W9-COU-354

COMPUTER AIDED MACHINE DESIGN

COMPUTER AIDED MACHINE DESIGN

Andrew D. Dimarogonas

W. Palm Professor of Mechanical Design,
Washington University, St Louis, USA

PRENTICE HALL

NEW YORK LONDON TORONTO SYDNEY TOKYO

First published 1988 by
Prentice Hall International (UK) Ltd,
66 Wood Lane End, Hemel Hempstead,
Hertfordshire, HP2 4RG
A division of Simon & Schuster International Group

Printed and bound in Great Britain at the
University Press, Cambridge.

Library of Congress Cataloging-in-Publication Data

Dimarogonas, Andrew, D., 1938–
Computer aided machine design.

Bibliography: p.
Includes index.
1. Machinery – Design – Data processing.
2. Computer aided design. 1. Title
TJ233.D5 1988 621.8′15′0285 87-29277
ISBN 0-13-166497-2

1 2 3 4 5 92 91 90 89 88

ISBN 0-13-166497-2
ISBN 0-13-166422-0 PBK

To
G.N. Sandor
on his 75th birthday

CONTENTS

Preface

1 MACHINE DESIGN METHODOLOGY

1.1 The art and science of machine design 1
1.2 Design strategies 2
1.2.1 Design continuation 2
1.2.2 Field of application 3
1.2.3 Design alternatives 4
1.2.4 Design inversion 4
1.2.5 Preliminary design: design synthesis 5
1.3 Design objectives 7
1.4 Design analysis and evaluation 7
1.5 Aesthetics in machine design 8
1.6 The seven stages in machine design 9
1.7 Systematic design 12
Examples 16
References and further reading 17
Problems 17

2 DESIGN DATABASES AND DESIGN STANDARDS

2.1 Design and artificial intelligence 21
2.2 Codes and standards 21
2.3 Dimensioning: preferred numbers 22
2.4 Dimensioning: fits and tolerances 23
2.5 Databases and the computer in design 28
2.5.1 Development of computers 29
2.5.2 Computer systems 29
2.5.3 Computer utilization 31
2.5.4 Computer programming 32
2.5.5 Computer aided machine design 34
2.6 Databases and data files 36
2.6.1 Sequential files 37
2.6.2 Random files 38
2.7 Program DESIGNDB 38
2.8 Expert systems 41
Examples 42
References and further reading 45
Problems 45
Appendix 2.A DESIGNDB: A database program for design 47

3 COMPUTER GRAPHICS

3.1 Organizing a design office 54
3.2 Computer aided versus manual drafting 55

3.3 Plane graphics 56
3.3.1 Graphics primitives 56
3.3.2 Drawing polygons 58
3.3.3 Drawing circular arcs 58
3.3.4 Clipping 59
3.3.5 Scaling 60
3.3.6 Translation 61
3.3.7 Rotation 61
3.3.8 Reflection 62
3.3.9 Zooming 63
3.4 Three-dimensional (3-D) graphics 63
3.4.1 Transformations 63
3.4.2 Mechanical drawings 65
3.4.3 Hidden line removal 68
3.4.4 Arc representation 72
3.5 Sectioned views 73
3.6 Automatic mesh generation 75
3.7 Solid modeling 79
3.8 Input 83
Examples 86
References and further reading 87
Problems 88
Appendix 3.A SOLID: A 3-D plotting program 91
Appendix 3.B AUTOMESH: An automatic mesh generation program 99

4 STRESS ANALYSIS IN MACHINE DESIGN

4.1 Modes of failure 107
4.1.1 Elastic deformation 108
4.1.2 Plastic deformation 108
4.1.3 Fracture 108
4.1.4 Wear 108
4.2 Modeling for stress analysis 109
4.3 Computer analysis of line members 111
4.3.1 The transfer matrix method 111
4.3.2 The finite-element method 118
4.3.3 The general prismatic element 121
4.3.4 Boundary conditions 123
4.3.5 System assembly and solution 124
4.4 Surface members 127
4.4.1 Membrane stresses 128
4.4.2 Plane stress – strain problems 129
4.4.3 Axisymmetric solids 133
4.4.4 System assembly and solution 135
4.5 Preprocessing and postprocessing 137
4.6 Commercial finite-element codes 138
Examples 139
References and further reading 147
Problems 148
Appendix 4.A TMSTAT: A transfer matrix analysis of beams program 155

Appendix 4.B PREFRAME: An input preparation program for FINFRAME 159
Appendix 4.C FINFRAME: A finite-element program for static analysis of space frames 165
Appendix 4.D FINSTRES: A finite-element program for plane stress analysis 170

5 MACHINE DESIGN OPTIMIZATION

5.1 The need for optimization 177
5.2 The general optimization problem statement 180
5.3 Analytical methods 180
5.4 Numerical methods 183
5.5 Search algorithms 185
5.6 Design for minimum cost 192
Examples 194
References and further reading 201
Problems 201
Appendix 5.A OPTIMUM: An optimization program 205

6 MATERIALS AND PROCESSES

6.1 Properties of materials for machine design 212
6.2 Material processing 221
6.2.1 Classification 221
6.2.2 Change of shape 222
6.2.3 Machining 224
6.2.4 Surface finish 225
6.2.5 Improvement of material strength 225
6.3 Material selection 228
6.3.1 High-strength materials in machine design 228
6.3.2 Designing with cast irons 231
6.3.3 Designing with structural steels and steel alloys 237
6.3.4 Designing with nonferrous alloys 248
6.3.5 Designing with engineering plastics 249
6.4 Factor of safety 250
6.5 Statistical character of strength: probablistic design 252
6.6 Strength theories 258
Examples 265
References and further reading 268
Problems 268
Appendix 6.A SAFFAC: A safety factor computation program 272
Appendix 6.B COMLOAD: A complex state of stress analysis program 273

7 FATIGUE AND FRACTURE

7.1 Stress concentration 277
7.2 Dynamic loads: fatigue 280
7.3 Combined loads 287
7.4 Cumulative fatigue damage 291
7.5 Low cycle fatigue 292
7.6 Brittle fracture 293
7.6.1 The transition temperature approach 294

7.6.2 Fracture mechanics approach 294
7.6.3 The congruency principle 296
Examples 297
References and further reading 303
Problems 303

8 DESIGN FOR STRENGTH

8.1 Simple states of stress 308
8.1.1 Tensile–compressive loading 308
8.1.2 Shear loading 309
8.1.3 Torsion 310
8.1.4 Bending of straight and curved elements 312
8.1.5 Design methodology for strength 315
8.2 Joint design 316
8.2.1 Rivets 317
8.2.2 Bolts 322
8.2.3 Welded joints 328
8.2.4 Interference fit joints 331
8.3 Stress analysis of joints 332
8.3.1 Interference stresses 332
8.3.2 Prestressed joints 336
8.3.3 Compound joints 341
8.3.4 Shear loaded joints 348
8.4 Joint stress concentration and safety factors in joints 353
8.5 Optimum design for strength 359
8.5.1 Optimum design of pin joints 359
8.5.2 Optimization of welds 359
Examples 359
References and further reading 366
Problems 367
Appendix 8.A RIVETS: An analysis of eccentric rivets program 381
Appendix 8.B SECTIONS: A program to calculate properties of complex sections 382
Appendix 8.C WELDS: A shear weld analysis program 385

9 DESIGN FOR RIGIDITY

9.1 Rigidity requirements in machine design 389
9.2 Rigidity of machine components 390
9.2.1 Rigidity material indices 390
9.2.2 Design rules for high rigidity 393
9.2.3 Rational sections 393
9.3 Stability of machine elements 395
9.4 Computer aided stability analysis 403
9.5 Machine frames 405
9.6 Design of springs 409
9.6.1 Design of torsional bars 415
9.6.2 Axially-loaded helical springs 416
9.7 Machine foundations 418
Examples 421
References and further reading 425
Problems 426

Appendix 9.A BUCKLING: Design of columns with Euler & Johnson formulas program 432
Appendix 9.B HELICSP: Design of helical springs program 433
Appendix 9.C TMSTABIL: Stability analysis of a general column with the transfer matrix method program 435

10 DESIGN OF FRICTION ELEMENTS

10.1 Sliding friction 438
10.2 Temperatures at sliding contacts 440
10.3 Wear owing to sliding friction 442
10.4 Clutches and brakes 443
10.4.1 Friction bands 443
10.4.2 Friction materials 444
10.4.3 Disk clutches and brakes 445
10.4.4 Conical clutches and brakes 448
10.4.5 Cylindrical radial air clutches 449
10.4.6 Cylindrical clutches and brakes 450
10.4.7 Safety and automatic clutches 453
10.4.8 CAD – optimum design of clutches and brakes 453
10.5 Friction belts 455
10.5.1 Mechanics of belt operation 456
10.5.2 Optimum design of belts 463
Examples 469
References and further reading 476
Problems 477
Appendix 10.A FLATBELTS: Design of flat friction belts program 485
Appendix 10.B VBELTS: Design of V-belts program 487

11 LUBRICATION

11.1 Friction and wear due to sliding 490
11.1.1 Mating surfaces 490
11.1.2 Wear mechanisms 490
11.2 Bearing materials 492
11.2.1 Copper alloys 493
11.2.2 Lead 493
11.2.3 White-metal bearing alloys—babbitts 493
11.2.4 Cadmium-base bearing alloys 494
11.2.5 Sintered bearings 494
11.2.6 'Dry' bearings 494
11.3 Fluid viscosity 494
11.4 Viscous flow 498
11.4.1 Viscous flow in a concentric bearing 498
11.4.2 Viscous flow in a pipe 499
11.4.3 Moving parallel plates 500
11.4.4 The Reynolds equation 500
11.5 Slider bearings 502
11.6 Journal bearings 506
11.6.1 Infinitely-long bearing 507
11.6.2 Short bearings 511
11.6.3 Frictional torque 512

11.6.4 Heat balance 513
11.6.5 Oil flow 514
11.6.6 Journal bearing design 515
11.6.7 Stability of journal bearings 518
11.6.8 Tilting pad bearings 520
11.7 Externally-pressurized (hydrostatic) bearing 521
11.7.1 The simple pad hydrostatic bearings 523
11.7.2 Externally-pressurized journal bearings 525
11.8 Computer methods 526
11.8.1 Numerical solution of the Reynolds equation 526
Examples 530
References and further reading 543
Problems 543
Appendix 11.1 FINLUB: A finite-element program for fluid lubrication 548

12 DESIGN OF CONTACT ELEMENTS

12.1 Dry contacts 555
12.2 Elastohydrodynamic lubrication 557
12.3 Rolling bearings 560
12.3.1 Rolling bearing types 561
12.3.2 Antifriction bearing design database 566
12.3.3 Fatigue load rating 566
12.3.4 Bearing loads 570
12.3.5 Combined loads 571
12.3.6 Fluctuating loads 577
12.3.7 Static loads 577
12.3.8 Lubrication of antifriction bearings 578
12.3.9 Load distribution within the bearing 580
12.4 Application of rolling bearings 582
12.4.1 Locating rolling bearings 582
12.4.2 Sealing 584
12.4.3 Lubricant application 587
12.5 Computer selection of antifriction bearings 587
Examples 589
References and further reading 592
Problems 592

13 DESIGN OF GEARING

13.1 General considerations 596
13.2 Geometry and kinematics of involute gearing 598
13.3 Mechanisms of failure 605
13.4 Materials and manufacture 606
13.5 Stress analysis of gear teeth 607
13.6 Design for surface strength 611
13.7 Computer aided graphics of spur gears 613
13.8 Computer aided design of spur gears 616
13.9 Helical gears 619
13.10 Bevel gears 623

13.11 Crossed gearing 627
13.12 Worm gears 629
13.13 Heat capacity and design of worm-gear drives 635
Examples 635
References and further reading 642
Problems 642
Appendix 13.A GEARPLOT: An involute gear teeth drawing program 647
Appendix 13.B GEARDES: A gear design program 648
Appendix 13.C WORMDES: A worm-gear design program 652

14 DESIGN OF AXISYMMETRIC ELEMENTS

14.1 Dynamics of rotary motion 655
14.2 Preliminary shaft design 661
14.2.1 Fundamentals 667
14.2.2 Shaft materials 664
14.2.3 Design criteria 665
14.2.4 Strength calculations 666
14.2.5 Fatigue strength 667
14.3 Design of shafts for rigidity 669
14.4 Vibration of rotating shafts 671
14.5 Computer aided design of shafts 673
14.6 Disks of revolution 676
Examples 679
References and further reading 687
Problems 687
Appendix 14.A ROTORDYN: A program for dynamic analysis of rotating shafts 689
Appendix 14.B SIMUL: A program for dynamic simulation of the rotation of a two-rotor system 695
Appendix 14.C SHRINKFIT: A program for the design of shrinkfits with compound hubs 697
Appendix 14.D SHAFTDES: A shaft design and antifriction bearing application program 698

Appendices

Appendix A Stress concentration factors 705
A.1 Theoretical stress concentration factors 705
A.2 Fatigue stress concentration factors 706
Appendix B Stress intensity factors for cracks 708
Appendix C Some standards for machine elements 710
C.1 Bolt dimensions 710
C.2 Key dimensions 712
C.3 Dimensions of slider bearings 713
C.4 Hot-rolled steel sections 715

Answers to selected problems 719

Index 723

PREFACE

Computer aided design (CAD) emerged in the 1960s out of the general acceptance of fast digital computers as tools to aid the design of complex systems. At that time, the most progressive industries used computer methods to aid their machine design efforts, using modern optimization techniques with the aid of computer graphics and computerized structural analysis.

Since then the capacity of computers has increased by many orders of magnitude while their price has decreased simultaneously, making them a generally available tool to the student and to the designer. The matching of computer techniques with the most traditional mechanical engineering courses such as machine design (or design of machine elements) came rather later. The purpose of this textbook is the introduction of computer methodology in the design of machines.

The need for rewriting of such a classical subject from a new point of view created the opportunity for a fresh look at the presentation of the subject.

All over the world, machine design textbooks have presented the material with component orientation, such as design of bolts, shafts, bearings, gears, etc. However, with the vast number of different elements of machines now available, such presentations were in any case incomplete. This led the author to try as for as possible a unifying approach, dividing the material from the point of view of design methodology rather than element function. Therefore, the material has been presented as, for example, design for strength, etc. The student has the opportunity to observe the common features of designing different elements with the same methods and with the same considerations. The aim is to guide the student into the design methodology rather than to teach how to design particular machine elements. Of course, this is not completely possible because the accumulated design experience is mostly component-oriented. Therefore, in certain cases such as gears, the presentation is component-oriented.

It has been natural to emphasize design methods which are particularly suitable for computer implementation.

Depending on the education philosophy and the students' background, teaching of CAD at the undergraduate level, in some schools involves utilization of application software only, while in others the students are required to do some programing of their own. For this reason, some software are developed in the book in a simplified way that students with a minimum experience of programing can follow and, in parallel, application of commercial software is introduced and encouraged.

The algorithms developed in the book are presented in a format which is language-independent. The student can program them to the high level language of his or her choice. For those who do not want to program the algorithms, the text is supplemented by appropriate software.

The software included in the book are not to substitute large commercial CAD programs. At the discretion of the instructor, the student can do the problems with such programs. The author suggests a parallel use of the programs of the book with available commercial packages. Even for the student who is interested only in commercial programs, this parallel study will enable an understanding of their potential and limitations. In this edition of the book, the software are written in BASIC. The author used the MICROSOFT Quick Basic 4.0 but most of the programs are made BASIC A compatible. An AT computer, preferably with an EGA card will suffice. Versions in FORTRAN and C are currently developed on floppy disks.

The first five chapters over the background of the CAD methods in machine design. Chapters 6 and 7 deal with the material selection in machine design. Chapters 8 and 9 present the strength considerations while Chapters 10, 11 and 12 deal with tribological aspects of machine design. Chapters 13 and 14 deal with the problems of rotary motion transmission.

Most case studies and problems are from the author's industrial experience and they have two aims:

1. To introduce the student to real engineering problems, yet simple enough to be understood and worked out.
2. To help the student reach computer literacy in an applied subject such as machine design.

The problems are designed to form five complete sets, so that the course can be repeated five times without repetition. Furthermore, there is a sufficient number of problems so that the instructor can emphasize design or programing, depending on the needs of the course.

Emphasis is placed in motivating the student to make his or her own complete design programs. In many cases however, complete programs are given which are beyond the student's reach at that level at a certain time, but which are necessary for further application and use, such as the optimization program.

All software included in the text have been carefully checked and used already by many students and designers. However, no warranty is given, expressed or implied, to the users. Their help in identifying bugs will be highly appreciated.

Andrew D. Dimarogonas
St Louis

ACKNOWLEDGEMENTS

The author expresses his great appreciation to Mr Bob Ambrose, Mechanical Engineer, for reading the manuscript and improving it in many ways, Professors G. Massouros, E. Gardner and J. Hochstein and many graduate and undergraduate students who read chapters of the manuscript and suggested improvements.

CHAPTER ONE

MACHINE DESIGN METHODOLOGY

1.1 THE ART AND SCIENCE OF MACHINE DESIGN

To an ever-increasing extent, our environment is dominated by machinery. They constitute the visible cultural landscape of everyday life, forming a complex pattern of function and meaning in which our perception of the world, our attitudes and sense of relationship with it, are closely interrelated.

The design process is a process of creation and therefore difficult to summarize in a simple design formula, a book, or a precise definition:

(i) It can be the work of a person.
(ii) It can be the effort of a team.
(iii) It may emanate from creative intuition or from executive decisions based upon market research.
(iv) It is generally constrained by resource availability, organizational, political, social and aesthetic considerations aiming, naturally, at an acceptance by the end user, the customer.

Machine design is a process of creation, invention and definition, involving an eventual synthesis of contributory and often conflicting factors into a three-dimensional form capable of multiple reproduction, at a marketable price, with acceptable quality of products, operating with a specified reliability and being socially acceptable.

Machine design is an applied science, drawing heavily from mechanical engineering science. Machines can neither defy Newton's Laws, nor the strength limits of their materials.

Machine design is also an art. One solution usually exists to a strength of materials' problem. In a machine design problem, an infinite number of solutions usually exists. In rough mathematical terms, the number of unknown parameters is usually orders of magnitude greater than available data, that is equations. This uncertainty can be eliminated to a small extent by optimization methods and to the full extent by good technical judgement of the designer. Now, what is 'good' and what is 'bad' technical judgement? Though experience has accumulated rules, the final judge is the end user and the quantitative measure of acceptance is the market success.

Competing designers have at their disposal about the same amount of information, with the contemporary dissemination of information. Proprietary information is usually much less, in reality, than advertized. Why is one product (and the manufacturer) successful and the others not? In most cases the main reasons are the decisions made at design stages, not those based on engineering science but those based on engineering (and business) judgement.

As stated above, this cannot be corrected with concrete rules. There are, however, guidelines which can help the designer with the decisions he has to make for a successful design. It is not certain that a machine designed 'by the book' will have guaranteed success. It is almost certain, however, that it will not be a total failure.

1.2 DESIGN STRATEGIES

When a designer is confronted with a design task, he has to start with some basic information. Sources of the information and the assignment might be:

(i) an assignment by a commercial organization or customer; such an assignment generally specifies the required parameters of the machine, and the field and conditions of its use;
(ii) a technical suggestion initiated by a designing organization or a group of designers;
(iii) a research work or an experimental model based on the research;
(iv) an invention proposition;
(v) an existing machine prototype which has to be developed with modifications and alterations.

The first case is more general and usually most convenient to the designer.

From the moment of the inception of a particular design, to the introduction of the machine into production a period elapses, which is longer for more complex machines. This period includes design, manufacture, finishing and adjustment of the prototype, industrial testing, introduction of necessary alterations to eliminate defects found when testing, state acceptance tests and final acceptance of the prototype. This is followed by the preparation of technical documents for the initial production series, its manufacture, and industrial tests, after which documents for mass production are prepared, and the production process modified to suit mass production starts.

Usually, this process takes from a few months to several years; sometimes from the beginning of design to the beginning of mass production two or three, or even more, years elapse.

The design stage can be considerably facilitated and shortened in time when the designer plans and executes a design strategy that takes into account certain established design practices.

1.2.1 Design continuation

The designer has to utilize the previous experience gained in the given branch of industry, as well as in the allied branches, and introduce into the particular design useful features of the existing machines.

Nearly every modern machine is the work of several generations of designers. The original model of a machine is gradually improved, equipped with new units and mechanisms, and made better as a result of new solutions. Some design solutions die due to the development of more rational ones, new technological procedures or higher operational demands, while some, proved exceptional, remain in force for very long periods, although at times being slightly modified. The Volkswagen Bug is an example.

When reviewing the development of any machine, it is seen that a huge number of miscellaneous design schemes were tried before. Many of them disappeared and became completely forgotten, then they were revived after many years on a new technological basis, and again returned to use. Such historical reviews help the designer to avoid errors and the repetitions of the stages traversed, and allow him to choose more prospective methods of machine development.

Particularly for new inventions, the inventor or designer must first conduct a thorough patent search to avoid reinventing the wheel or spending effort to solve problems other people have solved before.

It is useful to review the tendency of basic machine parameters over the years. Of particular importance is the study of the starting data when working on a new design. The selection of parameters must be preceded by a comprehensive study of all factors determining the machine's life. It is necessary to carefully study the experience gained in the operation of machines designed by others, correctly analyze their advantages and disadvantages, choose a correct prototype, and make clear the tendencies of the development and requirements of the branch of industry for which the machine is intended.

An important prerequisite for correct design is the availability of pertinent references on materials and manufacturing processes. Apart from archives of their own products, design organizations should have at their disposal design catalogues from manufacturers.

The concept of design continuation does not impose limitations on initiative. The designing of any machine presents an unlimited field of creativity to the designer. He must not invent anything already invented and must not forget the rule formulated early in the twentieth century by Gueldner: '*weniger erfinden, mehr konstruieren*' ('invent less, design more') (Orlov 1976).

The process of continuous machine improvement, caused by the ever-growing demands of the industry, results in the development of the design thought. The striving for a perfect design penetrates the flesh and blood of the designer, becomes his life. However, one should remember that the enemy of the 'good enough' is the 'perfect'.

The designer must constantly improve and enrich his store of design solutions. An experienced designer will invariably note and mentally 'snapshot' interesting solutions, even in machines foreign to him.

1.2.2 Field of application

The progress in mechanical engineering is linked with the development of the different branches of industry. Industrial development is accomplished by continuous improvements, including: growth of productivity; decrease of operational cycle; development of new technological processes; rearrangement of production lines; change in equipment layout; continuous expansion of mechanization; and automation of production. This accordingly results in higher requirements for machine parameters, their productivity and level of automation. As new technological innovations become available, some machines become obsolete. The necessity arises for new machines or for modifications of the old.

Occasionally such modifications may be very extensive and affect large categories of machines. Thus, the introduction of the continuous steel-casting process means the death or, in any case, reduction in the manufacture of such complex and large machines as blooming and slabbing mills – to name just one example.

Before designing machines intended for a specific branch of industry, it is necessary to examine the dynamics of the quantitative and qualitative development of that particular branch, the requirements for the given machine category and the possibility of the development of new technological processes and new methods of production.

1.2.3 Design alternatives

Generally, the final machine version is chosen from several which have been carefully evaluated and compared from all possible points of view, perfection of kinematic and force characteristics, cost of manufacture, energy consumption, cost of labor, reliable operation, size, durability, suitability for industrial production, safety, servicing, assembly–disassembly, inspection, setting and adjustment.

Not always, even during the most careful search, is a successful solution found, which fully answers the posed demands. Rarely is a solution a success in all respects. The matter is sometimes not the lack of inventiveness, but rather conflicting requirements. Under such circumstances, it is necessary to find a compromise solution, waiving some requirements which are not of prime importance in the given conditions of the machine application. Often a variant is chosen not because it incorporates more advantages, but rather because it has less disadvantages in comparison with others.

After selecting the scheme and basic parameters of the machine, a sketch is made, and then a general arrangement drawing, on the basis of which the initial, technical and working designs are composed.

The development of design alternatives is not a matter of individual preference or inspiration, but instead a regular design method aimed at seeking the most rational solution.

Design alternatives may also emanate from the study of nature itself, analogy with other branches of engineering and reassignment of functions among the several machine components.

1.2.4 Design inversion

Design inversion consists of inverting the function, shape and position of parts to yield a better design. In assemblies it is occasionally preferable to invert the functions of parts, e.g. to turn a driving element into a driven one, to turn a leader into a follower, to make a female component out of a male one, to convert a stationary part into a movable element, etc.

Furthermore, the shapes of parts are inverted, e.g. to change a female taper to a male one, a convex spherical surface to a concave one. In other cases it may be more suitable to shift some elements from one part to another.

Some typical examples of inversion in machine design are shown in Figure 1.1 (Orlov 1976).

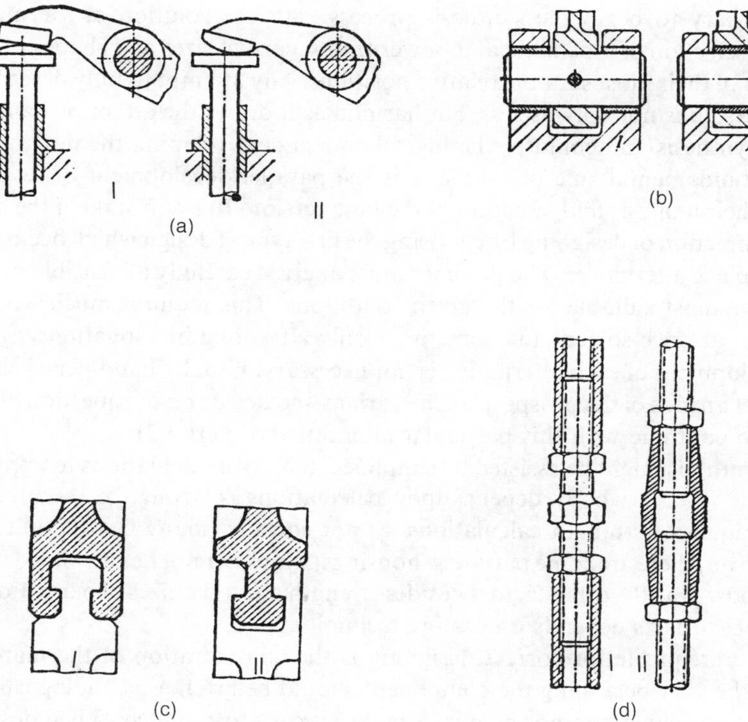

Figure 1.1 Examples of design inversion: (a) valve tarret; (b) connecting rod pin; (c) turbine bucket dovetail; (d) turnbuckle. (From Orlov, 1976, by permission of Mir Publishers, Moscow)

1.2.5 Preliminary design: design synthesis

During the design synthesis the main arrangement and general design of a given unit are elaborated (sometimes several versions). After discussing the draft, the working scheme is synthesized defining more accurately the design of the unit and serving as the initial material for furthering the project.

During synthesis it is important to separate the main items from the secondary. An attempt to conceive all elements of a design at a time is a typical error characteristic of inexperienced designers. After specifying the purposes and parameters of the projected design, the designer often begins to draw the complete design in all its details, depicting all the design elements like a picture. Such a method makes the design an irrational one, as it forms a mechanical string of constructional elements and units, poorly arranged. The reason is that the designer uses a heuristic process to evaluate the merits of the several possible parameter constellations and, naturally, leaves some of them out. The larger the number of elements considered, the larger the alternatives he is missing. He might then consider alternatives in items of minor importance and overlook better alternatives in the main features of the design.

It is necessary to begin the synthesis process with the solution of the main design problems, i.e. selection of kinematic and power trains, correct sizes and shapes of parts, and determination of their most suitable relative positions. Any attempt to fully describe parts at this stage of design is not only useless, but harmful as it draws the attention from the main problems of synthesis and confuses the logical course of developing the design.

Another fundamental rule of synthesis is the parallel development of several design alternatives, their analysis and selection of the best version. It is a mistake if the designer at once sets the direction of designing by choosing the first type of design which occurred to him, or a commonplace alternative. The designer must analyze carefully all feasible versions and choose the one most suitable for the given conditions. This requires much work and the problem is not at once solved, and sometimes only after long investigations.

Full development of each alternative is not necessary. Usually hand pencil sketches are sufficient to get an idea of the prospects of the variant and decide on the question of whether it is advisable to continue with this particular alternative (Figure 1.2).

Design synthesis might be assisted by simplified, tentative calculations to approximately size the system, but to wholly depend upon calculations is wrong. In the first place the existing techniques of strength calculations do not consider many factors influencing the design. Moreover, there are some parts (e.g. housings) which cannot be calculated except with tedious methods. Thirdly, other factors besides strength affect the sizes of parts. For example, the design of cast parts depends on casting technology.

Another prerequisite for correct designing is the consideration of the manufacturing problems; at the very beginning the component should be given a technologically rational shape. A skilled designer composing a part makes it easy to produce. When designing, the manufacturing process must be always considered.

Some units are not always successfully designed from start. The designer has to devise some 'tentative' alternatives and raise the design to the required level in the process of further activities. In such cases it is useful, as Italians say, *dare al tempo il tempo* ('to give time to time'), i.e. to take a breathing space, after which, as a result of subconscious thought, the problem is often solved. After a while the designer looks at the outline drawing in another light, and sees the mistakes made at the first stage of the development of the main design idea.

Sometimes the designer unintentionally loses his objectiveness and does not see the drawbacks of his favourite version or the potentialities of other version. In such cases,

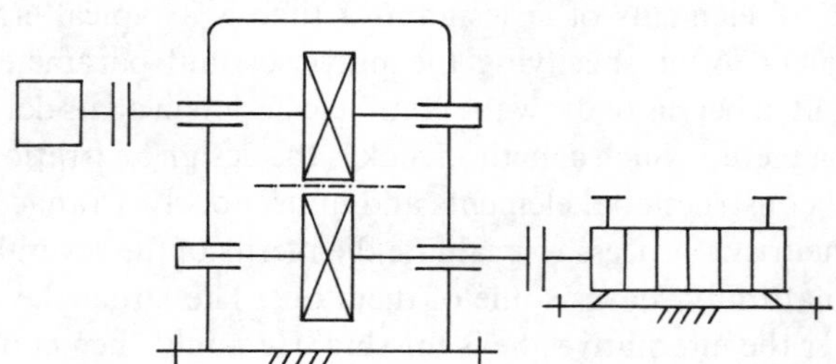

Figure 1.2 Simplified synthesis sketch for a motor gear box-pump assembly design

impartial opinions of outsiders, the advice of senior designers and co-workers, and even their fault-finding criticism turn out to be most helpful. Moreover, the sharper the criticism, the greater is the benefit derived by the designer. Of course, the rule 'perfect design is the main enemy of the good design' must be always observed.

1.3 DESIGN OBJECTIVES

A successful design effort indicates good planning, which starts with a clear definition of the design objectives. Such definition will provide guidance during the course of design, but it will also be eventually used to judge the result of the design effort.

The design objectives have the form of clearly-stated specifications of the machine to be designed or the product to be produced, or both. Specifications on the machine are imposed by marketing requirements but also for safety reasons, availability of certain raw materials, cooperation with other machines in production, etc. Specifications on the product are almost always imposed.

Usually, there is one main objective – the cost of product which is divided into running and capital cost. Many other objectives may be stated without an immediate cost interpretation, such as appearance, surface texture, noise, air pollution, etc.

The machine should not be overspecified, that is the design objectives should always be the minima of the acceptable limits. Overspecified machines have an unnecessary high cost. A friend of mine, a layman, once told me that he would place a lawsuit on Omega because his wristwatch was accurate only to one-tenth of a second while at the time of purchase they told him that it was accurate to one-hundredth of a second. Obviously in this case the watch was overspecified.

Overspecification also reduces creativity because it imposes restrictions and removes degrees of freedom from the designer's imagination and inventiveness.

1.4 DESIGN ANALYSIS AND EVALUATION

In a very few cases it is possible with preliminary calculations to design a final version of a machine or an element. In most cases, the designer, using his experiences and simple design calculations, makes a first draft of his design in the form of free-hand sketches with some basic dimensions. This is the product of his experience, creativity, imagination and calculations. It has to be thoroughly analyzed to verify that (a) it meets the design objectives and (b) it has adequate strength, stability, durability, etc. This process of design analysis is essential to any machine design process. Since most of this book is devoted to design analysis of the machines, this subject will come up quite often in the chapters to follow.

To evaluate a specific design, one must have a complete design analysis. Among other reasons, one has to verify not only that the machine designed meets the specifications but also that it is close to them and the machine is not overdesigned.

1.5 AESTHETICS IN MACHINE DESIGN

An aesthetically-pleasing appearance is by no means a luxury for a machine. Many times, for machines of established principle in particular, appearance is the only factor that counts. And for every machine it is a definite advantage.

Quality of a machine cannot be expressed by a simple number. There are several factors, often contradictory, which are included in the overall judgement of the quality, such as cost, production rate, reliability, durability, etc. Most times not all necessary information is available. Then, the judgement of quality becomes a relatively subjective matter which can be greatly influenced by the personal impression of the machine on the end user. Moreover, it is now verified that productivity and rate of accidents are greatly affected by the work environment, where the appearance of the work place is not of the least value.

The first machine designers were the blacksmiths of Ancient Greece and Rome, and of Medieval Europe. They did not know much about marketing but since they knew the users of their products, they had a continuous feedback and their products were in line with their contemporary ideas about aesthetics. Also, since conception and manufacture were the work of a single person, their design was simple and of human proportions. Mass production introduced the breaking of this process into many smaller ones, in particular design and manufacture. Mechanization brought about standardization and unification and isolated the personal impact of the designer to a large extent.

One basic design rule for the form, however not universally accepted, is Sullivan's rule – 'form follows function'. The first impression the machine or product has on the observer must be closely related to the machine function.

Sergio Pininfarina, the noted car designer, once wrote on the function versus appearance question for car design:

> Recent years have seen traditional design parameters joined by others arising out of matters of safety and the energy crisis, constraints which have brought with them a whole series of new functions.
>
> As one of the mind's main categories, Ancient Greek philosophy pointed to the Beautiful, meaning by this a concept inclusive of functional aesthetic (Useful) and even ethical (the Good) implications. The Beautiful was thus an extremely important achievement, laboriously or miraculously (with the help of the gods) arrived at by way of the convergence of varying and, at times, contrasting requirements.
>
> Modern philosophers, especially the Idealists and their followers, particularly Croce, have attempted to split the Beautiful and the Useful, and render them autonomous.
>
> In art, these two interpretations and classifications are readily observed in so-called classical and modern art: the eternal dilemma or contrast between Form and Content (Function) has set its imprint on almost all human production, both figurative and literary.
>
> But what is the situation in other not strictly-artistic fields? Can we speak of the Beautiful and the Useful, of Form and Content, of Appearance and Function?
>
> I believe that where these definitions concern creative activities the reply is positive, and the question or questions are legitimate.
>
> The design of a car is certainly creative in so far as it is a speculative mental activity: creativity applied to an object produced by a modern industrial society which is more complex precisely for that reason. ...
>
> By contrast, artistic creativity is and will always remain the work of just one or two

> people, and the resources he or they need to express their art; however much technological sophistication is deployed, will never be comparable with what we have been talking about, as the major contribution comes from the mind, eye, and hand. The very definition of Man, however, means that this invaluable contribution will not fall by the wayside; at the most it might become sublimated in a higher and more pregnant sphere than the one in which it has lived until now (Pininfarina 1978).

A very important aspect of form is style, which can be roughly defined as the generally accepted appearance at a certain time. A fast-back car would have looked ugly thirty years ago but today looks beautiful. An old Remington typewriter would not look right in a modern office – it is out of style.

Marketing demands greatly influence appearance. Producers for a large internal market do not put great emphasis on appearance. Industries which are mostly exporting, with greater competition, are very careful on form. A typical example again is USA automobiles on the one hand and Japanese and European on the other. Italian-made machinery has very carefully-designed appearance.

1.6 THE SEVEN STAGES IN MACHINE DESIGN

The machine design process is subject to a large number of variations. In every text on the subject of engineering design, a different division of the design process in distinct stages is proposed. They all make sense, although they seem very different from one another. This reflects the complex nature of the design process and the fact that every design problem requires a special treatment. This process cannot be exactly specified by an equation or an algorithm. A systematic approach is useful only to the extent that the designer is presented with a strategy that he can use as a base for planning the required design strategy for the problem at hand.

This strategy, proposed by Sandor (1964), is described in the flowchart of Figure 1.3. The flowchart is arranged in a Y-shaped structure:

1. The two upper branches of the Y represent, on one hand, the evolution of the design task, and on the other hand, the development of the available, applicable engineering background.
2. The junction of the Y stands for the merging of these branches: generation of design concepts.
3. The leg of the Y is the guideline toward the completion of the design, based on the selected concept.

The flowchart implies, but is not encumbered by, the feedbacks and iterations that are essential and inevitable in the creative process.

Stage 1

Stage 1A. Confrontation

The 'confrontation' is not a mere problem statement but rather the actual encounter of the

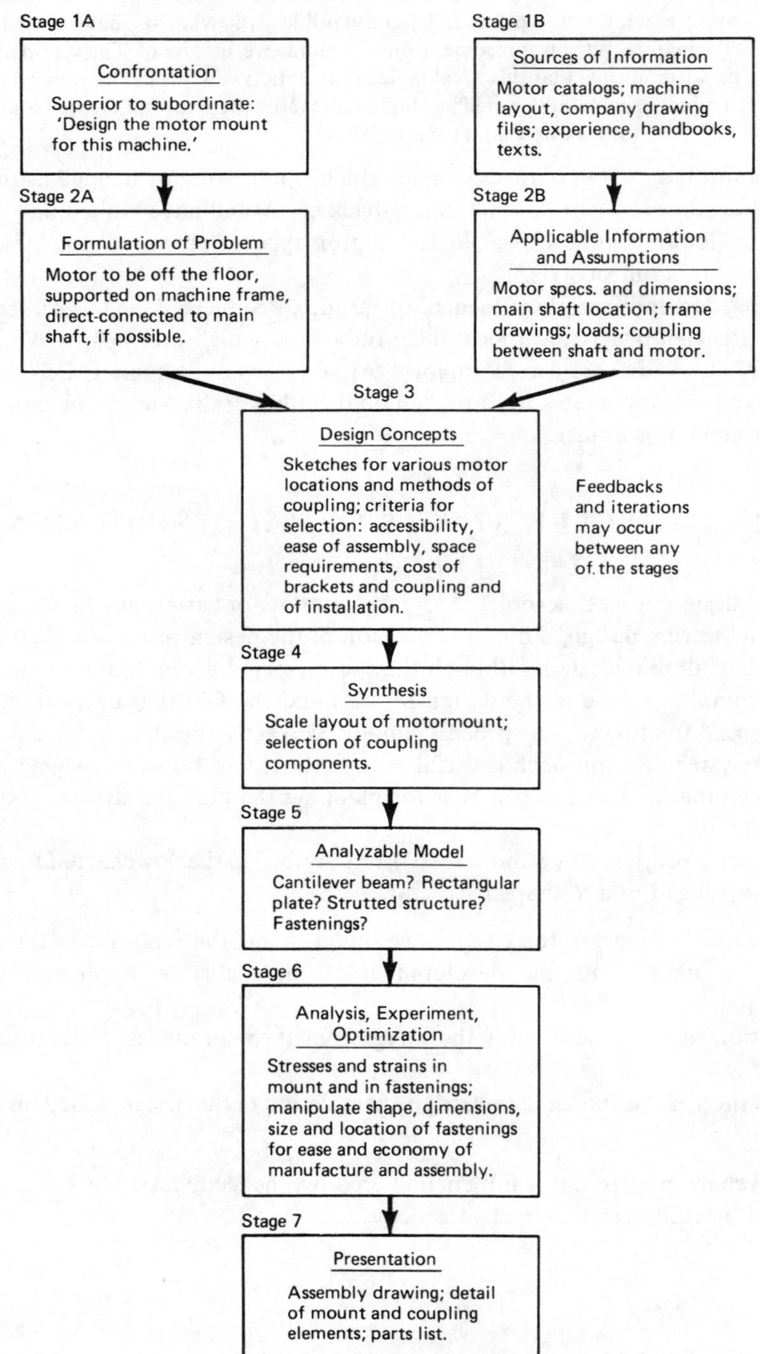

Figure 1.3 Sandor's Y structure of the design process. (Courtesy ASME)

engineer with a need to take action. It usually lacks sufficient information, and often demands more background and experience than the engineer possesses at the time. Furthermore, the 'real need' may not be obvious from this first encounter with an 'undesirable situation'.

Stage 1B. Sources of information

The sources of information available to the engineer encompass all human knowledge. Perhaps the best sources are other people in related fields.

Stage 2

Stage 2A. Formulation of problem

Since confrontation is often so indefinite, the engineer must clarify the problem that is to be solved: He must recognize and ferret-out the 'real need', and then define it in concrete, quantitative terms suitable for engineering action.

Stage 2B. Preparation of information and assumptions

From the vast variety of sources of information the designer must select the applicable areas, including theoretical and empirical knowledge, and, where information is lacking, fill the gap with sound engineering assumptions.

Stage 3. Generation and selection of design concepts

Here the background developed by the foregoing preparation is brought to bear on the problem as it was just formulated. All conceivable design concepts are prepared in schematic skeletal form, drawing on related fields as much as possible.

It should be remembered that creativity is largely a matter of diligence. If the designer lists all the ideas that he can generate or assimilate, workable design alternatives are bound to develop, and the most promising can be selected in the light of requirements and constraints.

Stage 4. Synthesis

The selected design concept is a skeleton. We must give it substance – fill in the blanks with concrete parameters with the use of systematic design methods guided by intuition. Compatibility with interfacing systems is essential. In many areas advanced analytical, graphical, and computer-aided methods have become available. However, intuition, guided by experience, is the traditional approach.

Stage 5. Analyzable model

Even the simplest physical system or component is usually too complex for direct analysis. It must be represented by a model amenable to analytic or empirical evaluation. In abstracting such a model, the engineer must strive to represent as many of the significant characteristics of the real system as possible, commensurate with the available time, methods, and means of analysis or experimental techniques. Typical models are simplified physical versions, free-body diagrams, and kinematic skeletal diagrams.

Stage 6. Experiment, analysis, optimization

Here the objective is to determine and improve the expected performance of the proposed design.

1. Design-oriented experiment, either on a physical model or on its analog, must take the place of analysis where the latter is not feasible.
2. Analysis or test of the representative model aims to establish the adequacy and responses of the physical system under the entire range of operating conditions.
3. In optimizing a system or a component, the engineer must decide three questions in advance:
   (a) With respect to what criterion or weighted combination of criteria should he optimize?
   (b) What system parameters can be manipulated?
   (c) What are the limits on these parameters?

Although systematic optimization techniques have been, and are, being worked out (see Chapter 5), this stage is largely dependent on the engineer's intuition and judgement. The amount of optimizing effort should be commensurate with the importance of the function or the system component and/or quantity involved.

'Experiment', 'analysis', and 'optimization' form one integral 'close-loop' stage in the design process. Their results may give rise to feedbacks and iterations involving any, or all, of the previous stages, including a possible switch to another design concept.

Stage 7. Presentation

No design can be considered complete until it has been presented to (and accepted by) two groups of people:

1. Those who will utilize it.
2. Those who will make it.

The engineer's presentation must therefore be understandable to the prospective user, and contain all the necessary details to allow manufacture and construction by the builder.

1.7 SYSTEMATIC DESIGN

Systematic design can be traced, perhaps, back to Leonardo da Vinci. His designs are the result of a systematic evolution from basic principles. Since the times of the Industrial Revolution and the rise of mechanization there have been many attempts to devise a coherent design theory. A qualitative description of the design process, the Y-structure proposed by Sandor, was presented in the previous section (1.6). The need to quantify this process was recognized in recent years, especially in view of the vast computing capacity now available.

A formalization of the machine design process (MDP) should be built upon certain generally-recognized principles, such as:

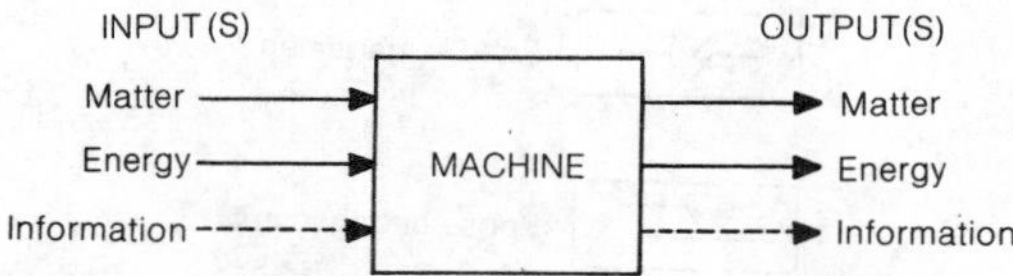

Figure 1.4 The machine as processor of matter, energy, information

(a) The MDP has a creative part performed by the designer and an algorithmic part, performed by machines.
(b) The creative part of the MDP is an heuristic one: the designer intuitively forms a great number of design variants and, rapidly, discards a great number of them to leave a small number for the application of rational design rules.
(c) A machine is a device which has a function of processing energy, matter and information (Figure 1.4).
(d) The function of a machine can be expressed as a structure of elementary functions, common to different machines.
(e) Each of these elementary function modules can be associated with a number of different physical effects. Each one of those can in turn be associated with a number of physical principles. Each one of the latter can yield several physical devices (Figure 1.5).
(f) There are five such elementary functions (Figure 1.6):

   (i) Fixed engagement, i.e. a gear.
   (ii) Loose engagement, i.e. a belt.
   (iii) Full constraint, i.e. a bolt.
   (iv) Partial constraint, i.e. a bearing.
   (v) House, i.e. a housing box.

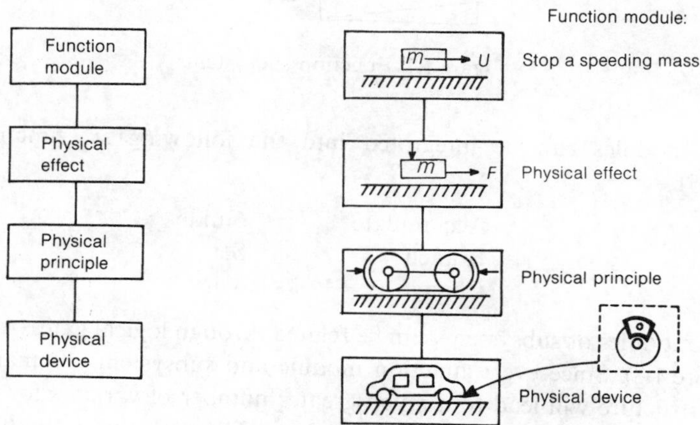

Figure 1.5 Evolution of a function module

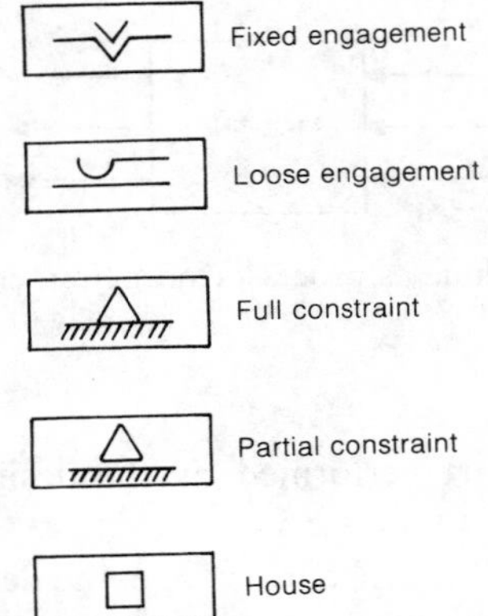

Figure 1.6 Elementary functions

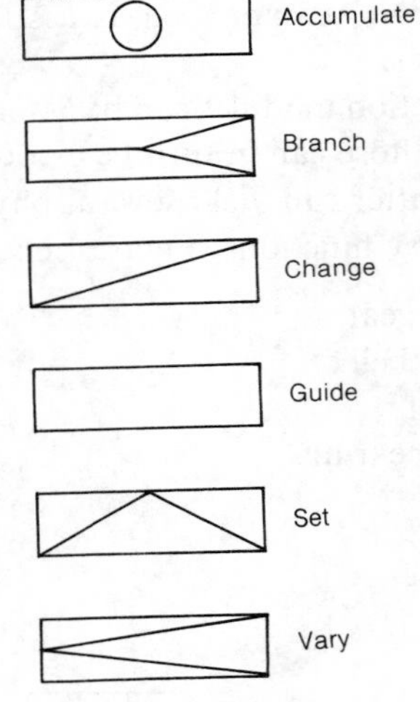

Figure 1.7 Function subsystems

(g) Function modules can be integrated into the following six function subsystems (Figure 1.7):

Accumulate	Guide
Branch	Set
Change	Vary

(h) Function modules or subsystems can be related through logical gates, AND, NOT, OR, etc. (Figure 1.8). Since every function module and subsystem has many variants, the function structure will lead to a much greater number of variants for evaluation.

(i) Every module or subsystem has a boundary. Through this boundary, the module exchanges matter, energy and information with the other modules.

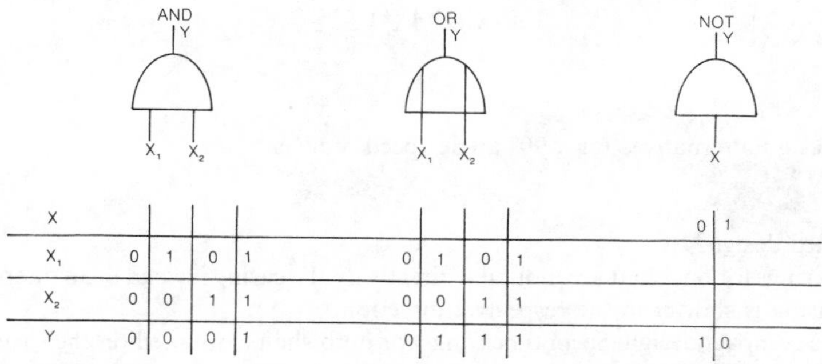

X	AND				OR				NOT	
X_1	0	1	0	1	0	1	0	1	0	1
X_2	0	0	1	1	0	0	1	1		
Y	0	0	0	1	0	1	1	1	1	0

Figure 1.8 Logical gates

The *creative part* of the MDP is the formulation of the function structure for the task at hand and the elimination of a number of the design variants without much computation effort. The *algorithmic part* of the design process takes over to yield the final design.

There are two hierarchies in the MDP:

(a) The flow hierarchy, as described by the Y-structure in the previous Section (1.6).
(b) The system hierarchy, i.e. the system–subsystems–modules structure.

Each design module corresponds to a particular physical realization of a function module and consists of:

(a) A set of information about standard sizes, available materials, processes, etc.
(b) A set of rules dictated by natural laws and engineering experience.
(c) A set of design parameters which must be determined so that the resulting physical element is fully specified.
(d) A set of parameters that the module communicates with the other modules, at higher, lower or parallel hierarchy.

Each design module operates at several levels, depending on the current level of the design flow, i.e. preliminary design, design evaluation, etc.

The function structure, once outlined by the designer, can be expressed in symbolic form using the elements of Figs 1.6–1.8. Since each of these elements can be realized by a number of physical elements, a symbolic function structure can yield a great number of design variants. The creative role of the designer is to define the symbolic function structure and to discard, intuitively, most of the resulting design variants. The role of the algorithm is to evaluate the remaining design variants, select the best one, perform its final and detailed design and communicate the results to the manufacturing operation, all under continuous monitoring and control by the designer.

The rest of this book will examine the knowledge on which the design modules are developed.

EXAMPLES

Example 1.1

Develop design alternatives for a 90° angle speed reducer.

Solution

(a) Standard design.

(b), (c) Bearings for one shaft are mounted directly on the casing instead of an intermediate sleeve. The casing is shorter in the respective direction.

(d) A more compact design. Second bearings for both shafts mounted on the casing without the use of sleeves. Smaller bearing reactions and smaller bearings.

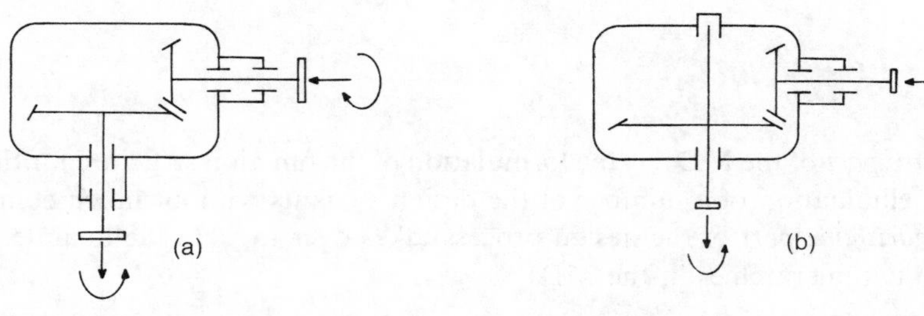

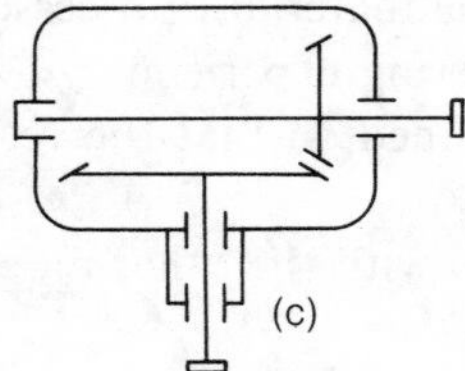

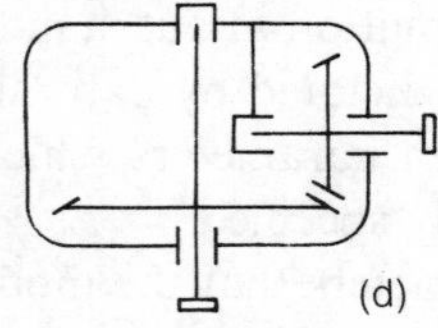

Figure E1.1

Example 1.2

Define the function structure and its symbolic form for the power train of an automobile.

Solution

A standard gear shift is assumed. The boundary of the system is B. The system is connected to the engine and to the ground and transmits and transforms energy between them. Also, it transmits and transforms information between the driver and the system (clutch and gearbox). The resulting function and symbolic structure are shown in Figure E1.2.

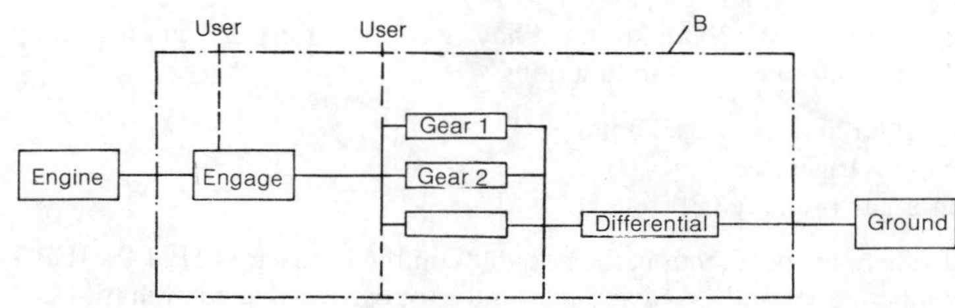

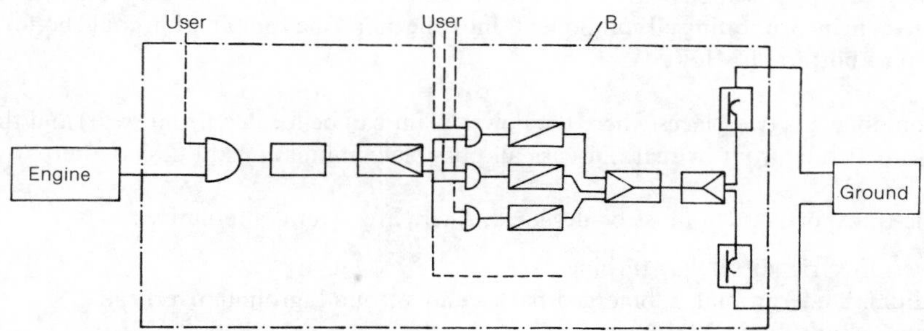

Figure E1.2

REFERENCES AND FURTHER READING

Heskett, J., 1980. *Industrial Design.* London: Thames & Hudson.

Hubka, Vl., 1982. *Principles of Engineering Design.* London: Butterworth Scientific. (Translated from German and edited by W. Eder).

Orlov, P., 1976. *Fundamentals of Machine Design.* Moscow: Mir.

Pahl, G. and Beitz, W., 1984. *Engineering Design.* London: The Design Council.

Pininfarina, S., 1978. 'Future trends', in *Function versus Appearance in Vehicle Design.* London: Inst. of Mech. Eng., Oct. 1978.

Rodenacker, W. G., 1970. *Methodisches Konstruieren*, 2nd Edn. Berlin: Springer Verlag.

Sandler, B.-Z., 1985. *Creative Machine Design*, Jamaica, NY: The Solomon Press.

Sandor, G. N., 1964. 'The seven stages of engineering design', *Mechanical Engineering*, Vol. 86, no. 4, pp. 21–5.

PROBLEMS

1.1 A shaft with axial and radial load is mounted on a casing by way of one axial and two radial bearings.

Show three design alternatives which will assure support of axial and radial loads and small axial freedom of motion to protect from thermal expansion of the shaft.

1.2 An LNG (liquified natural gas) tanker must have a container for the very low-temperature LNG. Therefore, the hull should have the functions:

To keep LNG and the vessel floating (F)
To insulate refrigerated LNG (I)
To sustain gas pressure (P)

From the outside to inside direction, the possible combinations are FIP, FPI, IFP, IPF, PIF, PFI. Discuss advantages and disadvantages and propose feasible combinations.

1.3 A microcomputer consists of main unit (U), keyboard (K), floppy-disk drive (F) and monitor screen (M). They can be assembled in 1, 2, 3 or 4 units. Propose possible combinations of design alternatives in incorporating all components into one unit. One combination could be, for example, from top to bottom, FMUK.

1.4 An automobile has two places where the power plant can be located (front–rear) and the driving wheels can also be front or rear. Discuss all possible combinations of design alternatives.

1.5 A public transport system must be designed. There are several alternatives:

1. Power: Electric, diesel, gas turbine.
2. Medium: Underground, submerged partly underground, ground, overhead.
3. Support: Rails, tyres, MHD support.

Organize and discuss design alternatives.

Problems 1.6 to 1.10

Invert the assemblies shown in Figures P1.6 to P1.10 respectively (Orlov 1976). In particular:

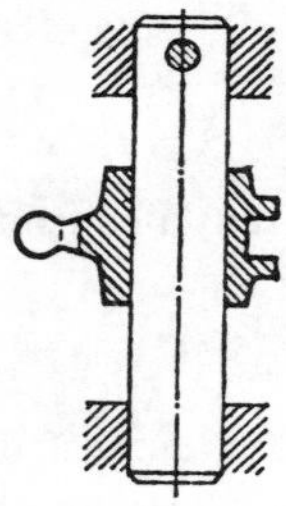

Figure P1.6

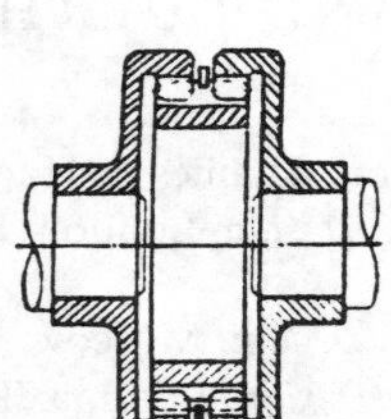

Figure P1.7

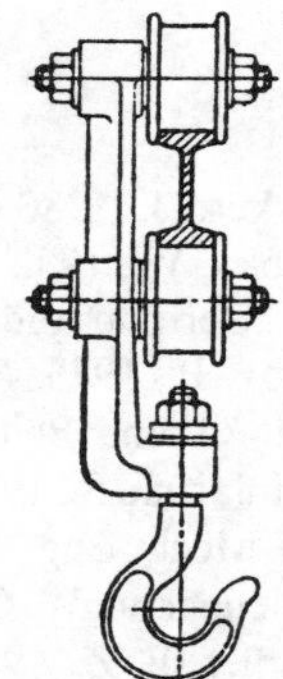

Figure P1.8

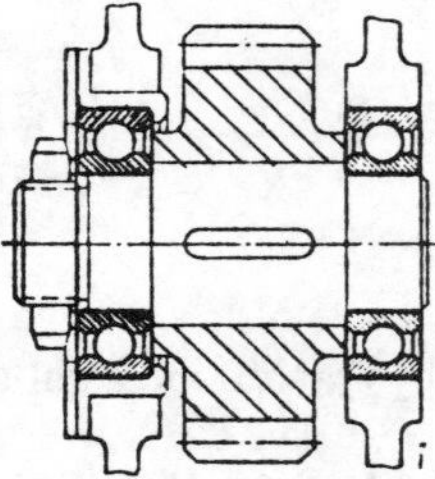

Figure P1.9

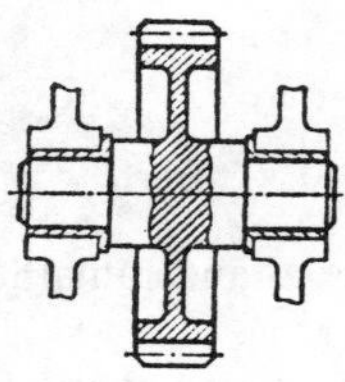

Figure P1.10

1.6 Relative motion, bracket–pin–support.
1.7 Internal–external gears.
1.8 Rail–wheel.
1.9 Gear–shaft–bearing.
1.10 Gear–shaft–bearing.

Problems 1.11 to 1.25

Design the machines respectively indicated. More specifically: using your everyday experiences and best judgement, write appropriate specifications, keeping in mind that these machines or gadgets will be manufactured for selling to the general public.

Plan a design strategy, outlining which of the seven stages that are applicable.

Execute the design plan and prepare a presentation to a prospective manufacturer who is interested to purchase your design.

1.11 A mouse trap for use in the house.
1.12 A parachute system for dropping an egg from 4 meters' height to the street without breaking.
1.13 A high-voltage mosquito killer for the backyard.
1.14 A machine that cuts almonds into halves.
1.15 A machine which places ten almonds in each chocolate bar while it is in the molten stage at a speed of one bar per second.
1.16 An automatic door-securing system which operates at night ten minutes after the lights go off.
1.17 A security lock of the car hood operated from the driver's seat.
1.18 A rififi mechanism for machining a hole in the side wall of a steel safe.
1.19 A mechanism which converts a small living-room table into an ironing table.
1.20 An artificial leg to simulate the walk of a regular person as if his leg were missing.
1.21 A machine to produce electricity out of sea waves.
1.22 A mechanism to provide orientation of a solar collector towards the sun during the day.
1.23 A welding table (1×1 meter) which can have adjusted height 0.5 to 1.2 meters and tilt up to 30° in any orientation and secured in the desired place.
1.24 A simple apparatus to incubate eggs. A heat source and a thermostatic device are essential.
1.25 A mechanism to pick up bricks from a moving transport belt and deposit them onto another one moving at 90° angle. Both belts are horizontal and they transport one brick per second.

Problems 1.26 to 1.40

Develop a function structure and a symbolic structure for the following systems:

1.26 A mouse trap for use in the house.
1.27 A parachute system for dropping an egg from 4 meters' height to the street without breaking.
1.28 A high-voltage mosquito killer for the backyard.
1.29 A machine that cuts almonds into halves.
1.30 A machine which places ten almonds in each chocolate bar while it is in the molten stage at a speed of one bar per second.
1.31 An automatic door-securing system which operates at night ten minutes after the lights go off.
1.32 A security lock of the car hood operated from the driver's seat.
1.33 A rififi mechanism for machining a hole in the side wall of a steel safe.
1.34 A mechanism which converts a small living-room table into an ironing table.

1.35 An artificial leg to simulate the walk of a regular person as if his leg were missing.

1.36 A machine to produce electricity out of sea waves.

1.37 A mechanism to provide orientation of a solar collector towards the sun during the day.

1.38 A welding table (1 × 1 meter) which can have adjusted height 0.5 to 1.2 meters and tilt up to 30° in any orientation and secured in the desired place.

1.39 A simple apparatus to incubate eggs. A heat source and a thermostatic device are essential.

1.40 A mechanism to pick up bricks from a moving transport belt and deposit them onto another one moving at 90° angle. Both belts are horizontal and they transport one brick per second.

CHAPTER TWO

DESIGN DATABASES AND DESIGN STANDARDS

2.1 DESIGN AND ARTIFICIAL INTELLIGENCE

Design has in it the element of creativity. No one design is exactly the same as another. Guidelines exist however which have been hammered out throughout decades of accumulated experience. Following those guidelines as much as possible will yield a rational and economical design in many aspects. Reduction of number of spare parts, reduction of number of special tools and materials, easy assembly and disassembly, compact design, and assurance of correct operation of the machine are some of the guiding targets that the designer of a machine must continuously consider. Modern manufacturing methods impose additional restraints: for example, assembly by robot arms is facilitated if all parts are assembled in a vertical direction. Last, but not least, national and international codes and standards have limited the choice of alternatives to a rigorously-defined set of acceptable dimensions and forms.

Limiting the possible number of acceptable design alternatives does not only improve on communication, reduction of parts in stock, required tools, drawings, etc. but also it helps in breaking down the design process into smaller and more universally-defined tasks. This is very important because it leads to a better understanding of the design process and the intelligence behind it. Indeed, the design process is an amalgam of many information-processing, decision-making and information-presentation talents, as in other human intelligence manifestations. An unambiguous understanding of this process is helped by breaking it down to smaller and better-defined tasks. If some of these tasks are fully and unambiguously defined, they might lead to machine automation of part of the design and manufacturing processes. This can eventually be implemented in computers, programed to the designer's intelligence, to a certain extent. This leads naturally to artificial intelligence, the study of ideas which enable computers to do the things that make people seem intelligent.

2.2 CODES AND STANDARDS

Long experience in machine design, manufacturing and operation has accumulated valuable data. Such data have been summarized and evaluated in many areas by government or industry standards' organizations.

Engineering standards have many purposes, such as:

Table 2.1 USA and international standards' organizations

Organization	Country	Address
AISI	USA	American Iron & Steel Institute, 1000 16th Street, N.W., Washington, D.C. 20036
ANSI	USA	American National Standards Institute, 1430 Broadway, New York, N.Y. 10018
ASME	USA	American Society of Mechanical Engineers, 345 East 47th Street, New York, N.Y. 10017
ASTM	USA	American Society for Testing and Materials, 1916 Race Street, Philadelphia, Penn. 19103
BSI	Great Britain	British Standards Institute, 2 Park Street, London, W1A 2BS, England
DNA	West Germany	Deutscher Normenausschuss (German Standards Organization), 4–7 Burggrafenstrasse, 1 Berlin 30, W. Germany
ISO	World	International Organization for Standardization, Central Secretariat, Case Postale 56, 1211 Geneva 20, Switzerland

(*a*) *Utilization of experience:* The designer should not reinvent the wheel every time he faces a problem with rotary motion.
(*b*) *Interchangeability of components:* This need became apparent at the first time of mass production of firearms and ammunition. Every bullet ought to fit properly with every firearm. Now, if you buy a ball-bearing of a certain type, it can replace a defective bearing of the same type regardless of its manufacturer.
(*c*) *Setting safety and quality standards for a range of products:* Such standards are adopted many times by governments to form mandatory codes.

The need for the designer to conform with such standards is apparent. However, he should not simply design 'to available standards' but use them with continuous criticism, wherever possible. The reason is that many times some standards are super-conservative, since in one simple standard they have to include a variety of different cases and the safety margins are set to the worst case, not always encountered.

Dimensioning standards are the most universally accepted. For this case, the International Organization for Standardization (ISO) has issued several standards which can be obtained from the national affiliate of ISO in every country. Some of the national and international standards' organizations mentioned in this book are listed in Table 2.1.

2.3 DIMENSIONING: PREFERRED NUMBERS

The old system of dimensioning in inch fractions was in effect some form of preferred numbers' system. Standard sizes however were used long ago. In Ancient Rome they used standard-size pipes for water supply. Much later, during the 1870s, Charles Renard, a French army captain, was successful in reducing considerably the number of dimensions of rope for military balloons, from 425 to 17 using known geometric series of numbers. The sequence of such numbers is obtained from one by multiplication by a constant number. In fact, for the series designated as R5, R10, R20 in accordance with ISO 3-1973, the constants are, respectively, $\sqrt[5]{10}$, $\sqrt[10]{10}$, $\sqrt[20]{10}$, giving:

R5	1				1.60				2.5	...
R10	1		1.25		1.60		2.00		2.5	...
R20	1	1.12	1.25	1.40	1.60	1.80	2.00	2.24	2.5	...
.										
.										
.										

Derived series are fractions of the standard series, such as R5/2, R10/3, R5/4,...

Using preferred numbers is very convenient in design because:

(a) it reduces inventory for many machine components, i.e. bolts, pipes, bearings, materials (for example, steels of various compositions and strength), manufacturing tools, such as drills and cutters, etc.;
(b) it results in product-line simplification when planning model sizes in respect to capacity, speed, power rating, etc.

2.4 DIMENSIONING: FITS AND TOLERANCES

In machine production, exact numbers are only the integer quantities, such as number of teeth of a gear, number of bolts of a coupling, etc.

Dimensions are never 'exact', for a variety of reasons. These are: (a) measuring errors; (b) manufacturing errors due to wear of tools, changes in temperature, etc.; and (c) intended deviations from the nominal dimension for functional purposes. For example, mass-produced shafts of diameter $d = 40$ mm have actual diameters very close to 40 mm but not exactly that. In accepting a shaft piece, one has to set some limits to the deviation from 40 mm, the 'nominal' dimension, because this shaft will have to work with other elements, such as bearings, gears, etc. or it will have to rotate within a confined annulus. For this reason, quality-control procedures set up limits to the deviations from the nominal diameter for acceptance. These limits have to be decided by the designer because many times they influence the function or the strength of a machine component.

Because interchangeability of parts imposes restrictions on the acceptable limits, one of the first standards was on those deviations called 'tolerances'. In particular, if two components must work together, their tolerances must be interrelated. In this case, the components form a 'fit'.

The tolerance range for a certain dimension depends on the manufacturing procedure used. ISO specifies 18 basic tolerance grades numbered as *IT01, IT0, IT1, IT2,...,IT16*, from finest to largest tolerances. Machine components used for fits are manufactured usually to grades *IT5* to *IT11*. The tolerance unit i is a function of the nominal diameter $D(i = 0.45\sqrt[3]{D} + 0.001D)$ and allowable tolerances for different qualities are shown in Table 2.2 as multiples of the tolerance unit i.

Table 2.2 Allowable tolerances and tolerance ranges (tolerance unit, 1 μm = 0.001 mm)

Basic hole system (D in mm)

Quality	Tolerance (μm)
IT 5	$7i$
IT 6	$10i$
IT 7	$16i$
IT 8	$25i$
IT 9	$40i$
IT 10	$64i$
IT 11	$100i$
IT 12	$160i$
IT 13	$250i$
IT 14	$400i$
IT 15	$640i$
IT 16	$1000i$

Limit dimensions (basic hole)

Hole

$D^{\mathrm{IT}}_{0.0}$

Shaft

$d^{-a_o}_{-a_o-it} = d^{-a_u+it}_{-a_u}$

Category	Minimum clearance a_o (μm)	Minimum interference a_u (μm)
a	$(265+1.3D)$, $D \leqslant 120$ $3.5D$, $D > 120$	
b	$(140+0.85D)$, $D \leqslant 160$ $1.8D$, $D > 160$	
c	$52D^{0.2}$, $D \leqslant 40$ $(95+0.8D)$, $D > 40$	
d	$16D^{0.44}$	
e	$11D^{0.41}$	
f	$5.5D^{0.41}$	
g	$2.5D^{0.34}$	
k		$-0.6\sqrt[3]{D}$, $3 < \mathrm{IT} < 8$ else, 0
m		$-\mathrm{IT7}+\mathrm{IT6}$
n		$-5D^{0.34}$
p		$-\mathrm{IT7}-0\ldots5$
r		Average of p and s
s		$-\mathrm{IT8}-1\ldots4$, $D \leqslant 50$ $-\mathrm{IT7}-0.4D$, $D > 50$
t		$-\mathrm{IT7}-0.63D$
u		$-\mathrm{IT8}-D$
v		$-\mathrm{IT7}-1.25D$
x		$-\mathrm{IT7}-1.6D$
y		$-\mathrm{IT7}-2.0D$
z		$-\mathrm{IT7}-2.5D$

Machining quality defines the tolerance range but it cannot give information on the location of this range in respect to the nominal dimension.

ISO has specified 27 fundamental deviations of the tolerance range from the nominal dimension. Each deviation is designated with a letter, lower case a, b, c,... for shaft dimensions, upper case A, B, C,... for hole dimensions. Figure 2.1 illustrates the relative position of tolerance ranges for the standard fundamental deviations (ISO 286) in respect to the nominal diameter (zero line) for shafts and holes. It can be seen that category H or h starts the tolerance range on the nominal diameter. It is apparent that the specific tolerance range and deviation for the shaft and hole will determine the operation of the fit. In fact, we usually distinguish three types of fits.

1. *Clearance, or running, fits:* All the tolerance range of the shaft is below the tolerance range of the hole (Figure 2.2a).
2. *Interference, or shrink fits:* All tolerance range of the shaft is above the tolerance range of the hole (Figure 2.2c). All hole diameters will be smaller than any of the shaft diameters and the fit should be 'forced'. Such fits are used to achieve solid connections transmitting force and torque. More about them will be discussed in Chapters 8 and 14.
3. *Transition fits:* If the two tolerance ranges overlap, that means that some pairs will have clearance fit, some others interference fit (Figure 2.2b). The same effect can be achieved for

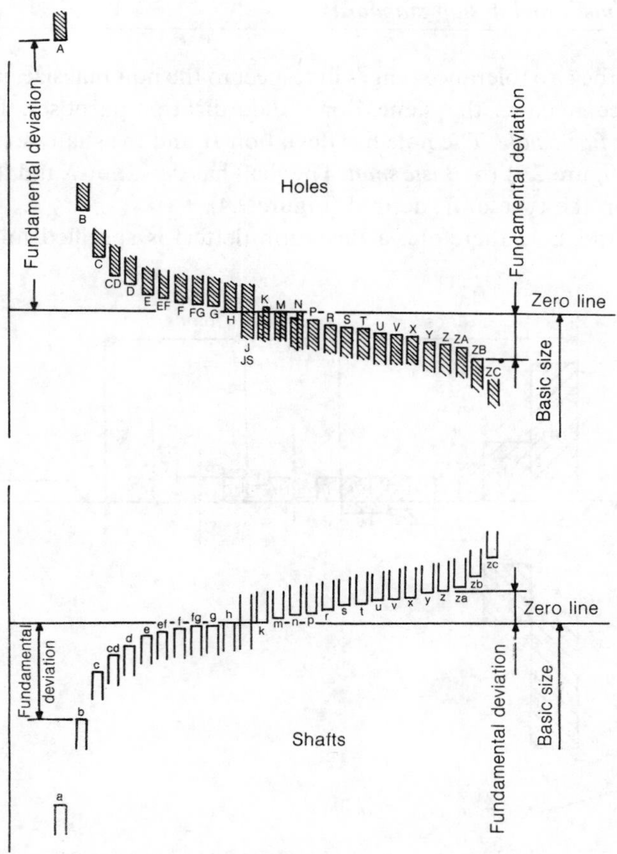

Figure 2.1 Relative deviation ranges for shafts and holes

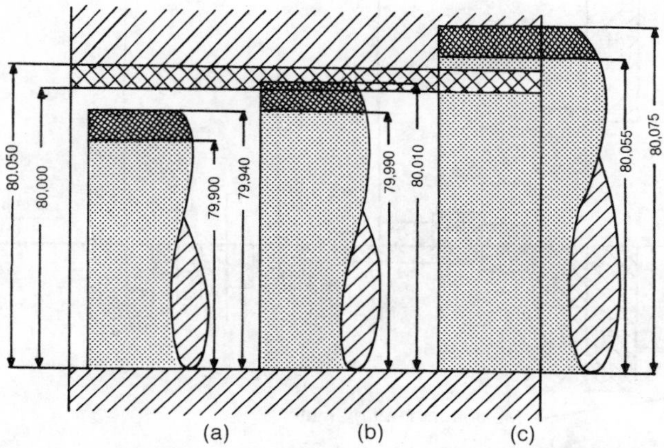

Figure 2.2 Clearance (a), transition (b) and interference (c) fits for 80-mm diameter shaft

any location of the two tolerance ranges in respect to the nominal size, provided that their relative distance remains the same. For standardization purposes, ISO specifies two alternatives: (a) *Basic hole*. The hole has deviation *H* and the shaft deviation depends on the desired fit (Figure 2.3). (b) *Basic shaft*. The shaft has deviation *h* and the deviation of the hole depends on the type of fit desired (Figure 2.4).

For shaft and hole therefore, a deviation (letter) is specified followed by a digit,

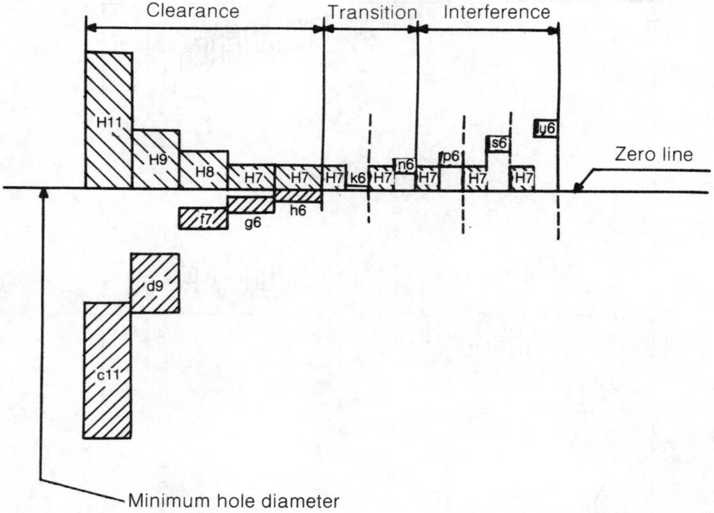

Figure 2.3 Standard fits, hole base (ANSI B4.2), to scale for 25 mm diameter

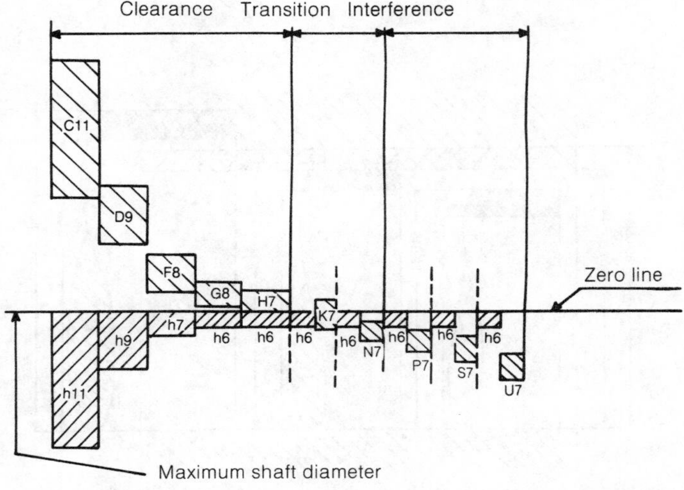

Figure 2.4 Standard fits, shaft base (ANSI B4.2), to scale for 25 mm diameter

Table 2.3 Description of preferred fits (ANSI B4.2)

	ISO symbol Hole basis	Shaft basis	Description	
Clearance fits	H11/c11	C11/h11	*Loose running* fit for wide commercial tolerances or allowances on external members.	More clearance
	H9/d9	D9/h9	*Free running* fit not for use where accuracy is essential, but good for large temperature variations, high running speeds, or heavy journal pressures.	
	H8/f7	F8/h7	*Close running* fit for running on accurate machines and for accurate location at moderate speeds and journal pressures.	
	H7/g6	G7/h6	*Sliding* fit not intended to run freely, but to move and turn freely and locate accurately.	
Transition fits	H7/h6	H7/h6	*Locational clearance* fit provides snug fit for locating stationary parts, but can be freely assembled and disassembled.	
	H7/k6	K7/h6	*Locational transition* fit for accurate location, a compromise between clearance and interference.	
	H7/n6	N7/h6	*Locational transition* fit for more accurate location where greater interference is permissible.	More interference
Interference fits	H7/p6	P7/h6	*Locational interference* fit for parts requiring rigidity and alignment with prime accuracy of location but without special bore pressure requirements.	
	H7/s6	S7/h6	*Medium drive* fit for ordinary steel parts or shrink fits on light sections, the tightest fit usable with cast iron.	
	H7/u6	U7/h6	*Force* fit suitable for parts which can be highly stressed or for shrink fits where the heavy pressing forces required are impractical.	

designating tolerance, such as $b8$, $H6$, $O7$, etc. A fit needs the specification of both shaft and hole, such as $H8/f7$, $P7/h6$, etc. Preferred fits per ANSI B4.2 (American National Standards Institute) are shown in Table 2.3 with suggested application. Figures 2.3 and 2.4 respectively show preferred base hole and base shaft fits according to ANSI B4.2. Values for minimum clearance or maximum interference for the basic hole system are shown in Table 2.2.

ISO and ANSI have published detailed tables with dimensions of different fits. These tables, for every combination of deviation and tolerance (i.e. $h6$, $D7, \ldots$) give maximum and minimum deviations from the nominal dimension A_u and A_o for the hole, and a_u and a_o for the shaft (Figure 2.5).

It is apparent that the maximum and minimum clearances between shaft and hole are

$$T_{\max} = a_u + A_u \tag{2.1a}$$

$$T_{\min} = a_o + A_o \tag{2.1b}$$

Negative clearances would mean interference fits.

The design procedure starts with selection of the machining method for shaft and hole. This yields the tolerance ranges IT and it. In usual applications, $IT = it + 1$, because machining holes is more difficult than machining shafts. Then, a basic hole or shaft system is selected. This is mostly arbitrary and machine manufacturers select one or the other on the basis of cost and availability of measurement tools. The basic hole system is more common.

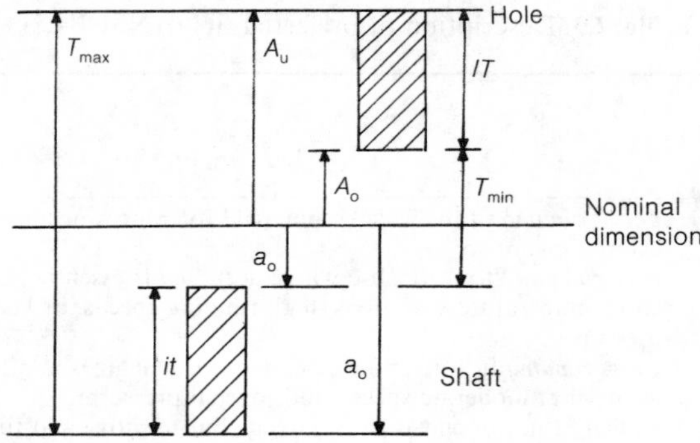

Figure 2.5 Terminology for fit features

Furthermore, from the preferred fits (depending on function) one is selected. Table 2.2 or the tolerance tables yield shaft and hole limits of tolerance a_o, a_u and A_o, A_u respectively. The dimensions N, n on the drawing are designated then as

$$N_{A_o}^{A_u}, n_{a_o}^{a_u} \quad \text{i.e.} \quad 80_0^{+30}, 80_{-7}^{+12}$$

for 80*H*7/*j*8. Nominal dimensions are in millimeters (mm) and tolerance limits in micrometers (μm).

Most important for the designer are the interference fits (or shrink fits) because they induce stresses on the fitted components which have to be taken into account in component design. This will be discussed later in Chapter 14.

In Example 2.1, a computer program is developed for the computation of the limit dimensions for a given fit, implementing Table 2.2.

2.5 DATABASES AND THE COMPUTER IN DESIGN

In the preceeding discussion on dimensioning and tolerances, one can observe certain common features:

(a) Long design experience and established standards have accumulated a vast amount of data in the form of design tables, which are essential to any design application.
(b) Dimensioning and tolerancing can be specified on the basis of clearly-defined procedures leaving no place for ambiguity. Such procedures are very simple forms of what is known as 'expert systems'.

These procedures lead naturally to computer implementation and they are simple but important applications. The first leads to what is known as a 'database', the second to an appropriate computer program. Both are some of the usual manifestations of what we define as 'computer aided machine design' (CAMD).

In turn, this is a part of computer aided design (CAD), the application of computer technology to the methods and procedures of planning, performing and implementing the design process.

2.5.1 Development of computers

The first digital computing device, the counting board known as the abacus, was used in the first century B.C. in eastern Mediterranean countries. The Antikythira computer of about the same time and place was the first device to perform extensive computations, namely the motion of the planets of our solar system.

The development of mathematics which followed the Industrial Revolution formed the basis for development of the need and the tools for the advent of the digital computer. It is not surprising that famous mathematicians like Pascal and Leibnitz are inventors of the first mechanical calculating machines.

The first device with storage and programing possibility was designed by Babbage in the nineteenth century, but not built at that time. This concept was materialized at the end of World War II in the Mark I computer in the USA and the Colossus in England, which were basically electromechanical devices.

The earliest of the vacuum-tube electronic computers, ENIAC, was developed at the University of Pennsylvania in 1946, while the first commercial computer was the UNIVAC in 1951.

The transistor, invented in 1948, was the basis for the first widely-used digital computers after 1959.

In the early 1960s a technique was developed for placing a number of transistors on a silicon chip forming complete electronic circuits. This led to the development of relatively small computers yet with substantial capacity, such as the PDP-8 minicomputer developed in 1965 by Digital Equipment Corporation.

In 1971 a new technology of large-scale integration, placing very complicated circuits on a silicon chip, led the Intel Corporation to the development of a complete computing device on such a chip, smaller than one square centimeter. Such a device is known as a 'microprocessor' and with addition of memory and input–output devices can perform full computer functions. This device was called a 'microcomputer' and resulted in the general availability of computing capacity. One of the first such microcomputers, the ALTAIR, was developed in 1975 by MTS Corporation.

2.5.2 Computer systems

The heart of a computer system is the central processing unit (CPU) (Figure 2.6), which manipulates the data. The data are presented to the CPU as electric pulses and pauses, corresponding to 1s and 0s. The communication is based on codes which make use of only these two binary digits. This smallest possible piece of information is called a 'bit'. To make communication faster, the system can send and receive a number of bits at the same time. Thus, one speaks about 8, 16, 32 and 64-bit systems. In an 8-bit system for example, the CPU receives and sends messages, called 'bytes', such as 10110001, 00010000, etc. one at a time, through several lines of communication. These lines form a communication bus. On this bus

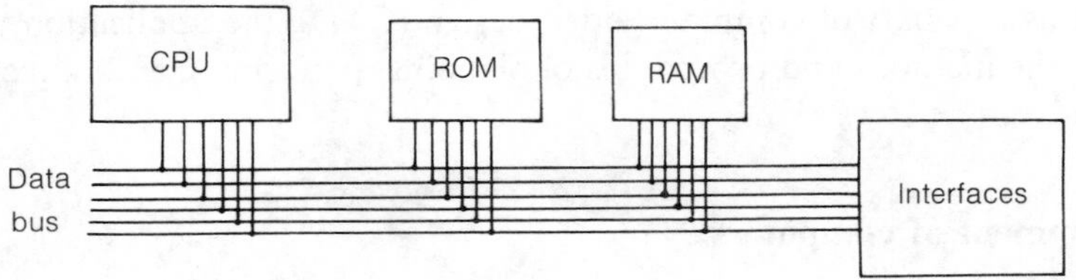

Figure 2.6 Computer architecture principle

several other devices are connected, such as memory modules, input–output devices, mass storage units, screens, printers, plotters, etc.

In microcomputers, in particular, all the computer programs which control the system are often stored in a memory device called ROM – read only memory – because it is protected from being written on by the computer user. It consists of modules which can store one bit, 0 or 1. Many bits are arranged in bytes and sometimes many bytes form words. This memory is not volatile, i.e. not erased when we turn the power off. It usually ranges in quantity between 8 and many hundred kbytes (1 kbyte is 1024 bytes).

Another piece of memory is connected to the bus: The read–write (random access) memory (RAM) where during the execution of a program all the data, results and the program itself are stored. When we turn the power off, this memory is usually erased.

Every computer can handle (address) a certain amount of memory. In 8-bit systems this maximum is approximately 64 kbytes. The 16- and 32-bit systems can address much more memory. Contemporary micros have RAM ranging from 256 kbytes to several megabytes.

Not all the RAM is available for storage to the user. Part of it is occupied by the graphics screen, operating system, etc., depending on the system and the high-level language used (see Section 2.5.3).

The bus is connected finally to interface devices which connect the computer to peripherals, such as disc drives, printers, plotters, etc.

The speed of the system depends on several factors. First is the CPU speed. It has a timing device which gives the tempo for the flow of information. In microcomputers this timing has a frequency from 5 to 20 MHz. One should not think that a 10-MHz processor has double the speed of a 5-MHz one. The reason is that memory access does not follow the speed of the CPU. It is advisable to check the speed of the system with a small program which should be representative for the intended use. Such programs are called 'benchmark tests' and the results differ for different benchmark tests for different systems. For the work described in this book, a quick representative test in BASIC will be

```
10 FOR I = 1 TO 1000
20 A = COS (I/1000)
30 NEXT I
```

This program in most 8- and 16-bit microcomputers takes about 30 seconds, the same as in an IBM-PC. In 16- and 32-bit machines, like IBM-AT and the APPLE McIntosh, it takes about 10 seconds. Compiled languages are several times faster. Minicomputers take about 0.01 second and mainframe computers take less than 0.001 second.

Computers are usually classified as mainframes, mini- or microcomputers. They differ in size, speed, cost and range of application.

There is no clear dividing line between them. In general, a microcomputer has the CPU on one silicon chip, the minicomputer on a printed circuit board and a mainframe in a more complex set of electronics. The difference in performance changes with time very rapidly. Technological advancement continuously changes the performance limits to the extent that a microcomputer today might be more powerful than a mainframe of fifteen years ago.

2.5.3 Computer utilization

It was said above that the computer communicates with 0 or 1 bits arranged in groups. Even the simplest arithmetic operation requires a number of such 8 (or more) bit messages (bytes), to be sent through the data bus. Programing directly with this language is a very complicated affair – quite hopeless. For this reason the CPU manufacturer has incorporated circuits which respond to certain combinations of bytes, corresponding to letters or numbers such as ADC, STA, etc. Each message of this type is equivalent to several one-byte messages. This language of communication is called ASSEMBLY and it is for experienced programmers only.

For the average user, software producers have written programs in the ASSEMBLY language which can respond to simpler commands, such as PRINT, $A = B + C$, etc., in plain English language or plain algebraic expressions. This feature makes programing easier for the user and the latter languages are called 'high-level languages' (HLL). They consist of a number of English words, between 50 and 200, with which one can give almost any relevant instruction to the computer. Such languages are BASIC, FORTRAN, PASCAL and C. The fundamental difference between BASIC and FORTRAN is that the first is interpretive (that is the computer executes the program translating instruction-by-instruction while executing the given program) while FORTRAN is compiled (that is the whole program is translated first and then executed). Another important difference is the program structure: A BASIC program is dedicated to a specific task; FORTRAN, PASCAL and C programs are structured to different levels, that is they consist of hierarchical modules which can be reassembled with little additional programing to form a different program.

Today the differences have been diminished by the advent of compiled versions of BASIC and interpretive versions of FORTRAN, while structured versions of BASIC are also available.

In general, BASIC is easier for inexperienced computer users to master computer literacy and for quick development of new programs, while FORTRAN, PASCAL and C are faster and more efficient, in general, and can be used for production programs.

Parts of high-level language programs, which must be executed at high speed, could be programed in ASSEMBLY to improve efficiency.

For specific purposes, problem-oriented languages have been developed, for example the ICES system for structural analysis. The problem here is programed in a technical language directly, such as:

```
TYPE SPACE FRAME
UNITS TON METER
JOINTS 12 MEMBERS 16
```

JOINT COORDINATES
1 Y 30. SUPPORT
.
.
.

2.5.4 Computer programing

Programing in high-level languages (HLL) involves a number of steps which end up to a specific code in the chosen HLL.

Solution of an engineering problem is achieved in a number of well-defined steps. These steps might be:

1. Specification of problem data
2. Selection of applicable formulas and constants
3. Decisions as to the applicable formulas
4. Decisions as to the applicable constants
5. Appropriate calculations
6. Presentation of the results

A set of instructions fully and exactly specifying the proper steps is an 'algorithm'. A simple algorithm might be expressed in plain language as a set of sentences describing the process. If the process is more complicated, a symbolic representation is used in the form of a block diagram. For example, Figure 2.7 shows a block diagram for the solution of quadratic equations.

A block diagram must unambiguously define the process and it must be complete by covering all possible alternatives. The more experienced the programer and the more simple the algorithm, the less need exists for a block diagram. However, a detailed block diagram will never be harmful.

The algorithms developed in this book will be presented in a simplified form, in algorithmic notation, in order that the reader will code them in the language of his choice. Block diagrams will be given when the complexity of the algorithm demands it. At the end of each chapter, programs coded in Advanced BASIC (compatible with QUICK BASIC 4.0) for the IBM-PC/AT systems will be given for the benefit of those who want to do the homework problems on their personal computer. The programs have been written upwards compatible to facilitate, as much as possible, transfer to minicomputers and mainframes. Most programs run well on an IBM-PC or AT. Larger graphics, modelling, optimization or finite-element programs will have to be transferred eventually to larger machines and more-efficient languages. It is expected that transferring the programs or programing directly in FORTRAN or PASCAL on larger machines will not be difficult. Moreover, FORTRAN and C language versions of the book software have been prepared.

An algorithm, or procedure, will have to include the following information (explanatory comments are indicated in italics):

(a) Name of the procedure, if possible a short-hand description of the procedure, i.e.

Procedure: QUADR.EQ (**Solution of a quadratic equation* $ax^2 + bx + c = 0$*)

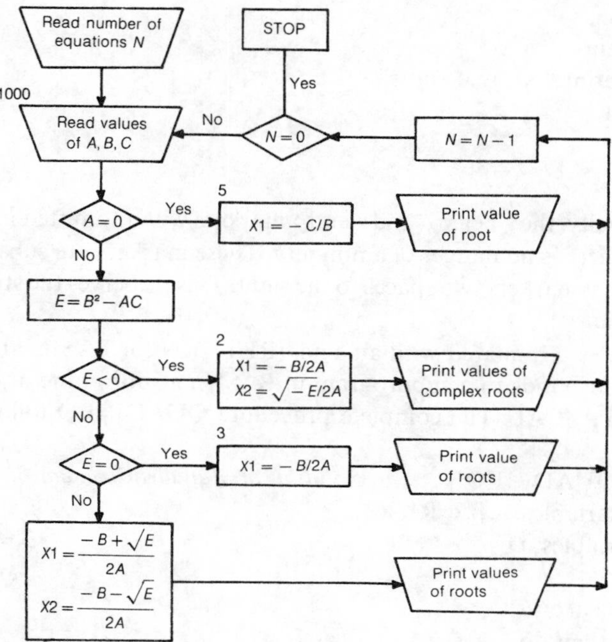

Figure 2.7 Algorithm to find all the roots of $ax^2 + bx + c = 0$

(b) Variables and constants which are global, that is communicated to main program:

GLOBAL variables: a, b, c, R1, R2

(c) Local variables and constants, used only within the procedure:

LOCAL variables: D (*$D =$ *the discriminant**)

(d) The main procedure. Marked with 'begin' and 'end' are the main procedure and all the elementary procedures that form the main procedure:

```
begin
   D:= b^2-4*a*c
   IF D:= 0 then do
      begin
         R1:= R2:= - b/(2*a)
      end
   else
      begin
         if D > 0 then do
            begin
               R1:= ( - b + SQR(D))/(2*a); R2:= ( - b - SQR(D))/(2*a)
            end
```

```
        else
          begin
            print "No real roots"
          end
      end
  end
```

For simplicity, the ⟨begin⟩ and ⟨end⟩ markers will be omitted in the algorithms to follow where there is no danger of ambiguity. These markers are substituted by moving the lines between markers two spaces to the right. This preserves the structured character of the algorithm.

(e) The procedure is terminated with an 'end' (RETURN for BASIC and FORTRAN). A procedure corresponds to a subprogram in BASIC, a subroutine in FORTRAN and a procedure in PASCAL. The complete procedure QUADR.EQ follows:

```
Procedure: QUADR.EQ        (*Solution of a quadratic equation ax^2 + bx + c = 0*)
GLOBAL variables: a, b, c, R1, R2
LOCAL variables: D                                   (*D = the discriminant*)
begin
  D:= b^2-4*a*c
  IF D:= 0 then do
    R1:= R2:= - b/(2*a)
  else if D > 0 then do
    R1:= (- b + SQR(D))/(2*a); R2:= (- b - SQR(D))/(2*a)
  else
    Print "No real roots"
end
```

Global variables in FORTRAN are the arguments or shared variables declared in a 'common' statement. In BASIC all variables are global. Dimensions must be declared whenever needed in PASCAL or FORTRAN in the procedure.

Wherever necessary, variable names will have a 'type' ending:

% for integer
$ for alphanumeric
# for double precision

The symbol := will be used to indicate an equation in the sense of replacement. A semicolon ; separates statements.

2.5.5 Computer aided machine design

Computers were used in machine design by only progressive industries in the '60s while in the '70s this became standard practice.

Computer aided machine design involves more than design calculations. As discussed previously in Section 2.4, a large number of data, standard tolerances, have been tabulated for use in design. Available parts in a large industry might be of the order 100,000. Extensive catalogs give properties, prices, etc. As outlined in the introduction to Section 2.5, a computer system for storage and retrieval of a large body of data is called a 'database'. In fact, accumulated design experience serves as a very large database from which the designer is obtaining continuously essential information.

There was always a link and a separating line between design and manufacturing – a set of drawings. The production of drawings can be facilitated by a computer with the aid of an output device called an automatic drafting machine or a plotter. But now, the communication between design and manufacturing is not only the drawings. The design program output can be transmitted to manufacturing in electronic form to directly program the machines for production, to initiate the material planning process, etc.

A preliminary design, once conceived by the designer but now possibly produced in a computer, needs to be analyzed, that is to find if this design satisfies the requirements set by the initial specifications. Design analysis usually involves lengthy calculations, especially when numerical methods such as finite element analysis are employed. Computers can facilitate this effort. Moreover, repeated calculations of the same nature can be performed safely in a short time with high reliability, once the calculation has been initially verified. (The author was once tempted to apply a computer-aided design analysis program to older machines designed the conventional way. This was easy because the data for each machine

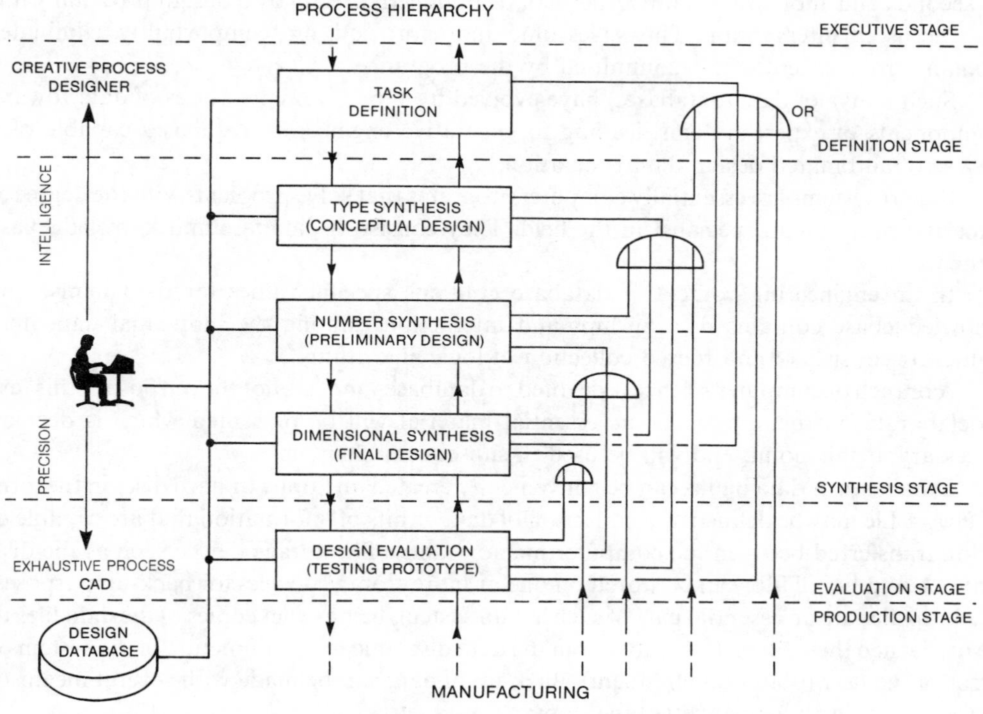

Figure 2.8 Design flow in CAD

was available in a database. The result was that more than 30% of the machines were designed with numerical mistakes in the hand calculations – fortunately none were critical.)

To assess the actual operating conditions of a machine, it is often necessary to approximate its operation by way of a mathematical model, expressed finally by a set of equations. Their solution is usually performed in the computer. The process is modeling. Furthermore, many times the design parameters can be improved to yield a better design satisfying some optimality requirements, such as lower cost, higher reliability, etc. This process is called 'optimization'.

Databases, computer drafting, design analysis, design synthesis, modeling and optimization, are the main manifestations of the computer aided machine design process.

2.6 DATABASES AND DATA FILES

In Section 2.4 the need to keep and look through tables with enormous amount of data became apparent. And this is not the only case in a design environment. A great number of tables must be kept and updated. Some examples are tolerances, materials properties, standard sizes of bolts, bearings, rivets, etc., steam tables and many others.

Computers can be of great help. For example, any one of the well-known mechanical engineering books with Steam Tables, can be included in a three page BASIC program, occupying no more than 4 kbytes of memory on a floppy disc. Properties then can be found in seconds and, more importantly, they can be obtained directly by a design program without the user's intervention. This saves time and effort and, most importantly, eliminates human errors, except errors committed by the programer.

Such banks of data (databases) have evolved from passive repositories of data toward components of expert systems, leading to generative engineering databases capable of a degree of automated design data generation.

Expert systems are essentially computer programs that solve problems with the degree of expertise of a human specialist in the field. They consist of database and knowledgebase systems.

In an engineering context, a database contains specific values for data items. The knowledgebase contains relationships and interactions among the individual data item values. It consists of an ordered collection of logic algorithms.

A branch of computer science is devoted to databases and it is not the purpose of this text to elaborate on the subject. Some essential material will be presented which is deemed necessary at this point, and will be used in subsequent chapters.

Databases or data banks can be conveniently created on floppy or hard discs in the form of files. A file may be defined as a collection of data or bits of information that are capable of being transferred between the computer memory and a file storage device such as the disk drive. A number of files can be stored on one or more storage devices for back-up purposes. Back-up files are an essential part of such a data system, being exact copies of the data files to be used when the information is lost from the data discs due to a malfunction of the system or error of the user (both quite frequent). Back-up of data can be made with several means of mass storage such as magnetic tapes, optical discs, etc.

Two main types of files are defined: program and data files.

1. Program files are the BASIC, FORTRAN, etc. programs that are also stored on discs.
2. Data files are collections of numbers, alphanumeric data, pictures, drawings, etc.

Unfortunately, the computer programs that operate the disc drive and transfer data back and forth between disc drive and computer memory are similar but not the same among different computer systems. These programs are called 'operating systems'. Transferring databases from one computer to another can be made under certain conditions.

Data files are reserved areas in the storage device where data are stored in an orderly way. Depending on the order of data storage, data files are distinguished as either sequential or random access.

2.6.1 Sequential files

A sequential disc file has the data arranged in the order of its input. To access one piece of data, we have to access sequentially all data up to the one we need. To change one piece of data, we have to transfer the file on memory, make the change and save again the file on disc. Only small data files can be efficiently stored as sequential data files. On the other hand, it is very easy to create and retrieve them.

To do anything with a file, we must OPEN it, assigning a name to it. Let us use the name FNAME$. Writing data in the file, numeric or alphanumeric, requires a WRITE command. Finally, the file must be closed to be saved with the CLOSE command. The file is now created and saved on disc for later use. The exact syntax of the commands depends on the system and language at hand (explanatory comments are indicated in italics):

```
Procedure: LSEQF                                  (*Load sequential file*)
GLOBAL variables:FNAME$, A1, A2,...
begin
  OPEN File FNAME$ for output
  WRITE Data A1, A2,...
  ...
  ...
  CLOSE file
end.
```

To retrieve data, the same sequence must be followed, READing or INPUTing the same, or smaller number of data. Data types must be maintained (integers, floating point, alphanumeric, etc.). Variable names need not be the same as in the writing statements.

```
Procedure: RSEQF                                  (*Retrieve sequential file*)
GLOBAL variables: FNAME$, A1, A2,...
begin
  OPEN File FNAME$ for input
  READ Data A1, A2,...
  ...
  ...
  CLOSE file
end.
```

2.6.2 Random files

A random file consists of several smaller sequential files, each assigned a numeral. They can be considered as numbered paragraphs in a longer text. Each such small sequential file is called a 'record' and has a specified length. These numbered records are arranged in the random file in a way convenient to the operating system and which does not interfere with the user. He can retrieve one or more records, by their numbers and use or modify them and save them back into the random file.

It is not necessary to read, modify and save the complete file, as in the case of sequential files. This saves time and very large files can be created exceeding the memory available to the computer.

Procedures for loading and retrieval of a random file record are as follow:

```
Procedure: LRNDF                    (*Loading random file FNAME$ with a record*)
GLOBAL variables: FNAME$, RECORD%, NCOL%, LEN%, C$(1..NCOL%)
begin
  OPEN file FNAME$ of record length LEN%:= length          (*Length in bytes*)
  FIELD description of the positions of variables on the
  record: 1 AS C$(1)$, 10 AS C$(2)$,...
  PUT or WRITE record RECORD% in the file (C$(I), I = 1 TO NCOL%)
  CLOSE the file
end.

Procedure: RRNDF                (*Retrieve a record No. RECORD% from
                                   a random file FNAME$ with NCOL% data*)
GLOBAL variables: FNAME$, NCOL%, RECORD%, LEN%, C$(1..NCOL%)
begin
  OPEN file FNAME$ of record length LEN%:= length in bytes
  FIELD description of the positions of variables on the
  record, i.e.: 1 AS C$(1), 10 AS C$(2),...
  GET or INPUT record RECORD% from the file (C$(I), I = 1 TO NCOL%)
  CLOSE the file
end.
```

In most systems, data stored in random files must be alphanumeric. Therefore, before loading and after retrieving records, appropriate conversions must be made.

In general, for small files one should use sequential files and for large ones use random files. Random files are also used when data are to be taken from the file frequently.

2.7 PROGRAM DESIGNDB

Databases for the storage and retrieval of data will be very frequently encountered in this book. Examples include physical and mechanical properties of materials, standard dimensions of screw threads, and catalogs of antifriction bearings. There are many commercial programs for this purpose. A simple program will be developed here suitable for design applications. For machine-design applications random files seem more suitable.

The program will consist of the following procedures:

(a) The heading file FNAME$ will contain data for the structure of the data file:

HEAD$, NREC%, NCOL%, C$[1], C$[2],..., C$[NCOL%]

where HEAD$ is the title of the table, NREC% is the number of records (one for each horizontal line in the table), NCOL% is the number of columns, and C$[I], I = 1, 2,..., NCOL% are the headings of the NCOL% columns.

The procedure HEADREC will create a new heading file or alter one previously created.

```
Procedure HEAD (File:FNAME$)
GLOBAL variables: HEAD$, NREC%, NCOL%, Array C$(1..NCOL%),
RECORD%
LOCAL variables INDEX%, I, CLOCAL$
begin
  enter INDEX%                                  (* = 1 for new heading*)
                                                (* = 2 to modify heading*)
  IF INDEX%:= 1 THEN
     enter HEAD$, NCOL%
  FOR I:= 1 to NCOL% do
     enter C$(I)
else
  if INDEX%:= 2 THEN
     enter file name FNAME$
     RECORD%:= 1
     CALL procedure RRNDF, file FNAME$
     FOR I:= 1 TO NCOL% do
        print C$(I); enter CLOCAL$
        IF CLOCAL$ <> " " THEN C$(I):= CLOCAL$
     CALL procedure LRNDF
  else print "try again"
end.
```

(b) A string variable XREC$[1..NCOL%] will be used for the NCOL% entries of each record. A procedure to create and add a new record of the data file follows:

```
Procedure ADDREC (File: FNAME$)
GLOBAL variables: HEAD$, RECORD%, NREC%, NCOL%, Arrays
REC$(1..NCOL%), C$(1..NCOL%)
LOCAL variable INDEX%, I, CLOCAL$
```

```
begin
  enter file name DFILE$
  RECORD%:= 1
  CALL procedure RRNDF, file DFILE$
  FOR I:= 1 to NCOL% do
    print C$(I), " = "; enter CLOCAL$
    XREC$(I):= CLOCAL$
  NREC%:= NREC% + 1
  CALL procedure LRNDF                (*Load new heading with new NREC%*)
  RECORD%:= NREC%
  FOR I:= 1 TO NCOL% do
    C$(I):= XREC$(I)
  CALL procedure LRNDF                                 (*Load new record*)
end.
```

(c) A procedure to examine and modify an existing record of number RECORD% follows:

```
Procedure MODREC (File: FNAME$)
GLOBAL variables: HEAD$, RECORD%, NREC%, NCOL%, Arrays
REC$(1..NCOL%), C$(1..NCOL%)
LOCAL variables INDEX%, I, CLOCAL$, Array TITLE$(1..NCOL%)
begin
  enter file name DFILE$
  RECORD%:= 0
  CALL procedure RRNDF
  FOR I:= 1 TO NCOL% do
    TITLE$(I):= C$(I)
  enter record number RECORD%
  CALL procedure RRNDF
  FOR I:= 1 TO NCOL% do
    print TITLE$(I), " = ", C$(I), " = ?"; enter CLOCAL$
    IF CLOCAL$ < > C$(I) THEN do
      C$(I):= CLOCAL$
    else continue
  CALL procedure LRNDF                            (*Load modified record*)
end.
```

A main program to call the appropriate procedure will be a simple menu program. Such a program appears in Appendix 2.A in BASIC. This program has some more features including a line printer tabulation of the data in the file. All the subsequent tables in this book have been compiled with the program DESIGNDB.

There is a wealth of excellent software for database systems. Such a program is SYMPHONY, developed by the Lotus Corporation.

2.8 EXPERT SYSTEMS

The problem of selecting the proper limit dimensions to achieve a desired fit, Section 2.4, is not merely one of application of a specific formula or a specific algorithm. It involves utilization of experience, rules of thumb, constraints, codes and standards, plus a large collection of data from which the proper information must be found. This is a very simple example for a class of problems which require specialized knowledge that experts in a particular field acquire from long experience. Such problems lead to interactive computer programs incorporating judgement, experience, rules of thumb, codes and standards and other expertise to provide knowledgeable advice to the designer. These programs are knowledgebases which together with databases form expert systems.

A knowledgebase consists of a number of knowledge modules. Some of these modules incorporate surface knowledge (empirical rules, codes, rules of thumb) while some others comprise deep knowledge (algorithmic procedures).

Expert systems cannot replace the designer's intelligence but they can approach it to varying degrees depending on the problem. Databases can incorporate almost any conceivable number of data in any design environment. Knowledgebases cannot replace the designer in most design endeavors. They can, however, do most of the work in many cases and therefore they are becoming an increasingly important part in the design process.

Computers play an increasing role in machine design today. In fact, a recent study revealed that computer aided design in mechanical engineering is greater than the similar activities in all the other engineering disciplines put together.

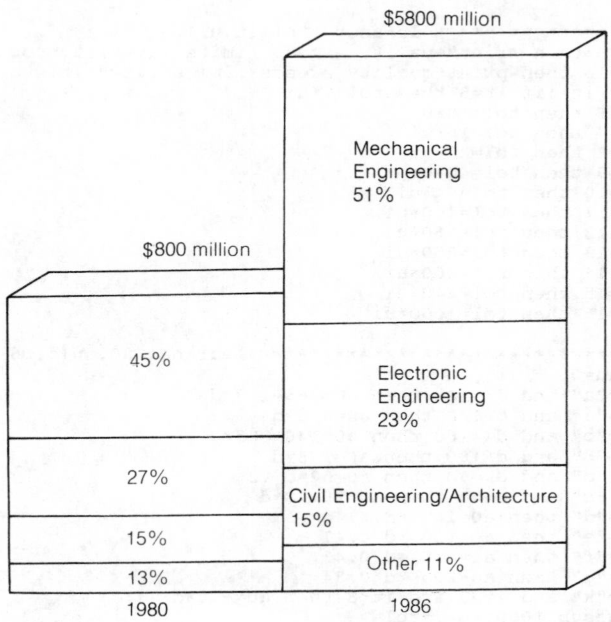

Figure 2.9 Use of computers in various professional fields, 1980 (left) and 1986 (right), expressed in $ million investment

EXAMPLES

Example 2.1

Write a program to compute limit dimensions of fits for qualities *IT6* to *IT11* using the values of Table 2.2, in the basic hole system.

Solution

The equations for the tolerance unit, the tolerance ranges and the minimum clearance and interference have been coded to materialize the data in Table 2.2. The resulting program and a test example for an 80 mm diameter, H8/*d*7 fit follows, Figure E2.1.

```
10 REM   Computation of dimension tolerances for the Basic Hole System
20 rem
30 rem
40 input"Enter Basic Dimension (mm)                ";d
50 input"Enter Quality of Shaft                  it";its
60 input"enter Quality of Hole                   IT";ith
70 input"Enter category of Shaft (Hole=H)         ";c$
100 rem ...................................tolerance unit, micrometers:
110 ui=.45*(d)^(1/3)
120 it=its:gosub 1000:stol=tol:rem ..............compute shaft tolerance
130 it=ith:gosub 1000:htol=tol:rem ..............compute hole tolerance
150 it=its:gosub 2000:rem .................compute minimum interference
160 print:print:print
170 dhupper=htol:dhlower=0
180 if a0>0 then dsupper=-a0:dslower=-a0-stol
190 if abs(au)>0 then dsupper=-au+stol:dslower=-au
192 print d;"H";ith;"/";c$;its;"  fit":print:print
200 print "SHAFT:  ";d+dslower/1000;"  <  d  <  ";d+dsupper/1000
210 print "HOLE:   ";d+dhlower/1000;"  <  D  <  ";d+dhupper/1000
300 end
1000 rem ***************tolerance  tol(it,ui)
1010 if it<5 then print"quality exeeds limits ,it=";it:stop
1020 if it>16 then print"quality exeeds limits ,it=";it:stop
1030 it=int(it):if it=5 then tol=7*ui
1040 if it=6 then tol=10*ui
1050 if it=7 then tol=16*ui
1060 if it=8 then tol=25*ui
1070 if it=9 then tol=40*ui
1080 if it=10 then tol=64*ui
1090 if it=11 then tol=100*ui
1100 if it=12 then tol=160*ui
1110 if it=13 then tol=250*ui
1120 if it=14 then tol=400*ui
1130 if it=15 then tol=640*ui
1140 if it=16 then tol=1000*ui
1150 return
2000 rem *******************************clearance a0,au(d,c$)
2010 ao=0:au=0
2020 if c$="a" and d<=120 then a0=265+1.3*d
2030 if c$="a" and d>120 then a0=3.5*d
2040 if c$="b" and d<=160 then a0=140+.85*d
2050 if c$="b" and d>160 then a0=1.8*d
2060 if c$="c" and d<=40 then a0=52*d^.2
2070 if c$="c" and d>40 then a0=95+.8*d
2080 if c$="d" then a0=16  *d^0.44
2090 if c$="e" then a0=11  *d^0.41
2100 if c$="f" then a0=5.5 *d^0.41
2110 if c$="g" then a0=2.5 *d^0.34
2120 if c$="k" and it>3 and it<8 then au=-.6*d^(1/3)
2130 it=7:gosub 1000:tol7=tol
2140 it=8:gosub 1000:tol8=tol
2150 it=6:gosub 1000:tol6=tol
2200 if c$="m" then au=-tol7+tol6
2210 if c$="n" then au=-5*d^.34
```

Figure E2.1 (*Contd.*)

```
2220 if c$="p" then au=-tol7-2.5
2230 if c$="s"  and d<=50 then au=-tol8-2.5
2240 if c$="s" and d>50 then au=-tol7-.4*d
2250 if c$="t" then au=-tol7-.63*d
2260 if c$="u" then au=-tol7-d
2270 if c$="v" then au=-tol7-1.25*d
2280 if c$="x" then au=-tol7-1.6*d
2290 if c$="y" then au=-tol7-2.*d
2300 if c$="z" then au=-tol7-2.5*d
2310 if a0=0 and au=0 then print"Category ;\";c$;" not programmed":stop
2320 return
```

```
Enter Basic Dimension (mm)           ? 80
Enter Quality of Shaft             it? 7
enter Quality of Hole              IT? 8
Enter category of Shaft (Hole=H)     ? d


 80 H 8 /d 7    fit


SHAFT:    79.85896    <  d  <   79.88998
HOLE:     80    <  D  <   80.04848
```

Figure E2.1

Example 2.2

Use DESIGNDB, Appendix 2.A, to generate a table of the first three powers of the integers between 1 and 10.

Solution

A heading defining file, POWERS is first created, selecting 'N' in the main menu. 10 byte long fields are used. Then, from the data file menu, 'N' is selected to create a new data file with the name DATAFILE. Finally, from the data file menu, 'L' is selected to list the results on the line printer, Figure E2.2. Other options in the data file menu are 'A' to add a new record, 'E' to edit an existing record, 'P' to print on the screen any records, 'F' to change data file. The procedure is tabulated in Figure E2.2.

```
                         DESIGNDB                    DESIGNDB
                          DESIGNDB                  DESIGNDB
                           DESIGNDB                DESIGNDB
                            DESIGNDB              DESIGNDB
                             DESIGNDB            DESIGNDB
                              DESIGNDB          DESIGNDB
                               DESIGNDB        DESIGNDB
   A random file development and maintainance program for design data bases
                                DESIGNDB  DESIGNDB
                                 DESIGNDBDESIGNDB
                                  DESIGNDESIGNDB
                                   DESIDESIGNDB
                                    DEDESIGNDB
                                     DESIGNDB

                       BASICA must be loaded with an extension:
                                 >A:BASICA/S:300
                       If you loaded BASICA properly, hit RETURN
           If NOT ,then hit S and RETURN to stop and then reload BASICA ?
```

Figure E2.2 (*Contd.*)

```
            INSTRUCTIONS
1. Database Defining file: The name of your choice for a file
   which will contain the top headings of your columns
   Examples: MYFILE , bolts.db , ME322, ...
2. Number of entries per record. Your Database consists of ta-
   bles with records, each record occupies one horizontal line.
   Each record (or line) has a number of columns-entries
3. Length,Name of entry: Each entry (column) has a heading
   Examples: 15,Diameter or 12,length or 20,age or 30,address,...
   All the database information are stored in the defining file
4. Each Database has many tables with the same headings
   For example, tables of different types of bolts. Give a file
   name to each of these tables.The structure will be:

                     DATABASE DEFINING FILE
  data file 1              data file 2              datafile 3    ...

  record 2                 record 2                 record2       --
  record 3                 record 3                 record 3      --
  --------                 --------                 --------      --
  Record 1 is reserved for file information
5. For a hard copy of a datafile, use the List File option
Hit return to continue, S to stop ?


                               DESIGNDB


         FILE DEFINITION MENU

    <N> Creation of new defining file
    <P> Use of previously defined file
    <Q> Quit

  Enter your selection ? n

  Enter name for the database defining file       ? powers
  Enter number of entries per record              ? 3
  Enter length, name of entry                   1 ? 10,number
  Enter length, name of entry                   2 ? 10,square
  Enter length, name of entry                   3 ? 10,cube
  DEFINING FILE powers CREATED

                     MENU

    <P>    Print records of file

    <E>    Edit records of file

    <A>    Add new records

    <L>    List of file

    <F>    Change file

    <Q>    Quit




    Enter your selection:  ? a

    New record number =  2

    number           ? 1
    square           ? 1
    cube             ? 1
```

Figure E2.2 (*Contd.*)

```
Edit (E) , Save (S),Add more records (A), Menu (M) ? a

New record number = 3

number          1   ? 2
square          1   ? 4
cube            1   ? 8
...
```

```
Defining file: POWERS    Data file:   DATAFILE
```

number	square	cube
1	1	1
2	4	8
3	9	27
4	16	64
5	25	125
6	36	216
7	49	343
8	64	512
9	81	729
10	100	1000

Figure E2.2

REFERENCES AND FURTHER READING

Clifford, J., Jarke, M., Vassiliou, Y., 1983. 'A short introduction to expert systems', *Database Engineering*, Vol. 6, pp. 3–16.

Date, C. G., 1981. *An Introduction to Database Systems.* Reading, Mass.: Addison-Wesley.

Encarnacao, J., Schlechtendahl, 1983. *Computer Aided Design.* Berlin: Springer-Verlag.

Horowitz, E., Sahni, S., 1984. *Fundamentals of Computer Algorithms.* Rockville, Maryland: Computer Science Press.

ISO, Standards ISO/R 286-1962, ISO/3-1973.

Keenan, J.H. *et al.*, 1969. *Steam Tables.* New York: John Wiley.

Knox, C. S., 1983. *CAD/CAM System, Planning and Implementation.* New York: Marcel Dekker.

Kverneland, K. O., 1978. *World Metric Standards for Engineering.* New York: Industrial Press.

Lang, J. C., 1982. *Design Dimensioning with Computer Graphics Applications*, New York: Marcel Dekker.

Pao, Y. C., 1984. *Elements of CAD/CAM.* New York: J. Wiley.

Rasdorf, W. J., Salley, G. C., 1985. 'Generative engineering databases – towards expert systems', *Computers and Structures*, Vol. 20, nos 1–3, pp. 11–15.

Rich, E., 1983. *Artificial Intelligence.* New York: McGraw-Hill.

Ullman, J.D., 1980. *Principles of Data Base Systems.* Potomac, Maryland: Computer Science Press.

Winston, P. H., 1977. *Artificial Intelligence.* Reading, Mass.: Addison-Wesley.

PROBLEMS

2.1 The crankshaft of an automobile engine revolves on main bearings of nominal diameter 42 mm. Find the diameter range of bearing and shaft if the fit is basic hole $H8/f7$.

2.2 Select proper fits for the following applications:

(a) Valve stem in the valve guide of an automobile engine.
(b) Head journal-bearing for a precision lathe.
(c) Interference fit of coupling and shaft, if the torque is transmitted through a key.
(d) A fit to locate a ball-bearing on a shaft.

2.3 Identify on a lawnmower:

(a) A clearance fit.
(b) A transition fit.
(c) An interference fit.

2.4 Draw a sketch of a sectional view along the axis of rotation of an electric motor, and identify all the fits that need to be specified.

2.5 Determine the *it* quality tolerance which describes a production lot of shafts of nominal diameter 30 and 72 mm with limit dimensions:

$30^{-0.007}_{-0.020}$ $72^{0.00}_{-0.019}$

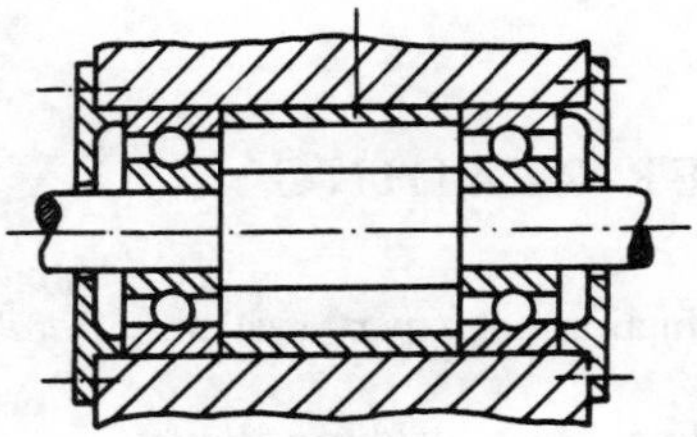

Figure P2.6

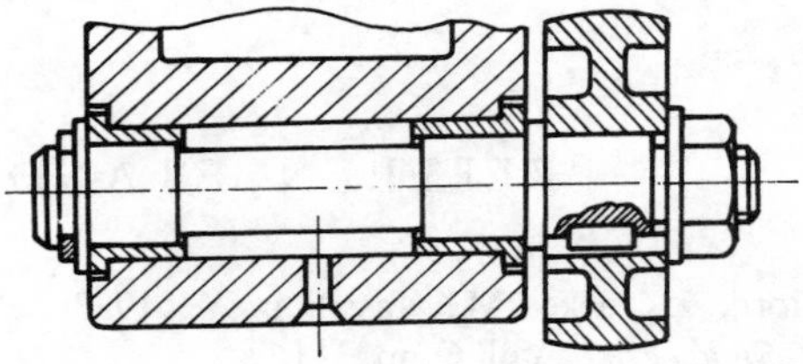

Figure P2.7

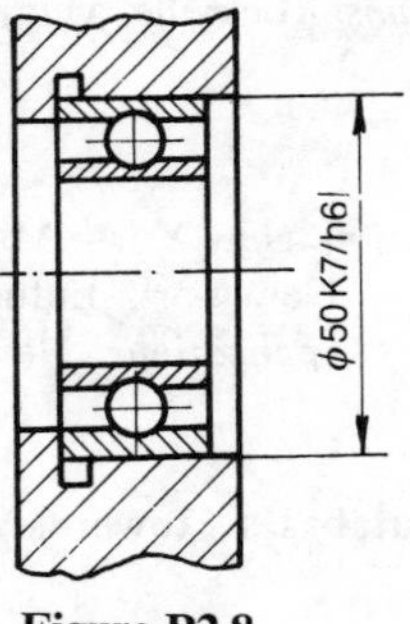

ϕ50 K7/h6

Figure P2.8

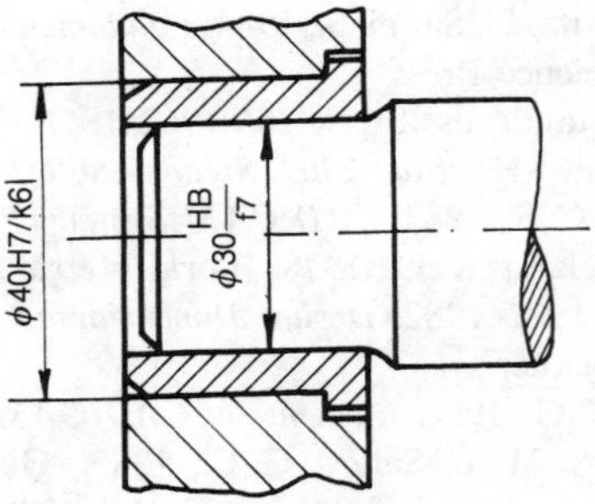

ϕ40H7/k6

ϕ30 HB/f7

Figure P2.9

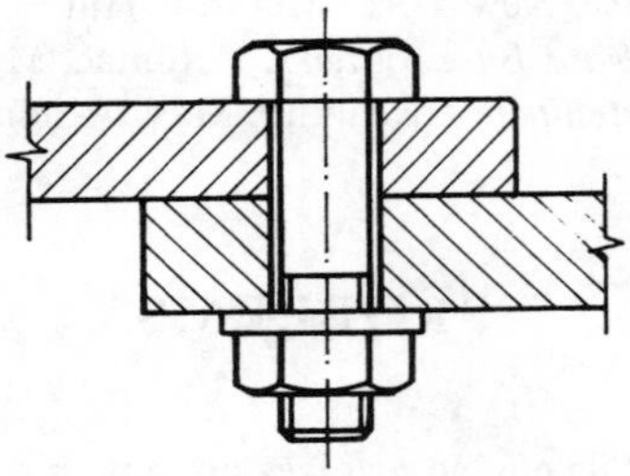

Figure P2.10

Problems 2.6 to 2.10

For the fits indicated in Figures P2.6 to P2.10 respectively, determine the limit dimensions for shaft and hole, selecting appropriate diameters and standard fits wherever it is applicable.

2.11 Using the DESIGNDB program compile a table of the integers from 1 to 30 with their squares, cubes and square roots.

2.12 Write a computer program to sum selected columns of a table of numbers in the form of a random text file and add a new record with the sums.

2.13 Write a computer program to add new columns to a table in the form of a random file which will be computed by addition, subtraction, multiplication and division of two other columns.

2.14 Use SYMPHONY, or any other database program available in your computer, to generate section areas and polar moments of inertia of circles between 10 mm and 100 mm diameter, in steps of 10 mm.

2.15 Repeat problem 2.14 for squares with sides 10 mm to 100 mm, in steps of 10 mm.

APPENDIX 2A
DESIGNDB: A DATABASE PROGRAM FOR DESIGN

```
10 REM ***************************************************
20 REM *                                                 *
30 REM *                   designdb                      *
40 REM *                                                 *
50 REM ***************************************************
60 REM          COPYRIGHT 1987
70 REM by Professor Andrew D. Dimarogonas, Washington University,St.Louis.
80 REM All rights reserved.  Unauthorized reproduction, dissemination or
90 REM selling is strictly prohibited.  This listing is for personal use.
100 REM
110 REM
120 REM
130 REM
140 REM A program to create and maintain random data files.
150 REM
160 locate 25,1:print" Dimarogonas,A.D., Computer Aided Machine
Design,Prentice-Hall,1988";
170 REM      ADD/11-20-85/revisions 12-22-86,5-29-1987
180 REM ***************************************************************
190 CLS:KEY OFF
200 FOR I=1 TO 14:COLOR I:LOCATE I,22+I:PRINT "DESIGNDB";
210 LOCATE I,50-I:PRINT "DESIGNDB";:NEXT I:COLOR 14
220 LOCATE 8,5:PRINT"A random file development and maintainance program
for";
230 PRINT " design data bases";
240 REM SN$:                              Name of defining file
250 REM NF$:                              Name of data file
260 DIM AA$(12,100 ):REM                  100 RECORDS, 12 ENTRIES PER RECORD
270 DIM N$(20):REM                        NAMES OF ENTRIES
275 DIM LS(20),L$(20):REM                 length of entry
280 DIM F1$(20),F2$(20),F3$(20):REM       AUX VARIABLES FOR PRINTING
290 LF = 300:REM                          RECORD LENGTH
300 PW = 135:REM                          PRINTER WIDTH, EPSON X-80, CONDENSED
MODE
310 M1=1:M2=1
320 FOR I=1 TO 20:LS(I)=1:L$(I)="1":NEXT I
```

```
330 IF X$="s" OR X$="S" THEN END
340 COLOR 14,0,0
350 LOCATE 16,21
360 PRINT"BASICA must be loaded with an extension:"
370 LOCATE 17,30:PRINT"     >A:BASICA/S:300 "
380 LOCATE 18,20:PRINT"If you loaded BASICA properly, hit RETURN":LOCATE
19,10
390 PRINT"If NOT ,then hit S and RETURN to stop and then reload BASICA ";
400 INPUT X$
410 IF X$="s" OR X$="S" THEN SYSTEM
420 LOCATE 23,21:PRINT"Do you want instructions? (Y/N) ";:INPUT X$
430 IF X$<>"y" AND X$<>"Y" THEN 460
440 CLS:GOSUB 3170
450 INPUT"Hit return to continue, S to stop ";X$
460 CLS:PRINT:PRINT"                         DESIGNDB"
470 PRINT:PRINT
480 REM FILE DEFINITION MENU
490 PRINT "      FILE DEFINITION MENU      "
500 PRINT
510 PRINT " <N> Creation of new defining file "
520 PRINT " <P> Use of previously defined file"
530 PRINT " <Q> Quit"
540 PRINT :INPUT "Enter your selection ";X$
550 IF X$="P" OR X$="p" THEN 770
560 IF X$="N" OR X$="n" THEN 590
570 IF X$="Q" OR X$="q" THEN END
580 GOTO 540
590 PRINT:INPUT "Enter name for the database defining file      ";SN$
600 INPUT       "Enter number of entries per record             ";NE
610 FOR IE = 1 TO NE
620 PRINT       "Enter length, name of entry                  ";IE;
630 INPUT LS(IE),N$(IE):NEXT IE
640 REM ********************************************* CREATING DEFINITION
FILE
650 OPEN SN$  AS #2 LEN=LF
660 FIELD #2,10 AS NE$,LS(1) AS F1$, LS(2) AS F2$, LS(3) AS F3$, LS(4) AS
F4$, LS(5) AS F5$, LS(6) AS F6$, LS(7) AS F7$, LS(8) AS F8$, LS(9) AS F9$,
LS(10) AS F10$, LS(11) AS F11$, LS(12) AS F12$
670 LSET NE$=MKS$(NE):LSET F1$=N$(1):LSET F2$=N$(2):LSET F3$=N$(3)
671 LSET F4$=N$(4):LSET F5$=N$(5):LSET F6$=N$(6): LSET F7$=N$(7)
672 LSET F8$=N$(8):LSET F9$=N$(9):LSET F10$=N$(10) :LSET F11$=N$(11):LSET
F12$=N$(12)
690 CODE%=1
700 PUT #2,CODE%
701 FIELD #2,10 AS LS1$,10 AS LS2$,10 AS LS3$,10 AS LS4$,10 AS LS5$,10 AS
LS6$,10 AS   LS7$ ,10 AS  LS8$ ,10 AS  LS9$ ,10 AS  LS10$,10 AS  LS11$ ,10
AS LS12$
702 LSET LS1$=STR$(LS(1)):LSET LS2$=STR$(LS(2)):LSET LS3$=STR$(LS(3)):LSET
LS4$=STR$(LS(4))
703 LSET LS5$=STR$(LS(5)):LSET LS6$=STR$(LS(6)):LSET LS7$=STR$(LS(7)):LSET
LS8$=STR$(LS(8))
704 LSET LS9$=STR$(LS(9)):LSET LS10$=STR$(LS(10)):LSET
LS11$=STR$(LS(11)):LSET LS12$=STR$(LS(12))
705 CODE%=2
706 PUT #2,CODE%
710 CLOSE #2
720 PRINT "DEFINING FILE ";SN$;" CREATED"
730 I=2:FOR IDELAY=1 TO 10000:NEXT IDELAY
740 GOTO 460
750 END
760 REM **************************************************** RETRIEVING
HEADING
770 INPUT "Enter defining file name ";SN$
780 OPEN SN$  AS #2 LEN=LF
784 FIELD #2,10 AS L$(1),10 AS L$(2),10 AS L$(3),10 AS L$(4),10 AS L$(5),10
AS L$(6),10 AS L$(7),10 AS L$(8),10 AS L$(9),10 AS L$(10),10 AS L$(11),10
AS L$(12)
786 GET #2,2
787 FOR I=1 TO 20:LS(I)=VAL(L$(I)):IF  LS(I)=0 THEN LS(I)=1
788 NEXT I
790 FIELD #2,10 AS NE$,LS(1) AS N$(1), LS(2) AS N$(2),LS(3) AS N$(3), LS(4)
```

```
AS N$(4),LS(5) AS N$(5), LS(6) AS N$(6), LS(7) AS N$(7), LS(8) AS
N$(8),LS(9) AS N$(9), LS(10) AS N$(10), LS(11) AS N$(11), LS(12) AS N$(12)
800 CODE%=1
810 GET #2, CODE%
830 NE=CVS(NE$)
831    FOR I=1 TO NE:L$(I)=N$(I):NEXT I
835 CLOSE #2
840 CLS
845    FOR I=1 TO NE:N$(I)=L$(I):NEXT I
850 M1=1:N1=1
860 REM
870 PRINT "                                          "
880 PRINT "             DESIGN DATA BASE             "
890 PRINT "                                          ":PRINT
900 PRINT "Defining file name:  ";:COLOR 28:PRINT SN$:COLOR 14
910 PRINT
920 PRINT "Creation and maintainance of data files"
930 PRINT:PRINT
940 PRINT "     DATA FILE MENU"
950 PRINT
960 PRINT " <E> Editing of existing data file"
970 PRINT
980 PRINT " <N> Creating a new data file"
990 PRINT
1000 PRINT " <Q> Quit"
1010 PRINT :PRINT :INPUT "Enter your selection";X$
1020 IF X$ ="Q" OR X$ ="q" THEN END
1030 IF X$="e" THEN X$="E"
1040 IF X$="n" THEN X$="N"
1050 IF X$ < > "E" AND X$ < > "N" THEN 1010
1060 PRINT
1070 INPUT "Enter data file name ";NF$
1080 IF X$ = "N" THEN 1230
1090 OPEN NF$  AS #1 LEN=LF
1100 FIELD #1, 10 AS N1$, 10 AS N2$
1110 CODE%=1
1120 GET #1,CODE%
1130 N1=CVS(N1$):N2=CVS(N2$)
1140 IF N1<1 THEN N1=1
1150 IF N2<1 THEN N2=1
1160 M1 = N1:M2 = N2
1170 CLOSE #1
1180 CLS
1190 PRINT "Defining file ";:COLOR 28:PRINT  SN$:COLOR 14 :PRINT
1200 PRINT "Data file     ";:COLOR 28:PRINT NF$:COLOR 14:PRINT
1210 IF M2>1 THEN M1=2
1220 PRINT :PRINT "Records already in file :";M1;" to ";M2
1230 PRINT
1240 PRINT "                  MENU"
1250 PRINT
1260 PRINT " <P>   Print records of file"
1270 PRINT
1280 PRINT " <E>   Edit records of file"
1290 PRINT
1300 PRINT " <A>   Add new records"
1310 PRINT
1320 PRINT " <L>   List of file"
1330 PRINT
1340 PRINT " <F>   Change file"
1350 PRINT
1360 PRINT " <Q>   Quit  ":PRINT
1370 PRINT
1380 INPUT "Enter your selection:  ";X$
1390 IF X$ = "F" OR X$="f" THEN 1070
1400 IF X$ = "Q" OR X$="q" THEN END
1410 IF X$ = "L" OR X$="l" THEN GOSUB 2540
1420 IF X$ = "E" OR X$="e" THEN 1520
1430 IF X$  =  "A" OR X$="a" THEN GOSUB 2150
1440 IF X$="P" OR X$="p" THEN 1460
1450 GOTO 1180
1460 INPUT " Record numbers n1,n2 ";N1,N2
```

```
1470 IF N1<2 THEN N1=2
1480 IF N2<200 THEN 1490
1490 GOSUB 1820
1500 GOSUB 2040
1510 GOTO 1180
1520 PRINT
1530 INPUT "Record numbers N1,N2   ";IR1,IR2
1540 IF IR1<2 THEN IR1=2
1550 IF IR2-IR1<200 THEN 1570
1560 COLOR 28:PRINT" 200 RECORDS MAX":COLOR 14:GOTO 1530
1570 N1 =IR1:N2=IR2:GOSUB  1820
1580 PRINT "To retain previous value, RETURN"
1590 PRINT "To change, type new value "
1600 PRINT:FOR IRG=IR1 TO IR2 :IR=IRG-IR1+1
1610 PRINT:PRINT
1620 PRINT "Record number     ";IRG
1630 FOR IE = 1 TO NE
1640 PRINT N$(IE);"      ";AA$(IE,IR);"    ";
1650 INPUT X$
1660 IF X$ > " " THEN AA$(IE,IR) = X$
1670 NEXT IE
1680 INPUT "Repeat record <R>, Next record <N> , Menu <M> ";X$
1690 IF X$="R" OR X$="r" THEN 1610
1700 IF X$="N" OR X$="n" THEN 1740
1710 IF X$="M" OR X$="m" THEN 1730
1720 GOTO 1680
1730 IRG=IR2
1740 NEXT IRG
1750 PRINT
1760 INPUT "(S) save, (N) new records, (M) menu: ";X$:IF X$ = "N" THEN 1530
1770 IF X$="M" OR X$="m" THEN 1180
1780 IF X$ = "S" OR X$="s" THEN 1800
1790 GOTO 1760
1800 GOSUB 2420
1810 GOTO 1180
1820 REM READ FROM FILE
1830 REM ******************************************** READING RECORD OF
FILE
1840 IF N1<2 THEN N1=2
1850 PRINT
1860 PRINT "Reading data: File ";NF$;" Record No: ";
1870 PRINT
1880 OPEN NF$ AS #1 LEN=LF
1890 FOR IGL = N1 TO N2 :I=IGL-N1+1
1900 FIELD #1, LS(1) AS A1$, LS(2) AS A2$, LS(3) AS A3$, LS(4) AS A4$,
LS(5) AS A5$, LS(6) AS A6$,LS(7) AS A7$, LS(8) AS A8$, LS(9) AS A9$, LS(10)
AS A10$, LS(11) AS A11$, LS(12) AS A12$
1910 CODE%=IGL
1920 GET #1, CODE%
1930 AA$(1,I) = A1$:AA$(2,I) = A2$
1940 AA$(3,I) = A3$:AA$(4,I) = A4$
1950 AA$(5,I) = A5$:AA$(6,I) = A6$
1960 AA$(7,I) = A7$:AA$(8,I) = A8$
1970 AA$( 9,I) = A9$:AA$(10,I) = A10$
1980 AA$(11,I) =A11$:AA$(12,I)=A12$
1990 PRINT IGL
2000 NEXT IGL
2010 CLOSE #1
2020 PRINT
2030 RETURN
2040 REM ****************************************************PRINTING OF
DATA
2050 FOR IGL = N1 TO N2 :I=IGL-N1+1
2060 PRINT " Record number      ";IGL:PRINT
2070 FOR IE = 1 TO NE
2080 PRINT "<";IE;"> ";N$(IE);"    ";AA$(IE,I)
2090 NEXT IE
2100 PRINT
2110 INPUT "RETURN for next, (M) for menu ";YY$
2120 IF YY$ = "M" OR YY$="m" THEN RETURN
2130 NEXT IGL
```

```
2140 RETURN
2150 REM ****************************************** INPUT DATA FOR NEW RECORD
2160 IF M2<1 THEN M2=1
2170 M2LAST=M2
2180 M2=M2+1
2190 PRINT:PRINT "New record number = ";M2:PRINT
2200 ILOC=M2-M2LAST
2210 IF M2<100 THEN 2230
2220 COLOR 28:PRINT"> 100 RECORDS":COLOR 14:M2=M2-1:GOTO 2260
2230 FOR IE = 1 TO NE
2240 PRINT N$(IE);"    ";AA$(IE,ILOC-1);"    ";:INPUT AA$(IE,ILOC)
2250 NEXT IE
2260 PRINT:PRINT:INPUT "Edit (E) , Save (S),Add more records (A), Menu (M)
";X$
2270 IF X$="M" OR X$="m" THEN 1180
2280 IF X$="E" OR X$="e" THEN 2230
2290 IR1=M2LAST+1:IR2=M2
2300 IF X$="S" OR X$="s" THEN 2340:M2LAST=M2LAST+1:IR2=M2
2310 IF X$<>"A" AND X$<> "a" THEN 2260
2320 IF X$="A" OR X$="a"  THEN 2180
2330 GOTO 2260
2340 REM**************************************** CHANGE RECORD 1 (NO OF
RECORDS)
2350 OPEN NF$ AS #1 LEN=LF
2360 FIELD #1, 10 AS N1$, 10 AS N2$
2370 LSET N1$=MKS$(M1):LSET N2$=MKS$(M2)
2380 CODE%=1
2390 PUT #1, CODE%
2400 CLOSE #1
2410 IR2=M2
2420 REM ***************************************************WRITE DATA IN
FILE
2430 OPEN NF$ AS #1 LEN=LF
2440  FIELD #1,LS(1) AS F1$, LS(2) AS F2$, LS(3) AS F3$, LS(4) AS F4$,
LS(5) AS F5$, LS(6) AS F6$, LS(7) AS F7$, LS(8) AS F8$, LS(9) AS F9$,
LS(10) AS F10$, LS(11) AS F11$, LS(12) AS F12$
2450 FOR IGL=IR1 TO IR2
2460 I=IGL-IR1+1
2470  LSET F1$=AA$(1,I):LSET F2$=AA$(2,I):LSET F3$=AA$(3,I):LSET
F4$=AA$(4,I):LSET F5$=AA$(5,I):LSET F6$=AA$(6,I)
2480  LSET F7$=AA$(7,I):LSET F8$=AA$(8,I):LSET F9$=AA$(9,I):LSET
F10$=AA$(10,I) :LSET F11$=AA$(11,I):LSET F12$=AA$(12,I)
2490 CODE%=IGL
2500 PUT #1, CODE%
2510 NEXT IGL
2520 CLOSE #1
2530 RETURN
2540 REM ********************************************LISTING OF FILE
RECORDS
2550 N = 0
2560 REM FILLING UP A DATA MATRIX
2570 PRINT
2580 REM READ FROM FILE
2590 PRINT
2600 PRINT "READING DATA FROM FILE"
2610  OPEN NF$ AS #1 LEN=LF
2620 FOR IGL = 2  TO M2  :I=IGL-M1+1
2630 FIELD #1,LS(1) AS A1$, LS(2) AS A2$, LS(3) AS A3$, LS(4) AS A4$, LS(5)
AS A5$, LS(6) AS A6$,LS(7) AS A7$, LS(8) AS A8$, LS(9) AS A9$, LS(10) AS
A10$, LS(11) AS A11$,LS(12) AS A12$
2640 CODE%=IGL
2650 GET #1, CODE%
2660 AA$(1,I) =A1$
2670 AA$(2,I) = A2$:AA$(3,I) = A3$
2680 AA$(4,I) = A4$:AA$(5,I) = A5$
2690 AA$(6,I) = A6$:AA$(7,I) = A7$
2700 AA$(8,I) = A8$:AA$(9,I) = A9$
2710 AA$(10,I) = A10$:AA$(11,I) = A11$
2720 AA$(12,I) = A12$
2730 PRINT I,AA$(1,I)
2740 NEXT IGL
```

```
2750 CLOSE  #1
2760 PRINT :PRINT:PRINT:PRINT
2770 PRINT "     TURN THE PRINTER ON..."
2780 PRINT:PRINT "Then, hit <RETURN> to continue"
2790 INPUT X$
2800 REM ********************************************** LIST ON LINE
PRINTER
2810 REM EPSON MX-80 IS ASSUMED
2820 LPRINT "Defining file: ";SN$;"   Data file:   ";NF$
2830 LPRINT :LPRINT:IPR=0
2840 LPRINT CHR$(15);: REM SET CONDENSED MODE
2850 WIDTH "LPT1:",PW:SP$="":FOR I=1 TO 50:SP$=SP$+" ":NEXT I
2860 LINEL=0:FOR I=1 TO NE:LINEL=LINEL+LS(I):NEXT I:DV=INT(LINEL/PW):IF
DV<1 THEN DV=1
2870 FOR I = 1 TO NE:N$(I)=N$(I)+SP$:F1$(I) = LEFT$ (N$(I),LS(I)/DV):NEXT I
2880 FOR I = 1 TO NE:F2$(I) = MID$ (N$(I),LS(I)/DV + 1,LS(I)/DV):NEXT I
2890 FOR I = 1 TO NE:F3$(I) = MID$ (N$(I),2 * LS(I)/DV + 2,LS(I)/DV):NEXT I
2900 FOR I = 1 TO NE:LPRINT F1$(I);SPC(LS(I)/DV+2-LEN(F1$(I)));:NEXT I
2901 LPRINT
2910 FOR I = 1 TO NE:LPRINT F2$(I);SPC( LS(I)/DV + 2 - LEN (F2$(I)));:NEXT
I
2911 IPR=0:FOR I=1 TO NE:IF F2$(I)>"a" THEN IPR=1
2912 NEXT I:IF IPR=1 THEN LPRINT
2920 FOR I = 1 TO NE:LPRINT F3$(I);SPC( LS(I)/DV + 2 - LEN (F2$(I)));: NEXT
I
2921 LPRINT
2940 FOR I=1 TO PW:LPRINT "_";:NEXT I:LPRINT
2950 FOR IGL = N1 TO N2 :I=IGL-N1+1
2960 FOR IE = 1 TO NE:F1$(IE) = LEFT$ (AA$(IE,I),LS(IE)/DV):NEXT IE
2970 FOR IE = 1 TO NE:F2$(IE) = MID$ (AA$(IE,I),LS(IE)/DV +
1,LS(IE)/DV):NEXT IE
2980 FOR IE = 1 TO NE:F3$(IE) = MID$ (AA$(IE,I),2 * LS(IE)/DV + 2,
LS(IE)/DV):NEXT IE
2990 EE = 0: FOR IE = 1 TO NE: IF F1$(IE) > " " THEN EE = 1
3000 NEXT IE
3010 IF EE = 0 THEN 3110
3020 FOR IE = 1 TO NE: LPRINT F1$(IE);SPC( LS(IE)/DV + 2 - LEN (F1$(IE));:
NEXT IE:LPRINT
3030 EE=0:FOR IE = 1 TO NE:IF F2$(IE) > " " THEN EE  = 1
3040 NEXT IE
3050 IF EE = 0 THEN 3110
3060 FOR IE = 1 TO NE:LPRINT F2$(IE);SPC(LS(IE)/DV + 2 - LEN
(F2$(IE)));:NEXT IE:LPRINT
3070 EE = 0:FOR IE = 1 TO NE: IF F3$(IE) > " " THEN EE = 1
3080 NEXT IE
3090 IF EE = 0 THEN 3110
3100 FOR IE = 1 TO NE:LPRINT F3$(IE);SPC(LS(IE)/DV + 2 - LEN
(F3$(IE)));:NEXT IE:LPRINT
3110 REM
3120 NEXT IGL
3125 INPUT"Hit ENTER to continue";X$
3130 RETURN
3140 PRINT "...ERROR.....TRY AGAIN"
3150 PRINT "FILE:   ";NF$
3160 STOP
3170 REM INSTRUCTIONS
3180 PRINT"          INSTRUCTIONS TO USE DESIGN.DBASE"
3190 PRINT"1. Database Defining file: The name of your choice for a file"
3200 PRINT"   which will contain the top headings of your columns"
3210 PRINT"   Examples: MYFILE , bolts.db , ME322, ..."
3220 PRINT"2. Number of entries per record. Your Database consists of ta-"
3230 PRINT"   bles with records, each record occupies one horizontal line."
3240 PRINT"   Each record (or line) has a number of columns-entries"
3250 PRINT"3. Length,Name of entry: Each entry (column) has a heading"
3260 PRINT"   Examples: 15,Diameter or 12,length or 20,age or
30,address,..."
3270 PRINT"   All the database information are stored in the defining file"
3280 PRINT"4. Each Database has many tables with the same headings"
3290 PRINT"   For example, tables of different types of bolts. Give a file"
3300 PRINT"   name to each of these tables.The structure will be:"
3310 PRINT:PRINT:
```

```
3320 PRINT"                    DATABASE DEFINING FILE"
3330 PRINT"  data file 1                data file 2              datafile 3
..."
3340 PRINT
3350 PRINT"  record 2                   record 2                 record2     --
"
3360 PRINT"  record 3                   record 3                 record 3    --
"
3370 PRINT"  --------                   --------                 --------    --
"
3380 PRINT"  Record 1 is reserved for file information"
3390 PRINT"5. For a hard copy of a datafile, use the List File option"
3400 RETURN
```

CHAPTER THREE
COMPUTER GRAPHICS

3.1 ORGANIZING A DESIGN OFFICE

A design office can be assumed to be:

(a) The room of a student who studies design.
(b) The office of a consulting engineer, involved in design.
(c) The office of a company, small or large, where a design team works.

Standard equipment of a design office must be the essentials of computing and drafting although the drafting room is usually separate.

A standard library should include manuals relevant to the design subject. Moreover, national and international standards and manufacturers' catalogs on the same subject are the source for much essential information on any design.

Today the design office is equipped with computer aided design (CAD) facilities. This, as discussed in Chapter 2, changes the design philosophy and the design practice.

The computer system employed depends very much on the scope and the amount of output of the design office.

Microcomputers are essential, regardless of the size of the design office. They can be operated at any time, day or night, by those not skilled in computers, and they encourage use through their simplicity. Contemporary microcomputers are equipped with main memory ranging between 256 kbytes and several megabytes (Mbytes), floopy-disk drives with storage capacity between 350 and 2000 kbytes and hard-disk storage devices with capacity between 20 and 400 Mbytes. Most systems have graphics screens with high-resolution graphics from 200×600 to 1024×1024 pixels (points on the screen). (It must be pointed out that these limits are constantly moving upwards at a pace that no textbook can keep up with.)

Printers are available for text 80 or 132 characters' wide and graphics with the dot matrix system or laser printers with high-quality text printing.

As a plotter device, one can use the printer either by directly transferring (dumping) the screen on paper or by directly programing the printer for plotting. The plot then consists of many dots with resolution varying with the spacing of the dots. There are, however, pen plotters for professional-quality plotting. Plotters can be drum type (Figure 3.1a), which can plot on a paper with the width of a drum and any length, or flat-bed type (Figure 3.1b), which can plot on a page (or multiple) size paper.

A pointing device is very useful in feeding geometric data to the computer, such as existing drawings. A mouse, a light pen or a digitizing tablet can be used for this.

Streaming tapes or floppy discs are used to store information for back-up purposes.

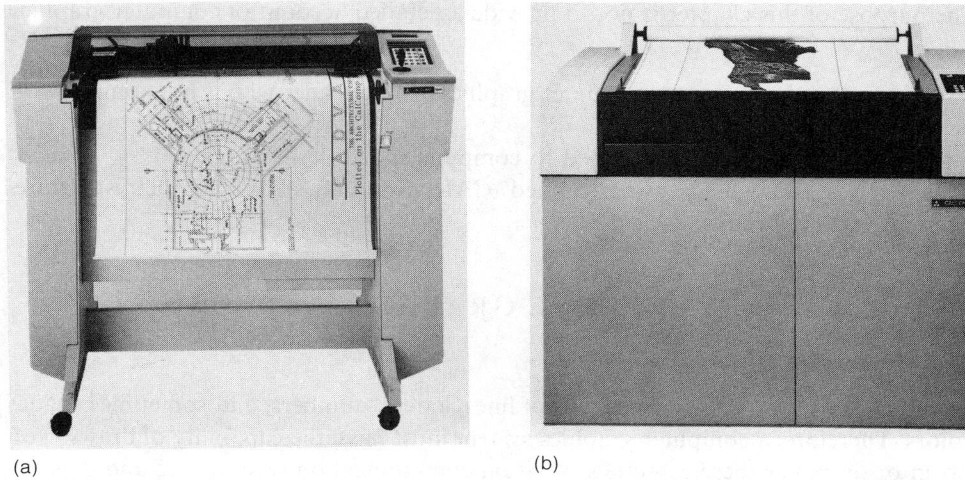

(a) (b)

Figure 3.1 (a) Drum plotter (courtesy California Computer Co.); (b) flat-bed x–y plotter (courtesy Hewlett-Packard Corp.)

3.2 COMPUTER AIDED VERSUS MANUAL DRAFTING

It is wrongly assumed by newcomers in this field that computer drafting will eventually make manual drafting obsolete. For the immediate future at least, this is not so. A recent study in one particular company revealed that for simple drawings, manual drafting was faster while for more complex drawings computer drafting was faster. Speed is not, therefore, the great advantage of computer drafting. The main advantage is in the fact that a computer-generated drawing can be stored away and recalled at any time. Furthermore, design is in many aspects an evolutionary process. Very few items are designed altogether new. Most designed components and machines are modifications and improvements of existing ones. In the manual mode, the drawing must be made from the very beginning. In a computer-drafting system, most of the time only the changes must be given and the new drawing is made by the system itself.

In many CAD systems basic drawings are available, called 'frames', that the designer can modify to produce the desired drawings. Such frames can also be combined to yield more complex parts and assemblies.

For most simple parts, computer programs can be written to make the drawing using the basic dimensions as input. Furthermore, if a design system is available in the computer, the input is merely the service conditions of the part.

Long experience and the need for simpler, faster and more economical drafting, resulted in standard drafting practices which in many cases deviate considerably from picturing a real object as a faithful photographic representation. Many conventional simplifications were merely for the designer's convenience and they could be waived in computer aided design. In representing gears, for example, the draftsperson never draws the gear teeth in detail. For the computer, this is an easily programable task. However, computer drafting has to comply in many aspects with conventional practices and therefore it needs special treatment.

The purpose of this chapter is not to provide a detailed account of computer graphics but,

(a) to serve as an introduction to computer graphics for the reader who is not familiar with the subject;
(b) to stress the points which are related to computer aided design; and
(c) to help the reader who is using 'canned' CAD programs to realize their structure, potential and limitations.

3.3 PLANE GRAPHICS

3.3.1 Graphics primitives

As in manual drafting, a drawing consists of lines, letters, numbers, and sometimes shades and colors. Therefore, a computer graphics system must have the capability of drawing on the screen or on paper these essentials of visual communication (Figures 3.2 and 3.3).

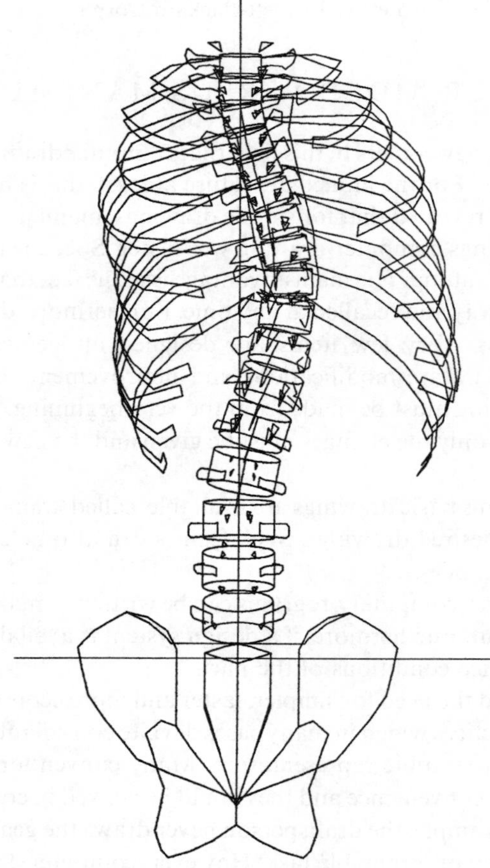

Figure 3.2 Axonometric wire drawing (courtesy ASME)

Figure 3.3 A graphics terminal

Unfortunately, graphics commands differ from one system to another and they have not been unified among computer manufacturers. For this reason, in the programs which appear in this book, graphics commands have been separated so that they can be changed to suit the system available to the reader. The commands used in this book are of the IBM-PC/XT/AT/BASIC system screen.

Each system has a set of simple graphics commands, called 'graphics primitives'.

(a) Pen control, up and down: Some systems use PEN UP and PEN DOWN commands while most screen-oriented systems specify the value of a parameter COLOR in visible colors for pen-down and an invisible color for pen-up.

(b) Moving the cursor pen or screen dot from one place to the other on absolute screen coordinates. A LINE $(x_1, y_1) - (x_2, y_2)$ will trace a straight line between points x_1, y_1 and x_2, y_2 in the screen or plotter coordinates and in the currently-specified color. The coordinates of the IBM-PC/XT/AT screen are X (the upper horizontal) and Y (the left vertical). To change the X-axis from upper to lower horizontal, a new Y-coordinate must be used, $y = a - Y$, where a is the lowermost position in screen coordinates.

(c) Some systems have commands to move the pen in a local coordinate system (one that is in respect to the last position of the cursor).

(d) Most systems can put letters and numbers (alphanumeric characters) on the graphics screen.

The following procedure PLOTPRIM will be used in the sequel, to provide the interface with the system available to the reader. To accommodate any particular system, the LINE statements of IBM should be properly modified.

```
Procedure PLOTPRIM                          (*Plots a line between x1, y1 and x2, y2*)
  GLOBAL variables: X1, Y1, X2, Y2, XCOLOR, XTYPE%
  IF XTYPE%:= 1 then LINE (X1, Y1) – (X2 – Y2)                  (*Straight line*)
  else
    LENGTH:= SQR((X2 – X1)*(X2 – X1) + (Y2 – Y1)*(Y2 – Y1))
    IF XTYPE%:= 2                                                (*Dashed line*)
      S:= 0
      repeat
        S:= S + 5
        X:= X1 + (S + 4)*(X2 – X1)/LENGTH
        Y:= Y1 + (S + 4)*(Y2 – Y1)/LENGTH
        LINE (X1, Y1) – (X, Y)
      until S > LENGTH
    else, IF XTYPE%:= 3 then do                                (*Symmetry line*)
      S:= 0
      repeat
        S:= S + 10
        X:= X1 + (S + 8)*(X2 – X1)/LENGTH
        Y:= Y1 + (S + 8)*(Y2 – Y1)/LENGTH
        LINE (X1, Y1) – (X, Y)
        XA:= X + (X2 – X1)/LENGTH
        YA:= Y + (Y2 – Y1)/LENGTH
        LINE (XA, YA) – (XA, YA)
      until S > LENGTH
    else, print message
end.
```

3.3.2 Drawing polygons

Most plane drawings can be plotted as polygons. Every vertex is called a 'node' and every side is called an 'edge'. Therefore one should specify for every polygon:

(i) Number of NNODES%
(ii) Coordinates of nodes X(1), Y(1), X(2), Y(2),...
(iii) Sequence of nodes defining the polygon in an array C$(PN, N1, N2,...). The first item is the number of nodes + 1, for closure, followed by the node numbers in sequence.

3.3.3 Drawing circular arcs

A circular arc is defined by its center, its radius, start and end angle. Sometimes, other data are available, such as three points, two points and the radius, etc. The arc is approximated as a polygon. For a full circle, such a polygon must have an adequate number of nodes for smooth representation of the circle. To give adequate resolution, a rule of thumb is that the number of nodes must be at least $s/5t$ for rough drawings and s/t for finer drawings, where s is the arc length and t the thickness of the plotted line or the screen pixel, but this depends also on the screen or plotter resolution.

A circular arc can be expressed in the usual parametric form, if the center is at x_0, y_0 and the radius is r,

$$\begin{aligned} x &= r\cos f + x_0 \\ y &= r\sin f + y_0 \end{aligned} \tag{3.1}$$

and the angle f varies between the given values of f_1 and f_2. Each step, however, requires computation of the sine and cosine of the angle, which is a slow process. To avoid this, let the increment of the angle be $\mathrm{d}f$. Then, for the n and $n+1$ steps

$$\begin{aligned} x_n &= r\cos f + x_0 \quad & x_{n+1} &= r\cos(f+\mathrm{d}f) + x_0 \\ y_n &= r\sin f + y_0 \quad & y_{n+1} &= r\sin(f+\mathrm{d}f) + y_0 \end{aligned} \tag{3.2}$$

Then, the $n+1$ step can be computed from the previous one as

$$\begin{aligned} x_{n+1} &= (x_n - x_0)\cos\mathrm{d}f - (y_n - y_0)\sin\mathrm{d}f + x_0 \\ y_{n+1} &= (x_n - x_0)\sin\mathrm{d}f + (y_n - y_0)\cos\mathrm{d}f + y_0 \end{aligned} \tag{3.3}$$

where $\cos\mathrm{d}f$ and $\sin\mathrm{d}f$ are constants and they are calculated only once.

The following procedure transforms a circular arc of radius R, center x_0, y_0 between polar angles $F1$ and $F2$ into an open polygon with 50 nodes and stores them into a record of a random data file.

```
Procedure CIRCLE
  GLOBAL R, XO, YO, F1, F2, array C$(1..101):var
  LOCAL XN, YN, DF, COSDF, SINDF, I, array X(1..50), array Y(1..50):var
  XN:= XO + R*COS(F1); YN:= YO + R*COS(F2)
  DF:= (F2 – F1)/49; COSDF:= COS(DF); SINDF:= SIN(DF)
  FOR I = 1 TO 50 do
    X(I):= XN; Y(I):= YN
    XN:= (X(I) – XO)*COSDF – (Y(I) – YO)*SINDF + XO
    YN:= (X(I) – XO)*SINDF + (Y(I) – YO)*COSDF + YO
    J:= 2*I
    C$(J):= string of X(I)
    C$(J + 1):= string of Y(I)
    C$(1):= string of 50
  enter RECORD% number and file name
  CALL procedure LRNDF
end.
```

A record is now placed in the data file containing the number of nodes (50) and the 50 pairs of X and Y of each node of the polygon replacing the circular arc.

3.3.4 Clipping

Many times, either accidentally or purposely, some node coordinates are outside the screen window. In other circumstances we want to plot part of a picture within a given window of

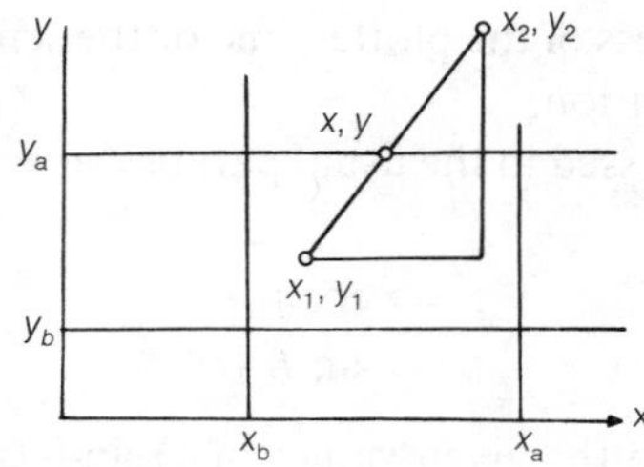

Figure 3.4 Straight line intersecting the window boundary

the plotting area. Therefore we must remove the lines, or part of lines, outside this window.

If a (straight) line is plotted between a point x_1, y_1, within the window, to an outside point x_2, y_2, the coordinates x_2, y_2 must be replaced with x, y, i.e. the intersection of the line with the screen window $x_b < x < x_a$, $y_b < y < y_a$. Simple geometry yields (Figure 3.4)

$$\begin{aligned} y &= y_a \\ x &= x_1 + (x_2 - x_1)(y_a - y_1)/(y_2 - y_1) \end{aligned} \qquad (3.4)$$

Similar results are obtained if there is intersection with the $x = x_a$, $x = x_b$, $y = y_b$ lines.

3.3.5 Scaling

A picture consists of a set of node definitions $x(i)$, $y(i)$ and set of polygon definitions C$(1...NP, 1...PNMAX). Scaling onto the x–y plane simply involves multiplication of the node coordinates by s_x and s_y, scale factors in the x and y directions, respectively (Figure 3.5). Representing the transformed coordinates with capital letters

$$\begin{aligned} X(i) &= x(i)s_x \\ Y(i) &= y(i)s_y \end{aligned} \qquad (3.5)$$

Usually $s_x = s_y$ and this transformation is called 'uniform'.

In matrix form, equations (3.5) can be written as

$$\begin{Bmatrix} X(i) \\ Y(i) \\ 1 \end{Bmatrix} = \begin{bmatrix} s_x & 0 & 0 \\ 0 & s_y & 0 \\ 0 & 0 & 1 \end{bmatrix} \begin{Bmatrix} x(i) \\ y(i) \\ 1 \end{Bmatrix} \qquad (3.5a)$$

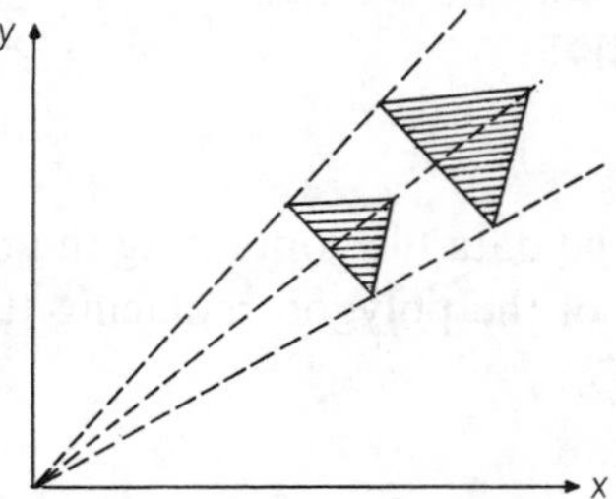

Figure 3.5 Plane scaling

The third equation is obviously an identity and it is used for computational convenience only.

3.3.6 Translation

Translation of a picture involves its physical parallel translation onto another location of the x–y plane (Figure 3.6).

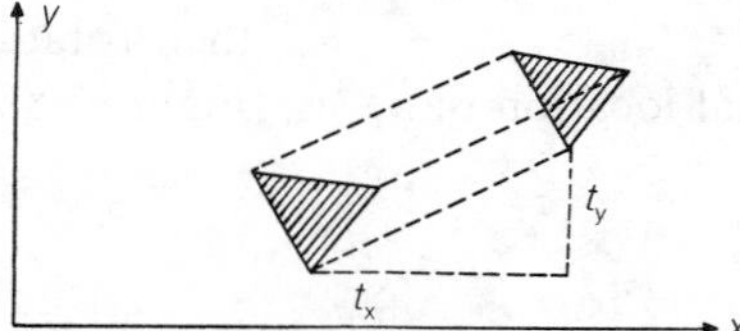

Figure 3.6 Plane translation

Only node coordinates are affected again:

$$X(i) = x(i) + t_x$$
$$Y(i) = y(i) + t_y \tag{3.6}$$

where t_x and t_y specify the magnitude of the translation in the x and y directions, respectively. In matrix form:

$$\begin{Bmatrix} X(i) \\ Y(i) \\ 1 \end{Bmatrix} = \begin{bmatrix} 1 & 0 & t_x \\ 0 & 1 & t_y \\ 0 & 0 & 1 \end{bmatrix} \begin{Bmatrix} x(i) \\ y(i) \\ 1 \end{Bmatrix} \tag{3.6a}$$

3.3.7 Rotation

The coordinates of a point x, y, after rotation by an angle θ about the origin, as shown in (Figure 3.7) become:

$$X = x\cos\theta + y\sin\theta$$
$$Y = -x\sin\theta + y\cos\theta \tag{3.7}$$

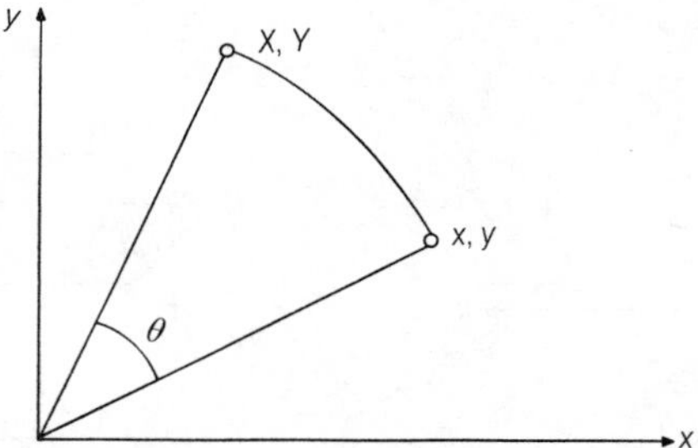

Figure 3.7 Point plane rotation

In matrix form:

$$\begin{Bmatrix} X(i) \\ Y(i) \\ 1 \end{Bmatrix} = \begin{bmatrix} \cos\theta & \sin\theta & 0 \\ -\sin\theta & \cos\theta & 0 \\ 0 & 0 & 1 \end{bmatrix} \begin{Bmatrix} x(i) \\ y(i) \\ 1 \end{Bmatrix} \tag{3.7a}$$

Rotation of the picture about the origin involves transformation of all the node coordinates.

Rotation about a general point x_0, y_0 involves first a translation, so that this point coincides with the origin, by $t_x = -x_0$, $t_y = -y_0$, then rotation, then translation $t_x = x_0$, $t_y = y_0$, i.e. back to the original location of x_0, y_0 (Figure 3.8).

3.3.8 Reflection

Most objects, both in nature and in engineering, have symmetry about one or more axes. It saves input information if only one half is given to the computer and a program plots the other half. This can be done with reflection (Figure 3.9).

The reflected coordinates are obtained in a very easy way. For example, for reflection about the x-axis:

$$\begin{aligned} X(i) &= x(i) \\ Y(i) &= -y(i) \end{aligned} \tag{3.8}$$

It is apparent that this is equivalent to a scaling of $s_x = 1$, $s_y = -1$.

For reflection about the y-axis:

$$\begin{aligned} X(i) &= -x(i) \\ Y(i) &= y(i) \end{aligned} \tag{3.9}$$

It is equivalent to a scaling of $s_x = -1$, $s_y = 1$.

Reflection about another axis can be obtained with translation and rotation first, until the desired axis of reflection coincides with one of the principal axes.

In mechanical engineering drawings, symmetry often refers to a section of an axisymmetric body of revolution. In this case, the definition of polygons must be appended with the definition of the new, reflected nodes (A′, B', C' in Figure 3.9).

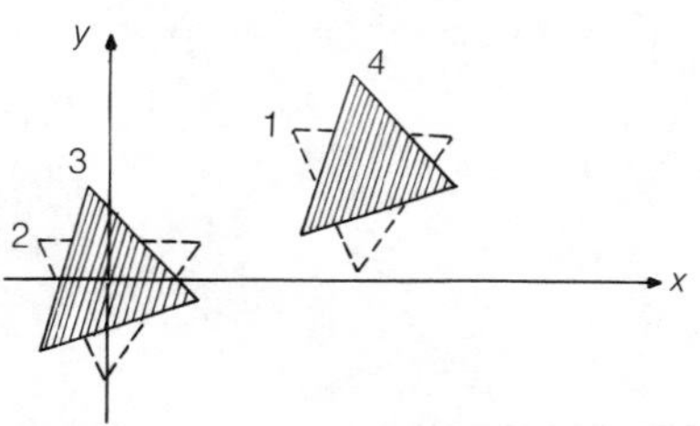

Figure 3.8 Successive steps in plane rotation

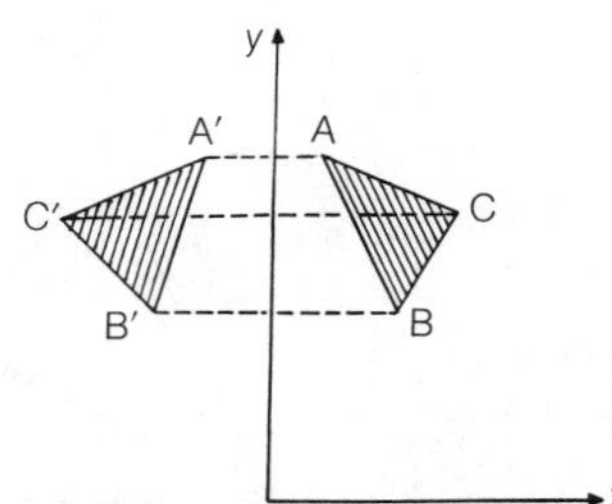

Figure 3.9 Mirror imaging

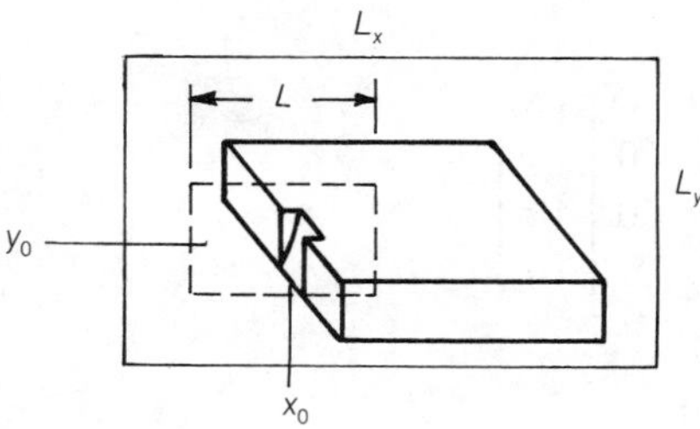

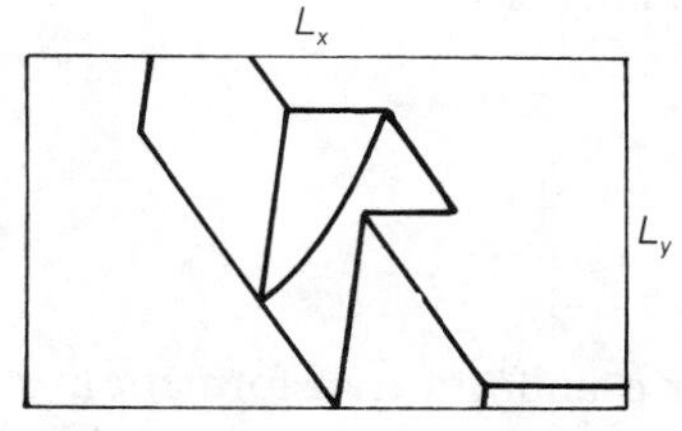

Figure 3.10 Zooming

3.3.9 Zooming

It is often necessary to provide an enlarged view of a picture detail (Figure 3.10). This can be achieved with a combination of scaling, translation and clipping, called 'zooming'. For example, for the detail shown in Figure 3.10, the transformations are:

Translation... $t_x = -x_0, t_y = -y_0$ (center of detail to origin)
Scaling... $s_x = s_y = L_x/L$
Translation... $t_x = L_x/2,\ t_y = L_y/2$ (center of detail to center of frame)
Clipping (to frame dimensions)

The same effect can be achieved by clipping the desired window end then performing the other operations.

3.4 THREE-DIMENSIONAL (3-D) GRAPHICS

3.4.1 Transformations

Since plotting devices are flat, only two-dimensional pictures can be plotted. One can, however, create three-dimensional objects and take several plane views, which is the standard practice in machine design. The description of the object requires the coordinates x, y, z of each node and the definition of polygons and curved lines.

As in plane graphics, transformations can also be applied on the node coordinates of the object while the polygon definition remains unchanged, except for reflection, in which case more nodes need to be defined.

The transformations can be expressed in matrix form, a notation very convenient. This method, however, is slow for microcomputer utilization. In this case, the direct algebraic expressions can be used.

Scaling can be accomplished by the following transformation on the node coordinates:

$$X = s_x x, \quad Y = s_y y, \quad Z = s_z z \tag{3.10}$$

In matrix form:

$$\begin{Bmatrix} X \\ Y \\ Z \\ 1 \end{Bmatrix} = \begin{bmatrix} s_x & 0 & 0 & 0 \\ 0 & s_y & 0 & 0 \\ 0 & 0 & s_z & 0 \\ 0 & 0 & 0 & 1 \end{bmatrix} \begin{Bmatrix} x \\ y \\ z \\ 1 \end{Bmatrix} \tag{3.10a}$$

For a uniform transformation, $s_x = s_y = s_z = s$.

Translation is performed by the transformation:

$$X = x + t_x, \quad Y = y + t_y, \quad Z = z + t_z \tag{3.11}$$

In matrix form:

$$\begin{Bmatrix} X \\ Y \\ Z \\ 1 \end{Bmatrix} = \begin{bmatrix} 1 & 0 & 0 & t_x \\ 0 & 1 & 0 & t_y \\ 0 & 0 & 1 & t_z \\ 0 & 0 & 0 & 1 \end{bmatrix} \begin{Bmatrix} x \\ y \\ z \\ 1 \end{Bmatrix} \tag{3.11a}$$

For reflection about the *xz* plane:

$$X = x, \quad Y = -y, \quad Z = z \tag{3.12}$$

For reflection about the *yz* plane:

$$X = -x, \quad Y = y, \quad Z = z \tag{3.13}$$

Reflection is equivalent to scaling with appropriate scale factors.

Rotation is somewhat more complicated. In plane transformations, rotation is about a point. In space, rotation is about an axis (Figure 3.11). The rotation about an axis passing through the origin is defined by way of the three direction cosines of the axis:

$$\begin{aligned} n_1 &= \cos(\varphi_x) \\ n_2 &= \cos(\varphi_y) \\ n_3 &= \cos(\varphi_z) \end{aligned} \tag{3.14}$$

where φ_x, φ_y, φ_z are the angles of the axis of rotation with the coordinate axes.

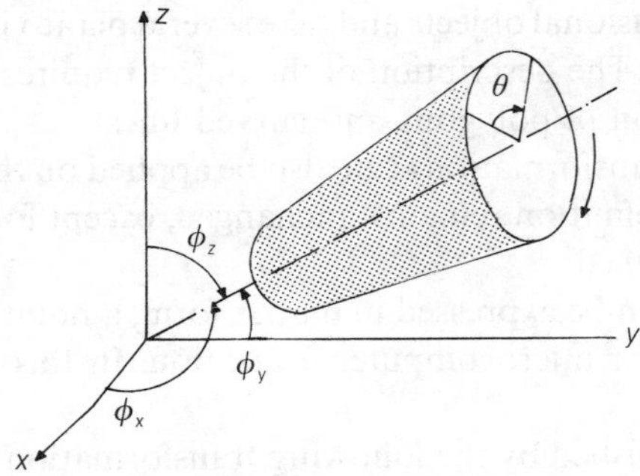

Figure 3.11 Solid rotation

Two of these need to be specified because:

$$n_1^2 + n_2^2 + n_3^2 = 1 \tag{3.15}$$

The relations given for the plane rotation are written in matrix form as

$$\begin{Bmatrix} X \\ Y \end{Bmatrix} = \begin{bmatrix} \cos\theta & \sin\theta \\ -\sin\theta & \cos\theta \end{bmatrix} \begin{Bmatrix} x \\ y \end{Bmatrix} \tag{3.16}$$

The general rotation matrix in three dimensions can be found in standard texts on analytic geometry, for an angle of rotation θ about an axis n_1, n_2, n_3 passing through the origin:

$$[R] = \begin{bmatrix} n_1^2 + (1 - n_1^2)\cos\theta & n_1 n_2(1 - \cos\theta) + n_3 \sin\theta & n_1 n_3(1 - \cos\theta) - n_2 \sin\theta & 0 \\ n_1 n_2(1 - \cos\theta) - n_3 \sin\theta & n_2^2 + (1 - n_2^2)\cos\theta & n_2 n_3(1 - \cos\theta) + n_1 \sin\theta & 0 \\ n_1 n_3(1 - \cos\theta) + n_2 \sin\theta & n_2 n_3(1 - \cos\theta) - n_1 \sin\theta & n_3^2 + (1 - n_3^2)\cos\theta & 0 \\ 0 & 0 & 0 & 1 \end{bmatrix} \tag{3.17}$$

For rotation about a general axis not passing through the origin, a translation is performed to bring the axis of rotation through the origin, then rotation, then translation again to the original position of the reference point.

Successive transformations can be performed by repeated application of the respective algorithms or by respective multiplication of the appropriate transformation matrices. For example, $\{P(i)\} = [R][T][S]\{p(i)\}$ means first scaling, then translation, then rotation. It must be emphasized that transformations of this type are not commutative, that this is if applied in different order they will yield, in general, different results.

3.4.2 Mechanical drawings

Mechanical drawings of an object are plane projections on the three planes *xy*, *yz*, *zx*, and axonometric projections.

A solid can be described by a number of plane polygons. Curves can be described also by way of polygons with a sufficient number of straight-line segments. Straight lines can also be considered as open polygons of two nodes. The following procedure reads the data for a solid object and creates a data file which can be subsequently retrieved for further processing. The following structure will be used:

Record No. 1: NNODES%, NARCS%, NPOLY%
Record No. 2: X, Y, Z coordinates of node 1
Record No. 3: X, Y, Z coordinates of node 2
. . . .
Record No. NNODES% + 1: X, Y, Z coordinates of node NNODES%
Record No. NNODES% + 2: for polygon 1 the number of nodes followed by the node numbers traced counterclockwise.
. . . .
Record No. NNODES% + NPOLY%: for polygon NPOLY%, data as above.
Record No. NNODES% + NPOLY% + 1: for circular arc No. 1, the numbers of the three defining nodes, beginning of arc, an intermediate and the end point of the circular arc.
. . . .

Record No. NNODES% + NPOLY% + NARCS%: for circular arc No. NARCS%, data as above.

```
Procedure MODEL
LOCAL NNODES%, NARCS%, NPOLY%,
        arrays X, Y, Z, C(1..NNODES%): var
GLOBAL array C$(1..NNODES%): var
begin
  enter NNODES%, NARCS%, NPOLY%
  enter file name FNAME$
  RECORD%:= 1; LEN:= 50
  C$(1, 2, 3):= NNODES%, NARCS%, NPOLY%, respectively
  LRNDF                                    (*Write record 1 of file*)
FOR I:= 1 TO NNODES% do
  enter X(I), Y(I), Z(I)
  C$(I):= string of X(I)
  RECORD%:= I + 1; LEN:= 30
  LRNDF           (*Load X, Y, Z coordinates of node I on record I + 1*)
FOR I:= NNODES% + 1 TO NNODES% + NPOLY% do
  IP:= I - NNODES%
  enter PNODES%
  C$(IP):= string of PNODES%
  FOR J:= 1 TO PNODES% do
    enter node numbers C(J)
    C$(IP + J):= string of C(J)
  RECORD%:= I; LEN:= 5*(PNODES% + 1)
  LRNDF                         (*Load polygon IP vertices on record I*)
NPR:= NNODES% + NPOLY%
FOR I:= NPR + 1 TO NPR + NARCS%
  FOR J:= 3*I - 2 TO 3*I
    enter C(J)
    C$(J):= string of C(J)
  RECORD%:= I; LEN:= 15; IC:= I - NPR
  LRNDF                                 (*Load arc IC data on record I*)
end.
```

Data can be recovered with the RMODEL procedure which involves the steps in procedure MODEL in reverse order, and with the use of the RRNDF procedure instead of the LRNDF.

To perform space transformations, only the coordinates of the nodes need to be transformed. The other data remain unchanged upon rotation, scaling, translation. Therefore, only the arrays X, Y, Z of records 2, 3, 4 of the model need to be transformed.

The procedures ROTATE, SCALE, TRANSL follow:

```
Procedure ROTATE
GLOBAL N1, N2, FI                          (*cos fx, cos fy, angle of rotation*)
        NNODES%, arrays X, Y, Z(1..NNODES%): var
LOCAL arrays C(1..NNODES%), R(1..6, 1..6): var
begin
  compute matrix R                                          (*Equation 3.17*)
  FOR I:= 1 TO NNODES% do
    matrix multiply C:= R*{X(I) Y(I) Z(I)}
    matrix set {X(I) Y(I) Z(I)}:= C
end.

Procedure SCALE
GLOBAL SX, SY, SZ
        NNODES%, arrays X, Y, Z(1..NNODES%): var
LOCAL arrays C(1..NNODES%), R(1..6, 1..6)
begin
  FOR I:= 1 TO NNODES% do
    X(I):= X(I)*SX
    Y(I):= Y(I)*SY
    Z(I):= Z(I)*SZ
end.

Procedure TRANSL
GLOBAL TX, TY, TZ
        NNODES%, arrays X, Y, Z(1..NNODES%): var
begin
  FOR I:= 1 TO NNODES% do
    X(I):= X(I) + TX
    Y(I):= Y(I) + TY
    Z(I):= Z(I) + TZ
end.
```

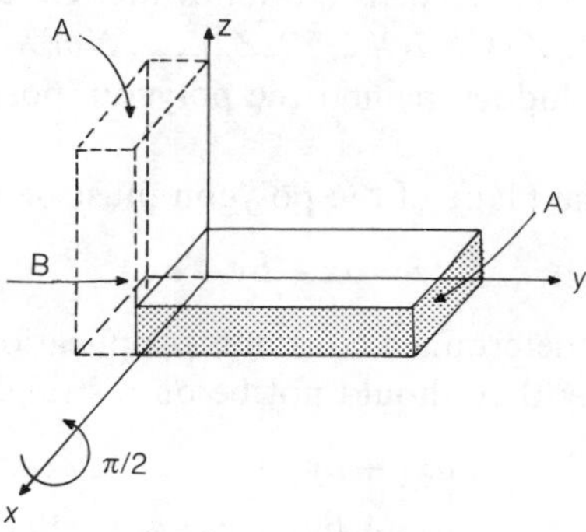

Figure 3.12 Rotation about the x-axis by angle $\pi/2$

A menu program using the above procedures can perform transformations of solid models.

If the description of an object by way of the node coordinates and line definition is given, the projection on the $z = 0$ plane is trivial. A plane plot is performed with the given values of the node coordinates x and y disregarding the z coordinate.

The projection on the $y = 0$ plane, view A, requires a rotation about the x-axis ($n_1 = 1$, $n_2 = 0$, $n_3 = 0$) by an angle $\theta = \pi/2$ and then plotting the new x, y coordinates (Figure 3.12).

The view B can be obtained with $\theta = -\pi/2$.

Axonometric views are obtained with proper selection of the axis and angle of rotation. An axonometric isometric view is obtained with a rotation such that the three orthogonal axes of symmetry of the solid, if such a situation exists, upon rotation are projected with equal angles 120° from one another. Such rotation is obtained with

$$n_1 = -0.707, \; n_2 = 0.707, \; \theta = \pi/4$$

3.4.3 Hidden line removal

In mechanical drawing, lines hidden behind visible planes are either not shown on the drawing or made with dashed lines. While plotting the projection of an object on the x–y plane by simply plotting all nodes and the connecting straight lines, the plotting resembles not a solid object but a wire mesh connecting the nodes. For drawings this is not acceptable and the hidden lines must be identified and removed or plotted as dashed lines. This procedure takes a lot of computer time and for microcomputers it can be considered a difficult task. Therefore, only in the final drawing can it be used because, for an average size object, it might take a long time to be completed.

Several changes have to be made to the plotting strategy to account for hidden lines.

(i) Lines cannot be plotted with one command (such as by one call of the LINE procedure for example) but instead they have to be plotted point-by-point or in small segments.
(ii) Before each point is plotted, it has to be tested against all bounding polygons for visibility. If it is found that the point is behind any one of them, then the point is not plotted.

Therefore, the problem is reduced to the following. Given a point X, Y and a plane polygon of nodes $N1(X1, Y1, Z1)$, $N2(X2, Y2, Z2), \ldots, Nm(Xm, Ym, Zm)$, find if the point, for an observer at $Z = \infty$, is hidden behind the polygon (positive check) or not (negative check).

First, the equation for the plane of the polygon must be determined in the form

$$z = ax + by + c \tag{3.18}$$

where a, b and c are as yet undetermined constants. Application of this equation at the first three consecutive nodes (*note*: they should not be on a straight line) yields:

$$\begin{aligned} ax_1 + by_1 + c &= z_1 \\ ax_2 + by_2 + c &= z_2 \\ ax_3 = by_3 + c &= z_3 \end{aligned} \tag{3.19}$$

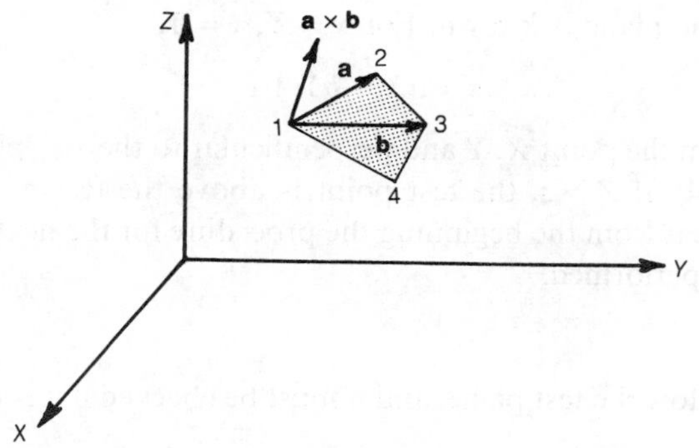

Figure 3.13 Determination of polygon orientation

Test 1
If the 3 × 3 coefficient matrix of the above system is singular, the plane is perpendicular to the x–y plane, and the check is negative. The next plane should then be tested. Moreover, almost half of the polygons can be eliminated from the test with a simple trick. Suppose that the node numbering was given counterclockwise when facing the surface from the outside. Upon projection on the x–y plane, the polygon facing up maintains the counterclockwise direction of node numbering. Otherwise, for a polygon facing down, the direction is reversed. The polygons facing down should not be plotted and should not be checked during the hidden line check. To this end, the cross-product is formed, **a** × **b**, where **a** and **b** are vectors connecting the first node to the second and third, respectively (Figure 3.13). If the z-component $a_x b_y - a_y b_x$ of the cross-product **a** × **b** is positive, then the polygon is facing up and should be plotted and checked for hidden lines.

Test 2
If check 1 is positive, the solution of the system (equation 3.19) yields the constants a, b and c

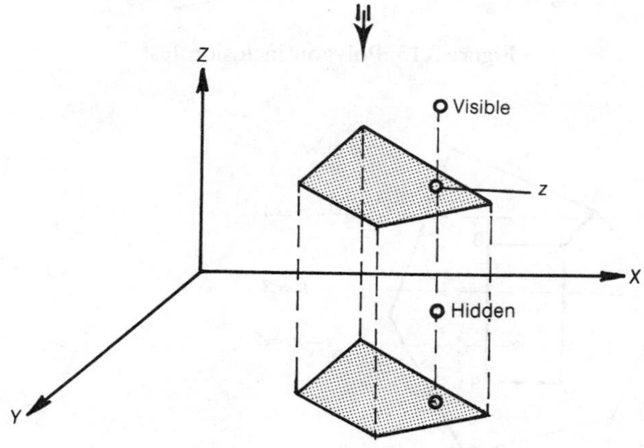

Figure 3.14 Hidden line detection

and the equation of the plane is known. For $x = X$, $y = Y$,

$$z = aX + bY + c \tag{3.20}$$

A straight line from the point X, Y and perpendicular to the x–y plane, meets the tested plane at z (Figure 3.14). If $Z > z$, the test point is above the test plane and the check is negative. Then we repeat from the beginning the procedure for the next plane. If $Z < z$, then the next test must be performed.

Test 3

The point X, Y, Z is below the test plane, and it must be checked if it is under the polygon or outside.

From the geometry of Figure 3.15 it is obvious that if a straight line emanating from the test point on the x–y plane meets the boundary of the polygon at an odd number of points, the point is inside the polygon. Otherwise, it is outside.

To perform this test, such a straight line must be selected. There is no loss in generality if one assumes a parallel to the x-axis. The straight line between points x_1, y_1 and x_2, y_2 is defined on the x–y plane by the equation

$$x = x_1 + (y - y_1)(x_2 - x_1)/(y_2 - y_1) \tag{3.21}$$

The intersection of this line with the line $y = Y$ is at

$$X = x_1 + (Y - y_1)(x_2 - x_1)/(y_2 - y_1) \tag{3.22}$$

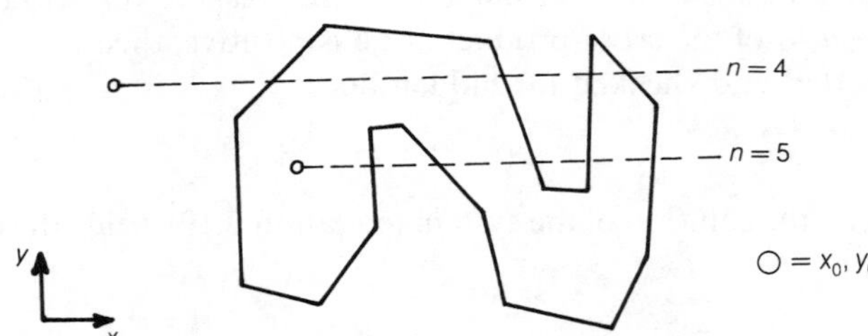

Figure 3.15 Polygon inclusion test

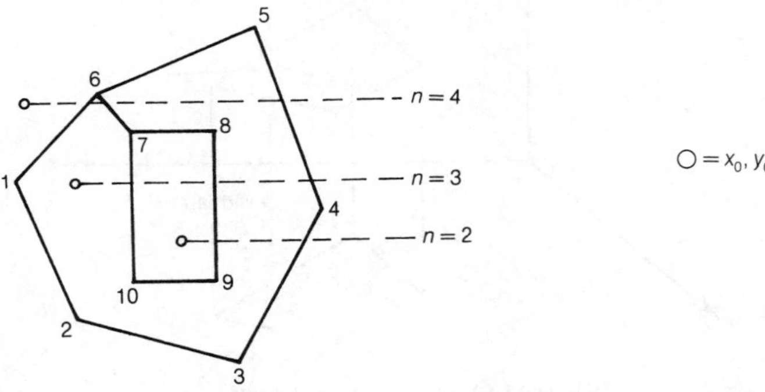

Figure 3.16 A multiple connected polygon

The point X, Y is between the points x_1, y_1, and x_2, y_2 if X is between x_1, x_2 and Y between y_1, y_2, since plotting is performed in the first quadrant.

Polygons inside other polygons can be treated by combining them as indicated in Figure 3.16. Instead of polygons 1–2–3–4–5–6–1 and 7–8–9–10–7, a new polygon is defined 1–2–3–4–5–6–7–8–9–10–7–6–1.

Obviously, if a test for any plane polygon is positive (point hidden), there is no need for testing against the other planes. Then, one must proceed to the next point of the line.

All the lines of the solid have to be checked and plotted point-by-point. For a speedier but less precise plot, one can decide and plot a small line segment at a time and consider it hidden or visible on the basis of the test of a point on it.

The algorithm HIDDEN, which follows, checks a point X, Y against a solid and returns the index HIND% with values 1 for hidden and 0 for visible. An index IND% = 0 is used at the first call and > 0 at the subsequent calls.

```
Procedure HIDDEN                                  (*Checks the visibility of a point*)
GLOBAL X, Y, Z, NNODES%, NPOLY%, HIND%, IND%, IPMAX%
arrays X, Y, Z(1..NNODES%), AP, BP, CP(1..NPOLY%), PVC(IPMAX%): var
LOCAL INDEX%: var
begin
  IF IND% = 0 then do
    FOR I:= 1 TO NPOLY% do
      solve eq. (3.19) for polygon i
      IF matrix singular then a = b = c = 0                (*Perpendicular to x–y*)
    AP(I):= a; BP(I):= b; CP(I):= c
else
  IP:= 1
  FOR I:= 1 TO NPOLY% do
    PNODES%:= INTEGER(PVC(IP))
    IF AP(I)^2+BP(I)^2+CP(I)^2:=0 then exit loop for I (*No. of nodes of polygon IP*)
    else
      z:= AP(I)*X + BP(I)*Y + BC(I)
      IF Z > z then exit loop for I
      else
        INDEX% = 0
  FOR J:= IP + 1 TO IP + PNODES%
    I1:= INTEGER(PVC(J)); X1:= X(I1)
    I2:= INTEGER(PVC(J + 1)); X2:= X(J2)
    Y2:= Y(I1); Y2:= Y(I2)
    IF x outside X1..X2 and y outside Y1..Y2, next J
      else
        IF x between X1 and X2 then INDEX% + INDEX% + 1
        else, next J
      IF INDEX% odd THEN HIND% = 1
      IF HIND%:= 1 then exit procedure
  IP:= IP + PNODES%
end.
```

A procedure SOLID will perform all the above transformations:

```
Procedure SOLID (NNODES%, NPOLY%, IPMAX%, X, Y, Z, CP)
GLOBAL NNODES%, NPOLY%, IPMAX%, HIND%
         arrays X, Y, X(1..NNODES%), int PVC(1..IPMAX%): var
begin
IP1:= 1
FOR I:= 1 to NPOLY% do
  PNODES%:= CP(IP1)
FOR N:= 1 TO PNODES% do
  I1:= PVC (IP1 + N), I2:= PVC (IP1 + N + 1)
X1:= X(I1); X2:= X(I2): Y1:= Y(I1); Y2:= Y(I2); Z1:= Z(I1); Z2 = (I2)
  LENGTH:= SQR((X2 - X1)^2 + (Y2 - Y1)^2 + (Z2 - Z1)^2)
  XP:= X1; YP:= Y1
  FOR J:= 1 TO INTEGER(LENGTH + 1) do
    X:= I*(X2 - X1)/LENGTH
    Y:= I*(Y2 - Y1)/LENGTH
    Z:= I*(Z2 - Z1)/LENGTH
    HIDDEN
    IF HIND%:= 0 THEN do
      LINE (XP, YP) - (X - Y)
      XP:= X; YP:= Y
    else
      XP:= X; YP:= Y
  IP1:= IP1 + PNODES%
end.
```

3.4.4 Arc representation

A circular arc can be defined in several ways. It is very convenient for computing algorithms to represent circular arcs by their center and three points – start, end, one intermediate. Upon general transformations these points will represent, in general, on the x–y plane the center and three points of an elliptical arc to which the circular arc has been transformed upon projection on the plane x–y.

The equation of an ellipse with major axis inclined by angle g in respect to the x-axis, has parametric equations

$$\begin{aligned} x &= (a\cos t\ \cos g) - (b\sin t\ \sin g) \\ y &= (a\cos t\sin g) + (b\sin t\cos g) \end{aligned} \tag{3.23}$$

where a and b are the same-axis of the ellipse (Figure 3.17).

Introducing the global angle $f = t + g$ and eliminating the angle t

$$\begin{aligned} x &= (a\cos^2 g + b\sin^2 g)\cos f + (a - b)\sin g\cos g\sin f \\ y &= (a - b)\sin g\cos g\cos f + (a\sin^2 g + b\cos^2 g)\sin f \end{aligned} \tag{3.24}$$

Since a, b and g have to be determined from the three points (which define the ellipse) on

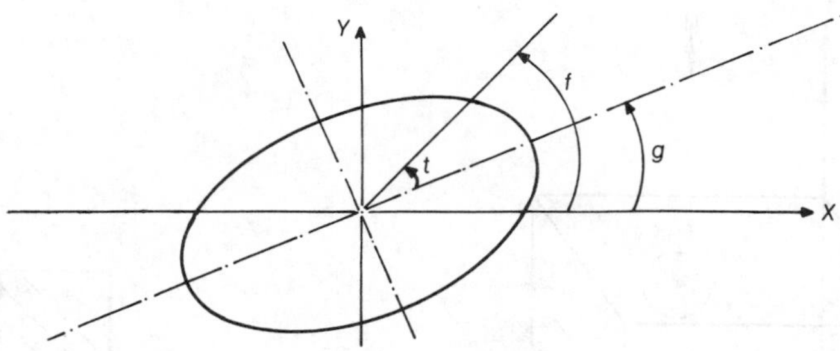

Figure 3.17 Ellipse geometry

the x–y plane, it is equivalent to the introduction of three new parameters for the parametric representation of the ellipse:

$$x = c_1 \cos f + c_2 \sin f \tag{3.25a}$$

$$y = c_2 \cos f + c_3 \sin f \tag{3.25b}$$

Parameters c_1 and c_2 are computed, applying equation (3.25a) to the first and second node. Parameter c_3 is then computed applying equation (3.25b) to the third node.

3.5 SECTIONED VIEWS

As mentioned above, machine design graphics deviates considerably from the geometric computer graphics because of established conventions and standards adopted during years of design practice. To the extent that some of these conventions were adopted to simplify the manual work, with the applications of computer methods they are gradually fading out. Many conventions however are useful for many other purposes, such as manufacturing, assembly, service, etc., and they have to receive special consideration. One such feature is the 'sectioned views'.

A sectioned view of a solid is essentially a view of the part of a solid which is below some plane $z = V$ (Figure 3.18). This solid is projected as usual on the x–y plane to obtain a sectional view of the solid in respect to the section plane. In practice, this means that:

(a) All lines which are above this plane are not plotted.
(b) Polygons which are entirely above this plane do not hide lines which are below the section plane.
(c) Lines which intersect with the section plane are plotted up to the intersection.
(d) For polygons which are intersecting with the section plane, their parts above the section plane have to be replaced with the line of intersection.

These requirements can be easily accommodated in the HIDDEN procedure.

Requirements (a) and (c) are satisfied if points with $Z > V$ are labeled hidden, setting HIND:= 1. Requirement (b) is satisfied if the index HIND:= 1 is cancelled for the hidden line tests if $z > V$.

For the requirement (d), the intersection of polygons with the plane $z = V$ can be found

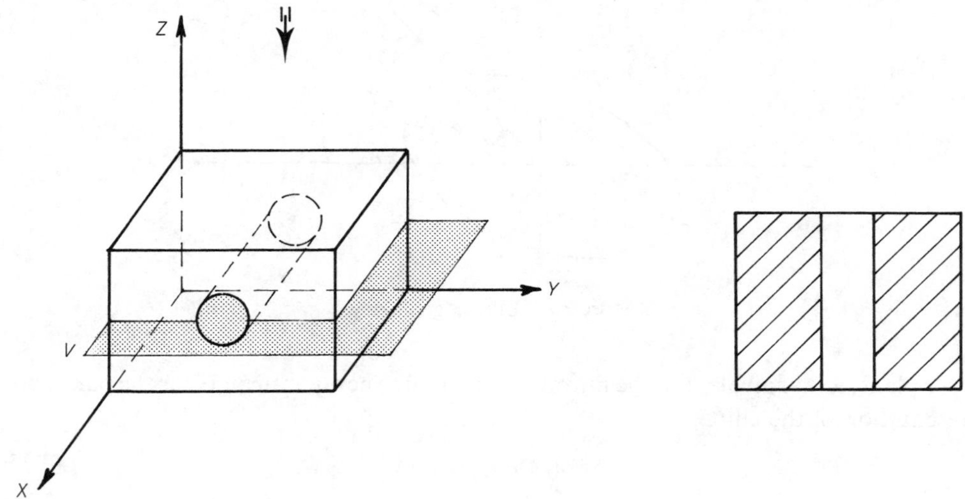

Figure 3.18 Sections of solids

from the equations of the polygon planes obtained already with the procedure HIDDEN where for each plane three constants a, b and c have been found defining the plane equation

$$z = aX + bY + c \tag{3.26}$$

Since on the intersection $z = V$,

$$aX + bY = V - c \tag{3.27}$$

Of this straight line only the parts that are inside the polygon must be plotted. This can be done with the method of Section 3.4. A procedure SECTION, which is an extension of HIDDEN, follows:

```
Procedure SECTION(V)                    (*Hidden line with section view below z = V*)
GLOBAL X, Y, Z, V, NNODES%, NPOLY%, HIND%, IND%, IPMAX%
  arrays X, Y, Z(1..NNODES%), AP, BP, CP(1..NPOLY%), PVC(IPMAX%)
begin
  IF IND% = 0 then do
    IP:= 1
    FOR I:= 1 TO NPOLY% do
      PNODES%:= INTEGER(PVC(IP))
      solve equation (3.19) for polygon i
  IF matrix singular then a = b = c = 0                    (*Perpendicular to x-y*)
  AP(I):= a; BP(I):= b; CP(I):= c                (*Intersection with the plane z = V*)
  if all vertices have z > V then next polygon I
  else
    FOR J:= IP + 1 TO IP + NNODES%
      NODE1:= INTEGER(PVC(J))
```

```
        NODE2:= INTEGER(PVC(J + 1))
        IF Z(NODE1) AND Z(NODE2)both < >V then next J
        else
          X1:= X(NODE1); X2:= X(NODE2)
          Y1:= (V - CP(NODE1) - AP(NODE1))/BP(NODE1)
          Y2:= (V - CP(NODE2) - AP(NODE2))/BP(NODE2)
          LINE (X1, Y1) - (X2, Y2)
else
  IP:= 1
  FOR I:= 1 TO NPOLY% do
    PNODES%:= INTEGER(PVC(IP))
    IF AP(I)^2 + BP(I)^2 + CP(I)^2:= 0 then next I          (*Vertical plane*)
    else
      IF Z > V THEN HIND%:= 1 and exit SECTION procedure
    else
      z:= AP(I)*X + BP(I)*Y + BC(I)
      IF Z > z then next I
      else
        INDEX = 0
    FOR J:= IP + 1 TO IP + NNODES%
      I1:= INTEGER(PVC(J)); X1:= X(I1)
      I2:= INTEGER(PVC(J + 1)); X2:= X(J2)
     Y2:= Y(I1); Y2:= Y(I2)
      IF x outside X1..X2 and y outside Y1..Y2, next J
      else
        IF x between X1 and X2 then INDEX + INDEX + 1
        else, next J
      IF INDEX odd THEN HIND% = 1
      IF HIND%:= 1 then exit procedure
  IP:= IP + NNODES%
end.
```

A typical sectioned view is shown in Figure 3.18.

3.6 AUTOMATIC MESH GENERATION

For many applications such as finite-element modeling and solid modeling, a plane or curved surface must be divided in a number of triangles or rectangles or other shapes (Figure 3.19). This follows simple rules of geometry if the bounding curve has a simple shape.

For a rectangle of dimensions L_x and L_y (Figure 3.20), divided into n columns and n rows, the node coordinates, line and polygon definitions in the notation of the SOLID procedure are, for division into rectangles:

$$\text{for } i = 1 \cdots n_x \quad \text{and} \quad j = 1 \cdots n_y$$

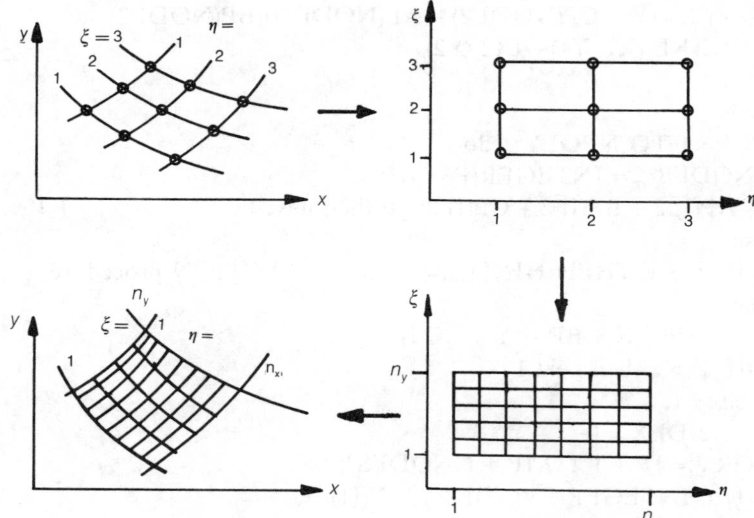

Figure 3.19 Curvilinear mesh transformations

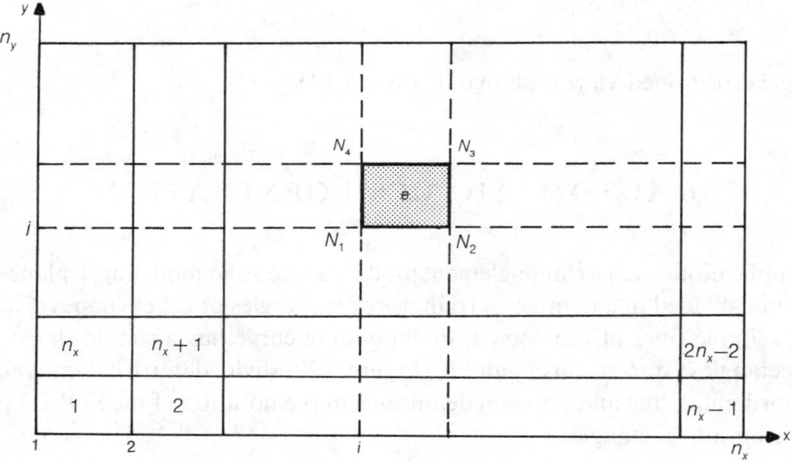

Figure 3.20 Rectangular mesh

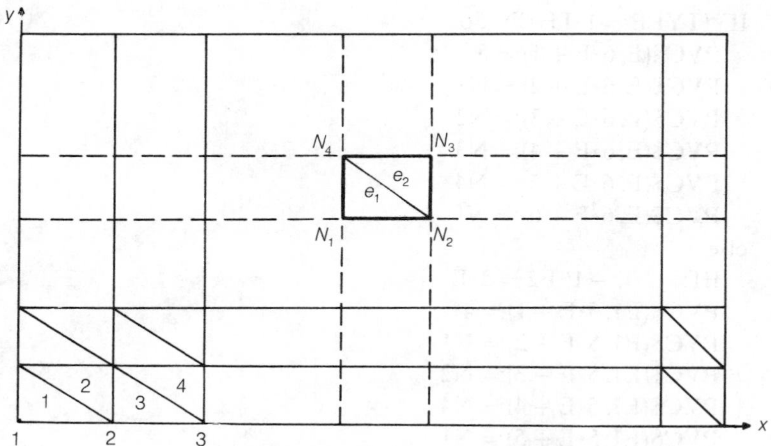

Figure 3.21 Triangular mesh

Nodes: $N_1 = ij,\ N_2 = ij + 1,\ N_3 = ij + n_x + 1,\ N_4 = ij + n_x + 1$

Node coordinates: $X(N_1) = iL_x/n_x,\ Y(N_1) = jL_y/n_y$

Polygon number: $e = (n_x - 1)(j - 1) + i$

Polygon vertices: $\text{PVC\$}(e, 6e + 1) = 5$

$\text{PVC\$}(e, 6e + 1 + k) = N_k,\ k = 1 \ldots 4$

$\text{PVC\$}(e, 6e + 6) = N_1$

For triangular facets (Figure 3.21), only the polygon definitions will change:

$e_1 = 2e - 1,\ e_2 = 2e$

$\text{PVC\$}(e_1, 5e_1 + 1) = 4$

$\text{PVC\$}(e_1, 5e_1 + 2) = N_1,\ \text{PVC\$}(e_1,\ 5e_1 + 3) = N_2,\ \text{PVC\$}(e_1, 5e_1 + 4) = N_4$

$\text{PVC\$}(e_2, 5e_2 + 1) = 4$

$\text{PVC\$}(e_2, 5e_2 + 2) = N_2,\ \text{PVC\$}(e_2, 5e_2 + 3) = N_3,\ \text{PVC\$}(e_2,\ 5e_2 + 4) = N_4$

The automesh procedure follows:

```
Procedure AUTOMESH(ITYPE, NX, NY, LX, LY, X, Y, PVC$)
GLOBAL var LX, LY, arrays X, Y(NX*NY)
          int NX, NY, ITYPE, array PVC$((NX - 1)*(NY - 1), 6)
begin
  FOR i:= 1 TO NX - 1 do
    FOR J:= 1 TO NY - 1 do
      N1:= (J - 1)*NX + I; N2:= N1 + 1; N3:= N2 + NX; N4:= N3 - N1
      X(N1):= I*LX/NX; Y(N1):= J*LY/NY
      E:= (NX - 1)*(J - 1) + I
```

```
    IF ITYPE:= 1 THEN do                                    (*Rectangles*)
      PVC$(E, 6*E + 1):= 5
      PVC$(E, 6*E + 2):= N1
      PVC$(E, 6*E + 3):= N2
      PVC$(E, 6*E + 4):= N3
      PVC$(E, 6*E + 5):= N4
      PVC$(E, 6*E + 6):= N1
    else                                                    (*Triangles*)
      E1:= 2*E - 1; E2:= 2*E
      PVC$(E1, 5*E + 1):= 4
      PVC$(E1, 5*E + 2):= N1
      PVC$(E1, 5*E + 3):= N2
      PVC$(E1, 5*E + 4):= N4
      PVC$(E1, 5*E + 5):= N1
      PVC$(E2, 5*E + 1):= 4
      PVC$(E2, 5*E + 2):= N2
      PVC$(E2, 5*E + 3):= N3
      PVC$(E2, 5*E + 4):= N4
      PVC$(E2, 5*E + 5):= N2
end.
```

The rectangular shape is of little use for most practical problems. In general, every plane shape can be divided into a number of curvilinear rectangles such as the one in Figure 3.19. It is assumed that the bounding curves can be expressed in parametric form, so that the opposite curves can be assigned the parameter values of 1 and 3 of the parameters η and ξ. Thus, in the parameter plane η, ξ the curvilinear rectangle will be transformed into a rectangle, which can be divided into triangles or rectangles with the AUTOMESH procedure. Therefore, the parametric equations relating x, y to η, ξ must be now developed.

For relatively smooth curves, a polynomial form is very practical. To this end, the parametric equations are set up, for polynomials of order, say 3, in the form:

$$x = (a_1\xi^2 + a_2\xi + a_3)\eta^2 + (a_4\xi^2 + a_5\xi + a_6)\eta + (a_7\xi^2 + a_8\xi + a_9)$$
$$y = (b_1\xi^2 + b_2\xi + b_3)\eta^2 + (b_4\xi^2 + b_5\xi + b_6)\eta + (b_7\xi^2 + b_8\xi + b_9)$$

There are nine transformation constants a_i and nine b_i to be determined. For this purpose, nine points are specified on the curvilinear rectangle corresponding to the intersections of the two families of three curves for $\eta, \xi = 1, (1+n)/2, n$. For each of these points there is a pair of values of η, ξ which correspond to known values of x, y. Therefore, they yield nine equations on the various a and nine equations on the various b. Solution of these equations will give the values of these a and b. Then the transformation is found.

Further, the rectangle in the η, ξ plane is divided with the AUTOMESH procedure in an $n_x \times n_y$ mesh. Each pair of known η and ξ is transformed into a point x, y. Therefore the mesh generation is completed, since the polygon definition does not change upon transformation. The corresponding procedure follows:

```
Procedure CURVEDMESH
global NX, NY, NEQ array PVC$((NX – 1)*(NY – 1), 6):int var
        arrays X, Y(NX*NY), A(NEQ, NEQ), FA, FB,
        ALFA, BETA(NEQ):real var
local I, J, IJ, NKSI, NETA, NX, NY:int var
begin
  input NX, NY, NKSI, NETA
  FOR I:= 1 TO NKSI do
    FOR J:= 1 TO NETA do
      IJ:= (J – 1)*NKSI + I
      input X(IJ), Y(IJ)                          (*Points defining the shape*)
  NEQ:= NKSI*NETA
                                          (*Defining system of linear equations*)
  FOR I:= 1 TO NKSI do
    FOR J:= 1 TO NETA do
      KSI:= 1 + (I – 1)*(NX – 1)/(NKSI – 1)
      ETA:= 1 + (J – 1)*(NY – 1)/(NETA – 1)
      IJ:= (J – 1)*NKSI + I
      FOR K:= 1 TO NEQ do
        A(K, J*NETA + I):= (KSI^I)*(ETA^J)
      FA(IJ):= X(IJ); FB(IJ):= Y(IJ)
                                  (*Solve linear equations by Gauss elimination*)
  GAUSSELIM(A, FA, NEQ, ALFA)
  GAUSSELIM(A, FB, NEQ, BETA)
                                                 (*Generate rectangular mesh*)
  AUTOMESH(2, NX, NY, NX, NY, KS, ET, PVC$)
                                               (*Back transformation to x–y plane*)
  FOR I:= 1 TO NX do
    FOR J:= 1 TO NY do
      IJ:= I*J
      X(IJ):= 0; Y(IJ):= 0
      FOR K1:= 1 TO NKSI do
        FOR K2:= 1 TO NETA do
          K12:= K1*K2
          X(IJ):= X(IJ) + (I^K1)*(J^K2)*ALFA(K12)
          Y(IJ):= Y(IJ) + (I^K1)*(J^K2)*BETA(K12)
end.
```

3.7 SOLID MODELING

The plane mesh generated in Section 3.6 can be supplemented by assigning z-coordinate values and then manipulated in space with the SOLID procedure. An example of this is shown in Figure 3.22, where the surface

$$z = \sin(R)/R$$

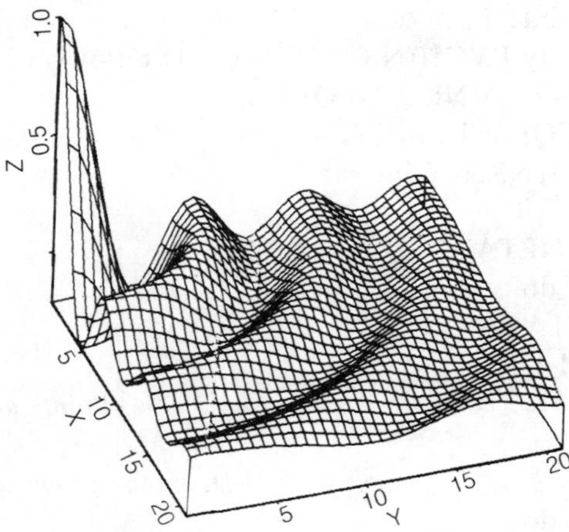

Figure 3.22 Description of a 3-D surface with rectangular patches

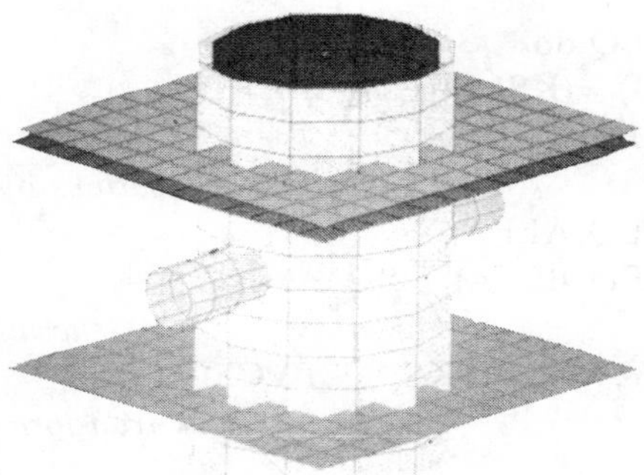

Figure 3.23 Solid modeling using cylinders and planes. (Courtesy Swanson Assoc., by permission)

is plotted in space. The rectangular facets of the resulting solid create an impression of solid surface. They are called 'surface patches' and they play an important role in solid surface presentation.

It is apparent that upon projection onto the x–y plane the density of the edges is proportional to the inclination of the patches in respect to the x–y plane. This creates a shading impression. The same thing can be achieved by painting the patches with a color of density proportional to the patch inclination. If the density of the mesh is high enough, a realistic three-dimensional picture can be created. Therefore, procedures to create solid primitives can be produced. Furthermore, the solid data can be appended with the desired intersecting solids. The result is, with the hidden line feature, that any complex solid can be modeled with the available solid primitives. In fact, this is essentially the way most commercial graphics packages operate. Examples of solid primitives are shown in Figure 3.23 while a complex solid is shown in Figure 3.24.

Figure 3.24 Solid modeled turbine blade. (Courtesy Swanson Assoc., by permission)

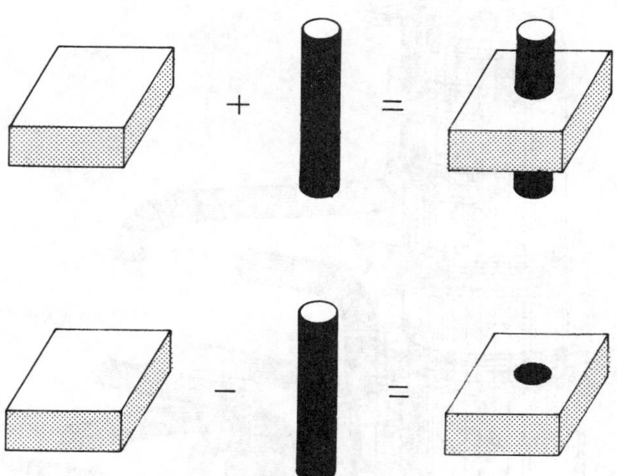

Figure 3.25 Solid operations

Solid primitives can be combined to yield complex solids with operations such as addition and subtraction (Figure 3.25), multiplication, and space transformations. Figure 3.26 shows a piston drawing made by solid modeling with a commercial package. The solid modeling can be extended to automatic mesh generation in three dimensions, such as the component shown in Figure 3.26. Other drawings made with commercial packages are shown in Figures 3.27 to 3.29.

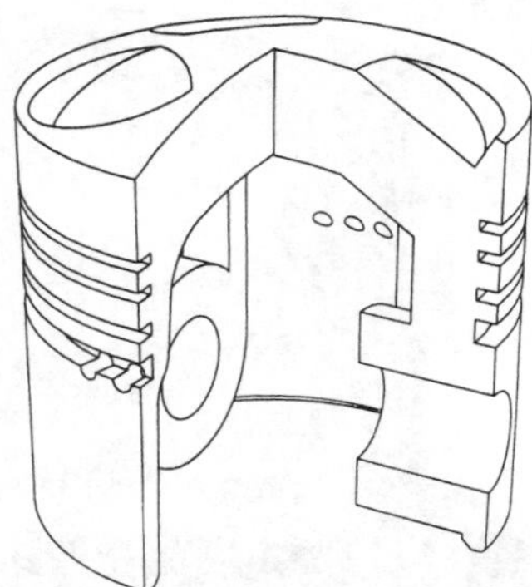

Figure 3.26 Complex components solid modeling: piston drawing made by solid modeling. (Courtesy ASME)

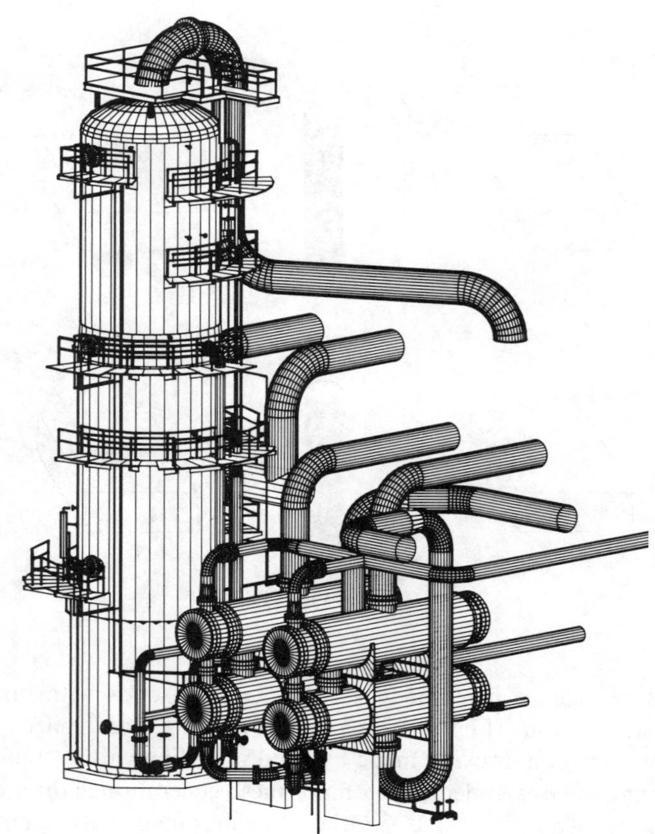

Figure 3.27 Solid modeling of a complex system. (Courtesy ASME)

Figure 3.28 Solid model of a gear drive. (Courtesy ASME)

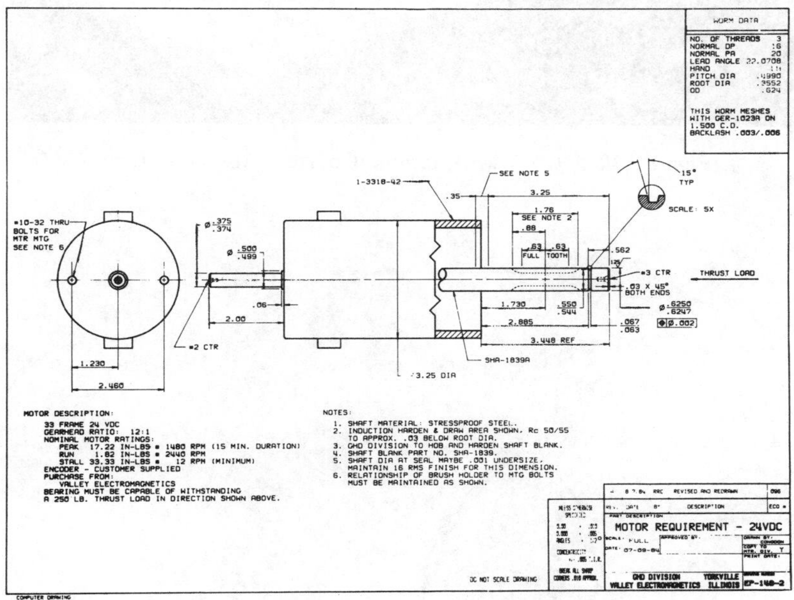

Figure 3.29 Plane drawing with a commercial package. (Courtesy ASME)

It must be pointed out that these procedures consume computer memory and time and can be efficiently implemented only on larger machines.

3.8 INPUT

Interactive computer graphics is a two-way communication. The output is presented to the user on a CRT, a printer or a plotter, perhaps stored also on a data file. The input information

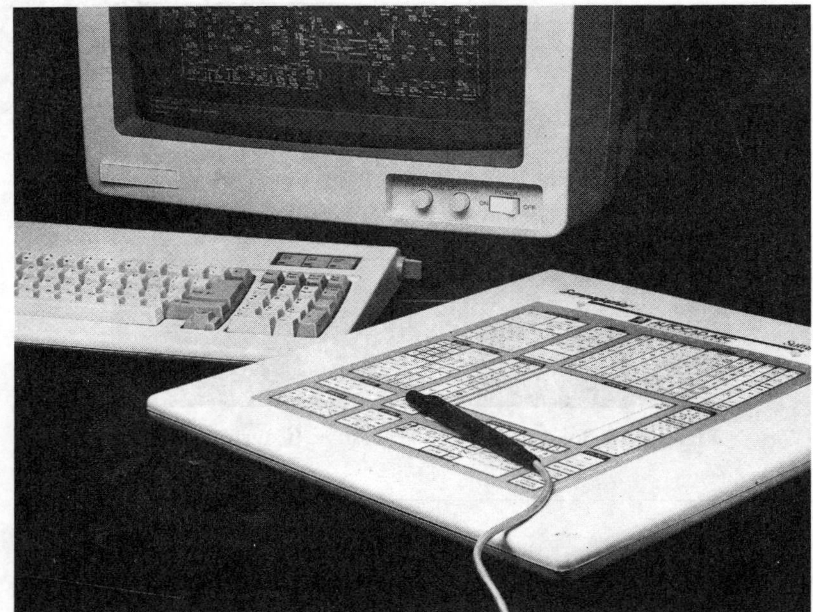

Figure 3.30 A CAD workstation. (Courtesy Autodesk, Inc.)

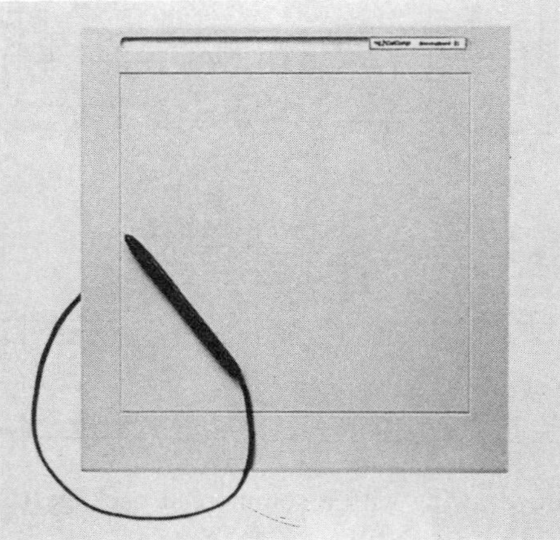

Figure 3.31 Digitizing tablet. (Courtesy California Computer Co.)

many times consists of a large amount of data that the user must transmit to the computer. There are several ways of doing it.

Tracing an existing drawing or a sketch can be done by a digitizing tablet (Figure 3.30). The drawing is placed on the tablet and a stylus or a cursor is traced over nodal points. Their x–y coordinates are transmitted to the computer. The z-coordinate can be transmitted through the keyboard or by tracing another view on the x–z or the y–z plane. Device-handling procedures are hardware-dependent and instructions are included in the appropriate user manuals. Robotic-type mouse heads are also used for both plotting and digitizing directly on the drawing board.

The general tendency is towards automatic data generation with solid modeling and the adoption of previously-defined solids (frames) which can be modified with a minimum of input data (Figure 3.32).

D
E
T HD
A
C
B
G
F
FAMILY NO.
33238
A
B
C
FAMILYNO.
30349
A
B DIA HOLE
D
E
C
F THD
G
CHM'F
I
J
K
FAMILY NO.
31235

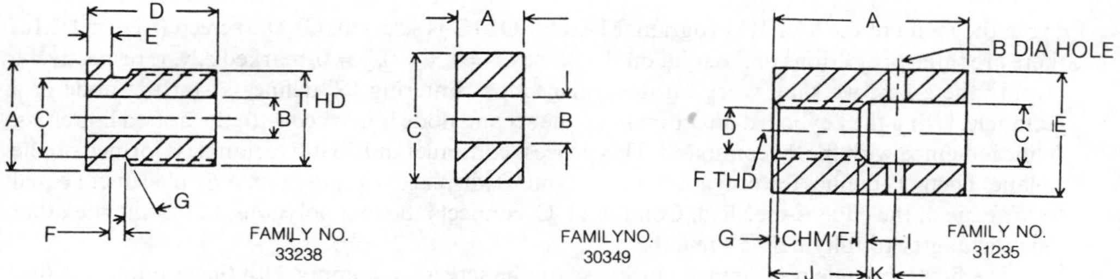

Figure 3.32 Drawing frames. (Courtesy Control Data Corporation)

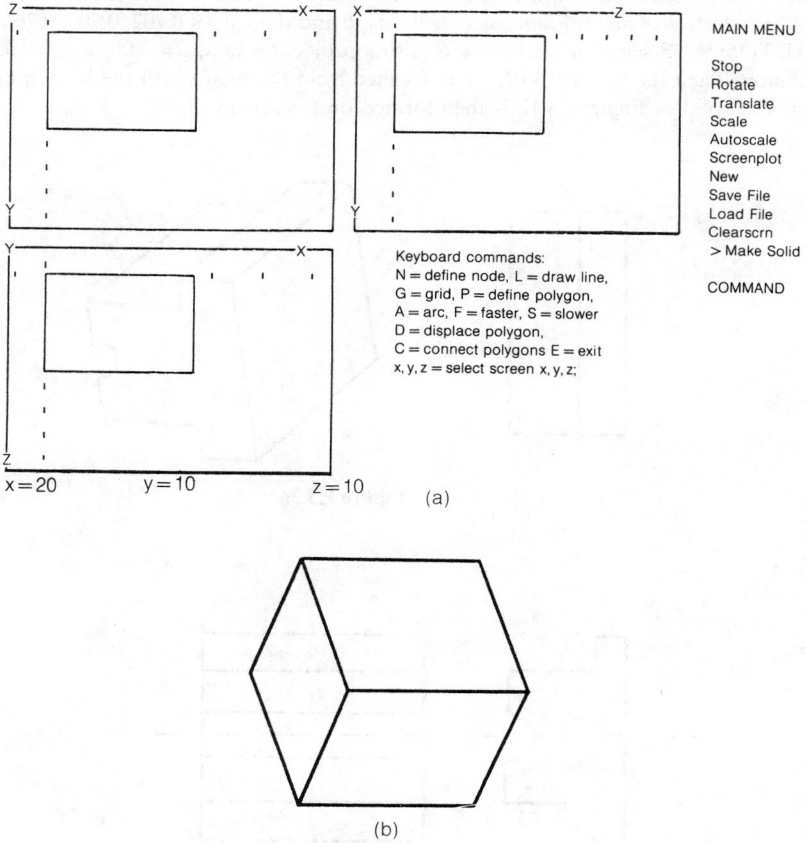

Z
X
Y
X
Z
Y
Y
X
Z
x=20
y=10
z=10
(a)
MAIN MENU
Stop
Rotate
Translate
Scale
Autoscale
Screenplot
New
Save File
Load File
Clearscrn
>Make Solid
COMMAND
Keyboard commands:
N = define node, L = draw line,
G = grid, P = define polygon,
A = arc, F = faster, S = slower
D = displace polygon,
C = connect polygons E = exit
x, y, z = select screen x, y, z;
(b)

Figure E3.1

EXAMPLES

Example 3.1

Using the program SOLID, draw an axonometric drawing of a cube with the hidden lines removed.

Solution

From the menu of the SOLID program, MAKE SOLID is selected. On the screen (Figure E3.1a) there are three projections of the solid on the planes $z = 0, x = 0, y = 0$, marked z, x, y, respectively. Hitting the z key, we shall work on the z projection. Entering 'N' defines a starting node of a rectangle. Using the keyboard direction keys, the second node is marked with the dotted line cross. A line is defined with the 'L' command. This process continues until a full rectangle is formed on the z plane. Then, 'P' defines the polygon 1. Command 'D' displaces the polygon. A displacement equal to the edge of the cube is specified. Command 'C' connects the two polygons and forms the other four rectangles to fully define the cube.

The first rectangle was defined clockwise on the screen, to comply with the requirement that its nodes are defined counterclockwise for an observer outside the solid.

Command 'E' returns to the main menu. The solid now can be saved with the SAVE FILE selection. The screen is cleared with the CLEARSCRN selection and the cube is rotated with the ROTATE selection. The program asks for n_1, n_2, θ and the values 0.707, 0.707, 0.79 are entered. The AUTOSCALE selection scales the resulting projection to $x_{max} = 200$, $y_{max} = 100$, in screen coordinates. Then the SCREENPLOT is selected from the menu with the hidden line removal option. The solid of Figure E3.1b is then formed on the screen.

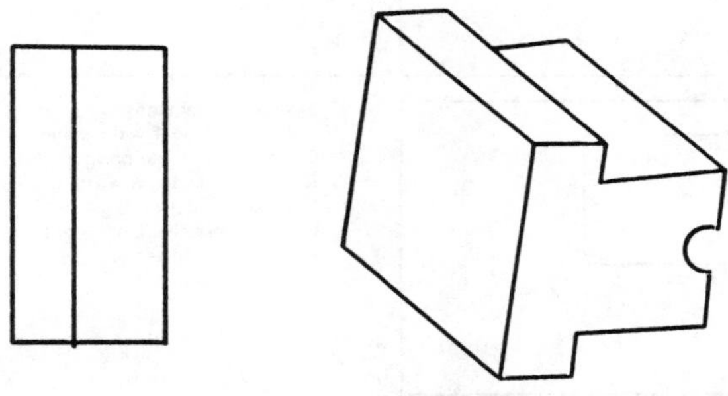

Figure E3.2a

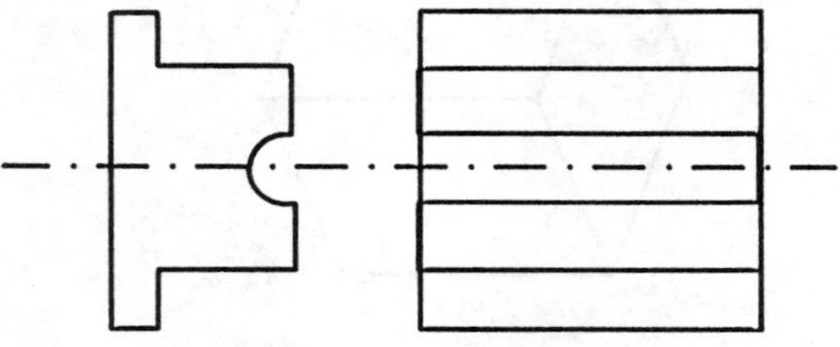

Figure E3.2b

Example 3.2

Using the program SOLID, draw axonometric and side views of a bearing support.

Solution

The input is prepared as in Problem 3.1 with the SOLID program. The plot is shown in Figure E3.2.

Example 3.3

Using AUTOCAD, or any similar package available in your system, draw the side views of a bracket shown.

Solution

The views shown in Figure E3.3, were made with AUTOCAD on an IBM AT, using the CALCOMP digitizing tablet and the 1044 drum plotter.

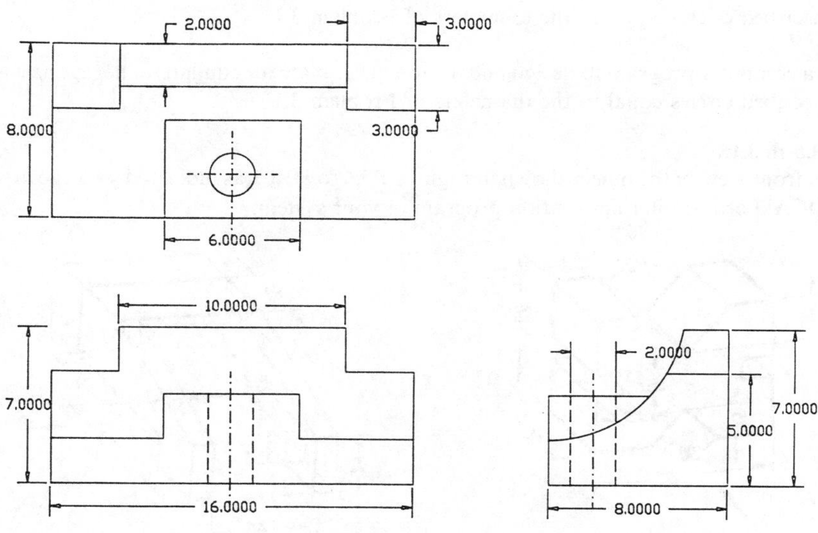

Figure E3.3

REFERENCES AND FURTHER READING

Angel, I. O., 1985. *Advanced Graphics with the IBM-PC.* London: Macmillan.

Encarnacao, J., Schlechtendahl, E. G., 1983. *Computer Aided Design.* Berlin: Springer-Verlag. (Contains extensive German bibliography.)

Faux, I. D., Pratt, M. J., 1981. *Computational Geometry for Design and Manufacture.* Chichester: Ellis Horward/Wiley. (Contains extensive bibliography.)

Foley, J. D., van Dam, A., 1982. *Fundamentals of Interactive Computer Graphics.* Reading, Mass.: Addison-Wesley.

Harrington, S., 1983. *Computer Graphics, A Programming Approach.* New York: McGraw-Hill.

Knox, C. S., 1983. *CAD/CAM System Planning and Implementation.* New York: Marcel Dekker.

Newman, W. M., Sproul, R. F., 1979. *Principles of Interactive Computer Graphics*, 2nd edn. New York: McGraw-Hill.

Pao, Y. C., 1984. *Elements of CAD/CAM.* New York: J. Wiley.

Rogers, D. F., Adams, I. A., 1976. *Mathematical Elements of Computer Graphics.* NY: McGraw-Hill.

PROBLEMS

3.1 Write a computer program to design and draw a guide plate of dimensions 120 × 240 mm with circles of diameters from 4 to 12 mm, in steps of 2 mm, and from 15 to 50 mm, in steps of 5 mm.

3.2 Write a computer program to design and draw a guide plate of rectangles with side length as the diameters of Problem 3.1.

3.3 Write a computer program to design and draw a guide plate of equilateral triangles with side length as the diameters of Problem 3.1.

3.4 Write a computer program to design and draw a guide plate for equilateral pentagons with radii of circumscribed circles equal to the diameters of Problem 3.1.

3.5. Write a computer program to design and draw a guide plate for equilateral hexagons with radii of circumscribed circles equal to the diameters of Problem 3.1.

Problems 3.6 to 3.15

Draw a front view of the object shown in Figures P3.6 to P3.15, as indicated with the arrow, using AUTOCAD or a similar application program of your system.

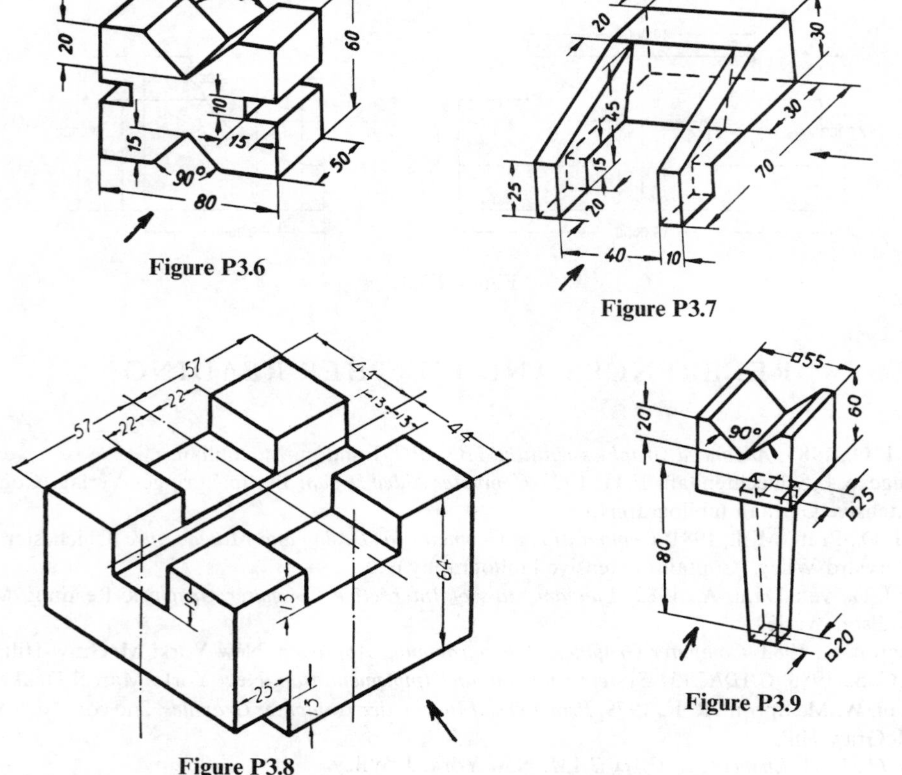

Figure P3.6

Figure P3.7

Figure P3.8

Figure P3.9

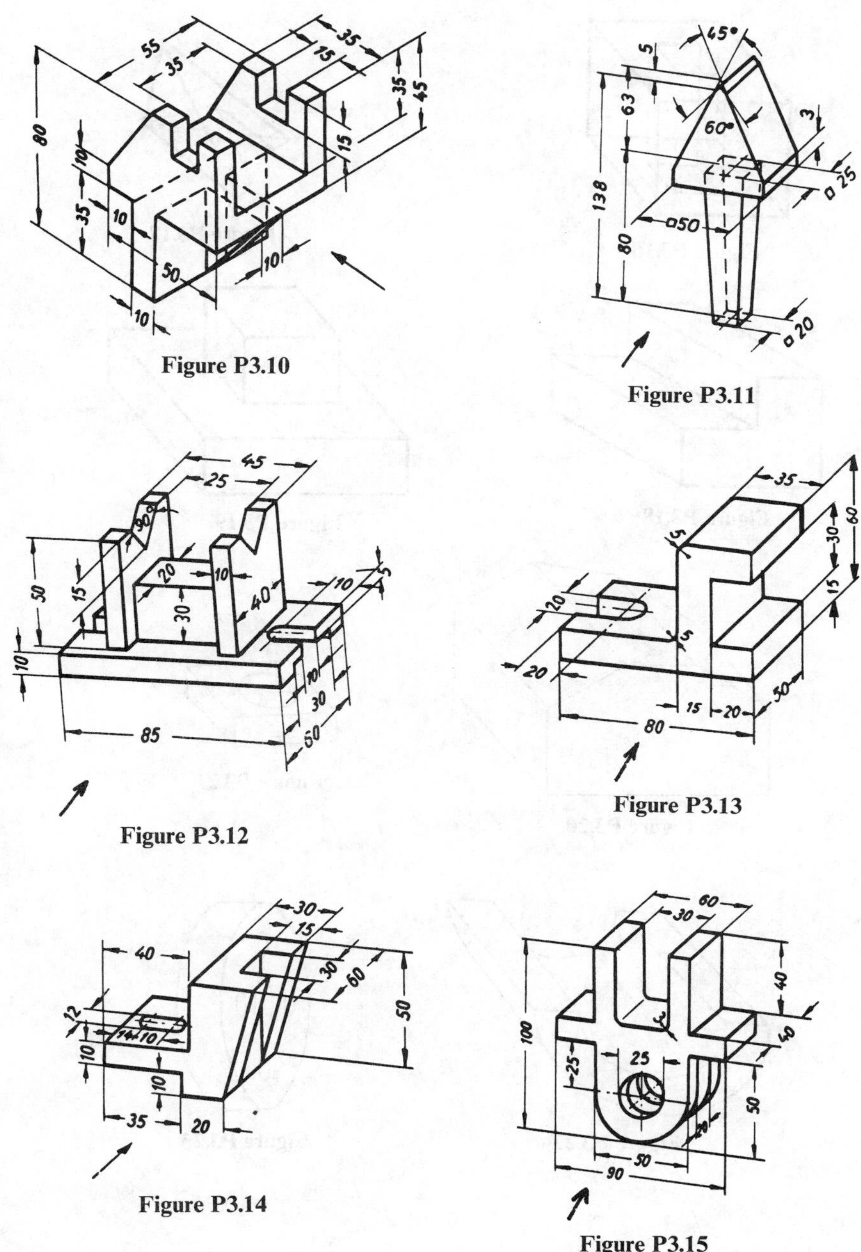

Figure P3.10

Figure P3.11

Figure P3.12

Figure P3.13

Figure P3.14

Figure P3.15

Problems 3.16 to 3.25

Use program SOLID to draw three views for the solids shown in Figures P3.16 to 3.25.

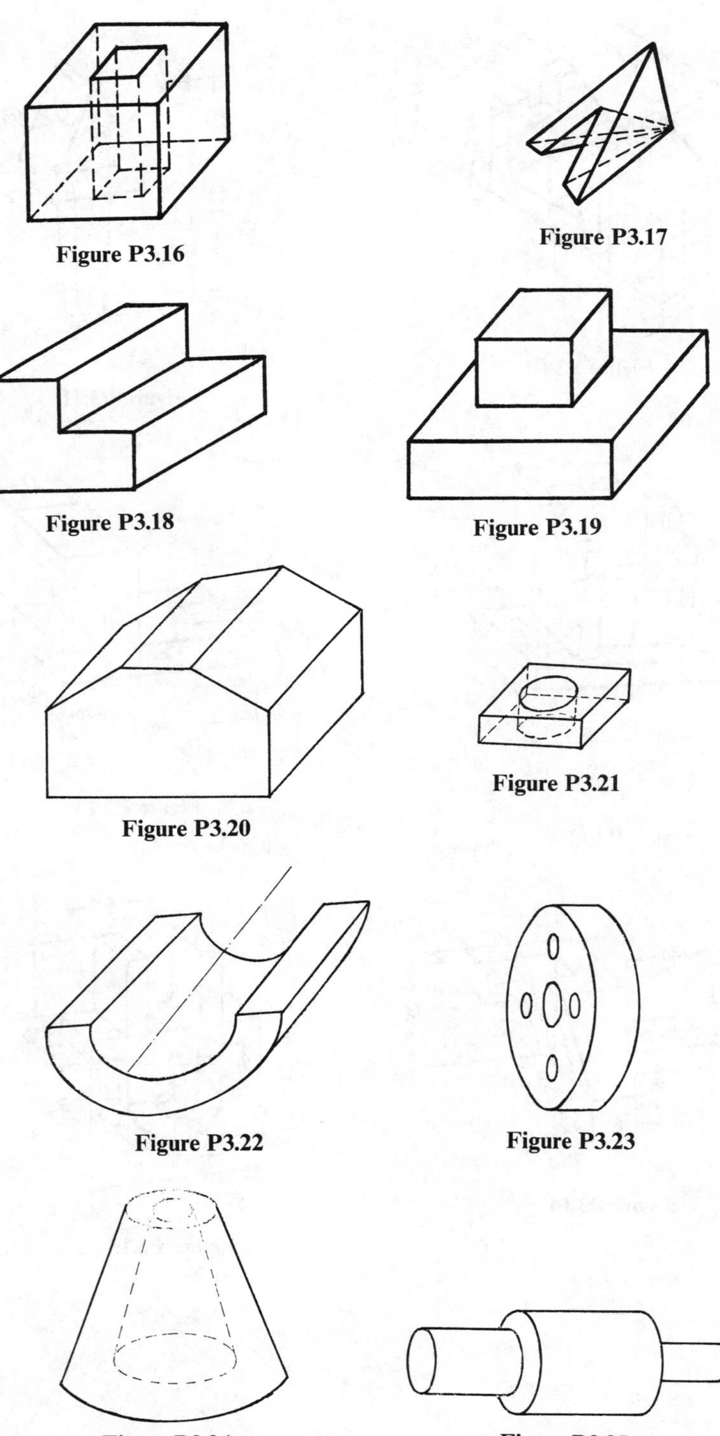

Figure P3.16

Figure P3.17

Figure P3.18

Figure P3.19

Figure P3.20

Figure P3.21

Figure P3.22

Figure P3.23

Figure P3.24

Figure P3.25

Problems 3.26 to 3.35

Using program SOLID, draw the objects shown in Figures P3.16 to 3.25, respectively, in axonometric view removing hidden lines.

APPENDIX 3.A
SOLID: A 3-D SOLID PLOTTING PROGRAM

```
10 DEFINT P
20 REM ***********************************************
30 REM *                                             *
40 REM *                    SOLID                    *
50 REM *                                             *
60 REM ***********************************************
70 REM          COPYRIGHT 1987
80 REM by Professor Andrew D. Dimarogonas, Washington University,St.Louis.
90 REM All rights reserved.  Unauthorized reproduction, dissemination or
100 REM selling is strictly prohibited.  This listing is for personal use.
110 REM
120 REM
130 REM
140 REM Plotting of 3-D line drawings with space transformations
150 REM Straight lines and circular arcs are allowed
160 REM Hidden line algorithm is included.
170 REM        ADD/11-20-85/revisions 12-22-86,5-29-87
180 REM ************************************************************
181 CLS:KEY OFF:SP=5:REM                   length of hidden line plotting
segment
182 FOR I=1 TO 14:COLOR I:LOCATE I,22+I:PRINT "SOLID";
183 LOCATE I,50-I:PRINT "SOLID";:NEXT I:COLOR 14
184 LOCATE 8,7:PRINT"A 3-D graphics package for solid development and";
185 PRINT " manipulation";
186 locate 25,1:print" Dimarogonas,A.D., Computer Aided Machine
Design,Prentice-Hall,1988";
190 LOCATE 18,10
191 INPUT"Enter xmax,ymax (600,200 for screen coordinates) ";XMAX,YMAX
196 XSC=3*600/XMAX:YSC=200/YMAX
200 DIM PD(100,70): REM                              Polygon definition
210 DIM D(100,4):REM                                 Plane coefficients
220 DIM X(100),Y(100),Z(100):REM                     Node coordinates
230 DIM R(3,3),XN(100),YN(100),ZN(100):REM           Storage matrices
240 SCREEN 2:XS=639:YS=199:ID=0
250 CLS:NMENUS=11
260 KEY OFF
270 DIM L$(100)
280 VIND=11:HIND=1
290 DIM MENU1$(25)
300 FOR IM=1 TO NMENUS:READ MENU1$(IM):NEXT IM
310 DATA "Stop      ","Rotate     ","Translate ","Scale      ","Autoscale "
320 DATA "Screenplot","New        ","Save File ","Load File ","Clearscrn "
330 DATA "Make Solid"
340 XS=0
350 LOCATE 1,70:PRINT "MAIN MENU"
360 FOR I=1 TO NMENUS:LOCATE I+2,70:PRINT MENU1$(I);:NEXT I
370 GOTO 470
375 ICOUNT=0
380 ICOUNT=ICOUNT+1:X$=INKEY$:IF ICOUNT<10 THEN 380
385 X$=INKEY$:XS=0
390 FOR I=17 TO 25:LOCATE I,66:PRINT SPC(14);:NEXT I
400 LOCATE 17,70:PRINT"COMMAND";
410 IF X$="" THEN 385
420 IF LEN(X$)=1 THEN XS=ASC(X$):GOTO 530
430 X$=RIGHT$(X$,1)
440 IF X$><"H" AND X$<>"P" THEN 500
450 IF X$="H" AND VIND>1 THEN VIND=VIND-1
460 IF X$="P" AND VIND<NMENUS THEN VIND=VIND+1
470 IF VIND>1 THEN LOCATE VIND+1,69:PRINT " ";MENU1$(VIND-1);
```

```
480 LOCATE VIND+2,69:PRINT ">";MENU1$(VIND);
490 IF VIND<NMENUS THEN LOCATE VIND+3,69:PRINT " ";MENU1$(VIND+1);
500 LOCATE 19,70:PRINT"          ";
510 LOCATE 21,65:PRINT"               ";
520 LOCATE 22,65:PRINT"               ";
530 IF XS=13 AND VIND=1 THEN SCREEN 0:END
540 IF XS=13 AND VIND=2 THEN GOSUB 1130:REM    rotation
550 IF XS=13 AND VIND=3 THEN GOSUB 1020:REM    translation
560 IF XS=13 AND VIND=4 THEN GOSUB 1080:REM    scaling
570 IF XS=13 AND VIND=5 THEN GOSUB 1290:REM    autoscale
580 IF XS=13 AND VIND=6 THEN GOSUB 1560:REM    screen plot
590 IF XS=13 AND VIND=7 THEN GOSUB  660:REM    start a new drawing
600 IF XS=13 AND VIND=8 THEN GOSUB 4000:REM    save file on disc
610 IF XS=13 AND VIND=9 THEN GOSUB 720:REM    load file from disc
620 IF XS=13 AND VIND=11 THEN GOSUB 5000:'     make a new file
630 IF XS=13 AND VIND=10 THEN CLS:GOTO 340:'   clear screen
640 GOTO 380
650 END
660 REM start a new drawing
670 LOCATE 22,65:INPUT"SURE (Y/N)";X$:IF X$="n" OR X$="N" THEN RETURN
680 ND=0:NP=0:NC=0:RETURN
690 REM************************************************************
700 REM              file retrieval
710 REM************************************************************
720 LOCATE 19,68:PRINT"          ";
730 LOCATE 18,70:PRINT"Data file  ":LOCATE 19,68:INPUT FILNA$
740 IF FILNA$="" THEN 730
750 ON ERROR GOTO 960
760 NDS=ND
770 OPEN FILNA$ FOR INPUT AS #1
780 INPUT#1,X$
790 IF X$="solend" OR X$="SOLEND" THEN 940
800 SYM$=""
810 ISYM=1
820 SYM$=MID$(X$,ISYM,1):ISYM=ISYM+1:IF SYM$="" AND ISYM<10 THEN 810
830 IF ISYM=10 THEN 780
840 IF SYM$="n" OR SYM$="N" THEN ND=ND+1:INPUT#1,X(ND),Y(ND),Z(ND):GOTO 780
850 IF SYM$="a" OR SYM$="A" THEN NC=NC+1:FOR I=1 TO 3:INPUT#1,SC(NC,I):NEXT
I:GOTO 780
860 IF SYM$="p" OR SYM$="P" THEN NP=NP+1:GOTO 890
870 IF X$="solend" OR X$="SOLEND" THEN 940
880 PRINT"FILE ERROR...":STOP
890 IP=IP+1
900 INPUT#1,PD(IP,1)
910 NN=PD(IP,1):FOR JP=1 TO NN
920 INPUT#1,NUNOD:PD(IP,JP+1)=NUNOD+NDS:NEXT JP
930 GOTO 780
940 CLOSE#1
950 GOTO 970
960 LOCATE 19,68:PRINT "BAD FILE     "
970 RETURN
1000 REM
***************************************************************
1010 REM COORDINATE TRANSFOMATION ROUTINES
1020 REM ****************************** SUBROUTINE TRANSLATE(XT,YT,ZT)
1030 LOCATE 21,65:PRINT "ENTER XT, YT, ZT":LOCATE 22,65:INPUT XT,YT,ZT
1040 FOR I=1 TO ND
1050 X(I)=X(I)+XT:Y(I)=Y(I)+YT:Z(I)=Z(I)+ZT
1060 NEXT I
1070 RETURN
1080 REM ****************************** SUBROUTINE SCALE(XS,YS,ZS) ****
1090 LOCATE 21,65:PRINT "ENTER SX,SY,SZ":LOCATE 22,65:INPUT SX,SY,SZ
1100 FOR I=1 TO ND
1110 X(I)=X(I)*SX:Y(I)=Y(I)*SY:Z(I)=Z(I)*SZ
1120 NEXT I:RETURN
1130 REM ****************************** SUBROUTINE ROTATE(IX,IY,IZ,fr)
1140 LOCATE 21,65:PRINT "ENTER IX, IY,fr":LOCATE 22,65:INPUT
IX,IY,FR:IZ=SQR (1-IX^2-IY^2)
1150 N1=IX:N2=IY:N3=IZ
1160 CF=COS (FR):SF=SIN (FR)
1170 R(1,1)=N1*N1+(1-N1*N1)*CF:R(1,2)=N1*N2*(1-CF)+N3*SF:F(1,3)=N1*N3*(1-
CF)-2*SF
```

```
1180 R(2,1)=N1*N2*(1-CF)-N3*SF:R(2,2)=N2*N2+(1-N2*N2)*CF:R(2,3)=N2*N3*(1-
CF)+N1*SF
1190 R(3,1)=N1*N3*(1-CF)+N2*SF:R(3,2)=2*N3*(1-CF)-N1*SF:R(3,3)=N3*N3+(1-
N3*N3)*CF
1200 FOR I=1 TO ND
1210 XN(I)=X(I)*R(1,1)+Y(I)*R(1,2)+Z(I)*R(1,3)
1220 YN(I)=X(I)*R(2,1)+Y(I)*R(2,2)+Z(I)*R(2,3)
1230 ZN(I)=X(I)*R(3,1)+Y(I)*R(3,2)+Z(I)*R(3,3)
1240 NEXT I
1250 FOR I=1 TO ND
1260 X(I)=XN(I):Y(I)=YN(I):Z(I)=ZN(I)
1270 NEXT I
1280 RETURN
1290 REM ******************************** AUTOMATIC SCALING *************
1300 REM AUTOMATIC POSITIONING
1310 XA=-1E+20:XB=1E+20:YA=-1E+20:YB=1E+20
1320 FOR I=1 TO ND
1330 IF X(I) < XB THEN XB=X(I)
1340 IF X(I) > XA THEN XA=X(I)
1350 IF Y(I) > YA THEN YA=Y(I)
1360 IF Y(I) < YB THEN YB =Y(I)
1370 NEXT I
1380 REM GIVEN WINDOW POSITION XB (LEFT), XA (RIGHT), YB (LOWER), YA
(UPPER)
1390 LOCATE 1,1
1400 LOCATE 21,65:PRINT "ENTER XMAX, YMAX":LOCATE 22,65:INPUT XS,YS
1410 IF XS >639 THEN XS=639
1420 IF YS >199 THEN YS=199
1430 FOR I=1 TO ND
1440 X(I)=(X(I)-XB)*XS/(XA-XB)
1450 Y(I)=(Y(I)-YB)*YS/(YA-YB)
1460 X=X(I):Y=Y(I):GOSUB 1490
1470 X(I)=X:Y(I)=Y:NEXT I
1480 RETURN
1490 REM ******************************** CLIPPING **********************
1500 REM IBM/PC screen at high resolution
1510 IF X <0 THEN X=0
1520 IF X>639 THEN X=639
1530 IF Y < 0 THEN Y=0
1540 IF Y > 199 THEN Y=199
1550 RETURN
1560 REM ******************************** SCREEN PLOTTING ***************
1570 ID=VIND-1
1580 IF ID < > 5 GOTO 1680
1590 FOR ILINE=21 TO 24:LOCATE ILINE,66:PRINT SPC(14);:NEXT ILINE
1600 LOCATE 21,68
1610 PRINT"Hidden Lines";:LOCATE 22,66:PRINT"1,full hidden";
1620 LOCATE 23,66:PRINT"2,convex solid";
1630 LOCATE 24,66:PRINT"3,wire...?";:Z$=INKEY$:IF Z$="" THEN 1630
1635 IF Z$<>"1" AND Z$<>"2" AND Z$<>"3" THEN 2000
1640 IH=VAL(Z$):PRINT Z$;
1650 IF IH=3 THEN IH=0
1660 IF IH = 1 THEN GOSUB 3020:REM find plane equations for polygons
1670 REM
1680 REM
1690 REM--------------------------------- PLOTTING STRAIGHT LINES
1700 FOR I= 1 TO NP:FOR JP=2 TO PD(I,1)
1710 I1=PD(I,JP):I2 = PD(I,JP+1)
1720 IF IH><2 AND I2<0 THEN 1980
1730 I1=ABS(I1):I2=ABS(I2):I3=ABS(PD(I,JP+2))
1740 X1=X(I1):X2 =X(I2):X3=X(I3)
1750 Y1=Y(I1):Y2 =Y(I2):Y3=Y(I3)
1760 Z1=Z(I1):Z2=Z(I2)
1770 IF IH=1 THEN 1830
1775 IF IH=0 THEN 1800:REM do not exclude hidden planes
1776 IF JP>2 THEN 1800
1780 DX1=X2-X1:DX2=X3-X1:DY1=Y2-Y1:DY2=Y3-Y1
1790 IF DX1*DY2-DX2*DY1>0 THEN  1990
1800 X=X1:Y=Y1:GOSUB 3490
1810 X=X2:Y=Y2:GOSUB 3580
1820 GOTO  1980
```

```
1830 DX=X2-X1:DY=Y2-Y1:DZ=Z2-Z1
1840 LOCATE 25,1:PRINT SPC(78);
1850 LOCATE 25,1:PRINT "Plotting line " ;JP-1;" of polygon ";I;
1860 LX=SQR (DX*DX+DY*DY)
1870 IF LX = 0 THEN 1980
1880 X = X1: Y = Y1
1890 FOR S =SP TO LX STEP SP
1900 XP = X: YP = Y
1910 X = X1 + S * DX / LX
1920 Y = Y1 + S * DY / LX
1930 Z = Z1 + S * DZ / LX
1940 GOSUB 3490
1950 NEXT S
1960 REM XP = X: YP= Y
1970 XP= X1 +DX: YP= Y1 + DY: GOSUB 3490
1980 NEXT JP
1990 NEXT I
2000 RETURN
2010 REM ******************************* ELLIPSE ***********************
2020 REM COMPUTATION OF THE PARAMETRIC EQUATIONS OF AN ELLIPSE WITH CENTER
AT XO, YO, GOING THROUGH 3 POINTS XJ, YJ,  J=1,2,3
2030 IF NC = 0 THEN 2160
2040 FX = X1 - X0: FY = Y1 - Y0: GOSUB 2170: F1 = F
2050 FX = X2 - X0: FY = Y2 - Y0: GOSUB 2170: F2 = F
2060 FX = X3 - X0: FY = Y3 - Y0: GOSUB 2170: F3 = F
2070 IF F3 = F1 THEN F3 = F3 + 3.14159 * 2
2080 IF S2 =0 THEN S2 = .0001
2090 IF F2 >F1 AND F3 <F2 THEN F3 =F3 + 6.283
2100 C1 = COS (F1): C2 = COS (F2): C3 = COS (F3)
2110 S1 = SIN (F1): S2 = SIN (F2): S3 = SIN (F3)
2120 DL = C1 * S2 - C2 * S1
2130 A1 = ((X1-X0) *S2 - S1 * (X2 - X0 )) / DL
2140 A2 = (C1 * (X2 - X0) - C2 * (X1 -X0)) / DL
2150 A3 = (Y2 - Y0- A2 * C2) /S2
2160 RETURN
2170 REM ******************************* FUNCTION ARCTAN(X) ************
2180 IF FX > < 0 THEN 2220
2190 IF FY > = 0 THEN F = 3.14159 / 2
2200 IF FY = < 0 THEN F = 3.14159 * 1.5
2210 GOTO 2260
2220 F = ATN (FY / FX)
2230 IF FX > = 0 THEN GOTO 2250
2240 F = F + 3.14159
2250 REM
2257 REM
2260 RETURN
3000 REM
3010 REM                                        Hidden line algorithms
3020 REM ******************************* PLANE EQUATIONS ***************
3030 REM   Z=D(IP,1)*X+D(IP,2)*Y+D(IP,3)
3040 FOR IP = 1 TO NP
3050 P1 = ABS(PD(IP,2)):P2 = ABS(PD(IP,3)):P3 = ABS(PD(IP,4))
3060 X1 = X(P1):X2 =X(P2):Y1 = Y(P1):Y2 = Y(P2)
3070 Z1 = Z(P1):Z2 = Z(P2):Z3 = Z(P3)
3080 X3 = X(P3):Y3 = Y(P3)
3090 DE =X1 * Y2 + Y1 * X3 + X2 * Y3 - X3 * Y2 - X1 * Y3 - X2 * Y1
3100 IF DE = 0 THEN DE = .001
3110 D(IP,1) = (Z1 * Y2 + Y1 * Z3 + Y3 * Z2 - Y2 * Z3 - Y3 * Z1 - Y1 * Z2)/
DE
3120 D(IP,2) = (X1 * Z2 + X3 * Z1 + X2 * Z3 - X3 * Z2 - X1 * Z3 - X2 * Z1)
/DE
3130 D(IP,3) = (X1 * Y2 * Z3 + Y1 * Z2 * X3 + Z1 * X2 * Y3 - X3 * Y2 * Z1 -
Y3 * Z2 * X1 - Z3 * X2 * Y1 ) / DE
3140 NEXT IP
3150 RETURN
3160 REM ******************************* HIDDEN LINE ALGORITHM *********
3170 REM
3180 REM ----------------------------FINDING PLANE INTERSECTION---
3190 HL=0
3200 FOR IP = 1 TO NP
3210 IH1=ABS(PD(IP,2)):IH2 = ABS(PD(IP,3))
```

```
3220 IH3=ABS(PD(IP,4))
3230 HX1=X(IH1):HX2 =X(IH2):HX3=X(IH3)
3240 HY1=Y(IH1):HY2 =Y(IH2):HY3=Y(IH3)
3250 LOCATE 25,42:PRINT "Hidden line test with polygon ";IP
3260 DX1=HX2-HX1:DX2=HX3-HX1:DY1=HY2-HY1:DY2=HY3-HY1
3270 IF DX1*DY2-DX2*DY1>0 THEN  3470
3280 ZP = D(IP,1) * X + D(IP,2) * Y + D(IP,3)
3290 IF Z > ZP-.001  THEN 3470
3300 REM ----------------------------CHECK IF INTERSECTION  IS IN THE
POLYGON-
3310 IGI=0
3320 PN=PD(IP,1)
3330 XZ=X:YZ=Y
3340 FOR LL=2 TO PN
3350 IG1=ABS(PD(IP,LL)):IG2=ABS(PD(IP,LL+1))
3360 EX=X(IG1):TX=X(IG2):EY=Y(IG1):TY=Y(IG2)
3370 IF EY>YZ AND TY>YZ THEN 3440
3380 IF EY<YZ AND TY<YZ THEN 3440
3390 IF TY=EY THEN EY=EY+9.999999E-21
3400 XX=EX+(YZ-EY)/(TY-EY)*(TX-EX)
3410 IF XX<XZ THEN 3440
3420 IF ABS(XX-EX)+ABS(TX-XX)>ABS(TX-EX) THEN 3440
3430 IGI=IGI+1
3440 NEXT LL
3450 IF IGI>2*INT(IGI/2) THEN HL=1
3460 IF HL=1 THEN IP=NP
3470 NEXT IP
3480 RETURN
3490 REM
3500 REM ****************************** DRAW LINE SEGMENT WITH HL TEST*
3510 GOSUB 1500: REM ------------------ SCREEN CLIPPING
3520 HL=0
3530 IF IH = 1 THEN GOSUB 3160:REM -----test for hidden line
3540 IF HL = 1 THEN GOTO 3570 :REM -----hidden, do not plot
3550 IF IH = 1 THEN LINE (XP*XSC/3,YP*YSC)-(X*XSC/3,Y*YSC):REM visible,
plot
3560 IF IH = 0 OR IH=2 THEN PSET(X*XSC/3,Y*YSC)
3570 RETURN
3580 REM ----------------------------- DRAW TO NEXT POINT
3590 GOSUB 1500: REM ------------------SCREEN CLIPPING
3600 LINE -(X*XSC/3,Y*YSC)
3610 RETURN
4000 REM ********************************STORING TRANSFORMED DATA ON DISC
4010 LOCATE 19,68:PRINT"            "
4020 LOCATE 18,70:PRINT"New file  ":LOCATE 19,68:INPUT FILNA$
4030 CLOSE
4040 OPEN FILNA$ FOR OUTPUT AS #1
4050 FOR I=1 TO ND
4060 PRINT#1,"NODE,";X(I);",";Y(I);",";Z(I):NEXT I
4070 REM **DATA** DEFINITION OF NC ARCS:NODES OF: CENTER, 3 POINTS COUNTER-
C-W
4080 IF NC=0 THEN GOTO 4120
4090 FOR I=1 TO NC:PRINT#1,"CIRCLE,";
4100 NP=NPOL
4110 FOR J=1 TO 4:PRINT#1,SC(I,J):NEXT J:NEXT I
4120 REM **DATA** DEFINITION OF POLYGONS
4130 FOR IP=1 TO NP
4140 PRINT#1,"POLYGON,";PD(IP,1);
4150 NN=PD(IP,1):FOR JP=1 TO NN
4160 PRINT#1,",";PD(IP,JP+1);:NEXT JP
4170 PRINT#1,"  " :NEXT IP
4180 PRINT#1,"SOLEND"
4190 CLOSE#1
4200 RETURN
5000 REM
*****************************************************************
5010 REM                    solid creation
5020 REM
*****************************************************************
5030 XSC=3:NPOL=NP
5040 LINE (0,0)-(261,91),,B:LOCATE 1,1:PRINT "Z";:LOCATE 1,80:PRINT"X";
```

```
5045 LOCATE 1,30:PRINT "X";:LOCATE 1,60:PRINT "Z";
5050 LINE (279,0)-(540,91),,B:LOCATE 1,36:PRINT "X";
5055 LOCATE 11,1:PRINT "Y";:LOCATE 11,36:PRINT "Y";
5060 LINE (0,99)-(261,191),,B:LOCATE 13,1:PRINT "Y";
5065 LOCATE 13,30:PRINT "X";:LOCATE 24,1:PRINT "Z";
5070 REM LINE (279,99)-(540,191),,B
5080 XSTEP=10:YSTEP=10:ZSTEP=10
5090 XHAIR=XSTEP:YHAIR=YSTEP:ZHAIR=ZSTEP
5100 X$="z":GOTO 5150
5110 X$=INKEY$:XS=0
5120 IF X$="" THEN 5110
5130 IF LEN(X$)>1  THEN 5260
5140 REM
5150 LOCATE 13,35
5160 PRINT " Keyboard commands:":LOCATE 14,35
5170 PRINT " N= define node, L= draw line,":LOCATE 15,35
5180 PRINT " G= grid, P= define polygon, ";:LOCATE 16,35
5190 PRINT " A= arc, F=faster, S=slower " ;
5200 LOCATE 17,35
5210 PRINT " D= displace polygon,":LOCATE 18,35
5220 PRINT " C= connect polygons ";"E= exit";
5230 LOCATE 19,35:PRINT " x,y,z= select screen x,y,z;
5235 IF X$="f"  AND XSTEP<10 THEN XSTEP=XSTEP+1:YSTEP=XSTEP:ZSTEP=XSTEP
5236 IF X$="s"  AND XSTEP>1 THEN XSTEP=XSTEP-1:YSTEP=XSTEP:ZSTEP=XSTEP
5237 IF X$="F"  AND XSTEP<10 THEN XSTEP=XSTEP+1:YSTEP=XSTEP:ZSTEP=XSTEP
5238 IF X$="S"  AND XSTEP>1 THEN XSTEP=XSTEP-1:YSTEP=XSTEP:ZSTEP=XSTEP
5240 IF X$<>"G" AND X$<>"g" THEN 5250
5245 FOR I= 10 TO 80 STEP 10:FOR J=9 TO 540 STEP 20:PSET(J,I):NEXT J:NEXT I
5246 FOR I=100 TO 190 STEP 10:FOR J=9 TO 260 STEP 20:PSET(J,I):NEXT J:NEXT
I
5250 IF LEN(X$)=1 THEN XS=ASC(X$):GOTO 5610
5260 X$=RIGHT$(X$,1)
5270 FOR I=5 TO 260 STEP 20:PSET(I,YHAIR),0:NEXT I
5280 FOR I=280 TO 539 STEP 20:PSET(I,YHAIR),0:NEXT I
5290 FOR I=5 TO 260 STEP 20:PSET(I,100+ZHAIR),0:NEXT I
5300 FOR I=6 TO 90 STEP 10:PSET(XSC*XHAIR,I),0:NEXT I
5310 FOR I=6 TO 90 STEP 10:PSET(280+XSC*ZHAIR,I),0:NEXT I
5320 FOR I=106 TO 190 STEP 10:PSET(XSC*XHAIR,I),0:NEXT I
5330 IF X$="H" AND PLANE=1 THEN YHAIR=YHAIR-YSTEP
5340 IF X$="H" AND PLANE=2 THEN YHAIR=YHAIR-YSTEP
5350 IF X$="H" AND PLANE=3 THEN ZHAIR=ZHAIR-ZSTEP
5360 IF X$="P" AND PLANE=1 THEN YHAIR=YHAIR+YSTEP
5370 IF X$="P" AND PLANE=2 THEN YHAIR=YHAIR+YSTEP
5380 IF X$="P" AND PLANE=3 THEN ZHAIR=ZHAIR+ZSTEP
5390 IF X$="K" AND PLANE=1 THEN XHAIR=XHAIR-XSTEP
5400 IF X$="K" AND PLANE=2 THEN ZHAIR=ZHAIR-ZSTEP
5410 IF X$="K" AND PLANE=3 THEN XHAIR=XHAIR-XSTEP
5420 IF X$="M" AND PLANE=1 THEN XHAIR=XHAIR+XSTEP
5430 IF X$="M" AND PLANE=2 THEN ZHAIR=ZHAIR+ZSTEP
5440 IF X$="M" AND PLANE=3 THEN XHAIR=XHAIR+XSTEP
5450 IF XHAIR<0 THEN XHAIR=0
5460 IF YHAIR<0 THEN YHAIR=0
5470 IF ZHAIR<0 THEN ZHAIR=0
5480 IF YHAIR>90 THEN YHAIR=90
5490 IF ZHAIR>90 THEN ZHAIR=90
5500 IF XHAIR>90 THEN XHAIR=90
5510 LOCATE 25,1:PRINT SPC(78);
5520 LOCATE 25,1:PRINT "x= ";XHAIR;:LOCATE 25,15:PRINT "y= ";YHAIR;
5530 LOCATE 25,30:PRINT "z= ";ZHAIR;" Nodes= ";ND;" Polygons= ";NP;
5540 FOR I=5 TO 260 STEP 20:PSET(I,YHAIR):NEXT I
5550 FOR I=280 TO 539 STEP 20:PSET(I,YHAIR):NEXT I
5560 FOR I=5 TO 260 STEP 20:PSET(I,100+ZHAIR):NEXT I
5570 FOR I=6 TO 90 STEP 10:PSET(XSC*XHAIR,I):NEXT I
5580 FOR I=6 TO 90 STEP 10:PSET(280+XSC*ZHAIR,I):NEXT I
5590 FOR I=106 TO 190 STEP 10:PSET(XSC*XHAIR,I):NEXT I
5600 GOTO 5110
5610 REM
5620 IF X$="x" OR X$="X" THEN PLANE=2
5630 IF X$="y" OR X$="Y" THEN PLANE=3
5640 IF X$="z" OR X$="Z" THEN PLANE=1
5650 IF X$><"n" AND X$<>"N" THEN 5760
```

```
5660 XPL=XHAIR:YPL=YHAIR:ZPL=ZHAIR
5670 ND=ND+1:X(ND)=XPL:Y(ND)=YPL:Z(ND)=ZPL
5680 N1=N2:N2=N3:N3=ND:LOCATE 25,1:PRINT SPC(78);
5690 LOCATE 25,1:PRINT "Start of a new polygon   (Y/N or Cancel) ? ";
5700 Z$=INKEY$:LOCATE 25,45:PRINT"...enter...";:IF Z$="" THEN 5700
5705 IF Z$="C" OR Z$="c" THEN PRINT "..cancelled..";:ND=ND-1:GOTO 5110
5710 IF Z$="y" OR Z$="Y" THEN PD(NP+1,2)=ND:ICPOL=1
5720 LOCATE 25,1:PRINT SPC(78);
5730 LOCATE 25,45:PRINT "new node= ";ND;
5740 PSET(XSC*XPL,YPL):PSET(280+XSC*ZPL,YPL):PSET(XSC*XPL,100+ZPL)
5750 GOTO 5110
5760 IF X$><"l" AND X$<>"L" THEN 5900
5761 LOCATE 25,1:PRINT SPC(78);
5762 LOCATE 25,1:PRINT "<1> line, <2> fantom line, <3> Cancel";
5763 L$="":L$=INKEY$:IF L$="" THEN 5763
5764 IF L$="1" OR L$="2" THEN 5770
5765 PRINT "..CANCELLED...";:GOTO 5940
5770 XPL=XHAIR:YPL=YHAIR:ZPL=ZHAIR
5780 LINE (XSC*X(ND),Y(ND)) -(XSC*XPL,YPL)
5790 LINE (280+XSC*Z(ND),Y(ND))-(280+XSC*ZPL,YPL)
5800 LINE (XSC*X(ND),100+Z(ND)) -(XSC*XPL,100+ZPL)
5810 INODE=ND+1:ION=0:FOR I=1 TO ND:D1=X(I)-XHAIR:D2=Y(I)-YHAIR:D3=Z(I)-
ZHAIR
5820 IF D1^2<1 AND D2^2<1 AND D3^2<1 THEN ION=1:INODE=I:I=ND
5830 NEXT I
5840 REM IF ION=1 and l$="1" THEN 5880:REM node already defined
5850 ND=ND+1:X(ND)=XPL:Y(ND)=YPL:Z(ND)=ZPL
5860 LOCATE 25,1:PRINT SPC(78);
5870 LOCATE 25,45:PRINT "new node= ";ND;
5880 ICPOL=ICPOL+1:PD(NPOL+1,1)=ICPOL:PD(NPOL+1,ICPOL+1)=INODE
5885 IF L$="2" THEN PD(NPOL+1,ICPOL+1)=-PD(NPOL+1,ICPOL+1)
5890 N1=N2:N2=N3:N3=INODE
5900 IF LEFT$(X$,1)><"p" AND X$<>"P" THEN 5940
5910 NPOL=NPOL+1:ICPOL=0
5920 NP=NPOL
5930 LOCATE 25,1:PRINT SPC(78);
5935 LOCATE 25,45:PRINT "new polygon no ";NPOL
5940  IF X$ <> "a" AND X$ <> "A" THEN 5960
5950 GOSUB 6430 :REM draw arc
5960  IF X$ = "e" OR X$ = "E" THEN 6010
5970 IF X$="d" OR X$="D" THEN GOSUB 6020
5980 IF X$="c" OR X$="C" THEN GOSUB 6240
5990 NP=NPOL
6000 GOTO 5110
6010 RETURN
6020 REM ************************************* Displacing last polygon
6030 LOCATE 25,1:PRINT SPC(78);
6035 DISPL$=""
6040 LOCATE 25,1:PRINT"enter displacement               ....C to cancel";
6041 LOCATE 25,1:PRINT"                          ";:P$=INKEY$:IF P$="" THEN 6040
6042 LOCATE 25,20+LEN(DISPL$):PRINT P$;
6043 IF P$="c" OR P$="C" THEN PRINT"..CANCELLED..";:RETURN
6044 DISPL$=DISPL$+P$:IF ASC(P$)<>13 THEN 6040
6045 DISPL=VAL(DISPL$)
6050 NPNODES=PD(NPOL,1):PD(NPOL+1,1)=NPNODES
6060 FOR I=1 TO PD(NPOL,1):NODE=ABS(PD(NPOL,I+1))
6070 ND=ND+1:IF PLANE=1 THEN
X(ND)=X(NODE):Y(ND)=Y(NODE):Z(ND)=Z(NODE)+DISPL
6080 IF PLANE=2 THEN X(ND)=X(NODE)+DISPL:Y(ND)=Y(NODE):Z(ND)=Z(NODE)
6090 IF PLANE=3 THEN X(ND)=X(NODE):Y(ND)=Y(NODE)+DISPL:Z(ND)=Z(NODE)
6100 XPR=X(ND-1):YPR=Y(ND-1):ZPR=Z(ND-1)
6110 XPL=X(ND):YPL=Y(ND):ZPL=Z(ND)
6120 PD(NPOL+1,I+1)=ND
6130 IF I=1 THEN 6170
6140 LINE (XSC*XPR,YPR) -(XSC*XPL,YPL)
6150 LINE (280+XSC*ZPR,YPR)-(280+XSC*ZPL,YPL)
6160 LINE (XSC*XPR,100+ZPR) -(XSC*XPL,100+ZPL)
6170 NEXT I
6180 NPOL=NPOL+1
6190 PD(NPOL,NPNODES+1)=PD(NPOL,2)
6200 REM inverse the sequence of the polygon
```

```
6210 FOR I=2 TO NPNODES+1:PD(NPOL+1,I)=PD(NPOL,I):NEXT I
6220 FOR I=2 TO NPNODES+1:PD(NPOL,I)=PD(NPOL+1,NPNODES+3-I):NEXT I
6230 RETURN
6240 REM ********************************** connect last two polygons
6241 LOCATE 25,1:PRINT SPC(79);
6242 LOCATE 25,1:PRINT "..press any key, ...C to cancel ";
6243 C$="":C$=INKEY$:IF C$="" THEN 6243
6244 IF C$="c" OR C$="C" THEN PRINT"..cancelled..";:RETURN
6250 NPNODES=PD(NPOL,1)
6260 FOR I=1 TO NPNODES-1
6270 I1=PD(NPOL-1,I+1):I2=PD(NPOL-1,I+2)
6280 J1=PD(NPOL,NPNODES+2-I):J2=PD(NPOL,NPNODES+2-I-1)
6290 PD(NPOL+I,1)=5
6300 PD(NPOL+I,2)=J1:PD(NPOL+I,3)=-J2
6310 PD(NPOL+I,4)=I2:PD(NPOL+I,5)=-I1
6320 PD(NPOL+I,6)=-J1
6330 KP(1)=ABS(J1):KP(2)=ABS(J2):KP(3)=ABS(I2):KP(4)=ABS(I1):KP(5)=ABS(J1)
6335 FOR J=2 TO 5
6340 XPR=X(KP(J-1)):YPR=Y(KP(J-1)):ZPR=Z(KP(J-1))
6350 XPL=X(KP(J)):YPL=Y(KP(J)):ZPL=Z(KP(J))
6360 LINE (XSC*XPR,YPR) -(XSC*XPL,YPL)
6370 LINE (280+XSC*ZPR,YPR)-(280+XSC*ZPL,YPL)
6380 LINE (XSC*XPR,100+ZPR) -(XSC*XPL,100+ZPL)
6390 NEXT J
6400 NEXT I
6410 NPOL=NPOL+NPNODES-1
6420 RETURN
6430 REM ******************************  circular arc on a plane
6440 LOCATE 25,1:PRINT SPC(78);
6450 LOCATE 25,1:PRINT "<1>,through last three nodes,<2> center,angle";
6460 PRINT " from last node,  1 or 2 ? ";
6470 X$=INKEY$:IF X$="" THEN 6470
6480 REM N1=ND-2:N2=ND-1:N3=ND
6490 IF PLANE=1 THEN X1=X(N1):X2=X(N2):X3=X(N3)
6500 IF PLANE=2 THEN X1=Z(N1):X2=Z(N2):X3=Z(N3)
6510 IF PLANE=3 THEN X1=X(N1):X2=X(N2):X3=X(N3)
6520 IF PLANE=1 THEN Y1=Y(N1):Y2=Y(N2):Y3=Y(N3)
6530 IF PLANE=2 THEN Y1=Y(N1):Y2=Y(N2):Y3=Y(N3)
6540 IF PLANE=3 THEN Y1=Z(N1):Y2=Z(N2):Y3=Z(N3)
6550 IF PLANE=1 THEN Z1=Z(N1):Z2=Z(N2):Z3=Z(N3)
6560 IF PLANE=2 THEN Z1=X(N1):Z2=X(N2):Z3=X(N3)
6570 IF PLANE=3 THEN Z1=Y(N1):Z2=Y(N2):Z3=Y(N3)
6580 IF X$="2" THEN 6650
6590 A11=X2-X1:A12=Y2-Y1:B1=(Y2^2-Y1^2+X2^2-X1^2)/2
6600 A21=X3-X2:A22=Y3-Y2:B2=(Y3^2-Y2^2+X3^2-X2^2)/2
6610 DELTA=A11*A22-A12*A21
6620 XC=(B1*A22-B2*A12)/DELTA
6630 YC=(B2*A11-B1*A21)/DELTA
6640 GOTO 6730
6650 XC=X3:YC=Y3:IF X$="1" THEN 6730
6660 LOCATE 25,1:PRINT SPC(78);:LOCATE 25,1
6670 PRINT"enter angle of the arc, degrees";
6680 Y$=INKEY$:IF Y$="" THEN 6680
6690 ANG$=ANG$+Y$:PRINT Y$;:IF ASC(Y$) <> 13 THEN 6680
6700 ANGLE=VAL(ANG$)*3.14159/180
6710 F0=ATN((Y2-YC)/(X2-XC+9.999999E-21)):IF X2<XC THEN F0=F0+3.14159
6720 F1=F0+ANGLE
6730 RADIUS=SQR((X2-XC)^2+(Y2-YC)^2)
6740 PI=3.14159:IF X$="2" THEN 6810
6750 F0=ATN((Y1-YC)/(X1-XC+9.999999E-21)):IF X1<XC THEN F0=F0+3.14159
6760 F1=ATN((Y2-YC)/(X2-XC+9.999999E-21)):IF X2<XC THEN F1=F1+3.14159
6770 F2=ATN((Y3-YC)/(X3-XC+9.999999E-21)):IF X3<XC THEN F1=F1+3.14159
6780 IF X1-XC=0 THEN IF Y1-YC>0 THEN F0=PI/2 ELSE F0=3*PI/2
6790 IF X2-XC=0 THEN IF Y2-YC>0 THEN F1=PI/2 ELSE F1=3*PI/2
6800 IF X3-XC=0 THEN IF Y3-YC>0 THEN F2=PI/2 ELSE F2=3*PI/2
6810 XPL=X2:YPL=Y2:ZPL=Z2
6820 IF F1<0 THEN F1=F1+3.14159*2
6830 IF F2<0 THEN F2=F2+3.14159*2
6840 IF F0<0 THEN F0=F0+3.14159*2
6850 IF F2>F0 THEN F=F0:FMAX=F2:IF F1>F2 OR F1<F0 THEN FMAX=F2-2*PI
6860 IF F2<F0 THEN F=F0:FMAX=F2:IF F1>F0 OR F1<F2 THEN FMAX=F2+2*PI
```

```
6870 DF=(FMAX-F)/10
6880 FOR IARC=1 TO 10
6890 F=F+DF
6900 XARC=XC+RADIUS*COS(F)
6910 YARC=YC+RADIUS*SIN(F)
6920 ZARC=Z3:P=PLANE
6930 XARC0=XC+RADIUS*COS(F-DF)
6940 YARC0=YC+RADIUS*SIN(F-DF):ZARC0=Z3
6950 IF P=1 THEN
XPL=XARC:YPL=YARC:ZPL=ZARC:XPREV=XARC0:YPREV=YARC0:ZPREV=ZARC0
6960 IF P=2 THEN
XPL=ZARC:YPL=YARC:ZPL=XARC:XPREV=ZARC0:YPREV=YARC0:ZPREV=XARC0
6970 IF P=3 THEN
XPL=XARC:YPL=ZARC:ZPL=YARC:XPREV=XARC0:YPREV=ZARC0:ZPREV=YARC0
6980 LINE (XSC*XPREV,YPREV) -(XSC*XPL,YPL)
6990 LINE (280+XSC*ZPREV,YPREV)-(280+XSC*ZPL,YPL)
7000 LINE (XSC*XPREV,100+ZPREV) -(XSC*XPL,100+ZPL)
7010 ND=ND+1:X(ND)=XPL:Y(ND)=YPL:Z(ND)=ZPL:ICPOL=ICPOL+1
7020 LOCATE 25,45:PRINT "new node= ";ND;:INODE=ND
7030 PD(NPOL+1,1)=ICPOL:PD(NPOL+1,ICPOL+1)=INODE
7040 N1=N2:N2=N3:N3=INODE
7050 NEXT IARC
7060 RETURN
```

APPENDIX 3.B
AUTOMESH: AN AUTOMATIC PLANE MESH GENERATION PROGRAM

```
10 REM *************************************************
20 REM *                                               *
30 REM *                   automesh                    *
40 REM *                                               *
50 REM *************************************************
60 REM          COPYRIGHT 1987
70 REM by Professor Andrew D. Dimarogonas, Washington University,St.Louis.
80 REM All rights reserved.  Unauthorized reproduction, dissemination or
90 REM selling is strictly prohibited.  This listing is for personal use.
100 REM
110 REM
120 REM
130 REM
140 REM Plane automatic mesh generation
150 REM Beam,Triangular and Quadrilateral elements
160 REM
170 REM        ADD/11-20-85/revision 12-22-86
180 REM ************************************************************
190 CLS:KEY OFF
200 FOR I=1 TO 14:COLOR I:LOCATE I,22+I:PRINT "AUTOMESH";
210 LOCATE I,50-I:PRINT "AUTOMESH";:NEXT I:COLOR 14
220 LOCATE 8,7:PRINT"An automatic mesh generation program for finite
element";
230 PRINT " analysis";
236 locate 25,1:print" Dimarogonas,A.D., Computer Aided Machine
Design,Prentice-Hall,1988";
240 LOCATE 16,I:PRINT "                        TYPES OF ELEMENTS";
250 LOCATE 18,I:PRINT "                        <1>         Beam";
260 LOCATE 19,I:PRINT "                        <2>   Triangular";
270 LOCATE 20,I:PRINT "                        <3> Quadrilateral";
280 LOCATE 22,I:INPUT "                      enter your selection...";E$
290 LOCATE 23,I:INPUT "                  enter screen window xmax,ymax ";XMAX,YMAX
300 XSC=550/XMAX:YSC=180/YMAX
310 DIM NI(200),NJ(200),NM(200),NK(200):REM           Element definitions
320 DIM X(200),Y(200),XP(200),YP(200):REM             Node coordinates
330 DIM XD(20,20),YD(20,20):REM                       Input nodal points
340 DIM GK(40,40),GF(40),DX(40),DY(40),D(40):REM      Auxiliary matrices
350 SCREEN 2:ID=0:TOLERANCE=YMAX/100
360 CLS:NMENUS=7
370 DIM L$(100)
```

```
380 VIND=5:IPLOT=0
390 DIM MENU1$(25)
400 FOR IM=1 TO NMENUS:READ MENU1$(IM):NEXT IM
410 DATA "Stop       ","Clear scrn","New        ","Save File ","Load File "
420 DATA "Automesh  ","Plot mesh"
430 XS=0
440 FOR I=1 TO NMENUS:LOCATE I+2,70:PRINT MENU1$(I);:NEXT I
450 ICOUNT=0
460 LOCATE 1,70:PRINT"MAIN MENU";
470 LINE(530,0)-(640,8),,B:GOTO 550
480 X$=INKEY$:XS=0
490 IF X$="" THEN 480
500 IF LEN(X$)=1 THEN XS=ASC(X$):GOTO 610
510 X$=RIGHT$(X$,1)
520 IF X$><"H" AND X$<>"P" THEN 580
530 IF X$="H" AND VIND>1 THEN VIND=VIND-1
540 IF X$="P" AND VIND<NMENUS THEN VIND=VIND+1
550 IF VIND>1 THEN LOCATE VIND+1,69:PRINT " ";MENU1$(VIND-1);
560 LOCATE VIND+2,69:PRINT ">";MENU1$(VIND);
570 IF VIND<NMENUS THEN LOCATE VIND+3,69:PRINT " ";MENU1$(VIND+1);
580 LOCATE 19,70:PRINT"           ";
590 LOCATE 21,65:PRINT"               ";
600 LOCATE 22,65:PRINT"               ";
610 IF XS=13 AND VIND=1 THEN SCREEN 0:END
620 IF XS=13 AND VIND=2  THEN CLS:GOTO 430:'        clear screen
630 IF XS=13 AND VIND=3 THEN GOSUB  710:REM         start a new drawing
640 IF XS=13 AND VIND=4 THEN GOSUB 1060:REM          save file on disc
650 IF XS=13 AND VIND=5 THEN GOSUB 770:REM          load file from disc
660 IF XS=13 AND VIND=6 THEN GOSUB 1250:'           generate mesh
670 IF XS=13 AND VIND=7 THEN GOSUB 2700:REM         plot mesh
680 FOR ICLEAR=14 TO 24:LOCATE ICLEAR,65:PRINT SPC(14);:NEXT ICLEAR
690 GOTO 480
700 END
710 REM start a new drawing
720 LOCATE 22,65:INPUT"SURE (Y/N)";X$:IF X$="n" OR X$="N" THEN RETURN
730 ND=0:NP=0:NC=0:RETURN
740 REM*****************************************************************
750 REM                file retrieval
760 REM*****************************************************************
770 LOCATE 19,68:PRINT"                ";
780 LOCATE 18,70:PRINT"Data file  ":LOCATE 19,68:INPUT FILNA$
790 IF FILNA$="" THEN 780
800 ICOUNT=0
810 OPEN FILNA$ FOR INPUT AS #1
820 ICOUNT=ICOUNT+1:INPUT#1,X$
830 IF X$="filend" OR X$="FILEND" THEN 990
840 REM
850 IF X$="beam" OR X$="BEAM" THEN E$="1"
860 IF X$="triangular" OR X$="TRIANGULAR" THEN  E$="2"
870 IF X$="quadrilateral" OR X$="QUADRILATERAL" THEN  E$="3"
880 SYM$=""
890 ISYM=1:IF ICOUNT>1000 THEN 1000
900 SYM$=MID$(X$,ISYM,1):ISYM=ISYM+1:IF SYM$="" AND ISYM<10 THEN 890
910 IF ASYM=10 THEN 820
920 IF SYM$="n" OR SYM$="N" THEN ND=ND+1:INPUT#1,I,X(ND),Y(ND):GOTO 820
930 IF SYM$<>"e" AND SYM$<>"E" THEN  820
940 NE=NE+1
950 IF E$="1" THEN INPUT#1,I,NI(NE),NJ(NE)
960 IF E$="2" THEN INPUT#1,I,NI(NE),NJ(NE),NM(NE)
970 IF E$="3" THEN INPUT#1,I,NI(NE),NJ(NE),NM(NE),NK(NE)
980 GOTO 820
990 CLOSE#1 :GOTO 1010
1000 PRINT"bad file";
1010 XMAX0=0:YMAX0=0:FOR I=1 TO ND:IF ABS(X(I))>XMAX0 THEN XMAX0=ABS(X(I))
1020 IF ABS(Y(I))>YMAX0 THEN YMAX0=ABS(Y(I))
1030 IF XMAX0>XMAX THEN XSC=550/XMAX0:LOCATE 25,1:PRINT"New xmax=";XMAX0;"
";
1040 IF YMAX0>YMAX THEN YSC=180/YMAX0:LOCATE 25,40:PRINT"New ymax=";YMAX0;"
";
1050 NEXT I:RETURN
1060 REM ********************************STORING TRANSFORMED DATA ON DISC
```

```
1070 LOCATE 19,68:PRINT"            "
1080 LOCATE 18,70:PRINT"New file  ":LOCATE 19,68:INPUT FILNA$
1090 CLOSE
1100 OPEN FILNA$ FOR OUTPUT AS #1
1110 IF E$="1" THEN PRINT#1,"BEAM"
1120 IF E$="2" THEN PRINT#1,"TRIANGULAR"
1130 IF E$="3" THEN PRINT#1,"QUADRILATERAL"
1140 FOR I=1 TO ND
1150 PRINT#1,"NODE,";I;",";X(I);",";Y(I):NEXT I
1160 REM **DATA** DEFINITION OF ELEMENTS
1170 FOR I=1 TO NE
1180 IF E$="1" THEN PRINT#1,"ELEMENT,";I;",";NI(I);",";NJ(I)
1190 IF E$="2" THEN PRINT#1,"ELEMENT,";I;",";NI(I);",";NJ(I);",";NM(I)
1200 IF E$="3" THEN
PRINT#1,"ELEMENT,";I;",";NI(I);",";NJ(I);",";NM(I);",";NK(I)
1210 NEXT I
1220 PRINT#1,"FILEND"
1230 CLOSE#1
1240 RETURN
1250 REM
***************************************************************
1260 REM                  mesh generation
1270 REM
***************************************************************
1280 LINE (530,0)-(640,8),0,B
1290 LOCATE 20,1
1300 PRINT"Using the N command create a grid of Ix by Iy nodes defining"
1310 PRINT"the geometry of the domain to be AUTOMESHed."
1320 PRINT"Then use A command to crate the mesh. Enter the desired Nx,Ny"
1330 PRINT"If you do a region in many patches,use R to remove overlapping
nodes"
1340 PRINT"Use F,S to change the step of the grid motion, P to move a
node";
1350 X$="":XSTEP=XMAX/10:YSTEP=YMAX/10
1360 XHAIR=XSTEP*5:YHAIR=YSTEP*5
1370 LOCATE 13,70:PRINT"COMMANDS:";
1380 LOCATE 14,70:PRINT"N-ode    ";
1390 LOCATE 15,70:PRINT"E-lement ";
1400 LOCATE 16,70:PRINT"S-lower  ";
1410 LOCATE 17,70:PRINT"F-aster  ";
1420 LOCATE 18,70:PRINT"W-rite   ";
1430 LOCATE 19,70:PRINT"G-rid    ";
1440 LOCATE 20,70:PRINT"A-utomesh";
1450 LOCATE 21,70:PRINT"R-edundant";
1460 LOCATE 22,70:PRINT"P-ush Node";
1470 LOCATE 23,70:PRINT"M-enu      ";
1480 IF X$="" THEN X$="P":GOTO 1630
1490 X$=INKEY$:XS=0
1500 LOCATE 25,1:PRINT ">> Command:  ";
1510 LOCATE 25,1:PRINT "               ";X$;SPC(10);
1520 IF X$="" THEN 1490
1530 IF LEN(X$)>1  THEN 1620
1540 REM
1550 IF X$="f" THEN XSTEP=XSTEP*2:YSTEP=XSTEP:ZSTEP=XSTEP
1560 IF X$="s" THEN XSTEP=XSTEP/2:YSTEP=XSTEP:ZSTEP=XSTEP
1570 IF X$="F" THEN XSTEP=XSTEP*2:YSTEP=XSTEP:ZSTEP=XSTEP
1580 IF X$="S" THEN XSTEP=XSTEP/2:YSTEP=XSTEP:ZSTEP=XSTEP
1590 IF X$<>"G" AND X$<>"g" THEN 1610
1600 FOR I= 10 TO 180 STEP 10:FOR J=9 TO 540 STEP 20:PSET(J,I):NEXT J:NEXT
I
1610 IF LEN(X$)=1 THEN XS=ASC(X$):GOTO 1790
1620 X$=RIGHT$(X$,1)
1630 FOR I=5 TO 500 STEP 20:PSET(I,YSC*YHAIR),0:NEXT I
1640 FOR I=6 TO 190 STEP 10:PSET(XSC*XHAIR,I),0:NEXT I
1650 IF X$="H" THEN YHAIR=YHAIR-YSTEP
1660 IF X$="P" THEN YHAIR=YHAIR+YSTEP
1670 IF X$="K" THEN XHAIR=XHAIR-XSTEP
1680 IF X$="M" THEN XHAIR=XHAIR+XSTEP
1690 IF XHAIR<0 THEN XHAIR=0
1700 IF YHAIR<0 THEN YHAIR=0
1710 IF YHAIR>190 THEN YHAIR=190
```

```
1720 IF XHAIR>500 THEN XHAIR=500
1730 LOCATE 25,3:PRINT SPC(75);
1740 LOCATE 25,35:PRINT "x= ";XHAIR;:LOCATE 25,45:PRINT "y= ";YHAIR;
1750 LOCATE 25,55:PRINT " Defining Nodes= ";NPD;
1760 FOR I=5 TO 500 STEP 20:PSET(I,YSC*YHAIR):NEXT I
1770 FOR I=6 TO 190 STEP 10:PSET(XSC*XHAIR,I):NEXT I
1780 GOTO 1490
1790 IF X$<>"W" AND X$<>"w" THEN 1900
1800 WORD$=""
1810 W$=INKEY$:IF W$="" THEN 1810
1820 IF ASC(W$)=13 THEN 1900
1830 IF ASC(W$)=8 THEN WORD$="":W$=""
1840 LOCATE YSC*YHAIR/8,XSC*XHAIR/7.875
1850 WORD$=WORD$+W$:PRINT WORD$;
1860 LOCATE YSC*YHAIR/8,(XSC*ZHAIR+280)/7.875
1870 PRINT WORD$;
1880 LOCATE (100+YSC*ZHAIR)/8,XSC*XHAIR/7.875
1890 PRINT WORD$;:GOTO 1810
1900 IF X$><"n" AND X$<>"N" THEN 1970
1910 XPL=XHAIR:YPL=YHAIR:ZPL=ZHAIR
1920 NPD=NPD+1:XP(NPD)=XPL:YP(NPD)=YPL
1930 N1=N2:N2=N3:N3=NPD:LOCATE 25,3:PRINT SPC(75);
1940 LOCATE 25,3:PRINT SPC(75);
1950 LOCATE 25,45:PRINT "new node= ";NPD;
1960 CIRCLE(XSC*XPL,YSC*YPL),3
1970 IF X$="a" OR X$="A" THEN GOSUB 2040
1980 IF X$="m" OR X$="M" THEN 2020
1990 IF X$="r" OR X$="R" THEN GOSUB 3840:REM  Remove redundant nodes
2000 IF X$="p" OR X$="P" THEN GOSUB 4070:REM  Push mode
2010 GOTO 1370
2020 LOCATE 25,1:PRINT SPC(78);:FOR I=13 TO 22:LOCATE I,68:PRINT
SPC(11);:NEXT I
2030 LINE (530,0)-(640,8),,B:RETURN
2040 REM ***********************************************
2050 REM *                                             *
2060 REM *                 AUTOMESH                    *
2070 REM *                                             *
2080 REM ***********************************************
2090 REM
2100 REM An automatic mash generation program.
2110 REM Curved quadrilateral is divided into triangular elements.
2120 REM The mesh is stored on disc for further processing.
2130 LINE (530,0)-(640,8),0,B
2140 LOCATE 23,64:PRINT SPC(15);
2150 LOCATE 23,64:PRINT "enter ix,iy";:INPUT IX,IY:LOCATE 23,68
2160 IF IX*IY><NPD THEN LOCATE 23,64:PRINT SPC(15);:GOTO 2150
2170 REM DEFINE BELOW NP CURVES, NP POINTS EACH
2180 XU = - 1E+20:XL = 1E+20 :  NPD=0
2190 YU = - 1E+20:YL = 1E+20
2200 FOR I = 1 TO IX
2210 FOR J = 1 TO IY
2220 XD(I,J)=XP((J-1)*IX+I):YD(I,J)=YP((J-1)*IX+I)
2230 IF XD(I,J) < XL THEN XL = XD(I,J)
2240 IF XD(I,J) > XU THEN XU = XD(I,J)
2250 IF YD(I,J) > YU THEN YU = YD(I,J)
2260 IF YD(I,J) < YL THEN YL = YD(I,J)
2270 NEXT J:NEXT I
2280 REM SCALE FACTOR:
2290 REM
2300 GOSUB 3160:REM FIND TRANSFORMATION COEFFICIENTS
2310 LOCATE 25,1:PRINT "COMPUTING MESH NODES...                    ";
2320 REM AUTOMATIC MESH GENERATION
2330 REM X-DIRECTION:NX NODES, Y-DIRECTION:NY NODES
2340 LOCATE 25,1:PRINT" nx,ny = mesh nodes in x,y directions";
2350 LOCATE 23,64:PRINT SPC(15);
2360 LOCATE 23,64:INPUT "Nodes nx,ny";NX,NY
2370 NEPN=4:IF E$="2" THEN NEPN=2
2380 IF E$="3" THEN NEPN=1
2390 NN = NX * NY:NPE=NE:NE =(NX - 1) * (NY - 1) * NEPN
2400 LOCATE 25,1:PRINT NN;"Nodes, ";NE;"elements                       ";
2410 REM INVERSE TRANSFORMATION AND ELEMENT DEFINITION
```

```
2420 FOR IP = 1 TO NX:FOR JP = 1 TO NY
2430 X = 0:Y = 0
2440 IN = (IP - 1) * NY + JP+ND
2450 LOCATE 25,40:PRINT "COMPUTING NODE ";IN;
2460 FOR I1 = 1 TO IX:FOR I2 = 1 TO IY
2470 IJ = (I1 - 1) * IY + I2
2480 I =1+(IX - 1) * (IP - 1) / (NX - 1)
2490 J =1+(IY - 1) * (JP - 1) / (NY - 1)
2500 X = X + DX(IJ) * I ^ (IX - I1) * J ^ (IY - I2)
2510 Y = Y + DY(IJ) * I ^ (IX - I1) * J ^ (IY - I2)
2520 NEXT I2:NEXT I1
2530 X(IN) = X:Y(IN) = Y
2540 IF JP = NY THEN 2630
2550 IF IP = NX THEN 2630
2560 E2=((IP-1)*(NY-1)+JP)*NEPN+NPE
2570 E1 = E2 - 1:NI1=IN:NJ1=IN+NY:NM1=IN+1:NI2=IN+NY:NJ2=NI2+1:NM2=IN+1
2580 IF E$="2" THEN NI(E1)=NI1:NJ(E1)=NJ1:NM(E1)=NM1
2590 NI(E2)=NI2:NJ(E2)=NJ2:NM(E2)=NM2:NK(E2)=IN:PRINT " ELEMENT  ";E2;
2600 IF E$<>"1" THEN 2630
2610 NI(E2-1)=NJ2:NJ(E2-1)=NM2:NI(E2-2)=NM2:NJ(E2-2)=NI1
2620 NI(E2-3)=NI1:NJ(E2-3)=NJ1
2630 NEXT JP:NEXT IP
2640 IF E2>0 THEN NE=E2:REM PLOT ELEMENTS
2650 GOSUB 2940
2660 IF IN>0 THEN ND=IN
2670 LOCATE 25,1:PRINT SPC(78);:FOR I=13 TO 22:LOCATE I,68:PRINT
SPC(11);:NEXT I
2680 LOCATE 23,64:PRINT SPC(15);
2690 RETURN
2700 REM ********************************************** PLOT MANIPULATION
2710 LINE (530,0)-(640,8),0,B
2720 ND1=1:ND2=ND
2730 FOR IP=14 TO 24:LOCATE IP,65:PRINT SPC(14);:NEXT IP
2740 LOCATE 16,68:PRINT"COMMANDS:";
2750 LOCATE 17,68:PRINT "R-otate";
2760 LOCATE 18,68:PRINT "T-ranslate";
2770 LOCATE 19,68:PRINT "S-cale";
2780 LOCATE 20,68:PRINT "M-enu ";
2790 LOCATE 21,68:PRINT "P-lot      ";
2800 LOCATE 22,68:PRINT "C-ross     ";
2810 LOCATE 25,1:PRINT"COMMAND>>  ";
2820 P$=INKEY$:IF P$="" THEN 2820
2830 PRINT "   ";P$;:IF P$="m" OR P$="M" THEN 2920
2840 IF P$="r" OR P$="R" THEN GOSUB 4650
2850 IF P$="t" OR P$="T" THEN GOSUB 4550
2860 IF P$="s" OR P$="S" THEN GOSUB 4600
2870 IF P$="p" OR P$="P" THEN GOSUB 2940
2880 IF P$="c" OR P$="C" THEN GOSUB 4220
2890 LOCATE 25,1:PRINT SPC(78);
2900 LOCATE 15,65:PRINT SPC(14);
2910 GOTO 2810
2920 LINE (530,0)-(640,8),,B
2930 RETURN
2940 REM **********************************************  MESH PLOTTING
2950 XS=XSC:YS=YSC
2960 FOR I = 1 TO NE
2970 I1 = NI(I):I2 = NJ(I):I3 = NM(I):I4=NK(I)
2980 X1 = XS * X(I1) + X0:Y1 = Y0 + Y(I1) * YS
2990 X2 = XS * X(I2) + X0:Y2 = Y0 + Y(I2) * YS
3000 X3 = XS * X(I3) + X0:Y3 = Y0 + Y(I3) * YS
3010 X4 = XS * X(I4) + X0:Y4 = Y0 + Y(I4) * YS
3020 IF X1>500 OR X2>500 OR X3>500 OR X4>500 THEN 3060
3030 IF Y1>180 OR Y2>180 OR Y3>180 OR Y4>180 THEN 3060
3040 IF E$="2" THEN LINE(X1,Y1)- (X2,Y2):LINE -(X3, Y3):LINE -(X1,Y1)
3050 IF E$="3" THEN LINE(X1,Y1)-(X2,Y2):LINE-(X3,Y3):LINE-(X4,Y4):LINE-
(X1,Y1)
3060 NEXT I
3070 REM PLOT DEFINING POINTS
3080 IF XT>0 OR YT>0 OR SX>0 OR SY>0 OR FR>0 THEN 3150
3090 FOR I = 1 TO IX:FOR J = 1 TO IY
3100 X = X0 + XS * XD(I,J)
```

```
3110 Y = Y0 + YS * YD(I,J)
3120 IF X>500 OR Y>180 THEN 3140
3130 LINE(X-2,Y-2)-(X+2,Y-2):LINE-(X+2,Y+2):LINE-(X-2,Y+2):LINE-(X-2,Y-2)
3140 NEXT J:NEXT I
3150 RETURN
3160 REM ********************************FINDING TRANSFROMATION INTO A
SQUARE
3170 REM FIND X-COEFFICIENTS
3180 REM DEFINE COEFFICIENT MATRIX
3190 LOCATE 25,1:PRINT "COMPUTING TRANSFORMATION COEFFICIENTS";
3200 PRINT "...FIRST CURVE FAMILY ";
3210 FOR I1 = 1 TO IX:FOR I2 = 1 TO IY:I = (I1 - 1) * IY + I2
3220 FOR J1 = 1 TO IX
3230 FOR J2 = 1 TO IY
3240 J = (J1 - 1) * IY + J2
3250 GK(I,J) = I1 ^ (IX - J1) * I2 ^ (IY - J2)
3260 NEXT J2:NEXT J1
3270 GF(I) = XD(I1,I2)
3280 NEXT I2:NEXT I1
3290 REM SOLVE FOR X-COEFFICIENTS
3300 GOSUB 3530:REM GAUSS ELIMINATION
3310 FOR I1 = 1 TO IY:FOR I2 = 1 TO IX:I = (I1 - 1) * IX + I2
3320 DX(I) = D(I):NEXT I2: NEXT I1
3330 REM FIND Y-COEFFICIENTS
3340 REM DEFINE COEFFICIENT MATRIX
3350 LOCATE 25,1:PRINT "SECOND CURVE FAMILY ";
3360 PRINT "COEFFICIENT MATRIX";
3370 FOR I1 = 1 TO IX:FOR I2 = 1 TO IY:I = (I1 - 1) * IY + I2
3380 FOR J1 = 1 TO IX
3390 FOR J2 = 1 TO IY
3400 J = (J1 - 1) * IY + J2
3410 GK(I,J) = I1 ^ (IX - J1) * I2 ^ (IY - J2)
3420 NEXT J2:NEXT J1
3430 GF(I) = YD(I1,I2)
3440 NEXT I2:NEXT I1
3450 REM SOLVE FOR Y-COEFFICIENT
3460 GOSUB 3530:REM GAUSS ELIMINATION
3470 FOR I1 = 1 TO IX:FOR I2 = 1 TO IY:I = (I1 - 1) * IY + I2
3480 DY(I) = D(I):NEXT I2:NEXT I1
3490 RETURN
3500 REM
3510 REM ************************************SOLUTION OF LINEAR EQUATIONS
3520 REM BY GAUSS ELIMINATION
3530 N = IY*IX:FOR I = 1 TO N:GK(I,N + 1) = GF(I):NEXT I:NN = N
3540 LOCATE 25,1:PRINT "Solving linear equations........."; SPC(30);
3550 FOR R = 1 TO NN - 1
3560 A = GK(R,R)
3570 T = R
3580 FOR I = R + 1 TO NN
3590 IF ABS(A) > = ABS (GK(I,R)) THEN 3620
3600 A = GK(I,R)
3610 T = I
3620 NEXT I
3630 IF T = R THEN 3690
3640 FOR S1 = R TO N + 1
3650 B1 = GK(R,S1)
3660 GK(R,S1) = GK(T,S1)
3670 GK(T,S1) = B1
3680 NEXT S1
3690 FOR I = R + 1 TO NN
3700 FOR J = R + 1 TO N + 1
3710 GK(I,J) = GK(I,J) - GK(I,R) * GK(R,J) / GK(R,R)
3720 NEXT J
3730 NEXT I
3740 NEXT R
3750 D(N) = GK(N,N + 1) / GK(N,N)
3760 FOR J = N - 1 TO 1 STEP -1
3770 S = 0
3780 FOR K = J + 1 TO N
3790 S = S + GK(J,K) * D(K)
3800 NEXT K
```

```
3810 D(J) = (GK(J,N + 1) - S) / GK(J,J)
3820 NEXT J
3830 RETURN
3840 REM **********************************REMOVING REDUNDANT NODES
3850 LOCATE 25,1:PRINT SPC(78);:ESIGN=1
3860 LOCATE 23,65:INPUT"tolerance";TOLERANCE
3870 LOCATE 23,65:PRINT"              ";
3880 FOR I=1 TO ND:FOR J=I+1 TO ND
3890 LOCATE 25,10:PRINT "Testing node ";I;" redundant node:";
3900 DELTA=(X(J)-X(I))^2+(Y(J)-Y(I))^2:IF DELTA>TOLERANCE THEN 4010
3910 FOR K=J TO ND-1:X(K)=X(K+1):Y(K)=Y(K+1):NEXT K
3920 FOR IE=1 TO NE
3930 IF NI(IE)=J THEN NI(IE)=I
3940 IF NJ(IE)=J THEN NJ(IE)=I
3950 IF NM(IE)=J THEN NM(IE)=I
3960 IF NI(IE)>J THEN NI(IE)=NI(IE)-1
3970 IF NJ(IE)>J THEN NJ(IE)=NJ(IE)-1
3980 IF NM(IE)>J THEN NM(IE)=NM(IE)-1
3990 NEXT IE
4000 ND=ND-1:PRINT J;" ";
4010 NEXT J
4020 LOCATE 25,70:PRINT SPC(9);
4030 IF ESIGN>0 THEN LOCATE 25,70:PRINT "working";
4040 ESIGN=ESIGN*(-1):NEXT I
4050 LOCATE 25,1:PRINT SPC(79);
4060 RETURN
4070 REM **********************************Push-relocate mode
4080 LOCATE 25,1:PRINT "Locate node with the CROSS and hit RETURN
"
4090 GOSUB 4240
4100 DN=9.999999E+24:XNODE=XHAIR:YNODE=YHAIR:FOR I=1 TO ND
4110 DELTA=(XNODE-X(I))^2+(YNODE-Y(I))^2:IF DELTA<DN THEN INODE=I:DN=DELTA
4120 NEXT I
4130 FOR I=5 TO 500 STEP 20:PSET(I,YSC*YHAIR),0:NEXT I
4140 FOR I=6 TO 190 STEP 10:PSET(XSC*XHAIR,I),0:NEXT I
4150 XHAIR=X(INODE):YHAIR=Y(INODE)
4160 LOCATE 25,1:PRINT INODE;"Move CROSS to new position, hit RETURN "
4170 GOSUB 4340
4180 X(INODE)=XHAIR:Y(INODE)=YHAIR
4190 FOR I=1 TO 24:LOCATE I,1:PRINT SPC(65);:NEXT I
4200 GOSUB 2940:REM plot mesh
4210 RETURN
4220 REM********************************************* move cross-hair
4230 LOCATE 25,1:PRINT" Move cross with keyboard arrows    ";
4240 X$=INKEY$:XS=0
4250 IF X$="" THEN 4240
4260 IF LEN(X$)>1  THEN 4330
4270 IF ASC(X$)=13 THEN 4500
4280 IF X$="f"  THEN XSTEP=XSTEP*2:YSTEP=XSTEP:ZSTEP=XSTEP
4290 IF X$="s"  THEN XSTEP=XSTEP/2:YSTEP=XSTEP:ZSTEP=XSTEP
4300 IF X$="F"  THEN XSTEP=XSTEP*2:YSTEP=XSTEP:ZSTEP=XSTEP
4310 IF X$="S"  THEN XSTEP=XSTEP/2:YSTEP=XSTEP:ZSTEP=XSTEP
4320 GOTO 4240
4330 X$=RIGHT$(X$,1)
4340 FOR I=5 TO 500 STEP 20:PSET(I,YSC*YHAIR),0:NEXT I
4350 FOR I=6 TO 190 STEP 10:PSET(XSC*XHAIR,I),0:NEXT I
4360 IF X$="H" THEN YHAIR=YHAIR-YSTEP
4370 IF X$="P" THEN YHAIR=YHAIR+YSTEP
4380 IF X$="K" THEN XHAIR=XHAIR-XSTEP
4390 REM
4400 IF X$="M" THEN XHAIR=XHAIR+XSTEP
4410 IF XHAIR<0 THEN XHAIR=0
4420 IF YHAIR<0 THEN YHAIR=0
4430 IF YHAIR>190 THEN YHAIR=190
4440 IF XHAIR>500 THEN XHAIR=500
4450 LOCATE 25,45:PRINT SPC(34);
4460 LOCATE 25,45:PRINT "x= ";XHAIR;:LOCATE 25,63:PRINT "y= ";YHAIR;
4470 FOR I=5 TO 500 STEP 20:PSET(I,YSC*YHAIR):NEXT I
4480 FOR I=6 TO 190 STEP 10:PSET(XSC*XHAIR,I):NEXT I
4490 GOTO 4240
4500 RETURN
```

```
4510 REM  ***************************************************************
4520 REM COORDINATE TRANSFOMATION ROUTINES
4530 REM ******************************* SUBROUTINE TRANSLATE(XT,YT)
4540 IF IROT=1 THEN 4560
4550 LOCATE 14,65:PRINT "ENTER XT, YT";:LOCATE 15,65:INPUT XT,YT
4560 FOR I=ND1 TO ND2
4570 X(I)=X(I)+XT:Y(I)=Y(I)+YT
4580 NEXT I
4590 IROT=0:RETURN
4600 REM ******************************* SUBROUTINE SCALE(sx,sy) ****
4610 LOCATE 14,65:PRINT "ENTER SX,SY";:LOCATE 15,65:INPUT SX,SY
4620 FOR I=ND1 TO ND2
4630 X(I)=X(I)*SX:Y(I)=Y(I)*SY
4640 NEXT I:RETURN
4650 REM ******************************* SUBROUTINE ROTATE(xrot,yrot,fr)
4660 XROT=0:YROT=0
4670 FOR I=ND1 TO ND2:XROT=XROT+X(I)/(ND2-ND1):YROT=YROT+Y(I)/(ND2-
ND1):NEXT I
4680 LOCATE 14,65:PRINT "ENTER ANGLE, deg";:LOCATE 15,65:INPUT FR
4690 IROT=1:XT=-XROT:YT=-YROT:GOSUB 4530:REM translate to origin
4700 FR=FR*3.14159/180:CF=COS (FR):SF=SIN (FR)
4710 FOR I=ND1 TO ND2
4720 XX=X(I):YY=Y(I)
4730 X(I)=XX*CF+YY*SF
4740 Y(I)=-XX*SF+YY*CF
4750 NEXT I
4760 IROT=1:XT=XROT:YT=YROT:GOSUB 4530:REM translate to original position
4770 RETURN
```

CHAPTER FOUR
STRESS ANALYSIS IN MACHINE DESIGN

4.1 MODES OF FAILURE

In most machines and structures each structural element must perform a certain task and at the same time it must retain its structural integrity under the operating conditions imposed on the element. In other words, under the influence of forces, thermal loads, friction, etc., it must not be damaged to such an extent that it will not be able to perform its function beyond some point in time. The term 'damage' does not always mean breakage. We shall use the term 'failure' for all unintended changes of the element which prevent it from continuing the performance of its duty.

Therefore, failure might mean fracture but also large plastic deformation, excessive wear, annealing owing to excessive temperatures, surface damage, local softening or hardening, etc.

The main purpose of design analysis is to find some relation between service conditions and mode of failure. Therefore, the first task is to identify the most probable mode of failure in the particular application. In general, there is some guidance in this respect by past experience or by codes or by standard design practice. And by 'mode of failure' we mean, *how*, *where* and *when* the element will fail. This simplifies the design work because it eliminates the need for many unnecessary calculations.

Take, for example, the design of a ball-bearing. As we shall see later, the probable mode of failure is at the rolling contact (*where*), by fatigue wear (*how*) and when the bearing makes a certain number of revolutions (*when*). Therefore, the designer must concentrate his efforts where the real need is and not, for example, to the calculations of the bulk strength of the outer ring. He must quantify the relation of the number of revolutions (*when*) to the failure of the ball race (*where*) by fatigue wear (*how*) and the service conditions.

The same type of element might have in different cases, different modes of failure, depending on the particular operating conditions, such as loads, temperatures, environment, etc. It also depends on the material of the element and the manufacturing process used. For example, a ductile material will lead an element to failure by large plastic deformation while with a brittle material, the mode of failure will be, probably, brittle fracture.

Static loads may result in plastic deformation or rupture. Dynamic loads might lead the same element to fatigue fracture.

The probable mode of failure will be the guide for the design analysis. In the following sections, some of the usual modes of failure will be discussed.

4.1.1 Elastic deformation

An element might not perform its duty if the elastic deformation exceeds certain limits. For example, in steam and gas turbines, air turbocompressors, jet engines, etc., there are close clearances around the rotating shafts. Excessive elastic deformation will lead to either shaft failure due to rubbing or opening the clearances and decreasing the efficiency of the machine. For this mode of failure, a deflection analysis must be performed.

For simple prismatic elements, well-known formulas for strength of materials are used, such as:

(i) Elongation of rod under axial loading: $\delta l = Pl/AE$
(ii) Torsion of rods of circular cross-section: $\delta\varphi = Tl/I_pG$
(iii) Deflection of beams due to bending: Several formulas exist for different loads and boundary conditions. For example, for a simply supported beam with a central load at mid-span, the maximum deflection is

$$y_{\max} = Pl^3/48EI$$

A systematic tabulation of design formulas for deformation will be given in Chapter 7. Because of the availability of computer methods, such formulas are used for very simple members or for a rough estimate before using computer techniques.

It must be noted that in the above equations, stresses do not appear explicitly. This suggests that one mode of failure might not occur together with the others at the same operating conditions. For example, a shaft might fail due to high elastic deformations while the stresses in the shaft are far below the safe limits.

4.1.2 Plastic deformation

Failure might occur in cases of general plastic deformation which will change the geometry of the elements. This is much more important in machine design than in structures because general plastic deformation might not affect the strength but it might prevent further operation due to geometric or other consequences.

As examples, a bolt deformed plastically is useless, a shaft with lateral plastic deformation will have excessive unbalance and vibration, and a gear tooth deformed plastically will not allow rotation.

4.1.3 Fracture

Fracture, obviously, will mean failure. By fracture we mean the physical separation of an element into two or more pieces. Fracture has, in general, three modes: Sudden fracture of brittle materials, progressive fracture of ductile materials, and creep rupture in elevated temperatures. Usually, this last is the most catastrophic mode of failure.

4.1.4 Wear

Elements with sliding surfaces might show excessive wear which will prevent them from further normal operation.

The above, and other, modes of failure, when identified, must be followed by design analyses to predict such failure and adjust the design parameters in order to prevent failure.

4.2 MODELING FOR STRESS ANALYSIS

Since most of the failure modes in machine design originate from high loads, a stress analysis is almost invariably needed for every machine design analysis. Stress analysis can be performed in three ways:

1. *Analytically*, using known principles of mechanics. Since analytical solutions exist only for very simple geometries, simplifying assumptions must be made in almost all cases. As the geometry of machines and components becomes more and more complicated, analytical methods become progressively more crude. However, no matter what computer programs are available, such analyses must always be performed for order-of-magnitude estimations.
2. *Experimentally*, using several methods of experimental stress analysis.
3. *Computationally*, using any of several existing computer codes, like those presented in this chapter. Such computer methods can be used for even very complicated geometries. The extreme caution however, in using these methods, will be emphasized continuously in this book, together with encouragement for their utilization.

The reader will recall that strengths of material methods are used in very specific and simple geometries, such as straight bars, cylindrical shafts with constant cross-section, perfect trusses and frames, etc. Such elements do not really exist in machines. Machine elements are usually much more complicated. Stress analysis in that geometry might be impossible analytically or difficult experimentally or computationally. The latter case might seem strange, since, in principle, contemporary methods can analyze the most complicated geometries. True. But one should always bear in mind that cost, time and reliability are essential ingredients in any machine design effort. The more complicated the computer analysis, the higher is the cost, the longer the time taken and accumulation of errors in the computer arithmetic makes the results progressively more unreliable. On one particular occasion when the author analyzed a thick pipe intersection by way of three-dimensional finite elements it took an experienced engineer six weeks only to prepare and verify the input. (By the way, it took a technician one afternoon to obtain experimental results!)

The need always exists, therefore, to substitute the real machine element with a model, which:

(a) for the mode of failure under examination, will be expected to yield results to an engineering accuracy close enough to the ones expected at the real element;
(b) will be manageable, that is, stress analysis can be performed with available resources (analytical methods, experimental capabilities, computer hardware and software); and
(c) will have stress analysis costs within budget and accuracy within acceptable limits.

The next question is, how? There are no equations for this. Experience, sound engineering judgement and the trial-and-error process will be the guides.

In general, elements which are manufactured in small quantities are analyzed with simpler models accompanied by higher safety margins. Elements in large quantities allow for detailed models with higher cost of stress analysis.

As an example, consider the element shown in Figure 4.1 which is loaded by opposing loads *F*. This particular form can be analyzed with the finite-element program FINSTRES which will be developed later in this chapter (p. 129). However, results for preliminary design can be obtained by modeling with two different methods, as indicated, depending on the relative dimensions. Model I assumes that the side members are very flexible and that they do not transmit moments to the members which bear the forces. Model II assumes the opposite. Therefore, the side members are beams loaded with eccentric loads.

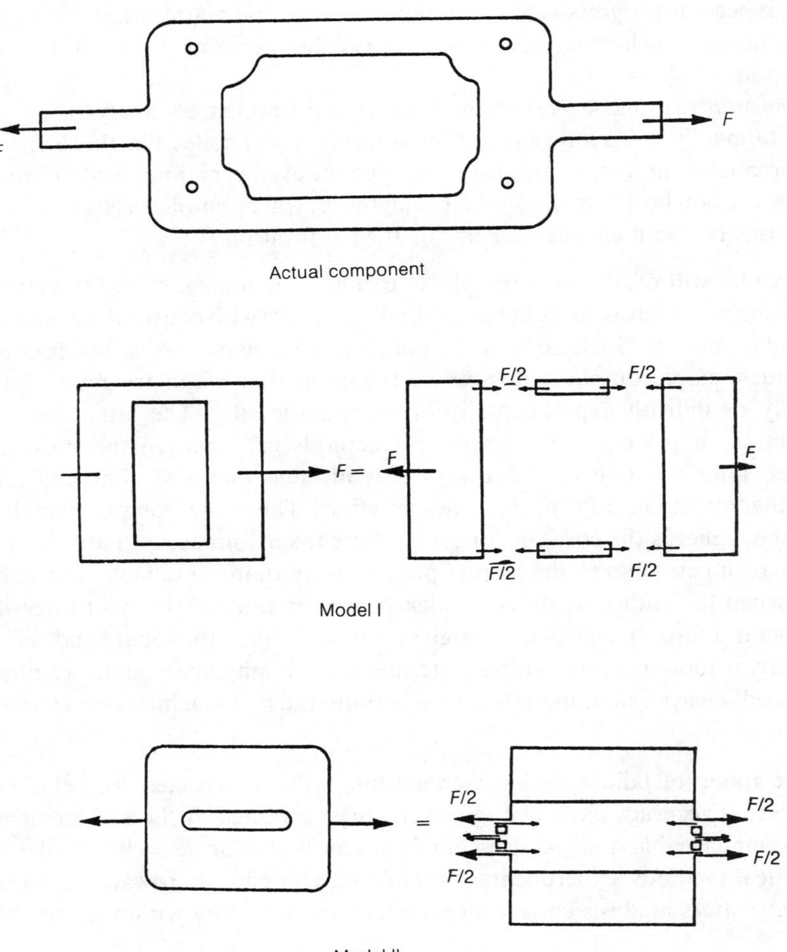

Figure 4.1 Modeling a complex component

4.3 COMPUTER ANALYSIS OF LINE MEMBERS

A generally-shaped machine member requires, in general, complicated procedures to produce the state of stress at each point owing to the static forces, thermal and dynamic loads, that the member sustains in service. This is a very complicated problem, in general, and the designer seeks always ways to achieve practical solutions to engineering accuracy.

Fortunately, most machine components have geometry which allows such solutions to be found by way of modeling these components approximately with simpler ones.

One such simplification is possible when the dimensions of the component along one direction are much greater than the ones perpendicular to it. We speak then of linear members which have the form of a rod when the length is several times greater than its maximum thickness. There is no specific rule to decide just when a long component can be considered a rod. It depends also on the form of loading and the accuracy required. In general, as a guide, a factor length/thickness of more than 10 can be considered as adequate. Such components are usually shafts, bolts, power screws, hooks, springs, torsion bars, etc. Some of them can be computed with strength of materials methods, as prismatic bars in tension/compression, bending, shear, torsion. Stress analysis then is relatively simple.

In many applications, however, machine components have more complicated geometry and the designer needs adequate solutions which cannot be obtained with strength of materials methods. Fortunately, computer methods come to the rescue and engineering solutions can be achieved for fairly complicated geometries. Examples are stepped shafts, machine frames, supports, steel structures, etc.

Some simple methods of computer aided design analysis will be discussed in the following sections.

4.3.1 The transfer matrix method

Stepped beams and shafts can be stress-analyzed with computer methods which make use of the one-dimensional nature of the component, under some simplifying assumptions.

In strength of materials, the problems have solutions in a more or less exact form. Of course, strictly speaking, perfectly elastic solids, uniform beams, uniform loads, etc., do not really exist in nature. But if one can conceive such systems, they can have exact solutions. Furthermore, in mechanics we study systems with distributed mass and elasticity. There again, idealizations of beams with constant properties, linearity, and boundary conditions of the rigid support type do not exist really. If one can conceive them, 'exact' solutions, in some form, exist and, in certain cases, they can be found.

This may have left us with a feeling of false security, however, because our solutions are, in any event, approximate. The first approximation is introduced in the modeling of the system itself as mentioned above. Moreover, only for very simple machine members with special and simple geometry, can an exact solution be found. For more-complicated systems, the very first difficulty is in their modeling. Finally, the associated equations must be solved.

In general, this has to be done numerically. Such situations exist for most moderately-complicated machine members. Therefore, for these problems we do eventually resort to numerical methods. In this case, it might be argued: why bother with analysis at all and not just work out the solution in the computer, from the beginning? The answer to this is that we should always try to pursue the analytical solution as far as possible and then resort to numerical methods. One should not expect the computer to do all the work, if analysis to any extent is not cumbersome. Reliable computer results are much more difficult to obtain and ascertain than the enthusiastic inexperienced user might think.

A rather simple method which is almost exclusively used for rotating shafts and also lends itself for application to microcomputers will be discussed in the following.

The *transfer matrix method*, was first introduced in an elementary form by Holzer, in the 1920s, for torsional vibration of rotating shafts. It makes use of the fact that in a large class of design problems, some structural member is designed along a line and the behavior at every point of the system is influenced by the behavior at neighboring points only. Typical examples are beams, shafts, piping systems, etc., Figure 4.2(a).

We shall start with the static lateral deflection of beams. We assume a beam, say simply supported, which consists of $n-1$ prismatic beam elements of different cross-section as in Figure 4.2(b).

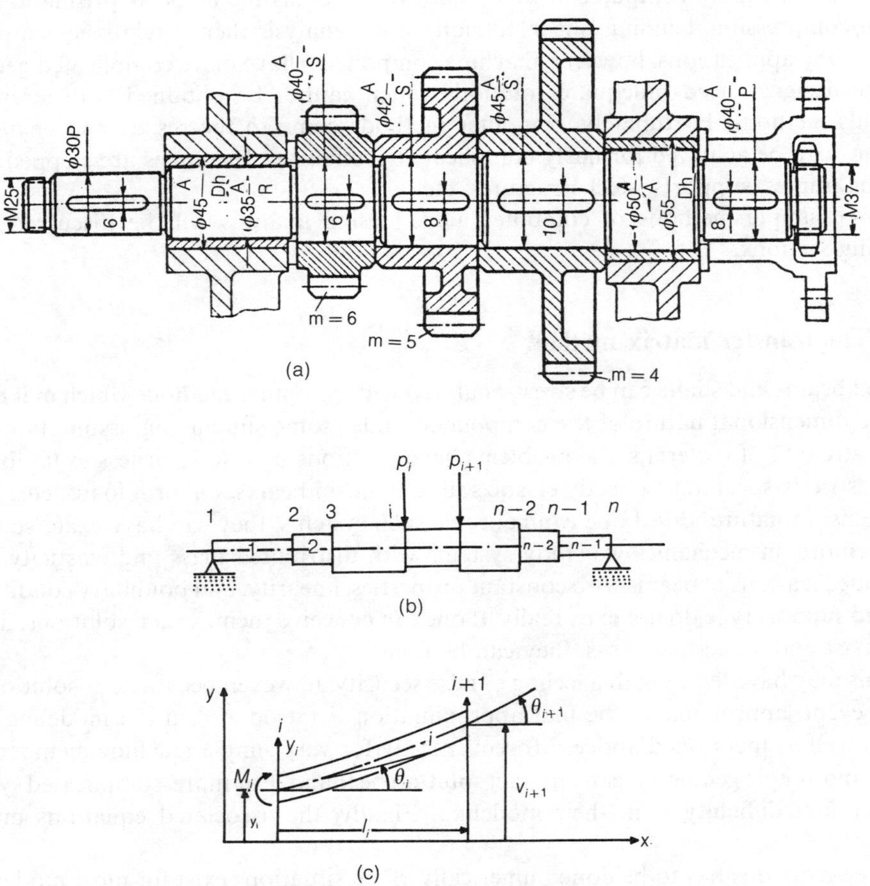

Figure 4.2 Modeling a stepped shaft (see text for explanation)

Thus the beam has $n-1$ elements with constant moments of inertia and n nodes, points (or planes) which define the beginning or the end of a uniform beam element.

To fully describe the situation at each node on the vertical plane, we need to know four quantities: The deflection y, the slope θ, the moment M and the shear force V. The element j of a beam between nodes j and $j+1$ is shown in Figure 4.2(c) with the sign conventions usual in statics.

These four quantities can be arranged in a vector form $z=\{y\theta MV\}$ which, because it describes the state of affairs at node j, is called 'state vector' and to designate the node j we shall use the state vector with a subscript 'j'.

Let us suppose that at the node 1 the state vector is

$$\{z_1\}=\{y_1\theta_1M_1V_1\} \tag{4.1}$$

but yet unknown. If no force is acting between nodes 1 and 2, the deflection, slope, moment and shear at node 2, from simple beam theory, will be

$$y_2=y_1+l_1\theta_1+\frac{l_1^2}{2EI_1}M_1+\frac{l_1^3}{6EI_1}V_1 \tag{4.2a}$$

$$\theta_2=\theta_1+\frac{l_1}{EI_1}M_1+\frac{l_1^2}{2EI_1}V_1 \tag{4.2b}$$

$$M_2=M_1+l_1V_1 \tag{4.2c}$$

$$V_2=V_1 \tag{4.2d}$$

$$1=1 \tag{4.2e}$$

Equations (4.2c) and (4.2d) express the equilibrium of the forces and moments on the beam and equations (4.2a) and (4.2b) give the deflections due to these moments and forces. Equation (4.2e) is an identity $1=1$ and it is added for computational convenience.

Equations (4.2) can be written in the following matrix form:

$$\{z_2\}=[L_1]\{z_1\} \tag{4.3}$$

where

$$[L_1]=\begin{bmatrix} 1 & l & l^2/2EI & l^3/6EI & 0\\ 0 & 1 & l/EI & l^2/2EI & 0\\ 0 & 0 & 1 & l & 0\\ 0 & 0 & 0 & 1 & 0\\ 0 & 0 & 0 & 0 & 1\end{bmatrix}_1 \tag{4.4}$$

$$\{z_1\}=\{y\theta MV\}_1$$
$$\{z_2\}=\{y\theta MV\}_2$$

The subscript 1 of the matrix indicates that the quantities l, E, I are properties of the element number 1.

Equation (4.3) tell us that the state vector at node 2 is the state vector at node 1 multiplied by a square 5×5 matrix L, which depends on the element properties only and it is well known. This matrix transferred the state from node 1 to node 2 and therefore it shall be called the 'transfer matrix'. For every element of the beam there exists one, known, transfer

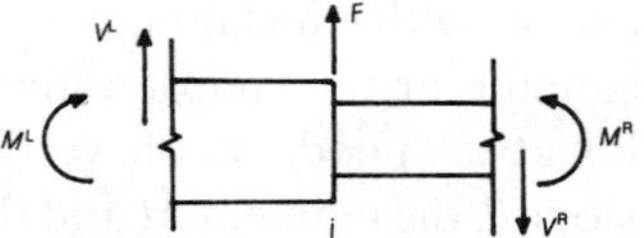

Figure 4.3 Node equilibrium

matrix $[L]$. We can repeat the procedure for elements 2, 3,... to obtain, also using the previous relations,

$$\left.\begin{aligned}\{z_2\} &= [L_1]\{z_1\}\\ \{z_3\} &= [L_2]\{z_2\} = [L_2][L_1]\{z_1\}\\ \{z_4\} &= [L_3]\{z_3\} = [L_3][L_2][L_1]\{z_1\}\end{aligned}\right\} \tag{4.5}$$

At the nodes, the state vector as we approach the node from left and right is the same. However, if at the node we have a static force F, this is not true. In Figure 4.3 we show the situation, that for a small length about the node, the deflection, slope and moment remain unchanged but, in order to maintain equilibrium, we must have $V^R = V^L + F$, where with superscript 'L' we designate the situation at the left of the node and 'R' refers to the situation at the right of the node. We can write

$$\left.\begin{aligned}y^R &= y^L\\ \theta^R &= \theta^L\\ M^R &= M^L\\ V^R &= V^L + F\\ 1 &= 1\end{aligned}\right\} \tag{4.6}$$

We can write this in matrix form as

$$\{z_j^R\} = [P_j]\{z_j^L\} \tag{4.7}$$

where

$$[P_j] = \begin{bmatrix} 1 & 0 & 0 & 0 & 0\\ 0 & 1 & 0 & 0 & 0\\ 0 & 0 & 1 & 0 & 0\\ 0 & 0 & 0 & 1 & F\\ 0 & 0 & 0 & 0 & 1 \end{bmatrix} \tag{4.8}$$

L matrices refer to a beam element and they are known as 'field matrices'. P matrices refer to a nodal point and they are called 'nodal' or 'point matrices'.

Finally, we can complete the sequence of equation (4.7) as follows:

$$\begin{aligned}\{z_2^R\} &= [P_2]\{z_2^L\} = \{P_2\}[L_1]\{P_1\}\{z_1^L\}\\ \{z_3^R\} &= [P_3]\{z_3^L\} = [P_3][L_2]\{z_2^R\} = \{P_3\}[L_2]\{P_2\}[L_1]\{z_1^R\}\end{aligned} \tag{4.9}$$

. . . . . .

$$\{z_n^R\} = [A]\{z_1^L\} \tag{4.10}$$

where the 5×5 matrix A is the product of all 5×5 matrices of the element and point matrices.

Equation (4.10) can be written as

$$\left.\begin{aligned} y_n &= a_{11}y_1 + a_{12}\theta_1 + a_{13}M_1 + a_{14}V_1 + a_{15} \\ \theta_n &= a_{21}y_1 + a_{22}\theta_1 + a_{23}M_1 + a_{24}V_1 + a_{25} \\ M_n &= a_{31}y_1 + a_{32}\theta_1 + a_{33}M_1 + a_{34}V_1 + a_{35} \\ V_n &= a_{41}y_1 + a_{42}\theta_1 + a_{43}M_1 + a_{44}V_1 + a_{45} \\ 1 &= 1 \end{aligned}\right\} \tag{4.11}$$

We have four equations with eight unknowns:

$$y_1,\ \theta_1,\ M_1,\ V_1;\ y_2,\ \theta_2,\ M_2,\ V_2$$

However, because of the boundary conditions we know four of these quantities. For example, for a simply-supported beam we shall have $y_1 = y_n = 0$ and $M_1 = M_n = 0$. Therefore, equations (4.11) have four unknowns: θ_1, θ_n, V_1, V_n.

After the computation of these unknowns, we can obtain the state vectors at the nodes, and thus the deflection of the beam, from equations (4.9), applied successively, from left to right.

What did we accomplish at this point? We have been able to compute the static deflection of the beam with only chain multiplications of 5 × 5 matrices and solution of a system of four algebraic equations. With the usual matrix inversion methods we would have to invert a large matrix and in addition to the much greater computation effort, much greater memory will be needed.

In general, for every element and node, appropriate matrices exist in the form

$$[L_j^{\mathrm{F}}] = \begin{bmatrix} & & & & .0 \\ & & & & . \\ & & & & .0 \\ & L_{4\times 4} & & & . \\ & & & & .0 \\ & & & & . \\ & & & & .0 \\ . & . & . & . & . \\ 0 & 0 & 0 & 0 & 1 \end{bmatrix}, \quad \{z\} = \begin{Bmatrix} y \\ \theta \\ M \\ V \\ 1 \end{Bmatrix}, \quad [P_j] = \begin{bmatrix} 1 & 0 & 0 & 0 & 0 \\ 0 & 1 & 0 & 0 & 0 \\ 0 & 0 & 1 & 0 & 0 \\ 0 & 0 & 0 & 1 & F \\ 0 & 0 & 0 & 0 & 1 \end{bmatrix} \tag{4.12}$$

where $[L_j^{\mathrm{F}}]$ is called element field matrix (for the element j), and $[P_j]$ point matrix for the node j.

In the product of equation (4.10), one can take into account any number of loads at the nodes by multiplying with all the respective point matrices.

It is as easy to account for linear springs at the nodes, such as bearings, supports, etc. Let such a spring react with a force proportional to deflection (linear spring) $-ky$, and a moment proportional to rotation (torsional spring) $-k\theta$, in addition to external forces, F_j and moments M_{0j} at the node. Force balance at this node yields:

$$M^{\mathrm{R}} = M^{\mathrm{L}} - k_{\mathrm{T}}\theta + M_{0j} \tag{4.13}$$

$$V^{\mathrm{R}} = V^{\mathrm{L}} - ky + F_j \tag{4.14}$$

Deflection and slope are the same before and after the spring. Therefore

$$\begin{Bmatrix} y \\ \theta \\ M \\ V \\ 1 \end{Bmatrix}^{R} = \begin{bmatrix} 1 & 0 & 0 & 0 & 0 \\ 0 & 1 & 0 & 0 & 0 \\ 0 & -k_T & 1 & 0 & M_0 \\ -k & 0 & 0 & 1 & F \\ 0 & 0 & 0 & 0 & 1 \end{bmatrix} \begin{Bmatrix} y \\ \theta \\ M \\ V \\ 1 \end{Bmatrix}^{L} \tag{4.15}$$

In a similar way, one can obtain the transfer matrix for a beam element with a distributed static load q per unit length (Figure 4.4).

In general, nodes are designated all points where some change takes place, such as changing sections, point forces, supports, springs, etc. For multisupported shafts and beams, it is easy to account for intermediate supports, treating them as linear springs with large spring constant k, two orders of magnitude, or more, higher than some representative spring constant of the shaft

$$k = 48EI/L^3 \tag{4.16}$$

at the nearest span using the maximum cross-section encountered within the span.

This procedure can easily be programed, as it was done in the following procedure, TMSTAT.

```
Procedure TMSTAT
GLOBAL NN, arrays PLOAD, SPRINGY, SPRINGT(1...NN)
    arrays QLOAD, EI, LENGTH(1...NN− 1)
    arrays FIELD, POINT, A(1...5, 1...5)
    array VECTOR(1...5): var
begin
input NN
FOR I:= 1 TO NN − 1 do
   input
   PLOAD(I), QLOAD(I), SPRINGY(I), SPRINGT(I), EI(I), LENGTH(I)
                                                    (*Right end data*)
input PLOAD(NN), SPRINGY(NN), SPRINGT(NN)
                                                    (*Forward sweep*)
set unit diagonal matrix A
FOR I:= 1 TO NN − 1
   MATPOINT(I)
   MATFIELD(I)
   MATMULT(POINT, A, ATEMP, 5, 5, 5)
   MATMULT(FIELD, ATEMP, A, 5, 5, 5)
                                                    (*End station*)
MATPOINT(NN)
MATMULT(POINT, A, ATEMP, 5, 5, 5)
                                        (*Apply boundary conditions*)
                                                    (*Free-free*)
```

```
DENOM:= A(3, 1)*A(4, 2) – A(3, 2)*A((4, 1)
V(1):= (A(3, 2)*A(4, 5) – A(3, 5)*A(4, 2))/DENOM
V(2):= (A(4, 1)*A(3, 5) – A(4, 5)*A(3, 1))/DENOM
V(3):= 0; V(4):= 0
                                              (*Computation of results*)
FOR I:= 1 TO NN – 1 do
   PRINT I, (V(J), J:= 1 TO 4)                (*Results at node I*)
   MATPOINT(I)
   MATFIELD(I)
   MATMULT(POINT, V, VTEMP, 5, 5, 1)
   MATMULT(FIELD, VTEMP, V, 5, 5, 1)
                                              (*End station*)
MATPOINT(NN)
MATMULT(POINT, TEMP, V, 5, 5, 1)
PRINT NN, (V(J), J:= 1 TO 4)                  (*Results at node NN*)
end.
```

Procedures MATPOINT(N), MATFIELD(N) compute the elements of the point and field matrices, respectively, at node N and section N (equation 4.15 and Figure 4.4). Procedure MATMULT(A, B, C, L, M, N) performs the matrix multiplication $C(L \times N) = A(L \times M) \times B(M \times N)$.

The beam has been assumed with free-free boundary conditions. Supports are designated in the input and treated as hard springs.

Members with continuously variable cross-section can be treated as in Figure 4.5 by taking enough nodes along the member and treating the in-between element as of constant section.

Certain caution must be exercised, however. Increasing the number of nodes beyond a point will cause the solution to be in large error because the computer cannot handle more than a certain number of digits for every number. In many cases, a quantity is computed as a difference of two large numbers. If the difference appears beyond the significant digits, it will be zero. Fortunately, this is usually apparent in the solution because it will be shown as a violation of the boundary conditions, erroneously large numbers, division by zero error message, etc.

In the program TMSTAT (Appendix 4.A), the above transfer matrix procedure was

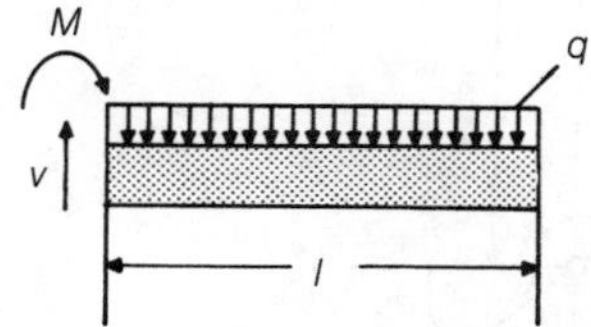

$$[L] = \begin{bmatrix} 1 & l & l^2/2EI & l^3/6EI & -ql^4/24EI \\ 0 & 1 & l/2EI & l^2/2EI & -ql^3/6EI \\ 0 & 0 & 1 & l & -ql^2/2 \\ 0 & 0 & 0 & 1 & -ql \\ 0 & 0 & 0 & 0 & 1 \end{bmatrix}$$

Figure 4.4 Matrix for element with distributed load

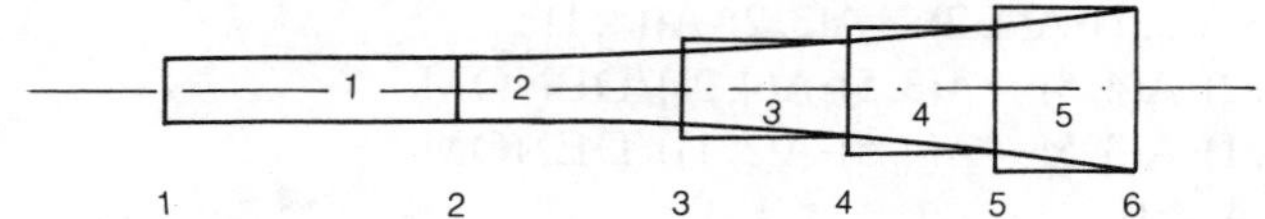

Figure 4.5 Modeling of shaft with variable diameter

coded. The user must observe the reactions calculated by the program. The accuracy of the computation can be easily tested by checking for vertical force balance.

The transfer matrix method is very convenient for design applications, especially on elements such as rotating shafts, and historically it has been one of the first applications in computer aided design of turbomachinery.

4.3.2 The finite-element method

The transfer matrix method has introduced the idea of an elastic member, the state of which can be approximated if the state at some points on its boundary is known. For a piece of a prismatic beam for example (Figure 4.6), if the conditions at the ends are known (deflection, slopes), the shape of the beam can be determined throughout this piece for static conditions.

This can be done with methods of strength of materials as it was done in Section 4.3.1. However, one can observe that the equations of the deflection of the beam are, by the simple beam theory,

$$\frac{d^2y}{dx^2} = \frac{1}{EI}M(x); \quad V(x) = \frac{dM}{dx}; \quad q(x) = \frac{d^2M}{dx^2} \tag{4.17}$$

If no load is assumed on the beam, $q(x) = 0$; differentiating twice the first equation and using the second, we obtain $d_4y/dx^4 = 0$. Its general solution is

$$y(x) = c_3x^3 + c_2x^2 + c_1x + c_0 \tag{4.18}$$

where c_0, c_1, c_2 and c_3 are constants to be determined from the boundary conditions $y_1, y_2, \theta_1, \theta_2$ $(= dy/dx)$. Therefore, if the end conditions are known, the deflections in between are functions only of these conditions and independent of the beam properties E and I. In fact, it can be shown very easily that

$$y(x) = f_1(x)y_1 + f_2(x)\theta_1 + f_3(x)y_2 + f_4(x)\theta_2 \tag{4.19}$$

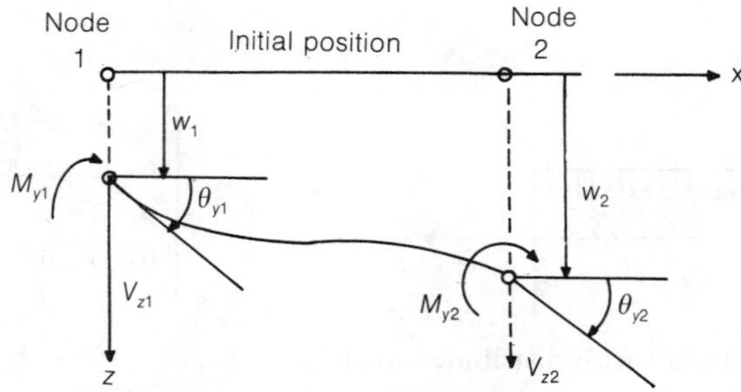

Figure 4.6 Deflected beam element

where, for $s = x/l$ and l the length of the element,

$$f_1(x) = 1 - 3s^2 + 2s^3; \quad f_3(x) = 3s^2 - 2s^3$$
$$f_2(x) = l(s - 2s^2 + s^3); \quad f_4(x) = l(-s^2 + s^3) \tag{4.20}$$

The slope $\theta(x) = \mathrm{d}y/\mathrm{d}x$ will be

$$\theta(x) = f'_1(x)y_1 + f'_2(x)\theta_1 + f'_3(x)y_2 + f'_4(x)\theta_2 \tag{4.21}$$

Equation (4.19) can also give the moments and shear forces along the beam, using equations (4.17). Therefore, the state at the distance x can be calculated once the end conditions are known. Therefore, instead of dealing with this beam segment, we can, in principle, deal only with the conditions at the two ends, already known as nodes. Every piece of a continuum which can be described by way of the conditions at a finite number of points along its boundary, will be called 'finite element'.

Another important property is that when using finite elements, the displacements at the nodes (deflections and slopes) are related to the generalized forces at these nodes (forces and moments). In the sequel we shall use the terms 'displacements' and 'forces' respectively. For a prismatic bar in flexure, we have already seen that the end conditions are related by way of a transfer matrix $[L]$. As we can see from equations (4.2), these can always be solved for $y_1, \theta_1, y_2, \theta_2$, in terms of M_1, V_1, M_2, V_2, and the reverse. The result is:

$$\{F\} = [K]\{Y\}, \quad \{Y\} = [K^{-1}]\{F\}$$

where (4.22)

$$\{F\} = \{V_1 M_1 V_2 M_2\}, \quad \{Y\} = \{y_1 \theta_1 y_2 \theta_2\}$$

$$[K] = EI\begin{bmatrix} 12/l^3 & & \text{Symmetric} & \\ 6/l^2 & 4/l & & \\ -12/l^3 & -6/l^2 & 12/l^3 & \\ 6/l^2 & 2/l & -6/l^2 & 4/l \end{bmatrix} \tag{4.23}$$

The matrix $[K]$ is the stiffness matrix and its inverse $[A]$ is the flexibility matrix.

The state of displacements at each node can have up to six components: three deflections and three slopes. Consequently, the forces at each node can have six components: three forces and three moments. Therefore, the stiffness and flexibility matrices can be of dimension 12×12, at the most, for two nodes. In general, for m nodes per element, the maximum dimension of these matrices will be $6m \times 6m$. In practical situations, we do not always use all the possible coordinates.

For example, if only lateral motion of a beam on a plane is considered with the state vector $\{Y\}$ and force vector $\{F\}$ above, the force equilibrium is

$$\{F\} = [K]\{Y\} \tag{4.24}$$

In general, this procedure can yield the desired solution for a continuous system. Using the fact that nodes connect elements we can write the force equilibrium equations for every element and use them as a system.

First however, mention must be made here about the nodal force F. This force is partly due to the external load, and partly due to the reaction on the node from the adjacent element. If we add the respective equations, only the sum of these nodal forces will appear, which is the known external force, because the internal forces at one node which belong to the two elements cancel each other upon addition. Therefore, in the equations of equilibrium, we have the external forces V and M. In a general system, this is called 'assembly of the elements' and it is shown in equation 4.25 where we can observe that element stiffness matrices overlap each other in places where each element has nodes common with the others. This means addition of the respective equations.

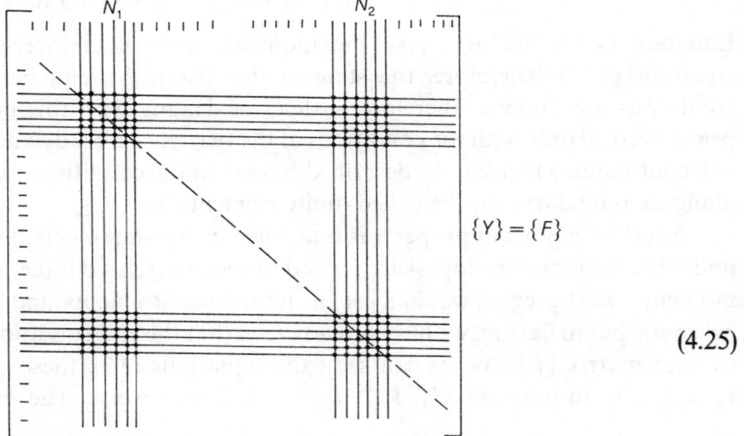

(4.25)

As pointed out above, element stiffness matrices can be computed from the transfer matrices, for simple beam elements. As an example, for the two-node beam with two degrees of freedom y and θ per node, the transfer matrix has the form

$$\begin{bmatrix} y \\ \theta \\ M \\ V \end{bmatrix}_{i+1} = \begin{bmatrix} L_{11} & L_{12} & L_{13} & L_{14} \\ L_{21} & L_{22} & L_{23} & L_{24} \\ L_{31} & L_{32} & L_{33} & L_{34} \\ L_{41} & L_{42} & L_{43} & L_{44} \end{bmatrix} \begin{bmatrix} y \\ \theta \\ M \\ V \end{bmatrix}_j \tag{4.26}$$

If we let $\{s_1\} = \{x\theta\}$ we can write

$$\begin{bmatrix} s_1 \\ s_2 \end{bmatrix}_{i+1} = \begin{bmatrix} L_1 & L_2 \\ L_3 & L_4 \end{bmatrix} \begin{bmatrix} s_1 \\ s_2 \end{bmatrix}_i \tag{4.27}$$

The matrix L has been partitioned to four 2×2 matrices L_1, L_2, L_3, L_4. If we solve for $(s_1)_i$ and $(s_1)_{i+1}$ in term of $(s_2)_i$ and $(s_2)_{i+1}$ we obtain

$$\begin{bmatrix} (s_1)_i \\ (s_1)_{i+1} \end{bmatrix} = \begin{bmatrix} F_1 & F_2 \\ F_3 & F_4 \end{bmatrix} \begin{bmatrix} (s_2)_i \\ (s_2)_{i+1} \end{bmatrix} \tag{4.28}$$

We observe that the left-hand side of the equation has the vector

$$\{Y\}_i = [x_i \theta_i x_{i+1} \theta_{i+1}]^{\mathrm{T}}$$

and the vector on the right-hand side is (note a difference with equation 4.22)

$$\{F\}_i = [M_i V_i M_{i+1} V_{i+1}]^{\mathrm{T}}$$

Therefore the matrix F is the flexibility matrix for the element.

In this way, given one of the matrices of the element – transfer, stiffness or flexibility matrices – we can compute the other two. The relations between these matrices are given below:

Flexibility matrix	*Transfer matrix*
$\begin{pmatrix} F_1 & F_2 \\ F_3 & F_4 \end{pmatrix}$	$\begin{pmatrix} F_4F_2^{-1} & F_3 - F_4F_2^{-1}F_1 \\ F_2^{-1} & -F_2^{-1} \end{pmatrix}$
Stiffness matrix	*Transfer matrix*
$\begin{pmatrix} k_1 & k_2 \\ k_3 & k_4 \end{pmatrix}$	$\begin{pmatrix} -k_2^{-1}k_1 & k_2^{-1} \\ k_3 - k_4k_2^{-1}k_1 & k_4k_2^{-1} \end{pmatrix}$
Transfer matrix	*Flexibility matrix* (4.29)
$\begin{pmatrix} L_1 & L_2 \\ L_3 & L_4 \end{pmatrix}$	$\begin{pmatrix} -L_1^{-1}L_2 & L_1^{-1} \\ L_4 - L_3L_1^{-1}L_2 & L_3L_1^{-1} \end{pmatrix}$
Transfer matrix	*Stiffness matrix*
$\begin{pmatrix} L_1 & L_2 \\ L_3 & L_4 \end{pmatrix}$	$\begin{pmatrix} -L_4^{-1}L_3 & L_4^{-1} \\ L_1 - L_2L_4^{-1}L_3 & L_2L_4^{-1} \end{pmatrix}$

4.3.3 The general prismatic element (Weaver 1967)

A beam element might have a more complex loading than the bending and shearing on one plane, as discussed in Section 4.3.2. Due to the assumed linearity, a stress analysis can be performed independently on two planes, vertical x–y and horizontal x–z (Figure 4.6), analyzing the external loads to their components on these two planes. Moreover, axial load and torsion can also be considered separately.

Structures and machine components often consist of interconnected prismatic elements, each under general space loading. Then, many components of stress and deformation are present and they may be related to one another. In this case, for the two-node prismatic element already discussed, 6 degrees of freedom per node are needed, with a total of 12 degrees of freedom per element, giving the displacement vector for the beam element, as

$$\{\delta^e\} = \{u_{x1}u_{y1}u_{z1}\theta_{x1}\theta_{y1}\theta_{z1}u_{x2}u_{y2}u_{z2}\theta_{x2}\theta_{y2}\theta_{z2}\}$$

with the force vector

$$\{F^e\} = \{F_{x1}F_{y1}F_{z1}M_{x1}M_{y1}M_{z1}F_{x2}F_{y2}F_{z2}M_{x2}M_{y2}M_{z2}\}$$

The forces and moments in the vertical plane x–y are related to the forces and moments on the same plane by way of equation (4.24). The same equation relates loads and displacements on the horizontal plane x–z.

An axial load F is related to the axial nodal displacements with

$$F_{x2} = (AE/L)u_{x2} - (AE/L)u_{x1} \tag{4.30}$$
$$F_{x1} = (AE/L)u_{x1} - (AE/L)u_{x2}$$

Torsion M is related to the twist at the nodes as

$$\left.\begin{aligned} M_{x2} &= (I_{\mathrm{p}}G/L)\theta_{x2} - (I_{\mathrm{p}}G/L)\theta_{x1} \\ M_{x1} &= (I_{\mathrm{p}}G/L)\theta_{x1} - (I_{\mathrm{p}}G/L)\theta_{x2} \end{aligned}\right\} \tag{4.31}$$

where A is the beam cross-section and I_{p} the polar area moment of inertia.

The force vector $\{\mathbf{F}^e\}$ is related to the displacement vector by way of equations (4.24), (4.30) and (4.31). This relation can be written in matrix form as

$$\{F^e\} = [K^e]\{\delta^e\} \tag{4.32a}$$

For such an element the coordinate and node-numbering system is indicated in Figure 4.6 with the y-axis normal to the plane of the paper. The stiffness matrix in equation (4.32b), obtained from equations (4.24), (4.30) and (4.31) by inspection, is

$$[K] = \left[\begin{array}{cccccc:cccccc}
\frac{EA_X}{L} & 0 & 0 & 0 & 0 & 0 & -\frac{EA_X}{L} & 0 & 0 & 0 & 0 & 0 \\
0 & \frac{12EI_Z}{L^3} & 0 & 0 & 0 & \frac{6EI_Z}{L^2} & 0 & -\frac{12EI_Z}{L^3} & 0 & 0 & 0 & \frac{6EI_Z}{L^2} \\
0 & 0 & \frac{12EI_Y}{L^3} & 0 & -\frac{6EI_Y}{L^2} & 0 & 0 & 0 & -\frac{12EI_Y}{L^3} & 0 & -\frac{6EI_Y}{L^2} & 0 \\
0 & 0 & 0 & \frac{GI_X}{L} & 0 & 0 & 0 & 0 & 0 & -\frac{GI_X}{L} & 0 & 0 \\
0 & 0 & -\frac{6EI_Y}{L^2} & 0 & \frac{4EI_Y}{L} & 0 & 0 & 0 & \frac{6EI_Y}{L^2} & 0 & \frac{2EI_Y}{L} & 0 \\
0 & \frac{6EI_Z}{L^2} & 0 & 0 & 0 & \frac{4EI_Z}{L} & 0 & -\frac{6EI_Z}{L^2} & 0 & 0 & 0 & \frac{2EI_Z}{L} \\
\hdashline
-\frac{EA_X}{L} & 0 & 0 & 0 & 0 & 0 & \frac{EA_X}{L} & 0 & 0 & 0 & 0 & 0 \\
0 & -\frac{12EI_Z}{L^3} & 0 & 0 & 0 & -\frac{6EI_Z}{L^2} & 0 & \frac{12EI_Z}{L^3} & 0 & 0 & 0 & -\frac{6EI_Z}{L^2} \\
0 & 0 & -\frac{12EI_Y}{L^3} & 0 & \frac{6EI_Y}{L^2} & 0 & 0 & 0 & \frac{12EI_Y}{L^3} & 0 & \frac{6EI_Y}{L^2} & 0 \\
0 & 0 & 0 & -\frac{GI_X}{L} & 0 & 0 & 0 & 0 & 0 & \frac{GI_X}{L} & 0 & 0 \\
0 & 0 & -\frac{6EI_Y}{L^2} & 0 & \frac{2EI_Y}{L} & 0 & 0 & 0 & \frac{6EI_Y}{L^2} & 0 & \frac{4EI_Y}{L} & 0 \\
0 & \frac{6EI_Z}{L^2} & 0 & 0 & 0 & \frac{2EI_Z}{L} & 0 & -\frac{6EI_Z}{L^2} & 0 & 0 & 0 & \frac{4EI_Z}{L}
\end{array}\right] \tag{4.32b}$$

This is not enough, however. The element was assumed on the x-axis, which is not always the case. In general, it is convenient to compute the stiffness matrix in a local coordinate system coinciding with the natural axes of symmetry of the beam element and then modify it for a general orientation of the element.

The vectors $\{\mathbf{F}^e\}$ and $\{\delta^e\}$ follow the rules of rotation discussed in Chapter 3. They have to be multiplied by a rotation matrix which accounts for:

(a) The three direction cosines C_x, C_y, C_z of the axis of the beam in respect to the global coordinate system x, y, z.
(b) The roll angle, i.e. the angle α of rotation of the beam section about the x-axis of the local coordinate system.

For each of the vectors $u_1, u_2, \theta_1, \theta_2, F_1, F_2, M_1, M_2$, the transformation matrix is

$$[R] = \begin{bmatrix} C_X & C_Y & C_Z \\ \dfrac{-C_X C_Y \cos\alpha - C_Z \sin\alpha}{(C_X^2 + C_Z^2)^{\frac{1}{2}}} & (C_X^2 + C_Z^2)^{\frac{1}{2}} \cos\alpha & \dfrac{-C_Y C_Z \cos\alpha + C_X \sin\alpha}{(C_X^2 + C_Z^2)^{\frac{1}{2}}} \\ \dfrac{C_X C_Y \sin\alpha - C_Z \cos\alpha}{(C_X^2 + C_Z^2)^{\frac{1}{2}}} & -(C_X^2 + C_Z^2)^{\frac{1}{2}} \sin\alpha & \dfrac{C_Y C_Z \sin\alpha + C_X \cos\alpha}{(C_X^2 + C_Z^2)^{\frac{1}{2}}} \end{bmatrix} \quad (4.32c)$$

Transformation of any one of the above vectors, elements of the total load or displacement vector, involves multiplication by the transformation matrix $[R]$. Since the load vector $\{F^e\}$ and the displacement vector $\{\delta^e\}$ consist of four three-component vectors each, the transformation of the 12-component vectors $\{F^e\}$ and $\{\delta^e\}$ from the position of the element along the x-axis (local) to the general space location (global), will be

$$\{F_G^e\} = [R_G]\{F_L^e\} \quad (4.33)$$

$$\{\delta_G^e\} = [R_G]\{\delta_L^e\} \quad (4.34)$$

where

$$[R_G] = \begin{bmatrix} [R] & 0 & 0 & 0 \\ 0 & [R] & 0 & 0 \\ 0 & 0 & [R] & 0 \\ 0 & 0 & 0 & [R] \end{bmatrix} \quad (4.35)$$

Equations (4.32) then relate loads and displacements in the local coordinate system. The vectors are referred to the global coordinate system upon multiplication by the matrix $[R]$. Then,

$$\{F^e\}[R_G] = [K^e]\{\delta^e\}[R_G] \quad (4.36)$$

Therefore the matrix relating loads and displacements in the global coordinate system is $\text{inv}[R_G][K^e][R_G]$. Inversion of matrix $[R_G]$ is not necessary because due to the reciprocity of rotation, the inverse of $[R_G]$ equals its transpose: $[K_G^e] = [R_G][K_L^e][R_G]$.

The stiffness matrices of the elements have to be combined to yield the system stiffness matrix relating the system nodal forces to the system nodal displacements (equation 4.25).

The way to assemble the stiffness matrix of the system from the element stiffness matrices has already been discussed and will be further exemplified later in this chapter.

4.3.4 Boundary conditions

The most usual boundary conditions are restraints to some of the degrees of freedom at certain nodes. For example, a completely fixed node i means that the six displacements associated with this node $\delta_{6(i-1)+j}$, $j = 1, 2, \ldots, 6$ are zero. There are many ways to take this into account in the equation:

$$[K]\{\delta\} = \{F\} \tag{4.37}$$

The best way, of course, is to eliminate all the constrained coordinates from the system and reduce accordingly the number of equations. It is much simpler however, to do a trick: The j component of the six coordinates of node i has coordinate number in the global system $p = (i-1)6 + j$. Then the p-row of matrix $[K]$ is eliminated and the diagonal term k_{pp} is set to 1. The corresponding component of the force vector is set to zero, if the coordinate is completely restrained (fixed), or to a specific value if an initial displacement is forced at this coordinate. Therefore, the boundary condition is always satisfied. A better utilization of computer time and memory would be to remove the superfluous coordinate, reducing accordingly the number of degrees of freedom. For simplicity, this procedure can be omitted.

4.3.5 System assembly and solution

To assemble the element stiffness matrices into the system stiffness matrix K, the procedure indicated in equation (4.25) must be followed. To this end we note that the element I has been defined by its node numbers $N1(I)$, $N2(I)$ and the coordinates $X(J)$, $Y(J)$, $Z(J)$ of the node $J = N1$ or $N2$. To each node, a number of degrees of freedom are assigned, in this case 6. Then, the element stiffness matrix will have dimension 12×12 in this case.

Six coordinates correspond to node j, $6(j-1)+k$, where $k = 1, \ldots, 6$. Therefore, to nodes N1, N2 correspond coordinates $6(\text{N1}-1)+k$ and $6(\text{N2}-1)+k$, respectively, $k = 1, \ldots, 6$. These coordinates are assigned numbers $I1$ to $I6$. Finally, each element of the 12×12 stiffness matrix K is added to the corresponding coordinates of the global stiffness matrix. The process is shown in equation (4.25). The 12×12 elements of the element stiffness matrix are denoted with black dots.

The process is repeated with each one of the elements filling up the stiffness matrix. This matrix thus obtained has some noticeable properties:

(a) It is symmetric (and of course real) due to the well-known reciprocity property of elastic systems.
(b) In most cases it is banded. All elements contribute to the main diagonal but also to a narrow band about the diagonal. This is not true to all problems however.

These two properties make the solution considerably easier.

The solution of the system $[K]\{\delta\} = \{F\}$ of linear equations is usually performed with two broad categories of methods: 'exact' and iterative. Here an 'exact' method, the Gauss elimination, will be used.

Take for example, the 3×3 system of linear equations

$$2x + 4y + 6z = 28 \tag{i}$$

$$3x + y + z = 8 \tag{ii}$$

$$x + 3y + 4z = 19 \tag{iii}$$

Multiply (i) by $-3/2$ and add to (ii), then multiply (i) by $-1/2$ and add to (iii). Then

$$\begin{aligned} 2x+4y+6z &= 28 \\ -5y-8z &= -34 \\ y+z &= 5 \end{aligned} \qquad \begin{bmatrix} 2 & 4 & 6 \\ 0 & -5 & -8 \\ 0 & 1 & 1 \end{bmatrix} \begin{Bmatrix} x \\ y \\ z \end{Bmatrix} = \begin{Bmatrix} 28 \\ -34 \\ 5 \end{Bmatrix} \quad \begin{matrix} \text{(i)} \\ \text{(ii)} \\ \text{(iii)} \end{matrix}$$

Multiply (ii) by 1/5 and add to (iii). The system becomes:

$$\begin{aligned} 2x+4y+6z &= 28 \\ -5y-8z &= -34 \\ -3/5z &= -9/5 \end{aligned} \qquad \begin{bmatrix} 2 & 4 & 6 \\ 0 & -5 & -8 \\ 0 & 0 & -3/5 \end{bmatrix} \begin{Bmatrix} x \\ y \\ z \end{Bmatrix} = \begin{Bmatrix} 28 \\ -34 \\ -9/5 \end{Bmatrix} \quad \begin{matrix} \text{(i)} \\ \text{(ii)} \\ \text{(iii)} \end{matrix}$$

This process is called 'elimination'. The solution follows immediately: From (iii) we find $z=3$; substituting in (ii) we find $y=2$ and substituting in (i) we find $x=1$ and the system is solved. The latter process is called 'back substitution'.

We shall not address ourselves at this point to the question of efficiency because in smaller computers the main problem is usually memory and time for the element computation rather than solution time. This is not always the case.

Solution of the linear system will yield the nodal displacements. For design calculations, the element stresses must be computed. To this end, the nodal forces for each element corresponding to the nodal displacements must be computed. Therefore, the displacement vector has to be multiplied with the element stiffness matrix (equation 4.32b). The resulting nodal forces are in the global coordinate system. Usually, these forces are needed in the local system so that strength of materials formulas can be applied for the determination of stresses. For this reason, the nodal displacements are first transformed into the local system and then the nodal forces in the local system are obtained multiplying by the local stiffness matrix:

$$\{\delta^e\}_{\text{local}} = [R^{\text{T}}]\{\delta^e\}_{\text{global}}$$

Therefore,

$$\{F^e\}_{\text{local}} = [K^e]_{\text{local}}[R^{\text{T}}]\{\delta^e\}_{\text{global}} = [H^e]\{\delta^e\}_{\text{global}}$$

The stress matrix $[H^e]$ is computed in the same procedure with the element stiffness matrix.

The FINFRAME procedure for the above algorithm follows:

```
Procedure FINFRAME
GLOBAL  var:int NNODes, NELements, arrays MBeginning, MEnd(NEL)
             NRestraints(NNOD, 6)
        real arrays node forces GF, coord NC(NNOD, 3)
          displ D(6*NNOD), ext forces DF(6*NNOD), element forces FE(12)
          element properties ARea, YModulus, SModulus, RAngle(NEL)
          stiffness arrays element KE(12, 12), global K(6*NNOD, 6*NNOD)
          stress H(12, 12)
begin
   input NNOD, NEL
   FOR I:= 1 TO NNOD do
      input node data:GF, NC, NR
```

```
FOR I:= 1 TO NEL do
  input element data: MB, ME, AR, YM, SM, RA
                                                  (*Stiffness matrix assembly*)
FOR I = 1 TO NEL do
  I1:= MB(I); I2 = ME(I)
  X1:= NC(I1, 1); X2:= NC(I2, 1); Y1:= NC(I1, 2); Y2:= NC(I2, 2)
  X3:= NC(I1, 3); Y3:= NC(I2, 3)
  BEAMELEMENT(I, 1)                               (*Element stiffness matrix*)
  FOR K1:= 1 TO 6 do
    FOR K2:= 1 TO 6 do
      IG1:= 6*(I1 - 1) + K1
      IG2:= 6*(I2 - 1) + K1
      JG1:= 6*(I1 - 1) + K2
      JG2:= 6*(I2 - 1) + K2
      K(IG1, JG1):= K(IG1, JK1) + KE(K1, K1)
      K(IG1, JG2):= K(IG1, JG2) + KE(K1, K2)
      K(IG2, JG1):= K(IG2, JG1) + KE(K2, K1)
      K(IG2, JG2):= K(IG2, JG2) + KE(K2, K2)
                                                  (*Assembly completed*)
                                                  (*Apply boundary conditions*)
FOR I:= 1 TO NNOD do
FOR J:= 1 TO 6 do
  IF NR(I, J):= 1 THEN do
    K1:= 6*(I - 1) + J
    GF(K1):= 0
    FOR K2:= 1 TO ^*NNOD do                       (*Eliminate column*)
      K(K1, K2):= 0
    K2:= 6*(I - 1) + J
    FOR K1:= 1 TO 6*NNOD do                       (*Eliminate row*)
      K(K1, K2):= 0
    K(K2, K2):= 1
    else
                                                  (*Boundary conditions applied*)
                                                  (*Solution of linear equations*)
NEQ:= 6*NNOD
GAUSSELIM(K, GF, D, NEQ)
                                                  (*Determination of stresses*)
INDEX%:= 2
FOR I:= 1 TO NEL do
  BEAMELEMENT(I, 2)
  FOR J:= 1 TO 6
    I1:= MB(I); I2:= ME(I)
    DE(J):= D(6*(I1 - 1) + J)
    DE(6 + J):= D(6*(I2 - 1) + J)
  MATMULT(H, DE, FE, 12, 12, 1)
```

```
    print element displacements, forces
end.

Procedure BEAMELEMENT(N, INDEX%)
GLOBAL var:int NNODES, NELements, arrays MBeginning, MEnd(NEL)
          real arrays node forces GF, coord NC(NNOD, 3)
          restraints NR(NNOD, 6)
          displ D, ext forces DF(6*NNOD), element forces FB, FE(NEL)
          element properties ARea, YModulus, SModulus, RAngle(NEL)
          stiffness matrices:element local LK, globalKE(12, 12)
          system stiffness GK(6*NOD, 6*NNOD)
begin
  I1:= MB(I); I2:= ME(I)
  compute X1, X2, Y1, Y2, Z1, Z2
  compute local stiffness matrix LK, eq. 4.32a
  compute rotation matrix R, eq. 4.35
  compute transpose of R, RT
  IF INDEX% = 1 THEN do
    compute [K] = [RT][LK][R]
    exit
  else
    compute [H] = [LK][RT]
end.
```

4.4 SURFACE MEMBERS

Structural members which at every point have dimensions along one direction much smaller than dimensions in the other directions, are called 'surface structures'. Usual cases are plates and shells (Figure 4.7). An additional property which many times exists in machines is symmetry about an axis. Axisymmetric plates and shells are very common, as rotating disks, pressure vessels, toroidal shells, etc.

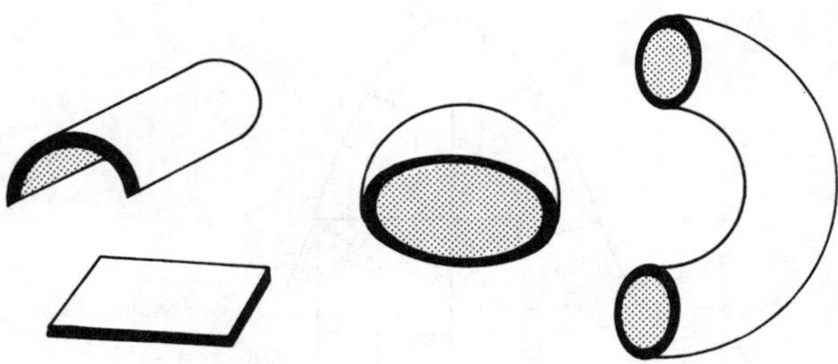

Figure 4.7 Surface structures

4.4.1 Membrane stresses

Shells are capable of sustaining very high loads mostly in the form of internal pressures. Therefore, they are often very thin and their resistance to bending is very small. The loads are sustained by tensile stresses in the shell material. Such stresses are called 'membrane stresses'. Considering a differential element $dS_1\,dS_2$ on the surface (Figure 4.8), and assuming an internal pressure p, summing up all forces along the direction perpendicular to the surface, the Laplace equation is obtained:

$$\frac{\sigma_t}{\rho_t} + \frac{\sigma_m}{\rho_m} = \frac{p}{h} \tag{4.38}$$

where σ_t and σ_m are circumferential and meridional stresses respectively, and ρ_t and ρ_m the corresponding radii of curvature, and h the thickness.

The Laplace equation is not enough to find the unknown stresses because there are two. In axisymmetric shells however, we can obtain a second equation in most cases by force equilibrium.

From Figure 4.9, summing up forces in the z-direction yields:

$$\sigma_m 2\pi rh \cos\theta = \pi r^2 p \tag{4.39}$$

and σ_t can be computed from equation (4.38).

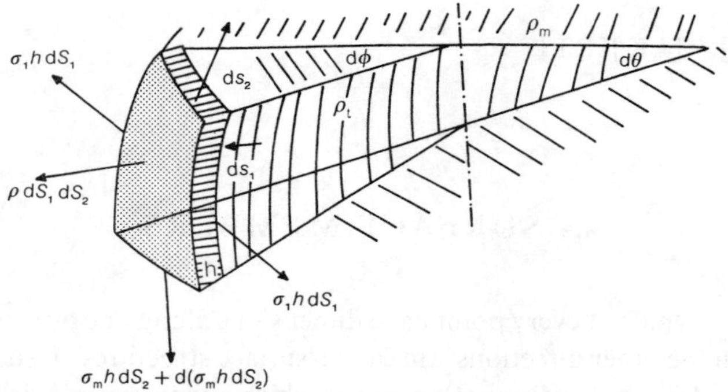

Figure 4.8 Equilibrium at a membrane element

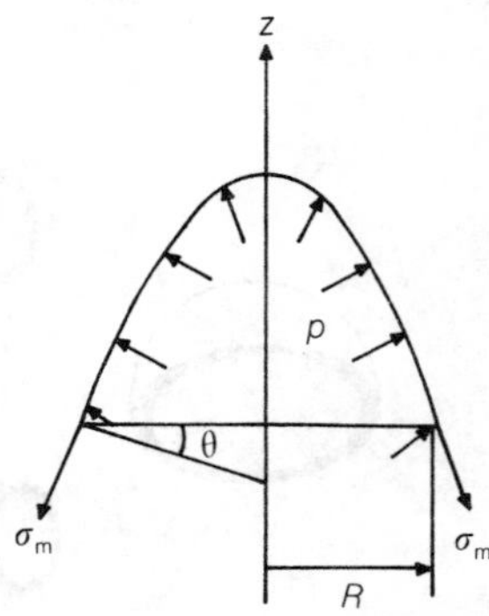

Figure 4.9 Membrane equilibrium along the axis of symmetry

4.4.2 Plane stress–strain problems (Rockey 1975)

Many machine components have plane form, like for example, the component shown in Figure 4.1. Analysis of complex plane geometric components with in-plane loading can be easily performed with the finite-element method.

As in the case of the frame structure which was broken down to prismatic beam elements, for which a stiffness matrix relating nodal forces to nodal displacement could be found, the plane structure is divided into a number of triangles, such as the one in Figure 4.10.

The vertices 1, 2, 3 will be used as nodes and the displacements u, v along the coordinates x and y, respectively will be used to describe the displacement within the triangle. This displacement will be a linear interpolation of the nodal displacements, which means that displacements will be linear functions of x, y.

Strains are derivatives of the displacements and therefore will be constant throughout the element. Consequently, the stresses within the element will be constant if linear elasticity is assumed.

This implies that the triangular element is small enough such that the stresses in the vicinity of the element are nearly constant. In general, these stresses will be normal stresses

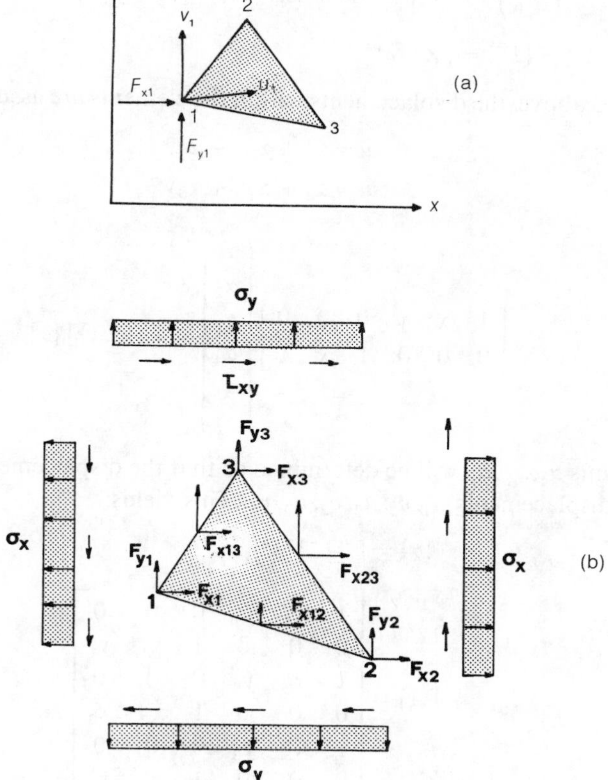

Figure 4.10 Free-body diagram of plane triangular element

σ_x, σ_y and shear stresses $\tau_{xy} = \tau_{yx}$. These stresses give forces on the edges of the triangle which will be assumed on the middle of each triangle side, as shown in Figure 4.10. These forces will be equally divided between the adjacent nodes.

Take node 1 for example. The nodal forces will be the sum of the contribution of stresses on the edges 1–2 and 1–3 for element thickness t:

$$\begin{aligned}
F_{x13} &= -\sigma_x(y_3 - y_1)t + \tau_{xy}(x_3 - x_1)t \\
F_{x12} &= -\sigma_x(y_1 - y_2)t - \tau_{xy}(x_2 - x_1)t \\
F_{x1} &= (F_{x13} + F_{x12})/2 = -\sigma_x(y_3 - y_2)t/2 - \tau_{xy}(x_2 - x_3)t/2 \\
F_{y1} &= (F_{y13} + F_{y12})/2 = -\sigma_x(x_2 - x_3)t/2 - \tau_{xy}(y_3 - y_2)t/2
\end{aligned}$$

In a similar way, the rest of the nodal forces can be related to the stresses. Writing these relations in matrix form, we obtain:

$$\begin{Bmatrix} F_1 \\ F_2 \\ F_3 \\ F_4 \\ F_5 \\ F_6 \end{Bmatrix} = -\frac{t}{2} \begin{bmatrix} y_3 - y_2 & 0 & x_2 - x_3 \\ 0 & x_2 - x_3 & y_3 - y_2 \\ y_1 - y_3 & 0 & x_3 - x_1 \\ 0 & x_3 - x_1 & y_1 - y_3 \\ y_2 - y_1 & 0 & x_1 - x_2 \\ 0 & x_1 - x_2 & y_2 - y_1 \end{bmatrix} \begin{Bmatrix} \sigma_x \\ \sigma_y \\ \tau_{xy} \end{Bmatrix} \tag{4.40}$$

or

$$\{F^e\} = [Z]\{\sigma^e\}$$

As proposed above, the displacements within the elements are assumed to have the form

$$\left.\begin{aligned}
u &= \alpha_1 + \alpha_2 x + \alpha_3 y \\
u &= \alpha_4 + \alpha_5 x + \alpha_6 y \\
&\begin{bmatrix} 1 & x & y & 0 & 0 & 0 \\ 0 & 0 & 0 & 1 & x & y \end{bmatrix} \begin{Bmatrix} \alpha_1 \\ \alpha_2 \\ \alpha_3 \\ \alpha_4 \\ \alpha_5 \\ \alpha_6 \end{Bmatrix} = [f(x, y)]\{\alpha\}
\end{aligned}\right\} \tag{4.41}$$

The constants $\alpha_1, \ldots, \alpha_6$ will be determined so that the displacements at the nodes 1, 2, 3 are the nodal displacements $u_1, v_1, u_2, v_2, u_3, v_3$. This yields:

$$\{\alpha\} = [A]^{-1}\{\delta^e\} \tag{4.42}$$

where

$$[A] = \begin{bmatrix} 1 & x_1 & y_1 & 0 & 0 & 0 \\ 0 & 0 & 0 & 1 & x_1 & y_1 \\ 1 & x_2 & y_2 & 0 & 0 & 0 \\ 0 & 0 & 0 & 1 & x_2 & x_2 \\ 1 & x_3 & y_3 & 0 & 0 & 0 \\ 0 & 0 & 0 & 1 & x_3 & y_3 \end{bmatrix}$$

The strains are now obtained from the displacements:

$$\left.\begin{aligned}\varepsilon_x &= \partial u/\partial x = \alpha_2\\ \varepsilon_y &= \partial u/\partial y = \alpha_6\\ \gamma_{xy} &= \partial u/\partial x + \partial u/\partial y\\ &= \alpha_3 + \alpha_5\end{aligned}\right\} \tag{4.43}$$

and in matrix form,

$$\begin{aligned}\{\varepsilon(x, y)\} &= \begin{Bmatrix} e_x \\ \varepsilon_y \\ \gamma_{xy}\end{Bmatrix}\\ &= \begin{bmatrix} 0 & 1 & 0 & 0 & 0 & 0\\ 0 & 0 & 0 & 0 & 0 & 1\\ 0 & 0 & 1 & 0 & 1 & 0\end{bmatrix}\{\alpha\}\\ &= [C]\{\alpha\}\\ &= [C][A^{-1}]\{\delta^e\}\end{aligned} \tag{4.44}$$

The stresses are now obtained from the strains by the generalized Hook's law:

$$\begin{aligned}\sigma_x &= [E/(1-\nu^2)][\varepsilon_x + \nu\varepsilon_y]\\ \sigma_y &= [E/(1-\nu^2)][\nu\varepsilon_x + \varepsilon_y]\\ \tau_{xy} &= [E/(1-\nu^2)](1-\nu)\gamma_{xy}/2\end{aligned} \tag{4.45}$$

In matrix form,

$$\begin{aligned}\{\sigma(x, y)\} &= \begin{Bmatrix} \sigma_x \\ \sigma_y \\ \tau_{xy}\end{Bmatrix}\\ &= \frac{E}{1-\nu^2}\begin{bmatrix} 1 & \nu & 0\\ \nu & 1 & 1\\ 0 & 0 & (1-\nu)/2\end{bmatrix}\{\varepsilon(x, y)\}\\ &= [D]\{\varepsilon(x, y)\}\end{aligned} \tag{4.46}$$

If the stresses are substituted into the equation (4.40), a matrix relation is obtained between nodal forces and nodal displacements, which yields directly the stiffness matrix of the element.

Therefore, the stiffness matrix is:

$$\begin{aligned}[K^e] &= [Z][D][C][A]^{-1}\\ &= [Z][D][B]\end{aligned} \tag{4.47}$$

It can be shown that $[Z] = [B]^{\mathrm{T}}$, therefore,

$$\{K^e\} = [B]^{\mathrm{T}}[D][B] \tag{4.48}$$

The stiffness matrix can be written in explicit form, Weaver (1967):

$$
[K^e] = \frac{t}{4\Delta}
\begin{bmatrix}
\begin{array}{l} d_{11}(y_2-y_3)^2 \\ \quad + d_{33}(x_3-x_2)^2 \end{array} &
\begin{array}{l} d_{12}(x_3-x_2)(y_2-y_3) \\ \quad + d_{33}(x_3-x_2)(y_2-y_3) \end{array} &
\begin{array}{l} d_{11}(y_2-y_3)(y_3-y_1) \\ \quad + d_{33}(x_3-x_2)(x_1-x_3) \end{array} &
\begin{array}{l} d_{12}(x_1-x_3)(y_2-y_3) \\ \quad + d_{33}(x_3-x_2)(y_3-y_1) \end{array} &
\begin{array}{l} d_{11}(y_1-y_2)(y_2-y_3) \\ \quad + d_{33}(x_2-x_1)(x_3-x_2) \end{array} &
\begin{array}{l} d_{12}(x_2-x_1)(y_2-y_3) \\ \quad + d_{33}(x_3-x_2)(y_3-y_1) \end{array} \\
\begin{array}{l} d_{21}(x_3-x_2)(y_2-y_3) \\ \quad + d_{33}(x_3-x_2)(y_2-y_3) \end{array} &
\begin{array}{l} d_{22}(x_3-x_2)^2 \\ \quad + d_{33}(y_2-y_3)^2 \end{array} &
\begin{array}{l} d_{12}(y_3-y_1)(x_3-x_2) \\ \quad + d_{33}(x_1-x_3)(y_2-y_3) \end{array} &
\begin{array}{l} d_{22}(x_3-x_2)(x_1-x_3) \\ \quad + d_{33}(y_2-y_3)(y_3-y_1) \end{array} &
\begin{array}{l} d_{21}(x_3-x_2)(y_1-y_2) \\ \quad + d_{33}(x_2-x_1)(y_2-y_3) \end{array} &
\begin{array}{l} d_{22}(x_2-x_1)(x_3-x_2) \\ \quad + d_{33}(y_1-y_2)(y_2-y_3) \end{array} \\
\begin{array}{l} d_{11}(y_2-y_3)(y_3-y_1) \\ \quad + d_{33}(x_1-x_3)(x_3-x_2) \end{array} &
\begin{array}{l} d_{12}(x_3-x_2)(y_3-y_1) \\ \quad + d_{33}(x_1-x_3)(y_2-y_3) \end{array} &
\begin{array}{l} d_{11}(y_3-y_1)^2 \\ \quad + d_{33}(x_1-x_3)^2 \end{array} &
\begin{array}{l} d_{12}(x_1-x_3)(y_3-y_1) \\ \quad + d_{33}(x_1-x_3)(y_3-y_1) \end{array} &
\begin{array}{l} d_{11}(y_1-y_2)(y_3-y_1) \\ \quad + d_{33}(x_1-x_3)(x_2-x_1) \end{array} &
\begin{array}{l} d_{12}(x_2-x_1)(y_3-y_1) \\ \quad + d_{33}(x_1-x_3)(y_1-y_2) \end{array} \\
\begin{array}{l} d_{21}(x_1-x_3)(y_2-y_3) \\ \quad + d_{33}(x_3-x_2)(y_3-y_1) \end{array} &
\begin{array}{l} d_{22}(x_1-x_3)(x_3-x_2) \\ \quad + d_{33}(y_2-y_3)(y_3-y_1) \end{array} &
\begin{array}{l} d_{12}(x_1-x_3)(y_3-y_1) \\ \quad + d_{33}(x_1-x_3)(y_3-y_1) \end{array} &
\begin{array}{l} d_{22}(x_1-x_3)^2 \\ \quad + d_{33}(y_3-y_1)^2 \end{array} &
\begin{array}{l} d_{21}(x_1-x_3)(y_1-y_2) \\ \quad + d_{33}(x_2-x_1)(y_3-y_1) \end{array} &
\begin{array}{l} d_{22}(x_1-x_3)(x_2-x_1) \\ \quad + d_{33}(y_1-y_2)(y_3-y_1) \end{array} \\
\begin{array}{l} d_{11}(y_1-y_2)(y_2-y_3) \\ \quad + d_{33}(x_2-x_1)(x_3-x_2) \end{array} &
\begin{array}{l} d_{12}(x_3-x_2)(y_1-y_2) \\ \quad + d_{33}(x_2-x_1)(y_2-y_3) \end{array} &
\begin{array}{l} d_{11}(y_1-y_2)(y_3-y_1) \\ \quad + d_{33}(x_1-x_3)(x_2-x_1) \end{array} &
\begin{array}{l} d_{12}(x_1-x_3)(y_1-y_2) \\ \quad + d_{33}(x_2-x_1)(y_3-y_1) \end{array} &
\begin{array}{l} d_{11}(y_1-y_2)^2 \\ \quad + d_{33}(x_2-x_1)^2 \end{array} &
\begin{array}{l} d_{12}(x_2-x_1)(y_1-y_2) \\ \quad + d_{33}(x_2-x_1)(y_1-y_2) \end{array} \\
\begin{array}{l} d_{21}(x_2-x_1)(y_2-y_3) \\ \quad + d_{33}(x_3-x_2)(y_1-y_2) \end{array} &
\begin{array}{l} d_{22}(x_2-x_1)(x_3-x_2) \\ \quad + d_{33}(y_1-y_2)(y_2-y_3) \end{array} &
\begin{array}{l} d_{12}(x_2-x_1)(y_3-y_1) \\ \quad + d_{33}(x_1-x_3)(y_1-y_2) \end{array} &
\begin{array}{l} d_{22}(x_1-x_3)(x_2-x_1) \\ \quad + d_{33}(y_1-y_2)(y_3-y_1) \end{array} &
\begin{array}{l} d_{21}(x_2-x_1)(y_1-y_2) \\ \quad + d_{33}(x_2-x_1)(y_1-y_2) \end{array} &
\begin{array}{l} d_{22}(x_2-x_1)^2 \\ \quad + d_{33}(y_1-y_2)^2 \end{array}
\end{bmatrix}
\tag{4.49}
$$

where 2Δ is the area of the triangular element and t the thickness.

Of particular importance is the relation between stresses and displacements:

$$\{\sigma^e\} = [H^e]\{\delta^e\} \tag{4.50}$$

where,

$$[H] = [D][B] \tag{4.51}$$

which will eventually yield the stresses after the determination of displacements from the equations

$$[K]\{\delta\} = \{F\} \tag{4.52}$$

The system, or global, equations (4.52), will be assembled from the element equations by assembling the system stiffness matrix out of the element stiffness matrices.

In this triangular element, with three nodes, for plane elasticity, the equations of equilibrium (equations 4.40) were obtained directly owing to the constant stresses assumed in the model. In other elements this is no longer possible and other methods are available. Without going into much detail, we can state that energy considerations lead to the following equation for the element stiffness matrix, instead of equation (4.48) (Zienkiewicz 1975):

$$[K^e] = \frac{1}{V}\int_V [B][D][B]\mathrm{d}V \tag{4.53}$$

where V is the element volume $= 2\Delta t$ and $\mathrm{d}V$ the differential volume $t\,\mathrm{d}x\mathrm{d}y$.

In this element, matrices $[B]$ and $[D]$ are constant within the element and the integration in equation (4.53) is trivial leading immediately to equation (4.48).

The algorithm to implement the above procedure is essentially the same as for the space frame. Only the definitions of the nodal displacement and the nodal force vector and the element stiffness matrix procedure will change. Input for the definition of the triangles will be obtained from the AUTOMESH procedure (Chapter 3).

This algorithm has been coded in the program FINSTRES, Appendix 4.C.

4.4.3 Axisymmetric solids

Stress analysis in axisymmetric solids can be performed similarly to plane stress–strain problems (Section 4.4.2). Assuming the geometry of Figure 4.10, let y be the axis of symmetry and x the radial direction. The coordinate axes are now shown in Figure 4.11.

The displacement and force vectors are the same:

$$\{F\} = \{F_{r1}F_{z1}F_{r2}F_{z2}F_{r3}F_{z3}\}$$
$$\{\delta\} = \{u_1 v_1 u_2 v_2 u_3 v_3\}$$

In addition to the previously-defined components of strain, there is a circumferential component $\varepsilon_t = u/r$, while the other terms are the same.

Therefore, another row and column are introduced between rows and columns 3 and 4 in the matrix $[C]$, and

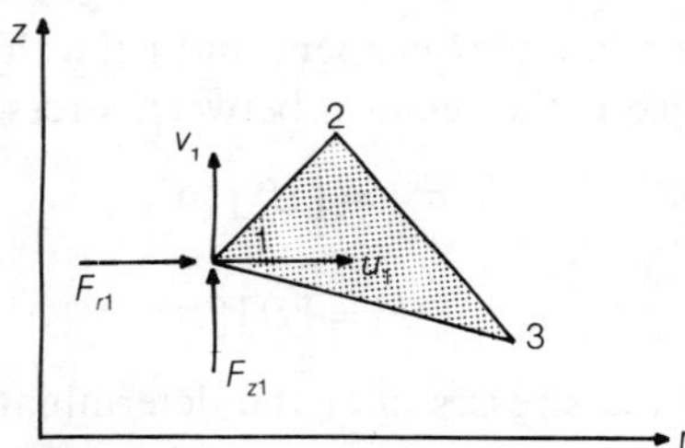

Figure 4.11 Axisymmetric triangular element

$$\{\varepsilon(r,z)\} = \begin{Bmatrix} \varepsilon_r \\ \varepsilon_z \\ \varepsilon_t \\ \gamma_{rz} \end{Bmatrix} = \begin{bmatrix} 0 & 1 & 0 & 0 & 0 & 0 \\ 0 & 0 & 0 & 0 & 0 & 1 \\ 1/r & 1 & z/r & 0 & 0 & 0 \\ 0 & 0 & 1 & 0 & 1 & 0 \end{bmatrix} \begin{Bmatrix} \alpha_1 \\ \alpha_2 \\ \alpha_3 \\ \alpha_4 \\ \alpha_5 \\ \alpha_6 \end{Bmatrix} \tag{4.54}$$

Therefore, the matrix $[B] = [C]\,\mathrm{inv}\,[A]$ will be:

$$[B] = \frac{1}{2\Delta} \begin{bmatrix} z_2 - z_3 & 0 & z_3 - z_1 & 0 & z_1 - z_2 & 0 \\ 0 & r_3 - r_2 & 0 & r_1 - r_3 & 0 & r_2 - r_1 \\ \dfrac{r_2 z_3 - r_3 z_2}{r} + (z_2 - z_3) + \dfrac{z}{r}(r_3 - r_2) & 0 & \dfrac{-r_1 z_3 + r_3 z_1}{r} + (z_3 - z_1) + \dfrac{z}{r}(r_1 - r_3) & 0 & \dfrac{r_1 z_2 - r_2 z_1}{r} + (z_1 - z_2) + \dfrac{z}{r}(r_2 - r_1) & 0 \\ r_3 - r_2 & z_2 - z_3 & r_1 - r_3 & z_3 - z_1 & r_2 - z_1 & z_1 - z_2 \end{bmatrix} \tag{4.55}$$

$$2\Delta = \det \begin{vmatrix} 1 & r_1 & z_1 \\ 1 & r_2 & z_2 \\ 1 & r_3 & z_3 \end{vmatrix}$$

With the introduction of the circumferential stress, the stress–strain relation can be written in the form (now it is always plane strain);

$$\begin{Bmatrix} \sigma_r \\ \sigma_z \\ \sigma_\theta \\ \tau_{rz} \end{Bmatrix} = \frac{E(1-v)}{(1+v)(1-2v)} \begin{bmatrix} 1 & \dfrac{v}{1-v} & \dfrac{v}{1-v} & 0 \\ \dfrac{v}{1-v} & 1 & \dfrac{v}{1-v} & 0 \\ \dfrac{v}{1-v} & \dfrac{v}{1-v} & 1 & 0 \\ 0 & 0 & 0 & \dfrac{1-2v}{2(1-v)} \end{bmatrix} \begin{Bmatrix} \varepsilon_r \\ \varepsilon_z \\ \varepsilon_\theta \\ \gamma_{rz} \end{Bmatrix} \tag{4.56}$$

which also defines the new matrix $[D]$. Finally,

$$[K^e] = \frac{1}{V}\int_V [B][D][B]2\pi r \, dr dz \tag{4.57}$$

Matrix B is a function of r, and the integration in equation (4.57) has to be performed numerically. However, for small enough elements, it is possible to evaluate B at the centroid of the element,

$$\bar{r} = (r_1 + r_2 + r_3)/3, \quad z = (z_1 + z_2 + z_3)/3 \tag{4.58}$$

with small effect on accuracy. Then, integration of equation (4.57) is trivial and

$$[K^e] = [B]^T[D][B] \tag{4.59}$$

Matrix $[H^e]$ is again $[D][B]$.

The program FINSTRES can be modified changing the definition of matrices B and D and their smaller dimension from 3 to 4. In fact, this has been included as an option in the program.

4.4.4 System assembly and solution

The procedure indicated in equation (4.25) must be followed to assemble the element stiffness matrices into the system stiffness matrix $[K]$. To this end, we note that the element I has been defined by its node numbers $NI(I)$, $NJ(I)$, $NM(I)$ and each node J by the coordinates $X(J)$, $Y(J)$. For the element I, node J is $NI(I)$ or $NJ(I)$ or $NM(I)$. To each node, a number of degrees

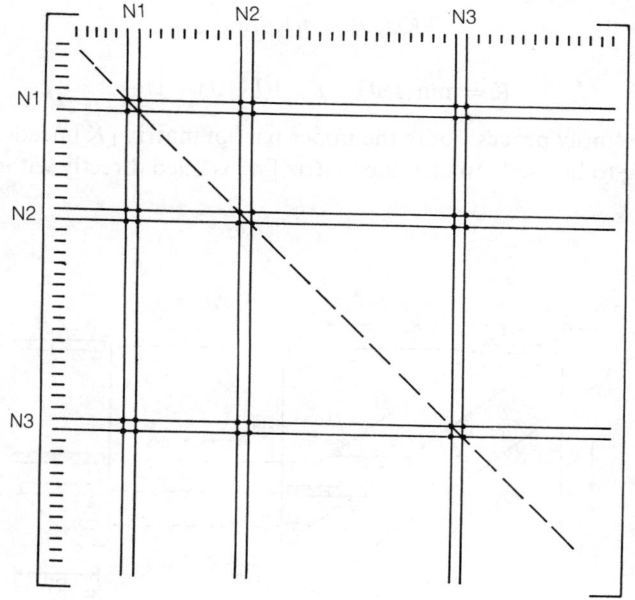

Figure 4.12 Assembly of the stiffness matrix

of freedom are assigned. For example, in plane elasticity, two degrees of freedom per node are assigned. Then, in this case the element stiffness matrix will have dimension 6×6.

Two coordinates, the $2J$-1 and the $2J$, correspond to node J. Therefore, for nodes NI, NJ and NM, the corresponding coordinates are $2NI$-1,$2NI$, $2NJ$-1,$2NJ$ and $2NM$-1,$2NM$. These coordinates are assigned numbers $I1$ to $I6$. Finally, each element Ip,Iq of the 6×6 stiffness matrix KE is added to the corresponding coordinates p,q of the global stiffness matrix. The process is shown in Figure 4.12. The 6×6 elements of the element stiffness matrix are denoted with black dots. The process is repeated with each one of the elements filling up the stiffness matrix. This matrix thus obtained has two noticeable properties:

(a) It is symmetric (and of course real) due to the well-known reciprocity property of elastic systems.
(b) In most cases it is banded. All elements contribute to the main diagonal but also to a narrow band about the diagonal. This is not true in all problems however.

These two properties make the solution considerably easier.

The solution of the system

$$[K]\{\delta\} = \{F\} \tag{4.60}$$

will be performed in the way as explained in Section 4.3.5, using the Gauss elimination method. We shall use the two properties mentioned above to assist compute memory, namely the symmetry and banded form of the matrix. We define a matrix A with a number of rows equal to the number of rows of matrix K but number of columns equal to, roughly, half bandwidth including the diagonal (Figure 4.13).

The required computer memory ($N \times BW$) is much less than $N \times N$, usually in the order of 10%. The relationship between the stiffness matrix K and the band storage matrix A is:

$$K(I,J) = A(K,L) \tag{4.61}$$

where

$$K = \min(I,J), \quad L = |(I-J)+1| \tag{4.62}$$

During the assembly process, only the upper part of matrix $[K]$ needs to be filled in, if matrix $[A]$ is going to be used . In fact, the matrix $[A]$ is filled directly without using $[K]$ at all.

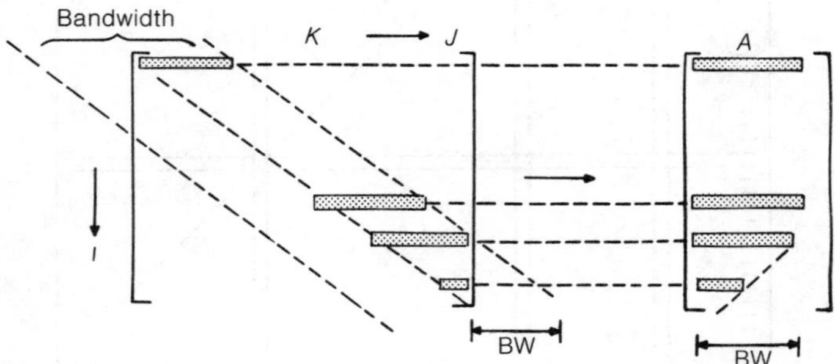

Figure 4.13 Banded matrix

4.5 PREPROCESSING AND POSTPROCESSING

The numerical methods of stress analysis widely employed today in machine design, involve an enormous amount of input data and output results. This prompted the development of preprocessing methods for automatic input generation and postprocessing to automatically

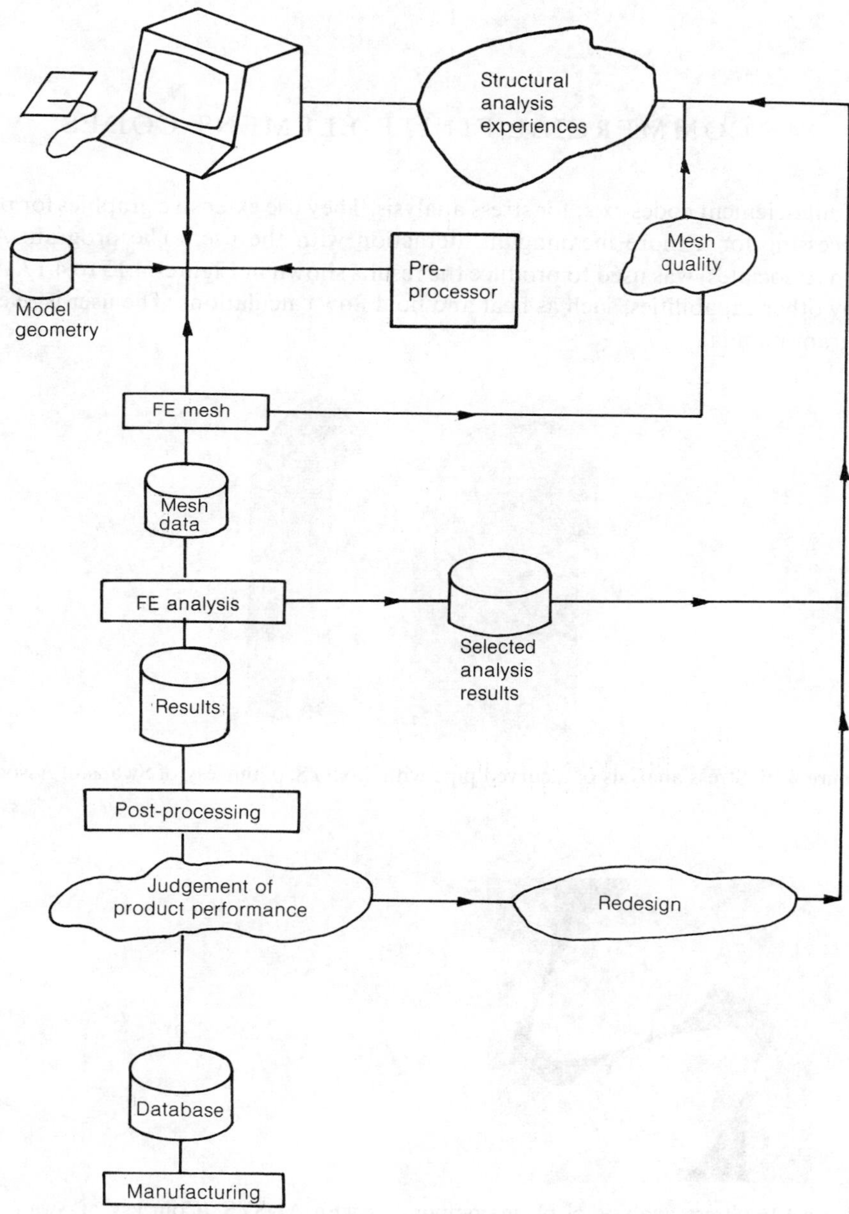

Figure 4.14 Finite-element analysis of a machine element

sort the important results for identification and presentation in an effective way.

The automatic mesh generation already presented in Chapter 3 is used to prepare inputs for finite-element analyses in the form of grids with triangular or rectangular elements. The design procedure is shown in Figure 4.14. The FINSTRES program has a postprocessor which generates on a color screen the deflected shape and a stress distribution map, if an EGA card and a color monitor are available, by assigning different colors to different stress levels.

4.6 COMMERCIAL FINITE-ELEMENT CODES

Several finite-element codes exist for stress analysis. They use extensive graphics for pre- and post-processing for a more meaningful interaction with the user. The program ANSYS (Swanson Associates) was used to produce the results shown in Figures 4.15 to 4.17. ANSYS has many other capabilities, such as heat and fluid flow calculations. The user is referred to the program manual.

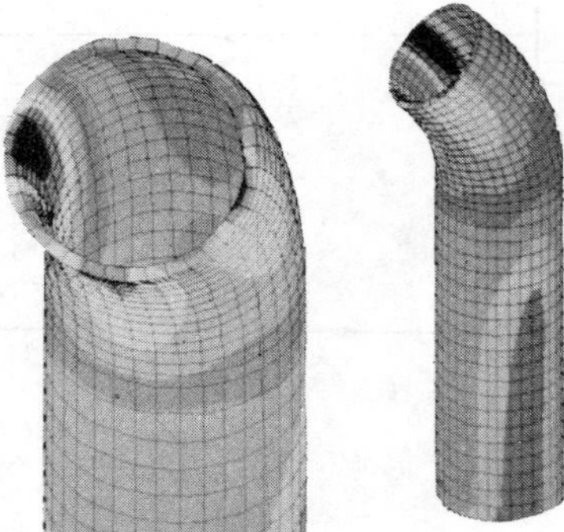

Figure 4.15 Stress analysis of a curved pipe with ANSYS. (Courtesy of Swanson Assoc.)

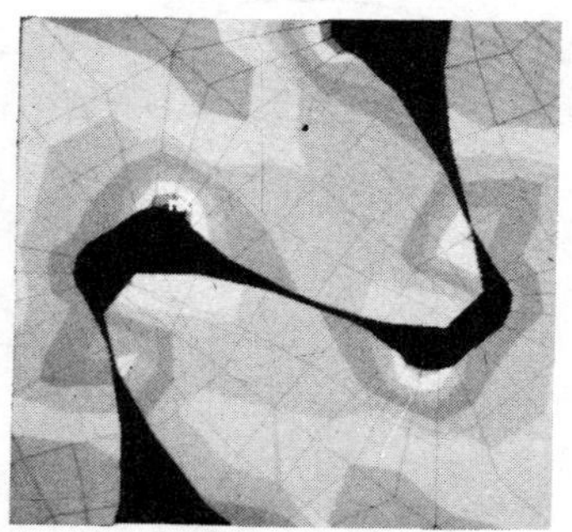

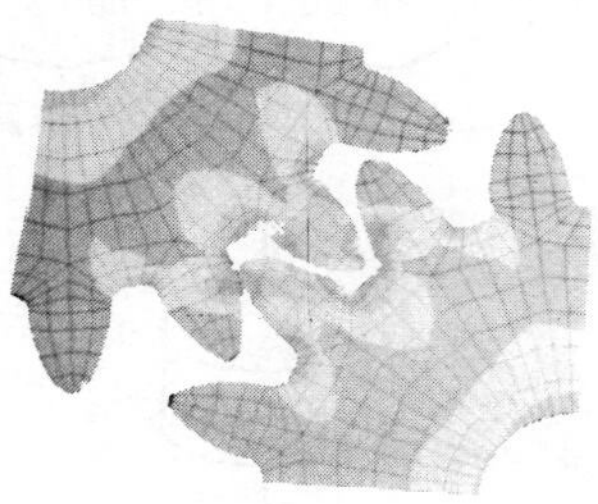

Figure 4.16 Stress analysis of plane components with ANSYS. (Courtesy of Swanson Assoc.)

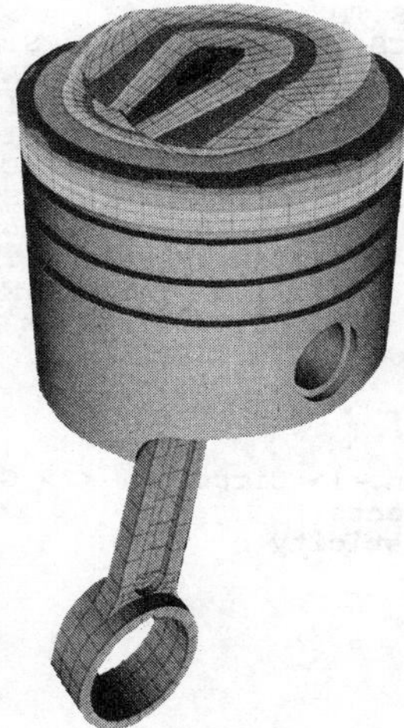

Figure 4.17 Solid modeling and stress analysis of three-dimensional solids with ANSYS. (Courtesy of Swanson Assoc.)

Optimizing finite-element codes are now available, which have integral tests for the computation accuracy. Such a code, the PROBE (Noetic Technologies Corp.) developed by Dr B. Szabo, is using new techniques for obtaining results to the desired accuracy.

EXAMPLES

Example 4.1

A shaft is loaded as shown in Figure E4.1a, dimensions in millimeters.

The Young's modulus $E = 2.1 \times 10^{11}$ N/m^2. Compute maximum deflection, reactions, shear forces and bending moments.

Solution

Hard copy of the input for the program TMSTAT (in SI units Nm) follows, together with the output:

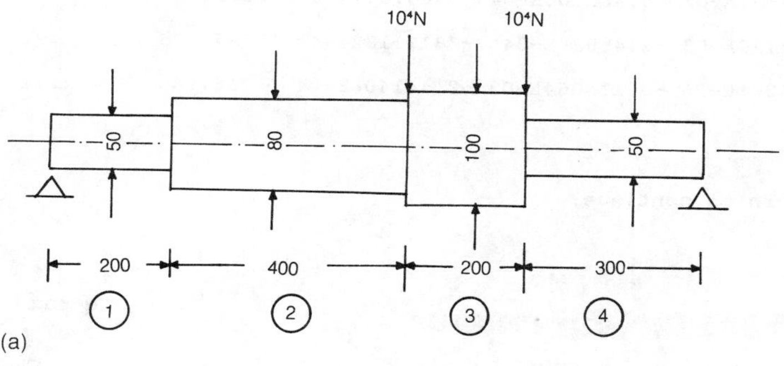

Figure E4.1 (*Contd.*)

```
                    STATICS.TM                         STATICS.TM
                     STATICS.TM                      STATICS.TM
                      STATICS.TM                   STATICS.TM
                       STATICS.TM                STATICS.TM
                        STATICS.TM             STATICS.TM
                         STATICS.TM          STATICS.TM
                          STATICS.TM       STATICS.TM
A Transfer Matrix Program for Continuous Beam  analysis
                            STATICS.TMSTATICS.TM
                             STATICS.STATICS.TM
                              STATICSTATICS.TM
                               STATSTATICS.TM
                                STSTATICS.TM
                                 STATICS.TM

   Type of section,<1> Circular, <2> General ? 1
   Number of Elements:                        ? 4
   Modulus of Elasticity                      ? 2.1e11


Element data                                  node data
El/No
   length    Diameter      distr load      force          spring
 1  ? .2         ? .05          ?            ?               ?
 2  ? .4         ? .08          ?            ?               ?
 3  ? .2         ? .1           ?            ? 1e4           ?
 4  ? .3         ? .05          ?            ? 1e4           ?
 right end....                               ?               ?
Enter no of solid supports? 2
Enter numbers of support nodes:

Support  1   at node ? 1
Support  2   at node ? 5

Are data correct (Y/N)   ? y
```

SOLUTION ...
======================

Node	Deflection	Slope	moment	Shear	Reaction
1	1.470575E-06	5.561773E-03	0	0	7272.728
2	9.633144E-04	3.304113E-03	-1454.546	-7272.728	
3	1.825637E-03	5.481806E-04	-4363.637	-7272.728	
4	1.854139E-03	-2.45528E-04	-3818.182	12727.273	
5	2.573361E-06	-9.135064E-03	-2.441406E-04	.7167969	12726.56

print return to continue? _

q-max= 0

Figure E4.1 (*Contd.*)

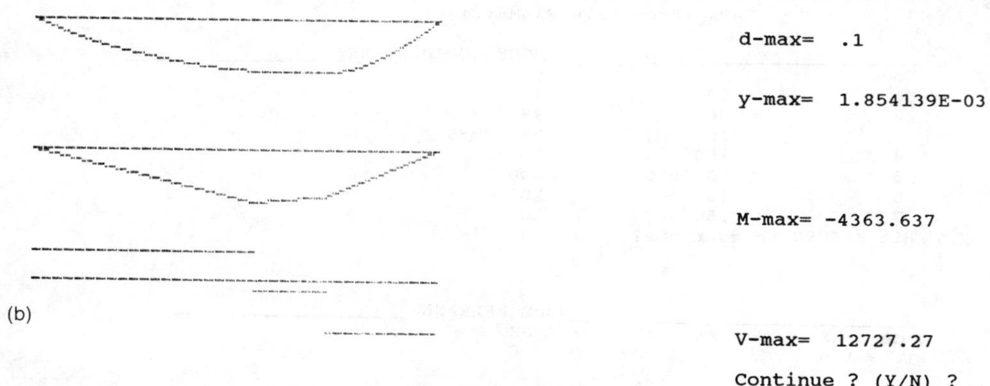

Figure E4.1 (a) Model for static analysis of a shaft E4.1; (b) program TMSTAT run

Maximum deflection is 0.00185 m (1.85 mm). It could be somewhat greater between stations. One can either draw the deflection curve or use more nodes along the beam to achieve more precision in the parameter tabulation. That, of course, depends on the particular problem and the required accuracy. If an EGA card is available, the program plots deflection, moment and shear diagrams on the screen.

Example 4.2

Using the program FINFRAME (Appendix 4.B) calculate displacements and member forces for a double crane hook shown in Figure E4.2(a, b) made of 4-cm thick plate.
Data: $F = 5$ ton; $t = 4$ cm; $b = 15$ cm; $E = 2.1 \times 10^{11}$ Pa; $\nu = 0.3$; material, steel.

Solution

Figure E4.2(b) shows a model, plane frame, which can be analyzed as a space frame setting all z-coordinates of nodes equal to zero.

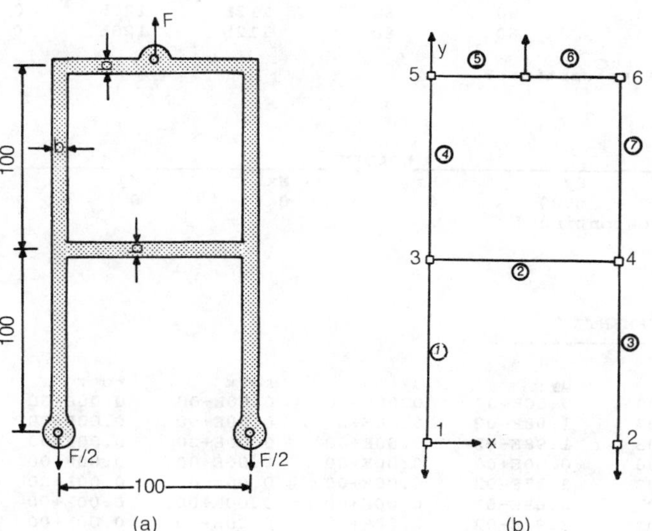

Figure E4.2 (*Contd.*)

DATA (From file FRAME42)

NODE COORDINATES

Node	x	y	z
1	50	0	0
2	50	99.99999	0
3	150	99.99999	0
4	150	0	0
5	50.00001	200	0
6	100	200	0
7	150	200	0

Hit RETURN to continue?

NODE RELEASES

Node	RX	RY	RZ	FX	FY	FZ
1	0	0	1	0	0	0
4	0	0	1	0	0	0

Hit RETURN to continue?

ELEMENT INCIDENCES

Member	Node 1	Node 2
1	1	2
2	2	3
3	3	4
4	2	5
5	5	6
6	6	7
7	7	3

Hit RETURN to continue?

ELEMENT PROPERTIES

No	Material	Area	Iy	Iz	Iyz	Roll angle
1	1	60	80	1125	1205	0
2	1	60	80	1125	1205	0
3	1	60	80	1125	1205	0
4	1	60	80	1125	1205	0
5	1	60	80	1125	1205	0
6	1	60	80	1125	1205	0
7	1	60	80	1125	1205	0

Hit RETURN to continue?

LOADING

Node	Fx	Fy	Fz	Mx	My	Mz
6	0	5000	0	0	0	0

Hit RETURN to continue?

NODE DISPLACEMENTS

Element	ux	uy	uz	rotx	roty	rotz
1	0.00E+00	0.00E+00	0.00E+00	0.00E+00	0.00E+00	0.00E+00
2	2.60E-03	1.98E-02	0.00E+00	0.00E+00	0.00E+00	-9.41E-04
3	-2.60E-03	1.98E-02	0.00E+00	0.00E+00	0.00E+00	9.41E-04
4	0.00E+00	0.00E+00	0.00E+00	0.00E+00	0.00E+00	0.00E+00
5	-2.10E-03	3.97E-02	0.00E+00	0.00E+00	0.00E+00	4.77E-03
6	-1.61E-07	2.69E-01	0.00E+00	0.00E+00	0.00E+00	4.66E-10
7	2.10E-03	3.97E-02	0.00E+00	0.00E+00	0.00E+00	-4.77E-03

Figure E4.2 (*Contd.*)

```
ELEMENT FORCES
------------------

Element   Fx          Fy          Fz          Mx          My          Mz
 1    1.26E+02   -2.50E+03   0.00E+00    0.00E+00    0.00E+00   -4.08E+03
 2    6.55E+02    1.53E-04   0.00E+00    0.00E+00    0.00E+00   -4.45E+03
 3    1.26E+02    2.50E+03   0.00E+00    0.00E+00    0.00E+00    8.53E+03
 4   -5.29E+02   -2.50E+03   0.00E+00    0.00E+00    0.00E+00    1.30E+04
 5   -5.29E+02   -2.50E+03   0.00E+00    0.00E+00    0.00E+00   -4.00E+04
 6   -5.29E+02    2.50E+03   0.00E+00    0.00E+00    0.00E+00    8.50E+04
 7   -5.29E+02    2.50E+03   0.00E+00    0.00E+00    0.00E+00   -4.00E+04

Normal termination. Press any key.
```

(c)

Figure E4.2 Analysis of a plane frame with FINFRAME

Moments of inertia:

$$IY = bt^3/12 = 80\,\text{cm}^4$$

$$IZ = tb^3/12 = 1125\,\text{cm}^4$$

$$IJ = IY + IZ = 1205\,\text{cm}^4$$

Section area $A = bt = 60\,\text{cm}^2$.

For numerical convenience ton, cm units will be used:

Young's modulus, $E = 2100\,\text{ton/cm}^2$.
The model with node and element numbers is shown in Figure E4.2(b).
Nodes 1 and 2 will be restrained: node 1 in x, y, z, and node 2 in y, z. This situation is statically equivalent with the given loading.
The results are shown in Figure E4.2.

Example 4.3

A communications satellite has the dimensions shown in Figure E4.3(a), axisymmetric geometry, thickness $h = 10$mm and internal pressure 0.2 MPa. Outside pressure is zero. Draw a stress map in the section shown.

Solution

Point C: From equation (4.38), $\sigma_t = \sigma_m = pr/2h = 0.2E6 \times 0.5/0.02 = 0.5E7\text{ Pa} = 5\text{ MPa}$

Point F: Similarly $\sigma_t = \sigma_m = pr/2h = 3\text{ MPa}$

Point D: $\tan\theta = (250 - 150)/800$, $\theta = 7.12°$

$$\sigma_m = pr/2h\cos\theta = 0.2E6 \times 0.15/2 \times 0.01 \times \cos 7.12 = 1.51\text{ MPa}$$

$$r_t = 0.150 \times \cos 7.12 = 0.149\text{ m}$$

$$\sigma_t = (p/h - \sigma_m/r_m)r_t = 2.98\text{ MPa},\ r_m = \infty$$

Point A: $\sigma_m = 0.2E6 \times 0.250/2 \times 0.01 \times \cos 7.12 = 2.52\text{ MPa}$

$$r_t = 0.250\cos 7.12 = 0.248\text{ m}$$

$$\sigma_t = 0.2 \times 0.248/0.01 = 4.96\text{ MPa}$$

The stress map is shown in Figure E4.3(b).

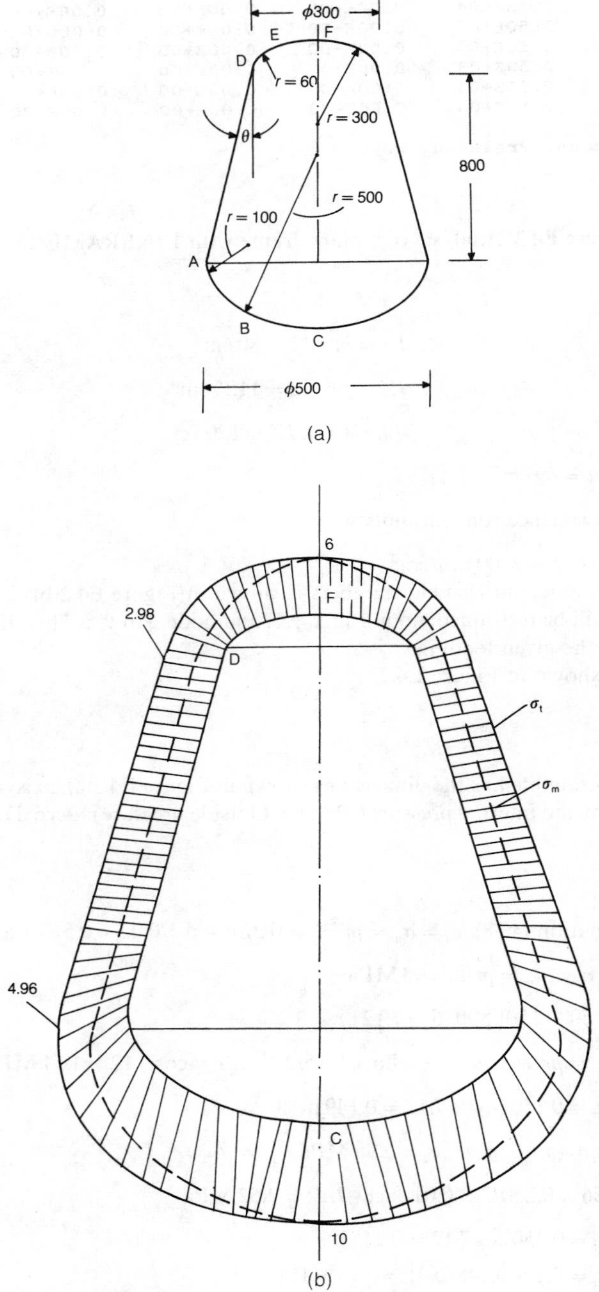

Figure E4.3 Stress analysis of a communications satellite vessel

Example 4.4

A steel bracket as shown in Figure E4.4 supports a load of 70,000 N uniformly distributed on the upper end of the bracket. Determine the stresses in the bracket, the maximum stress and the stress concentration factor in the smallest section. Thickness = 100, $b_0 = 100$, $b = 200$ mm, height = $2h = 400$ mm.

Solution

The program FINSTRES (Appendix 4.C), will be used.

To simplify the problem (although it is not necessary) we note that only one-quarter needs to be considered (Figure E4.4b) with half the load on the upper half surface and restraining the horizontal motion on the center line (x-axis). One point will be fully supported to have a determinate system.

An 8 × 6 grid of triangular elements is generated with program AUTOMESH, Appendix 3.B. From the main menu, the AUTOMESH is selected. Using the dotted cross with the keyboard arrows, nine nodes are specified, the four corners and five points in-between, Figure E4.4c. The points are defined with the 'N' command (Node), in a row- or column-wise sequence. Then the 'A' command is used to automatically generate the mesh. The specified nine nodes were on a 3 × 3 mesh and the generated mesh is 6 × 8. Note that points 2, 5 and 6 were selected closer to point 3 where stress concentration is expected. This creates a more dense mesh in the high stress area. Then, command 'M' returns control to the main menu and with the keyboard arrows the SAVE FILE option is used. The name EXAMPLE.44 is entered and the mesh is stored on floppy disk.

Now, the FINSTRES program, Appendix 4.C, is used. From the main menu, LOAD FILE is selected and file EXAMPLE.44 is loaded. Then, PLOT is selected and a screen plot of the mesh

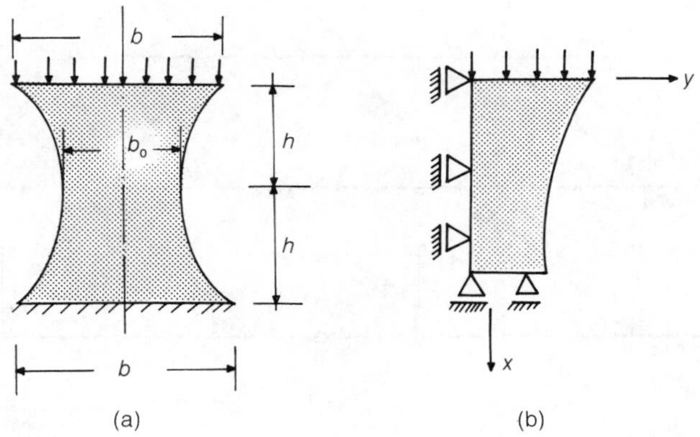


```
FINSTRES                         FINSTRES
 FINSTRES                       FINSTRES
  FINSTRES                     FINSTRES
   FINSTRES                   FINSTRES
    FINSTRES                 FINSTRES
     FINSTRES               FINSTRES
      FINSTRES             FINSTRES
A 3-node finite element program for plain or axisymmetric stress analysis
             FINSTRES  FINSTRES
              FINSTRESFINSTRES
               FINSTRFINSTRES
                FINSFINSTRES
                 FIFINSTRES
                  FINSTRES
```

Figure E4.4 (*Contd.*)

```
                          TYPE OF PROBLEM

<1>Plane stress                         <2>     Plane strain
<3> Axisymmetric, plain stress          <4> Axisymmetric, plain strain
               enter your selection...? 1

Enter:  Modulus of Elasticity, Poisson Ratio,  thickness ? 2.1e6,.3,100
```

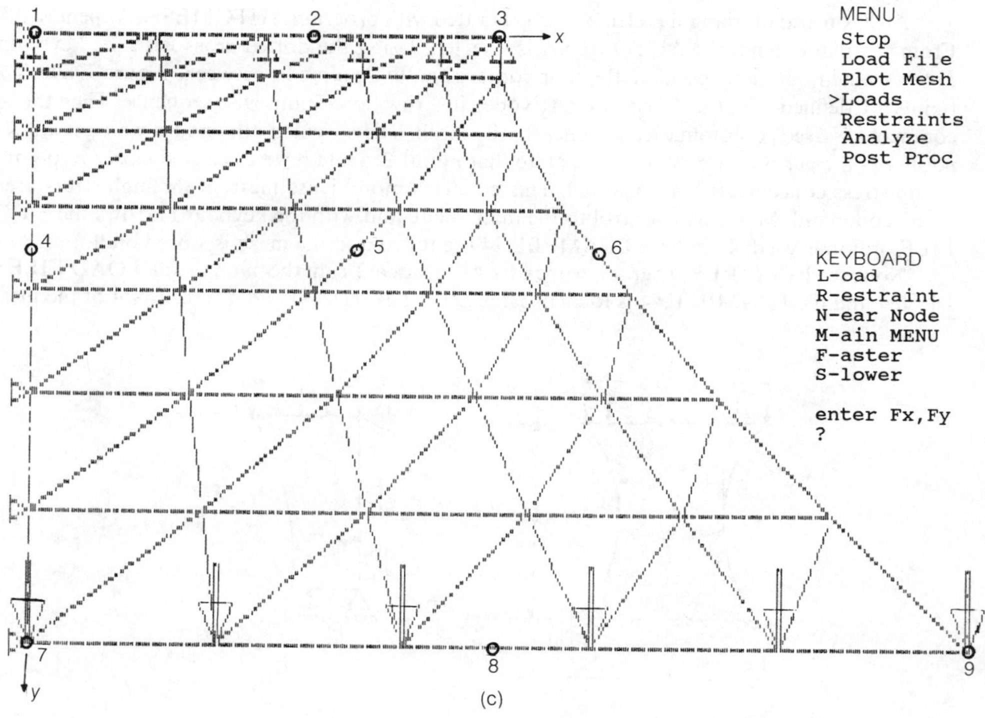

(c)

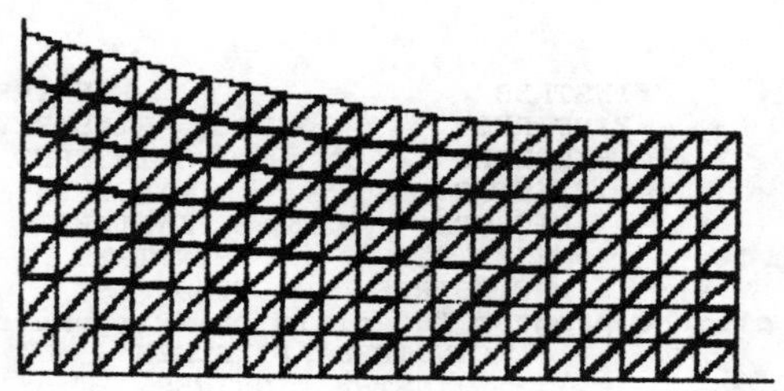

Screen dump of finite element mesh.

(d)

Figure E4.4 Stress analysis of a curved column with FINSTRES

appears. The LOADS command is selected from the main menu and the keyboard commands appear on the screen. On the upper nodes, the lower on the screen, loads $70{,}000/(2 \times 5) = 7000$ N for the four inner nodes in the vertical direction, and half that, 3500 N on the outer nodes, represent half of the total load of 70,000 N, because half of the bracket width is used. Finally, the 'M' command returns control to the main menu. The Analyze command is selected then and displacements and stresses are printed on the line printer, at the nodes and area center of the elements, respectively.

The maximum stress, in the vertical (y) direction is 8.087 N/mm^2 and the nominal stress is $70{,}000/(100 \times 100) = 7$ N/mm^2. Therefore, the stress concentration factor is $K_T = 8.087/7 = 1.15$.

The Post Processing selection in the main menu yields, if a color monitor and an EGA card are available, the deflected shape of the mesh and the distribution of stresses.

Example 4.5

A steel spur gear (Figure E4.5) is secured on the shaft by a shrink fit. It was found that the hoop stress σ_{max} at the outer surface was 6.5 kgf/mm^2 by strain gage measurements. Find the pressure between shaft and gear. Dimensions in millimeters are: Inner radius = 50, width = 100, outer radius = 100. Material properties for carbon steel. Plane stress.

Solution

A 3×3 node grid is made with eight triangular elements with the AUTOMESH program, Appendix 3.B. An arbitrary internal pressure of 1 Pa is used. On node 4 it generates force $2\pi r_i p \times 50$ and on nodes 1 and 7 force $2\pi r_i p \times 25$. The maximum stress is found with the FINSTRES program, Appendix 4.D, $\sigma_{t\max} = 1.66$ Pa. The actual pressure between shaft and gear will be

$$p_{\max} = 1 \times (65 \times 10^6/1.66) = 39 \text{ MPa}$$

With the analytical methods of Chapter 14, we would obtain almost identical results with the ones obtained here with a very crude model of eight elements.

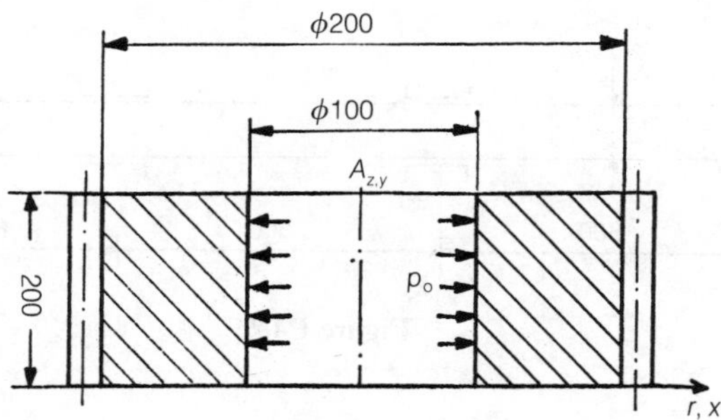

Figure E4.5 Stress analysis of an axisymmetric hub with FINSTRES

REFERENCES AND FURTHER READING

Dimarogonas, A. D., 1976. *Vibration Engineering.* St. Paul: West Publishers.

Fadeeva, V. N., 1959. *Computation Methods in Linear Algebra.* New York: Dover. (Translated from the Russian by C. Benster.)

Pestel, E. C., Leckie, F. A., 1963. *Matrix Methods in Elastomechanics.* New York: McGraw-Hill.
Rockey, K. E., *et al.* 1975. *The Finite Element Method.* New York: Halsted Press/Wiley.
Weaver, W., Jr 1967. *Computer Programs for Structural Analysis.* New York: Van Nostrand.
Zienkiewicz, O. C., 1977. *The Finite Element Method.* London: McGraw-Hill.

PROBLEMS

Problems 4.1. to 4.5

Using the program TMSTAT, determine reactions and draw deflection, moment and shear diagrams to a proper scale, for the systems shown in Figures P4.1 to P4.5 respectively. Material, steel; dimensions in millimeters; forces in N. Assume simple supports.

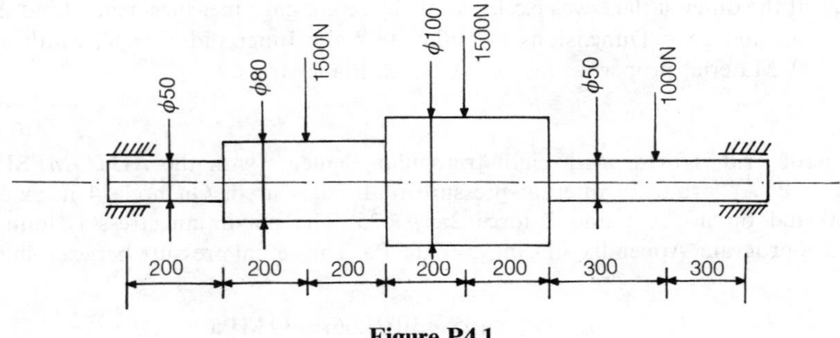

Figure P4.1

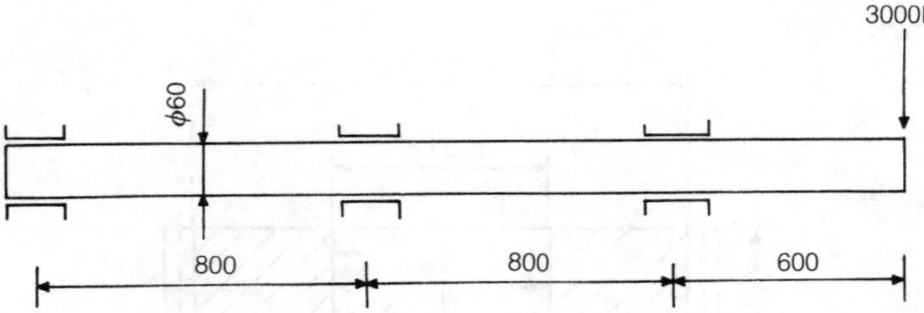

3000N

ϕ60

800

800

600

Figure P4.2

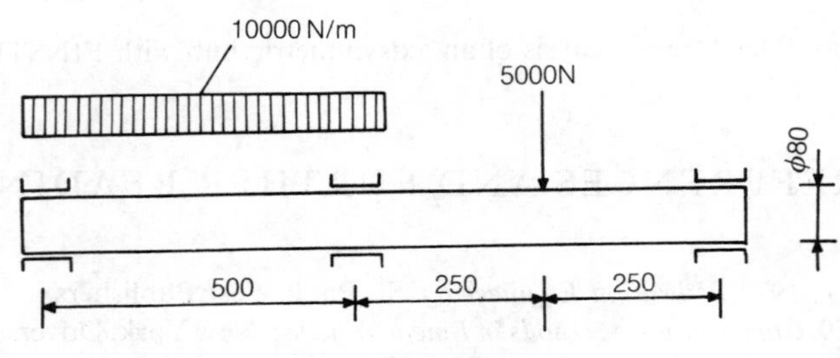

Figure P4.3

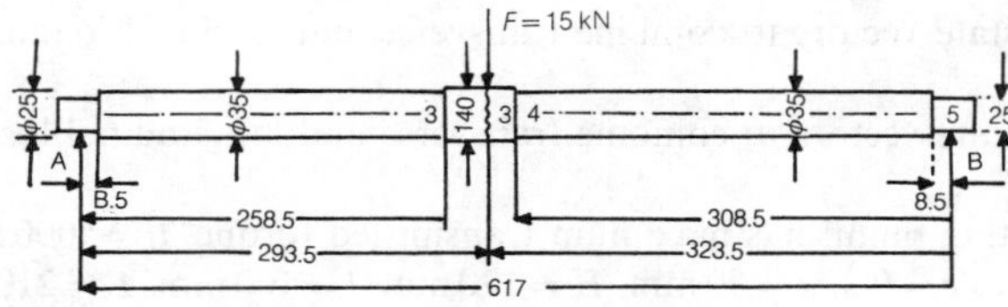

Figure P4.4

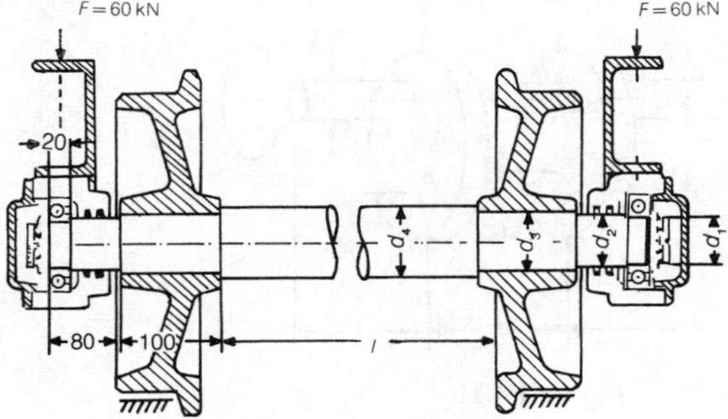

$l = 1.2\,\mathrm{m}$
$d_1 = 80\,\mathrm{mm}$
$d_2 = 100\,\mathrm{mm}$
$d_3 = 110\,\mathrm{mm}$
$d_4 = 130\,\mathrm{mm}$

Figure P4.5

4.6 Derive transfer matrices for torsional loading of a cylindrical shaft of uniform diameter between nodes. Torques are applied at the nodes and the shaft is clamped at one end, free at the other. Write then a program similar to TMSTAT but with two-dimensional vectors (angle of twist, torque) to solve torsion of shafts problems.

4.7 Do the same as in Problem 4.6, but for axial loading of shafts and beams. The state vector is (axial motion, axial force).

4.8 For a beam where at some node there is change of direction of the center line by some angle, the slope and moment remain the same but deflection and shear force change by rotation. Derive the relation of the state vector before and after the node and write in a matrix form, thus deriving a change of direction point matrix. You have to include now in all transfer matrices and state vectors the axial force because rotation gives a shear component.

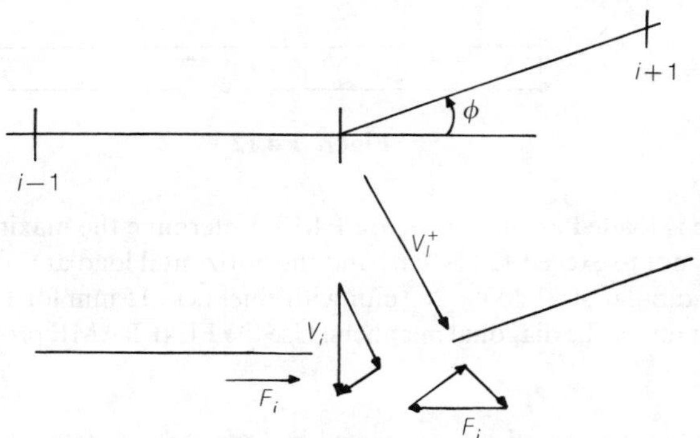

Figure P4.8

4.9 Derive transfer matrices and state vectors to combine transverse and torsional loading.

4.10 Derive transfer matrices and state vectors to combine transverse and longitudinal loading.

4.11 The crankshaft of a reciprocating pump has maximum transmitted torque $T = 5000$ Nm. The material is steel, $E = 2.1E11$ Pa, $v = 0.3$, $d = 30$ mm, $H = 120$ mm, $l = 100$ mm, $L = 210$ and the section of the transverse members is orthogonal of thickness 30 mm and width 60 mm. Determine the maximum stresses and the rotation of the one end relative to the other, using the FINFRAME program.

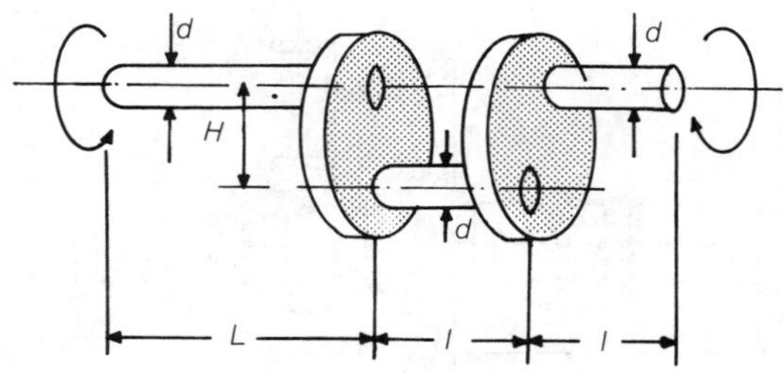

Figure P4.11

4.12 A steel coil compaction press works in a frame of tubular steel, as shown, with a rectangular cross-section 160 × 160 mm and thickness 10 mm. The 65-ton maximum force acts between points A and B (Figure P4.12). Using the program FINFRAME find the member forces and the change of the distance (AB) after loading. With proper boundary conditions, only one-eighth of the frame needs to be considered (dotted line).

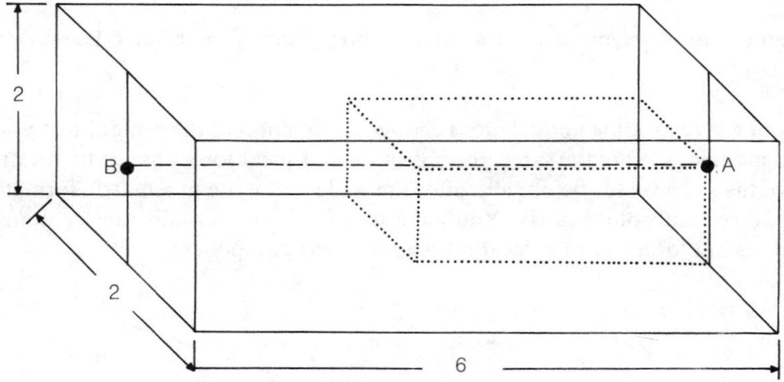

Figure P4.12

4.13 A portal crane is loaded as shown Figure P4.13. Determine the maximum load if the maximum tensile stress is not to exceed 12 kN/cm^2 and the horizontal load at point A is 20% of the vertical. The section is tubular steel 200 × 200 mm with thickness 15 mm for the columns and beam and strip 100 × 20 mm for the diagonal members. Use the FINFRAME program with simple supports at the rails.

4.14 A marine gear has a hub and rim connected by four rods of rectangular cross-section $b \times w$. At point A there are three forces: radial F_r, axial F_a and tangential F_t. Assuming that hub and rim are

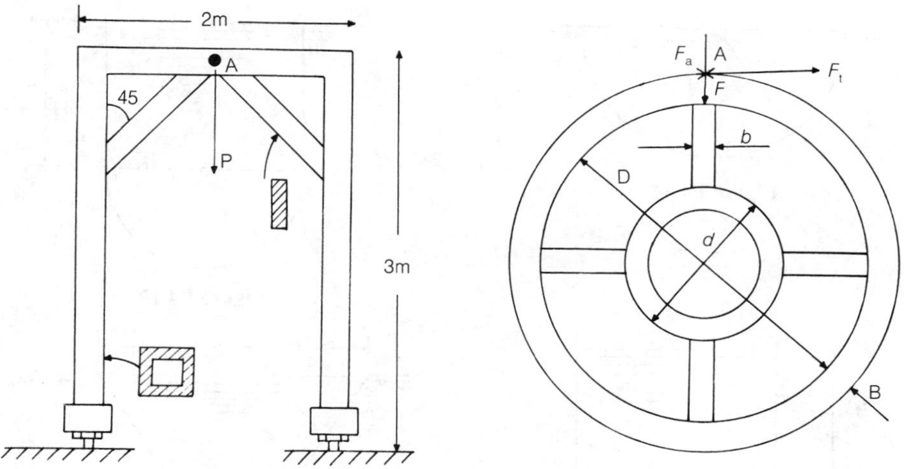

Figure P4.13

Figure P4.14

absolutely rigid, determine the stresses in the rods using program FINFRAME.

Data: $F_a = 1500$ N; $F_r = 2000$ N; $F_t = 3000$ N; $b = 100$ mm; $w = 200$ mm; $d = 400$ mm; $D = 1000$ mm; $E = 1.9 \times 10^{11}$ N/m^2; $\nu = 0.3$.

(*Hint*: Assume clamped connection to the hub (ground) and model the rim as a rectangle with beams of cross-section one order of magnitude greater than the one of the rods.)

4.15 A car bumper is a beam connected to the body by way of two springs, in the simplest version. If we assume force P and a rigid car frame, the deflection at point A of the bumper due to P will be y. The energy of elastic deformation will be $Py/2$. Suppose now that the car moves with a speed V and bumps on a solid object at point A. Since at maximum deflection, when the car stops, the elastic energy equals kinetic energy $mV^2/2$ of the car, find the maximum speed V so that for a bumper that you can measure on some car, the maximum stress at point A will not exceed the yield strength (say 22 kN/cm^2). Assume mass and dimensions for any car.
(*Hint*: Analyze first with FINFRAME for a force $P = 1000$ N and find the deflection and the maximum stress. Calculate then the maximum allowable P and from the elastic energy find V.)

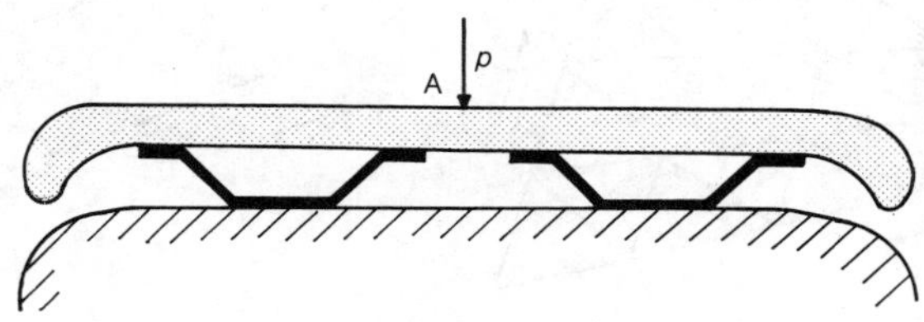

Figure P4.15

Problems 4.16 to 4.20

Determine the stresses of the pressure vessels shown with the membrane method. Thickness everywhere is t and when a fluid is indicated, it is water.

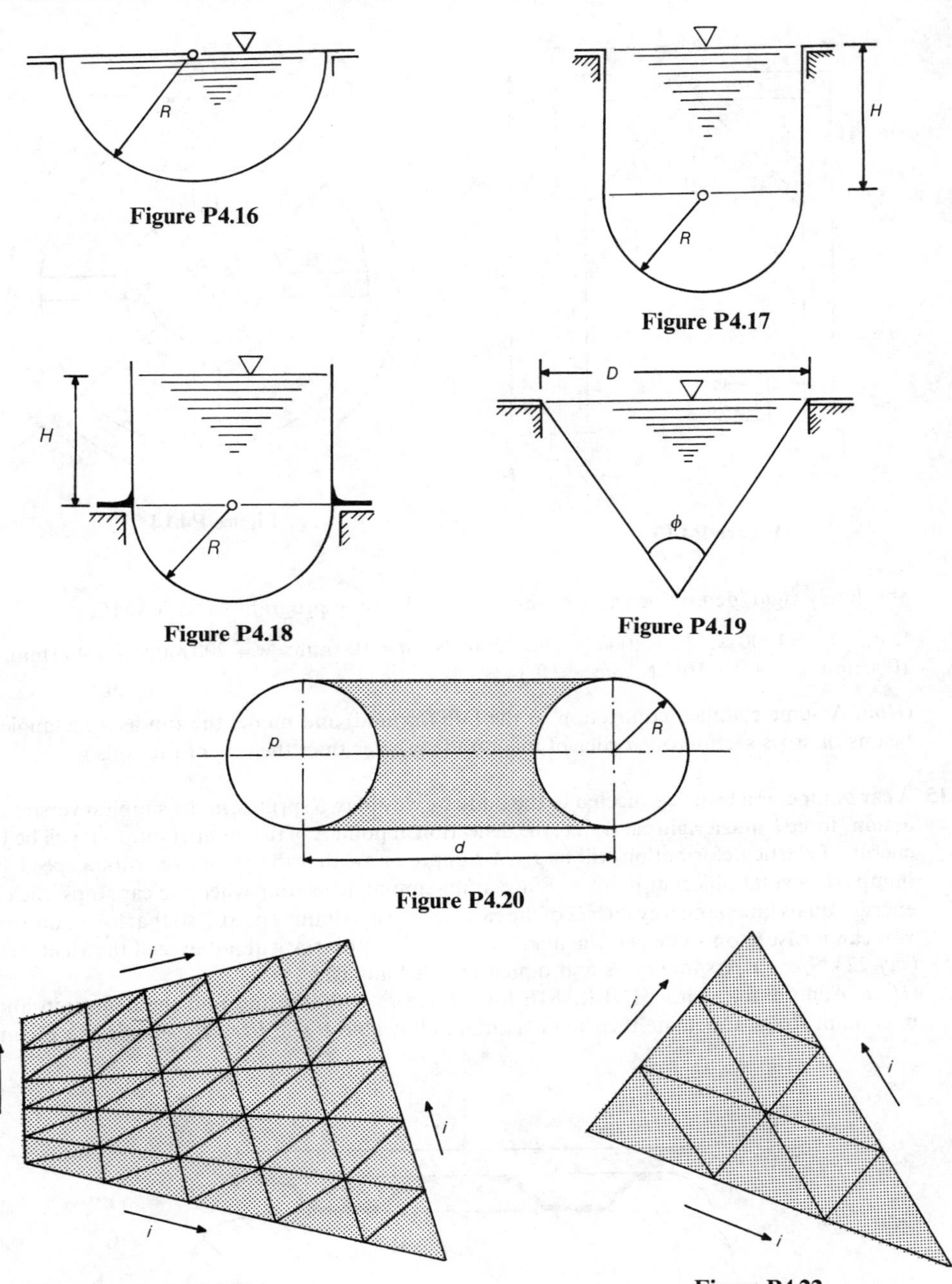

Figure P4.16

Figure P4.17

Figure P4.18

Figure P4.19

Figure P4.20

Figure P4.21

Figure P4.22

Problems 4.21 to 4.25

Write a program of automatic triangular mesh generation for the shapes shown in Figures P4.21 to P4.25 respectively using as a guide the one in program AUTOMESH for a rectangle. The number of nodes on each side is specified and also the respective node numbers. Given also is the

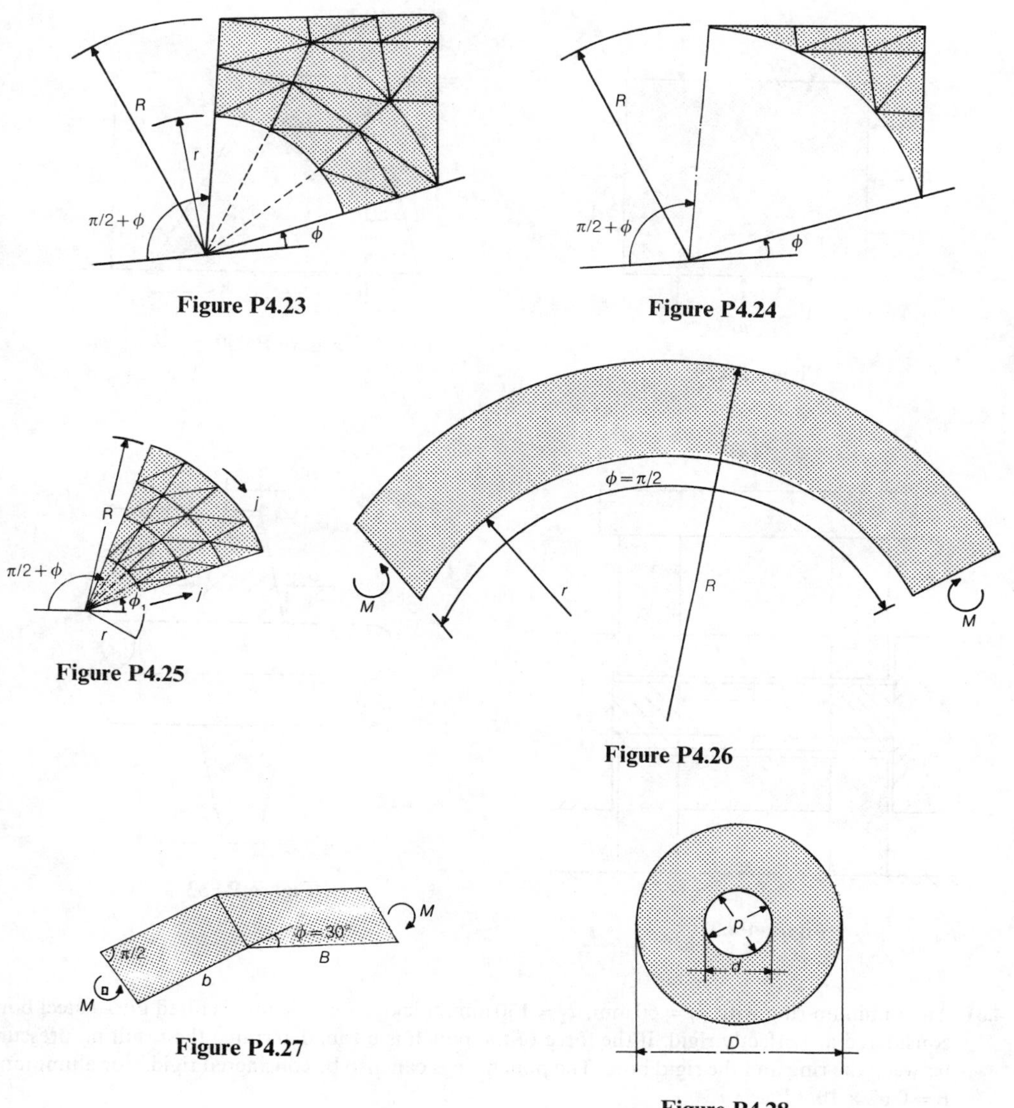

Figure P4.23

Figure P4.24

Figure P4.25

Figure P4.26

Figure P4.27

Figure P4.28

lower node number and lower element number to be used. All nodes and elements should be numbered sequentially above these starting numbers.

Problems 4.26 to 4.29

For the plates shown of thickness 1 cm and loading as shown, find maximum stresses. Data (wherever applicable): $M = 1000$ Nm; $R = 300$ mm; $r = 180$ mm; $a = 50$ mm; $b = B = 200$ mm; $d = 60$ mm; $D = 200$ mm; pressure, $p = 100$ bar; $a = 200$ mm. Use ANSYS (See Section 4.6.)

4.30 A water dam with $W = 4$ m, $B = 16$ m, $H = 30$ m is made of concrete with density 2,700 kg/m^3. Find the stresses at the bearing surface of the dam.

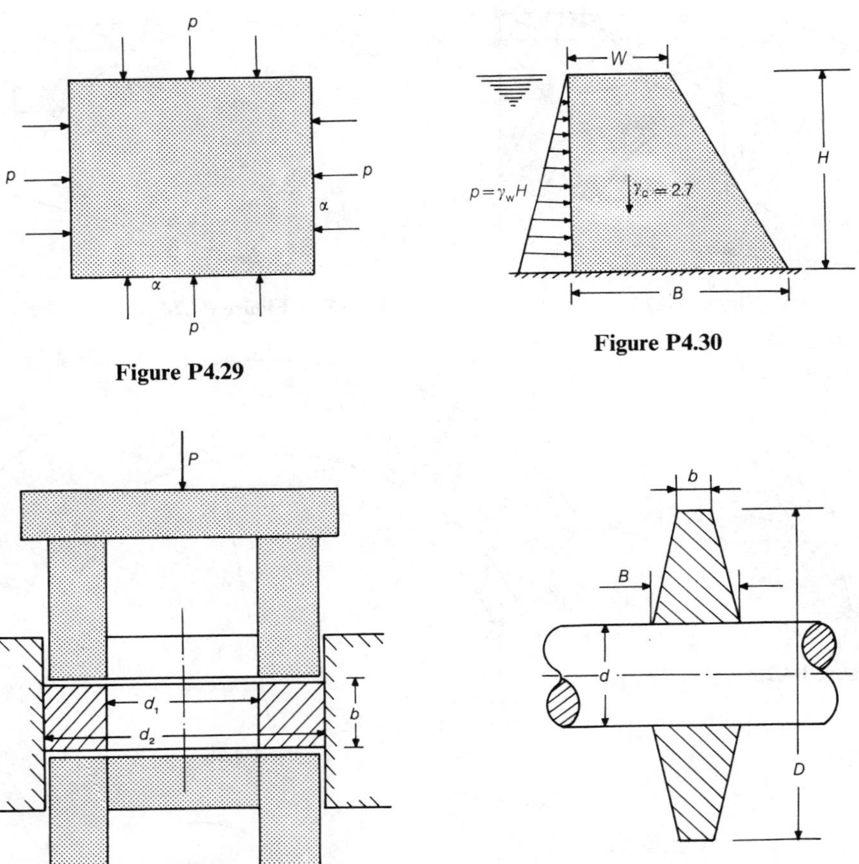

Figure P4.29

Figure P4.30

Figure P4.31

Figure P4.33

4.31 An aluminum ring with $d_1 = 60$ mm, $d_2 = 130$ mm, thickness $b = 30$ mm is fitted into a steel bore considered as perfectly rigid. If the force of the punch is 6 ton, determine the resulting pressure between the ring and the rigid bore. The punch rings can also be considered rigid. For aluminum, $E = 0.62 \times 10^{11}$ Pa.

4.32 In Problem 4.31, suppose that the steel ring of the bore is not rigid but it has outer diameter $D = 180$ mm, width $B = 60$ mm and the inner ring is symmetrically located. Find the resulting pressure.

4.33 A turbine disk with $d = 300$ mm, $D = 900$ mm, $B = 100$ mm, $b = 60$ mm rotates at 3,600 rpm. Use the axisymmetric program FINSTRES with node loads from inertia (centrifugal) dividing equally on the three nodes the load of each element. Find the maximum stress and the increase of the inner diameter d. Neglect interference pressure (no restraints of the inner diameter). Material, steel.

4.34 The revolving head of a machine to atomize the insulation plastic which covers the lower part of the large Coca-Cola bottles consists of a conical steel disk rotating at 12,000 rpm. Assuming no

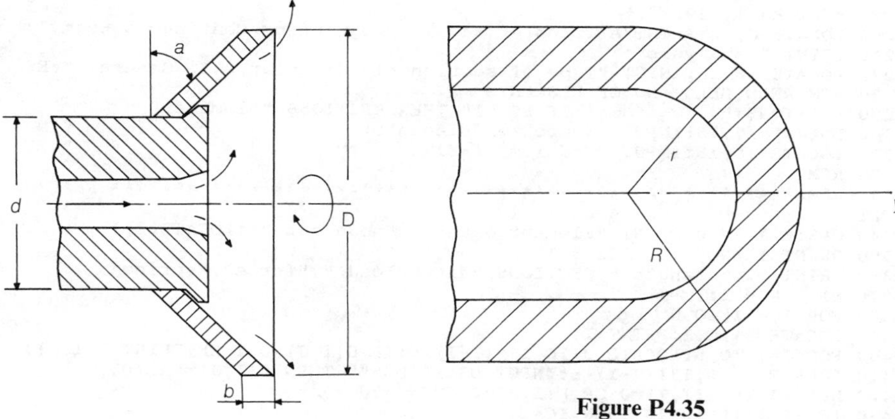

Figure P4.34

Figure P4.35

interference between shaft and disk, determine the maximum stress, with FINSTRES as in Problem 4.33.

Data: $D = 300$ mm; $d = 100$ mm; $b = 10$ mm; $a = 30°$.

4.35 The end of a high-pressure test tube is hemispherical, as shown in Figure P4.35. Using the axisymmetric FINSTRES program determine the maximum stresses. Take the straight part length equal to one diameter and boundary conditions on the left restraints of displacement along the y-axis. Data: $r = 100$ mm, $R = 70$ mm, $p = 10$ MPa.

APPENDIX 4.A
TMSTAT: A TRANSFER MATRIX ANALYSIS OF BEAMS PROGRAM

```
10 KEY OFF:CLS
20 REM ***************************************************
30 REM *                                                 *
40 REM *                 PROGRAM TMSTAT                  *
50 REM *                                                 *
60 REM ***************************************************
70 REM          COPYRIGHT 1987
80 REM by Professor Andrew D. Dimarogonas, Washington University,St.Louis.
90 REM
100 REM
110 REM
120 REM
130 REM
140 REM STATIC ANALYSIS OF CONTINUOUS BEAM BY THE TRANSFER MATRIX METHOD
150 REM
160 REM               BY ADD 6/11/84     Revised 5/17/87
170 REM
180 DIM D(200,5):REM PARAMETER MATRIX
190 DIM P(5,5),S(5,5),A(5,5),G(5,5):REM        AUXILIARY TRANSFER MATRICES
200 DIM V(5,1),W(5,1):REM                      STATE VECTORS
210 DIM IS(20)  :REM                           SUPPORT MATRIX
220 DIM RES(200,4):REM                         RESULTS MATRIX
230 FOR I=1 TO 14:COLOR I:LOCATE I,22+I:PRINT "TMSTAT";
240 LOCATE I,50-I:PRINT "TMSTAT";:NEXT I:COLOR 14
245 LOCATE 25,1:PRINT"Dimarogonas,A.D.,Computer Aided Machine Design,
```

```
Prentice-Hall, 1988";
250 LOCATE 8,12:PRINT"A Transfer Matrix Program for Continuous Beam ";
260 PRINT " analysis";
270 LOCATE 16,15:INPUT "Type of section,<1> Circular, <2> General ";E$
280 REM READ BELOW NO OF ELEMENTS
290 SECTION$="  Iy        " :IF E$="1" THEN SECTION$="Diameter"
300 LOCATE 17,15:INPUT "Number of Elements:                          ";N
310 LOCATE 18,15:INPUT "Modulus of Elasticity                        ";EY
320 REM
330 REM ****************************************************** READING
DATA
340 CLS:SCREEN 0:PRINT "Element data",,,"node data";:LOCATE 2,1
350 PRINT"El/No";:LOCATE 3,1
360 PRINT "     length ",SECTION$,"distr load ","force","spring";
370 N1 = N - 1:IC=1
380 FOR I = 1 TO N
390 LOCATE 3+IC,1:PRINT I;
400 FOR J=1 TO 5:LOCATE 3+IC,13*(J-1)+6:IF D(I,J)>0 THEN PRINT D(I,J);
410 LOCATE 3+IC,13*(J-1)+5:INPUT D$:IF D$>"" THEN D(I,J)=VAL(D$)
420 NEXT J:IF D(I,1)=0 OR D(I,2)=0 THEN 390
430 IC=IC+1:IF IC=10 THEN IC=1
440 NEXT I
450 LOCATE 3+IC:PRINT" right end....";
460 FOR J=4 TO 5 :LOCATE  3+IC,13*(J-1)+6:IF D(N+1,J)>0 THEN PRINT
D(N+1,J);
470 LOCATE 3+IC,13*(J-1)+5:INPUT D$:IF D$>"" THEN D(N+1,J)=VAL(D$)
480 NEXT J
490 REM ******************************** SET "STIFFENESS" OF RIGID SUPPORTS
500 REM
510 N2=INT((N+1)/2):EI=EY*D(N2,2):IF E$="1" THEN EI=D(N2,2)^4/64*EY*3.14
520 LL = N * D(N2,1):KS=(48 * EI) / LL ^ 3 * 1000
530 PRINT
540 INPUT"Enter no of solid supports";NS
550 PRINT"Enter numbers of support nodes:":PRINT
560 FOR I = 1 TO NS:PRINT "Support "; I;"  at node ";:INPUT IS(I):NEXT I
570 PRINT:INPUT"Are data correct (Y/N) ";X$
580 IF X$="N" OR X$="n" THEN 340
590 REM ************************************** FINDING MAXIMUM VALUES
600 CLS:LTOT=0:DMAX=0:QMAX=0:FMAX=0
610 FOR I=1 TO N+1
620 DIA=D(I,2):IF E$="2" THEN DIA=(64*D(I,2)/3.14)^.25
630 LTOT=LTOT+D(I,1):IF DIA>DMAX THEN DMAX=DIA
640 IF D(I,3)>QMAX THEN QMAX=D(I,3)
650 IF D(I,4)>FMAX THEN FMAX=D(I,4)
660 NEXT I
670 GOSUB 1780:REM  Plot beam and loading
680 LOCATE 23,1:INPUT"Hit return to continue";CON$
690 SCREEN 2:WIDTH 80
700 LOCATE 23,1:PRINT "COMPUTING ....PLEASE WAIT ...";
710 REM
720 REM ************************************FORWARD SWEEP OF TRANSFER
MATRICES
730 REM
740 REM                                                    SET A(5,5) TO
IDENTITY
750 FOR I=1 TO 5:FOR J=1 TO 5:A(I,J)=0:NEXT J:A(I,I) =1:NEXT I
760 REM                          COMPUTE AND MULTIPLY TM FROM LEFT TO
RIGHT
770 FOR M = 1 TO N + 1
780 F1 = D(M,4):K1=D(M,5)
790 IF M > N THEN GOTO 820
800 L1=D(M,1):D1=EY*D(M,2):Q1=D(M,3):IF E$="1" THEN
D1=3.14159*D(M,2)^4*EY/64
810 GOSUB 1510:REM                                     fetch field matrix
S
820 GOSUB 1620:REM                                     fetch point matrix
P
830 FOR I = 1 TO 5: FOR J = 1 TO 5:G(I,J) = A(I,J): NEXT J: NEXT I
840 GOSUB 1690:REM
A=P*G
850 IF M>N THEN GOTO 880
```

```
860 FOR I = 1 TO 5:FOR J = 1 TO 5:G(I,J) = A(I,J):NEXT J: NEXT I
870 GOSUB 1740: REM
A=S*G
880 NEXT M
890 FOR I = 1 TO 5:V(I,1) = 0:NEXT I
900 REM        SOLVE FOR UNKNOWN END CONDITIONS,HERE, FREE-FREE BEAM IS
ASSUMED
910 REM                              WITH V0=VL=0, MO=VL=0, UNKNOWN
XO,XL,FO,FL
920 D9 = A(3,1) * A(4,2) - A(4,1) * A(3,2)
930 V(1,1) = (-A(4,2) * A(3,5) + A(3,2) * A(4,5)) / D9
940 V(2,1) = (-A(3,1) * A(4,5) + A(4,1) * A(3,5)) / D9
950 V(5,1) = 1
960 REM              RESULTS, PROCEED FROM LEFT TO RIGHT MULTIPLYING WITH
TM'S
970 REM                              AND PRINTING STATE VECTOR AT EACH
STATION
980 CLS
990 PRINT: PRINT: PRINT "SOLUTION ..."
1000 PRINT "======================"
1010 PRINT:PRINT
1020 PRINT"Node";TAB(6);"Deflection";TAB(19);"Slope";TAB(33);"moment";
1030 PRINT TAB(47);"Shear";TAB(61);"Reaction"
1040
PRINT"_______________________________________________________________________
_______"
1050 PRINT
1060 FOR M = 1 TO N + 1
1070 IF M > N THEN GOTO 1160
1080 PRINT M;TAB(5);V(1,1);TAB(19);V(2,1);TAB(33);V(3,1);
1090 FOR IRES=1 TO 4:RES(M,IRES)=V(IRES,1):NEXT IRES
1100 PRINT TAB(47);V(4,1);TAB(61);
1110 FOR IS = 1 TO NS: IF IS(IS) < > M THEN 1140
1120 SR = KS * V(1,1)
1130 PRINT SR;:RES(M,4)=RES(M,4)-SR
1140 PRINT
1150 NEXT IS:RES(M,4)=RES(M,4)+D(M,4)-V(1,1)*D(M,5)
1160 F1 = D(M,4):K1 = D(M,5)
1170 IF M > N THEN GOTO 1200
1180 L1=D(M,1):D1=EY*D(M,2):Q1=D(M,3):IF E$="1" THEN
D1=3.14159*EY*D(M,2)^4/64
1190 GOSUB 1510
1200 GOSUB 1620
1210 FOR I = 1 TO 5: W(I,1) = V(I,1):NEXT I
1220 FOR I = 1 TO 5: V(I,1) = 0: FOR K = 1 TO 5
1230 V(I,1) = P(I,K) * W(K,1) + V(I,1): NEXT K:NEXT I
1240 IF M > N THEN GOTO 1280
1250 FOR I = 1 TO 5: W(I,1) = V(I,1):NEXT I
1260 FOR I = 1 TO 5: V(I,1) = 0:FOR K = 1 TO 5
1270 V(I,1) =S(I,K) * W(K,1) + V(I,1):NEXT K: NEXT I
1280 NEXT M
1290 M = M-1
1300 REM                                         PRINT RESULTS AT END
STATION
1310 PRINT M;TAB(5);V(1,1);TAB(19);V(2,1);TAB(33);V(3,1);
1320 PRINT TAB(47);V(4,1);TAB(61);
1330 FOR IRES=1 TO 4:RES(M,IRES)=V(IRES,1):NEXT IRES
1340 FOR IS= 1 TO NS: IF IS(IS) < > N + 1 THEN 1370
1350 SR = KS * V(1,1)
1360 PRINT SR
1370 NEXT IS:FOR I=1 TO N+1
1380 IF ABS(RES(I,1))>ABS(YMAX) THEN YMAX=RES(I,1)
1390 IF ABS(RES(I,2))>ABS(FMAX) THEN FMAX=RES(I,2)
1400 IF ABS(RES(I,3))>ABS(MMAX) THEN MMAX=RES(I,3)
1410 IF ABS(RES(I,4))>ABS(VMAX) THEN VMAX=RES(I,4)
1420 NEXT I : LOCATE 24,1
1430 INPUT "print return to continue";X$
1440 CLS:GOSUB 2050:LOCATE 24,60:INPUT "Continue ? (Y/N) ";X$
1450 IF X$="n" OR X$="N" THEN 1470
1460 GOTO 340
1470 SCREEN 0:CLS:COLOR 14:END
```

```
1480 REM
1490 REM *****************************************TRANSFER MATRIX
SUBROUTINES
1500 REM
1510 REM *****************************************FIELD MATRIX - EULER BEAM
1520 FOR I = 1 TO 5: S(I,I) = 1: NEXT I
1530 EI = D1:S(1,2) = L1
1540 S(3,4) =L1:S(1,3) = L1 ^ 2 / (2 * EI)
1550 S(2,4) =S(1,3):S(2,3) = L1 / EI
1560 S(1,4) =S(1,3) * L1 /3
1570 S(1,5) =Q1 * L1 ^ 4 / (24 * EI)
1580 S(2,5) =S(1,5) * 4 / L1
1590 S(3,5) = Q1 * L1 ^ 2 / 2
1600 S(4,5) = S(3,5) * 2 /L1
1610 RETURN
1620 REM ******************POINT MATRIX:FORCE, DISTRIBUTED LOAD AND SPRING
1630 FOR I = 1 TO 5: FOR J= 1 TO  5: P(I,J) = 0: NEXT J: NEXT I
1640 FOR I = 1 TO 5: P(I,I) = 1:NEXT I
1650 FOR I = 1 TO NS:SI = IS(I)
1660 IF M = SI THEN K1 = K1+KS
1670 NEXT I
1680 P(4,5) = F1: P(4,1) = - K1: RETURN
1690 REM A = P * G
1700 FOR I = 1 TO 5: FOR J = 1 TO 5
1710 C = 0
1720 FOR K = 1 TO 5:C = C + P(I,K) * G(K,J):NEXT K:A(I,J) = C: NEXT J:NEXT
I
1730 RETURN
1740 REM A = S * G
1750 FOR I = 1 TO 5: FOR J = 1 TO 5: A(I,J) = 0
1760 FOR K = 1 TO 5: A(I,J) = A(I,J) + S(I,K) * G(K,J)
1770 NEXT K:NEXT J: NEXT I:RETURN
1780 REM ******************************************* PLOTTING BEAM AND
LOADS
1790 REM
1800 SCREEN 2:CLS:XSC=300/LTOT:YSC=XSC/2.5:LSTART=0
1810 FOR I=1 TO N+1
1820 DIA=D(I,2):IF E$="2" THEN DIA=(64*DIA/3.14)^.25
1830 IF DIA=0 THEN DIA=DIAPR
1840 DIAPR=DIA
1850 X1=10+LSTART:Y1=50-DIA*YSC/2:X2=10+LSTART+D(I,1)*XSC:Y2=50+DIA*YSC/2
1860 LINE(X1,Y1)-(X2,Y2),7,BF:IF D(I,4)=0 THEN 1890:REM              plot
beam
1870 LINE (X1-1,Y1)-(X1+1,Y1-20),3,BF:REM                             plot
force
1880 LINE (X1,Y1)-(X1+3,Y1-5),3:LINE -(X1-3,Y1-5),3:LINE -(X1,Y1),3:REM
same
1890 FOR II=1 TO NS:IF IS(II)<>I THEN GOTO 1910
1900 LINE(X1,Y2)-(X1-3,Y2+5):LINE -(X1+3,Y2+5):LINE -(X1,Y2):REM    plot
support
1910 NEXT II
1920 IF D(I,3)>0 THEN LINE(X1,20)-(X2,20-D(I,3)*10/QMAX),1,BF:REM   plot q
1930 IF D(I,5)>0 THEN LINE(X1,Y2)-(X1+10,Y2+5):LINE -(X1-10,Y2+10):'plot
spring
1940 IF D(I,5)>0 THEN LINE -(X1,Y2+15):REM                            plot
spring
1950 LSTART=LSTART+D(I,1)*XSC:NEXT I
1960 FOR II=1 TO NS:IF IS(II)><N+1 THEN  1980
1970 LINE(X2,Y2)-(X2-3,Y2+5):LINE -(X2+3,Y2+5):LINE -(X2,Y2):REM    plot
support
1980 NEXT II
1990 LOCATE 3,60:PRINT "q-max= ";QMAX;
2000 LOCATE 7,60:PRINT "d-max= ";DMAX;
2010 LOCATE 10,60:PRINT "y-max= ";YMAX;
2020 LOCATE 16,60:PRINT "M-max= ";MMAX;
2030 LOCATE 22,60:PRINT "V-max= ";VMAX
2040 RETURN
2050 REM ********************************************** plot results
2060 GOSUB 1780
2070 LSTART=0
2080 FOR I=1 TO N
```

```
2090 X1=10+LSTART:Y1=70+RES(I,1)*20/YMAX:TSC=20/YMAX/XSC
2100 X2=10+LSTART+D(I,1)*XSC:Y2=70+RES(I+1,1)*20/YMAX
2110 SL1=RES(I,2)*TSC:SL2=RES(I+1,2)*TSC:PSET(X1,Y1)
2120 V=-RES(I,4):MOM=RES(I,3):VMAX=ABS(VMAX):MMAX=ABS(MMAX)
2130 FOR X=X1 TO X2:S=(X-X1)/(X2-X1):F1=1-3*S^2+2*S^3:F2=3*S^2-2*S^3
2140 F3=(X2-X1)*(S-2*S^2+S^3):F4=(X2-X1)*(-S^2+S^3)
2150 Y=(F1*Y1+F2*Y2+F3*SL1+F4*SL2)
2160 LINE -(X,Y),3:NEXT X:FOR X=X1 TO X2
2170 VPR=V:V=V-1*D(I,3)/XSC
2180 MPR=MOM:MOM=MOM-1*V/XSC
2190 PSET (X,-V*20/VMAX+170),3:PSET (X,-MOM*20/MMAX+120),3
2200 NEXT X
2210 LINE (X1,70)-(X2,70),7
2220 LINE (X1,120)-(X2,120),7
2230 LINE (X1,170)-(X2,170),7
2240 LSTART=LSTART+D(I,1)*XSC
2250 NEXT I
2260 RETURN
```

APPENDIX 4.B
PREFRAME: AN INPUT PREPARATION PROGRAM FOR FINFRAME

```
10 REM ***********************************************
20 REM *                                             *
30 REM *                PREFRAME                     *
40 REM *                                             *
50 REM ***********************************************
60 REM          COPYRIGHT 1987
70 REM by Professor Andrew D. Dimarogonas, Washington University,St.Louis.
80 REM All rights reserved.  Unauthorized reproduction, dissemination or
90 REM selling is strictly prohibited.  This listing is for personal use.
100 REM
110 REM
120 REM
130 REM Development of frame models for structural analysis
140 REM with the FINFRAME program
150 REM
160 REM        ADD 5-30-1987
170 REM *********************************************************
180 CLS:KEY OFF
190 FOR I=1 TO 14:COLOR I:LOCATE I,22+I:PRINT "PREFRAME";
200 LOCATE I,50-I:PRINT "PREFRAME";:NEXT I:COLOR 14
210 LOCATE 8,7:PRINT"A 3-D graphics package to create frame models for";
220 PRINT " structural analysis";
230 LOCATE 18,20
240 INPUT"Enter xmax,ymax (600,200 for screen coordinates) ";XMAX,YMAX
250 XSC=600/XMAX:YSC=200/YMAX
260 DIM PLOAD(100,8):REM                              Node loads
270 DIM PD(10,50):REM                                 Polygon definition
280 DIM PROP(100,8):REM                               Element properties
290 DIM ELEM(100,3):REM                               Element definitions
300 DIM X(100),Y(100),Z(100):REM                      Node coordinates
310 DIM IDREL(100,8):REM                              Joint releases
320 SCREEN 2:XS=639:YS=199:ID=0
330 CLS:NMENUS=7
340 KEY OFF
350 DIM L$(100)
360 VIND=1:HIND=1:FOR I=1 TO 100:FOR J=1 TO 6:IDREL(I,J)=1:NEXT J:NEXT I
370 DIM MENU1$(25)
380 FOR IM=1 TO NMENUS:READ MENU1$(IM):NEXT IM
390 DATA "Stop       "
400 DATA "Screenplot","New        ","Save File ","Load File ","Make model"
410 DATA "Clear scrn"
420 XS=0
430 LOCATE 1,70:PRINT "MAIN MENU"
440 FOR I=1 TO NMENUS:LOCATE I+2,70:PRINT MENU1$(I);:NEXT I
450 GOTO 570
460 ICOUNT=0
```

```
470 ICOUNT=ICOUNT+1:X$=INKEY$:IF ICOUNT<10 THEN 470
480 X$=INKEY$:XS=0
490 FOR I=17 TO 25:LOCATE I,66:PRINT SPC(14);:NEXT I
500 LOCATE 17,70:PRINT"COMMAND";
510 IF X$="" THEN 480
520 IF LEN(X$)=1 THEN XS=ASC(X$):GOTO 630
530 X$=RIGHT$(X$,1)
540 IF X$><"H" AND X$<>"P" THEN 600
550 IF X$="H" AND VIND>1 THEN VIND=VIND-1
560 IF X$="P" AND VIND<NMENUS THEN VIND=VIND+1
570 IF VIND>1 THEN LOCATE VIND+1,69:PRINT " ";MENU1$(VIND-1);
580 LOCATE VIND+2,69:PRINT ">";MENU1$(VIND);
590 IF VIND<NMENUS THEN LOCATE VIND+3,69:PRINT " ";MENU1$(VIND+1);
600 LOCATE 19,70:PRINT"         ";
610 LOCATE 21,65:PRINT"              ";
620 LOCATE 22,65:PRINT"              ";
630 IF XS=13 AND VIND=1 THEN SCREEN 0:END
640 IF XS=13 AND VIND=2 THEN GOSUB 3750:REM    screen plot
650 IF XS=13 AND VIND=3 THEN GOSUB  720:REM    start a new drawing
660 IF XS=13 AND VIND=4 THEN GOSUB 1010:REM    save file on disc
670 IF XS=13 AND VIND=5 THEN GOSUB 780:REM    load file from disc
680 IF XS=13 AND VIND=6 THEN GOSUB 1390:'     make a new file
690 IF XS=13 AND VIND=7  THEN CLS:GOTO 420:'   clear screen
700 GOTO 470
710 END
720 REM start a new drawing
730 LOCATE 22,65:INPUT"SURE (Y/N)";X$:IF X$="n" OR X$="N" THEN RETURN
740 ND=0:NP=0:NC=0:RETURN
750 REM*************************************************************
760 REM                 file retrieval
770 REM*************************************************************
780 LOCATE 19,68:PRINT"            ";
790 LOCATE 18,70:PRINT"Data file  ":LOCATE 19,68:INPUT FILNA$
800 IF FILNA$="" THEN 790
810 REM ON ERROR GOTO 990
820 NDS=ND:ICOUNT=0
830 OPEN FILNA$ FOR INPUT AS #1
840 ICOUNT=ICOUNT+1:INPUT#1,X$
850 IF X$="framend" OR X$="FRAMEND" THEN 970
860 SYM$=""
870 ISYM=1:IF ICOUNT>1000 THEN 990
880 SYM$=MID$(X$,ISYM,1):ISYM=ISYM+1:IF SYM$="" AND ISYM<10 THEN 870
890 IF ASYM=10 THEN 840
900 IF SYM$="n" OR SYM$="N" THEN ND=ND+1:INPUT#1,X(ND),Y(ND),Z(ND):GOTO 840
910 IF SYM$="e" OR SYM$="E" THEN NE=NE+1:GOTO 940
920 GOTO 840
930 PRINT"FILE ERROR...":STOP
940 FOR J=1 TO 3
950 INPUT#1,ELEM(NE,J):NEXT J
960 GOTO 840
970 CLOSE#1
980 GOTO 1000
990 LOCATE 19,68:PRINT "BAD FILE     "
1000 RETURN
1010 REM ********************************STORING TRANSFORMED DATA ON DISC
1020 LOCATE 19,68:PRINT"            "
1030 LOCATE 18,70:PRINT"New file  ":LOCATE 19,68:INPUT FILNA$
1040 CLOSE
1050 OPEN FILNA$ FOR OUTPUT AS #1
1060 FOR I=1 TO ND
1070 PRINT#1,"NODE,";X(I);",";Y(I);",";Z(I):NEXT I
1080 REM **DATA** DEFINITION OF ELEMENTS
1090 FOR I=1 TO NE
1100 PRINT#1,"ELEMENT ";
1110 FOR J=1 TO 3
1120 PRINT#1,",";ELEM(I,J);:NEXT J
1130 PRINT#1,"  " :NEXT I
1140 REM                                    **DATA** element properties
1150 FOR I=1 TO NE
1160 PRINT#1,"PROPERTY ,";I;
1170 FOR J=1 TO 6
```

```
1180 PRINT#1,",";PROP(I,J);:NEXT J
1190 PRINT#1,"  " :NEXT I
1200 REM                                 **DATA** node loads
1210 FOR I=1 TO ND:INDEX=0:FOR J=1 TO 6:IF PLOAD(I,J)>0 THEN INDEX =1
1220 NEXT J:IF INDEX=0 THEN 1270
1230 PRINT#1,"LOADING @ ,";I;
1240 FOR J=1 TO 6
1250 PRINT#1,",";PLOAD(I,J);:NEXT J
1260 PRINT#1,"  "
1270 NEXT I
1280 REM                                 **DATA** node releases
1290 FOR I=1 TO ND:INDEX=0:FOR J=1 TO 6:IF IDREL(I,J)=0 THEN INDEX =1
1300 NEXT J:IF INDEX=0 THEN 1350
1310 PRINT#1,"RELEASE @ ,";I;
1320 FOR J=1 TO 6
1330 PRINT#1,",";IDREL(I,J);:NEXT J
1340 PRINT#1,"  "
1350 NEXT I
1360 PRINT#1,"FRAMEND"
1370 CLOSE#1
1380 RETURN
1390 REM
**********************************************************************
1400 REM                    frame creation
1410 REM
**********************************************************************
1420 NPOL=NP
1430 LINE (0,0)-(261,91),,B:LOCATE 1,1:PRINT "Z";:LOCATE 1,80:PRINT"X";
1440 LOCATE 1,30:PRINT "X";:LOCATE 1,60:PRINT "Z";
1450 LINE (279,0)-(540,91),,B:LOCATE 1,36:PRINT "X";
1460 LOCATE 11,1:PRINT "Y";:LOCATE 11,36:PRINT "Y";
1470 LINE (0,99)-(261,191),,B:LOCATE 13,1:PRINT "Y";
1480 LOCATE 13,30:PRINT "X";:LOCATE 24,1:PRINT "Z";
1490 REM LINE (279,99)-(540,191),,B
1500 XSTEP=10/XSC:YSTEP=XSTEP:ZSTEP=XSTEP
1510 XHAIR=XSTEP:YHAIR=YSTEP:ZHAIR=ZSTEP
1520 LOCATE 13,35
1530 PRINT " Keyboard commands:":LOCATE 14,35
1540 PRINT " N= define node, E= element    ":LOCATE 15,35
1550 PRINT " L= load          R= release   ";:LOCATE 16,35
1560 PRINT " F= faster        S= slower    " ;
1570 LOCATE 17,35
1580 PRINT " P= polygon       D= displace ":LOCATE 18,35
1590 PRINT " C= connect       M= Menu";
1600 LOCATE 19,35
1610 PRINT " X-Y-Z screens    W= Write";
1620 X$="z":GOTO 1790
1630 X$=INKEY$:XS=0
1640 IXS=IXS+1:IF IXS<10 THEN LOCATE 25,1:PRINT">";
1650 IF IXS>10 THEN LOCATE 25,1:PRINT" ";
1660 IF IXS>20 THEN IXS=0
1670 PRINT X$;
1680 IF X$="" THEN 1630
1690 IF LEN(X$)>1  THEN 1790
1700 REM
1710 IF X$="f"  THEN XSTEP=XSTEP*2:YSTEP=XSTEP:ZSTEP=XSTEP
1720 IF X$="s"  THEN XSTEP=XSTEP/2:YSTEP=XSTEP:ZSTEP=XSTEP
1730 IF X$="F"  THEN XSTEP=XSTEP*2:YSTEP=XSTEP:ZSTEP=XSTEP
1740 IF X$="S"  THEN XSTEP=XSTEP/2:YSTEP=XSTEP:ZSTEP=XSTEP
1750 IF X$<>"G" AND X$<>"g" THEN 1780
1760 FOR I= 10 TO 80 STEP 10:FOR J=9 TO 540 STEP 20:PSET(J,I):NEXT J:NEXT I
1770 FOR I=100 TO 190 STEP 10:FOR J=9 TO 260 STEP 20:PSET(J,I):NEXT J:NEXT
I
1780 IF LEN(X$)=1 THEN XS=ASC(X$):GOTO 2140
1790 X$=RIGHT$(X$,1)
1800 FOR I=5 TO 260 STEP 20:PSET(I,YHAIR*YSC),0:NEXT I
1810 FOR I=280 TO 539 STEP 20:PSET(I,YHAIR*YSC),0:NEXT I
1820 FOR I=5 TO 260 STEP 20:PSET(I,100+ZHAIR*YSC),0:NEXT I
1830 FOR I=6 TO 90 STEP 10:PSET(XSC*XHAIR,I),0:NEXT I
1840 FOR I=6 TO 90 STEP 10:PSET(280+XSC*ZHAIR,I),0:NEXT I
1850 FOR I=106 TO 190 STEP 10:PSET(XSC*XHAIR,I),0:NEXT I
```

```
1860 IF X$="H" AND PLANE=1 THEN YHAIR=YHAIR-YSTEP
1870 IF X$="H" AND PLANE=2 THEN YHAIR=YHAIR-YSTEP
1880 IF X$="H" AND PLANE=3 THEN ZHAIR=ZHAIR-ZSTEP
1890 IF X$="P" AND PLANE=1 THEN YHAIR=YHAIR+YSTEP
1900 IF X$="P" AND PLANE=2 THEN YHAIR=YHAIR+YSTEP
1910 IF X$="P" AND PLANE=3 THEN ZHAIR=ZHAIR+ZSTEP
1920 IF X$="K" AND PLANE=1 THEN XHAIR=XHAIR-XSTEP
1930 IF X$="K" AND PLANE=2 THEN ZHAIR=ZHAIR-ZSTEP
1940 IF X$="K" AND PLANE=3 THEN XHAIR=XHAIR-XSTEP
1950 IF X$="M" AND PLANE=1 THEN XHAIR=XHAIR+XSTEP
1960 IF X$="M" AND PLANE=2 THEN ZHAIR=ZHAIR+ZSTEP
1970 IF X$="M" AND PLANE=3 THEN XHAIR=XHAIR+XSTEP
1980 IF XHAIR<0 THEN XHAIR=0
1990 IF YHAIR<0 THEN YHAIR=0
2000 IF ZHAIR<0 THEN ZHAIR=0
2010 IF YHAIR>YMAX THEN YHAIR=YMAX
2020 IF ZHAIR>YMAX THEN ZHAIR=YMAX
2030 IF XHAIR>XMAX THEN XHAIR=XMAX
2040 LOCATE 25,3:PRINT SPC(75);
2050 LOCATE 25,10:PRINT "x= ";XHAIR;:LOCATE 25,20:PRINT "y= ";YHAIR;
2060 LOCATE 25,30:PRINT "z= ";ZHAIR;" Nodes= ";ND;" Elements= ";NE;
2070 FOR I=5 TO 260 STEP 20:PSET(I,YHAIR*YSC):NEXT I
2080 FOR I=280 TO 539 STEP 20:PSET(I,YHAIR*YSC):NEXT I
2090 FOR I=5 TO 260 STEP 20:PSET(I,100+ZHAIR*YSC):NEXT I
2100 FOR I=6 TO 90 STEP 10:PSET(XSC*XHAIR,I):NEXT I
2110 FOR I=6 TO 90 STEP 10:PSET(280+XSC*ZHAIR,I):NEXT I
2120 FOR I=106 TO 190 STEP 10:PSET(XSC*XHAIR,I):NEXT I
2130 GOTO 1630
2140 IF X$<>"W" AND X$<>"w" THEN 2250
2150 WORD$=""
2160 W$=INKEY$:IF W$="" THEN 2160
2170 IF ASC(W$)=13 THEN 2250
2180 IF ASC(W$)=8 THEN WORD$="":W$=""
2190 LOCATE YSC*YHAIR/8,XSC*XHAIR/7.875
2200 WORD$=WORD$+W$:PRINT WORD$;
2210 LOCATE YSC*YHAIR/8,(XSC*ZHAIR+280)/7.875
2220 PRINT WORD$;
2230 LOCATE (100+YSC*ZHAIR)/8,XSC*XHAIR/7.875
2240 PRINT WORD$;:GOTO 2160
2250 IF X$="x" OR X$="X" THEN PLANE=2
2260 IF X$="y" OR X$="Y" THEN PLANE=3
2270 IF X$="z" OR X$="Z" THEN PLANE=1
2280 IF X$><"n" AND X$<>"N" THEN 2420
2290 XPL=XHAIR:YPL=YHAIR:ZPL=ZHAIR
2300 ND=ND+1:X(ND)=XPL:Y(ND)=YPL:Z(ND)=ZPL
2310 N1=N2:N2=N3:N3=ND:LOCATE 25,3:PRINT SPC(75);
2320 LOCATE 25,3:PRINT "Start of a new polygon   (Y/N or Cancel) ? ";
2330 Z$=INKEY$:LOCATE 25,45:PRINT"...enter...";:IF Z$="" THEN 2330
2340 IF Z$="C" OR Z$="c" THEN PRINT "..cancelled..";:ND=ND-1:GOTO 1630
2350 INODE=ND:IF ND=1 THEN 2356
2352 ION=0:FOR I=1 TO ND-1:D1=X(I)-XHAIR:D2=Y(I)-YHAIR:D3=Z(I)-ZHAIR
2354 TOL=XSTEP^2:IF D1^2+D2^2+D3^2<TOL THEN ION=1:INODE=I:I=ND
2355 NEXT I
2356 IF Z$="y" OR Z$="Y" THEN PD(NP+1,2)=INODE:ICPOL=1
2360 LOCATE 25,3:PRINT SPC(75);:IF ION=1 THEN ND=ND-1
2370 IF ION=0 THEN LOCATE 25,45:PRINT "new node= ";ND;
2380
CIRCLE(XSC*XPL,YSC*YPL),3:CIRCLE(280+XSC*ZPL,YSC*YPL),3:CIRCLE(XSC*XPL,100+
YSC*ZPL),3
2390 LOCATE 23,40:PRINT"Node data (Y/N) ";:INPUT N$
2400 IF N$="y" OR N$="Y" THEN GOSUB 3520
2410 GOTO 1630
2420 IF X$><"e" AND X$<>"E" THEN 2660
2430 LOCATE 25,2:PRINT SPC(75);
2440 LOCATE 25,3:PRINT "Hit any key, C to Cancel";
2450 L$="":L$=INKEY$:IF L$="" THEN 2450
2460 IF L$="c" OR L$="c" THEN PRINT "..CANCELLED...";:GOTO 2710
2470 XPL=XHAIR:YPL=YHAIR:ZPL=ZHAIR
2480 NN1=INODE:NN2=ND+1
2490 LINE (XSC*X(NN1),YSC*Y(NN1)) -(XSC*XPL,YSC*YPL)
2500 LINE (280+XSC*Z(NN1),YSC*Y(NN1))-(280+XSC*ZPL,YSC*YPL)
```

```
2510 LINE (XSC*X(NN1),100+YSC*Z(NN1)) -(XSC*XPL,100+YSC*ZPL)
2520 INODE=ND+1:ION=0:FOR I=1 TO ND:D1=X(I)-XHAIR:D2=Y(I)-YHAIR:D3=Z(I)-
ZHAIR
2530 SS=XSTEP^2:IF D1^2<SS AND D2^2<SS AND D3^2<SS THEN ION=1:INODE=I:I=ND
2540 NEXT I
2550 IF ION=1  THEN N2=INODE:GOTO 2620:REM                node already defined
2560 ND=ND+1:X(ND)=XPL:Y(ND)=YPL:Z(ND)=ZPL
2570
CIRCLE(XSC*XPL,YSC*YPL),3:CIRCLE(280+XSC*ZPL,YSC*YPL),3:CIRCLE(XSC*XPL,100+
YSC*ZPL),3
2580 LOCATE 25,3:PRINT SPC(75);
2590 LOCATE 25,45:PRINT "new node= ";ND;
2600 LOCATE 23,40:PRINT"Node data (Y/N) ";:INPUT N$
2610 IF N$="y" OR N$="Y" THEN GOSUB 3520
2620 ICPOL=ICPOL+1:PD(NPOL+1,1)=ICPOL:PD(NPOL+1,ICPOL+1)=INODE
2630 NE=NE+1:ELEM(NE,1)=NE:ELEM(NE,2)=PD(NPOL+1,ICPOL):ELEM(NE,3)=INODE
2640 GOSUB 3290
2650 N1=N2:N2=N3:N3=INODE
2660 IF LEFT$(X$,1)><"p" AND X$<>"P" THEN 2710
2670 NPOL=NPOL+1:ICPOL=0
2680 NP=NPOL
2690 LOCATE 25,3:PRINT SPC(75);
2700 LOCATE 25,45:PRINT "new polygon no  ";NPOL
2710 REM
2720  IF X$ = "m" OR X$ = "M" THEN 2770
2730 IF X$="d" OR X$="D" THEN GOSUB 2780
2740 IF X$="c" OR X$="C" THEN GOSUB 3080
2750 NP=NPOL
2760 GOTO 1630
2770 RETURN
2780 REM ************************************** Displacing last polygon
2790 LOCATE 25,3:PRINT SPC(75);
2800 DISPL$=""
2810 LOCATE 25,3:PRINT"  enter displacement....C to cancel";
2820 LOCATE 25,3:PRINT"                                   ";:P$=INKEY$:IF P$="" THEN
2810
2830 DISPL$=DISPL$+P$:LOCATE 25,45+LEN(DISPL$):PRINT P$;
2840 IF P$="c" OR P$="C" THEN PRINT"..CANCELLED..";:RETURN
2850 IF ASC(P$)<>13 THEN 2810
2860 DISPL=VAL(DISPL$)
2870 NPNODES=PD(NPOL,1):PD(NPOL+1,1)=NPNODES
2880 FOR I=1 TO PD(NPOL,1):NODE=ABS(PD(NPOL,I+1))
2890 ND=ND+1:IF PLANE=1 THEN
X(ND)=X(NODE):Y(ND)=Y(NODE):Z(ND)=Z(NODE)+DISPL
2900 IF PLANE=2 THEN X(ND)=X(NODE)+DISPL:Y(ND)=Y(NODE):Z(ND)=Z(NODE)
2910 IF PLANE=3 THEN X(ND)=X(NODE):Y(ND)=Y(NODE)+DISPL:Z(ND)=Z(NODE)
2920 LOCATE 23,40:INPUT"Node data (Y/N) ";N$:IF N$="y" OR N$="Y" THEN GOSUB
3520
2930 XPR=X(ND-1):YPR=Y(ND-1):ZPR=Z(ND-1)
2940 XPL=X(ND):YPL=Y(ND):ZPL=Z(ND)
2950
CIRCLE(XSC*XPL,YSC*YPL),3:CIRCLE(280+XSC*ZPL,YSC*YPL),3:CIRCLE(XSC*XPL,100+
YSC*ZPL),3
2960 IF I>1 THEN NE=NE+1:ELEM(NE,1)=NE:ELEM(NE,2)=ND-1:ELEM(NE,3)=ND
2970 IF I>1 THEN GOSUB 3290:REM              enter element properties
2980 PD(NPOL+1,I+1)=ND
2990 IF I=1 THEN 3030
3000 LINE (XSC*XPR,YPR) -(XSC*XPL,YPL*YSC)
3010 LINE (280+XSC*ZPR,YPR)-(280+XSC*ZPL,YPL*YSC)
3020 LINE (XSC*XPR,100+ZPR) -(XSC*XPL,100+ZPL*YSC)
3030 NEXT I
3040 NPOL=NPOL+1
3050 PD(NPOL,NPNODES+1)=PD(NPOL,2)
3060 ELEM(NE,3)=PD(NPOL,2):ND=ND-1
3070 RETURN
3080 REM ******************************** connect last two polygons
3090 IF NPOL<2 THEN LOCATE 25,3:PRINT" No polygons. Define them..";:RETURN
3100 LOCATE 25,3:PRINT SPC(75);
3110 LOCATE 25,3:PRINT "..press any key, ...C to cancel ";
3120 C$="":C$=INKEY$:IF C$="" THEN 3120
3130 IF C$="c" OR C$="C" THEN PRINT"..cancelled..";:RETURN
```

```
3140 NPNODES=PD(NPOL,1)
3150 FOR I=1 TO NPNODES-1
3160 I1=PD(NPOL-1,I+1):I2=PD(NPOL-1,I+2)
3170 J1=PD(NPOL,I+1):J2=PD(NPOL,I+2)
3180 NE=NE+1:ELEM(NE,1)=NE:ELEM(NE,2)=I1:ELEM(NE,3)=J1
3190 GOSUB 3290:REM              enter element properties
3200 FOR J=1 TO 2:L1=ELEM(NE-2+J,2):L2=ELEM(NE-2+J,3)
3210 XPR=X(L1):YPR=Y(L1):ZPR=Z(L1)
3220 XPL=X(L2):YPL=Y(L2):ZPL=Z(L2)
3230 LINE (XSC*XPR,YPR*YSC) -(XSC*XPL,YPL*YSC)
3240 LINE (280+XSC*ZPR,YPR*YSC)-(280+XSC*ZPL,YPL*YSC)
3250 LINE (XSC*XPR,100+ZPR*YSC) -(XSC*XPL,100+ZPL*YSC)
3260 NEXT J
3270 NEXT I
3280 RETURN
3290 REM ****************************************input of element
properties
3300 LOCATE 21,40:PRINT"Enter element properties..........";
3310 LOCATE 22,40:PRINT "element number                ";NE;"  ";
3320 LOCATE 23,40:PRINT SPC(39);
3330 LOCATE 23,40:INPUT "New data or previous (N/P)      ";X$
3340 LOCATE 23,40:PRINT SPC(39);
3350 IF X$="n" OR X$="N" THEN 3390
3360 LOCATE 23,40:INPUT "No of previous same element     ";NSE
3370 LOCATE 23,40:PRINT SPC(39);
3380 FOR IEQV=1 TO 6:PROP(NE,IEQV)=PROP(NSE,IEQV):NEXT IEQV:GOTO 3510
3390 LOCATE 23,40:INPUT"Material type               ";X$:PROP(NE,1)=VAL(X$)
3400 LOCATE 23,40:PRINT SPC(39);
3410 LOCATE 23,40:INPUT"Area of cross section  ";X$:PROP(NE,2)=VAL(X$)
3420 LOCATE 23,40:PRINT SPC(39);
3430 LOCATE 23,40:INPUT"Moment of inertia Iy   ";X$:PROP(NE,3)=VAL(X$)
3440 LOCATE 23,40:PRINT SPC(39);
3450 LOCATE 23,40:INPUT"Moment of inertia Iz   ";X$:PROP(NE,4)=VAL(X$)
3460 LOCATE 23,40:PRINT SPC(39);
3470 LOCATE 23,40:INPUT"Moment of inertia Ip   ";X$:PROP(NE,5)=VAL(X$)
3480 LOCATE 23,40:PRINT SPC(39);
3490 LOCATE 23,40:INPUT"Roll angle of section  ";X$:PROP(NE,6)=VAL(X$)
3500 LOCATE 23,40:PRINT SPC(39);
3510 X$="":RETURN
3520 REM ****************************************input of node data
3530 LOCATE 21,40:PRINT"Enter node data                   ";
3540 LOCATE 22,40:PRINT "node number           ";ND;"          ";
3550 LOCATE 23,40:PRINT SPC(39);
3560 LOCATE 23,40:INPUT "Enter number of loads (0=none)   ";X$
3570 IF X$="" THEN 3630
3580 LOADN=LOADN+1
3590 NLOADS=VAL(X$):FOR ILOAD=1 TO NLOADS
3600 LOCATE 23,40:PRINT SPC(39);
3610 LOCATE 23,40:PRINT "Fx,Fy,Fz,Mx,My,Mz (1-6),value";:INPUT IT,XLOAD
3620 PLOAD(ND,IT)=XLOAD:NEXT ILOAD
3630 LOCATE 23,40:PRINT SPC(39);
3640 LOCATE 23,40:INPUT"Enter number of releases (0=none) ";X$
3650 IF X$="" THEN 3710
3660 NREL=NREL+1
3670 NR=VAL(X$):FOR IREL=1 TO NR
3680 LOCATE 23,40:PRINT SPC(39);
3690 LOCATE 23,40:PRINT "Release Rx,Ry,Rz,Fx,Fy,Fz (1-6) ";:INPUT IT
3700 IDREL(ND,IT)=0:NEXT IREL
3710 X$=""
3720 LOCATE 21,40:PRINT SPC(38);
3730 LOCATE 22,40:PRINT SPC(38);
3740 RETURN
3750 REM ************************************************** plotting frame
3760 FOR I=1 TO NE
3770 N=ELEM(I,2):M=ELEM(I,3)
3780 LINE (XSC*X(N),YSC*Y(N)) -(XSC*X(M),YSC*Y(M))
3790 LINE (280+XSC*Z(N),YSC*Y(N))-(280+XSC*Z(M),YSC*Y(M))
3800 LINE (XSC*X(N),100+YSC*Z(N)) -(XSC*X(M),100+YSC*Z(M))
3810 NEXT I
3820 RETURN
```

APPENDIX 4.C
FINFRAME: A FINITE-ELEMENT PROGRAM FOR STATIC ANALYSIS OF SPACE FRAMES

```
10 CLS
20 REM **************************************************
30 REM *                                                *
40 REM *                  FINFRAME                      *
50 REM *                                                *
60 REM **************************************************
70 REM             COPYRIGHT 1985
80 REM by Professor Andrew D. Dimarogonas, University of Patras, Greece.
90 REM   All rights reserved. Unauthorized reproduction, dissemination,
100 REM  selling or use is strictly prohibited.
110 REM  This listing is for reference purpose only.
120 REM
130 REM
140 REM A finite element program for analysis of space frames
150 REM  with prismatic beam elements.
160 REM A banded matrix storage scheme with Gauss elimination is used.
170 REM
180 REM   ADD/10-12-84
190 REM   Revised,4-19-87
200 REM
210 REM
220 REM ************************************************************
230 CLS:KEY OFF
240 FOR I=1 TO 14:COLOR I:LOCATE I,22+I:PRINT "FINFRAME";
250 LOCATE I,50-I:PRINT "FINFRAME";:NEXT I:COLOR 14
260 LOCATE 8,12:PRINT"A 3-D frame structural analysis program for static";
270 PRINT " loading";
275 locate 25,1:print" Dimarogonas,A.D., Computer Aided Machine
Design,Prentice-Hall,1988";
280 REM ****************************************** data ******************
290 READ NN,NM,BW,N: DATA 20,20,30,10
300 REM NN=NO OF NODES, nM=NO OF ELEMENTS, BW=BANDED MATRIX HALF BAND WIDTH
310 REM =(MAX DIFF OF NODE NUMBERS AT ELEMENT ENDS+1)*3
320 REM N = NO OF MATERIALS
330 M = NN
340 DIM ND(M):REM ..............................NODE NUMBER
350 DIM NG(M,6):REM ............................COORDINATE NO
360 DIM NC(M,3):REM ............................NODE COORDINATES
370 DIM NR(M,6):REM ............................NODE RESTRAINTS
380 DIM DD(M,6):REM ............................NODE INITIAL DISPLACEMENTS
390 DIM D(6*M): REM ............................NODE DISPLACEMENTS
400 DIM MB(M),ME(M):REM ........................ELEMENT NODES, BEGINNING,
END
410 DIM NM(M):REM ..............................ELEMENT MATERIAL
420 REM SECTION PROPERTIES:....... AR=AREA, YM=YOUNG MODULUS, SM=SHEAR
MODULUS
430 DIM AR(NM),YM(NM),SM(NM):REM ..............ELEMENT PROPERTIES
440 DIM RA(NM): REM ............................ROLL ANGLE
450 DIM DF(M,6):REM ............................APPLIED NODE FORCES
460 DIM GF(6 * M): REM .........................FORCE VECTOR
470 DIM FB(NM,6),FE(NM,6): REM ................ELEMENT FORCES
480 DIM C(3): REM ..............................ELEMENT DIRECTION COSINES
490 DIM LK(12,12): REM .........................LOCAL STIFFNESS MATRIX
500 DIM KEL(12,12): REM ...............UNROTATED ELEMENT STIFFNESS MATRIX
510 DIM A(6*M,BW): REM ................BANDED STIFFNESS MATRIX (GLOBAL)
520 DIM KN(12):REM ....................POINTER CONNECTING LOCAL ELEMENT
530 DIM IY(NM),IZ(NM), IJ(NM):REM ..............ELEMENT INERTIAS
540 DIM R(12,12): REM ..........................ROTATION MATRIX
550 DIM T(12,12): REM ..........................AUXILIARY MATRIX
560 KEY OFF
570 REM ********* ZERO NODE RELEASE MATRIX ***************
580 FOR I = 1 TO NN: FOR J = 1 TO 6: NR(I,J) = 1: NEXT J: NEXT I
590 GOSUB 2950:REM  *********  load data file  *********
600 CLS:LOCATE 1,1:PRINT"              DATA (From file ";FILNA$;")":PRINT
610 D$="__________"
620 PRINT D$;"____________________NODE COORDINATES____________________"
630 PRINT"Node";TAB(16);"x";TAB(31);"y";TAB(46);"z"
```

```
640 FOR I = 1 TO NN: PRINT ND(I) ;TAB(15);NC(I,1);
650 PRINT TAB(30) ;NC(I,2) ;TAB(45) NC(I,3):NEXT I
660 INPUT"Hit RETURN to continue";XR$:CLS
670 PRINT D$;"____________________NODE RELEASES____________________"
680 PRINT "Node ";TAB(8);"RX   RY   RZ   FX   FY   FZ"
690 FOR I=1 TO NN:INDEX=0:FOR J=1 TO 6:IF NR(I,J)=0 THEN INDEX=1
700 NEXT J:IF INDEX=0 THEN 730
710 PRINT I;TAB(8);
720 FOR J = 1 TO 6:PRINT NR(I,J); TAB(8+5*J);: NEXT J
730 PRINT:NEXT I
740 INPUT"Hit RETURN to continue";XR$:CLS
750 PRINT D$;"____________________ELEMENT INCIDENCES____________________"
760 PRINT "Member           Node 1          Node 2"
770 FOR I=1 TO NM:PRINT I,MB(I),ME(I):NEXT I
780 INPUT"Hit RETURN to continue";XR$:CLS
790 REM ELEMENT PROPERTIES
800 PRINT D$;"__________________ELEMENT PROPERTIES__________________"
810 PRINT "No         Material  Area        Iy          Iz          Iyz        Roll
angle"
820 FOR I = 1 TO NM
830 PRINT I;" ";TAB(10);NM(I);TAB(20);AR(I);TAB(30);IY(I);
840 PRINT TAB(40);IZ(I);TAB(50);IJ(I);TAB(60);RA(I):NEXT I:PRINT
850 FOR I = 1 TO NM
860 IF NM(I)> 10 THEN 890
870 YM(I) = 210000!:SM(I) = 105000!
880 GOTO 910
890 REM
900 PRINT "NO MATERIAL IN PROGRAM": END
910 NEXT I
920 INPUT"Hit RETURN to continue";XR$:CLS
930 REM EXTERNAL LOADS, FORCES AND MOMENTS
940 REM NUMBER OF FORCED NODES NF
950 PRINT D$;"______________________LOADING________________________"
960 PRINT "Node Fx            Fy           Fz           Mx          My
Mz"
970 FOR I = 1 TO NN:INDEX =0:FOR J=1 TO 6:IF DF(I,J)>0 THEN INDEX =1
980 NEXT J
990 IF INDEX=0 THEN 1010
1000 PRINT I;: FOR J = 1 TO 6: PRINT TAB(12*(J-1)+5);DF(I,J);: NEXT J:PRINT
1010 NEXT I
1020 GOSUB 1210
1030 END:REM *********************END OF MAIN PROGRAM*******************
1040 REM
1050 REM **********************************FINDING ELEMENT BETWEEN NODES
AA,BB
1060 MN = 0
1070 FOR I = 1 TO NN
1080 IF AA = ND(I) THEN A = I
1090 IF AA = ND(I) GOTO 1140
1100 NEXT I
1110 PRINT "NUMBERING ERROR  ";AA; " NODE"
1120 MN = 1
1130 GOTO 1200
1140 FOR I = 1 TO NN
1150 IF BB = ND(I) THEN B = I
1160 IF BB = ND(I) GOTO 1200
1170 NEXT I
1180 PRINT "NUMBERING ERROR AT  ";BB; " NODE"
1190 MN = 1
1200 RETURN
1210 REM *******FINITE ELEMENT ANALYSIS *************"
1220 INPUT"Hit RETURN to continue";X$
1230 REM BOUNDARY CONDITIONS
1240 FD=0: FOR I = 1 TO NN
1250 FOR J = 1 TO 6
1260 IF NR(I,J) = 0 GOTO 1280: REM  ...........(1-ACTIVE; 0-NONACTIVE)
1270 FD=FD+1:NG(I,J)=FD:GF(FD) = DF(I,J)
1280 NEXT J:NEXT I
1290 REM
1300 REM ....................................STIFFNESS MATRIX FORMATION
1310 REM
```

```
1320 PRINT "FORMING STIFFNESS MATRIX..."
1330 FOR IM = 1 TO NM
1340 AA = MB(IM):BB = ME(IM):  GOSUB 1050
1350 FOR K = 1 TO 6
1360 KN(K) = NG(A,K):KN(K + 6) = NG(B,K)
1370 NEXT K
1380 DX = NC(B,1) - NC(A,1):DY = NC(B,2) -NC(A,2): DZ = NC(B,3) - NC( A,3)
1390 Q = SQR (DX * DX + DY * DY + DZ * DZ)
1400 C(1) = DX / Q:C(2) = DY / Q: C(3) = DZ /Q
1410 REM
1420 REM ..................................ELEMENT STIFFNESS MATRIX
1430 GOSUB 1780
1440 REM ..................................ASSEMBLY INTO THE SYSTEM MATRIX
1450 FOR U = 1 TO 12: FOR V = U TO 12
1460 N1 = KN(U): N2 = KN(V)
1470 IF N1 = 0 GOTO 1520
1480 IF N2 = 0 GOTO 1520
1490 A1 = N1: IF N2 < N1 THEN A1 = N2
1500 A2 = ABS (N1 - N2) + 1
1510 A(A1,A2)= A(A1,A2) + LK(U,V)
1520 NEXT V
1530 NEXT U
1540 NEXT IM
1550 REM ..................................SOLVE LINEAR EQUATIONS
1560 REM ..................................BANDED MATRIX, GAUSS ELIMINATION
1570 GOSUB 2680
1580 REM ..................................SOLUTION
1590 PRINT "Computing nodal displacements..."
1600 FOR K = 1 TO FD:FOR I = 1 TO NN:FOR J =1 TO 6
1610 IF NG(I,J) = K THEN DD(I,J) = D(K)
1620 NEXT J:NEXT I:NEXT K
1630 REM ..................................ELEMENT FORCES; SIGMA = H*DELTA
1640 PRINT "ELEMENT FORCES"
1650 FOR IM = 1 TO NM
1660 AA = MB(IM):BB = ME (IM):GOSUB 1050
1670 DX = NC(B,1) - NC(A,1):DY = NC(B,2) - NC(A,2):DZ = NC(B,3) - NC (A,3)
1680 Q = SQR (DX * DX + DY * DY +DZ * DZ)
1690 C(1) = DX / Q: C(2) = DY / Q: C(3) = DZ / Q
1700 GOSUB 1780
1710 FOR U = 1 TO 6
1720 FOR V = 1 TO 6
1730 FB(IM,U) = FB(IM,U) + LK(U,V) * DD(A,V) + LK(U,V+6) * DD(B,V)
1740 FE(IM,U) = FE(IM,U) + LK(U + 6,V) * DD(A,V) + LK(U + 6,V + 6) *
DD(B,V)
1750 NEXT V:NEXT U:NEXT IM
1760 GOSUB 1880
1770 RETURN
1780 REM ***********************************ELEMENT  LOCAL STIFFNESS MATRIX
1790 PRINT "COMPUTING ELEMENT ";IM
1800 FOR K = 1 TO NM
1810 IF MM$(IM) = CM$(K) THEN 1830
1820 NEXT K
1830 GOSUB 2060
1840 GOSUB 2250
1850 GOSUB 2520
1860 GOSUB 2600
1870 RETURN
1880 REM ***********************************************************RESULTS
1890 PRINT : PRINT "NODE DISPLACEMENTS"
1900 PRINT "------------------------":PRINT
1910 PRINT"Element  ux        uy          uz          rotx        roty
rotz"
1920 FOR I = 1 TO NN
1930 PRINT ND(I);TAB(5);
1940 FOR J=1 TO 6:PRINT USING "##.##^^^^ ";DD(I,J);:IF J<6 THEN PRINT
TAB(6+12*J);
1950 NEXT J:PRINT
1960 NEXT I
1970 PRINT :PRINT "ELEMENT FORCES"
1980 PRINT "------------------":PRINT
1990 PRINT"Element  Fx        Fy          Fz          Mx          My
```

```
Mz"
2000 FOR I = 1 TO NM
2010 PRINT I;TAB(5);
2020 FOR J=1 TO 6:PRINT USING "##.##^^^^ ";FB(I,J);:IF J<6 THEN PRINT
TAB(6+12*J);
2030 NEXT J:PRINT
2040 NEXT I
2050 RETURN
2060 REM ******************************************ELEMENT STIFFNESS IN
L.C.S
2070 AE = AR(IM) * YM(K) / Q
2080 ZY = IY(IM) * YM(K) / (Q ^ 3)
2090 ZZ = IZ(IM) * YM(K) / (Q ^ 3)
2100 ZJ = IJ(IM) * SM(K) / Q: Q2 = Q ^ 2
2110 KEL(1,1) = AE: KEL(2,2) = 12 * ZZ:KEL(3,3) = 12 * ZY:KEL(4,4) = ZJ
2120 KEL(5,3) = - 6 * ZY * Q
2130 KEL(5,5) = 4 * ZY * Q2:  KEL(6,2) = 6*ZZ*Q: KEL(6,6) = 4 * ZZ * Q2
2140 KEL(7,1) = - AE:         KEL(7,7) = AE:      KEL(8,2) = - 12 * ZZ
2150 KEL(8,6) = - 6 * ZZ * Q: KEL(8,8) = 12 * ZZ:KEL(9,3) = - 12 * ZY
2160 KEL(9,5) = 6 * ZY * Q:   KEL(9,9) = 12 * ZY:KEL(10,4)=-
ZY:KEL(10,10)=ZJ
2170 KEL(11,3) = - 6 * ZY * Q:KEL(11,5) = 2 * ZY * Q2: KEL(11,9) = 6 * ZY *
Q
2180 KEL(11,11) = 4 * ZY * Q2:KEL(12,2) = 6 * ZZ * Q:KEL(12,6) = 2 * ZZ *
Q2
2190 KEL(12,8) = - 6 * ZZ * Q:KEL(12,12) = 4 * ZZ * Q2
2200 REM ............................................SYMMETRY
2210 FOR U = 1 TO 12: FOR V = 1 TO 12
2220 KEL(U,V) = KEL(V,U)
2230 NEXT V:NEXT U
2240 RETURN
2250 REM **********************************ELEMENT STIFFNESS MATRIX IN
G.C.S
2260 QQ = SQR (C(1) * C(1) + C(3) * C(3))
2270 A5 = (RA(IM)) *3.14159/ 180
2280 CS = COS (A5):SS = SIN (A5)
2290 REM CHECK FOR PARALLEL TO Z ELEMENT
2300 IF C(1) = 0 AND C(3) = 0 GOTO 2390
2310 R(1,1) = C(1):R(1,2) = C(2):R(1,3) = C(3)
2320 R(2,1) = - C(1) * C(2) * CS /QQ - C(3) * SS / QQ
2330 R(2,2) = QQ * CS
2340 R(2,3) = - C(2) * C(3) * CS / QQ + C(1) * SS / QQ
2350 R(3,1) = C(1) * C(2) * SS / QQ - C(3) * CS / QQ
2360 R(3,2) = - QQ * SS
2370 R(3,3) = C(2) * C(3) * SS / QQ + C(1) * CS / QQ
2380 GOTO 2420
2390 R(1,1) = 0:R(1,2) = C(2):R(1,3) = 0
2400 R(2,1) = - C(2) * CS:R(2,2) = 0: R(2,3) = C(2) * SS
2410 R(3,1) = SS:R(3,2) = 0:R(3,3) = CS
2420 REM
2430 FOR U = 1 TO 3: FOR V = 1 TO 3
2440 U3 = U + 3:V3 = V + 3
2450 U6 = U + 6:V6 = V + 6
2460 U9 = U + 9:V9 = V + 9
2470 R(U3,V3) = R(U,V)
2480 R(U6,V6) = R(U,V)
2490 R(U9,V9) = R(U,V)
2500 NEXT V:NEXT U
2510 RETURN
2520 REM T= UK * R
2530 FOR U = 1 TO 12: FOR V = 1 TO 12
2540 T(U,V) = 0
2550 FOR W = 1 TO 12
2560 T(U,V) = T(U,V) + KEL(U,W) * R(W,V)
2570 NEXT W
2580 NEXT V:NEXT U
2590 RETURN
2600 REM ..................................LK = TRANSPOSE (R) * T
2610 FOR U = 1 TO 12: FOR V = 1 TO 12
2620 LK(U,V) = 0
2630 FOR W = 1 TO 12
```

```
2640 LK(U,V) = LK(U,V) + R(W,U) * T(W,V)
2650 NEXT W
2660 NEXT V:NEXT U
2670 RETURN
2680 REM *******BANDED SYMMETRIC MATRIX SYSTEM OF LINEAR EQUATION.GAUSS
METHOD
2690 MS = BW
2700 DIM DI(MS)
2710 N1 = FD - 1
2720 FOR K = 1 TO N1
2730 C = A(K,1):K1 = K + 1: NI = K1 + MS - 2
2740 L = NI: IF FD < NI THEN L = FD
2750 FOR J = 2 TO MS: DI(J) = A(K,J):NEXT J
2760 FOR J = K1 TO L: K2 = J - K + 1: A(K,K2) = A(K,K2) / C: NEXT J
2770 GF(K) = GF(K) / C
2780 FOR I = K1 TO L
2790 K2 = I - K1 + 2: C = DI(K2)
2800 FOR J = I TO L
2810 K2 = J - I + 1:K3 = J - K +1
2820 A(I,K2) = A(I,K2) - C * A(K,K3):NEXT J
2830 GF(I) = GF(I) - C * GF(K)
2840 NEXT I:NEXT K
2850 GF(FD) = GF(FD) / A(FD,1)
2860 FOR I = 1 TO N1:K = FD - I
2870 K1 = K + 1: NI = K1 + MS - 2
2880 L = NI: IF NI > FD THEN L = FD
2890 FOR J = K1 TO L
2900 K2 = J - K + 1
2910 GF(K) = GF(K) - A(K,K2) * GF(J)
2920 NEXT J:NEXT I
2930 FOR I = 1 TO FD: D(I) = GF(I):NEXT I
2940 RETURN
2950 REM***********************************************************
2960 REM                  file retrieval
2970 REM***********************************************************
2980 NN=0:NM=0:NJ=0:NF=0
2990 LOCATE 20,45:PRINT"           ";
3000 LOCATE 20,30:PRINT"Data file  ":LOCATE 20,45:INPUT FILNA$
3010 IF FILNA$="" THEN 3000
3020 OPEN FILNA$ FOR INPUT AS #1
3030 ICOUNT=ICOUNT+1:INPUT#1,X$
3040 IF X$="framend" OR X$="FRAMEND" THEN 3230
3050 SYM$=""
3060 ISYM=1:IF ICOUNT>1000 THEN 3250
3070 SYM$=MID$(X$,ISYM,1):ISYM=ISYM+1:IF SYM$="" AND ISYM<10 THEN 3070
3080 IF ISYM=10 THEN 3030
3090 IF SYM$<>"n" AND SYM$<>"N" THEN GOTO 3110
3100 NN=NN+1:ND(NN)=NN:INPUT#1,NC(NN,1),NC(NN,2),NC(NN,3):GOTO 3030
3110 IF SYM$="e" OR SYM$="E" THEN NM=NM+1:GOTO 3200
3120 IF SYM$<>"r" AND SYM$<>"R" THEN 3140
3130 NJ=NJ+1:INPUT#1,NOD:FOR J=1 TO 6:INPUT#1, NR(NOD,J):NEXT J
3140 IF SYM$<>"l" AND SYM$<>"L" THEN 3160
3150 NF=NF+1:INPUT#1,NOD:FOR J=1 TO 6:INPUT#1,DF(NOD,J):NEXT J
3160 IF SYM$<>"p" AND SYM$<>"P" THEN 3180
3170 INPUT#1,NOD,NM(NOD),AR(NOD),IY(NOD),IZ(NOD),IJ(NOD),RA(NODE)
3180 GOTO 3030
3190 PRINT"FILE ERROR...":STOP
3200 INPUT#1,NDUM
3210 INPUT#1,MB(NM),ME(NM)
3220 GOTO 3030
3230 CLOSE#1
3240 GOTO 3270
3250 LOCATE 19,68:PRINT "BAD FILE      "
3260 STOP
3270 RETURN
```

APPENDIX 4.D
FINSTRES: A FINITE-ELEMENT PROGRAM FOR PLANE STRESS ANALYSIS

```
10 REM ****************************************************
20 REM *                                                  *
30 REM *                    FINSTRES                      *
40 REM *                                                  *
50 REM ****************************************************
60 REM
70 REM by Professor Andrew D. Dimarogonas, Washington Univeristy.
71 REM All rights reserved.  Unauthorized reproduction, dissemination,
72 REM selling or use is strictly prohibited.
73 REM This listing is for reference purpose only.
80 REM
90 REM
100 REM  A program for finite elements analysis of plane or axisymmetric
101 REM solids.  Triangular elements are used and a banded matrix scheme.
110 REM
120 REM           ADD/18/12/84, Revised 4-30-87
130 REM
140 N =200:REM                              MAX NUMBER OF NODES
150 DIM X(N),Y(N):REM                       NODE COORDINATES
160 DIM NI(N),NJ(N),NM(N),NK(N):REM         ELEMENT NODES
170 DIM NF(N),XF(N),YF(N):REM               NODES WITH FIXED DISPLACEMENTS
180 DIM TT(N):REM                           FIXED DISPLACEMENTS AT ABOVE NODES
190 REM A(N,BW):'                           STIFFNESS MATRIX OF HALF BANDWIDTH
BW
200 DIM GF(N):'                             FORCE VECTOR
210 DIM D(N),DA(N):'                        SOLUTION VECTOR
220 DIM SS(N):REM                           ELEMENT EQ STRESS
230 DIM S(10),ED(10):REM                    AUXILIATY MATRICES
240 CLS:KEY OFF
250 AX=0
260 FOR I=1 TO 14:COLOR I:LOCATE I,22+I:PRINT "FINSTRES";
270 LOCATE I,50-I:PRINT "FINSTRES";:NEXT I:COLOR 15
280 LOCATE 8,1:PRINT"A 3-node finite element program for plain or
axisymmetric";
290 PRINT " stress analysis";
295 locate 25,1:print" Dimarogonas,A.D., Computer Aided Machine
Design,Prentice-Hall,1988";
300 LOCATE 16,1:PRINT "                         TYPE OF PROBLEM";
310 LOCATE 18,1
320 PRINT "<1>Plane stress                      <2>     Plane strain";
330 PRINT "<3> Axisymmetric, plain stress       <4> Axisymmetric, plain
strain"
350 INPUT "              enter your selection...";E$
360 LOCATE 22,1:PRINT"Enter:  Modulus of Elasticity, Poisson Ratio, ";
370 PRINT " thickness ";:INPUT EY,NU,TH
380 IF E$="1" OR E$="3" THEN IX=1
390 IF E$="2" OR E$="4" THEN IX=2
400 IF E$="3" OR E$="4" THEN AX=2
410 SCREEN 2:XS=639:YS=199:ID=0:TOLERANCE=1
420 CLS:NMENUS=7
430 XSC=3:YSC=1
440 DIM L$(100)
450 VIND=2
460 DIM MENU1$(25)
470 FOR IM=1 TO NMENUS:READ MENU1$(IM):NEXT IM
480 DATA "Stop      ","Load File ","Plot Mesh","Loads     ","Restraints"
490 DATA "Analyze   ","Post Proc "
500 XS=0
510 FOR I=1 TO NMENUS:LOCATE I+2,70:PRINT MENU1$(I);:NEXT I
520 ICOUNT=0
530 LOCATE 1,70:PRINT"MAIN MENU";
540 LINE(530,0)-(640,8),,B:GOTO 620
550 X$=INKEY$:XS=0
560 IF X$="" THEN 550
570 IF LEN(X$)=1 THEN XS=ASC(X$):GOTO 680
580 X$=RIGHT$(X$,1)
590 IF X$><"H" AND X$<>"P" THEN 650
```

```
600 IF X$="H" AND VIND>1 THEN VIND=VIND-1
610 IF X$="P" AND VIND<NMENUS THEN VIND=VIND+1
620 IF VIND>1 THEN LOCATE VIND+1,69:PRINT " ";MENU1$(VIND-1);
630 LOCATE VIND+2,69:PRINT ">";MENU1$(VIND);
640 IF VIND<NMENUS THEN LOCATE VIND+3,69:PRINT " ";MENU1$(VIND+1);
650 LOCATE 19,70:PRINT SPC(9);
660 LOCATE 21,65:PRINT SPC(14);
670 LOCATE 22,65:PRINT SPC(14);
680 IF XS=13 AND VIND=1 THEN SCREEN 0:END
690 IF XS=13 AND VIND=2 THEN GOSUB 2770:REM          load file from disc
700 IF XS=13 AND VIND=4 THEN GOSUB 3730:REM        define loads on nodes
710 IF XS=13 AND VIND=5 THEN GOSUB 3730:REM        define boundary
conditions
720 IF XS=13 AND VIND=6 THEN GOSUB  910:REM          solve
730 IF XS=13 AND VIND=7 THEN GOSUB 3070:REM          post processing
740 IF XS=13 AND VIND=3 THEN GOSUB  800:REM          plot mesh
750 IF XS=13 THEN XS=0:VIND=VIND+1:GOTO 510
760 FOR ICLEAR=14 TO 24:LOCATE ICLEAR,65:PRINT SPC(14);:NEXT ICLEAR
770 IF XS=13 AND VIND>6 THEN VIND=1:GOTO 510
775 LOCATE 25,1:PRINT"Back to the main menu. Use arrows to select,then
RETURN";
780 GOTO 550
790 END
800 REM ******************************************************* PLOT
ELEMENTS
810 FOR I = 1 TO NE
820 I1 = NI(I):I2 = NJ(I): I3 = NM(I)
830 X1 =XSC* X(I1) :Y1 = Y(I1)*YSC
840 X2 =XSC* X(I2) :Y2 = Y(I2)*YSC
850 X3 =XSC* X(I3) :Y3 = Y(I3)*YSC
860 LINE(0,0)-(0,190):LINE (0,0)-(550,0)
870 LINE(X1,Y1)-(X2,Y2):LINE-(X3,Y3):LINE-(X1,Y1)
880 NEXT I
890 RETURN
900 REM
910 REM ******************************************* FINITE ELEMENT
ANALYSIS
920 CLS
930 NN=ND
940 W1 = BW + 2
950 N = 2 * NN
960 DIM A(N,W1)
970 LOCATE 1,1:PRINT "COMPUTING ......";
980 REM STIFFNESS MATRIX
990 LOCATE 1,1:PRINT "FORMING STIFFNESS MATRIX...";
1000 DIM B(4,6),KE(6,6),DB(4,6),BT(6,4),DD(4,4)
1010 FOR I = 1 TO NE
1020 LOCATE 1,1:PRINT "COMPUTING ELEMENT NO: ";I;"       ";
1030 NI = NI(I):NJ = NJ(I):NM = NM(I)
1040 XI = X(NI):YI = Y(NI)
1050 XJ = X(NJ):YJ = Y(NJ)
1060 XM = X(NM):YM = Y(NM)
1070 DL = XJ * YM + XI * YJ + YI * XM - YI * XJ - YJ * XM - XI * YM
1080 DL=DL / 2
1085 IF DL=0 THEN PRINT"zero area element no :";I:STOP
1090 BI = YJ - YM:CI = XM - XJ
1100 IA = I: IF J < I THEN IA = J
1110 BJ = YM - YI:CJ = XI - XM
1120 BM = YI - YJ:CM = XJ - XI
1130 REM FETCH ELEMENT MATRIX
1140 GOSUB 2000: REM H=DD*B
1150 GOSUB 2340: REM BT*DD*B
1160 REM SYSTEM ASSEMBLY
1170 II(1) = 2*NI - 1:II(2) = 2 * NI
1180 II(3) = 2 *NJ - 1:II(4) = 2 * NJ
1190 JA = ABS (NI - NJ) + 1
1200 II(5) = 2 * NM - 1:II(6) = 2 * NM
1210 FOR I1 = 1 TO 6: FOR J1 = I1 TO 6
1220 IE = II(I1):JE = II(J1)
1230 MA = ABS(IE - JE) + 1:JA = IE: IF JE < IE THEN JA = JE
1240 A(JA,MA) = A(JA,MA) + KE(I1, J1)
```

```
1250 NEXT J1:NEXT I1
1260 NEXT I
1270 FOR I =1  TO NF
1280 ISU = NF(I):XX = XF(I):YY = YF(I)
1290 S1 = 2 * ISU - BW: IF S1 < 1 THEN S1 = 1
1300 S2 = 2 * ISU + BW: IF S2 > N THEN S2 = N
1310 FOR S = S1 TO S2
1320 I1 = 2 * ISU - 1: I2 = 2 * ISU
1330 T1 = I1: IF S < I1 THEN T1 = S
1340 T2 = ABS (I1 - S) + 1
1350 T3 = I2: IF S < I2 THEN T3 = S
1360 T4 = ABS (I2 - S) + 1
1370 IF XX > 0 THEN 1400
1380 GF(I1) = GF(I1) - A(T1,T2) * XX
1390 A(T1,T2) =0
1400 IF YY > 0 THEN 1430
1410 A(T3,T4) = 0
1420 GF(I2) = GF(I2) - A(T3,T4) * YY
1430 NEXT S
1440 IF XX > 0 THEN 1460
1450 A(I1,1) = 1: GF(I1) = 0
1460 IF YY > 0 THEN 1480
1470 A(I2,1) = 1000000!:GF(I2) = 0
1480 NEXT I
1490 GOSUB 2450
1500 E0 = 0:CLS
1510 LPRINT "SOLUTION - DISPLACEMENTS*1000"
1520 LPRINT "Node   x-displacement   y-displacement":LPRINT
1530 LPRINT "----------------------------------":LPRINT
1540 XMAX=0:YMAX=0
1550 FOR I = 1 TO ND
1560 X = D(2*I-1) * 1000
1570 Y = D(2*I) * 1000
1580 LPRINT I;TAB(5);X;TAB(20);Y
1590 IF ABS(X)>XMAX THEN XMAX=ABS(X)
1600 IF ABS(Y)>YMAX THEN YMAX=ABS(Y)
1610 NEXT I
1615 LOCATE 25,1:LPRINT "Maximum nodal desplacements, xmax=";XMAX;",
ymax=";YMAX;
1620 LPRINT
1630 LOCATE 24,1:PRINT "HIT RETURN TO CONTINUE WITH STRESSES":INPUT X$
1640 REM CALCULATIONS OF STRESSES AT THE CENTROID OF EACH ELEMENT
1650 PRINT :PRINT "COMPUTING STRESSES..."
1660 LPRINT:LPRINT
1670 LPRINT"No   x          y            sx           sy            txy";
1675 LPRINT "          sth";
1680 LPRINT
1690 FOR I = 1 TO NE
1700 NI = NI(I):NJ = NJ(I):NM = NM(I)
1710 XI = X(NI):YI = Y(NI)
1720 XJ = X(NJ):YJ = Y(NJ)
1730 XM = X(NM):YM = Y(NM)
1740 DL = XJ * YM + XI * YJ + YI * XM - YI * XJ - YJ * XM - XI * YM
1750 DL = DL / 2
1760 BI = YJ - YM:CI = XM - XJ
1770 IA = I: IF J < I THEN IA = J
1780 BJ = YM - YI:CJ = XI - XM
1790 BM = YI - YJ:CM = XJ - XI
1800 REM COMPLETE ELEMENT MATRIX H
1810 GOSUB 2000:REM  DB=DD*B
1820 REM S = (DD*B)*DISPLACEMENTS
1830 REM ELEMENT DISPLACEMENT VECTOR ED
1840 ED(1) = D(2 * NI - 1): ED(2) = D(2 * NI)
1850 ED(3) = D(2 * NJ - 1): ED(4) = D(2 * NJ)
1860 ED(5) = D(2 * NM - 1): ED(6) = D(2 * NM)
1870 FOR ISU = 1 TO 4
1880 S(ISU) = 0
1890 FOR JS = 1 TO 6: S(ISU) = S(ISU) + DB(ISU ,JS) * ED(JS):NEXT JS
1900 NEXT ISU
1910 LPRINT I;TAB(5);
1920 LPRINT (XI + XJ + XM) / 3;TAB(15);(YI +YJ + YM) / 3 ;TAB(25);
```

```
1930 LPRINT S(1);TAB(39);S(2);
1940 LPRINT TAB(53);S(4);TAB(67);S(3)
1950 SSS=SQR((S(1)-S(2))^2+4*S(4)^2):IF SSS>SMAX THEN SMAX=SSS:IEMAX=I
1960 SS(I)=SSS
1970 NEXT I
1975 LPRINT:LPRINT"Max equivalent stress= ";SMAX;"    at element ";IEMAX
1980 RETURN
1990 REM ************************************** FINITE ELEMENT ROUTINES
2000 REM ELEMENT STIFFNESS MATRIX
2010 B(1,1) = BI:B(1,3) = BJ:B(1,5)=BM
2020 B(2,2) = CI:B(2,4) = CJ:B(2,6)=CM
2030 B(4,1) = CI:B(4,2) = BI:B(4,3)=CJ
2040 B(4,4) = BJ:B(4,5) = CM:B(4,6)=BM
2050 IF AX = 0 THEN 2100:REM NON AXISYMMETRIC
2060 XX = (XI + XJ + XM) / 3:YY = (YI + YJ + YM) / 3
2070 B(3,1) = (XJ * YM - XM * YJ) / XX + BI + YY * CI / XX
2080 B(3,3) = ( - XI * YM + XM * YI) / XX +BJ + YY * CJ / XX
2090 B(3,5) = (XI * YJ - XJ * YI) / XX + BM + YY * CM / XX
2100 FOR Z = 1 TO 4: FOR J = 1 TO 6:B(Z,J) = B(Z,J) / (2 * DL): NEXT J:NEXT
Z
2110 IF AX = 2 THEN IX = 2
2120 IF IX = 2 THEN 2200
2130 REM PLANE STRESS
2140 DD(1,1) = EY / (1 - NU * NU)
2150 DD(2,2) = DD(1,1)
2160 DD(1,2) = NU * DD(1,1):DD(2,1) = DD(1,2)
2170 DD(3,3) = 1
2180 DD(4,4) = EY / (2 + 2 * NU)
2190 GOTO 2290
2200 REM PLAIN STRAIN
2210 DD(1,1) = EY * (1 - NU) / (1 + NU) / (1 - 2 * NU)
2220 DD(1,2) = NU * EY / (1 + NU) / (1 - 2 * NU)
2230 DD(2,2) = DD(1,1):DD(2,1) = DD(1,2)
2240 DD(3,3) = 1
2250 IF AX = 0 THEN 2280:REM NON AXISYMMETRIC
2260 DD(1,3) = DD(1,2):DD(2,3) = DD(2,1)
2270 DD(3,1) = DD(2,1):DD(3,2) = DD(3,1)
2280 DD(4,4) = EY / (2 + 2 * NU)
2290 REM DB=D*B
2300 FOR Z = 1 TO 4: FOR J = 1 TO 6
2310 DB(Z,J) = 0:FOR K = 1 TO 4:DB(Z,J) = DB(Z,J) + DD(Z,K) * B (K,J):NEXT
K
2320 NEXT J:NEXT Z
2330 RETURN
2340 REM BTRANSPOSE
2350 FOR Z = 1 TO 4: FOR J = 1 TO 6
2360 BT(J,Z) = B(Z,J):NEXT J:NEXT Z
2370 REM KE=BT*D*B*DELTA*T
2380 FOR Z = 1 TO 6: FOR J = 1 TO 6
2390 KE(Z,J) = 0:FOR K = 1 TO 4
2400 KE(Z,J) = KE(Z,J) + BT(Z,K) * DB(K,J)
2410 NEXT K
2420 KE(Z,J) = KE(Z,J) * DL * TH
2430 NEXT J:NEXT Z
2440 RETURN
2450 REM BANDED SYMMMETRIC MATRIX SYSTEM OF LINEAR EQUATIONS.GAUSS METHOD
2460 PRINT "SOLVING LINEAR EQUATIONS..."
2470 FD = N
2480 MS = BW
2490 DIM DI(MS)
2500 N1 = FD - 1
2510 FOR K = 1 TO N1
2520 C = A(K,1):K1 = K  + 1:NI = K1 + MS - 2
2530 L = NI: IF FD < NI THEN L = FD
2540 FOR J = 2 TO MS: DI(J) = A(K,J):NEXT J
2550 FOR J = K1 TO L: K2 = J - K + 1: A(K,K2) = A(K,K2) / C: NEXT J
2560 GF(K) = GF(K) / C
2570 FOR I =K1 TO L
2580 K2 = I - K1 + 2:C = DI(K2)
2590 FOR J = I TO L
2600 K2 = J - I + 1: K3 = J - K + 1
```

```
2610 A(I,K2) = A(I,K2) - C * A(K,K3):NEXT J
2620 GF(I) = GF(I) - C * GF(K)
2630 NEXT I:NEXT K
2640 GF(FD) = GF(FD) / A(FD,1)
2650 FOR I = 1 TO N1:K = FD - I
2660 K1 = K + 1:NI = K1 + MS -2
2670 L = NI: IF NI > FD THEN L = FD
2680 FOR J = K1 TO L
2690 K2 = J - K + 1
2700 GF(K) = GF(K) - A(K,K2) * GF(J)
2710 NEXT J: NEXT I
2720 FOR I = 1 TO FD:D(I) = GF(I):NEXT I
2730 RETURN
2740 REM************************************************************
2750 REM              file retrieval
2760 REM************************************************************
2770 LOCATE 19,68:PRINT"          ";
2780 LOCATE 18,70:PRINT"Data file  ":LOCATE 19,68:INPUT FILNA$
2790 IF FILNA$="" THEN 2780
2800 ICOUNT=0:XMAX=0:YMAX=0
2810 OPEN FILNA$ FOR INPUT AS #1
2820 ICOUNT=ICOUNT+1:INPUT#1,X$
2830 IF X$="filend" OR X$="FILEND" THEN 3040
2840 IF X$="beam" OR X$="BEAM" THEN E$="1"
2850 IF X$="triangular" OR X$="TRIANGULAR" THEN  E$="2"
2860 IF X$="quadrilateral" OR X$="QUADRILATERAL" THEN  E$="3"
2870 SYM$="":IF X(ND)>XMAX THEN XMAX=X(ND)
2875 IF Y(ND)>YMAX THEN YMAX=Y(ND)
2880 ISYM=1:IF ICOUNT>1000 THEN 3050
2890 SYM$=MID$(X$,ISYM,1):ISYM=ISYM+1:IF SYM$="" AND ISYM<10 THEN 2880
2900 IF ASYM=10 THEN 2820
2910 IF SYM$="n" OR SYM$="N" THEN ND=ND+1:INPUT#1,I,X(ND),Y(ND):GOTO 2820
2920 IF SYM$<>"e" AND SYM$<>"E" THEN  2820
2930 NE=NE+1 :ES=VAL(E$)
2940 IF E$="1" THEN INPUT#1,I,NI(NE),NJ(NE)
2950 IF E$="2" THEN INPUT#1,I,NI(NE),NJ(NE),NM(NE)
2960 IF E$="3" THEN INPUT#1,I,NI(NE),NJ(NE),NM(NE),NK(NE)
2970 IF ABS(NI(NE)-NJ(NE))>NDIF THEN NDIF=ABS(NI(NE)-NJ(NE))
2980 IF ES>1 AND ABS(NI(NE)-NM(NE))>NDIF THEN NDIF=ABS(NI(NE)-NM(NE))
2990 IF ES>1 AND ABS(NJ(NE)-NM(NE))>NDIF THEN NDIF=ABS(NJ(NE)-NM(NE))
3000 IF E$="3" AND ABS(NI(NE)-NK(NE))>NDIF THEN NDIF=ABS(NI(NE)-NK(NE))
3010 IF E$="3" AND ABS(NJ(NE)-NK(NE))>NDIF THEN NDIF=ABS(NJ(NE)-NK(NE))
3020 IF E$="3" AND ABS(NM(NE)-NK(NE))>NDIF THEN NDIF=ABS(NM(NE)-NK(NE))
3030 GOTO 2820
3040 CLOSE#1 :GOTO 3060
3050 PRINT"bad file";
3060 BW=2*NDIF+2:XSC=500/XMAX:YSC=160/YMAX
3065 RETURN
3070 REM ****************************************************** PLOT
ELEMENTS
3090 FOR I = 1 TO NE
3100 I1 = NI(I):I2 = NJ(I): I3 = NM(I)
3110 X1 =XSC* X(I1):Y1 = Y(I1)*YSC
3120 X2 =XSC* X(I2):Y2 = Y(I2)*YSC
3130 X3 =XSC* X(I3):Y3 = Y(I3)*YSC:XG=(X1+X2+X3)/3:YG=(Y1+Y2+Y3)/3
3140 LINE(0,0)-(0,190):LINE (0,0)-(300,0)
3150 LINE(X1,Y1)-(X2,Y2):LINE-(X3,Y3):LINE-(X1,Y1)
3160 DMAX=SQR(XMAX^2+YMAX^2)/1000
3170 XD1=X1+D(2*I1-1)*XSC*10/DMAX:YD1=Y1+D(2*I1)*YSC*10/DMAX
3180 XD2=X2+D(2*I2-1)*XSC*10/DMAX:YD2=Y2+D(2*I2)*YSC*10/DMAX
3190 XD3=X3+D(2*I3-1)*XSC*10/DMAX:YD3=Y3+D(2*I3)*YSC*10/DMAX
3200 LINE(XD1,YD1)-(XD2,YD2),1:LINE-(XD3,YD3),1:LINE-(XD1,YD1),1
3210 NEXT I
3220 LOCATE 25,1:INPUT"Hit RETURN to continue ";CON$
3230 CLS:SCREEN 8
3231 CLS:FOR IC=1 TO 8:ICOL=IC:IF IC=2 OR IC=3 THEN ICOL=IC+2
3232 IF IC=3 OR IC=4 THEN ICOL=IC-2
3233 LOCATE 2*IC,73:PRINT INT(IC/8*100);:LOCATE 2*IC,78:PRINT"%";
3235 LINE(560,16*IC+6)-(570,16*IC+16),ICOL,BF
3238 NEXT IC
3240 FOR I = 1 TO NE
```

```
3250 I1 = NI(I):I2 = NJ(I): I3 = NM(I)
3260 X1 =XSC* X(I1):Y1 = Y(I1)*YSC
3270 X2 =XSC* X(I2):Y2 = Y(I2)*YSC
3280 X3 =XSC* X(I3):Y3 = Y(I3)*YSC:XG=(X1+X2+X3)/3:YG=(Y1+Y2+Y3)/3
3290 LINE(0,0)-(0,190):LINE (0,0)-(600,0)
3300 ICOL=8
3305 IF SS(I)>.125*SMAX THEN ICOL=1
3310 IF SS(I)>.25*SMAX THEN ICOL=4
3315 IF SS(I)>.375*SMAX THEN ICOL=5
3320 IF SS(I)>.5*SMAX THEN ICOL=2
3325 IF SS(I)>.675*SMAX THEN ICOL=3
3330 IF SS(I)>.75*SMAX THEN ICOL=6
3335 IF SS(I)>.825*SMAX THEN ICOL=7
3340 LINE(X1,Y1)-(X2,Y2),1:LINE-(X3,Y3),1:LINE-(X1,Y1),1
3350 IF ICOL<8 THEN PAINT (XG,YG),ICOL
3360 NEXT I
3370 LOCATE 25,1:INPUT"Hit return to continue";CON$
3380 WIDTH 80:VIND=2:CLS
3390 RETURN
3400 REM ************************************ NODAL FORCES
3410 FOR IER=22 TO 24:LOCATE IER,68:PRINT SPC(11);:NEXT IER
3420 X=XSC*X(INODE):Y=Y(INODE)*YSC
3430 LOCATE 22,68:PRINT "enter Fx,Fy";
3440 LOCATE 23,68:INPUT GF(2*INODE-1),GF(2`INODE)
3450 IF GF(2*INODE-1)<>0 THEN LINE(X,Y)-(X+40,Y+1),,B
3460 IF GF(2*INODE-1)<>0 THEN LINE(X+10,Y-3)-(X,Y):LINE -(X+10,Y+3)
3470 IF GF(2*INODE)<>0 THEN LINE(X-1,Y)-(X+1,Y-20),,B
3480 IF GF(2*INODE)<>0 THEN LINE(X+8,Y-10)-(X,Y):LINE -(X-8,Y-10)
3490 RETURN
3500 REM ************************************ NODE BOUNDARY CONDITIONS
3510 FOR IER=22 TO 24:LOCATE IER,68:PRINT SPC(11);:NEXT IER
3520 NFNODES=NFNODES+1:NF(NFNODES)=INODE
3530 X=XSC*X(INODE):Y=Y(INODE)*YSC
3540 LOCATE 25,1:PRINT "Direction of restraint: <1> -x, <2> -y, <3> both
";
3550 LOCATE 22,68:PRINT "direction:";
3560 LOCATE 23,68:INPUT DIR$:IF DIR$<>"1" AND DIR$<>"2" AND DIR$<>"3" THEN
3560
3570 IF DIR$="1"  THEN XF(NFNODES)=0:YF(NFNODES)=1
3580 IF DIR$="2" OR DIR$="3" THEN LINE(X,Y)-(X-6,Y+6):LINE -(X+6,Y+6):LINE-
(X,Y)
3590 IF DIR$="1" OR DIR$="3" THEN LINE(X,Y)-(X-9,Y+3):LINE -(X-9,Y-3):LINE-
(X,Y)
3600 IF DIR$="2" THEN XF(NFNODES)=1:YF(NFNODES)=0
3610 IF DIR$="3" THEN XF(NFNODES)=0:YF(NFNODES)=0
3620 NF=NFNODES
3630 RETURN
3640 REM ******************************LOCKING TO THE NEAREST NODE
3650 DIST=0:PREDIST=1E+30::FOR I=1 TO ND
3660 DIST=SQR((X(I)-XHAIR)^2+(Y(I)-YHAIR)^2)
3670 IF DIST<PREDIST THEN INODE=I:PREDIST=DIST
3680 NEXT I
3690 LOCATE 25,1:PRINT SPC(78);
3700 LOCATE 25,1:PRINT "Nearest Node is  ";INODE;
3710 XLOCK=X(INODE):YLOCK=Y(INODE)
3720 RETURN
3730 REM********************************************** move cross-hair
3740 LOCATE 25,1:PRINT" Move cross with keyboard arrows   ";
3750 LOCATE 14,68:PRINT"KEYBOARD:";
3760 LOCATE 15,68:PRINT"L-oad     ";
3770 LOCATE 16,68:PRINT"R-estraint";
3780 LOCATE 17,68:PRINT"N-ear Node";
3790 LOCATE 18,68:PRINT"M-ain MENU";
3800 LOCATE 19,68:PRINT"F-aster   ";
3810 LOCATE 20,68:PRINT"S-lower   ";
3820 XSTEP=10/XSC:YSTEP=10/YSC:XHAIR=1/XSC:YHAIR=1/YSC
3830 X$="aa":GOTO 3960
3840 X$=INKEY$:XS=0
3850 IF X$="" THEN 3840
3860 IF LEN(X$)>1  THEN 3960
3865 IF X$="f" OR X$="F" OR X$="s" OR X$="S" THEN LOCATE 25,1:PRINT "step
```

```
x,y =";XSTEP,YSTEP;"          ";
3870 IF X$="f"  THEN XSTEP=XSTEP*2:YSTEP=XSTEP:ZSTEP=XSTEP
3880 IF X$="s"  THEN XSTEP=XSTEP/2:YSTEP=XSTEP:ZSTEP=XSTEP
3890 IF X$="F"  THEN XSTEP=XSTEP*2:YSTEP=XSTEP:ZSTEP=XSTEP
3900 IF X$="S"  THEN XSTEP=XSTEP/2:YSTEP=XSTEP:ZSTEP=XSTEP
3910 IF X$="l" OR X$="L" THEN GOSUB 3400
3920 IF X$="R" OR X$="r" THEN GOSUB 3500
3930 IF X$="n" OR X$="N" THEN GOSUB 3640:GOTO 3960
3940 IF X$="m" OR X$="M" THEN GOTO 4150
3950 GOTO 3840
3960 X$=RIGHT$(X$,1)
3970 FOR I=5 TO 500 STEP 20:PSET(I,YHAIR*YSC),0:NEXT I
3980 FOR I=6 TO 190 STEP 10:PSET(XSC*XHAIR,I),0:NEXT I
3990 IF XLOCK>0 OR YLOCK>0 THEN XHAIR=XLOCK:YHAIR=YLOCK
4000 XLOCK=0:YLOCK=0
4010 IF X$="H" THEN YHAIR=YHAIR-YSTEP
4020 IF X$="P" THEN YHAIR=YHAIR+YSTEP
4030 IF X$="K" THEN XHAIR=XHAIR-XSTEP
4040 REM
4050 IF X$="M" THEN XHAIR=XHAIR+XSTEP
4060 IF XHAIR<0 THEN XHAIR=0
4070 IF YHAIR<0 THEN YHAIR=0
4080 IF YHAIR*YSC>190 THEN YHAIR=190/YSC
4090 IF XHAIR*XSC>500 THEN XHAIR=500/XSC
4100 LOCATE 25,55:PRINT SPC(23);
4110 LOCATE 25,35:PRINT "x= ";XHAIR;:LOCATE 25,45:PRINT "y= ";YHAIR;
4120 FOR I=5 TO 500 STEP 20:PSET(I,YHAIR*YSC):NEXT I
4130 FOR I=6 TO 190 STEP 10:PSET(XSC*XHAIR,I):NEXT I
4140 GOTO 3840
4150 RETURN
```

CHAPTER FIVE
MACHINE DESIGN OPTIMIZATION

5.1 THE NEED FOR OPTIMIZATION

In machine design, as for almost any design problem, the designer has to determine a number of design parameters, to unambiguously specify a machine or even the simplest machine element. Looking at any drawing one can observe many independent design decisions which were not obvious, but instead required rational selection processes. In addition, materials, surface finish, heat treatment and other manufacturing processes must be specified. On the other hand, the machine or component has to meet certain operational requirements so that it will perform its duty satisfactorily without failure. Finally, the part must be appealing to the user, have low price, or some other feature the user wants.

Take, for example, the case of the design of a simple rod to resist an extensional axial force F. A number of rectangular shapes of dimensions $a \times b$ are available for a number of materials, each having a certain strength S and price c_m per unit mass. The surface finishing of the rod also has a cost c_s per unit of surface.

The cost (C) for a length L is:

$$C = Labdc_m + 2L(a+b)c_s$$

where d is the material density.

The selection of the parameters a, b, material (cost, density) is not altogether arbitrary since the rod must sustain an axial load F without failing. This means that the following design equation must be satisfied:

$$S \geqslant F/ab$$

Engineering common sense dictates that the equals sign in the design equation will yield the minimum rod section and the minimum cost. Therefore, there is a relation between material, a and b which can, in this case, be introduced in the cost equation:

$$b = F/aS; \quad C = LFdc_m/S + 2L\left[a + \frac{F}{aS}\right]c_s$$

The cost now depends on many parameters which can be grouped into four categories:

(a) Functional requirements (F)
(b) Material properties (S, d, c_m)
(c) Geometry (a, L)
(d) Manufacturing processes (c_s)

It must be noted that some of the parameters might be specified beforehand, such as the length and surface finish, due to additional functional requirements or availability which cannot be expressed mathematically, while other parameters are related, such as density, strength and material cost. It seems, then, that the material and the rod width a are free for the designer to select. However, he wants to minimize the cost. To this end, he plots the cost for the available materials I, II, III as a function of the width a, as in Figure 5.1. The lower cost is for material I and rod width a_0 and the resulting minimum cost is C_0.

It turns out that although in the beginning it was apparent that the solution of this design problem had some ambiguity, this was cleared up by the process of obtaining minimum cost.

We can draw some more general conclusions for the analysis of this simple design problem.

In general, a design problem might have a number of design parameters much greater than the number of available design equations. Therefore, the design problems seem to have many solutions. In fact, if n is the number of design parameters, p the number of design equations ($n > p$), for every arbitrary choice of $n - p$ design parameters, the rest p design parameters can be determined from the p design equations. Therefore, the design problem has an $(n - p)$ order infinity of solutions, not counting multiple solutions of the design equations which might exist.

The designer, especially the student, in a situation like this has the feeling that his design is arbitrary. All of the solutions would be formally 'correct'. Not all, however, would be equally successful. This is apparent in the fact that although all the designers have at their disposal, in general, the same equations, the same manufacturing facilities, the same materials, etc., they end up with designs which do not have all the same commercial success. The experience, ingenuity and creativity of the designer usually account for the better selection of the arbitrary design parameters. Now, what is a 'better' design? Or what feature distinguishes a 'good' from a 'bad' design? There is no simple answer to this question, beyond that, it depends on the design: For example, for steel construction – the cost; for a ship – the operation cost; for a spacecraft component – the weight, for a function generator mechanism – the maximum error. Furthermore, in many cases the quality of the design depends on more than one parameter. Sometimes it is difficult to combine quantitatively the advantages and disadvantages of several parameters.

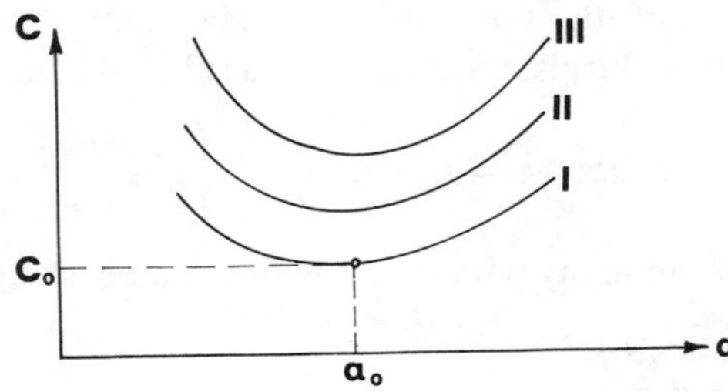

Figure 5.1 Cost of the rod for three different materials, I, II, III

In most cases, the designer can identify a single criterion or a combination of criteria in the form of a function of the design parameters which better satisfies his objectives as it becomes smaller. Cost is a typical example. This function is called an 'objective function'. For every arbitrary combination of the design parameters, this function yields a single value which the designer has to minimize. To the aid of the designer come modern methods, like computer methods, which can yield an intelligent selection of the design parameters which will give a better or even the best possible design, called an 'optimum design'. This procedure is called 'optimization'.

A group of machine design students was asked once to design a beam from a list of materials with specific loading and length. One design equation, the strength of the beam. The outcome was interesting: beam costs ranged from $320 to $1500.

Sometimes, there are no restrictions on the n design parameters and the problem can be stated as follows:

Minimize the objective function $\mathrm{f}(x)$

where

$$x = \{x_1 x_2, \ldots, x_n\}$$

and $x_1, x_2, \ldots, x_n$ are yet unknown design parameters.

This problem is known as an 'unconstrained optimization' problem. Such a problem is not usual in machine design. Its discussion, however, is important because more complex problems can be transformed to an unconstrained optimization problem.

In most cases the design parameters are not independent:

(a) There are restrictions on the range of the design parameters. For example, dimensions cannot be negative and some of them have upper limits due to space limitations.
(b) There are equations or inequalities that the design parameters must satisfy. They are called, respectively, equality and inequality constraints. 'Equality constraints' are, for example, the speed, capacity, transmitted force, torque, moment of a machine or member, etc. 'Inequality constraints' are, for example, total weight limitations, limiting stresses and deflections, space limitations, etc. The associated problem is called a 'constrained optimization' problem.

If there are only p equality constraints, it is possible to use them in order to reduce the number of design parameters by a number p, by elimination. Then, the problem is reduced to an unconstrained one with $(n - p)$ unknown design parameters called 'decision variables'. In fact, this was done in the example of the tension rod when the dimension b was eliminated by way of the application of an equality constraint, the limiting stress.

In the design example of the tension rod, all practically feasible solutions, that is combinations of the design parameters satisfying the design equations, have been computed and the resulting designs were sorted to find the optimum one for minimum cost. This kind of search is called 'exhaustive' and for small problems it is common. For larger problems however, the computation effort becomes prohibitive and the exhaustive search is too exhaustive. For such problems intelligent methods have been developed to yield an optimum design with minimum computation effort.

5.2 THE GENERAL OPTIMIZATION PROBLEM STATEMENT

In most design situations, the optimization problem can be stated as follows:

Find an appropriate x which minimizes f(x) (objective function) subject to:

$$g_i(x) = 0, \quad i = 1, 2, \ldots, m \tag{5.1}$$

$$h_j(x) > 0, \quad j = 1, 2, \ldots, p \tag{5.2}$$

where

$$x = \{x_1 x_2, \ldots, x_n\} \tag{5.3}$$

and the functions g and h are respectively the equality and inequality constraints.

When the functions f, g, and h are linear we call the problem 'linear programing'. The definition of the nonlinear programing problem will follow logically.

The parameters $x_1, x_2, \ldots,$ are the design parameters of the system and, in general, their range is restricted as follows:

$$x_{j\min} < x_j < x_{j\max}, \quad j = 1, 2, \ldots, n \tag{5.4}$$

In general, there are two methods to attack the optimization problem:

1. In cases where we possess simple mathematical expressions for the objective function and the constraints, it is sometimes possible to obtain explicit expressions for the minima, even the global minimum.
2. In cases lacking the mathematical expressions or having very complicated equations, one has to resort to numerical methods.

One special feature of the machine design optimization is that usually there are many objectives which have to be combined in one objective function.

In many engineering applications, there is a common denominator – the cost. For example, capital cost and operating cost of a machine can be easily combined into one equation. There is no general rule for the selection of the objective function. It just depends on the problem. We shall return to the question of the selection of the objective function in Section 5.5.

5.3 ANALYTICAL METHODS

As stated in Section 5.1, the general optimization problem consists of the following:

$$\text{Minimize f}(x), \quad x = \{x_1, x_2, \ldots, x_n\}$$

subject to equality and inequality constraints.

Well-known methods for maxima and minima of functions can be used.

Case 1. Unconstrained minimum of a function

The conditions for relative minimum of a function f(x) in n variables:

$$x_1, x_2, \ldots, x_n$$

are

(a)
$$\frac{\partial f}{\partial x_i} = 0, \quad i = 1, 2, \ldots, n \tag{5.5}$$

or

$$\nabla f(x) = 0$$

(b) For the minimum, some additional conditions must be met because some of the solutions of equation (5.5) might yield maxima or inflection points. There are mathematical criteria to distinguish the minima but in most computer solutions it is more convenient to just compare the solutions for the relative minimum.

The conditions in equation (5.5) provide us with n equations in n unknowns. Therefore, we can generally solve this system and determine the design variables. This system does not necessarily have a unique solution. Every solution of the problem is a local extremum. As pointed out above, comparison will isolate the lowest minimum and no further consideration should be given to the other solutions. There is no general method to determine the lowest minimum – the global minimum.

Case 2. Equality constraints

We want the minimum of a function with the equality constraints,

$$g_i(x) = 0, \quad i = 1, 2, \ldots, m \tag{5.6}$$

We form the Lagrangian function:

$$P = f + \sum_{i=1}^{m} \lambda_i g_i(x) \tag{5.7}$$

In optimization theory, it is proven that an extremum of this function corresponds to a minimum of the objective function subject to the equality constraints. In fact, at a minimum

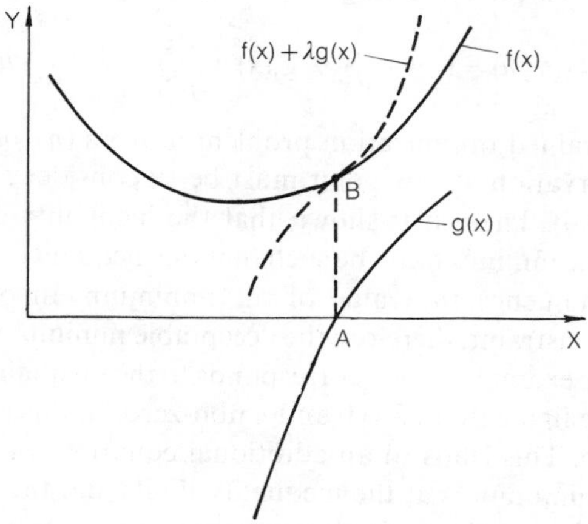

Figure 5.2 Geometric interpretation of a Lagrange multiplier

of the objective function, the Lagrangian function has an inflection point. This can be seen in Figure 5.2 for a function f(x) of one variable subject to the constraint g(x) = 0.

Function $y = \mathrm{g}(x)$ meets the x-axis at A and the minimum of f(x) subject to g(x) = 0 is simply point B. With dashed line (Figure 5.2), the Lagrangian function $\mathrm{P} = \mathrm{f}(x) + \lambda \mathrm{g}(x)$ has been plotted. This function apparently has an inflection point at B.

The constants λ are called 'Lagrange multipliers'. The necessary conditions for minima are:

$$\frac{\partial \mathrm{P}(x, \lambda)}{\partial x_i} = 0, \quad i = 1, 2, \ldots, n \tag{5.8}$$

We possess $m + n$ equations (5.6) and (5.8) with $m + n$ unknowns

$$x_1, x_2, \ldots, x_n, \lambda_1, \lambda_2, \ldots, \lambda_M$$

Therefore, the problem can be formally solved.

Case 3. Inequality constraints

This is the case when we want the minimum of a function subject to the following inequality constraints:

$$\mathrm{h}_j(x) \geqslant 0, \quad j = 1, 2, \ldots, p \tag{5.9}$$

We introduce at this point p new variables, called 'slack variables' to be determined,

$$u = \{u_1, u_2, \ldots, u_p\}$$

and we transform the inequalities to the following equalities:

$$\mathrm{h}_j(x) - u_j = 0, \quad j = 1, 2, \ldots, p \tag{5.10}$$

We thus have a problem of equality constraints as before with p additional variables u. Then, it is like Case 2 and can be solved with the method of Lagrange multipliers. The objective function will be a part of a Lagrangian function:

$$\mathrm{P}(x, \lambda, u) = \mathrm{f}(x) + \sum_{i=1}^{m} \lambda_j \mathrm{g}_j(x) + \sum_{j=m+1}^{m+p} \lambda_j(\mathrm{h}_j - u_j) \tag{5.11}$$

This is an unconstrained optimization problem in n x's $(p + m)$ λ's and p u's.

An important observation at this point must be to consider some results observed in Example (5.1) (see p. 194). There it is shown that the inequality constraint is not always active. In other words, the minimum can be such that the inequality constraint is satisfied and therefore it does not influence the value of the minimum. In other circumstances, the minimum violates the constraint, therefore the acceptable minimum will be at the inequality limits, a hyperbola in the example, which corresponds to the inequality becoming an equality. This corresponds, in the first case to $\lambda = 0$ and u non-zero. In the second case $u = 0$ and the multiplier λ is non-zero. This leads to an additional equation: $\{u\}^{\mathrm{T}}\{\lambda\} = 0$. This equation means that either the minimum is at the inequality limits and the slack variables are zero, $\{u\} = 0$, or not, in which case the multipliers are zero, $[\lambda] = 0$.

In analytical solutions, one has to try all possible combinations to be sure of finding the true minimum. In large problems this is very difficult and we resort to numerical methods which automatically find the true local minimum.

5.4 NUMERICAL METHODS

There are numerous difficulties in applying analytical methods for design optimization except for cases when the design equations are fairly simple. The equality constraints are mostly nonlinear and elimination of design parameters sometimes is impractical. Many times, solutions are obtained with numerical methods and the constraints do not have analytical expressions, and further, they are not continuously differentiable. Therefore, one has to resort to numerical methods. Moreover, even if analytical formulation of the problem is feasible, solution of the system of equations for the determination of the design parameters, usually highly nonlinear, has to be done with numerical methods.

Most numerical methods start with the definition of a penalty function which includes the objective function and the constraints and is subject to unconstrained optimization. It comes from the objective function f(x) with proper development to include also equality and inequality constraints, as was discussed in the Section 5.3. This function will be used in the form

$$P(x) = f(x) + K \sum_{i=1}^{m} [g_i(x)]^2 + L \sum_{i=1}^{p} \langle h_i(x) \rangle^2 \qquad (5.12)$$

where $\langle h(x) \rangle = h(x)$ for $h(x) < 0$ and zero otherwise.

The penalty multipliers K and L (not to be confused with the Lagrange multipliers, although they are very similar) have to be selected in such a way that will assure the proper contribution of the penalty terms and the objective function in the penalty function. The

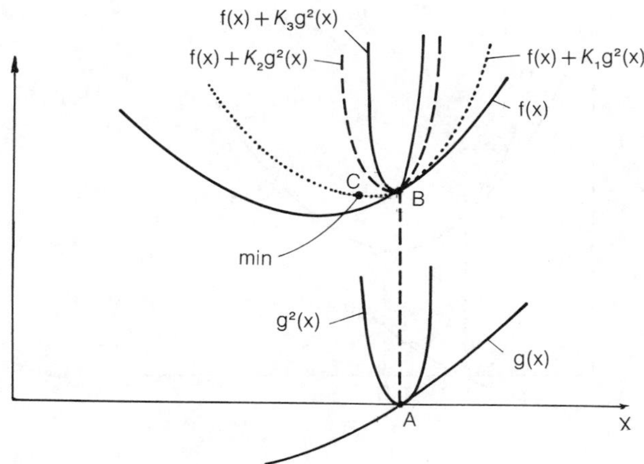

Figure 5.3 Penalty function, equality constraint

optimization procedure starts with small values of K and L and it is repeated from the last optimum point with progressively higher values of K and L.

'Hard' constraints, i.e. constraints which must be kept absolutely, require large multipliers. 'Soft' constraints, i.e. constraints which accept small violation, need only small multipliers. In general, the multiplier is a weighting function expressing the relative importance of each penalty term in the penalty function. The squaring of the constraint functions is for smooth boundaries which, as it will be apparent later, help for numerical convenience.

A geometric interpretation of the penalty function can be obtained again for one variable x and objective function f(x) (Figure 5.3). There is an equality constraint g(x) = 0 which yields a constrained minimum at point B. Then, the penalty function is plotted for three different values of the parameter K. We can observe that, indeed, for $K = K_3$ and K_2, the minimum of the penalty function is at B. There are values of K however which can shift the minimum, to C in this case, $K = K_1$.

A similar geometric interpretation of the inequality constraints is shown in Figure 5.4. An inequality constraint h(x) > 0 is imposed. If the limit point A is on the right of the unconstrained minimum, which is at D (Figure 5.4a), the penalty function is plotted for three

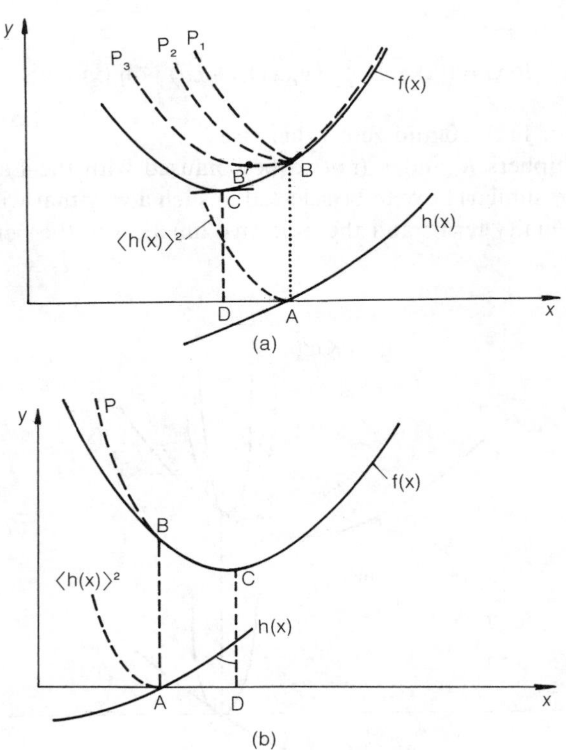

Figure 5.4 Penalty function, inequality constraint. (a) limit point A on right situation; (b) limit point A on left situation

different values of L. It is observed that for $L = L_1$ or L_2, the point B is, indeed, the constrained minimum. For a smaller L, however and penalty function L_3, the minimum is shifted towards C at point B′.

If the limit point A is on the left (Figure 5.4b) of the unconstrained minimum at D, then the minimum is still at C regardless of the value of L and the inequality constraint is inactive.

Several versions of the method, some having advantages over others, exist in the literature.

This method looks more complicated than elimination of parameters by way of the equality and, perhaps, inequality constraints. It has certain advantages, however:

1. Many times, the equations have a form that does not allow for elimination.
2. Using inequality constraints as equalities for parameter elimination does not always lead to an optimum, because, as shown in Example (5.1) (see p. 194), this is not always the case. One, then, has to obtain two solutions – one without the inequality constraint and the other using it as equality. This procedure is often tedious. On the other hand, the method of the penalty function takes care of the inequalities automatically. For larger problems, the process of elimination is out of the question anyway.
3. An analytical method with the Lagrange multipliers is possible in principle, but requires the solution of a large number of nonlinear algebraic equations, usually a tedious task.

5.5 SEARCH ALGORITHMS

(a) First-order methods: steepest descent

In an optimization problem we might look for either maxima or minima, depending on the nature of the problem. We do not lose generality if we assume, from this point, that we look for minima, because, when a function f(x) has a maximum, the function $-$f(x) has a minimum.

Most of the methods for seeking minima with a step-by-step algorithm are really extensions of the steepest descent method, introduced by Cauchy in 1847. (The reader must be careful when using the many methods because there is a tendency to underestimate their limitations and to generalize their range of application.)

In order to visualize the search strategy we start with a two-dimensional, unconstrained model (Figure 5.5). The penalty function (objective in this case) is $f(X_1, X_2)$, subject to optimization.

Suppose that we have a guess at the decision variables $A(X_1, X_2)$. We want to change the values X_1 and X_2 in order to obtain a better (smaller) value of the function $f(X_1, X_2)$. As a first approximation we assume that the objective function is linear with respect to X_1 and X_2 about the point A. In other words we substitute the surface $f(X_1, X_2) = C$ with a plane aAa tangent to it at point A. Then, for small variations ΔX_1 and ΔX_2 the value of the function will be, for a sufficiently small step, at point B,

$$f(X_1 + \Delta X_1, X_2 + \Delta X_2) \cong f(X_1, X_2) + \frac{\partial f}{\partial X_1}\Delta X_1 + \frac{\partial f}{\partial X_2}\Delta X_2 \tag{5.13}$$

The length (AB) of the step will be:

$$\delta^2 = \Delta S^2 = \Delta X_1^2 + \Delta X_2^2 \tag{5.14}$$

The change of the function BB′, approximated by the tangent plane, will (from equation 5.13) be:

$$\Delta f = \frac{\partial f}{\partial X_1}\Delta X_1 \pm \frac{\partial f}{\partial X_2}\Delta X_2 \tag{5.15}$$

In order that this change be a maximum,

$$\frac{\partial}{\partial X_1}(\Delta f) = 0, \quad \frac{\partial}{\partial X_2}(\Delta f) = 0 \tag{5.16}$$

which yields, with equation (5.14),

$$\frac{\Delta X_1}{\Delta X_2} = \pm\left(\frac{\partial f}{\partial X_1}\right)\Big/\left(\frac{\partial f}{\partial X_2}\right) \tag{5.17}$$

Equations (5.14) and (5.17) give the lengths ΔX_1 and ΔX_2. The new point B′ $(X_1 + \Delta X_1, X_2 + \Delta X_2)$ can be written in an operational form

$$X_B = X_A - \delta \nabla f_A(X) \tag{5.18}$$

where δ is the step length ($= \Delta S = AB$).

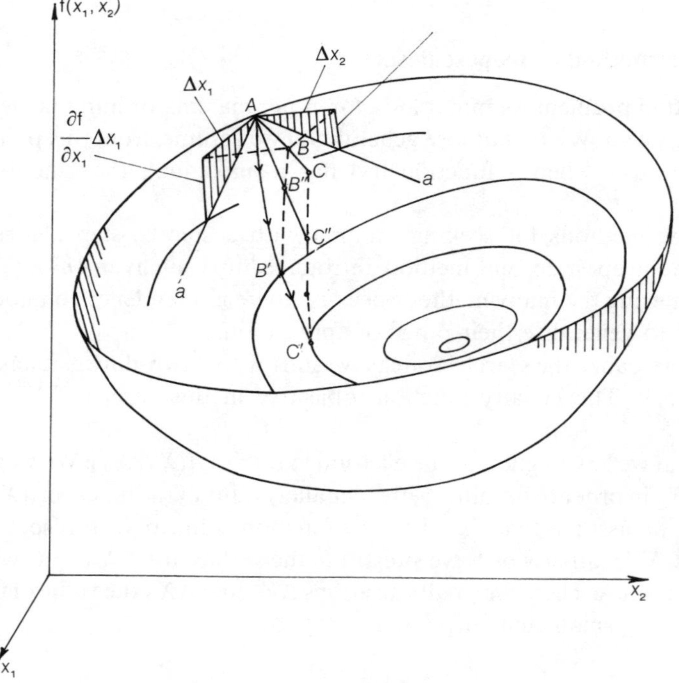

Figure 5.5 Steepest descent on a surface

The derivatives $\partial f/\partial X_1$ and $\partial f/\partial X_2$ can be calculated explicitly or numerically, depending on the form of the objective function. We note that the vector $\nabla \mathbf{f}$ is the one which is perpendicular to the isocontour line which passes through point A (Figure 5.5).

We note that further steps along the same direction give smaller values of the objective function, up to some point C, which means that the step length was not the proper one. Therefore, equation (5.18) has to be modified as follows:

$$\{X_C\} = \{X_A\} - \lambda_C\{\nabla f_A(X)\} \tag{5.19}$$

We can find point C and the value of $\lambda_C = AC$ by several methods. For example, we can proceed with successive steps until we observe an increase of the objective function. Then we can apply locally some numerical method to locate C. Such methods will be discussed later on (p. 188). In equations (5.18) and (5.19), from equations (5.14) and (5.17),

$$\{\nabla f_A(X)\} = \left\{\frac{\partial f}{\partial X_1}\frac{\partial f}{\partial X_2}\right\}\left[\left(\frac{\partial f}{\partial X_1}\right)^2 + \left(\frac{\partial f}{\partial X_2}\right)^2\right]^{-1/2} \tag{5.20}$$

We can generalize the method for n decision variables $X_1, X_2, \ldots, X_n$ applying the same formula

$$\{\nabla f_A(X)\} = \left\{\frac{\partial f}{\partial X_1}\frac{\partial f}{\partial X_2}\cdots\frac{\partial f}{\partial X_n}\right\}\left[\left(\frac{\partial f}{\partial X_1}\right)^2 + \left(\frac{\partial f}{\partial X_2}\right)^2 + \cdots + \left(\frac{\partial f}{\partial X_n}\right)^2\right]^{-1/2} \tag{5.21}$$

This method requires the computation of the function and its first derivative. It belongs to the class of first-order methods. In computer applications however, the derivative is computed numerically using small increments of the design variables and computing the change of the function.

(b) Zero-order methods: the Monte Carlo method

The previous method and similar ones converge very fast in some cases but they are very sensitive to discontinuities and singularities which very often exist in machine design. For

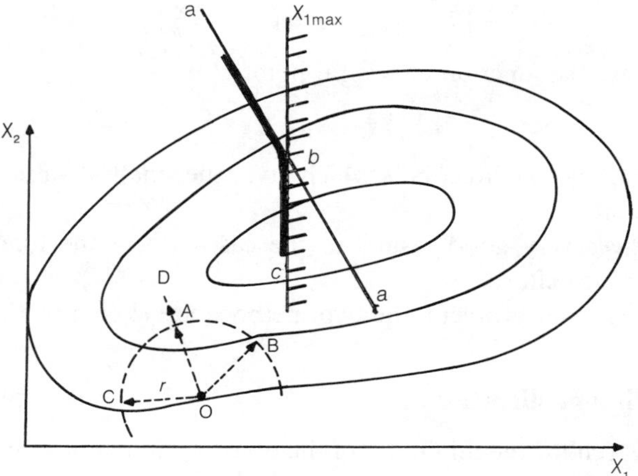

Figure 5.6 Random search in two directions

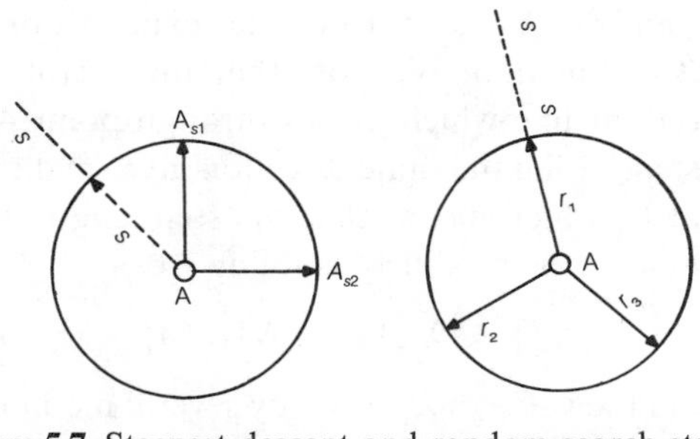

Figure 5.7 Steepest descent and random search strategies

example, in applications related to steam properties, problems are encountered when crossing the saturation line which appears as a steep ridge.

A very simple and efficient method is the Monte Carlo method. As the name suggests, it is based on random selection of directions. In the two-dimensional cases, for example, we start from point O in Figure 5.6.

We select a number of arbitrary directions, say three, and we calculate the value of the function at the points A, B, and C on a circle with center O and a radius r, the length of the step. To pick-up an arbitrary direction we take two random numbers r_1 and r_2 and form the vector $s(X_{01} + r_1, X_{02} + r_2)$ of length r where X_{01} and X_{02} are the coordinates of the point O.

We compare then, the values of the objective function at points A, B and C. Suppose that the value of the function at A is the smallest among the three points. Then we proceed along the direction of point A until we find point D, where the random search is repeated. We have to test at least two directions at every point. We can now generalize the method as follows:

Start from point O_0 $(X_1, X_2, \ldots, X_n)$ (Figure 5.6), and construct random directions:

$$\{s_j^*\} = \{r_{j1}, r_{j2}, \ldots, r_{jn}\} \tag{5.22}$$

where $j = 1, 2, \ldots, m \geqslant n$ and r_{ji}, $i = 1, 2, \ldots, n$ are $m \times n$ arbitrary or random numbers selected with a random-number generator. We normalize vector s to a desired length as follows:

$$\{s_j\} = [\{s_j^*\}[r_{j1}^2 + r_{j2}^2 + \cdots + r_{jn}^2]^{-\frac{1}{2}}]\delta \tag{5.23}$$

Then we calculate the function at several points:

$$\{X_j\} = \{X_0\} + \{s_j\} \tag{5.24}$$

We then move along the direction s which gives the smallest value of the function, to determine the value of λ_m.

This method gives very good results in the cases where the function $f(X)$ can be calculated without much effort.

For a two-dimensional problem, the two methods are shown in Figure 5.7.

(c) Optimization in one direction

In any numerical search for the minimum of the objective function, one always encounters the problem of optimization along the optimum direction. The lowest point along this direction has to be found in order to start the new search.

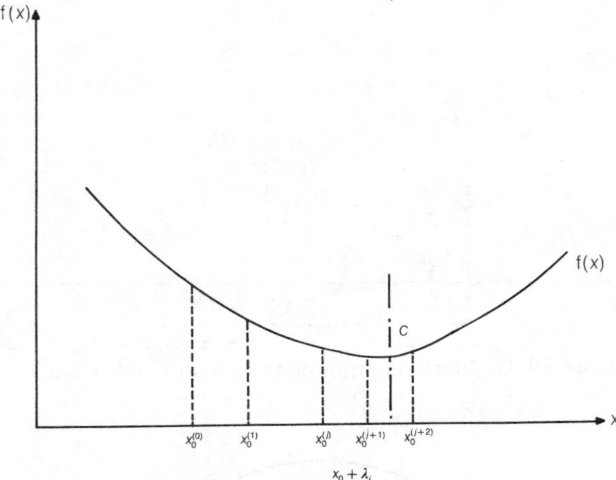

Figure 5.8 Undirectional optimization with interpolation

It seems easy, since the minimum in one variable can be found with a differentiation. This is not usually the case because the directional derivative of the objective function is usually not available. Only values of the objective function can usually be computed. In this case, one proceeds with a preselected step, until the objective function starts increasing again, for example, at the fourth step (Figure 5.8). Two strategies will be described here for the location of point C, the minimum along the direction s.

I. Lagrangian interpolation

As the function is evaluated at successive steps in the s direction, it is checked to see when it starts increasing, thus having passed the minimum. The last three points are used in a three-point Lagrangian interpolation function (F)

$$F = \frac{(x - x^{j+1})(x - x^{j+2})}{(x^j - x^{j+1})(x^j - x^{j+2})} f_j + \frac{(x - x^j)(x - x^{j+2})}{(x^{j+1} - x^j)(x^{j+1} - x^{j+2})} f_{j+1}$$

$$+ \frac{(x - x^j)(x - x^{j+1})}{(x^{j+2} - x^j)(x^{j+2} - x^{j+1})} f_{j+2} \tag{5.25}$$

where x is the coordinate along the direction s, f_j the value of the objective function at point x_j and $j = 1, 2, 3$, correspond to the last three steps of the unidirectional search.

This is a second-order polynomial in respect to x. Upon differentiation, it yields the local minimum of this function. The new point is used with the nearest two points for further iteration with the same procedure.

Higher-order interpolation has also been used. The appropriate Lagrange interpolation function must be found and a local minimum is obtained with one of the unidirectional optimization methods, such as the Golden Section method, which is discussed below.

II. The Golden Section method (Figure 5.9)

From the last three points, when the functions start increasing, the extreme two, A and B, are

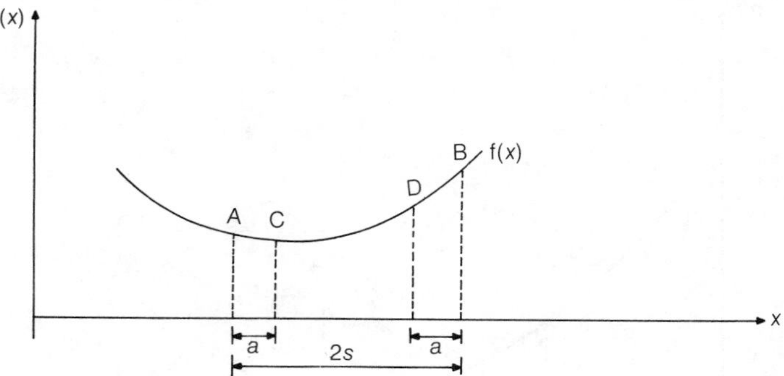

Figure 5.9 Undirectional optimization with Golden Section

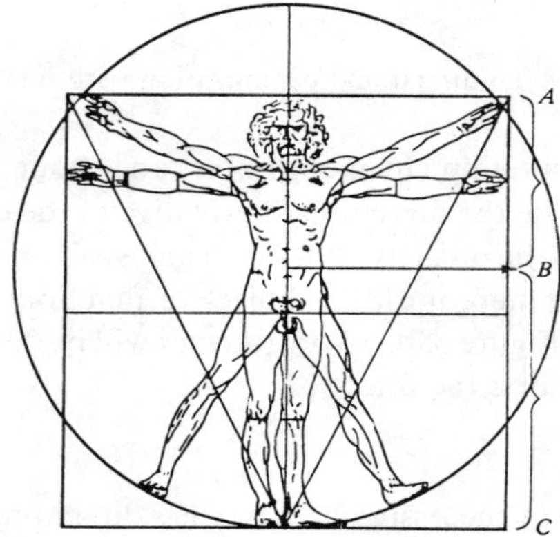

Figure 5.10 Human proportions and Golden Section (after Leonardo da Vinci). $BA/AC =$ the Golden Section ratio

kept. Two new points, C and D, are calculated at distances $a(2s)$ from the two ends where a is the Golden Section ratio 0.382 or some convenient fraction, usually around 0.3. The lower of the two points, say C, is found. The procedure is repeated between points A and D. This procedure converges rapidly to the required minimum.

The Golden Section principle known from ancient times, was demonstrated by Leonardo da Vinci in his studies of human proportions (Figure 5.10).

The method stops when a sufficient accuracy has been achieved or the number of steps exceed a predetermined number.

(d) Constraints on the design variables

The constraints

$$X_{lj} < X_j < X_{uj}$$

on the design variables can be simply observed during advance along a unidirectional search by setting the value of X to the limit which is violated. For example, consider a two-dimensional objective function with a search along line a–a. At point b, X_1 becomes greater than its higher limit $X_{1\max}$ (Figure 5.6). While the values of λ change normally, the values of $X_{1\max}$ are kept equal to $X_{1\max}$ until the lower point c is detected. A new search must be conducted for an optimum direction at this point.

(e) Interior point algorithms

The form of the penalty function of equation (5.12) belongs to a broad category of algorithms called 'exterior point algorithms', because the search can start from any initial point, even if the inequality constraints are violated. The penalty terms will force the search to gradually approach the feasible region, that is the region where the inequality constraints are not violated.

If we have enough information for the behavior of our system to select an initial guess in the feasible region, there are penalty functions which will keep the search within the feasible region imposing very high values to the penalty terms when the boundaries of the region are approached. For example, functions of the form

$$f(x) = \ln[g(x)] \quad \text{or} \quad 1/g(x) \tag{5.26}$$

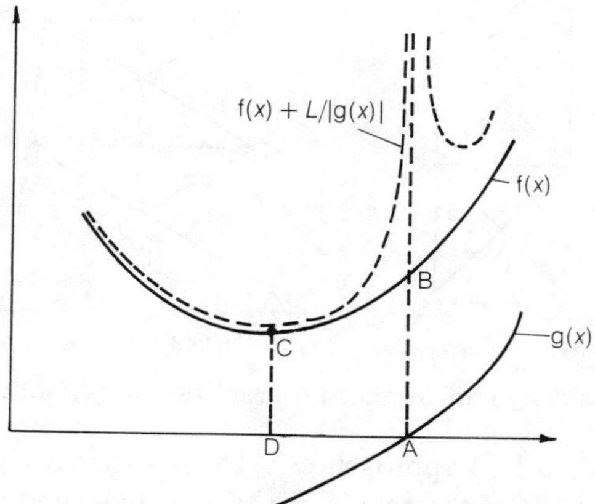

Figure 5.11 Interior point algorithms

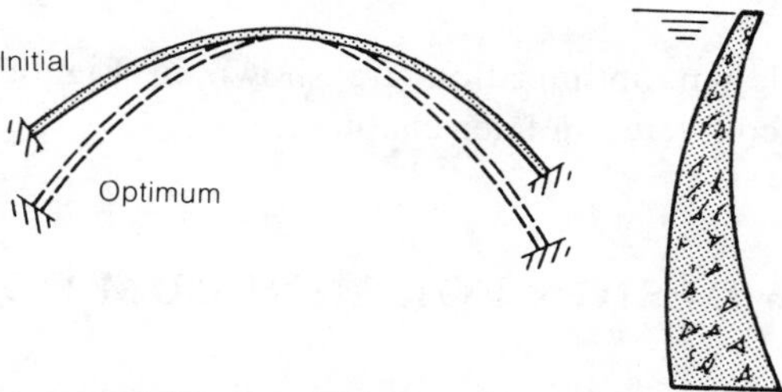

Figure 5.12 Optimum design of an arch and a dam (Gottfried and Weisman 1973)

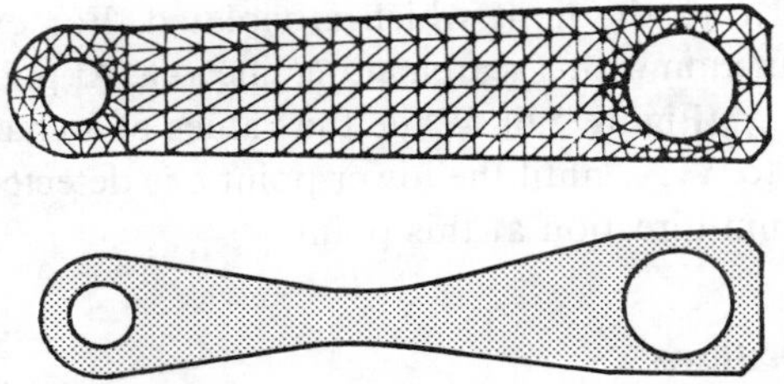

Figure 5.13 Optimum automotive arm design using the finite-element method (Gottfried and Weisman 1973)

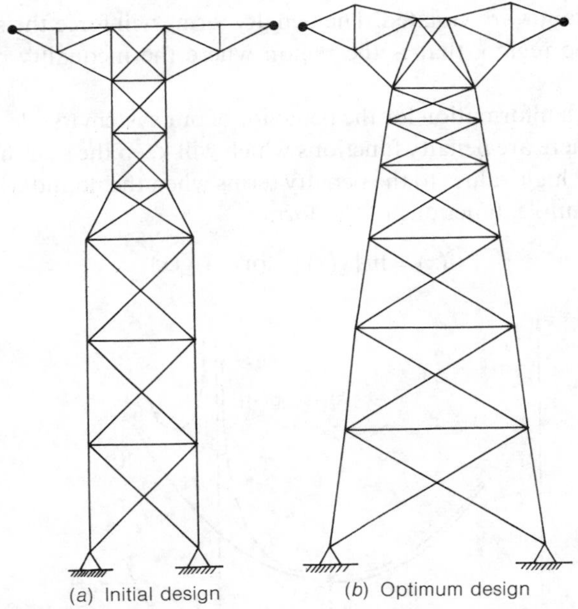

Figure 5.14 Optimum truss design for overhead transmission lines (Gottfried and Weisman 1973)

have very high values when h(x) approaches 0. The associated algorithms are known as interior point ones. The form of the penalty function is indicated in Figure 5.11.

All the above outlined search techniques, and some more, have been included in the program OPTIMUM (Appendix 5.A) and they are menu-selectable. The user has to supply a penalty function incorporating the objective function and equality and inequality constraints.

Some results of design optimization are shown in Figures 5.12 to 5.14. Further applications will be encountered in later chapters.

5.6 DESIGN FOR MINIMUM COST

In any society, especially in a free-enterpise economy, the engineer designs under cost control. For most machines and industrial products, the objective function is the profit which much be maximized. Almost invariably, this is synonymous with minimizing the cost. Then, constraints are imposed on certain design features such as stresses, deflections, etc. as a part of applicable design rules. Furthermore, additional constraints can be imposed on the environmental impact, safety etc.

The cost of an industrial product consists of several components which can be differentiated into two fundamental groups: *fixed costs* and *variable costs*.

Fixed costs are not related to the quantity of products manufactured. They are related to the cost of having the manufacturing facility in operation. Such costs are:

(a) Capital costs: The manufacturing facility requires an investment which should be repaid within a certain schedule, because the invested capital has a financial cost and the facility has a finite life during which the cost must be repaid (depreciation).
(b) Labor costs: In some cases, the labor costs are independent of the production and they have to be accounted as fixed costs, such as supervision.
(c) Maintainance costs: These include materials and labor to maintain the facility.
(d) Fixed operating costs: These include consumables, energy, taxes, etc., not related to the production volume.

Variable costs are directly related to the amount of products manufactured. Such costs are:

(a) Materials needed for the product itself.
(b) Materials for the facility operation which are related to the production volume.
(c) Labor costs directly related to the production volume.

Usually, the design engineer does not have to go through the optimization of the total production process when he designs an improved part or a new machine in the production process. With the planned production of the facility, *unit costs* are available for the estimation of the cost of machine or component, which include fixed and variable costs.

Unit costs include material and machining costs:

(a) Material cost: This is computed on the basis of the available size of the stock from which the final dimensions will be made and not the dimensions of the finished product. This cost varies between location and time. In a particular locality, at a certain time, some representative costs (in \$/kg) are

Low carbon steel	= 0.30
Alloy steel	= 0.5–2.0
Cast iron (product)	= 0.20
Aluminum	= 2.0
Copper	= 3.0
Copper alloys	= 3.0–6.0

(b) Machining cost: For some operations, this depends on the amount of machined volume (e.g. drilling), for others it depends on the amount of machined surface (e.g. grinding), while for others it depends on the length of machining (e.g. keyways). Some are on a per-operation basis (e.g. quenching).

Some representative machining costs based on time and locality are:

Drilling ($\$/cm^3$)	$= 0.1$
Machining ($\$/cm^2$)	$= 0.02$
Grinding ($\$/cm^2$)	$= 0.01$
Welding, hand ($/cm/pass)	$= 0.03$
Welding, machine ($/cm/pass)	$= 0.01$
Cleaning, deburring ($/cm)	$= 0.005$
Painting ($\$/cm^2$)	$= 0.005$
Gear cutting ($\$/cm^3$)	$= 0.20$

Therefore, the cost function will have the form

$$C = c_m Vd + \sum (\text{process cost} \times \text{process quantity}) \tag{5.27}$$

where V is the volume of the material stock, c_m the unit cost of the material and d its density.

The cost function will be supplemented by the constraints to yield the penalty function for optimization.

EXAMPLES

Example 5.1

We want to design a rectangular container with the following specifications:

Contained volume $= V_0 = 1\ \text{m}^3$.
Maximum base area $= A_0 = 1\ \text{m}^2$.

Design objective: minimum cost of the container's material (thickness given).

The design parameters will be:

$$x_1 = \text{length } L; \quad x_2 = \text{width } B; \quad x_3 = \text{height } H$$

The objective function is the container surface which must be minimized:

$$\text{f} = 2(LB + LH + BH) \tag{a}$$

The equality constraint (volume $= V_0$)

$$\text{g}_1 = LBH - V_0 = 0 \tag{b}$$

The inequality constraint is on the base area $\leqslant A$

$$\text{h}_1 = A_0 - LB \geqslant 0 \tag{c}$$

In order to meet the design objective we must find a set of values L, B and H to give the minimum function f(total area) and satisfy the constraints of the problem. We must take into account some physical limitations on the design parameters. For example:

$$0 < L, B, H < M \tag{d}$$

In order to visualize the optimization process we substitute the value of H from equation (b) into the equation (a)

$$\text{f} = 2[LB + (L + B)V_0/LB] \tag{e}$$

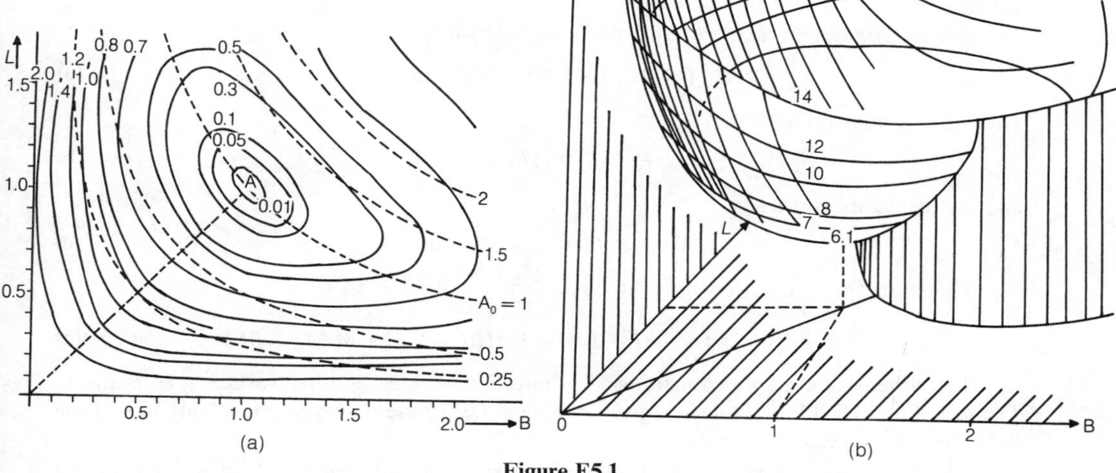

Figure E5.1

We eliminate one parameter using one equality constraint. This suggests that, in general, we can eliminate some of the design parameters using an equal number of equality constraints. This is not always possible, or even wise (p. 183).

The remaining parameters, the decision variables, L and B, are now subject to optimization, under the remaining inequality constraint. Now, we plot the function f in an (L–B) Cartesian plane (Figure E.5.1a), in the form:

$$f = C \tag{f}$$

for the given volume $V_0 = 1$.

For several values C we will have different contours. We note that from the L–B plane, we can select only values (L, B) which are below the hyperbola of equation (c),

$$LB = A_0 \tag{g}$$

These hyperbolas, for several values of A_0 have been plotted with dashed lines. We see that, if we disregard the inequality constraint, there is a minimum of the function at the point A (1, 1) where the function equals six. Therefore, the container will have the shape of a cube as one should expect from well-known geometric rules. This type of minimum is called an 'unconstrained minimum'. If we add the constraint (c) we note that for values $A_0 \geqslant 1$, the minimum is valid and the inequality constraint is ineffective. However, for $A_0 < 1$ the unconstrained minimum is not acceptable because it violates the inequality constraint.

The new minimum will be found as the point of contact of a contour line and a hyperbola. For example, for $A_0 = 0.25$ the minimum is at point B (0.5, 0.5) and the value of the function is about nine. This minimum is called a 'constrained minimum'. In this case, the inequality constraint appears as an equality.

Because of the small number of parameters involved and the form of the objective function we have only one minimum in the permissible range of variation of the decision variables. In other cases, however, we might have several minima. We shall call them 'local minima' and the absolute minimum will be called the 'global minimum'.

Example 5.2

Formulate Problem 5.1 by way of the penalty function. Apply program OPTIMUM with data:

Volume $V_0 = 10\,\mathrm{m}^3$

Minimum base area $A_0 = 4\,\mathrm{m}^2$

Solution

The design variables are H, B, L. The objective function:

$$f(H, B, L) = 2(HB + BL + LH) \tag{a}$$

The equality constraint

$$g(H, B, L) = HBL - V_0 = 0 \tag{b}$$

and the inequality constraint

$$h(B, L) = A_0 - BL > 0 \tag{c}$$

Therefore, the penalty function will be

$$P(H, B, L) = 2(HB + BL + LH) + K(HBL - V_0)^2 + M\langle A_0 - BL\rangle^2 \tag{d}$$

Therefore, we have an unconstrained optimization problem in three variables, H, B and L. The program OPTIMUM, solves just this problem. Let us use it now as a tool only to find the minimum of the penalty function P.

The program (see Appendix 5.A for details) expects a penalty function F definition after line 10,000, where the design variables are the elements of an array $H(1)$, $H(2)$,.... The program then yields the solution:

$$H = 2.5,\ B = 2,\ L = 2 \text{ and } f(H, B, L) = 24$$

Obviously, $BL = A$, which means that the obtained optimum is on the boundary of the inequality constraint, $A - BL > 0$. On the other hand, use of $A = 6\,\text{m}^2$, results in a cube,

$$H = B = L = 2.44$$

In this case the inequality constraint is ineffective and the obtained minimum is the global one.

Example 5.3

A torque arm is to be designed for an automobile application, such as in Figure E5.3. A vertical force P on the right produces a torque PL on the shaft on the left. Simple strength of materials formulas should be used:

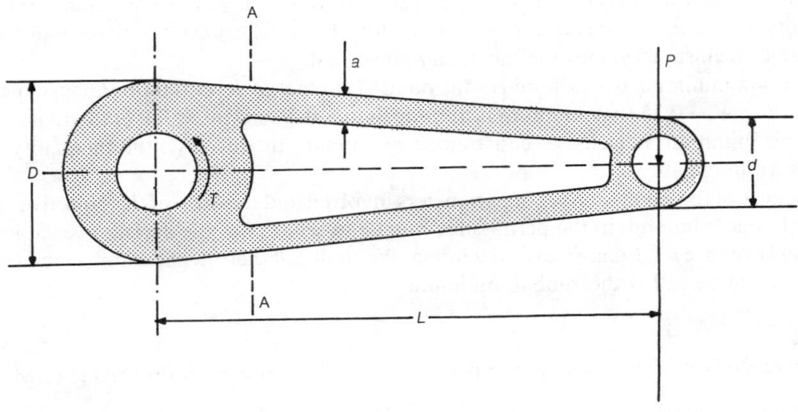

Figure E5.3

(a) The maximum stress at section A–A should not exceed some value:

$$\sigma_{\max} = \frac{M}{I} y_{\max} = \frac{P(L - D/2)}{I} \frac{D}{2} \leqslant s_y/N, \quad g(a, D) = 1 - \frac{P(2L - D)D}{4I} \frac{N}{s_y} = 0$$

$$I = 2a^2\left(\frac{D - a}{2}\right)^2 + \frac{2a^4}{12} = \frac{a^2(D - a)^2}{2} + \frac{a^4}{6}$$

(b) The axial load on the two connecting rods is, approximately,

$$F = PL/(D - a)$$

and it should not exceed the buckling load:

$$F \leqslant \frac{4\pi^2 EI}{l^2}, \quad h(a, D) = \frac{2\pi^2 Ea^4}{3(2L - D - d)} \frac{(D - a)}{PL} - 1 > 0$$

Determine the optimum D and a (thickness) for minimum volume of the arm.
Data: $E = 2.1 \times 10^{11}$ N/mm^2; $L = 200$ mm; $d = 30$ mm; $P = 5000$ N; $s_y/N = 100$ N/mm^2.

Solution
The volume is

$$V = \frac{\pi D^2}{4} a + 2\left(L - \frac{D}{2} - \frac{d}{2}\right) a^2$$

The stress equation is used as an equality constraint

$$g(a, D) = 1 - \frac{P(2L - D)}{4I} \frac{N}{s_y} = 0$$

The buckling equation is used as an inequality constraint

$$h(a, D) = \frac{4\pi^2 Ea^4}{3(2L - D - d)} \frac{(D - a)}{P} - 1 > 0$$

Then, the penalty function is assumed in the form

$$P(a, D) = V + K_1[g(a, D)]^2 + K_2\langle h(a, D)\rangle^2$$

The program OPTIMUM was used with $K_1 = 10^6$ and $K_2 = 10^4$. The optimum is at $a = 17.8$ mm and $D = 46.8$ mm. At this point, $g = -0.047$ and $h = 13{,}704$. This means that the error in satisfaction of the stress equation is approximately 5% and the buckling equation is inactive. The first attempt was with $K_1 = K_2 = 10$. It yielded smaller volume but great value of the equality constraint which means that the stress equality was not satisfied. Progressively higher values for K were used until the equality constraint was approximately satisfied.

Example 5.4

Optimization of a solar collector orientation linkage.

(The reader should be familiar with kinematic synthesis of linkages at an elementary level to follow this example.)

Solar tracking by means of linkages is a simple, reliable and economical solution to improve efficiency of solar collectors. An optimization method for this purpose is presented in this example.

The apparent altitude and azimuth of the sun changes with the hour of the day, the season and the latitude of the station. A diagram for a specific data and locality is shown in Figure E5.4.1. The

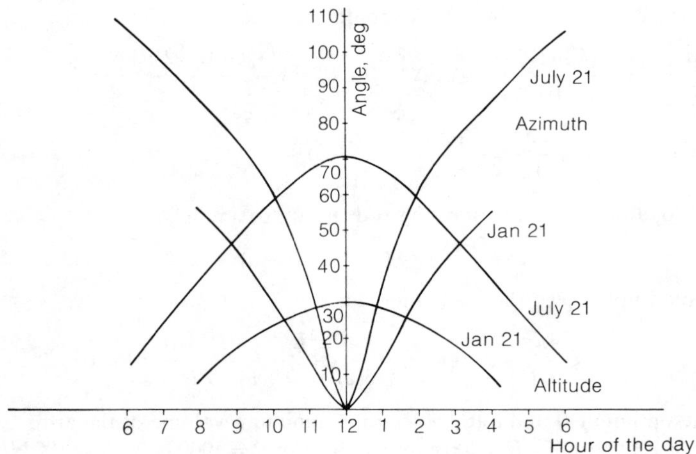

Figure E5.4.1 Azimuth and altitude of sun, versus hour of the day for a latitude 40°N at 21 January and 21 July

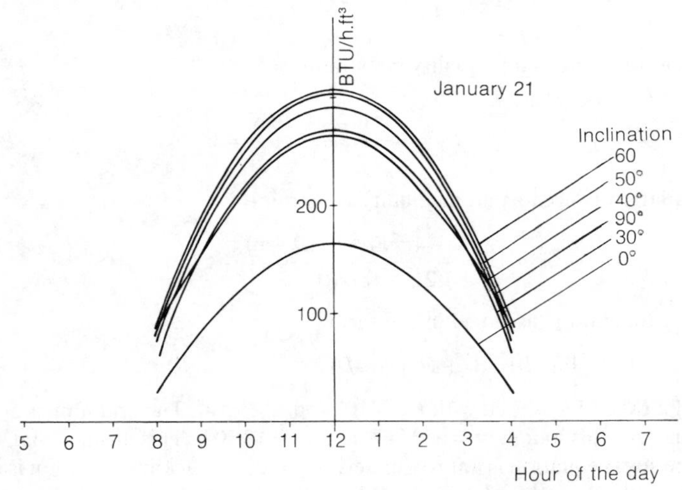

Figure E5.4.2 Total surface insolation versus hour of the day for 40°N latitude at 21 January for different surface inclinations

azimuth is plotted against the hour of the day for January 21 at a latitude of 40° North. For the same locality, the altitude plotted against the hour of the day is also shown in Figure E5.4.1 and the total surface insolation versus hour of the day and surface inclination for a south-facing surface is shown in Figure E5.4.2.

An ideal tracking should provide for a space motion to track both altitude and azimuth. Though this can be achieved with spatial mechanisms, in this example only the azimuth tracking is considered. It is evident from Figure E5.4.1 that one should approximate the azimuth functions in a range of angles $-60°$ to $60°$.

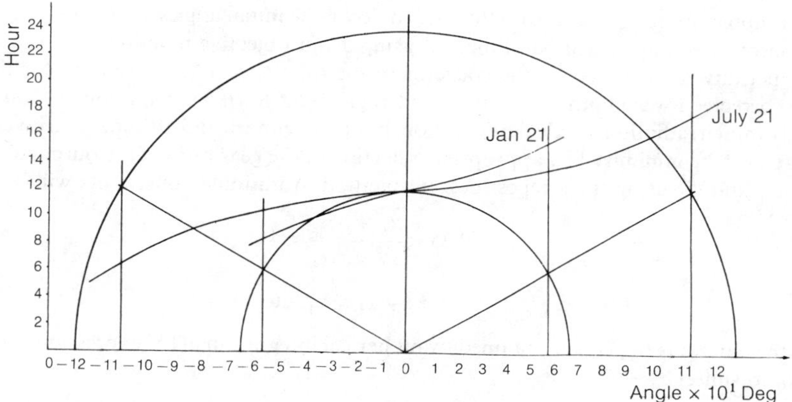

Figure E5.4.3 Spacing of accuracy points (Tschebytschev) for the function to be generated, angle versus hour of the day

The corresponding ranges of the hours will be 8 a.m. to 4 p.m. The algebraic method of synthesis will be utilized with three accuracy points selected by means of the Tschebytschev polynomials in a way shown in Figure E5.4.3. Finally, the following accuracy points were used in a form of couples of coordinate points for the system of Figure E5.4.3 (21 January):

Hours	Degrees
$x_1 = 8.4$	$y_1 = -52$
$x_2 = 12$	$y_2 = 0$
$x_3 = 15.6$	$y_3 = 52$

Range $E = 100°$

The mechanism used is shown in Figure E5.4.4. The input angle is provided by way of a geared motor delivering 1 revolution per 24 hours. The output angle ψ corresponds to the azimuth angle y. Because the range of y is very wide, the range of ψ is reduced by way of a gear couple attached to the link 3, having a gear ratio π/ES so that the range of φ is 0 to π and the range of ψ is $-\pi/S$ to π/S, leaving S as an arbitrary parameter. Therefore the transformation from (x, y) to (φ, ψ) will be

$$\varphi = \varphi_1 + \frac{x - x_1}{24} 2\pi; \quad \psi = \psi_1 + \frac{y - y_1}{2ES} 2\pi \tag{a}$$

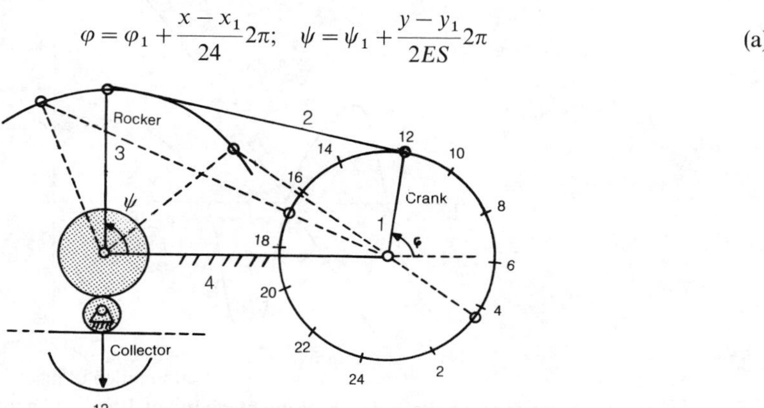

Figure E5.4.4 Optimum mechanism configuration for 21 January

It is apparent from equation (a) that the choice of the initial angles φ_1 and ψ_1 is arbitrary. The selection of these angles will be optimized using as an objective function the ratio of the total energy per day (U) received by the collector to the total energy (U_α) available for the collector ideally corrected for azimuth only with a fixed angle to the horizontal at noon ω_0, while U_i will be the daily total irradiation for ideal orientation, both for azimuth and altitude. The inverse function $f = U_\alpha/U_i$ will be minimized by appropriate selection of the gear ratio of the output and the initial angles φ_1 and ψ_1 using the steepest descent method. Additional constraints will be

$$0.25 < \frac{\alpha_1}{\alpha_4}, \frac{\alpha_2}{\alpha_4}, \frac{\alpha_3}{\alpha_4} < 4$$

$$\alpha_1 + \alpha_2 + \alpha_3 > \alpha_4, \text{ etc.}$$

where the functions U_α, U_c are obtained by numerical integration of the expressionos from sunrise time x_0 to sunset $x_0 + E_x$.

$$U_\alpha = \int_{x_0}^{x_0+E_x} q_i \cos(a - \omega_0) \mathrm{d}x$$

$$U_c = \int_{x_0}^{x_0+F_x} q_i \cos(a - \omega_0) \cos(\psi - \psi^*) \mathrm{d}x$$

Where q_i is the instantaneous irradiation rate, ψ is the output angle delivered by the linkage and ψ^* the ideal output angle corresponding to the correct azimuth; in other words $\psi - \psi^*$ is the absolute linkage error.

The optimization procedure used the program OPTIMUM (Appendix 5.A) and yielded a variety of mechanisms having very low error in the energy function U_c owing to the stationary character of the cosine function about zero. For example, the following linkage was obtained.

Data:
January 21. Sun rise time = 8 a.m.; sun set time = 4 p.m.
$U_\alpha = 2127.45$ BTU/ft^2; $U_i = 2182$ BTU/ft^2.

Results:
$\alpha_1 = 0.3766$; $\alpha_2 = 1.057$; $\alpha_3 = 0.6007$; $\alpha_4 = 1$ (link lengths)

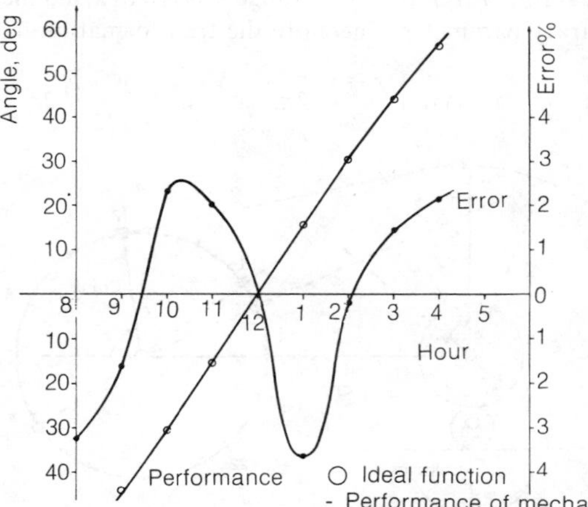

Figure E5.4.5 Performance of solar tracking mechanism. Ideal and generated angle versus hour of the day. Error per cent versus hour of the day

Input angle at 12 noon = 82.65°.
Output gear ratio = 0.19.
Surface daily total energy, $U_c = 2127.36$ BTU/ft².

It is observed that U_c is almost identical with the ideal value of $U_\alpha = 2127.45$ BTU/ft². Turning now to the accuracy of the output angle, which is most important for focusing collectors, in Figure E5.4.5 the correct function and the linkage output angle are plotted against the hour of the day. There is observed a maximum error of 3.62% which is tolerable for most focusing collector applications. The mechanism is shown in Figure E5.4.4.

REFERENCES AND FURTHER READING

Dimarogonas, A. D., 1972. *Machine Design Optimization.* General Electric, Technical Information Series. Schenectady, N.Y.

Fiacco, A.V., McCormic, G. P., 1968. *Nonlinear Programing: Sequential Unconstrained Optimization Techniques.* New York: J. Wiley.

Gottfried, S. B., Weisman, J., 1973. *Introduction to Optimization Theory.* New York: Prentice-Hall.

Johnson, R. C., 1961. *Optimum Design of Mechanical Elements.* New York: J. Wiley.

Mischke, C. R., 1968. *An Introduction to Computer Aided Design.* New York: Prentice-Hall.

Reclaitis, G. V., Ravindran, A., Ragsdell, K. M., 1983. *Engineering Optimization: Methods and Application.* New York: J. Wiley.

Siddal, J. N., 1982. *Optimal Engineering Design.* New York: Marcel Dekker.

PROBLEMS

5.1 A circular rod of diameter d will be machined to a rectangular shape to give a beam with maximum resistance moment in bending $W = bh^3/6$. Given d, find analytically the optimum value of b. Then, for $d = 10$ cm, define the objective function and the equality constraint of the problem and define the penalty function.

5.2 A cyclindrical container (20 m³) is filled with water. The stress near the lower end is pr/t, where p the pressure, r the radius $d/2$ and t the thickness. The material has allowable stress 100 N/mm². Analytically, find the diameter and height which gives minimum weight of the container. The thickness t is uniform.

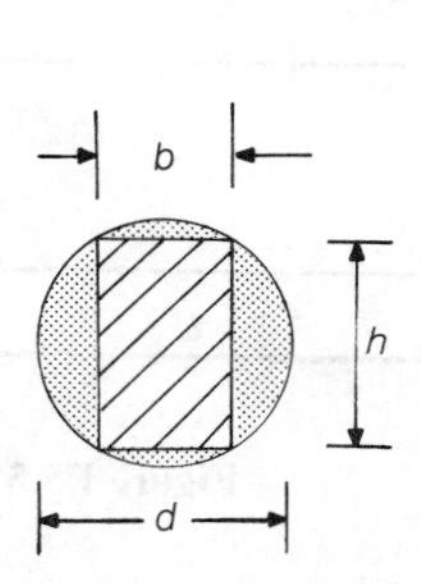

Figure P5.1

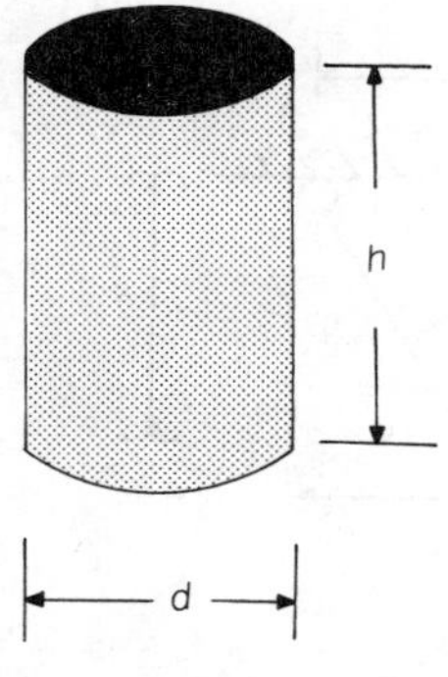

Figure P5.2

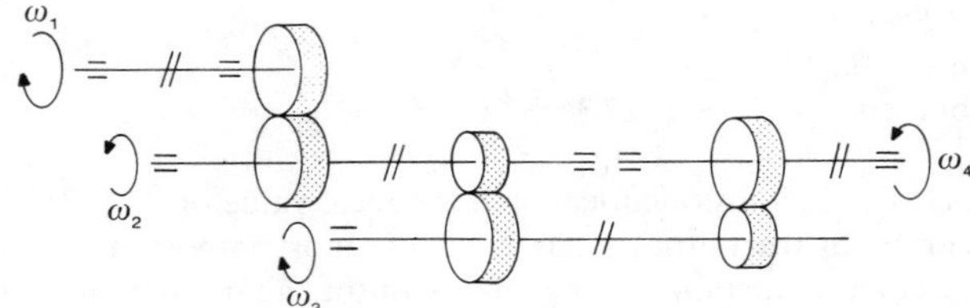

Figure P5.3

5.3 A three-stage gear transmission has overall gear into 10 or $\omega_1/\omega_4 = 10$. On the motor at $\omega = \omega_1$, the equivalent moment of inertia for gear i which has $\omega = \omega_i$ is $I_{eq} = (\omega/\omega_1)^2\, I_p$ where I_p is the polar moment of inertia of the particular gear. Assuming that I_p is $1/2mr^2 = (\pi/2)\rho br^4$ where r is the radius, ρ the density and b the constant width. Find analytically the three gear ratios i_1, i_2, i_3 to have:

(a) Total gear ratio $i_1 i_2 i_3 = 10$.
(b) Minimum equivalent moment of inertia on the motor shaft, if $r_1 = r_2 = r_3$.

[From Gottfried and Weisman *Introduction to Optimization Theory*]

5.4 A hook carrying a 10-ton load is supported from a roof through two circular rods AB and AC. If the load does not always have vertical direction, there is also a maximum horizontal force of 5-ton in either direction. Calculate analytically the distance BC (x) for the minimum diameter of the rods, where $h = 1$ m if the allowable stress is 8 kN/cm^2. Do not consider buckling.

5.5 A pressure vessel has internal pressure $p = 10$ bar and a volume $V = 4$ m^3. The tangential stress $pd/2t$ should not exceed 10 kN/cm^2. Find analytically the values of L and d for minimum weight. The thickness t is constant throughout the vessel.

5.6 The vessel shown is filled with water. The thickness t is calculated based on tensile strength at point A which should not exceed 12 kN/cm^2. Find analytically the values of d and h for minimum vessel weight if the volume is 6 m^3 and the thickness is constant.

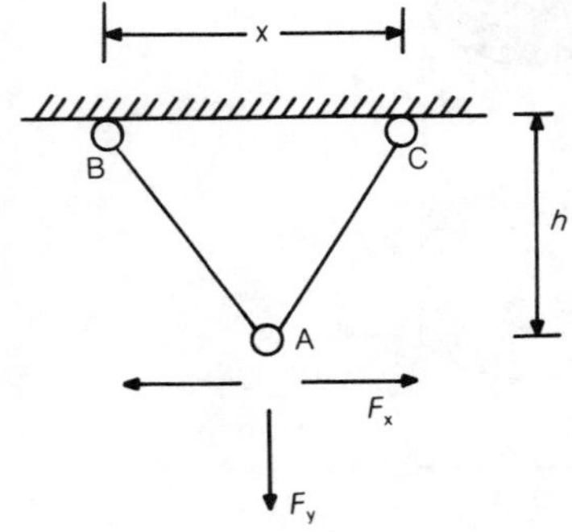

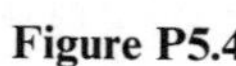

Figure P5.4

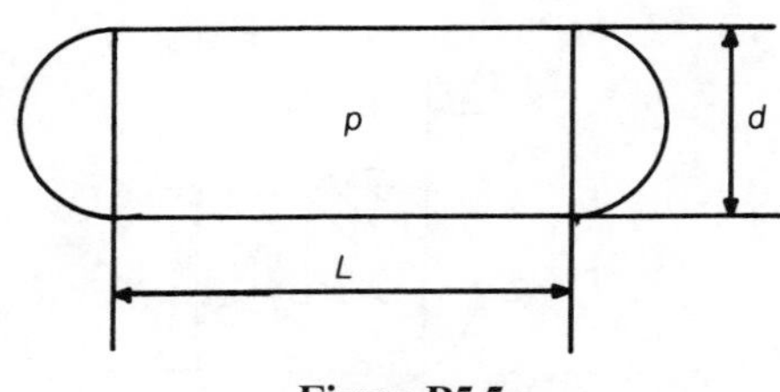

Figure P5.5

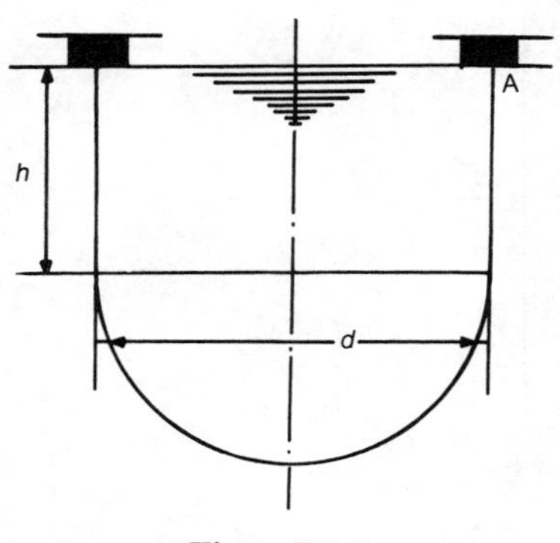

Figure P5.6

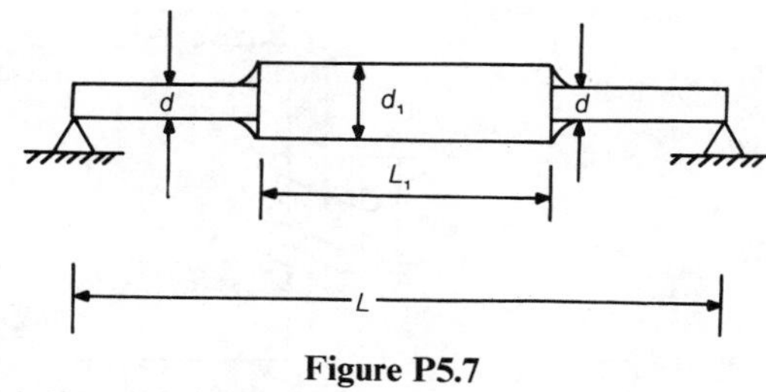

Figure P5.7

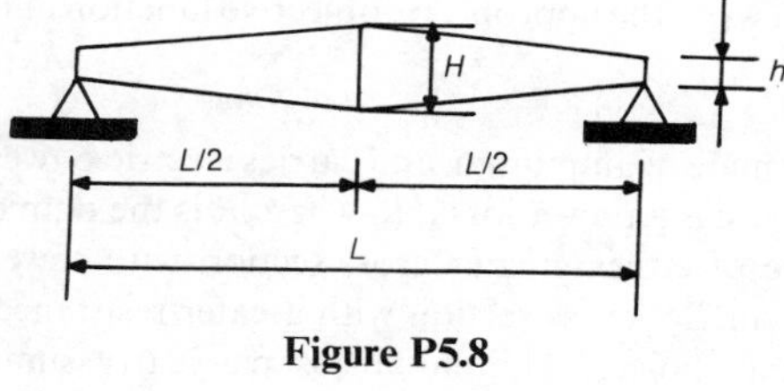

Figure P5.8

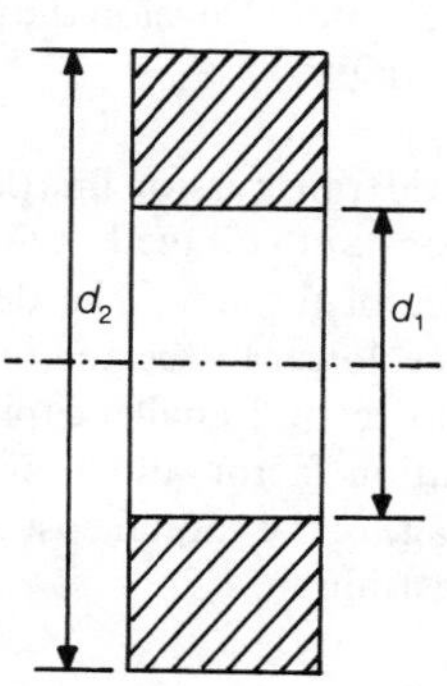

Figure P5.9

5.7 The shaft shown is made of steel with $E = 2.1 \times 10^5$ MPa, density $7800\,\text{kg/m}^3$, and allowable tensile stress 120 MPa. The shaft is loaded by its own weight. Find analytically the optimum values of the ratios L_1/d, d_1/d for maximum L.

5.8 For maximum L as in Problem 5.7, same material, find the optimum value of h/H for a tapered beam, loaded by its own weight.

5.9 A rotating disc as shown is used for storage of kinetic energy. If $\omega = 300$ rad/sec $= 40$ cm, the maximum allowable stress is $30\,\text{N/mm}^2$ and the density is $2200\,\text{kg/m}^3$, find t_1, t_2, for maximum stored kinetic energy per unit mass of the disk, see page 679.

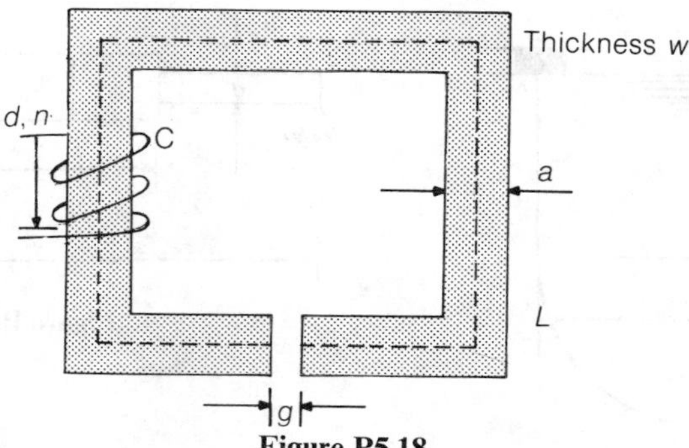

Figure P5.18

Problems 5.10 to 5.17

Solve the Problems 5.1 to 5.8 with the program OPTIMUM, forming a proper penalty function. Use all four minimization methods of the menu and tabulate the computer time required to reach the same level of accuracy.

5.18 Figure P5.18 shows a typical electromagnetic circuit. A coil C with copper wire of diameter d and n turns is fed with electric current to produce a certain magnetic flux in the core. The designer has the choice of using more copper winding or more iron core. To optimize this selection, formulate the optimization problem of minimizing the cost of the materials for the given flux, current, material cost, width a and thickness w of the core at the lower section, gap g, and mean length of the magnetic path L. Develop the design equations, then write the appropriate objective function and the constraints.

5.19 An electric transmission line (length l, diameter d) is made of aluminum, and carries electric power W at a constant voltage V. Owing to the current i, there is a power loss i^2R, where R is the ohmic resistance of the line. The designer has the choice of either greater cross-section with lower resistance, lower losses and greater capital cost, or smaller cross-section with greater resistance, greater losses and smaller capital cost. To optimize this choice, for minimum cost per year, assume a utilization factor and year energy flow constant, cost of line per kg, cost per kWh, fixed percentage of the capital cost as interest and depreciation. Then formulate the objective functions and constraints.

5.20 Discuss the optimization of the selection of the actual power to pump a constant flow of water to a certain distance and height. Tabulate the data you need and formulate the objective function and the boundary conditions.

Problems 5.21 to 5.25

Optimization methods are used for curve fitting, that is, given n pairs of data, $n \times x$ and the corresponding $n \times y$, find the values of $a, b, c, \ldots$ which make the function $f(x; a, b, c, \ldots)$ deviate from the given data to the least possible extent. To this end we select the function f to a convenient form intuitively, or from experience, to a form which contains undetermined coefficients $a, b, c, \ldots$ Then assuming an optimization criterion, usually the sum of the squares of the differences:

$$F(a, b, c, \ldots) = \sum_{i=1}^{n} [y_i - f(x_i; a, b, c, \ldots)]^2$$

This is then the objective function to be minimized. Now, given the following table of six pairs of points x, y, find the best fit for the equations indicated next to each problem number:

x	y
0.5	0.256
1.0	0.480
1.5	1.045
2.0	1.843
2.5	2.240
3.0	2.566

5.21 $ax^2 + bx + c$

5.22 $a\cos(bx + c)$

5.23 $ae^{bx} + c$

5.24 $ax^b + c$

5.25 $ae^{-bx} + ce^{-dx}$

APPENDIX 5.A
OPTIMUM: AN OPTIMIZATION PROGRAM

```
10 CLS:KEY OFF
20 REM **************************************************
30 REM *                                                *
40 REM *                   OPTIMUM                      *
50 REM *                                                *
60 REM **************************************************
70 REM          COPYRIGHT 1987
80 REM by Professor Andrew D. Dimarogonas, Washington University,St.Louis.
81 REM All rights reserved.  Unauthorized reproduction, dissemination or
82 REM selling is strictly prohibited.  This listing is for personal use.
90 REM
100 REM
110 REM Unconstrained optimization of a function.  Steepest descent,
Conjugate directions,
120 REM Monte Carlo and Random coordinate methods are used.
121 REM       ADD/11-20-85/revision 12-22-86
122 REM ***********************************************************
123 CLS:KEY OFF
124 FOR I=1 TO 14:COLOR I:LOCATE I,22+I:PRINT "OPTIMUM";
125 LOCATE I,50-I:PRINT "OPTIMUM";:NEXT I:COLOR 14
126 LOCATE 8,7:PRINT"An Optimization Program for a user-supplied penalty";
127 PRINT " function";
128 locate 25,1:print" Dimarogonas,A.D., Computer Aided Machine Design,
Prentice-Hall,1988";
129 locate 23,28:input"Hit RETURN to continue";HR$
130 REM
140 REM BOUNDS OF VARIABLES BL<VE<VU
150 DIM VU(20): REM UPPER BOUND
160 DIM VL(20): REM LOWER BOUND
170 DIM H(20),HA(20), BB(20):REM STATE VECTORS
180 DIM SR(20,20),HR(20,200),S(20),F(20),DF(20):REM AUX MATRICES
190 DIM IN(10),CO(10), VE(10)
200 REM N=NUMBER OF VARIABLES
210 N = 2
220  REM PRINTOUT:FULL = 1 , RESULTS ONLY=2
230 IN(1) = 1
240 REM IN(20 = NO OF RANDOM DIRECTONS>=N
250 IN(2) = 2 * N
260 REM CONSTANTS:STEP,MIN.STEP,MIN. OF OBJECTIVE FUNCTION, ACCELERATION
FACTOR, DECELERATION FACTOR
270 FOR I = 1 TO 5: READ CO(I):NEXT I
280 DATA  .01,.001, .1, 1.5,.5
290 REM UPPER BOUNDS
300 FOR I = 1 TO N: READ VU(I):NEXT I
310 DATA 5, 20
320 REM LOWER BOUNDS
```

```
330 FOR I = 1 TO N: READ VL(I):NEXT I
340 DATA 0, 0
350 REM INITIAL GUESS OF VARIABLES
360 FOR I = 1 TO N: READ VE(I):NEXT I
370 DATA 1,1
380 CLS
390     PRINT"********************************************"
400     PRINT"*                                          *"
410     PRINT"*                OPTIMUM                   *"
420     PRINT"*                                          *"
430     PRINT"********************************************"
440 PRINT:PRINT"     OPTIMIZATION PARAMETERS"
450       PRINT"  RETURN for default parameters"
460 PRINT"                               Default:"
470 PRINT"Number of variables             ";N;TAB(60);:INPUT X$
480 IF X$>"" THEN N=VAL(X$)
490 PRINT"Printout, full=1, results=2   ";IN(1);TAB(60);:INPUT X$
500 IF X$>"" THEN IN(1)=VAL(X$)
510 PRINT"Search step length              ";CO(1);TAB(60);:INPUT X$
520 IF X$>"" THEN CO(1)=VAL(X$)
530 PRINT"Minimum step length             ";CO(2);TAB(60);:INPUT X$
540 IF X$>"" THEN CO(2)=VAL(X$)
550 PRINT"Minimum of penalty function   ";CO(3);TAB(60);:INPUT X$
560 IF X$>"" THEN CO(3)=VAL(X$)
570 PRINT"Acceleration Factor             ";CO(4);TAB(60);:INPUT X$
580 IF X$>"" THEN CO(4)=VAL(X$)
590 PRINT"Deceleration Factor             ";CO(5);TAB(60);:INPUT X$
600 IF X$>"" THEN CO(5)=VAL(X$)
610 FOR I=1 TO N
620 PRINT"Upper bound of variable ";I;"   ";VU(I);TAB(60);:INPUT X$
630 IF X$>"" THEN VU(I)=VAL(X$)
640 NEXT I
650 FOR I=1 TO N
660 PRINT"Lower bound of variable ";I;"   ";VL(I);TAB(60);:INPUT X$
670 IF X$>"" THEN VL(I)=VAL(X$)
680 NEXT I
690 FOR I=1 TO N
700 PRINT"Initial guess of variable";I;"  ";VE(I);TAB(60);:INPUT X$
710 IF X$>"" THEN VE(I)=VAL(X$)
720 NEXT I
730 PRINT:PRINT"Are the data correct? (Y/N)   ";:INPUT X$
740 IF X$="n" OR X$="N" THEN 380
750 CLS
760 CLS
770 PRINT "   OPTIMIZATION METHOD"
780 PRINT :PRINT "                  MENU"
790 PRINT "--------------------------"
800 PRINT
810 PRINT "STEEPEST DESCENT           1"
820 PRINT "MONTE CARLO                2"
830 PRINT "RANDOM COORDINATES         3"
840 PRINT "CONJUGATE DIRECTIONS       4"
850 PRINT :PRINT "ENTER SELECTION:  ":INPUT IN(3)
860 IF IN(3)<1 OR IN(3)>4 THEN 750
870 PRINT :PRINT: PRINT
880 PRINT "UNIDIRECTIONAL SEARCH MODE:"
890 PRINT "            MENU"
900 PRINT "-----------------------------"
910 PRINT "BEST VALUE              0"
920 PRINT "GOLDEN SECTION          1"
930 PRINT "INTERPOLATION           2"
940 PRINT : PRINT "ENTER SELECTION ":INPUT IN(4)
950 IF IN(4)<0 OR IN(4)>2 THEN 750
960 CLS:color 14
970 IF IN(3)=1 THEN METHOD$="Steepest Descent"
980 IF IN(3)=2 THEN METHOD$="Monte Carlo     "
990 IF IN(3)=3 THEN METHOD$="Random Coordinates"
1000 IF IN(3)=4 THEN METHOD$="Conjugate Directions"
1010 IF IN(4)=0 THEN UNIDIR$="Best Value         "
1020 IF IN(4)=1 THEN UNIDIR$="Golden Section     "
1030 IF IN(4)=2 THEN UNIDIR$="Interpolation      "
```

```
1040 LOCATE 1,1:PRINT"Direction Method: ";:COLOR 15:PRINT METHOD$;:COLOR 14
1050 LOCATE 1,41:PRINT"Line Search Mode: ";:COLOR 15:PRINT UNIDIR$;:COLOR
14
1052 LOCATE 8,1:PRINT"Initial Guess Vector                Initial Penalty
Function";
1053 FOR I=1 TO N
1054 LOCATE 8+I,5:PRINT I;"    ";VE(I);
1055 NEXT I
1060 GOSUB 3010:REM CALL OPTIMUM
1062 FOR I=1 TO 5:LOCATE 19+I,1:PRINT"
";
1063 NEXT I
1070 LOCATE 8,58
1080 PRINT "FINAL OPTIMUM VECTOR:  "
1090 FOR I = 1 TO N:LOCATE 8+I,60: PRINT I;"    ";VE(I);"   ";:NEXT I
1100 LOCATE 20,1
1110 PRINT "Final Penalty Function     ";F;:LOCATE 22,1
1120 PRINT "Optimization completed in  ";IN(10);"directions";
1130 END
2000 REM Subroutine  PENALTY FUNCTION [H(N)]
2010 REM *************************************************************
2020 REM    OBJECTIVE OR PENALTY FUNCTION DEFINITION (LINES 2000-2999)
2030 REM *************************************************************
2040 REM
2050 REM DEFINE BETWEEN LINES 2000-2999 OR MERGE AFTER LINE 10000
2060 GOSUB 10000
2070 RETURN
3000 REM **********************
3010 REM        OPTIMUM
3020 REM **********************
3030 REM
3040 REM STEP LENGTH
3050 DL = CO(1)
3060 REM MINIMUM STEP LENGTH
3070 DM = CO(2)
3080 REM FUNCTION TOLERANCE
3090 FT = CO(3)
3100 REM ACCELERATION FACTOR
3110 AC = CO(4)
3120 REM DECELERATION FACTOR
3130 DC = CO(5)
3140 NI = N
3150 REM VECTOR INITIATION
3160 FOR I= 1 TO N: HS(I) = VE(I):NEXT I
3170 IP = IN(1):M = IN(2):IM = IN(3):ID = IN(4)
3180 L1 = DL: IF IM = 1 THEN M = N
3190 IF IM = 4 THEN M = N
3200 REM DEFINE STARTING POINT
3210 FOR I = 1 TO NI:HA(I) = HS(I): NEXT I
3220 REM FIND OBJ FUN AT A, BOX 6
3230 ND = 1: FOR I = 1 TO N: H(I) = HA(I): NEXT I
3240 GOSUB 2020:REM CALL SYNTAN
3250 FA = F
3260 IF INFSTART<1 THEN LOCATE 10,40:PRINT F;:INFSTART=1
3270 REM SET DIRECTION COUNTER
3280 IC = 0
3290 IC = IC + 1
3300 LOCATE 3,1:PRINT"STATUS: ";:LOCATE 3,10:COLOR 15
3310 IF IP = 1 THEN PRINT "FINDING OPTIMUM DIRECTION AT POINT ";IC;:LOCATE
5,1
3320 IF IP=1 THEN PRINT "POINT ";IC;"  CURRENT STEP="DL;" FUNCTION=";FA;"
";
3330 IF IP = 0 THEN 4170
3340 LOCATE 8,60
3350 PRINT "CURRENT STATE VECTOR"
3360 COLOR 15:FOR I=1 TO N
3370 LOCATE 8+I,60:PRINT I;"    ";HA(I);:NEXT I
3380 COLOR 14
3390 IF IM = 2 THEN 3470
3400 IF IM = 3 THEN 3470
3410 FOR I = 1 TO NI: FOR JM = 1 TO NI
```

```
3420 SR(I,JM) = 0
3430 NEXT JM:SR(I,1) = 1:NEXT I
3440 GOTO 3500
3450 REM FIND RANDOM NUMBERS
3460 RANDOMIZE
3470 FOR J = 1 TO M: FOR I = 1 TO M
3480 SR(I,J) = RND:NEXT I:NEXT J
3490 REM DEFINE COEFFICIENTS
3500 FOR J =  1 TO M: SA = 0: FOR I = 1 TO NI
3510 SA = SA + SR(I,J) * SR(I,J) :IF SA=0 THEN SA=1E-30
3520 NEXT I: S(J) = DL / SQR (SA): NEXT J
3530 REM DEFINE DIRECTIONS
3540 FOR J = 1 TO M: FOR I = 1 TO NI
3550 SR(I,J) = S(J) * SR(I,J)
3560 NEXT I:NEXT J
3570 FOR J = 1 TO M
3580 FOR I = 1 TO NI:H(I) = HA(I):NEXT I:GOSUB 2020:FA = F
3590 NEXT J
3600 FO = F
3610 REM FIND POINTS ABOUT A
3620 FOR  J = 1 TO M: FOR I = 1 TO NI
3630 HR(I,J) = HA(I) + SR(I,J)
3640 NEXT I:NEXT J
3650 REM FIND OBJECTIVE FUNCTION
3660 FOR J = 1 TO M: FOR I = 1 TO NI
3670 H(I) = HR(I,J)
3680 NEXT I
3690 GOSUB 2020:REM CALL SYNTAN
3700 F(J) = F
3710 REM DIFFERENTIALS
3720 DF(J) = FA - F(J):NEXT J
3730 REM FIND MAX AMONG DF
3740 DD = - 1E+20
3750 FOR J = 1 TO M
3760 IF DD > DF(J) THEN 3780
3770 DD = DF(J):KK = J
3780 NEXT J
3790 REM BOX 24
3800 IF IN(3) > < 2 THEN 3820
3810 IF DD < 0 THEN 4790
3820 IF IM = 2 THEN 3980
3830 REM FIND OPTIMUM DIRECTION
3840 FOR I = 1 TO NI
3850 DH(I) = 0
3860 FOR J = 1 TO M
3870 DH(I) = DH(I) + SR(I,J) * DF(J)
3880 NEXT J: NEXT I
3890 REM NORMALIZE, BOX 32
3900 DH = 0: FOR I = 1 TO NI
3910 DH = DH + DH(I) * DH(I)
3920 NEXT I
3930 DH = DL / SQR (DH)
3940 REM VECTOR OF LENGTH DL, BOX 34
3950 FOR I = 1 TO NI
3960 DH(I) = DH * DH(I): NEXT I
3970 GOTO 4010
3980 FOR I = 1 TO NI
3990 DH(I) = SR(I,KK)
4000 NEXT I
4010 REM
4020 REM BOXES 36,38
4030 FOR I = 1 TO NI
4040 HB(I) = HA(I): NEXT I
4050 FB = FA
4060 IE = 0: LL = 0
4070 REM D3=DISTANCE ALONG OPTIMUM DIRECTION FROM START
4080 D3 = DL: NS = 0
4090 LOCATE 20,1:IF IP = 1 THEN PRINT "OPTIMUM DIRECTION WAS FOUND    ";
4100 LOCATE 21,1:IF IP = 1 THEN PRINT "------------------------------" ;
4110 REM ADVANCE ONE STEP ALONG O.D.
4120 IF IM > < 4 THEN GOTO 4170
```

```
4130 IJ = IJ + 1
4140 IF IJ > M THEN GOTO 4170
4150 FOR  I = 1 TO N: SR(I,IJ) = DH(I)
4160 NEXT I
4170 REM
4180 D0 = DL
4190 FOR I = 1 TO NI
4200 BB(I) = HB(I) + DH(I)
4210 NEXT I
4220 GOSUB 5770: REM CHECK BOUNDS
4230 NS = NS:D1 = D2:D2 = D3
4240 D0 = D0 * AC
4250 D3 = D3 + D0
4260 IF IE > 10 AND F1 = F2 AND F2 = F3 THEN 4860
4270 F1 = F2: F2 = F3
4280 REM BOXES 41,42,44
4290 ND = 1
4300 FOR I = 1 TO N:H(I) = BB(I):NEXT I
4310 GOSUB 2020:REM CALL SYNTAN
4320 LOCATE 22,1:COLOR 14
4330 IF IP = 1 AND  IE = 0 THEN PRINT "STEP  DISTANCE  FUNCTION";
4340 BF = F:LOCATE 23,1
4350 IF IP = 1 AND IE = 0 THEN PRINT  "--------------------------------";
4360 LOCATE 24,1:IF IP = 1 THEN  PRINT IE+1;"     ";D3;"     ";F;"     ";
4370 FO = FB - BF
4380 F3 = BF
4390 IF IE = 0 AND FO < DD THEN LL = 1
4400 IF LL = 0 THEN 4460
4410 FOR I = 1 TO NI
4420 DH(I) = SR(I,KK):NEXT I
4430 FO = DD
4440 BF = F(KK)
4450 IE = 1: LL = 0
4460 IF FO < 0 THEN 4570
4470 IE = IE + 1
4480 IF IE > 20 THEN 3290
4490 REM ADVANCE ONE STEP ALONG O.D.
4500 FOR I = 1 TO NI
4510 HB(I) = BB(I): NEXT I
4520 FB = BF
4530 FOR I = 1 TO NI
4540 DH(I) = DH(I) * AC
4550 NEXT I
4560 GOTO 4190
4570 IF IP = 1 THEN LOCATE 20,1
4580 IF IP = 1 THEN PRINT "END OF UNIDIRECTIONAL SEARCH";
4590 IF ID = 1 THEN GOSUB 5020: REM CALL GOLDSEC
4600 FOR I = 1 TO NI:HA(I) = BB(I)
4610 H(I) = HA(I):NEXT I
4620 IF ID = 2 THEN GOSUB 5410: REM CALL INTER
4630 FOR I = 1 TO NI
4640 IF ID = 1 OR D = 2 THEN HA(I) = BB(I):H(I) = HA(I)
4645 NEXT I
4650 FA = FB
4660 IF ID = 1 OR ID = 2 THEN GOSUB 2020: REM CALL SYNTAN
4670 FA = F
4680 IF ID >0 THEN 4750
4690 FOR I = 1 TO N
4700 FA = F
4710 IF HA(I) > = VU(I) THEN 4740
4720 IF HA(I) < = VL(I) THEN 4740
4730 HA(I) = HA(I) - DH(I)
4740 NEXT I
4750 IF ABS (F0 - F) < FT THEN 4860
4760 IF FA < FT THEN 4860
4770 GOTO 3290
4780 GOTO 4840
4790 IF IP = 1 THEN LOCATE 24,40:PRINT"Search failed...Decreasing Step
"
4800 IC = IC - 1
4810 DL = DL * DC
```

```
4820 IF DL < DM THEN 3870
4830 GOTO 3290
4840 AA = 1
4850 REM
4860 IN(10) = IC
4870 FOR I = 1 TO N
4880 VE(I) = HA(I):NEXT I
4890 RETURN
4900 REM ******************************
4910 REM                STEP
4920 REM ******************************
4930 FOR I = 1 TO NI:H1(I) = BB(I):H(I) = H1(I) + DN(I) * EL
4940 NEXT I
4950 GOSUB 5770: REM CHECK BOUNDS
4960 GOSUB 2020: REM CALL SYNTAN
4970 FOR I = 1 TO N:BB(I) = H1(I)
4980 NEXT I
4990 RETURN
5000 REM *****************************
5010 REM              GOLDSEC
5020 REM *****************************
5030 FOR I = 1 TO N:DN(I) = DH(I) / (D3 - D2)
5040 IF BB(I) = > VU(I) THEN 5070
5050 IF BB(I) = < VL(I) THEN 5070
5060 BB(I) = BB(I) - DN(I) * (D3 - D1)
5070 NEXT I
5080 GOSUB 5770: REM CHECK BOUNDS
5090 ST = .375
5100 D7 = D3 - D1:COLOR 14:LOCATE 20,40
5110 IF IP = 1 THEN PRINT "GOLDEN SECTION PROCEDURE..                    "
5120 IF IP > < 1 THEN 5170
5130 LOCATE 21,40
5140 PRINT "STEP          DISTANCE FROM START";:LOCATE 22,40
5150 PRINT "----------------------------------  "
5160 LOCATE 23,40:PRINT"                                       "
5170 DI = D1
5180 NC = 10:REM MAX ITERATIONS
5190 S1 = ((D3 - D1) * ST + D1)
5200 S2 = (( - D3 + D1) * ST + D3):LOCATE 23,40
5210 IF IP = 1 THEN PRINT IT;"    ";D1;"  ";D3;"    "
5220 EL = S1 - D1
5230 GOSUB 4910:REM CALL XSTEP
5240 Y1 = F
5250 EL = S2 - D1: GOSUB 4910: Y2 = F
5260 IF Y1 < Y2 THEN 5320
5270 FOR I = 1 TO N:BB(I) = BB(I) + DN(I) * (S1 - D1)
5280 NEXT I
5290 GOSUB 5770: REM CHECK BOUNDS
5300 D1 = S1
5310 GOTO 5330
5320 D3 = S2
5330 IT = IT + 1
5340 IF IT > NC THEN 5360
5350 GOTO 5190
5360 FOR I = 1 TO N:BB(I) = BB(I) + DN(I) * (S1 - S2) / 2
5370 NEXT I
5380 GOSUB 5770: REM CHECK BOUNDS
5390 IT = 0
5400 RETURN
5410 REM *********************************
5420 REM                 INTER
5430 REM *********************************
5440 IF IP  < > 1 THEN 5480
5450 LOCATE 20,40
5460 PRINT "INTERPOLATION PROCEDURE...         ";:LOCATE 21,40
5470 PRINT "-------------------------         "
5480 IX = 8: REM  MAX NO ITERATIONS
5490 IF IP < 1 THEN 5520
5500 LOCATE 22,40:PRINT "STEP       DISTANCE FROM START    ";
5510 LOCATE 23,40:PRINT "---------------------------       ";
5520 TM = .00001: IT = 0
```

```
5530 FOR I = 1 TO NI
5540 DN(I) = DH(I) / (D3 - D2)
5550 IF BB(I) > = VU(I) THEN 5580
5560 IF BB(I) < = VL(I) THEN 5580
5570 BB(I) = BB(I) - DN(I) * (D3 - D1)
5580 NEXT I
5590 GOSUB 5770: REM CHECK BOUNDS
5600 D6=D1-D2:D7=D1-D3:D8=D2-D3:LOCATE 24,40
5610 IF IP = 1 THEN PRINT IT;"          ";D1;"          ";D2;"          ";D3;
5620 P6 = D1 + D2: P7 = D1 + D3: PS = D2 + 3
5630 G8 = 1! / (D6 * D6):G7 = 1 / (D6 * D8): G6 = 1 / (D7 * D8)
5640 D = ( - P8 * G8 * F1 + P7 * G6 * F2 - F6 * G6 * F3) / (2! * (G8 * F1 -
G7 * F2 + G6 * F3 + .0001))
5650 D = - D
5660 EL = D - D1: GOSUB 4900:Y = F
5670 TO = ABS (D - D2)
5680 IF D < D2 THEN 5710
5690 D3 = D2:D2 = D:F3 = F3:F2 = Y
5700 GOTO 5740
5710 FOR I = 1 TO NI:BB(I) = BB(I) + DN(I) * (D2 - D1): NEXT I
5720 GOSUB 5770: REM CHECK BOUNDS
5730 D1 = D2:D2 = D:F1 = F2:F2 = Y
5740 IT = IT +1
5750 IF IT < IX AND TO > TM THEN 5600
5760 RETURN
5770 REM
5780 REM BOUNDS CHECK ON VARIABLES
5790 FOR I = 1 TO N
5800 IF HA(I) < VL(I) THEN HA(I) = VL(I)
5810 IF BB(I) < VL(I) THEN BB(I) = VL(I)
5820 IF HA(I) > VU(I) THEN HA(I) = VU(I)
5830 IF BB(I) > VU(I) THEN BB(I) = VU(I)
5840 IF H(I) < VL(I) THEN H(I) = VL(I)
5850 IF H(I) > VU(I) THEN H(I) = VU(I)
5860 NEXT I
5870 RETURN
10000 REM SUBROUTINE DEFINING PENALTY FUNCTION F[H(N)]
10010  F = (H(1) - 10) ^ 2 + (H(2) - 10) ^ 2
10020 RETURN
```

CHAPTER SIX

MATERIALS AND PROCESSES

6.1 PROPERTIES OF MATERIALS FOR MACHINE DESIGN

Capacity of a machine component is related to the most severe service condition it can sustain without a change which will prevent the component from continuing its intended function. In most cases, loads are the main manifestations of capacity. To assess the load-carrying capacity of a machine component, the maximum unit load (stress) has to be compared with the appropriate material property. Finding stresses in a component under service loading has been discussed in Chapter 4 and will be further exemplified in Chapters 7, 8, 14. Here, we shall discuss the capacity of engineering materials to sustain loads.

A most informative material test is the simple tensile test. A specimen in the form of a cylindrical bar, machined to a certain specification, is slowly loaded in a tensile testing machine and load P and displacement (extension) Δl are recorded. The resulting load–displacement curve is shown in Figure 6.1 for a typical mild steel. Assuming constant cross-section of the rod, the same curve relates stress and strain. The curve of this figure does not correspond exactly to the real stress–strain relationship because of the way it is made; that is, we measure force and displacement and interpret them as stress and strain based on the initial length and the initial cross-section of the test specimen. These properties continuously change during the experiment therefore the results of such a test will have only a formal value.

For most materials used in machine design from this diagram we observe a linear stress–strain relationship which extends up to some point, and shortly thereafter one can observe an increasing deformation without a proportional increase in the load and the stress. This roughly corresponds to the point that we have a substantial yielding of the material and we call this the 'yield point'. The corresponding stress (S_y) is called the 'yield strength' of the material. We have been intentionally vague in the precise definition of the yield point owing

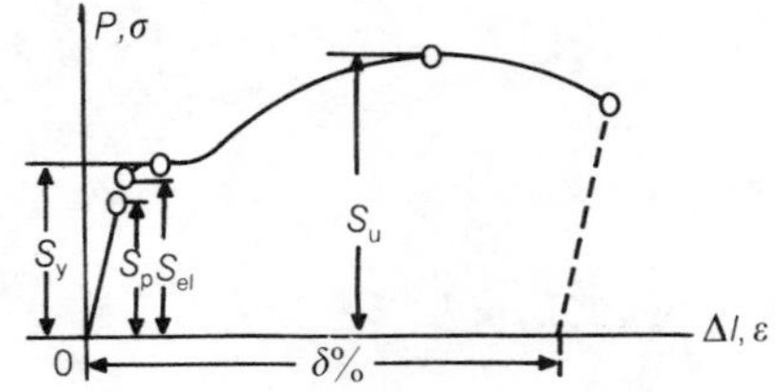

Figure 6.1 Stress–strain diagram for mild steel

to the fact that in most materials it is not possible to identify a single point where we have the transition from elastic to plastic behavior. In fact, in most engineering materials this transition is not abrupt and it is a matter of definition to specify the yield point. Usually we define the yield point as the point where a certain percentage of plastic deformation remains after loading.

Ultimate tensile strength (S_u) is the maximum nominal stress which can be observed into the stress–strain diagram, which corresponds to the maximum nominal stress that the material can sustain, the ratio of maximum load to the original cross-sectional area.

In ductile materials there is a substantial difference between yield and tensile strength. In high-strength materials, the difference between these two values decreases.

In most materials the strength is the same in tension and in compression. Some materials however, have very different values of strength in tension and compression, such as cast iron for example. As we shall see later (p. 293) micro-cracks exist in the structure of this material which give rise to high stress concentration during tensile loading while in compressive loading for geometrical reasons these micro-cracks are ineffective. Therefore the material can sustain much higher loads. In such materials the strength in tension and in compression have to be recorded independently.

In material property tables, especially for design purposes, one will also observe different strengths in tension and bending. Similar differences might be observed in shear and torsion. Although it appears that no matter what loads cause the stresses, the strength to a particular type of stress must be the same, and this is not always the case. Take for example the strength in pure tension and in bending. In both cases the direction of stress is the same, namely tensile stress. In flexure however, only the outer fibres of the material have high stress while the stresses diminish as we move towards the neutral line. Since, as shall be seen later (p. 293), the micro-cracks already mentioned are uniformly distributed in the material, the probability of having a micro-crack in the area of high stress is smaller in flexure than in pure tension, therefore the strength in flexure is, in general, greater than that in pure tension. A similar situation exists for strengths in torsion and shear.

Another deviation of the stress–strain curve from reality must be pointed out: The horizontal scale is usually arbitrarily nonlinear since elastic deformation at small strains is very small compared with plastic deformation. For this reason, the strain scale is enlarged for small strains.

This yield point in ductile materials is usually well defined. In cases where there is no pronounced yield point in the diagram the yield strength is defined as the stress at which the permanent set $\varepsilon_{pl} = 0.002$ or 0.2% (Figure 6.2). In some cases the yield strength is established for $\varepsilon_{pl} = 0.5\%$.

This yield point in ductile materials is usually well defined. In cases where there is no permanent set. To distinguish between the yield point in tension and in compression, an additional subscript 't' or 'c' is introduced in the notation when it is necessary in some materials. Thus we obtain the symbols S_{yt} and S_{yc} for the yield point.

The yield point is one of the main mechanical characteristics of a material in machine design where yielding is mostly unwanted, even if the structural integrity of the component is maintained, because of geometric and/or other implications.

As noted above, the ratio of the maximum force that the specimen is capable of sustaining to its original cross-sectional area is termed the 'ultimate tensile strength' and is

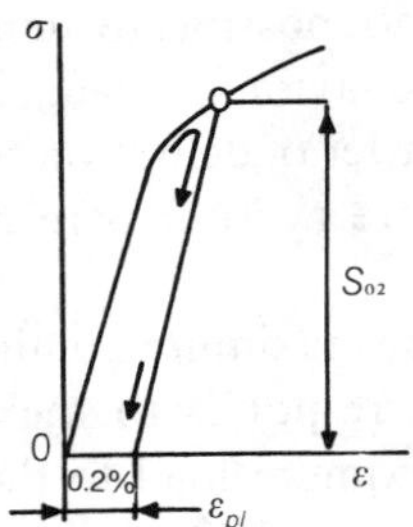

Figure 6.2 Stress–strain diagram for high strength steel

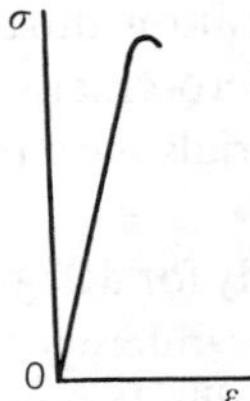

Figure 6.3 Stress–strain diagram for a brittle material

denoted by S_u. It is important to note that S_u is not the stress at which the specimen fractures. If the tensile force is referred to the minimum section at a given point in time rather than to the original cross-sectional area of the specimen, it may be observed that the average stress on the narrowest section of the specimen before rupture is appreciably greater than S_u. Thus the ultimate tensile strength is also a conventional quantity. Owing to the convenience and simplicity of its determination, the ultimate tensile strength is widely used in design practise as a basic comparative characteristic of the strength properties of materials.

Another important characteristic of a material for design is determined from tension tests, the percentage elongation at rupture, which is the average permanent deformation produced in a specified standard length of the specimen at the moment of rupture.

The ability of a material to acquire large permanent deformations without fracture is known as 'ductility'. The property of ductility is of prime importance in such manufacturing processes as extrusion, drawing, bending, etc. The measure of ductility δ is the percentage elongation at rupture. The greater δ, the more ductile is the material. Highly ductile materials include annealed copper, aluminum, brass, low-carbon steel, etc. Duralumin and bronze are less ductile. Slightly ductile materials include many alloy steels.

A property opposite to ductility is brittleness, i.e. the tendency of a material to fracture without any appreciable permanent deformation. Materials possessing this property are called 'brittle'. For such materials, the amount of elongation at rupture does not exceed 2 to 5%, and in some cases it is expressed by a fraction of 1%. Brittle materials include cast iron, high-carbon tool steel, glass, brick, stone, etc. The tension test diagram for brittle materials has no yield point or strain hardening zone (Figure 6.3).

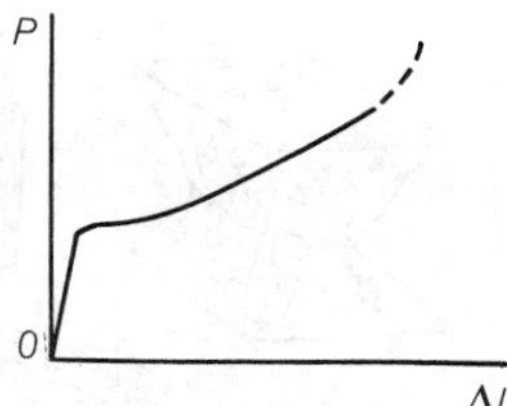

Figure 6.4 Stress–strain diagram for a ductile material in compression

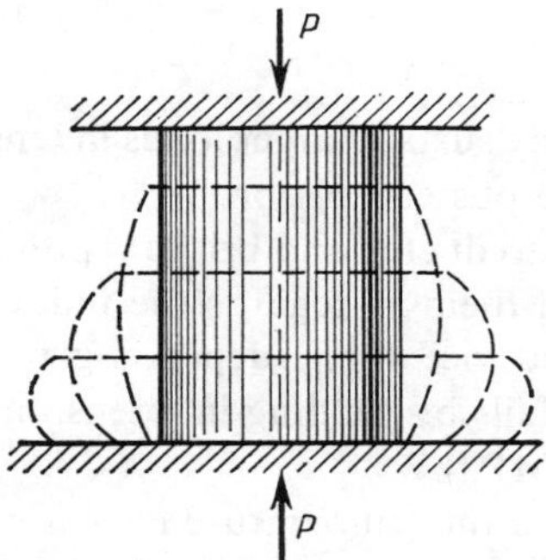

Figure 6.5 Compression of ductile materials. (After Feodosyev 1973)

Ductile and brittle materials behave differently in compression tests as well. The compression test is conducted on short cylindrical specimens placed between parallel plates. The compression test diagram for low-carbon steel has a shape such as represented in Figure 6.4. Here, as for tension, the yield point can be observed with subsequent transition to the zone of strain hardening. Thereafter, however, the load does not fall, as in tension, but increases abruptly. This is owing to the fact that the cross-sectional area of the compressed specimen increases; the specimen takes a barrel-like shape owing to friction at the ends (Figure 6.5). It is practically impossible to bring the specimen of a ductile material to fracture. The test cylinder is compressed into a thin disk (see Figure 6.5) and further testing is limited by the capacity of the machine. Hence the ultimate compressive strength cannot be found for materials of this kind.

Brittle materials behave in a different way in compression tests. The compression test diagram for these materials retains the qualitative features of the tension test diagram. The ultimate compressive strength of a brittle material is determined in the same way as in tension. The fracture of the specimen occurs with cracks forming on inclined or longitudinal planes (Figure 6.6).

A comparison of the ultimate strength S_{ut} and the ultimate compressive strength S_{uc} of brittle materials shows that these materials possess, as a rule, higher strength indices in compression than in tension. The magnitude of the ratio

$$k = S_{ut}/S_{uc}$$

for cast iron ranges from 0.2 to 0.4. For ceramic materials, $k = 0.1$ to 0.2. For ductile materials, usually $S_{ut} \approx S_{uc}$.

Figure 6.6 Compression of brittle materials. (After Feodosyev 1973)

Fiber-reinforced plastics may sustain larger loads in tension than in compression. Some metals such as magnesium also possess this property.

The division of materials into ductile and brittle is purely conventional because there is no sharp dividing line between them as regards the index k. Many brittle materials may behave as ductile materials and vice versa, depending on the conditions of testing. For example, a cast-iron specimen fails by yielding in a tension test under high pressure of the surrounding medium ($p > 400$ MPa).

The duration of loading and the temperature have a very great effect on ductility and brittleness. Under rapid loading, brittleness is displayed more sharply, while under prolonged loading ductility is more pronounced. For example, brittle glass is capable of developing permanent deformations under sustained loading at normal temperature. Ductile materials, such as low-carbon steel, exhibit brittle properties under sudden impact loading.

Tension and compression tests give an objective assessment of material properties. In industry, however, this method of testing is often very inconvenient where the quality of manufactured parts is desired. For example, it is difficult to measure the results of heat treatment on finished parts by tension and compression tests.

It is therefore common practice to resort to a comparative assessment of material properties by hardness tests. 'Hardness' is the capacity of material to resist mechanical penetration of sharp objects. The most commonly used are the Brinell and Rockwell hardness tests. In the first case a steel ball 10 mm in diameter and in the second case a pointed diamond indenter is pressed against the surface of the part. The hardness of the material is determined by measuring the resulting indentation. Experiments have provided empirical conversion tables for estimating the ultimate tensile strength of a material from hardness indices, as in Figure 6.7 for example. Thus non-destructive hardness tests provide means of measuring the strength indices of the material.

The foregoing discussion of material properties applies to tests under normal conditions, i.e. at room temperature and at relatively small rates of change of loads and elongations.

The range of temperatures over which structural materials actually operate extends far beyond the limits of the above normal conditions. In some structures the material is subject to extremely high temperatures, as, for example in the walls of combustion chambers of aircraft engines. In other structures, in contrast, working temperatures are extremely low, such as in elements of refrigerating plants and tanks containing liquefied natural gases.

The speeds of loading and the duration of external forces also vary over a wide range. There are loads varying very slowly and other loads varying very rapidly. Some loads act for

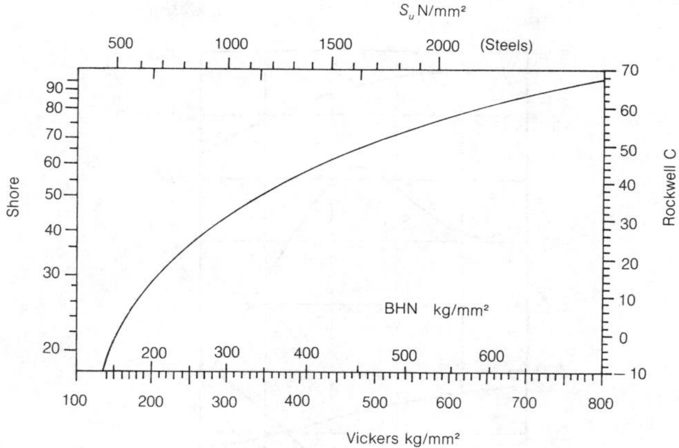

Figure 6.7 Hardness scales

years and others for a fraction of a second. It is clear that the mechanical properties of materials will differ depending on the environment and the loading conditions.

A general analysis of material properties taking into account the temperature and time effects is very complicated and cannot be confined to simple experimental curves similar to tension test diagrams. Therefore, temperature and time effects are treated at present with reference to particular types of problems.

Many machine loads are slowly varying, or static, loads. The rate of change of such loads in time is so small that the kinetic energy which is acquired by moving particles of a deformed body is a negligible fraction of the work done by external forces. In other words, the work done by external forces is transformed only into elastic potential energy and also into irreversible thermal energy as a result of plastic deformation of the material. The testing of materials in the so-called normal conditions is performed under static loads.

Figure 6.8 shows the relation between temperature and the modulus of elasticity E, the yield point S_y, the ultimate tensile strength S_u and the percentage elongation at rupture δ for a low-carbon steel over a range from 0 to 500 °C. As can be seen from these curves, the modulus of elasticity practically does not vary with temperature up to 300 °C. The quantities S_{yt} and especially δ undergo more substantial changes; so-called embrittlement of the steel takes place, i.e. the percentage elongation at rupture is reduced. With further increase of temperature the ductile properties of the steel are recovered while the strength indices decrease rapidly.

Embrittlement at elevated temperatures is encountered mostly in low-carbon steel. Alloy steels and non-ferrous alloys mainly exhibit a monotonic increase of δ and a similar monotonic reduction of S_{yt} and S_{ut} with increase in temperature. Figure 6.9 represents the corresponding curves for a chrome–manganese steel.

The variation in time of strains and stresses induced in a loaded part is called 'creep'. A particular case of creep is a growth of irreversible strains under a constant stress as on the disks and blades of a gas turbine subjected to large centrifugal forces and high temperatures. This increase in dimensions is irreversible and usually occurs after long operation. A result of creep is relaxation – a redistribution of stresses in time under a constant strain. An example

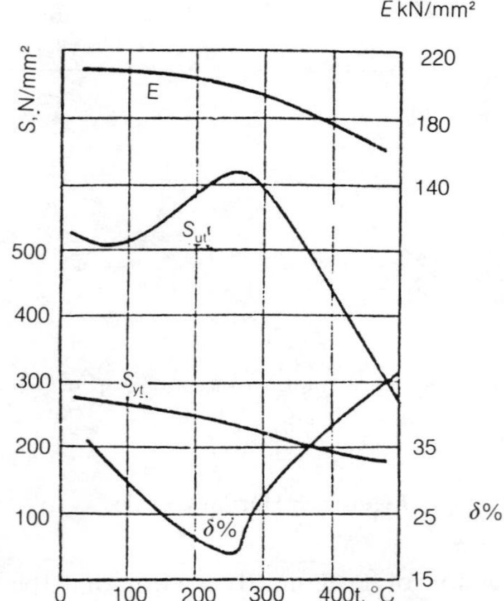

Figure 6.8 Properties of a mild steel at elevated temperatures

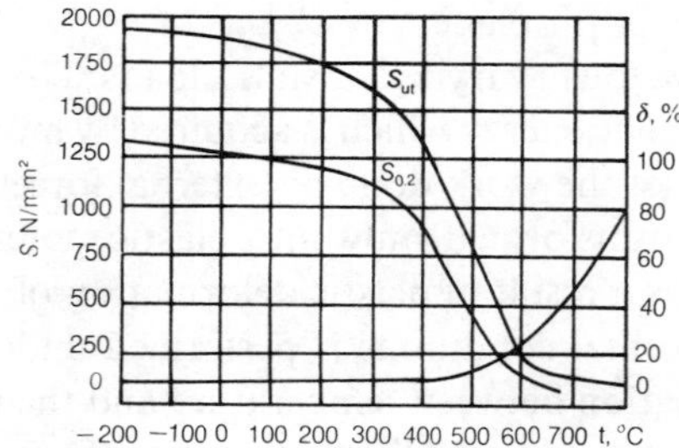

Figure 6.9 Properties of a high strength steel at elevated temperatures

of relaxation is the loosening of bolt connections operating under high-temperature conditions.

Creep lends itself most readily to experimental studies. If a specimen is loaded with a constantly acting force (Figure 6.10) and the variation of its length at a fixed temperature is observed, it is possible to obtain creep diagrams (Figure 6.11) giving the strain versus time relationship at various values of stress. As can be seen from these curves, the strains grow very rapidly at the start. Then the process settles down and strains increase at a constant rate. As in ordinary testing, the specimen begins to neck down through time. Shortly before rupture a rapid increase in local strains occurs as a consequence of the reduction in cross-sectional area. At higher temperatures, the variation of strains with time takes place more rapidly.

For a given material, it is possible to convert creep diagrams to relaxation diagrams by using the methods of the theory of creep. Relaxation diagrams can, however, be obtained

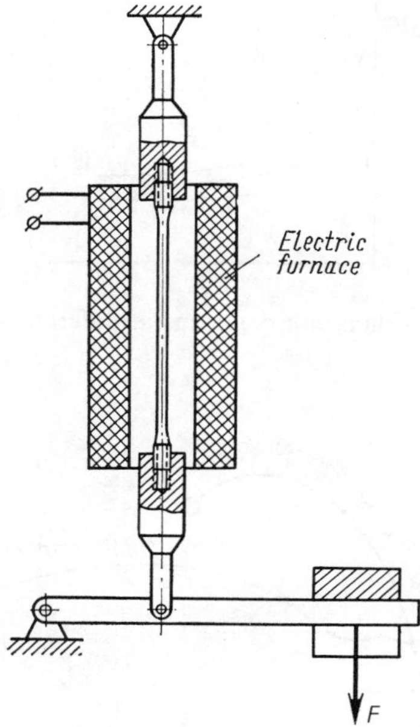

Figure 6.10 Testing of steel at elevated temperatures. (After Feodosyev 1973)

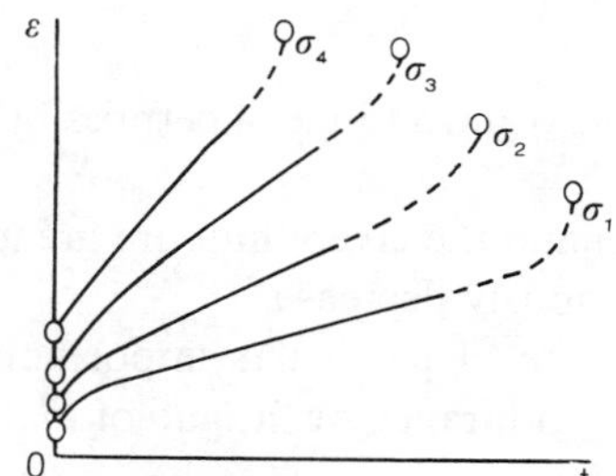

Figure 6.11 Creep at different stress levels

experimentally. This demands more complex apparatus as it is necessary to measure changes in magnitude of the tensile force while maintaining the elongation of the specimen. Relaxation diagrams giving the stress–time relation are shown in Figure 6.12.

The basic mechanical characteristics of a material under creep deformation are the creep–rupture strength and the creep limit. The 'creep–rupture' strength is defined as the ratio of the load, at which a tension specimen fails in a given length of time, to the original cross-sectional area. Thus, the creep–rupture strength depends on a given time to rupture. The latter is chosen equal to the service life of a part and ranges from minutes in rockets to 30–40 years in large steam turbines.

The 'creep limit' is defined as the stress at which plastic strain reaches a given value in a given time. To determine the creep limit it is necessary to assign a time interval (which

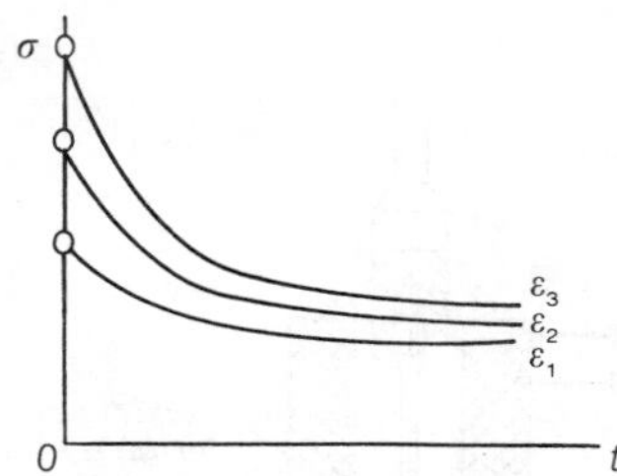

Figure 6.12 Relaxation diagrams at different strain levels

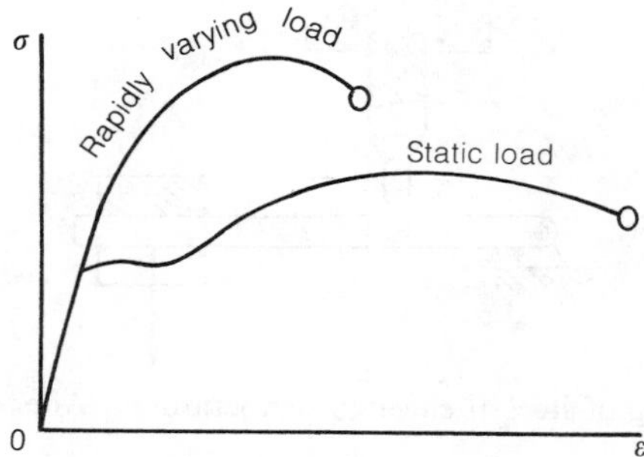

Figure 6.13 Effect of rapid loading

depends on the service life of a part) and a range of permissible strains (which depends on the service conditions of a part).

The creep–rupture strength and the creep limit are greatly affected by temperature. As temperature increases, they obviously decrease.

Periodically varying loads are of particular importance in machine design and are associated with the concepts of endurance or fatigue of materials. These problems will be discussed in detail in Chapter 7.

Some loads vary very rapidly producing appreciable velocities of the particles of a deformed body, these velocities being so high that the overall kinetic energy of moving masses is now a considerable fraction of the total work done by external forces. On the other hand, the rate of change of a load may be related to the rate of development of plastic deformations. A load may be considered as rapidly varying if plastic deformations cannot fully develop during the process of loading. This materially affects the character of observed stress–strain relations.

It is quite apparent that since the development of plastic deformations cannot be fully accomplished under rapid loading the material becomes more brittle (Figure 6.13). Strain rate effects become significantly more important at high temperatures.

The last of the three kinds of load under consideration are loads varying very rapidly in time. Their rate of change is so great that the work done by external forces is almost completely transformed into kinetic energy of moving particles of a body while the elastic and plastic strain energy is relatively small.

Very rapidly varying loads are produced by impact of bodies moving with velocities of several hundred meters per second and higher. These loads are dealt with in the study of problems of armour piercing, in the assessment of the destructive action of a blast, and in the investigation of the penetrating power of interplanetary dust encountered by spacecraft.

Since the strain energy is relatively small under conditions of very high rates of loading, the elastic properties of a solid are of minor importance in the present case. However, problems involving very high rates of loading lie outside the framework of machine design.

6.2 MATERIAL PROCESSING

6.2.1 Classification

Materials are commercially available in standard forms and almost invariably, the material will be processed to the designed form. Not all materials, however, are processed in the same way. That will depend on the material, the type and size of the element to be manufactured, the particular service conditions of the element, and the available manufacturing facilities; last, but not least, is the production cost. Therefore, the designer has to be aware of the available methods of material processing and their main features. Many times, the manufacturing requirements for a component have decisive influence on the design. For example, wall thickness of a cast part cannot be less than castability requires for the particular material and size, even if strength is assured with much smaller thickness. With the advent of manufacturing automation, many of our design philosophies have to change, to facilitate the application of New Technology.

The selection of the proper processing method in the process of designing a given machine element requires a knowledge of all possible production methods. Factors that must be considered are volume of production, quality of the finished product, and the advantages and limitations of the various types of equipment capable of doing the work. Most parts can be produced by several methods, but one of them will be the most economical.

Metal working may be classified according to various types of process, many of which, with some modifications, can be applied to most nonmetallic materials (De Garmo 1974):

A. Processes used to change the shape of material:

1. Casting
2. Hot and cold working
3. Power metallurgy forming
4. Plastics molding

B. Processes used for machining parts to a fixed dimension:

1. Traditional machining, chip removal
2. Nontraditional machining

C. Processes for obtaining a surface finish:

1. Metal removal
2. Polishing
3. Coating

D. Processes used for joining parts or materials (welding).

E. Processes used to change the physical properties (heat treatment).

6.2.2 Change of shape

Most metal products originate as an ingot casting or continuous casting. Molten metal is poured into metal or graphite molds to form ingots of convenient size and shape for further processing.

Processes used primarily to change the shape of metals include the following:

1. Casting
2. Forging
3. Extruding
4. Rolling
5. Drawing
6. Squeezing
7. Crushing
8. Piercing
9. Swaging
10. Bending
11. Shearing
12. Spinning
13. Stretch forming
14. Roll forming
15. Torch cutting
16. Explosive forming
17. Electrohydraulic forming
18. Magnetic forming
19. Electroforming
20. Powder metal forming
21. Plastics molding

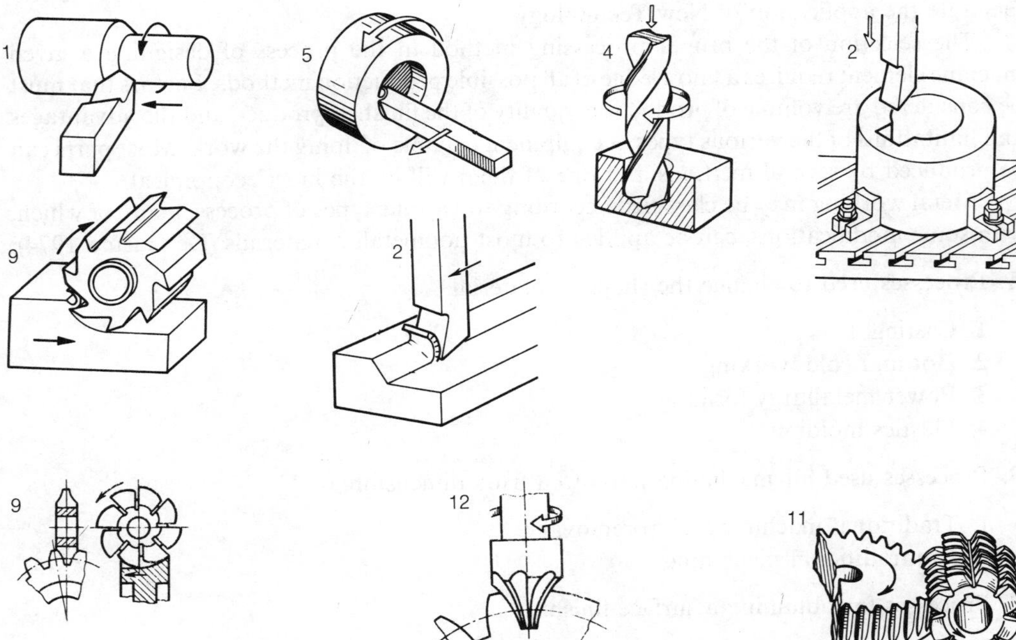

(a)

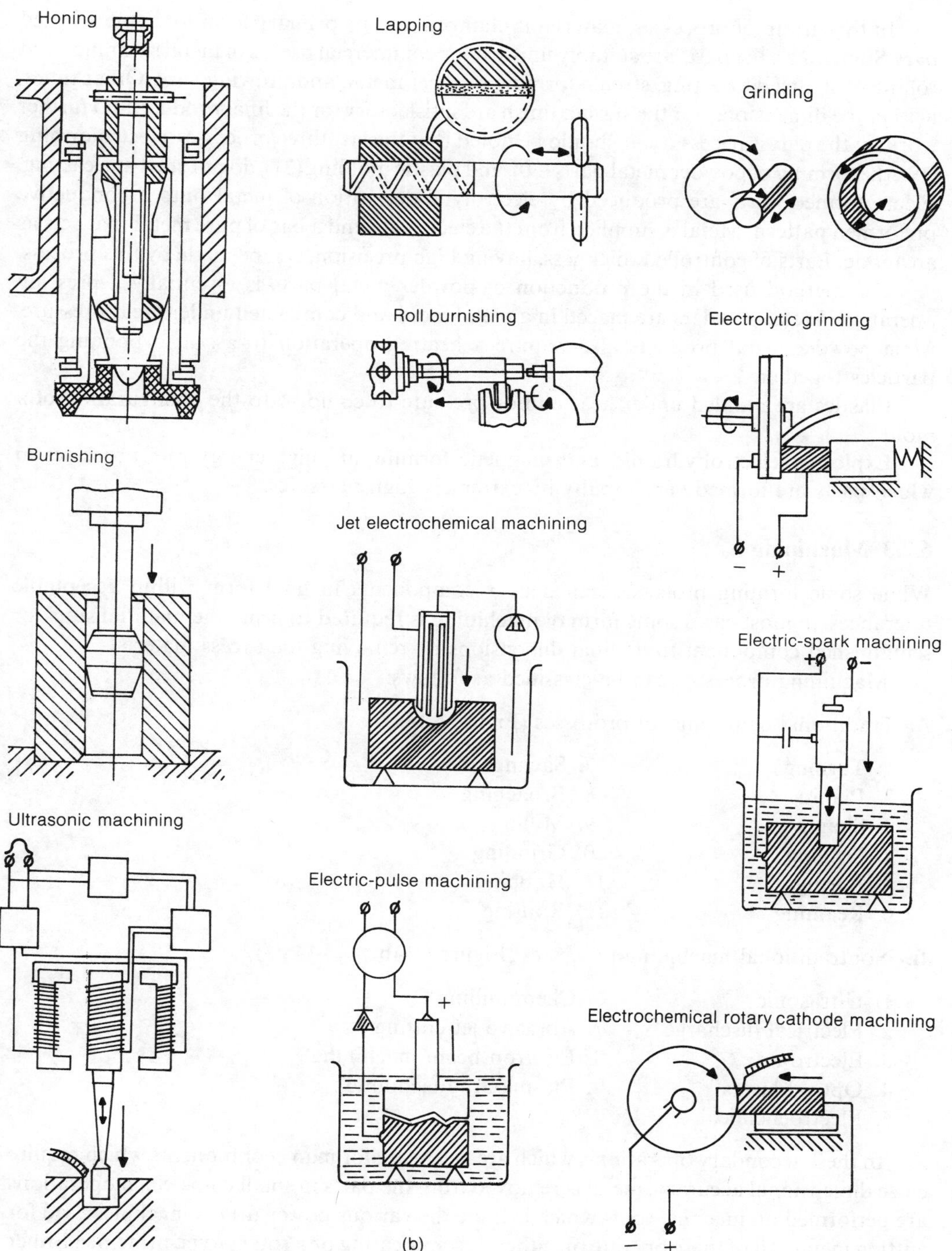

Figure 6.14 (a) Traditional chip-removal processes; (b) nontraditional machining processes: fine machining and polishing

In this group of processes, material is changed into its primary form for some selected part. Sometimes, the parts are suitably finished for commercial use, as in metal spinning, cold rolling of shafts, die casting, stretch forming of sheet metal, and drawing wire. Other times, neither the dimensions nor the surface finish are satisfactory for the final product, and further work on the part is necessary. It should be noted that the last three processes, electroforming (19), the forming of powder metal parts (20), and plastic molding (21), do not start as a casting. Electroformed parts are produced by electrolytic deposition of metal onto a conductive preformed pattern. Metal is supplied from the electrolyte and a bar of pure metal that acts as an anode. Parts of controlled thickness, having high precision, can be made by this process.

The method used in the production of powder metal parts is essentially a pressing operation. Metal powders are placed in a metal mold and compacted under great pressure. Most powder metal products also require a heating operation to assist in bonding the particles together.

Plastics are molded under heat and/or pressure to conform to the configuration of a mold.

Explosive, electrohydraulic, and magnetic forming are high-energy rate processes in which parts are formed very rapidly by extremely high pressures.

6.2.3 Machining

While some forming processes can deliver components in final form within acceptable tolerances, in most cases some form of machining is required to bring the material stock or semifinished component to its final dimensions by removing the excess material.

Machining processes can be classified as follows:

A. Traditional chip-removal processes (Figure 6.14a):

1. Turning
2. Planing
3. Shaping
4. Drilling
5. Boring
6. Reaming
7. Sawing
8. Broaching
9. Milling
10. Grinding
11. Hobbing
12. Routing

B. Nontraditional machining processes (Figure 6.14b):

1. Ultrasonic
2. Electrical discharge
3. Electro-arc
4. Optical lasers
5. Electrochemical
6. Chem-milling
7. Abrasive jet cutting
8. Electron beam machining
9. Plasma-arc machining

In these secondary operations, which are necessary for many components which require close dimensional accuracy, metal is removed from the parts in small chips. Such operations are performed on machine tools which include the various power-driven machines used for cutting metal. All of these operate on either a reciprocating or a rotary-type principle: Either the tool or the work reciprocates or rotates, as indicated in Figures 6.14(a, b). The planer is an example of a reciprocating machine, since the work reciprocates past the tool, which is held in a stationary position. In other machines, such as the shaper, the work is stationary and the tool reciprocates. The rotary machine is the lathe, which has the work rotating and the tool stationary. In the drill press it is the tool that rotates.

In ultrasonic machining, metal is removed by abrasive grains which are carried in a liquid and attack the work surface at high velocity by means of an ultrasonic generator. For electrical discharge and electro-arc machining, special arcs are generated that can be used to machine any conducting material. The optical laser is a strong beam of photons that can be used to generate extremely high temperatures and thus cut or weld metal. Chemical machining is done by either attacking the metal chemically or by using a reverse plating process.

6.2.4 Surface finish

Surface finishing operations are used to ensure a smooth surface, great accuracy, esthetic appearance, or a protective coating.

Processes used are:

1. Polishing
2. Abrasive belt grinding
3. Barrel tumbling
4. Electroplating
5. Honing
6. Lapping
7. Superfinishing
8. Metal spraying
9. Inorganic coating
10. Parkerizing
11. Anodizing
12. Sheradizing

In this group there are processes that cause little change in dimension and result primarily in finishing the surface. Other processes, such as grinding, remove some metal to the designed dimension in addition to giving it a good finish. Processes such as honing, lapping, and polishing consist in removing small scratches with little change in dimension. Superfinishing is also a surface-improving process that removes undesirable fragmented metal, leaving a base of solid crystalline metal. Plating and similar processes – used to obtain friction, wear and corrosion-resisting surfaces or just to give a better apearance – do not change dimensions materially.

6.2.5 Improvement of material strength

There are a number of processes in which the physical properties of the material are changed by the application of an elevated temperature or by rapid or repeated stressing of the material. Processes in which properties are changed include:

1. Heat treatment
2. Hot working
3. Cold working
4. Shot peening

Heat treating includes a number of processes that result in changing the properties and structure of metals. Although both hot and cold working are primarily processes for changing the shape of metals, these processes have considerable influence on both the

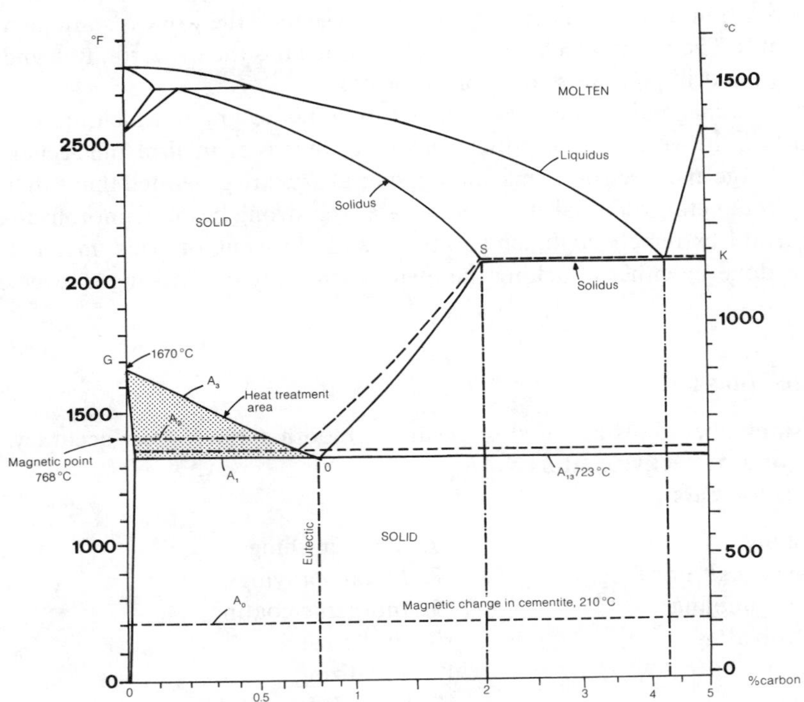

Figure 6.15 Iron–carbon equilibrium diagram for steel

structure and the properties of the metal. Shot peening renders many small parts, such as springs, resistant to fatigue failure.

Because strength of metals and alloys can be drastically improved by heat treatment, it has become one of the most important and commonly used processes, in particular for steel. Steel and cast iron consist basically of iron and carbon in varying composition and structure. The crystal and chemical structure of these two elements in equilibrium, even for fixed composition, largely depends on the temperature history. Therefore, heat treatment is mostly the application of a temperature scenario suitable for the particular purpose.

The several heat-treatment methods can be identified on the phase equilibrium diagram of steel which shows the phases and structure of the iron–carbon equilibrium for various compositions and temperatures (Figure 6.15).

Since ancient times, it was known that hot iron immersed in water would become hard and brittle and this process was used to make stronger weapons. This process, known as 'quenching', improves strength and hardness and requires heating to about 20 to 50 °C above the line GOSK in Figure 6.15 and rapid cooling rate. Schematically, it is shown in Figure 6.16 together with some of the other important heat treatments. All processes are described in terms of the two temperatures corresponding to the intersection of the carbon content vertical and the curves A1 and A3 of Figure 6.15, which bound the transformation range.

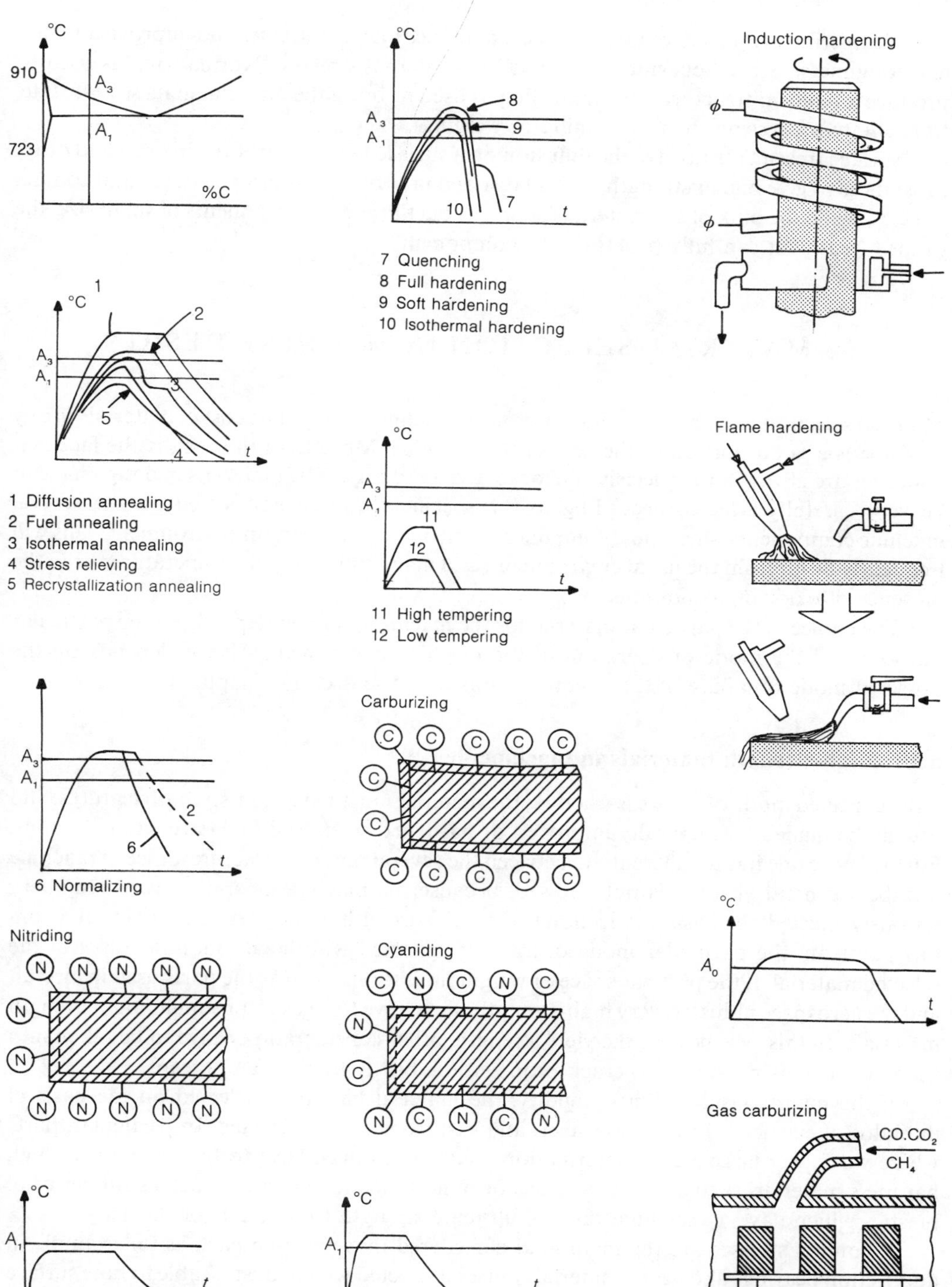

Figure 6.16 Heat treatment processes for steel

'Tempering' is applied to quenched steel to reduce internal stresses and improve ductility and toughness. 'Stress relieving' in turn relieves internal stresses. 'Normalizing' is used to produce a uniform structure, and 'annealing' is used to bring the material in its softest state, to facilitate processing such as forming or welding.

Surface treatments involve the diffusion on a surface layer of carbon, nitrogen and other substances to give higher strength to this layer and improve resistance to fatigue and contact stresses, while the bulk of the material retains its ductility. For components of small size, the surface treatment can fully penetrate the component.

6.3 MATERIAL SELECTION IN MACHINE DESIGN

Material selection plays a very important role in machine design. The cost of materials in any machine is a good portion of the cost of the machine. More than the cost, is the fact that materials are always a very decisive factor for a good design. The materials in design have to be very carefully selected according to the specific requirements posted on the several machine components, since these components will operate in a certain environment. This will frequently be far from the usual environment, and many times at high temperatures or in the presence of oxidizing conditions.

The choice of the particular material for the machine member depends on the particular purpose and the mode of operation of the machine component. Also it depends on the expected mode of failure of this component as was discussed in Chapter 4.

6.3.1 High-strength materials in machine design

If the expected mode of failure is fracture, the material must have high-strength features, and usually be made of structurally improved or hardened steels and high-strength cast irons. Furthermore one has to differentiate between the several modes of fracture. Since in machine members general yielding is not allowed, because the machine operation is usually very seriously affected, the basis of selection is the yield strength of the particular material. From this point on, the particular mode of fracture expected will dictate more features of the selected material. If the part has space or weight limitations, such as the parts used in aircraft and the aerospace industry, very high-strength steels have to be used, or similar high-strength materials. In this case besides the yield strength, other considerations have to be taken into account, such as resistance to crack propagation, brittle fracture and creep.

If the main consideration is rigidity, the material has to be selected on the basis of modulus of elasticity. In this case steels and cast irons can be selected. In particular, parts where we expect large elastic deformations, such as springs, have to be made out of high hardness materials such as hardened steel or of non-metallic materials such as rubber or of plastics which have a very high ratio of ultimate strength to modulus of elasticity.

If contact stresses are the main load considered in a machine part, as for example in antifriction bearings and gears, materials must be selected which must be able to take surface hardening and similar treatments. Structurally-improved steel or hardened steel must be utilized.

Many machine parts move relative to others and their wear characteristics are very important. In this case one of the two materials has to be made of very hard steel or cast iron and the other material must be a softer one with good antifriction properties. The harder material is for the larger part which is not easily replaceable while the antifriction material is for the part which can be replaced easier. Antifriction materials such as bronzes, babbitt metals and other non-ferrous alloys, antifriction plastics, etc., have low coefficient of friction, high wear resistance, good resistance to sizing, good running-in properties and low wear on the mating part. Sliding couples have to be made from very dissimilar materials because similar materials have a tendency to form local bonds which increase wear and friction. In such couples where lubrication is also present, the effect this might have on oil circulation and on the disposal of the wear debris should be considered.

In other cases where we want a high coefficient of friction, for example in brakes and clutches, friction materials are used. These materials have high coefficient of friction and wear rate, and also present high resistance to heat. They must also have a coefficient of friction which does not change very much with temperature and environmental conditions, and further they have to have low wear on the mating parts under dry or lubricated conditions. In modern machinery operating in high temperatures such as aircraft engines and gas turbines, materials must have very high wear resistance and in this case heat-resistant superalloys are used.

Many times when we design for high-contact stresses or for friction and wear resistance, it is possible to use surface hardening or surface-coating techniques. In this case we do not have to use a very strong and usually very expensive material for the whole part, and instead can use a material which gives adequate strength for the body of the part and can be surface-strengthened or coated with the proper harder or softer (depending on the case) layer.

As noted above, cost of the material is a primary consideration. Therefore in selecting a material, the mere strength value can not be the absolute criterion. For this purpose we use specific material indices, such as strength divided by the unit price, strength divided by density if the weight of the material is of importance, and other indices of this type depending on particular situations. Many times, of course, material selection is limited due to the material availability or manufacturing limitations in the particular locality.

Due to the fact that stocks are expensive, every manufacturing facility tries as hard as possible to reduce the number of available materials on these. This is a usual limitation on the available materials for design and many times the designer has to look through such a list of available materials for his design. For machine parts which are going to be produced in a very large quantity, this is not a limitation because the materials will be purchased for the production of the particular component.

When the material does not contribute greatly to the cost of the machine, or for a new design of the machine, materials of high quality must always be selected. This is because we usually design machines and their members based on the major modes of expected failure without considering secondary effects (which in many cases are much greater than we think) and having a good material in many aspects will help in this direction. Later, when the machine has to be redesigned, based on the experience gained during the period of its initial use, economizing with the use of lower-cost materials can be considered.

Materials frequently impose technological limitations in developing larger and more powerful machines and structures. For example, we can design a simply-supported beam

Table 6.1 Physical properties of engineering materials

Material number	Material name	Specific weight	Young's modulus (GPa)	Shear modulus (GPa)	Thermal coeff. $(\times 10^6)/^\circ C$	Sp. heat (kJ/kg °C)	Thermal cond. (kJ/mh °C)	Electrical resistance $\mu(\Omega m)$	Poisson ratio
1	Aluminum	2.70	62.1	23.3	22.2	0.921	775	0.027	0.34
2	Wrought Al alloys	2.72	74	28	22	0.921	500	0.045	0.34
3	Cast Al alloys	2.7	68	28	23	0.921	560	0.053	0.34
4	Structural steels	7.85	210	85	11.45	0.477	190	0.17	0.27
5	Alloy steels	7.85	210	84	11.4	0.510	120	0.7	0.27
6	Stainless steels	7.7	200	86	18	0.5	45	0.7	0.29
7	Heat res. steels	7.83	210	82	11.45	0.4	45	0.8	0.35
8	Copper	8.97	117	50	16	0.385	1400	0.017	0.295
9	Bronze	8.5	112	41	17	0.385	600	0.045	0.295
10	Brass	8.5	109	40	17	0.377	245	0.08	0.295
11	Cast iron	7.5	66–170	9.6–28	10	0.586	180	0.9	0.2
12	Cast steel	7.83	207	77	12.5	0.48	134	1	0.31
13	Mg alloys	1.8	45	16.6	26	1.05	300	0.14	0.3
14	Titanium	4.51	107	41	8.5	0.469	50	0.12	0.34

with some load up to a certain span. Beyond that span if we try to increase the strength of the beam by strengthening the section of the beam, the additional weight we put on the beam is higher than the additional strength of the beam. To make a beam with longer span one must use a material with a higher strength to density ratio or a different design of the section of the beam to increase the strength without simultaneously increasing the weight per unit length. Similar limitations by the materials are posed on the development of more powerful aircraft engines. To have more power in an engine one has to have longer blades in the compressor and turbine section. Longer blades mean higher loads and therefore higher stresses as a result. To make an engine with higher power one has to use a material with higher strength over density ratio. There is a limit however to this ratio for the available materials, beyond which at the present state of technology we cannot go. A breakthrough in material development is usually followed by other technological breakthroughs in areas where this material can be utilized.

Physical properties of the materials are not influenced substantially by small variations in composition and can be represented as in Table 6.1 which was compiled with the program DESIGNDB (Chapter 2). The reader can add or modify data using this program. Strength, on the other hand, is greatly influenced by composition. Therefore, for every type of material, a separate table needs to be compiled.

It must be emphasized that the material properties given are average values or ranges. Some wide ranges in strength properties indicate the influence of heat treatment. Fatigue strength given is discussed in Chapter 7.

6.3.2. Designing with cast irons

Irons with carbon content above 1.2% are usually referred to as 'cast irons'. The carbon appears mostly as graphite inclusions in spherical or other forms which impose on cast iron certain unique properties, such as a great difference between tensile and compressive strengths, high internal friction, low coefficient of friction and high rigidity.

Gray cast iron

Gray cast iron is the principal material for casting larger shapes. It has very good castability, average strength, small elongation at fracture (which means limited impact strength), good wear resistance and high internal friction with no sensitivity to heating. Gray cast iron has low tensile strength because of the graphite inclusions which result in local stress concentrations which in turn lower the strength of the material. In compression this factor does not appear, therefore this strength of cast iron is comparable with the strength of steel.

Gray cast iron is used mainly for parts of relatively complicated shape which makes casting easier than fabricating. They are used mostly for stationary parts of machines like machine housing, and have a good resistance to wear; for this purpose they are used for parts of machines over which other parts are sliding.

According to the ISO, gray cast irons are identified by two letters followed by two numbers representing the tensile strength and the bending strength in kilograms per square millimeter.

The modulus of elasticity of cast iron increases with the tensile strength.

Many times the size of the machine component made out of gray cast iron is determined not on the basis of strength but on the minimum thickness that a castable part can have because of casting limitations. The higher the percentage of carbon, the lower is the strength and the better is the castability, therefore the smaller the permissible dimensions of the component.

1. Low-strength castings are employed for parts which are subject to low loads and no severe sliding wear.
2. Medium-strength castings are used for parts which have medium loads and work at low sliding speeds and low pressures. These castings are the ones most extensively used, as for example in most machine housing and supporting parts.
3. High-strength castings are used for parts which are subject to high loads and stresses or work at high speeds and pressures such as crankshafts, drive components, heavily-loaded guideways, etc.

The properties of cast iron can be improved with alloying or adding particular forms of carbon before casting. For extremely thin-wall castings of cast iron we use a high carbon content, up to 3.6%, and silicon up to 2.8% also with increased phosphorus content.

For good antifriction properties we alloy the cast iron with nickel (0.3 to 0.4%) and chromium (0.2 to 0.35%). Nickel and chromium together with low percentage of silicon and phosphorus are used for heat- and wear-resistance castings. In high-strength cast irons, magnesium and other materials are added, causing the graphite to precipitate as the cast iron solidifies in the form of spherical nodules and having the effect of reducing the internal stress concentrations. The modulus of elasticity of such cast irons ranges from 160 to 190 GPa, while the endurance limit for cross-sections of medium size is approximately equal to that of medium-carbon steel. Such cast irons are frequently used instead of steel because although they might have somewhat lower strength, very complicated parts can be made and strength improved by making the parts lighter.

White cast irons

White cast irons are very hard and they can be machined only with special tools. For this reason they have a very high wear and heat resistance and also a high resistance to corrosion. They are used:

(a) as parts with high wear such as brake shoes, grinding and crushing parts, pumps for abrasive particle-carrying liquids etc.;
(b) parts which are subject to flame and heat environments;
(c) parts operating in chemical environments.

White cast irons can be also alloyed with the addition of elements such as nickel and boron for wear resistance, chromium for wear and heat resistance, and silicon for acid resistance.

Malleable cast irons

Malleable cast irons are used for parts of somewhat complicated shape but with certain impact loads. The word "malleable' is only used by convention since these cast irons cannot be subject to plastic working as the name suggests.

The properties of the several types of cast irons and steels are shown in Tables 6.2 to 6.4.

Table 6.2 Material properties, cast iron and steel

Material number	Commercial name	ISO designation	USA standard	USA designation	DIN designation	Ultimate strength (MPa)	Yield strength (MPa)	Fatigue strength (MPa)	Elongation (%)	Notes
1	Cast iron	185 GR 10	ASTM A159	G1800	GG-10	100	100	70	0.37	Machine frames 3-30 mm thick
	Cast iron									
	Cast iron									
2	Cast iron	185 GR 15	ASTM A159	G2500	GG-15	550–700	150–200	70	0.37	Above
3	Cast iron	185 GR 20	ASTM A159	G3000	GG-20	600–850	200–250	90	0.37	Above
4	Cast iron	185 GR 25	ASTM A159	G3500	GG-25	700–1000	250–300(T)	120	0.37	Above
5	Cast iron	185 GR 30	ASTM A159	G4000	GG-30	800–1200	300–350(T)	140	0.33	Above
6	Cast iron	185 GR 35	ASTM A48	CLASS 50	GG-35	950–1400	340–400(T)	150	0.33	Above
7	Cast iron	185 GR 40	ASTM A48	CLASS 55	GG-40	1100–1400	400–450(T)	160	0.33	Above
8	Cast iron/ steel	1083 GR 38-17	ASTM A536	GR 60-40-18	GGG-38	370	230	160	17	Cast parts with impact loads.
9	Cast iron/ steel	1083 GR 42-12	ASTM A536	GR 60-45-12	GGG-40	410	270	170	12	Above
10	Cast iron/ steel	1083 GR 50-7	ASTM A536	GR 80-55-06	GGG-50	490	340	200	20	Above
11	Cast iron/ steel	1083 GR 60-2	NA	NA	GGG-60	590	390	230	23	Above
12	Cast iron/ steel	1083 GR 70-2	ASTM A536	GR 100-70-03	GGG-70	690	440	260	26	Above
13	Cast iron/ steel	NA	ASTM A536	GR 120-90-02	GGG-80	800	450	270	27	Above
14										

Table 6.3 Material properties, austenitic cast iron

Material number	Commercial name	ISO designation	USA standard	USA designation	DIN designation	Ultimate strength (MPa)	Yield strength (MPa)	Fatigue strength (MPa)	Elongation (%)	Notes
1	Austenitic cast iron	S-NiMn13 7			GGG-NiMn13 7	390	210	NA	15	
2	Austenitic cast iron	S-NiCr20 2	ASTM A439	TYPE D-2	GGG-NiCr20 2	370	210	NA	7	
3	Austenitic cast iron	S-NiCr20 3	ASTM A439	TYPE D-2B	GGG-NiCr20 3	390	210	NA	7	
4	Austenitic cast iron	S-NiSiCr-2 0 5 2			GGG-NiSiCr 20 4 2	370	210	NA	10	
5	Austenitic cast iron	S-Ni22	ASTM A439	TYPE D-2C	GGG-Ni 22	370	170	NA	20	
6	Austenitic cast iron	S-NiMn23 4			GGG-NiMn23 4	440	210	NA	25	
7	Austenitic cast iron	S-NiCr30 1	ASTM A439	TYPE D-3A	GGG-NiCr30 1	370	210	NA	13	
8	Austenitic cast iron	S-NiCr30 3	ASTM A439	TYPE D-3	GGG-NiCr30 3	370	210	NA	7	
9	Austenitic cast iron	S-NiSiCr-3 0 5 5	ASTM A439	TYPE D-4	GGG-NiSiCr 30 5 5	390	240			
10	Austenitic cast iron	S-Ni35	ASTM A439	TYPE D-5	GGG-Ni35	370	210	NA	20	
11	Austenitic cast iron	S-NiCr35 3	ASTM A439	TYPE D-5B	GGG-NiCr35 3	370	210		7	
12	Austenitic cast iron	L-NiCu-Cr1 5 6 2	ASTM A436	TYPE 1	GGT-NiCuCr 15 6 2	170				Laminated gray
13	Austenitic cast iron	L-NiCu-Cr1 5 6 3	ASTM A436	TYPE 1b	GGL-NiCuCr 15 6 3	190				Laminated gray Laminated gray
14	Austenitic cast iron	L-NiCr-20 2	ASTM A436	TYPE 2	GGL-NiCr20 2	170				
15	Austenitic cast iron	L-NiCr20 3	ASTM A436	TYPE 2b	GGL-NiCr20 3	190				
16	Austenitic cast iron	L-NiCr-30 3	ASTM A436	TYPE 3	GGNiCr30 3	190				
17	Austenitic cast iron	L-NiSi-Cr3 0 5 5	ASTM A436	TYPE 4	GGL-NiSiCr 30 5 5	170				
18	Austenitic cast iron	L-Ni35	ASTM A436	TYPE 5	GGL-Ni35	120				
19										

Table 6.4 Material properties, steel castings

Material number	Commercial name	ISO designation	USA standard	USA designation	DIN designation	Ultimate strength (MPa)	Yield strength (MPa)	Fatigue strength (MPa)	Elongation (%)	Notes
1	Mall iron/BL	943 GR C				290			6	Small parts/ Light loads
2	Mall iron/BL	943 GR B	ASTM A47	GR 32510		310	190	NA	10	Above
3	Mall iron/BL	943 GR A	ASTM A47	GR 35018	GTS-35	340	210	180	12	Engine parts up to 100 kg thick
4	Mall iron/BL	942 GR B			GTW-35	310	NA	NA	4	Whitehart CI/ Parts up to 50 kg
5	Mall iron/ Perl		ASTM A220	GR 40010		410	270	NA	10	
6	Mall iron/ Perl	944 GR E	ASTM A220	GR 45006	GTS-45	440	270	NA	7	As 4 above
7	Mall iron/ Perl	944 GR D	ASTM A220	GR 50005		490	310	NA	5	
8	Mall iron/ Perl	944 GR C	ASTM A220	GR 60004	GTS-55	540	350	NA	4	See 3 above
9	Mall iron/ Perl	944 GR B	ASTM A220	GR 80002	GTS-65	640	420	NA	3	See 3 above
10	Mall iron/ Perl	944 GR A	ASTM A220	GR 90001	GTS-70	690	540	NA	2	

Table 6.4 (*Contd.*)

Material number	Commercial name	ISO designation	USA standard	USA designation	DIN designation	Ultimate strength (MPa)	Yield strength (MPa)	Fatigue strength (MPa)	Elongation (%)	Notes
11	Steel castings	3755 GR 20-40	ASTM A27	GR 60-30	GS-38	400	200	NA	25	
12	Steel castings	3755 GR 26-52	ASTM A27	GR 70-40	GS-52	520	260	18	18	
13	Steel castings	3755 GR 26-52	ASTM A27	GR 70-40	GS-52	520	260	NA	18	
14	Steel castings	3755 GR 30-57	ASTM A148	GR 80-50	GS-60	570	300	NA	15	
15	Steel castings		ASTM A148	GR 90-60	GS-62	620	410	NA	20	
16	Steel castings		ASTM A148	GR 105-85	GS-70	720	590	NA	17	
17	Steel castings		ASTM A148	GR 120-95		830	650		14	
18	Steel castings		ASTM A148	GR 150-125		1030	860	NA	9	
19										
20	Steel castings	944 GR D	ASTM A220	GR 50005		490				
21	Steel castings	944 GR C	ASTM A220	GR 60004	GTS-55	540				
22	Steel castings	944 GR B	ASTM A220	GR 80002	GTS-65	640				
23	Steel castings	944 GR A	ASTM A220	GR 90001	GTS-70	690				
24	Steel castings	3755 GR 20-40	ASTM A27	GR 60-30	GS-38	400				
25	Steel castings	3755 GR 23-45	ASTM A27	GR 65-35	GS-45	450				
26	Steel castings	3755 GR 26-52	ASTM A27	GR 70-40	GS-52	520				
27	Steel castings	3755 GR 30-57	ASTM A148	GR 80-50	GS-60	570				
28	Steel castings		ASTM A148	GR 90-60	GS-62	620				
29	Steel castings		ASTM A148	GR 105-85	GS-70	720				
30	Steel castings		ASTM A148	GR 120-95		830				
31	Steel castings		ASTM A148	GR 150-125		1030				
32										
33										

6.3.3 Designing with structural steels and steel alloys

Steel is by far the most widely used material in machine construction. In general, it has very high strength per unit price. Beyond that it has many other advantages over other structural materials.

Structural steels are classified on the bases of their carbon content and alloying elements. With respect to their carbon content and type of heat treatment, steels are classified as:

1. low-carbon carburizing steels with a carbon content up to 0.25%;
2. medium-carbon structurally improvable and hardening steels with the carbon content between 0.25 and 0.6%;
3. high-carbon hardening steels with a carbon content over 0.6%.

Castings are also made out of steel with relatively high content of carbon. Such steels are somewhat inferior to rolled or forged steels and have lower machineability. Structural section steels are supplied in various types and size ranges as rounds, squares, and several other forms. Steels of ordinary quality are used for parts which are not to be heat-treated.

Uses of carbon steels are as follows:

1. Low-carbon structural steels are used for parts that during manufacturing will have to undergo plastic deformations or machining which is based also on plastic deformation.
2. Medium-carbon steels are used for parts which have low loads and are not going to be heat-treated after machining. Such steels are in general not used for sliding parts.
3. High-carbon structural steels are used for parts which have medium stresses and are subject to heat treatment.

Carbon steels with manganese content are used for larger parts to improve strength and wear resistance. They are known for their hardenability with special heat treatments. Carbon steels with an increased sulphur content are called 'free-cutting steels' and they are used for parts where smooth surfaces and high machineability are required.

As mentioned above, medium-carbon steels can be heat treated and hardened with special methods over their surface to assume good friction properties while their body can take high loads, especially impact loads. When high strength and special surface properties are required, alloyed steels are used which can also in general be heat treated. In particular, chromium steels have high strength and high resistance to wear and corrosion. However, since they can be hardened near the surface they are generally used for parts with small cross-section.

Under high-temperature conditions, chromium–nickel steels can be used and they have good machineability, a property which makes them especially useful for gears. Critical parts with large cross-sections can be made out of chromium–nickel steels which have high hardenability, high strength and wear resistance and high toughness. For even higher mechanical and processing properties, steels with molybdenum or tungsten additions are used in very critical parts of machines like gas turbines and aircraft engines.

In small quantities, titanium has advantageous effects on the grain size of the steel, raising its hardenability. It is also used for critical parts of machines.

Construction steel is a low-alloyed steel with a carbon content up to 0.18% and having addition of manganese, silicon, chromium, nickel and copper.

Table 6.5 Material properties, structural steel

Material number	Commercial name	ISO designation	USA standard	USA designation	DIN designation	Ultimate strength (MPa)	Yield strength (MPa)	Fatigue strength (MPa)	Elongation (%)	Note
1	Structural steel	630 Fe 37-A	ASTM A284	GRADE D	USt37-1	360	230	170	26	General use
2	Structural steel	630 Fe 42-A	ASTM A570	GRADE D	USt42-1	410	250	190	23	Impact and dynamic loads
3	Structural steel	630 Fe 44-A	ASTM A470	GRADE E	RSt46-2	430	270	210	23	
4	Structural steel	630 Fe 52-B	ASTM A572	GRADE 50	St52-3	490	350	265	22	High strength crane components
5	Structural steel	1052 Fe 50-1	ASTM A572	GRADE 50	St52-3	690	290	265	20	
6	Structural steel	1052 Fe 60-1	ASTM A572	GRADE 50	St60-1	590	330	290	15	High strength power transmissions
7	Structural steel	1052 Fe 70-2	ASTM A572	GRADE 55	St70-2	690	360	320	11	High local stresses/ dies/rolls

Table 6.6 Material properties, carbon steels

Material number	Commercial name	ISO designation	USA standard	USA designation	DIN designation	Ultimate strength (MPa)	Yield strength (MPa)	Fatigue strength (MPa)	Elongation (%)	Notes
1	Carbon steels		AISI	1010	C10	460	355	250	25	Cold drawn/ General use in machinery
2	Carbon steels		AISI	1015	C15	527	430	273	25	As above
3	Carbon steels		AISI	1020	C22	500–650	300–420	190	20	As above
4	Carbon steels	C25	AISI	1025		540–690	360–	NA	18	As above
5	Carbon steels	C30	AISI	1030		580–730	390–	NA	17	As above
6	Carbon steels	C35	AISI	1035	C35	620–760	420–500	290	17	As above
7	Carbon steels	C40	AISI	1040		660–800	450–	NA	16	
8	Carbon steels	C45	AISI	1045	C45	700–840	480–500	330	14	
9	Carbon steels	C50	AISI	1050		720–880	510–	NA	13	
10	Carbon steels	C55	AISI	1055	C55	780–930	540–	NA	12	
11	Carbon steels	C60	AISI	1060	C60	830–980	570–600	400	11	
12										

Table 6.7 Material properties, stainless steels, ferritic/martensitic

Material number	Commercial name	ISO designation	USA standard	USA designation	DIN designation	Ultimate strength (MPa)	Yield strength (MPa)	Fatigue strength (MPa)	Elongation (%)	Notes
1	Stainless FERR/MART	683/13 GRADE 1	AISI	403	X7CR13	440–640	250	NA	20	
2	Stainless FERR/MART	GRADE 2	AISI	405	X7CRAL13	410–610	250	NA	20	
3	Stainless FERR/MART	GRADE 8	AISI	430	X8CR17	440–640	250	NA	18	
4	Stainless FERR/MART	GRADE 8A	AISI	430F	X8CR17	440–640	250	NA	15	
5	Stainless FERR/MART	GRADE 8B	AISI	—	X8CRTI17	440–640	250	NA	18	
6	Stainless FERR/MART	GRADE 9C	AISI	436	X6CRMO17	440–640	250	NA	18	
7	Stainless FERR/MART	GRADE 3	AISI	410	X10CR13	590–780	410	NA	16	
8	Stainless FERR/MART	GRADE 7	AISI	416	X12CRS13	640–830	440	NA	12	
9	Stainless FERR/MART	GRADE 4	AISI	420	X20CR13	690–880	490	NA	14	
10	Stainless FERR/MART	GRADE 9	AISI	431	—	830–1030	640	NA	10	
11	Stainless FERR/MART	GRADE 9B	AISI	—	X22CINI17	880–1130	690	NA	9	
12	Stainless FERR/MART	GRADE 5	AISI	420 FSE	X30CR13	780–980	590	380	11	
13										

Table 6.8 Material properties, alloy, direct hardening steels

Material number	Commercial name	ISO designation	USA standard	USA designation	DIN designation	Ultimate strength (MPa)	Yield strength (MPa)	Fatigue strength (MPa)	Elongation (%)	Notes
1	Alloy direct hardening steels	R683 PART 2 GR 1	AISI	4130	25CrMo4	880–1080	690–	NA	12	Large parts heavily stressed
2	Alloy direct hardening steels	R683 PART 2 GR 2	AISI	4135	34CrMo4	980–1080	690–850	450	11	As above
3	Alloy direct hardening steels	R683 PART 2 GR 3	AISI	4140	43CrMo4	1080–1270	880–1000	500	10	As above
4	Alloy direct hardening steels	R683 PART 5 GR 1	AISI	1527	28Mn6	780–930	880–1000	350	13	Low stresses/ automotive use
5	Alloy direct hardening steels	R683 PART 6 GR 1			32CrMo12	1080–1270	880–1050	630	10	Large parts highly stressed
6	Alloy direct hardening steels	R683 PART 7 GR 1	AISI	5132	34Cr4	880–1080	500–700	NA	12	As above
7	Alloy direct hardening steels	R683 PART 7 GR 2	AISI	5135	37Cr4	930–1130	740–	NA	11	As above
8	Alloy direct hardening steels	R683 PART 7 GR 3	AISI	5140	41Cr4	980–1240	780–	NA	11	As above
9	Alloy direct hardening steels	R683 PART 8 GR 1	AISI	8740		1030–1230	830–	NA	10	
10	Alloy direct hardening steels	R683 PART 8 GR 2	AISI	9840		1030–1230	830–	NA	10	
11	Alloy direct hardening steels	R683 PART 8 GR 3	AISI	4340	34CrNiMo6	1180–1370	980–	NA	9	
12	Alloy direct hardening steels	R683 PART 8 GR 5			30CrNiMo8	1230–1420	1030–	NA	9	
13	Alloy direct hardening steels	R683 PART 8 GR 6				1230–1420	1030–	NA	9	

Table 6.9 Material properties, free cutting steels

Material number	Commercial name	ISO designation	USA standard	USA designation	DIN designation	Ultimate strength (MPa)	Yield strength (MPa)	Fatigue strength (MPa)	Elongation (%)	Notes
1	Free cutting steels	R683/9 GR 1	AISI	1211	9S20	490–780	390	NA	8	Non hardened use in machines
2	Free cutting steels	R683/9 GR 2	AISI	1212	9SMn28	510–800	410	NA	7	Above
3	Free cutting steels	R683/9 GR 2Pb	AISI	12L13	9SMnPb28	510–800	410	NA	7	Above
4	Free cutting steels	R683/9 GR 3	AISI	1214	9SMn36	540–830	430	NA	7	Above
5	Free cutting steels	R683/9 GR 3Pb	AISI	12L14	9SMnPb36	540–830	430	NA	7	Above
6	Free cutting steels	R683/9 GR 4	AISI	1108	10S20	490–780	390	NA	8	Case hardened
7	Free cutting steels	R683/9 GR 4Pb	AISI	11L08	10SPb20	490–780	390	NA	8	Above
8	Free cutting steels	R683/9 GR 5	AISI	1117		510–800	410	NA	7	Above
9	Free cutting steels	R683/9 GR 6	AISI	1115		540–830	410	NA	7	Above
10	Free cutting steels	R683/9 GR 7	AISI	1138	35S20	570–760	390	NA	14	Hardened
11	Free cutting steels	R683/9 GR 8	AISI	1140		620–810	420	NA	14	Above
12	Free cutting steels	R683/9 GR 9	AISI	1137		740–930	510	NA	12	Above
13	Free cutting steels	R683/9 GR 10	AISI	1146	45S20	650–840	450	NA	11	Above

Table 6.10 Material properties, nitriding, case hardening steels

Material number	Commercial name	ISO designation	USA standard	USA designation	DIN designation	Ultimate strength (MPa)	Yield strength (MPa)	Fatigue strength (MPa)	Elongation (%)	Notes
1	Nitriding steel	R683 PART 10 GR 1			31CrMo12	1080–1270	880	NA	10	Components resistant to wear
2	Nitriding steel	R683 PART 10 GR 2			39CrMoV139	1270–1470	1080	NA	8	Above
3	Nitriding steel	R683 PART 10 GR 3	ASTM CLASS D	A355	34CrAlMo5	780–930	590	NA	14	Above
4	Nitriding steel	R683 PART 10 GR 4	ASTM CLASS A	A355	41CrAlMo7	930–1130	740	NA	12	Above
5	Case hardening steel	R683 PART 11 GR 1	AISI	1010	C10	490–830	290	250	13	
6	Case hardening steel	R683 PART 11 GR 2	AISI	1015	C15	590–930	340	170	12	
7	Case hardening steel	R683 PART 11 GR 3	AISI	1016		640–980	390	NA	10	
8	Case hardening steel	R683 PART 11 GR 4	AISI	5120	12Cr3	830–1180	540	320	10	
9	Case hardening steel	R683 PART 11 GR 5	AISI	5115	16MnCr5	930–1270	640	440	9	
10	Case hardening steel	R683 PART 11 GR 7				1030–1370	690	NA	8	
11	Case hardening steel	R683 PART 11 GR 8	AISI	4118	20MoCr4	930–1270	640	NA	9	
12	Case hardening steel	R683 PART 11 GR 9	AISI	4718		980–1320	640	NA	8	
13	Case hardening steel	R683 PART 11 GR 10			15CrNi6	1030–1370	690	450	8	
14	Case hardening steel	R683 PART 11 GR 11				980–1320	640	NA	8	
15	Case hardening steel	R683 PART 11 GR 12	AISI	8620		980–1320	640	NA	8	
16	Case hardening steel	R683 PART 11 GR 13	AISI	4320		1080–1420	740	NA	8	
17	Case hardening steel	R683 PART 11 GR 14	AISI	9310		1130–1470	780	NA	8	
18	Case hardening steel	R683 PART 11 GR 15				1270–1620	880	NA	7	
19										

Table 6.11 Material properties, stainless steels, austenitic

Material number	Commercial name	ISO designation	USA standard	USA designation	DIN designation	Ultimate strength (MPa)	Yield strength (MPa)	Fatigue strength (MPa)	Elongation (%)	Notes
1	Stainless steels-austenitic	683/13 GR 10	AISI	304L	X2CrNi18 9	440	180	NA	40	
2	Stainless steels-austenitic	683/13 GR 15	AISI	321	X12CrNiTi1 8 9	490	210	NA	35	
3	Stainless steels-austenitic	683/13 GR 16	AISI	347	X10CrNiNb1 8 9	490	210	NA	35	
4	Stainless steels-austenitic	683/13 GR 11	AISI	304	X5CrNi18 9	490–690	200	NA	40	
5	Stainless steels-austenitic	683/13 GR 12	AISI	302	X12CrNi 18 8	490–690	210	40		S
6	Stainless steels-austenitic	683/13 GR 17	AISI	303	X12CrNiS 18 8	490–690	210	NA	35	
7	Stainless steels-austenitic	683/13 GR 13	AISI	305	X5CrNi19 1 1	490–690	180	NA	40	
8	Stainless steels-austenitic	683/13 GR 14	AISI	301	DIN 17224 X12CrNi 17 7	590–780	220	NA	NA	
9	Stainless steels-austenitic	683/13 GR 19	AISI	316L	X2CrNiMo 18 10	440–660	200	270	40	
10	Stainless steels-austenitic	683/13 GR 20	AISI	316	X5CrNiMo 18 10	490–690	210	270	40	
11	Stainless steels-austenitic	683/13 GR 21			X10CrNiMoT i18 10	490–690	220	270	35	
12	Stainless steel	ISO683/13 GR 23			X10CrNiMoN b18 10	490–690	220	270	35	
13	Stainless austenitic steel	GRADE 19A	AISI	316L	X2CrNiMo 18 12	440	200	270	40	
14	Stainless austenitic steel	GRADE 20A	AISI	316	X5CrNiMo 18 12	490	210	270	40	
15	Stainless austenitic steel				X10CrNiMoT i18 12	490–690	220	270	35	
16	Stainless austenitic steel	GRADE 23A				490–690	220	NA	35	
17	Stainless austenitic steel	GRADE 24	AISI	317	X2CrNiMo 18 16	490–690	200	NA	35	

18	Stainless austenitic steel	GRADE A-3	AISI	202	X8CrMnNi 18 9	640–830	300	NA	40
19									
20	Stainless steels-austenitic	683/13 GR 15	AISI	321	X12CrNiTi1 8 9	490			
21	Stainless steels-austenitic	683/13 GR 16	AISI	347	X10CrNiNb1 8 9	490			
22	Stainless steels-austenitic	683/13 GR 11	AISI	304	X5CrNi18 9	490			
23	Stainless steels-austenic	683/13 GR 12	AISI	302	X12CrNi 18 8	490			
24	Stainless steels-austenitic	683/13 GR 17	AISI	303	X12CrNiS 18 8	490			
25	Stainless steels-austenitic	683/13 GR 13	AISI	305	X5CrNi19 1 1	490			
26	Stainless steels-austenitic	683/13 GR 14	AISI	301	DIN 17224 X12CrNi 17 7	590			
27	Stainless steels-austenitic	683/13 GR 19	AISI	316L	X2CrNiMo 18 10	440			
28	Stainless steels-austenitic	683/13 GR 20	AISI	316	X5CrNiMo 18 10	490			
29	Stainless steels-austenitic	683/13 GR 21			X10C2NiMoT i18 10	490			
30	Stainless steel	ISO683/13-GR 23			X10CrNiMoN b18 10	490–690	220	270	35
31	Stainless austenitic steel	GRADE 19A	AISI	316L	X2CrNiMo 18 12	440	200	270	40
32	Stainless austenitic steel	GRADE 20A	AISI	316	X5CrNiMo 18 12	490	210	270	40
33	Stainless austenitic steel				X10CrNiMoT i18 12	490	220	270	35
34	Stainless austenitic steel								
35									
36									

Table 6.12 Material properties, spring steels

Material number	Commercial name	ISO designation	USA standard	USA designation	DIN designation	Ultimate strength (MPa)	Yield strength (MPa)	Fatigue strength (MPa)	Elongation (%)	Notes
1	Spring steels	683 PART 1 4 GR 1	AISI	1074 1080	D75-2	1180	880	NA	6	
2	Spring steels	683 PART 1 4 GR 2				1180	880	NA	6	
3	Spring steels	683 PART 1 4 GR 3			46Si7	1270	1080	NA	6	
4	Spring steels	683 PART 1 4 GR 4			51Si7	1320	1130	550	6	
5	Spring steels	683 PART 1 4 GR 5	AISI	9255	55Si7	1320	1130	550	6	
6	Spring steels	683 PART 1 4 GR 6	AISI	9260		1370	1180	NA	5	
7	Spring steels	683 PART 1 4 GR 7			60Cr7	1370	1180	650	5	
8	Spring steels	683 PART 1 4 GR 8	AISI	5155	55Cr3	1370	1180	650	6	
9	Spring steels	683 PART 1 4 GR 9	AISI	5160		1370	1180	650	5	
10	Spring steels	GRADE 10	AISI	51B60		1370	1180	650	6	
11	Spring steels	GRADE 11	AISI	—		1370	1180	650	6	
12	Spring steels	GRADE 12	AISI	4161		1370	1180	650	6	
13	Spring steels	GRADE 13	AISI	6150	50CrV4	1370	1180	650	6	
14	Spring steels	GRADE 14	AISI	—	51CrMoV4	1370	1180	650	6	
15										

Table 6.13 Material properties, flame, induction hardening steels

Material Number	Commercial Name	ISO designation	USA standard	USA designation	DIN designation	Ultimate strength (MPa)	Yield strength (MPa)	Fatigue strength (MPa)	Elongation (%)	Notes
1	Flame and induction hardening	683/12 GR 1	AISI	1035	Cf35	620–760	420	NA	17	
2	Flame and induction hardening	683/12 GR 2	AISI	1040		660–800	450	NA	16	
3	Flame and induction hardening	683/12 GR 3	AISI	1045	Cf45	700–840	400	NA	14	
4	Flame and induction hardening	683/12 GR 4	AISI	1050		740–880	510	NA	13	
5	Flame and induction hardening	683/12 GR 5	AISI	1055	Cf53	740–880	510	NA	12	
6	Flame and induction hardening	683/12 GR 6	AISI	5145	45Cr2	880–1080	640	NA	12	
7	Flame and induction hardening	683/12 GR 7	AISI	5135	38Cr4	930–1130	740	NA	11	
8	Flame and induction hardening	683/12 GR 8	AISI	5140	42Cr4	980–1180	780	NA	11	
9	Flame and induction hardening	683/12	AISI	4140	41CrMo4	1080–1270	880	NA	10	
10	Flame and induction hardening	683/12 GR 10	AISI	8640 8740		1030–1230	830	NA	10	
11	Flame and induction hardening	683/13 GR 11				1030–1230	830	NA	10	
12										

Corrosion-resisting, stainless, and acid-resisting steels are alloyed with cromium, nickel, and manganese.

Heat-resisting steels are used for temperatures above 700 °C and consist of low- and medium-alloy chromium steels.

Strength properties of steels are shown in Tables 6.5 to 6.13, which can be created and maintained with program DESIGNDB (Chapter 2).

6.3.4 Designing with nonferrous alloys

Copper alloys are very widely used in machine construction, primarily for two main advantages: good friction properties and high resistance to corrosion.

Brasses are copper alloys with zinc as the main alloying element (up to 50%). All other copper alloys are called 'bronzes'. With respect to the main component besides copper, bronzes are classified as tin, lead, aluminum, beryllium, silicon and others. In general, bronzes have high antifriction properties, high corrosion resistance and they are conducive to several processing methods, such as casting or machining. Because of the above properties, bronzes are used for bearings, guides, gears, nuts of power screws, and for fittings and parts in corroding environments.

Bronzes with tin contents between 4 and 12% are widely used with smaller contents of lead, zinc and phosphorus resulting in high corrosion resistance. Because tin is expensive, these bronzes are also expensive.

Lead bronzes contain 27 to 32% lead and have good antifriction properties, being used mostly for bearings. Because of their low hardness they require that the mating material be surface-hardened and also have a good surface finish.

Aluminium bronzes are used also as antifriction material at high pressures but at only low and medium sliding speeds. The requirements on the mating parts are as for lead bronzes.

Brasses have good resistance to corrosion, high electrical conductivity, sufficient strength, and especially good processing properties. Brasses can be casted in the foundry but also they can be subject to cold working and rolled into thin sheets and wires. With the exception of higher grades, brasses can be machined at high speeds and they give a high-class surface finish. Because of these properties, brasses are widely used for tubing, sheets, wires, fittings, instruments, and electrical machinery.

Babbitt metals, or white metals, are alloys with soft metals such as tin, lead, and calcium. They are very good antifriction materials and also have good running-in properties in bearings operating at high speeds and high pressures.

Light-weight alloys are based on aluminum or magnesium and they have specific weights below 3.5. They can be made both in castings or rolled stock. These alloys have high strength to weight ratio and they are used for parts which have high speed and intermittent motion such as reciprocating machines and in general where dynamic loads are high. They are also used for rapidly rotating parts to reduce the centrifugal forces and for housings of engines and machines which are used in the aircraft industry.

In mass production they are particularly useful because they can be machine-casted to precise dimensions, therefore eliminating the need for expensive machining.

6.3.5 Designing with engineering plastics

Plastics are materials consisting of high molecular-weight organic compounds mostly synthetic, and less-frequently natural materials. To improve mechanical properties fillers are used and small amounts of additives. The fillers must be in the form of cloth fabric, paper, glass or graphite fibers and small particles.

Due to their good properties, plastics have been developed to a very high degree during the last few decades. Their use is very economical because usually their price is low and they can be manufactured to a great variety of forms with a very small amount of material loss during manufacturing.

Generally, plastics have low density, high heat and electrical insulation, chemical stability, very high damping capacity, good appearance and, last but not least, high strength. In fact, there are some plastics which have strength comparable with that of very high-quality steels. Limitations of plastics are their low heat resistance and some tendency to change dimensions and shape and sometimes degrade with time or exposure to heat or water, etc.

With respect to the resins used, plastics are classified as thermosetting or thermoplastic. Thermosetting materials undergo a chemical change owing to high temperature during manufacturing. They cure to an infusible shape not permitting any reforming. Thermoplastic materials soften with heat and harden on cooling. They can be resoftened again by heating to be formed to a different shape.

Plastic parts are produced by hot- or cold-pressure molding, injection molding, transfer molding and machining.

Table 6.14 Mechanical properties of some nonferrous materials

Material name	S_u(MPa)	S_y(MPa)	S'_n(MPa)	delta (%)	Use
Titanium	500	340	280	18	Aircraft-turbine
Titanium alloys	1000	950	500	12	as above
Al-Mg alloys	220	110	70	20	Automotive-chemical
Cast Mg alloys	200	160	70	5	Aircraft-automotive
Wrought Mg alloys	270	170	160	15	High strength parts
Al-Zn-Mg alloys	240	130	120	12	as above
Cast Al-Ti-Cu alloys	400	300	80	10	as above
Cast Al-Si-Mg alloys	180	160	110	3	as above
Copper	200	40	—	35	Electric-heat conductors
Cu Sn bronze	320	160	150	40	Friction parts
Cu Ni alloys	310	120	110	35	Corrosion resistant
Cu Al alloys	450	180	150	20	as above
Wrought Cu Zn brass	370	180	155	25	as above
Cast Cu Zn brass	450	170	150	12	as above
Plastics: Acetal	60	—	—	—	Structural and mechanical parts
Glass reinforced	133				
Nylon 6/12	66				—
Glass reinforced	150				
Polyester	58				
Glass reinforced	133				

Thermosetting plastics can be laminated with a fabric base with a filler of cotton cloth in sheets, plates and fibers. Moreover, hardened paper, laminated asbestos fabric, wood laminate and fiberglass laminate can be used. The last produces a plastic which has very high strength, elasticity, low notch sensitivity, high heat resistance and good electrical insulating properties. It is among materials with highest strength per unit mass.

Thermoplastic materials are also very extensively used; they have somewhat lower mechanical properties than thermosetting plastics but they have good manufacturing properties. Such thermoplastic materials are plexiglass, polyethylene and polyvinyl chloride.

Mechanical properties of some nonferrous materials are shown in Table 6.14.

6.4 FACTOR OF SAFETY

As a result of tension and compression tests we obtain basic data on the mechanical properties of a material. Let us now see how the results so obtained can be used in machine design.

As already stated in Chapter 4, the fundamental and most commonly used method of analysis is the method based on stresses. According to this method the design is based on the maximum stress developed at some point of a loaded structure. This stress is called the maximum working stress. It must not exceed a certain value, characteristic of a given material and the service conditions of the structure.

Design based on stresses uses a relation of the form

$$\sigma_{max} = S_L/N \tag{6.1}$$

where S_L is a certain limiting stress for a given material, and N is a number greater than unity called the 'factor of safety'. The dimensions of a structure are usually already known and assigned, for example, from service or technological considerations. In this case the value of σ_{max} is calculated to determine the existing factor of safety

$$N = S_L/\sigma_{max} \tag{6.2}$$

If this factor of safety satisfies the designer, it is concluded that the design is acceptable.

When a structure is in the design stage and some characteristic dimensions are to be assigned directly from strength requirements, the magnitude of N is prescribed beforehand. The required dimension is obtained from a relation of the form

$$\sigma_{max} \leqslant S_L/N \tag{6.3}$$

This quantity is called the 'allowable stress'. It remains to decide what stress is to be taken as a limiting stress S_L and how to assign N.

In order to avoid appreciable permanent deformation of a functioning structure it is customary in machine design to take the yield strength as S_L for ductile materials. Then the maximum working stress is S_y/N (Figure 6.17). Here the factor of safety is denoted by N and is called the factor of safety with respect to yielding. For brittle materials and in some cases for moderately ductile materials, S_L is taken to be the ultimate tensile strength S_u. We then

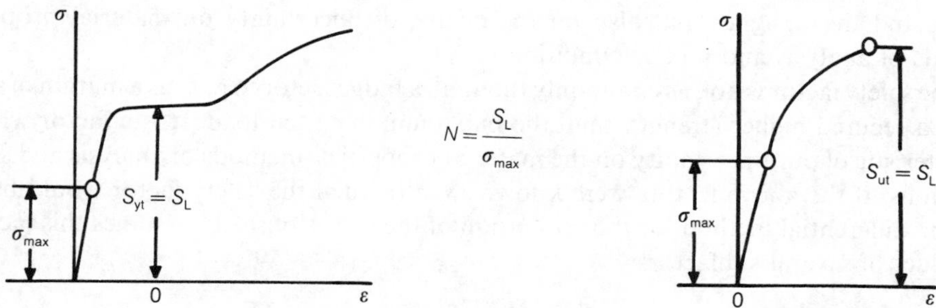

Figure 6.17 Definition of the safety factor

obtain

$$N_u = S_u/S_{max} \tag{6.4}$$

where N_u is the factor of safety with respect to fracture.

As stated in Chapter 4, the design is based on the limiting load. It is possible to introduce similarly the factor of safety based on the limiting load:

$$N = P_L/P_W \tag{6.5}$$

where P_L and P_W are the limiting and working loads. In design based on stiffness

$$N = \delta_L/\delta_W \tag{6.6}$$

where δ_L and δ_W are the limiting and working displacements, respectively.

The choice of N is made in two ways: (a) analytically, based on reliability analysis as shown in Section 6.5; (b) empirically, as is shown below.

The factor of safety assigned is based on experience with the specific service conditions for the structure being designed. The factor N is virtually determined by past practice and by the state of the art at the particular moment. Each field of engineering has its own traditions, requirements, methods and finally, specificity of designs, and the factor of safety is assigned accordingly. Thus, for example, in the designing of stationary engineering structures intended for prolonged service, the factors of safety are rather large ($N = 2$–5). In aircraft engineering where severe weight restrictions are imposed on the structures, they are in the range 1.5 to 2. In view of the high reliability requirements it has become the practise in this field to conduct obligatory static tests on individual components and complete aircraft for direct determination of the limiting loads.

The choice of the factor of safety depends on the methods of stress analysis, the degree of accuracy of these methods, and the graveness of consequences that the failure of a part may entail. The factor of safety depends also on the properties of the material. In the case of a ductile material, the factor of safety with respect to yielding may be lower than for a part made of brittle material. This is quite evident since a brittle material is more sensitive to accidental damage and unexpected manufacturing defects. Moreover, any accidental increase of stresses may cause only small permanent deformations in a ductile material, whereas for a brittle material this may result in failure. The proper choice of the factor of safety depends to a considerable extent on the judgement, experience and ingenuity of the analyst and the designer, but also on the degree of uncertainty on material properties, methods of analysis and service condition.

The safety factor is not, as commonly thought, a fudge factor to give us a margin of safety, that is a secured higher strength than the maximum expected load. It is a factor which is characteristic of our uncertainty on the material properties, methods of analysis and service conditions. If the above factors were known *exactly*, then the safety factor would be 1.

The differential method for the estimation of the factor of safety assumes this factor as a product of several subfactors:

$$N = N_1 N_2 N_3 N_4 \ldots \tag{6.7}$$

where:

N_1 reflects the reliability with which design loads can be determined. It takes values between 1 and 1.5. $N_1 = 1$, when rated loads are determined with unquestionable accuracy, or where safety devices protect the machine from overloading, for example pressure vessels with relief valves; $N_1 = 1.5$, when the load is determined from questionable data, for example wind generators.

N_2 reflects the reliability of the material properties. It is taken to be 1.2 to 1.5 for rolled or forged steel and 1.5 to 2.5 for cast iron and brittle material partly owing to easier inspection of the rolled steel and smaller effect of imperfection on ductile materials.

N_3 depends on the consequences of a failure. It is taken to be 1 when failure will not affect anything else, or $\geqslant 1.5$ when failure results in total loss of the machine, environmental damage, or danger to operators.

N_4 refers to starting and accidental overloads, if the calculations are based on the rated load of the machine, to frequent starts, to operation with or without shocks, etc. This subfactor, also called 'service factor', depends on the particular application and can sometimes reach high values. N_4 can be taken up to a value of 2 for smooth operation and low starting torque, to 5 for rough operation, high starting torque and frequent start-ups. Of course it is taken to be 1 when the exact operating loads have been used in the calculations. Design is based on the yield strength.

As a guide only, and if design is based on both the yield and fatigue strengths, N_4 is taken (see K-H. Decker, *Maschinenelemente*, 1973) to be:

1.3 to 1.5 for static loads, turbines and electric machines;
1.4 to 1.6 for reciprocating machines and machine tools;
1.5 to 1.8 for punching and pressing machines; and
1.6 to 2.0 for impact machines and steel mills.

6.5 STATISTICAL CHARACTER OF STRENGTH: PROBABILISTIC DESIGN

The foregoing discussion might have left the reader with the impression that the purpose of the safety factor is to assure that the strength of the structure is greater than the expected loading by a fixed percentage. This, of course is not true. If the expected loads were known exactly (loading) and the material properties (strength) were known exactly, then we have already explained that the safety factor would be exactly 1. There would be no reason at all for it to be otherwise. However, loads and material properties have a certain degree of uncertainty. Take, for example, the design of the shaft for an elevator motor. The maximum load is usually specified by the maximum allowed number of passengers. For the designer, this is translated into load, multiplying the number of passengers by the average weight of each person. However, people's weights are different. If we have the weights of a great number of people, we can plot the fraction p(L) of the number of people with weight within a specified step range L, say between five successive kilogram integers (60–65, 65–70, 70–75, etc.) (Figure 6.18).

Such a histogram represents the distribution of weights of people of a certain type (locality, sex, age, etc.). If the sample is large enough and the spacing of steps small, this distribution appears as a continuous curve (Figure 6.19), the probability of the load curve. This curve has the property that the area under the curve has a value 1 because the sum of the loads which belong to all load ranges equals the total sample.

There is a point $\bar{L}$which is the average of the sample population. This point is called the 'mean' and can be computed as

$$\bar{L} = \sum_{i=1}^{n} L_i/n \tag{6.8}$$

where L_i is the value of L of the sample i and n the total number of sample loads. The closer the distribution curve is to the mean, the less is the deviation of the sample from the mean. To

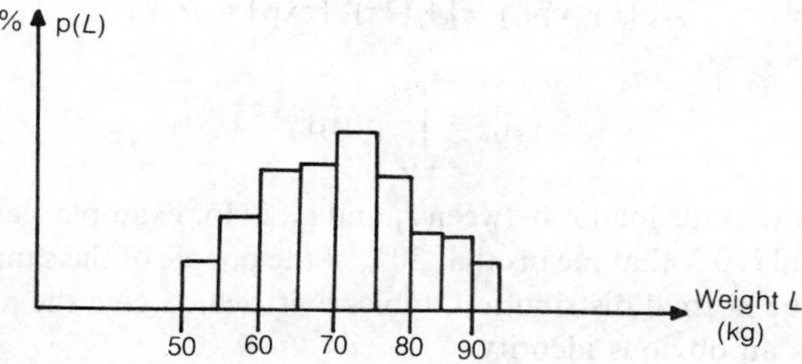

Figure 6.18 Histogram of weight distribution of people

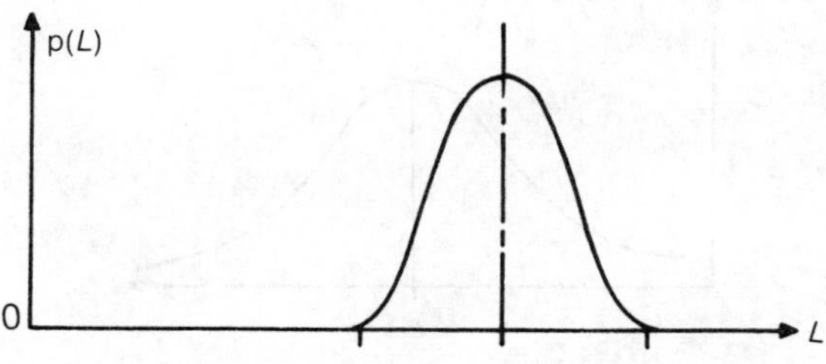

Figure 6.19 Load distribution

quantify that, we sum all the squares of the differences between each step range and the mean:

$$S_L^2 = \sum_{i=1}^{n} (L_i - \bar{L})^2/(n-1) \tag{6.9}$$

We call this quantity 'variance' and its square root (S) the 'standard deviation' and denote it by σ if n is very large. Many physical quantities closely follow a specific type of curve, called the 'normal distribution' (Figure 6.20) which is defined by

$$\mathrm{p}(L) = [1/S_L(2\pi)^{\frac{1}{2}}]\exp[-(L-\bar{L})^2/2S_L^2] \tag{6.10}$$

Introducing the new variable

$$t = (L - \bar{L})/S_L$$

equation (6.8) gives

$$0 = \sum_{i=1}^{n} (L_i - \bar{L})/n = S_L \sum_{i=1}^{n} (L_i - \bar{L})/nS_L = S_L\bar{t}$$

which means that $\bar{t} = 0$. Equation (6.9) now gives

$$S_L^2 = \sum_{i=1}^{n} (L_i - \bar{L})^2/(n-1) = \sum_{i=1}^{n} t_i^2 S_L^2/(n-1)$$

and

$$\sum_{i=1}^{n} t_i^2/(n-1) = 1, \quad S_t = 1$$

which means that the standard deviation of t is unity. The normal distribution becomes

$$S_L\mathrm{p}(L) = \bar{\mathrm{p}}(t) = [1/(2\pi)^{\frac{1}{2}}]\exp[-t^2/2] \tag{6.11}$$

The integral

$$A_{t_1}^{t_2} = \int_{t_1}^{t_2} \mathrm{p}(t)\,\mathrm{d}t$$

gives the probability that the load is between t_1 and t_2. If, for example, between loads $L = 60$ and 70 kg the integral is 0.3, that means that 30% of the people of the sample weigh between 60 and 70 kg. For the normal distribution, tables exist which give the above integral. The following relation is an obvious identity:

$$A_{-\infty}^{t_1} + A_{t_1}^{t_2} + A_{t_2}^{\infty} = 1 \tag{6.12}$$

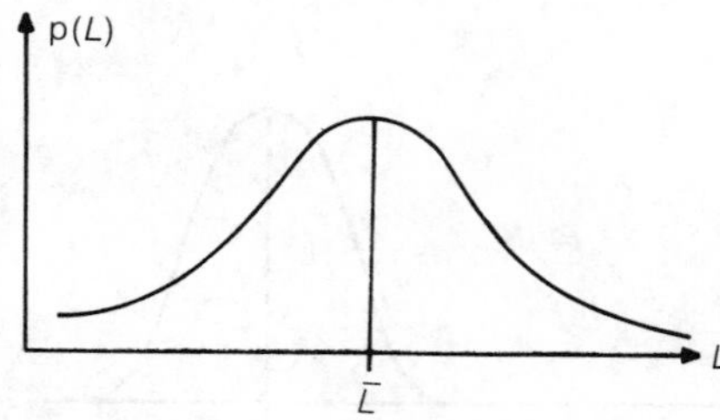

Figure 6.20 Normal distribution

The same behavior is exhibited by the material properties. The tabulated properties are actually mean values and it is expected that during repeated tensile tests, half the specimens will fail below the mean strength and half above it. The strength has also a standard deviation S which with most engineering standards should not exceed 8% of the value of the mean.

Suppose now that we plot in the same diagram the load, in the form of stress L on a particular section and the material capacity, in the form of the strength c (Figure 6.21). It is obvious that if overlap exists, some of the designs will fail, though the mean of the capacity (material strength) is greater than the mean of the load (stress).

The ratio

$$N = \bar{c}/\bar{L} \tag{6.13}$$

is, by what was indicated in Section 6.4, the safety factor, expressing the degree of uncertainty of the expected loads and material strengths. The greater the safety factor, the less probability of failure (load exceeds capacity).

To quantify this, let us form the difference for each design $D = c - L$. It is known from probability theory that

$$\begin{gathered} \bar{D} = \bar{c} - \bar{L} \\ S_D^2 = S_c^2 + S_L^2 \end{gathered} \tag{6.14}$$

Therefore, if the statistical properties, mean and standard deviation, of the load and the

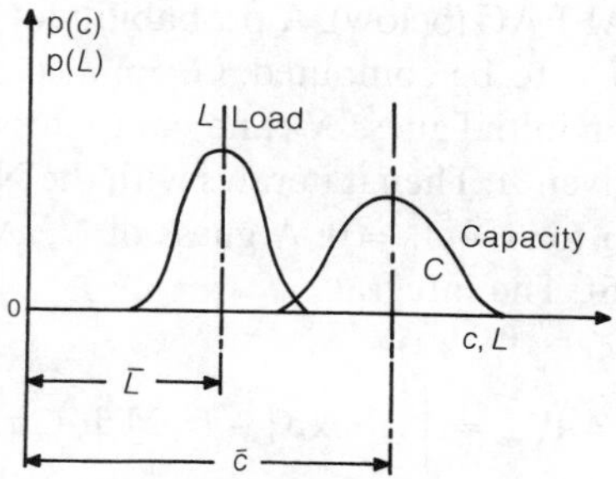

Figure 6.21 Distributions of load and capacity

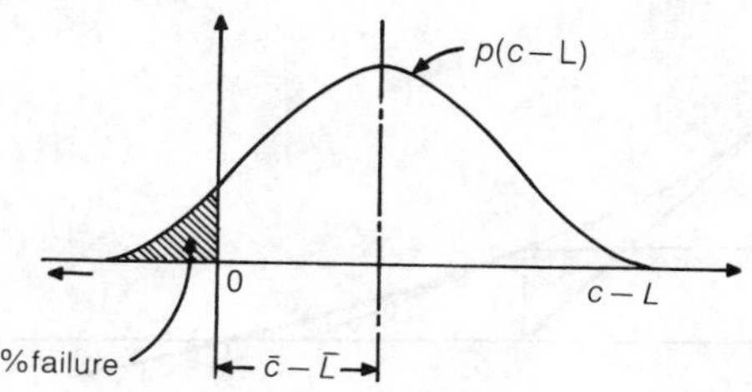

Figure 6.22 Distribution of the over capacity (capacity − load)

capacity are known, so are the statistical properties of the difference D. The distribution of D will also be normal (Figure 6.22). We shall have failure when $D < 0$. Therefore, the probability of failure will be

$$A^{0}_{-\infty} = \int_{-\infty}^{0} \exp[-t^2/2]\,dt/(2\pi)^{\frac{1}{2}} \tag{6.15}$$

where $t = [0 - (\bar{c} - \bar{L})]/S_D$, or

$$t = (-\bar{c} + \bar{L})/[S_c^2 + S_L^2]^{\frac{1}{2}}$$

Since $N = \bar{c}/\bar{L}$, dividing by $\bar{c}$,

$$t = (1 - N)/[S_c/\bar{c})^2 + (S_L/\bar{L})^2/N^2]^{\frac{1}{2}} \tag{6.16}$$

Therefore, if the relative standard deviations $S_c/\bar{c}$ and $S_L/\bar{L}$ are known, from a certain value of the safety factor N one can compute t and from a table of the standard distribution compute the probability of failure. Therefore, a function $A = f(N)$ can be plotted (Figure 6.23). Since such function has been established for a given probability of failure, the appropriate safety factor N can be found.

The probability of failure is not purely a technical matter. For the Apollo mission, for example, it must have been close to zero. For home appliances, experience shows that it is higher than zero. In general, customer demands and economic factors determine the allowable probability of failure. Because the penalty for lower probability of failure is a higher safety factor, which leads to heavier sections and more expensive product. N is computed in the procedure SAFFAC (below). A probability of failure A is assigned and the corresponding safety factor N is to be computed. Given also are the ratios $S_L/\bar{L}$ and $S_c/\bar{c}$.

The program starts with an initial guess N_0 and computes the probability of failure A_0, not equal in general with the given A. Then it iterates with the Newton–Raphson method on N until it comes close enough to $A - A_0 = 0$: A guess of N, (N_0), is assumed and t is then computed with equation (6.16). The integral

$$A = A^{t}_{-\infty} = \int_{-\infty}^{t} \exp[-t^2/2]\,dt/(2\pi)^{\frac{1}{2}} \tag{6.17}$$

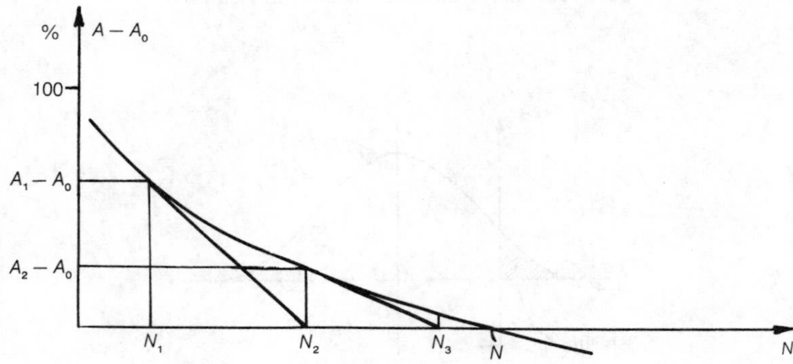

Figure 6.23 Probability of failure versus safety factor

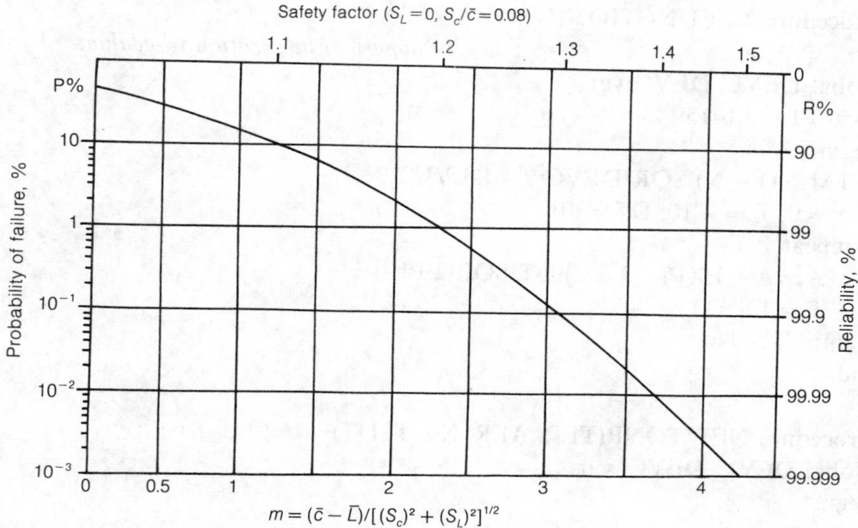

Figure 6.24 Nomogram for probability of failure

is then evaluated numerically from a large negative value of t to t_1. This can be done conveniently with the trapezoidal rule.

Let A_1 be the value of this integral for $N = N_1$. Given a small variation in N, say ΔN, this results in a change, ΔA. A numerical value of the derivative $\mathrm{d}A/\mathrm{d}N$ is $\Delta A/\Delta N$. From geometry of Figure 6.23,

$$N_2 = N_1 - (A_1 - A_0)/(\Delta A/\Delta N)$$

In the same way the sequence of points $N_3, N_4, \ldots$, are found which converge, in general, to the desired point N.

A useful nomogram is shown in Figure 6.24, made by the program SAFFAC. It shows the probability of failure on the left as a function of the ratio

$$m = \frac{\bar{c} - \bar{L}}{[S_c^2 + S_L^2]^{\frac{1}{2}}} = \frac{N - 1}{[(S_c/\bar{c})^2 N^2 + (S_L/\bar{L})^2]^{\frac{1}{2}}}$$

Considering material only ($S_L = 0$), $m = (N - 1)/0.08N$ (for standard deviation of the material 8%), the safety factor is shown in the upper horizontal.

The procedures SAFFAC and NEWTONR follow (explanations are given in italics):

```
Procedure SAFFAC(A, N)
global DEVL, DEVC:real
begin
   input probability of failure A
   input standard deviations DEVL, DEVC
   NER:= .01; AER:= .001; N0:= 1.5; DN:= .001
   NEWTONR(NER, AER, N0, N, DN)
   print A, N, message
end.
```

```
Procedure NLFUNCTION(N, A)
                        (*Numerical integration to evaluate A from N*)
global DEVL, DEVC:var
local PI:= 3.14159
begin
  TM:= (1 - N)/SQR(DEVC^2 + DEVL^2/N^2)
  A:= 0; T:= - 10; DT:= .01
  repeat
    A:= A + EXP(-T^2/2)*DT/SQR(2*PI)
    T:= T + DT
  until T > TM
end.

Procedure NEWTONR(TER, AER, X0, T, DT)
global DEVL, DEVC: var
begin
  repeat
    TC:= T; AC:= A                          (*Save current T and A*)
    NLFUNCTION(X, F)
    NLFUNCTION(X + DX, F1)                  (*Compute A for current T*)

    DERIV:= (F1 - F)/DX
    X:= X - F/DERIV                         (*New value of safety factor*)
  until abs(X - XC) < TER or abs(F - FC) < AER or 100 iterations
end.
```

6.6 STRENGTH THEORIES

The tensile test results need to be interpreted in cases where there are stresses other than tensile stresses in one direction, because otherwise we would need an infinite number of tests with every possible combination of loads. To this end, strength theories have been developed for

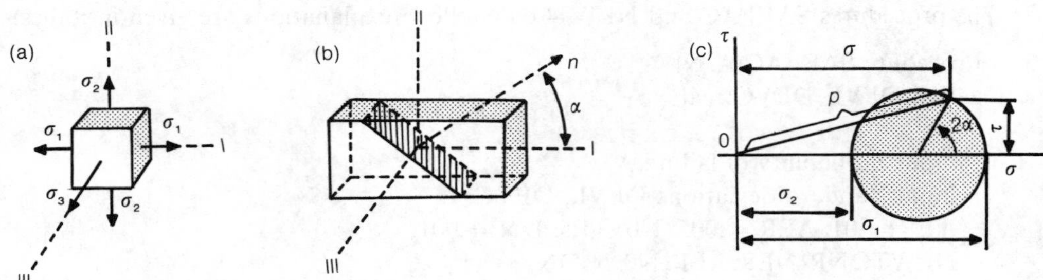

Figure 6.25 Stress state at planes parallel to axis III

design which usually leads to a single stress, equivalent to the actual loading from the point of view of failure, which then is compared with the allowable stress

$$\sigma_{eq} < S_y/N \tag{6.18}$$

A solid in a general state of stress, has in general three principal non-zero stresses: $\sigma_1, \sigma_2, \sigma_3$ (Figure 6.25) along the faces of an elementary cube taken near the point being considered in the solid. The planes they act upon (which are free from shearing stresses) are called the 'principal planes of stress'. The axes (I, II, III) perpendicular to these planes are called the 'principal axes of stress'.

For inclined sectional planes the normal, shearing and resultant stresses p are determined for equilibrium of a prismatic element as in Figure 6.25 separated in two by the inclined plane. For example, for a plane parallel to axis III (Figure 6.25)

$$\left.\begin{aligned} \sigma &= \sigma_1 \cos^2 \alpha + \sigma_2 \sin^2 \alpha \\ \tau &= \frac{\sigma_1 - \sigma_2}{2} \sin 2\alpha \\ p &= [\sigma^2 + \tau^2]^{\frac{1}{2}} = [\sigma_1^2 \cos^2 \alpha + \sigma_2^2 \sin^2 \alpha]^{\frac{1}{2}} \end{aligned}\right\} \tag{6.19}$$

For other orientations, the stress computations are similar.

These stresses can be found graphically, using the Mohr diagram in Figure 6.25, which is merely a graphical representation of equations 6.19. The general Mohr diagram is as shown in Figure 6.26, if we use all three principal stresses.

The extremal shearing stresses are equal to the radii of the Mohr circles:

$$\left.\begin{aligned} \tau_1 &= \pm\frac{\sigma_2 - \sigma_3}{2} \\ \tau_2 &= \pm\frac{\sigma_1 - \sigma_3}{2} \\ \tau_3 &= \pm\frac{\sigma_1 - \sigma_2}{2} \end{aligned}\right\} \tag{6.20}$$

of which τ_2 is the maximum (in magnitude).

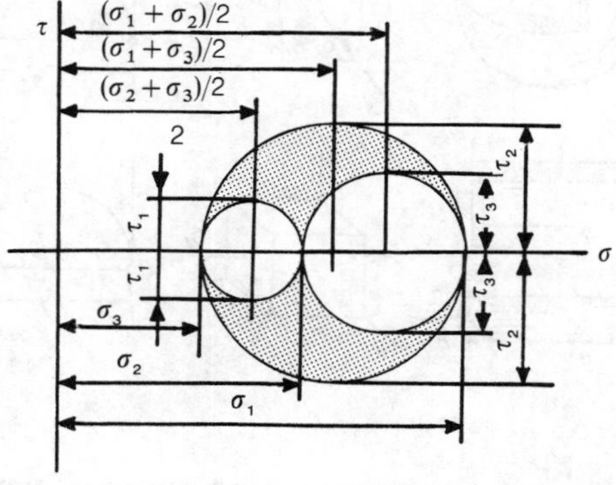

Figure 6.26 Mohr circles for general state of stress

These stresses occur in planes inclined at 45° to the direction of the principal stresses: τ_1 in two mutually perpendicular planes parallel to axis I; τ_2 in two mutually perpendicular planes parallel to axis II; and τ_3 in two mutually perpendicular planes parallel to axis III.

All the equations relating to the general state of stress can also be applied to the planar state of stress, if one principal stress is equated to zero, as well as to the linear state of stress, if two principal stresses are equated to zero.

The various strength theories propose criteria which determine the strength of an element of a material subject to complex stress conditions. According to these criteria, equivalent, or reduced, stresses ($\sigma_1, \sigma_2, \sigma_3$) are established, i.e. stresses due to uniaxial tension of an element of a material which makes its stressed state equivalent for failure to a given complex state of stress. In other words, since only uniaxial tests are available, the strength

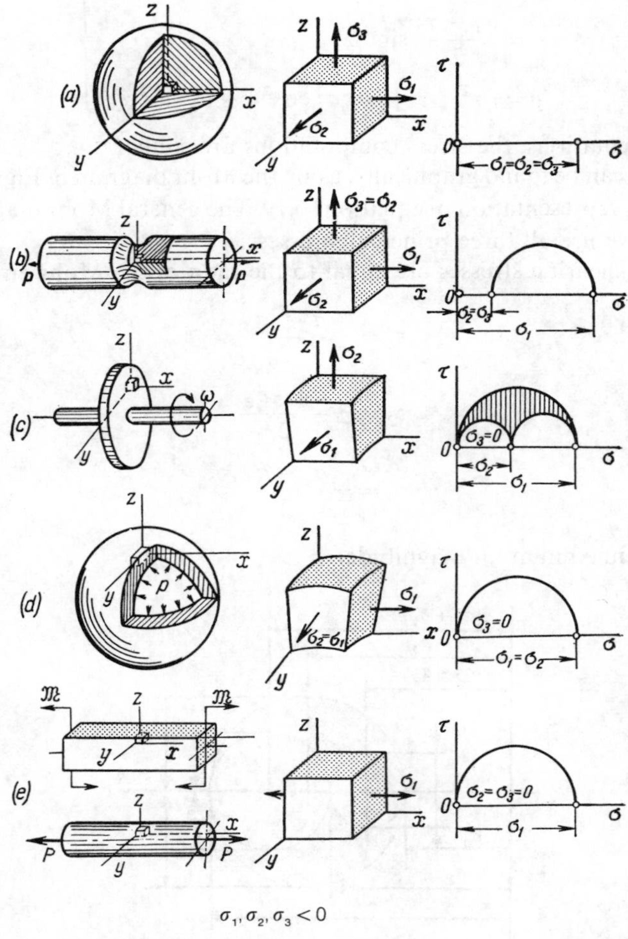

Figure 6.27 Complex states of stress–tension, shear, bending. (From Feodosyev, *Strength of Materials*, by permission of Mir Publishers, Moscow)

theories interpret the failure under general state of stress to failure under uniaxial stress which is better understood.

For the general state of stress of an element, the equivalent stresses have the following values:

I. According to the maximum normal stress theory, failure occurs when the maximum principal stress equals the uniaxial stress of failure, therefore,

$$\sigma_{\text{eqI}} = \max(\sigma_i), \quad i = 1, 2, 3 \tag{6.21}$$

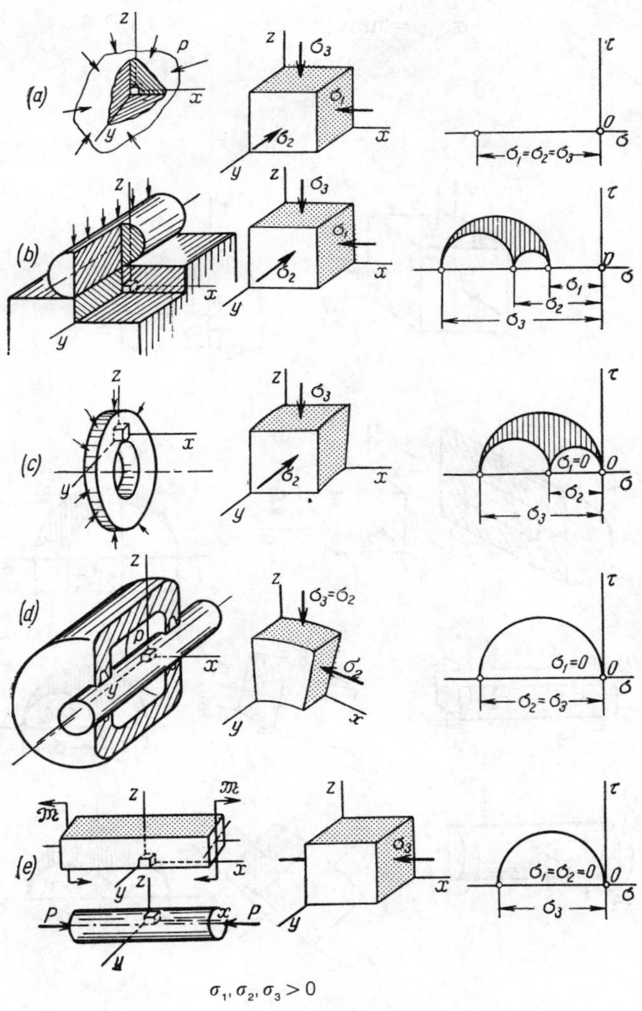

Figure 6.28 Complex states of stress–torsion. (From Feodosyev, *Strength of Materials,* by permission of Mir Publishers, Moscow)

II. According to the maximum shearing stress theory, failure happens when the maximum shear stress reaches the shear stress at failure for the uniaxial tension test. Since this is half the tension stress,

$$\sigma_{\text{eqII}} = \max|\sigma_i - \sigma_j|, \quad i, j = 1, 2, 3 \tag{6.22}$$

III. According to the distortion energy theory, failure occurs when the strain energy due to distortion equals the same energy for the uniaxial test, therefore,

$$\sigma_{\text{eqIII}} = [\tfrac{1}{2}[(\sigma_1 - \sigma_2)^2 + (\sigma_2 - \sigma_3)^2 + (\sigma_3 - \sigma_1)^2]^{\frac{1}{2}} \tag{6.23}$$

IV. The theory of limiting states of stress is a modification of the shear stress theory for materials with different tension and compression strengths. Then,

$$\sigma_{\text{eqIV}} = \max|\sigma_i - k\sigma_j|, \quad i, j = 1, 2, 3 \tag{6.24}$$

where $k = S_{ut}/S_{uc}$

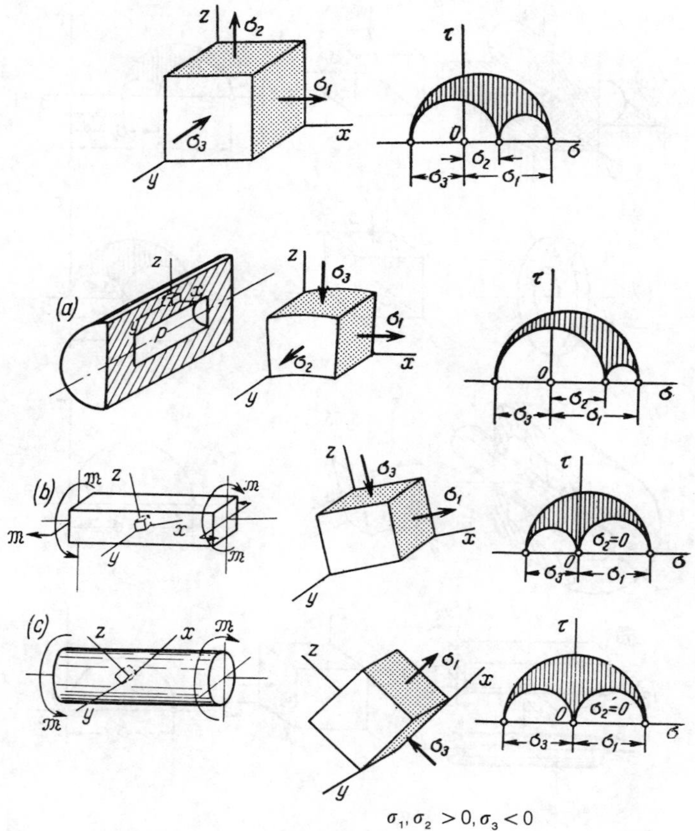

Figure 6.29 Complex states of stress, general loading. (From Feodosyev, *Strength of Materials*, by permission of Mir Publishers, Moscow)

In machine design, the theories II and III are usually applied, namely the maximum shear stress theory and the theory of the distortion strain energy.

The most usual case in machine design involves normal and shearing stresses σ and τ respectively on a plane. The equivalent stress is then

$$\sigma_{eq} = [\sigma^2 + a\tau^2]^{\frac{1}{2}} \tag{6.25}$$

where $a = 3$ for the theory of distortion strain energy and $a = 4$ for the theory of maximum shear stress. Therefore, the latter theory usually is more conservative and widely used for ductile materials while the method of distortion strain energy approaches better the experimental results. For highly brittle materials, the theory of maximum normal stress should be used. Finally, for brittle materials exhibiting different strength in tension and compression, the theory of limiting states of stress should be used.

In machine design, sometimes there are complex states of stresses. For some usual design cases, the stresses and Mohr circles are shown in Figures 6.27 to 6.29. From these figures it is apparent that complex states of stress occur which do not facilitate immediate application of the failure criteria which require that the principal stresses are known. This can be done, in principle, graphically with Mohr circles but it is rather tedious. It is easier to do it algebraically.

To this end, suppose that the state of stress in respect to three mutually perpendicular planes forming the Cartesian system Oxyz is known, given by the three normal and three shear stresses, components of the stress tensor at point O. At a plane inclined in respect to Oxyz by way of three direction cosines l, m, n, the stress vector has components along x, y, z

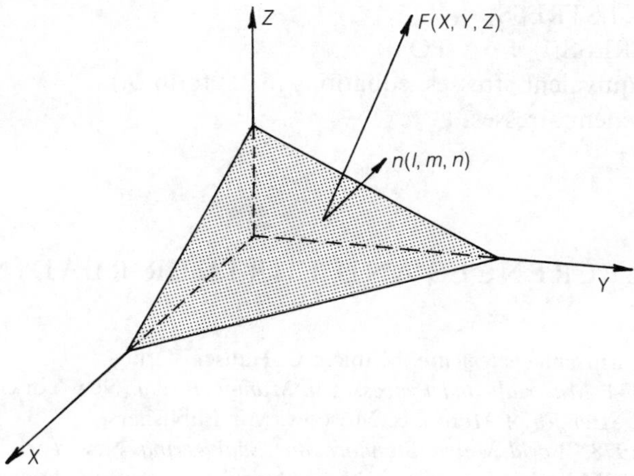

Figure 6.30 Stresses at a diagonal section

(Figure 6.30),

$$\begin{Bmatrix} X \\ Y \\ Z \end{Bmatrix} = \begin{bmatrix} S_{xx} & S_{xy} & S_{xz} \\ S_{yx} & S_{yy} & S_{yz} \\ S_{zx} & S_{zy} & S_{zz} \end{bmatrix} \begin{Bmatrix} l \\ m \\ n \end{Bmatrix} \tag{6.26}$$

where the same indices designate normal stresses.

This plane is principal if the stress vector is perpendicular to the plane, that is the vectors F and n are collinear:

$$X = Sl \quad Y = Sm, \quad Z = Sn \tag{6.27}$$

where S is the (unknown) principal stress. Therefore, equations (6.26) and (6.27) form a homogeneous system of linear algebraic equations in the unknown direction cosines l, m, n. Solution exists if

$$\begin{vmatrix} S_{xx} - S & S_{xy} & S_{xz} \\ S_{yx} & S_{yy} - S & S_{yz} \\ S_{zx} & S_{zy} & S_{zz} - S \end{vmatrix} = 0 \tag{6.28}$$

Equation (6.28) is, in general, a cubic algebraic equation in the unknown principal stress S admitting, in general, three solutions, the three principal stresses.

A general method for the solution of the above problem, the Eigenvalue problem, is the Jacobi method which is used in the program COMLOAD (Appendix 6.B). Detailed description of the method is beyond the scope of this book. The procedure follows:

```
Procedure COMLOAD
global arrays S(3, 3), PSTRESS(3):var
begin
   input S(1, 1), S(2, 2), S(3, 3)          (*Normal stresses SXX, SYY, SZZ*)
   input S(1, 2), S(1, 3), S(2, 3)          (*Shear stresses TXY, TXZ, TYZ*)
   fill symmetric terms of matrix S
   JACOBI(S, PSTRESS)
   print (PSTRESS(I), I = 1 TO 3)
   compute equivalent stresses, equations (6.21) to (6.24)
   print equivalent stresses
end.
```

REFERENCES AND FURTHER READING

Decker, K.-H., 1973. *Maschinenelemente.* Munich: C. Hanser Verlag.

De Garmo, E. P., 1974. *Materials and Processes in Manufacturing.* New York: Macmillan.

Feodosyev, V., 1973. *Strength of Materials.* Moscow: Mir Publishers.

Kverneland, K. O., 1978. *World Metric Standards for Engineering.* New York: Industrial Press.

Machine Design, 1981. Materials Issue v. 53, No. 6, New York: Penton Publishers.

Metals Handbook. 1974. Ohio: American Society of Metals.

Popov, E P., 1976. *Mechanics of Materials.* New Jersey: Prentice-Hall.
Reshetov, D. N., 1978. *Machine Design.* Moscow: Mir Publishers.
Shigley, J. E., 1977. *Mechanical Engineering Design.* New York: McGraw-Hill.

EXAMPLES

Example 6.1

A carbon steel AISI 1020 rod is subject to a compressive axial impact. Calculate the elastic energy up to yielding and the plastic energy up to rupture, absorbed per unit volume of the material. Neglect dynamic effects.

Solution

From the appropriate tables, we obtain $S_y = 340\ \text{N/mm}^2$, $\delta = 39\%$, $E = 2.1 \times 10^5\ \text{N/mm}^2$.

In the elastic region, the elastic energy is the area under the straight portion of the stress–strain curve. The plastic energy is the area under the yielding curve at (approximately) constant stress S_y up to the point of rupture $\varepsilon_p = \delta$.

$$U_e = \tfrac{1}{2}(S_y A)(\varepsilon_y L) = S_y^2 V/2E$$

$$U_p = (S_y A)\Delta l = S_y V\delta/100$$

where A is the cross-section, L the length, and V is the volume of the bar. Therefore, per unit volume

$$u_e = S_y^2/2E = 0.275\ \text{N mm/mm}^3$$

$$u_p = S_y/100 = 123.6\ \text{N mm/mm}^3$$

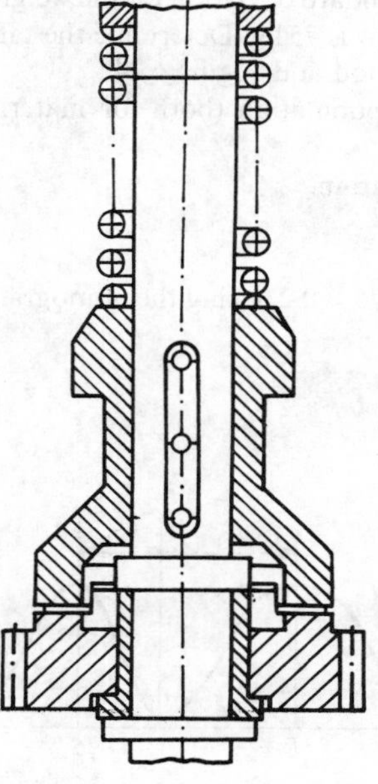

Figure E6.2

Example 6.2

The indexing mechanism shown has a rotating gear with a cam profile on its top horizontal surface which pushes the slider, also having a cam profile, upwards once per revolution. The slider moves along a stationary shaft and it is pressed downwards by a helical spring. Select the appropriate materials.

Solution

The slider has sliding friction against the shaft and the gear. A good selection for them is a grade of surface-hardened steel. The slider has to be made from a very dissimilar material, such as cast iron or phosphor-bronze, with good antifriction properties. Similar material is selected for the bushing between gear and shaft. One of the spring steels grade is selected for the springs.

Design calculations are needed to establish the particular material grade for each component.

Example 6.3

Select an empirical safety factor for the shaft of an electric motor driving an air compressor for paint spray, if the design is based on the rated load, with the differential method.

Solution

Rated loads are very accurately determined because there is electrical load limit switches and pressure relief valves on the air side. Therefore, assuming that safety devices operate 10% above the rated load, $N_1 = 1.1$.

Rolled steel used for shafts is reliable and therefore $N_2 = 1.2$. For the application envisaged, $N_3 = 1$.

Smooth compressor operation and frequent start-ups are expected, requiring a rather high subfactor $N_4 = 1.6$. Finally $N = 1.1 \times 1.2 \times 1 \times 1.6 = 2.11$.

Example 6.4

The cable for a high-rise building elevator is made of high strength steel with standard deviation 8% of the yield strength. The standard deviation for the weight of the town's population per capita is 15 kg while the average weight is 75 kg. Determine the safety factor for a reliability of 99.99%:

(a) With the statistical method and Figure 6.24.
(b) Assuming that all the population (both for materials and passengers) is 100% within 3 standard deviations.
(c) With the SAFFAC program.

Solution

(a) It is $S_c/\bar{c} = 0.08$, $S_L/\bar{L} = 15/75 = 0.2$. Using the nomogram of Figure 6.24, $P = 0.01\%$, $m = 3.7$.

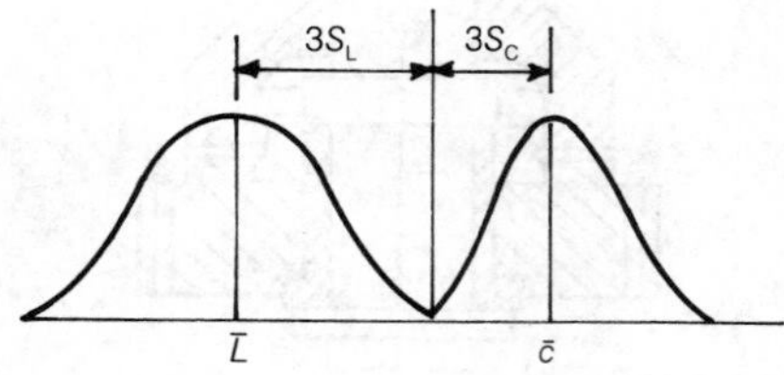

Figure E6.4(a)

Therefore,

$$3.7 = \frac{N-1}{[(S_c/\bar{c})^2 N^2 + (S_L/\bar{L})^2]^{\frac{1}{2}}} = \frac{N-1}{[0.08^2 N^2 + 0.2^2]^{\frac{1}{2}}}$$

Solution of the resulting quadratic equation yields $N = 1.936$.

(b) The situation is shown in the Figure E6.4(a). Obviously

$$\bar{c} - \bar{L} = 3(S_L + S_c)$$

Dividing by L,

$$N - 1 = 3\left(\frac{S_L}{\bar{L}} + \frac{S_c N}{\bar{c}}\right)$$

$$N = \frac{1 + 3\dfrac{S_L}{\bar{L}}}{1 - 3\dfrac{S_c}{\bar{c}}} = \frac{1 + 3 \times 0.2}{1 - 3 \times 0.08} = 2.1$$

(c) The RUN of the SAFFAC program follows.

```
Enter probability of failure     %      ? .01
Enter a guess for the safety factor     ? 1
Standard deviation of load DL/L         ? .2
Standard deviation of capacity DC/C     ? .08


Iterating.....: N= 1.270442        Function= .113956
Iterating.....: N= 1.412884        Function= .0360767
Iterating.....: N= 1.529622        Function= 1.186078E-02
Iterating.....: N= 1.632812        Function= 3.941402E-03
Iterating.....: N= 1.726521        Function= 1.301396E-03
Iterating.....: N= 1.810809        Function= 4.137386E-04
Iterating.....: N= 1.880927        Function= 1.15984E-04
Iterating.....: N= 1.926562        Function= 2.119408E-05
Iterating.....: N= 1.941211        Function= 4.437752E-07

RESULT:   Safety factor= 1.941211

Ok
```

Figure E6.4(b)

Example 6.5

A thin tube of inner diameter $d = 30$ mm, thickness $t = 5$ mm, is subject to internal pressure $p = 30\ \text{N/mm}^2$ and a torque $T = 200$ Nm. Calculate the stresses at a point far from the ends, the principal stresses and the equivalent stresses with (a) the maximum shear stress theory, (b) the equivalent distortion strain energy theory.

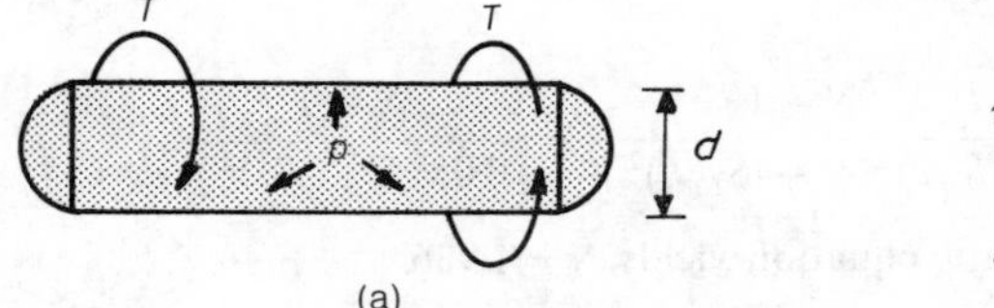

(a)

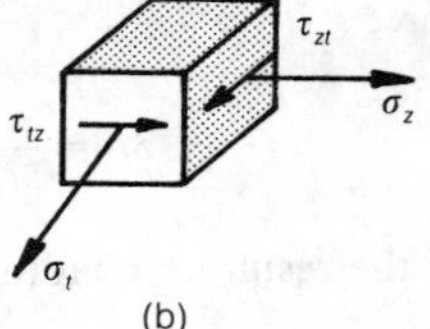

(b)

(c)

```
                    Input normal stresses   Sx, Sy, Sz   ? 90,0,45
                    Input shear stresses    Txy,Txz,Tyz ? 0,28,0

RESULTS:
----------------

----------Principal stresses-----------
 0            31.57996        103.4201
-------------eigenvectors----------------
 0            1               0
 .4322089     0              -.9017736
 .9017736     0               .4322089

Do you want the limiting states of stress theory? (Y/N) ? n
Maximum shear stress theory,              Seq= 103.42
Distortion strain energy theory,          Seq= 91.79
Ok
```

Figure E6.5(a), (b), (c)

Solution

The circumferential stress $\sigma_t = pR/t = 30 \times 15/5 = 90\,\text{N/mm}^2$
The longitudinal stress $\sigma_z = pR/2t = 45\,\text{N/mm}^2$
The shear stress

$$\tau_{tz} = \frac{T}{W} = \frac{200 \times 10^3}{2\pi \times 15^2 \times 5} = 28\,\text{N/mm}^2$$

The program COMLOAD is used to determine the principal stresses and the equivalent stresses. The RUN of the program is shown in Figure E6.5(c).

PROBLEMS

6.1 Determine the elastic energy up to yielding and the plastic energy up to rupture for a simply-supported beam loaded by axial impact at mid-span, per unit volume of the beam. The cross-section is circular and the material is ASTM A570 structural steel.

6.2 Solve Problem 6.1 for a tapered square cross-section of side a at mid-span and $0.8a$ at the supports.

6.3 Solve Problem 6.1 if the impact is torsional at the one end and the other end is fixed.

6.4 Use the program DESIGNDB to tabulate the elastic and plastic energies of materials in the available files for Example 6.1.

6.5 Use the program DESIGNDB to tabulate the tensile strengths per unit mass of materials on file.

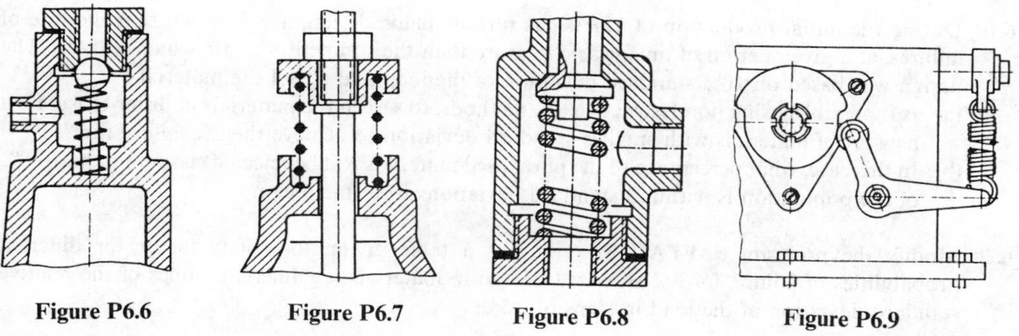

Figure P6.6 Figure P6.7 Figure P6.8 Figure P6.9

Problems 6.6 to 6.8
For the fluid (steam) valves shown, select the proper material for each component.

6.9 For the indexing mechanism shown, operating in normal environment, select the proper element materials.

Problems 6.10 to 6.15
Select the appropriate safety factors by the empirical method for the following elements:

6.10 Main shaft of a lawnmower.
6.11 Valve stem for an automotive engine.
6.12 Head bolt for a diesel engine.
6.13 Anchoring cables for a TV tower.
6.14 Main shaft of a steel mill.
6.15 Main propeller shaft of a helicopter.

6.16 On a small gasoline engine, measurements were conducted at full throttle and at rated speed. The 8% standard deviation materials, for 99% reliability

1	2	3	4	5	6	7	8	9	10
3.5	3.65	3.32	3.35	3.74	3.61	3.6	3.52	3.56	3.41

If the computations are based on the rated horsepower 3.5 HP, find the proper safety factor for 8% standard deviation materials, for 99% reliability.

6.17 An air compressor of 10 kW was designed for standard deviations of the load 15% and the material 10% for a 95% reliability. Subsequently, the marketing department suggested that 99% reliability was essential. It was decided to lower the rated power in order to achieve the required reliability. Calculate the reduced rated power.

6.18 Measurements of stresses owing to pressure were conducted at rated load on a boiling-water reactor pressure vessel. It was found that standard deviation was 10%. Thermal stresses could not be measured with accuracy. Temperature measurements were made, showing a standard deviation of 18%. It was then assumed that thermal stresses would have a similar behavior. Assuming that mean values are additive while standard deviations have a square root of sum of the squares law, determine the safety factor for standard material 8% if the design stresses are the sum (pressure + thermal) stresses.

6.19 During the initial production of a new gas turbine blade design, it was found that the rate of failures, at a given period of time, was 3%, more than the competitor's rate which was 2%. The design was based on 10% standard deviation of the load and 8% of the material.

(a) Is it possible, using nondestructive test methods, to screen the materials in order to isolate a new lot of materials with smaller standard deviation to achieve the 2% failure rate?

(b) In this case, what percentage of the purchased materials will be rejected, assuming that 100% of the population is within 3 standard deviations from the mean?

6.20 Modify the program SAFFAC to print out a table giving the safety factor for different probabilities of failure, for 8% standard deviation materials for different values of the relative standard deviation of the load in steps of 2.5%.

Problems 6.21 to 6.25

For the elements and the loads shown, determine the most dangerous stress state, the normal stresses, the maximum shear stresses and the equivalent stress with the failure theories. Draw also the appropriate Mohr circles (T is torque, M is bending moment) for Problems 6.21 to 6.24. $T = 100$ Nm, $M = 80$ Nm, $d = 80$ mm, $t = 5$ mm, $p = 1.2$ MPa, $F = 1000\ N$, length $= 500$ mm.

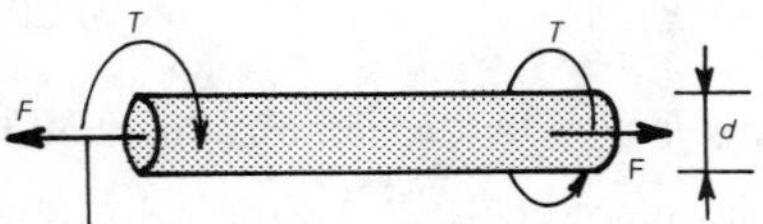

Figure P6.21

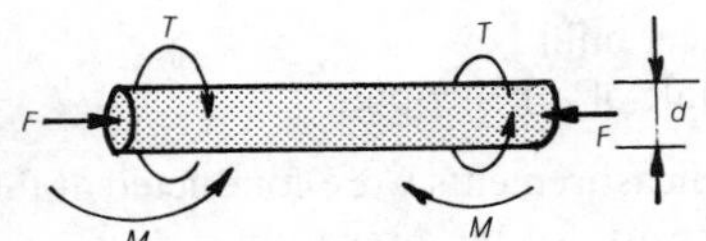

Figure P6.22

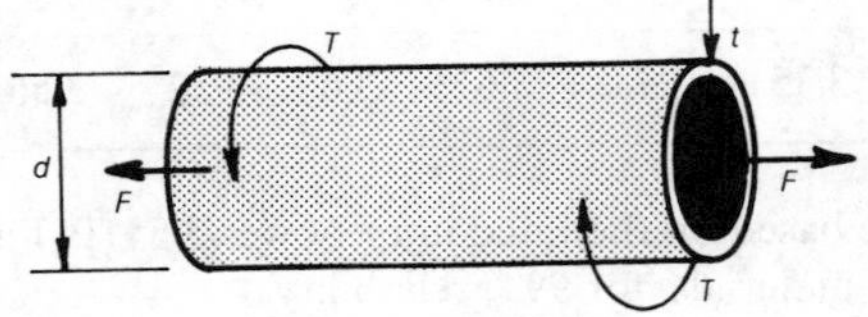

Figure P6.23

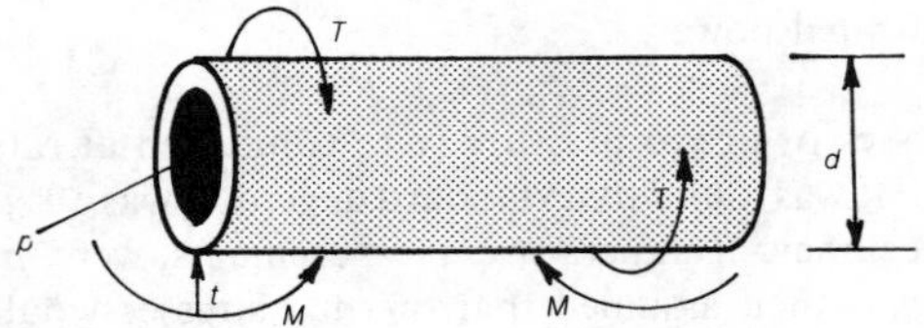

Figure P6.24

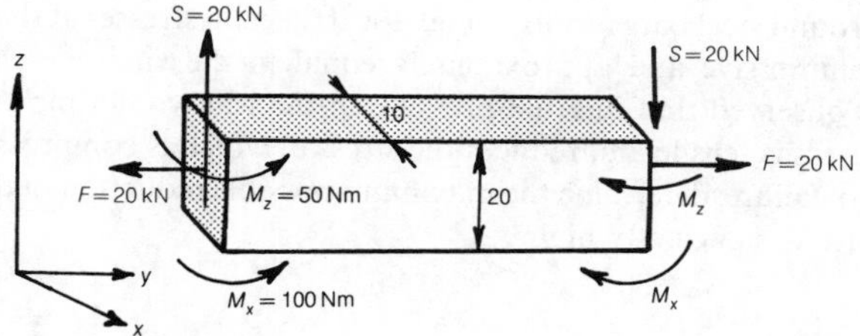

Figure P6.25

6.26 For the road sign shown and the indicated loading, determine what appears to you as the most dangerous stress state, the principal stresses and the equivalent stresses with the strength theories.

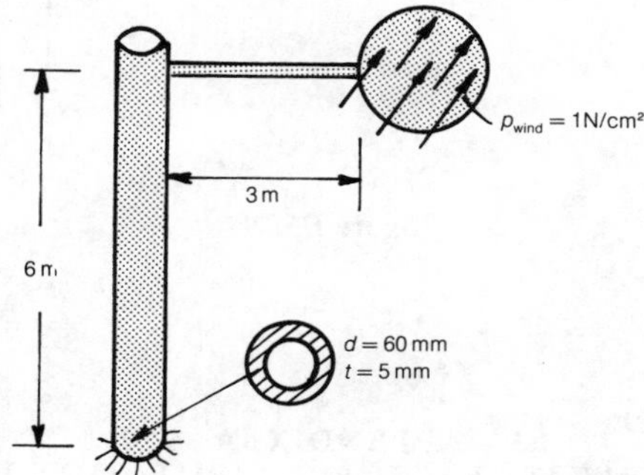

Figure P6.26

6.27 A turbine wheel is loaded as shown. Find the state of stress near the wheel, at the most dangerous point, the principal stresses and the equivalent stress with the available strength theories.

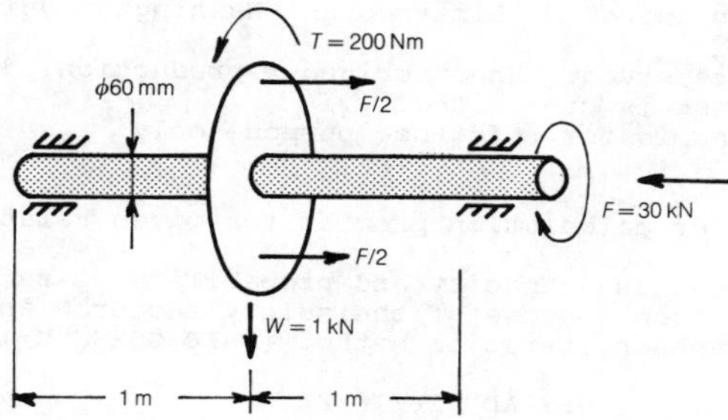

Figure P6.27

6.28 A rotating shaft of diameter 50 mm for water pumps rotates at 200 rpm and transmits power of 10 kW. No other loads are acting on the shaft. After a number of failures an investigation was conducted and found that bending moments of 100 Nm maximum were loading the shaft due to unbalance. It was decided to operate the pumps at lower loads to achieve the design safety factor. Find the new load rating.

6.29 Hot rolling of round steel bars results in high local thermal stresses at the surface of the roll. These stresses are compressive and approximately equal in the circumferential and axial direction $\sigma_t = \sigma_z$. It was observed that internal pressure in the roll would improve roll strength because the resulting stress is tensile and reduces the stress σ_t which is compressive. Using the distortion energy theory of failure, determine the maximum possible reduction of the equivalent stress if the internal pressure is sufficiently high.

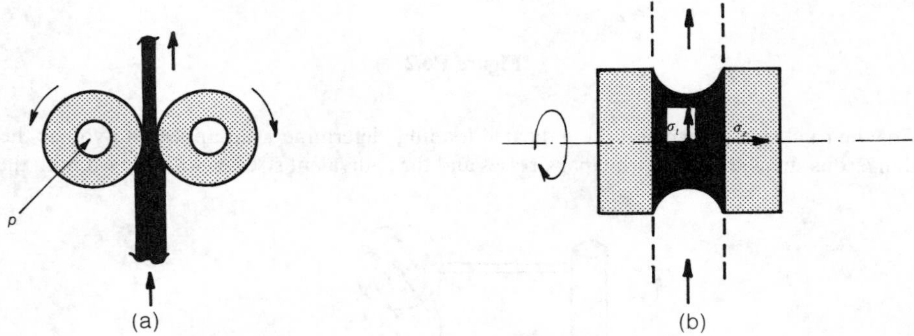

Figure P6.29

APPENDIX 6.A
SAFFAC: A SAFETY FACTOR COMPUTATION PROGRAM

```
10 REM ***********************************************
20 REM *                                             *
30 REM *                  SAFFAC                     *
40 REM *                                             *
50 REM ***********************************************
60 REM                 COPYRIGHT 1985
70 REM by Professor Andrew D. Dimarogonas, Washington Univ., St. Louis,
Mo.
80 REM All rights reserved.  Unauthorized reproduction, dissemination,
90 REM selling or use is prohibited.
100 REM This listing is for reference purpose only.
110 REM
120 REM
130 REM Safety factor calculation program for given relative standard
deviation
140 REM  DL/L of load and capacity and probability of failure PA.
150 REM  It starts from a guess of the safety factor N and uses the
160 REM   Newton-Raphson iteration method (SAFE GUESS N=1)
170 REM
180 REM                     BY ADD 26/10/84
190 CLS:KEY OFF
200 FOR I=1 TO 14:COLOR I:LOCATE I,22+I:PRINT "SAFFAC";
210 LOCATE I,50-I:PRINT "SAFFAC";:NEXT I:COLOR 14
220 LOCATE 8,7:PRINT"A safety factor calculation program for
probabilistic";
230 PRINT " design";
232 locate 25,1:print" Dimarogonas,A.D., Computer Aided Machine  Design,
Prentice-Hall,1988";
235 LOCATE 23,28:INPUT"Hit ENTER to continue";X$:CLS
240 LOCATE 1,15
250 INPUT"Enter probability of failure   %     ";PA:PA=PA/100
```

```
260 LOCATE 2,15
270 INPUT"Enter a guess for the safety factor  ";N
280 LOCATE 3,15
290 INPUT"Standard deviation of load DL/L      ";D1
300 LOCATE 4,15
310 INPUT"Standard deviation of capacity DC/C  ";D2
320 PI = 3.14159
340 GOSUB 500
350 REM ...............................NEWTON ITERATION STARTS HERE
360 DN = .01:REM ....................STEP FOR NUMERICAL DERIVATIVE
COMPUTATION
380 N0=N:N = N + DN:Y0 = Y: GOSUB 500:DE = (Y - Y0) / DN
390 IF DE < > 0 THEN 420
400 PRINT "ENTER A NEW GUESS OF SAFETY FACTOR N(SAFE GUESS IS N=1)":INPUT N
410 GOTO 350
420 N = N0 - Y0 /DE
430 GOSUB 500:Y0 = Y
440 LOCATE 15,1:PRINT  "Iterating.....: N=";N, "Function=";Y;SPC(20);
450 IF ABS (Y0 / DE) < .001 THEN GOTO 470:REM ............ACCURACY OF N IS
.001
460 GOTO 350
470 LOCATE 20,1: PRINT"RESULT:";
480 PRINT "  Safety factor=";N;"                ":PRINT
490 END
500 REM **************************FUNCTION DEFINITION FOR NEWTON-RAPHSON
METHOD
510 X = (N - 1) / SQR (D2^ 2 * N ^ 2 + D1^ 2)
520 GOSUB 550
530 Y = A - PA
540 RETURN
550 REM ************************************************INTETRAP
560 REM
570 REM ............................INTEGRATION ROUTINE BY THE TRAPEZOIDAL
RULE
580 REM
590 REM  .........INTEGRATION OF THE NORMAL DISTRIBUTION FUNCTION FROM X TO
INF
600 X0 = X
610 A = 0: DX = .05 :ER = .000001
620 X = X0: GOSUB 710: W0 = W
630 X = X + DX
640 WP = W
650 GOSUB 710
660 A = A + W * DX
670 IF ABS (W - WP) > ER THEN 630
680 A = A + W0 / 2 * DX
690 REM              PRINT "RESULT:INTEGRAL=",A,"FROM ";X0;" TO ";X
700 RETURN
710 REM **********************FUNCTION DEFINITION:NORMAL DISTRIBUTION
FUNCTION
720 W = EXP ( - X * X / 2) / SQR (2 * PI)
730 RETURN
```

APPENDIX 6.B
COMLOAD: A COMPLEX STATE OF STRESS ANALYSIS PROGRAM

```
10 REM **********************************
20 REM *                                *
30 REM *            COMLOAD             *
40 REM *                                *
50 REM **********************************
60 REM           COPYRIGHT 1985
70 REM by Professor Andrew D. Dimarogonas, Washington Univ., St. Louis, Mo.
80 REM  All rights reserved.  Unauthorized reproduction, disssemination,
90 REM selling or use is strictly prohibited.
100 REM  This listing is for reference purpose only.
```

```
110 REM
120 REM
130 REM General state of stress analysis
140 REM and equivalent stresses
150 REM
160 REM          ADD / 1-12--84
170 REM
180 CLS:KEY OFF
200 FOR I=1 TO 14:COLOR I:LOCATE I,22+I:PRINT "COMLOAD";
210 LOCATE I,50-I:PRINT "COMLOAD";:NEXT I:COLOR 14
220 LOCATE 8,20:PRINT"Stress analysis for general state of stress";
222 locate 25,1:print" Dimarogonas,A.D., Computer Aided Machine Design,
Prentice - Hall,1988";
230 REM  Equivalent stresses are computed with:
240 REM  a) The maximum shear stress theory
250 REM  b) The distortion strain energy theory
260 REM  c) The Mohr's limiting states of stress theory
270 REM
280 REM
290 DIM A(3,3),B(3),U(3,3)
300 REM B..............................diagonal eigenvalue matrix
310 REM U..............................eigenvector matrix
320 LOCATE 17,20
330 MX = - 1E+10:MN = 1E+10:N=3
340 INPUT "Input normal stresses  Sx, Sy, Sz  ";A(1,1),A(2,2),A(3,3)
350 LOCATE 18,20
360 INPUT "Input shear stresses   Txy,Txz,Tyz ";A(1,2),A(1,3),A(2,3)
380 S1=3:REM .........................3 significant digits
390 A(2,1) = A(1,2):B(1)=1
400 A(3,1) = A(1,3):B(2)=1
410 A(3,2) = A(2,3):B(3)=1
420 CLS
430 PRINT :PRINT "RESULTS:":PRINT "-----------------":PRINT
440 GOSUB 1000:REM ......................compute solution
450 PRINT"----------Principal stresses-----------"
460 PRINT B(1),B(2),B(3)
470 PRINT"-------------eigenvectors-----------------"
480 FOR I=1 TO 3:FOR J=1 TO 3:PRINT U(J,I),:NEXT J:PRINT
490 NEXT I
500 MX = - 1E+20:MN = 1E+20
510 PRINT
520 FOR I = 1 TO 3
530 IF B(I) < MN THEN MN = B(I)
540 IF B(I) > MX THEN MX = B(I)
550 NEXT I
560 M1 = ABS (MX - MN)
570 M1 = INT (100 * M1) / 100
580 M2 = SQR (((B(1)-B(2))^2+(B(2)-B(3))^2+(B(3)-B(1))^2)/2)
590 M2 = INT (100 * M2) / 100
600 INPUT "Do you want the limiting states of stress theory? (Y/N) ";X$
610 IF X$ > < "Y" AND X$<> "y" THEN 660
615 PRINT
620 INPUT "Enter ratio k=Sut/Suc                        ";K
630 PRINT:PRINT
640 M3 = ABS (MX - K * MN)
650 M3 = INT(100 * M3) / 100
655 PRINT"-----------Equivalent stresses-------------":PRINT
660 PRINT "Maximum shear stress theory,               Seq=";M1
670 PRINT "Distortion strain energy theory,           Seq=";M2
680 IF X$ > < "Y" AND X$ <> "y" THEN 700
690 PRINT "Limiting states of stress theory (Mohr),Seq=";M3
700 END
1000 REM ********************************JACOBI(A,B,U,S1,N)
1010 REM ==CYCLIC JACOBI METHOD==
1020 REM A..........................NxN stiffness matrix
1030 REM B..........................Nx1 diagonal mass matrix
1040 REM                            and eigenvalue matrix
1050 REM S1.........................No of significant figures
1060 REM ==INITIALIZE==
1070 GOSUB 1110:REM ................INITIALIZE
1080 GOSUB 1180:REM ................SOLVE
1090 RETURN
```

```
1100 REM **********************INITIALIZE TOLERANCE AND MAX No  OF
ROTATIONS
1110  Z=2*S1
1120  T1=1/(10^Z)
1130  R=5*N*N
1140  R1=0
1150  T2=.1
1160  N1=N-1
1170 RETURN
1180 REM *******************************************EIGENPROBLEM
SOLUTION==
1190 GOSUB 1380
1200 REM ==PERFORM ONE CYCLE OF ROTATIONS==
1210 GOSUB 1540
1220 REM ==CHECK TOLERANCE==
1230 IF X1<T1 THEN GOTO 1360
1240 REM CHECK No OF ROTATIONS
1250 IF R1>R THEN GOTO 1280
1260  T2=.1*X1
1270 GOTO 1210
1280 PRINT
1290 PRINT "******  ERROR  ******"
1300 PRINT
1310 PRINT "NO CONVERGENCE ATTAINED"
1320 PRINT
1330 PRINT "WITH ";R1;"ROTATIONS"
1340 PRINT "......END......"
1350 STOP
1360 GOSUB 2140
1370 RETURN
1380 FOR I=1 TO N
1390 FOR J=1 TO N
1400  U(I,J)=0
1410 NEXT J
1420  U(I,I)=1
1430 NEXT I
1440 FOR I=1 TO N
1450  B1=SQR (B(I))
1460  B(I)=1/B1
1470 NEXT I
1480 FOR I=1 TO N
1490 FOR J=1 TO N
1500  A(I,J)=B(I)*A(I,J)*B(J)
1510 NEXT J
1520 NEXT I
1530 RETURN
1540  X1=0
1550 FOR K=1 TO N1
1560  K1= K+1
1570 FOR L=K1 TO N
1580  A1=A(K,K)
1590  A2=A(K,L)
1600  A3=A(L,L)
1610  X=A2*A2/(A1*A3+9.999999E-21)
1620 REM ==CHECK IF ROTATION IS NEEDED==
1630 IF X>X1 THEN GOTO 1650
1640 GOTO 1660
1650  X1=X
1660 IF X<T2 THEN GOTO 2100
1670  R1=R1+1
1680 REM ==COMPUTE ANGLE==
1690 IF A1=A3 THEN GOTO 1740
1700  Z=.5*(A1-A3)/(A2+9.999999E-21)
1710  Z1=1+1/(Z*Z)
1720  T=-Z*(1+SQR(Z1))
1730 GOTO 1750
1740  T=1
1750  C=1/SQR (1+T*T)
1760  S=C*T
1770  S2=S*S
1780  C2=C*C
1790  A(K,L)=0
```

```
1800 REM ==TRANSFORM DIAGONAL ELEMENTS==
1810  A0=2*A2*C*S
1820  A(K,K)=A1*C2+A0+A3*S2
1830  A(L,L)=A1*S2-A0+A3*C2
1840 REM ==TRANSFORM OFF DIAGONAL ELEMENTS==
1850 FOR I=1 TO N
1860 IF I<K THEN GOTO 1890
1870 IF I>K THEN GOTO 1930
1880 GOTO 2030
1890  AO=A(I,K)
1900  A(I,K)=C*A0+S*A(I,L)
1910  A(I,L)=-S*A0+C*A(I,L)
1920 GOTO 2030
1930 IF I<L THEN GOTO 1960
1940 IF I>L THEN GOTO 2000
1950 GOTO 2030
1960  A0=A(K,I)
1970  A(K,I)=C*A0+S*A(I,L)
1980  A(I,L)=-S*A0+C*A(I,L)
1990 GOTO 2030
2000  A0=A(K,I)
2010  A(K,I)=C*A0+S*A(L,I)
2020  A(L,I)=-S*A0+C*A(L,I)
2030 NEXT I
2040 REM ==TRANSFORM MATRIX U TO GENERATE EIGENVECTORS==
2050 FOR I=1 TO N
2060  U0=U(I,K)
2070  U(I,K)=C*U0+S*U(I,L)
2080  U(I,L)=-S*U0+C*U(I,L)
2090 NEXT I
2100 NEXT L
2110 NEXT K
2120 RETURN
2130 REM ==NORMALIZE EIGENVECTORS==
2140 FOR I=1 TO N
2150 FOR J=1 TO N
2160  U(I,J)=U(I,J)*B(I)
2170 NEXT J
2180 NEXT I
2190 FOR I=1 TO N
2200  B(I)=A(I,I)
2210 NEXT I
2220 REM ==ORDER EIGENSOLUTION==
2230 FOR I=1 TO N1
2240  I1=I+1
2250  Z=B(I)
2260  M=I
2270 FOR J=I1 TO N
2280 IF Z<B(J) THEN GOTO 2310
2290  Z=B(J)
2300  M=J
2310 NEXT J
2320  B(M)=B(I)
2330  B(I)=Z
2340 FOR J=1 TO N
2350  Z=U(J,I)
2360  U(J,I)=U(J,M)
2370  U(J,M)=Z
2380 NEXT J
2390 NEXT I
2400 RETURN
```

CHAPTER SEVEN

FATIGUE AND FRACTURE

7.1 STRESS CONCENTRATION

In strength of materials analysis, one usually assumes uniform distribution of stresses in tensile members or linear distribution of stresses and strains in beams. However, this is usually an approximation and it is a reasonable approximation only under certain conditions. For example, stresses have to be considered far from point forces, supports, etc. Moreover, it is known that elastic materials present non-uniformity of stress distribution near geometric disturbances such as notches, holes, keyways, etc. Such occurrences are quite usual in machine members, which almost invariably are not simple cylinders, beams or plates. In order to perform their function they have to have such field disturbances and the designer must consider the stress concentration near them. Such a situation exists, for example, if we have a strip under uniform tension. As we know from strength of materials far from the point of application of the forces, the distribution of the stress everywhere in the section is uniform and equal to the applied stress. If, however, there is a hole in the strip, then near the hole there is increased stress as shown in Figure 7.1. For this case an analytical solution exists if the size of the hole is much smaller than the width of the strip and the hole has the shape of an ellipse with semi-axes *a* and *b* oriented as shown, and the maximum stress is given by equation 7.1.

$$\sigma_{\max} = \sigma_0(1 + 2a/b) \tag{7.1}$$

The ratio

$$K_t = \sigma_{\max}/\sigma_0 \tag{7.2}$$

is called the theoretical 'stress concentration factor' and expresses the intensity of the stress concentration due to the presence of the hole. In curved beams, the usual beam formula does

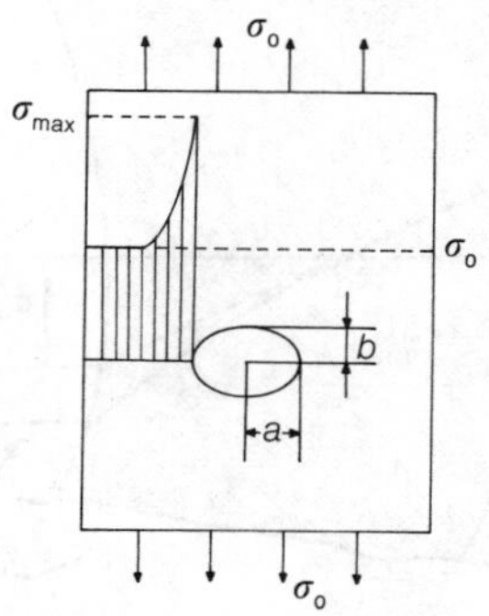

Figure 7.1 Stress concentration near an elliptic hole

not apply because at the inner circle the stress is greater than that at the outer circle, while in a straight beam these stresses are equal (Figure 7.2). Geometric disturbances which cause stress concentration are also referred to as 'stress raisers'.

There are several methods to assess the stress concentration. Many cases can be solved analytically and there is a vast amount of literature on this subject. In more complicated cases one can apply numerical and experimental methods to estimate the stress fields and consequently the stress concentration factors. The finite-element method presented in Chapter 4 yields a very convenient tool to assess the stress field in geometries encountered in engineering problems. For most practical situations, stress concentration factors are tabulated in the literature. In Appendix A.1, we have listed some results pertinent to design application.

Tabulated stress concentration factors (SCFs) are usually obtained for ideal elastic and homogeneous materials and for this reason they are called theoretical or geometric SCFs.

It would seem that design on the basis of the material strength has to be performed considering the maximum stresses at stress concentrations. This is not always the case. The reason is that in certain materials with high degree of ductility, the maximum stress may exceed the yield strength of the material. The local yielding in the area of stress concentration results in a redistribution of stresses such that a more uniform distribution will finally be achieved. This phenomenon is called 'stress redistribution' and it is very common in ductile materials.

In brittle materials however, stress concentration can be very dangerous because the yield point is almost indistinguishable from the point of fracture, and yielding can be accompanied by the development of a crack which might propagate and lead to final fracture. Therefore, for static loading, in general, stress concentrations have to be considered only in brittle materials. For dynamic loading the situation is quite different. This will be discussed later in this chapter (p. 292). For some cases of static loading of brittle materials, stress concentration is not considered. This is the case when internal discontinuities such as microcracks exist and which produce very high stress concentrations – higher than the sharpest geometrical stress raisers. Such a material is cast iron in tension where this

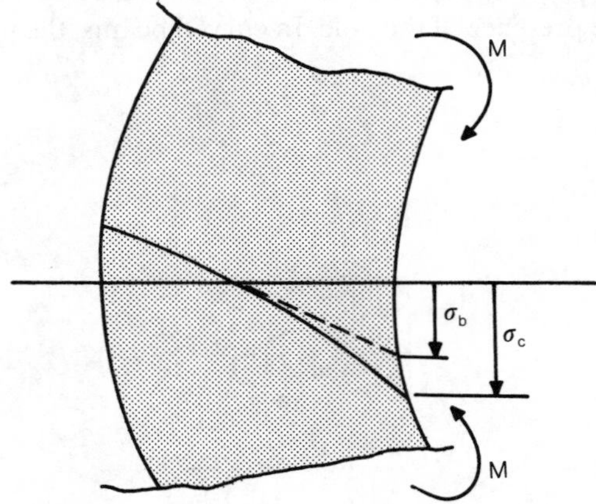

Figure 7.2 Stress concentration on a curved beam

(a)

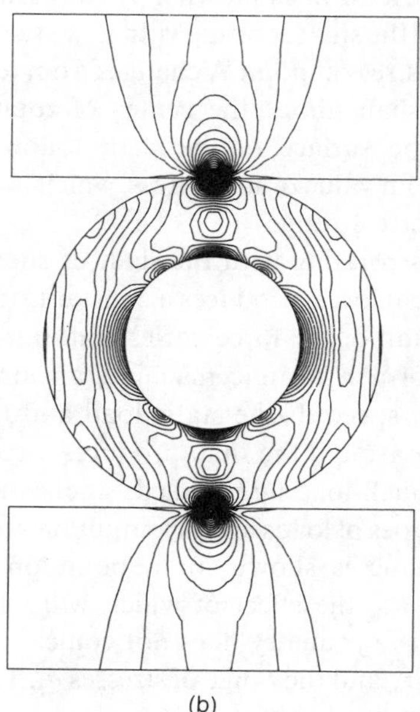

(b)

Figure 7.3 (a) Photoelastic demonstration for stress concentration; (b) Finite-element analysis for stress concentration. (Courtesy ASME)

phenomenon is taken into account by assigning less strength in tension than in compression. Therefore, stress concentration for static loading must be considered only in compression and not in tension, in materials like this.

Local yielding near stress raisers in ductile materials results in residual stresses after unloading, a phenomenon common to most problems where partial yielding takes place.

In designing machine members one has to be careful where to put field disturbances like holes, notches, etc. If such disturbances have to be placed on a machine member (through the function of the machine), the designer must seek locations of low stress levels – if he has any choice at all. For example, in a member which is loaded by bending, holes and other disturbances must be located, if possible, on the neutral line and not near the outer fibers of the beam. A handbook of stress concentration factors is invariably a very good companion for every design office.

Photoelasticity is a useful method to demonstrate and measure the stress concentration. Details can be found in any textbook on experimental stress analysis. Figure 7.3 shows a case of stress raiser. High fringe concentration indicates stress concentration.

7.2 DYNAMIC LOADS: FATIGUE

Since most machine members move, many loads are not constant through time. Further, some constant loads, as we shall see later (p. 666), can produce stresses which vary throughout time in a moving machine member. Take for example a rotating shaft which is loaded at mid-span with a static load W as shown in Figure 7.4(a). If we measure the stress at a fixed point A on the surface of the shaft rotating with it, we see that although the load on the shaft is constant, the produced stress at point A changes from tensile to compressive stress as the point A moves with the shaft about the center of rotation. Therefore, in time the maximum tensile stress on the surface of the shaft follows a harmonic curve, called 'alternating stress'. The maximum value of such stress, which is equal to the semi-width of the stress variation, is called the 'stress range'.

When an eccentric mass is rotating with the shaft as shown in Figure 7.4(b), then the resulting inertia force (centrifugal force) produces a constant stress at the point A of the shaft as shown, although the direction of the force varies continuously.

Next, consider the case of a hydraulic machine, an aircraft engine compressor for example, rotating at very high speed. If the static load and the unbalance give negligible effects, the turbulent flow of air around the shaft produces forces which are very irregular in time and appear as random. Such loading is called 'stochastic' (Figure 7.4c).

In most machines, all the types of loads appear simultaneously, as shown in Figure 7.4(d), where the stress history in time is shown for a point on the periphery of the shaft. Disregarding the stochastic load, the effect of which will be discussed later (p. 292), we have a periodic force. Its axis of symmetry does not coincide with the time axis. Thus, we separate the mean static stress σ_m and the range of stresses σ_r. This is the most usual situation in machines, particularly in rotating shafts.

When discussing static strength of materials, the assumption is that in the strength test, the force is applied slowly and only once. Early this century, the German engineer Woeler,

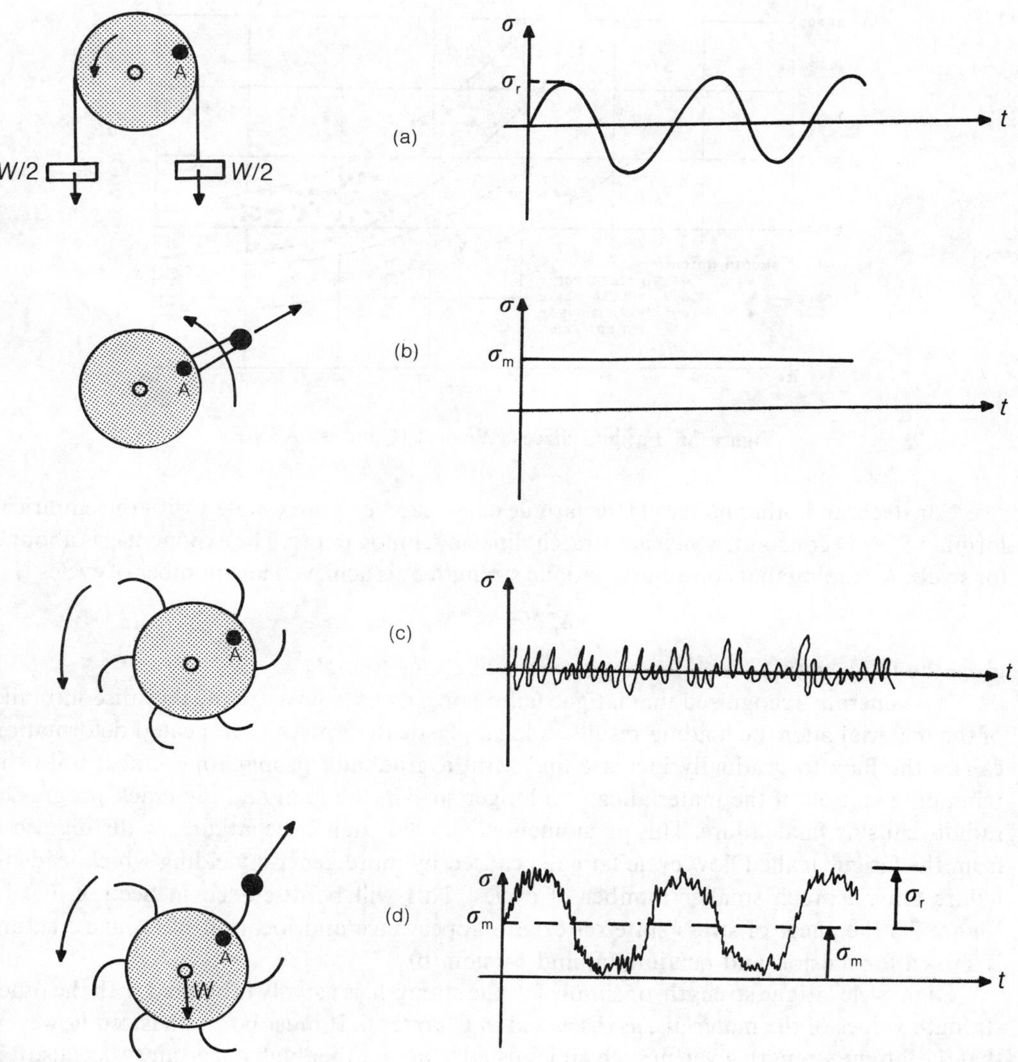

Figure 7.4 Modes of dynamic loading

experimenting with railroad axles, discovered that when he loaded them with forces producing stresses below the ultimate strength of the material, failure was observed after a number of full reversals of the stresses (revolutions in this case). Moreover, when the load was decreased, the number of cycles to failure increased. There was a stress level at which the material did not fail after any number of revolutions – stress reversals. This was called 'continuous strength' or 'fatigue strength'. The situation is shown in Figure 7.5, in a stress–time (number of cycles) diagram. The fatigue curve (called also a 'Woeler curve') levels are usually between 5 and 10 million cycles. The asymptote to this curve would indicate the fatigue strength.

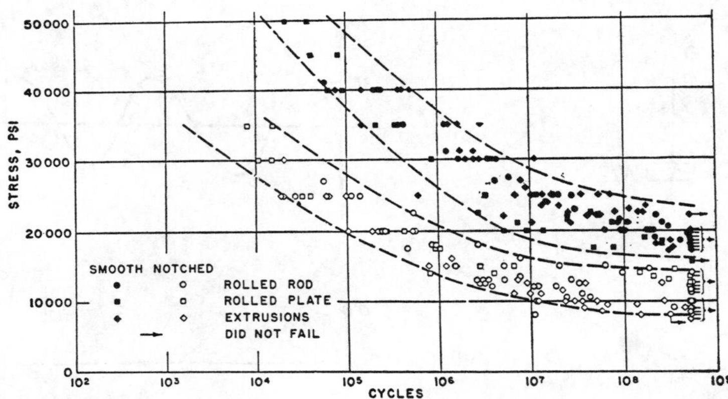

Figure 7.5 Fatigue curves (Woeler). (Courtesy ASME)

For steels and other materials the fatigue curve can be approximated with the empirical formula $S^m N = \text{constant}$, which is a straight line on semilog paper. The exponent m is about 9 for steels. Assuming that continuous fatigue strength S_n' is achieved in a number of cycles N_0,

$$S_n^m N = S_n'^m N_0$$

gives the fatigue strength S_n for any number of cycles $N < N_0$.

It is generally recognized that fatigue failure originates from a minute flaw discontinuity of the material after the loading results in local plastic deformation. Repeated deformation causes the flaw to gradually increase and form a gradually propagating crack, until the remaining section of the material can no longer sustain the load and the crack progresses rapidly causing final failure. This phenomenon is called 'high cycle fatigue', to distinguish it from the fatigue (called 'low cycle fatigue') caused by more general yielding which leads to failure after a much smaller number of cycles. This will be discussed in Section 7.5. In Figure 7.6 the effect of stress state on origin, appearance and location of fatigue fracture is shown for tension and bending (a) and torsion (b).

High cycle fatigue strength, or simply fatigue strength, is usually tabulated with the other strength values of the materials, as discussed in Chapter 6. It must be emphasized however, that the fatigue strength given in such tables usually has a rather high uncertainty because it is influenced by a great number of factors. From what was said for the procedure leading to fatigue failure, it is clear that fatigue strength depends on the size, orientation and density of the material flaws in relation to the element size and the mode of application of the load. For this reason, the tabulated values of the fatigue strength refer to very carefully performed experiments with a standard specimen. It has been found in practice that in actual machine elements the fatigue strength differs from the tabulated values, depending on the operating conditions and the geometry of the element. To obtain the design fatigue strength for the particular application, the theoretical fatigue strength is usually multiplied by a number of derating factors. Each factor reflects a different mechanism of deviation from the ideal laboratory test. Here we shall use the expression

$$S_e = C_F C_S C_R C_H C_L S_n' / K_f \tag{7.3}$$

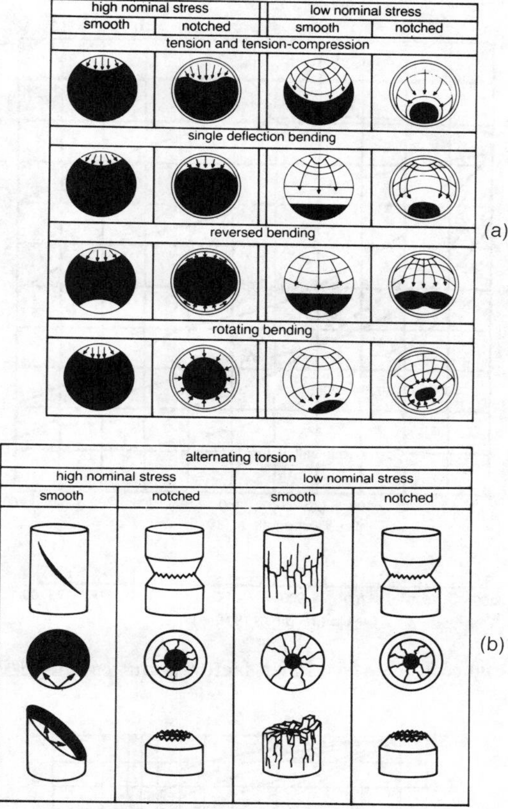

Figure 7.6 Effect of stress state on fatigue fracture origin, location and appearance. (From Engel, L. and Klingele, K., *An Atlas of Metal Damage*, Hansen Verlag, Munich, 1981)

C_F is used to express the fact that different surface conditions result in different fatigue strength. This is due to the fact that rough surfaces have micro-cracks which with repeated loading might propagate and result in fracture. For structural steels this factor is given in Figure 7.7, for different levels of surface finishing.

Fatigue tests are performed with standard diameters, usually 10 mm, and the variation of the diameter has an effect on fatigue strength. This is due to the fact (similar to the surface finish) that surface anomalies have smaller effect on the bulk strength of larger parts than in smaller parts. On the other hand, if the dimensions become smaller then the probability of existence of internal cracks and other irregularities in the material is minimized. This is accounted for with the factor C_S (Figure 7.8).

Published values of the fatigue strength are mean values of strength, as was explained in Chapter 6. Therefore, the reliability of the component will be 50%. Higher reliability can be obtained by a proper safety factor (Chapter 6). If however, this is not accounted for in the safety factor, the fatigue strength must be derated by a reliability factor, C_R, which is the inverse $1/N$, of the safety factor obtained in the upper horizontal scale of Figure 6.24. If

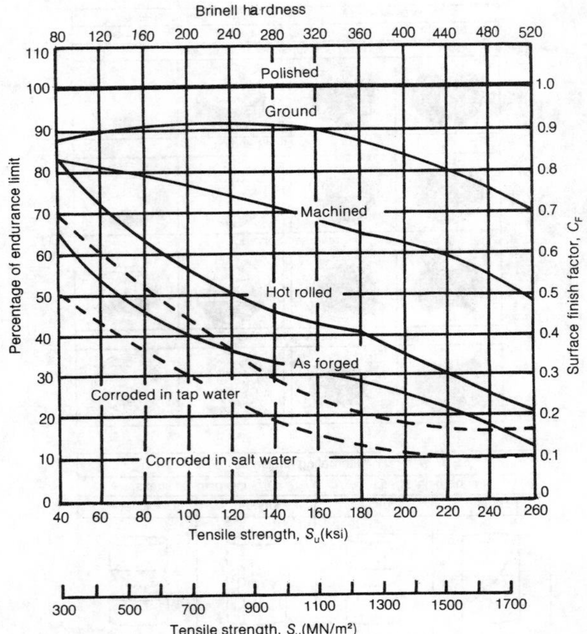

Figure 7.7 Surface finish factor. (Courtesy ASME)

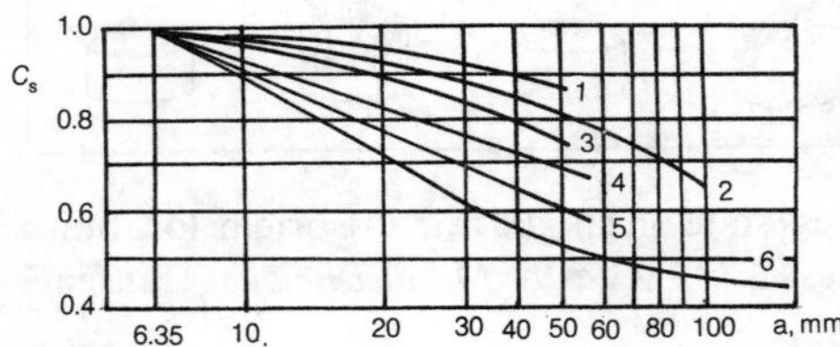

1–carbon steel, smoothly polished; 2–carbon steel, smoothly ground; 3–alloy steel, smoothly polished; 4–alloy steel, smoothly ground; carbon steel with stress concentration; 5–alloy steel with moderate stress concentration; 6–structural steel ($S_u < 65\,\text{kgf/mm}^2$ or $S_u < 650\,\text{MN/m}^2$), shaft with press-fitted part made of un-pressworked steel; at $d < 60$ mm–alloy steel with intense stress concentration.

Figure 7.8 Size factor. (From Mirolyubov *et al.*, Mir Publishers, Moscow)

reliability is accounted for in the safety factor used in the design equations $\sigma_{max} = S_e/N$, then $C_R = 1$.

The fatigue strength is influenced by the nature and density of the material flaws. It is then apparent that the fatigue strength of elements under pure tension–compression loading, must be lower than that in reversed bending because for the latter high stress occurs only at and near the outer fibers, therefore the probability of existence of a critical flaw in the highly stressed region is smaller. It is therefore expected, and experimentally verified, that the fatigue strength in the tension–compression test is less than the values of fatigue strength which are obtained for cylindrical specimens in reversed (rotating) bending. For this case, the load

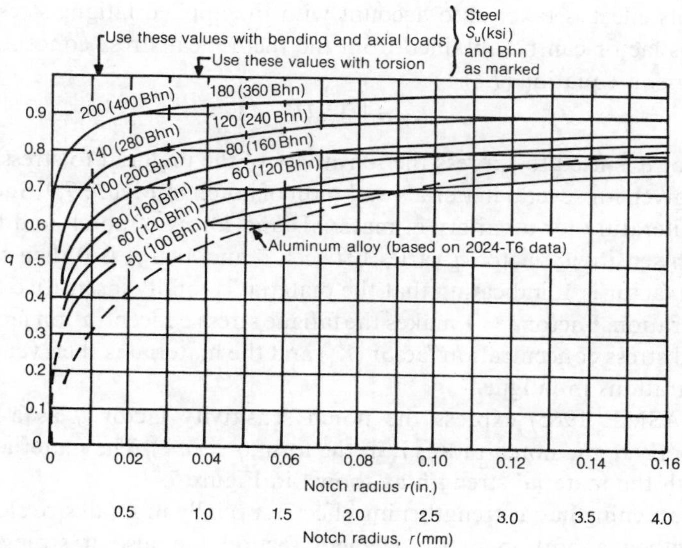

Figure 7.9 Notch sensitivity factor. (Courtesy ASME)

derating factor C_L is taken to be 0.9, while for reversed bending and torsion (which has similar stress distribution to bending) it is taken to be 1.

It was pointed out earlier that the stress concentration factors do not have the same influence on ductile and brittle materials. In fatigue, the situation is a little more complicated. This is because even ductile materials under repeated loadings and unloadings, and therefore repeated yielding, might also fail. Experiments show that the proportion of the stress concentration that must be applied for design analysis is more or less a property of the

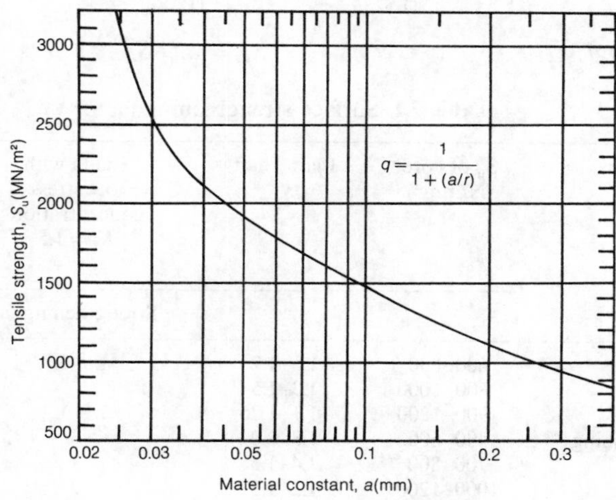

Figure 7.10 Notch sensitivity factor according to Neuber. (Courtesy ASME)

material and its effect is taken into account with the applied fatigue stress concentration factor K_f. This factor can be obtained from the theoretical stress concentration factor K_t using the Peterson equation (7.4).

$$K_f = 1 + q(K_t - 1) \tag{7.4}$$

Here, the factor q is used to express the sensitivity of the material to stress concentrations. This factor is given for several materials and radii of notch in Figure 7.9. Such figures can be found in the literature for a variety of materials and forms of notch and loading.

The notch sensitivity factor q varies between 0 and 1. If it is 0 then the fatigue stress concentration factor is 1, indicating that the material is totally insensitive to fatigue due to stress concentration. Factor $q = 1$ makes the fatigue stress concentration factor (K_f) equal to the theoretical stress concentration factor (K_t) and the material is then very sensitive to the stress concentrations in fatigue.

Neuber (ASME 1965) express the notch sensitivity factor q as a function of the material property a, the notch radius r, in the form $q = f(a/r)$. The material constant a was correlated with the material strength as shown in Figure 7.10.

Surface hardening has a strengthening effect, especially in small size elements. Not only does it strengthen a portion of the element section but also it strengthens the stress concentration areas near the surface where a fatigue crack is most likely to originate. Therefore, an overrating factor C_H must be used, greater than 1. For the most common surface-hardening processes, C_H is given in Table 7.1.

In the material files presented in Chapter 6 (p. 233), fatigue strength is tabulated for materials from rotating beam tests. In absence of this information, sometimes good estimates can be obtained from the ultimate strength based on experimentally obtained factors *ff* (after Haenchen and Decker 1967) multiplying the ultimate strength S_u.

Loading		$ff = S'_n/S_u$	
	$S_u < 500$	$S_u = 500–700$	$S_u > 700\ \text{N/mm}^2$
Bending	0.6	0.5	0.4
Axial	0.5	0.48	0.45
Shear (torsion)		$S'_{ns} = 0.58S'_n$	

Table 7.1 Surface strengthening factor

Surface hardening procedure	S_u of core (N/mm²)	Plain shafts	Shafts with low stress concentration $K_f = 1.5$	Shafts with high stress concentration $K_f = 1.8$ to 2
		Strengthening factor C_H		
Induction surface hardening	600–800	1.5–1.7	1.6–1.7	2.4–2.8
	800–1000	1.3–1.5	–	–
Nitriding	900–1200	1.1–1.25	1.5–1.7	1.7–2.1
Carburizing and quenching	400–600	1.8–2.0	3	–
	700–800	1.4–1.5	–	–
	1000–1200	1.2–1.3	2	–
Work hardening by shot-peening	600–1500	1.1–1.25	1.5–1.6	1.7–2.1
Roll burnishing	–	1.2–1.3	1.5–1.6	1.8–2.0

7.3 COMBINED LOADS

In Chapter 6 we have already seen that for static loading the results of the tensile tests and the associated material properties can be extended to the more general loading of a material. An equivalent stress is defined as a function of the stresses acting on the material, and this equivalent stress is compared with the material strength. The related strength theories refer to the multiaxial static load on the material. With dynamic loading, the designer has to assess the strength of a component which is loaded by different mechanisms, both static and dynamic, and he has then to come up with some equivalent load which will be subsequently compared with the material strength.

Assume uniaxial loading of a component with a combination of a static stress and an alternating stress as shown in Figure 7.11. The strength has to be evaluated based on the static and dynamic properties of the given material. To this end we note that for two combinations of static/dynamic load, the solution is already known. For zero dynamic load the stress should not exceed the yield strength while for zero static load the stress cannot exceed the fatigue strength of the material. If we now make a failure diagram on the (σ_m, σ_r) plane, Figure 7.12, there are two known points, the point S_e on the σ_r axis and the point S_y on the σ_m axis. The Soderberg Hypothesis is that if these two points are connected with a straight line, this line separates the plane into a safe and an unsafe region. Below the Soderberg Line the combination of static and dynamic load is assumed to be safe. Above the

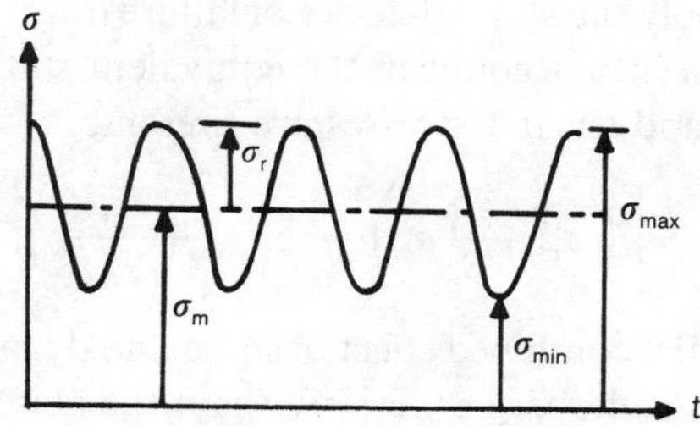

Figure 7.11 Static and alternating stress

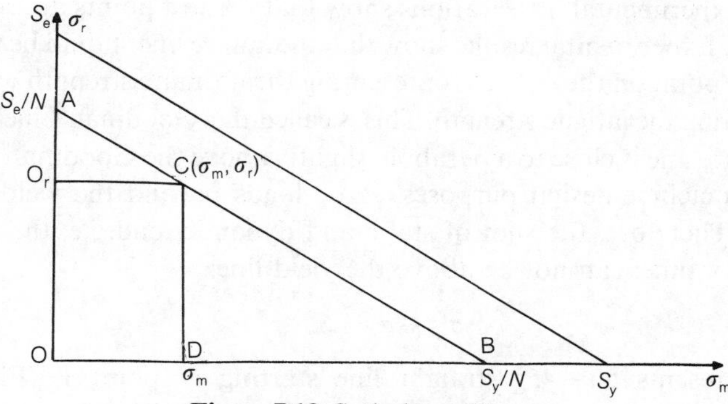

Figure 7.12 Soderberg Diagram

line the material will fail. Furthermore, if also a safety factor N is taken into account, the Soderberg Line moves towards the origin by a factor of N giving the line AB (Figure 7.12). In this diagram every loading situation is represented by a point on the plane. In the limiting state the point C is on the line AB. For this case, from the similar triangles AOB and CDB we obtain

$$\frac{(S_y/N) - \sigma_m}{\sigma_r} = \frac{S_y}{S_e} \tag{7.5}$$

$$\frac{S_y}{N} = \sigma_m + \sigma_r \frac{S_y}{S_e} \tag{7.6}$$

It is apparent from equation (7.5) that we can replace the combination of static and dynamic loads by an equivalent load given by equation (7.6). The result can then be compared with the yield strength of the material:

$$\sigma_{eq} = \frac{S_y}{N} = \sigma_m + \sigma_r \frac{S_y}{S_e} \tag{7.6a}$$

The effective safety factor is

$$N = \frac{S_y}{\sigma_{eq}} = \frac{1}{(\sigma_m/S_y) + (\sigma_r/S_e)} \tag{7.6b}$$

In the above, only uniaxial static and dynamic loading is considered. In the case of more general loading one may apply the static theories of failure (see Chapter 6) since the effect of dynamic loads is now taken into account in the equivalent static loads. We can write, for example, if only normal σ and shear τ stresses are present,

$$\sigma_{eq} = \left[\left(\sigma_m + \frac{S_y}{S_e}\sigma_r\right)^2 + a\left(\tau_m + \frac{S_{sy}}{S_{se}}\tau_r\right)^2\right]^{\frac{1}{2}} \tag{7.7}$$

Equation 7.7 expresses the combined effect of static and dynamic loads acting as normal and shear stresses, with the Soderberg criterion. Again the factor a is 4 or 3 depending on application of the equivalent shear stress or the equivalent strain energy criteria.

The Soderberg Diagram is very convenient for design, because of the facility of expressing the dynamic loads as equivalent static loads. It is a rather conservative criterion, however, because experimental observations show that the test points are somewhat above the Soderberg Line. Experimental results show that the failure line should be taken as the one which connects the point on the σ_m axis representing the ultimate strength with the point on the σ_r axis representing the fatigue strength. This is called the 'Goodman Line'. The tests show that the actual failure line is close to a parabola slightly above the Goodman Line. It must be observed that for machine-design purposes, static loads beyond the yield point are very rarely acceptable. Therefore, the sum of static and dynamic load, i.e. the maximum load which occurs at any time, can not be above the yield line:

$$\sigma_m + \sigma_r = S_y$$

This equation represents a $-45°$ straight line starting at point E (Figure 7.13). The intersection with the Goodman Line completes the Goodman Diagram: ADE is the failure

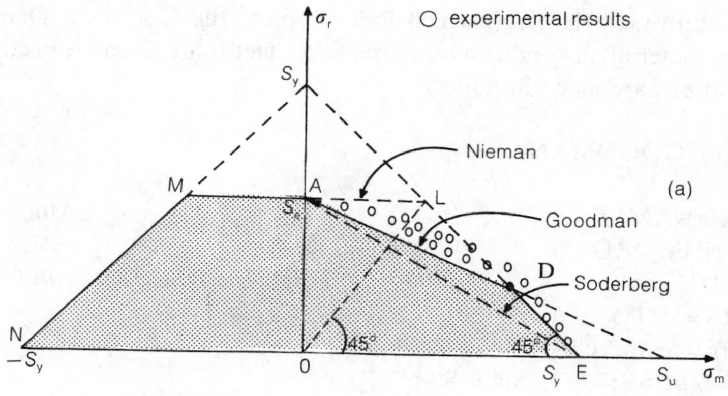

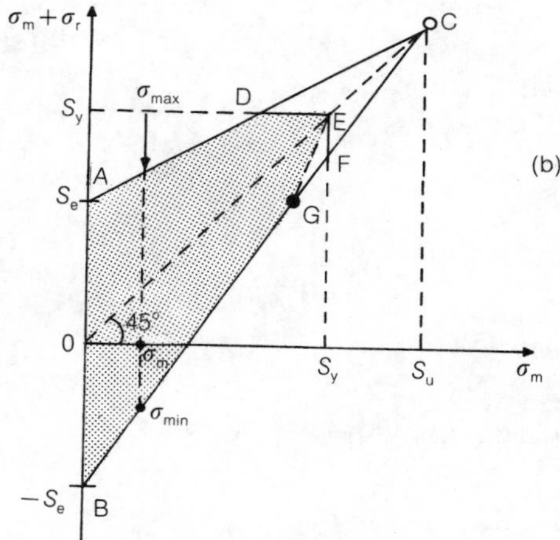

Figure 7.13 Goodman Diagrams

line (Figure 7.13a). This situation may be also expressed in a diagram of maximum total stress versus static stress. Then the failure line is again ADE (Figure 7.13b).

On the left of the Diagram for negative static stresses (compression), the situation is a little different. Experiments show that most materials in static compression are rather insensitive to additional static loads. Therefore, the Goodman Line continues from point A to point M and then follows to the point N, the safe static yielding line. This concludes the Diagram.

There have been many attempts to define more realistic lines ADE, which better express the experimental data. Since the points A and E are known, the in-between corner point has been the subject of different definitions. One such definition, very usual in West Germany, is the pulsating strength of the material, which consists of subjecting the material to equal static stress and stress range. Obviously this point on the Goodman Diagram must be on the 45° dichotomous of the AOE angle, and, according to Nieman, on the yield line.

The procedure GOODMAN, which follows, plots the Goodman Diagram for any material on the material files and computes the safety factor for given mean stress and stress range (explanations are given in italics).

```
procedure GOODMAN
begin
  input FNAME$                                  (*Material file name*)
  input RECORD                                  (*Material number*)
  RRNDF                          (*Retrieve RECORD from file FNAME$*)
  FOR i:= 1 TO 10 do
    S(I):= real of C$(I)
  SU:= S(6); SY:= S(7); SN:= S(8)
                                                (*Plotting the axes*)
  LINE (0,0 TO 0,100)
  LINE (0,0 TO 100,0)
                                                (*Finding point D*)
  X:= SU*(SN-SY)/(SN-SU)
  Y:= SN*(SY-SU)/(SN-SU)
  LINE (0, SN TO SY, 0)                         (*Soderberg Line*)
  LINE (0, SN TO X, Y)
  LINE (X, Y TO SU, 0)                          (*Goodman Line*)
  input SM, SR
  RATIO:= SR/SM
  RATIOD:= Y/X
  if RATIO > RATIOD then do
    NFACTOR:= SY/(SM + SR*SU/SN)
  else
    NFACTOR:= SY/(SM + SR)
  PRINT NFACTOR Goodman
  NFACTOR:= SY/(SM + SR*SU/SN)
  PRINT NFACTOR Soderberg
end.
```

This procedure has been included in the program SHAFTDES, for the design of rotating shafts, Chapter 14, Appendix 14.D.

This equivalent static stress and the effective safety factor are, from Figure 7.13(a), with similar reasoning as in the Soderberg Diagram:

(i) Line ED

$$e_{eq} = \sigma_m + \sigma_r < \frac{S_y}{N}$$

$$N = \frac{S_y}{\sigma_m + \sigma_r}$$

(ii) Line DA

$$\sigma_{eq} = \sigma_m + \frac{S_u}{S_e}\sigma_r$$

$$N = \frac{S_y}{\sigma_m + \dfrac{S_u}{S_e}\sigma_r}$$

The foregoing discussion was based on the fatigue strength for a number of cycles greater than 10 million (continuous strength). Similar diagrams can be made for strengths at lower number of cycles and the corresponding fatigue strength at the particular number of cycles must be obtained from the Woeler curve, if available.

7.4 CUMULATIVE FATIGUE DAMAGE

In Section 7.3, it was assumed that the loads are constant during the life of the machine member. This might not be the case in an actual machine operation.

The machine by its nature might not operate all the time at the same load. On other occasions it is possible that the machine operates accidentally for a period at loads higher than the design load. In this case we need to know if we can have safe operation of the machine or component for the same number of operating cycles by way of giving an appropriate reduction to the machine load.

Palmgren, in the 1920s, made the assumption, which was re-stated later by Miner (ASME 1965), that operation at a certain level of dynamic loads for certain number of cycles, has a cumulative effect of damage which, when summed up to a certain level, leads to the eventual failure of the material.

If we assume that the material is subjected to different loads, $\sigma_1, \sigma_2, \ldots,$ at $n_{r1}, n_{r2}, \ldots,$ cycles respectively, for each loading the ratio of the number of cycles over the number of cycles for failure $n_1, n_2, \ldots,$ at that load (Figure 7.14), is a quantitative measure of fatigue damage occurred on the material. When the sum of all these damages reaches a certain limit, the material will fail. This is expressed as:

$$\frac{n_{r1}}{n_1} + \frac{n_{r2}}{n_2} +, \ldots, + \frac{n_{rn}}{n_n} \geqslant 1 \tag{7.8}$$

In equation (7.8) this limit is one.

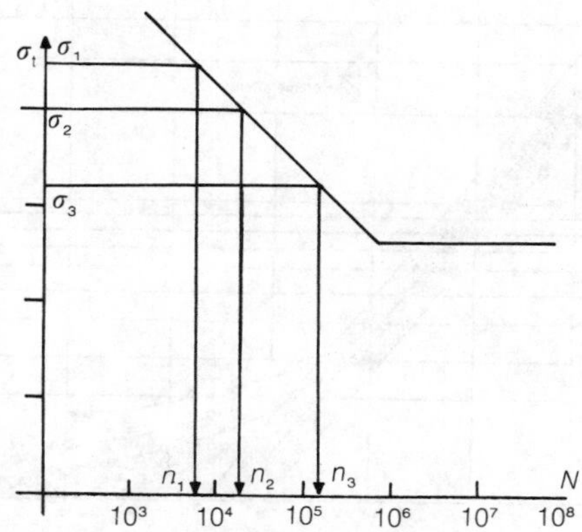

Figure 7.14 Miner's Rule

Subsequent experiments showed that this value differs from one for many materials and for many loading situations. There is a vast number of experimental data for this sum. For softer steels it is above one and can reach the value of three. For high-strength steels it might fall slightly below one. It is a good practice to use the value of one, which results in more conservative design for lower-strength steels.

It must be pointed out that the above results are based on very strict assumptions, such as sinusoidal loading, constant load over a period of time, etc. Machine members usually sustain loads which are of the type shown in Figure 7.4(d). This suggests that the above results are only approximate and further experiments and assessment of the field-service experience are necessary for every relevant machine-design effort.

7.5 LOW CYCLE FATIGUE

In many service conditions the material will only sustain a few thousand cycles of severe loading for which plastic strains are developed. Such situations exist many times in thermal machines where, during start-up, the thermal stresses developed lead to plastic flow. In other cases such yielding comes from impact loads during start-up and in general where yielding is not present during normal operation of the machine, so that for the life of the machine the total number of occasions where yielding occurs is relatively low. Several investigators have shown that for a finite life of less than about 100,000 cycles it is apparently the cyclic strain which is important for failure rather than the cyclic stress. For such situations it is usually impossible or irrelevant to measure or compute the stress under practical conditions of operation due to the plastic behavior of the material. For a number of metals such as aluminum alloys, magnesium, low-carbon steels and alloy steels, a fairly narrow scatterband was obtained when the test data were plotted in terms of maximum fiber strain versus endurance life in cycles, as shown in Figure 7.15. As a further refinement it has been observed

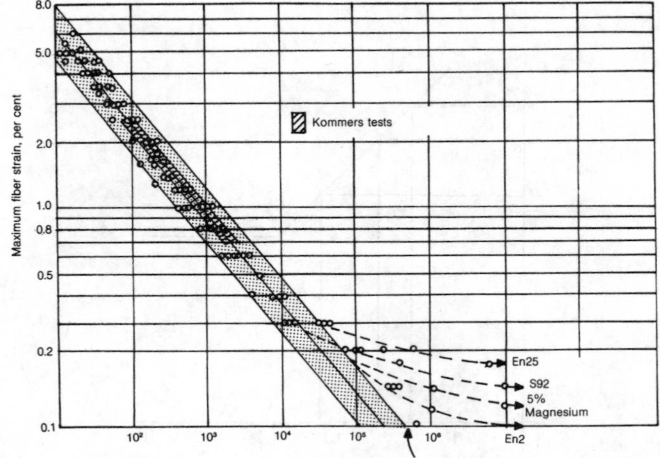

Figure 7.15 Low cycle fatigue tests. (Courtesy ASME)

that if the elastic strain is subtracted from the total strain measured in tests, a good straight-line relationship exists between the logarithm of plastic strain and the logarithm of cycles to failure. It has been shown that the experimental data are expressed with a relationship

$$N^{\frac{1}{2}} \Delta\varepsilon_p = C \tag{7.9}$$

where $\Delta\varepsilon_p$ is taken as the cyclic plastic strain, N is the fatigue life in cycles and C is a constant of the material. The value of ε_p is determined by subtracting the elastic strain from the total observed strain. On log paper, strain lines are obtained such as in Figure 7.15.

It was found that the factor $C \cong 1/2$, the natural strain at fracture, which is defined as the logarithm of the ratio of the initial and final cross-sectional areas at fracture

$$C = \ln(A_0/A) \tag{7.10}$$

For most steels $C = 0.3$ to 0.8.

7.6 BRITTLE FRACTURE

For many years, machine parts and engineering structures subjected to steady loads at moderate and low temperatures have been designed primarily on the basis of the yield strength. This method of design does not always produce satisfactory results as large structures frequently fail at stresses below the yield strength, with disastrous consequences.

For example, the tanker SS 'Schenectady' failed while in dock, when the nominal stress in the deck was only 60 kN/mm^2. There was also the failure of a liquid natural gas storage tank in Cleveland, Ohio, on October 20, 1944. Brittle fracture of one large tank allowed the escape of gas into a heavily populated industrial area. Over 130 people lost their lives. More recently, a number of large steam turbine and generator rotors have burst.

The aircraft and missiles industry is frequently facing similar problems. During the initial operation of the B-29 airplane, the landing-gear assemblies failed with monotonous regularity, sometimes when the plane was sitting on the runway. These landing gears were made on high-strength (1400 to 1550 MPa) tensile strength steel forgings welded and sometimes plated.

Further failures occurred on welded pressure vessels used in the missiles program. These vessels must generally operate at very high stresses and sometimes at very low temperature. Elevated-temperature 'brittle fracture' may be basically the same phenomenon as that observed at lower temperatures, for it has many features in common with low-temperature 'brittle fracture'.

'Brittle fracture' is the term that is commonly applied to the sudden failures described above. The usage of the word 'brittle' arises largely from the absence of any noticeable deformation, preceding or accompanying the failure.

Although the term 'brittle fracture' is most commonly defined as a failure that was not preceded or accompanied by any noticeable plastic flow, actually it is not this brittleness which is the primary cause for concern.

The terms 'brittle' and 'brittle crack propagation' both apply to failure under 'static' load, as opposed to 'fatigue' loading. For a very large number of completely reversed fatigue loading cycles (zero mean stress), the failure is always brittle; that is, there is no visible plastic deformation preceding crack initiation.

The following is a list of some of the more common criteria of brittleness (see ASME 1965):

1. Failure without any plastic flow preceding the initiation of a crack in an unnotched element.
2. Failure without any noticeable plastic flow accompanying the propagation of the crack. Such failure could also be called 'failure with low crack propagation resistance'.
3. Failures characterized by a 'granular' or 'crystalline' appearance, as contrasted with a 'fibrous' or 'silky' texture.
4. Failure characterized by flat fracture surfaces normal to the direction of stress, as contrasted with fracture surfaces that are inclined at 45° to the direction of stress.
5. Sudden failures, as contrasted with gradual tearing.
6. Failures in which the total energy absorbed is small compared to the energy that would be absorbed at a higher temperature.
7. Failures that occur before the load–deflection curve has deviated appreciably from a straight line.
8. Failures characterized by cleavage of the grains, as ascertained by a microsection through the fracture surface.
9. Failure at a load significantly less than that required for plastic flow to spread across the entire cross-section. For example, it is common practise in the case of annular grooves in cylindrical tension test bars to describe the behavior as 'notch brittle' if the notch strength is less than the unnotched strength and 'notch ductile' if the notch strength is greater than the unnotched strength.

All of the currently used or proposed approaches to engineering design are based on the principle that if one of the necessary conditions for fracturing is eliminated, or sufficiently reduced, failure can be avoided.

Some approaches to the brittle-fracture design problem are:

7.6.1 The transition temperature approach

This approach is based on the concept that each material possesses a characteristic transition temperature, below which brittle fracture will not occur. Sufficient experimental data on the material at hand are necessary for this approach (which is purely empirical).

7.6.2 Fracture mechanics approach

According to Griffith (ASME 1965), rapid crack propagation will commence whenever a small extension of the crack releases as much elastically-stored energy as the energy used to form the additional crack surface.

The energy-release rate for the specimen is a function of nominal stress. It must be determined either by calculation or by measurement. For example, for bending of a notched

Table 7.2 Critical energy release rate for some materials (from Sih 1972)

Material	G_c(lb/in)
Dural	1.60×10^3
Key steel	5.71×10^2
Brass	3.43×10^2
Teak wood	68.5
Cast iron	45.7
Cellulose	22.8
Polystyrene	11.4
Polymethyl methacrylate	5.71
Epoxide resin	3.77
Polyester resin	2.51
Graphite	0.571–1.14
Alumina	0.457
Magnesia	0.114
Glass	4.57×10^{-2}

beam, the energy-release rate G is

$$G = \frac{2h(1 - \nu^2)}{E} \mathrm{f}\left(\frac{a}{d}\right) < G_c \tag{7.11}$$

where, h is the net beam depth, (a/d) is the ratio of notch depth to gross beam depth, and 'f' is the function of (a/d).

The critical energy release rate is a material property and it is shown for some materials exhibiting brittle behavior in Table 7.2. For the particular geometry and loading the energy release rate can be computed for most design problems.

We note that equation (7.1), for the stress concentration near an elliptical hole, gives infinite stress when the minor semi-axis tends to zero, as we can model a crack. In general, Irwin (ASME 1965) has shown that near the tip of a crack the stress field has the form

$$\sigma = K\mathrm{f}(\theta)[2\pi r]^{-\frac{1}{2}} \tag{7.12}$$

where r is the distance of the point from the crack tip, $\mathrm{f}(\theta)$ is a function of the angle with the crack plane. K is a factor which depends on loading and geometry. For example, for a strip in uniform tension, perpendicular to the crack plane,

$$K = \sigma_0[\pi a]^{\frac{1}{2}}\mathrm{F}(a) \tag{7.13}$$

where $2a$ is the crack length and $\mathrm{F}(a)$ is a function which depends on the geometry. For an infinite strip, that is if the strip boundary is very far from the crack, $\mathrm{F}(a) = 1$.

The utility of the factor K, called the 'stress intensity factor' (SIF), follows the observation that for every material when this factor reaches a critical value, brittle fracture will follow. The critical value of the SIF is then a material property called 'fracture toughness' (Table 7.3) which is known to be related to the critical energy release rate as

$$S_c = (1 + \nu)(1 - 2\nu)K_c^2/2\pi E \tag{7.14}$$

Therefore, for a given material with given either critical strain energy release rate or

Table 7.3 Fracture toughness for some materials (from Sih 1972)

Material	Ultimate strength, S_u(ksi)	Critical stress-intensity factor, K_c ksi(in)$^{1/2}$
A517F Steel (AM)	120	170
AISI 4130 Steel (AM)	170	100
AISI 4340 Steel (VAR)	300	40
AISI 4340 Steel (VAR)	280	40
AISI 4340 Steel (VAR)	260	45
AISI 4340 Steel (VAR)	240	60
AISI 4340 Steel (VAR)	220	75
300M Steel (VAR)	300	40
300M Steel (VAR)	280	40
300M Steel (VAR)	260	45
300M Steel (VAR)	240	60
300M Steel (VAR)	220	75
D6AC Steel (VAR)	240	40–90
H-ll Steel (VAR)	320	30
H-ll Steel (VAR)	300	40
H-ll Steel (VAR)	280	45
12Ni–5Cr–3Mo Steel (VAR)	190	220
18Ni(300) Maraging Steel (VAR)	290	50
18Ni (250) Maraging Steel (VAR)	260	85
18Ni (200) Maraging Steel (VAR)	210	120
18Ni (180) Maraging Steel (VAR)	195	160
9Ni–4Co-0.3C Steel (VAR)	260	60
Al 2014-T651	70	23
Al 2024-T851	65	23
Al 2219-T851	66	33
Al 2618-T651	64	32
Al 7001–T75	90	25
Al 7075-T651	83	26

fracture toughness, an expression is needed giving the stress intensity factor or the energy release rate for the particular geometry and loading. The SIF has been tabulated in recent years in many publications for a variety of problems. Some useful results are given in Appendix B at the end of the book.

7.6.3 The congruency principle

Failure criteria that are based purely on nominal stress are highly empirical in nature. They stem from the idea that fracturing is a local phenomenon and therefore is not influenced by conditions remote from the location of the fracture origin. For example, let us compare the behavior of an unnotched beam specimen with that of a rotating disk with a hole in the center (Figure 7.16). Suppose the beam depth is such that when the stress at the tension surface of the beam equals the bore stress in the disk, the stress gradient at the tension surface of the beam also equals the stress gradient at the bore of the disk. Now two elements of volume, one at the bore of the disk and the other at the tension surface of the beam, encounter the same stress situation, and so should fail at the same stress and strain.

This same concept, that failure will occur in two different objects when the local conditions are the same, can also be applied to notches.

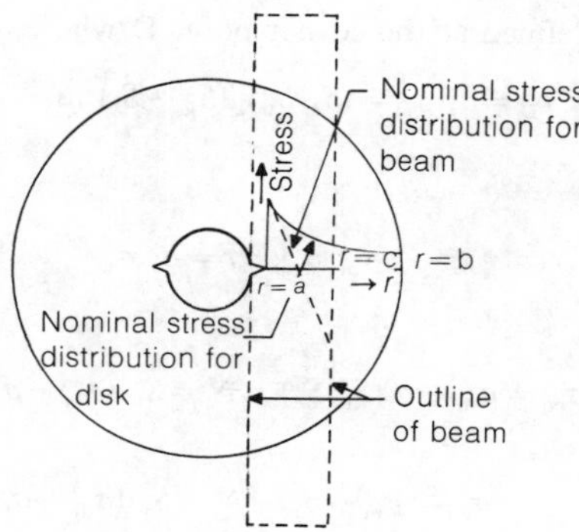

Figure 7.16 Congruency principle

REFERENCES AND FURTHER READING

ASME, *Metals Engineering-Design.* 1965. New York: McGraw-Hill.

Dimarogonas, A. D., *Vibration Engineering.* 1976. St. Paul: West.

Duggan, T. V., Byrne, J., 1977. *Fatigue as a Design Criterion.* London: Macmillan.

Haenchen, R., Decker, K.-H., 1967. *Neue Festigkeitsberechnung fuer den Maschinenbau.* Munich: K. Hansen Verlag.

Mirolyubov, I. N., *et al.*, 1974. *An Aid to Solving Problems in Strength of Materials.* Moscow: Mir Publishers.

Rooke, D. P., Cartwight, D. J., 1976. *Stress Intensity Factors.* London: Ministry of Defence.

Shigley, J. E., *Mechanical Engineering Design.* 1972. New York: McGraw-Hill.

Sih, G. C., *et al.*, 1972. *Applications of Fracture Mechanics to Engineering Problems.* Bethlehem, Pa: Lehigh University.

Tada, H., Paris, P., Irwin, G., 1973. *The Stress Analysis of Cracks Handbook*, Hellortown, Pa: Del Res. Corp.

EXAMPLES

Example 7.1

Determine the effective safety factor on the modified Goodman Diagram.

Solution

The safety factor based on the Soderberg Line is given by equation (7.6):

$$N = S_y/\sigma_{eq} = 1/(\sigma_m/S_y + \sigma_r/S_e)$$

The Goodman Line ADE consists of two segments

$$\text{DE:} \quad \sigma_m + \sigma_r = S_y \qquad \text{AD:} \quad \sigma_m/S_u + \sigma_r/S_e = 1$$

The coordinates of point D is the solution of the above system

$$\sigma_m = S_u(S_e - S_y)/(S_e - S_u)$$
$$\sigma_r = S_e(S_y - S_u)/(S_e - S_u)$$

The asymmetry factor is defined at the corner point D where

$$r_D = \sigma_r/\sigma_m = (S_e/S_u)[(S_u - S_y)/(S_y - S_e)]$$

Therefore, given

$$(\sigma_m, \sigma_r), \quad r = \sigma_r/\sigma_m$$

For $r > r_D$,

$$\sigma_{eq} = \sigma_m + \sigma_r(S_u/S_e), \quad N = S_u/(\sigma_m + \sigma_r S_u/S_e)$$

For $r < r_D$,

$$\sigma_{eq} = \sigma_m + \sigma_r, \quad N = S_y/(\sigma_m + \sigma_r)$$

Example 7.2

A circular shaft of diameter $d = 40$ mm is made of AISI 1020 carbon steel, normalized, with $S_u = 440$, $S_y = 340$, $S_n' = 226$ N/mm^2. It has a normally machined surface and it is integral with a wheel weighing 100 kg. There is also a constant thrust force of 2 ton. The shaft is connected to the wheel with a fillet of radius 5 mm. The shaft surface is not heat-treated. Find the effective safety factor for 99% reliability with the Soderberg and the Goodman Diagrams.

Solution

The effective fatigue strength

$$S_e = C_F C_S C_R C_H S_n'/K$$

where $C_F = 0.82$, from Figure 7.7; $C_L = 1$, bending is dominant; $C_S = 0.72$, from Figure 7.8, curve 4; $C_R = 1/1.2 = 0.83$, Figure 6.24 (p. 257) for $R = 99\%$; $C_H = 1$, no surface hardening; $K_t = 1.7$, $r/d = 5/40$, $D/d \gg 1$; $q = 0.7$ for $S_u = 440$ N/mm^2 = 62 ksi, $r = 5$ mm; and $K_t = 1 + q\ (K_t - 1) =$ 1.49.

Therefore,

$$S_e = 0.82 \times 0.72 \times 0.83 \times 1 \times 226/1.49 = 74.3 \text{ N/mm}^2$$

The axial force produces constant stress

$$\sigma_m = 20{,}000/(\pi \times 40^2/4) = 15.9 \text{ N/mm}^2$$

The lateral force produces alternating stresses

$$\sigma_r = M/W = (1000 \times \tfrac{500}{4})/(\pi \times 40^3/32) = 19.9 \text{ N/mm}^2$$

Soderberg formula, equation (7.6b) gives

$$N = 1/[(\sigma_m/S_y) + (\sigma_r/S_e)] = 1/[(15.9/340) + (19.9/74.3)] = 3.17$$

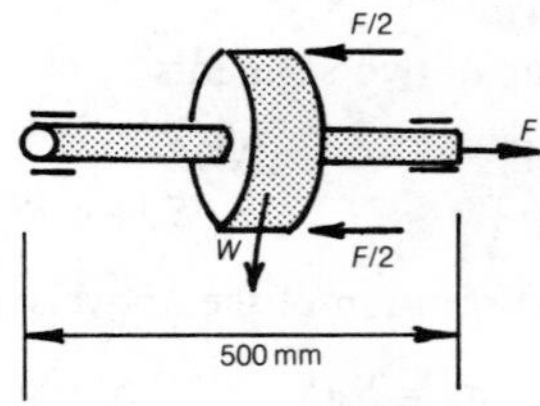

Figure E7.2

To apply the Goodman formula, as in Example 7.1, we compute the asymmetry factor

$$r = \sigma_r/\sigma_m = 19.9/15.9 = 1.25$$

The corner point

$$r_D = (S_e/S_u)[(S_u - S_y)/(S_y - S_e)] = 0.0635$$

and $r > r_D$, therefore

$$N = S_u/[\sigma_m + (\sigma_r S_u/S_e)] = 440/[15.9 + (19.9 \times 440/74.3)] = 3.28$$

As expected, the Soderberg formula yields a smaller factor of safety, thus more conservative design. One more iteration is needed, to take into account factors which depend on the results (K_t for example).

Example 7.3

Compute the shaft diameter in Example 7.2 if all data remain the same and the safety factor must be 2.5. Use the Soderberg criterion.

Solution

From Example 7.2, if d the unknown diameter $\neq 40$, $\sigma_m = 15.9(40/d)^2$, $\sigma_r = 19.9(40/d)^3$. Therefore, Soderberg equation (7.6b) gives

$$2.5 = 1/\{[15.9(40/d)^2/S_y] + [19.9(40/d)^2/S_e]\}$$

The equation can be solved by iteration. Solve for d^3:

$$d^3 = \{40^3 \times 19.9/[(1/2.5) - (15.9(40/d)^3/S_y)]\}/S_e$$

```
10 D = 50
20 DD = (1/2.5 - 15.9*(40/D)^2/340)/(40^3*19.9)*74.3      RUN
30 D = (1/DD)^.333                                        35.7847485
40 PRINT D                                                36.7526236
40 GOTO 20                                                36.6444461
                                                          36.6560516
                                                          36.6548009
                                                          36.6549356
```

The diameter is 36.65 mm. The next standard diameter must be selected.

Example 7.4

Measurements were conducted on a mechanism component to determine the service conditions. A purely alternating stress was measured at a frequency of 40 cpm. The distribution of stresses and the relative frequency for 60 operations recorded were

σ_r(N/mm^2)	140–150	130–140	120–130	110–120	100–110	90–100
N_r	3	7	9	12	15	14

The material had an effective fatigue strength of 90 N/mm^2 reached at 10^7 cycles, while the strength at 1000 cycles was 440 N/mm^2. Assuming a straight line of the Woeler curve between the two points on (log – time) scale, determine the fatigue life of the component.

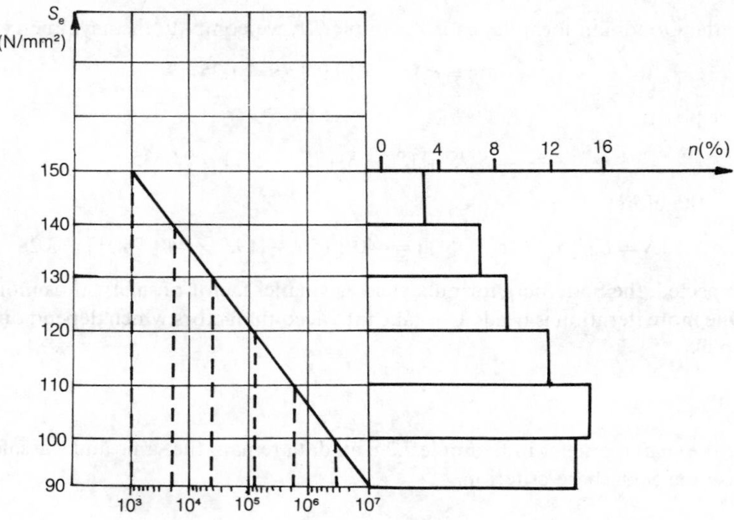

Figure E7.4

Solution

From Figure E7.4 where the effective fatigue life curve is plotted with the component stress distribution, the fatigue life at each stress level is obtained:

σ_r	140–150	130–140	120–130	110–120	100–110	90–100
n	10^3	7×10^3	3×10^4	1.2×10^5	8×10^5	4×10^6

If N is the fatigue life, the number of cycles at each load is $n_r = N(N_r/100)$. Miner's rule gives

$$N(N_{r1}/n_1 + N_{r2}/n_2 + \cdots) = 1 \times 100$$

$$N = 100/[(3/10^3) + (7/7 \times 10^3) + (9/3 \times 10^4) + (12/1.2 \times 10^5) + (15/8 \times 10^5) + (14/4 \times 10^6)]$$

$N = 22612$ cycles which corresponds to 9.42 hours of operation.

Example 7.5

For a typical test of mild steel for ductility, a strip of thickness $a = 10$ mm is bent successively about two cylinders of diameter $2R = 200$ mm. It is assumed that plastic deformation prevails and elastic deformation can be neglected. The material is AISI 1015 steel with area reduction at failure 60%. Determine the number of deformation cycles to failure.

Solution

The total strain at the outer fibers of the strip will be, at some arc length S,

$$\varepsilon = (S_2 - S_1)/S_1 = (R + a)/R - 1 = 0.1$$

It is assumed that elastic deformation is negligible, therefore $\Delta\varepsilon_p = 0.1$.

The low cycles fatigue constant C is

$$C = \ln(A_0/A) = \ln(1/0.6) = 0.51$$

Equation (7.9) then yields

$$N = (C/\Delta\varepsilon_p)^2 = (0.51/0.1)^2 = 26 \text{ cycles}$$

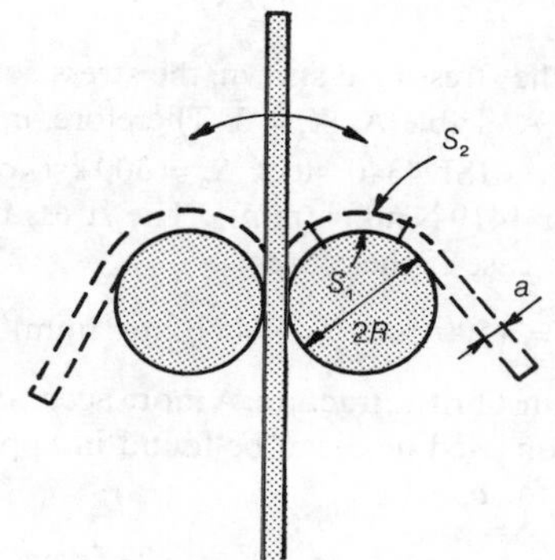

Figure E7.5

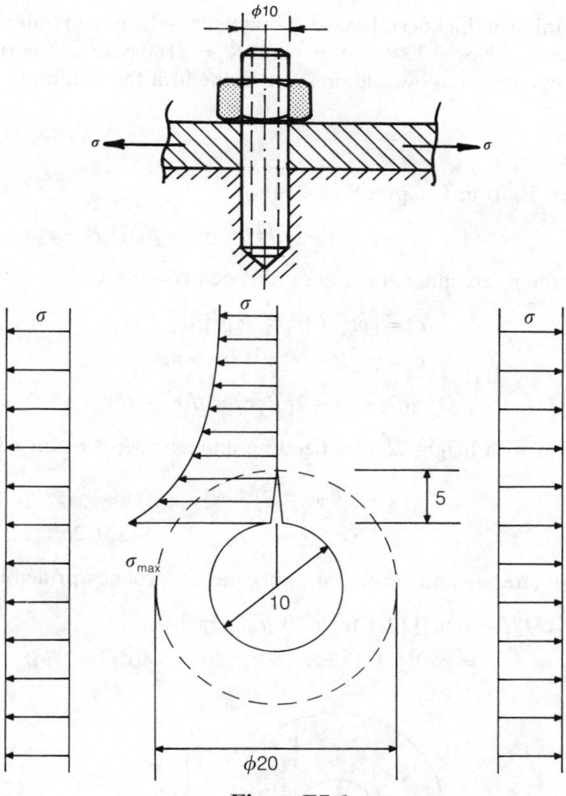

Figure E7.6

Example 7.6

A wide strip of AISI 4340 steel with $S_u = 300\,\text{ksi}$ is under uniform tension of $50\,\text{N/mm}^2$. It is secured on the machine frame by a 10-mm diameter bolt and a 20-mm diameter nut. If a crack develops at the hole, it cannot be detected until it propagates outside the nut external diameter, that is, it has a length of 5 mm. Assuming the crack to behave as an edge crack of a strip as shown with the same tension, accounting for stress concentration at the hole, find out if the strip is secured against brittle fracture when the crack has the length of 5 mm. Find also the critical crack length.

Solution

With the approximation for the stress field shown, the stress concentration factor for the 10-mm hole will be, from Appendix A, Table A, $K_t = 3$. Therefore, $\sigma_{max} = 3 \times 50 = 150\ \text{N/mm}^2$. From Table 7.3 it is found that for AISI 4340 steel, $S_u = 300$ ksi, or $2112\ \text{N/mm}^2$ and the fracture toughness $K_c = 40$ ksi $(\text{in})^{\frac{1}{2}}$ or $1419\ \text{N mm}^2\ (\text{mm})^{\frac{1}{2}}$. The stress intensity factor is $K = \sigma\ (\pi a)^{\frac{1}{2}}$ for $f(a/b) = 1$, $a \ll b$, Appendix B, case 2, therefore

$$K = 150(\pi \times 5)^{\frac{1}{2}} = 594\ \text{N/mm}^2\ (\text{mm})^{\frac{1}{2}} < K_c$$

Therefore, the strip is safe against brittle fracture. A more accurate value for $f(a/b)$ without the edge crack of a strip approximation used here can be found in appropriate handbooks. The critical crack length will be, solving for a,

$$a = (K_c/\sigma_{max})^2/\pi = 28.5\ \text{mm}$$

Example 7.7

A turbine wheel of uniform thickness, has inner diameter 160 mm and outer 400 mm. It is made of 12 Ni–5Cr–3Mo steel with $S_u = 1340\ \text{N/mm}^2$ and $K_c = 7180\ \text{N/mm}^2\ (\text{mm})^{\frac{1}{2}}$. It is press-fitted on the shaft. Find the maximum allowable pressure of the fit if the minimum observable crack has length 15 mm.

Solution

The tangential stress is, from Chapter 8 (p. 334)

$$\sigma_t = [(p_1 r_1^2 - p_2 r_2^2) + (r_1^2 r_2^2/r^2)(p_1 - p_2)]/(r_2^2 - r_1^2)$$

where r_1 and r_2, p_1 and p_2 are inner and outer radii and pressures respectively. Since $p_1 = p$, $p_2 = 0$

$$\sigma_t = [pr_1^2 + (r_1^2 r_2^2/r^2)p]/(r_2^2 - r_1^2)$$
$$\sigma_t = pr_1^2(1 + r_2^2/r^2)/(r_2^2 - r_1^2)$$
$$\partial\sigma_t/\partial r = (-2pR_1^2 r_2^2/r^3)/(r_2^2 - r_1^2)$$

The equivalent beam with height $2b$ and bending moment M, per unit width, would produce

$$\sigma_{max} = 3M/2b^2$$
$$\partial\sigma/\partial x = -\sigma_{max}/b = -3M/2b^3$$

Equating maximum stresses and slopes, to apply the congruency principle,

$$3M/2b^2 = pr_1^2[[1 + (r_2^2/r_1^2)]/(r_2^2 - r_1^2)]$$
$$= p80^2[(1 + (200^2/80^2)/(200^2 - 80^2)] = 1.38p$$

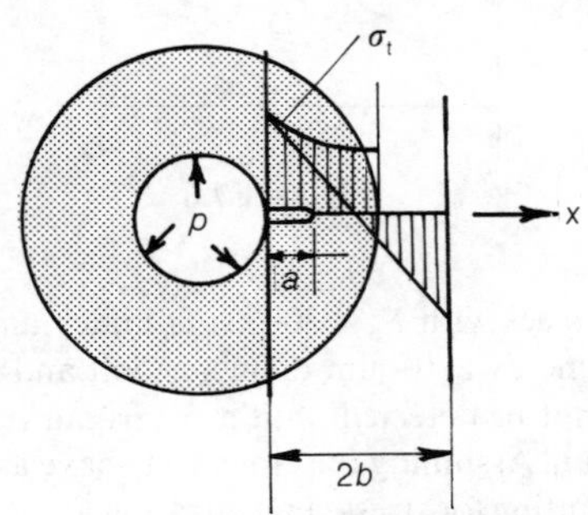

Figure E7.7

and

$$-3M/2b^3 = -2pr_2^2/(r_2^2 - r_1^2)r_1 = -2.97 \times 10^{-2}p$$

Dividing the last two equations,

$$b = 1.38/2.97 \times 10^{-2} = 46.5\,\text{mm}$$

Then, $a/b = 15/46.5 = 0.32$. Then from Appendix B, for an edge crack on a strip, $f(a/b) = 0.67$ and the stress intensity factor

$$K = f(a/b)M/(b-a)^{\frac{3}{2}} = K_c$$

Replacing the moment $M = 1.38p2b^2/3$,

$$K_c = f(a/b)5.52pb^2/(b-a)^{\frac{3}{2}}$$

Solving for p

$$p = K_c(b-a)^{\frac{3}{2}}/5.52\,f(a/b)b^2 = 272\,\text{N/mm}^2$$

PROBLEMS

Problems 7.1 to 7.10

For the elements shown, determine the maximum allowed alternating load, as indicated. No heat treatment, reliability 99%, safety factor $N = 1$.

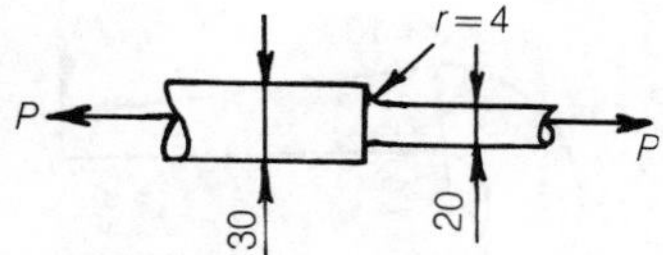

Figure P7.1 and P7.11

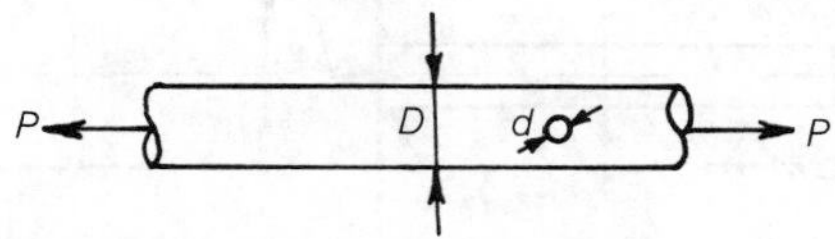

Figure P7.2 and P7.12

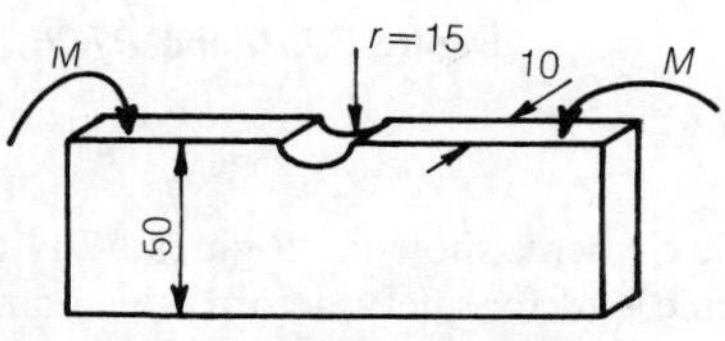

Figure P7.3 and P7.13

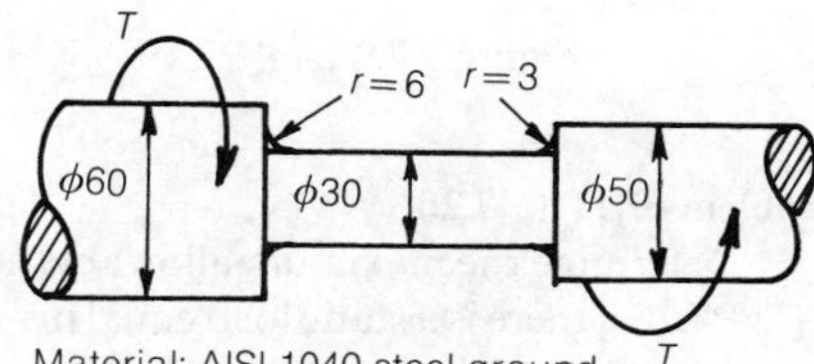

Figure P7.4 and P7.14

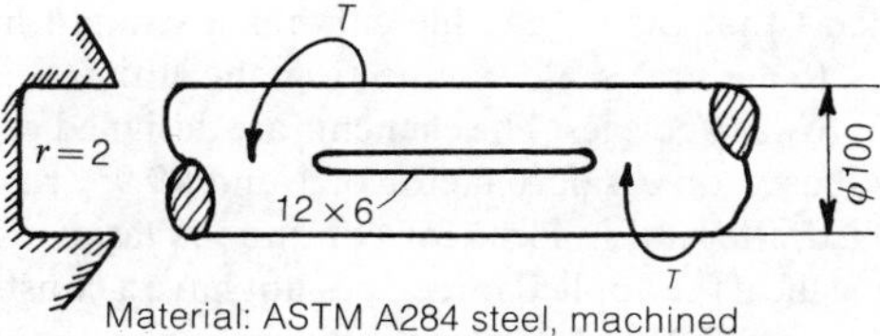

Figure P7.5 and P7.15

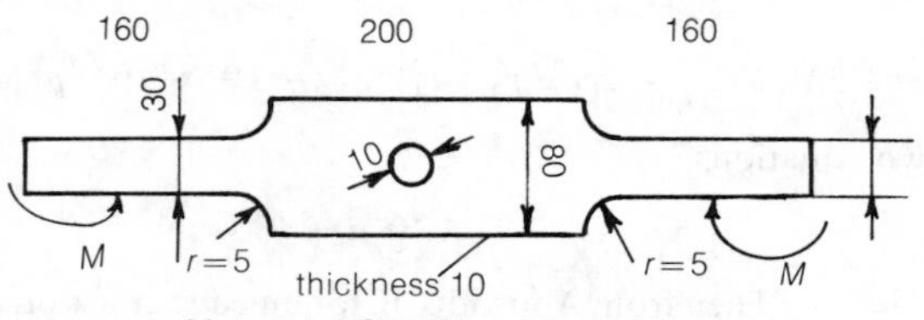

Figure P7.6 and P7.16

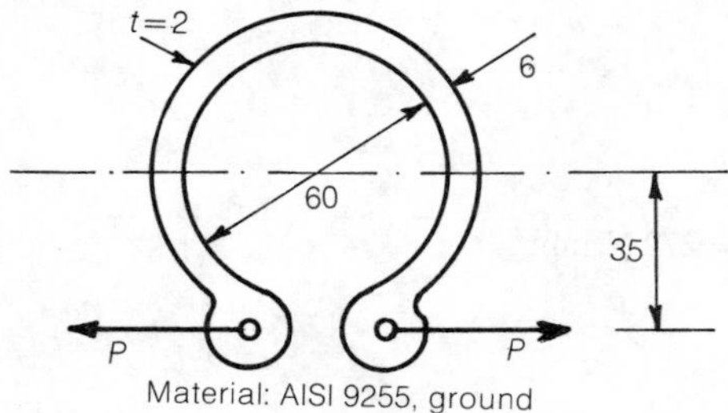

Figure P7.7 and P7.17

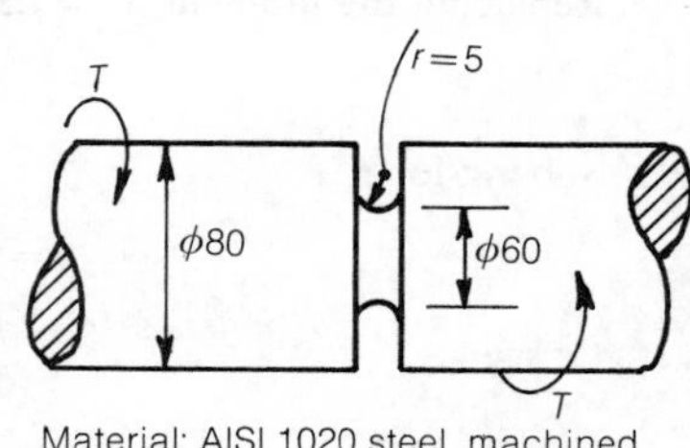

Figure P7.8 and P7.18

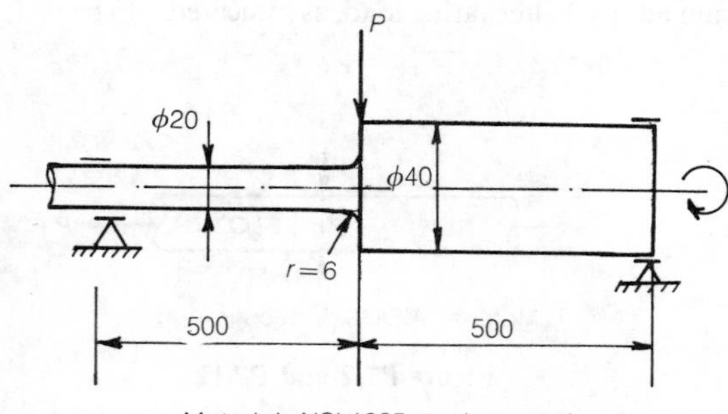

Figure P7.9 and P7.19

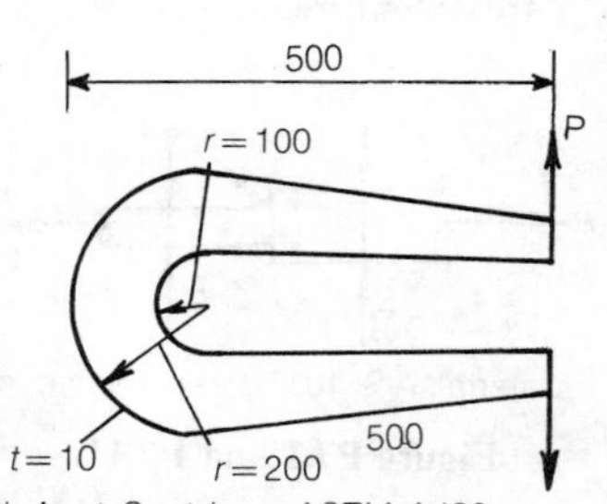

Figure P7.10 and P7.20

Problems 7.11 to 7.20

Determine the maximum allowable alternating loads for the elements shown in Figures P7.11 to P7.20 if there is a static load equal in magnitude to the dynamic load, for safety factor 1.8 including reliability.

Problems 7.21 to 7.24

Assume for these problems that the fatigue life curve is a straight line on semi-log paper with logarithmic time (cycles to failure) scale, connecting the ultimate strength at 1 cycle to the continuous fatigue strength at 10^7 cycles. The elements are designed with applied alternating force giving life of 10^6 cycles, based on a safety factor of 1 and 99.9% reliability.

(a) Determine the applied alternating force for continuous fatigue strength.

(b) Determine the fatigue life if the applied force does not have a constant amplitude but it has the form indicated in Figures P7.21 to P7.25, respectively, as percentages of the design alternating load for the above life of 10^6 cycles. Assume the Miner's Rule sum to be 1.

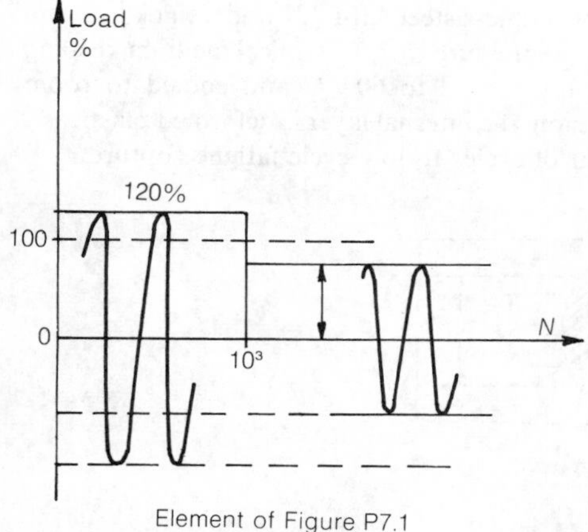

Figure P7.21

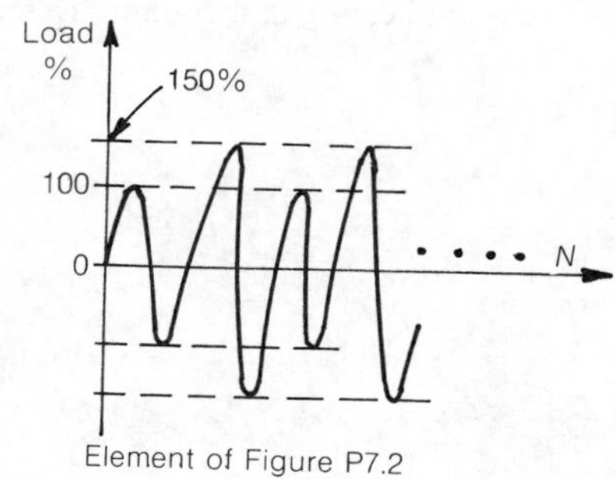

Figure P7.22

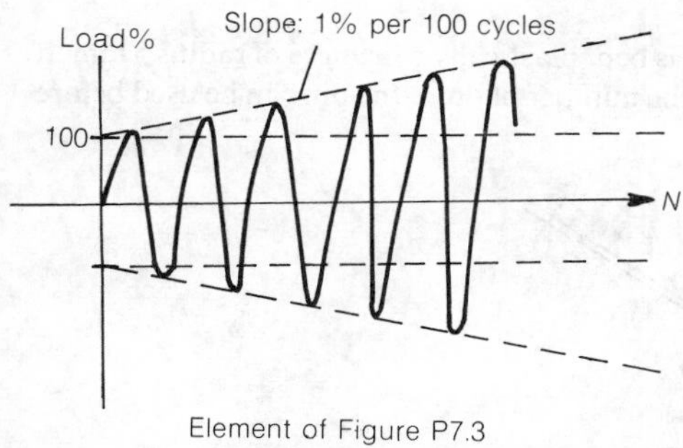

Figure P7.23

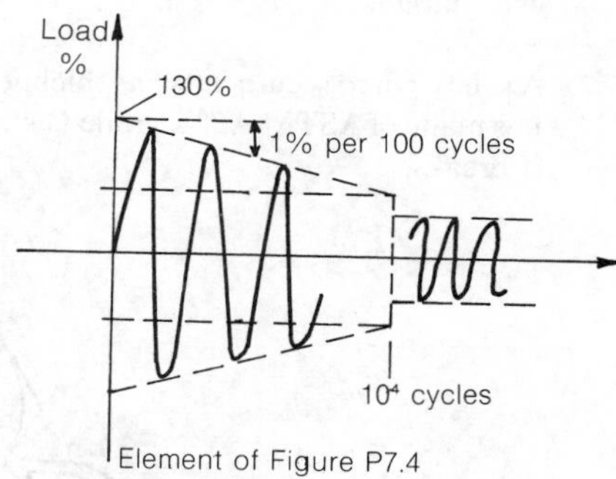

Figure P7.24

7.25 On the element shown in Figure P7.8, with a load designed for 10^6 cycles for $N = 1$, $R = 99.9\%$, it was found by measurements that the actual load had the design mean value but it had a standard distribution with deviation 5% of the mean, within two standard deviations. Under the same assumptions as in Problems 7.21–7.24, determine the number of cycles to failure.

7.26 An AISI 1025 steel pipe is carrying a hot liquid at 480 °C. Every time the hot liquid passes through the pipe, the pipe undergoes plastic deformation owing to thermal expansion, because the pipe is fixed at the ends to a structure of constant temperature and very high rigidity. Plastic deformation occurs also at the cooling of the pipe. Determine the number of cycles up to the low cycle fatigue rupture of the pipe. Room temperature is 30 °C.

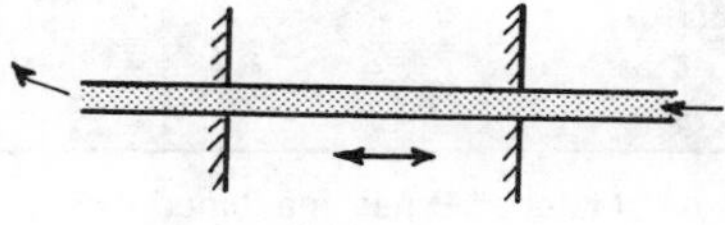

Figure P7.26

7.27 A compound pipe consists of a thin internal layer of stainless steel AISI 321 and a thick external layer of AISI 4130 steel. The pipe is under high internal pressure with a chemical medium keeping the two layers in contact. The medium is gradually heated to 600 °C and cooled to room temperature. Owing to differences in thermal expansion, the internal layer is deformed plastically during heating and cooling. Determine the number of cycles to low cycle fatigue rupture.

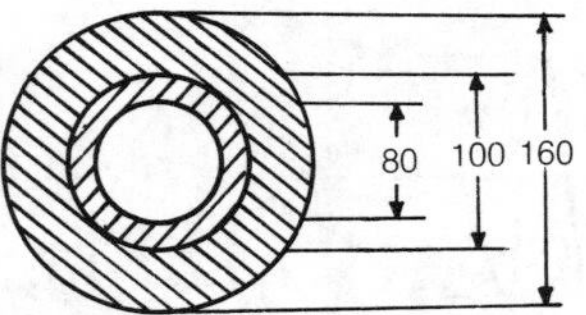

Figure P7.27

7.28 Measurements on a turbine rotor revealed that during start-up the surface of the rotor is heated faster than the core and a plastic flow of 0.008 was measured. The rotor is made of an AISI 5135 alloy steel and it has normally one start-up per week. Determine its expected life considering low cycle fatigue.

7.29 A safety pin to secure nuts has thickness 2 mm and it is bent plastically to a curve of radius 5 mm. If it is made of ASTM A284, grade D steel, determine the number of times the pin can be used before it breaks.

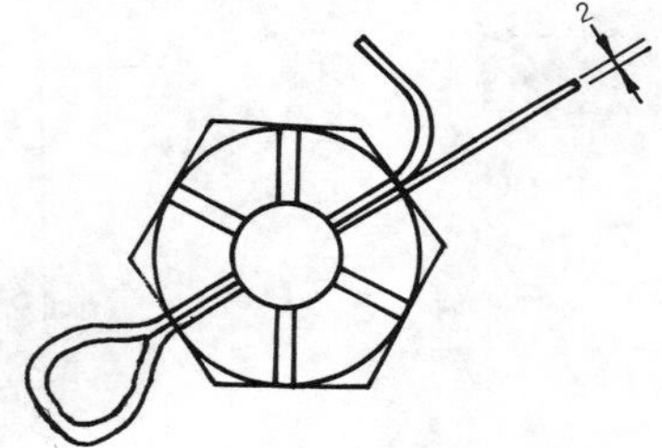

Figure P7.29

7.30 A large electric generator shaft undergoes plastic deformation approximately 0.007 every time it is synchronized out of phase with the electric grid. If the shaft is made of carbon steel AISI 1025, determine its expected life if one incident of that type is expected every month.

Problems 7.31 to 7.33

For the cracked elements shown, determine the critical crack length.

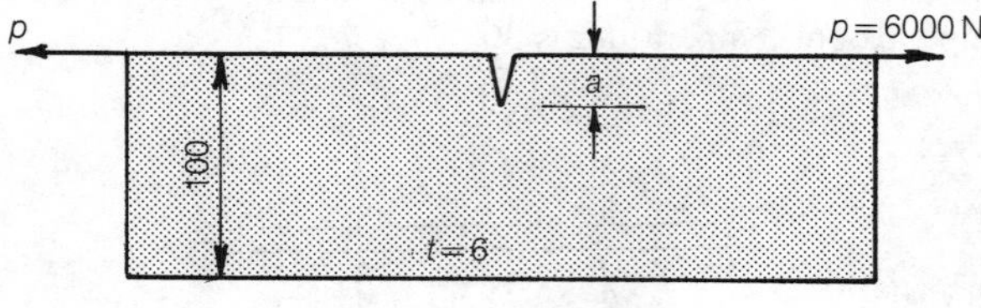

Figure P7.31

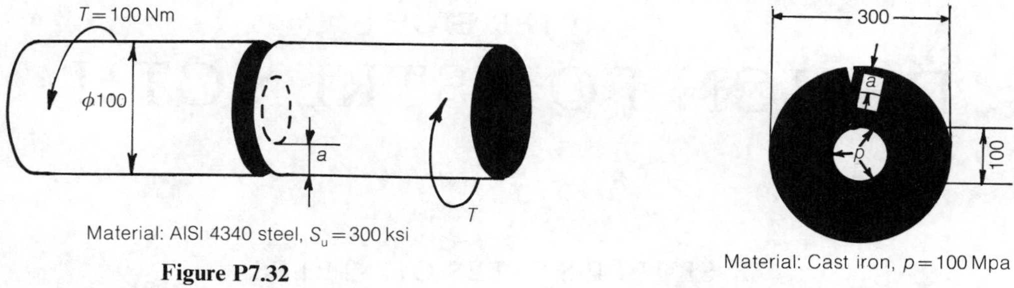

Figure P7.32

Figure P7.33

7.34 A cylindrical pressure vessel is made of A517F steel. It has a diameter of 2 m, thickness of 20 mm. In the longitudinal welds, the inspection method can detect cracks down to 2 mm length. Determine the maximum allowable pressure to avoid brittle fracture.

7.35 In Problem 7.31, the plate is heated so that the upper side is kept at constant temperature 500 °C while the lower is cooled to 20 °C. The fracture toughness of the material is $K_c = 100 + 1.2T$, where the temperature T varies linearly along the width of the plate. Determine the expected length where the crack will stop propagating once it started (crack arrest) when it will reach a value where $K = K_c$, because K_c varies as the crack propagates.

CHAPTER EIGHT
DESIGN FOR STRENGTH

8.1 SIMPLE STATES OF STRESS

Almost any realistic loading of a machine element results in a general distribution of stresses and strains within the element. Detailed analysis is in principle possible, especially in view of the computer analysis methods discussed in Chapter 4. However, engineers are always trying to isolate in a problem the most important features for the specific requirements and disregard factors which, to their best judgement, do not contribute to the design goal in proportion to the effort needed for taking these factors into account. It seems strange that, although sufficient computation facilities exist to analyze most conceivable loading situations, we still concentrate on the important parts of the problem disregarding minor factors. But, the following have to be considered:

(a) Many times, analytical or computational effort to include a certain factor in design might cost much more than the additional cost that will result from accepting a somewhat greater safety factor instead.
(b) Calculations of high complexity sometimes lead to a loss of the general understanding of the problem.
(c) Even in the case of a justified complete computation effort, simplified analyses are needed for preliminary design, to establish design solutions which subsequently might be further analyzed and optimized.
(d) Long design practise has accumulated invaluable experience. Such experience is difficult to be transferred and incorporated in a short time into the new methods.

Simplified and modern methods therefore have to co-exist for a long time to come.

Now, when can a certain factor be considered as unimportant? There is no simple answer to such a question. The designer is not simply a computer – experience and judgement cannot be easily replaced.

8.1.1 Tensile–compressive loading

This is the simplest, yet the most useful, in machine design loading situations because direct use of the results of the material tests can be made. In fact, most material properties are tabulated for such loading.

Therefore, the design equation for members which are loaded significantly only by tension–compression is

$$\sigma = F/A < S_y/N \tag{8.1}$$

where F is the axial force and A the section of the member.

This relation is mostly applicable to long prismatic elements, but approximations for other shapes can be effectively made. Usually, the force is given, a material is selected and equation (8.1) yields the required section. For certain brittle materials, stress concentration in local notches has to be considered as was discussed in Chapter 7.

Equation (8.1) is simple but requires care in its application. It presumes a uniform distribution of stresses over the cross-section which might not be true in certain cases, such as the following:

(a) Near the point of application of forces and the points of supports: For this reason, axial members such as bolts, rivets, etc., are designed with an increased cross-section near such points.
(b) Eccentric application of load or eccentric reaction of the support: In such cases, bending moments are introduced which result in bending stresses which might be much higher than pure tensile stresses.
(c) Misalignment of axially-loaded members: This again can introduce shear or bending loads and can be avoided with proper design and manufacturing to ensure alignment.
(d) Holes, notches, stress raisers in general: Such field disturbances are most dangerous in pure tensile loading because all the section of the element is under maximum stress.

In addition, limiting the stress below the allowable limits does not necessarily mean a good design. Other requirements must be met, such as rigidity and stability, resistance to fatigue and fracture, surface strength, resistance to wear, etc.

8.1.2 Shear loading

Pure shear is a state of stress in which the faces of an isolated element are acted upon only by shearing stresses. Such a state of stress is characterized by linear displacements of two parallel faces of the element relative to each other. Total shear is the magnitude of linear displacement s (Figure 8.1a).

The ratio of the total shear s between two adjacent faces to the distance l between these faces

$$\frac{s}{l} = \tan \gamma \approx \gamma \tag{8.2}$$

gives the unit shearing deformation or the angle of shear.

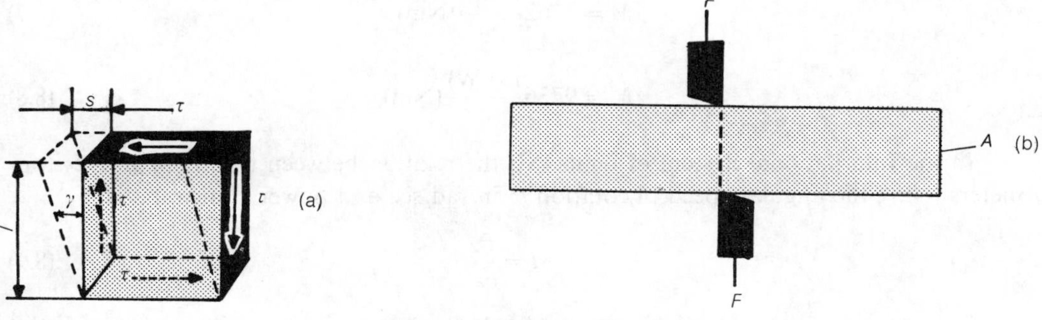

Figure 8.1 Shear loading

The angles of shear $\gamma_1, \gamma_2, \gamma_3$ (i.e. the angular shift in the right angles between the planes of action of the external shearing stresses τ_1, τ_2, τ_3, which are equal in magnitude but opposite in sign) are determined according to Hooke's Law:

$$\left.\begin{aligned}\gamma_1 = \varepsilon_2 - \varepsilon_3 = \frac{\tau_1}{G}\\ \gamma_2 = \varepsilon_1 - \varepsilon_3 = \frac{\tau_2}{G}\\ \gamma_3 = \varepsilon_1 - \varepsilon_2 = \frac{\tau_3}{G}\end{aligned}\right\} \tag{8.3}$$

where G is the shear modulus of elasticity.

If the shearing stresses are considered to be uniformly distributed over the area A of their action, then the shearing force is

$$F = A\tau \tag{8.4}$$

and the design equation for pure shear

$$\tau = F/A \leqslant S_{sy}/N \tag{8.5}$$

where S_{sy} is the yield strength in shear, which is related to the tensile yield strength by the theories of failure under combined loading (see Chapter 6) and

$$\begin{aligned}S_{sy} &= 0.5S_y \text{ (equivalent shear stress criterion)}\\ &= 0.577S_y \text{ (equivalent strain energy criterion)}\end{aligned} \tag{8.6}$$

The same reservations for application of the design equation (8.5) exist as for pure axial loading. In particular, non-aligned application of the punch (Figure 8.1b) results in non-uniform shear stress over the section A. This is more pronounced in brittle materials while in ductile materials it might be partly relieved by local yielding.

8.1.3 Torsion

Prismatic elements are many times mainly loaded by torque, such as in shafts, torsional springs, etc. Torque due to power transmitted is computed from the rotational speed n (rpm) and the power P:

$$M = 7162\frac{P(\text{hp})}{n}(\text{Nm}) \tag{8.7}$$

$$M = 9736\frac{P(\text{kW})}{n}(\text{Nm}) \tag{8.8}$$

In the International System of Units (SI) the relation between torque M in Newton-meters (Nm), the angular speed of rotation ω in rad/sec and power P in watts is

$$M = \frac{P}{\omega} \tag{8.9}$$

For a round cylindrical rod of diameter $d = 2r$, the shearing stress τ, at an arbitrary point at distance r from the center in a cross-section is

$$\tau = M_T r / J_p \tag{8.10}$$

in which $J_p = \pi d^4/32 \approx 0.1d^4$ is the polar moment of inertia of a circular cross-section.

The maximum shearing stresses, at points most remote from the center, are

$$\tau_{max} = M_T/W_p \leqslant S_{sy}/N \tag{8.11}$$

in which $W_p = J_p/r \approx 0.2d^3$ is the polar section modulus of a circular cross-section.

Table 8.1 Torsional loading of sections

Section	Maximum stress	Twist angle per unit length
 Solid ellipse (2b, 2h; point A)	$\tau_A = \dfrac{2T}{\pi h b^2}$	$\phi = \dfrac{h^2+b^2}{\pi h^3 b^3}\cdot\dfrac{T}{G}$
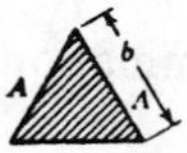 Equilateral triangle (side b; point A)	$\tau_A = \dfrac{20T}{b^3}$	$\phi = \dfrac{80}{b^4\sqrt{3}}\cdot\dfrac{T}{G} = \dfrac{46.2}{b^4}\cdot\dfrac{T}{G}$
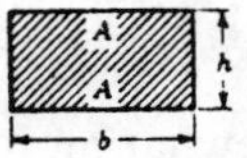 Rectangle (b, h; point A)	$\tau_A = \dfrac{T}{\alpha b h^2}$	$\phi = \dfrac{1}{\beta b h^3}\cdot\dfrac{T}{G}$
	$\alpha = 8/(3a+1.8b), \beta = 16/3 - 3.36b/a[1 - b^4/12a^4]$	
Split thin tube (r, t)	$\tau = \dfrac{3T}{2\pi r t^2}$	$\phi = \dfrac{3}{2\pi r t^3}\cdot\dfrac{T}{G}$
	$= \dfrac{3r}{t}\cdot\dfrac{T}{2\pi r^2 t}$	$= \dfrac{3r^2}{t^2}\cdot\dfrac{1}{2\pi r^3 t}\cdot\dfrac{T}{G}$
Closed thin tube (r, t)	$\tau = \dfrac{Tr}{J}$	$\phi = \dfrac{T}{JG}$
	$= \dfrac{T}{2\pi r^2 t}$	$= \dfrac{1}{2\pi r^3 t}\cdot\dfrac{T}{G}$
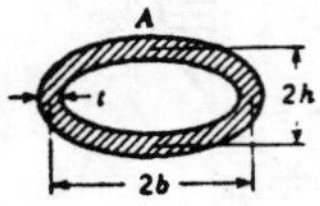 Hollow thin ellipse (2b, 2h, t; point A)	$\tau_A = \dfrac{T}{2\pi b h t}$	$\phi = \dfrac{\sqrt{2(b^2+h^2)}}{4\pi b^2 h^2 t}\cdot\dfrac{T}{G}$
Hollow thin rectangle (b, h, t, t_1; points A, B)	$\tau_A = \dfrac{T}{2bht_1}$	
	$\tau_B = \dfrac{T}{2bht}$	$\phi = \dfrac{bt + ht_1}{2tt_1 b^2 h^2}\cdot\dfrac{T}{G}$

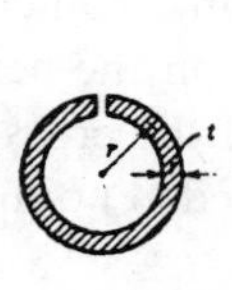

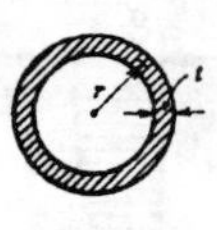

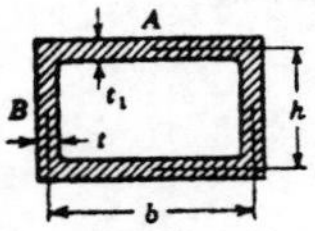

For rods of noncircular cross-section the maximum shearing stress is

$$\tau_{max} = \frac{M_T}{W_t} \tag{8.12}$$

in which W_t is section modulus for torsion (given for various cross-sections in handbooks and textbooks on the strength of materials). In Table 8.1, design equations for the maximum stress and the angle of twist have been compiled.

The angle of twist is, for a prismatic element of length L, in torsion,

$$\Delta\varphi = M_T L / J_p G \tag{8.13}$$

The polar moment of inertia J_p for a rod of non-circular cross-section is given in textbooks on the strength of materials. It can also be computed by program SECTIONS (see Section 8.3.3).

If the rod has several portions over which M varies in accordance with some law, the total angle of twist (angle through which the end sections of the rod turn with respect to each other) is found from the expression

$$\varphi = \sum \int M_T \, dx / GJ_p \tag{8.14}$$

Integration is carried out over the length of each portion; then summation, over all the portions of the rod.

8.1.4 Bending of straight and curved elements

Design of prismatic elements when bending is predominant, is somewhat more complex.

Since the design equations for bending of straight and curved elements are very similar, we shall consider the more general case of a beam element under bending moment where the longitudinal axis of symmetry is curved. The curvature is measured on the neutral axis z, which for curved beams does not pass through the center of area of the cross-section.

The bending moment M determines the normal stress σ_m developed at points of a cross-section of a curved bar.

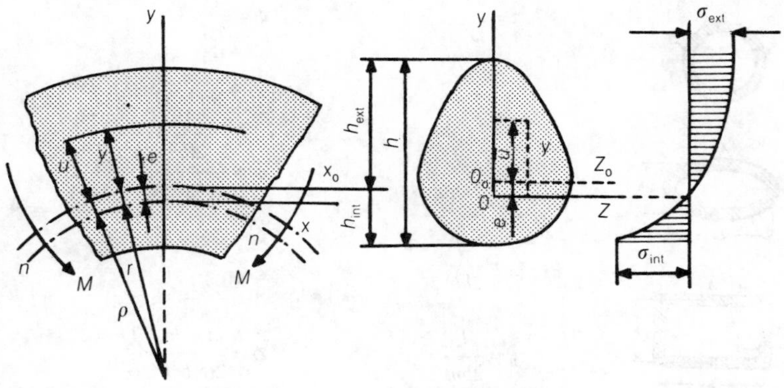

Figure 8.2 Bending of a curved beam

Stresses σ_m are distributed over the cross-sectional area A, according to a hyperbolic law:

$$\sigma_m = \frac{M}{S} \frac{y}{(r+y)} \tag{8.15}$$

in which S is the static moment of area F about the neutral axis z, r is the radius of curvature of the neutral line nn and y is the coordinate of the point being considered from the z-axis (Figure 8.2).

The neutral line nn is displaced with respect to the geometric axis of the beam towards the center of its curvature by the distance $e = \rho - r$, in which ρ is the radius of curvature of the geometric axis of the beam.

The radius of curvature of the neutral line of the beam for any shape of its cross-section is found from the equation

$$r = \frac{F}{\displaystyle\int_F \frac{dF}{(\rho + u)}} \tag{8.16}$$

in which u is the coordinate of the point being considered in the cross-section from the centroidal axis z_0 (Figure 8.2) and F the cross-sectional area.

For a rectangular cross-section,

$$r = h/\ln(r_{ext}/r_{int}) \tag{8.17}$$

For a round cross-section,

$$r = \{\rho + [\rho^2 - (d/2)^2]^{\frac{1}{2}}\}/2 \tag{8.18}$$

For a trapezoidal cross-section,

$$r = F/\{[b_{ext} + r_{ext}(b_{int} - b_{ext})/h]\ln(r_{ext}/r_{int}) - (b_{int} - b_{ext})\} \tag{8.19}$$

in which r_{ext}, r_{int}, b_{ext} and b_{int} are the radii of curvature and the widths of the external and internal fibers of the cross-section, respectively.

For certain other shapes of cross-sections, the values of r are given in textbooks on the strength of materials.

For beams of not very large curvature, the value of e can also be found by using the approximate equation

$$e = i^2/\rho \tag{8.20}$$

in which $i = (I/F)^{\frac{1}{2}}$ is the radius of gyration of the beam cross-section about the centroidal axis z_0, F the cross-sectional area and I the moment of inertia of the cross-section about the centroidal axis z_0.

The closest approximation to an accurate solution is given by equation (8.20) for beams with cross-sections symmetrical with respect to axis z_0.

Since

$$S = Fe = \frac{I}{\rho}; \quad y = u + e = u\left(1 + \frac{i^2}{\rho u}\right) \tag{8.21}$$

and

$$r + y = \rho + u = \rho\left(1 + \frac{u}{\rho}\right)$$

in which u is the distance of the point being considered in the cross-section from the centroidal axis z_0, then equation (8.15) can be written as

$$\sigma_m = (Mu/I)(1 + i^2/\rho u)/(1 + u/\rho) = Mu\alpha/I \tag{8.22}$$

where

$$\alpha = (1 + i^2/\rho u)/(1 + u/\rho) \tag{8.23}$$

characterizes the nonlinear law of distribution of σ_m over the cross-section and is dependent on the shape of the latter and on the initial curvature of the beam. If $h/\rho \leqslant 0.1$, then α differs only slightly from unity and σ_m can be calculated from the equation for a straight beam, $\sigma = Mu/I$.

The maximum and minimum normal stresses are obtained in the extreme fibers of the beam at $u = h_{ext}$ and $u = -h_{int}$. They are equal to

$$\sigma_{int}^{ext} = \pm M\alpha_{int}^{ext}/W_{int}^{ext} \tag{8.24}$$

Table 8.2 Geometric properties of sections

Cross section		W_{int}^{ext}	e	α_{ext}	α_{int}
1	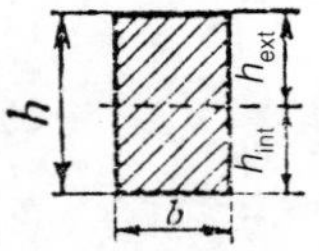	$\frac{bh^2}{6}$ $\frac{bh^2}{6}$	$\frac{h^2}{12}$	$\dfrac{1+\frac{h}{6\rho}}{1+\frac{h}{2\rho}}$	$\dfrac{1-\frac{h}{6\rho}}{1-\frac{h}{2\rho}}$
2	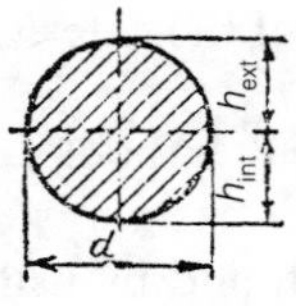	$\frac{\pi d^3}{32}$ $\frac{\pi d^3}{32}$	$\frac{d^2}{16\rho}$	$\dfrac{1+\frac{d}{8\rho}}{1+\frac{d}{2\rho}}$	$\dfrac{1-\frac{d}{8\rho}}{1-\frac{d}{2\rho}}$
3	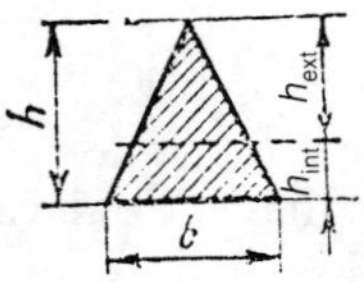	$\frac{bh^2}{24}$ $\frac{bh^2}{24}$	$\frac{h^2}{18\rho}$	$\dfrac{1+\frac{h}{12\rho}}{1+\frac{2h}{3\rho}}$	$\dfrac{1-\frac{h}{6\rho}}{1-\frac{1}{3}\frac{h}{\rho}}$
4	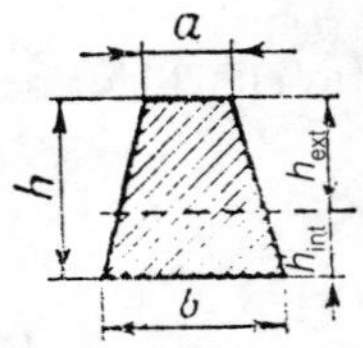 $\frac{a}{b} = n$	$\frac{n^2+4n+1}{2n+2}\cdot\frac{bh^2}{12}$ $\frac{n^2+4n+1}{2n+1}\cdot\frac{bh^2}{12}$	$\frac{n^2+4n+1}{(n+1)^2}\cdot\frac{h^2}{18\rho}$	$\dfrac{1+\frac{n^2+4n+1}{(n+1)(n+2)}\cdot\frac{h}{6\rho}}{1+\frac{n+2}{n+1}\cdot\frac{h}{3\rho}}$	$\dfrac{1-\frac{n^2+4n+1}{(n+1)(2n+1)}\cdot\frac{h}{6\rho}}{1-\frac{2n+1}{n+1}\cdot\frac{h}{3\rho}}$

Table 8.3 Design equations for strength

Loading	Material	Static loads	Dynamic loads	Combined loads
Tension–Compression	Ductile	$\sigma = F/A < S_y/N$	$\sigma = F/A < S_e/N$	$\sigma_{eq} < S_y/N$
	Brittle	$\sigma = FK_t/A < S_u/N$	$\sigma = F/A < S_e/N$	$\sigma_{eq} < S_y/N$
Crushing/bearing		$\sigma = F/A < (2-2.5)S_u/N$	$\sigma = F/A < (2-2.5)S_e/N$	$\sigma_{eq} < (2-2.5)S_y/N$
Bending	Ductile	$\sigma = Mc/I < S_y/N$	$\sigma = Mc/I < S_e/N$	$\sigma_{eq} < S_y/N$
	Brittle	$\sigma = McK_t/I < S_u/N$	$\sigma = Mc/I < S_e/N$	$\sigma_{eq} < S_y/N$
Shear	Ductile	$\tau = F/A < 0.577S_y/N$	$\tau = F/A < 0.577S_e/N$	$\sigma_{eq} < S_y/N$
	Brittle	$\tau = FK_t/A < 0.577S_u/N$	$\tau = F/A < 0.577S_e/N$	$\sigma_{eq} < S_y/N$
Torsion	Ductile	$\tau = TR/I_p < 0.577S_y/N$	$\tau = TR/I_p < 0.577S_e/N$	$\sigma_{eq} < S_y/N$
	Brittle	$\tau = TRK_t/I_p < 0.577S_u/N$	$\tau = TR/I_p < 0.577S_e/N$	$\sigma_{eq} < S_y/N$

$S_e = C_F C_S C_R C_H C_L S_n'/K_f$ Soderberg theory $\sigma_{eq} = [(\sigma_m + \sigma_r S_y/S_e)^2 + 3(\tau_m + \tau_r S_y/S_e)^2]^{\frac{1}{2}}$

Goodmann theory:

Asymmetry factor

$r_D = \sigma_r/\sigma_m$, for $r > r_D$, $\sigma_{eq} = \sigma_m + \sigma_r S_u/S_e$; $r < r_D$, $\sigma_{eq} = \sigma_m + \sigma_r$

$$r = \frac{1}{\dfrac{S_y S_u - S_e}{S_e S_u - S_y} - 1}$$

$r_T = \tau_r/\tau_m$, for $r > r_T$, $\tau_{eq} = \tau_r$; for $r < r_T$, $\tau_{eq} = \tau_m + \tau_r$

$\sigma_{eq} = [\sigma_{eq}^2 + 3\tau_{eq}^2]^{\frac{1}{2}}$

in which W is the equatorial section modulus of the cross-section which is $W_{ext} = I/h_{ext}$ for the external fiber and $W_{int} = I/h_{int}$ for the internal fiber and

$$\begin{aligned}\alpha_{int}^{ext} &= \alpha_{u=\pm h_{int}^{ext}} \\ &= (1 \pm i^2/\rho h_{int}^{ext})/(1 \pm h_{int}^{ext}/\rho)\end{aligned} \tag{8.25}$$

The magnitudes of the coefficients α_{int}^{ext} for certain cross-sectional shapes are given in Table 8.2.

The resultant normal stress at an arbitrary point of a cross-section of a curved beam can be found by the equation

$$\sigma = \frac{M}{I} u\alpha \tag{8.26}$$

Again, certain caution must be exercised in using the design equation (8.26). Beyond the reasons presented in Section 8.1.1, the usual compound sections of beams in bending must be checked for local stability of webs, and shear and normal forces which might be present.

If the material of the beam has different tensile and compressive strengths, the strength conditions at the most dangerous cross-section should comply with the requirements for both the internal and external fibers in accordance with the allowable stress values.

A summary of design equations for strength is included in Table 8.3.

8.1.5 Design methodology for strength

On the basis of element strength, the design procedure is to apply the equations of Table 8.3 for comparison of the maximum stresses occurring on the element to the applicable material properties. Many times, more than one test must be made at different locations because of different loading at different areas of the element.

The stresses are computed with simple strength of materials equations if possible. For complex geometry, numerical methods are usually employed.

The comparisons indicated in Table 8.3 presume the knowledge of elementary geometry. If the design stresses are substantially higher or lower than the material strength, the geometry has to be adjusted and the computation repeated.

Many times it is possible to leave the unknown dimensions in algebraic form and use the design equations to determine the unknowns. For simple problems this can be done explicitly. For more complicated situations, optimization methods are employed, as in Chapter 5. The corresponding computational procedure follows:

```
procedure DESIGN.FOR.STRENGTH
begin
    Select materials. Basis: engineering judgement                    (Step 1)
    Identify unknown parameters defining geometry                     (Step 2)
    Express design requirements (material limits, operating features, etc.)
    in the form of design equations                                   (Step 3)
    Compute parameters of design equations which can be derived from
    data                                                              (Step 4)
    Based on engineering experience and judgement, select superfluous
    parameters to leave undetermined as many design parameters as
    design equations                                                  (Step 5)
    Solve design equations to determine unknowns                      (Step 6)
    Check results against common sense, experience                    (Step 7)
    Optimize by an educated better selection of superfluous parameters at
    step 5                                                            (Step 8)
    Make detailed sketches of final design                            (Step 9)
    Perform final design analysis to verify design.                   (Step 10)
end.
```

As we have seen in Chapter 5, steps 5–8 can be replaced by an optimization algorithm.

8.2 JOINT DESIGN

The several machine elements required in order to form a working machine frequently have to be connected together. In other cases it is advantageous to form a complicated machine element out of more than one component which have to be connected together to give the final element. Some reasons have been presented previously, such as producibility, components made out of dissimilar materials, serviceability with parts that have high wear rate, etc. Since in most cases joints have the purpose of carrying loads, we shall use them as examples of how to apply methods of design for strength.

Since joint elements are used in a variety of cases they have been standardized to a very great extent. The standardization of such elements must be examined first.

A joint element, such as a bolt for example, can be used in a variety of tasks. The same bolt can be used in an automobile, in an aircraft engine or in a paper mill. On all occasions its purpose is to carry a certain load in a certain mode. We can get an idea of the frequency of use of joint elements if we consider the fact that an automobile sometimes consists of more than 10,000 parts, a machine tool might have up to 20,000 parts, a rolling mill might reach 1,000,000 parts.

Joints may be permanent or removable. The selection is based on the purpose and on the economy of the joint. Permanent connections cost less in general but they have the disadvantage that to disconnect the parts, the joints have to be destroyed and they cannot be used again. On the other hand, permanent joints are safer especially in parts which sustain dynamic loads. Therefore, for permanent joints the main consideration is the strength. For separable joints permitting assembly and disassembly, additional security from accidental separation is necessery.

Permanent joints are held together either by the forces of molecular cohesion, such as the welded joints, or by mechanical means such as riveted and bolted joints.

The main consideration in the design of joints is to make them function as closely as possible as if the connected elements were a single solid part and to comply with the condition of equal strength of the joint and connected elements. Otherwise the material of the connecting element will not be utilized to its full capacity.

There are occasions when the joints are part of fluid-carrying vessels under pressure. In this case the joint has the additional condition of fluid tightness. For this purpose the contacting surfaces must be held together by a pressure exceeding the pressure of the fluid. If high accuracy under the application of the load is required, the joint must in addition have sufficient rigidity.

Several types of joints are categorized below:

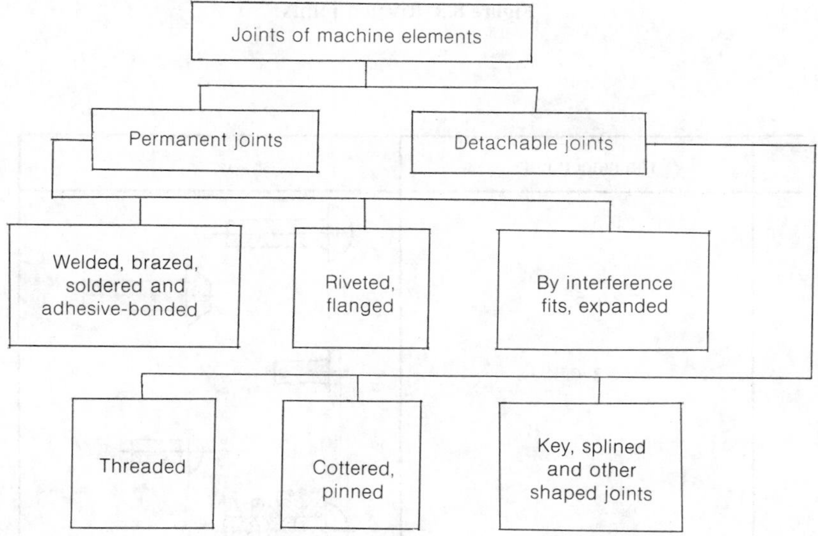

8.2.1 Rivets

A rivet is a short round bar with heads at each end. One of the heads is made beforehand on the body of the rivet and the second head, called the 'closing head', is formed during riveting. A riveted joint is made by inserting rivets into holes in the elements to be connected while holding the elements together in some way so that the holes coincide. Then the closing head is formed by riveting, which also expands the shank of the rivet as is shown in

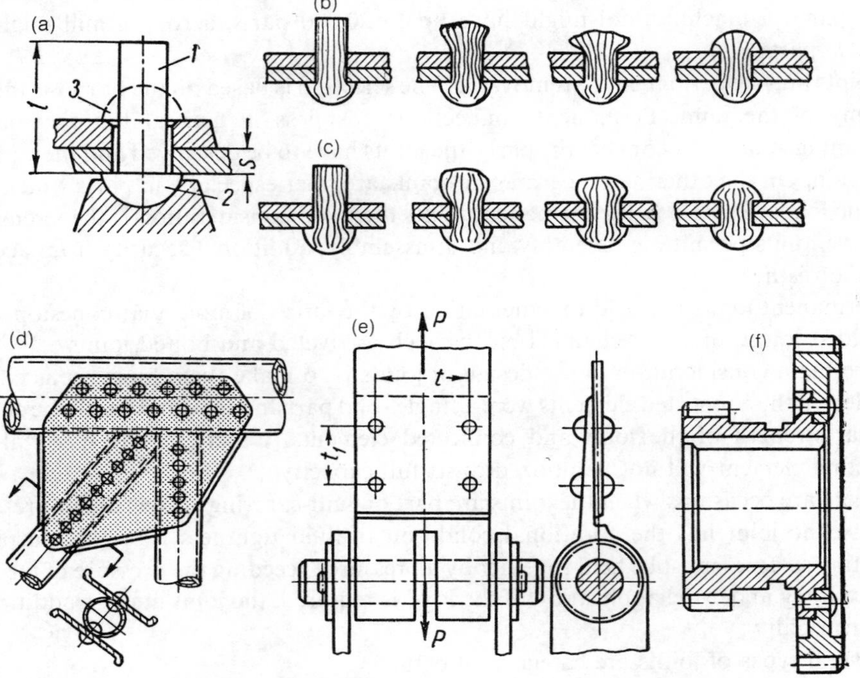

Figure 8.3 Riveted joints

Diameter (mm)	Shape
1–37	
2.6–6	
2.3–6	
2–7	
8	
4–20	
4–20	

Figure 8.4 Types of rivet

Table 8.4 Dimensions of rivets

Rivet	d	10	12	(14)	16	(18)	20	22	24	27	30	(33)	36
	d_1	11	13	15	17	19	21	23	25	28	31	34	37
Bolt (eq.)		M10	M12	—	M16	—	M20	—	M24	—	M30	—	M36
Pressure vessels	D	18	22	25	28	32	36	40	43	48	53	58	64
	k	7	9	10	11.5	13	14	16	17	17	21	23	25
	R	9.5	11	13	14.5	16.5	18.5	20.5	22	24.5	27	30	33
	r	1	1.6	1.6	2	2	2	2	2.5	2.5	3	3	4
Steel construction	D	16	19	22	25	28	32	36	40	43	48	53	58
	k	6.5	7.5	9	10	11.5	13	14	16	17	19	21	23
	R	8	9.5	11	13	14.5	16.5	18.5	20.5	22	24.5	27	30
	r	0.5	0.6	0.6	0.8	0.8	1	1	1.2	1.2	1.6	1.6	2
Sunken heads	D	14.5	18	21.5	26	30	31.5	34.5	38	42	42.5	46.5	51
	k	3	4	5	6.5	8	10	11	12	13.5	15	16.5	18
	r	27	41	58	85	113	124.5	75.5	91	111	114	136	164

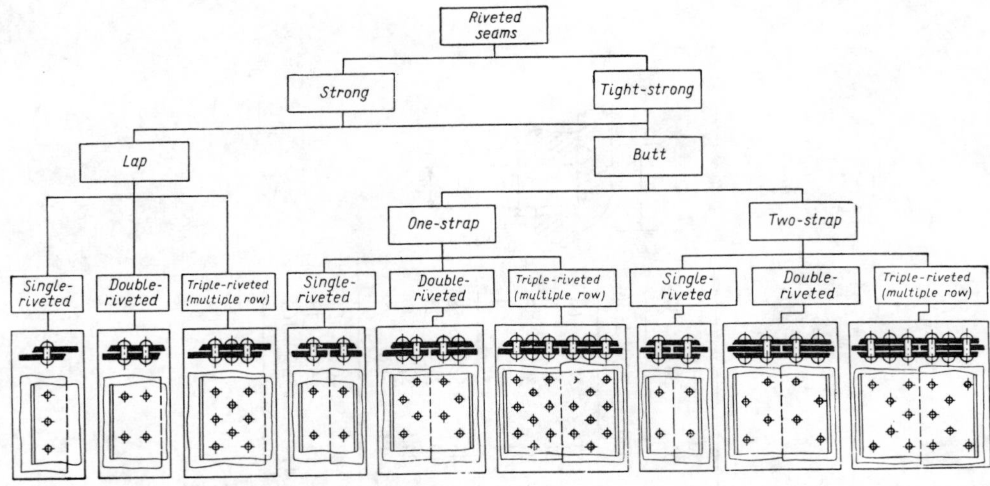

Figure 8.5 Types of riveted joint (Reshetov 1980)

Table 8.5 Allowable stresses in riveted joints (MPa): (a) plate material; (b) rivet material

(a)

		ASTM					ASTM A159*			ASTM A48*
	Loading	A284 GC	A284 GD	A570 GD	A572 G50	A572 G55	G1800	G3000	G4000	Class 55
Tension	static	120	140	160	180	220	35	65	100	135
	pulsating	85	100	120	140	170	25	40	75	100
	alternating	70	85	95	110	130	20	35	50	70
Bending	static	170	195	225	250	310	50	90	140	190
	pulsating	95	110	130	155	185	28	45	80	110
	alternating	75	95	100	120	145	20	40	55	80
Shear	static	240	280	320	360	410	65	130	200	270
	pulsating	170	200	240	280	340	45	85	130	170
	alternating	140	170	190	220	260	35	65	100	130

(b)

	Shear				Bearing		Tension		
	A284 GrC	A570 GrD	A572 Gr50	A572 Gr60	A570 GrD	A572 Gr50	A284 GrC	A570 GrD	A572 Gr50
static	140	180	225	280	360	440	70	90	110
pulsating	100	140	170	200	280	340	50	70	85
alternating	85	110	130	170	220	260	40	55	65

*Cast irons

Figure 8.3(a, b, c). Some applications are shown in Figure 8.3(d, e, f) and some types of rivets are shown in Figure 8.4. Rivets are perhaps the oldest known type of machine element. Today they are used in many applications:

1. When the material is such that welding might temper the heat-treated components or may warp them and change their shape.
2. In materials which are very dissimilar or non-weldable.
3. In the case of very heavy repeated impact and vibrational loads.
4. In cases where maximum safety is required, such as in aircraft structures.

Standard dimensions of rivets are shown in Table 8.4 and the usual types of riveted connections are shown in Figure 8.5. The type of riveting selected depends on the loads that riveting will carry.

Materials used for machinery riveting are shown in Table 8.5, with allowable stresses for static and dynamic loads. For steel constructions, local codes are mandatory and they must be consulted.

For pressure-vessel and boiler riveting, the ASME or other applicable pressure vessel codes may be applied.

Riveting for light and aircraft structures need special treatment and they will be discussed later in this chapter (p. 336).

A riveting might fail with a number of failure mechanisms (Figure 8.6), as of course any pin-type joint such as bolting or spot welding. In Figure 8.6 (below) each mode of failure diagram is accompanied by its design equation.

(*a*) *By shear failure of the rivet*

$$\tau = 4F/\pi d^2 t \leqslant S_{sy}/N$$

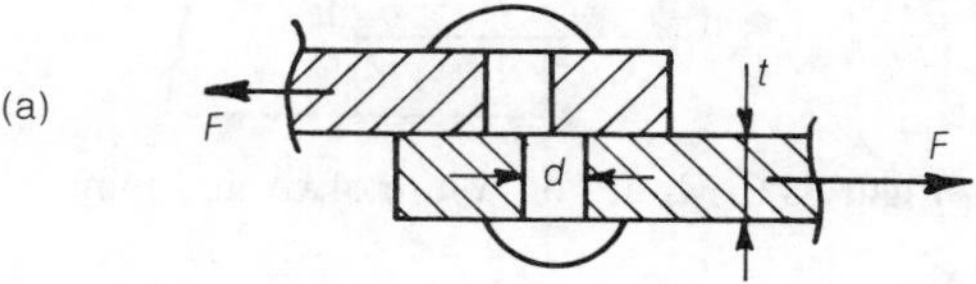

(*b*) *By tensile failure of the plate along the riveting*

$$\sigma = F/t(w - d) \leqslant S_y/N$$

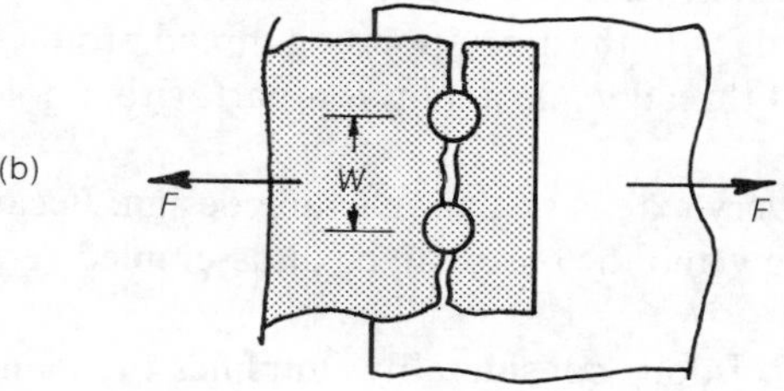

(c) *By crushing of the rivet or the plate owing to excess bearing pressure*

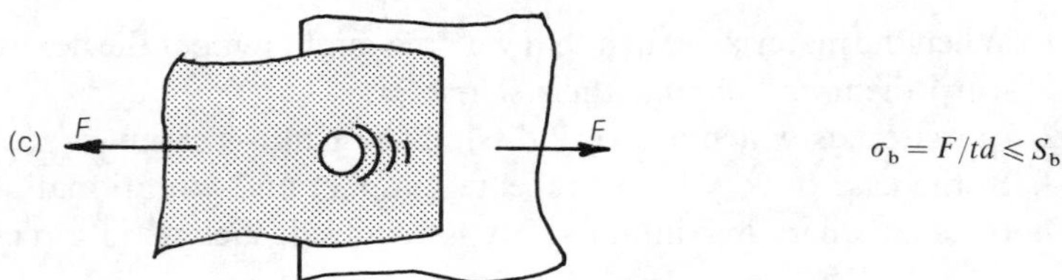

$$\sigma_b = F/td \leqslant S_b$$

(d) *By double shear of the plate*

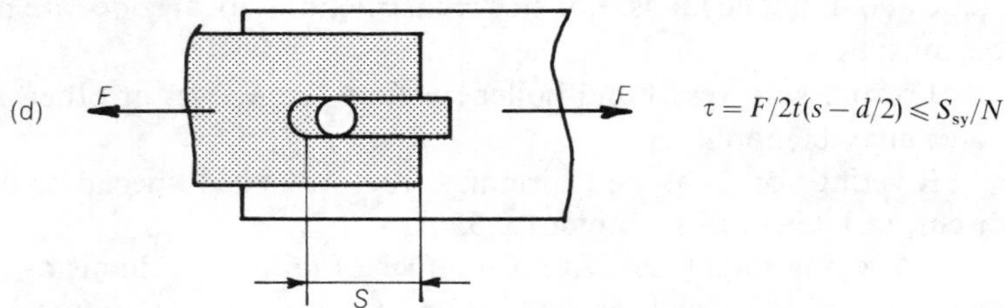

$$\tau = F/2t(s - d/2) \leqslant S_{sy}/N$$

(e) *By shear tear-out of the plate*

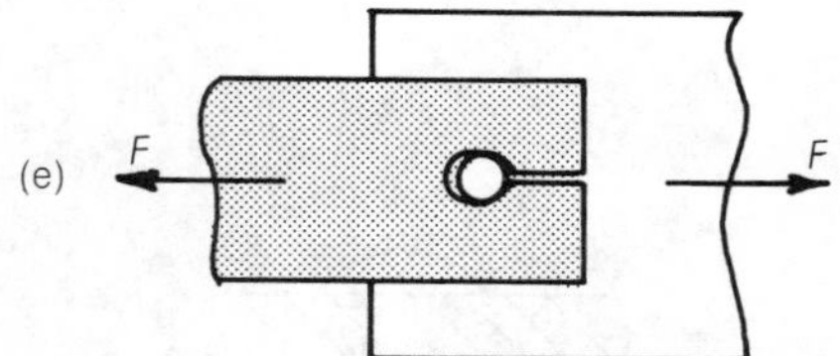

Figure 8.6 Modes of failure of riveted joints

8.2.2 Bolts

Bolt or screw joints are separable joints held together by screw fastening, such as screws, bolts, studs and nuts or by thread cut on the parts to be joined (Figure 8.7).

The thread is formed as a helical groove into the surface of a cylindrical bar or hole and has a cross-section complying with the corresponding thread profile. The term 'screw' is used with any part with external thread while a 'nut' is a part with a hole which has an internal thread.

Screw joints are used very extensively in machine design. Because of their standardization, their cost is usually low and the parts can be disassembled very easily usually without destroying the bolts.

In steel structures, use of bolts considerably simplifies the manufacture and assembly. According to their purpose, screw threads are classified as follows:

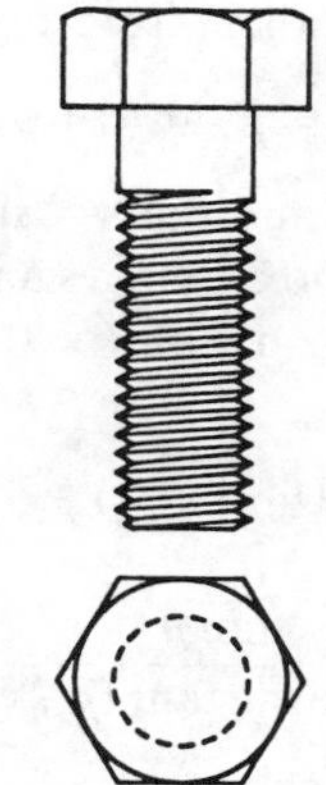

Figure 8.7 A computer-generated bolt drawing

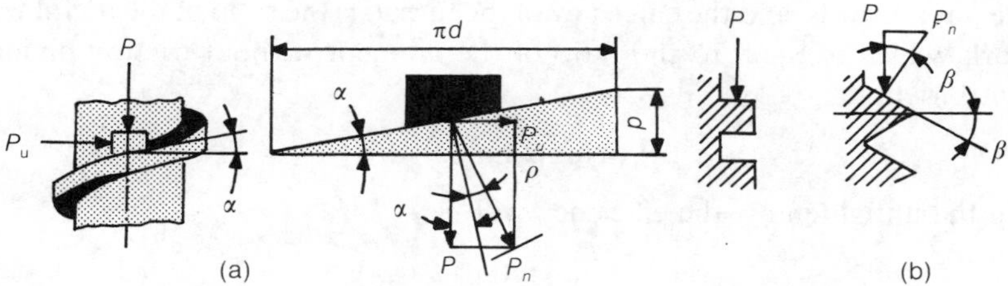

Figure 8.8 Geometry and forces on the thread

1. *Fastener threads*: These threads are usually very heavily loaded and they can develop high fastening forces with relatively small torque owing to the wedge effect of their thread.
2. *Fastening and sealing threads*: These are for both fastening parts together and preventing the leakage of fluids at high pressure.
3. *Power threads for transmitting motion*: These reduce friction and are of a special shape such as trapezoidal or orthogonal.

Accordingly, the international standards have specified different kinds of threads some of which are shown in Appendix C.1.

A bolt is loaded in general by an axial force P which is transferred to the bolt from the nut (Figure 8.8). The contact plane of the bolt with the nut is not on a plane perpendicular to the bolt axis but it forms an angle β' in respect to it and is determined by the angle of the helix α and another angle β in respect to the axis, which corresponds to the inclination of the thread (Figure 8.8b).

To move the nut we have to apply a peripheral force P_u. This force must balance the load and the friction force $P_n f$, where P_n is the normal force at the contact and f the coefficient of friction between the bolt and the nut. The friction force has a positive or a negative sign depending on the direction of motion.

If ρ is the friction angle, $\tan\rho = f$, then, from Figure 8.8a,

$$P_u = P\tan(\alpha \pm \rho) \tag{8.27}$$

where the '+' sign refers to tightening and the '−' sign to untightening.

When, as usual, $\beta > 0$, the normal force changes and becomes $P/\cos\beta$. It is equivalent to the normal force for an orthogonal thread with $\beta = 0$ when the effective friction angle ρ' is introduced and then,

$$P_u = P\tan(\alpha \pm \rho') \tag{8.28}$$

where

$$\tan\rho' = \frac{\tan\rho}{[1+\cos^2\alpha\tan^2(\beta/2)]^{\frac{1}{2}}} \cong \frac{\tan\rho}{\cos(\beta/2)} \tag{8.29}$$

The required moment at the nut is

$$M_T = P_u d_2/2 = P\tan(\alpha \pm \rho')d_2/2 \tag{8.30}$$

where d_2 is the pitch diameter of the thread.

Because of the friction the work applied to the nut is not all transformed into the work pushing the nut upwards, and the efficiency of the thread is the ratio of the useful work to the applied work, which is equal to the ratio of the moment without friction divided by the moment with friction,

$$\eta_t = \tan\alpha/\tan(\alpha + \rho') \tag{8.31}$$

During the untightening, the efficiency will be

$$\eta_u = \tan\alpha/\tan(\alpha - \rho') \tag{8.32}$$

The above equation means that if $\alpha - \rho'$ is less than 0, we have to apply moment in order to untight the nut. If, however, the efficiency is positive, it means that the nut has the tendency to untight itself freely. In order for this to happen we must have

$$\tan(\alpha - \rho') > 0 \tag{8.33}$$

or in other words $\alpha > \rho'$. This suggests that the thread angle α must be smaller than the friction angle for stable fastening. In other words, the smaller the angle α, the more steady is the bolt. For this reason, for bolts which are loaded with dynamic load and there is a danger of self-untightening of the nut, we use special threads with small pitch, and consequently small angle α.

When the nut is tightened against a stationary part there is friction between the nut and the part during rotation of the nut. In this case the applied torque has to overcome this friction as well. If we assume that the mean diameter of the application of the friction forces on the nut is d_A (generally the average of the two diameters of bolt and nut), f_A the coefficient of friction between the nut and the stationary part, there will be friction from the thread and the nut seating:

$$M = M_T + M_A = P[\tan(\alpha \pm \rho')d_2/2 + f_A d_A/2] \tag{8.34}$$

In Table 8.6, coefficients of friction are tabulated for normal practises in bolt design and utilization.

Table 8.6 Friction coefficients for nuts and bolts

Material		Lubrication		
Bolt	Nut	None	Oil	MoS_2
Surface treatment				
Mn–Ph	None	0.14...0.18	0.14...0.15	0.10...0.11
Zn–Ph	None	0.14...0.21	0.14...0.17	0.10...0.12
8 μm Zn	None	0.125...0.18	0.125...0.17	
7 μm Cd	None	0.08...0.12	0.08...0.11	
8 μm Zn	5 μm Zn	0.125...0.17	0.14...0.19	
7 μm Cd	6 μm Cd	0.08...0.12	0.14...0.15	

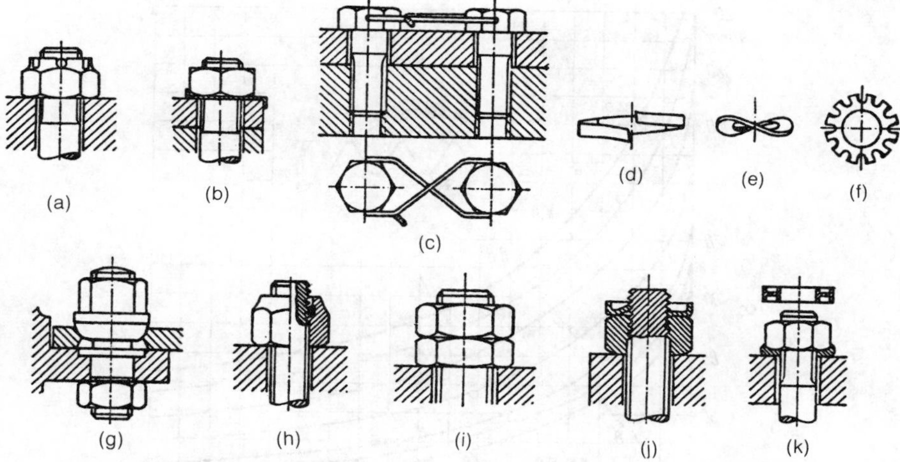

Figure 8.9 Different bolts, nuts and securing washers (Niemann 1965)

Nut friction contributes towards a more stable bolt. In this case the condition for stability of the bolt is

$$\tan(\alpha - \rho')d_2/2 < f_A d_A/2$$

To increase the friction between the nut and the stationary parts, thus making a more stable bolting, special safety parts are utilized (Figure 8.9).

In principle, bolts can be made from a variety of materials; in fact for special applications they are indeed made of many different materials. Steel bolts are the most widely used and they have been standardized.

The ISO has specified twelve categories of bolt strength as shown in Table 8.7, where the number designation means approximately tensile strength in kg-f/mm^2.

The Society of Automotive Engineers uses the standard shown in Table 8.8 where in the left column is a sign which is pressed on the head of the bolt.

Bolts are made also from the following materials, amongst others:

1. Aluminum 2024-T4 + 2011-T3 for machine bolts.
2. Copper, brass, bronze for general use, especially for chemical industries.

Table 8.7 Allowable static and dynamic loads for bolts (ISO)

		SAE/ISO Class											
Bolt		3.6	4.6	4.8	5.6	5.8	6.6	6.8	6.9	8.8	10.9	12.9	14.9
S_u	N/mm²	340	400	400	500	500	600	600	600	800	1000	1200	1400
S_y	N/mm²	200	240	320	300	400	360	480	—	—	—	—	—
$S_{y0,2}$	N/mm²	—	—	—	—	—	—	—	540	640	900	1080	1260
δ_u	%	25	25	14	20	10	16	8	12	12	9	8	7
Nut		4			5		6			8	10	12	14
S_u	N/mm²	400			500		600			800	1000	1200	1400

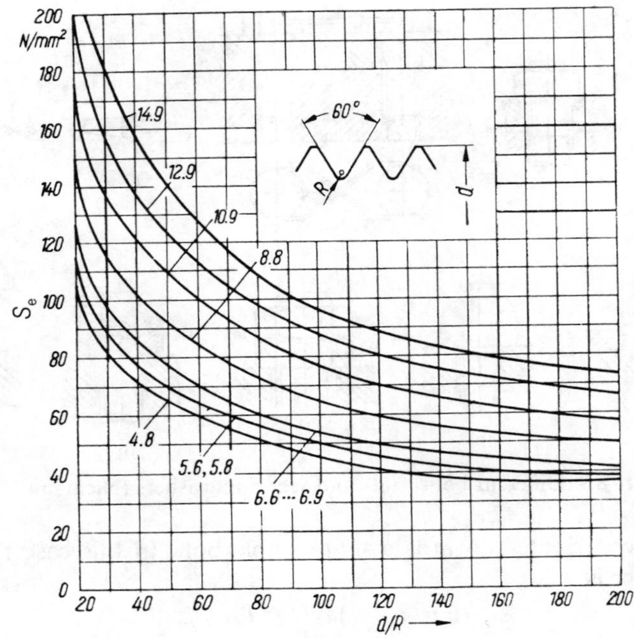

3. Nickel alloys for extremely low and high temperatures and acid environment.
4. Inconel nickel alloy for high temperatures and acid environments.
5. High alloys for magnetic applications and in demand of high hardness and high resistance to oxidation.
6. Nylon for resistance to temperature, impact, vibration, chemicals and for electrical insulation.
7. Polyvinyl chloride for acid environments.
8. Teflon for resistance to oxidation and relatively high temperatures.
9. Zinc, cadmium, nickel, and chromium coatings or plating are used for protection in acid or humid environments.

Table 8.8 SAE standards of materials and strengths for bolts

Identification mark	Grade designation	Fastener description	Material	Is Mfr's identification symbol required?	Nominal size range (in.)	Mechanical properties: Proof load. (psi)	Mechanical properties: Yield strength (min. psi)	Mechanical properties: Tensile strength (min. psi)	Hardness: Brinell	Hardness: Rockwell	Remarks
No mark	SAE grade 0	Bolts and screws	Steel	Yes	$\frac{1}{4}$ to $1\frac{1}{2}$	—	—	—	—		
No mark	SAE grade 1	Bolts and screws	Carbon steel	Yes	$\frac{1}{4}$ to $1\frac{1}{2}$	—	—	55,000	207 max.	B95 max.	Equivalent to ASTM A307, grade A
	GM 255-M	Bolts and screws	Carbon steel	Optional							
No mark	SAE grade 2	Bolts and screws	Carbon steel	Yes	$\frac{1}{4}$ to $\frac{1}{2}$	55,000	—	69,000	241 max.	B100 max.	
					over $\frac{1}{2}$ to $\frac{3}{4}$	52,000	—	64,000	241 max.	B100 max.	
	GM 260-M	Bolts and screws	Carbon steel	Optional	over $\frac{3}{4}$ to $1\frac{1}{2}$	28,000	—	55,000	207 max.	B95 max.	
	SAE Grade 3	Bolts and screws	Medium carbon steel	Yes	$\frac{1}{4}$ to $\frac{1}{2}$	85,000	—	110,000	207/269	B95/104	
					over $\frac{1}{2}$ to $\frac{5}{8}$	80,000	—	100,000	207/269	B95/104	
	SAE Grade 5	Bolts and screws	Medium carbon steel, heat treated	Yes	$\frac{1}{4}$ to $\frac{3}{4}$	85,000	—	120,000	241/302	C23/32	Equivalent to ASTM A449
					over $\frac{3}{4}$ to 1	78,000	81,000	115,000	235/302	C22/32	
	GM 280-M	Bolts and screws		Optional	over 1 to $1\frac{1}{2}$	74,000	77,000	105,000	223/285	C19/30	
	SAE grade 5.1	Sems	Carbon steel, heat treated	Yes	$\frac{3}{8}$ and smaller	85,000	—	120,000	241/375	C23/40	
	GM 275-M	Sems		Optional							
	SAE grade 6	Bolts and screws	Medium carbon steel, heat treated	Yes	$\frac{1}{4}$ to $\frac{5}{8}$	110,000	—	140,000	285/331	C30/36	
					over $\frac{5}{8}$ to $\frac{3}{4}$	105,000	—	133,000	269/331	C28/36	
	SAE grade 7	Bolts and screws	Low alloy steel, heat treated	Yes	$\frac{1}{4}$ to $1\frac{1}{2}$	105,000	110,000	133,000	269/321	C28/34	Threads rolled after heat treatment
	GM 290-M	Bolts and screws		Optional							
	SAE grade 8	Bolts and screws	Low alloy steel, heat treated	Yes	$\frac{1}{4}$ to $1\frac{1}{2}$	120,000	125,000	150,000	302/352	C32/38	Equivalent to ASTM A354, grade BD
	GM 300-M	Bolts and screws		Optional							
	GM 455-M	Bolts and screws	Corrosion resistant steel	Optional	$\frac{1}{4}$ to $1\frac{1}{2}$	40,000	—	55,000	143 min.	B79 min.	

Source: Camcar

8.2.3 Welded joints

Welded joints are permanent joints formed by localized heating with the addition of a welding element or by welding together the two mating parts.

Welded joints are the most advanced type of permanent joints because the properties of the welded components and the weldment are closest to those of the solid member under certain conditions. Furthermore, very complicated members can be fabricated by welding.

All structural steels including high alloys and also nonferrous alloys and plastics can be efficiently welded.

Metal arc welding is performed with an electric arc which is formed between a metal electrode and the mating parts. The heat, which is generated by the arc, melts both the electrode and the surface of the mating parts. This process can be automated – called 'submerged arc welding' – and can give a very high quality of joints. It can be applied to a wide range of thicknesses of the welded parts and it can join parts of ordinary structural steels as well as of high-strength alloys. *Robots* have been very effectively used in welding processes, reducing the cost and improving quality.

To avoid oxidation due to the high temperatures, either a paste is used with the electrode (the paste is melting at relatively low temperature and covering the weld, preventing the oxygen from coming into contact with the hot metal) or an inert gas is supplied (which again prevents the oxygen from coming in contact with the hot metal).

Resistance welding is based on heating the conducting surfaces of the joint by passing an electric current through them and applying a pressure on the seam. Resistance welding is mostly used in large lot and mass production.

Friction welding is performed by making use of the heat generated in the relative motion of the parts being joined.

There are a very great number of different methods of welding which suggests the wide use of these methods of joining machine elements.

With respect to the mutual position of the welded components, welded joints can be classified into the following categories:

(1) *Butt joints*: One of the parts is a continuation of the other. The end faces are welded together.
(2) *Lap joints*: The side surfaces of the welded parts overlap partly. Welding is performed at one or both the edges over the side of the other part.
(3) *Tee joints*: The joint parts are perpendicular to each other and less frequently at some other angle. The end face of one part is welded to the side of the other.
(4) *Corner joints*: The joint parts are perpendicular to each other or at some other angle and are welded together along the adjacent edges.

Butt joints are closest in behavior to solid components and their application is gradually increasing at the expense of lap joints.

Because of their geometry, welded joints require less base metal for the components, in comparison with riveted joints (usually between 15 and 20%).

In comparison to castings, welded parts are much lighter than steel castings (by up to 50%) and cast-iron castings (by up to 30%).

One disadvantage of welded joints is for welding performed by hand – reliability depends very much on the skill of the welder. There are however methods to test the quality of the weld and improve reliability. Another disadvantage of welded joints is that, due to the high temperatures in the area of the weld as compared with the base metal, very high thermal stresses are developed which deform the welded parts and leave behind internal stresses. Corrective measures and processing are available for reducing or preventing distortion from welding, such as the symmetry of the welds, reduction of amount of weld metal, and proper application of the welding sequence. There are also methods of relieving stresses by heating after the welding. The metal near the weld is annealed.

In particular, welded parts to be manufactured after welding to a high standard of accuracy should be tempered to prevent subsequent warping due to the residual stresses.

A special type of welded joint is when an intermediate element is used, an element with a strength much less than that of the basic materials. These joints are called 'soldered', 'brazed' or 'adhesive joints'.

Soldering and *brazing* are processes for joining parts by means of a material with low melting point as compared with the basic material. These processes are mainly for joining metal parts. Soldering is known from ancient times and recently has been applied to much more critical joints than previously. Most widely used are the lap joints. In comparison to welding, soldering and brazing have a number of advantageous features: The soldering or brazing temperature is low and causes no change in the properties of the metals being joined. In a lap joint, soldering or brazing is carried out along a surface, showing more uniform stress distribution than in welds and enabling shorter joints to be used. The influence of defects on soldered or brazed joints is relatively small. Further, owing to the fact that the surface carrying the load can be very great, as compared with welding, soldered and brazed joints can reach the strength of welding.

For joining of various materials, *adhesives* based on synthetic resins have found extensive application in recent years. These adhesives have very high strength and have been used in critical machines and structures such as aircraft and bridges. They have also the advantage that they can be used for joining parts of very dissimilar materials. The main disadvantage of such adhesives is their low resistance to higher temperatures.

Adhesives applied in mechanical design are divided into two groups. Most widely used are the *adhesives based on organic polymeric resins* with a heat resistance not over the range 300 to 350 °C. The shear strength of such adhesives at normal temperatures reaches 18 to 20 N/mm^2 and at 200 to 300 °C it is reduced drastically. The second group includes *adhesives based on organic silicon compounds and inorganic polymers* which have a heat resistance up to 1000 °C but also have higher brittleness. The shear strength of such adhesives at normal temperatures is between 7 and 8 N/mm^2 and at 1000 °C is between 2 and 3 N/mm^2.

Adhesives have been very efficiently used to strengthen the fastening between parts which are joined by other methods.

Some of the more usual welded joints are shown in Figure 8.10.

For welding electrodes, the American Welding Society (AWS) specifies the symbol E*yyxx*, where *yy* is the tensile strength in ksi and *xx* is the type of welding (Table 8.9).

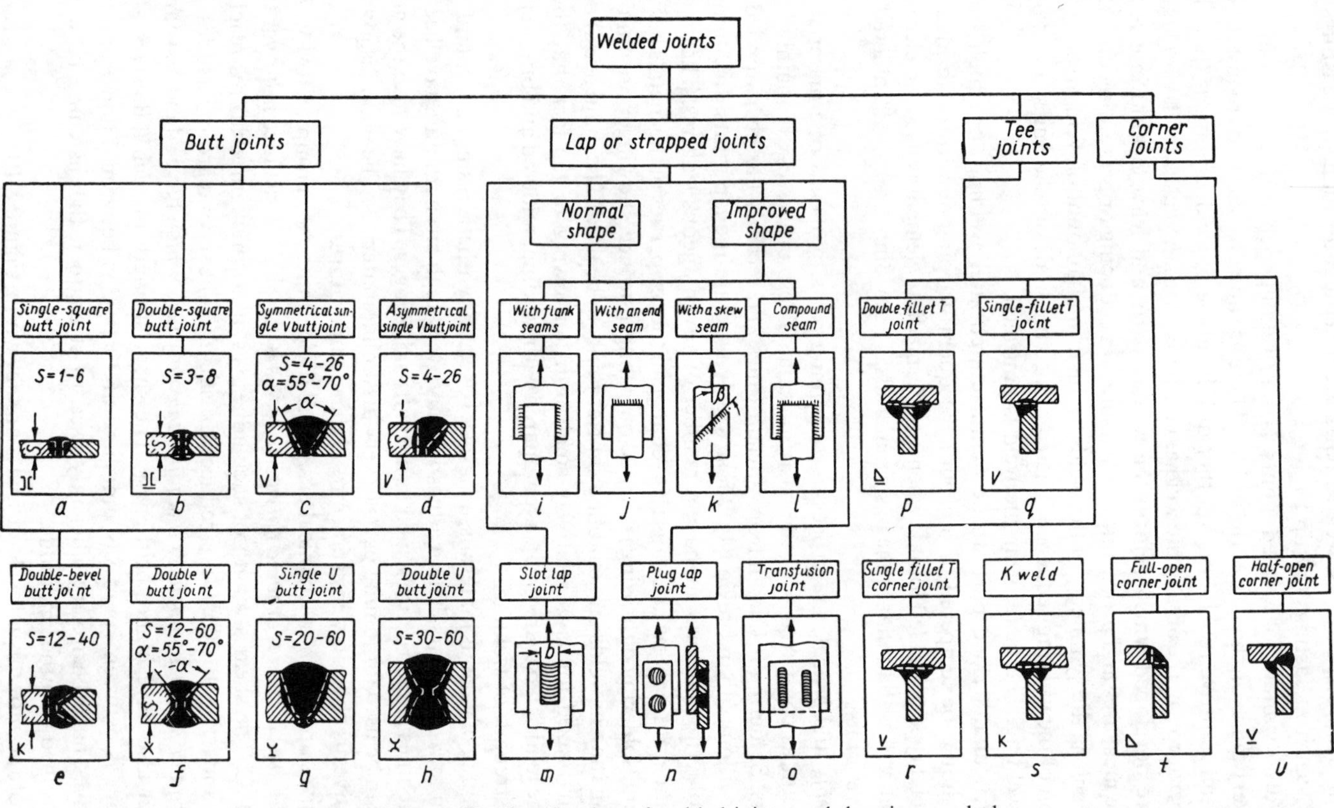

Figure 8.10 Classification of welded joints and drawing symbols

Table 8.9 AWS standard welding electrodes

AWS class	S_u (psi)	S_y (psi)
E60*xx*	62,000	50,000
E70*xx*	70,000	57,000
E80*xx*	80,000	67,000
E90*xx*	90,000	77,000
E100*xx*	100,000	87,000
E120*xx*	120,000	107,000

8.2.4 Interference fit joints

Interference fit joints (as has been seen already in Chapter 2) are formed by a fit where the shaft has a diameter larger than the hole of the hub. The joining is performed by applying pressure or by heating or cooling one or other part so that the shaft will enter freely into the hole and by equalizing the temperatures afterwards, the fit will be formed. During the contraction of the hub, pressure is developed between the shaft and the hole. Thus, mutual displacement of the joined parts is prevented by frictional forces at the contact surfaces.

Their chief advantage is easy manufacturing and assembly and the heavy loads they can carry. The disadvantage is that they have some difficulty in disassembly and during that process the surfaces might be damaged.

Cylindrical interference fit joints have very wide application for heavy loads and when there is no need for frequent assembly and disassembly.

Typical examples of parts joined by interference fit are cranks, wheels and bands for railway rolling stock, rims of gears and worm gears, turbine disks, rotors of electric motors, and ball and roller bearings (Figure 8.11).

The tightness of the joint is determined by the amount of interference which is selected from the fit established by the existing standard system of tolerances and fits, as presented in Chapter 2.

The most commonly used interference fits of grade 6, 7 and 8 tolerances are, in the order of decreasing interference, H7/u7, H7/s6, H7/r6, H7/p6.

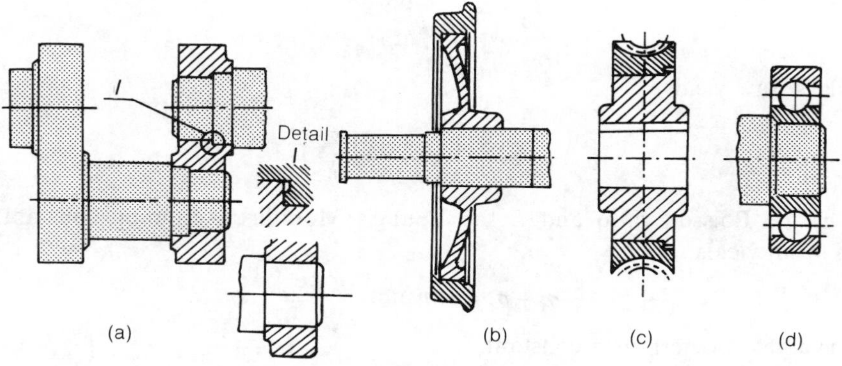

Figure 8.11 Shrink-fitted connections

Press fitted joints are used for lighter loads and thermal expansion fit joints are used for heavier loads.

The limitation of the surface damage possible during disassembly, can be overcome by tapered fit joints. These joints are obtained by a taper usually of 1:50. Critical joints are assembled by shrinkage or expansion or by hydraulic forcing. The same method is used for disassembly.

8.3 STRESS ANALYSIS OF JOINTS

8.3.1 Interference stresses

Stress analysis in interference fit joints is usually performed by the assumption that a hub of orthogonal cross-section is press-fitted around a rigid cylindrical shaft. For other geometries, computer methods are used.

Between the hub and the shaft there is uniform pressure p and the friction coefficient is f. The resulting friction moment is $M = fp \cdot Ld(d/2)$, where L is the hub length. The axial friction force is $F = fp\pi Ld$. The pressure p to secure the fit will then be

$$p = [(2M/f\pi d^2 L)^2 + F/f\pi Ld)^2]^{\frac{1}{2}} \tag{8.35}$$

The design requirements are:

1. For the fit with the minimum interference (see Chapter 2), the pressure must be at least p, as given in equation (8.35).
2. At maximum interference, the hub must not fail.

Looking at an elementary segment of the hub, Figure 8.12, summing up the forces on the half ring, Figure 8.12c, gives

$$2\sigma_t \, d\rho = -2\rho_r \, d\rho + 2\rho \, d\sigma_r + 2d\rho \, d\sigma_r \tag{8.36}$$

Neglecting the product of differentials, as a higher order one,

$$\sigma_t = \sigma_r + \rho(d\sigma_r/d\rho)$$

$$\sigma_t - \sigma_r = \rho \frac{d\sigma_r}{d\rho} \tag{8.37}$$

Hooke's Law yields

$$\varepsilon_z = -\nu\left(\frac{\sigma_r}{E}\right) - \nu\left(\frac{\sigma_t}{E}\right) \tag{8.38}$$

where ν is the Poisson ratio and E the Young's Modulus. Assuming constant strain, equation (8.38) yields

$$\sigma_t + \sigma_r = \text{constant} = -2a \tag{8.39}$$

where a is a yet-undetermined constant.

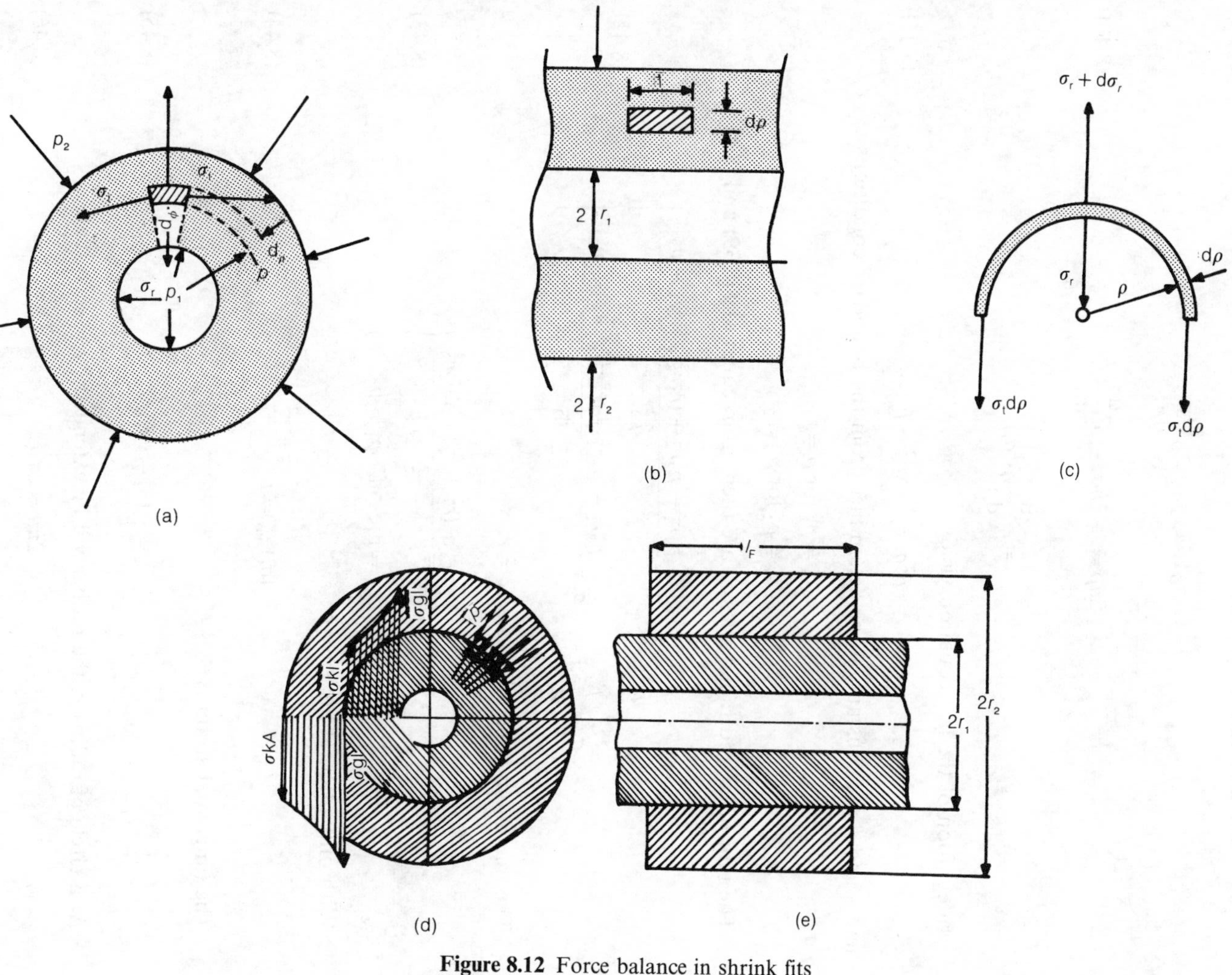

Figure 8.12 Force balance in shrink fits

Equations (8.37) and (8.39) yield

$$2a = -2\sigma_r - \rho\frac{d\sigma_r}{d\rho}$$

$$2a\rho = -2\rho\sigma_r - \rho^2\frac{d\sigma_r}{d\rho} \tag{8.40}$$

$$= -\frac{d}{d\rho}(\rho^2\sigma_r)$$

Integration of the last equation yields

$$\rho^2\sigma_r = -\alpha\rho^2 + \beta \tag{8.41}$$

where α and β are constants to be determined from the boundary conditions:

$$\sigma_r = p_1 \text{ for } \rho = r_1$$
$$\sigma_r = p_2 \text{ for } \rho = r_2$$

The tangential and radial stresses are then, from equations (8.39) and (8.41),

$$\left.\begin{aligned} \sigma_r &= \frac{\beta}{\rho^2} - \alpha, \quad \sigma_t = \frac{p_1r_1^2 - p_2r_2^2 + (r_1^2r_2^2/\rho^2)(p_1 - p_2)}{r_2^2 - r_1^2} \\ \sigma_t &= \frac{\beta}{\rho^2} + \alpha, \quad \sigma_r = \frac{p_2r_2^2 - p_1r_1^2 + (r_1^2r_2^2/\rho^2)(p_1 - p_2)}{r_2^2 - r_1^2} \end{aligned}\right\} \tag{8.42}$$

The maximum value of σ_t is at $\rho = r$:

$$\left.\begin{aligned} (\sigma_t)_{max} &= \frac{p_1(r_1^2 + r_2^2) - 2p_2r_2^2}{r_2^2 - r_1^2} \\ (\sigma_r)_{max} &= p_1, \quad \text{if} \quad p_1 > p_2 \end{aligned}\right\} \tag{8.43}$$

With only internal pressure p

$$(\sigma_t)_{max} = p\frac{r_2^2 + r_1^2}{r_2^2 - r_1^2} \tag{8.44}$$

The tangential strain is

$$\varepsilon_t = \frac{1}{E}(\sigma_t - \nu\sigma_r) \tag{8.45}$$

At a circle of radius ρ, the radius will be extended by u, such that

$$2\pi u = 2\pi\rho\varepsilon_t \tag{8.46}$$

From equations (8.45) and (8.46), the radial displacement is

$$u = \frac{\rho}{E}(\sigma_t - \nu\sigma_r) \tag{8.47}$$

Equations (8.42) and (8.47) yield

$$u(\rho) = \frac{(1+\rho)(p_1 r_1^2) - (p_2 r_2^2)\rho + (1-\nu)(p_1 - p_2) r_1^2 r_2^2/\rho}{E(r_2^2 - r_1^2)} \tag{8.48}$$

At the internal hub diameter, $\rho = r_1$,

$$u(r_1) = u_1 = \alpha_{11} p_1 + \alpha_{12} p_2 \tag{8.49}$$

where

$$\left.\begin{aligned} \alpha_{11} &= [(1+\nu) r_1^3 + (1-\nu) r_1 r_2^2]/[E(r_2^2 - r_1^2)] \\ \alpha_{12} &= [-(1+\nu) r_1 r_2^2 - (1-\nu) r_1 r_2^2]/[E(r_2^2 - r_1^2)] \end{aligned}\right\} \tag{8.50a}$$

At the external hub diameter, $\rho = r_2$, similarly,

$$u(r_2) = u_2 = \alpha_{21} p_1 + \alpha_{22} p_2$$

where

$$\left.\begin{aligned} \alpha_{21} &= [(1+\nu) r_1^2 r_2 + (1-\nu) r_1^2 r_2]/[E(r_2^2 - r_1^2)] \\ \alpha_{22} &= [-(1+\nu) r_2^3 - (1-\nu) r_1^2 r_2]/[E(r_2^2 - r_1^2)] \end{aligned}\right\} \tag{8.50b}$$

Again, if for the fit $p_1 = p$, $p_2 = 0$,

$$u_1 = \frac{r_1 p}{E}\left(\frac{r_2^2 + r_1^2}{r_2^2 - r_1^2} + \nu\right) \tag{8.51}$$

Recapitulating, an interference fit must be designed for:

1. *Minimum interference e_{min}.* The minimum pressure is given by equation (8.35). Under this pressure p, the shaft diameter is reduced by

$$u_s = -\frac{(1-\nu_s) r_2 p}{E_s} \tag{8.52}$$

assuming zero internal diameter and only external pressure. The minimum interference $e_{\min}$ must consist of the reduction of the shaft diameter plus the extension of the internal hub diameter, equation (8.51), plus the roughness of the shaft R_s and hub R_h.

Therefore

$$e_{\min} = \frac{p r_1}{E}\left(\frac{r_2^2 + r_1^2}{r_2^2 - r_1^2 + \nu}\right) + \frac{(1-\nu_s) r_1 p}{E_s} + R_h + R_s \tag{8.53}$$

2. *Maximum interference e_{max}.* This is computed from the allowable stress of the hub material S_y/N. The pressure needed to result in tangential stress S_y/N is given by equation (8.54)

$$p_{\max} = \left(\frac{r_2^2 - r_1^2}{r_2^2 + r_1^2}\right)\frac{S_y}{N} \tag{8.54}$$

Then, the corresponding extension of the internal hub diameter will be

$$e_{\max} = \frac{r_1 p_{\max}}{E}\left(\frac{r_2^2 + r_1^2}{r_2^2 - r_1^2} + \nu\right) \tag{8.55}$$

Equations (8.53) and (8.55) are the design equations for the interference fit.

The above figures assume 100% reliability. Since only a small percentage of fits have maximum interferences near e_{max}, if $e_m = (e_{max} + e_{min})/2$, the ranges $t_d = d_{max} - d_{min}$, $t_D = D_{max} - D_{min}$, analysis based on normal distribution yields the maximum interference, for $d = 2r_1$, $D = 2r_2$,

$$e_{max} = e_m + 0.5c[t_d^2 + t_D^2]^{\frac{1}{2}} \tag{8.56}$$

where the factor c is related to the desired reliability P as:

P	0.99	0.95	0.9	0.5
c	0.78	0.55	0.43	0.0

For $P = 1$, $c = 1$. For lower reliability, the maximum interference will be substantially lower and so will be the maximum design pressure.

Compound hubs, wheels, etc., will be discussed in detail in Chapter 14 where the procedure and computer program SHRINKFIT is presented which also can be used instead of the above analysis.

After the computation of e_{max} and e_{min}, the proper fit should be selected with the methods discussed in Chapter 2 (see also Example 2.2).

8.3.2 Prestressed joints

Bolts and rivets may, in certain load situations, be axially loaded, with the tensile loading being predominant.

Bolts are loaded by tightening the nut while rivets are loaded by thermal stresses during cooling.

During cooling, steel rivets with hot riveting almost invariably reach thermal stresses higher than the material yield stress, because the rivet is hot and the joined elements relatively cold. Therefore, in general, after hot riveting there is remaining stress equal to the rivet material yield stress. In fluid-tight joints this is desirable because the axial force developed presses the two parts together and the force might also be sufficient to support the joint load by friction.

For cold riveting, heat is produced during the plastic deformation of the rivet head which increases the rivet temperature. Upon cooling, thermal stresses are also developed, usually below the yield strength of the material.

The situation is shown in Figure 8.13. At the beginning of the hemispherical head formation, the hammering force is P_b, the stress S_y, and the diameter is d. At the end of the hammering, the force is P_e corresponding to a diameter approximately $2d$ and a stress equal to tensile strength S_u. Therefore

$$\begin{aligned} P_b &= \pi d^2 S_y/4 \\ P_e &= \pi d^2 S_u \end{aligned} \tag{8.57}$$

The change in height is computed for invariable volume, $\Delta h = 5d/3$ and assuming a linear law of variation of the hammering force, the mechanical work during the riveting process is

$$W = (P_b + P_e)\Delta h/2 \tag{8.58}$$

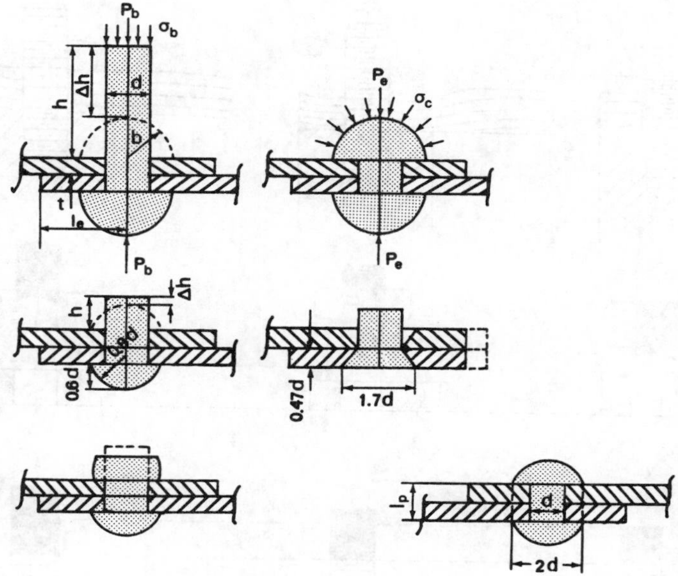

Figure 8.13 Cold riveting

Assuming that most of the produced heat enters the rivet, the rise of temperature will be

$$\Delta T = W/c\rho V \tag{8.59}$$

where c is the specific heat, ρ the density and V the volume of the rivet. Therefore

$$\Delta T = (5\pi/6)d^3[(S_y/4) + S_u]/c\rho V \tag{8.60}$$

The thermal stress during cooling

$$\sigma = \alpha E \Delta T = (5\pi/6)(\alpha E/c\rho)(d^3/V)[(S_y/4) + S_u] \tag{8.61}$$

where α is the coefficient of thermal expansion.

For bolts, the axial force is known from the applied torque on the nut and equations (8.27) to (8.30).

For both bolts and rivets, applications exist where loads are expected in service in addition to the initial loads imposed with tightening the nut or riveting. Such an example is the flanges of pressure vessels, engine heads, etc. (Figure 8.14). The head bolts are tightened initially to develop an axial force F_V (prestress), without internal pressure. When such pressure p is applied, a force is applied on the head, F_B. During prestressing, the bolt is elongated by δ_s while the flange is compressed by δ_f (subscript 'f' denotes flange and 's' bolt). Upon application of pressure p, with a head force F_B per bolt, the bolt is further elongated while the compression of the flange decreases. Since the elements are elastic, there is a linear relationship between forces and deformations on the bolt and flange (Figure 8.14a, b). At equilibrium, without pressure, the forces on both bolt and flange are equal to the prestress force F_V. Approaching the two force-deformation diagrams, for common point

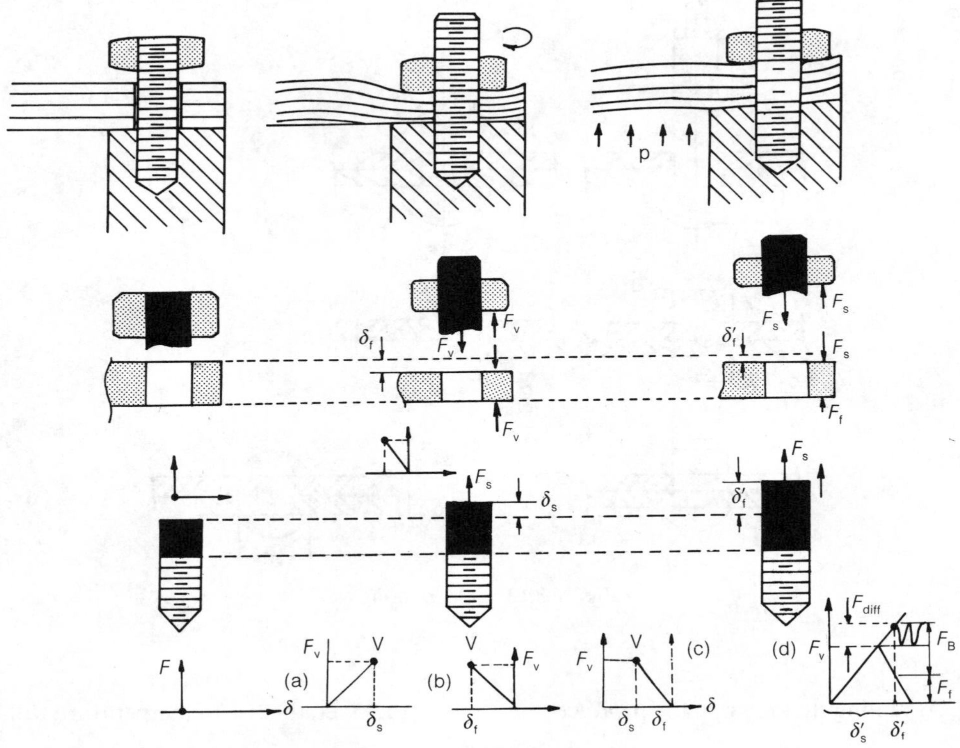

Figure 8.14 Prestressing of bolts

V, we obtain the diagram of Figure 8.14(c). There, the bolt and flange force–displacement relationships are described by the two straight lines. Geometry demands that the sum $\delta_s + \delta_f$ remains constant. External load F_B moves the equilibrium point to the right and the bolt load is F_s while the flange load is F_f. Obviously $F_B = F_s - F_f$. The increase of the bolt load is F_{diff}.

At point V, the prestress force

$$F_V = A_f E_f \delta_f / l = A_s E_s \delta_s / l; \frac{\delta_s}{\delta_f} = \frac{A_f}{A_s} \cdot \frac{E_f}{E_s} \tag{8.62}$$

because for a prismatic bar $\delta = Fl/AE$. Here A is the cross-section, l the length, F the force, and E the Young's modulus. The geometry of Figure 8.16(b) yields

$$\frac{F_{diff}}{F_B} = \frac{\delta_f}{\delta_f + \delta_s}, \quad F_{diff} = F_B \frac{1}{1 + (\delta_s/\delta_f)}$$

$$F_{diff} = C_K F_B, \quad C_K = \frac{1}{1 + (\delta_s/\delta_f)} \tag{8.63}$$

The factor C_K is computed from equations (8.62) and (8.63). The former equation assumes a prismatic flange, which is not very realistic because only a portion around the bolt is

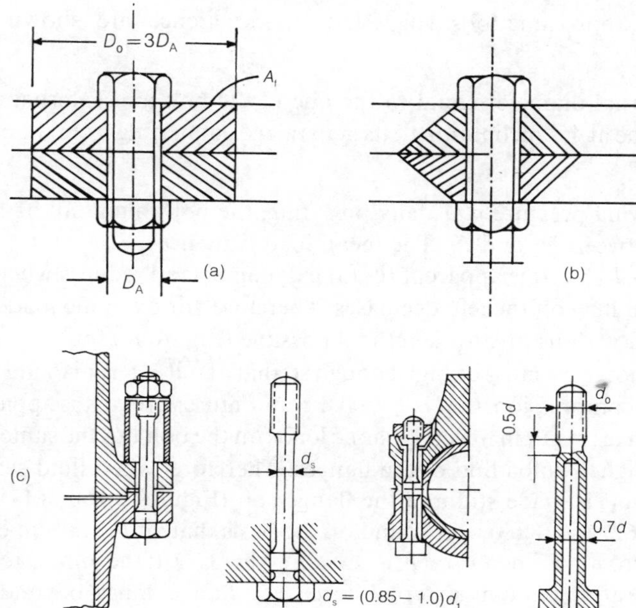

Figure 8.15 Bolts for dynamic loads

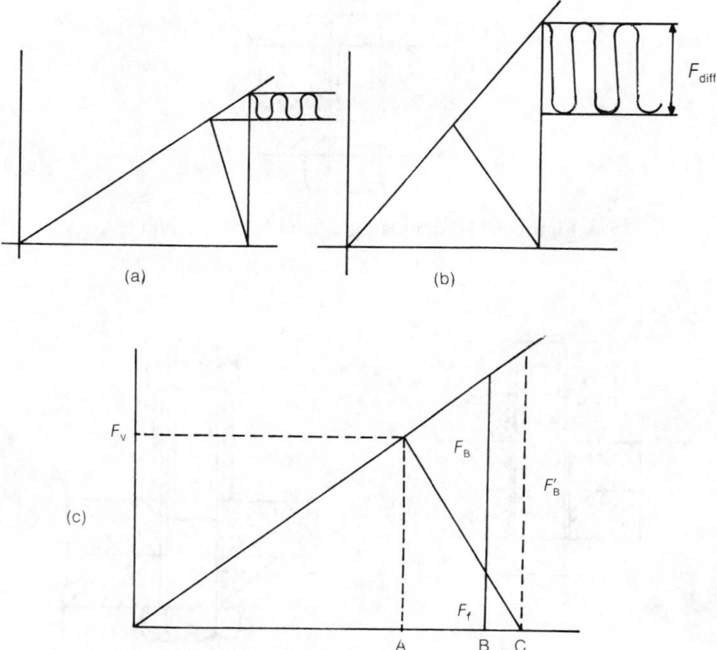

Figure 8.16 Prestressing for fluid-tight joints

deformed. Some approximations suggested by experience are shown in Figure 8.15, assuming:

(a) A ring of internal diameter equal to the one of the bolt and external three times that.
(b) A frustrum cone at 45° inclination to its generatrix, emanating from the external diameter of the nut.

For time-varying pressure p, usually pulsating, the bolt might fail by fatigue. The bolt load fluctuates between F_V and F_s. The mean load is then $F_m = (F_V + F_s)/2$ while the load range is $F_r = (F_s - F_V)/2$. It is apparent that the dynamic load reduces when the slope of the force–deformation line, on the left, decreases. Therefore, for dynamic loaded joints, slender bolts are designed with as great a length as possible (Figure 8.15c).

The prestress force F_V must be high enough so that at full external load F_B, the flange, will not have negative compression, that is $F_f > 0$. From Figure 8.16(a) it is apparent that slender bolts and rigid flanges result in lower dynamic loads on the bolt but the same external load F_B results in much higher unloading of the flanges. Therefore, when fluid sealing is of prime importance, the bolt is made stiff and the flange soft (Figures 8.16b, 8.17 and 8.18).

Equations (8.62) and (8.63) were found on the basis that, at maximum external load, the flange will have zero load. Then $F_B = F_s - F_f = F_V/(1 - C_K)$. If the joint is to be fluid-tight, as usually happens, at maximum external force the flange must be under compression. Therefore, it must be designed for a greater force F'_B. The situation is shown in Figure 8.16(c) where F_f is the remaining flange load. Denoting the ratio $AC/AB = N_P$, from the triangle

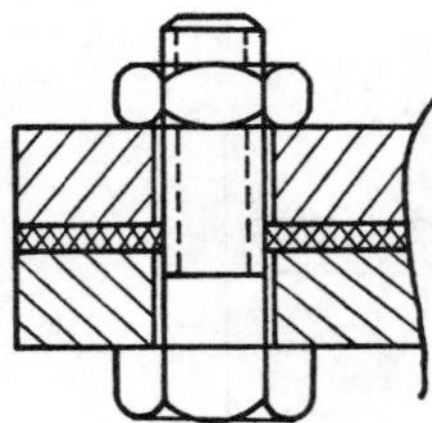

Figure 8.17 Increasing elasticity of the flange

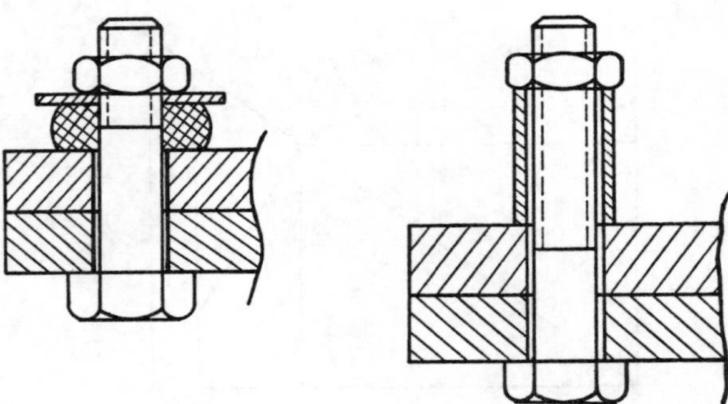

Figure 8.18 Increasing elasticity of the bolt

analogy it is found that

$$F_V = (1 - C_K)N_P F_B$$
$$F_{diff} = [1 - (1 - C_K)N_P]F_B = C_K N_P F_B \tag{8.64}$$

When $N_P = 1$ equation (8.64) is equivalent to equation (8.63). The value of N_P depends on the machining on the flange surfaces and the bolt spacing. It is suggested that $N_P = 2$ for static and 4 for dynamic loading must be selected. The ratio N_P is called the 'preload factor'.

For the computation of the steady and alternating forces on the bolt, we note that $F_m = (F_s + F_V)/2, F_r = (F_s - F_V)/2$.

8.3.3 Compound joints

In many instances, a joint might consist of several simple elements forming a compound joint and it requires special treatment. Such joints are, for example, compound riveting, bolting and welding with any loading condition, in particular bending and torsion.

Take, for example compound riveting (it could also be bolting or spot welding), Figure 8.19. It is loaded by shear force Q and bending moment M in the same plane. The shear force can be assumed equally distributed among the rivets. Each rivet has a shear force Q/n. The shear forces $P_1, P_2, \ldots, P_n$ owing to the bending moment are not equal in magnitude or direction.

Assuming that the plate is rigid and the rivets elastic, the shear deformation of each rivet, thus the shear force of each rivet owing to bending, is proportional to its distance from the area centroid. This is

$$\frac{P_1}{a_1} = \frac{P_2}{a_2} = \frac{P_3}{a_3} =, \ldots, = \frac{P_n}{a_n} \tag{8.65}$$

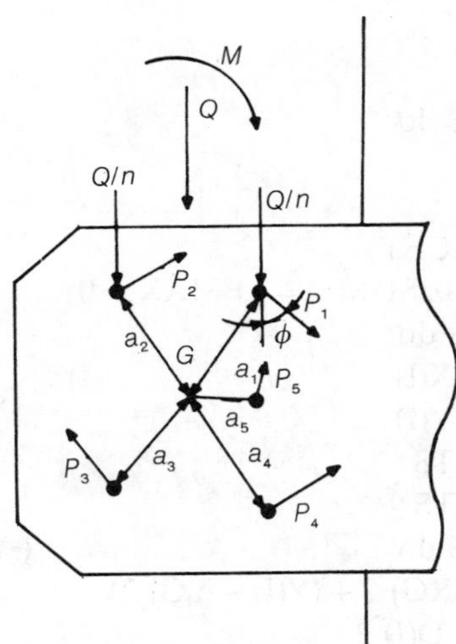

Figure 8.19 Compound joint

Balance of moments requires

$$M = P_1 a_1 + P_2 a_2 + , \ldots , + P_n a_n \tag{8.66}$$

Equations (8.65) and (8.66) yield

$$P_i = M a_i / \sum_j a_j^2 \tag{8.67}$$

P_1 here will apparently be the highest shear load owing to the torque. The total shear force on rivet 1 is then

$$F_1 = [P_1^2 + (Q/n)^2 + 2P_1(Q/n)\cos\varphi]^{\frac{1}{2}} \tag{8.68}$$

More precisely, the force F_i is perpendicular to the line connecting each rivet with the center of gravity of the riveting which has coordinates:

$$x_G = \left(\sum_i x_i\right)/n, \quad y_G = \left(\sum_i y_i\right)/n$$

The components of the force F are therefore,

$$F_{yi} = F_i/[1 + \lambda_i^2]^{\frac{1}{2}}, \quad F_{xi} = \lambda_i F_{yi}$$
$$\lambda_i = (y_i - y_G)/(x_i - x_G)$$

Program RIVETS, Appendix 8.A, performs this analysis. This program can also be used for bolting and spot welding. The corresponding procedure follows (explanations are in italics):

```
Procedure RIVETS
global FMAX: real var
local FORCEX, FORCEY, MOMENT, SUMX, SUMY, SUMD, XG, YG, P, PX
     PY, PXTOT, PYTOT arrays X, Y, D(NRIVETS): real var
     NRIVETS: int var
begin
  input NRIVETS                                  (*Number of rivets*)
  for I:= 1 to NRIVETS do
    input X(I), Y(I)                             (*Rivet coordinates*)
  input MOMENT                                   (*External moment*)
  input FORCEX, FORCEY                           (*External force*)
  SUMX:= 0; SUMY:= 0; SUMD:= ; FMAX:= 0
  for I:= 1 to NRIVETS do                        (*Finding center of area*)
    SUMX:= SUMX + X(I)
    SUMY:= SUMY + Y(I)
  XG:= SUMX/NRIVETS
  YG:= SUMY/NRIVETS
  for I:= 1 to NRIVETS do                        (*Finding distances, sum 1/d^2*)
    D(I):= SQR((X(I) - XG)^2 + (Y(I) - YG)^2)
    SUMD:= SUMD + D(I)^2
  for I:= 1 to NRIVETS do                        (*Net rivet forces*)
```

```
    P:= MOMENT*D(I)/SUMD
    RATIO:= (Y(I) – YG)/(X(I) – XG)
    PY:= P/SQR(1 + RATIO^2)
    PX:= PY*RATIO
    PXTOT:= PX + FORCEX/NRIVETS
    PYTOT:= PY + FORCEY/NRIVETS
    RIVETF:= SQR(PXTOT^2 + PYTOT^2)
    if FMAX < RIVETF then do
      FMAX:= RIVETF                          (*Maximum rivet force*)
  end.
```

For other compound joints without uniformity and symmetry of rivets of equal diameter, compound section properties must be considered. For this case and, generally, in the further study of strength, some geometrical characteristics of a section will be dealt with: static moments, moments of inertia, and section moduli. The static moment of an area with respect to an axis x (Figure 8.20) is the quantity defined by an integral of the form

$$S_x = \int_F y \, \mathrm{d}F \tag{8.69}$$

over the area F, where y is the distance of the elementary area $\mathrm{d}F$ from the x-axis.

The static moment of an area may be positive, negative or zero. The last defines the area centroid which has a distance from the arbitrary x-axis

$$y_c = S_x/F \tag{8.70}$$

Similarly, the static moment with respect to the y-axis is

$$S_y = \int_F x \, \mathrm{d}F \tag{8.71}$$

and the distance of the centroid from the y-axis

$$x_c = S_y/F \tag{8.72}$$

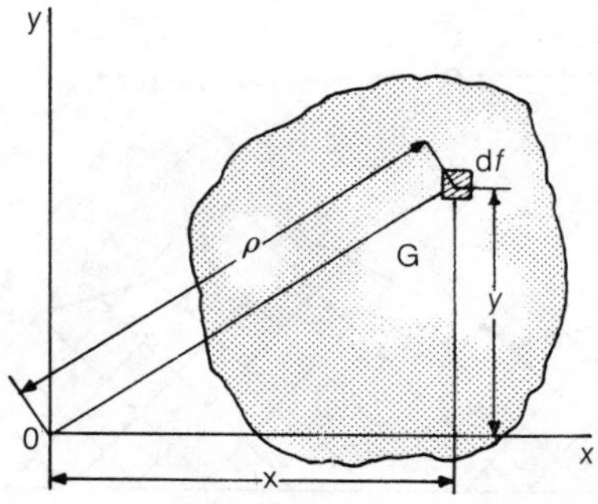

Figure 8.20 Element of a plane section

It can be seen that if the x and y axes pass through the centroid of an area, the static moment of the area with respect to these axes is equal to zero.

If a figure can be presented in the form of separate simple figures (squares, triangles, etc.) for which the positions of the centroids are known, the static moment of the area of the entire figure may be obtained as the sum of the static moments of the areas of these simple figures. Area computation is performed by summing up the areas of the trapezia formed by each side projected on the x-axis. The area of the trapezium under the side between nodes i and $i + 1$ is (Figure 8.21),

$$F_i = (x_i - x_{i+1})(y_i + y_{i+1})/2 \tag{8.73}$$

Note that this area is negative when $x_{i+1} > x_i$. If the nodes are numbered counterclockwise, then the area below the polygon will be negative. Therefore, summing up all trapezia will yield the polygon area.

The axial (or equatorial) moment of inertia of the cross-section area is equal to the integral:

$$I_x = \int_F y^2 \, dF, \quad I_y = \int_F x^2 \, dF \tag{8.74}$$

The product of inertia of the area of a figure is a geometric characteristic of the section defined by an integral of the form

$$I_{xy} = \int_F xy \, dF \tag{8.75}$$

The static moments needed for the computation of the centroid can be found for polygons by summing up the static moments of the trapezia (Figure 8.21),

$$\begin{aligned} A &= \sum (x_i - x_{i+1})(y_i + y_{i+1})/2 \\ \bar{y} &= \sum (x_i - x_{i+1})(y_i^2 + y_i y_{i+1} + y_{i+1}^2)/6A \\ \bar{x} &= \sum (y_{i+1} - y_i)(x_i^2 + x_i x_{i+1} + x_{i+1}^2)/6A \end{aligned} \tag{8.76}$$

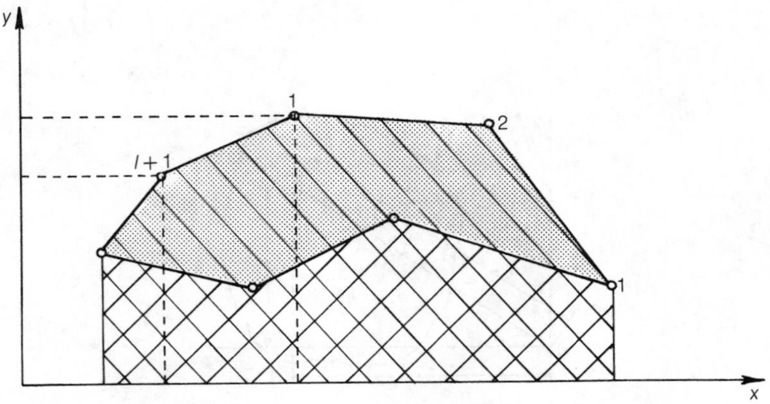

Figure 8.21 Numerical evaluation of plane-section properties

Similarly, the moments of inertia may be numerically computed for polygons by summing up the moments of inertia of the trapezia formed by the polygon sides

$$
\begin{aligned}
I_{XX} &= \sum(x_i - x_{i+1})(y_i^3 + y_i^2 y_{i+1} + y_i y_{i+1}^2 + y_{i+1}^3)/12 \\
I_{XY} &= \sum(x_i - x_{i+1})[x_i(9y_i^2 + 6y_i y_{i+1} + 3y_{i+1}^2) + x_{i+1}(3y_i^2 + 6y_i y_{i+1} + 9y_{i+1}^2)] \\
I_{YY} &= \sum(y_{i+1} - y_i)(x_i^3 + x_i^2 x_{i+1} + x_i x_{i+1}^2 + x_{i+1}^3)/12
\end{aligned} \tag{8.77}
$$

The product of inertia I_{XY} may be positive, negative and, in a particular case, zero.

If orthogonal axes x and y, or one of them, are axes of symmetry of a figure, the product of inertia with respect to such axes is equal to zero. A Cartesian system of axes x, y rotated by an angle α in respect to axes x, y, such that

$$\tan 2\alpha = 2I_{xy}/(I_y - I_x) \tag{8.78}$$

determines the position of two axes about one of which the axial moment of inertia is a maximum and about the other a minimum, while the product of inertia is zero. Such axes are called the 'principal axes'. Moments of inertia with respect to the principal axes are called the 'principal moments of inertia'.

The magnitude of the principal moments of inertia may be found as the principal stresses discussed in Chapter 6:

$$I_{max/min} = (I_x + I_y)/2 \pm 0.5[(I_x + I_y)^2 + 4I_{xy}^2]^{\frac{1}{2}} \tag{8.79}$$

The corresponding procedure follows:

```
Procedure SECTIONS
begin
  input NPOLY                                          (*Number of polygons*)
    for I:= 1 to NPOLY do
      input NNODE(I)                                   (*Nodes of polygon I*)
      for J:= 1 to NNODE(I) do
      input X(I, J), Y(I, J)                           (*Node coordinates*)
  for i:= 1 to NPOLY do                                (*Computing area*)
    for J:= 1 to NNODE(I) – 1 do
      AREA:= AREA + (X(I, J) – X(I, J + 1))*(Y(I, J) + Y(I, J + 1))/2
  for I:= 1 to NPOLY do                                (*Computing moments*)
    for J:= 1 to NNODE(I) – 1 do
      XI:= X(I, J); XIP1:= X(I, J + 1)
      YI:= Y(I, J); YIP1:= Y(I, J + 1)
      YBAR:= YBAR + (XI – XIP1)*(YI^2 + YI*YIP1 + YIP1^2)/(6*AREA)
      XBAR:= XBAR + (YI – YIP1)*(X1^2 + XI*XIP1 + XIP1^2)/(6*AREA)
      IXX:= IXX + (XI – XIP1)*(YI^3 + YI^2*YIP1 + YI*YIP1^2 + YIP1^3)/12
      IXY:= IXY + (XI – XIP1)*(XI*(9*YI^2 + 6*YI*YIP1 + 3*YIP1^2)
              + XIP1*(3*YI^2 + 6*YI*YIP1 + 9*YIP1^2))
      IYY:= IYY + (YIP1 – Y1)*(XI^3 + XI^2*XIP1 + XI*XIP1^2 + XI^3)/12
  ALFA:= ATAN(2*IXY/(IYY – IXX))/2                     (*Principal axes*)
  A1:= (IXX + IYY)/2
```

```
    A2:= .5*SQR((IXX + IYY)^2 + 4*IXY^2)
    I1:= A1 + A2; I2:= A1 - A2              (*Principal moments of inertia*)
  end.
```

This procedure has been coded in the program SECTIONS, Appendix 8.B.

The program is in the interactive mode. It considers the following:

(a) Polygons defined by their number of corner nodes n and their coordinates: According to the preceding discussion, if the definition of node coordinates is made counterclockwise, the section properties will be positive, representing solid sections. Section openings must be defined clockwise.
(b) Circular segments between an arc and the beginning and end radii: The coordinates of the center, the radius and the angles of the beginning and end radii in respect to the x-axis must be specified. Again, the sequence of the two angles will define positive or negative properties, that is solid segments or openings.

The program transforms the circular segment into a polygon dividing the arc into thirty straight segments. The error is of the order of 0.1%.

With this program, fairly complicated sections can be computed.

```
SECTION PROPERTIES PROGRAM

POLYGONS:
NO  1:0  0  50  0  50  100  0  100
NO  2:20  20  20  80  30  80  30  20

CIRCULAR SEGMENTS:
NO  1  X0 = 0   Y0 = 0     Radius = 15  Fbeg 90   Fend = 0
NO  2  X0 = 50  Y0 = 0     Radius = 15  Fbeg 180  Fend = 90
NO  3  X0 = 50  Y0 = 100   Radius = 15  Fbeg 270  Fend = 180
NO  4  X0 = 0   Y0 = 100   Radius = 15  Fbeg 360  Fend = 270
```

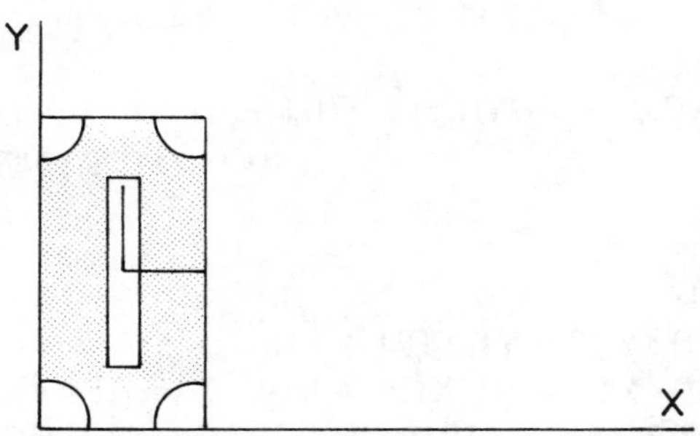


```
Area        3693.48784
Centroid    X0 = 25.0000016   Y0 = 50.0000001
Inertias:   Ix = 2630334.36
            Iy = 780209.581
            Ixy = 3410543.94
```

Principal Inertias: I1 = 2630334.36
I2 = 780209.58
Principal angle: – 1.39102712E-05 deg
MAX distances from principal axes:
HX:25.0000698
HY:50.0000422

A similar program which performs stress analysis of compound welded sections, WELDS, is presented in Appendix 8.C. Data are entered in the program interactively. This program treats welds with the simplifying assumption that weld width is very small compared with the other dimensions. The input–output follow.

Scale: 1 cm × 1 cm grid

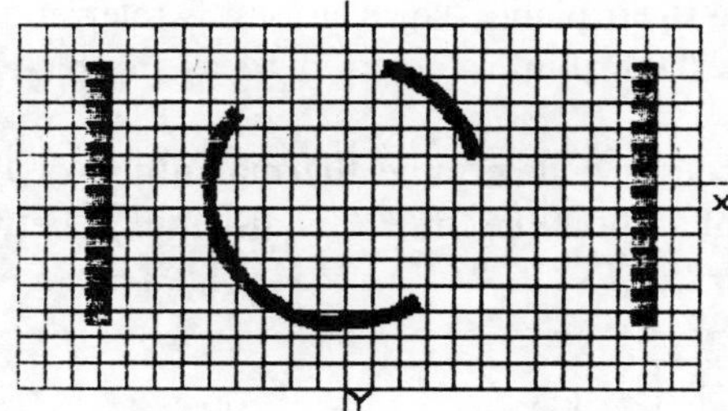

Scale: 1 cm × 1 cm grid

LOADING:

FX = 1000
FY = 2000
FZ = 3000
RX = 1000000
RY = 1000000
RZ = 5000000

AREA = 2016.58061
AREA CENTER:XG = .41033016 YG = 0

MOMENTS OF INERTIA:

IX = 16823.0904
IY = 20042622.3
IR = 20059445.4

RESULTS:

MAXIMUM EQUIVALENT STRESS = 54.6336206
AT THE STRAIGHT SEGMENT
NUMBER 1
AT POINT 2

8.3.4 Shear loaded joints

When a transverse force Q is acting on cross-sections of a bar and the other internal forces are equal to zero, this type of loading is called 'shear'. In this case only shearing stresses arise at the section.

In most practical problems a transverse force acts simultaneously with a bending moment and a longitudinal force so that normal stresses as well as shearing stresses usually act over the cross-sections. However, in cases where shearing stresses are considerably larger than normal stresses, shearing strength analysis only needs to be done.

A typical example of such a simplified but, as experience shows, quite reliable analysis is the calculation of the shear strength of riveted, bolted and welded joints.

Figure 8.22(a) shows two sheets joined by means of rivets (lap joint). A possible failure of a riveted joint is shown in Figure 8.22(b). The joint fails due to shear of rivets along the line of contact of the sheets.

If failure of each rivet occurs along a single shear plane, the rivet joint is referred to as single-shear (Figure 8.22a), if along two planes the joint is referred to as double-shear (Figure 8.22c), and so on.

Shearing stresses act over rivet shear planes. To determine the magnitude of these shearing stresses, it is necessary to know first of all how the force P is distributed between

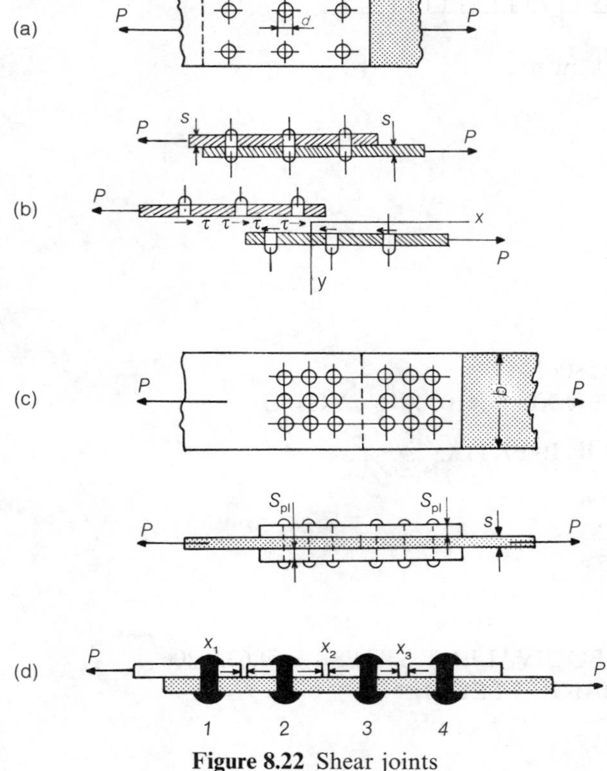

Figure 8.22 Shear joints

individual rivets. To do this the riveted joint should be calculated as a statically indeterminate system.

The redundant unknowns may be taken to be the forces between rivets and one of the sheets (Figure 8.22d). (Other variants of the basic system are also possible.) Equating to zero the displacements at the places where the sheets are cut, we obtain the equations of deformations necessary for the determination of the unknown forces.

After determining the forces in the sheets we find the shearing forces in the rivets as the difference of the forces adjacent to the rivet; for example, the shearing force in the third rivet (Figure 8.22d) is $Q_3 = X_3 - X_2$, etc.

The results of the solution of this problem for a lap joint with the same cross-sectional area of the sheets being riveted are presented in Table 8.10. It can be seen that unequal loading conditions for rivets become more distinct with increase in number of rivets in a row.

With six rivets the shearing forces in the outer rivets (first and sixth) are about 2.5 times larger than in the middle rivets (third and fourth).

In cases where elements of different cross-sectional area are connected, unequal loading conditions for rivets are encountered. The most overstressed rivets are on the side of the sheet having the smaller cross-sectional area.

Experiments show, however, that for ductile materials the rivets fail simultaneously under static loading. This is due to the fact that the forces in the rivets are equalized by the moment of failure as a result of ductility of the material and the clearances between the rivets and the sheets.

When impact and vibratory loads are in question, unequal loading conditions for rivets must be taken into consideration, in the form of stress concentration. Thus, under the action of a static load it may be taken that the shearing force in each rivet is

$$Q = P/n \tag{8.80}$$

where P is the force acting on the joint, and n is the number of rivets.

In the case of a riveted joint in double shear (see, for instance, Figure 8.22c), n denotes the number of rivets located on one side of the joint of the sheets being connected.

In addition, the assumption is made that shearing stresses are distributed uniformly over the shear plane though actually, as experimental observations show, their distribution is not uniform. However, a strict theoretical solution of this problem is rather involved, as there are clearances between rivets and sheets, frictional forces between sheets, etc. Furthermore, rivets are usually made of the most ductile steels, and hence irregularity in the distribution of

Table 8.10 Load sharing in multiple riveting

Number of rivets	Q_1	Q_2	Q_3	Q_4	Q_5	Q_6	Q_m
	expressed as fractions of P						
3	0.353	0.294	0.353				0.333
4	0.29	0.21	0.21	0.29			0.25
5	0.26	0.17	0.14	0.17	0.26		0.20
6	0.24	0.15	0.11	0.11	0.15	0.24	0.166

shearing stresses due to the occurrence of plastic deformations disappears by the moment of failure, by stress relaxation, as discussed above.

Having assumed the shearing stresses to be uniformly distributed over the cross-section of the rivet, we can easily find their magnitude.

Deriving the equation of equilibrium of the cut portion of the joint, for instance, of the upper portion (Figure 8.22b), we obtain

$$\sum X = 0, \quad -P + n\tau F = 0$$

whence

$$\tau = P/nF \tag{8.81}$$

where $F = \pi d^2/4$ is the cross-sectional area of a rivet with diameter d.

By using equation (8.80) we have from equation (8.81),

$$\tau = Q/F \tag{8.82}$$

The design equation for the shearing strength of rivets is then

$$\tau = P/Fn = Q/F \leqslant S_{sy}/N \tag{8.83}$$

Here S_{sy}/N is the allowable shearing stress.

From equation (8.83) the necessary number of single-shear rivets can readily be determined:

$$n = P/(\pi d^2 S_{sy}/4N) \tag{8.84}$$

With double-shear or multishear rivet joints, the total number of rivet shear planes located on one side of the joint should multiply n in equation (8.83), $2n$, $3n$, ...

The magnitude of allowable shearing stresses is usually established experimentally to reveal the influence of irregularity in stress distribution on the strength of a joint, the influence of frictional forces, of clearances, etc. In the design of rivets it is assumed that $S_{sy}/N = (0.6\text{–}0.8)S_y/N$ (Table 8.5).

Shearing stresses act (Figure 8.23) not only over the shear areas BC and AD but also, as follows from the law of the conjugate shearing stresses, over the areas CD and AB perpendicular to the former.

Both normal and shearing stresses act on inclined sections. The maximum normal stresses occur on the principal planes. Their magnitude and direction can be determined with the method presented in Chapter 6 and procedure COMLOAD.

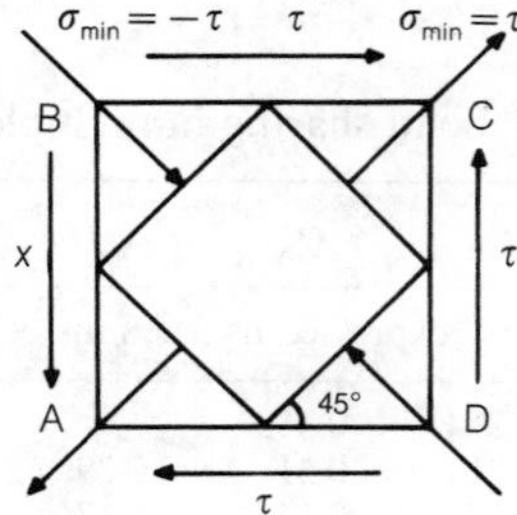

Figure 8.23 Shear loading

Table 8.11 Riveting joints for pressure vessels (after Decker)

No.	Figure	Rows	$P_{1\,mm}$ (kP/mm)	V_{max}	d_1 (mm)	t (mm)	e	e_1	e_2	s_L
1	a	1	50	0.6	$\sqrt{50s}-4$	$2d_1+8$	$1.5d_1$	—	—	—
2	a	2	40–95	0.7	$\sqrt{50s}-4$	$2.6d_1+15$	$1.5d_1$	$0.6t$	—	—
3	a	3	70–135	0.75	$\sqrt{50s}-4$	$3d_1+22$	$1.5d_1$	$0.5t$	—	—
4	b	1	35–95	0.65	$\sqrt{50s}-5$	$2.6d_1+10$	$1.5d_1$	—	$1.35d_1$	$0.6\cdots0.7s$
5	c	2	85–160	0.8	$\sqrt{50s}-6$	$5d_1+15$	$1.5d_1$	$0.4t$	$1.5d_1$	$0.8s$
6	d	2	85–160	0.8	$\sqrt{50s}-6$	$5d_1+15$	$1.5d_1$	$0.4t$	$1.5d_1$	$0.8s$
7	b	2	85–135	0.75	$\sqrt{50s}-6$	$3.5d_1+15$	$1.5d_1$	$0.5t$	$1.35d_1$	$0.6\cdots0.7s$
8	c	8	130–230	0.85	$\sqrt{50s}-7$	$6d_1+20$	$1.5d_1$	$0.4t$	$1.5d_1$	$0.8s$
9	b	3	110–240	0.7	$\sqrt{50s}-7$	$3d_1+10$	$1.5d_1$	$0.6t$	$1.5d_1$	$0.8s$
10	c	4	190–320	0.85	$\sqrt{50s}-8$	$6d_1+20$	$1.5d_1$	$0.4t$	$1.5d_1$	$0.8s$
11	b	4	180–320	0.7	$\sqrt{50s}-8$	$3d_1+10$	$1.5d_1$	$0.6t$	$1.5d_1$	$0.8s$

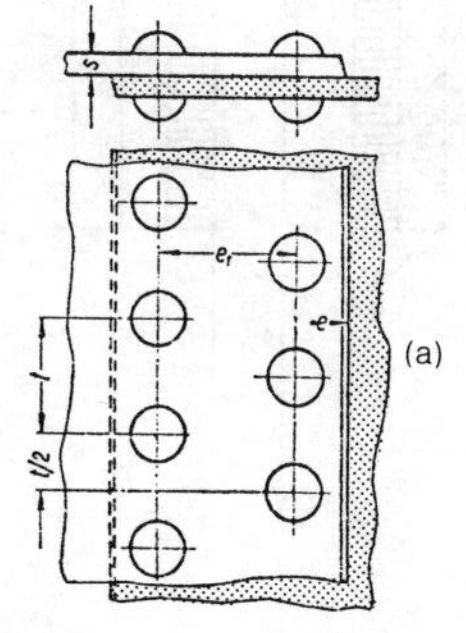

(a)

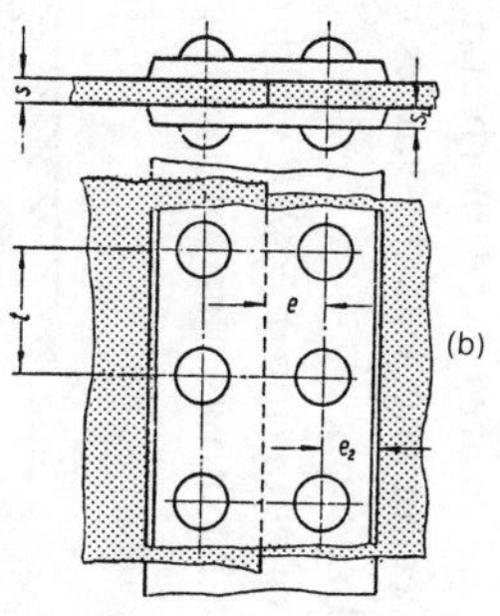

(b)

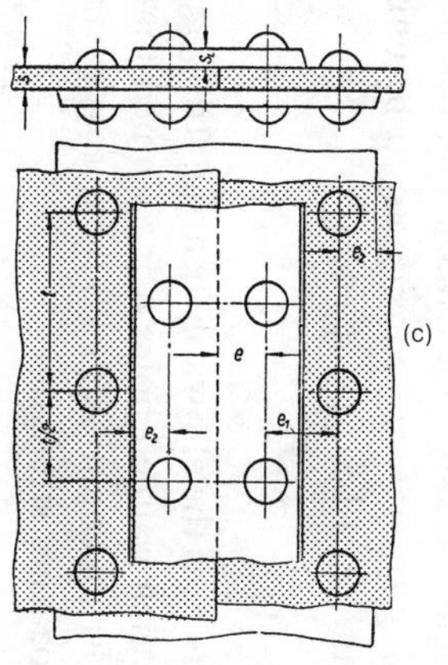

(c)

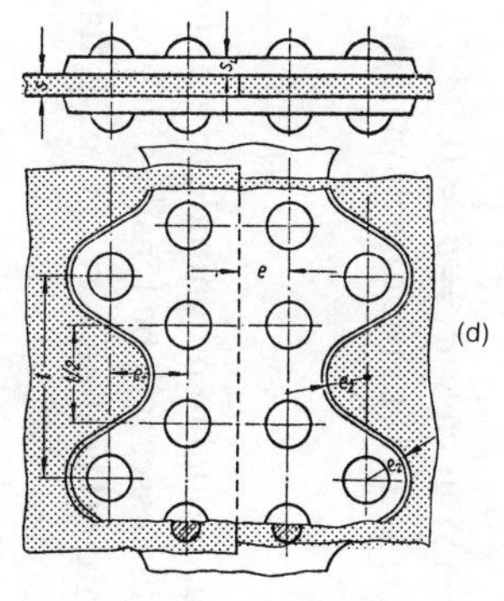

(d)

Since in our case, $\sigma_1 = \sigma_2 = 0$, $\tan 2\alpha = \infty$. Consequently, the principal areas are inclined at an angle $\alpha = 45°$ to the direction of the shear planes (see Figure 8.23). The principal stresses are

$$\sigma_{1,2} = \pm\tau \tag{8.85}$$

i.e. they are equal in magnitude to the shearing stresses acting on the shear planes; one of the principal stresses will be tensile, the other compressive (see Figure 8.23). Since two principal stresses are other than zero, shear, as has been pointed out above, is a particular case of the plane (biaxial) state of stress.

In addition to shear, bearing strength also has to be provided when rivet joints are designed. Bearing stresses are checked over the contact area between the rivets and sheets being connected. The projections of the bearing areas on the plane of the drawing are shown in Figure 8.22 by thick lines. The bearing area of one rivet is assumed to be $F = td$. The bearing stresses are considered to be uniformly distributed over the bearing area and the condition of bearing strength is

$$\sigma_b = P/n'F_b \leqslant S_b \tag{8.86}$$

where S_b is the allowable bearing stress and n' is the number of rivets. For drilled or punched and then reamed holes it is assumed that $S_b = 2S_u$.

From equation (8.87) the necessary number of rivets can be determined on the basis of the bearing strength:

$$n' = P/tdS_b \tag{8.87}$$

Of the two quantities n and n', the greater is taken.

In general, the characteristic load for a riveting is the load per unit width. Depending on this figure, the proper type of riveting is selected which, experience shows, will be near optimum. Decker suggests the values given in Table 8.11, which also gives some details for typical steel riveting.

By use of equation $\tau = Q/F$ welded joints loaded in shear can also be designed.

Figure 8.24 shows a lap joint of two sheets by means of transverse and side fillet welds. It

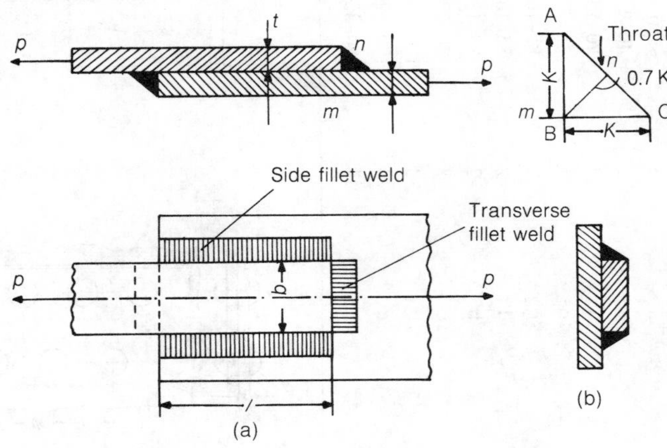

Figure 8.24 Weld geometry

is assumed that the dangerous section in the weld coincides with the plane passing through the bisector *mn* of the right angle ABC (Figure 8.24). Thus, for one transverse fillet weld the area of the dangerous section is $b(0.7k)$, and for one side fillet weld $l(0.7k)$ where k is the leg of the weld; in the case represented in Figure 8.24 the leg of the weld is equal to the thickness t of the upper sheet. The shearing stresses are assumed to be uniformly distributed over the area of the dangerous section. Based on the above assumptions, the allowable load for a transverse fillet weld is

$$P_{\mathrm{tr}} = 0.7kb(S_{\mathrm{sy}}/N) \tag{8.88}$$

The allowable load for one side fillet weld is

$$P_{\mathrm{s}} = 0.7lk(S_{\mathrm{sy}}/N) \tag{8.89}$$

Strength considerations obviously require that the total allowable resistance of the welds should not be less than the force acting on the joint:

$$P = \left(\sum_i a_i L_i\right)(S_{\mathrm{sy}}/N) \tag{8.90}$$

where a_i is the effective weld section width ($0.7k$ in the above case) and L_i the length of each segment.

By using this equation, it is possible, given the dimension k, to determine the required length of welds.

8.4 JOINT STRESS CONCENTRATION AND SAFETY FACTORS IN JOINTS

Stress concentration can be present in single or compound joints for many reasons.

Bolts have inherent stress concentration due to their geometry. The thread of the bolt is a succession of notches with substantial stress concentration. For fatigue loading, the effective stress concentration factor (K_{f}), for standard ISO threads, might be taken as follows for three bolt materials:

	Stress concentration factor, K_{f}	
Bolt material	Cut threads	Rolled threads
Carbon steels	3.0–4.0	2.4–3.2
Alloy steels	4.0–6.0	3.2–4.8
Titanium alloys	4.5–5.5	3.6–4.4

Lower values are for lower grades. Stress concentration appears also in the first threads which are heavier loaded than the distant ones, as one can see from Figure 8.25. This results in a non-uniform stress distribution (Figure 8.26) on the threads of both the bolt and the nut.

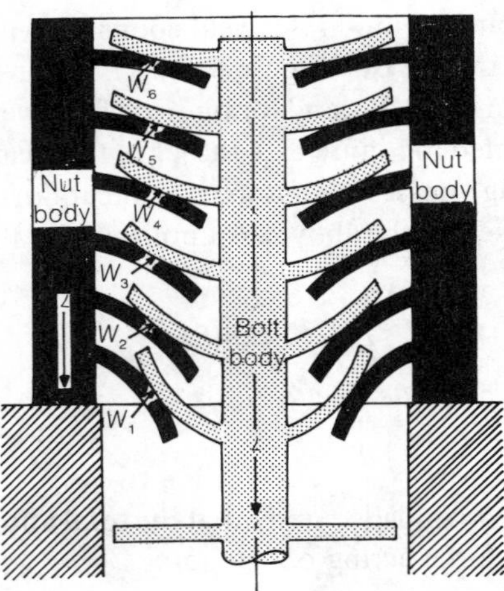

Figure 8.25 Stress concentration in the threads (ASME 1965)

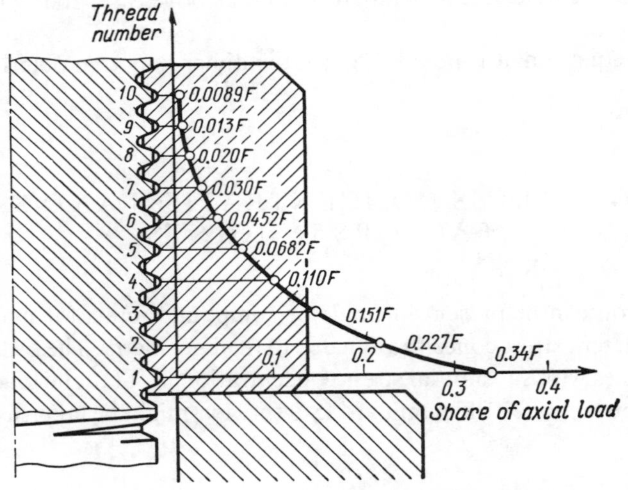

Figure 8.26 Stress concentration in a nut. (From Reshetov, by permission of Mir Publishers, Moscow)

Because of this, increase of the nut length beyond a point, does not add to the strength of the thread. The associated stress concentration factor K_f is between 1.3 and 1.8; the lower value is for standard coarse pitch metric threads and the higher value is for finer threads. This value must be applied for both static and dynamic loading because substantial yielding of the threads is not allowed for geometric reasons.

Safety factors for bolts, if no rational estimate of the safety factor is available, are taken between 1.5 and 3, depending on the control of the tightening torque. Lower values are for tightening with torque measuring wrenches and higher values for frequent tightening without accurate control of the tightening torque.

Fatigue strength of bolts can be improved by mechanical working of threads (rolling, shot peening) by 20–30%, surface hardening by 40% and cold rolled stock by 60%.

Stress concentration in bolt heads is also present. Special designs have been devised to allow for uniform transmission of the bolt load to the connected elements through the head and nut (Figure 8.27). Designs in Figure 8.27(c), (d) and (f) reduce substantially the stress concentration in the thread to almost 1.

Similar reasoning can be applied for stress concentration on welds.

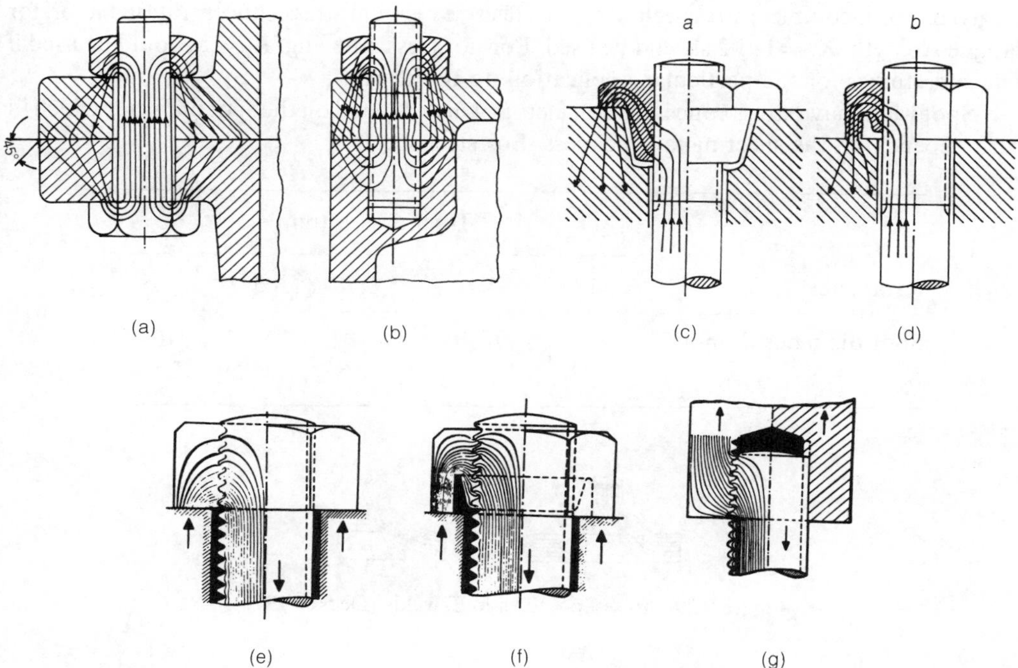

Figure 8.27 Stress flow lines between bolt and nut. (From Niemann, by permission of Springer Verlag, Berlin)

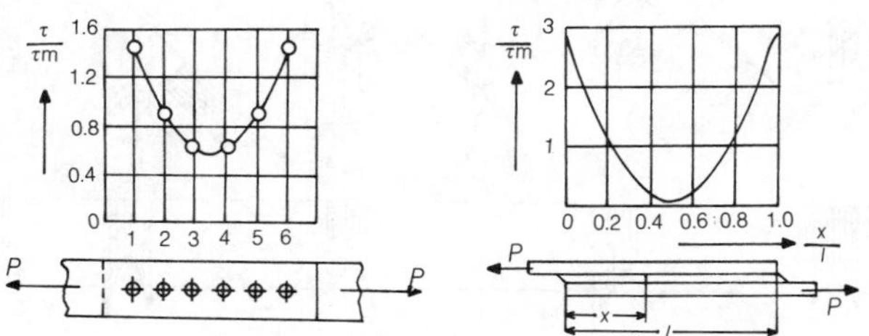

Figure 8.28 Stress concentration in long welds and riveting

Transverse welds show the same behavior as multiple riveting and the threads: The end parts of the weld are higher stressed, as shown in Figure 8.28. For ductile materials, yielding is acceptable and thus it is not a problem for static loads. For dynamic loads and brittle materials, long transverse seams must be avoided, since the stress concentration is difficult to estimate and a long weld is in this case not economical.

Figure 8.29 shows the transmission of load between the welded elements, while Figure 8.30 shows stress concentrations across the weld.

For different types of welds, effective stress concentration factors K_f are shown in Figure 8.31.

Welds are associated with internal stresses owing to the local yielding during the welding process. If full stress relieving is applied to the welded parts, no further consideration should be given. For incomplete stress-relieving, an effective residual stress concentration factor, for fatigue strength, $K_f = 1$–1.2 should be used. For no stress relieving, $K_f = 2$ should be used if relevant studies for the particular application do not exist.

Spot welds have stress concentration factors which depend on the spot diameter which in turn depends on the sheet metal thickness. For steel sheets:

	Thickness, *s* (mm)				
Parameter	1	1.5	2	3	5
Spot diameter *d* (mm)	4–8	6–10	8–10	10–12	10–14
K_f	1–2	1.5–2.5	2–2.5	2.5–2.9	2.5–4

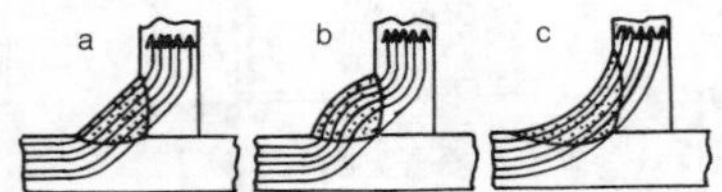

Figure 8.29 Stress flow lines in T-welds (Decker 1973)

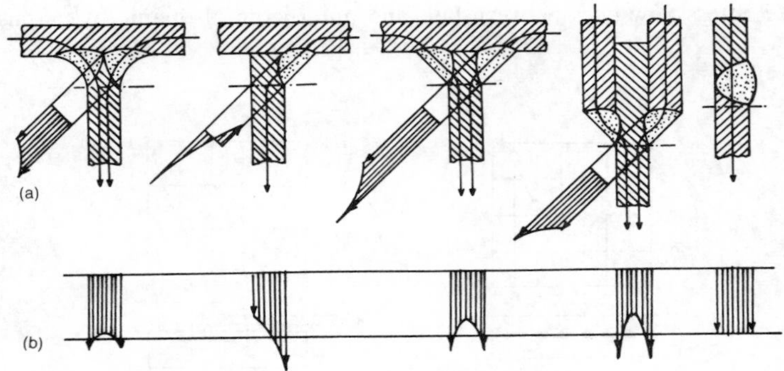

Figure 8.30 Stress concentration across the weld (Decker 1973)

Weld type	Design width, $t =$	Loading: Tension compression	Bending	Torsion
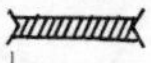	s	1	0.83	1.25
	s	2	1.66	2.38
	s	1.43	1.19	1.78
	s	1.09	0.91	1.37
	s	1.43	1.19	1.78
	s	1.25	1.02	1.54
	$2a$	3.12	1.45	3.12
	$2a$	2.85	1.43	2.86
	$2a$	2.44	1.15	2.44
	a	4.55	9.1	4.55
	s	1.59	1.25	2.0
	s	1.79	1.25	2.22
	s	1.43	1.19	1.79
	a	4.55	9.9	4.55
	$2a$	3.33	1.67	3.33
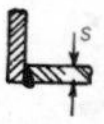	s	2.22	1.81	2.10
	s	1.67	1.33	2.0
	$2a$	2.87	1.43	2.86
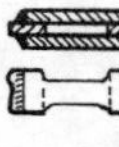	$2a$	4.55		
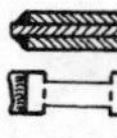	$2a$	4.0		
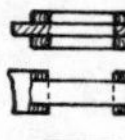	$2a$	4.0		
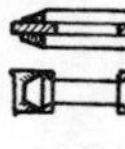	$2a$	2.08		

Figure 8.31 Stress concentration for different weld types (Niemann 1965)

For the end rivets in multiple row riveting, stress concentration occurs as for the end threads. The associated theoretical stress concentration factors K_t are given below and it depends on the number of rows:

	Number of rows				
	2	3	4	5	6
K_t	1	1.06	1.16	1.3	1.44

Stress concentration exists also in the riveted part due to the hole which from the photoelastic studies of Heywood (ASME 1965) is curve (b) in Figure 8.32,

$$K_{tn} = \sigma_{max}/[F/(D-d)t] = 2 + [1 - d/D]^3$$

where the subscript n indicates that the stress concentration factor was computed with the average stress over the net area.

The rivet in the hole increases the stress concentration factor, curve (a) in Figure 8.32.

For ductile materials, and static loads, only the reduction in the available section is accounted for owing to stress relieving. The section reduction factor is then

$$v = A/A_0$$

where A_0 is the theoretical section between adjacent rivets Dt and A the actual section $(D-d)t$. For pressure-vessel riveting, section reduction factors and design details are given in Table 8.12.

Safety factors for most riveting applications, such as construction and pressure vessels, are dictated by local codes. Usually, for steel construction factors of safety for rivetings are taken between 1.5 and 2, while for pressure vessels and steam boilers up to 5.

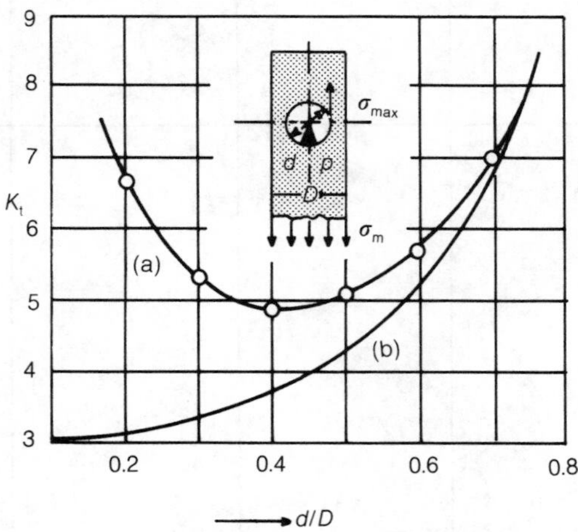

Figure 8.32 Stress concentration in a riveted connection

8.5 OPTIMUM DESIGN FOR STRENGTH

The objective for any design for strength effort is usually the optimum distribution of the load-carrying capacity of a structure in order to achieve the smallest possible weight and the lowest cost, or usually both since they are in close relationship. No general rules can be discussed, however, because the corresponding procedures are problem-oriented and the general ideas described in Chapter 5 apply, together with the methods of analysis discussed in this chapter and Chapter 4. Some pertinent problems will also be discussed here.

8.5.1 Optimum design of pin joints

Multiple pin joints, such as riveting and bolting, under shear loading, have an inherent deficiency in that the stress distribution within the elements is not uniform. For practical purposes, it is not technically sound to use fasteners of different size for the same joint, even for different joints of the same structure. Therefore, the design objective is to find the optimum distribution of the fasteners so that for a given load the minimum number will be used. Therefore, the formulation of the optimization problem should be:

(a) *Objective function*: number of fasteners of given size (cost).
(b) *Design variables*: spacing of fasteners, i.e. their location.
(c) *Constraints*: strength, space limitations.

The procedure OPTIMUM can be used in conjunction with the RIVETS procedure with a properly formulated penalty function.

8.5.2 Optimization of welds

Substantial savings in the amount of weld needed to carry a given load can be achieved by optimum distribution of the weld. The problem can be formulated as follows:

(a) *Objective function*: total weld length for given weld thickness.
(b) *Design variables*: the beginning and end points defining the several (perhaps) weld segments.
(c) *Constraints*: strength, maximum available length.

The procedure OPTIMUM can be coupled to the procedure WELDS.

EXAMPLES

Example 8.1

Design a riveted joint connecting two sheets of the same thickness $t = 16$ mm having two cover plates (Figure 8.22c) if $P = 60$ tonnes. Allowable stress $S_y/N = 160\ \text{N/mm}^2$, $S_{sy}/N = 100\ \text{N/mm}^2$, $S_b = 320\ \text{N/mm}^2$.

Solution

In this case the rivets are in double shear because in order for a joint to fail there must be a failure in

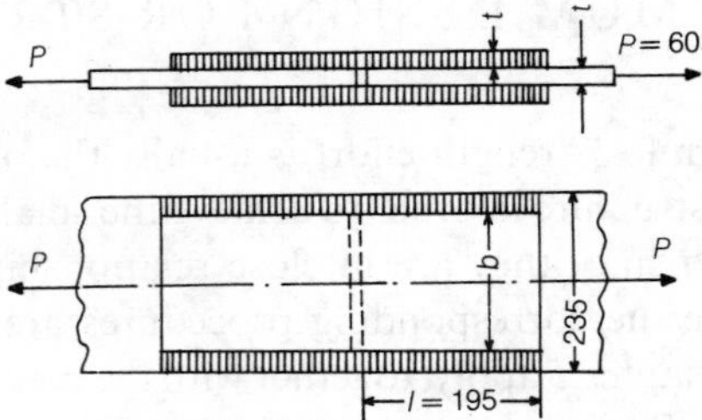

Figure E8.2

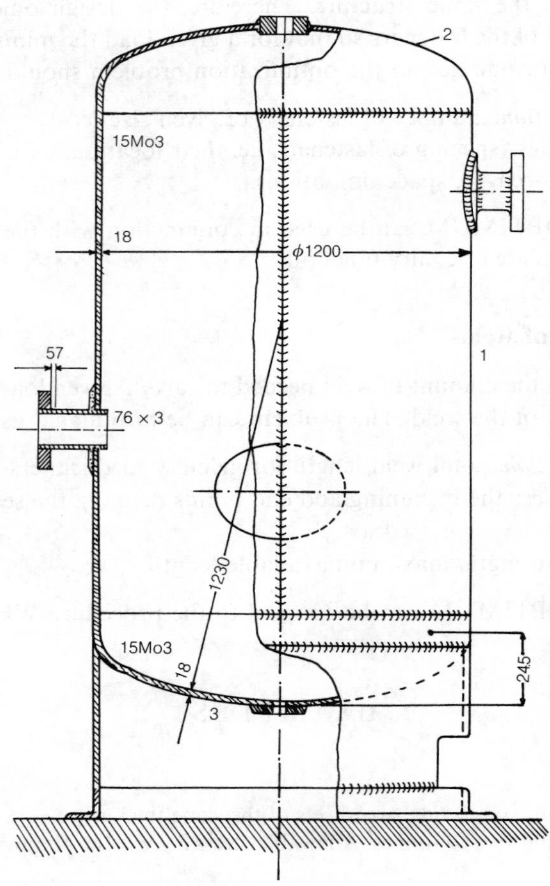

Figure E8.3 (after Decker)

each rivet by shearing along two planes. Take the diameter of a rivet $d = 20$ mm. The necessary number of shear planes is found according to equation (8.84):

$$n = \frac{P}{\frac{\pi d^2}{4}(S_{sy}/N)} = \frac{4 \times 600{,}000}{3.14 \times 20^2 \times 100} = 19.1 \text{ shear planes}$$

Consequently, it is necessary to take 10 rivets. The required number of rivets based on bearing is given by equation (8.87).

$$n' = \frac{P}{tdS_b} = \frac{600{,}000}{16 \times 20 \times 320} = 5.85$$

The design against resistance to shearing has proved to be decisive. We take 10 rivets on each side of the joint in three rows with 3 rivets in a row. Choose the section F of the sheet based on the tension on the sheet:

$$F = \frac{P}{(S_y/N)} = \frac{600{,}000}{160} = 3750 \text{ mm}^2$$

Hence for the thickness $t = 16$ mm we find the width of the sheet

$$b_1 = \frac{F}{t} = \frac{3750}{16} = 235 \text{ mm}$$

The width of the holes $3d = 60$ mm must be added to this working width to obtain the total width of the sheet $b = 235 + 60 = 295$ mm. This width is quite enough to take three rivets. (The distance between the centers of rivets is taken to be $3d$.) The thickness of each cover plate must not be less than the thickness of the sheet; we take $t_c = 16$ mm.

Example 8.2

Design a welded joint according to the data of the preceding example (Figure E8.2). The allowable shearing stress for welds is $S_{sy}/N = 110 \text{ N/mm}^2$.

Solution

In order to leave a place for laying out side fillet welds take the width of the cover plate b_c somewhat narrower than the width of the sheet b_s, i.e. $b_c = b_s - 2t = 235 - 32 = 203$ mm. According to the condition of equal strength, the cross-sectional area of the two cover plates must not be less than the cross-sectional area of the sheet, i.e. $2b_c t_c \geqslant F_s$. Hence the thickness of the cover plate is

$$t_c \geqslant \frac{16 \times 235}{2 \times 203} = 9.2 \text{ mm}$$

Assume $t_c = 10$ mm. The necessary working length of the side fillet welds, L, can be determined from the condition $S_{sy}/N > \Sigma P/L(0.7k)$. Hence $L = 600{,}000/0.7 \times 40 \times 110 = 195$ mm.

Example 8.3

The pressure vessel shown is welded and made of a hot-rolled steel grade with $S_y = 480 \text{ N/mm}^2$ at the operating temperature of 200 °C. A local code requires that $N = 2$. For this type of welding, the weld derating factor is $v = 0.8$. Calculate the wall thickness of the vessel for an internal pressure $p = 3.2$ MPa.

Solution

(A) Main vessel. The tangential stress

$$\sigma_t = pr/t \leqslant S_y/N, \quad t = pr/(S_y/N)$$

The thickness will be computed on the basis of tangential stress only, because at points of the peripheral seam near the tangential weld, the axial stress is very small because it is relieved by the welding process.

Therefore, $t = 3.2 \times 600/(480/2) = 8$ mm

(B) The curved end elements have the longitudinal stress

$$\sigma_z = K_t pr/2t \leqslant S_{sy}/N, \quad t = K_t pr/2(S_y/N)$$

The stress concentration factor $K_t = 2.0$ (Figure 8.31). Therefore,

$$t = 2.0 \times 3.2 \times 600/2(480/2) = 8 \text{ mm}.$$

Here, the tangential stress is very small because it is relieved by the end cover.

Example 8.4

The node point shown in a welded steel structure is loaded by axial loads $V = -30$ kN, $D = 50.75$ kN, $U_1 = 85$ kN, $U_2 = 40$ kN. The weld thickness is $a = 5$ mm. Check the weld for strength if the local building code requires allowable shear strength of the weld 105 N/mm^2.

Solution

(i) For the force V: $\tau = V/(\Sigma \alpha L) = 30{,}000/5(75 + 30) = 57.14$ N/mm^2.
(ii) For the force D: $\tau = D/(\Sigma \alpha L) = 50{,}750/5(90 + 40) = 78.1$ N/mm^2.
(iii) For the force U: the flange transmits the difference $U = U_1 - U_2 = 85 - 40 = 35$ kN. Therefore

$$\tau = U/(\Sigma \alpha L) = 35{,}000/5(90 + 45 + 45) = 38 \text{ N/mm}^2$$

All welds are safe, but overdesigned. Weld thickness or length for every member can be reduced.

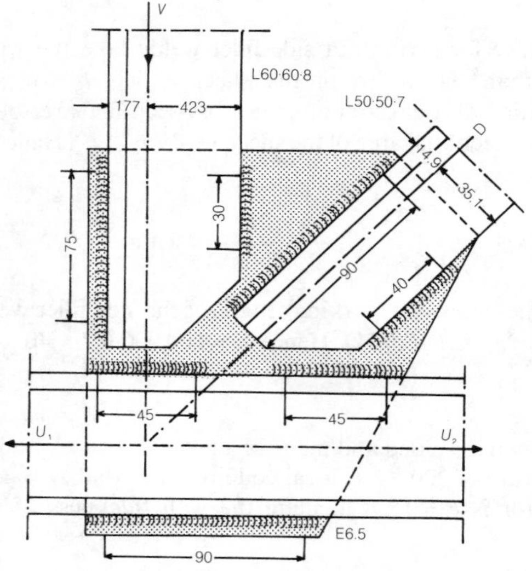

Figure E8.4 (after Decker)

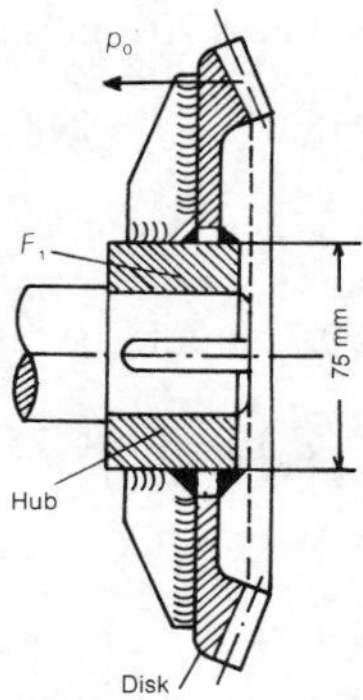

Figure E8.4 (after Decker)

Example 8.5

For the compound welded bevel gear shown, transmitting a torque of 10 kNm, and an axial force $P_a = 7500$ N, calculate the weld if the allowable shear stress is $S_{sy}/N = 120\ \text{N/mm}^2$.

Solution

(A) The torque is transmitted mainly through the peripheral weld of the disk to the hub. The section more severely loaded is the one facing the disk. If the thickness there is *s*, the section modulus, for small *s*,

$$W = \pi 2ds(d/2)^2/(d/2) = 2\pi d^2 s/2$$

The shear stress $\tau = M/W = S_{sy}/N$. Therefore

$$120 = 10{,}000 \times 10^3 \times 2/2 \times (\pi \times 75^2 \times s/2)$$

$$s = 10{,}000 \times 10^3 \times 2/2(\pi \times 75^2 \times 120) = 4.7\ \text{mm}$$

(B) The axial force (F_a) will be loading primarily the reinforcing fins and the force is transmitted through the four welds of the two fins at the hub. If *L* is the length at each weld and the same weld thickness *s* is used, the shear stress will be, for $a = s/(2)^{\frac{1}{2}} = 3.5$ mm,

$$\tau = F_a/(\Sigma \alpha L) = F_a/4aL = 120$$

Therefore $L = F_a/4 \times 120a = 7500/(4 \times 120 \times 3.5) = 4.46$ mm.

Example 8.6

The pressure vessel shown in Figure E8.3 will now be made riveted. Calculate the wall thickness and the longitudinal riveting, if the material has $S_y = 480\ \text{N/mm}^2$ and the safety factor must be $N = 3$.

Solution

The longitudinal seam transmits force per unit length *pt*:

$$P_{1\,\text{mm}} = PD/2 = 1200 \times 3.2/2 = 1920\ \text{N/mm}$$

Riveting of type 9 is selected (Table 8.11), with weakening factor $v = 0.7$. Therefore, the wall thickness will be

$$t = Dp/2(S_y v/N) = 17.5\ \text{mm}.$$

Allowing 1 mm for corrosion, wall thickness 20 mm is selected. As for the rivets, from Table 8.11

$$d_1 = (50s)^{\frac{1}{2}} - 7 = 24.6 = 25\,\text{mm}$$

spacing $w = 3d_1 + 10 = 85\,\text{mm}$, $e = 1.5d_1 = 38\,\text{mm}$, $e_1 = 0.6w = 51\,\text{mm}$, $e_2 = 1.5d_1 = 38\,\text{mm}$, and $s_L = 0.8t = 16\,\text{mm}$.

The weakening factor, taken as 0.7, should be tested

$$v = (s - d_1)/s = (85 - 25)/85 = 0.706$$

which is acceptable.

Example 8.7

A riveted joint for steel construction consists of two hot-rolled profiles $\llcorner 60 \times 60 \times 6\,\text{mm}$ and gosset plate. The member transmits force $U_1 = 154\,\text{kN}$. The two $\llcorner$ sections are joined with the 100-mm gosset plate with two 15-mm diameter rivets. Check the joint for riveting strength, if

$$S_{sy} = 120\,\text{N/mm}^2, \quad S_y = 160\,\text{N/mm}^2$$

Solution

For transmission of force there are two 177-mm^2 rivets (four sections),

$$\tau_a = U_1/mf_1n = 154{,}000/2 \times 177 \times 4 = 108.8\,\text{N/mm}^2$$

The acceptable safety factor is $N = S_{sy}/\tau_a = 1.1$. The gosset plate has tensile stress (bearing)

$$\sigma_1 = U_1/wd_1n = 154{,}000/10 \times 15 \times 4 = 256.6\,\text{N/mm}^2$$

The axial member is a double angle. It has a section of 691 mm^2 (see Appendix C.4). The section is reduced by $6 \times 15 = 90\,\text{mm}^2$, owing to the rivet. Therefore, the net section is $A = 691\text{–}90 = 601\,\text{mm}^2$. Therefore, because there are two loaded members, the average tensile stress is

$$\sigma = U_1/2A = 154{,}000/2 \times 601 = 128\,\text{N/mm}^2.$$

The acceptable safety factor is

$$N_s = S_y/\sigma = 1.25$$

Example 8.8

The electromagnetic coupling shown in Figure E8.8 has the friction material riveted on the coupling plate. The rivet material is ASTM A572 Grade 50 and the maximum transmitted torque is 500 Nm. Find the safety factor of the riveting. Rivet spacing is 90°.

Solution

The total transmitted force $P = T/r = 500/0.185 = 2710\,\text{N}$. This is divided into $3 \times 4 = 12$ rivets. Therefore, per rivet,

$$P_n = P/n = 2710/12 = 226\,\text{N}$$

The rivet section is $A = 32.17\,\text{mm}^2$. The average shear stress

$$\tau_a = P_n/mA = 226/1 \times 32.17 = 7\,\text{N/mm}^2$$

For the rivet material, $S_y = 225\,\text{N/mm}^2$. The yield strength in shear is, then, with the strain

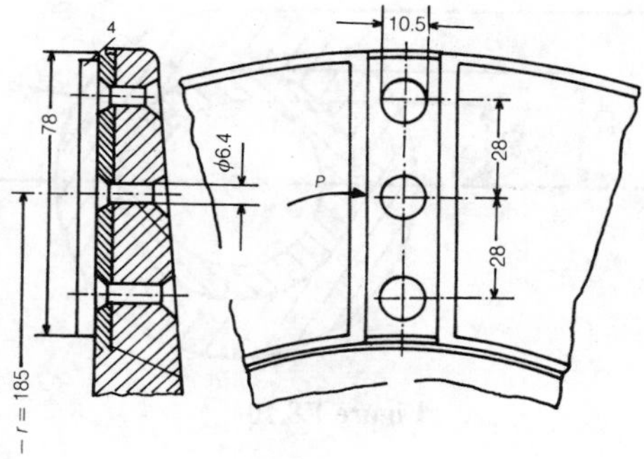

Figure E8.8

energy criterion $S_{sy} = 0.577S_y = 133\ \text{N/mm}^2$. Therefore, the safety factor is

$$N_s = S_{sy}/\tau_a = 133/7 = 19$$

The bearing load is

$$\sigma_b = P_n/sd = 226/4 \times 6.4 = 8.83\ \text{N/mm}^2$$

The bearing strength is $S_c = 2S_y = 2 \times 230 = 460\ \text{N/mm}^2$. The associated safety factor is

$$N_c = S_c/\sigma_c = 52$$

The smaller safety factor is to be applied, which is 19.

Example 8.9

A gear hub has inner diameter $d = 100$ mm, outer diameter $d = 200$ mm and width $b = 40$ mm. It transmits 15 hp at 140 rpm. The coefficient of friction between hub and shaft is $f = 0.4$, the hub material has yield strength 220 N/mm², the steel is ASTM A284 Grade D, and the desired safety factor is 3. Find the limit dimensions of the shrink fit. The surfaces can be machined to a roughness of 0.6 μm.

Solution

The minimum pressure required is

$$P_{\min} = 2M/f\pi b d_1^2 = 2 \times 71620 \times 15/140 \times 0.4\pi \times 4 \times 100 = 300\ \text{N/mm}^2$$

Minimum interference

$$e_{\min} = (\pi d_1/2E)[(d_2^2 + d_1^2)/(d_2^2 - d_1^2) - v] = 1\ \mu\text{m}$$

Maximum pressure

$$P_{\max} = (d_2^2 - d_1^2)/(d_2^2 + d_1^2)S_y/N = 40\ \text{N/mm}^2$$

Maximum interference

$$e_{\max} = (d_1 P_{\max}/2E)[(d_2^2 + d_1^2)/(d_2^2 - d_1^2) - v] = 20\ \mu\text{m}$$

Therefore, the fit should be, taking into account the roughness of shaft and hub,

$$d_1 = 100^{-0.0022}_{-0.0222}$$

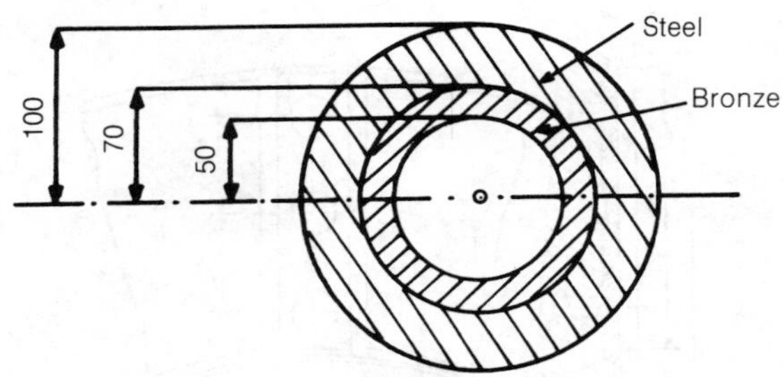

Figure E8.10

Example 8.10

A bearing consists of a bronze sleeve with $E_b = 1.1E11$ N/mm^2 and $S_{yb} = 70$ N/mm^2. It is shrink fitted inside a steel bore of $S_y = 220$ N/mm^2. Find the maximum allowable interference with a safety factor 1. Then, find the temperature of the steel ring necessary for the initial insertion of the bronze sleeve, if the coefficient of linear thermal expansion of steel is $\alpha = 12E - 6/°C$ and the Poisson ratio for both materials is 0.3. The surfaces are machined to a roughness 0.8 μm.

Solution

With the dimensions shown in Figure E8.10,

$$\sigma_t = -(2p \times 70^2)/(70^2 - 50^2) = S_{yb} = 70 \text{ N/mm}^2; \quad p = 17.1 \text{ N/mm}^2$$

The contraction of the bronze ring will be for $p_1 = 0$, $p_2 = 17.1$ N/mm^2

$$u_b = [(1 + 0.3)(-17.1 \times 70) \times 70 + (1 - 0.3)(-17.1)50^2 \times 70^2/70]/1.1E11 \times (70^2 - 50^2) = 0.05 \text{ mm}$$

Similarly, the expansion of the steel sleeve will be for $p_1 = 17.1$ N/mm^2, $p_2 = 0$

$$u_s = 0.015 \text{ mm}$$

The maximum interference will be the sum

$$e_{max} = u_b + u_s + 2(Ra_b + Ra_s) = 50 + 15 + 2 \times 2 \times 0.8 = 68.2 \ \mu\text{m}$$

The radial strain $\varepsilon_r = e_{max}/r_2 = \alpha T$, therefore

$$T = E_{max}/\alpha r_2 = 0.068/12E - 6 \times 70 = 77.4 \text{ °C}$$

over the room temperature.

REFERENCES AND FURTHER READING

ASME, 1965. *Metals Engineering-Design.* New York: McGraw-Hill.

Decker, K.-H., 1973. *Maschinenelemente*, Munich: Carl Hanser Verlag.

Duggan, T. V., Byrne, J., 1977. *Fatigue as a Design Criterion.* London: Macmillan.

Haenchen, R., Decker, K.-H., 1967. *Neue Festigkeitsberechnung fuer den Maschinenbau.* Munich: K. Hanser Verlag.

Kverneland, K. O., 1978. *World Metric Standards in Engineering.* New York: Industrial Press.

Machine Design, 1980. Fastening and joining issue. Cleveland: Penton/IPC.

Niemann, G., 1965. *Maschinenelemente.* Berlin: Springer Verlag.
Reshetov, D. N., 1980. *Machine Design.* Moscow: Mir Publishers.
Shigley, J. E., 1972. *Mechanical Engineering Design.* New York: McGraw-Hill.

PROBLEMS

8.1 The riveted steel construction node shown for a crane structure is loaded by a horizontal load $F = 47.5\,\text{kN}$. The thickness of the gosset plate equals the thickness of the web of the horizontal member. Steel is of the ASTM A572 grade 42 and the rivets are made of the same materials. Compute the rivet diameter for the riveting of the horizontal beam, if the safety factor is 1.5.

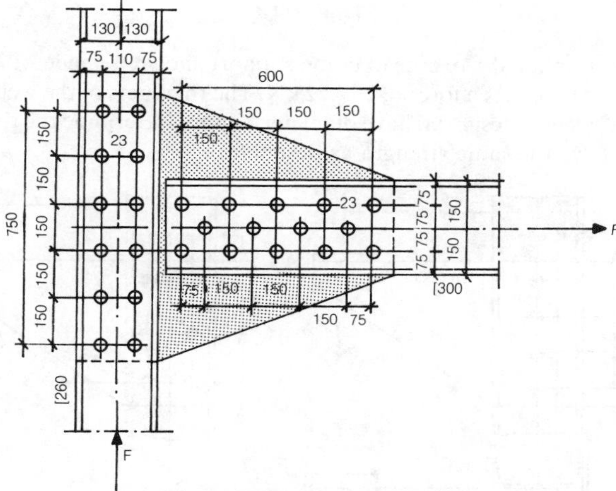

Figure P8.1 (after Decker)

8.2 In the riveting of Problem 8.1, compute the diameter of the rivets for the riveting on the vertical column.

8.3 A plate of 8-mm thickness is joined with a vertical steel beam of thickness 10 mm by way of two rivets and supports an eccentric weight $F = 12\,\text{kN}$. The rivets are 16 mm in diameter. If the material for all components is construction hot-rolled steel ASTM A572 grade 42 and the rivets are made from the same material, determine which one of the two solutions has higher safety factor.

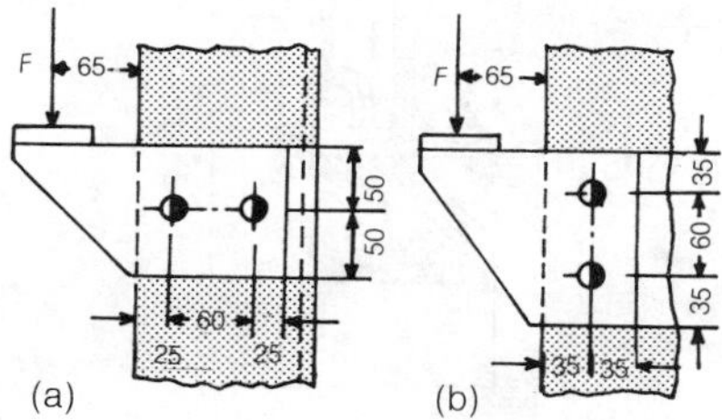

Figure P8.3

8.4 Solve Problem 8.3, except that the riveting is as in Figure P8.4, the thickness of the column web is 12 mm, the force $F = 35$ kN and the material is ASTM A570 grade D steel. Rivet diameter is 16 mm, and plate thickness is 8 mm.

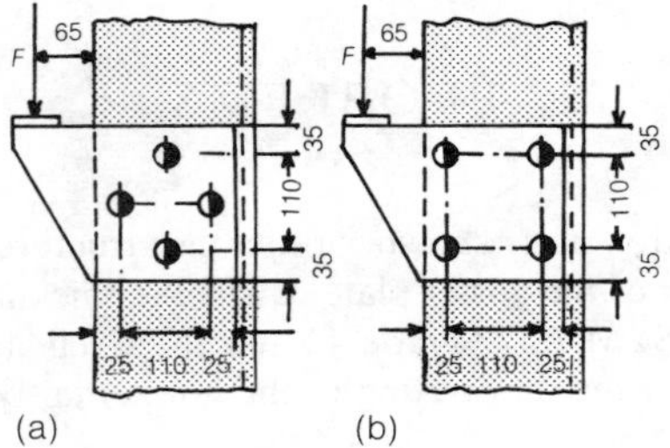

Figure P8.4

8.5 The steel construction for the overhead crane support shown is made of ASTM A284 grade D structural steel, and supports a force of $F = 72$ kN. The thickness of the web and the two plates is 14 mm. The eight rivets of design (a) have diameter 20 mm. Determine the rivet diameter of design (b) (12 rivets) to have the same strength as design (a).

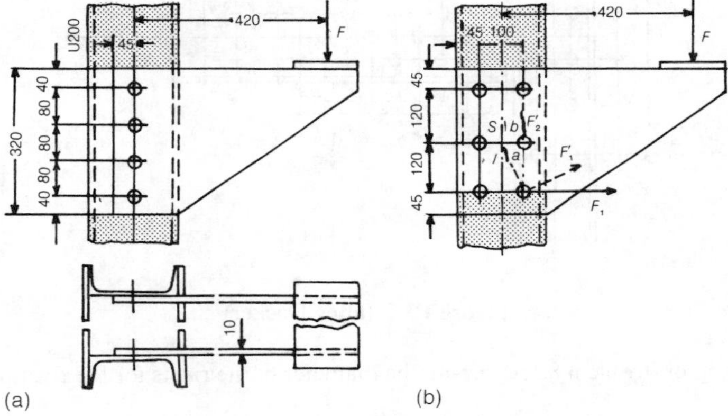

Figure P8.5

8.6 The chain gear shown is fixed on the plate of a toothed coupling with six rivets 6.4-mm diameter. It runs at 2,000 rpm with power 43 kW. The material of the gear and the nut is carbon steel SAE 1018. Determine the safety factor of the joint.

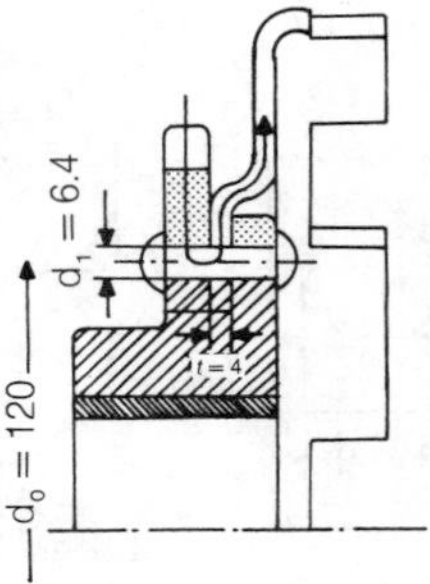

Figure P8.6 (Decker)

8.7 The blade of a centrifugal blower is attached to the rotating disk by four rivets of 8.4-mm diameter. Their material is ASTM A284 grade D steel and the material of the blade and disk is ASTM A570 grade D steel. The blade weighs 0.3 kg and its mass center is at 175-mm radius. Determine the necessary rivet diameter if the safety factor must be $N = 1.8$ and $n = 3000$ rpm.

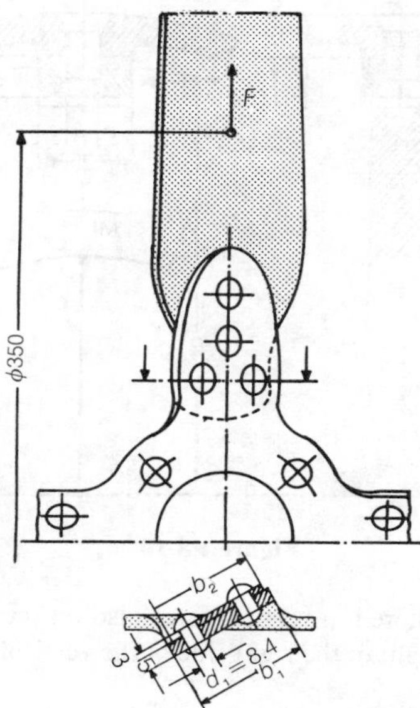

Figure P8.7 (Decker)

8.8 For a disk clutch, the plate is fixed on the disk by eight rivets of material carbon steel 1018 while the material of the disk and the plate is carbon steel 1025. Determine the necessary rivet diameter for a safety factor $N = 2$. The friction moment is generated by an axial force 80 kN at a diameter 130 mm with a coefficient of friction $f = 0.3$.

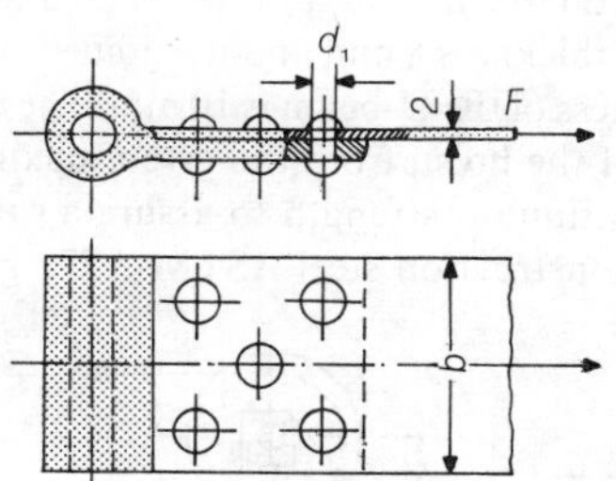

Figure P8.9

8.9 The strip band of a band brake is transforming a force $F = 12$ kN. It is fixed on the pulling bracket by five rivets. The material of the bracket and the rivet is ASTM A284-D structural steel while the band is made of a spring steel with $S_u = 400$ N/mm^2, and $S_y = 320$ N/mm^2. Determine the width of the band and the rivet diameter for a safety factor $N = 4$.

8.10 A brake drum is fixed on a disk rim with eight rivets as shown. Assuming material steel SAE 1020 for rivets, rim and drum, brake moment 1000 Nm, safety factor $N = 1.7$, determine the rivet diameter.

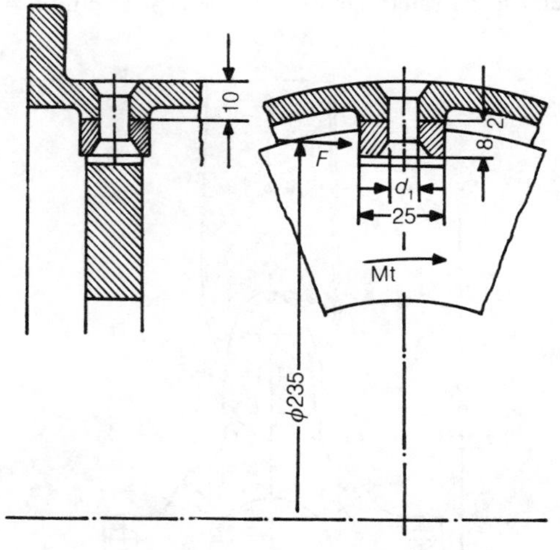

Figure P8.10

8.11 In the node of Problem 8.1, there is a vertical force also on the horizontal member of 50 kN at a distance of 2,000 mm to the right of the right side of the vertical column. Solve the problem using the RIVETS program.

8.12 Solve Problem 8.7 if there is also an axial force owing to the flow at the center of mass perpendicular to the plane of the paper, downwards, of magnitude 1000 N.

8.13 Solve Problem 8.9 for fatigue strength of the rivet and the band.

8.14 Solve Problem 8.10 if the drum is heated owing to abnormal braking to 150 °C while the temperature of the rim remains at 60 °C.

8.15 Two steel plates, riveted on two U-beams, support the shaft and pulley system of a hoist. The cable force is 6500 N. The plate has thickness 8 mm and it is joined by the U-beam by three rivets of 16-mm diameter. The web thickness of the U-beam is 10 mm. The cable force varies a great number of times during the operation of the hoist, from the given maximum value to zero. Find a proper material for the rivets for continuous strength to assure a safety factor $N = 2.5$. The beam and plate material is hot rolled construction steel ASTM A572 grade 42.

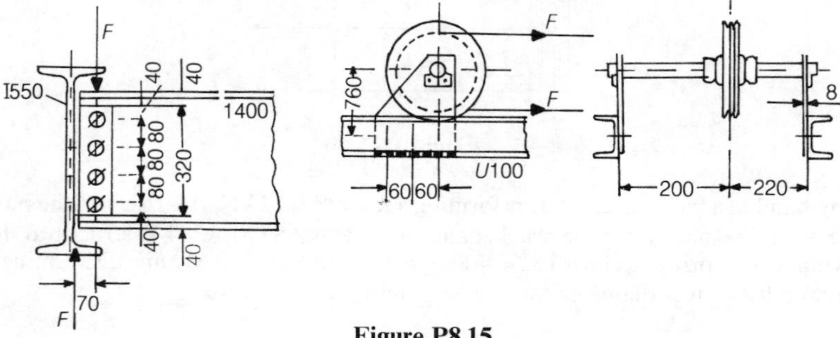

Figure P8.15

8.16 The turn-buckle shown consists of two bolts with opposite threads with the nuts on one piece. When the common nut is rotated, the two threaded studs are pulled towards one another. Design a 2-ton pulling force turn-buckle of class 4.6. The turn-buckle is rotated under load and the safety factor is taken as $N = 3$. Assuming length $L = 10$ mm, design the section between nuts if the turning is performed by a 10-mm diameter rod inserted between the turn-buckle rods which have circular cross-section and their centers are 40-mm apart.

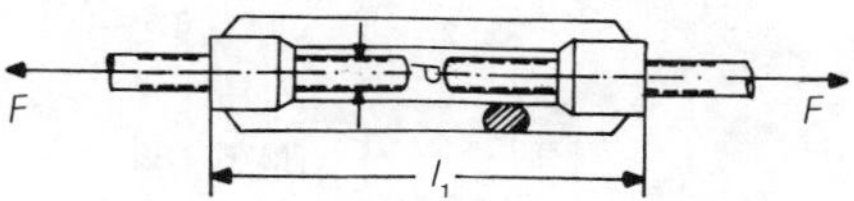

Figure P8.16

8.17 A hook is made of AISI 1020 steel and it is rated at 100 kN. Find the necessary nut length and the safety factor for the bolt.

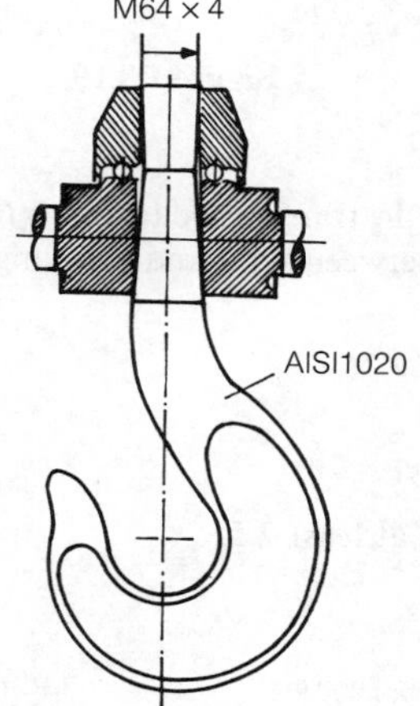

Figure P8.17

8.18 In the steel construction shown, the load is 45 kN. Determine the smallest safety factor for either one of the two bolts.

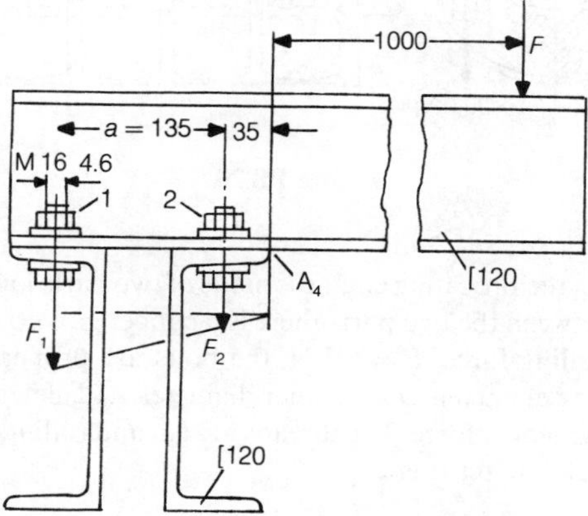

Figure P8.18

8.19 A coupling is joined by bolts as shown. The coupling material is ASTM A159 grade 3000, and the bolts are of Category 8.8. The transmitted torque is 10 kNm and the desired safety factor is 2.2. The torque must be transmitted with friction, supplied with sufficient bolt preload. The coefficient of friction is 0.2 between coupling faces. For the case of loosening the bolts, their shank must be able to support the torque by shearing. Determine the necessary number of bolts and the tightening torque.

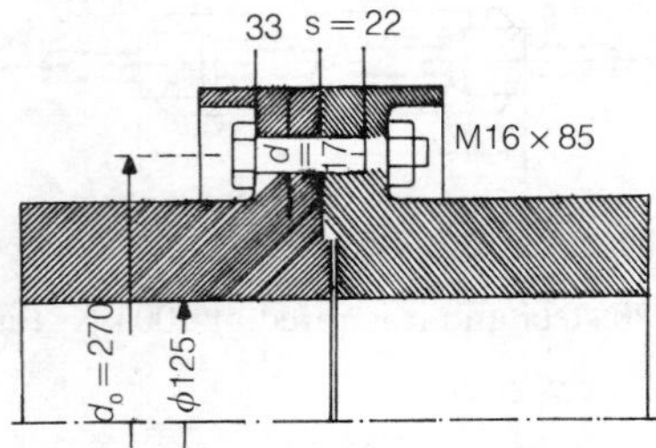

Figure P8.19

8.20 In the friction coupling shown, the transmitted torque is 600 Nm, the shaft diameter is $D = 60$ mm, and the coefficient of friction between shaft and coupling is 0.2. The coupling is held together by four bolts on each side.
Determine:

(a) The bolt diameter.
(b) The tightening torque.

The safety factor must be at least 2.5.

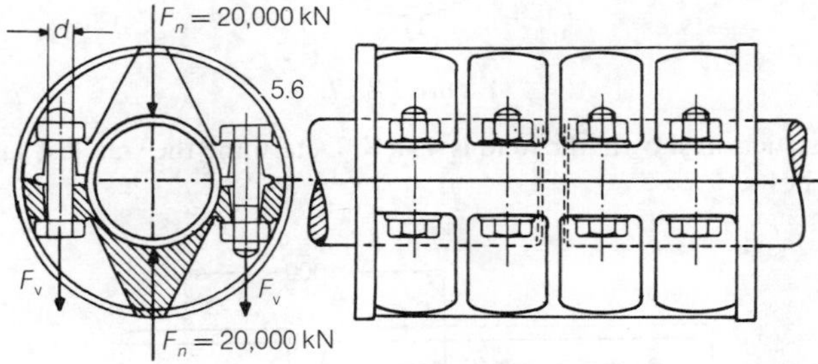

Figure P8.20

8.21 The follower bearing of a reciprocating engine is made of two parts joined together by two bolts of ISO/SAE Class 10.9. Between the two parts there is a copper gasket. To maintain contact during the operation of the pulling force, $P = 30$ kN, the bolts are properly preloaded. Assuming an associated flange of outer diameter $2d$ and inner diameter $1.1d$, determine the bolt diameter, the preload force and the tightening force. The threads are cut and rolling-hardened. The safety factor must be 1.8 and the preload factor is 2.

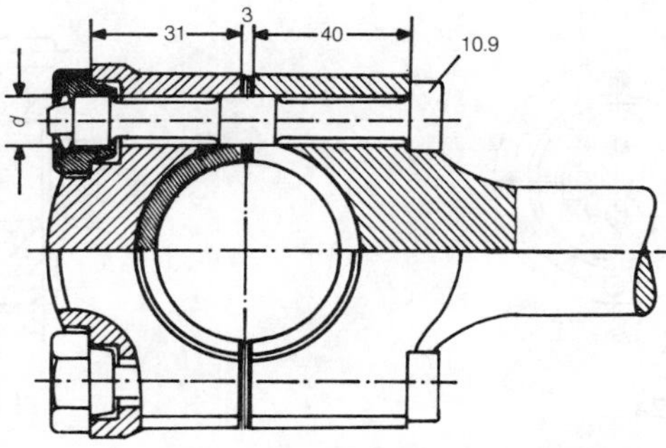

Figure P8.21

8.22 The head flange of a small diesel engine is secured on the engine body by six bolts, Category 8.8. The maximum pressure is 400 N/mm^2. The threads are cut without other treatment. The safety factor is $N = 2$. Determine the bolt diameter and the tightening torque if the preload factor is 1.5. Assume that the gasket compression is negligible.

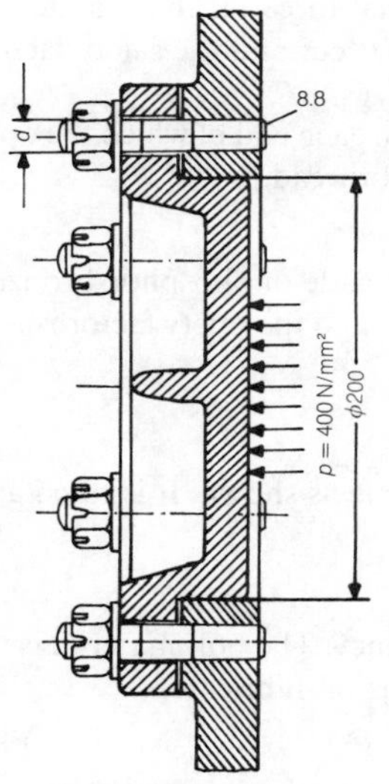

Figure P8.22 (Decker)

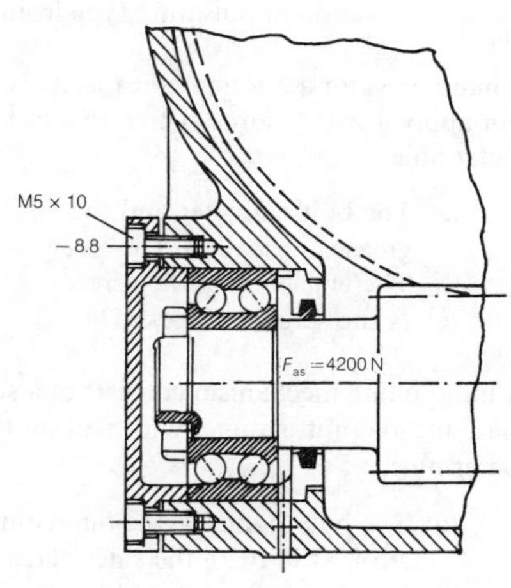

Figure P8.23

8.23 The left end bearing of the worm for a worm gear box is secured by a flange with four bolts. From experience it is known that $\delta_f/\delta_s = 0.6$, the safety factor is $N = 1.5$ and the preload factor is 1.1. Determine the bolt diameter for cut threads without further treatment. The bolt is Class 8.8.

8.24 A gear pump delivers pressure at 200 N/mm^2, assumed uniform over the flange. It is fixed by six bolts SAE 5.6.
Determine:

(a) The preload and the tightening torque for preload factor 1.5.
(b) The bolt diameter for rolled and heat-treated threads with safety factor $N = 1.8$.
(c) The safety factor if a 3-mm thick copper gasket is used with the previously-designed bolts.

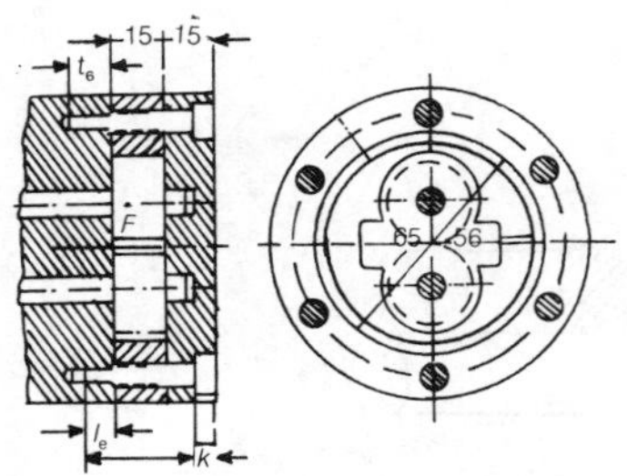

Figure P8.24

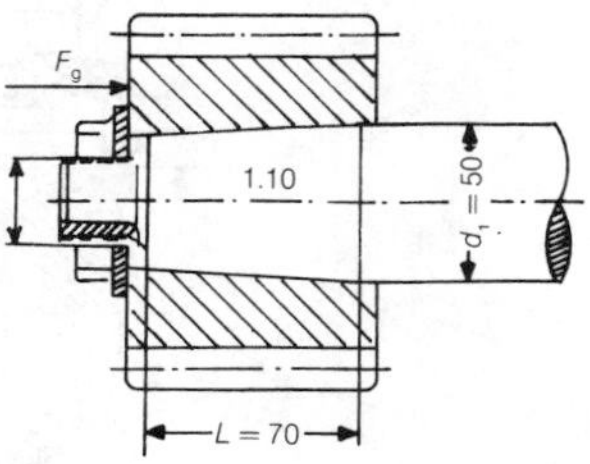

F_g
1.10
$d_1 = 50$
$L = 70$

Figure P8.25

8.25 A pinion gear is secured on a shaft with a tapered fit with a 10% taper. The torque transmitted is 500 Nm and the coefficient of friction between shaft and gear is 0.25.
Determine:

(a) The required axial force.
(b) The relation of axial force versus radial force versus radial movement versus axial movement.
(c) The necessary preload.
(d) The bolt diameter, if during operation, there is an axial force on the gear to the left direction of pulsating type from 0 to 5 kN. The thread is cut and the safety factor is 2.

8.26 A hand press for a 2-ton rated capacity is operated with a lever on each end of which the operator can apply a 200-N force. The screw has a standard trapezoidal thread.
Determine:

(a) The bolt diameter and the applied torque if the nut is made of phosphor-bronze, with grease lubrication. The screw material is AISI 1020 steel and the safety factor is $N = 2.2$.
(b) The efficiency of the screw.
(c) Is the screw self-locked?

8.27 A hub pulling mechanism consists of a screw and a holding frame as shown. It is rated at 5-kN load and the nut is directly formed on the cast iron flange.
Determine:

(a) The bolt diameter, the tightening torque and the efficiency. The bolt material is carbon steel AISI 1020, the safety factor $N = 2$, and the bolt is oil-lubricated.
(b) The maximum allowable travel L of the screw.

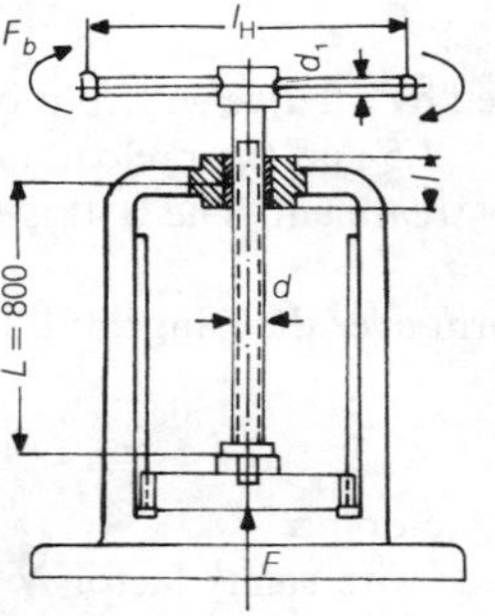

Figure P8.26

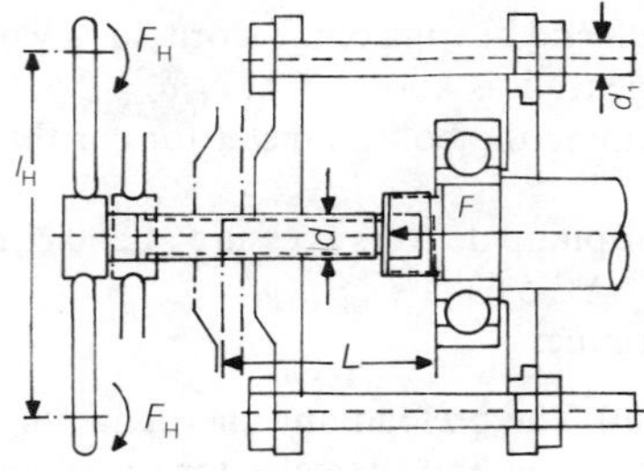

Figure P8.27

8.28 A screw jack has an oil-lubricated phosphor-bronze nut and an AISI 1015 steel screw with standard square thread 36 × 6. It is rated at 2-ton load.
Determine:

(a) The torque required at maximum load.
(b) The efficiency of the screw.
(c) The safety factor.

Is the bolt self-locked?

8.29 The steam valve shown has a standard ISO screw and a phosphor-bronze nut. Assuming 500-N peripheral force at a 100-mm radius at the valve wheel,

(a) determine the screw diameter and efficiency, for a safety factor $N = 2$;
(b) check for self-locking.

8.30 A 200-kN press consists of a power screw with square threads, phosphor-bronze nut, grease lubricated, operated by a worm gear system. Determine the screw diameter for a factor of safety $N = 2$, the efficiency, and the loading and unloading torque.

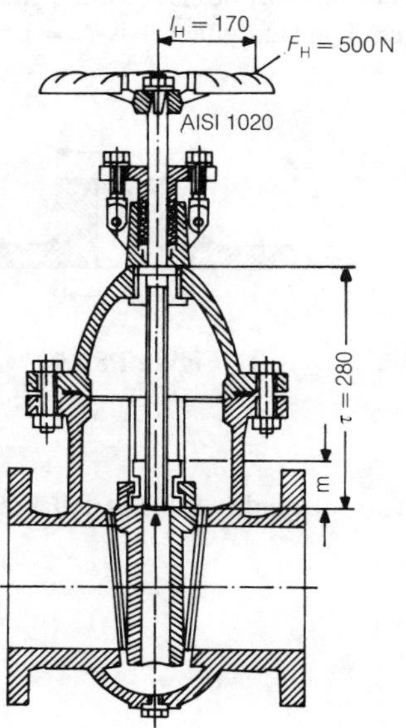

Figure P8.29

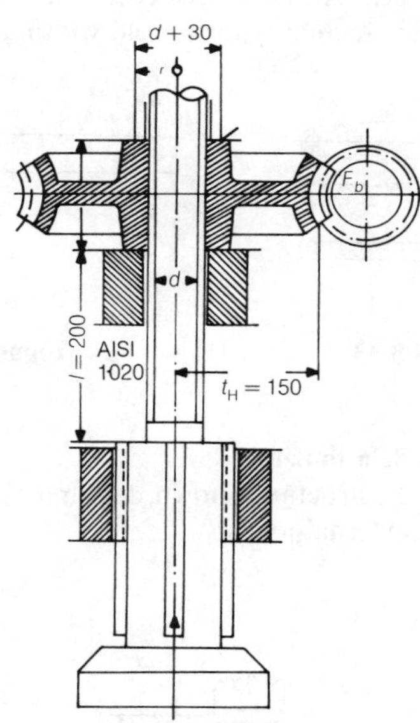

Figure P8.30

8.31 The I-beam shown is welded on a vertical steel column by welds of 6-mm width with a E60 electrode. The beam is loaded as shown by a 6-ton force. Assuming that the shear force is uniformly distributed, determine the safety factor of the welded structure.

8.32 A supporting structure is made of 80-mm diameter pipe with 3-mm thickness, welded on a vertical angle section. The welding has width $a = 5$ mm and the load F is static. Assuming a weld around the pipe determine, on the basis of the welds on the vertical column, the maximum permissible force F if the weld is made of quality E70 electrode.

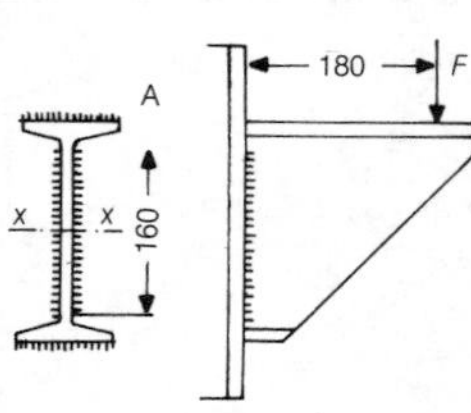

Figure P8.31

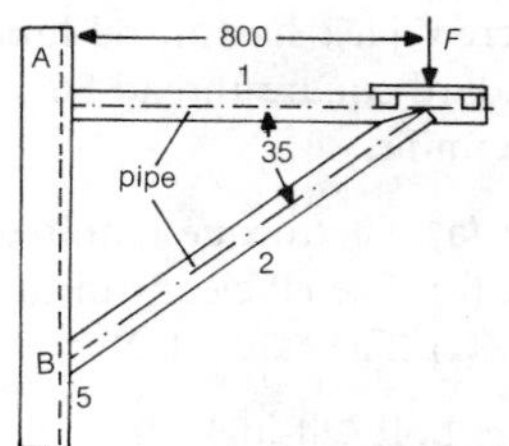

Figure P8.32

8.33 The bracket of Figure P8.33 is welded, as shown, on a vertical column. If the load F is 60 kN and the thickness of the plates is 12 mm, determine the weld width with E70 electrodes, for static load, with safety factor $N = 2$.

8.34 The lever shown is welded with E60 electrode and loaded as shown. For a safety factor $N = 1.5$, determine the weld width a.

8.35 The gear shown is welded on the hub and transmits a 500-Nm torque. Determine the proper electrode quality for a weld width $a = 5$ mm on both sides. The safety factor is $N = 1.6$.

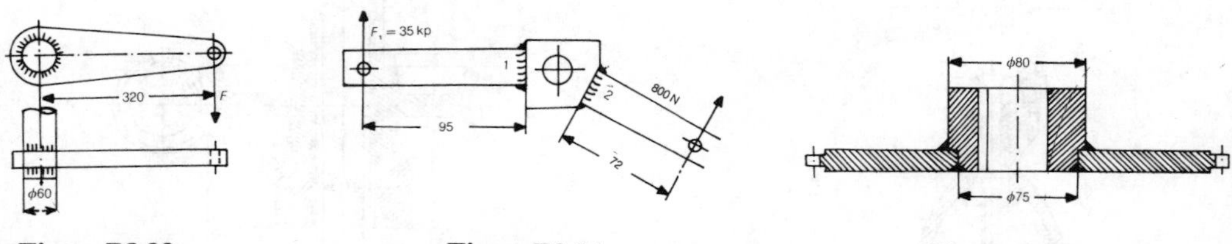

Figure P8.33 Figure P8.34 Figure P8.35

Problems 8.36 to 8.40

For the structures shown, determine the weld width a for electrode quality E70, safety factor $N = 2$ and external loads.

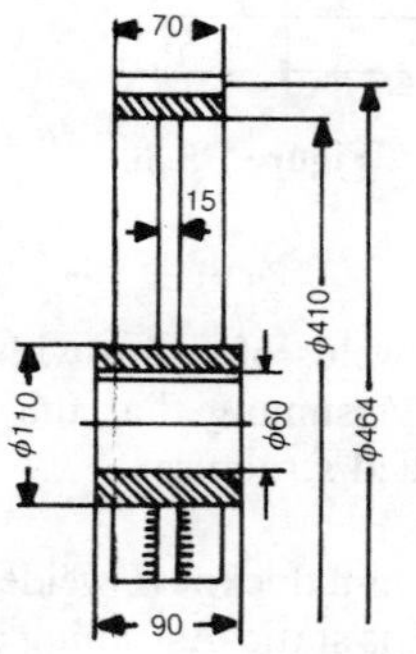

Figure P8.36

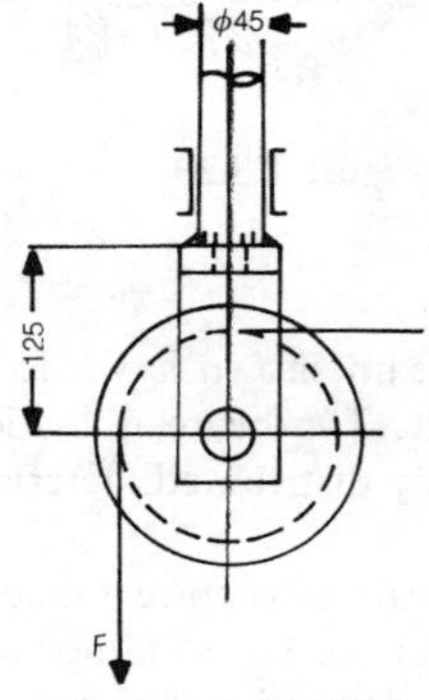

Figure P8.37

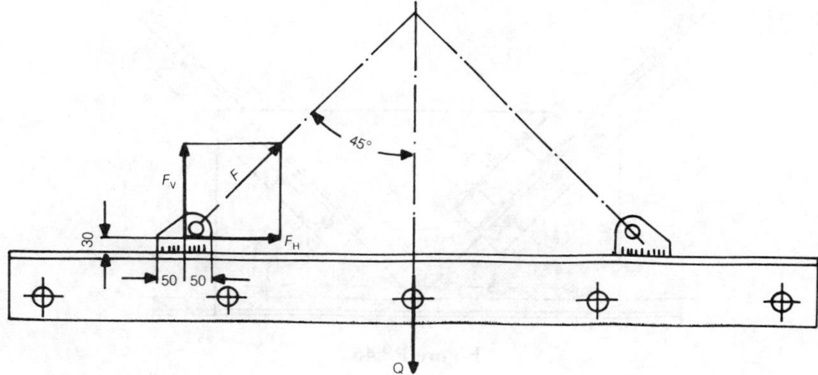

Figure P8.38 (Decker)

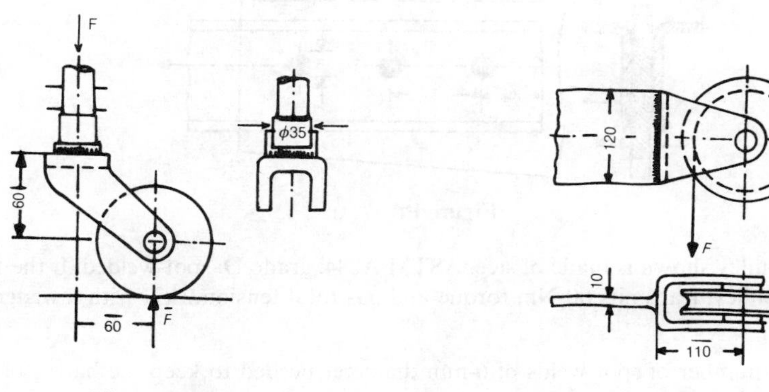

Figure P8.39

Figure P8.40

8.36 Power 11 kW at 100 rpm.

8.37 Force $F = 6\,\text{kN}$

8.38 Weight $Q = 50\,\text{kN}$.

8.39 Force $F = 3\,\text{kN}$.

8.40 Force $F = 5\,\text{kN}$.

Problems 8.41 to 8.45

Solve Problems 8.36 to 8.40 if the applied load is pulsating, that is, varies between zero and the given maximum value, for fatigue strength.

8.46 The node shown of a steel construction is made with spot welding and loaded as shown. Find the safety factor if the material is ASTM A284, grade D. $1\,\text{kp} \approx 10\,\text{N}$.

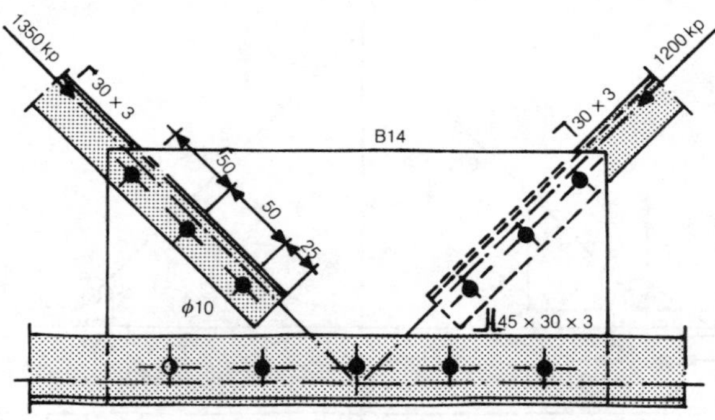

Figure P8.46

8.47 For the light-steel construction detail shown, with material steel ASTM A284, grade D, find the maximum of the loading force F if the safety factor must be $N = 2$.

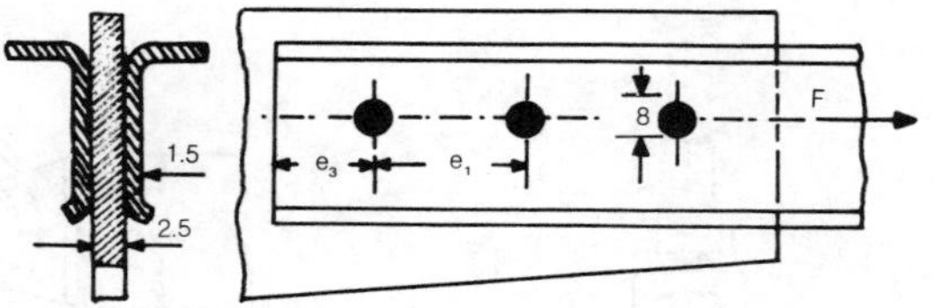

Figure P8.47

8.48 The V-belt pulley shown is made of steel ASTM A284, grade D, spot-welded. If the belt is 180° around the pulley, transmits 60-Nm torque and has total tension 2 kN with a wedge angle 30° determine:

(a) The number of spot welds of 6-mm diameter needed to keep the halves of the pulley together.
(b) The number of spot welds between the left half of the pulley and the hub, of 6-mm diameter.

The material is steel ASTM A284, grade D and the safety factor $N = 2.5$.

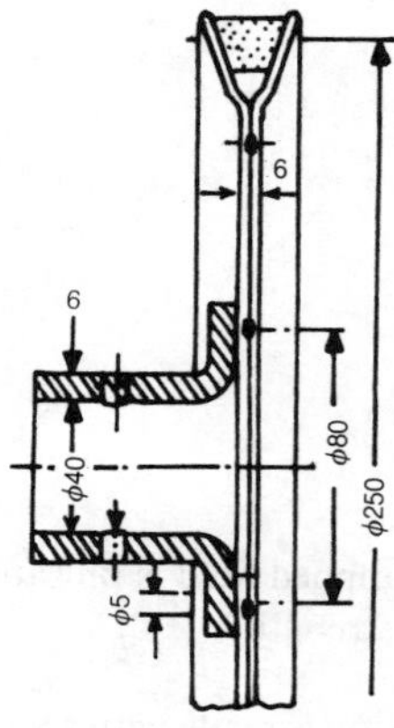

Figure P8.48 (Decker)

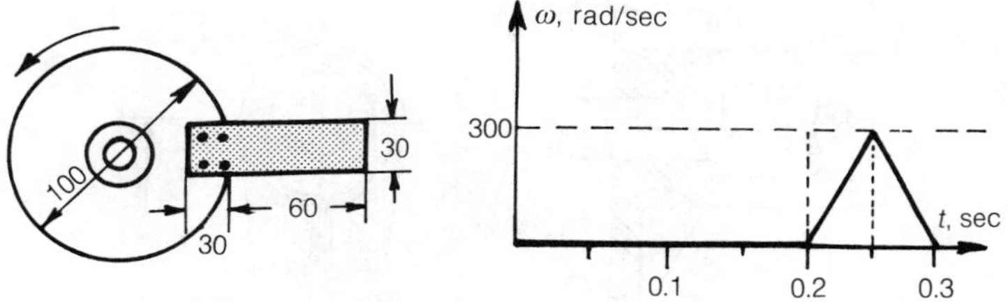

Figure P8.49

8.49 A control device has a strip of thickness 2 mm fixed on a disk with four spot welds. The material of disk and strip is carbon steel AISI 1020. The motion of the disk is intermitted in cycles of 0.3 sec as shown. With a safety factor $N = 2.2$, determine the diameter of the spot welds.

8.50 A disk clutch consists of a hub and a disk fixed by spot welds of 5-mm diameter on the hub. If the maximum transmitted torque is 50 Nm, the disk and hub material is steel ASTM A572, grade 50, the safety factor is 1.5, determine the number of spot welds required.

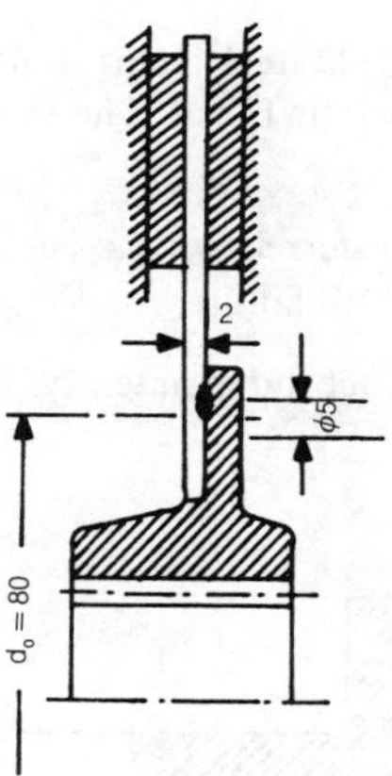

Figure P8.50

8.51 A pinion gear with inner diameter = 50 mm, outer diameter = 80 mm, width = 60 mm, and material carbon steel AISI 1030, is press-fitted on a shaft from carbon steel AISI 1015 with fit H7t6

Determine, for precision-machined surfaces:

(a) The maximum transferred torque.
(b) The safety factor against material yielding. (The shaft and bore surfaces are precision machined.)
(c) The preheating temperature required for assembly.

8.52 A control arm has a maximum activating force 150 N in the position shown. It is press-fitted on a 15 mm diameter shaft and the hub has 37-mm outer diameter and 22-mm width. The material of the shaft and hub is carbon steel AISI 4140. Determine the required fit for a safety factor against slipping 1.5 and against material failure 2. The fit surfaces are ground to a roughness of 10 μm.

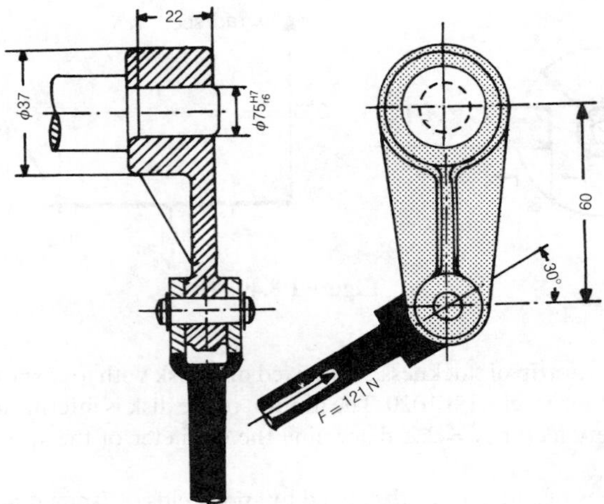

Figure P8.52

8.53 A disk and hub of ASTM A284, grade D steel, is shrink-fitted on a shaft of material AISI 1015 steel. The shaft diameter is 80 mm and the fit H7u6. The surfaces are normally machined. Determine:

(a) The maximum torque transferred with a safety factor against slipping 2.2.
(b) The safety factor against hub failure.

Assume that the disk and the hub act independently, transferring a certain torque each.

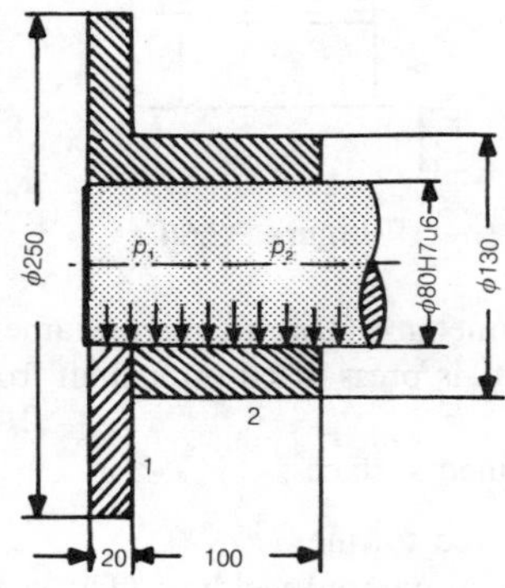

Figure P8.53

8.54 The crank arm shown is press-fitted on a shaft. The shaft material is ASTM A572 grade 65 steel and the crank is of grade 50 steel. Determine the required fit for safety factors against yielding 1.5 and against slipping 2.5. If the crank is to be press-fitted, determine the required axial force. Shaft diameter 30 mm, hub width 60 mm, length 40 mm.

8.55 The flywheel of an internal combustion engine is shrink-fitted on the crankshaft as shown. The moment of inertia of the flywheel is $4\,\text{kg/m}^2$ and the maximum expected angular acceleration is $100\,\text{rad/sec}^2$. Determine the required fit if the factor of safety against slipping is 1.8, against yielding 2. The material of the shaft is AISI 1030 steel, that of the flywheel is cast iron ASTM A159, grade G3000, and the surfaces are ground.

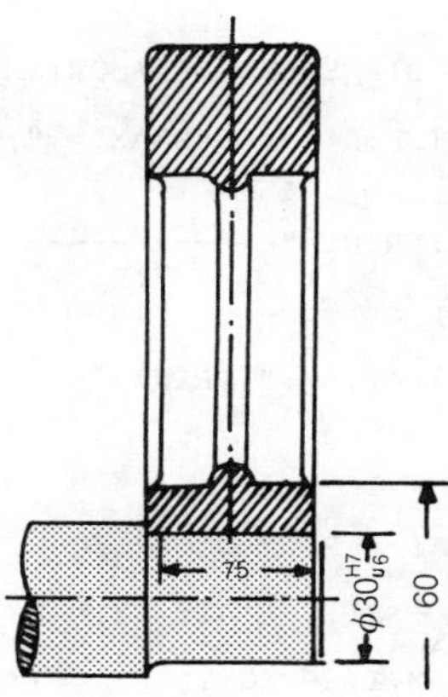

Figure P8.55

APPENDIX 8.A
RIVETS: AN ANALYSIS OF ECCENTRIC RIVETS PROGRAM

```
10 REM ***********************************************
20 REM *                                             *
30 REM *                   RIVETS                    *
40 REM *                                             *
50 REM ***********************************************
60 REM                   COPYRIGHT 1985
70 REM by Professor Andrew D. Dimarogonas, Washington Univ., St. Louis, Mo.
80 REM
90 REM
100 REM A rivet calculation program for eccentrically loaded riveting.
110 REM
120 REM
130 REM          ADD 27/10/84
140 REM
150 REM
160 CLS:KEY OFF
170 FOR I=1 TO 14:COLOR I:LOCATE I,22+I:PRINT "RIVETS";
180 LOCATE I,50-I:PRINT "RIVETS";:NEXT I:COLOR 14
190 LOCATE 8,12:PRINT"Stress analysis of eccentrically loaded riveted";
200 PRINT " joints";
202 locate 25,1:print" Dimarogonas,A.D., Computer Aided Machine Design,
Prentice-Hall,1988";
210 LOCATE 16,20:INPUT"Allowable shear stress            ";SZ
220 LOCATE 17,20:INPUT"Total number of rivets            ";N
230 LOCATE 18,20:INPUT"Force in the x direction          ";FX
240 LOCATE 19,20:INPUT"Force in the y direction          ";FY
250 LOCATE 20,20:INPUT"Torque in the z direction          ";MR
260 LOCATE 22,20:INPUT"Are the data correct  (Y/N)        ";D$
270 IF D$="n" OR D$="N" THEN 210
280 DIM X(10), Y(10), XB(10), YB(10), W(10)
```

```
290 CLS
300 PI=3.14159
310 DATA 1200, 4, 100, 200, 50000
320 FOR I = 1 TO N
330 LOCATE 3+I,20:PRINT"x,y coordinates of rivet ";I;:INPUT"    ";X(I),Y(I)
340 NEXT I
350 LOCATE 4+I,20:INPUT"Enter a guess for the rivet diameter";D
360 IT = 0
370 IF D < 0 THEN GOTO 350
380 PRINT
390 REM CORRECTION OF GUESS DIAMETER: D=D*SQR(SMAX/SALLOWABLE)
400 GOSUB 530
410 PRINT "WITH GUESS DIAMETER:D =";D,"SMAX = ";SM
420 D = D * SQR (SM / SZ)
430 GOSUB 530
440 PRINT :PRINT "RESULTS:":PRINT "------------------"
450 PRINT :PRINT "Diameter=                 ";D
460 PRINT :PRINT "Max. shear stress=    ";SM
470 PRINT :PRINT
480 PRINT "Area center of riveting:":PRINT
490 PRINT "XG=";XG
500 PRINT "YG=";YG
510 PRINT
520 END
530 A = N * PI * D ^ 2 / 4:AI = PI * D ^ 2 /4
540 SX = 0: SY = 0: SW = 0: SM = 0
550 FOR I = 1 TO N: SX = SX + S(I) * AI: SY = SY + Y(I) * AI: NEXT I
560 SG = SX / A: YG = SY / A
570 FOR I = 1 TO N: XB(I) = X(I) - XG:YB(I) = Y(I) - YG: W(I) = SQR (XB(I)
^ 2 + YB(I) ^ 2): NEXT I
580 FOR I = 1 TO N: SW = SW + W(I) ^ 2: NEXT I
590 FOR I = 1 TO N: TX(I) = MR * YB(I) / SW: TY(I) = MR * XB(I) / SW
600 PX = TX(I) + FX / N: PY = TY(I) + FY / N: PR = SQR (PX ^ 2 + PY ^ 2):
SI = PR / (PI * D ^ 2 / 4)
610 IF SI > SM THEN SM = SI
620 NEXT I
630 FU = SM - SZ
640 RETURN
```

APPENDIX 8.B
SECTIONS: A PROGRAM TO CALCULATE PROPERTIES OF COMPLEX SECTIONS

```
10 REM **************************************************
20 REM *                                                *
30 REM *                 SECTIONS                       *
40 REM *                                                *
50 REM **************************************************
60 REM           COPYRIGHT 1987
70 REM by Professor Andrew D. Dimarogonas, Washington University,St.Louis.
80 REM All rights reserved.  Unauthorized reproduction, dissemination or
90 REM selling is strictly prohibited.  This listing is for personal use.
100 REM
110 REM
120 REM
130 REM
140 REM A section properties calculation program
150 REM Sectionas are created with the SOLID program
160 REM
170 REM       ADD/11-20-85/revision 12-22-86
180 REM ************************************************************
190 CLS:KEY OFF
200 FOR I=1 TO 14:COLOR I:LOCATE I,22+I:PRINT "SECTIONS";
210 LOCATE I,50-I:PRINT "SECTIONS";:NEXT I:COLOR 14
```

```
220 LOCATE 8,3:PRINT"A section properties calculation program. Sections crea";
230 PRINT "ted with SOLID";
232 locate 25,1:print" Dimarogonas,A.D., Computer Aided Machine Design, Prentice-Hall,1988";
240 NS = 20:REM ................................... MAX NO OF POLYGONS
250 DIM PD(NS,35):REM ............................POLYGONS
260 DIM X(99),Y(99),Z(99):REM ....................NODE COORDINATES
270 DIM AR(NS): REM ..............................POLYGON AREA
280 DIM IX(NS):REM ...............................POLYGON INERTIA
290 DIM IY(NS):REM ...............................POLYGON INERTIA
300 DIM P(NS):REM ................................POLYGON INERTIA/CROSS
310 DIM MY(NS):REM ...............................POLYGON AREA MOMENT
320 GOSUB 1440:REM     ...........................RETRIEVE FILE
330 GOSUB 1730:REM ...............................FIND MAXIMA AND MINIMA
340 REM PLOTTING .................................SECTION
350 SCREEN 2
360 REM ..........................................SCALING
370 XSC =110/ (X2 - X1)
380 YSC =100 / (Y2 - Y1)
390 FOR IP = 1 TO NP
400 NN = PD(IP,1):N1 = PD(IP,2)
410 V1 = (X(N1)) *XSC:V2 =  (Y(N1)) *YSC
420 V1 = V1
430 PSET(V1,V2)
440 FOR J = 2 TO NN + 1
450 K = PD(IP,J)
460 X = (X(K)) *XSC:Y = (Y(K)) *YSC
470 LINE -(X,Y)
480 NEXT J:NEXT IP
490 LOCATE 3,50:PRINT"Xmax=";X2
500 LOCATE 4,50:PRINT"Xmin=";X1
510 LOCATE 5,50:PRINT"ymin=";Y1
520 LOCATE 6,50:PRINT"ymax=";Y2
530 LOCATE 20,1
540 PRINT "Computing ...please wait."
550 FOR IP = 1 TO NP:I = IP
560 NN = PD(IP,1):AR(I)=0:P(I)=0:IX(I)=0:IY(I)=0:MX(I)=0:MY(I)=0
570 FOR J = 2 TO NN
580 N1 = PD(IP,J):N2 = PD(IP,J + 1)
590 X1 = X(N1):X2 = X(N2):Y1 = Y(N1):Y2 = Y(N2)
600 GOSUB 650:REM ..........................PROPERTIES FOR EACH SEGMENT
610 NEXT J:NEXT IP
620 GOSUB 870:REM ..........................SUM SEGMENTS AND PRINT RESULTS
630 END
640 REM ********************************************SEGMENT PROPERTIES
650 AA = X2 - X1:BB = Y1:CC = Y2 - Y1:DD = X1
660 PRINT ".";
670 AR(I) = AR(I) - AA * (BB + .5 * CC)
680 I1 = AA * BB * BB * BB / 3
690 I2 = AA * CC * CC * CC / 36
700 I3 = .5 * AA * CC * (BB + CC / 3) * (BB + CC / 3)
710 IX(I) = IX(I) - I1 - I2 - I3
720 I1 = CC * DD * DD * DD / 3
730 I2 = CC * AA * AA * AA / 36
740 I3 = .5 * CC * AA * (DD + AA / 3) * (DD + AA / 3)
750 IY(I) = IY(I) + I1 + I2 + I3
760 P1 = (DD + AA / 2) * (BB / 2) * (AA * BB)
770 P2 = AA * AA * CC * CC / 72
780 P3 = AA * CC * .5 * (DD + 2 * AA / 3) * (BB + CC / 3)
790 P(I) = P(I) - P1 - P2 - P3
800 A1 = AA * BB * BB * .5
810 A2 = AA * CC * .5*(BB + CC / 3)
820 MX(I) = MX(I) - A1 - A2
830 B1 = CC * DD * DD *.5
840 B2 = AA * CC * .5 * (DD + AA / 3)
850 MY(I) = MY(I) + B1 + B2
860 RETURN
870 REM ************************************SUM SEGMENT PROPERTIES
880 FOR I = 1 TO NP
890 TA = TA + AR(I)
```

```
900 XO = XO + MY(I): YO = YO + MX(I)
910 XI = XI + IX(I)
920 YI = YI + IY(I)
930 PI = PI + P(I)
940 NEXT I
950 XO = XO / TA:YO = YO / TA
960 XI = XI - TA * YO * YO
970 YI = YI - TA * XO * XO
980 JI = XI + YI
990 PI = PI - TA * XO * YO
1000 I1 = (XI + YI) * .5 + SQR ((XI - YI) * (XI - YI) * .25 + PI * PI)
1010 I2 = (XI + YI) * .5 - SQR ((XI - YI) * (XI - YI) * .25 + PI * PI)
1020 IF XI = YI GOTO 1050
1030 PA = .5 * ATN ( - 2 * PI / (XI - YI))
1040 GOTO 1060
1050 PA = 0
1060 HX = 0:HY = 0
1070 FOR I = 1 TO NP
1080 FOR II = 2 TO PD(I,1)+1
1090 N = PD(I,II)
1100 XR = X(N) - XO:YR = Y(N) - YO
1110 AA = ABS (XR * COS(PA) + YR * SIN (PA))
1120 IF AA > HX THEN HX = AA
1130 AA = ABS (XR * SIN (PA) - YR * COS (PA))
1140 IF AA > HY THEN HY = AA
1150 NEXT II
1160 NEXT I
1170 CA = COS (PA):SA = SIN(PA)
1180 X1=XMIN-10:X2=XMAX+10:Y1=YMIN-10:Y2=YMAX+10
1190 O1=X1*XSC:O2=(YO+(XO-X1)*TAN(PA))*YSC:O3=X2*XSC:O4=(YO-(XO-
X1)*SA/CA)*YSC
1200 V2=Y1*YSC:V1=(XO-(YO-Y1)*TAN(PA))*XSC:V4=Y2*YSC:V3=(XO+(YO-
Y1)*SA/CA)*XSC
1210 PRINT
1220 PRINT "PLOTTING PRINCIPAL AXES"
1230 LINE( O1,O2)-( O3,O4 )
1240 LINE( V1,V2)-(V3,V4)
1250 INPUT "Hit RETURN for  results ";X$
1260 SCREEN 0 :COLOR 14,1
1270 PA = 180 * PA / 3.14159
1280 PRINT
1290 PRINT "Area          ";TA
1300 PRINT "Centroid  XO=";XO;" Y0=";YO
1310 PRINT
1320 PRINT "Inertias:  Ix=";XI
1330 PRINT "           Iy=";YI
1340 PRINT "           Ixy=";JI
1350 PRINT
1360 PRINT "Principal Inertias:  I1=";I1
1370 PRINT "                     I2=";I2
1380 PRINT
1390 PRINT "Principal angle:     ";PA;"  deg"
1400 PRINT :PRINT "Max distances from principal axes:"
1410 PRINT "  HX:  ";HX
1420 PRINT "  HY:  ";HY
1430 RETURN
1440 REM*************************************************************
1450 REM                 file retrieval
1460 REM*************************************************************
1470 LOCATE 19,35:PRINT"                ";
1480 LOCATE 18,35:PRINT"Data file  ":LOCATE 19,35:INPUT FILNA$
1490 IF FILNA$="" THEN 1480
1500 rem ON ERROR GOTO 1710
1510 NDS=ND
1520 OPEN FILNA$ FOR INPUT AS #1
1530 INPUT#1,X$
1540 IF X$="solend" OR X$="SOLEND" THEN 1690
1550 SYM$=""
1560 ISYM=1
1570 SYM$=MID$(X$,ISYM,1):ISYM=ISYM+1:IF SYM$="" AND ISYM<10 THEN 1570
1580 IF ISYM=10 THEN 1530
```

```
1590 IF SYM$="n" OR SYM$="N" THEN ND=ND+1:INPUT#1,X(ND),Y(ND),Z(ND):GOTO
1530
1600 IF SYM$="a" OR SYM$="A" THEN NC=NC+1:FOR I=1 TO
3:INPUT#1,SC(NC,I):NEXT I:GOTO 1530
1610 IF SYM$="p" OR SYM$="P" THEN NP=NP+1:GOTO 1640
1620 IF X$="solend" OR X$="SOLEND" THEN 1690
1630 PRINT"FILE ERROR...":STOP
1640 IP=IP+1
1650 INPUT#1,PD(IP,1)
1660 NN=PD(IP,1):FOR JP=1 TO NN
1670 INPUT#1,NUNOD:PD(IP,JP+1)=NUNOD+NDS:NEXT JP
1680 GOTO 1530
1690 CLOSE#1
1700 GOTO 1720
1710 LOCATE 19,68:PRINT "BAD FILE        " :CLOSE
1720 RETURN
1730 REM ************************************FINDING XMIN,XMAX,YMIN,YMAX
1740 X1 = 1E+10:X2 = - 1E+10:Y1 = X1:Y2 = X2
1750 FOR I = 1 TO ND
1760 IF X1 > X(I) THEN X1 = X(I)
1770 IF X2 < X(I) THEN X2 = X(I)
1780 IF Y2 < Y(I) THEN Y2 = Y(I)
1790 IF Y1 > Y(I) THEN Y1 = Y(I)
1800 NEXT I
1810 XMIN=X1:YMIN=Y1:XMAX=X2:YMAX=Y2
1820 RETURN
```

APPENDIX 8.C
WELDS: A SHEAR WELD ANALYSIS PROGRAM

```
10 REM ************************************
20 REM *                                  *
30 REM *             WELDS                *
40 REM *                                  *
50 REM ************************************
60 REM          COPYRIGHT 1985
70 REM by Professor Andrew D. Dimarogonas, Washington Univ., St. Louis, Mo.
80 REM  All rights reserved.  Unauthorized reproduction, disssemination,
90 REM selling or use is strictly prohibited.
100 REM  This listing is for reference purpose only.
110 REM
120 REM
130 REM CALCULATIONS OF STRESSES AT A COMPLEX PLANE WELD
140 REM WITH GENERAL LOADING.
150 REM
160 REM          ADD / 1-12--84
170 REM
180 CLS:KEY OFF
190 LOCATE 25,1:PRINT" Andrew Dimarogonas, Computer Aided Machine Design,
Prentice-Hall, London 1988";
200 FOR I=1 TO 14:COLOR I:LOCATE I,22+I:PRINT "WELDS";
210 LOCATE I,50-I:PRINT "WELDS";:NEXT I:COLOR 14
220 LOCATE 8,15:PRINT"Stress analysis of welded joints on the x-y plane";
230 LOCATE 16,20:INPUT"Enter xmax,ymax                    ";XMAX,YMAX
240 LOCATE 17,20:INPUT"No of line segments                ";NL
250 LOCATE 18,20:INPUT"No of arc  segments                ";NA
260 XSC=600/XMAX:YSC=200/YMAX:SC=XSC:IF YSC<XSC THEN SC=YSC
270 DIM X(10),X2(10),Y1(10),Y2(10),A(10)
280 DIM XO(10),YO(10),F1(10),F2(10),R(10),B(10)
290 CLS:PRINT"                                WELD DATA":PRINT
300 IF NL=0 THEN 370
310 FOR I=1 TO NL
320 PRINT"line segment No ";I:PRINT
330 INPUT "Beginning, x,y=     ";X1(I),Y1(I)
340 INPUT "End         x,y=     ";X2(I),Y2(I)
350 INPUT "weld width a         ";A(I) :PRINT
360 NEXT I
```

```
370 IF NA = 0 THEN 470
380 FOR I=1 TO NA
390 PRINT "arc segment No                          ";I
400 INPUT "arc center Xc,Yc                        ";XO(I),YO(I)
410 INPUT "arc angles f1,f2 from x,ccw (deg)       ";F1,F2
420 INPUT "arc radius                              ";R(I)
430 INPUT "weld width                              ";B(I)
440 F1(I)=F1*3.14159/180
450 F2(I)=F2*3.14159/180
460 NEXT I:PRINT:PRINT"                  LOADING:":PRINT
470 INPUT"Force Fx          ";FX$:FX=VAL(FX$)
480 INPUT"Force Fy          ";FY$:FY=VAL(FY$)
490 INPUT"Force Fz          ";FZ$:FZ=VAL(FZ$)
500 INPUT"Moment mx         ";MX$:MX=VAL(MX$)
510 INPUT"Moment my         ";MY$:MY=VAL(MY$)
520 INPUT"Moment mz         ";MZ$:MZ=VAL(MZ$)
530 SCREEN 2
540 REM ...........................................PLOTTING THE
WELD
550 REM SCALE FACTOR SC TO KEEP -140<X<140 AND -75<Y<75
560 XSC=SC:YSC=SC/2
570 REM ...........................................PLOTTING THE
GRID
580 LAB$="10":GSTEP=10*SC:IF SC>10 THEN GSTEP=SC:LAB$="1"
590 IF SC<1 THEN GSTEP=100*SC:LAB$="100"
600 FOR I = 0 TO 200 STEP GSTEP/2
610 LINE(10, I)-( 400,I):NEXT I
620 FOR I =0 TO 400 STEP GSTEP
630 LINE(I, 10)-(I, 200): NEXT I
640 LINE( 1,100)-( 500,101),,BF:LINE( 199, 1)-( 201, 200),,BF
650 PRINT LAB$;" x ";LAB$;" GRID---X=HORIZ, ---Y=VERT":PRINT
660 REM PLOT ...........................................STRAIGHT
SEGMENTS
670 FOR I = 1 TO NL
680 DX = X2(I) - X1(I): DY = Y2(I) - Y1(I)
690 DL = SQR (DX * DX + DY * DY)
700 CF = DX / DL:SF = DY / DL
710 IA = A(I):FOR J = -IA / 2 TO IA / 2 STEP 1/SC
720 JX = - J * SF:JY = J * CF
730 X1=X1(I)+200/XSC:X2=X2(I)+200/XSC:Y1=Y1(I)+100/YSC:Y2=Y2(I)+100/YSC
740 LINE((X1+JX) *XSC,(Y1+JY) *YSC )-((X2+JX) *XSC,(Y2+JY) *YSC )
750 NEXT J:NEXT I
760 IF NA = 0 THEN 890
770 REM ....................................PLOTTING ARC SEGMENTS
780 FOR I = 1 TO NA
790 R1 = R(I) - B(I) / 2:R2 = R(I) + B(I) / 2
800 FOR R = R1 TO R2 STEP B(I)/10:DF=(F2(I)-F1(I))/20
810 FOR F = F1(I) TO F2(I) STEP DF
820 Y = YO(I) *YSC +100 + R * SIN(F)* YSC
830 X = XO(I) *XSC + 200 + R * COS(F) *XSC
840 IF F = F1(I) THEN PSET (X,Y)
850 IF F<> F1(I) THEN LINE-(X,Y)
860 NEXT F
870 NEXT R
880 NEXT I
890 REM
900 NF = 10: REM ..............................NO OF INTERGRATION
SEGMENTS
910 REM ..............................CALCULATION OF AREA, CENTER OF
AREA
920 IF NL = 0 THEN 980
930 REM ...................................................LINE
SEGMENTS
940 FOR I = 1 TO NL:S = SQR((X2(I) - X1(I)) ^ 2 + (Y2(I) - Y1(I)) ^ 2)
950 AI = S * A(I):AR = AR + AI: MX = MX + AI * (Y2(I) + Y1(I)) / 2
960 MY = MY + (AI * (X2(I) + X1(I)) / 2)
970 NEXT I
980 IF NA = O THEN 1150
990 REM ...............................................ARC SEGMENTS
1000 FOR I = 1 TO NA
1010 PI = 3.14159
```

```
1020 F1 = F1(I) * PI / 180: F2 = F2(I) * PI / 180
1030 REM
1040 F1(I) = F1:F2(I) = F2
1050 DF = (F2 - F1) / 10
1060 FOR F = F1 TO F2 STEP DF
1070 EP = 1: IF F = F1 THEN EP = .5
1080 IF F > .99999 * F2 THEN EP = .5
1090 X = XO(I) + R(I) * COS (F): Y = YO(I) + R(I) * SIN (F)
1100 AR = AR + R(I) * DF * EP * B(I)
1110 MX = MX + Y * R(I) * DF * ER * B(I)
1120 MY = MY + X * R(I) * DF * ER * B(I)
1130 NEXT F
1140 NEXT I
1150 XG = MY / AR: YG = MX / AR
1160 LOCATE 3,60:PRINT "Area=";AR
1170 LOCATE 4,60:PRINT "Area Center:";
1180 LOCATE 6,60:PRINT "Xg=";XG;
1190 LOCATE 7,60:PRINT "Yg=";YG;
1200 REM ........................................MOMENTS OF INERTIA
1210 IF NL = 0 THEN 1310
1220 REM LINE SEGMENTS
1230 FOR I = 1 TO NL
1240 X1 = X1(I) - XG:Y1 = Y1(I) - YG: X2 = X2(I) - XG: Y2 = Y2(I) -YG
1250 HX = X2 - X1:HY = Y2 - Y1:S = SQR (HX ^ 2 + HY ^ 2): AI = A(I) * S
1260 IF HX = 0 THEN HX = A(I)
1270 IF HY = 0 THEN HY = A(I)
1280 IX = IX + A(I) * HY ^ 2 / 12 + ((Y1 + Y2) / 2) ^2 * AI
1290 IY = IY + A(I) * HX ^ 2 / 12 + ((X1 + X2) / 2) ^ 2 * AI
1300 NEXT I
1310 IF NA = 0 THEN 1440
1320 REM ARCS
1330 FOR I = 1 TO NA
1340 F1 = F1(I):F2 = F2(I): DF = (F2 - F1) / NF
1350 FOR F = F1 TO F2 STEP DF
1360 EP = 1: IF F = F1 THEN EP = .5
1370 IF F > .999 * F2 THEN EP = .5
1380 XO = XO(I) - XG:YO = YO(I) - YG
1390 X = XO + R(I)*COS (F):Y = YO + R(I) * SIN (F)
1400 IX = IX + Y * Y * R(I) * DF*EP * B(I)
1410 IY = IY + X * X * R(I) * DF * EP * B(I)
1420 NEXT F
1430 NEXT I
1440 LOCATE 9,60:PRINT "Moments of inertia:";
1450 LOCATE 11,60:PRINT "Ix=  ";IX;
1460 LOCATE 12,60:PRINT "Iy=  ";IY;
1470 IR = IX + IY
1480 LOCATE 13,60:PRINT "Ir=  ";IR;
1490 REM ................................FINDING MAXIMUM EQUIVALENT
STRESS
1500 REM ................................LINE SEGMENTS
1510 IF NL = 0 THEN 1610
1520 FOR I = 1 TO NL
1530 X = X2(I) - XG:Y = Y1(I) - YG
1540 GOSUB 1890
1550 IF ABS (EQ) < SM THEN 1570
1560 SM = EQ:ID$ = "STRAIGHT":IN = 1:ID = 1
1570 X = X2(I) - XG:Y = Y2(I) - YG: GOSUB 1890
1580 IF ABS (EQ) < SN THEN 1600
1590 SM = EQ:ID$ = "STRAIGHT":IN = 2: ID = 1
1600 NEXT I
1610 IF NA = 0 THEN 1770
1620 REM ..................................................ARC
SEGMENTS
1630 FOR I = 1 TO NA
1640 F1 = F1(I):F2 = F2(I):DF = (F2 - F1) / NF
1650 FOR  F = F1 TO F2 STEP DF
1660 X = XO(I) - XG +R(I) * CS(F)
1670 Y = YO(I) - YG +R(I) * SIN (F)
1680 GOSUB 1890
1690 IF ABS (EQ) < SM THEN GOTO 1710
1700 SM = EQ: ID$ = "ARC":ID = I:FF = F
```

```
1710 NEXT F
1720 NEXT I
1730 REM ......................STRESS ARE CALCULATED ON THE BASIS OF
STENGTH OF
1740 REM ......................MATERIALS WITH AVERAGING OF DIRECT SHEAR.
FOR E-
1750 REM ......................QUIVALENT STRESS THE MAX SHEAR STRESS
CRITERION
1760 REM ......................FAILURE IS ASSUMED.
1770 LOCATE 15,60:PRINT "RESULTS:";
1780 LOCATE 17,53:PRINT "Smax-eq=";SM;
1790 LOCATE 18,53:PRINT "at the ";ID$;"segment";
1800 LOCATE 19,53:PRINT "number ";ID;
1810 IF ID$ = "STRAIGHT" THEN 1870
1820 X=XO(ID)-XG+R(ID)*COS (FF):Y = YO(ID)-YG + R(ID) * SIN (FF):GOSUB 1890
1830 LOCATE 21,53:PRINT "at the point";
1840 LOCATE 22,53:PRINT "x=";X;
1850 LOCATE 23,53:PRINT "y=";Y;
1860 GOTO 1880
1870 LOCATE 21,53:PRINT "at the point no ";IN;
1880 LOCATE 23,1:END
1890 REM ...................................................STRESS
CALCULATIONS
1900 SZ = FZ /AR + MX * Y / IX + MY * X / IY
1910 TY = FY /AR + RZ * X / IR + MZ* Y/(IX+IY)
1920 TX = FX /AR + RZ * Y / IR - MZ * X/(IX+IY)
1930 TC = SQR (TX * TX + TY * TY)
1940 EQ = SQR (SZ * SZ + 4 * TC * TC)
1950 RETURN
```

CHAPTER NINE

DESIGN FOR RIGIDITY

9.1 RIGIDITY REQUIREMENTS IN MACHINE DESIGN

Rigidity is the capacity of machines to sustain loads without appreciable change of their geometry. The deformation owing to loading might not be high enough, many times, to geometrically disturb the operation of the machine. However, it might cause other effects such as instability, friction, vibration, etc., which in turn might influence proper machine operation.

Some requirements imposed on the rigidity of machine parts are:

1. Strength during dynamic instabilities or shock loads.
2. Performance of the parts in connection with mating parts.
3. Easy manufacturing.
4. Satisfactory operation of the machine as a whole.

The conditions based on the performance of mating parts are of prime importance. For example, the proper rigidity of shafts determines the satisfactory behavior of bearings, gears, worm and other drives whose components are mounted on the shafts. The forces exerted by drives bend shafts and lead to interference in bearings and gears. Permissible deflection and angles of inclination are determined on the basis of rigidity requirements.

Requirements for rigidity on the basis of manufacturing processes are of prime importance for various parts, especially with mass production.

Rigidity requirements based on conditions of satisfactory operation of the machine as a whole are usually quite specific. Here, design proceeds for the working loads and the accuracy specifications of the workpiece.

The rigidity of machine components is determined by their inherent rigidity when dealt with as beams, plates or shells with idealized supports and by their contact rigidity, i.e. the rigidity of the surface layers at the areas of contact.

Of prime importance for most parts subject to considerable loads is their inherent deformation. But in precision machinery at relatively low loads, the contact deformations in unsecured joints (between mutually moveable mating parts) play an essential and even predominant role in the balance of displacements.

The contact of parts may be under conditions of:

1. Initial contact in a point or line, as in pressing two balls or cylinders together.
2. A large nominal area of contact.

In both cases, the contact displacements are significant due to the small actual area of contact. In the first case, this is associated with the nominal shape of the contacting surfaces, and in the second, with surface microirregularities and wavyness. The load is carried by the microirregularities at the peaks of the macrowaves. Experiments show that the actual contact area is usually only a very small part of the nominal value (discussed in Chapter 11).

Stability is a criterion determining the dimensions of the following:

(a) Long thin parts in compression.
(b) Thin plates subjected to compression acting on opposite edges.
(c) Shells subjected to external pressure.
(d) Thin-walled hollow shafts.

The most common parts checked for buckling are the screws of jacks, lead screws, piston rods and compression springs. Many components are checked for stability in the steel structures of materials handling machinery.

Sometimes, low rigidity in the form of high flexibility is wanted in many machine members, like springs. This property permits constant loads to be maintained in pressure-tight components and mechanisms, vibration isolation, etc.

Since the dynamic characteristics of mechanical systems depend on rigidity (or flexibility), this property is frequently carefully controlled in order to maintain proper dynamic properties of the machine. This will be discussed in Chapter 14.

9.2 RIGIDITY OF MACHINE COMPONENTS

9.2.1 Rigidity material indices

High strength of a material is not necessarily associated with high rigidity. One can assert the opposite: Improvement of strength properties (S_u, S_y etc.) has little effect on elastic properties (E, G, ν). Since deflection, in general, is influenced by loads, geometry and elastic properties, improvement of material strength results in smaller sections, for the same load, thus higher deflections and smaller rigidity. Therefore, a representative index expressing material rigidity would be, for example, E/S_y. For machines where the major part of the load is determined by the mass of the machine, the material density (or specific weight) γ is equally important. The associated index would be $E\gamma/S_y$ in this case.

For the tension rod, rigidity is defined as the ratio of axial load to resulting axial deformation, also called the 'spring constant'

$$k = P/\delta L = EA/L$$

where A is the cross-section, L the length, and E Young's Modulus. For constant k, the weight of the part γLA, where γ is the specific weight, is minimized when the quantity

$$1/G = 1/A\gamma L = (1/L^2 k)(E/\gamma)$$

is maximized. For given length and rigidity, the material of minimum weight is the one with maximum E/γ index.

For maximum rigidity at given material strength S_u, load P and length L,

$$k = EA/L = (P/L)(E/S_u)$$

and the index E/S_u must be maximum, or E/S_y, depending on the design method.

Similarly, maximum strength and maximum rigidity for minimum weight is achieved for given geometry A, L and load P, when,

$$\frac{1}{G}kN = \frac{1}{AL\gamma}\cdot\frac{EA}{L}\cdot\frac{S_y}{P/A} = \left(\frac{A}{LP_{\max}}\right)\left(\frac{ES_y}{\gamma}\right)$$

is maximized. The general index ES/γ applies therefore generally for cases where tensile stresses are the main material load. For flexure and torsion, similar results can be obtained.

For the basic material categories, rigidity indices are shown in Table 9.1.

Let us compare the rigidity, strength and weight of structures when changing over from castings of pig iron to castings of aluminum and carbon steels without altering the parts' shape. Assuming the rigidity, strength and weight of grey cast iron structures as equal to unity, we obtain:

Parameter	Aluminum alloys	Carbon steel
Rigidity (E)	0.9	2.6
Strength (S)	0.72	2.3
Weight (γ)	0.39	1.1

Thus, the transfer to aluminum alloy castings hardly affects rigidity, somewhat lowers strength (approximately 30%) and significantly reduces the weight of the structure (2.5 times). Changing to steel castings increases rigidity and strength approximately 2.5 times and the weight is practically the same.

The above-listed relations are based on the fact that Young's Modulus has a constant value and only slightly depends on the presence (in normal amounts) of alloying elements, heat treatment and strength characteristics of alloys of the given metal. For example, in respect of steels, beginning from low carbon and up to high-alloy ones, Young's Modulus varies from 190,000 to 220,000 MPa, and the shear modulus, from 79,000 to 82,000 MPa. With regard to aluminum alloys, $E = 70{,}000$–$75{,}000$ MPa and $G = 24{,}000$–$27{,}000$ MPa.

Consequently, for the manufacture of identically-shaped components when rigidity is the first requirement and the stress level is low, the use of weakest materials is recommended (low carbon steels). This will not affect the rigidity of the construction but will enable the cost to be reduced.

The above does not hold when the strength of the construction is as important as the rigidity. Thus, for instance, identically-shaped structures, one made from low-carbon steel, and the other from alloy steel, will have the same rigidity but the load capacity of the first structure will be as many times less as the tensile strength of carbon steel is less than that of alloy steel.

For equally rigid parts the greatest strength advantage is with materials having the highest S_u/E ratio (high-strength steels, titanium alloys and wrought aluminum alloys). In terms of weight advantage (this is, in this case, proportional to the γ/E factor) the above-listed

Table 9.1 Strength and rigidity characteristics of structural materials (Orlov 1976)

Material	Specific weight, γ, kgf/dm^3	Ultimate tensile strength, S_u kgf/mm^2	Yield limit, $S_{0.2}$, kgf/mm^2	Modulus of elasticity, E, kgf/mm^2	Modulus of shear, G, kgf/mm^2	Rigidity characteristics			Generalized factor, $\frac{S_{0.2}E}{\gamma}\cdot 10^{-5}$
						$\frac{E}{\gamma}\cdot 10^{-3}$	$\frac{E}{S_u}\cdot 10^{-2}$	$\frac{E}{S_{0.2}}\cdot 10^{-2}$	
Carbon steels	7.85	35–80	21–48	21000	8000	2.67	2.6	1.56	1.3
Alloy steels		100–180	80–145				1.17	0.94	3.8
High-strength steels		250–350	225–315				0.6	0.54	8.4
Grey cast irons	7.2	20–35	14–25	8000	4500	1.1	2.3	1.6	0.3
High-strength cast irons	7.4	45–80	32–56	15000	7000	2	1.9	1.3	1.1
Aluminum alloys, cast	2.8	18–25	13–17.5	7200	2500	2.67	2.9	2.04	0.45
Aluminum alloys, wrought		40–60	28–42				1.2	0.83	1.1
Magnesium alloys, cast	1.8	12–20	8–13	4500	1500	2.3	2.1	1.35	0.32
Magnesium alloys, wrought		25–30	18–21				1.4	1.2	0.52
Structural bronzes	8.8	40–60	32–48	11000	4200	1.25	1.85	2.3	0.6
Titanium alloys	4.5	80–150	70–135	12000	4200	2.66	0.8	0.72	3.6

materials possess approximately equivalent values. Worse weight characteristics are held by bronzes and grey cast irons.

In practice, the choice of material is determined not only by the strength–rigidity characteristics, but also by some other properties. That is why preference is given to those design features which enable strength and rigidity to be enhanced even when using materials of low strength and rigidity.

9.2.2 Design rules for high rigidity

Some design methods to increase rigidity are:

1. Avoidance of bending which is weak from the viewpoint of rigidity and strength, changing the force-carrying mechanism to compression and tension.
2. For parts working in bending, rational support positioning.
3. Rational increase of section inertia moments, reinforcing joints and the parts transferring forces from one section to another.
4. For thin-walled box-shaped parts, using optimum shapes.

9.2.3 Rational sections

It is important that the increase in rigidity is not accomplished by increasing weight of the machine. In general, the solution of the problem involves strengthening sections, which under the given loads are subjected to the highest stresses, and removing weight from the unloaded or slightly loaded areas. In flexure, the stressed sections are those farthest from the neutral axis. In torsion the external fibers are mostly stressed; moving radially and inwards, the stresses become weaker and at the center are zero. Consequently, for these cases it is more rational to concentrate material at the periphery and remove it from the center.

Generally, the greatest rigidity and strength characteristics with smallest weight are found in components with thin-walled sections, i.e. parts such as box sections, tubes and shells.

Table 9.2 gives rigidity and strength comparisons for differently-shaped sections. The base of the comparison depends upon similar weight conditions of parts, expressed as similar cross-sectional areas. The strength and rigidity improvements are obtained by successful application of the material-distribution principle in the regions of the highest acting stresses.

For cylindrical sections the moment of inertia I and the section modulus W of a solid round section are taken as the basis of comparison – with respect to the other parts, a solid square shaped section.

Figure 9.1 shows reinforcements of beams by transversal ribs and stiffening boxes: (a) by stiffening partitions; (b) by stiffening boxes; (c) by semi-circular stiffening elements.

Diagonal stringers in the form of webs strongly improve rigidity (Figures 9.2a, b), and also load section stiffening (Figure 9.3). Thus, constructions with longitudinally-formed stiffening rib angles at the transition points where vertical walls change to horizontal ones (Figure 9.3b) have greater rigidity than the original construction (Figure 9.3a) in spite of the formal reduction in inertia moment. The rigidity parameter increases also when the longitudinal rib has transverse stiffening ribs spaced over the part length (Figures 9.3c, d).

Table 9.2 Rigidity/strength of sections of the same weight (from Orlov, 1976, Mir Publishers, Moscow, by permission)

Section	Ratios		Inertia I/I_0	W/W_0
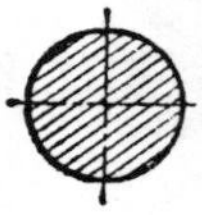		—	1	1
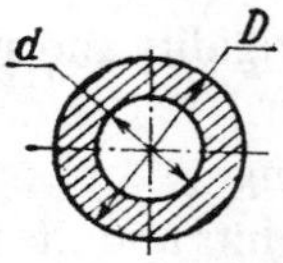	$\frac{d}{D}$	0.6	2.1	1.7
		0.8	4.5	2.7
		0.9	10	4.1
		—	1	1
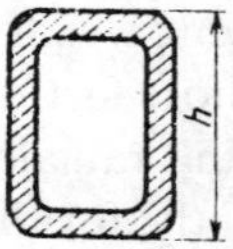	$\frac{h}{h_0}$	1.5	3.5	2.2
		2.5	9	3.7
		3.0	18	5.5
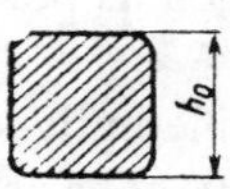		—	1	1
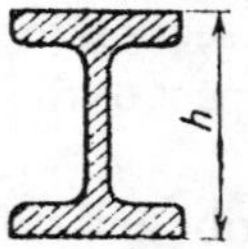	$\frac{h}{h_0}$	1.5	4.3	2.7
		2.5	11.5	4.5
		3.0	21.5	7.0

Table 9.3 illustrates how longitudinally-arranged webs affect the rigidity of profiles during flexure and torsion. One diagonal stringer will suffice; another stringer will enhance the rigidity but to a small degree.

Ribbing is used to improve rigidity, particularly of cast housing-type components (Figures 9.4 and 9.5).

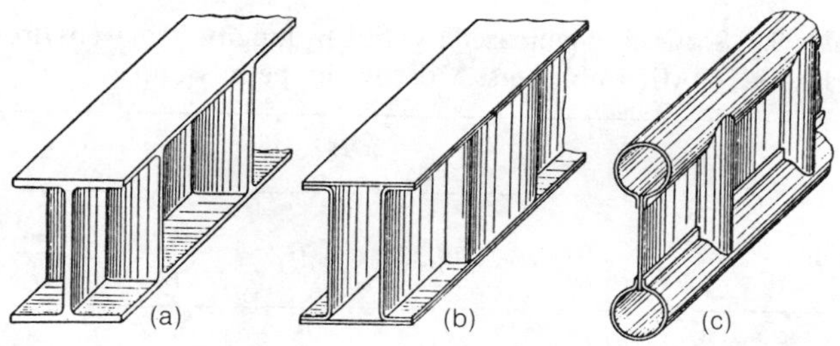

Figure 9.1 Beams with high rigidity; (a) by stiffening partitions; (b) by stiffening boxes; (c) by semi-circular stiffening elements. (From Orlov, 1976, by permission of Mir Publishers, Moscow)

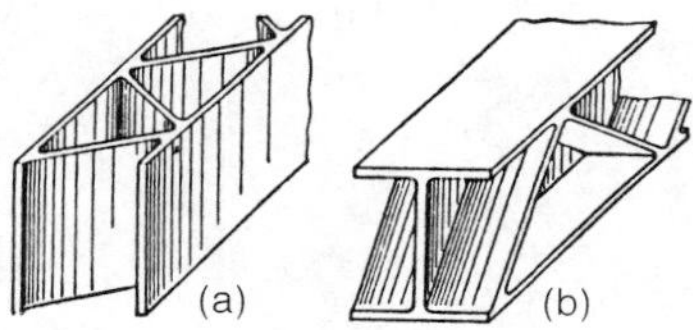

Figure 9.2 Reinforced beams. (From Orlov, 1976, by permission of Mir Publishers, Moscow)

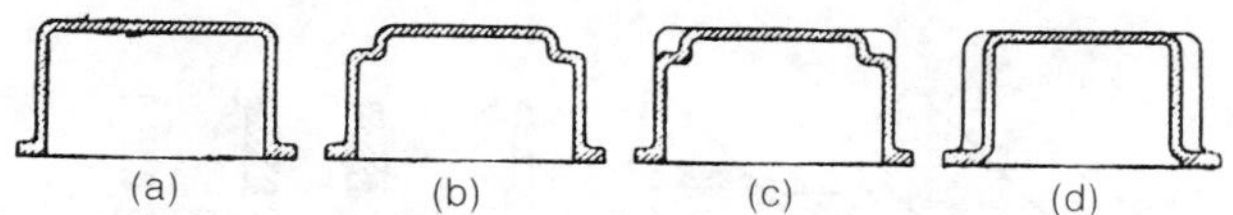

Figure 9.3 Local section stiffening. (From Orlov, 1976, by permission of Mir Publishers, Moscow)

9.3 STABILITY OF MACHINE ELEMENTS

Stability, loosely defined, is the property of a system to return to its original state after if has been displaced from the position of equilibrium. Otherwise, the system is unstable.

A system that has lost stability may behave in different ways. There is usually a transition to a new position of equilibrium, this being commonly accompanied by large displacements, the development of large plastic strains or complete collapse. In some cases a structure continues to operate properly and performs its basic functions after loss of stability, as for example, thin-walled panels in aircraft structures. Finally, a system that has lost stability may set into continuous oscillatory motion. The simplest case is the buckling of a centrally-compressed bar (Figure 9.6). At a sufficiently-large force the bar cannot maintain the straight-line configuration and deflects laterally.

Table 9.3 Sections of increased rigidity, by longitudinal webs (from Orlov, 1976, Mir Publishers, Moscow, by permission)

Profile	Factors				
	I_{flex}	I_{tors}	G	$\dfrac{I_{\text{flex}}}{G}$	$\dfrac{I_{\text{tors}}}{G}$
	1	1	1	1	1
	1.17	2.16	1.38	0.85	1.56
	1.55	3	1.26	1.23	2.4
	1.78	3.7	1.5	1.2	2.45

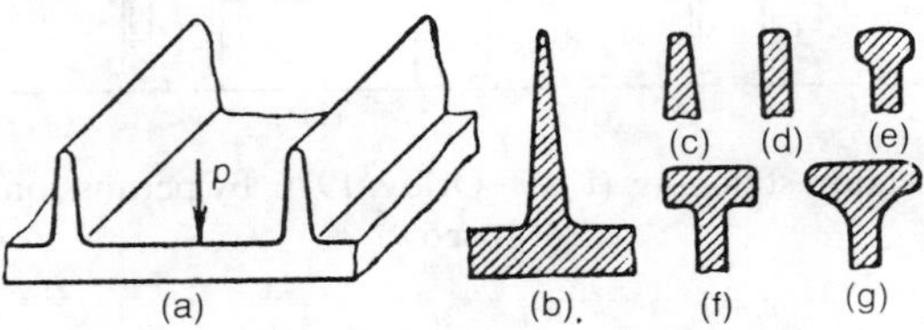

Figure 9.4 Ribbed sections (in order of increasing strength). (From Orlov, 1976, by permission of Mir Publishers, Moscow)

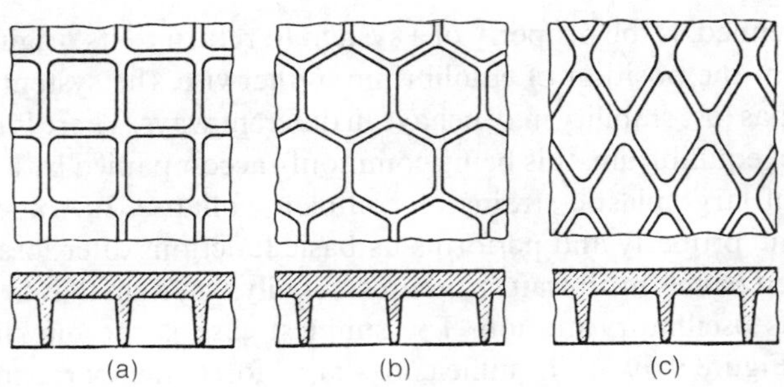

Figure 9.5 Rib forms. (From Orlov, 1976, by permission of Mir Publishers, Moscow)

Figure 9.6 Buckling of a column

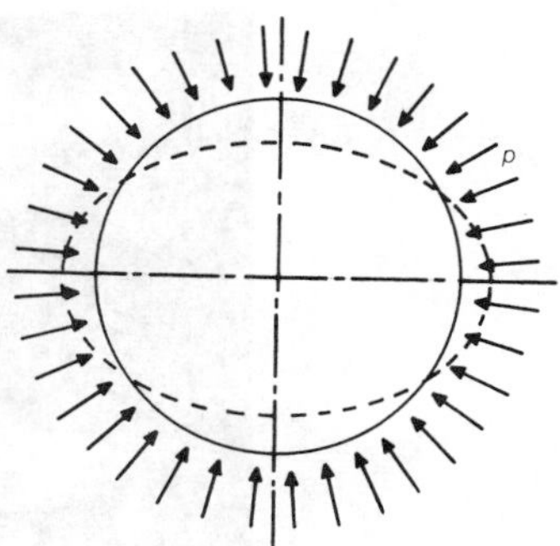

Figure 9.7 Buckling of a ring or pipe under external pressure

Figure 9.8 Collapse of a cylindrical shell under axial compressive force

A thin-walled tube (Figure 9.7) under external pressure may lose stability. The circular shape of the section then changes into an elliptical one and the tube is completely flattened though the stresses in the walls are far from reaching the yield point at the moment of buckling.

The same tube may lose stability under axial compression (Figure 9.8). A similar phenomenon takes place when a tube is subjected to torsion (Figure 9.9).

It is apparent that instability is most pronounced in light thin-walled structures, such as compressed shells and thin walls. Therefore, in designing such elements stability analysis of both separate components and the system is required in addition to the strength analysis.

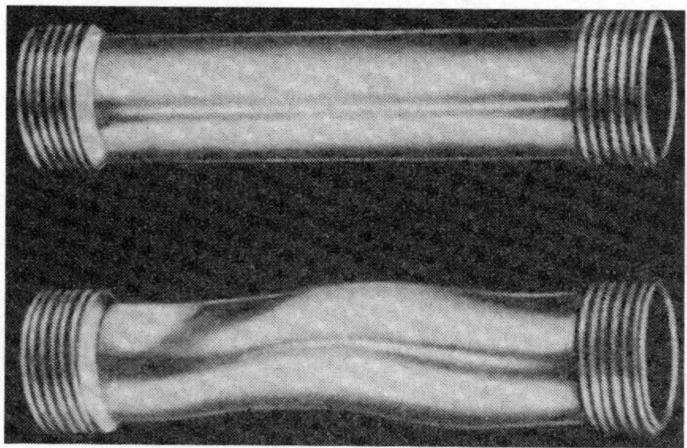

Figure 9.9 Collapse of a pipe under torsion

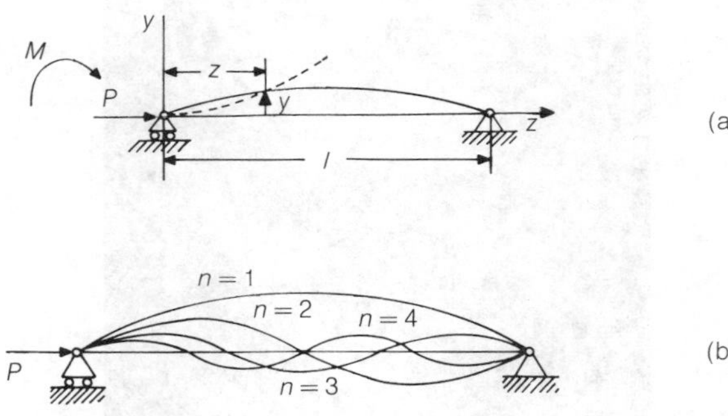

Figure 9.10 Modes of buckling of compressed columns

For the analysis of stability it is assumed that the system is ideal, i.e. if it is a compression member, its axis is perfectly straight, the material is homogeneous, and the forces are applied centrally.

The simplest elastic element is a bar compressed by central forces P (Figure 9.10a). This problem was first formulated and solved by Euler in the middle of the 18th century. Hence the expressions 'Euler problem' or 'stability of a bar after Euler' are often used when reference is made to the stability of a compressed bar.

Suppose that the bar has deflected slightly for some reason.

The coordinates of the elastic curve of the bar are denoted by z and y. For small deflections

$$EJy'' = M \tag{9.1}$$

The bending of the bar occurs in the plane of minimum rigidity, and so the quantity J is understood to be the minimum moment of inertia of the section.

The bending moment M is equal to $-Py$.

$$EJy'' = -Py \tag{9.2}$$

Let

$$P/EJ = k^2 \tag{9.3}$$

Equation (9.2) then becomes

$$y'' + k^2 y = 0 \tag{9.4}$$

whence

$$y = C_1 \sin kz + C_2 \cos kz \tag{9.5}$$

The constants C_1 and C_2 must be chosen so as to satisfy the boundary conditions: when $z = 0$, $y = 0$ and when $z = l$, $y = 0$. From the first condition it follows that $C_2 = 0$, and from the second condition

$$C_1 \sin kl = 0 \tag{9.6}$$

This equation has two possible solutions: either $C_1 = 0$ or $\sin kl = 0$.

In the first case the displacements y (equation 9.5) are identically zero for $C_1 = C_2 = 0$, and so the bar maintains the straight-line configuration. This case is of no interest to us. In the second case $kl = \pi n$, where n is an arbitrary integer. Taking into account expression (9.3), we obtain $P = \pi^2 n^2 EJ/l^2$. This means that, for the bar to maintain a curvilinear configuration, the force P must take a definite value. This minimum force P is when $n = 1$

$$P_{cr} = \pi^2 EJ/l^2 \tag{9.7}$$

This force is termed the 'first critical load' or 'Euler's load'. When $n = 1$, $kl = \pi$ and the elastic curve, equation (9.5), becomes

$$y = C_1 \sin \pi z/l \tag{9.8}$$

The bar buckles in the half-wave of a sine curve with a maximum deflection C_1.

For any whole-number value of n

$$y = C_1 \sin \pi n z/l \tag{9.9}$$

and the elastic curve of the bar is represented by a curve in the form of n half-waves (Figure 9.10b).

Our solution however, does not say what happens if $P > P_{cr}$, $kl \neq \pi$. Then it follows from equation (9.6) that $C_1 = C_2 = 0$ since $\sin kl \neq 0$. This means that the function y (equation 9.5) is identically zero and the bar remains straight. Thus at $P = P_{cr}$ the bar assumes a curvilinear shape and becomes straight again at a value somewhat greater than P_{cr}, this contradicting the accepted physical concepts of mechanics of buckling.

These predicaments can readily be overcome if we take into account that the differential equation (9.2) is approximate and applicable only for small deflections. If this equation is written accurately, we obtain

$$EJ\frac{1}{\rho} = \frac{EJy''}{(1 + y^2)^{\frac{3}{2}}} = -Py \tag{9.10}$$

With a force P greater than critical, the displacements grow so rapidly that the quantity y in the denominator cannot be neglected.

Solution of the nonlinear differential equation (9.10) can be obtained numerically. This solution will yield the maximum deflection, at the mid-span, as a function of the applied load (Figure 9.11, curve a).

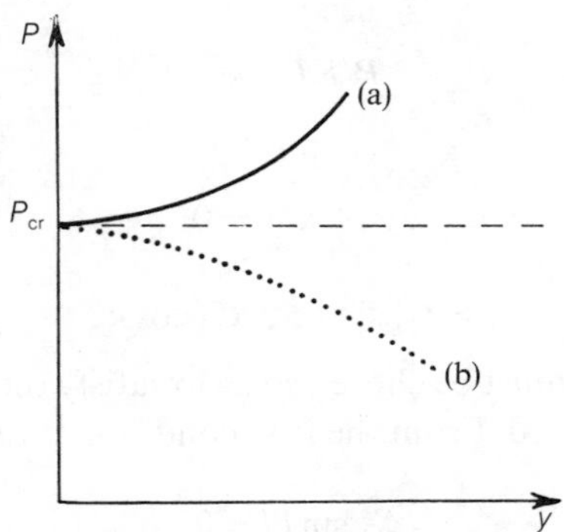

Figure 9.11 Postbuckling behavior of beams (a) and shells (b)

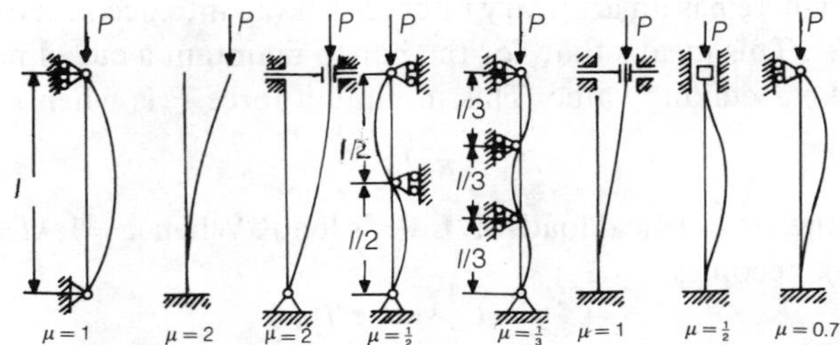

Figure 9.12 Buckling of columns with different boundary conditions

Non-zero maximum deflection occurs for $P > P_{cr}$ and the critical load P_{cr} coincides with the one calculated with linear analysis (equation 9.7).

Such a procedure is called 'post-buckling analysis'.

It can be seen from Figure 9.11 that post-buckling behavior of the rod is relatively smooth, although for small increases of the load above the critical value, very high deflections occur, which might lead to exceeding the yield point and unwanted permanent damage.

There are instability cases, however, in which the situation is worse. For example, in Figure 9.11(b) the dotted line indicates the post-buckling behavior of certain curved plates. When the critical load is reached, the plate moves to another equilibrium point with a sudden jump.

Equation 9.7 was obtained for a bar with guided hinges (simple supports) at the two ends. Other end conditions will yield similar results. In general

$$P_{cr} = \pi^2 EJ/(\mu l)^2 \tag{9.11}$$

where μ takes the values indicated in Figure 9.12.

Some useful results for critical loads of simple structural members are shown in Table 9.4.

In the preceding analysis, 'small' deflections were assumed, that is small in respect to the column length. For relatively short columns, the Euler critical load deviates from

Table 9.4 Design equations for stability of structural members

Geometry	Description	Critical load
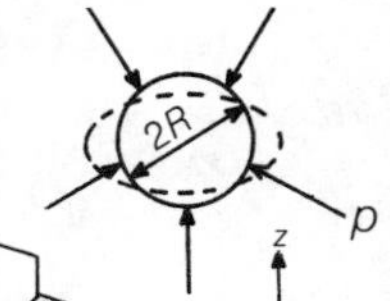	Buckling of a ring or tube of thickness t under external pressure p_{cr}	$p_{cr} = 3EI/R^3$ $I = t^3/12$
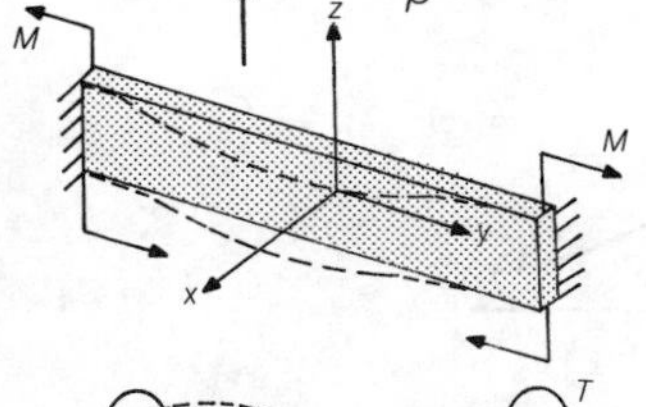	Lateral buckling of a thin beam under a moment on its plane $I_p = I_x + I_z$, Length L	Hinged ends: $M_{cr} = \frac{\pi}{L}(EI_z GI_p)^{\frac{1}{2}}$ Fixed ends: $M_{cr} = \frac{2\pi}{L}(EI_z GI_p)^{\frac{1}{2}}$
	Torsional buckling of a thin rod in torsion	$T_{cr} = 2\pi EI_p/L$
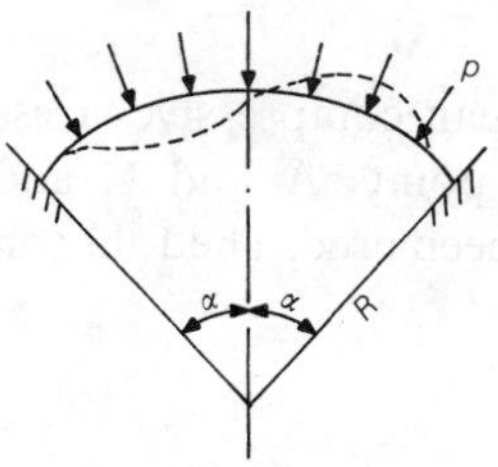	Arch or shell clamped at both ends	$p_{cr} = EI(k^2 - 1)/R^3$ $k \tan\alpha \cot k\alpha = 1$
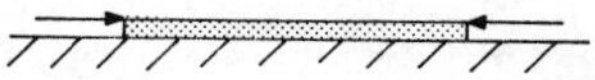	Column on elastic foundation constant $\beta = \bar{\beta}L^4/\pi^4 EI$ $\bar{\beta}$: soil modulus (Table 9.6)	B.C. — $P_{cr} =$ pinned–pinned — $2(\beta EI)^{\frac{1}{2}}$ free–free — $(\beta EI)^{\frac{1}{2}}$
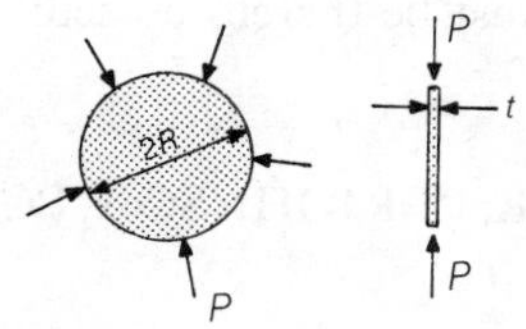	Circular plate force P per unit arc length	Boundary — $P_{cr} =$ free — $2.88t^2 E/R^2$ fixed — $9.79t^3 E/R^2(1 - \nu^2)$
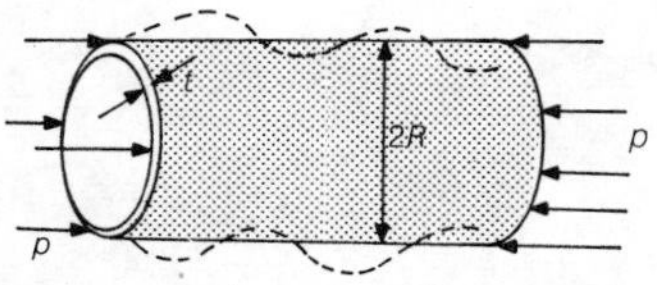	Long tube under axial thrust pressure p	$p_{cr} = \frac{2tE}{R[3(1-\nu^2)]^{\frac{1}{2}}} \cdot \frac{t}{R}$
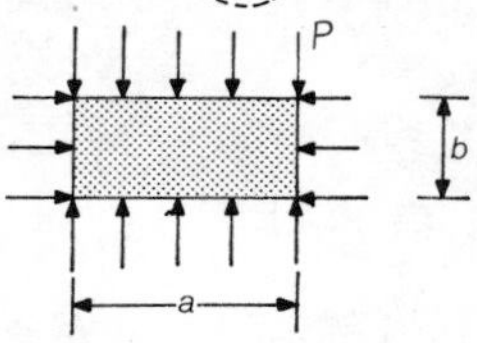	Rectangular plate of thickness t under uniform force per unit length P	$P_{cr} = \frac{\pi^2 Eh^2}{3(1-\nu^2)}\left(\frac{1}{a^2} + \frac{1}{b^2}\right)$

experimental results (open circles in Figure 9.13). The Euler critical load is in satisfactory agreement with experiments up to a compressive stress equal to one-half of the yield strength.

The column might undergo considerable plastic deformation before becoming unstable. In machine design this is usually not acceptable because geometric limitations will be

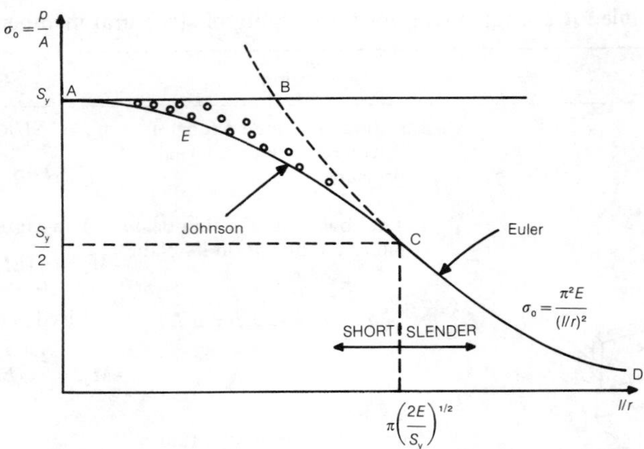

Figure 9.13 Buckling near the yield strength of the material

violated by the resulting permanent deformation. Therefore, the mean compressive stress should not exceed the yield strength, point A. To connect the points A and C and approximate the experimental results many empirical relations have been used. The Johnson formula is very convenient, a second-order curve

$$\sigma_0 = \frac{P_{cr}}{A} = S_y\left[1 - \frac{S_y(l/r)^2}{4\pi^2 E}\right]$$

where r is the radius of gyration of the section $r^2 = I/A$.

To design a column for stability, the following procedure must be therefore used:

```
Procedure BUCKLING
local   A, I, LENGTH, SY, EYOUNG, RGYR, PCREULER, PCRJOHNSON, PCR:
real var
    PI:= 3.14159: real constant
begin
  input data: A, I, LENGTH, SY, EYOUNG
  input boundary condition coefficient MI
  RGYR:= I/A
  STRESS:= P/A
  if STRESS < SY/2 then do
    PCREULER:= A*PI^2*EYOUNG/(MI*LENGTH^2)
    PCR:= PCREULER
    print "SLENDER COLUMN"
  else
    PCRJOHNSON:= A*SY
                  *[1 - SY*MI*(LENGTH/RGYR)^2/(4*PI*EYOUNG)]
    PCR:= PCRJOHNSON
    print "SHORT COLUMN"
  print PCR
end
```

The procedure can be used in the inverse way to determine the section properties from a given load. This is done in the program BUCKLING, Appendix 9.A.

9.4 COMPUTER AIDED STABILITY ANALYSIS

With the computer algorithms and programs presented in Chapter 4, stability questions cannot be answered because the associated mechanisms have not been included. For example, in the transfer matrix and stiffness matrix analysis with straight beam elements, axial force and the resulting moment (as in equation 9.2) do not appear. This indicates the proper course of action.

Recall the beam equations which were used to obtain transfer and stiffness matrices

$$\left.\begin{aligned} EIy'' &= M \\ M'' &= \mathrm{q}(x) \end{aligned}\right\} \tag{9.12}$$

In view of the additional moment due to the axial force P, equation (9.12) takes the form

$$\left.\begin{aligned} EIy'' &= M - Py \\ M'' &= \mathrm{q}(x) \end{aligned}\right\} \tag{9.13}$$

Combining in one equation

$$EIy^{\mathrm{IV}} + Py'' = \mathrm{q}(x) \tag{9.14}$$

The general solution of this equation is not a third order polynomial as in equation (4.18) but (for $k = (P/EI)^{\frac{1}{2}}$),

$$y(x) = C_1 \cos hkx + C_2 \sin hkx + C_3 \cos kx + C_4 \sin kx + \frac{1}{EI}\iiiint \mathrm{q}\, \mathrm{d}x \tag{9.15}$$

For stability analysis we assume, at the moment, $\mathrm{q}(x) = 0$.

The transfer matrix for the beam element with axial force P will be obtained with application of the same boundary conditions. Then

$$L = \begin{bmatrix} 1 & \sin kl/k & (1-\cos kl)/EIk^2 & (kl - \sin kl)/EIk^3 & 0 \\ 0 & \cos kl & \sin kl/EIk & (1-\cos kl)/EIk^2 & 0 \\ 0 & (P \sin kl)/k & \cos kl & \sin kl/k & 0 \\ 0 & 0 & 0 & 1 & 1 \end{bmatrix} \tag{9.16}$$

This transfer matrix can be utilized to yield the solution with the boundary conditions at the end of the beam. Recalling the conditions in Chapter 4, we note that the unknown end conditions are obtained as solution of the algebraic equations (4.11) (see p. 115). If the loading is zero, so is the right-hand side vector. Buckling means non-zero solution which is possible only if the determinant of the coefficient matrix is zero.

$$\det[A(P)] = 0 \tag{9.17}$$

Since the elements of the matrix A are functions of the axial force P, the proper design approach is to seek the lowest value of P which satisfies equation (9.17).

Sometimes, due to numerical inaccuracies this process might be slow or even impossible. For this reason we usually seek with methods of Chapter 5 (program OPTIMUM, see Appendix 5A) the minimum of the function det $[A(P)]$ and test if it is close enough to zero.

Another method is to assume a small lateral load and obtain some response quantity, such as deflection at mid-span, for example. Increasing gradually the axial load P, when the critical load is reached then the response quantity will increase rapidly.

Exactly the same approach is used with the stiffness matrix formulation.

One word of caution: transfer matrix (9.16) must be used very carefully and only when stability analysis is required, although it appears that it could replace the matrix in equation (9.16) making static analysis programs versatile and usable also for stability analysis. This is not the case because some of the elements of the transfer matrix in equation (9.16) are differences of two almost equal members, for example $1 - \cos kL$. Small errors in the evaluation of the $\cos kL$ for small P will have great effect on the results. For this reason, equation (9.16) must be used only when necessary and with caution. For example, very short beams will make kL very small and this will amplify the above unwanted effects.

The procedure TMSTABIL and the corresponding program TMSTABIL given in Appendix 9.C use the transfer matrix approach to compute the function, deflection versus axial load. It includes non-zero static loads and a given distribution of axial loads. Then it increments by a constant factor all axial loads and calculates the response at a selected point. By inspection, one can find the critical load at a point where there is a sudden jump in the increment rate of the deflection.

The procedure TMSTABIL follows (with explanations in italics). It uses the procedure TMSTAT, Chapter 4 (p. 116). A maximum load range must be specified up to which the procedure will check the system for stability. A maximum deflection must be specified beyond which the system will be considered unstable.

```
Procedure TMSTABIL
GLOBAL NN, arrays PLOAD, SPRINGY, SPRINGT(1..NN)
     arrays QLOAD, EI, LENGTH(1..NN – 1)
     arrays FIELD, POINT, A(1..5,1..5)
     array VECTOR(1..5):var
begin                                              (*Input section*)
  input MAXLOAD                               (*Maximum load range*)
  input YMAX                          (*Maximum y for stable shape*)
  input NN
  For I:= 1 TO NN – 1 do
  input
PLOAD(I), QLOAD(I), SPRINGY(I), SPRINGT(I), EI(I), LENGTH(I)
                                                  (*Right end data*)
  input PLOAD(NN), SPRINGY(NN), SPRINGT(NN)
  repeat
                                                   (*Forward sweep*)
  set unit diagonal matrix A
  For I:= 1 TO NN – 1 do
     MATPOINT(I)
     MATFIELD(I)
```

```
      MATMULT(POINT, A, ATEMP, 5, 5, 5)
      MATMULT(FIELD, ATEMP, A, 5, 5, 5)
                                                  (*End station*)
   MATPOINT(NN)
   MATMULT(POINT, A, ATEMP, 5, 5, 5)
                                    (*Apply boundary conditions*)
                                                     (*Free-free*)
   DENOM := A(3, 1)*A(4, 2) - A(3, 2)*A((4, 1)
   V(1) := (A(3, 2)*A(4, 5) - A(3, 5)*A(4, 2))/DENOM
   V(2) := (A(4, 1)*A(3, 5) - A(4, 5)*A(3, 1))/DENOM
   V(3) := 0; V(4) := 0
                                         (*Computation of results*)
   FOR I := 1 TO NN - 1 do
     MATPOINT(I)
     MATFIELD(I)
     MATMULT(POINT, V, VTEMP, 5, 5, 1)
     MATMULT(FIELD, VTEMP, V, 5, 5, 1)
                                                  (*End station*)
     MATPOINT(NN)
     MATMULT(POINT, TEMP, V, 5, 5, 1)
     IF VMAX > YMAX then do
       print "system is unstable at load LOAD"
       exit
     else
     LOAD := LOAD + MAXLOAD/100
   until LOAD > MAXLOAD
     if LOAD > MAXLOAD then do
       print "the system is stable up to maximum given load"
     else
end.
```

9.5 MACHINE FRAMES

Machine frames support the basic units of a machine, ensuring their proper location with respect to one another, and resisting all the principal forces acting in the machine. Baseplates support machines and their drives, ordinarily consisting of a number of separate units, as well as machines of vertical design. Housings and other box-type parts enclose or support various mechanisms of machines.

The design of these components has a considerable influence on the performance and reliability of machines, vibration stability, operating accuracy under load and service life (when these components have slideways or other surfaces subject to wear).

According to their purpose, housing-type components can be divided into the following groups: (1) beds, mainframes and housings, (2) bases, baseplates and foundation plates, and (3) housing-type components of the units or assemblies, Figures 9.14a–l.

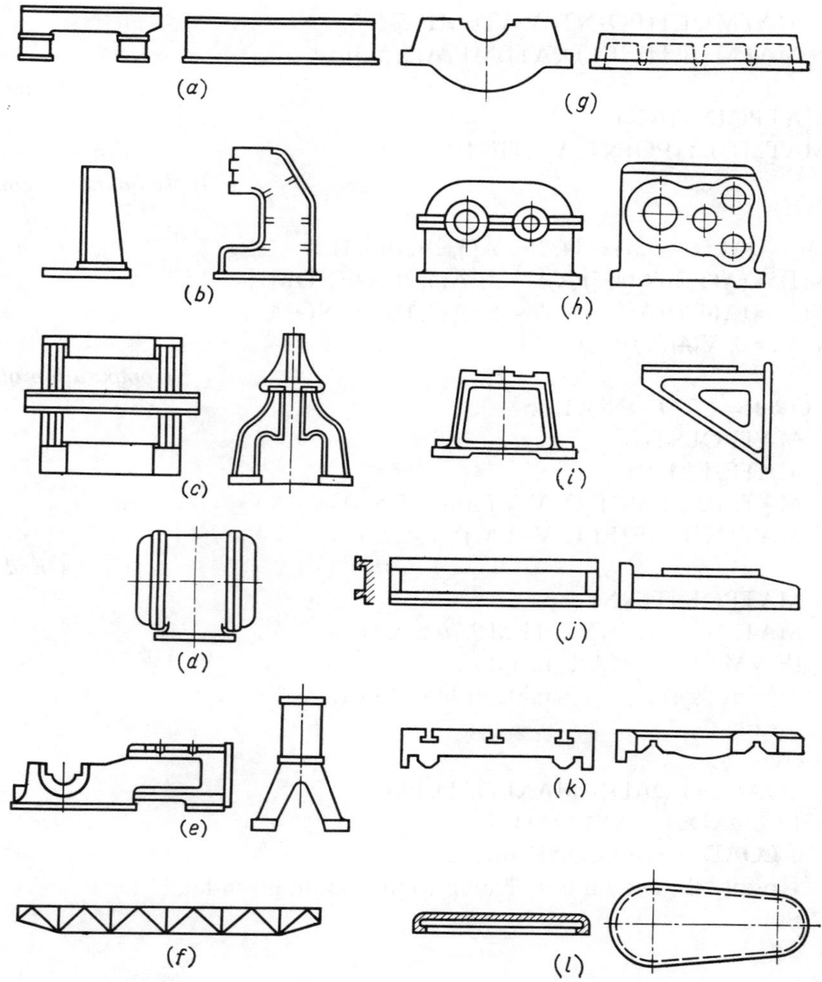

Figure 9.14 Machine housing components. (From Orlov, 1976, by permission of Mir Publishers, Moscow)

Criteria of performance and reliability of housing-type components are strength, rigidity and durability. *Strength* is the basic criterion for components subject to heavy loads, mainly impact (shock) and variable loads. *Rigidity* is the main criterion of performance for the great majority of housing-type components. Increased elastic deflections in such components usually lead to faulty operation of the mechanisms, inaccuracy and the initiation of vibrations. When high rigidity is desirable, the components are made of materials with a high modulus of elasticity, i.e. cast iron and steel used without heat treatment.

The *wear life* is important for components with sliding friction.

Housing-type components of transportation machinery (for instance, engine crankcases, as well as components subject to heavy inertia forces), are frequently made of light alloys. Such alloys possess increased strength per unit mass.

Most housing-type components are of cast iron. This is based on the feasibility of obtaining intricate geometric shapes, and the comparatively low cost in lot production, in which the cost of the die is distributed over a large number of castings.

Housing-type components of welded design are employed to reduce the mass and overall dimensions, and, in job and small-lot production, also to lower production costs and shorten the lead time (time required to begin production).

Welded components are made of: (a) elements of simple shapes in poorly equipped and job and small-lot production, (b) fabricated elements in sufficiently well-equipped lot production, and (c) press-worked elements of refined, streamline shapes in large-lot and mass production.

Weldments made up of cast elements considerably simplify the required castings.

Housing-type components which should be of minimum mass, but are not subject to appreciable loads and do not require dimensional stability, can be efficiently made of plastics. Such components include the housings of portable and hand-held machines and tools, instruments, covers, hoods, etc.

It is good practise to design housing-type components subject to bending and torsion with thin walls whose thickness is usually determined from the condition of castability. Components subject to torsion should, wherever possible, be of closed box section, and those subject to bending should have the main part of the material placed as far as possible from the neutral axis. If opening or hatches are required to enable the inner cavity of the casting to be utilized, they should not coincide in location in opposite walls of the casting. The weakening of the casting by such openings should be compensated for by the provision of flanges or rigid covers. The most effective method of economizing on materials in machine manufacture is, as a rule, the reduction of the wall thicknesses. The required rigidity of the walls can be provided by the proper ribbing.

Recommended wall thicknesses for iron castings are listed in Table 9.5 on the basis of the so-called equivalent overall size of the casting. The equivalent overall size, i.e. the size of a box-like casting of cubic shape which is equivalent, with respect to boundary conditions, to the casing being designed, can be estimated as

$$N = \frac{2L + B + H}{4}$$

where L, B and H are the length, breadth and height of the casting.

Inner walls and ribs cool more slowly than do outer ones, and therefore, to ensure

Table 9.5 Recommended wall thickness for iron castings

Equivalent overall size, m	Thickness of outer walls, mm	Thickness of inner walls, mm	Equivalent overall size, m	Thickness of outer walls, mm	Thickness of inner walls, mm
0.4	6	5	2.0	16	12
0.75	8	6	2.5	18	14
1.0	10	8	3.0	20	16
1.5	12	10	3.5	22	18
1.8	14	12	4.5	25	20

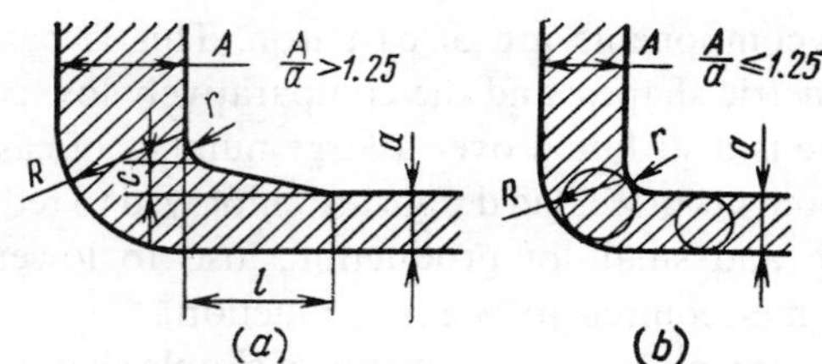

Figure 9.15 Casting details

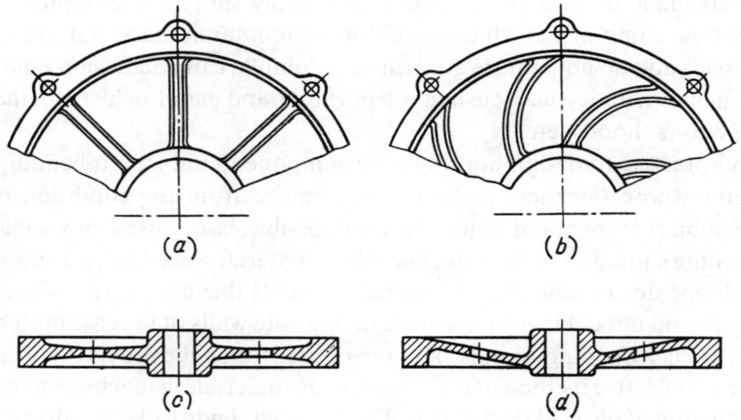

Figure 9.16 Casting of pulleys

simultaneous cooling, they should be 80% the thickness of the outer walls and ribs. The height of ribs or fins should not be more than five times their thickness. To comply with conditions of castability, walls of steel castings are designed from 20 to 40% thicker than those of iron castings. Nonferrous casting alloys allow substantially thinner walls than iron castings. Thick-walled castings are used when the overall size of the component is strictly limited.

The walls and sections should, as far as possible, be of the same thickness. If a constant thickness cannot be maintained, there should be a gradual blending of the thinner into the thicker sections (Figure 9.15).

Castings should be designed so that in cooling the members are free to shrink without developing excessive residual stresses. For this reason, the ribs of round cast plates are designed with a curved shape (Figure 9.16b); a special network of ribbing is provided, etc.

To prevent an accumulation of nonmetallic inclusions and the formation of blowholes, it is best to avoid large flat surfaces if, according to the mounding conditions, they are to be poured in the horizontal position (Figures 9.16c, d).

To simplify pattern-making, the geometric shapes of elements of castings should be ones easy to machine, i.e. they should be bounded by planes, and cylindrical and conical surfaces.

Special efforts should be made to design castings so that they are easy to mold. It should

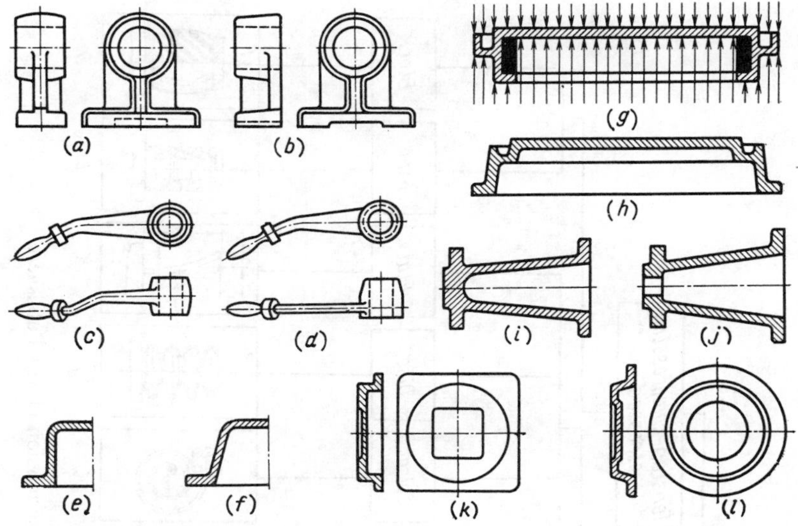

Figure 9.17 Casting forms. (From Reshetov, by permission of Mir Publishers, Moscow)

be feasible to mold simple castings in one half-mold (Figure 9.17b) or with a single flat parting line (Figure 9.17a). To facilitate withdrawing of the pattern from the mold, sidewalls should be designed with a slight inclination from the vertical (Figure 9.17f), as otherwise it will be necessary to provide foundry draft on both the outside and inside surfaces.

Wherever possible, the shapes of castings should be designed so that no cores or loose pieces on the patterns are required (Figure 9.17h). The possibility of withdrawing the pattern from the mold (without employing loose pieces or cores) is checked as follows: An imaginary stream of rays, perpendicular to the parting plane, should not produce shadowed portions (Figure 9.17g). If cores are an absolute necessity, then holes should be provided (for production purposes) to reliably anchor the covers (Figure 9.17j). Cores should be unified wherever possible.

Provision should be made in job-lot production for the sweep-molding of large castings (Figure 9.17l) using a strickleboard or sweep template.

9.6 DESIGN OF SPRINGS

In Chapter 8 we saw several types of joint elements with the general main purpose of connecting as rigidly as possible other larger elements. This is not always the case. Many times joining two machine elements needs to be done in a flexible way.

Such joints are used when we need the following:

(a) To produce constant forces, even if the two joining elements move to some extent in respect to one another.
(b) To adjust machine members when clearances exist or can develop.

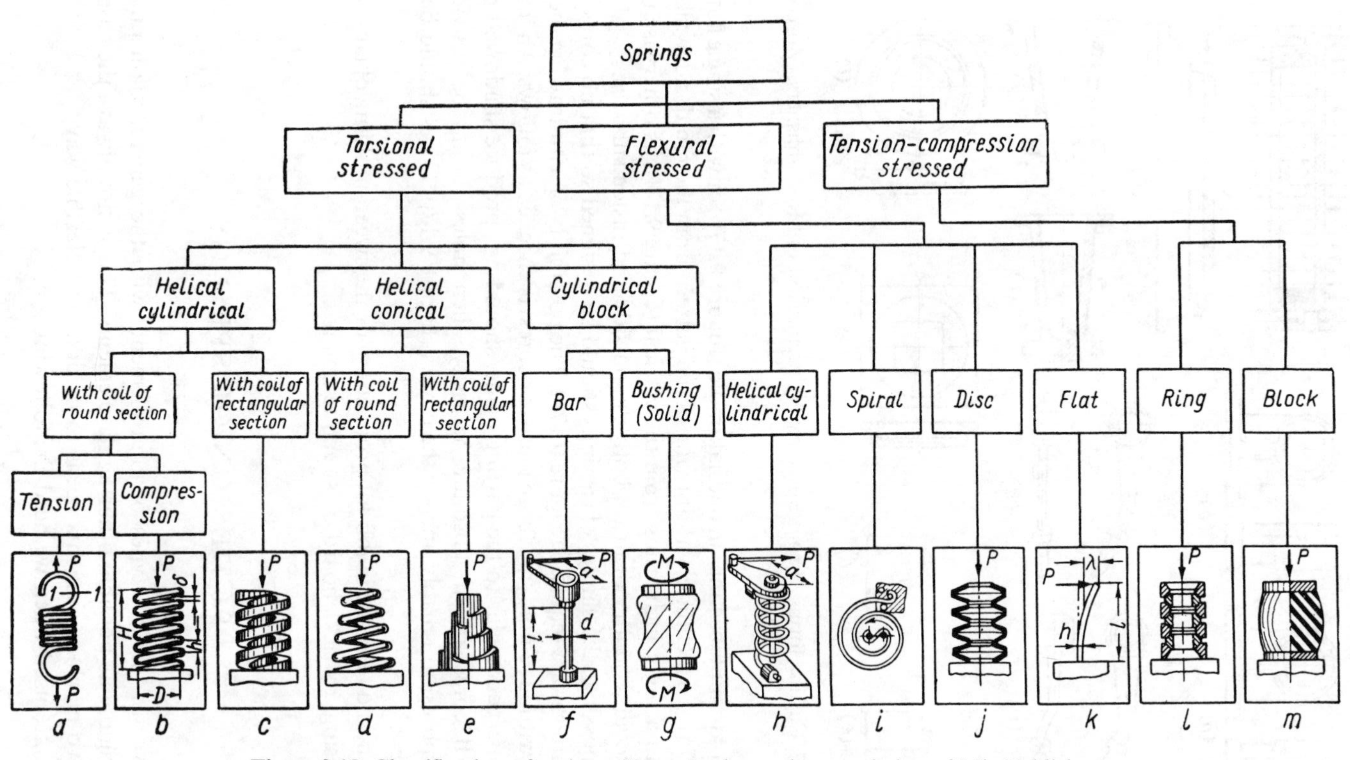

Figure 9.18 Classification of springs. (From Reshetov, by permission of Mir Publishers, Moscow)

(c) To accumulate elastic strain energy, which subsequently is used to drive an element into motion.
(d) For vibration isolation, when we expect dynamic loads and we want to minimize maximum stresses on the machine.
(e) To absorb impact energy, when impact loads are expected in a machine, in order to keep the peak impact stresses as low as possible.

Linear elements are loaded by forces in one direction and they are deformed in the same direction. In most cases there is a linear relationship between force and displacement,

$$F = k\delta \tag{9.18}$$

where k is a constant, depending on the material and geometry of the spring and is called the 'spring constant'. Such springs are helical springs (Figure 9.18) and other types which are shown in Figures 9.19 to 9.22.

Torsion springs are loaded with a torsional moment M and there is a relative twist between the two ends of the spring which are again related linearly

$$M_T = k_T \Delta\varphi \tag{9.19}$$

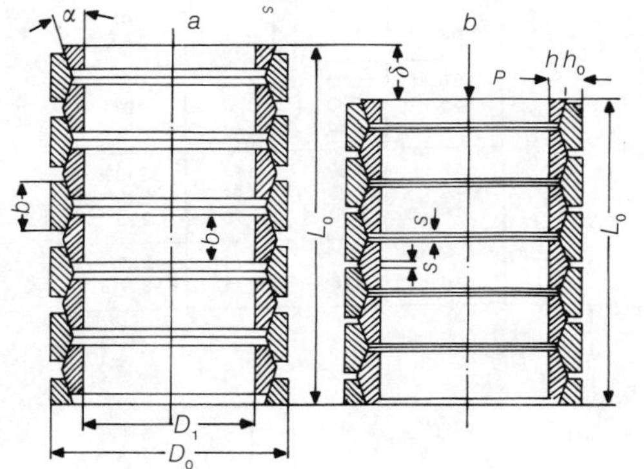

Design equations
Outer ring
$\sigma_1 = P/\pi b \tan(a + \rho)h$
Inner ring
$\sigma_1 = P/\pi b \tan(a + \rho)h$
$\delta = (N-1)P(D_0/h_0 + D/h)/2\pi bE \tan\alpha \tan(\alpha+\rho)$
$k = P/\delta = 2\pi bE \tan\alpha \tan(\alpha+\rho)/(N_a - 1)(D_0/h_0 + D_1/h_1)$
$\tan\rho = f$ – (friction coefficient)

Figure 9.19 Design equations for ring springs. (Decker)

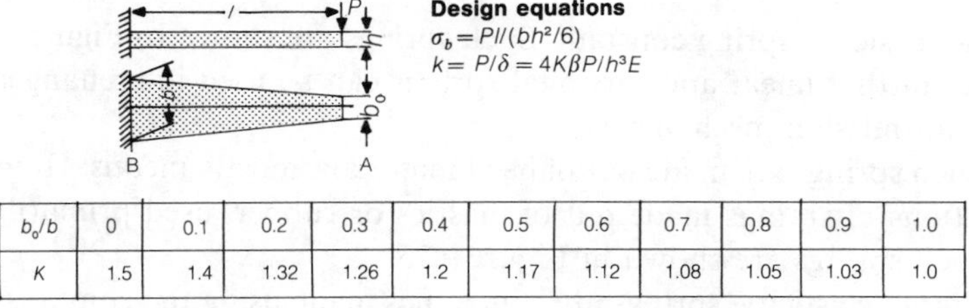

b_0/b	0	0.1	0.2	0.3	0.4	0.5	0.6	0.7	0.8	0.9	1.0
K	1.5	1.4	1.32	1.26	1.2	1.17	1.12	1.08	1.05	1.03	1.0

Figure 9.20 Design equations for tapered beam springs

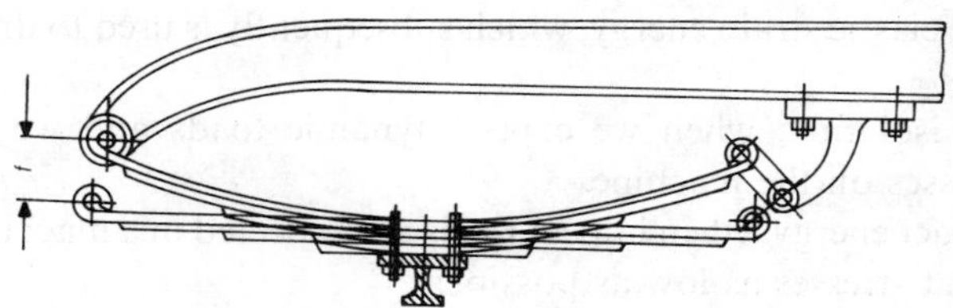

Figure 9.21 Leaf springs

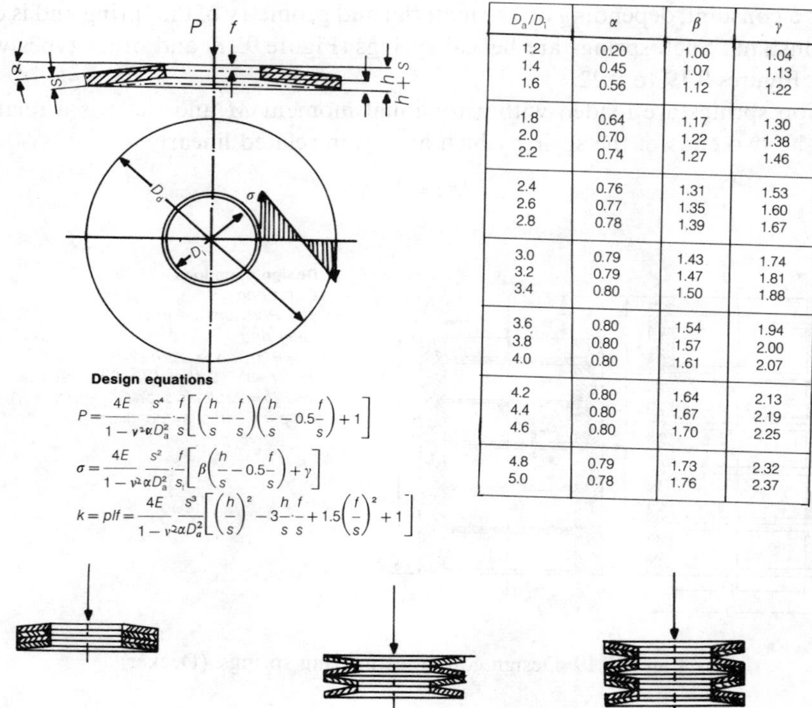

D_a/D_i	α	β	γ
1.2	0.20	1.00	1.04
1.4	0.45	1.07	1.13
1.6	0.56	1.12	1.22
1.8	0.64	1.17	1.30
2.0	0.70	1.22	1.38
2.2	0.74	1.27	1.46
2.4	0.76	1.31	1.53
2.6	0.77	1.35	1.60
2.8	0.78	1.39	1.67
3.0	0.79	1.43	1.74
3.2	0.79	1.47	1.81
3.4	0.80	1.50	1.88
3.6	0.80	1.54	1.94
3.8	0.80	1.57	2.00
4.0	0.80	1.61	2.07
4.2	0.80	1.64	2.13
4.4	0.80	1.67	2.19
4.6	0.80	1.70	2.25
4.8	0.79	1.73	2.32
5.0	0.78	1.76	2.37

Figure 9.22 Design equations for Belleville springs

where k_T is the torsional spring constant. Such springs are shown in Figures 9.23 and 9.24.

It is apparent that linear and torsional springs can be used interchangeably, by using proper load transmission mechanisms.

Usually such springs are made out of hard materials, mostly metals. However there are many applications of springs made out of plastics or rubber, used primarily for machine mountings. Such springs are shown in Figure 9.25.

For even more elasticity, springs are sometimes made using the compression of a large volume of air (Figure 9.26).

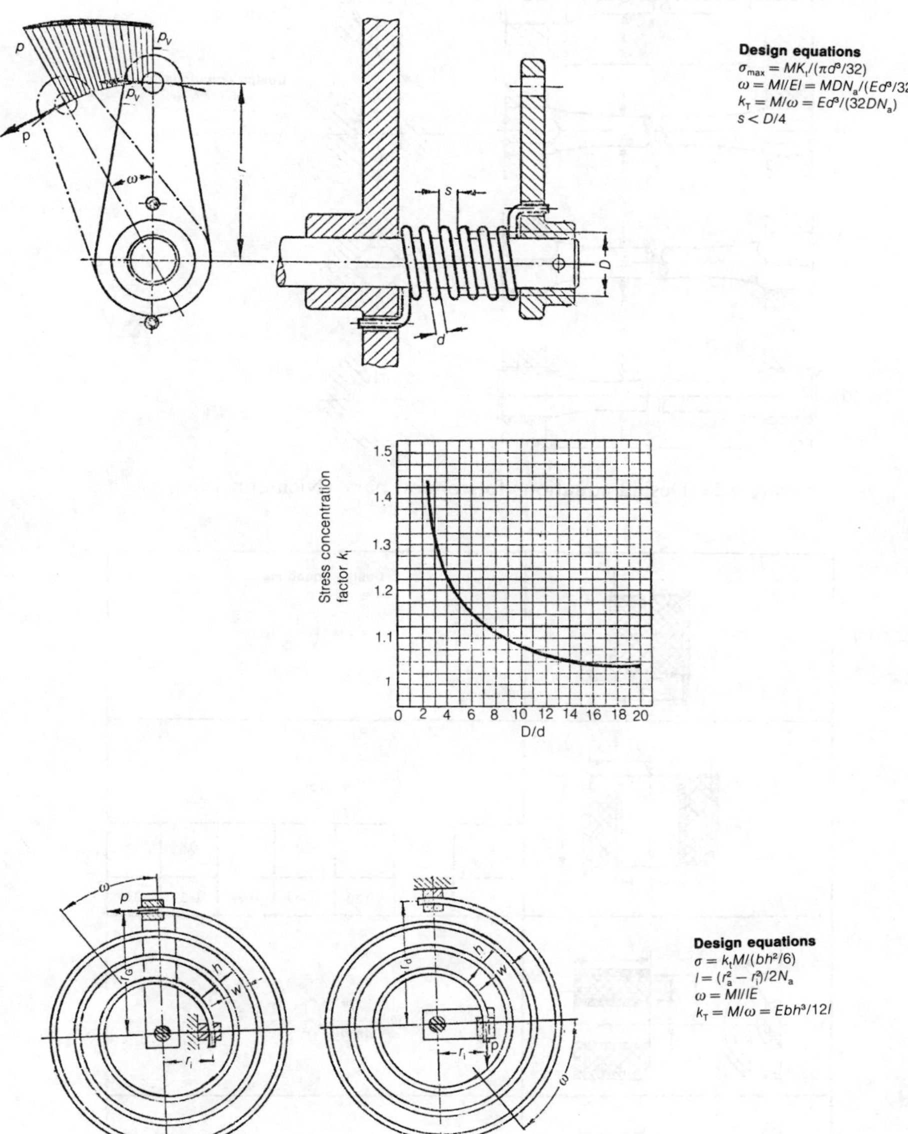

Figure 9.23 Design equations for torsional helical springs. (Decker)

Materials for springs must have high elastic properties, high strength and they must be stable with time. Because springs made out of weaker materials must have large dimensions, for dynamic loadings which are common in spring applications, there are high inertia forces which must be avoided. Since the modulus of elasticity does not change considerably amongst various types of steel, high-strength steels are mostly used, as discussed in Chapter 6.

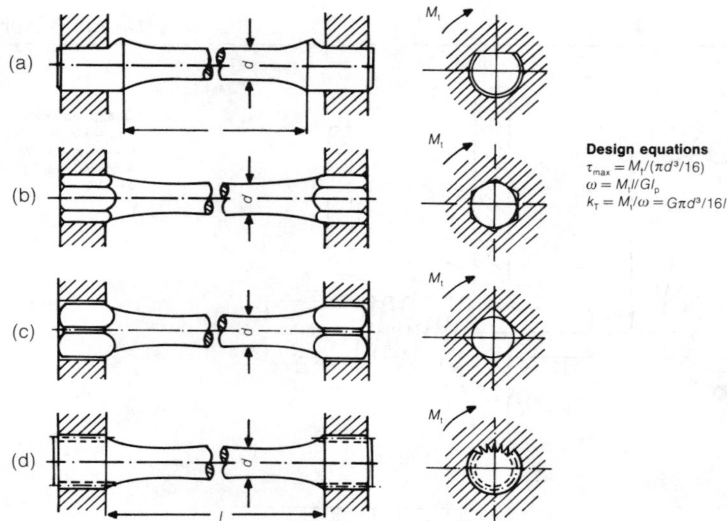

Figure 9.24 Design equations for torsion bars. (Niemann 1965)

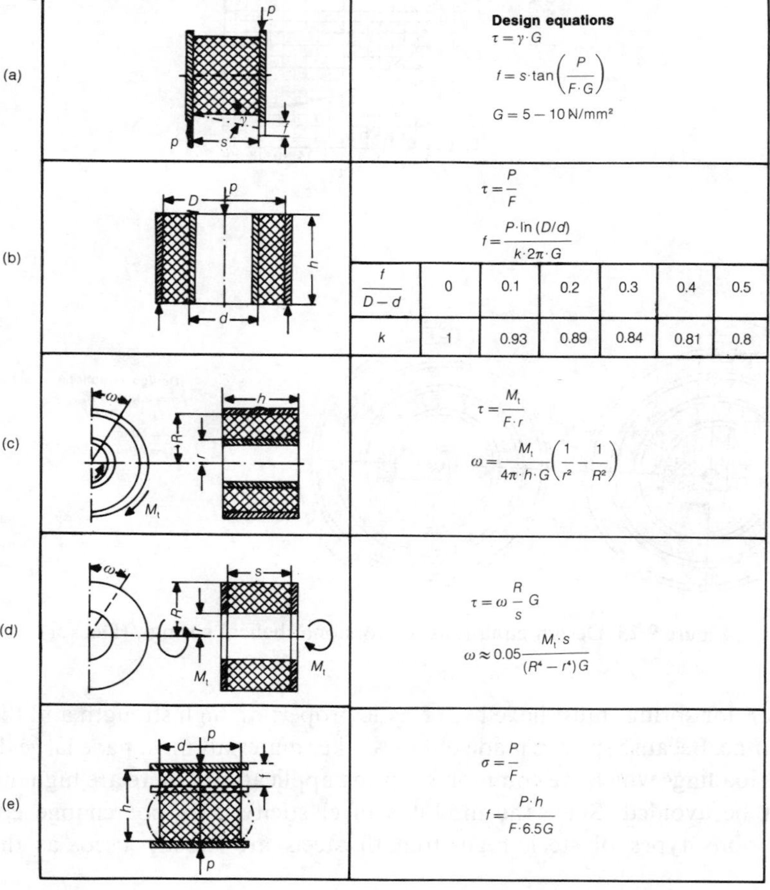

Figure 9.25 Design equations for rubber springs. (Niemann 1965)

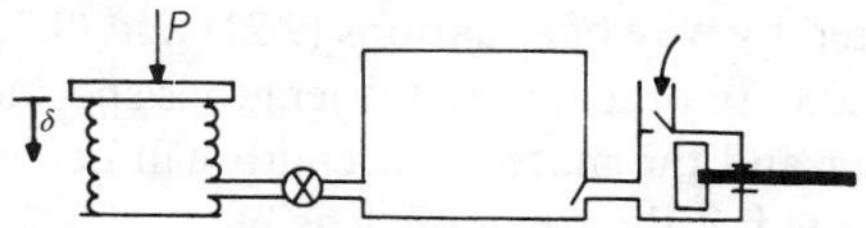

Figure 9.26 Air springs

The main materials for springs are high-carbon steels, manganese steel, chromium steel, chromium–vanadium steel, etc. However, the most widely used are carbon steels because they have high strength and low price.

Manganese, silicon and chromium–manganese steels have higher strength and can be hardened, enabling them to be used for springs of small cross-sections. Chromium–vanadium steel has high mechanical strength and also high endurance limit, heat resistance and good mechanical processing properties; therefore, it is used for critically-loaded springs and in particular for repeated loading such as in valve springs for engines. Springs operating in corrosive environments are made of nonferrous alloys such as several types of bronzes. Steel springs in the same environment can resist oxidation when coated with cadmium and other coatings.

For critical applications, springs are processed with methods such as shot peening or plastic-free stress. The latter is achieved by way of application of a load greater than design load to produce plastic strains. Upon release the remaining stresses act in a way opposite to the applied stresses, therefore enabling the spring to take as much as 25% higher loads.

9.6.1 Design of torsion bars

A torsion bar (Figure 9.24), has a well-known relation between torque and angle of twist

$$\Delta\varphi = M_T L / G I_P \tag{9.20}$$

which considers the torsion bar as a perfect cylinder under the action of the torque.

By definition, the spring constant is then

$$k_T = G I_P / L \tag{9.21}$$

Maximum stresses at the outer fiber owing to torsion are

$$\tau_{max} = \frac{M_T}{I_P} R \tag{9.22}$$

if pure torsion is assumed and bending and shear are not present.

As in the case of buckling (that is, stability of thin rods under axial compressive loads), long twisted rods can undergo torsional buckling, assuming a helical form. According to the theory of elastic stability, there is a critical torsion

$$T_{cr} = 2\pi EI/L = 2\pi^2 ED^4/32L \tag{9.23}$$

To design a torsional spring in the form of a torsion bar, the maximum torque T and a specific value (or a maximum value or an allowable range) of the spring constant is given. Depending on other service conditions, such as environment, temperature etc., a material is selected and then its strength properties are known. Design parameters are the length and the diameter which are calculated by way of equations (9.21) and (9.22), and a stability check is made by way of equation (9.23). In other cases, the length is specified and then by way of the same equations the diameter and the material strength will be determined from which the proper material will be selected, if the design is feasible.

For dynamic loads, equivalent stresses must be computed with the method presented previously.

9.6.2 Axially-loaded helical springs

In axially-loaded helical springs the wire at any section is twisted by a torque equal to $PD/2$, where P is the axial force acting on the center-line of the coil and D is the mean coil diameter. If we consider a small part of the wire as a torsion bar, for the angle of twist, equation (9.20) applies. This twist gives deflection in the center of the spring equal to $\Delta\varphi D/2$. Therefore, the length is πD, the deflection will be

$$\delta = \frac{8N_a PD^3}{Gd^4} \tag{9.24}$$

where N_a is the number of the active turns of the spring, D the mean diameter of the spring, d the wire diameter, and G the shear modulus. Therefore, the torsional spring constant will be

$$k_T = \frac{P}{\delta} = \frac{Gd^4}{8N_a D^3} \tag{9.25}$$

The same twisting torque of the wire produces maximum shear stresses at the outer fiber owing to torsion

$$\tau = \frac{8PD}{\pi d^3}\frac{C - 0.25}{C - 1}; \quad C = \frac{D}{d} \tag{9.26}$$

where $C = (2R/d)$ is a correction factor because it is a torsion of a curved bar.

Helical springs in compression are also subject to buckling. In reality, it is torsional buckling of the wire but appears as a column buckling. Applying equation (9.23) for the torsional buckling of the wire, we can use the Euler equation with an equivalent moment of inertia

$$I_{eq} = \frac{Ld^4}{64N_a D(1 + v/2)} \tag{9.27}$$

where v is the Poisson ratio. Then the Euler equation can be applied

$$P_{cr} = \frac{\pi^2 EI_{cr}}{(\mu L)^2} \tag{9.28}$$

The coefficient μ, as previously, characterizes the boundary conditions at the point of support of the springs. The same values apply as for the buckling of columns.

To design a helical spring in axial loading, the maximum load and the spring constant are usually given. A material is selected and its mechanical properties determined. Design parameters are the mean diameter, the wire diameter and the number of acting turns of the coil. In principle, the three design equations can yield the unknown design parameters. Since there are other design considerations, it is more rational to use the computer to calculate, for a number of active turns (say from 5 to 50 in steps of 5), the diameters d and D by the two non-linear algebraic equations

$$f_1(d, D) = \frac{Gd^4}{8N_a D^3} - k = 0 \tag{9.29}$$

$$f_2(d, D) = \frac{8\pi D}{\pi d^3}\left(\frac{C-0.5}{C-1}\right) - \frac{S_{sy}}{N} = 0 \tag{9.30}$$

The solution of these equations is performed with the Newton–Raphson method and each solution is checked for buckling. Finally, the feasible solutions are tabulated and the designer must pick up the most suitable solution based on some other requirements.

Such obvious requirements might be, for example, that the spring must be allowed to be compressed by a certain distance before the subsequent turns of the coil come in contact. Also for practical purposes the number of active turns cannot be a very small one. Other times there are space limitations for the external diameter of the coil. In cases where it is necessary that the requirement of stability be waived, by providing an external guide to the spring with suitable lubrication, friction and wear will be minimized and the spring will be supported, avoiding lateral buckling. In this case care must be exercised with high-speed engines because heating of the spring or the pipe and rapid deterioration might occur.

For relatively thick wire, the shear stress owing to the axial force must also be included in the calculation. On the occasions (rather rare), where the coil is made out of ductile material and for static loading, the correction factor owing to curvature can be neglected (there is redistribution of stresses because of local yielding). For brittle materials and especially for dynamic loading this factor has to be always taken into account.

The procedure HELICAL SPRINGS and the related program (HELICSP) in Appendix 9.B perform the above calculations.

It appears that at least two parameters in a helical spring design are arbitrary. This naturally leads to optimization. Depending on the problem at hand, other limitations might be imposed, such as space, maximum inclination of the spiral, etc., while the most obvious objective function is the weight. The program HELICSP (Appendix 9.B) uses an exhaustive search method to tabulate the results for a parameter study with given limits in terms of the material and the number of turns. A typical computation follows the program. Since the number of design parameters may be larger than the design equations, spring design can be optimized. For helical springs, for example, the formulation of the optimization problem follows:

Data:	Free length L, travel δ, load P, spring constant k, material S, E, G
Design variables:	N, D, d
Objective function:	$f(N, D, d) = \pi^2 D d^2 N/4$, the volume of the spring
Constraints:	
Equality constraints:	
a. maximum stress:	$g_1(N, D, d) = S_{sy} - (16PR/\pi d^3)(C - 0.25)/(C - 1) = 0$, $C = d/D$, $R = D/2$

b. spring constant: $g_2(N, D, d) = k_T - Gd^4/64NR^3 = 0$

Inequality constraints:

a. buckling load: $h_1(N, D, d) = \pi^2 EI\mu/L^2 - P > 0$

b. travel: $h_2(N, D, d) = L - Nd - \delta > 0$

c. helix angle 10°: $h_3(N, D, d) = 0.176 - L/\pi ND > 0$, because $\tan\alpha = L/\pi ND$, $\tan 10° = 0.176$

d. maximum diameter: $D_{\max} > D$, $h_4(N, D, d) = D_{\max} - D > 0$

Penalty function:

$$P(N, D, d) = f(N, D, d) + K(g_1^2 + g_2^2) + L(\langle h_1\rangle^2 + \langle h_2\rangle^2 + \langle h_3\rangle^2 + \langle h_4\rangle^2)$$

9.7 MACHINE FOUNDATIONS

Machines which are expected to transmit substantial static or dynamic forces through their pedestal are installed on foundations. A typical arrangement of this kind is shown in Figure 9.27. The machine, of mass m mounted on a massive foundation of mass M, rests directly on soil or some other elastic material such as cork, rubber, springs, etc. It can therefore be represented as a one degree of freedom system. The spring constant k can be determined from the dimensions and properties of the elastic material. If the mass M rests on soil, the spring constant will be

$$k = Ak_s \tag{9.31}$$

where A is the footing surface and k_s is a constant called the 'coefficient of compression of the soil' (lb/in^3 or equivalent SI units). Typical properties of relevant materials and soils are given in Table 9.6.

Since foundation affects directly the machine operation and the environmental effects of machine vibration and noise, its design is a part of the machine design effort.

Usually, the machine rests on a heavy base (which is resiliently supported by some elastic substance) or directly on the soil. The purpose of this construction is to keep at a minimum

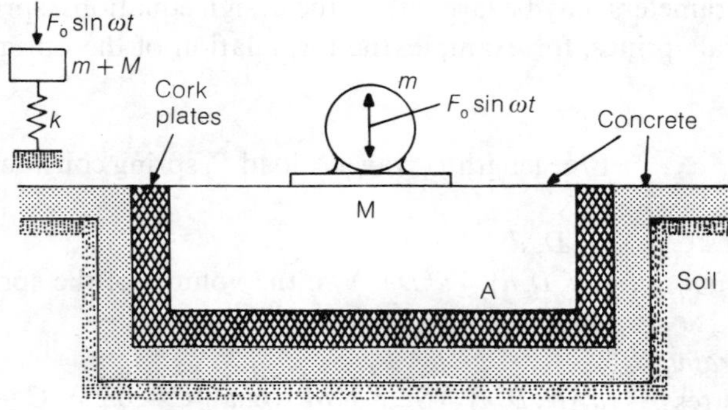

Figure 9.27 Machine foundation

Table 9.6 Load-carrying capacity of soils

Soil Type	Permissible load psi/(N/mm²)	$k_s = \bar{\beta}$ Vertical lb/cu. in N/mm³(×2×10⁵)	Horizontal lb/cu. in N/mm³
Grey plastic silty clay with sand and organic silt	15/10	75/200	110/300
Brown, saturated silty clay with sand	22/15	110/300	170/460
Dense silty clay with sand	75/50	190/520	315/850
Medium moist sand	30/20	110/300	200/550
Dry sand with gravel	30/20	110/300	200/550
Fine saturated sand	35/25	150/400	340/925
Medium sand	35/25	150/400	340/925
Loesial, natural moisture	45/30	170/460	370/1000
Gravel	90/62	375/1000	1000/2700
Sandstone	150/10	350/950	1000/2700
Limestone	165/115	400/1100	1200/3250
Granite, partly decomposed	600/415	1000/2700	3500/9500
Granite, sound	850/590	1050/2850	3600/9800

the force transmitted through the foundation to the surroundings. We are, therefore, interested in the force transmitted. This is the force carried through the springs, which is, accounting also for damping,

$$f = kx + c\dot{x} \tag{9.32}$$

x is the displacement which can be computed from the imposed force $F \sin \omega t$ and the properties of the system, $(m + M)$ and k. The amplitude of this force will be

$$F_T = kX + ic\omega X = \frac{F_0[1 + (2\zeta\omega/\omega_n)^2]^{\frac{1}{2}}}{\{[1 - (\omega/\omega_n)^2]^2 + (2\zeta\omega/\omega_n)^2\}^{\frac{1}{2}}} \tag{9.33}$$

where ω_n is the natural frequency $(k/m)^{\frac{1}{2}}$ and ζ the fraction of critical damping $c/4(km)^{\frac{1}{2}}$, being 2.7–3.5% for concrete frames, 0.5–2.5% for steel frames, 5–6% for heavy mass resting on soil, 0.5–1.0% for steel spring support, and 1 – 2% for cork or rubber support.

Therefore, the ratio of transmitted to imposed force (T_R) is

$$T_R = \frac{F_T}{F_0} = \{[1 - (\omega/\omega_n)^2]^2 + (2\zeta\omega/\omega_n)^2\}^{-\frac{1}{2}} \tag{9.34}$$

This ratio is called 'transmissibility' and is given in Figure 9.28. For quick calculations, we sometimes use the static deflection δ_{st} of the system owing to its own weight. This can readily be measured by loading the foundation with a static weight w and multiplying the resulting static deflection by $(M + m)g/w$. The natural frequency is

$$\omega_n = \left[\frac{kg}{(m + M)g}\right]^{\frac{1}{2}} = \frac{(g)^{\frac{1}{2}}}{(\delta_{st})^{\frac{1}{2}}} \tag{9.35}$$

The following formula is commonly used for the natural frequency:

$$f_n = 188(\delta_{st})^{-\frac{1}{2}} \tag{9.36}$$

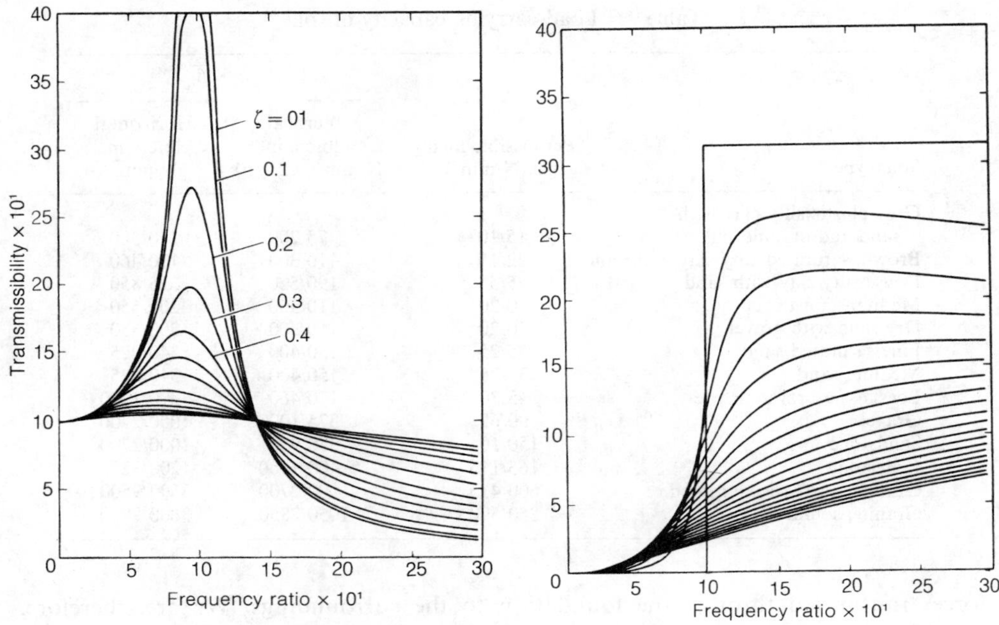

Figure 9.28 Response spectrum of a one degree of freedom system

where n is in counts per minute (cpm) and static deflection is in inches. In terms of the frequency $f = \omega \times 60/2\pi$ (cpm), for $\zeta = 0$, we obtain

$$T_R = \left(\frac{(2\pi f)^2 \delta_{st}}{g} - 1\right)^{-1} \tag{9.37}$$

From equation (9.34) we observe that for $\zeta = 0$, we have effective isolation ($T_R < 1$) only if $(\omega/\omega_n)^2 - 1 > 1$ or $\omega > \omega_n(2)^{\frac{1}{2}}$. From Figure 9.28, however, we observe that for $\omega/\omega_n = 3.0$ the transmissibility is about 0.1 and decreases very slowly after that. The value $\omega = 3\omega_n$ is very often used for foundation design.

In Figure 9.28, we observe also that damping has no effect on the point where effective isolation ($T_R < 1$) starts. This point is $\omega/\omega_n = (2)^{\frac{1}{2}} = 1.41$. For frequencies $\omega < 1.41\omega_n$, damping reduces the transmissibility ratio, and thus the transmitted force. For frequencies $\omega > 1.41\omega_n$, however, damping increases the transmissibility ratio and the transmitted force. This observation shows that damping is not always a desirable feature in engineering systems.

The spectrum of Figure 9.28 can also be used to study the transmissibility of motion in the case of base excitation. Thus, if the base has a motion $y = Y\cos\omega t$, the mass will have a motion $x = X\cos(\omega t - \varphi)$. The ratio of resulting to imposed motion (output/input) is $X/Y = H(\omega)$, where $H(\omega)$ is exactly the function plotted in Figure 9.28.

Physically, the situation might represent the isolation of an instrument, for example, mounted on a vibrating floor. The engineer very often has to design a flexible mounting of the instrument in order to minimize its vibration if the floor or the base vibrates. Here the motion transmissibility ratio equals the force transmissibility ratio in the case of an harmonic force imposed on the mass. They are both plotted in Figure 9.28. From this we again observe that effective motion isolation is incurred for $\omega > 1.41\omega_n$ only, regardless of damping. For higher exciting frequencies, damping has an adverse effect. As a general rule, therefore, for a given ω a flexible mounting (low k, thus high ω/ω_n) with light damping must be provided.

EXAMPLES

Example 9.1

Determine the critical load for an aluminum alloy pipe with inner diameter 30 mm, outer diameter 40 mm, Young's Modulus $E = 71{,}000\ \text{N/mm}^2$, yield strength $S_y = 180\ \text{N/mm}^2$ and length 1.2 m, fixed at one end and hinged on the other.

Solution

The moment of inertia and the cross-section area are

$$I = \pi(D^4 - d^4)/64 = \pi(256 - 81)10^4/64 = 8.59 \times 10^4\ \text{mm}^4$$

$$A = \pi(D^2 - d^2)/4 = \pi(1600 - 900)/4 = 549\ \text{mm}^2$$

For the given boundary conditions, $\mu = 0.7$ (Figure 9.12). The slenderness ratio is

$$\lambda = \mu L/(I/A)^{\frac{1}{2}} = 67.2$$

The limiting value is

$$\lambda_0 = \pi(E/S_y)^{\frac{1}{2}} = \pi(71{,}000/180)^{\frac{1}{2}} = 62$$

Therefore the Euler formula is valid. Therefore, further,

$$P_{cr} = \pi^2 EI/(\mu L)^2 = \pi^2 \times 71{,}000 \times 85{,}900/(0.7 \times 1200)^2 = 85{,}308\ \text{N}$$

$$\sigma_{cr} = P_{cr}/A = 155\ \text{N/mm}^2$$

Example 9.2

Using the BUCKLING program, design a solid round bar for the data in Example 9.1.

Solution

The run of the program follows:

```
INPUT LOAD, LENGTH, YIELD POINT, YOUNG MODULUS
285300, 1200, 180, 71000
BOUNDARY CONDITIONS:
-----------------------------------
PIN-PIN            M = 1
CLAMPED-FREE           2
  >> - CLAMPED         .5
  >> - PIN             .7
ENTER PROPER       M
?.7
SOLUTION:
-----------------------------------
DIAMETER       = 28.011119
COMPRESSIVE STRESS = 138.419921
SHORT COLUMN
```

The section is $\pi \times 28^2/4 = 616\,\text{mm}^2$, heavier than the hollow section of Example 9.1, as expected. It must be noted that the column is now short and the Johnson formula applies.

Example 9.3

The supporting column for a power-plant installation is loaded with 350 kN and it is fabricated with two channel sections with welded cross-plates. The column material is structural steel ASTM A284 grade D with $S_y = 225\,\text{N/mm}^2$, $E = 210{,}000\,\text{N/mm}^2$. The safety factor must be at least 1.4. Select the proper channel section and find the spacing of the plates and the distance of the channel sections.

Solution

For buckling in a direction perpendicular to the figure plane, the plates do not contribute and the channel sections take all the buckling load. Therefore, by the Euler formula:

$$2I_z = P(\mu L)^2/\pi E = 350{,}000(6{,}000)^2/\pi^2 210{,}000 = 6.1E6\,\text{mm}^4$$

The channel section will then have

$$I_z = 3.05E6\,\text{mm}^4$$

The section with width 120 mm is selected with

$$I_z = 3.04E6\,\text{mm}^4, \quad I_x = 0.31E6\,\text{mm}^4, \quad A = 1330\,\text{mm}^2.$$

In the x-direction, the section must be also stable. The moment of inertia of the section in the x-direction is, if x_0 is the distance of the centroid, equal to 15.4 mm,

$$I'_x = 2I_x + 2A[(a/2) + x_0]^2 = 2 \times 0.31E6 + 2 \times 1330[(a/2) + 15.4]^2$$

This must equal $2I_z$ to have equal buckling resistance in both directions. Therefore

$$[(a/2) + 15.4]^2 = (6.1E6)/2 \times 1330 = 2060$$

Therefore $a = 60\,\text{mm}$.

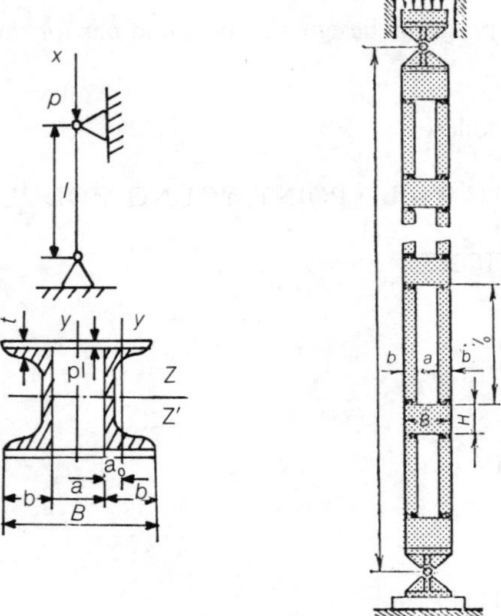

Figure E9.3

Each section between two adjacent plates at distance l must resist local buckling. In the z-direction, this condition is obviously satisfied. In the x-direction it must, from the Euler formula, be:

$$I_x = 0.31E6 > (P/2)l^2/\pi^2 E$$

Therefore, for fixed-fixed mounting,

$$l_o^2 = 0.31E6 \times \pi^2 E/(\mu P/2) = 0.31E6 \times \pi^2 \times 210{,}000/(2 \times 350{,}000/2)$$
$$l_o = 1354\text{ mm.}$$

Example 9.4

Find the bending moments, stresses and deflection for a beam member with axial and transverse load (beam–column), Figure E9.4.

Solution

The problem can be solved by adding to the right-hand side of equation (9.2) the induced bending moment by the transverse load and solving accordingly. For relatively symmetrical transverse loads, as is usually the case in machine design, an approximate method yields adequately accurate results.

To this end, we assume that the shape of the deflected beam is similar to one without transverse forces, that is sinusoidal, as in equation (9.9).

$$y = \delta_{\max} \sin \pi x/L \quad \text{(a)}$$

Equation (9.2) with the addition of the moment by the transverse forces becomes, if y_t is the deflection due to these forces acting alone,

$$EIy'' = -Py + EIy_t'' \quad \text{(b)}$$

If δ_T is the total maximum deflection and δ_t the one due to the transverse forces only,

$$EI\delta_T\pi^2/L^2 = EI\delta_t\pi^2/L^2 + P\delta_T \quad \text{(c)}$$

Therefore

$$\delta_T = \delta_t/[1 - (P/P_{cr})] \quad \text{(d)}$$

where P_{cr} is the Euler critical load without the transverse forces. Since deflections are proportional to bending moments,

$$M = M_t/[1 - (P/P_{cr})] \quad \text{(e)}$$

The corresponding maximum stresses are proportional to bending moments

$$\sigma_{\max} = \sigma_{t\max}/[1 - (P/P_{cr})] \quad \text{(f)}$$

Equations (d), (e) and (f) can be used for designing under any type of loading and boundary conditions. For the particular transverse loading the maximum deflection δ_t, the bending moment

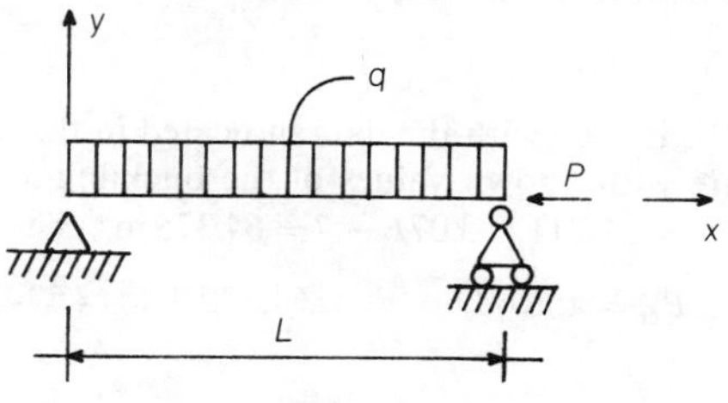

Figure E9.4

$M_t(x)$ and the maximum stress $\sigma_{t_{max}}(x)$ are computed. Then, the Euler load P_{cr} is computed for the column without transverse loads. Finally, equations (d), (e) and (f) yield the deflection, bending moment and maximum stresses along the beam–column for the combined loading.

Example 9.5

The suspension spring for a car is of the form of Figure 9.21, made of AISI 9255 steel with $S_y = 1130\,\text{N/mm}^2$, has strip width 50 mm, thickness 5 mm, length 1200 mm and consists of seven strips. The total weight of the car is $G = 10\,\text{kN}$. The load is distributed 45% front and 55% rear. Determine:

(a) The safety factor for the spring strength.
(b) The spring constant.
(c) The natural frequency of the car on these springs.

Solution

Each strip behaves as a beam independent from the others supporting an equal portion ($\frac{1}{7}$th) of the load.

The section modulus

$$W = bh^2/6 = 50 \times 5^2/6 = 208.3\,\text{mm}^3$$

(a) The stresses in the front spring

$$\sigma_F = (0.45G/4)L/nW = (0.45 \times 1000/4) \times 600/7 \times 208.3 = 463\,\text{N/mm}^2$$

The stresses in the rear spring

$$\sigma_R = (0.55G/4)L/nW = (0.55 \times 1000/4) \times 600/7 \times 208.3 = 566\,\text{N/mm}^2$$

The corresponding safety factors are

$$N_F = 1130/463 = 2.44, \quad N_R = 1350/566 = 2.00$$

(b) The spring constant is, for half spring of length 600 mm (Figure 9.20), with n layers,

$$k = P/\delta = 3nEI/K_t L^3$$

The stress concentration factor $K_t = 1.36$, the section moment of inertia $I = bh^3/12 = 50 \times 5^3/12 = 520.8$ mm. Therefore,

$$k = 3 \times 7 \times 21{,}000 \times 520.8/1.36 \times 600^3 = 7.82\,\text{N/mm}$$

(c) The natural frequency is, for eight half springs supporting the car

$$\omega_n = (k/m)^{\frac{1}{2}} = [(8 \times 7.8 \times 1000)/1000]^{\frac{1}{2}} = 7.9\,\text{rad/sec}$$

Example 9.6

Using the program TMSTABIL (Appendix 9.C) find the limiting axial load of the shaft shown of diameter, uniform, $d = 50$ mm, material carbon steel with $E = 2.1E5\,\text{N/mm}^2$.

Solution

The procedure TMSTABIL is used with the data indicated in the output and Figure E9.6a. The bar is uniform to compare with known values of the buckling load. The moment of inertia is $I = \pi d^4/64 = 3.07E-7$, $EI = 2.1E11 \times 3.07E-7 = 64{,}375\,\text{m}^4$. The Euler critical load is

$$P_{cr} = \pi^2 EI/L^2 = \pi^2 \times 64{,}375/1.1^2 = 524{,}000\,\text{N}$$

for pinned ends.

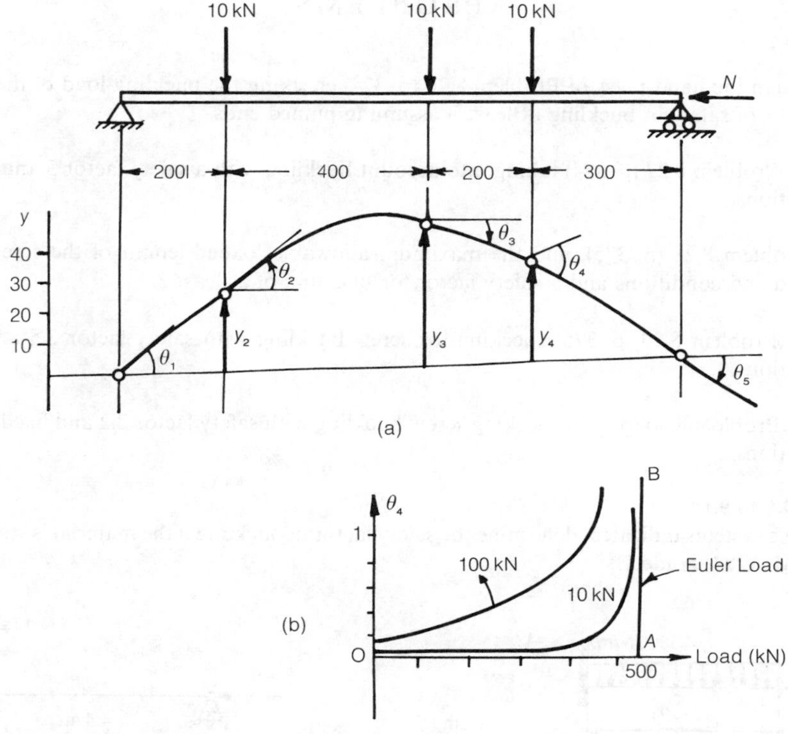

Figure E9.6

Successive runs are performed with increasing axial load and one of the deflection parameters, say θ_4 – the slope at the right end, is recorded. At 525 kN, there is a rapid increase, indicating instability. As expected, this coincides with the Euler critical load of 524 kN. At that load, the deflections and slopes determine the buckling mode, plotted in Figure 9.6b. As expected, it has a sinusoidal shape. The lateral loads cause the deviation from the theoretical Euler deflection curve OAB. At 100 kN lateral loads, this deviation is more noticeable.

REFERENCES AND FURTHER READING

Decker, K.-H., 1973. *Maschinenelemente.* (1st edn. 1962, Leipzig: VEB), 6th edn. Munich: C. Hanser.
Dimarogonas, A. D., 1976. *Vibration Engineering.* St. Paul: West.
Johnson, R. C., 1961. *Optimum Design of Mechanical Elements.* New York: J. Wiley.
Niemann, G., 1965. *Maschinenelemente.* Berlin: Springer.
Orlov, P., 1976. *Fundamentals of Machine Design.* Moscow: Mir Publishers.
Roark, R. J., Young, W. C., 1975. *Formulas for Stresses and Strain.* New York: McGraw-Hill.
Simitsis, G., 1976. *Elastic Stability of Structures.* Englewood Cliffs, N. J.: Prentice-Hall.
Timoshenko, S., Gere, J. M., 1961. *Elastic Stability.* New York: McGraw-Hill.
Veodosyev, V., 1973. *Strength of Materials.* Moscow: Mir Publishers.

PROBLEMS

9.1 Redesign the hand press of Problem 8.26 (p. 374) checking the buckling load of the screw with a factor of safety in buckling at least 3 assuming pinned ends.

9.2 Solve Problem 8.27 (p. 374) taking into account buckling with a safety factor 3 and pinned end conditions.

9.3 In Problem 8.28 (p. 375), find the maximum allowable loaded length of the screw assuming pinned end conditions and a safety factor for buckling of at least 3.

9.4 Solve Problem 8.29 (p. 375) checking for screw buckling with safety factor 2.5 and fixed-free conditions.

9.5 Solve Problem 8.30 (p. 375) checking screw buckling with safety factor 2.2 and fixed-pinned end conditions.

Problems 9.6 to 9.10

For the systems indicated, determine the safety factor in buckling if the material is structural steel ASTM A284, grade D.

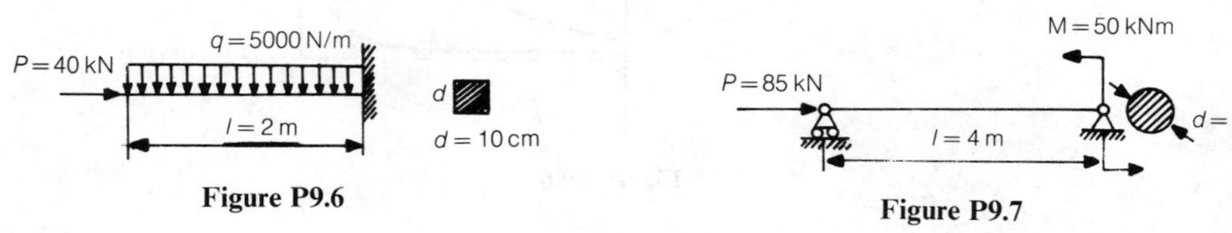

Figure P9.6

Figure P9.7

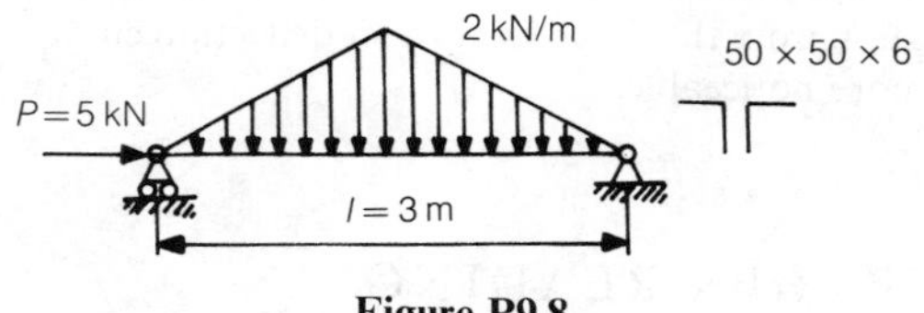

Figure P9.8

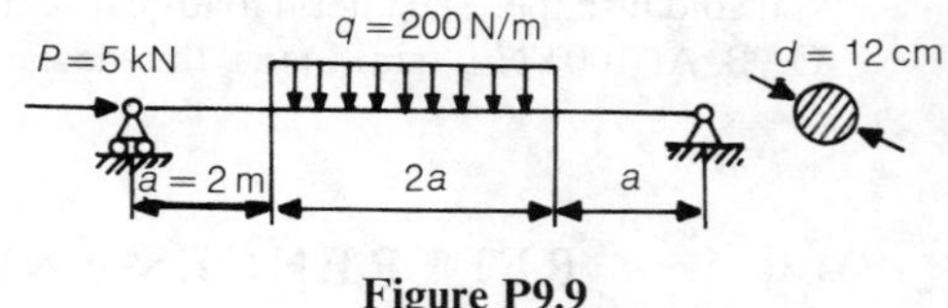

Figure P9.9

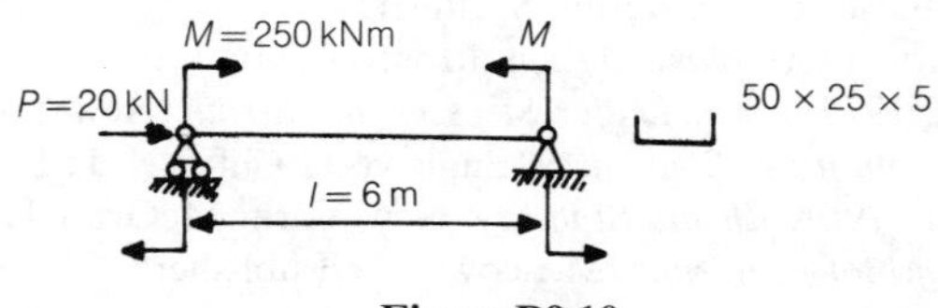

Figure P9.10

Problems 9.11 and 9.12
The shaft shown is made of carbon steel SAE 1020. Determine the maximum allowable axial load F for factor of safety $N = 3.5$.

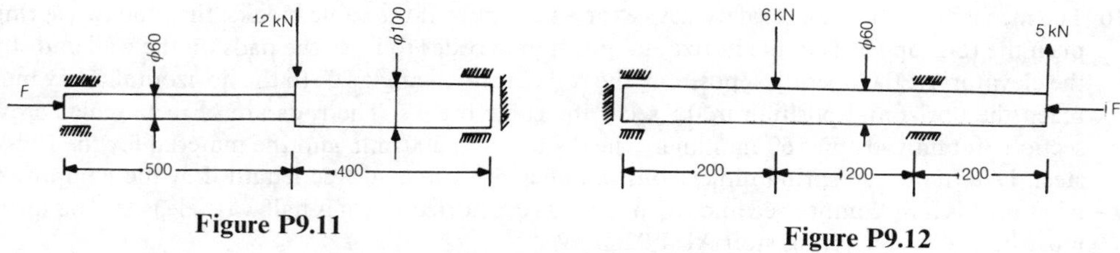

Figure P9.11

Figure P9.12

9.13 The bar shown is elastically supported on coil springs with equal stiffness $k = 1000$ N/cm. Determine the maximum axial load with a safety factor $N = 3$. The material is steel SAE 1020.

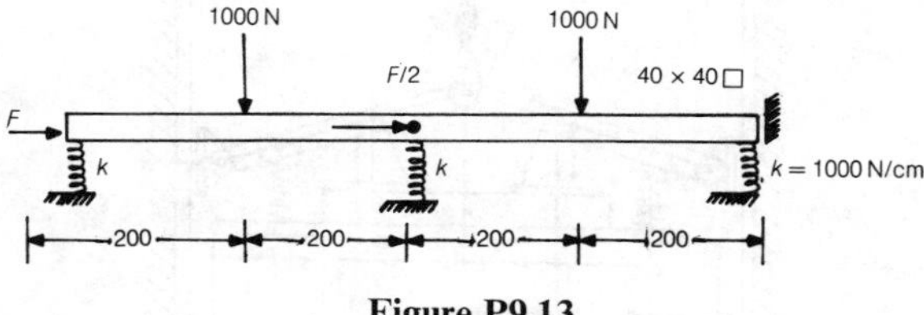

Figure P9.13

9.14 The shaft shown was designed with an axial load F giving safety factor in buckling $N = 3.5$ without lateral loads. After the addition of the lateral loads, determine the safety factor. The material is carbon steel SAE 1020.

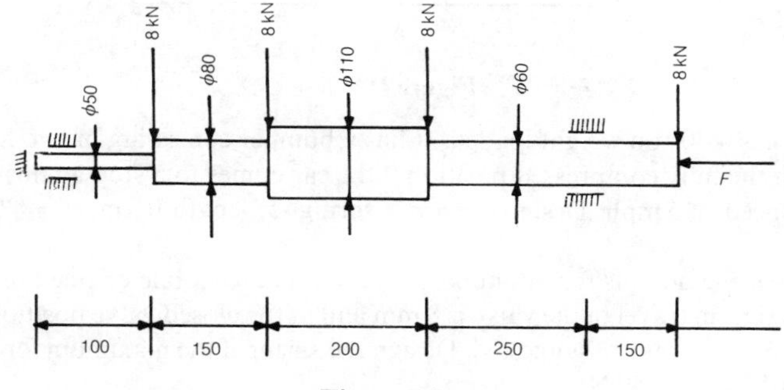

Figure P9.14

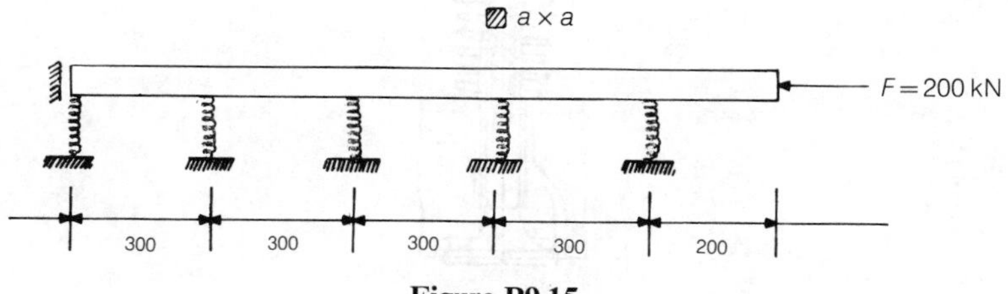

Figure P9.15

9.15 The bus-bar of an electric generator is made of pure copper. Determine the section size a if the safety factor in buckling must be at least 2.5. What will be the safety factor with the center spring removed? The spring constants are all equal to 5 kN/cm.

9.16 The mechanism shown is a safety device for an elevator. If the cable breaks, then the two springs push the rods upwards to the horizontal position in order to force the pads on the wall and stop the elevator. At the operation position, the rods are at an angle 20° to the horizontal. They must reach the horizontal position in 0.5 sec if the cable breaks. The rods are of rectangular cross-section 30-mm wide and 60-mm high. The distance $l = 300$ mm, and the material for the rods is steel. Determine the spring dimensions, number of turns and free length if at the compressed position it is fully compressed and when the rods are horizontal, it is half way released. The spring must be made of a spring steel AISI 9260.

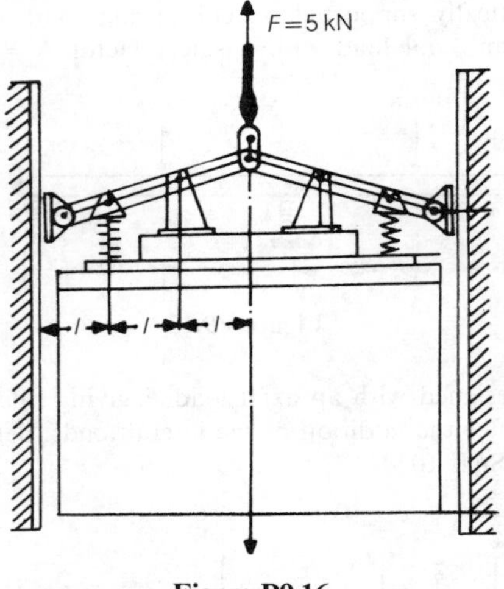

Figure P9.16

9.17 A railroad car has a 40-ton weight loaded. It has a bumper consisting of two helical springs and they must be in the fully compressed position if the car comes to a stop against the end wall at a maneuvering speed of 5 mph. Design a proper spring, of length 0.5 m, $N_a = 30$.

9.18 The return of seating actions for an intake valve of an automobile engine are secured by a coil spring. The maximum travel of the valve is 8 mm and at the closed valve position it must deliver a force of 150 N; 5 to 8 turns are suggested. Design the spring if the maximum force, when the valve opens, must not exceed 250 N.

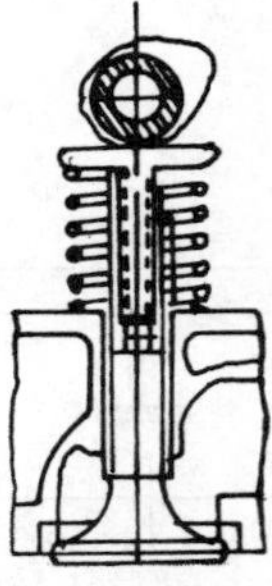

Figure P9.18

9.19 The pressure relief valve shown uses a spring made of AISI 9260 spring steel. At the position shown the spring was compressed to half of the maximum compression. Determine:

(a) The pressure at which the relief valve opens.
(b) The pressure required to force the spring at the fully compressed position.

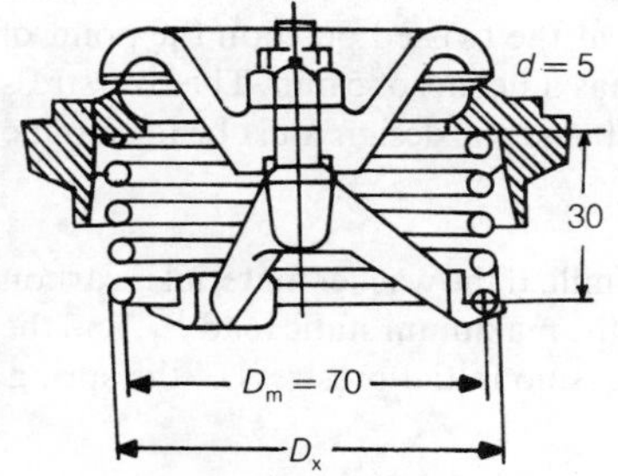

Figure P9.19

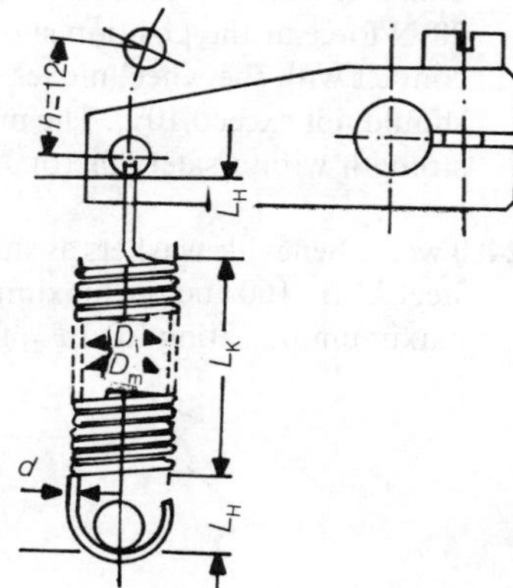

Figure P9.20

9.20 The return spring of a control arm must pull the arm with a 20-N force at the position shown and 25-N force at maximum extension which is 12 mm. Design a spring of AISI 9260 spring steel for fatigue strength (large number of operations) with a safety factor of $N = 2$. The total length of the spring $L + 2L_H$ must not exceed 50 mm.

9.21 A torque wrench consists of a torsion rod of diameter d and length l. A pointer fixed on the lower end indicates at the upper end the total angle of twist, proportional to the applied moment. The wrench is rated at 100 Nm and the angle of twist at that torque is 20°. Design the torsion rod of AISI 1035 carbon steel with a factor of safety $N = 1.5$.

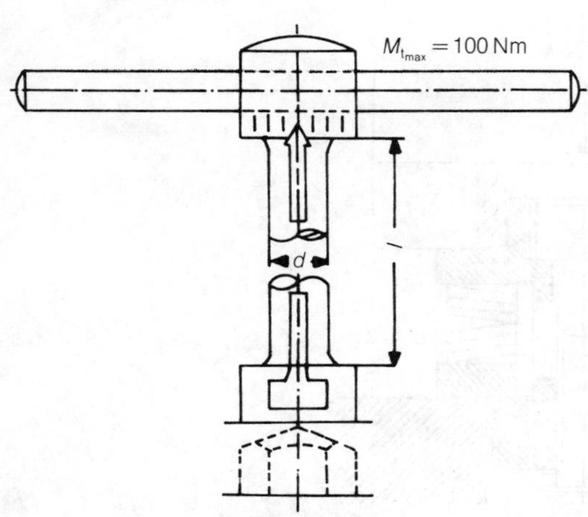

Figure P9.21

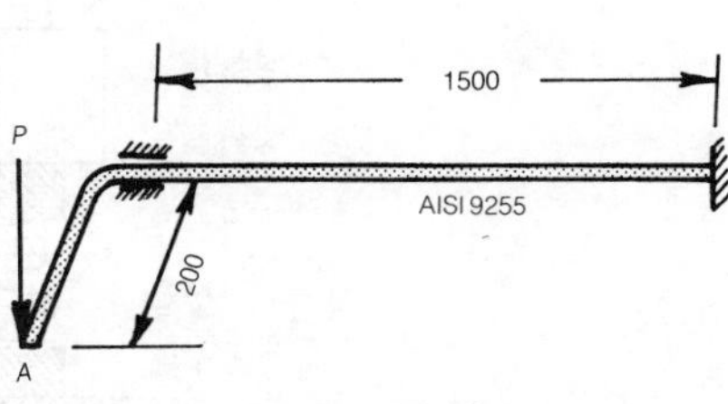

Figure P9.22

9.22 A torsion bar for an automobile, supports a 3-kN load (approximately one-quarter of the automobile mass) as shown. The travel of point A from the dead load (2 kN) to the full load (3 kN) must not exceed 50 mm. The natural frequency with the dead load only must not exceed 1 Hz. Is the design feasible under a safety factor for strength $N = 1.6$?

9.23 The ratchet mechanism for a toothed wheel consists of a helical spring around a pin of 20-mm diameter which forces the ratched arm on the wheel as shown. The spring is prestressed to deliver 20-N force at the position shown. Design the spring so that at the twisted position the point of contact with the wheel moves to the tip of the tooth which has a height of 5 mm. The overstress should not exceed 10%. The material is spring steel AISI 9260 and the design must be for fatigue strength with a safety factor $N = 2$ for strength.

9.24 Twelve Belleville washers as shown form a spring for a rolling mill. If the washer material is carbon steel AISI 5160 and the maximum travel is 15 mm, determine the maximum static load F_1, and the maximum pulsating load F_2, for safety factor $N = 1.5$. There is no initial prestress of the spring.

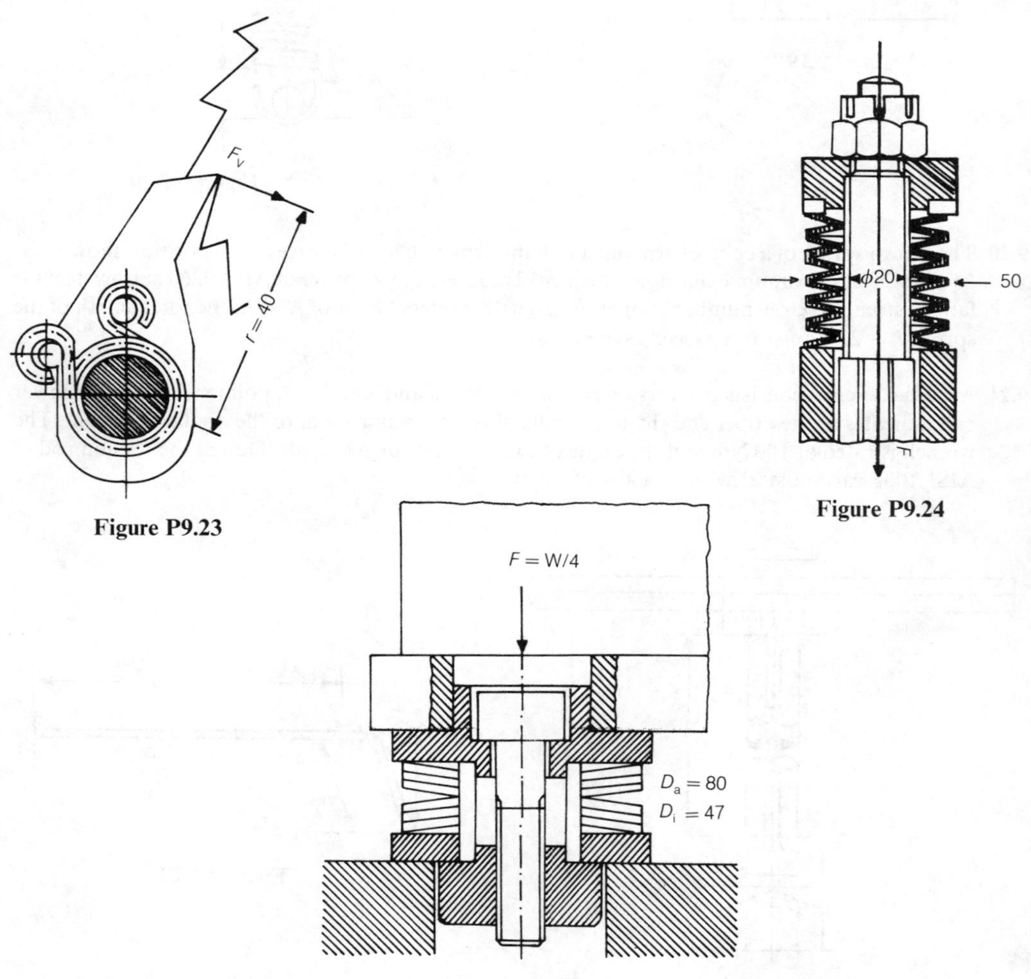

Figure P9.23

Figure P9.24

Figure P9.25

9.25 Six Belleville washers as shown form a support spring for vibration isolation of a heavy machine. The machine weight is 6 ton, equally divided among four springs. The natural frequency for vertical vibration should not exceed 20 Hz.
Determine:

(a) The required spring constant.
(b) The washer dimension for fatigue strength if the dynamic load is 15% of the static load, the material is AISI 1035 steel and the safety factor 1.5.

9.26 Design the spring of Problem 9.25 with a rubber spring.

9.27 A rubber spring is used to reduce impact in a highly elastic machine. The machine has weight 1000 N equally divided among four rubber springs. It is expected that the rubber spring will take impact of the above weight at speed 1 m/sec. During impact, the rubber spring must not be compressed more than 20%. Determine the diameter d. $S_u = 5$ MPa.

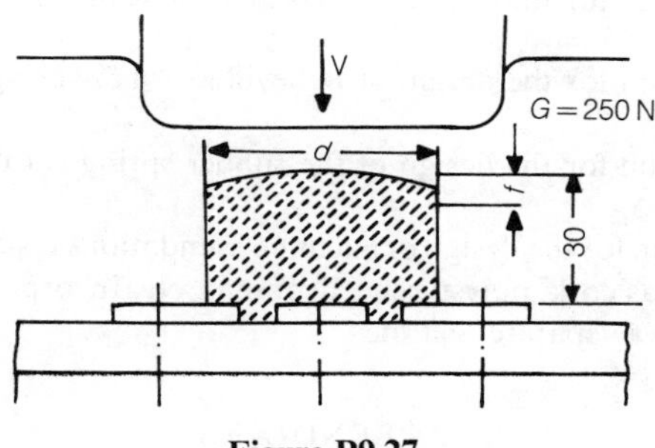

Figure P9.27

9.28 An electronic instrument aboard a destroyer ship weighs 100 N and it is mounted on four rubber springs of the type shown in Figure 9.25a. The dominant vibration frequency in the vicinity of the instrument is 30 Hz. Design the rubber spring for effective vibration isolation, that is with natural frequency of the instrument on the spring no greater than 10 Hz.

9.29 An automobile engine weighs 2300 N and it is mounted on two rubber springs of the type of Figure 9.25e. The modulus of elasticity of the particular rubber is 30 N/mm^2. The rubber spring has the shape of a rectangular pad 60 × 60 mm and thickness 30 mm. Determine the natural frequency of the engine on the springs.

9.30 The torsion bar of Problem 9.22 is to be replaced with two rubber springs, at the support points, of the type in Figure 9.24c. Design the rubber springs.

9.31 It was decided that the system in Problem 9.25 should be founded directly on a sandy soil. Determine the dimensions of the base plate.

9.32 A printing press has a length of 15 m, weighs 60 ton and operates at 2000 rpm. It must be isolated to avoid transferring its vertical vibration and noise to the surroundings. To this end, a concrete base plate is made of thickness 1 m and density of 2700 kg/m^3. This plate rests on gravel. Determine the necessary base-plate dimensions for effective isolation. Determine the transmissibility ratio at 2000, 4000, and 10,000 rpm rotating speeds of the press for $\omega/\omega_n = 6$.

9.33 A reciprocating refrigeration compressor operates at 3000 rpm and consequently it vibrates in the vertical and horizontal direction. It weighs 10 kN and it is installed on a concrete base with cork support as in Figure 9.27. Determine the dimensions of the concrete base and the required cork plate thickness if cork has properties $G = 20\ \text{N/mm}^2$, $E = 40\ \text{N/mm}^2$, and $S_u = 1\ \text{N/mm}^2$.

9.34 An air compressor operating at 1500 rpm was installed with an arrangement such as in Figure 9.27. The concrete base was 1 × 1 m, 300-mm high, weighing 8.1 kN, while the compressor weight was 5 kN. The cork plate had $E = 40\ \text{N/mm}^2$. Find the cork plate thickness if the transmissibility ratio is required to be below 0.1, assuming fraction of critical damping $\zeta = 0.05$.

9.35 From a local store, get some samples of base-plate support materials, such as cork, polystyrene plates and neoprene rubber, and determine in a materials laboratory their elastic properties G and E and the compression strength S_u. Is the stress–strain relation linear?

9.36 Write a computer aided machine design (CAMD) program for the design of torsion bars.

9.37 Write a CAMD program for the design of helical springs in torsion.

9.38 Write a CAMD program for the design of Belleville washer springs.

9.39 Write a CAMD program for the design of the rubber springs in Figure 9.25.

9.40 Write a CAMD program for the design of machine foundations resting: (a) directly on soil and (b) on soft materials, such as cork, polystyrene, neoprene, etc. Incorporate soil and plate properties into the program or in a separate text file.

APPENDIX 9.A
BUCKLING: DESIGN OF COLUMNS WITH EULER AND JOHNSON FORMULAS PROGRAM

```
10 REM *********************************************
20 REM *                                           *
30 REM *                BUCKLING                   *
40 REM *                                           *
50 REM *********************************************
60 REM
70 REM By Prof. Andrew Dimarogonas, Washington Univ., St. Louis, Mo.
80 REM
90 REM
100 REM Diameter of a column for buckling with Euler and Johnson equations.
110 REM
120 REM
130 CLS:KEY OFF
140 FOR I=1 TO 14:COLOR I:LOCATE I,22+I:PRINT "BUCKLING";
150 LOCATE I,50-I:PRINT "BUCKLING";:NEXT I:COLOR 14
160 LOCATE 8,7:PRINT"Design of a column for buckling with Euler and
Johnson";
170 PRINT " equations";
175 locate 25,1:print" Dimarogonas,A.D., Computer Aided Machine Design,
Prentice-Hall,1988";
180 LOCATE 18,20
190 LOCATE 18,20:INPUT"Enter axial load                ";P
200 LOCATE 19,20:INPUT"Enter column length             ";L
210 LOCATE 20,20:INPUT"Enter yield strength            ";SY
220 LOCATE 21,20:INPUT"Enter Young modulus             ";E
230 PI = 3.14159
240 CLS:LOCATE 5,1
250 PRINT "Boundary conditions     m":PRINT "______________________________"
260 PRINT
```

```
270 PRINT "pinned-pinned            1"
280 PRINT "clamped-free             2"
290 PRINT " >>      -clamped       .5"
300 PRINT " >>      -pinned        .7"
310 PRINT:INPUT "Enter m";M
320 PRINT:PRINT
330 LR = M * L
340 D = 4 * SQR (P *LR ^ 2 / PI ^ 3 / E)
350 D = SQR (D)
360 SO = 4 * P / (PI * D ^ 2)
370 IF SO < SY / 2 THEN GOTO 400
380 D = SQR (4 * P / (PI * SY) + SY * LR ^ 2 / (PI ^ 2 * E))
390 SO = 4 * P / (PI * D ^ 2)
400 PRINT "SOLUTION:": PRINT "______________________________"
410 PRINT "Diameter=.........................";D
420 PRINT "Compressive stress=...............";SO
430 PRINT "Equivalent minimum moment of inertia=";PI*D^4/64
440 X = LR / (D / 2): XC = PI * SQR (2 * E / SY)
450 PRINT:PRINT
460 IF X < XC THEN PRINT "SHORT COLUMN"
470 IF X > XC THEN PRINT "SLENDER COLUMN"
480 PRINT
490 END
```

APPENDIX 9.B
HELICSP: DESIGN OF HELICAL SPRINGS PROGRAM

```
10 REM ************************************
20 REM *                                  *
30 REM *         HELICSP                  *
40 REM *                                  *
50 REM ************************************
60 REM            COPYRIGHT 1985
70 REM by Professor Andrew D. Dimarogonas, Washington Univ., St. Louis, Mo.
80 REM  All rights reserved.  Unauthorized reproduction, disssemination,
90 REM selling or use is strictly prohibited.
100 REM  This listing is for reference purpose only.
110 REM
120 REM
130 REM The program computes coil and wire diameter of a helical spring
140 REM for given spring constant and material strength.
150 REM Performs also stability check.
160 REM
170 REM
180 REM          ADD / 1-12--84
190 REM
200 CLS:KEY OFF
210 LOCATE 25,1:PRINT" Andrew Dimarogonas, Computer Aided Machine Design,
Prentice-Hall, London 1988";
220 FOR I=1 TO 14:COLOR I:LOCATE I,22+I:PRINT "HELICSP";
230 LOCATE I,50-I:PRINT "HELICSP";:NEXT I:COLOR 14
240 LOCATE 8,15:PRINT"Design of helical springs and parameter studies";
250 LOCATE 23,25:INPUT"Hit return to run the program";X$
260 CLS:PRINT "        ENTER DATA"
270 PRINT "----------------------"
280 PRINT "Design data:"
290 INPUT "Spring constant         ";K
300 INPUT "Maximum load            ";PM
310 INPUT "Number of turns         ";NA
320 PRINT
330 PRINT "Material properties:"
340 INPUT "Shear modulus           ";G
350 INPUT "Young modulus           ";YM
360 INPUT "Poisson ratio           ";PO
370 IF PO > .5 THEN 360
380 INPUT "Allowable shear stress";SA
390 DX = .001:DY = .001:REM ......................DIFFERENIALS FOR
DERIVATIVES
```

```
400 GOSUB 1080:REM ...............................SOLVE FOR d, R
410 PRINT
420 PRINT "        RESULTS"
430 PRINT "--------------------"
440 PRINT "Coil Diameter  :  ";2 * Y
450 PRINT "Wire Diameter  :  ";X
460 PRINT "Max. Deflection:  ";PM / K
470 FL = PM / K + X * (NA + 2)
480 PRINT "Free length    :  ";FL
490 PRINT "Min free length:  ";(NA + 2) * X
500 HA = ATN ((PM / (K * NA) + X) / (6.28 * Y)) * (180 / 3.14)
510 PRINT "Helix angle    :  ";HA;" deg";
520 IF HA < 30 THEN 540
530 COLOR 28 :PRINT ">>>>";: COLOR 14
540 PRINT
550 REM ..................................................BUCKLING
LOAD
560 IE = FL * X ^ R / (128 * NA * Y * (1 + PO / 2))
570 PC = 4 * (3.14) ^ 2 * YM * IE / FL ^ 2
580 PRINT "Buckling load  :  ";
590 IF PC < PM THEN COLOR 28
600 PRINT PC;
610 IF PC < PM THEN PRINT "<DESIGN LOAD"
620 COLOR 14
630 PRINT
640 INPUT "HIT RETURN TO CONTINUE ";X$
650 S1 = SA:S2 = SA:S3 = 1
660 N1 = NA:N2 = NA:N3 = 1
670 CLS
680 PRINT "          MENU"
690 PRINT
700 PRINT " <D>   NEW DESIGN"
710 PRINT " <P>   PARAMETER STUDY"
720 PRINT " <Q>   QUIT"
730 PRINT :PRINT
740 INPUT "Enter your selection:   ";X$
750 IF X$ = "Q" OR X$="q" THEN END
760 IF X$ = "D" OR X$="d" THEN 250
770 IF X$ = "P" OR X$="p" THEN 790
780 GOTO 670
790 CLS
800 PRINT "   PARAMETER STUDY"
810 PRINT:PRINT
820 INPUT "Parameter study on number of turns (Y/N)";N$
830 IF N$ = "N" OR N$="n" THEN 870
840 PRINT :PRINT "Enter ";
850 INPUT "NMIN, NMAX, STEP";N1,N2,N3
860 PRINT:PRINT
870 INPUT "Parameter study on strength (Y/N)     ";S$
880 IF S$ = "N" OR S$="n" THEN 910
890 PRINT :PRINT "Enter ";
900 INPUT "SMIN,SMAX,SSTEP ";S1,S2,S3
910 CLS
920 PRINT "FEASIBLE DESIGNS:  ":PRINT
930 PRINT " N  Strength    wire dia     coil dia        volume"
940 PRINT "_____________________________________________________":PRINT
950 FOR SA = S1 TO S2 STEP S3
960 FOR NA = N1 TO N2 STEP N3
970 GOSUB 1080
980 FL = PM / K + X * (NA + 2)
990 IE = FL * X ^ 4 / (128 * NA * Y * (1 + PO / 2))
1000 PC = 4 * (3.14) ^ 2 * YM * IE / FL ^ 2
1010 VOLUME=3.14159^2*X^2*2*Y*NA/4
1020 IF PC < PM THEN 1040
1030 PRINT NA; TAB( 5);SA; TAB( 17);X; TAB( 32);2 * Y;TAB(47);VOLUME
1040 NEXT NA:NEXT SA
1050 PRINT:INPUT "HIT RETURN TO CONTINUE ";X$
1060 GOTO 670
1070 END
1080 REM *********************************************NEWTON-RAPHSON
SOLUTION
```

```
1090 X = (G / (64 * NA * K) * ( 16 * PM / 3.14 / SA) ^ 3) ^ .2
1100 Y = 3.14 * X ^ 3 * SA / 16 /PM
1110 REM A(I,J) = DFI/DXJ
1120 GOSUB 1270
1130 FA = F1:FB = F2
1140 X = X + DX:GOSUB 1270
1150 A(1,1) = (F1 - FA) / DX:A(2,1) = (F2 - FB) / DX
1160 X = X - DX
1170 Y = Y + DY:GOSUB 1270
1180 A(1,2) = (F1 - FA) / DY:A(2,2) = (F2 - FB) / DY
1190 Y = Y - DY
1200 DD = A(1,1) * A(2,2) - A(1,2) * A(2,1)
1210 SX = (FB * A(1,2) - FA * A(2,2)) / DD
1220 SY = (FA * A(2,1) - FB * A(1,1)) / DD
1230 X = X + SX:Y = Y + SY
1240 IF ABS(SX) + ABS(SY) < .001 THEN 1260
1250 GOTO 1120
1260 RETURN
1270 REM ***********************************COMPUTE FUNCTIONS
F1(X,Y),F2(X,Y)
1280 C = 2 * Y / X
1290 F1 = G * X ^ R / (64 * NA * Y ^ 3) - K
1300 F2 = 16 * PM * Y / (3.14159 * X ^ 3) * (C - .5) / (C - 1) - SA
1310 F2 = F2 / SA
1320 RETURN
```

APPENDIX 9.C
TMSTABIL: STABILITY ANALYSIS OF A GENERAL COLUMN WITH THE TRANSFER MATRIX METHOD PROGRAM

```
10 REM ************************************************
20 REM *                                              *
30 REM *                tmstabil                      *
40 REM *                                              *
50 REM ************************************************
60 REM          COPYRIGHT 1987
70 REM by Professor Andrew D. Dimarogonas, Washington University,St.Louis.
80 REM All rights reserved.  Unauthorized reproduction, dissemination or
90 REM selling is strictly prohibited.  This listing is for personal use.
100 REM
110 REM
120 REM
130 REM ************************************************************
140 CLS:KEY OFF
150 FOR I=1 TO 14:COLOR I:LOCATE I,22+I:PRINT "TMSTABIL";
160 LOCATE I,50-I:PRINT "TMSTABIL";:NEXT I:COLOR 14
170 LOCATE 8,12:PRINT"Stability of a compound column with variable";
180 PRINT " cross-section";
190 LOCATE 25,1:PRINT" Dimarogonas,A.D., Computer Aided Machine Design,
Prentice-Hall, London 1988";
195 LOCATE 18,30
200 INPUT"Hit RETURN to continue";X$
210 REM Stability of a column with the transfer matrix method.
220 REM                          D= DATA MATRIX
230 REM                          P=POINT MATRIX
240 REM                          S=FIELD MATRIX
250 REM                          A,G,W=AUXILIARY MATRICES
260 DIM D(10,6), P(5,5), S(5,5), A(5,5), V(5,1)
270 DIM G(5,5), W(5,1)
280 COLOR 14,0:CLS:LOCATE 25,1:PRINT"for each data entry, or 0, hit
RETURN";
290 LOCATE 1,1:PRINT"                    ENTER DATA ":PRINT:PRINT
300 INPUT"Number of Elements     ";N
310 INPUT"Modulus of Elasticity ";E
320 PRINT
330 COLOR 15
340 PRINT"length";TAB(9);"Inertia";TAB(18);"Force-axial";TAB(30);" spring
";:COLOR 7,0:PRINT " (at left end)"
350 COLOR 14,0
```

```
360 N1 = N - 1: FOR I = 1 TO N
370 FOR J=1 TO 4:LOCATE 8+I,10*(J-1)+1:INPUT D(I,J):NEXT J
375 IF D(I,1)=0 OR D(I,2)=0 THEN PRINT"repeat..";:GOTO 370
376 NEXT I
380 PRINT" right end....";
390 FOR J=3 TO 4:LOCATE 8+I,10*(J-1)+1:INPUT D(N+1,J):NEXT J
400 PRINT:PRINT
410 DMIN=1E+20
420 FOR I=1 TO N
430 IF D(I,2)>DMAX THEN DMAX=D(I,2)
440 IF D(I,2)<DMIN THEN DMIN=D(I,2)
450 LTOT=LTOT+D(I,1)
460 NEXT I
470 NMS=INT(N/2+.5):KSHAFT=48*E*D(NMS,2)/LTOT^3
480 PRINT "DATA"
490 PRINT "=========="
500 PRINT "   NUMBER OF SECTIONS ";N
510 PRINT "   YOUNG MODULUS=      ";E
520 PRINT
530 PRINT "LENGTH          I                   FORCE      SPRING"
540 PRINT "-------------------------------------------------------"
550 FOR I = 1 TO N: FOR J = 1 TO 4
560 PRINT TAB(12 * J - 8);D(I,J);
570 NEXT J: PRINT: NEXT I:PRINT "END STATION:";
580 PRINT TAB( 28);D(N + 1,3); TAB( 40);D(N + 1,4)
590 PRINT:INPUT"Enter number of rigid supports";NSUP
600 FOR ISUP=1 TO NSUP
610 PRINT"Node number for support no ";ISUP;:INPUT INODE
620 D(INODE,4)=D(INODE,4)+100*KSHAFT
630 NEXT ISUP
640 PRINT:PRINT
650 PRINT"Hit return to continue.."
660 INPUT Y$
670 CLS
680 COLOR 0,14
690 PRINT"*******************************************************"
700 PRINT"*                                                     *"
710 PRINT"*            COLUMN STABILITY ANALYSIS                *"
720 PRINT"*                                                     *"
730 PRINT"*******************************************************"
740 COLOR 14,0
750 PCRMAX=3.14^2*E*DMAX/LTOT^2
760 PCRMIN=3.14^2*E*DMIN/LTOT^2
770 PRINT:PRINT"For simple supports at the two ends, the range"
780 PRINT"of  the Euler critical load= max:";PCRMAX;" /  min:";PCRMIN
790 PRINT"Enter Pmin,Pmax,Pstep        ";
800 INPUT PM1,PM2,PSTEP
810 REM
820 PRINT:PRINT"Computing Characteristic Determinant...."
830 PRINT:FOR PAX=PM1 TO PM2 STEP PSTEP
840 PRINT "Additional Axial load=";PAX;
850 FOR I = 1 TO 5: FOR J = 1 TO 5: A(I,J) = 0: NEXT J: NEXT I
860 FOR I = 1 TO 5: A(I,I) = 1: NEXT I
870 FOR M = 1 TO N : FAX = D(M,3):D1 = D(M,2)
880 L1 = D(M,1):K1 = D(M,4)
900 GOSUB 1220
910 GOSUB 1310
920 FOR I = 1 TO 5: FOR J = 1 TO 5: G(I,J) = A(I,J): NEXT J: NEXT I
930 GOSUB 1340: REM A = P * G
950 FOR I = 1 TO 5: FOR J = 1 TO 5: G(I,J) = A(I,J): NEXT J: NEXT I
960 GOSUB 1390: REM A = S * G
970 NEXT M
980 FAX = D(M,3):K1 = D(M,4): GOSUB 1310
990 FOR I = 1 TO 5: FOR J = 1 TO 5: G(I,J) = A(I,J): NEXT J: NEXT I
1000 GOSUB 1340
1010 FOR I = 1 TO 5: V(I,1) = 0: NEXT I
1020 DPREV=D9
1030 D9 = A(3,1) * A(4,2) - A(4,1) * A(3,2)
1040 V(1,1) = - A(3,2) / A(3,1)
1050 V(3,1)=0:V(4,1)=0:V(5,1)= 0
1060 REM
```

```
1070 IF X$="R" THEN 1160
1080 PRINT "Det= ";D9
1090 IF DPREV*D9>=0 THEN 1180
1100 PCR=PAX+D9/(DPREV-D9)*PSTEP:ICR=ICR+1
1110 PRINT:PRINT"Critical load      ";ICR;" is:  ";PCR
1120 PAX=PCR
1130 D9=0
1140 INPUT"hit RETURN for next critical load, Q to quit ";X$
1150 IF X$="q" OR X$="Q" THEN 1210
1160 PRINT"Iterating for next critical load ...    ":PRINT:PRINT
1180 NEXT PAX
1190 PRINT"Hit return to continue....";:INPUT Y$
1200 GOTO 670
1210 END
1220 REM FIELD MATRIX - EULER BEAM
1230 FOR I = 1 TO 5: S(I,I) = 1: NEXT I
1240 I1 = D1:IF PAX+FAX <0 THEN PRINT"No solution..high load";:END
1250 KX=SQR((PAX+FAX)/(E*I1)):SK=SIN(KX*L1):CK=COS(KX*L1)
1260 S(3,4)=SK/KX:S(1,3) =(1-CK)/(E*I1*KX^2):S(3,3)=CK
1270 S(1,2) = SK/KX:S(2,2)=CK
1280 S(2,4) = S(1,3):S(2,3) = SK/(E*I1*KX)
1290 S(1,4) = (KX-SK)/(E*I1*KX^3):S(3,2)=-E*I1*KX*SK
1300 RETURN
1310 REM POINT MATRIX: FORCE AND SPRING
1320 FOR I = 1 TO 5:FOR J=1 TO 5:P(I,J)=0:NEXT J:P(I,I) = 1: NEXT I
1330 P(3,2)=-KT: P(4,5) = F1: P(4,1) = - K1 : RETURN
1340 REM A = P * G
1350 FOR I = 1 TO 5: FOR J = 1 TO 5
1360 C = 0
1370 FOR K = 1 TO 5: C = C + P(I,K) * G(K,J):NEXT K: A(I,J) = C: NEXT
J:NEXT I
1380 RETURN
1390 REM A=S*G
1400 FOR I = 1 TO 5: FOR J = 1 TO 5: A(I,J) = 0
1410 FOR K = 1 TO 5: A(I,J) = A(I,J) + S(I,K) * G(K,J): NEXT K: NEXT J:
NEXT I: RETURN
```

CHAPTER TEN

DESIGN OF FRICTION ELEMENTS

10.1 SLIDING FRICTION

It has already been mentioned in Chapter 2 that real surfaces are always irregular and if amplified they would look as in Figure 10.1. Curve b is the actual geometry and curve a shows the surface with amplification in the vertical direction much greater than the one in the horizontal direction.

It appears that the surface profile consists of two waveforms, one with large wavelength, called 'waviness' and another with a much smaller wavelength, called 'roughness'.

When two solids with real surfaces as in Figure 10.1 approach each other under a normal force, the tips of the surface irregularities come first in contact. For most engineering situations, they deform elastically first and then plastically. The approach of the two surfaces stops when the interaction forces between surface irregularities sum up to the normal load.

The real area of contact A_r is the sum of the contact surfaces of the surface asperities, usually much smaller than the apparent area of contact A_a (Figure 10.2).

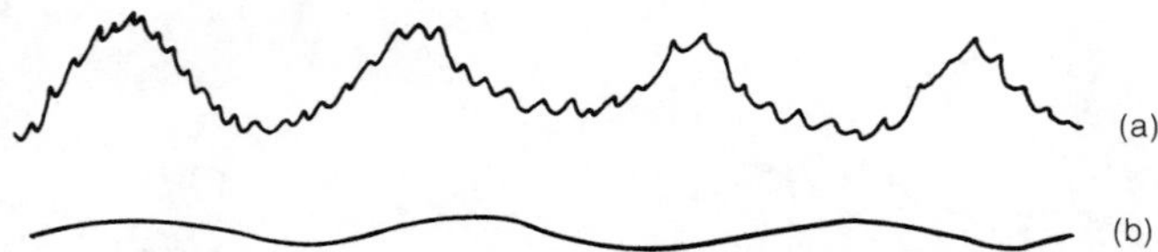

Figure 10.1 Geometry of machined surfaces

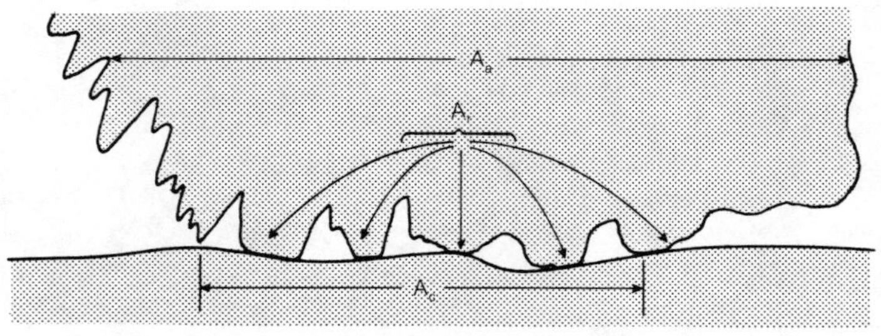

Figure 10.2 Real area of contact

At contact points there is almost always plastic deformation accompanied by elastic deformation of the neighboring areas (Figure 10.3).

Herz has studied the contact mechanics (Kragelskii 1965) macroscopically and proved that when a sphere is pressed against a plane, the apparent area of contact will be a circle (Figure 10.4), with diameter

$$d = 1.75\left[Wr\left(\frac{1}{E_1} + \frac{1}{E_2}\right)\right] \tag{10.1}$$

where E is the elasticity modulus of the solids 1 and 2.

The stress distribution has an elliptic shape and the maximum interface pressure is

$$p_{\max} = 0.42W^{\frac{1}{3}}r\left[\left(\frac{1}{E_1} + \frac{1}{E_2}\right)\right]^{-\frac{2}{3}} \tag{10.2}$$

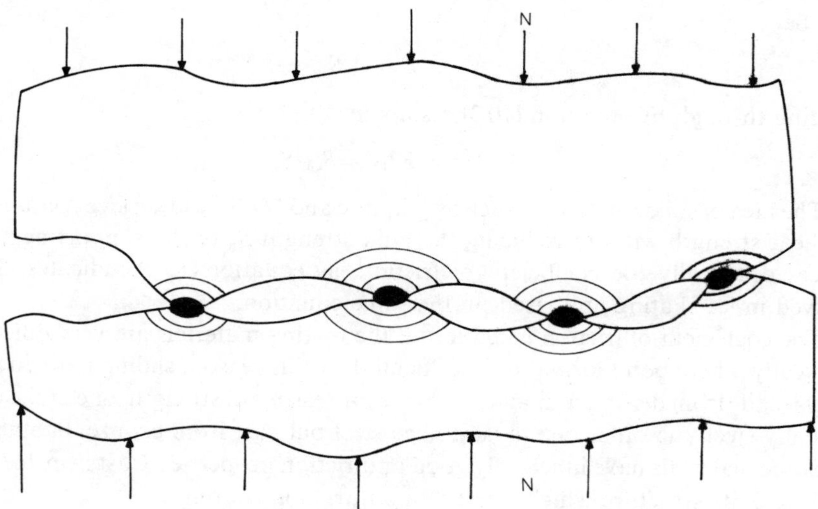

Figure 10.3 Formation of friction junctions

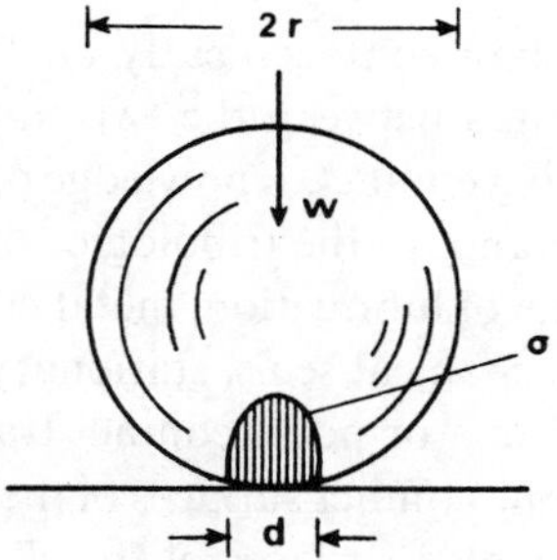

Figure 10.4 Herz contacts

Plastic deformation of the surface asperities is so sudden that the heat of the plastic deformation raises the interface temperatures and local bonds may be created. In most engineering cases, the force necessary to break these bonds during sliding is the main contribution of the frictional force.

Molecular forces, microcutting, and elastic hysteresis are some of the other mechanisms contributing to the frictional force.

In metal contacts, the prevailing mechanism for dry friction is usually the bond formation and this helps in the estimation of the coefficient of friction.

Ernst and Merchant (Kragelskii 1965) suggested that the real area of the contact has magnitude determined by the assumption that the external load N is balanced by general yielding of the softer material

$$N = A_r S_u \tag{10.3}$$

The frictional force F to break the bonds of surface A_r with ultimate shear strength S_{su} must be

$$F = A_r S_{su} \tag{10.4}$$

Dividing through by equation (10.3) results in

$$f = F/N = S_{su}/S_u \tag{10.5}$$

The idea of solid lubricants, such as graphite and MoS_2 and surface coatings is to reduce the shear strength without reducing the bulk strength S_u of the supporting material. This reduces drastically the coefficient of friction, as equation (10.5) indicates. The concepts involved in ice skating are based on the same equation.

The coefficient of friction is reduced if the mating materials are very different physico-chemically – here bond formation is difficult. For this reason, sliding pairs in machines are always made from dissimilar materials. Since for reasons of strength, one of the two materials is usually steel, the other one must not be steel but cast iron, bronze, babbitt, etc.

Some materials have inherently good antifriction properties. Cast iron, for example, has graphite in its structure which acts as an antifriction coating.

10.2 TEMPERATURES AT SLIDING CONTACTS

When two bodies come into sliding contact nearly all the energy dissipated by friction appears as heat, which is distributed between the two bodies and raises appreciably their temperature in the area of the sliding contact. Knowledge of these temperatures at the sliding interface is of fundamental importance to the tribological behavior of the materials and has immediate application in the fields of lubrication, metal cutting, grinding and the design of gears, bearings, forming tools, mechanical seals, stationary or moving electric contacts etc.

Steady rubbing over a given area or point contact tracing a closed path at high speed results in high temperatures over the contact surfaces of the stationary and the moving body. Heat is transferred through the cooled surfaces of the sliding bodies by convection.

The average surface temperature depends on the amount of heat produced, the product

Table 10.1 Heat resistance coefficients (h is the heat transfer coefficient)

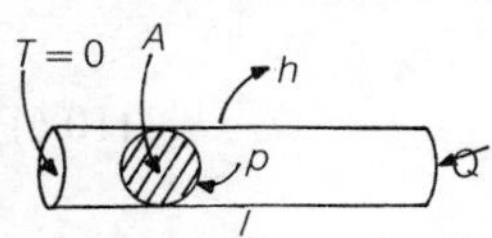	Cylinder, heat input at one end, the other at constant temperature	$H = \frac{A}{l^2 \tan h(N)} \frac{N}{}, N = l\left(\frac{hp}{kA}\right)^{\frac{1}{2}}$ $H = l\left(\frac{hp}{kA}\right)^{\frac{1}{2}}$ for $h = 0$
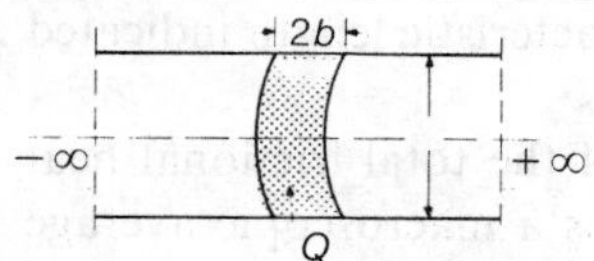	Rotating cylinder of infinite length, radius l, friction on cylindrical surface band of width $2b$	$H = \frac{2}{S_1}\pi^2\beta^2, \beta = \frac{b}{l}$ with $S_1 = \int_0^\infty \frac{\sin^2(\beta\gamma)\,d\gamma}{\gamma^2\left[\gamma\frac{I_1(\gamma)}{I_0(\gamma)} + \frac{lh}{k}\right]}$
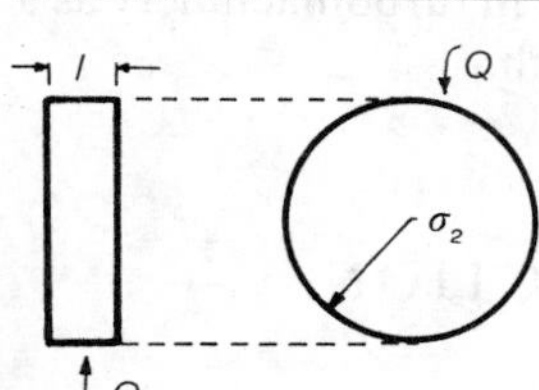	Full cylindrical ring of radius a_2 and thickness l. Friction on the cylindrical surface of width l side surface cooled with heat transfer coefficient H	$H = 2\pi m a_2 \frac{I_1(ma_2)}{I_0(ma_2)}, m^2 = \frac{2h}{kl}$ I(x) modified Bessel Functions of the first kind
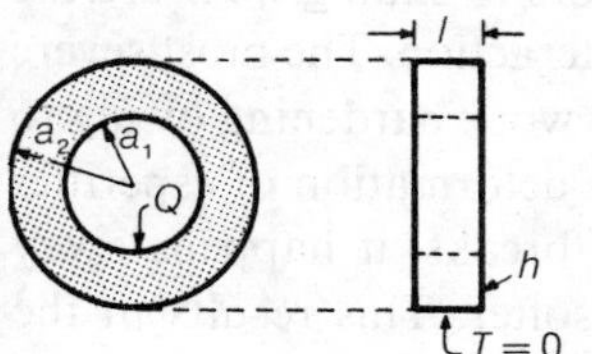	Circular ring with inner radius a_1, outer radius a_2, thickness l. Friction on the inner surface. Outer surface insulated. Side surface at heat transfer coefficient H	$H = 2\pi a_1 m\psi$ with $m^2 = \frac{2h}{kl}$ and $\psi = \frac{I_1(ma_2)k_1(ma_1) - I_1(ma_1)k_1(ma_2)}{I_0(ma_1)k_1(ma_2) + I_1(ma_2)k_0(ma_1)}$ $k(x)$ modified Bessel Functions of the second kind
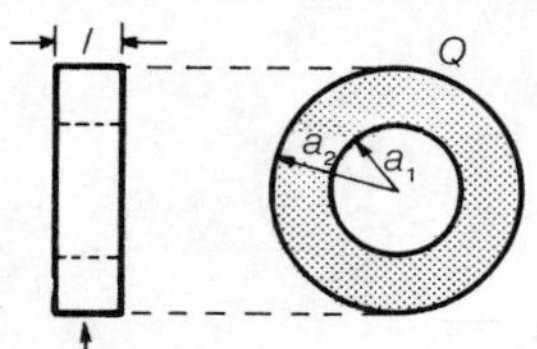	Circular ring with inner radius a_1, outer radius a_2, thickness l. Friction on the outer surface, inner surface insulated. Side surface at heat transfer coefficient H	$H = 2\pi a_2 m g$ with $g = \frac{I_1(ma_1)k_1(ma_2) - I_1(ma_2)k_1(ma_1)}{I_0(ma_2)k_1(ma_1) + I_1(ma_1)k_0(ma_2)}$
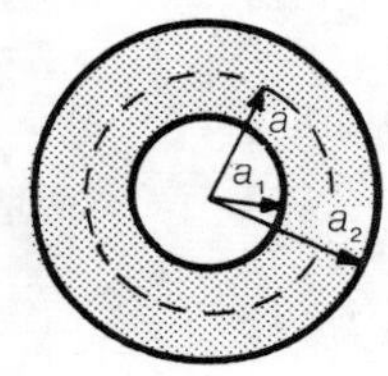	Circular ring with inner radius a_1, outer radius a_2. Friction on the side surface at radius a. Inner and outer surfaces insulated	$H = 2\pi a_2 m(X + Y)$ with $X = \frac{I_1(ma_1)k_1(ma) - I_1(ma)k_1(ma_1)}{I_0(ma)k_1(ma_1) + I_1(ma_1)k_0(ma)}$ $Y = \frac{I_1(ma)k_1(ma_2) - I_1(ma_2)k_1(ma)}{I_0(ma)k_1(ma_2) - I_1(ma_2)k_0(ma)}$

of frictional force and sliding speed, and the resistance of the two elements to the flow of heat. This has been expressed in the form:

$$T_0 = \frac{Q}{l_1 k_1 H_1 + l_2 k_2 H_2} \tag{10.6}$$

where $Q = UF$ (the rate of heat generation), U is the sliding speed, F the frictional force, H the dimensionless resistance of one of the mating parts to the heat flow (Table 10.1), k is the thermal conductivity of the material (Table 6.1, p. 240), l is a characteristic length indicated in Table 10.1, and indices 1, 2 are for the two mating parts.

This yields the average surface temperature as a function of the total frictional heat generated and the properties of the sliding bodies. Although it is a macroscopic average temperature it can be used for all practical design purposes as maximum contact temperature. This temperature is used in selecting the friction, brake and seal materials and calculating the capacity of these elements to carry the pressure loads. High temperatures owing to friction frequently cause yielding of the steam packing teeth in turbomachinery as a result of the pressure difference between the two faces of each tooth.

10.3 WEAR OWING TO SLIDING FRICTION

Wear is not a process with a single mechanism. Major contributors to sliding wear are the interaction of asperities, fatigue, oxide removal, and molecular interaction. The most severe wear mechanism is, where it exists, the tearing in depth caused by work hardening owing to plastic deformation. During the approach of the surfaces, plastic deformation of asperities might lead to their hardening. When, during sliding, the bond breaks, it happens away from the bond interface, at some depth where the material is softer. This results in the removal of large amounts of debris and rapid surface deterioration. For this reason, the softer material should never be steel which in most forms is work hardening

The wear rate can be expressed as volume wear,

$$V/L = KN/S_y$$

Table 10.2 Wear rates

Material	Hardness (10^{12} N/m^2)	$K \times 10^4$
Soft steel on soft steel	18.6	70
Hard steel on:		
Brass 60/40	9.5	6
Teflon	0.5	0.25
Brass 70/3	6.8	1.7
Plexiglass	2.0	0.07
Hardened steel	85	1.3
Stainless steel	25	0.47
Polyethylene	0.17	0.0043

where V is the volume of wear debris, L is the sliding distance, N is the normal force, and S_y is the yield pressure of the material. K is a wear coefficient having the values in Table 10.2.

Friction-exposed materials must have a high coefficient of friction with low wear rate. For metallic materials, this combination is very difficult. For this reason, while one of the mating parts is usually steel for strength purposes, the other is usually non-metallic like asbestos, leather, plastics, etc.

10.4 CLUTCHES AND BRAKES

Couplings with controlled engagements, i.e. those that during operation can be activated to connect two rotating members and the opposite, are called 'clutches'. Brakes are clutches which connect one rotating member to non-rotating ones, usually the machine frame. Clutches and brakes have very similar operating characteristics and can be conveniently studied together. Almost invariably, torque is transmitted via sliding friction between two mating surfaces.

10.4.1 Friction bands

Perhaps the simplest brake/clutch consists of a flexible band around a cylindrical drum (Figure 10.5). The drum is fixed on one rotating member while the band is fixed on the other (clutch) or on the stationary frame (brake). One end is fixed on a pivot point A and the other on a mechanical or hydraulic forcing device. During operation, the two ends of the band are stressed by forces P_1 and P_2. The friction forces between drum and band have total moment about the center of rotation equal to the shaft torque. Along the contact zone the band tension P is variable.

An elementary part of the band corresponding to an angle $\mathrm{d}\theta$ will be acted upon by forces P and $P + \mathrm{d}P$ at the two ends, normal force $\mathrm{d}N$ and frictional force $f\,\mathrm{d}N$ from the drum, where f is the coefficient of friction. Summing up forces in the radial direction

$$(P + \mathrm{d}P)\sin\frac{\mathrm{d}\theta}{2} - P\sin\frac{\mathrm{d}\theta}{2} - \mathrm{d}N = 0 \qquad (10.7)$$

$$\mathrm{d}N = \mathrm{d}P$$

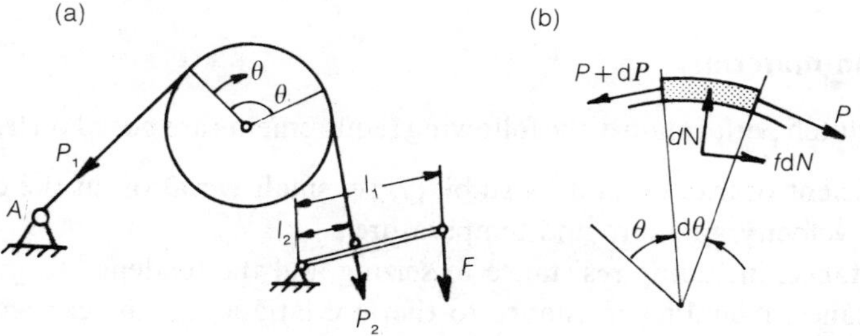

Figure 10.5 Friction band brake

In the tangential direction

$$(P+\mathrm{d}P)\cos\frac{\mathrm{d}\theta}{2}-P\cos\frac{\mathrm{d}\theta}{2}-f\,\mathrm{d}N=0 \tag{10.8}$$

$$\mathrm{d}P\pm f\,\mathrm{d}N=0$$

(upper sign '+' for clockwise, lower sign '−' for counterclockwise rotation).

Therefore, eliminating dN,

$$\mathrm{d}P/P=\pm f\,\mathrm{d}\theta \tag{10.9}$$

Assuming a constant coefficient of friction and integrating along the contact arc θ,

$$\int_{P_1}^{P_2}\mathrm{d}P/P=\pm f\int_0^{\theta}\mathrm{d}\theta,\quad \frac{P_1}{P_2}=e^{\pm f\theta} \tag{10.10}$$

This relation was obtained by Euler. The friction torque will be

$$T=(P_1-P_2)R=P_2(e^{\pm f\theta}-1)R \tag{10.11}$$

Therefore, for given torque T, the required actuating force P_2 will be

$$P_2=T/(e^{f\theta}-1)R \tag{10.12}$$

The normal force between band and drum is of importance owing to surface pressure limitations of the lining materials.

For width b of the band and pressure $p(\theta)$,

$$\mathrm{d}N=pbR\,\mathrm{d}\theta \tag{10.13}$$

Using equation (10.7),

$$p=P/bR \tag{10.14}$$

and the maximum value is for $\theta=\theta_0$, $p=P_1$

$$p_{\max}=P_1/bR \tag{10.15}$$

$$p_{\max}=(T/bR^2)[e^{\pm f\theta_0}/(e^{\pm f\theta_0}-1)] \tag{10.16}$$

Design is based on the required transmitted torque and the allowable maximum pressure of the lining material. Control parameters are the arc θ, radius R, width b while the coefficient of friction f depends on the selected materials and environment.

10.4.2 Friction materials

For adequate clutch performance, the following requirements are posed on friction materials:

(1) High coefficient of friction and its stability, i.e. small variation in the coefficient upon changes in velocity, pressure and temperature.
(2) Wear resistance, including resistance to seizing and the tendency to grab.
(3) Heat resistance, including resistance to thermal fatigue, i.e. the capacity to withstand elevated temperatures without failure, retaining the necessary properties of the material for a prolonged period.

Dry clutches and brakes usually have friction pairs consisting of steel or cast iron on a lining of some asbestos-base friction material.

Thermosetting (phenol–cresol–formaldehyde) resins, natural or synthetic rubbers, or both resin and rubber together are used as the bond for the friction material.

Friction linings may be:

(a) A fabric woven of asbestos and cotton fibres and metal wire, molded at high temperature.
(b) Molded in press molds from short-fibered asbestos.

Band stock for shoe-type clutches and brakes is rolled off the same materials.

At working temperatures up to 300 °C, for asbestos friction materials, the allowable pressure is up to 0.6 to 1.0 N/mm^2.

It is better practise to glue on friction linings than to rivet them. Glueing allows wear to a greater depth and increases the effective area. This doubles the service life.

It proves expedient in clutches manufactured in large lots to use disks with 'Cer-met' (metal–ceramic) linings made by sintering. Up-to-date Cer-met friction materials contain the following components: copper or iron which constitutes the base and provides for heat dissipation, graphite and lead which serve as the lubricant, and asbestos and quartz sand which increase the friction. Cer-met friction materials have a higher wear resistance and thermal conductivity than ordinary asbestos materials. Their properties change less when they are heated.

The Cer-met materials are applied on the steel disks or shoes and they are joined by sintering under pressure. First the steel surfaces are copper-plated. The thickness of a disk with a Cer-met coating is from 30 and 40% less than the glued-on friction lining. This means that the axial overall size of the clutch is correspondingly reduced.

The friction elements of clutches that are to run in oil are made of steel with subsequent hardening or sulphocyaniding, as well as of combinations of materials: hardened steel on a friction plastic, or hardened steel on a Cer-met in the form of a lining (for large-lot production).

10.4.3 Disk clutches and brakes

Such elements are extensively employed. They have working surfaces of the simplest shape. Even with a small overall size, they can have a large friction surface. The force required for engagement is not very large because it consecutively applies pressure on the friction surfaces and is not distributed among them.

Clutches may be of single-disk (with two friction surfaces) and multiple-disk clutch. The disk is keyed to one shaft and is compressed for engagement between two flanges keyed to the other shaft. These clutches are widely used in automobiles for which clear-cut disengagement is a desirable feature (Figure 10.6).

A multiple-disk clutch consists of a housing, a sleeve, a set of disks linked to the housing, a set of disks linked to the sleeve and a pressure mechanism.

The valuable features of multiple-disk clutches are:

(a) High load capacity in conjunction with small overall size, especially diametral size, which is of prime importance for high-speed drives.
(b) Smooth engagement.

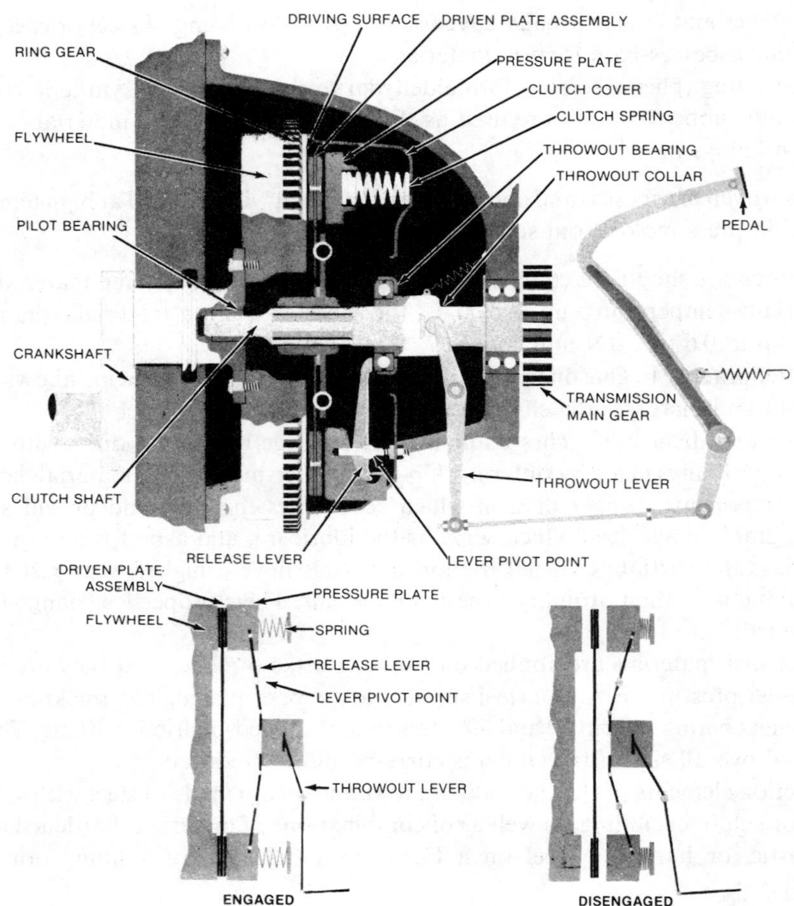

Figure 10.6 Automobile disk clutch. (Courtesy Mobil Oil Corp.)

(c) The possibility of varying the number of disks, which is an essential advantage in limiting the number of components of different sizes in standard clutches.

One drawback of multiple-disk clutches is poor disengagement, especially for clutches mounted on vertical shafts.

Hardened steel or metal–ceramic disks are commonly used for clutches operating with lubrication. In dry clutches, one set of disks (usually the outside disks) has friction linings.

Disks are linked to the housing and sleeve by means of parallel-sided or involute spline joints. It should be noted that the joint between the disks and the sleeve, and even the housing, is subject to very high stresses. Frequently, the disks' grooves wear on the sides of the splines which prevent smooth engagement by interfering with axial motion of the disks. For this reason, the surfaces of the splines must be properly hardened.

As a rule, not more than 8 to 12 disks are used. The use of more disks leads to nonuniform pressure between them, owing to friction on the splines, and the poor disengagement.

The following clearances are to be provided between the disks when the clutch is released:

(a) *For metal disks*, from 0.5 to 1 mm for single- and two-disk clutches, and 0.2 to 0.5 mm for multiple-disk clutches.
(b) *For non-metallic disks*, from 0.8 to 1.5 mm for single- and two-disk clutches, and 0.5 to 1 mm for multiple-disk clutches.

Disk disengagement can be improved by providing disk-type springs on one-half of the disks (for instance, the inside disks). These are not flat in the free state, but are tapered or wavy). Such disks are also called 'sine disks' and behave like springs themselves.

It is more difficult to ensure proper disk disengagement on vertical shafts. Disks are sometimes made with different outside diameters for such installations providing steps on which they rest. When such a clutch is disengaged, each disk drops by gravity onto its own step.

Multiple-disk disengaged clutches, normally used in speed gearboxes, are frequently of twin design. This enables two gears to be alternately engaged to the shaft to change speeds and to reverse rotation.

The torque (T) that can be transmitted by a disk clutch is

$$T = \frac{1}{S} F R_{\mathrm{m}} i f = \frac{\pi}{S}(R^2 - r^2) R_{\mathrm{m}} i p_{\mathrm{a}} f \tag{10.17}$$

where F = force pressing the disks together; S = margin of frictional engagement; R and r = outside and inside radii of the annular friction surfaces respectively. The ratio r/R usually ranges from 0.5 to 0.7.

From this, for selected radial dimensions, we find the required number of pairs of friction surfaces (i):

$$i = \frac{ST}{\pi(R^2 - r^2) R_{\mathrm{m}} p_{\mathrm{a}} f} \tag{10.18}$$

where f = coefficient of friction, p_{a} = allowable pressure on lining, and i = number of pairs of friction surfaces, equal to the number of outside and inside disks together with the end flanges minus one.

The mean radius of the friction surface is:

$$R_{\mathrm{m}} = \frac{R + r}{2}$$

The force required to press the disks together is

$$F = \frac{ST}{R_{\mathrm{m}} i f} \tag{10.19}$$

An especially critical mechanism of friction clutches is the control device. Lever-cam pressure mechanisms are quite extensively used for controlling clutches for transmitting low and medium torques if automatic controls are not required.

Motion is transmitted from the manual control lever to a collar which is shifted along the axis of the shaft. Most pressure mechanisms have operating levers. When the collar is shifted, its bevelled surface engages one end of the operating levers whose other end applies pressure to the friction components of the clutch (Figure 10.6).

Pressure mechanisms must have a considerable mechanical advantage. This is obtained by properly selecting the lengths of the arms of the operating levers, the angle of the bevel on the collar, etc.

10.4.4 Conical clutches and brakes

One of the members of an ordinary cone clutch has an internal conical working surface and the other member has a mating external cone. The clutch is engaged and disengaged by axially shifting one of the members. The conical friction surfaces enable considerable normal pressure and frictional forces to be produced with a relatively small engaging force (principle of a wedge mechanism) (Figure 10.7).

To avoid self-engagement and to facilitate disengagement, the cone angle (angle between an element of the conical surface and the shaft axis) is taken greater than the angle of static friction. For metallic friction surfaces the cone angle is taken from 8 to 10° or more, while for asbestos-base linings, it is taken from 12 to 15° or more.

Advantages of cone clutches are their ready release and comparatively simple design. But the clutches have essential shortcomings as well: their considerable radial overall size and the strict requirements made to the coaxiality of the shafts being joined.

For these reasons, cone clutches, previously extensively employed in the engineering industries, have only restricted application today.

The torque that can be transmitted by a cone clutch with a mean radius R_m of the friction surfaces with width b is

$$T=\frac{2}{S}\pi R_m^2 b p_a f \tag{10.20}$$

Hence, the width b of the friction surface for a selected value of R is

$$b=\frac{ST}{2\pi R_m^2 p_a f} \tag{10.21}$$

Usually $b/R_m=\psi=0.3$ to 0.5.

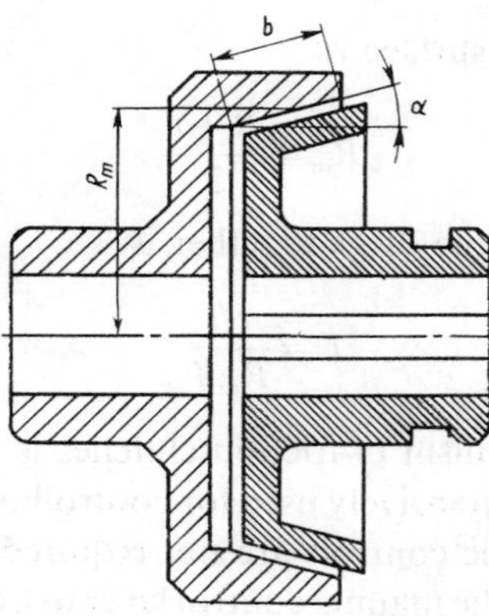

Figure 10.7 Conical clutch

Another procedure in design is to first assign the value of ψ. Then

$$R_m = \left[\frac{ST}{2\pi\psi p_a f}\right]^{\frac{1}{3}} \tag{10.22}$$

The force required to engage the clutch is

$$F \approx \frac{ST\sin\alpha}{R_m f} \tag{10.23}$$

where α is the angle between an element of the cone surface and the shaft axis (cone angle).

In deriving equation (10.23), it is assumed that the force of friction on the working surfaces (in accordance with the direction of the sliding velocity) acts in the peripheral direction and does not empede clutch engagement.

10.4.5 Cylindrical radial air clutches

In these clutches friction is developed between the shoes of a rubber tire, secured to one clutch member, and the cylindrical rim of the second member (drum). To engage the clutch, air under pressure is admitted into the tire which expands so that the shoes are pressed uniformly against the drum.

The tire shown in Figure 10.8 transmits a torque. It consists of: (a) an elastic rubber inner tube which holds the air, (b) multiple-ply load-carrying lining of tough rubberized fabric (cord), and (c) outer tread of rubber.

The tire is manufactured by hot vulcanization and constitutes an integral unit.

The shoes are secured to the tire by plain pins which pass through holes in the tread and are kept from falling out by a wire threaded through holes in the pins. The tire is heat-insulated from the shoes with lining which also protects it against the products of wear.

The shoes are coated by a friction lining held by glue. This lining is usually made of an asbestos fabric band impregnated with phenolic resin.

The merits of radial tyre-type air clutches are:

(a) convenient controls;

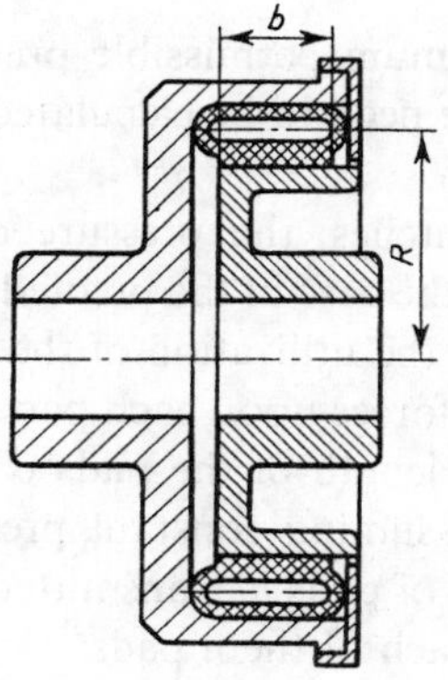

Figure 10.8 Radial pneumatic clutch, outwards expanding (Decker)

(b) possibility of regulating the limiting torque and the rate of engagement so that the clutch can be a safety device;
(c) compensation for axial, radial and angular displacements of the shafts being jointed without subjecting them to appreciable radial or axial loads; in practise, radial displacements up to 2 or 3 mm can be observed and even considerably larger axial displacements;
(d) self-compensation for wear so that periodic regulation is unnecessary;
(e) noise attenuation, shock cushioning and damping of torsional vibrations.

The drawbacks of these clutches include: high cost of the tire, ageing of the rubber, and the liability to damage by oil, alkalis or acids getting on the rubber.

These clutches operate well in the temperature range from −20 to +50 °C.

Tire-type air clutches are employed mainly in heavy engineering: in oil-well winches, in drives from a marine engine to the propeller, in excavators and other similar machinery.

10.4.6 Cylindrical clutch and brakes

Such elements consist of a drum and one or more friction pads pressed against the inner or outer surface of the drum by a variety of force mechanisms.

On the basis of friction interface on the inner or outer surface, these are classified as inner or outer clutches (brakes). On the basis of forcing mechanism they are classified as follows:

(a) *Centrifugal clutches.* The friction pads are on the driving shaft and when the speed exceeds some value, the centrifugal force is enough to provide the normal force required to drive the drum and the driven shaft.
(b) *Pneumatic clutches.* The friction material is glued on the outer (or inner) surface of a rubber tire which when inflated creates enough normal pressure between the mating surfaces to drive the drum.
(c) *Mechanical clutches.* This is where a mechanism provides the force on the friction pad necessary to drive the drum.
(d) *Hydraulic clutches.* This is where the force is supplied by one or two hydraulic actuators.

For a given torque and maximum permissible pressure of the friction material, the dimensions and the actuating force need to be calculated. Figure 10.9 shows and inwardly expanding radial pneumatic clutch.

Except for the pneumatic clutches, the pressure distribution between the rubbing surfaces is not exactly uniform. In the case of the pivoted pads, the pressure depends on the distance from the pivot point and the utilization of the friction material is poor. For this reason, the tendency is to use two forces upon each pad to assure almost uniform pressure between the mating surfaces. The length of the pads is chosen as a fraction of the circle periphery for the same reason. Assuming constant pressure $p_{\max}$, constant coefficient of friction f, radius of drum R, width of pads b, transmitted torque T, the total actuating pad force will be (Figure 10.10b), for each of the n pads:

$$F = T/fRn \tag{10.24}$$

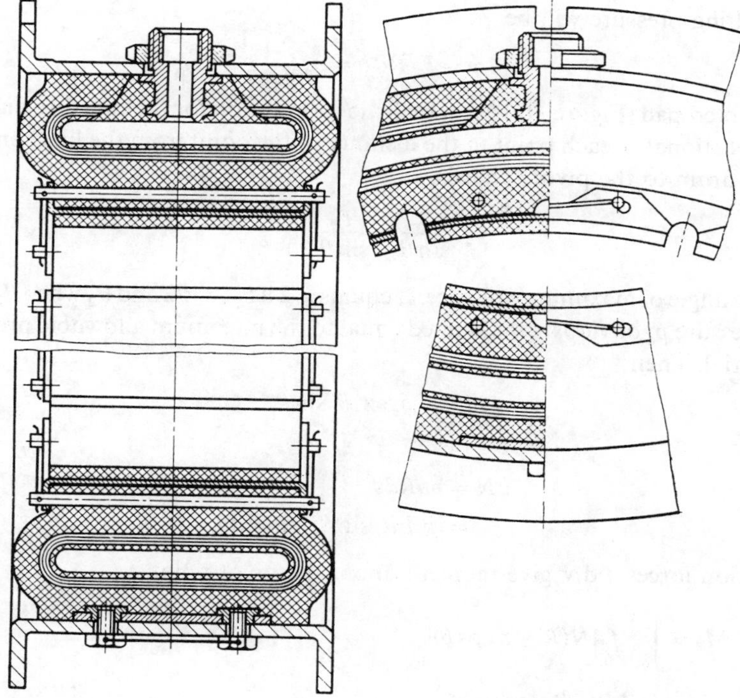

Figure 10.9 Radial pneumatic clutch, inwards expanding (Decker)

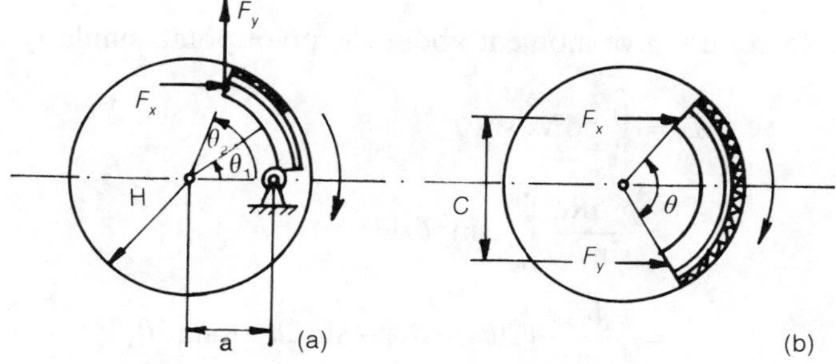

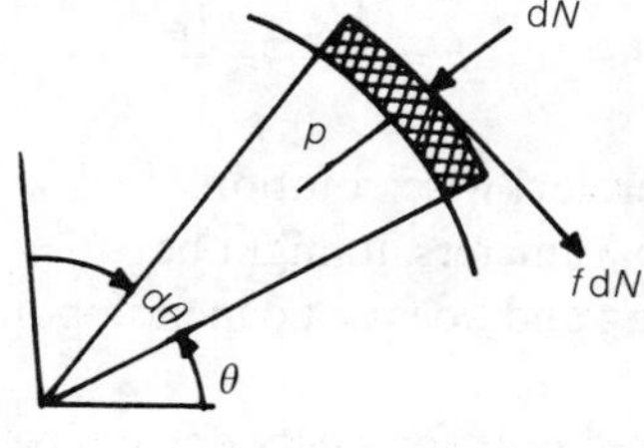

Figure 10.10 Drum brake

The resulting pressure will be

$$p = F/R\varphi b = p_{max} \tag{10.25}$$

For a pivoted pad (Figure 10.10a), a realistic assumption for the pressure distribution is that it is proportional at each point to the distance of the point from the line connecting the center of the drum to the pivot. That is

$$\frac{p}{\sin\theta} = \frac{p_a}{\sin\theta_a} \tag{10.26}$$

where θ_a is the angle of maximum pressure. It equals $\pi/2$ if $\theta_2 \geqslant \pi/2$ and θ_2 when $\theta_2 < \pi/2$. The maximum pressure p_a obviously is assumed equal to the maximum allowable pressure of the friction material. Then

$$p(\theta) = p_a \sin\theta/\sin\theta_a \tag{10.27}$$

and

$$\begin{aligned} dN &= pbRd\theta \\ &= (p_a bR/\sin\theta_a)\sin\theta\,d\theta \end{aligned} \tag{10.28}$$

The friction forces $f\,dN$ give moment about the pivot point

$$\begin{aligned} M_f &= \int_{\theta_1}^{\theta_2} f\,dN(R - a\cos\theta) \\ &= \frac{fp_a bR}{\sin\theta_a}\int_{\theta_1}^{\theta_2} \sin\theta(R - a\cos\theta)\,d\theta \\ &= fp_a bR[a(\cos 2\theta_2 - \cos 2\theta_1)/4 - R(\cos\theta_2 - \cos\theta_1)]/\sin\theta_a \end{aligned} \tag{10.29}$$

The normal forces dN give moment about the pivot point, similarly,

$$\begin{aligned} M_n &= \int_{\theta_1}^{\theta_2} dNa\sin\theta \\ &= \frac{p_a bRa}{\sin\theta_a}\int_{\theta_1}^{\theta_2} \sin^2\theta\,d\theta \\ &= \frac{p_a bRa}{4\sin\theta_a}[2(\theta_2 - \theta_1) - \sin 2\theta_2 + \sin 2\theta_1] \end{aligned} \tag{10.30}$$

The actuating force must supply the balance moment. This yields

$$F = \frac{M_n \pm M_f}{c} \tag{10.31}$$

The '+' sign stands for counterclockwise rotation.

With proper choice of the parameters, it might be $M_n = M_f$, and then $F = 0$. This means that the coupling is self-actuating and does not depend upon the magnitude of the actuating force.

This feature, however, is used only for safety devices because coupling is rather sudden and uncontrollable, accompanied usually by high impact forces.

The coupling torque will be

$$T = \int_{\theta_1}^{\theta_2} fR\,\mathrm{d}N = \frac{fp_a bR^2}{\sin\theta_a}\int_{\theta_1}^{\theta_2}\sin\theta\,\mathrm{d}\theta$$

$$= \frac{fp_a bR^2}{\sin\theta_a}(\cos\theta_1 - \cos\theta_2) \tag{10.32}$$

10.4.7 Safety and automatic clutches

Self-acting (self-controlled) clutches release automatically under certain operating conditions. The operating conditions of certain classes of machinery – crushers, excavators, cultivators, for the press-forming of metals, etc. – include systematic overloads. In machinery not subject to impacts and in machining homogeneous workpieces, for instance machine tools, overloads may be caused by selecting excessively high machining variables (speeds, feeds or depths of cut), by dulling or breakage of the cutting tools, etc.

Overloads occurring in the operation of mechanisms may be caused by lubrication failures, seizing, jamming, etc.

As to its action, an overload may build up gradually (for instance, in the dulling of a tool) or it may be of the impact type.

The function of a safety link can be performed, in addition to special safety clutches, by other elements of the drive which permit slipping upon overloads: friction clutches, hydraulic clutches, hydraulic and pneumatic drives. In hydraulic drives, for instance, overloads are prevented by safety valves. In pneumatic drives, any appreciable overload is impossible owing to the elasticity of the working medium – air.

10.4.8 CAD – optimum design of clutches and brakes

The general idea behind computer aided design and optimization of clutches and brakes is to achieve optimum performance with minimum dimensions and cost.

The design objectives depend on the particular application and no general formulation can be made at this point. By way of an example we shall see some aspects of this class of problems (Figure 10.11).

An automobile brake is to be designed of the disk type. The brake consists of a disk of diameter D and two friction pads between radii r_1 and $r_2 = D/2$ and angle θ. The car has design speed v and mass m per wheel. There is a linear relationship between car speed v and angular velocity ω of the shaft:

$$v = L\omega \tag{10.33}$$

where L is a constant parameter in terms of length. The pad area is

$$A = \pi(r_2^2 - r_1^2)/4 \tag{10.34}$$

The average pad pressure (p), equal to the maximum allowed pressure p_a, is

$$p = p_a = P/A = 2P/\theta(r_2^2 - r_1^2)$$

where

$$P = p_a\theta(r_2^2 - r_1^2)/2 \tag{10.35}$$

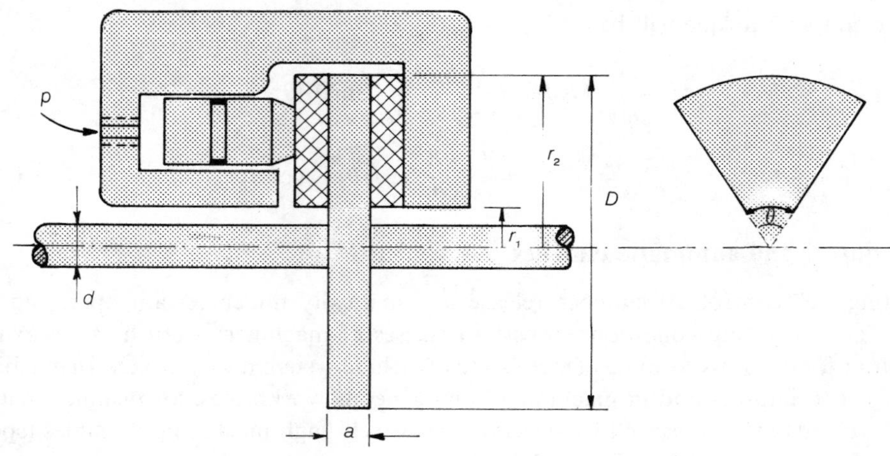

Figure 10.11 Disk brake

Table 10.3 Material properties for clutch friction materials

Friction pair materials	Coefficient of friction (f)	Allowable pressure p_a (N/mm^2)	
		With several friction surfaces (disk clutches)	With one friction surface (cylindrical clutches)
Running in oil			
Hardened steel on hardened steel	0.06	0.6 to 0.8	–
Cast iron on cast iron or hardened steel	0.08	0.6 to 0.8	1.0
Cer-met on hardened steel	0.1	0.8 to 0.10	–
Running dry			
Asbestos-base pressed material on steel or cast iron	0.3	0.2 to 0.3	0.3
Cer-met on hardened steel	0.4	0.3 to 0.4	–
Cast iron on cast iron or hardened steel	0.15	0.2 to 0.3	0.3

Notes: 1. Lower values of pressure are for a larger number of friction surfaces and vice versa.
2. If no thermal calculations are to be carried out, then, at high velocities (measured at the middle of the width of the friction surface) and a large number of engagements per hour, the pressure should be reduced somewhat, especially for multiple-disk clutches with a great number of disks. For the latter this reduction should be 15% at $v = 5$ m/sec, 30% at $v = 10$ m/sec, and 35% at $v = 15$ m/sec.

where P is the braking force on the pad. The moment of the friction forces on both sides of the disk is

$$M = Pf(r_1 + r_2)/2 = p_a f\theta(r_1 + r_2)(r_2^2 - r_1^2)/4 \tag{10.36}$$

To find a moment of inertia J equivalent to the mass per wheel m, the kinetic energy is used

$$mv^2/2 = J\omega^2/2, \quad \text{or} \quad J = mL^2 \tag{10.37}$$

Momentum balance gives

$$M = J\dot{\omega} = mL^2\gamma/L = mL\gamma, \quad \gamma = M/mL \tag{10.38}$$

The time to stop the car from a speed v is

$$t = v/\gamma = umL/M = (umL/4p_a f\theta)[1/(r_1 + r_2)(r_2^2 - r_1^2)] \tag{10.39}$$

The kinetic energy of the car $mv^2/2$ will be transformed to heat which will increase the temperature of the disk by

$$T = (8mv^2/\pi\rho c)(1/ar_2^2) \tag{10.40}$$

Equation (10.39) gives the objective function $t(r_1, r_2, a)$ which must be minimized under certain constraints:

(a) The maximum disk temperature should not exceed the temperature allowed by the brake material, T_0.
(b) The radius r_1 has a minimum $d/2$ and r_2 has a maximum allowed by the space available.

The constraint (a) can be waived since it can be used for the determination of a after the radii have been optimized. In this case, the design variables are the two radii r_1 and r_2.

10.5 FRICTION BELTS

A belt drive consists of a driving and a driven pulley and a belt which is mounted around the pulleys, with a certain amount of preload. It transmits peripheral force from one pulley to the other.

Some arrangements of belt drives are shown in Figure 10.12. More than one driven pulley can be used and also in limited applications the pulleys can be in different planes.

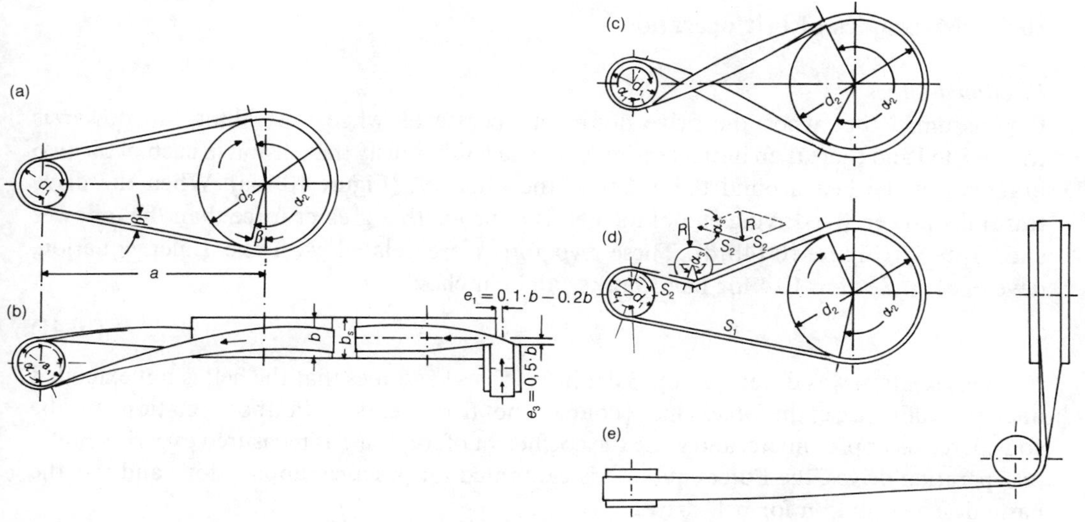

Figure 10.12 Belt drive geometry

Figure 10.13 Belt sections: (a) Flat, (b) V-belts, (c) round, (d) multiple V-belts

The usual form of belts is indicated in Figure 10.13. Flat belts have a narrow rectangular cross-section (Figure 10.13a). The V-belts have trapezoidal cross-section (Figure 10.13b) and multiple V-belts consist of several V-belts in parallel arrangement (Figure 10.13d). Round belts are used also with a circular cross-section.

In many applications, instead of using one large belt, several belts of smaller thickness are utilized, to reduce bending stresses and fatigue loading, as will be shown later (p. 465).

Belts have certain advantages which make their use very wide:

1. They have the capability of transmitting motion at considerable distances.
2. Their operation is smooth and they can absorb impact loads on one drive and not transmit them to the other drive.
3. They can slip, preventing overloading of motors during start-up of the machine with high inertial loads.
4. They operate at very high speeds.
5. Their cost is usually low.

Disadvantages of belt drives are:

1. Considerable overall size, much greater than other types of drives, such as gear drives.
2. Because of slipping, which is not constant, the relation of the two speeds of rotation of driven and driving pulley is not constant in time.
3. Frequent maintenance is required, because the belt needs frequent tensioning and its service life is usually short.

10.5.1 Mechanics of belt operation

Frictional forces

It is assumed that when the drive does not operate or when it is idling, no power is transmitted and there is an initial tension of the belt. Obviously the tension in each of the two branches of the belt around the pulley is the same, S_0 (Figure 10.14a). When the drive transmits power, the driving end of the belt is pulled with a greater force than the follower end, $S_1 > S_2$ (Figure 10.14b, d). These two forces are related with the Euler equation, developed in Section 10.4 for band brakes and clutches:

$$S_1/S_2 = e^{f\theta} \tag{10.41}$$

This relation was developed by Euler in 1775, and assumes that the belt is not extended under the influence of the force. This of course is not true, therefore the above relation must be considered as approximate, and since the coefficient of friction f is measured experimentally on operating drives, the Euler equation is calibrated for practical applications and it is the basic design equation for belt drives.

In Figure 10.14a a belt is shown around a pulley in idle operation. As mentioned above,

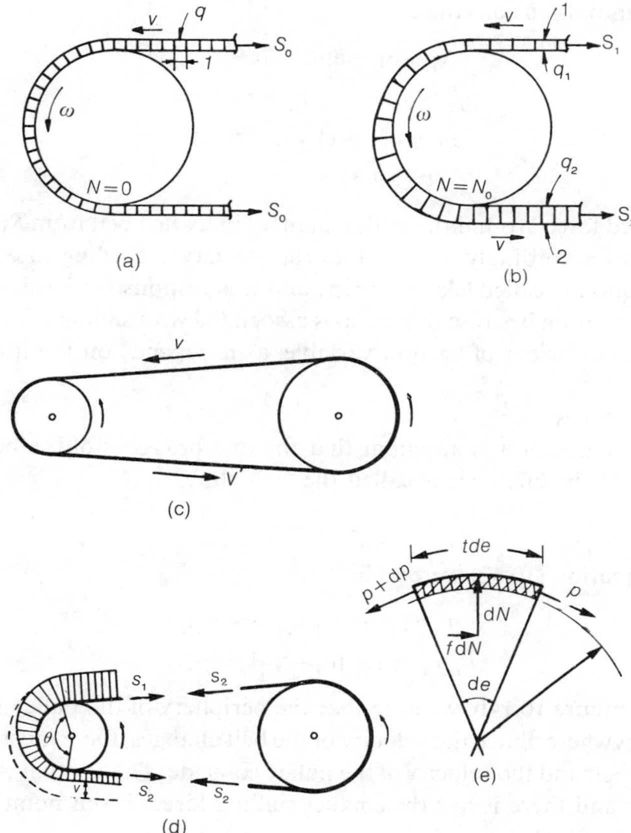

Figure 10.14 Belt kinematics

both branches of the belt have the same tension. Let us draw lines perpendicular to the belt in equal distances, say at unit lengths. Their distances during rotation remain constant and equal to the unit length.

Suppose now that the belt transmits a certain power N. This requires that the tension in the two branches of the belt is different and their difference produces the power. Let these two forces be S_1 and S_2. It is apparent that $S_1 > S_0$ and $S_2 < S_0$. That means that the branch 1 of the belt is elongated, and the branch 2 of the belt is contracted. Therefore the lines previously traced will have greater distances on branch 1 and smaller distances on branch 2.

Let q be the mass per unit length, between two lines on the belt. If we assume that the deformation of the belt does not change its volume, then since the mass rate entering the pulley and the one leaving the pulley must be equal, the principle of Conservation of Mass requires that q_u must be constant, where q is the said mass per unit length. Let v be the velocity of the belt in each position. It is obvious now that in the loaded belt the mass per unit length in the two branches is different and it is $q_2 > q_0 > q_1$. Therefore it must be $v_1 > v_0 > v_2$.

If q_0 refers to belt with initial tension S_0 and q refers to an unloaded belt, and if e is the strain during the initial tension, then

$$S_1 > S_0 \quad \text{and} \quad S_2 < S_0$$
$$q_2 > q_0 > q_1 \quad v_1 > v_0 > v_2$$
$$q_0 = q(1+e) = \text{constant}$$
$$v/(1+e) = \text{constant} \tag{10.42}$$

The transmitted force around the pulley changes in loaded belt from S_1 to S_2. Therefore the strain e changes accordingly and so does the velocity according to equation (10.42).

This phenomenon is called 'elastic creep', and it is responsible for deviations from the theoretical Euler equation because this creep is associated with sliding of the belt around the pulley making the coefficient of friction variable, as it depends on the friction velocity.

Kinematics of belt drives

From the above discussion it is apparent that the two branches of the belt have different strains e_1 and e_2. Their difference is called the 'belt slip',

$$s = e_1 - e_2 \tag{10.43}$$

Because of equation (10.42) we obtain

$$v_1/(1+e_1) = v_2/(1+e_2)$$
$$v_1/v_2 \approx 1 + (e_1 - e_2) = 1 + s \tag{10.44}$$

Returning to Figure 10.14b, we note that the periphery of the pulley has of course the same velocity everywhere. Since the velocity of the belt changes, there must be a point where the velocity of the belt and the velocity of the pulley coincide. This is point A where the belt is leaving the pulley and there it has the smaller pulling force. From point A to the driving branch, there is a gradual slip.

With this assumption, which is experimentally substantiated, if ω_1 and ω_2 are the angular velocities of the driving and the driven pulleys respectively and d_1 and d_2 are the respective diameters, we obtain

$$\left.\begin{aligned} v_1 = \omega_1 d_1/2, \quad v_2 = \omega_2 d_2/2 \\ i = \omega_1/\omega_2 = (1+s)d_2/d_1 \end{aligned}\right\} \tag{10.45}$$

Equation (10.45) gives the transmission ratio i which is related not only to the diameters of the pulleys but also to the relative belt slip which depends on the operating conditions and it is not constant.

This is a serious shortcoming of friction belts and they cannot be used in applications where we require that the two pulleys are in constant speed relation with one another.

The transmission ratio for flat belts is up to 5, using a tension pulley up to 10 and with the V-belt drives with transmission ratio up to 15 can be effectively used.

Pull factor

As mentioned before, the belt at idle operation has a preload force S_0 and assumes forces S_1 and S_2 during operation. The corresponding strains will be respectively e_0, e_1, and e_2. If the volume is constant, the contraction of one branch must equal the elongation of the other. If L

is the length of each branch of the belt, it must be

$$\left.\begin{aligned}(e_1 - e_2)L &= (e_0 - e_2)L\\ e_1 + e_2 &= 2e_0\end{aligned}\right\} \qquad (10.46)$$

Because the strain $e = S/EA$, equation (10.46) yields the Ponçelet equation

$$S_1 + S_2 = 2S_0 \qquad (10.47)$$

If the driving peripheral force is U, the peripheral velocity is v and the transmitted power is N,

$$N = Uv \qquad (10.48)$$

$$U = S_1 - S_2 \qquad (10.49)$$

Now the initial tension and the operation tensions can be obtained using equations (10.49) and (10.41):

$$\begin{aligned}&S_2 = U/(m-1), \quad S_1 = Um/(m-1)\\ &S_0 = U(m+1)/2(m-1), \quad m = e^{f\theta}\end{aligned} \qquad (10.50)$$

The ratio

$$d = U/2S_0 = (m-1)/(m+1) \qquad (10.51)$$

is called the 'pull ratio' and for flat belts is, for efficient operation, between 0.4 and 0.6 and for V-belts is between 0.7 and 0.9.

The pull ratio is an important factor for the belt operation. The slip factor and the efficiency of the belt are functions of the pull factor as is shown in Figure 10.15. Efficiencies η of belts are usually high, around 95%. Near the point where there is a rapid increase in the slip factor it is approximately the maximum of the efficiency. Before that point, at low pull factor, the belt is heavy and there are many losses which reduce efficiency. After that optimum point sliding friction increases rapidly and so do the losses, lowering the efficiency.

Experimental observations show that the optimum value of the pull factor is 0.6 for elastic and leather belts, 0.5 for cotton and 0.4 for wool belts while for plastic belts is about 0.5.

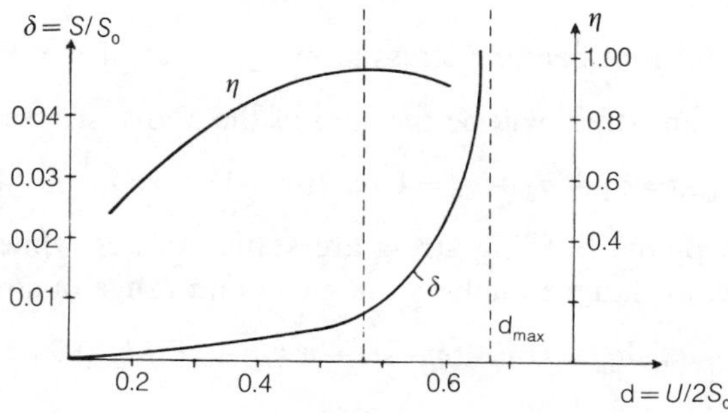

Figure 10.15 Performance of belts

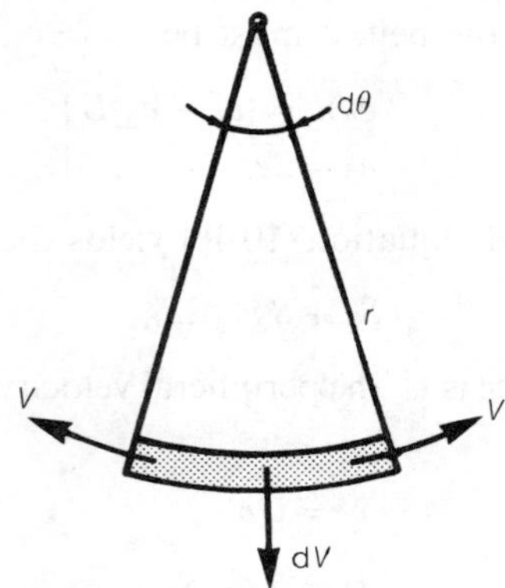

Figure 10.16 Effect of centrifugal force

Stresses in belts

When the belt operates at high speeds, there are considerable inertial forces due to the circular motion (centrifugal forces). Assuming an elementary part of the belt (Figure 10.16), on which the tension force due to the centrifugal force is V while the elementary centrifugal force is dV, we observe that the centrifugal force is

$$dV = v^2 \, dm/r = qv^2 \, d\theta/g \tag{10.52}$$

Equilibrium of the elementary part of the belt (Figure 10.16), requires that $dV = 2V \, d\theta/2$. With equation (10.52),

$$V = qv^2/g \tag{10.53}$$

The belt is stressed due to the following:

1. *Static stresses* due to the maximum tension S_1 over the cross-sectional area $A = bt$

$$\sigma_1 = S_1/A = Um/A(m+1) \tag{10.54}$$

2. *Tensile stresses* due to the centrifugal forces

$$\sigma_u = V/A = qv^2/Ag = v^2\gamma/g \tag{10.55}$$

3. *Bending stresses* due to the curvature of the belt around the pulleys

$$\sigma_b = Et/d_{1,2} \tag{10.56}$$

where there is a different bending stress at every pulley, if the diameters are different.

The total maximum stress will be the sum of the above stresses (1–3), all tensile:

$$\sigma_{max} = \sigma_1 + \sigma_u + \sigma_b = Um/A(m-1) + v^2\gamma/g + Et/d_{1,2} \tag{10.57}$$

All stresses except the bending stress are static stresses while the bending stress is pulsating. Therefore, for fatigue analysis the mean and range of stress are:

$$\sigma_m = Um/A(m-1) + v^2\gamma/g + (Et/d_{min})/2 \tag{10.58}$$

and

$$\sigma_r = (Et/d_{min})/2 \tag{10.59}$$

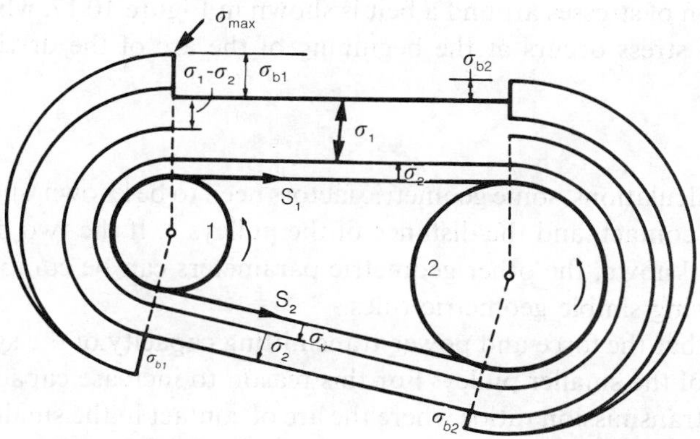

Figure 10.17 Stresses on belts

D1, D2, α, A	D1, D2, α, A	D1, D2, A	D1, Dp, C, Ap, α, E, γ, A, D2
$\varphi \approx 180^0 - \frac{D_2 - D_1}{A} 60^0$	$\varphi \approx 180^0 + \frac{D_2 + D_1}{A} 60^0$	$\varphi \approx 180^0 + \frac{D_1}{A} 60^0$	$\varphi \approx 180 - \frac{D_2 - D_1}{2A} 60 + \frac{(D_1 + D_2 - 2E)}{2A_p}$
$L = 2A + \frac{\pi}{2}(D_2 + D_1) + \frac{(D_2 - D_1)^2}{4A}$	$L = 2A + \frac{\pi}{2}(D_1 + D_2) + \frac{(D_1 + D_2)^2}{4A}$	$L = 2A + \frac{\pi}{2}(D_1 + D_2) + \frac{D_1^2 + D_2^2}{2A}$	$L = (A + A_p + C) + \frac{\pi}{2}(D_1 + D_2) + \frac{(D_2 - D_1)^2}{8A} + \frac{(D_2 + D_p)^2}{8A} + \frac{(D_1 + D_p)^2}{8C} - \frac{E}{A_p}\left(\frac{D_1 + D_p}{2}\right) - \frac{E}{C}\left(\frac{D_2 + D_p}{2}\right)$

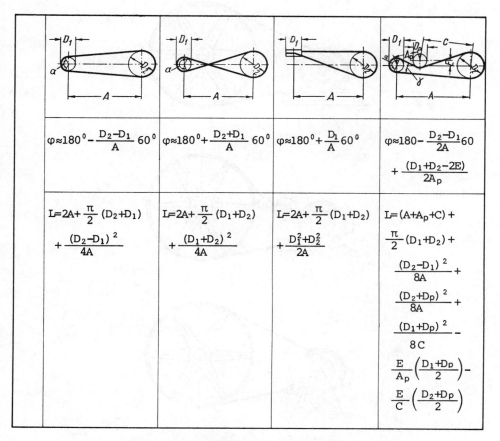

Figure 10.18 Geometric relations in belt drives

The distribution of stresses around a belt is shown in Figure 10.17, where it is indicated that the maximum stress occurs at the beginning of the arc of the driving pulley, at the entering branch.

Geometry

For all the above calculations, some geometric factors need to be known, such as the length of the belt, the arc of contact, and the distance of the pulleys *a*. If the two diameters and the pulley distance are known, the other geometric parameters can be computed as shown in Figure 10.18 following simple geometric rules.

It is apparent that the force and power-transmitting capacity of the system depends on the arc of contact of the smaller pulley. For this reason to increase capacity of belt drives, especially for high transmission ratios where the arc of contact in the smaller pulley is small, auxiliary pulleys are used to increase the arc of contact. Such a system can also provide for the initial tension. This and other systems for pretensioning belts are shown in Figure 10.19.

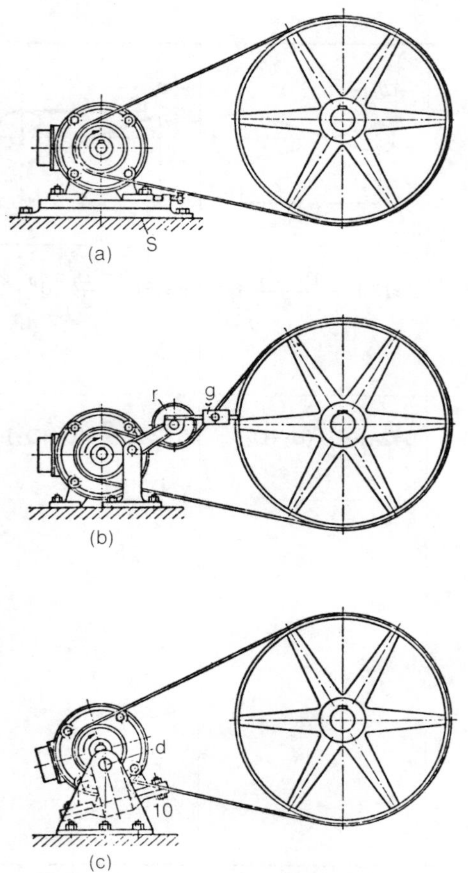

Figure 10.19 Belt tension (Dobrovolski 1975)

10.5.2 Optimum design of belts

For most belting applications the process of designing is merely a process of application engineering, selecting belts according to manufacturing catalogs. Each manufacturer gives also a suggested design procedure for his belts. This of course is associated with the material properties that are given by such manufacturers, in a way that is related with a method of design. Many manufacturers however, are supplying information about the basic material properties for their belts. In this case it is suggested to proceed with a rational design method to check and substantiate the manufacturer's application rules.

In most design applications, the power transmitted N_p and the speeds of the two pulleys, n_1 and n_2, are given and in some cases the distance between the two pulleys. Design should begin from the determination of the diameter of the driving pulley which is the more critical. There are some empirical relations which however are all based on the idea of optimum selection of the driving pulley diameter. This is associated with the centrifugal stresses on the belt which, if the diameter of the driving pulley is large enough, lead to such values that produce tension on the belt greater than the initial tension S_0, which means that there is no normal force between the belt and the pulley and the transmitted power is 0. To quantify this, let us return to Figures 10.5 and 10.16 and rewrite the equation of equilibrium of an elementary part of the belt also taking into account the centrifugal forces. Following the analysis presented in Section 10.4, we obtain:

$$P\,\mathrm{d}\theta = \mathrm{d}N + q\frac{v^2}{g}\mathrm{d}\theta \tag{10.60}$$

$$\mathrm{d}P = f\,\mathrm{d}N \tag{10.61}$$

$$\left(P - q\frac{v^2}{g}\right)\mathrm{d}\theta = \frac{1}{f}\mathrm{d}P \tag{10.62}$$

$$\int_0^\theta f\,\mathrm{d}\theta = \int_{S_1}^{S_2} \frac{\mathrm{d}P}{P - q\left(\dfrac{v^2}{g}\right)} \tag{10.63}$$

$$\frac{S_1 - q\left(\dfrac{v^2}{g}\right)}{S_2 - q\left(\dfrac{v^2}{g}\right)} = e^{f\theta} = m \tag{10.64}$$

$$S_1 - S_2 = \frac{m-1}{m}\left(S_1 - q\frac{v^2}{g}\right) = U = \frac{75N_p}{v} \tag{10.65}$$

$$N_p = \frac{m-1}{75m}v\left(S_1 - q\frac{v^2}{g}\right) \tag{10.66}$$

Equation (10.66) implies that the belt can transmit power which depends on the velocity. For zero velocity we have transmitted power zero and for some value of the velocity:

$$v = (S_1 g/q)^{\frac{1}{2}} \tag{10.67}$$

the quantity in parentheses in equation (10.66) becomes zero and therefore the belt does not transmit power. In between those two points there is a third-degree curve which has a maximum when

$$\frac{\mathrm{d}}{\mathrm{d}v}\left[v\left(S_1 - q\frac{v^2}{g}\right)\right] = S_1 - 3\frac{q}{g}v^2 = 0$$

$$v_{\mathrm{opt}}^2 = \frac{S_1 g}{3q} = \frac{(S_y/N)Ag}{3A\gamma} = \frac{(S_y/N)g}{3\gamma} = v^2 = \left(\frac{\pi n d_1}{60}\right)^2 \tag{10.68}$$

This determines the diameter of the small pulley

$$d_1 = \frac{60}{\pi n}\left(\frac{S_y/Ng}{3\gamma}\right)^{\frac{1}{2}} \tag{10.69}$$

This equation can be used only qualitatively, because it does not take into account many other factors influencing optimum design, such as the cost of the pulley itself. In general, it yields high values for the pulley diameters for the best use of the belt materials but the pulley and the whole system become larger and heavier. This is the reason for using empirical relations which take into account more factors than the centrifugal force in the form

$$d_1 = (1100 \text{ to } 1300)(N_p/n_1)^{\frac{1}{3}} \tag{10.70}$$

where d_1 is in millimeters, N_p in kW, and n_1 in rpm.

The determination of the diameter of the driven pulley is made with equation (10.45), in the form (10.71).

$$d_2 = d_1 i/(1 + s) \approx 0.985 d_1 i \tag{10.71}$$

At this point a thickness of the belt must be selected, based on available sizes or suggested diameter to belt thickness ratio, where they are available. Then the width of the belt can be computed from the relation

$$b = N_p C/N_0 \tag{10.72}$$

where C is a total service factor, and N_0 is the power capacity of the belt per unit width. This capacity is given in most manufacturers' catalogs and if not available it can be computed from equations (10.48 and 10.57) in the form

$$N_0 = (S_c/N - v^2\gamma/g - Et/d)vt(m-1)/m \tag{10.73}$$

In this relation, S_c is the service strength, much lower than the tensile strength obtained with static tests. This is due to the fact that at a point far below the static strength, the creep increases rapidly and the belts lose their tension. Repeated retensioning can be done but it requires continuous attention and service and the operation will be less reliable. For this reason, experimental values of the creep strength S_c are used, given in Table 10.4 for some belt materials, instead of the static strength S_u given also for reference purposes only.

A similar situation exists for the elasticity modulus E used for the computation of the bending stresses. Repeated bending results in local creep in the tensile stress area which is apparent when we cut a belt and let it free on a plane surface. It will take a curved form owing

Table 10.4 Material properties for belts

Material	E (N/mm²)	S_u (N/mm²)	γ (N/m³ × 10⁻³)	v_{max} (m/sec)	E_c (N/mm²)	S_c (N/mm²)	S_n (N/mm²)	m (exp)	d_{min}
Rubber	200–350	45–60	11–12	40	10	2.5	6	5	
Leather	150–250	25–30	10–11	30–50	30	2.9	6	6	
Fabric	200–350	30–50	9–10	20–25	15	2.1	3	5	
Plastic	600	200	10–11	60	60	6.1	6	6	
V-belts O	250–400	60–70	11–12	25	10	2.3	9	8	63
A					12	2.5			90
B					18	2.8			125
C					21	3.0			200
D					28	3.2			315
E					35	3.2			500
F					44	3.2			800

to unequal creep along the thickness. For this reason an apparent value of the elasticity modulus E is used which corresponds to correct bending stresses when using equation (10.69). Such values are given also in Table 10.4.

Therefore, wherever strength and elasticity modulus appear in the design equations, S_c and E_c from Table 10.4 must be used.

Again, N_0 is the power capacity of the belt per unit length and for service factor equal to 1. The service factor depends on the particular application and can be broken down to five subfactors

$$C = C_1 C_2 C_3 C_4 C_5 \tag{10.74}$$

Values of the five subfactors are given in Table 10.5. For V-belts, two more factors are used, C_6 and C_7 given also in Table 10.5.

The belt must now be checked for fatigue. Fatigue tests on friction belts have not revealed specific endurance limits. The point where the deterioration of belt starts has a great uncertainty and it seems that after a number of reversals of the stress, deterioration starts independently of the magnitude of stress range. Therefore the endurance limit must be considered as a statistical quantity with a great deviation. In that respect, service life calculation based on fatigue strength, is (using the expression for the fatigue curve)

$$\sigma_{max}^m N = \text{constant} \tag{10.75}$$

Where N is the number of cycles and σ_{max} is the maximum stress. If the fatigue strength is assumed at a number of cycles N_0, usually taken to be 10^6, and this strength S_n, then

$$\sigma_{max}^m N = S_n^m N_n \tag{10.76}$$

The number of cycles to failure of the belt is then

$$N = N_n(S_n/\sigma_{max}^m) = 3600\, Hzv/L \tag{10.77}$$

where H is the life in hours, z is the number of bends per one full cycle of the belt (equal to

Table 10.5 Design factors for belts

C_1, overload factor, 1.0–1.5
C_2, environment factor, 1.0–1.3
C_3, continuous operation factor, if fatigue test not performed,

Hours per day	C_3
3–4	1.45
8–10	1.5
16–18	1.9
24	2.0

C_4, contact angle factor (=1 if equation (10.73) used)

θ	80°	100°	140°	180°
C_4	1.5	1.35	1.1	1

C_5, tension factor,
$C_5 = 1$, tension with bolts
$= 1.2$, with belt shortening (flat belts)
$= 0.8$, self-tensioned system
C_6, pulley selection factor (V-belts)
$C_6 = 1$, if d_1 computed by equation (10.69)
$= d_1/d_0$, d_0 the value from equation (10.69)
C_7, multiple belt factor (V-belts)
$C_7 = 1$, 1 V-belt
$= 1.25$, multiple V-belts
f, friction coefficient, 0.25 to 0.40

the number of pulleys), and v/L is the number of full revolutions of the belt per second. On this basis the service life of the belt is

$$H = N_n(S_n/\sigma_{max})^m/(3600\, zv/L) \tag{10.78}$$

In the above equations σ_{max} is supposed to be the range of the alternating stress. Since most experiments of that type to determine the endurance limit are not made on a regular material testing machine because the belt cannot be tested very easily in compression, belts are tested in an asymmetrical cycle, similar to the real operation in service. Therefore, unless detailed values are given for the endurance limit under similar conditions as for metals, the range σ_{max} in equation (10.78) must be taken equal to the maximum stress which appears on the belt, given by equation (10.57) The values in Table 10.4 for endurance limits are given on that basis. Finally, the pull factor and the slip coefficient have to be computed and checked against experience.

The V-belts can be designed exactly the same way except that in this case the size of the belt is selected and the number of belts is computed. To this end, we first note that the coefficient of friction for an equivalent flat belt, for a given normal force giving the same friction force as for flat belts (Figure 10.20), is given by equation (10.79).

$$f = f/\sin\theta \approx 3f \tag{10.79}$$

Dimensions of V-belts are given in manufacturers' catalogs.

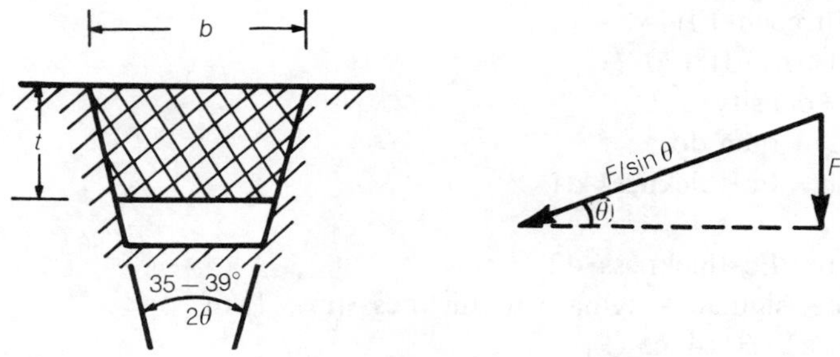

Figure 10.20 V-belt wedge effect

From that point on, the calculations proceed as for the case of flat belts with a modified coefficient of friction. Eventually instead of using equation (10.72), a similar equation can be used giving the number of V-belts needed to transmit the given power.

$$z = N_p C/N_0$$

Again C is the service factor and N_0 is the power transmitted for the given conditions by one V-belt.

The unit power transmitted N_0 is given usually in manufacturers' catalogs or in the absence of this can be computed by way of equation (10.57), in the form of equation (10.81) because here the section are A is known, with the values given in Table 10.4.

$$N_0 = (S_e/N - v^2\gamma/g - Et/d)Au(m-1)/m \tag{10.81}$$

If more than one V-belt is used, their rating has to be reduced (C_7 factor) because of the fact that after a certain operation the tensions of the belts are not equal and therefore there is no equal distribution of the load.

The procedures FLATBELTS and V-BELTS (programs are Appendices 10.A and 10.B, respectively) for the design of flat and V-belts, respectively, follow (explanations are given in italics):

```
Procedure FLATBELTS
local N, n1, distance, iratio, density
     Ec, thickness, Scall, frcoef, Snall, zeta: var
     Nn:= 1E7 :const
global bwidth, wratio, Hlife :var
begin
  input N, n1, distance, iratio, density, thickness, zeta
  input Scall, frcoef, Snall
                                                    (*Geometry*)
  d1:= 120*(N/n1)^(1/3)
  d2:= .985*d1*iratio
  alfa:= arctan(d2 – d1)/2/SQR(distance^2 – (d2 – d1)^2/4))
  FI:= 3.14 – 2*alfa
                                                    (*Strength*)
```

```
     m:= exp(frcoeff*FI)
     u:= 2*3.14*n1*d1/120
     sigmau:= density*u^2
     if iratio > 1 then do
        sigmab:= Ec*thickness*d1
     else
        sigmab:= Ec*thickness*d2
     No:= (Sc - sigmau - sigmab)*u*thickness*(m - 1)/m
     input c1, c2, c3, c4, c5
     c:= c1*c2*c3*c4*c5
     bwidth:= N*c/No
                                                   (*Endurance calculations*)
     U:= N/u
     Fo:= U*(m + 1)/2/(m - 1)
     F1:= Fo + U/2
     F2:= Fo - U/2
     sigmao:= Fo/(thickness*bwidth)
     sigman:= U/(thickness*bwidth)
     smax:= sigmao + sigman + sigmau + sigmab
     Hours:= Nn*(Sn/smax)^n/(3600*zeta*U*length)
     PULL:= U/(2*Fo)
     print bwidth, b/d1, PULL, Hours
  end
```

For optimization, the above procedure can be used with variable belt width and thickness. The objective function will consist of the belt cost plus the cost of the two pulleys. Most constraints have been used already in the design procedure to eliminate unknown quantities. Additional constraints will be:

(a) The service life Hours > Hmin
(b) The pull and thickness/width ratios must be within allowable limits.

A similar formulation can be applied for V-belts. Unknown here are the number of belts and the type of sections, both discrete quantities. They have to be used by way of an exhaustive search algorithm to determine the required optimum.

```
  Procedure VBELTS
  local N, n1, distance, iratio, density
       Ec, section, thickness, Scall, frcoef, Snall, zeta: var
       Nn:= 3.14159, PI:= 3.14159: const
  globalHlife, number: var
  begin
     input N, n1, distance, iratio, density, thickness, section, zeta
     input Scall, frcoef, Snall
                                                          (*Geometry*)
```

```
    d1:= 120*(N/n1)^(1/3)
    d2:= .985*d1*iratio
    alfa:= arctan(d2 - d1)/2/SQR(distance^2 - (d2 - d1)^2/4))
    FI:= 3.14 - 2*alfa
                                                    (*Strength*)
    m:= exp(frcoeff*FI)
    u:= 2*3.14*n1*d1/120
    sigmau:= density*u^2
    if iratio > 1 then do
      sigmab:= Ec*thickness*d1
    else
      sigmab:= Ec*thickness*d2
    No:= (Sc - sigmau - sigmab)*u*thickness*(m - 1)/m
    input c1, c2, c3, c4, c5, c6, c7
    c:= c1*c2*c3*c4*c5*c6*c7
    number:= N*c/No
                                        (*Endurance calculations*)
    U:= N/u
    Fo:= U*(m + 1)/2/(m - 1)
    F1:= Fo + U/2
    F2:= Fo - U/2
    sigmao:= Fo/section
    sigman:= U/section
    smax:= sigmao + sigman + sigmau + sigmab
    Hours:= Nn*(Sn/smax)^n/(3600*zeta*U*length)
    PULL:= U/(2*Fo)
    print number, Hours, PULL
  end
```

EXAMPLES

Example 10.1

A disk brake has dimensions $d_1 = 30$ mm, $d_2 = 250$ mm and the brake pads operate at radius $a = 180$ mm. The disk is made of carbon steel. The coefficient of friction was measured $f = 0.3$. The brake force applied by a hydraulic cylinder on the pads is 1000 N. Find the maximum interface temperature, assuming that practically all the heat goes into the disk, because of the low thermal conductivity of the asbestos pad, and that at the surface of the disk the convection heat transfer coefficient is 40 W/m^2 °C h and the speed of rotation is 100 rpm.

Solution

For carbon steel, thermal conductivity $k = 190$ kj/m h°C = 52.7 j/m s°C. Therefore, from Table 10.1, $m = (2h/kl)^{\frac{1}{2}} = (2 \times 40/52.7 \times 0.09)^{\frac{1}{2}} = 4.1$. From tables of Bessel Functions (see Spiegel, M., 1968. *Mathematical Handbook*, Shaum) we obtain for the disk with side friction, Table 10.1.

$$H = 2a_2 m(X + Y) = 2\pi \times 0.125 \times 4.1 \times 0.23 = 0.74$$

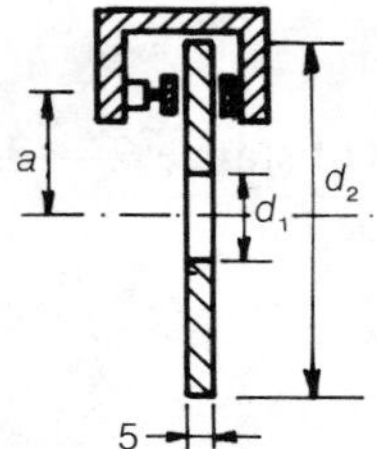

Figure E10.1

From equation (10.15) if all the heat is entering the disk,

$$T_0 = Q/lkH = fP\omega a/akH = fP2\pi n/60kH$$
$$= 0.3 \times 1000 \times 2\pi \times 100/60 \times 52.7 \times 0.74 = 80°\text{C}$$

Example 10.2

The collector rings of an electric generator are made of hard steel and the brushes of 70/30 brass. The brushes, of section 40 × 40 mm, are pressed against the rings by a force of 10 N and have a length of 40 mm. The useful length is 30 mm. The yield strength of the brass is $S_y = 80\ \text{N/mm}^2$. Determine the life of the brushes, in hours, if the machine rotates at 3600 rpm.

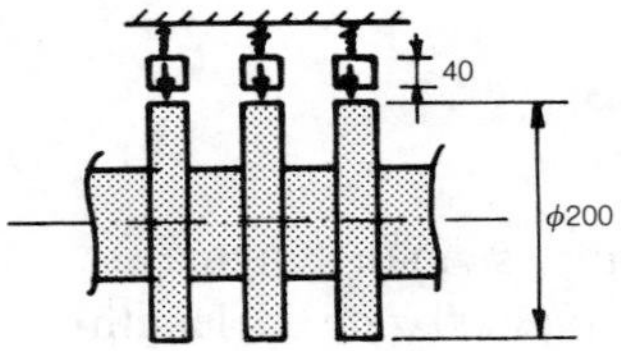

Figure E10.2

Solution

From Table 10.2, the wear coefficient is $k = 1.7 \times 10^{-4}$. Equation (10.16) yields the sliding distance

$$L = VS_y/kN = 40 \times 40 \times 30 \times 80/1.7 \times 10^{-4} \times 10 = 2.25 \times 10^9\ \text{mm}$$
$$t = (L/\pi d)/(n/60) = (2.25 \times 10^9/\pi \times 200)/(3600/60)$$
$$= 59{,}713\ \text{sec}$$
$$= 16.5\ \text{hours.}$$

Example 10.3

The band brake for a construction lifting crane is hand operated by a lever arm as shown in Figure 10.5(a). The operator force is 300 N and the lever has $l_2 = 100$ mm, $l_1 = 400$ mm. The band is perpendicular to the lever at the position shown and the contact arc is $\theta = 120°$. The diameter of the drum is 250 mm, lined with asbestos friction material. Determine:

(a) The braking moment for both directions of rotation.
(b) The steel bandwidth.

Solution

From Table 10.3, for asbestos lining on steel, the coefficient of friction is 0.3 and the allowable pressure is $0.3\ \text{N/mm}^2$.

(a) *For clockwise rotation*, the force P_2 is, from the lever arm equilibrium

$$P_2 = Fl_1/l_2 = 300 \times 400/100 = 1200\,\text{N}$$

Equation (10.32) yields the braking torque

$$T = P_2R(e^{f\theta} - 1) = 1200 \times 0.125[\exp(0.3 \times 2\pi/3) - 1] = 131\,\text{Nm}$$

For counterclockwise rotation,

$$T = P_2R(e^{f\theta} - 1) = 1200 \times 0.125[\exp(-0.3 \times 2\pi/3) - 1] = 70\,\text{Nm}$$

(b) *For clockwise rotation*, $P = 1200$ N and

$$P_1 = P_2e^f = 1200\exp(0.3 \times 2/3) = 2249\,\text{N}$$

The maximum pressure is given by equation (10.25). The bandwidth is then,

$$b = P_1/p_{\max}R = 2249/0.3 \times 125 = 60\,\text{mm}$$

For counterclockwise rotation

$$b = P_2/p_{\max}R = 1200/0.3 \times 125 = 33\,\text{mm}$$

because $P_1 < P_2$.

Example 10.4

A single disk clutch for an automobile transmits maximum torque at 4100 rpm where it delivers 92 kW power. The two friction surfaces are lined with asbestos friction material rubbing against the cast iron flywheel. The external diameter of the disk lining is 300 mm and the internal is 100 mm. Determine:

(a) the factor of safety against clutch slipping;
(b) the required actuating force.

Solution

From Table 10.3, $f = 0.3$ and maximum allowable pressure is 0.3 N/mm². The maximum torque is

$$T_{\max} = 9550P/n = 9{,}550 \times 92/4100 = 214\,\text{Nm}$$

The area (A) of the lining is

$$A = \pi(d_1^2 - d_2^2)/4 = \pi(300^2 - 200^2)/4 = 39{,}250\,\text{mm}^2$$

The actuating force is

$$F = Ap_{\max} = 39{,}250 \times 0.3 = 11{,}775\,\text{N}$$

The maximum capacity of transmitting torque is

$$T_c = FR_mif = 35{,}325 \times (400 + 100)/2 \times 2 \times 0.3 = 1766.25\,\text{Nm}$$

The margin of safety against slipping

$$N_s = T_c/T_{\max} = 1766.25/214 = 8.25$$

Example 10.5

For safety reasons against control malfunction, a typical motor drive for an overhead crane consists of a conical rotor pressed axially against a conical brake by a helical spring. When electrical voltage is applied, the rotor moves to the left and the brake is released, and rotation starts. The motor is designed for a maximum torque of 400 Nm. Design the conical brake if the

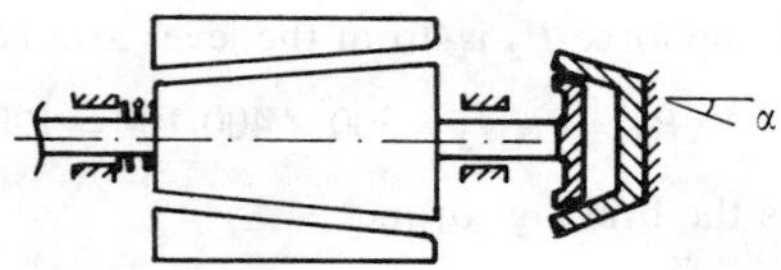

Figure E10.5

friction surfaces are asbestos lining against cast iron. The inner diameter of the cone must be at least 100 mm, the cone angle must be $a = 10°$ with a safety factor $N = 2$ against slipping.

Solution

From Table 10.3, $f = 0.3$ and $p_{max} = 0.3\ \text{N/mm}^2$. If b is the unknown width of the friction surface, the outer diameter d_1 is (if $d_2 = 150$ mm, the inner diameter),

$$d_1 = d_2 + b \tan a$$

and the mean radius

$$R_m = (d_1 + d_2)/2 = d_2 + b \tan a/2$$

Equation (10.20) yields

$$N_s T = 2R_m^2 b p_{max} f = 2(d_2 + b \tan a/2)^2 b p_{max}$$

$$(d_2 + b \tan a/2)^L b = N_s T / 2\pi P_{max}$$

$$(100 + b \tan 10°/2)^2 b = 2 \times 400{,}000/2 \times 0.3$$

Integrating on

$$b = 2 \times 400{,}000[2\pi \times 0.3(100 + b \tan 10°/2)^2]^{-1}$$

we find $b = 40$ mm.

The required axial force that the helical springs must provide will be

$$F = \frac{TN_s}{R_m} \tan a = \frac{400{,}000 \times 2}{100 + 40 \tan a/2} \sin 10° = 1362\ \text{N}$$

Example 10.6

A drum brake is shown in Figure E10.6 with two centrally pivoted brake pads. The pad angle $\theta = \theta_2 - \theta_1 = \pi/2$, the diameter of the drum is $d = 220$ mm, and the lever geometry is determined by $l_2 = 200$, $l_1 = 200$, $e = 60$ mm. Asbestos friction material will be used in a cast iron drum. Determine, for maximum braking torque 180 Nm:

(a) The required actuating force F.
(b) The width of the pads.

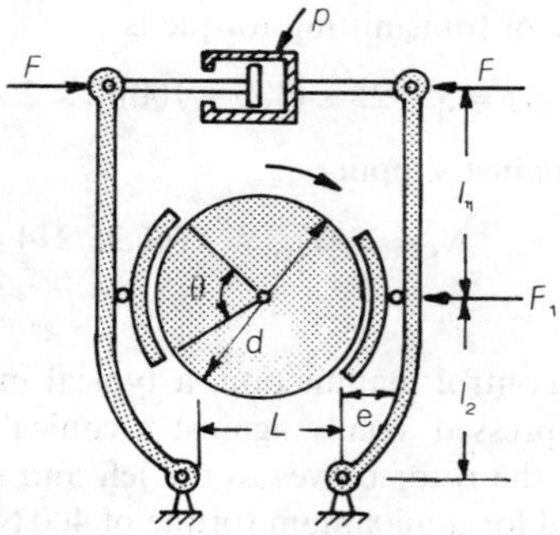

Figure E10.6

Solution
For asbestos on cast iron, $f = 0.3$, $p_a = 0.3\ \text{N/mm}^2$ from Table 10.3. It is $-\theta_1 = \theta_2 = \pi/4$, therefore $\theta_a = \theta_2 = \pi/4$. Equation (10.32) gives, for two pads,

$$b = \frac{-T\sin\theta_a/2}{fp_aR^2(\cos\theta_1 - \cos\theta_2)} = \frac{180{,}000(\sin\theta/4)/2}{0.3\times 0.3\times 110^2(2\cos\pi/2)} = 41\ \text{mm}$$

The normal and friction forces have horizontal elements $dN\cos\theta$ and $f dN\sin\theta$, respectively. Using equation (10.21) and integrating, yields the total pivot force

$$F_1 = \int_{\theta_1}^{\theta_2}\cos\theta\,dN + \int_{\theta_1}^{\theta_2} f\sin\theta\,dN = \frac{p_abR}{\sin\theta_a}(\sin\theta\cos\theta + f\sin^2\theta)\,d\theta$$

$$= \frac{p_abR}{\sin\theta_a}[-\cos 2\theta_1/2 + f(\theta/2 - \sin 2\theta/4)]|_1^2$$

$$F_1 = \frac{P_abR}{\sin\theta_a}[(\cos 2\theta_1 - \cos 2\theta_2)/2 + f(\theta_2/2 - \theta_1/2 - (\sin 2\theta_2)/4 + (\sin 2\theta_1)/4)]$$

Because $-\theta_1 = \theta_2 = \pi/4$,

$$F_1 = \frac{P_abR}{\sin\theta_a}\left[f\left(\frac{\pi}{4} - \frac{1}{2}\right)\right] = \frac{0.3\times 41\times 110\times 0.3}{\sin\frac{\pi}{4}}\left(\frac{\pi}{4} - \frac{1}{2}\right) = 164\ \text{N}$$

$$F = F_1l_2/(l_1 + l_2) = 164\times 200/400 = 82\ \text{N}.$$

Example 10.7
A 30 kW, 1450 rpm electric motor drives an ammonia compressor at 500 rpm. Design a flat belt for this transmission if the distance of the two pulleys is 1500 mm.

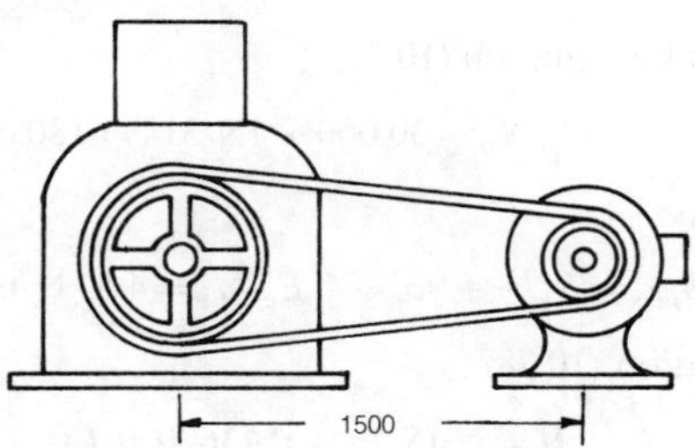

Figure E10.7

Solution
A plastic belt is selected with (Table 10.4): $E_c = 60\ \text{N/mm}^2$, $\gamma = 11{,}000\ \text{N/m}^3$, $S_c = 6.1\ \text{N/mm}^2$.
The motor pulley diameter d is selected with equation (10.70):

$$d_1 = 1200(N/n)^{1/3} = 1200(30/1450)^{1/3} = 330\ \text{mm}$$

The driven pulley diameter by equation (10.71):

$$d_2 \approx 0.985d_1(n_1/n_2) = 0.985\times 330\times(1450/500) = 940\ \text{mm}.$$

The peripheral velocity

$$v = (2\pi n_1/60)R_1 = 2\pi\times 1450\times 0.165/60 = 25\ \text{m/sec} < v_{max}.$$

The contact angle

$$\cos\frac{a_1}{2} = \frac{d_2 - d_1}{2a} = \frac{940 - 330}{2 \times 1500}, \quad a_1 = 1560 = 2.74\,\text{rad.}$$

From Table 10.5, the coefficient of friction $f = 0.3$.
The belt tension forces

$$m = S_1/S_2 = \exp(fa_1) = \exp(0.3 \times 2.74) = 2.27$$
$$S_1 - S_2 = u = N/v = 30{,}000/25 = 1200\,\text{N.}$$

Therefore,

$$S_2 = U/(m - 1) = 1200/(2.27 - 1) = 945\,\text{N.}$$
$$S_1 = mS_2 = 2.27 \times 945 = 2145\,\text{N}$$
$$S_0 = U(m + 1)/2(m - 1) = 1200 \times 3.27/2 \times 1.27 = 1544\,\text{N}$$

Select belt thickness 5 mm to compute the width b. Specific belt capacity, equation (10.73), for $b = 1$ mm

$$N_0 = (S_G - \sigma_u - \sigma_b)t(m - 10)/\text{m}$$
$$= [S_G - (\gamma v^2/g) - (Et/d_1)]t(m - 1)/m$$

For unit width b,

$$N_0 = [6.1 - (11E - 6 \times 25{,}000^2/9810) - (10 \times 5/330)]25 \times 5(2.27 - 1)/2.27 = 314\,\text{W/mm}$$

The service factor

$$C = C_1C_2C_3C_4C_5$$

where, from Table 10.5, $C_1 = 1.5$ (reciprocating machine), $C_2 = 1$ (normal environment), $C_3 = 1$, $C_4 = 1.06(\varphi = 156°)$, $C_5 = 1.2$ (tension by adjustment),

$$C = 1.5 \times 1 \times 1.06 \times 1.2 = 1.9$$

Therefore the belt width is by equation (10.72)

$$b = NC/N_0 = 30{,}000 \times 1.9/314 = 180\,\text{mm}$$

The maximum stress is

$$\sigma_{max} = S_1/A + \gamma v^2/g + E_c t/d_1 = 4.53\,\text{N/mm}^2$$

Service life, from equation (10.78):

$$H = N_n(S_n/\sigma_{max})^m/(3600zu/L)$$

$$\text{Length } L = 2a4\pi(d_1 + d_2)/2 + (d_2 - d_1)^2/4a$$
$$= 5056\,\text{mm.}$$

Number of bends per full pass, $z = 2$

$$H = 10^7(6/4.53)^6/(3600 \times 2 \times 25{,}000/5056)$$
$$= 1516\text{h}$$

Loads on the shaft:

$$R = 2S_0 \sin a/2 = 2 \times 1544 \times \sin 78°$$
$$= 3020\,\text{N.}$$

Example 10.8

Design a V-belt for the auxiliary drive system of an automobile engine. The crankshaft pulley 1 drives the alternator 2, water pump 3 and air-conditioning compressor 4, rated at $n_1 = 4000$ rpm, $n_2 = 5000$ rpm, $N_2 = 0.5$ kW, $n_3 = 3500$ rpm, $N_3 = 0.4$ kW, $n_4 = 4000$ rpm, $N_4 = 1$ kW.

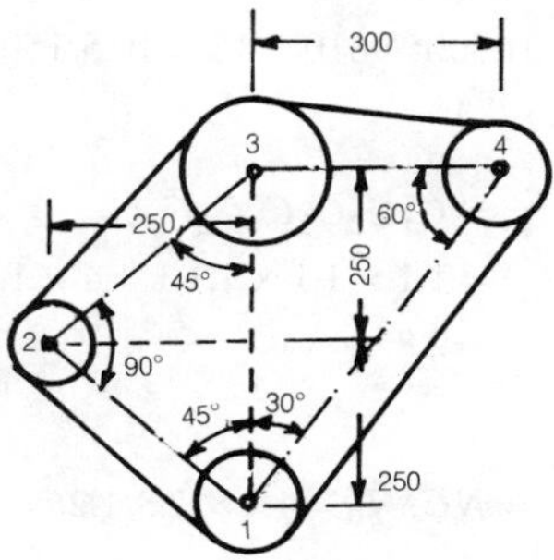

Figure E10.8

Solution

For V-belts, the coefficient of friction is $f = 0.30$. The wedge angle is 40°, therefore the apparent coefficient of friction is

$$f = f'/\sin(\alpha/2) = 0.3/\sin 20° = 0.877$$

Taking approximately the contact angles equal to the angles of the lines of centers indicated in Figure E10.8,

$$m_1 = \exp(f_1\varphi_1) = 3.15$$

The total power transmitted by the driving pulley

$$N = N_2 + N_3 + N_4 = 0.5 + 0.4 + 1 = 1.9\,\text{kW}$$

Selection of driving pulley diameter

$$d_1 = 1200(P_1/n_1)^{1/3} = 1200(3.3/4000)^{1/3} = 112\,\text{mm}.$$

$$d_2 = 0.985d_1(n_1/n_2) = 0.985 \times 112(4000/5000) = 88\,\text{mm}$$

$$d_3 = 0.985d_1(n_1/n_3) = 0.985 \times 112(4000/3500) = 126\,\text{mm}$$

$$d_4 = 0.985d_1(n_1/n_4) = 0.985 \times 112(4000/4000) = 110\,\text{mm}.$$

The peripheral velocity is

$$(2\pi n_1/60)R_1 = (2\pi \times 4000/60) \times 56/1000 = 23.5\,\text{m/sec}.$$

Belt forces for pulley 1,

$$S_1/S_2 = m = 3.15$$

$$S_1 - S_2 = U = P/v = 3300/23.5 = 140\,\text{N}$$

$$S_2 = U/(m-1) = 140/(3.15-1) = 65\,\text{N}$$

$$S_1 = mS_2 = 3.15 \times 65 = 205\,\text{N}$$

$$S_0 = (S_1 + S_2)/2 = 135\,\text{N}$$

Section A is selected with $b = 13\,\text{mm}$, $t = 8\,\text{mm}$, $A_u = 81\,\text{mm}^2$. Specific belt capacity (equation 10.73)

$$N_0 = (S_c - \sigma_u - \sigma_b)A_u(m-1)/m = (S_c - \gamma v^2/g - E_c t/d)vt(m - 10/m)$$

For V-belts, $S_c = 2.3\,\text{N/mm}^2$, $E_c = 20\,\text{N/mm}^2$, $\gamma = 12{,}000\,\text{N/m}^3$. Therefore,

$$N_0 = (2.8 - 12{,}000 \times 10^{-9} \times 31{,}500^2/9810 - 18 \times 10.5/150)31.5 \times 138 \times 2.15/3.15 = 2740\,\text{W}$$

The service factor

$$\begin{aligned} C &= C_1C_2C_3C_4C_5C_6C_7 \\ &= 1.1 \times 1.1 \times 1.1 \times 1.4 \times 1.2 \times 1.25 \\ &= 2.8 \end{aligned}$$

The number of belts required

$$j = NC/N_0 = 1.9 \times 2.8/1.26 = 4.22$$

The number of belts is not acceptable. There are now two alternatives: To use steel cord V-belts or heavier section. Let us try the second.

B-section is selected with $E_c = 18\,\text{N/mm}^2$, $S_c = 2.8\,\text{N/mm}^2$, $A = 138\,\text{mm}^2$, $t = 10.5\,\text{mm}$. Therefore, increasing d_1 to 150 mm,

$$N_0 = (2.8 - 12{,}000 \times 10^{-9} \times 31{,}500^2/9810 - 18 \times 10.5/150)31.5 \times 138 \times 2.15/3.15 = 2740\,\text{W}$$

Therefore $j = NC/N = 3.3 \times 2.8/2.7 = 1.92$ belts. Two belts are selected. The diameters now become,

$$d_1 = 150\,\text{mm}, \quad d_2 = 118, \quad d_3 = 168, \quad d_4 = 148\,\text{mm}.$$

The service life will be computed from equation (10.78). The maximum stress (from bending stresses) is at pulley 2. Therefore,

$$\sigma_{max} = S_1/A + \gamma v^2/g + E_c t/d_1 = 3.95\,\text{N/mm}^2$$

Number of bends per full pass is 4. Therefore,

$$H = 10^7(6/3.95)^6/(3600 \times 4 \times 31{,}500/1600) = 433\,\text{h}.$$

REFERENCES AND FURTHER READING

Dobrovolsky, V. *et al.*, 1975. *Machine Elements.* Moscow: Mir Publishers.

Dimarogonas, A. D., Michalopoulos, D., 1981. 'A compilation of heat distribution coefficients at sliding contacts'. *Tribology*, August 1981.

Halling, J., 1976. *Introduction to Tribology.* London: Wykeham Publishers.

Kragelskii, I. V., 1965. *Friction and Wear.* London: Butterworths.

Machine Design, Mechanical Drives Reference Issue. 1981. Penton Publ. Vol. 53, No 6.

Niemann, G., 1965. *Maschinenelemente.* Berlin: Springer Verlag.

Oliver, L. R., Johnson, C. O., Breig, W. F., 1976. 'V-belt life prediction and power rating'. *Trans. ASME J. Eng. for Industry*, p. 340.

Shigley, J. E., 1972. *Mechanical Engineering Design.* New York: McGraw-Hill.

Spiegel, M., 1968. *Mathematical Handbook*, New York: Shaum Outline Series, McGraw-Hill.

Ummen, A. (ed.), 1965. *Keilriemen.* Essen: Verlag Ernst Heyer.

PROBLEMS

10.1 For the disk of Example 10.1, find the average temperature in the disk, for a short application of the brake to bring the car from a 60 mph speed to complete stop. The mass of the car is 1200 kg and the heat convection over the surface of the disk can be neglected. All heat can be assumed to enter the disk because of the low thermal conductivity of the lining material.

10.2 The drum of Example 10.6 is shown in Figure P10.2. Determine the drum temperature, assumed uniform, if the brake is applied for 15 rev. Neglect heat loses of the drum and assume all heat enters the drum.

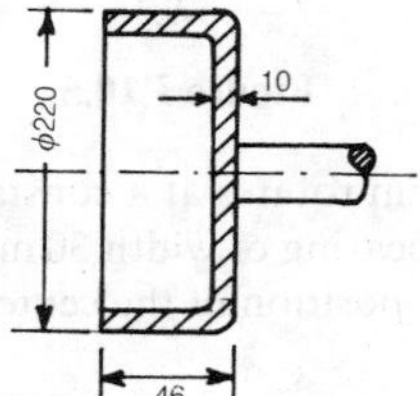

Figure P10.2

10.3 A cam and follower mechanism consists of a nearly circular cam 1 of diameter $d_1 = 120$ mm and a follower disk 2 of diameter $d_2 = 40$ mm, pressed against the cam by a force $F = 100$ N by a lever and a helical spring. The width of the disks is 15 mm and the coefficient of rolling friction was found by measurement to be $f = 0.05$. Assuming convection heat-transfer coefficient $h = 50\ \mathrm{W/m^2\ {}^\circ C\ h}$, determine the interface temperature of the rolling surfaces, if the cam rotates at 3000 rpm.

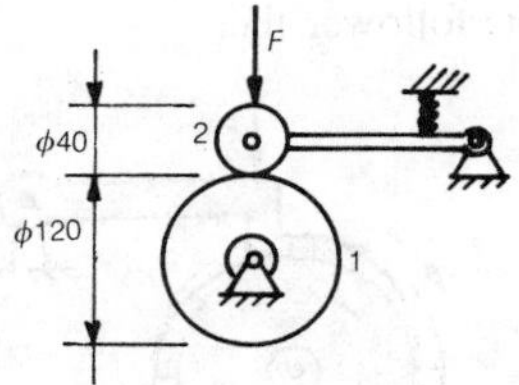

Figure P10.3

10.4 In a friction welding operation of two circular carbon steel shafts of diameter $d = 100$ mm, one piece is stationary while the other revolves with a constant speed n. At the same time, an axial force F is applied. If the coefficient of friction is $f = 0.35$, determine the required speed of rotation to reach a temperature of 1300 °C required. The axial force is $F = 200$ kN and the surface heat-transfer coefficient for the stationary shaft is $30\ \mathrm{W/m^2\ {}^\circ C\ h}$ and for the rotating one $60\ \mathrm{W/m^2\ {}^\circ C\ h}$. Assume infinite length.

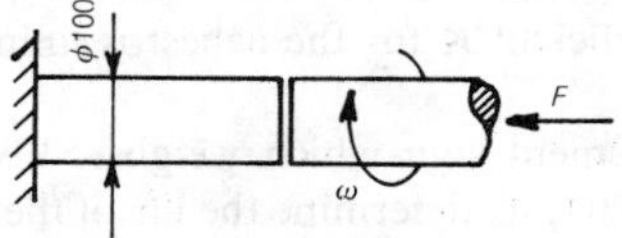

Figure P10.4

10.5 Solve Problem 10.4 for the friction welding of two carbon steel disks, one circular the other annular with a slight taper of angle $a = 5°$. The axial force is 10 kN and the heat transfer coefficient is as in Problem 10.4.

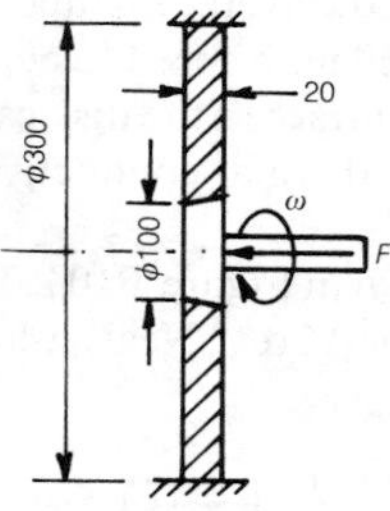

Figure P10.5

10.6 A steel shaft of diameter $d = 40$ mm rotates at a constant speed of 125 rpm loaded by a 1500-N force. It is supported by a teflon bearing of width 30 mm. Determine after 20 hours of operation the bearing wear and the vertical position of the center of the shaft in respect to the initial one.

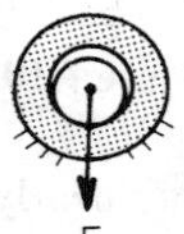

Figure P10.6

10.7 An indexing mechanism consists of a toothed wheel of diameter $d = 60$ mm of hard steel and a follower of 60/40 brass pressed against the wheel by a force $F = 180$ N. The sliding section of the follower is 20 × 20 mm and the operation is without lubrication. If the limit of wear for the follower is 1 mm, determine the follower life.

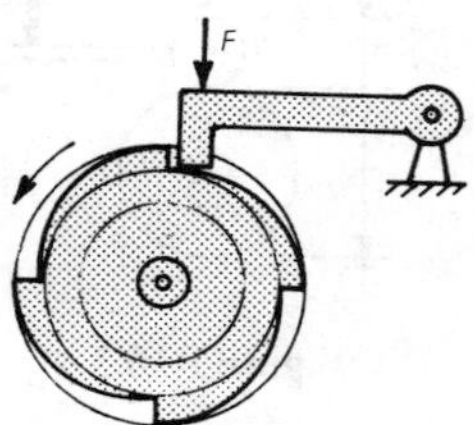

Figure P10.7

10.8 A disk brake such as the one of Example 10.1 was tested for wear. The asbestos pads had friction area 2000 mm^2 and they were pressed against the rotating disk for 10 sec every 1 min. The speed of rotation was 4000 rpm and the braking force was 1000 N. After 100 hours of testing, the wear of the pads was 3 mm. The yield strength of the asbestos lining was measured $S_y = 2$ N/mm^2 by a static test. Determine the coefficient K for the asbestos lining.

10.9 A revolving pole of advertisement sign which weighs 5 kN is supported by a polyethylene bearing. If the pole rotates at 20 rpm, determine the life of the bearing if the linear wear allowed, by geometry, is 10 mm.

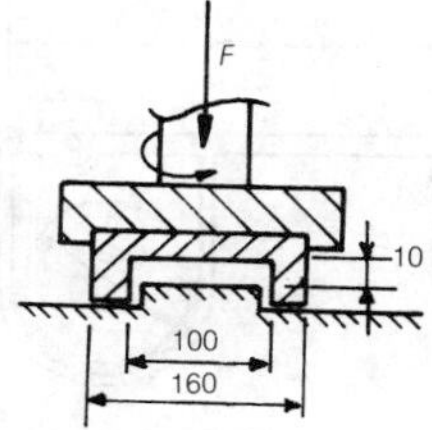

Figure P10.9

10.10 A differential expansion detector for a steam turbine rotor consists of a brass 70/30 slider and a helical spring prestressed to a force $F = 150$ N applied on the slider. When the rotor expands to the right, it contacts the slider and the 150-N force is then applied between rotor and slider. The sliding surface is $10 \times 10\,\text{mm}^2$. During an overhaul, the slider was inspected and 8-mm wear was measured. Determine the total duration of the differential expansions of the rotor, revolving at 3600 rpm. If between overhauls, the turbine made 150 start-ups, determine the average duration of the differential expansion, given that such expansion is expected during every start-up.

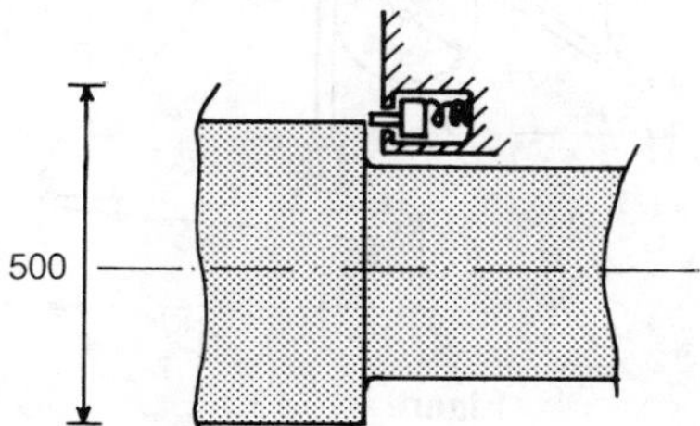

Figure P10.10

10.11 On a differential band brake, the force applied is $P = 300$ N and the braking torque is $T =$ 1000 Nm. The drum diameter is 280 mm and the contact angle 230°. The one end of the band is hinged at a distance of 60 mm from the lever pivot as shown. Determine the proper position for the hinge of the other end to support the braking torque. Both ends of the band are perpendicular to the lever. The lining is asbestos friction material and the drum is made of cast iron. Then, design the band of carbon steel SAE 1025 with a safety factor $N = 3$.

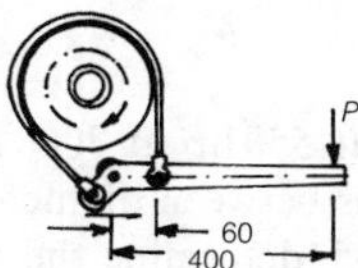

Figure P10.11

10.12 The band brake shown, is activated by force $F = 250$ N. The lining is asbestos friction material and the drum is cast iron. Determine the braking torque and the lining width for operation of the brake with both directions of rotation.

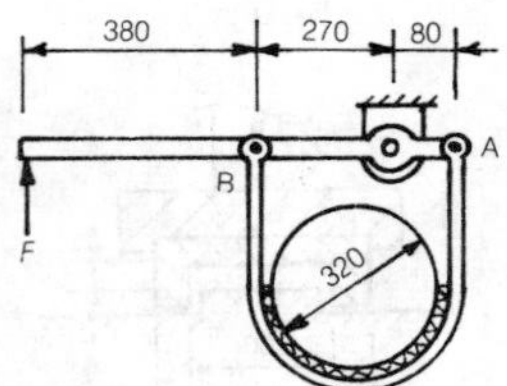

Figure P10.12

10.13 A double drum hoisting machine has a band brake with common band as shown. If the braking torque required for each drum is 800 Nm, determine the required activating force P and the width of the asbestos lining. The drums are made of cast iron.

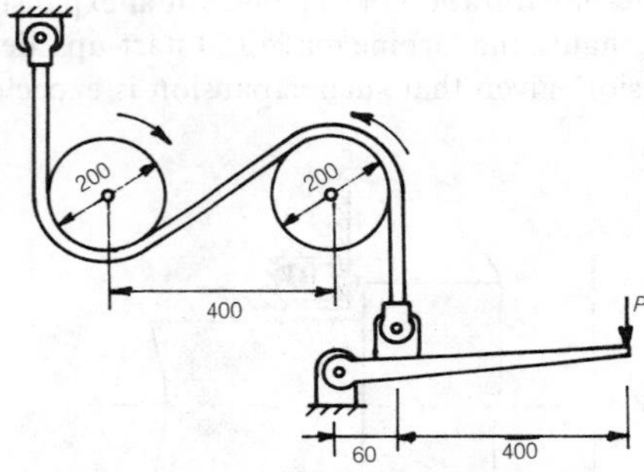

Figure P10.13

10.14 For the brake of Problem 10.12, determine the braking torque for both directions of rotation if the left end of the band is hinged with the other end at point B.

10.15 For the system of the Problem 10.13, determine the activating force if the direction of the drum on the left is reversed. Determine also the tension of each part of the band.

10.16 A single disk automotive clutch is designed for 80 kW rated power at 3600 rpm. Both sides of the disk have lining material with minimum diameter 140 mm. The lining is asbestos friction material. For a factor of safety against slipping $N = 3$, determine the external lining diameter and the actuating force.

10.17 An engine for a bulldozer is rated at 550 hp at 3000 rpm. It uses a multiple disk clutch with asbestos lining. The friction surface is between diameters $d_1 = 200$ mm and $d_2 = 300$ mm. If the factor of safety against slipping is 2.5, determine the required number of clutch disks.

10.18 An automobile weighs 1100 kg and the four brakes must bring it to a complete stop from a speed of 90 mph at a 60-m distance. The braking force is distributed 60% in the front and 40% in the rear wheels and it is assumed constant during braking.

Design a disk brake for the front wheels with asbestos lining with one disk and two pads. The distance of the area center of pads from the axis of rotation should be approximately 120 mm while the wheel diameter is 500 mm.

10.19 In Problem 10.18, design the rear wheel brakes as drum brakes with two 120° pads each and internal drum diameter 360 mm. The pads are forced at both ends by hydraulic cylinders. (Shigley 1972.)

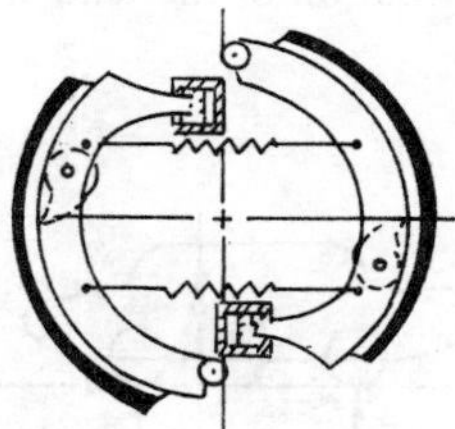

Figure P10.19

10.20 Solve Problem 10.19 for pad activation with one cylinder and the other pad end hinged to the opposite one. (Shigley 1972.)

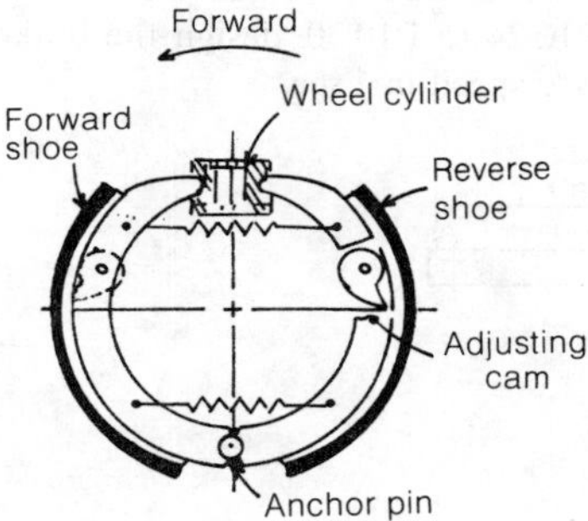

Figure P10.20

10.21 The inflated rubber clutch shown in Figure 10.7 is rated at 40 kW, 3600 rpm. The coefficient of friction between rubber and steel drum is $f = 0.35$ and the shear strength of the rubber is 6 N/mm^2. The available air pressure is 1 N/mm^2. Determine the main dimensions of the clutch trying to keep the ratio b/R between 0.3 and 0.8.

10.22 A centrifugal clutch consists of four pads free to move outwards against the inner cylindrical surface of the drum. Coupling is achieved by frictional forces between pads and drum. The pads are forced to rotate with the inner rotor.

Such a clutch was rated at 10 hp, 1000 rpm. The pads had asbestos lining. The inner diameter of the drum was 300 mm. Determine the size and mass of the pads. Find pad size.

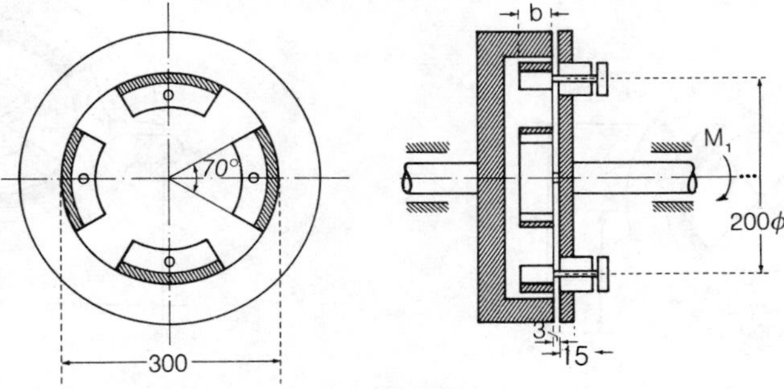

Figure P10.22

10.23 An automobile has a 1200 kg mass with a center of mass as shown, with $l_1 = 1000$, $l_2 = 1200$, $h = 900$ mm. When braking, without slipping, frictional forces develop, equal to deceleration × mass. Determine the distribution of the load G to the front and rear wheels,

(a) at constant speed,
(b) at deceleration 0.6 g.

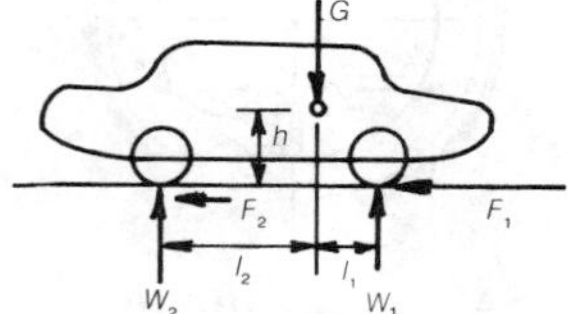

Figure P10.23

10.24 to 10.29

On a hoisting crane, a load of 1 ton is lowered with a cable on a 200-mm diameter drum. For the brakes shown in Figures P10.24 to P10.30, design the brake so that the load will come to a complete stop from a 1 m/sec speed in 1 sec.

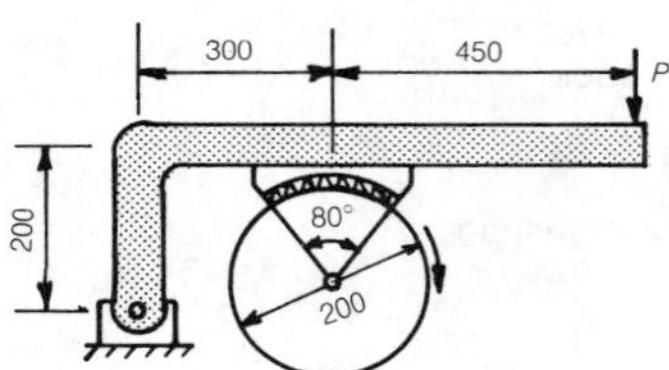

Figure P10.24

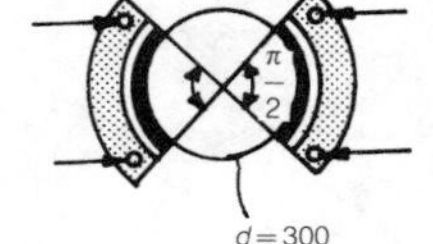

Figure P10.25

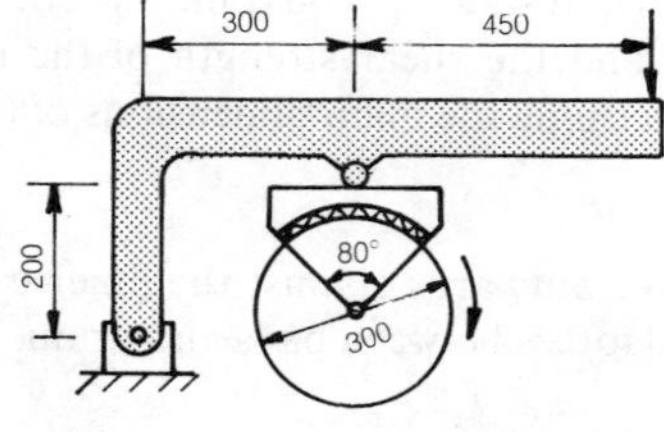

Figure P10.26

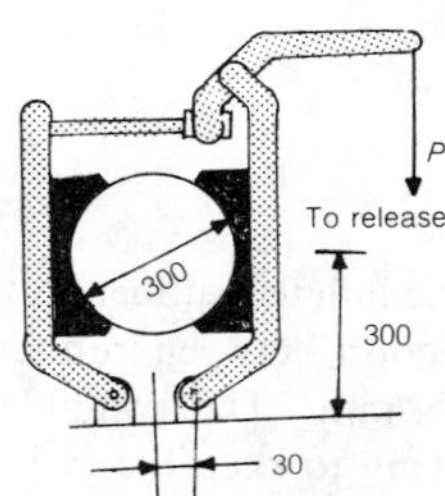

Figure P10.27

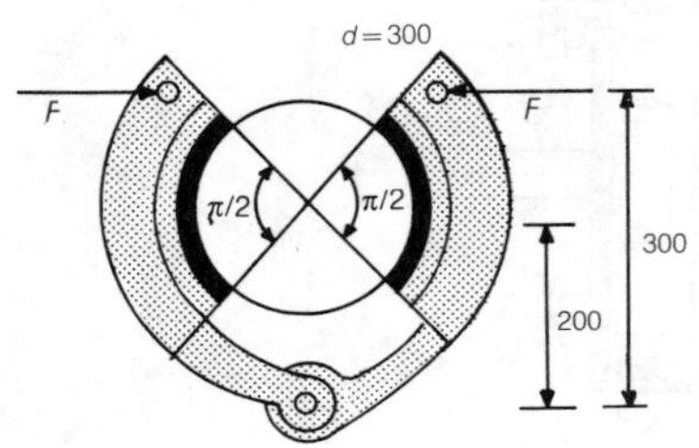

Figure P10.28

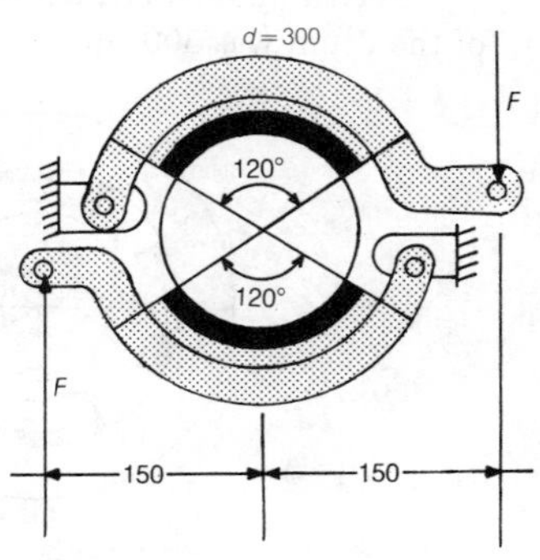

Figure P10.29

10.30 The main shaft of a milling machine is powered by an electric motor at $N = 20\,\text{kW}$ at 1750 rpm, with a 160-mm pulley while the driven pulley has $d_2 = 300\,\text{mm}$. The distance of the two shafts is 1400 mm. The selected leather belt is 3 mm-thick and 160-mm wide. Check the belt for strength and determine the fatigue life.

10.31 In Problem 10.30, design a proper rubber belt and determine the fatigue life.

10.32 A 10-hp, 750-rpm motor drives a mixing machine at 150 rpm. A plastic flat belt is used. Determine:

(a) The belt thickness, width and the pulley diameters.
(b) The service life.
(c) The proper tightening weight.

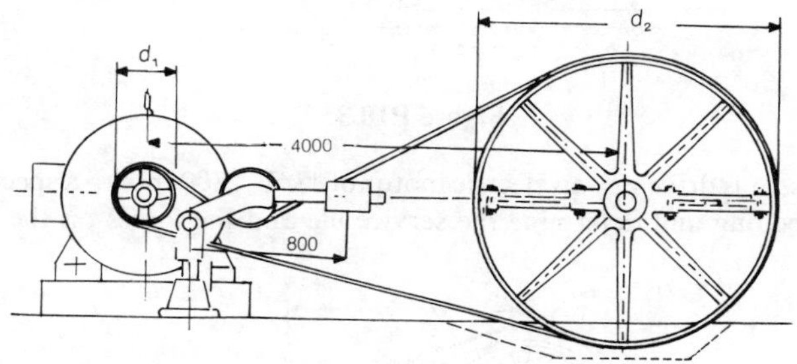

Figure P10.32

10.33 A 10-kW motor drives a press with 350 rpm while the diameter of the motor pulley and the tightening pulley is 125 mm and the driven pulley 1000 mm. The tightening weight $G = 60\,\text{N}$. A

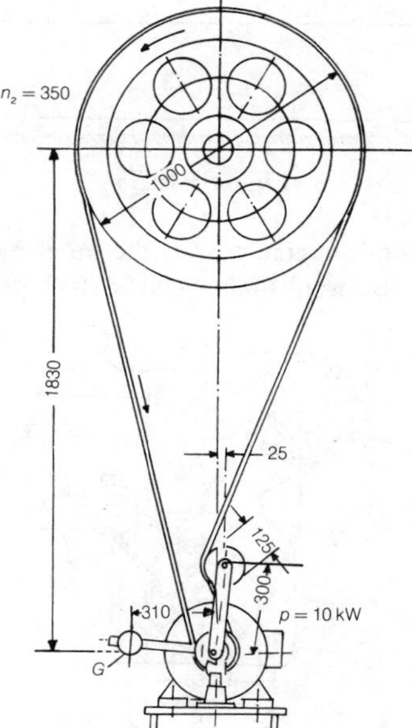

Figure P10.33

rubber belt is selected of 4-mm thickness and 140-mm width. Check the strength of the belt, and determine the fatigue life and the forces on the shaft owing to the belt tension.

10.34 In Problem 10.33, design a plastic belt and determine the service life.

10.35 A paper machine is driven by an 8-kW, 1700 rpm motor at equal speed. Design the proper V-belting and determine the service life.

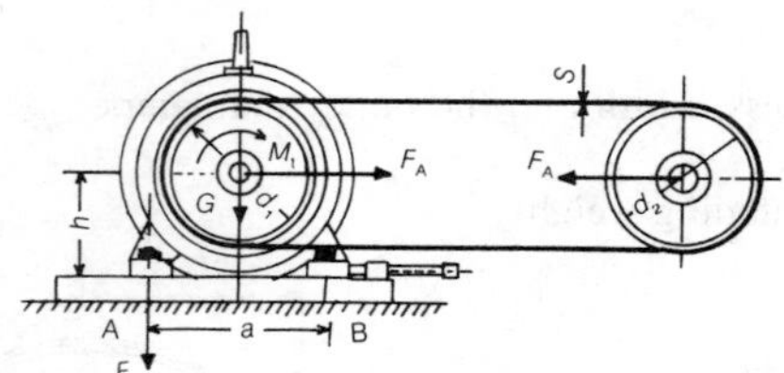

Figure P10.35

10.36 An air compressor is driven by an electric motor of 3 kW, 2800 rpm to a speed of 710 rpm. Design the proper V-belting and determine the service life and the forces on the shafts.

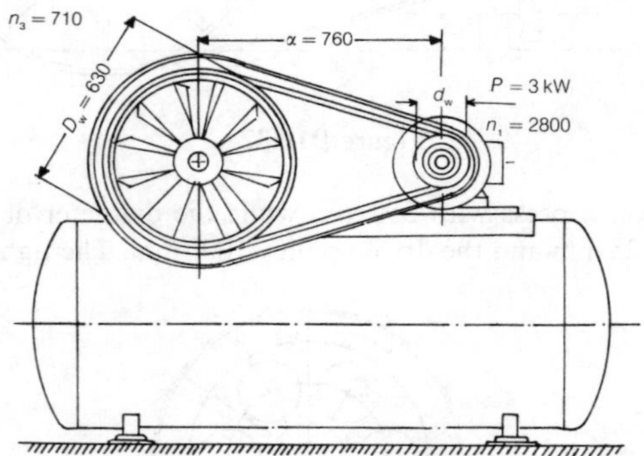

Figure P10.36

10.37 The crankshaft of an automobile engine powers the water pump and ventilator of 6 hp and the electric generator of 0.3 hp. The crankshaft speed is 1000 rpm. Determine the V-belts required.

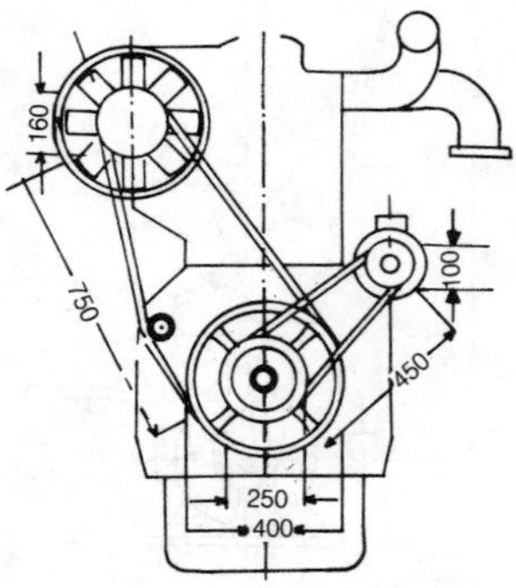

Figure P10.37

APPENDIX 10.A
FLATBELTS: DESIGN OF FLAT FRICTION BELTS PROGRAM

```
10 REM ************************************
30 REM *                                  *
40 REM *             FLATBELTS            *
50 REM *                                  *
60 REM ************************************
70 REM            COPYRIGHT 1987
80 REM by Professor Andrew D. Dimarogonas, Washington Univ., St. Louis, Mo.
90 REM  All rights reserved.  Unauthorized reproduction, disssemination,
100 REM selling or use is strictly prohibited.
110 REM  This listing is for reference purpose only.
120 REM
130 REM
140 REM DESIGN OF FLATBELTS
150 REM
160 REM
170 REM          ADD / 6-2-1987
180 REM
190 CLS:KEY OFF
200 LOCATE 25,1:PRINT"   Dimarogonas,A.D., Computer Aided Machine Design,
Prentice-Hall, 1988";
210 FOR I=1 TO 14:COLOR I:LOCATE I,22+I:PRINT "FLATBELTS";
220 LOCATE I,50-I:PRINT "FLATBELTS";:NEXT I:COLOR 14
230 LOCATE 8,30:PRINT"Design of flatbelts ";
240 LOCATE 22,25:INPUT"Hit RETURN to run the program";X$
250 DIM S(4),B(4),D(4),E(4),Y(4),K(4),Z(4)
260 FOR I= 1 TO 4
270 READ S(I),B(I),D(I),E(I)
280 NEXT I
290 DATA 0.44,25,20,45,0.44,10,25,35,0.39,5,30,25,2,80,80,55
300 FOR I=1 TO 4
310 READ Y(I),K(I),Z(I)
320 NEXT I
330 DATA 0.75,5,2,0.75,6,2,0.75,7,2,3,55,3
340 COLOR 14:CLS:PRINT:PRINT"                    FLAT BELT DESIGN":PRINT:PRINT
350 PRINT"INPUT DATA:":PRINT
360 INPUT "Power transmitted (hp)                    ";N
370 INPUT "Speed of driving shaft (rpm)              ";N1
380 INPUT "Speed of driven shaft (rpm)               ";N2
390 INPUT "Distance between shafts (mm)              ";A
400 INPUT "Service factors:  C1                      ";C1
410 INPUT "                  C2                      ";C2
420 INPUT "                  C3                      ";C3
430 INPUT "                  C4                      ";C4
440 INPUT "                  C5                      ";C5
450 PRINT
460 INPUT "Are the data correct (Y/N)                ";X$
470 IF X$="n" OR X$="N" THEN 340
480 CLS:C=C1*C2*C3*C4*C5
490 NITER=0
500 LOCATE 3,1:PRINT TAB(15);"leather(ag)";TAB(30);"leather g";TAB(45);
510 PRINT "leather f or s";TAB(60);"plastic a,b,c";
520 LOCATE 9,1:PRINT "belt width (mm)=";
530 LOCATE 10,1:PRINT "initial belt length (mm)=";
540 LOCATE 11,1:PRINT "static overload,% ";
550 LOCATE 12,1:PRINT "fatigue overload,% ";
560 LOCATE 13,1:PRINT "CROSS-LOOP OPERATION:----------------------------------
-------------------------"
570 LOCATE 14,1:PRINT"belt width";
580  LOCATE 15,1:PRINT "initial belt length (mm)";
590  LOCATE 16,1:PRINT "static overload ,% ";
600 LOCATE 17,1:PRINT "fatigue overload ,% ";
610 LOCATE 8,1
620 PRINT "OPEN LOOP OPERATION:-----------------------------------------------
--------------"
630 FOR I= 1 TO 4
640 IF I=1 THEN LOCATE 2,1:PRINT "material No=";
650 LOCATE 2,5+15*I:PRINT I
660 X=0
```

```
670 D1=90*SQR(2*(D(I)+X))*(N*C/(S(I)*N1))^.33+DDIA
680 D2=.985*D1*N1/N2
690 S=D1/(2*(D(I)+X))
700 D1=INT(D1)+1
710 D2=INT(D2)+1
720 IF I>3 THEN GOTO 800
730 S=INT(S)+1
740 IF X>0 THEN GOTO 890
750 IF S<8 THEN GOTO 890
760 IF S<14 THEN GOTO 780
770 IF S>13 THEN GOTO 790
780 X=5:GOTO 670
790 X=15:GOTO 670
800 IF S<1 THEN GOTO 860
810 IF S<1.4 AND X=0 THEN GOTO 840
820 IF S<1.4 AND X>0 THEN GOTO 870
830 IF S>=1.4 AND X=0 THEN GOTO 850 ELSE GOTO 880
840 X=10 :GOTO 670
850 X=20 :GOTO 670
860 S=(INT(S/.5)+1)*.5 :GOTO 890
870 S=(INT(S/.7)+1)*.7 :GOTO 890
880 S=(INT(S/.9)+1)*.9
890 IF I=1 THEN LOCATE 5,1:PRINT "driving pulley dia (mm)=";
900 LOCATE 5,20+(I-1)*15:PRINT D1
910 IF I=1 THEN LOCATE 6,1:PRINT "driven  pulley dia (mm)=";
920 LOCATE 6,20+(I-1)*15:PRINT D2
930 LOCATE 7,1:PRINT"belt thickness (mm)";
940 LOCATE 7,20+(I-1)*15:PRINT S
950 U=D1*N1*75/(1.43*10^6)
960 LOCATE 8,1
970 L=2*A+1.57*(D2+D1)+((D2-D1)^2)/(4*A)
980 DL=Y(I)*L/(100+Y(I))
990 LO=L-DL
1000 F=3.14-3.14*(D2-D1)/(3*A)
1010 FF=F*(.3+U/100)
1020 KM=(2.7183^FF-1)/2.7183^FF
1030 H=150
1040 SK=1.43*10^6*N*C1/(D1*N1*H*S)
1050 NO=(S(I)-SK-K(I)*S/D1)*U*S*KM/75
1060 B=N*C/NO
1070 NITER=NITER+1:IF NITER>100 THEN 1610
1080 IF ABS(B-H)<50 THEN GOTO 1110
1090 IF (B-H)<0 THEN H=100 :GOTO 1040
1100 H=H+50 :GOTO 1040
1110 B=INT(B)+1
1120 LOCATE 9,20+(I-1)*15:PRINT B
1130 LO=INT(LO)+1
1140 LOCATE 10,20+(I-1)*15:PRINT LO
1150 T=Y(I)*E(I)/100+.5*SK+K(I)*S/D1
1160 T1=S(I)-T
1170 IF T1>0 THEN GOTO 1210
1180 T2=(S(I)-T)/S(I)
1190 T3= ABS(T2)*100
1200 LOCATE 11,20+(I-1)*15:PRINT T3
1210 BB=1000*Z(I)*U/L
1220 B1=B(I)-BB
1230 IF B1>0 THEN GOTO 1270
1240 B2=(B(I)-BB)/B(I)
1250 B3=ABS(B2)*100
1260 LOCATE 12,20+(I-1)*15:PRINT B3
1270 LOCATE 13,1
1280 L=2*A+1.57*(D1+D2)+((D1+D2)^2)/(4*A)
1290 DL=Y(I)*L/(100+Y(I))
1300 LO=L-DL
1310 F=3.14+3.14*(D2+D1)/(3*A)
1320 FF=F*(.3+U/100)
1330 KM=(2.7183^FF-1)/2.7183^FF
1340 H=150
1350 SK=1430000!*N*C1/(D1*N1*H*S)
1360 NO=(S(I)-SK-K(I)*S/D1-E(I)*(H/A)^2)*U*S*KM/75
1370 B=N*C/NO
```

```
1380 IF ABS(B-H)<50 THEN GOTO 1420
1390 IF (B-H)>=0 THEN GOTO 1420
1400 IF (B-H)<0 THEN H=H-50
1410 GOTO 1580
1420 B=INT(B)+1
1430 LOCATE 14,20+(I-1)*15:PRINT B
1440 LO=INT(LO)+1
1450 LOCATE 15,20+(I-1)*15:PRINT LO
1460 T=Y(I)*E(I)/100+.5*SK+K(I)*S/D1
1470 T1=S(I)-T
1480 IF T1>0 THEN GOTO 1520
1490 T2=(S(I)-T)/S(I)
1500 T3=ABS(T2)*100
1510 LOCATE 16,20+(I-1)*15:PRINT T3
1520 BB=1000*Z(I)*U/L
1530 B1=B(I)-BB
1540 IF B1>0 THEN GOTO 1580
1550 B2=(B(I)-BB)/B(I)
1560 B3=ABS(B2)*100
1570 LOCATE 17,20+(I-1)*15:PRINT B3
1580 IF I=1 THEN LOCATE 20,1:PRINT "belt speed,(m/s) ";
1590 LOCATE 20,20+(I-1)*15:PRINT U
1600 LOCATE 19,1:PRINT"-----------------------------------------------------
------------------------";
1610 NEXT I
1620 LOCATE 22,1:INPUT"Any change (D-iameter,C-enter distance,Q-uit)";C$
1630 LOCATE 22,1:PRINT SPC(78);
1640 IF C$><"d" AND C$>< "D" THEN 1670
1650 LOCATE 22,1:PRINT "enter increment of the";
1660 PRINT" diameter of the driving pulley";:INPUT DDIA:GOTO 480
1670 IF C$><"c" AND C$>< "C" THEN 1690
1680 LOCATE 22,1:PRINT "new center distance (old=";A;")";:INPUT A:GOTO 480
1690 END
```

APPENDIX 10.B
VBELTS: DESIGN OF V-BELTS PROGRAM

```
10 REM ************************************
20 REM *                                  *
30 REM *                                  *
40 REM *              VBELTS              *
50 REM *                                  *
60 REM ************************************
70 REM          COPYRIGHT 1987
80 REM by Professor Andrew D. Dimarogonas, Washington Univ., St. Louis, Mo.
90 REM  All rights reserved.  Unauthorized reproduction, disssemination,
100 REM selling or use is strictly prohibited.
110 REM  This listing is for reference purpose only.
120 REM
130 REM
140 REM DESIGN OF V-BELTS
150 REM
160 REM
170 REM          ADD / 6-2-1987
180 REM
190 CLS:KEY OFF
200 LOCATE 25,1:PRINT"   Dimarogonas,A.D., Computer Aided Machine Design,
Prentice-Hall, 1988";
210 FOR I=1 TO 14:COLOR I:LOCATE I,22+I:PRINT "VBELTS";
220 LOCATE I,50-I:PRINT "VBELTS";:NEXT I:COLOR 14
230 LOCATE 8,30:PRINT"Design of V-belts ";
240 LOCATE 22,25:INPUT"Hit RETURN to run the program";X$
250 DIM T(49),B(11),S(11),D(11),K(11),L(11)
260 FOR I=1 TO 49
270 READ T(I)
280 NEXT I
```

```
290 DATA 20,22,25,28,32,36,40,45,50,56,63,71,80,90,100,112
300 DATA 125,140,160,180,200,224,250,280,315,355,400,450
310 DATA 500,560,630,710,800,900,1000,1120,1250,1400,1600
320 DATA 1800,2000,2240,2500,2800,3150,3550,4000,4500,5000
330 FOR I=1 TO 11
340 READ B(I),S(I),D(I)
350 NEXT I
360 DATA 5,3,40,6,4,50,8,5,63,10,6,80,13,8,100,17,11,132
370 DATA 20,12.5,180,25,16,236,32,20,315,40,25,450,50,32,600
380 FOR I=1 TO 11
390 READ K(I),L(I)
400 NEXT I
410 DATA 150,860,212,1262,296,1916,420,2820,585,4275
420 DATA 832,6332,1100,9540,1650,14050,2303,18063
430 DATA 3230,18080,4620,18100
440 CLS:COLOR 13:PRINT A1$:PRINT A2$:PRINT A3$:PRINT A4$:PRINT A5$
450 COLOR 14:PRINT:PRINT"                              V - BELT DESIGN":PRINT:PRINT
460 PRINT"INPUT DATA:":PRINT
470 INPUT "Power transmitted (hp)                            ";N
480 INPUT "Speed of driving shaft (rpm)                      ";N1
490 INPUT "Speed of driven shaft (rpm)                       ";N2
500 INPUT "Distance between shafts (mm)                      ";A
510 INPUT "Service factor C1                                 ";C1
520 INPUT "c2                                                ";C2
530 INPUT "c3                                                ";C3
540 INPUT "c4                                                ";C4
550 INPUT "c5                                                ";C5
560 PRINT
570 COLOR 12:INPUT"Are the data correct? (Y/N)                 ";X$
580 COLOR 13:CLS:PRINT A1$:PRINT A2$:PRINT A3$:PRINT A4$:PRINT A5$
590 COLOR 14:PRINT"                              results:"
600 IF X$="n" OR X$="N" THEN 440
610 C=C1*C2*C3*C4*C5*1.25
620 I=1
630 D=D(I)+1
640 J=1
650 K=D-T(J)
660 IF K<0 THEN GOTO 680
670 J=J+1 :GOTO 650
680 T=(T(J+1)-T(J))/2
690 IF K>T THEN GOTO 710
700 D1= T(J):GOTO 720
710 D1=T(J+1)
720 D2=D1*.985*N1/N2
730 P=D2-T(J+1)
740 IF P<0 THEN GOTO 760
750 J=J+1 :GOTO 730
760 T=(T(J+2)-T(J+1))/2
770 IF P>T THEN GOTO 790
780 D2=T(J+1) :GOTO 800
790 D2=T(J+2)
800 A1=1.5*D2
810 IF A1>A THEN GOTO 830
820 I=I+1 :GOTO 630
830 IF D2>A THEN GOTO 1060
840 L=2*A+1.57*(D2+D1)+((D2-D1)^2)/(4*A)
850 DL=.75*L/(100.75)
860 L0=L-DL
870 U=D1*N1/19100
880 L0=INT(L0)+1
890 LI=L-3.14*S(I)
900 IF LI<K(I) THEN GOTO 820
910 IF LI>L(I) THEN GOTO 820
920 B=2000*U/L
930 R=B/40
940 PRINT "B/Bmax =                          ";R
950 PRINT "belt width (mm)=                  ";B(I)
960 PRINT "belt thickness (mm)=              ";S(I)
970 PRINT "initial belt length (mm)=         ";L0
980 Q=N1*D1/D(I)
990 PRINT "n1 * d1/dm =                      ";Q
```

```
1000 INPUT "unit power tranmission N0=     ";N0
1010 O=N*C/N0
1020 O=INT(O)+1
1030 PRINT "number of belts=                ";O
1040 INPUT "want the next size (Y/N)        ";A$
1050 IF A$="y" OR A$="Y" THEN GOTO 820 ELSE GOTO 1070
1060 PRINT "design not feasible"
1070 END
```

CHAPTER ELEVEN

LUBRICATION

11.1 FRICTION AND WEAR DUE TO SLIDING

11.1.1 Mating surfaces

When the surfaces of machine elements are in contact, they rest on each other in a way discussed in Chapter 10. The resistance to the sliding of one surface over the other, that is the frictional force, is a result of the resistance offered by the interaction of the projections on both surfaces. During the process of sliding, very high local temperatures will be generated. If the surfaces are only lightly pressed against each other, the tendency will be for the projections to be smeared over and this will be facilitated by the local high temperatures and possible local melting of the metals.

The process of smearing over the projections of both surfaces is considerably facilitated if there is a lubricant in the space between them. Local melting of the projections is followed by rapid cooling and the ultimate result is a very highly polished extremely hard amorphous layer on the surface known as the 'Beilby layer'. The process of forming the Beilby layer is known as 'running-in'. (This is generally true, but it must not be assumed that a Beilby layer must always exist. Examples of successful lubrication are known where it is absent.)

Another important factor is that when metals come into contact with a lubricant, an absorbed film of lubricant will be formed on its surface. There are indications that the nature of this film, formed with the first oil to contact the surface, is of importance. The process of running-in can be facilitated by the use of solid lubricants as graphite and molybdenum bisulphide.

11.1.2 Wear mechanisms

Wear will be apparent after two surfaces have been rubbing together for a considerable period. A good design assures that the maximum amount of wear is confined to those parts that can be readily replaced; as an example, journals which cannot readily be replaced are made of hard materials, while mating bearing 'brasses' or shells are made of soft material so that most of the wear takes place in the softer bearing material.

If good fluid film lubrication could always be maintained, that is if the mating surfaces could always be separated by a film of lubricant, extremely little, if any, cutting wear would result, and this would only take place during the starting and stopping periods when full fluid film lubrication is not possible.

Abrasive wear results from the intrusion of some abrasive such as grit, and this can be reduced to a minimum by the use of filters in lubrication systems (Figure 11.1).

Wear due to scuffing can only be reduced by improvement in lubrication conditions.

Corrosion wear results from chemical action which is often a result of acidic

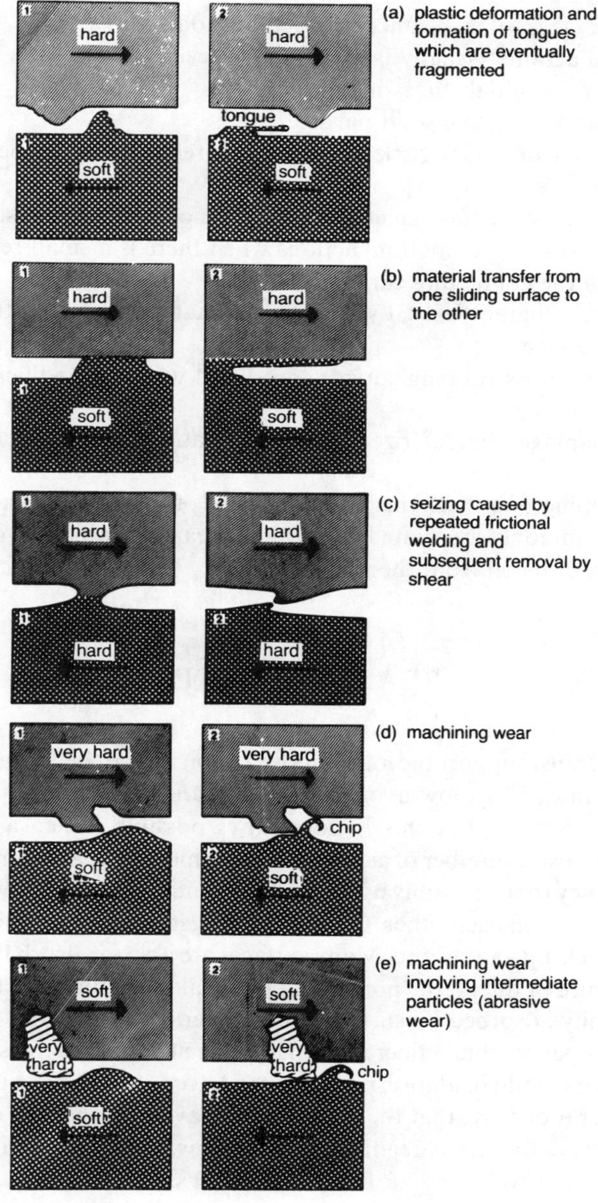

Figure 11.1 Wear mechanisms. (From Klingele and Engel.)

deterioration products in the lubricating oil, though it could arise from naturally-occurring free acids in fatty oils, or residual acid from bad refining.

The following definitions of terms used in connection with wear were agreed upon at the Conference of Lubrication and Wear arranged jointly by ASME and the Institution of Mechanical Engineers in October, 1957.

Wear Progressive loss of substance from the surface of a body brought about by mechanical action. (Usually it reduces the serviceability of the body, but it can be beneficial in its initial stages in running-in)

Abrasion Wear caused by fine solid particles

Embedding Inclusion of solid particles, abraded or foreign matter, in the surfaces of wearing parts

Fretting corrosion Destruction of metal surfaces by a combination of sliding and corrosive (usually oxidation) actions when there is a small relative movement at the contacting surfaces

Pitting Local wear characterized by removal of material to a depth comparable with surface damage

Scoring Scratches across rubbing surfaces produced without modification of the general form

Scuffing Gross damage characterized by the formation of local welds between sliding surfaces

Seizure The stopping of a mechanism as the result of interfacial friction

Spalling Wear, commonly associated with rolling bearings, which involves the separation of flakes of metal from the surface

11.2 BEARING MATERIALS

A bearing material must support the load imposed upon it. The coefficient of friction between shaft and journal must be as low as possible. This can be achieved by a film of lubricant between shaft and journal, but this is not always possible. When a continuous film of lubricant does not exist, a number of factors assume importance. Some metals have a natural 'lubricity' or tendency to slide readily over each other and this is generally more marked when dissimilar surfaces rub on each other. Others weld together more readily, a property which leads to scuffing, pick-up and seizure. Many surfaces are able to adsorb lubricants because of their chemical nature or due to the porosity of physical constitution of their material. Fatty oils are more readily adsorbed by surfaces than mineral oils and they also have far better surface-wetting properties than mineral oils. Even the adventitious presence of microscopic amounts of water or of mild oxidation products can have a similar effect to that of a lubricant.

Other considerations are: that the alloy should have high thermal conductivity so that the heat generated by friction is readily conducted away; that it should be durable, i.e. it should have a comparatively long life in service; that it should be easily replaced and should have appreciable resistance to corrosion. Bearing metals which are to be cast should flow freely when molten and have minimum contraction on cooling.

11.2.1 Copper Alloys

Phosphor-bronze is a popular bearing metal. The phosphorus is added to make the alloy more molten for casting purposes, to give a sounder casting by thus removing the oxygen and to form some phosphide of copper, the presence of which improves wear-resisting properties of the alloy.

The so-called aluminum-bronzes fall into two groups: up to 7.5% aluminum gives the alpha bronze, while from 8 to 14% aluminum gives an alloy with a duplex structure.

Bronzes are used for average bearing pressures up to 30 MPa.

11.2.2 Lead

Lead has a low coefficient of friction, provides a soft 'bedding' surface and yields sufficiently to take up small misalignment. It will not dissolve in copper and hence does not form a solid solution but disperses as very small droplets throughout the metal.

Lead can be added to the normal bronzes and brasses to improve machineability, or in larger amounts to provide a sound bearing material. It can also be added to copper in proportions of up to 30% to give 'leaded bronzes'. This material is used in the copper–lead big-end bearings for aircraft and other heavy-duty internal combustion engines.

11.2.3 White-metal bearing alloys – Babbitts

Popularly known as the white-metals are alloys of lead, tin, antimony and a little copper. They are divided into two groups: those in which the major constituent is tin (referred to as tin-based) and those in which lead is the principal constituent (lead-based). All are relatively soft and are used as bearing linings, i.e. as surface coatings of harder materials forming the structural component of the step or 'brass'.

Tin-base alloys are often called 'Babbitt metals' after their inventor, but Isaac Babbitt's original metal consisted of a mixture of 24 lb tin, 8 lb antimony, 4 lb copper, to every pound of which 2 lb of tin was added. This gives a percentage composition of tin 88.9, antimony 7.4, and copper 3.7%. In addition to these constituents, present-day tin-base alloys may contain some lead, nickel or cadmium. Small amounts of copper combining with tin tend to form a fibrous network which prevents segregation of the hard particles that may form, thus giving a more uniform structure. Lead tends to increase resistance to deformation, antimony to increase hardness, and cadmium increases hardness and tensile strength.

Lead-base alloys, in general, are less expensive than tin-base alloys, but they will not stand up to their performance. Antimony increases the tensile strength of the alloy, but in excess of 15% it can cause brittleness. Tin increases hardness and tensile strength.

The thickness of the babbitt layer is determined from the maximum allowable bearing wear during running-in and normal operation. It usually is 1% of the bearing diameter. In large diameters, the babbitt is cast in place with the liner and then machined. For mass production and similar diameters, the antifriction material is bonded on strip stock and pressed to the final shape. In this case, the thickness of the strip stock is 1.5 to 2.5 mm and the antifriction layer is 0.2 to 0.3 mm thick.

Babbitts are used for pressures up to 20 MPa and temperatures up to 110 °C.

11.2.4 Cadmium-base bearing alloys

These are heavy-duty bearing alloys that are much superior in mechanical strength to tin-base white metals. Nickel, copper and silver may be present in small proportions. Their drawback is susceptibility to corrosion.

11.2.5 Sintered bearings

Porous bearings, mainly bushes, can be produced by the methods of powder metallurgy, and having a porous structure they can be impregnated with oil. Such sintered bearings generally contain, in powdered form, the same materials as ordinary bearings with the addition of powdered graphite of the order of 4.5 to 5.5%.

11.2.6 'Dry' bearings

Ordinary sleeve-type bearings require a continuous supply of lubricant. A possible alternative is the oil-impregnated porous bearing. Such a bearing contains a limited amount of lubricant and its life may be short if it is subjected to heat or the action of solvents.

Where the complete absence of any lubricant is highly desirable or it is necessary to eliminate periodical servicing, 'dry' bearings may be used. These bearings are made of a material that inherently possesses the property of being able to slide smoothly over other materials with minimum frictional resistance. Such a material is polytetrafluoroethylene, generally known as PTFE. It is, however, mechanically weak, dimensionally unstable and has low thermal conductivity. A very good bearing material can be made by impregnating sintered bronze with PTFE and adding a small amount of lead. Alternatively, PTFE may be used as a very thin shell bonded to a steel backing.

Other unusual bearing materials are carbon, which has a rather specialized application in the case of clutch thrust washers and nylon which is meeting with some success when used in inaccessible positions. Nylon is tough, hard-wearing, easy to mold and machine, and any lubrication difficulties can be overcome by the use of molybdenum bisulphide powder.

11.3 FLUID VISCOSITY

The viscosity of a liquid can be simply defined as its resistance to flow, and is a measure of the friction between the liquid molecules when moving past each other. In his **Law of Viscous Flow**, Newton first stated the relation between the mechanical forces involved as follows: the internal friction (i.e. viscosity) of a fluid is constant with respect to the rate of shear.

Absolute or dynamic viscosity can be defined in terms of the simple model shown in Figure 11.2, with two parallel flat plates, distance h apart, the upper one moving and the lower one stationary, separated by a film of oil. In order to move the upper plate, of area A, at a constant velocity U across the surface of the oil and cause adjacent layers to flow past each other, a tangential force F must be applied. Since oil will adhere to the two surfaces, the lowest layer of molecules will remain stationary, the uppermost layer will move with the

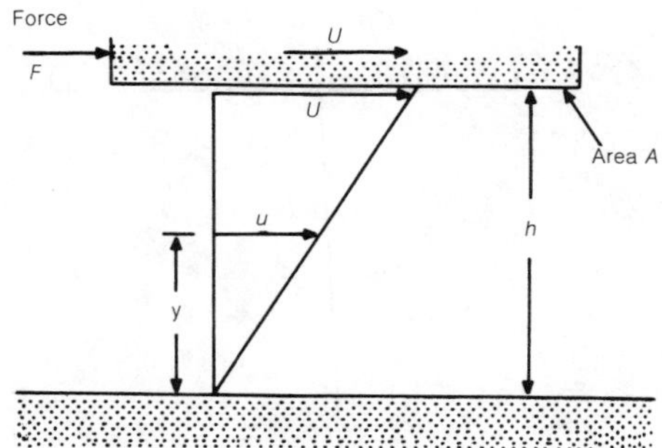

Figure 11.2 Lubricated plate in parallel motion

velocity of the upper plate and each intermediate layer will move with a velocity directly proportional to its distance y from the stationary plate.

The shear stress on the oil causing relative movement of the layers is equal to F/A. The rate of shear R of a particular layer, sometimes called the 'velocity gradient', is defined as the ratio of its velocity u to its perpendicular distance from the stationary surface y, and is constant for each layer, i.e. $R = u/y = U/h$. Newton deduced that the force F required to maintain a constant velocity U of the upper plate was proportional to the area A and to the rate of shear U/h, or

$$F = \mu A U/h$$

where μ is the coefficient of viscosity or absolute viscosity.

Absolute viscosity is thus defined by

$$\mu = \frac{\text{Shear stress}}{\text{Rate of shear}} = \frac{F/A}{U/h}$$

Since the dimensions of shear stress are $[\mathrm{MLT^{-2}L^{2}}]$ and of rate of shear $[\mathrm{T^{-1}}]$, it can be seen that the units of absolute viscosity are mass divided by length times time, i.e. [M/LT], or force-seconds divided by length squared.

Two units of absolute viscosity have been used, the poise and the reyn, the former based on the centimeter-gram-second (cgs) system and the latter on the inch-pound-second system of units. Both fundamental units of viscosity are too large for practical problems or presentation of data and they are therefore subdivided into the centipoise (1 cP = 0.01 poise) and microreyn ($1\,\mu\mathrm{R} = 10^{-6}$ reyn).

In the SI system, the unit of absolute viscosity is $\mathrm{Nsec/m^2}$ or Pa.s.

For conversion of absolute viscosity units,

$$1\ \text{reyn} = \frac{1\ \text{lb fs}}{\text{in}} = \frac{454 \times 981}{2.54^2}\frac{\text{dyn sec}}{\text{cm}^2} = 6.895 \times 10^6\ \text{cP}$$

or approximately,

$$1\,\mu\mathrm{R} = 7\ \mathrm{cP}$$
$$1\ \mathrm{cP} = 1\ \mathrm{mPa.s}$$

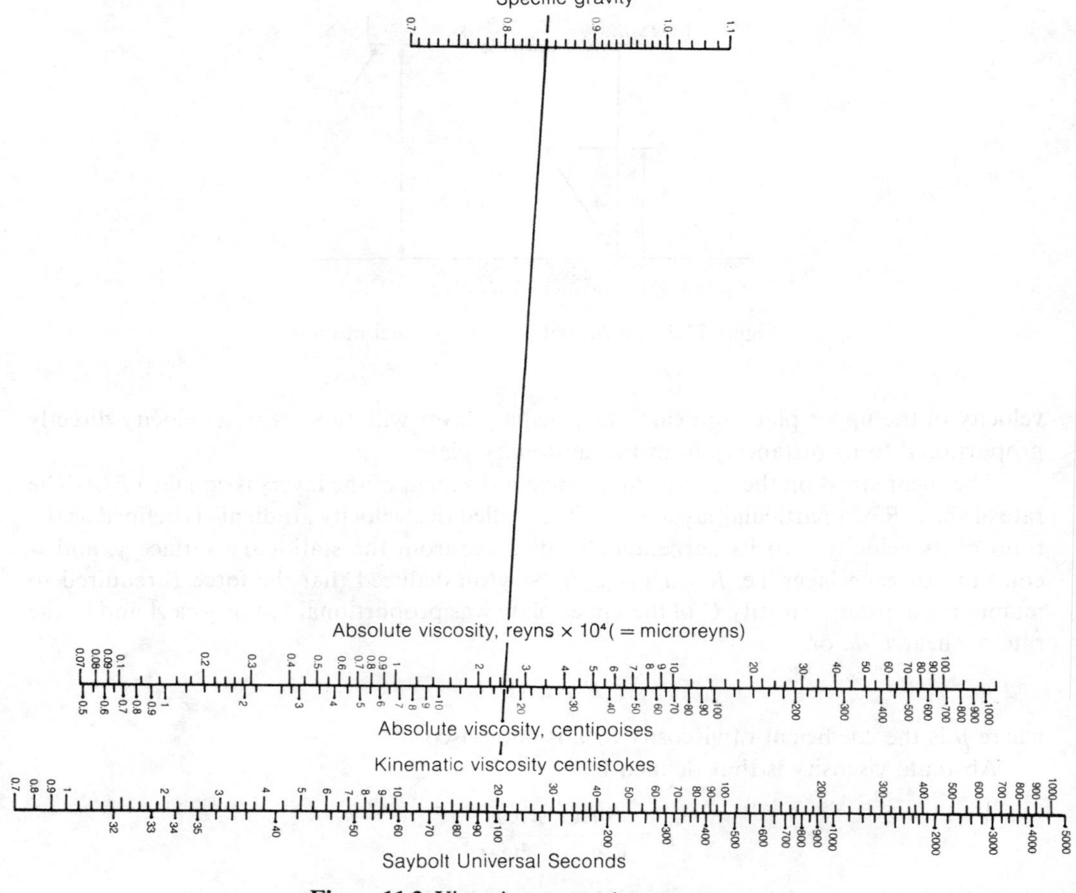

Figure 11.3 Viscosity conversion nomogram

The most traditional methods of measuring viscosity were based on measuring the time needed for a certain oil quantity to flow through a small pipe with gravity, a principle known for measurement of time since Ancient Egypt. For this reason, the kinematic viscosity was introduced $\nu = \mu/\rho$ with units L^2/T where ρ is the oil density. In the cgs system, the unit was cm^2/s, called Stokes (St), with the most widely used unit being the centistoke (cSt). The SI unit is then m^2/s equal to 10^4 St or 100 cSt. Therefore, 1 cSt is 10^{-2} (m^2/s). Oil density depends on oil type and it is 850 to 930 kg/m^3 for most mineral oils with an average value of 890.

A nomogram for viscosity units conversion is shown in Figure 11.3, including some of the empirical scales used in the USA and Europe.

An increase in temperature or a decrease in pressure weakens the intermolecular bonds in a fluid and leads to a reduction in viscosity. To be precise therefore, viscosity should always be quoted at a specified temperature and pressure; if the pressure is omitted it is understood to be atmospheric. Some oils are less sensitive to changes in viscosity with temperature than others; these are said to have a high viscosity index. Figure 11.4 shows this dependence for some oils.

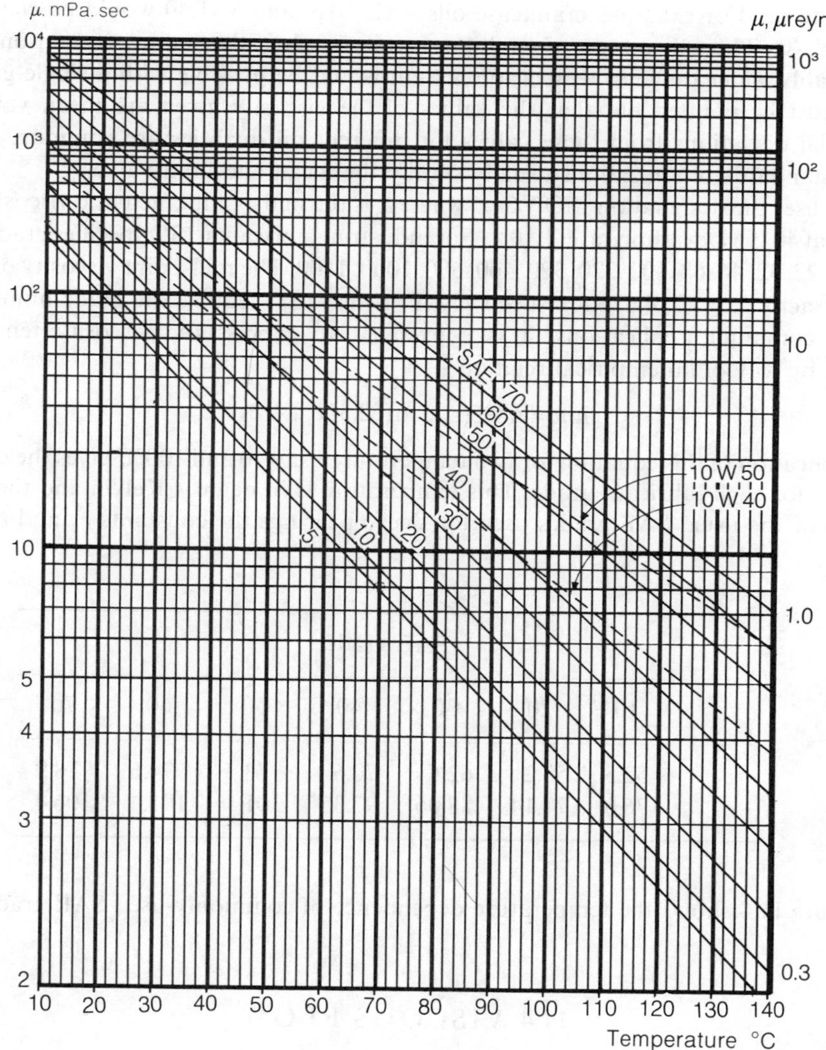

Figure 11.4 Viscosity of SAE oil grades

Pressure affects viscosity much less. For hydrodynamic lubrication, this effect is usually negligible. An approximate expression for the dynamic viscosity μ at pressure p as function of the viscosity μ_0 at atmospheric pressure, is

$$\mu = \mu_0 \exp(ap)$$

where $a = (1.3 \text{ to } 3.5)10^{-4}\,\text{mm}^2/\text{N}$ for most mineral oils.

Since the SAE crankcase oil specification is based on viscosity requirements at two temperatures only (0°F and 210°F), it becomes possible by suitable formulation to make an oil meeting the requirements of two SAE grades simultaneously. Such an oil is described as a multigrade oil. For example, crankcase oils SAE 10W and SAE40 would be classified as SAE 10W/40. The advantage of multigrade oils is that they allow operation over a considerably wider range of temperatures than would be possible with a single-grade oil.

It must be pointed out that the values of the viscosity given are mean values and substantial deviations from these values for different oil qualities are common. For this reason, standards organizations usually specify the mean and limit values.

ISO uses the designation ISO VG x, where x is the mid-point kinematic viscosity in cSt (mm^2/s) at 40 °C. Deviations are $\pm 10\%$. For industrial lubricants, ISO specifies grades 2, 3, 5, 7, 10, 15, 22, 32, 46, 68, 100, 220, 320, 460, 680, 1000, 1500. The mid-point viscosity of the first four is exactly 2.2, 3.2, 4.6, 6.8 respectively. For the rest, it equals the grade number.

For computer applications, it is convenient to express the viscosity–temperature function by way of an empirical relation

$$\ln\mu = 6895(c_1 \exp(63.3/T_K) + c_2)$$

Application of this equation for two temperatures, say 40 and 100°C gives the constants c_1 and c_2 for each lubricant grade. This temperature is in degrees Kelvin and then μ is in mPa.s. For the usual SAE grades used for sleeve bearings the constants c_1 and c_2 are:

	SAE grade						
	10	20	30	40	50	60	70
c_1	52.5	53.2	62.6	70.6	73.3	74.6	78.9
$-c_2$	76.6	77.14	88.06	97.35	101.1	101.4	106.4

Figure 11.4 shows the temperature dependence of commonly-used SAE grades.

11.4 VISCOUS FLOW

11.4.1 Viscous flow in a concentric bearing

Petroff applied Newton's Law of Viscous Flow to calculate the frictional torque and power loss in certain types of journal bearings which tend to run concentrically since they carry little or no transverse loads. Figure 11.5 shows a vertical guide bearing of this type. Oil completely fills the clearance space, which is small compared with the shaft diameter; it is also assumed that negligible axial flow of oil takes place. The resisting torque is

$$T = \tau R\pi DL \tag{11.1}$$

where the shear stress is

$$\tau = \mu U/c = \mu\pi DN/c \tag{11.2}$$

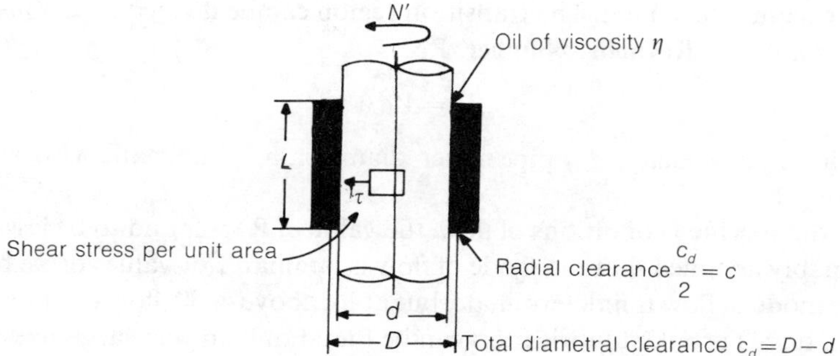

Figure 11.5 Lubricated concentric cylinders

Then, the **Petroff Equation** is obtained

$$T = \pi^2 \mu N L D^3 / 2c \tag{11.3}$$

where N is the rotating speed (revolutions per second). The power loss is

$$P = \omega T = \pi^2 N^2 L D^3 \mu / c \tag{11.4}$$

We can define a coefficient of friction:

$$f = T/W = T/pLD = (\pi^2 D^2 / 2c)(\mu N / p) \tag{11.5}$$

where W is the bearing load, p the bearing mean pressure W/LD. It is apparent that the coefficient of friction depends on two groups of parameters:

Geometry: $\pi^2 D^2 / 2c$

Operation: $\mu N / p$

Petroff's equation is of particular value in estimating the power losses in journal bearings that are lightly loaded and operate nearly concentric at high rotational speeds, e.g. turbine bearings. Even in the case of moderately loaded bearings running with a considerable degree of eccentricity, Petroff's equations may be applied as a first approximation to assess the minimum losses to be expected. Note that the power loss and hence the frictional heat developed becomes of increasing significance in large diameter bearings since T is proportional to D^3.

11.4.2 Viscous flow in a pipe

When a liquid flows in a parallel-sided pipe it loses energy in friction between adjacent layers of the liquid and also between boundary layers of the liquid and the pipe walls. Provided that rate of flow does not exceed a certain critical velocity, the mode of flow is described as viscous or laminar flow and is governed by a fundamental relationship known as **Poiseuille's Law**. The chief characteristics of viscous flow are that the velocity is independent of the internal roughness of the pipe and the velocity distribution across a diameter of the pipe is parabolic in form with maximum velocity along the pipe axis. At flow velocities above the critical, the flow pattern becomes turbulent. The transition region can be defined by a non-dimensional criterion known as the **Reynolds Number**, Re.

$$\mathrm{Re} = Vd/v$$

where V = the fluid velocity, d = pipe inner diameter, v = kinematic viscosity (μ/ρ) and ρ = density of oil.

If, under the specified conditions of flow, the value of Re is found to be less than 2000, it can be reasonably assumed that the mode of flow is laminar. For values of Re between 2000 and 4000, the mode of flow is indeterminate, but at Re above 4000, flow is almost certainly of a turbulent nature. Turbulent lubrication can be found only in very large sizes or very high speeds.

Fox and McDonald showed that below the critical velocity the volume of liquid flowing through a narrow, parallel-sided channel in time t is given by

$$q = \Delta p h^3/12\mu L \tag{11.6}$$

where q = volume of liquid per unit time, Δp = difference in the pressure causing the flow, h = channel height, μ = absolute viscosity, and L = length of channel, unit width.

11.4.3 Moving parallel plates

A flat plate, moving parallel to a stationary plate with a velocity u, (Figure 11.2), is accompanied by a flow

$$q = wUh/2 \tag{11.7}$$

assuming a linear variation of velocity across the plate distance h, constant along the width w.

11.4.4 The Reynolds equation

Consider now a flat surface stationary and a curved surface, nearly parallel to the stationary one moving in respect to it in the x-direction with velocity U. It also moves away from the stationary surface with a velocity $V = \dot{h}$, where $h(x, y; t)$ is the distance between the two surfaces at location x, y in the z-direction. The pressure is, in general, a function of x, y and t.

The flow in a control volume of sides Δx and Δy and height h (variable), as in Figure 11.6(a), is considered constant, for incompressible liquid.

Flow in the x-direction:
Owing to pressure difference (Figure 11.6b)

$$Q_x = \frac{h^3}{12\mu}\frac{\partial p}{\partial x}\Delta y,$$

from equation (11.6).
Owing to motion of the plate, (Figure 11.6c)

$$Q_u = (Uh/2)\Delta y,$$

from equation (11.7).

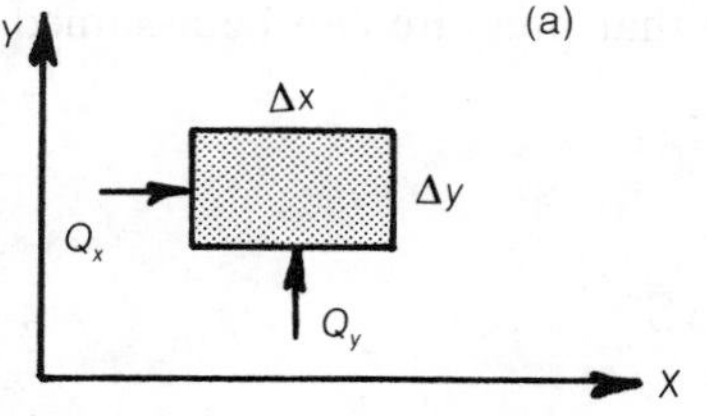

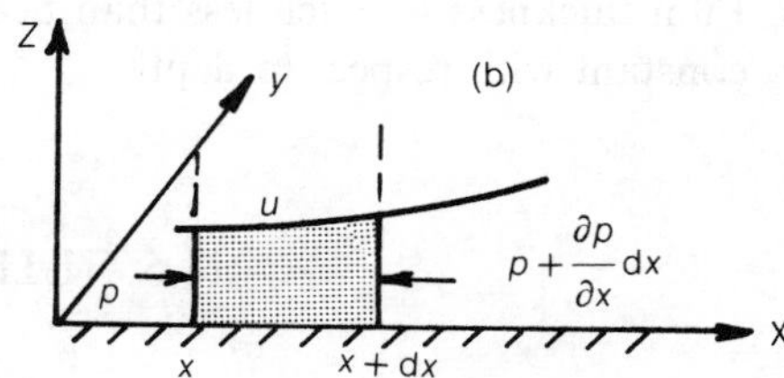

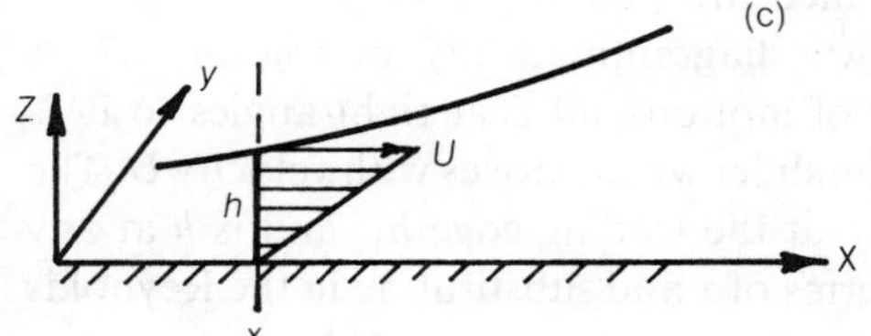

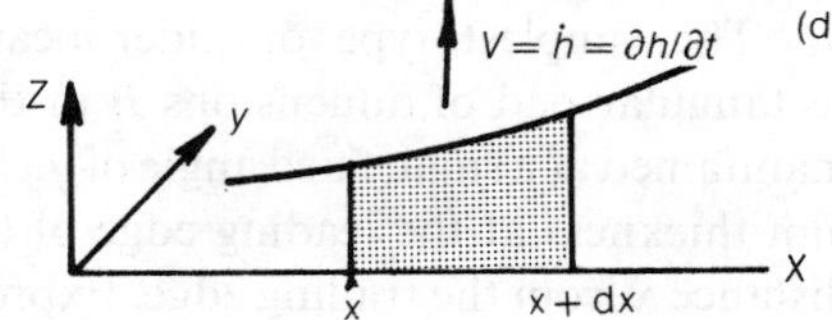

Figure 11.6 Elementary lubricated area

Flow in the y-direction:

$$Q_x = \frac{h^3}{12\mu}\frac{\partial p}{\partial y}\Delta x,$$

from equation (11.6), Figure 11.6(b).
The total flow will be

$$Q = \frac{Uh}{2}\Delta y - \frac{h^3}{12\mu}\frac{\partial p}{\partial x}\Delta y - \frac{h^3}{12\mu}\frac{\partial p}{\partial y}\Delta x \tag{11.8}$$

Because of continuity, the derivative of the flow will equal the change of volume due to the upwards motion of the moving plate with velocity $-\dot{h} = V$.

$$\Delta Q = \frac{\partial Q}{\partial x}\Delta x + \frac{\partial Q}{\partial y}\Delta y \tag{11.9}$$

$$\Delta Q = \frac{U}{2}\frac{\partial h}{\partial x}\Delta x\Delta y - \frac{\partial}{\partial x}\left(\frac{h^3}{12\mu}\frac{\partial p}{\partial x}\right)\Delta x\Delta y - \frac{\partial}{\partial y}\left(\frac{h^3}{12\mu}\frac{\partial p}{\partial y}\right)\Delta x\Delta y = -\dot{h}\Delta x\Delta y \tag{11.10}$$

This yields the general form of the Reynolds equation

$$\frac{\partial}{\partial x}\left(\frac{h^3}{12\mu}\frac{\partial p}{\partial x}\right) + \frac{\partial}{\partial y}\left(\frac{h^3}{12\mu}\frac{\partial p}{\partial y}\right) = \frac{U}{2}\frac{\partial h}{\partial x} + \dot{h} \tag{11.11}$$

From the foregoing discussion, it is apparent that the Reynolds equation is based on the following assumptions:

1. Laminar flow of lubricant within the bearing clearance.
2. Constant viscosity; temperature rise due to shearing of the oil and effects of pressure on viscosity are neglected.

3. Film thickness is much less than bearing dimensions so that pressure can be assumed constant with respect to depth.

11.5 SLIDER BEARINGS

Slider bearings consist of two plane-surfaced members, separated by a wedge-shaped film of oil. The moving member, called the 'slider' or 'runner', moves in the direction of convergence of the oil film; the other, stationary member, is called the 'pad' or 'shoe'.

The simplest type of slider bearing is shown diagrammatically in Figure 11.7. A rectangular pad of dimensions B in the direction of motion and L at right-angles to it, is maintained at a small fixed angle of inclination to the slider which moves with velocity U. The film thickness at the leading edge of the pad is h_0, at the trailing edge h_1, and is h at any distance x from the trailing edge. Expressing h in terms of x and substituting in the Reynolds

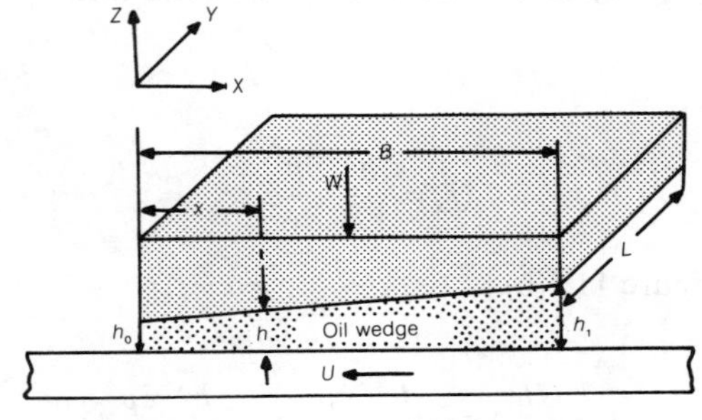

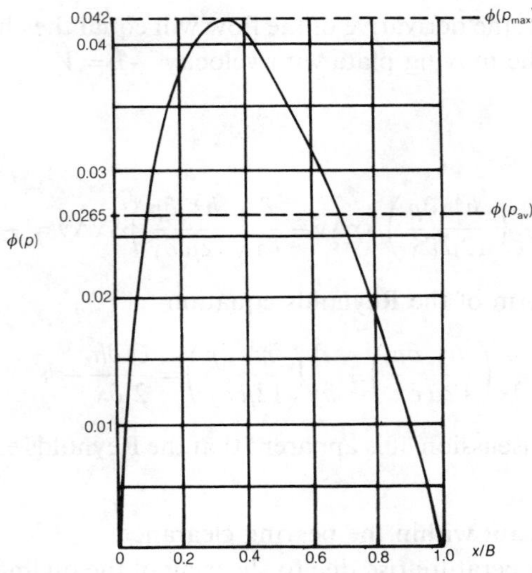

Figure 11.7 Bearing oil wedge pressure distribution. (Courtesy BP 1969)

equation, it can be shown that for a pad in which $L \gg B$, $\partial p/\partial y = 0$, (i.e. a pad of 'infinite' width), the pressure distribution in the direction of motion is given by integrating twice equation (11.11):

$$p = \frac{6\mu UB}{h_0^2} \cdot \phi(p) \tag{11.12}$$

where $\phi(p)$ is a non-dimensional pressure function whose value depends on the angle of tilt of the pad. Evaluation of $\phi(p)$ enables a pressure profile to be plotted across the face of the pad as shown in Figure 11.7. The average pressure across the pad, p_{av}, is found by integration to be:

$$p_{av} = \frac{6\mu UB}{h_0^2} \cdot \phi(p_{av}) \tag{11.13}$$

From this equation, the minimum film thickness at the trailing edge is given by:

$$h_0 = \left(\frac{6\mu UB}{p_{av}} \cdot \phi(p_{av})\right)^{\frac{1}{2}} \tag{11.14}$$

A typical basis for design is to incline the pad so that the film thickness at the leading edge is twice that of the trailing edge (i.e. $h_1/h_0 = 2$). In this case the pressure function $\phi(p)$ as plotted in Figure 11.7 shows a maximum value of about 0.042 and an average value of 0.0265. Calculations show that $\phi(p_{av})$ is not greatly affected by small changes in pad inclination provided h_1/h_0 lies within the range 1.5 to 3. For design purposes, therefore, it is sufficiently accurate to take $\phi(p_{av})$ as equal to 0.026.

The load-carrying capacity W of the simple rectangular pad is equal to the product of the average pressure and the pad area ($B \times L$):

$$W = \frac{6\mu UB^2L}{h_0^2} \cdot \phi(p_{av}) \tag{11.15}$$

It appears that by progressively reducing the clearance, the load capacity could be steadily increased and the coefficient of friction decreased. However, practical problems, such as the accurate machining of truly convergent planes of good surface finish, and the mechanical and thermal distortion of the surfaces under working conditions, limit the extent to which this objective can be pursued. A modified arrangement of the slider, the 'pivoted pad' bearing, developed by Michell (1905) in Australia and independently by Kingsbury in the USA, eliminates this problem.

In the pivoted-pad bearing illustrated in Figure 11.8 the pad rests on a pivot through which the bearing load is transmitted. The pivoted pad is inherently stable but at the same time free to adjust its inclination to varying operating conditions. If the pivot is at the center of the pad, it is possible to run the slider in either direction without seriously affecting the performance of the bearing. For unidirectional running, the optimum position of the pivot is slightly beyond the middle of the pad towards the trailing edge. Five-ninths of the way along is often chosen in the design of such a bearing.

The equations developed for the fixed-pad bearing apply equally well to the pivoted-pad bearing, provided the pivot is located at the centre of pressure of the pad. Given the bearing load, the minimum film thickness and the pad angle are found.

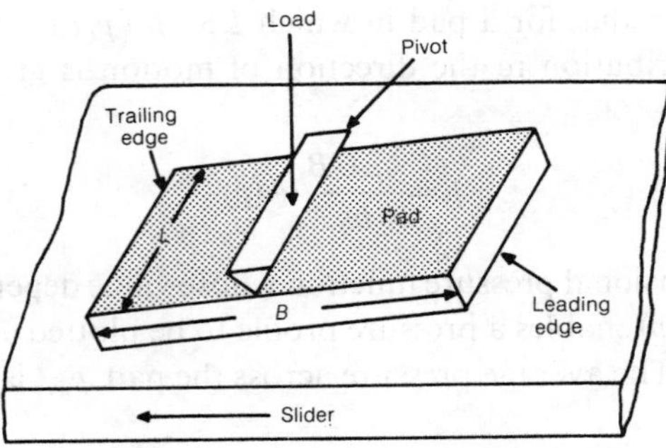

Figure 11.8 Principle of a tilting pad

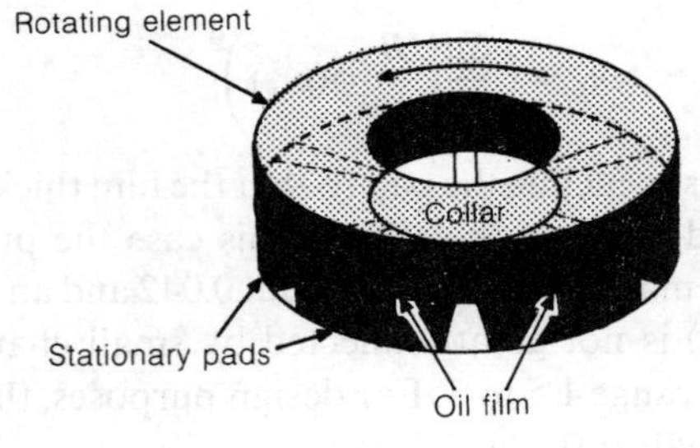

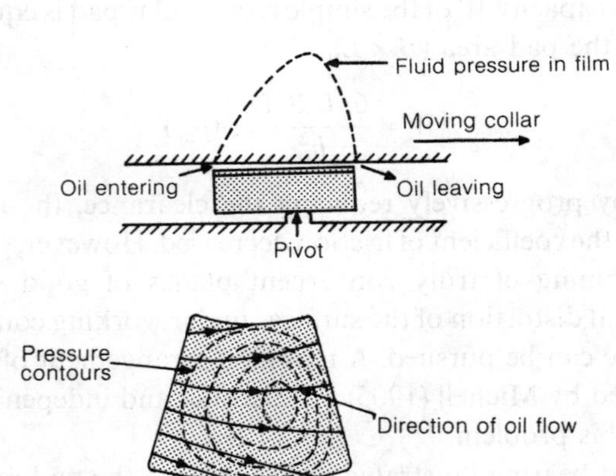

Figure 11.9 Tilting, thrust pad bearing. (Courtesy BP 1969)

A number of pivoted pads can be assembled circumferentially to form a complete bearing capable of sustaining an axial load between rotating surfaces. The pattern of oil flow and the pressure distribution in the form of pressure contours are superimposed on the plane view of one of the sector-shaped pads. It will be observed that the flow of oil in a radial direction (i.e. the side leakage) is appreciable when the length of a pad is approximately equal to its width (Figure 11.9).

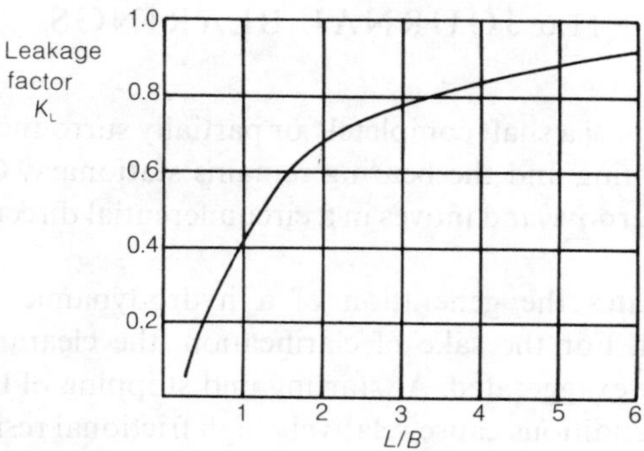

Figure 11.10 Leakage factors for bearings of finite width

The foregoing equations have been derived on the assumption that there is no side leakage of oil and consequently that the pressure remains constant in that direction. In fact, this could be true only for bearing pads having infinite length. However, the equations may be applied to real bearings with reasonable accuracy provided the ratio of L/B is greater than about 4. If L/B is less than 4, Kingsbury and Needs (Pugh 1970) have derived a series of 'leakage factors' (K_L) which when introduced into the above equations for infinitely long bearing pads enables satisfactory solutions to be obtained for pads with L/B ratios of less than 4 (Figure 11.10).

Then,

$$W = \frac{6\mu U B^2 L}{h_0^2} \cdot \phi(p_{av}) \cdot K_L \tag{11.16}$$

$$h_0 = \left[\frac{6\mu U B}{p_{av}} \cdot \phi(p_{av}) \cdot K_L \right]^{\frac{1}{2}} \tag{11.17}$$

The most important criterion in design is the minimum film thickness, which should not be less than a certain critical amount. This figure is governed by such factors as the efficiency of filtration in the oil system, the surface finishes of the bearing parts and the surface speed. Typically h_0 is of the order of 0.1 to 0.4 mm and the pad dimensions are chosen so that the average oil film pressure is between 2 and 3 N/mm^2, depending on the application. Integrating the shear stress over the slider surface we obtain the friction force

$$F = \frac{\mu U L B}{h_0} \cdot \phi(F_R) \tag{11.18}$$

where $\phi(F_R)$ is a non-dimensional function. Calculations show that the value of $\phi(F_R)$ is relatively insensitive to changes in pad inclination and, for bearings in which h_1/h_0 is in the region of 2, may be taken as equal to 0.77.

The power loss is equal to the product of the frictional drag F and the velocity U:

$$\text{Power loss} = \frac{\mu U^2 L B}{h_0} \cdot \phi(F_R) \tag{11.19}$$

11.6 JOURNAL BEARINGS

A journal bearing consists of a shaft completely or partially surrounded by a bearing surface. Usually the shaft is rotating and the bearing remains stationary. Oil enters the clearance space through a hole or groove and moves in a circumferential direction, leaving the bearing by leakage at its ends.

Figure 11.11 illustrates the generation of a hydrodynamic oil film under various conditions of operation. For the sake of clarification, the clearance between sleeve and journal has been greatly exaggerated. At starting and stopping of the shaft (Figure 11.11a) boundary or thin-film conditions cause relatively high frictional resistance, but as soon as a sufficient speed of rotation is exceeded, journal and bearing surfaces separate and hydrodynamic conditions prevail, greatly reducing the frictional resistance (Figure 11.11b). If the load is raised further or the speed reduced, the eccentricity increases and the 'attitude' angle, ϕ, decreases (ϕ is defined as the angle between the direction of the load and the line joining journal and bearing centers, measured in the direction of rotation). This reduces the minimum film thickness (Figure 11.11c). Increasing load or decreasing rotational speed will cause complete or partial breakdown of hydrodynamic conditions. At very light loads and

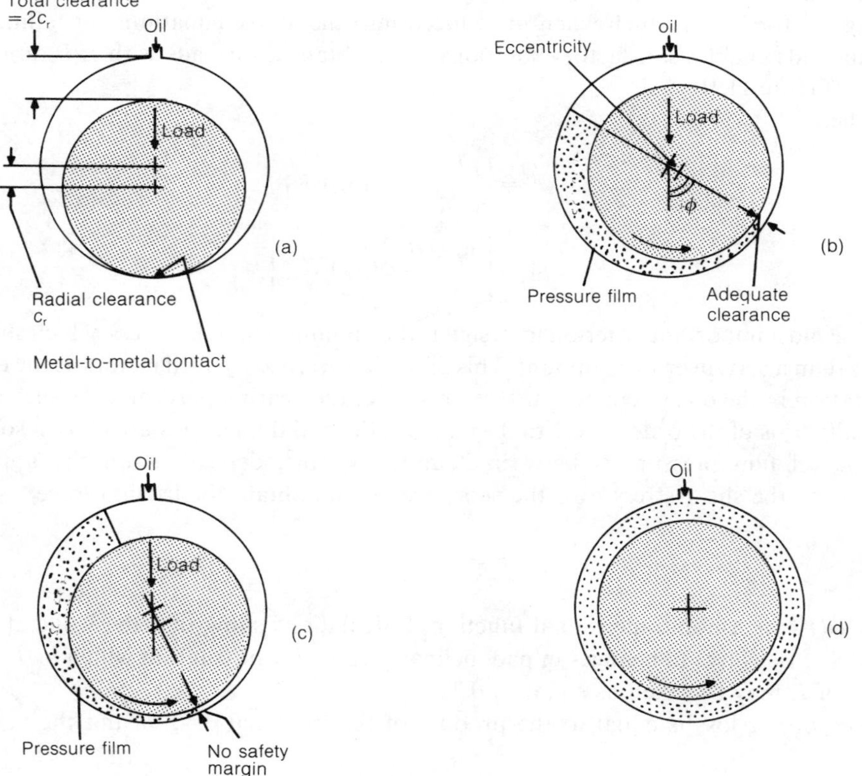

Figure 11.11 Unloading of a cylindrical bearing

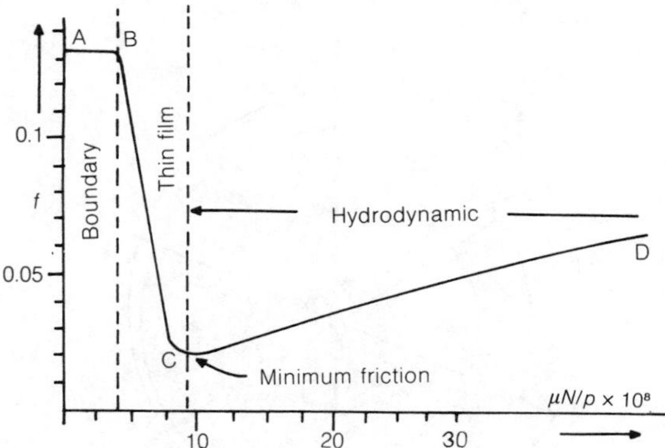

Figure 11.12 Bearing operating regimes

high rotational speeds, on the other hand, the journal will rotate almost concentrically within the bearing, i.e. eccentricity, $e = 0$ (Figure 11.11d). In this case frictional torque and power losses due to viscous shear of the oil film can be readily estimated by application of Petroff's equation. Figure 11.12 illustrates how the state of lubrication and frictional resistance in a journal bearing are influenced by changes in the three main parameters that govern its performance, as was found in Petroff's equation.

The short section AB of the curve indicates a region of operation in which a state of boundary lubrication prevails. The fact that AB is nearly horizontal confirms that in this region frictional drag is almost uninfluenced by changes in speed, load or viscosity. Section BC indicates a rapid fall in friction as $\mu N/p$ is increased, for example by a significant increase in the rotational speed; in this region thin-film or mixed conditions of lubrication exist. At C transition from thin-film to full hydrodynamic lubrication occurs, further increase in $\mu N/p$ causing a gradual rise in friction owing to the additional viscous shear.

11.6.1 Infinitely-long bearing

Figure 11.13 shows a cross-section of a 360° journal bearing with exaggerated clearance in which the geometrical relationships between the bearing and journal are illustrated. Taking as a datum the line O′O joining bearing and journal centers, the film thickness h at an angular position θ in the direction of rotation is given by

$$h = c_r(1 + \varepsilon \cos \theta)$$

where c_r is the radial clearance (i.e. half the total clearance) and the eccentricity ratio is defined by $\varepsilon = e/c_r$.

The journal is capable of supporting a load equal to the resultant of the oil pressures acting on its surface. In order to predict the actual load capacity, the circumferential pressure distribution from beginning to end of the load-bearing oil film must first be determined. The Reynolds equation for pressure distribution in a hydrodynamic bearing (equation 11.11), may be re-written in polar form by putting $\partial x = r\,\partial\theta$, then if U is surface velocity and neglecting

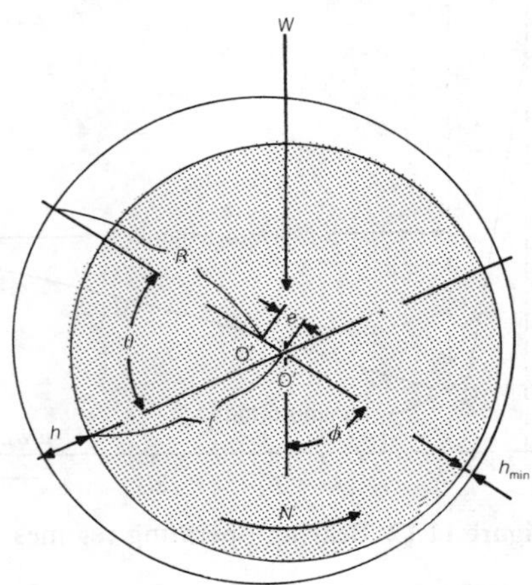

Figure 11.13 Geometry of a cylindrical bearing

$\partial/\partial y$ terms and $\dot{h}$, and integrating once

$$\frac{dp}{d\theta} = \frac{6\mu Ur(h - h')}{h^3} \tag{11.20}$$

By neglecting $\partial p/\partial y$ terms, it was assumed a long bearing (in theory, 'infinitely long'); then the above equation will correctly express the circumferential pressure gradient in any plane perpendicular to the bearing axis, since leakage of oil and change of pressure in an axial or y-direction can be neglected. Substituting for h the relation for film thickness gives:

$$\frac{dp}{d\theta} = 6\mu UR\left[\frac{c_r(1 + \varepsilon\cos\theta) - c_r(1 + \varepsilon\cos\theta')}{c_r^3(1 + \varepsilon\cos\theta)}\right] \tag{11.21}$$

where θ' is the position at which $\mathrm{d}p/\mathrm{d}\theta = 0$ or $h' = c_r(1 + \varepsilon\cos\theta')$. θ' is a constant to be determined by the boundary conditions that define the beginning and end of the generated pressure film. The pressure film can reasonably be expected to commence at the thickest part of the film (i.e. $p = 0$ when $\theta = 0$), and continue for slightly more than half the journal circumference (i.e. $p = 0$ when $\theta = +$ some angle θ' and $\mathrm{d}p/\mathrm{d}\theta = 0$). These boundary conditions, illustrated in the top diagrams of Figure 11.14, are physically realistic and generally confirmed by experimental evidence, but their exact adoption adds greatly to the complication of further mathematical analysis. Reynolds was only partially successful in solving the above equation for pressure distribution, which was subsequently integrated by Sommerfeld in 1904.

In his solution, Sommerfeld suggested that the pressure again falls to zero when $\theta = 2\pi$, with the implication that a continuous pressure film from 0 to 2π extends round the whole of the bearing circumference. In this case the oil-film pressure p at any angular position θ is given by

$$p = \frac{6\mu Ure}{c_r^2}\frac{(2 + \varepsilon\cos\theta)\sin\theta}{(2 + \varepsilon^2)(1 + \varepsilon\cos\theta)^2} \quad \text{and} \quad \varphi = \cos^{-1}\left(\frac{-3}{2 + \varepsilon^2}\right) \tag{11.22}$$

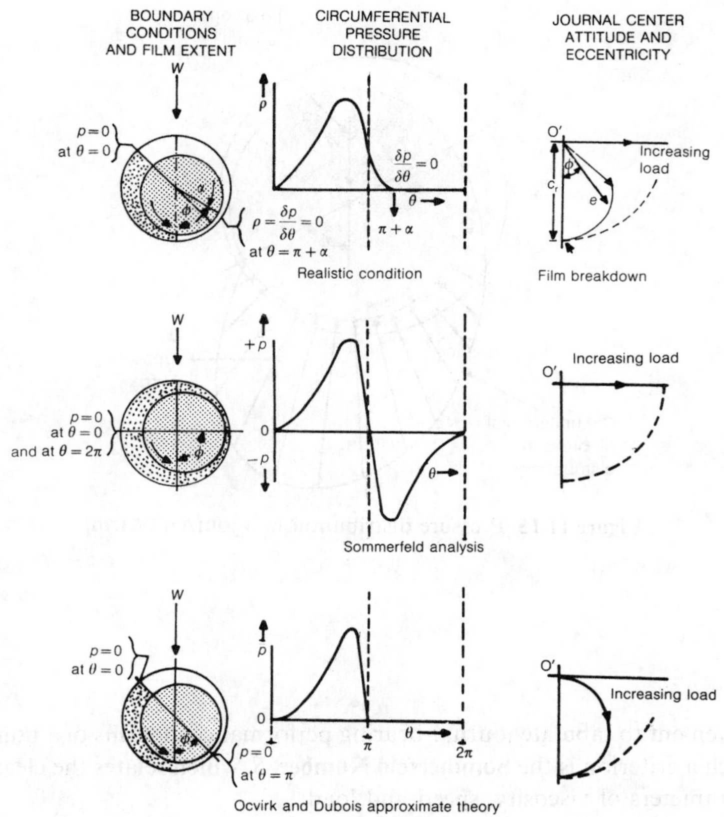

Figure 11.14 Boundary conditions for cylindrical bearings. (Courtesy BP 1969)

In Figure 11.14, the pressure distribution is shown and the bearing locus for the representative sets of boundary conditions. Realistic is the first, where the film breaks down when negative pressures are encountered (cavitation). The original Sommerfeld analysis for a full bearing accepts negative pressures. Ocvirk and Dubois boundary conditions (Cameron 1966) are a very good approximation as experiments have shown. This approximation facilitates analytical and numerical treatment of the problem.

Figure 11.15 shows the forces acting on a journal bearing under load. At equilibrium the load, acting through the center of the bearing, must be equal and opposite to the resultant of the fluid pressures in the film surrounding the shaft. By resolving and equating these, Sommerfeld showed that the load capacity W of a full, infinitely-long journal bearing is given by:

$$W = \mu UL\left(\frac{r}{c_r}\right)^2 12\pi\varepsilon/[(2+\varepsilon^2)(1-\varepsilon^2)]^{\frac{1}{2}} \tag{11.23}$$

where, μ is the oil viscosity, U is the peripheral speed of the journal, L is axial length of the bearing, c_r is the radial clearance and c_r/r the clearance ratio.

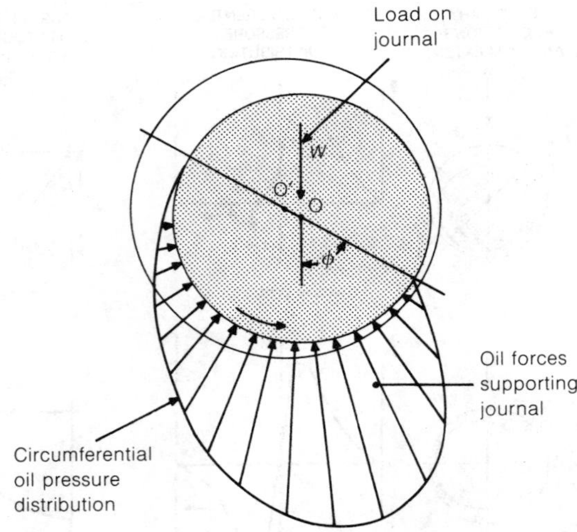

Figure 11.15 Pressure distribution in a journal bearing

It is convenient to tabulate journal-bearing performance in terms of a non-dimensional criterion. Such a criterion is the **Sommerfeld Number**, S, which relates the clearance ratio to the other parameters of viscosity, speed and load:

$$S = \frac{\mu N}{p}\left(\frac{r}{c_r}\right)^2 \tag{11.24}$$

where μ = absolute viscosity, N = speed (rev/sec), p = load per unit projected bearing area = W/LD, r = journal radius, and c_r = radial clearance.

Equation (11.23) assumes a dimensionless form

$$S = \frac{(2+\varepsilon^2)(1-\varepsilon^2)^{\frac{1}{2}}}{12\pi^2\varepsilon}, \quad \varphi = \pi/2 \tag{11.25}$$

A full 360° journal bearing, for which equation (11.25) holds, encounters problems of cavitation and stability. Partial bearings have been used. The 180° bearing (the bottom half) is very widely used. Solution of the Reynolds equation for infinitely-long bearing yields

$$S_{180^\circ} = \frac{(2+\varepsilon^2)(1-\varepsilon^2)^{\frac{1}{2}}}{12\pi^2\varepsilon[1/4 + \varepsilon^2/\pi^2(1-\varepsilon^2)]^{\frac{1}{2}}} \quad \tan\varphi = -\pi(1-\varepsilon^2)^{\frac{1}{2}}/2\varepsilon \tag{11.26}$$

For finite length full 360° bearings, numerical solutions yield the pressure distribution for finite bearings, Figure 11.16. Upon integration of the pressures, the load can be related to the bearing eccentricity as in Figure 11.17, giving the eccentricity as function of the Sommerfeld Number.

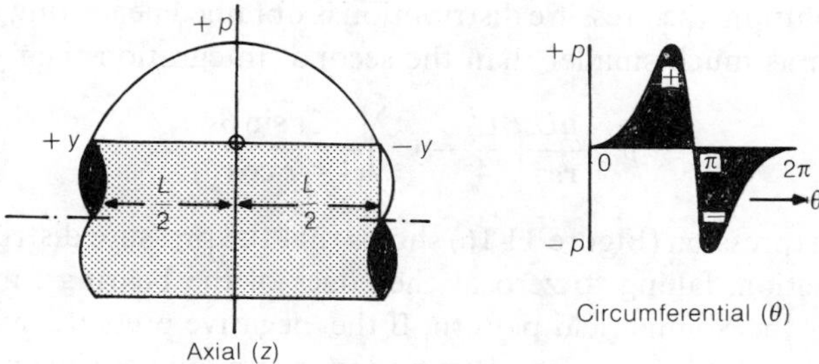

Figure 11.16 Axial pressure distribution

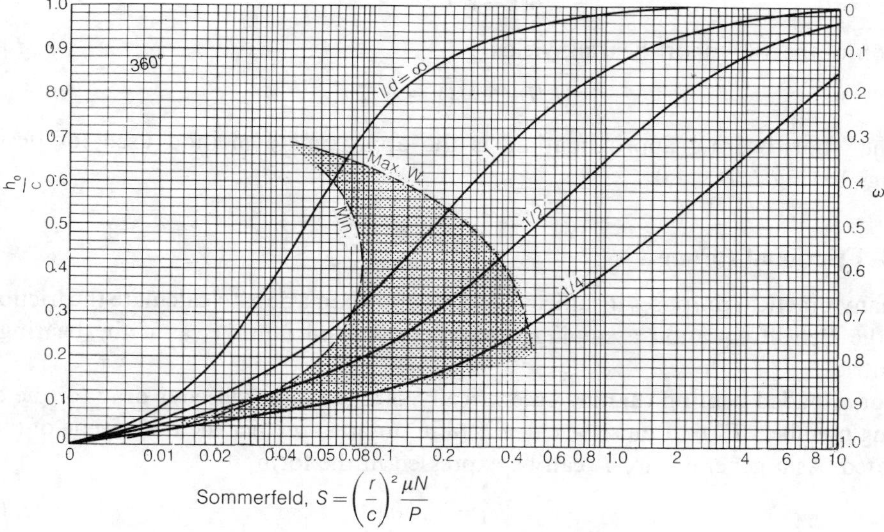

Figure 11.17 Minimum oil film thickness. (After Raimondi and Boyd 1958; courtesy ASLE)

11.6.2 Short bearings

The present trend in industrial practice is to utilize narrow bearings in which the length-to-diameter ratio is less than 1. This reduced L/D ratio emphasizes the practical consequences of side flow of oil and introduces a factor of unreality into design procedures based on Sommerfeld's solution of the Reynolds equation.

An alternative solution which includes the axial oil flow and also retains a considerable part of the circumferential flow was proposed by Michell in 1929 and later extended by Ocvirk and Dubois in 1952 (Cameron 1966) as a basis of a chart design procedure for narrow bearings. This latter analysis supplements the Sommerfeld analysis for narrow bearings and the two may be considered as the halves of the same type of solution.

In Michell's solution, the pressure distribution is obtained neglecting the first term in the Reynolds equation as much smaller than the second. Integration then yields

$$p = \frac{\mu U}{rc_r^2}\left(\frac{L^2}{4} - y^2\right)\frac{3\varepsilon \sin\theta}{(1 - \varepsilon\cos\theta)^3} \tag{11.27}$$

A plot of this expression (Figure 11.16) shows that the pressure distribution is parabolic in the axial y-direction, falling to zero at the edges of the bearings; in a circumferential direction it follows the Sommerfeld pattern. If the 'negative pressure' region from $\theta = \pi$ to $\theta = 2\pi$ is omitted, the overall pressure distribution is in good agreement with experimental results. Integration of the pressure gives,

$$1/(L/D)^2 S = \frac{\varepsilon\pi}{4(1-\varepsilon^2)^2}[\pi^2(1-\varepsilon^2) + 16\varepsilon^2]^{\frac{1}{2}} \tag{11.28}$$

Solution of the 180° short bearing yields

$$(L/D)^2 S = (1-\varepsilon)^{\frac{3}{2}}/2\pi^2\varepsilon$$

The 'short-bearing approximation' is the approximate method used for design of bearings having $L/D \leqslant 1.0$.

11.6.3 Frictional torque

For many practical purposes it suffices as a first approximation, to calculate the friction loss from the Petroff equation assuming a parallel-sided oil film, i.e. as if the bearing were concentric.

For a loaded bearing running eccentrically, it is the larger viscous drag torque on the rotating member, T_J, that must be calculated in order to predict the magnitude of the heat generated. As a general rule, T_J can be expressed in the form

$$T_J = T_P \phi(\varepsilon) \tag{11.29}$$

where T_P is the torque predicted by Petroff's equation and $\phi(\varepsilon)$ is a friction factor involving the degree of eccentricity (at concentricity, $\phi(0) = 1$). As the eccentricity increases, the calculated value of $\phi(\varepsilon)$ rises gradually but still remains near unity until eccentricity ratios of the order of 0.5 to 0.6 are reached. However at higher eccentricities, $\phi(\varepsilon)$ increases more rapidly and for very high eccentricities approaches a value of 3.

For the infinitely long and the short bearings, function $\phi(\varepsilon)$ can be readily computed. To this end, the power loss per unit time for the portion of the film at an angle d will be

$$d\dot{W} = Qdp = q\frac{\partial p}{\partial \theta}\mathrm{d}\theta + Q\frac{\partial p}{\partial y}\mathrm{d}y \tag{11.30}$$

Integrating

$$\dot{W} = \int_{\theta_1}^{\theta_2}\int_0^L Q(\theta, y)\left[\frac{\partial p}{\partial \theta}\mathrm{d}\theta + \frac{\partial p}{\partial y}\mathrm{d}y\right] = T_J\omega = T_P\omega\phi(\varepsilon) \tag{11.31}$$

where the flow Q is given by equation (11.8). Integration under the assumptions of infinitely

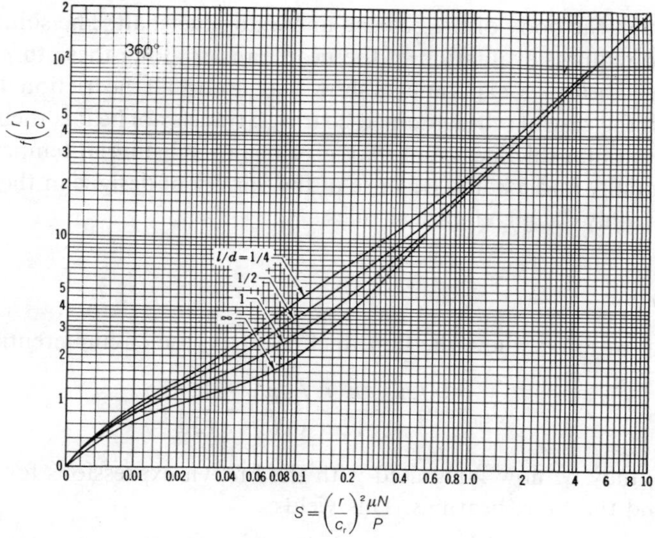

Figure 11.18 Friction factor. (After Raimondi and Boyd 1958; courtesy ASLE)

long and short, and 180° and 360°, bearing possibilities, yields the following results:

(i) 360° bearing, infinitely long:

$$\phi_1 = 2(1+2\varepsilon)^2/[(2+\varepsilon^2)/(1-\varepsilon^2)^{\frac{1}{2}}] \tag{11.32}$$

(ii) 180° bearing, infinitely long:

$$\phi_2 = (1+2\varepsilon^2)/[(2+\varepsilon^2)(1-\varepsilon^2)^{\frac{1}{2}}] \tag{11.33}$$

(iii) 360° bearing, short:

$$\phi_3 = [1+\varepsilon^2/2(1-\varepsilon^2)](L/D)^2/(1-\varepsilon^2)^{\frac{1}{2}} \tag{11.34}$$

(iv) 180° bearing, short:

$$\phi_4 = \phi_3/2 \tag{11.35}$$

Numerical solutions for finite bearings yielded friction-coefficient functions given in Figure 11.18.

11.6.4 Heat balance

In order to estimate the actual oil temperature and hence the working viscosity, a balance must be established between the frictional heat generated within the bearing clearance and the heat dissipated from the bearing itself. At thermal equilibrium, the frictional heat is partly carried away in the oil flowing out from the ends of the bearing and partly transmitted by conduction through the body of the bearing and along the shaft. The solution of the heat-balance equation requires evaluation of the heat transfer along these three paths.

For forced-feed bearings, experiments show that almost all the frictional heat is removed by the oil flow. In the case of self-contained bearings, transfer of heat is then localized to the region of minimum film thickness and may give rise to substantial temperature gradients both circumferentially and axially. Because of the variety of designs in these bearings, it is possible to apply only rough, empirical rules.

11.6.5 Oil flow

At any angular position θ where the film thickness is h, the circumferential flow Q_c is

$$Q_c = \frac{ULh}{2} - \frac{h^3 L}{12\mu r}\left(\frac{\partial p}{\partial \theta}\right) - \frac{h^3}{12\mu}\frac{\partial p}{\partial y} \tag{11.36}$$

Equation (11.36) may be now integrated with the known expressions for pressure for the infinitely long and the short bearings. This yields:

$$\textit{Infinitely long: } Q_L = \pi c_r DLN(1-\varepsilon^2)/2(2+\varepsilon^2) \tag{11.37}$$

$$\textit{Short: } Q_S = \pi c_r DLN[5-(1+24\varepsilon^2)^{\frac{1}{2}}]/16 \tag{11.38}$$

Oil flows for the 360° finite bearing have been obtained numerically and given in Figure 11.19 in the form of a flow factor Q/r_cNL.

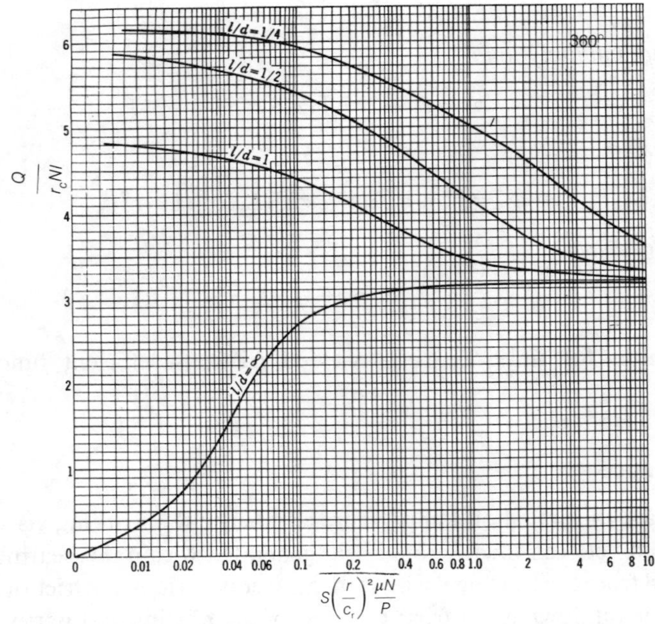

Figure 11.19 Oil flow. (After Raimondi and Boyd 1958; courtesy ASLE)

11.6.6 Journal bearing design

From practical experience in bearing design, a number of design rules have emerged that provide general guidance in choosing bearing dimensions and acceptable operating conditions.

(a) Heat generation

The product pU of the average pressure and the peripheral velocity is proportional to the generated heat FU, where F is the frictional force, in proportion to p. Since the capacity to absorb and transmit the generated heat depends on the overall design and operating conditions, it is suggested for automotive engines that pU is up to 25–35 Nm/mm^2 s, high-speed internal combustion engines up to 30–50, rolling-mill bearings 40–200, and steam turbines up to 100 Nm/mm^2 s. This experience has been included in Figure 11.20(a, b).

For thermal calculations, the heat transfer coefficient at the bearing surface can be taken $H = 10(1 + u^{\frac{1}{2}})$ W/m^2 °C, where u (m/s) is the velocity of the air flow around the bearing housing. If no forced flow is available, natural convection can be assumed with $H = 14$ W/m^2 °C.

For bearings without oil circulation, all generated heat has to be transmitted to the environment by the bearing enclosure surface A. The equilibrium temperature will be at the point where the generated heat, given by equation (11.30), equals the heat removed by convection over the surface A:

$$T_P\omega\phi(\varepsilon) = HA(T_b - T_e) \tag{11.39}$$

where T_b is the bearing temperature and T_e the environment temperature.

T_b must not exceed, for usual oils, 70 °C for normal conditions and 90 °C for heavily-loaded bearings. If T_b is substantially higher than the environment temperature, it affects

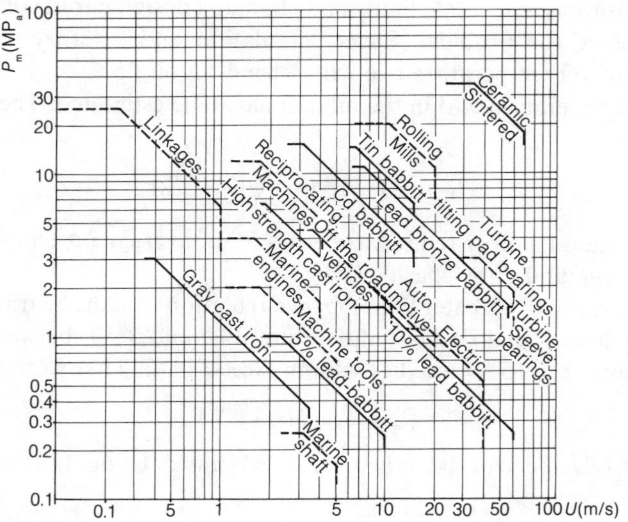

Figure 11.20(a) Design experience with fluid bearings

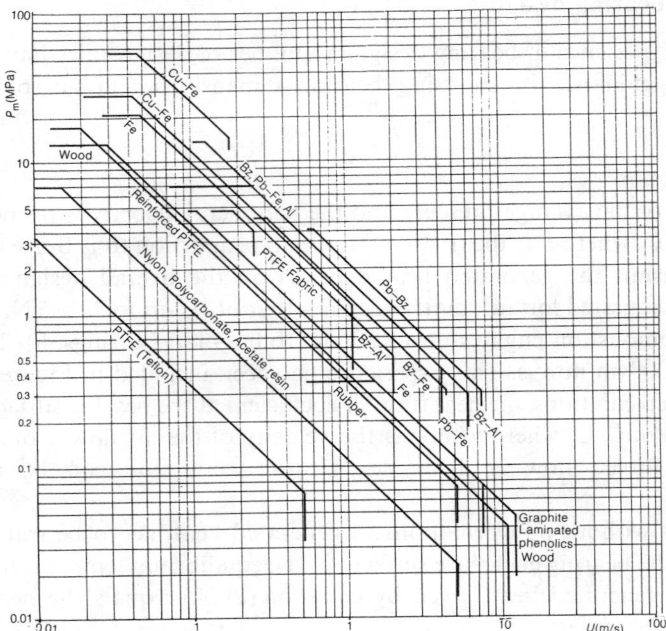

Figure 11.20(b) Materials for fluid bearings

viscosity and the bearing calculations have to be repeated until the difference between assumed viscosity and the one at the resulting bearing temperature are close enough.

If the temperature is above safe limits and changing design parameters within practical limits cannot resolve the problem, forced circulation and external oil cooling can be provided. Then, the oil temperature rise will depend on oil flow.

First, the heat generation within the oil film has to be estimated. Then, energy balance demands that

$$T_P \omega \phi(\varepsilon) = \dot{W} = c\rho Q(T - T_0) \tag{11.40}$$

where T_0 is here the oil supply temperature and T an average oil temperature, Q the oil supply to the bearing and c the specific heat.

The oil supply is usually greater than the oil circulation Q in the bearing in order to keep exit temperatures low. Usual design-temperature differences ΔT between oil supply and return are in the range 8 to 15 °C and they seldom exceed 20 °C. The external oil supply will be

$$Q_e = T_P \omega \phi(\varepsilon)/c\rho \Delta T \tag{11.41}$$

For mineral oils, ρc can be taken as 1.36 N/mm^2 °C or 110 lb/in^2 °F at normal temperatures.

Self-contained bearings without external oil circulation have been successfully operated in the following range:

Diameter (mm)	Maximum speed (rpm)
75	3600
200	1000
600	200

(b) Clearance ratio (c_r/r)
The bearing performance is particularly sensitive to changes in this ratio. Clearance ratios are usually within the range of 0.001 to 0.0005. The lower figure applies to slow speed bearings and it is usually necessary to increase the ratio with increasing speed to avoid undue heating. A diametral clearance of 0.001 in per inch of shaft diameter is commonly accepted as standard practice, i.e. $c_r/r = 0.001$. Vogelpohl suggests $r/c_r = 1000 + 0.4(U)^{\frac{1}{2}}$ within U m/s.

It is suggested that calculations are performed for a range of clearance ratios and the results tabulated. Then, the optimum value is selected by an overall evaluation of the results obtained.

(c) Minimum film thickness (h_{min})
Production tolerances, surface finish of journal and bearing, and the effects of possible contaminants in the oil supply must all be considered. For small bearings (less than 50-mm diameter) having high-quality surface finishes and running at slow to medium speeds, h_{min} is usually designed to be not less than about 2.5 μm. Trumpler suggested minimum film thickness (μm) equal to $5 + 0.4D$, where D is in millimeters. Limiting factors for the minimum film thickness are the journal and bearing surface roughnesses R_j and R_b respectively. The minimum film thickness should account also for the maximum expected size of solid particles in the oil d_f usually dictated by the filter type. For usual industrial quality filters this size is 50 μm while with special filters particle size can be limited to 5 to 10 μm. Finally

$$h_{min} > S(R_j + R_b) + d_f$$

where S should be at least 2. This should be the minimum film thickness for the running fit selected.

(d) Radial clearance (c_r)
Allowances may have to be made for wear and for possible shaft misalignment or deflection. For example, if the minimum clearance ratio desired is 0.001, allowance for wear could be 0.0005 and production tolerance 0.0005. Calculations of load capacity and performance must be made at both possible extremes of clearance resulting from these allowances.

(e) Length-to-diameter ratio (L/D)
There is no obvious advantage in making the bearing much longer than the shaft diameter since reduced loading per unit projected area of bearing may be offset by increased frictional torque; the chances of edge loading owing to misalignment are also increased. If no special restrictions are imposed, an L/D ratio of 1.0 is a reasonable basis to adopt.

In general, length L is computed for the selected oil, clearance and minimum film thickness.

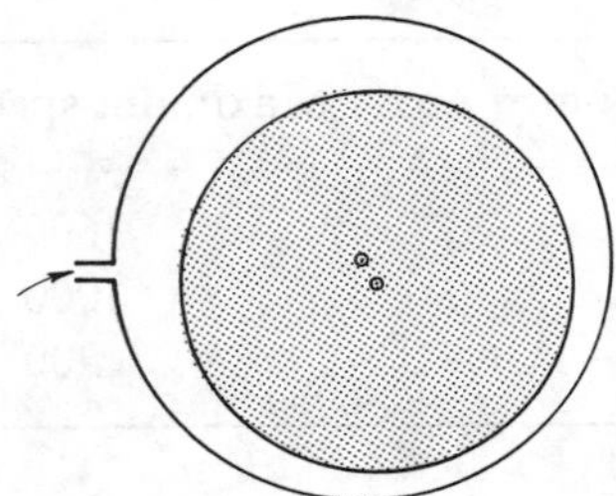

Figure 11.21 Geometry of a cylindrical journal bearing

(f) Oil grooves

The purpose of oil grooves is to distribute the lubricant in the bearing so that an effective pressure film may be formed as soon after the point of oil entry as possible (Figure 11.21). Grooves that are incorrectly placed, for example within or close to the loaded area, reduce the load-carrying capacity.

11.6.7 Stability of journal bearings

Hydrodynamically-lubricated journal bearings can exhibit a form of instability known as 'oil-film whirl'. This most readily occurs in high-speed, lightly-loaded bearings such as those of some turbines and gives rise to the transmission of vibrations throughout the equipment, which is most undesirable because of the possible fatigue effects on components. There are two types of oil film whirl.

In 'half-speed' or 'half-frequency' whirl, the journal performs an orbital path in the bearing. The locus of the journal center usually encloses the bearing center; its direction is in that of journal rotation and its speed is half the speed of rotation of the journal.

In 'resonant' whirl, or 'oil whip', the journal center describes a similar path, but its frequency is that of the critical speed for mechanical whirling of the shaft alone. Resonant whirl usually occurs only at speeds above twice the critical whirling speed. In resonant whirl the stiffness of the rotating shaft is seen therefore to be a controlling factor; the larger the diameter of the shaft relative to its length, the less likely is instability in the bearing.

This unstable interaction of rotating shaft with hydrodynamic oil film is to be avoided in the design of bearings. The following factors, some of which cannot be taken into account by the bearing designer because they are not under his control, all reduce the likelihood of oil-film whirl:

1. Reduced bearing clearance
2. Increased oil supply pressure
3. Lower running temperature
4. Higher oil viscosity
5. Lower running speed
6. Lower L/D ratio of journal and shaft
7. Higher bearing load

8. Absence of vibrations or shocks, including load fluctuations, that could cause an initial displacement of the journal from its equilibrium position.

Oil grooves in the bearing give greater stability, as does an 'anti-whirl' bearing design such as a slightly elliptical bore (Figures 11.21 and 11.22).

Tilting pad bearings usually resolve oil-film whirl problems as they follow the motion of the shaft without generation of lateral forces which could initiate an unstable shaft motion.

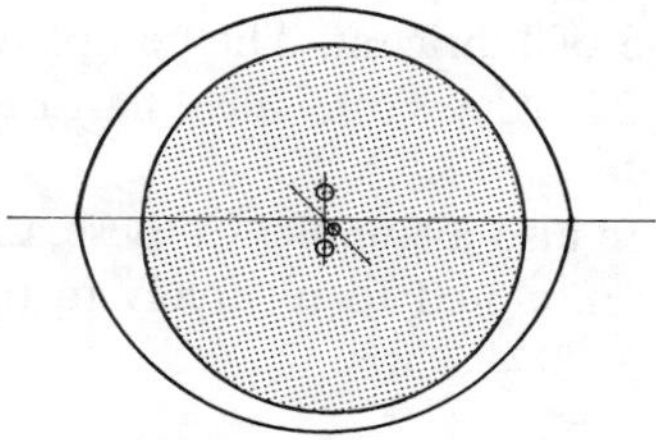

Figure 11.22 Geometry of an elliptical bearing

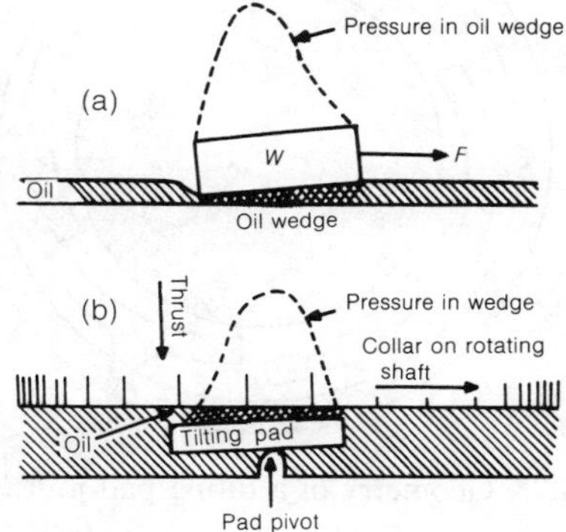

Figure 11.23 Pressure distribution on a tilting pad

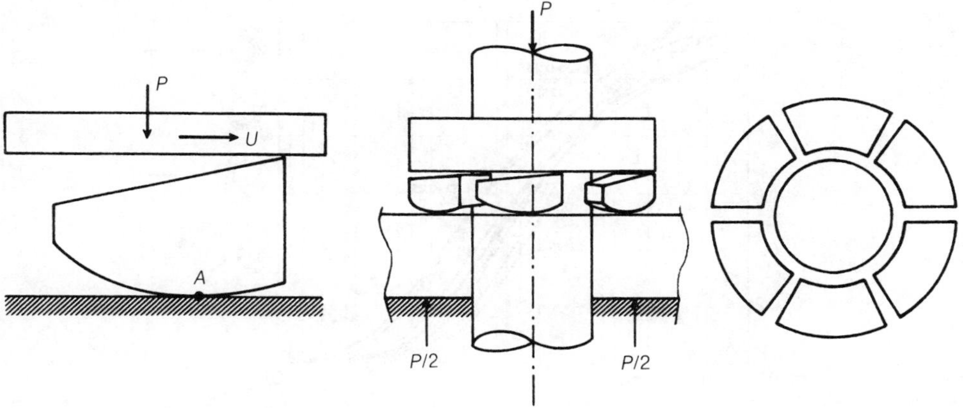

Figure 11.24 Geometry of an axial tilting pad bearings

11.6.8 Tilting pad bearings

For axial bearings used for ship propulsion there was a problem of wedge formation which was solved by Michell, dividing the thrust surface into a number of individual pads with an inclination in respect to the direction of rotation, such as in Figure 11.23a. It was further found that improved operation could be achieved by non-fixed pads, free to turn about a pivot (Figure 11.23b).

The pads were so pivoted that they could take up the angle of approach necessary for the formation of the wedge of lubricant. The complete thrust ring for such a thrust block is shown in Figure 11.24. The same idea has been used for journal bearings (Figure 11.25).

Michell journal bearings are also available. In these, tilting pads replace the ordinary fixed bearing, and are suitably mounted to tilt freely to the angle necessary to form the

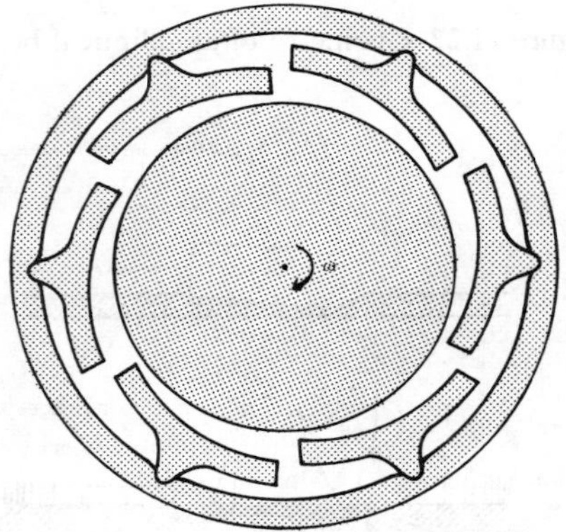

Figure 11.25 Geometry of a tilting pad journal bearing

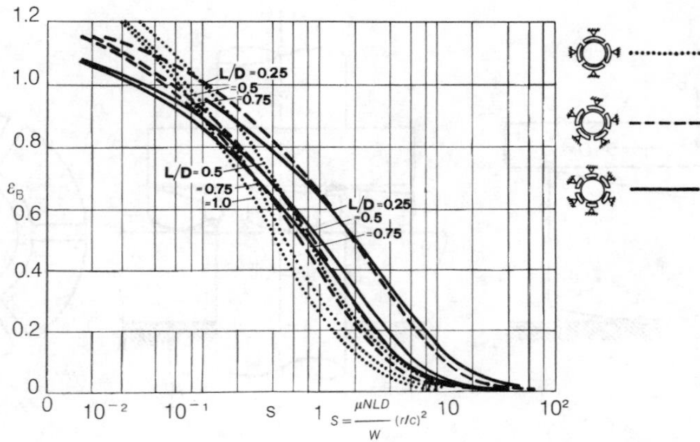

Figure 11.26 Eccentricity ratio for tilting pad journal bearings

wedge of lubricant, but they cannot move circumferentially with the shaft. Much higher bearing pressures then become possible and better vibration attenuation. Load-carrying capacity for usual types of pad bearing can be found from Figure 11.26. Power loss and oil flow can be taken equal to the ones for the full journal bearing with the same eccentricity ratio, if appropriate design data are not available.

11.7 EXTERNALLY-PRESSURIZED (HYDROSTATIC) BEARINGS

To keep two solid surfaces separated, fluid pressure must be applied at the interface. Consider a block of area A subjected to a load W and separated from a plane by a film of fluid held at a constant pressure p_s, as in Figure 11.27. The film can be sustained provided that

$$W = p_s A \tag{11.42}$$

If p_s is not constant at all points, so that each element $\mathrm{d}A$ carries a load $\mathrm{d}W$,

$$\mathrm{d}W = p\,\mathrm{d}A \tag{11.43}$$

For the total load

$$W = \int_A p\,\mathrm{d}A \tag{11.44}$$

where the integral is taken over the total area.

For flow along a uniform pipe of diameter d, (Figure 11.28), the Poiseuille flow (equation 11.6), can be written as

$$q = \frac{\pi d^4}{128\mu}\left(\frac{\mathrm{d}p}{\mathrm{d}x}\right) = \frac{\pi d^4}{128\mu}\left(\frac{p_2 - p_1}{L}\right) \tag{11.45}$$

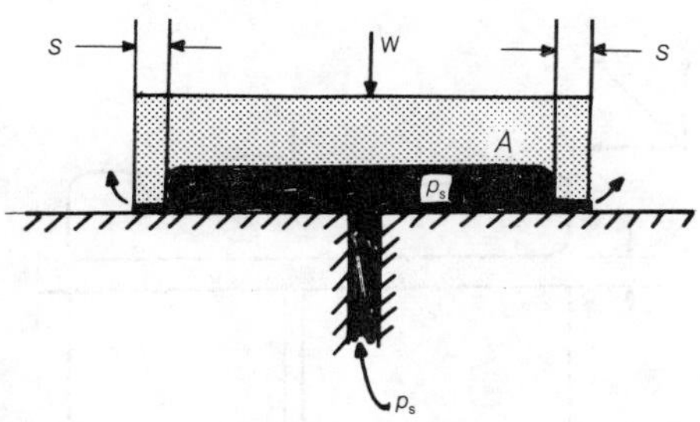

Figure 11.27 Externally pressurized (hydrostatic) bearing

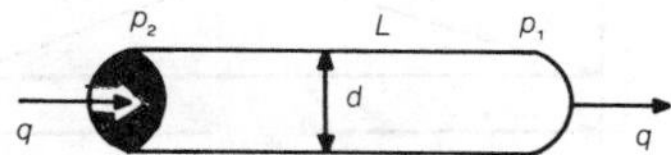

Figure 11.28 Pipe flow

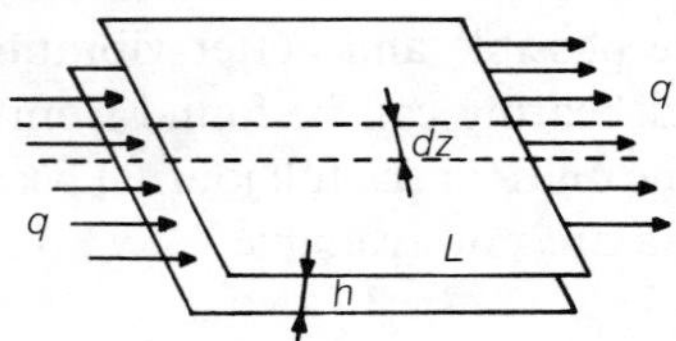

Figure 11.29 Plate flow

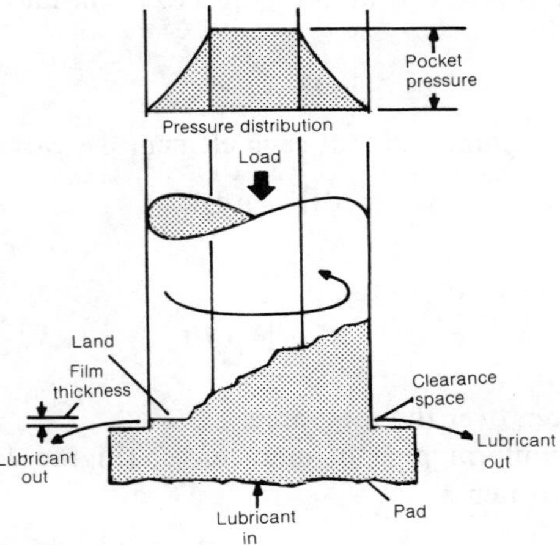

Figure 11.30 Hydrostatic bearing, uniform pressure

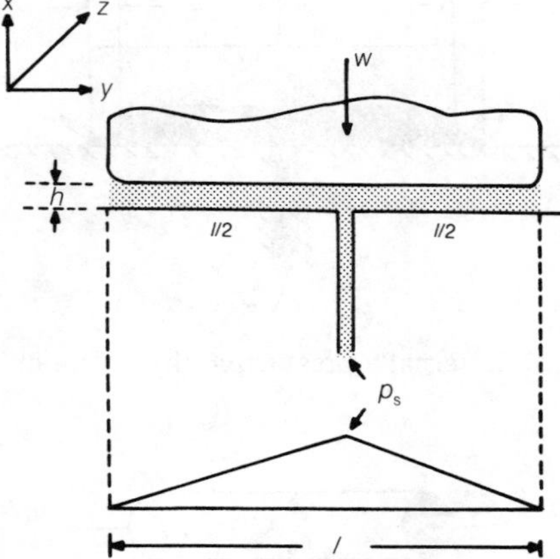

Figure 11.31 Hydrostatic bearing, uniform oil film thickness

For flow between parallel surfaces (Figure 11.29), equation (11.6) can be written as

$$dq = \frac{h^3 dz}{12\mu}\left(\frac{\mathrm{d}p}{\mathrm{d}x}\right) = \frac{h^3 dz}{12\mu}\left(\frac{p_2 - p_1}{L}\right) \tag{11.46}$$

The pressure drop along the large section is very small in relation to the other section, and for such large sections we can therefore assume constant pressure (Figure 11.30).

For the geometry shown in Figure 11.31, for constant flow,

$$\frac{\mathrm{d}p}{\mathrm{d}x} = -\frac{12\mu dq}{h^3 dz} = \frac{k}{h^3} \tag{11.47}$$

where k is a constant.

11.7.1 The simple pad hydrostatic bearing

A simple hydrostatic pad is supplied by constant pressure p_s along the central line and the x-direction. Flow is then assumed along the y-direction only.

The load per unit width in the z-direction is

$$W = 2\left(\frac{p_s - 0}{2}\right)l/2 = p_s l/2 \tag{11.48}$$

If we now increase the load we obviously have to increase p_s to maintain equilibrium.

There are two possible solutions. We may use not a constant pressure but a constant flow supply, or we may use a constant pressure supply together with a 'compensating' element.

(a) Constant flow supply

Suppose a pump supplies fluid at a constant flow rate q. Neglecting flow in the z-direction, the flow from the bearing in the y-direction is from the center to the two edges. Applying equation (11.46) for parallel-walled channels to one-half of the bearing gives (Figure 11.31).

$$1/2q = \frac{h^3}{12\mu}\left(\frac{p_s - 0}{l/2}\right) \tag{11.49}$$

Thus,

$$p_s = \frac{3q\mu l}{h^3} \tag{11.50}$$

and

$$W = \frac{p_s l}{2} = \frac{3q\mu l^2}{2h^3} \tag{11.51}$$

Since q is constant, any change in W is accommodated by a change in the film thickness h, and the supply pressure p_s now varies to ensure a constant flow rate q. An improvement in the bearing design will significantly increase its load capacity (Figure 11.32). Since the film thickness in the recess is relatively large with respect to the film thickness at the lands, the pressure in the recess is essentially constant. The load capacity of such a bearing is

$$W = p_s b + (p_s/2)l \tag{11.52}$$

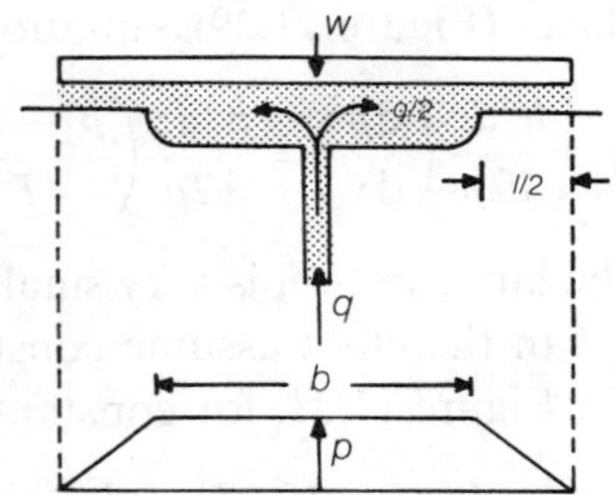

Figure 11.32 Hydrostatic bearing, mixed

Considering the constant flow supply through the lands given by equation (11.49) as before, the load capacity is now

$$W = \frac{3q\mu l^2}{2h^3}[1 + (2b/l)] \tag{11.53}$$

This is an increased load capacity by a factor $1 + (2b/l)$, and this type of geometry is that most commonly employed in externally-pressurized bearing designs.

However, the use of a constant flow type of pump implies that every bearing would require its own separate pump, a very expensive arrangement. The advantage of the other alternative, a constant pressure pump, is that the same pump may supply a whole series of bearings provided that its flow capacity can meet their total requirements.

(b) Constant pressure supply

We have seen that a direct connection of a constant pressure supply to the bearing is not practical since we would have to adjust the supply pressure to accommodate changes in load. Automatic adjustment can be achieved by a restricting element. The simplest compensating element is a length of capillary tube. Consider the system shown in Figure 11.33a, where a constant supply pressure p_s drops to a pressure p_r at the bearing supply point. From the flow through a tube we note, from equation (11.44), that

$$q = \frac{\pi d^4}{128\mu}\left(\frac{p_s - p_r}{l_c}\right) \tag{11.54}$$

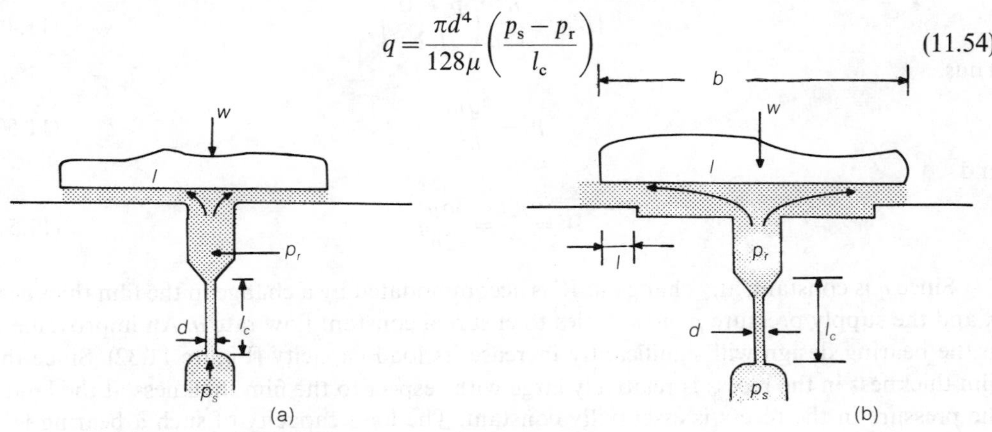

Figure 11.33 Hydrostatic bearing, constant pressure supply, capillary compensated

The linear pressure drop along each half of the bearing (through which flows half of the fluid) must give flow

$$\frac{q}{2}=\frac{h^3 w}{12\mu}\frac{p_r}{l/2} \tag{11.55}$$

where w is the width of the bearing in the z-direction. Thus eliminating q one obtains,

$$p_r=\frac{p_s}{1+(h^3/k)} \tag{11.56}$$

where

$$k=\frac{3\pi d^4 l}{128 l_c w}$$

The load carried is then given by

$$W=\frac{p_r l w}{2}=\frac{P_s l w}{2[1+(h^3/k)]} \tag{11.57}$$

This equation relates the load to the film thickness for a given geometry of bearing. The flow through such a bearing is then found from equations (11.54) and (11.55),

$$q=\frac{k(p_s l w-2w)}{1+(h^3/k)} \tag{11.58}$$

Thus, although the supply pressure is constant, changes in load cause a change in the film thickness and in the flow rate. For a constant p_s any increase in load increases p_r and reduces flow rate. The operation is stable. A more useful geometry, in practise, is to have a large central recess together with narrow lands (Figure 11.33b). The load capacity for such a bearing then becomes

$$W=\frac{p_s(b+l/2)w}{1+(h^3/k)} \tag{11.59}$$

11.7.2 Externally-pressurized journal bearings

Figure 11.34 shows a single-pad and a multi-pad arrangement for such a bearing. Each pad is supplied separately from a constant pressure supply via appropriate compensating elements. The oil then flow over the lands into the drainage channels and the calculations of Section 11.7.1 apply.

Such bearings are used in slow speed applications where the speed is not high enough to sustain a hydrodynamic film. It is used also in heavy machinery, such as large steam turbines for start-up until they reach a speed high enough to assure an adequate hydrodynamic oil film thickness. The flow for each pad is:

$$q=\frac{k[p_s(b+l/2)w-W]}{3\mu l(b+l/2)} \tag{11.60}$$

For cases with oil-flow in the z-direction or with more complex geometries, the basic behavior is similar.

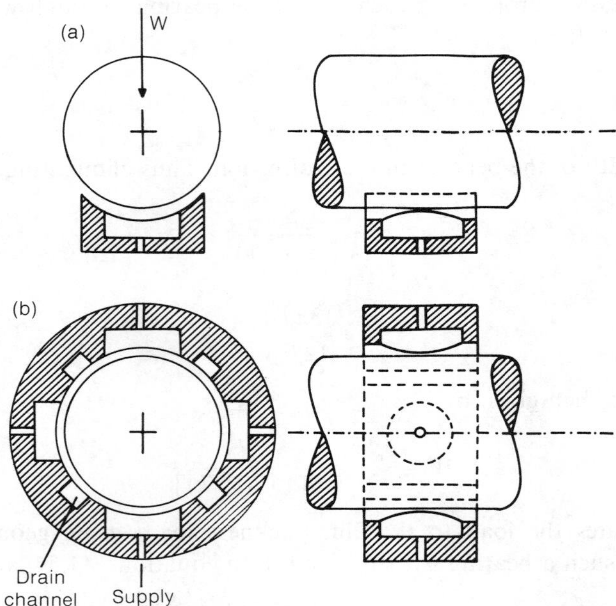

Figure 11.34 Hydrostatic journal bearing

11.8 COMPUTER METHODS

11.8.1 Numerical solution of the Reynolds equation

The hydrodynamic lubrication theory has a strong analytical tool in the form of Reynolds equation to predict bearing behavior. Solutions of this equation in analytical form have been obtained only in special cases such as the infinitely long and the short bearing. In the ranges $0.5 < L/D < 2$ and $0.6 < \varepsilon < 0.9$ which are the most important in applications, such solutions deviate considerably from reality. For this reason, numerical methods have been employed, implemented in digital computers.

The finite-element method, introduced in Chapter 4, is widely used today for bearing analysis and design. The direct method used in Chapter 4 is not applicable here and one has to work directly with the Reynolds equation.

The bearing area in the form of a rectangle of dimensions $2\pi RL$ is divided in triangular elements as in elasticity problems of Chapter 4.

Each node is assigned one degree of freedom, the oil pressure at the node. Within the element, an interpolation function will be used. The pressure within the element will be then a function of $x = R\theta$, z.

$$\bar{p}^{(e)}(x, z) = N_1 p_1 + N_2 p_2 + N_3 p_3 = \sum_{i=1}^{3} N_i p_i \tag{11.61}$$

where p_i are the pressures at nodes $i = 1, 2, 3$, and N_i are functions of x and z such that at the nodes the above expression gives the nodal pressures p_i. The shape or interpolation functions

N_i are the same as in the plane elasticity problem (Chapter 4).

$$\begin{aligned} N_1 &= (a_1 + b_1 x + c_1 z)/2\Delta \\ N_2 &= (a_2 + b_2 x + c_2 z)/2\Delta \\ N_3 &= (a_3 + b_3 x + c_3 z)/2\Delta \end{aligned} \tag{11.62}$$

where 2Δ is the area of the triangular element.

$$\Delta = \frac{1}{2}\begin{bmatrix} 1 & x_1 & z_1 \\ 1 & x_2 & z_2 \\ 1 & x_3 & z_3 \end{bmatrix} \tag{11.63}$$

The constants a, b, c are determined from the conditions that, at the nodes, the pressures are equal to the node pressure. This yields,

$$a_1 = x_2 z_3 - x_3 z_2, \quad b_1 = z_2 - z_3, \quad c_1 = x_3 - x_2 \tag{11.64}$$

The other coefficients are obtained by circular permutation.

The Reynolds equation (11.11) is

$$\frac{\partial}{\partial x}\left[h^3(x)\frac{\partial p}{\partial x}\right] + \frac{\partial}{\partial z}\left[h^3(x)\frac{\partial p}{\partial z}\right] = 6\mu U\frac{\mathrm{d}h}{\mathrm{d}x} + 12\frac{\mathrm{d}h}{\mathrm{d}t} \tag{11.65}$$

assuming that the oil film thickness h is a function of x (or θ) only. The right-hand side is a function of x only, $-q(x)$.

The **Galerkin Principle** (Huebner 1975) states that multiplying the equation

$$\frac{\partial}{\partial x}\left(h^3\frac{\partial \bar{p}^{(e)}}{\partial x}\right) + \frac{\partial}{\partial z}\left(h^3\frac{\partial \bar{p}^{(e)}}{\partial z}\right) + q = 0 \tag{11.66}$$

by functions $N_i(x, z)$ and integrating over the problem surface, if the integrals are zero, the greater the number of such functions, the greater the proximity of the function $\bar{p}(x, z)$ to the exact solution. Let us use three such functions, the shape functions N_i:

$$\iint_{(e)} N_i\left[\frac{\partial}{\partial x}\left(h^3\frac{\partial p^{(e)}}{\partial x}\right) + \frac{\partial}{z}\left(h^3\frac{\partial \bar{p}^{(e)}}{\partial z}\right) + q\right]\mathrm{d}x\,\mathrm{d}z = 0, \quad i = 1, 2, 3 \tag{11.67}$$

Recall now the **Green's Theorem** for integration by parts of double integrals

$$\iint_\Omega \varphi\frac{\partial\psi}{\partial x}\mathrm{d}x\,\mathrm{d}y = -\iint_\Omega \frac{\partial\varphi}{\partial x}\psi\,\mathrm{d}x\,\mathrm{d}y + \Phi_\Gamma \varphi\psi n_x d\Gamma$$

$$\iint_\Omega \varphi\frac{\partial\psi}{\partial y}\mathrm{d}x\,\mathrm{d}y = -\iint_\Omega \frac{\partial\varphi}{\partial y}\psi\,\mathrm{d}x\,\mathrm{d}y + \Phi_\Gamma \varphi\psi n_y d\Gamma \tag{11.68}$$

where n_x, n_y are the direction cosines of the angles between a normal on the boundary and the x and y axes, respectively.

Application to the integrals in equation (11.67) yields

$$-\iint_{(e)}\left(h^3\frac{\partial p^{(e)}}{\partial x}\cdot\frac{\partial N_i}{\partial x}+h^3\frac{\partial p^{(e)}}{\partial z}\frac{\partial N_i}{\partial x}\right)\mathrm{d}x\,\mathrm{d}z+\iint_{(e)}N_i q\,\mathrm{d}x\,\mathrm{d}y$$

$$+\Phi_\Gamma\left(h^3\frac{\partial p^{(e)}}{\partial x}n_x+h^3\frac{\partial p^{(e)}}{\partial z}n_z\right)N_i d\Gamma^{(e)}=0 \tag{11.69}$$

Introducing the value of $p^{(e)}=\Sigma_i N_i p_i$ and writing the three equations (for $i=1,2,3$) in vector-matrix form, for $\{p^e\}=\{p_1p_2p_3\}$

$$\left(-\iint_{(e)}(h^3)\left[\frac{\partial N_i}{\partial x}\frac{\partial N_j}{\partial x}\right]\{p^{(e)}\}+h^3\left[\frac{\partial N_i}{\partial z}\frac{\partial N_j}{\partial z}\right]\{p^{(e)}\}\right)\mathrm{d}x\,\mathrm{d}z$$

$$+\iint_{(e)}\{N\}q\,\mathrm{d}x\,\mathrm{d}z=0 \tag{11.70}$$

Because of the term $h^3(x)$, numerical integration must be performed. However, experience shows that for even a rough finite element mesh of, say 5×5 nodes which is already very coarse, no more than 2–3% error is encountered if the film thickness h is considered constant within the element. For finer mesh, which is needed for adequate accuracy, this assumption does not introduce a measurable error. Also,

$$\frac{\partial N_i}{\partial x}=\frac{b_i}{2\Delta},\quad \frac{\partial N_i}{\partial z}=\frac{c_i}{2\Delta},\ \text{etc.}$$

therefore,

$$[h^3b_ib_j+h^3c_ic_j]/4\Delta\{p^{(e)}\}=-\frac{\Delta q}{3}\begin{Bmatrix}1\\1\\1\end{Bmatrix} \tag{11.71}$$

and the element 'stiffness' matrix, correlating pressure to flow is

$$k_{ij}=h^3(b_ib_j+c_ic_j/4\Delta \tag{11.72}$$

and the force vector

$$f_i=-\Delta q/3 \tag{11.73}$$

Element assembly follows the method presented in Chapter 4. As boundary conditions, zero pressure will be assumed at the boundary nodes. The program FINSTRES was modified to accommodate the above stiffness matrix and force vector. Then, the program FINLUB was devised. The program starts with data: $DL=D/L$, $RA=R$, $EC=c$, $MU=\mu$, $OM=\omega$, $E=\varepsilon$, $FF=\varphi$ angles for several fixed pads, the bearing angle $FI=\theta$, the attitude angle.

The program yields the pressure distribution. Then, it integrates the horizontal and vertical components of the pressure to yield the bearing forces, horizontal and vertical, and the Sommerfeld Number.

A 15 × 7 mesh usually gives accurate results. Because banded matrices have been used, a moderate personal computer can accommodate more than 100 nodes and 200 elements, adequate for most applications. On the IBM-PC a 20 × 7 mesh with 140 nodes and 228 elements took about 1 min to run.

Integrating for the forces, the program neglects negative pressures.

The program can be easily modified to include such factors as misalignment $h = H(x, z)$, prescribed pressure or flow at some part of the boundary, etc.

It must be made clear here that a partial bearing, that is $\varphi < 360°$, is symmetrically located about the vertical. In general, for attitude angle $\theta = 0$ there are vertical and horizontal forces. Usually the loads are vertical. Therefore, the attitude angle must be found which makes the horizontal force zero. This can be done by the Newton–Raphson method. To this end, an initial solution is pursued with $\theta = 0$ which yields a horizontal force X_0. An increment in attitude angle $\Delta\theta$ yields a horizontal force $X_0 + \Delta X$. An estimate of the new attitude angle is then

$$\theta_1 = \theta_0 - \frac{X_0}{(\Delta X/\Delta\theta)}$$

This is repeated until the horizontal force becomes small enough. This yields the attitude angle which with the eccentricity ratio gives the bearing equilibrium point.

The program can be used in two ways:

(a) For a particular bearing configuration, it can be used to find the equilibrium position ε, θ for given load and speed.

(b) For the given bearing type to give design diagrams such as the ones presented in Section 11.6 to be subsequently used for bearing design.

A simple run for a full 360° bearing follows. The program produces also colored pressure distribution on the screen and the mesh utilized, created with the AUTOMESH program of Chapter 4.

```
BEARING DATA-hit ENTER for default data

Bearing Radius                 50?
Bearing Ang. Velocity          377?
Bearing Clearance              .2?
Oil Dynamic Viscosity          2.76E – 07?
Bearing Eccentricity           .1?
Bearing Att Angle, deg.      – 62.9?
Number of arcs                 1?

Arc no 1
Entrance Angle, deg.     – 180?
Exit Angle, deg.           180?
Length                     100?
Preload                    0?
Are the data correct (Y/N)?
```

```
Iterating: for equilibrium
FORCES: FX = 41194.6    FY = 4352.675   Fa = - 62.9
FORCES: FX = 41193.2    FY = 4360.398   Fa = - 62.89
FORCES: FX = 41557.44   FY =  210.343   Fa = - 68.53625
FORCES: FX = 41557.44   FY =  210.343   Fa = - 68.53625
for design data
FORCES: FX = 41607.03   FY = 195.7998   Fa = - 68.53625
FORCES: FX = 41480.95   FY = 232.7317   Fa = - 68.53625
Ix..SOLUTION..node pressures.....Iy >>>>>>
ARC 1
0.0E+00   0.0E+00   0.0E+00   0.0E+00  0.0E+00
0.0E+00  -.5E+01   -.6E+01   -.5E+01   0.0E+00
0.0E+00  -.8E+01   -.1E+02   -.8E+01   0.0E+00
0.0E+00   0.4E+01   0.5E+01   0.4E+01  0.0E+00
0.0E+00   0.9E+01   0.1E+02   0.9E+01  0.0E+00
0.0E+00   0.5E+01   0.7E+01   0.5E+01  0.0E+00
0.0E+00   0.2E+01   0.3E+01   0.2E+01  0.0E+00
0.0E+00   0.8E+00   0.1E+01   0.8E+00  0.0E+00
0.0E+00  -.1E+00   -.1E+00   -.1E+00   0.0E+00
0.0E+00   0.0E+00   0.0E+00   0.0E+00  0.0E+00
Hit ENTER?
RESULTS:

Sommerfeld Number            .2496406
Vertical Load                41480.95
Horizontal Load              232.7317
Spring Constant Kxx          247929.7
Spring Constant Kxy         -382460.9
Spring Constant Kyx         -72715.76
Spring Constant Kyy          111943.5
Relative Eccentricity         4994346
Attitude Angle, deg,        -68.53625
Length/Diameter ratio        1
Hit ENTER?
```

EXAMPLES

Example 11.1

An automobile manufacturer specifies SAE 30 oil for an engine. The design temperature of the oil during normal engine operation is 80 °C. Determine the dynamic and kinematic viscosities of the oil at that temperature, if it is known that the density of the oil is 880 kg/m^3.

Solution

Figure 11.4 shows dynamic viscosity at 80 °C for SAE 30 oil:

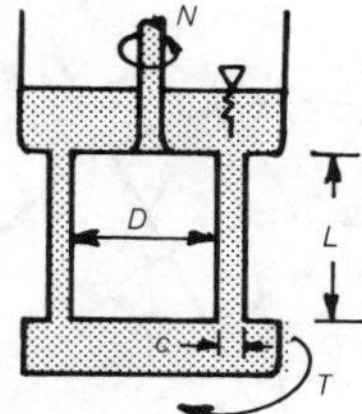

Figure E11.2

$$\mu = 12.2 \text{ MPa. s} = 12.2 \text{ cP} = 1.77 \text{ reyn}$$

The kinematic viscosity,

$$\nu = \mu/\rho = 12.2E - 3/880 = 1.38E - 5 \text{ m}^2/s = 13.8 \text{ cSt}$$

From Figure 11.3, 13.8 cSt correspond to 73 Saybolt Universal Seconds.

Example 11.2

On a viscosity measuring instrument, a cylinder of diameter $D = 30$ mm and length $L = 30$ mm rotates with $N = 12$ rps in a cylindrical cavity with a radial clearance of $c = 1$ mm. The torque on the stationary cylinder is measured $T = 0.5$ Nmm. Find the absolute oil viscosity.

Solution

The Petroff equation (11.3) yields

$$\mu = 2cT/\pi^3 NLD^3 = 2 \times 0.001 \times 0.5E - 3/(\pi^2 \times 12 \times 0.003 \times 0.003^3) = 0.0104 \text{ Pa.s}$$

or

$$\mu = 10.4 \text{ mPa.s} = 10.4 \text{ cP} = 1.51 \text{ reyn.}$$

Example 11.3

A lightly loaded sleeve bearing is split and the upper half has a wide groove as shown in Figure E11.3. The radial clearance is 0.05 mm, the journal rotates at 3600 rpm and oil SAE 10 is used. The bearing temperature is 60 °C. Find the power loss, assuming that the journal rotates in a concentric position.

Solution

Petroff's equation will be used. To this end, the lower half is one-half of the loss of a full bearing and the upper lands of width 10 mm are one-half of a full bearing each 10-mm long. The viscosity is from Figure 11.4, $\mu = 12.5$ mPa.s. Therefore the total power loss will be, for $N = 3600/60$ rps,

$$W = W_1 + 2W_2 = (\pi^3 \times 60^2 \times 0.04 \times 0.03^3 \times 12.5E - 3/5E - 5) \times 0.5$$
$$+ (\pi^3 \times 60^2 \times 0.01 \times 0.03^3 \times 12.5E - 3/5E - 5) \times 0.5 \times 2 = 15.06 + 7.53 = 22.6 \text{ W}.$$

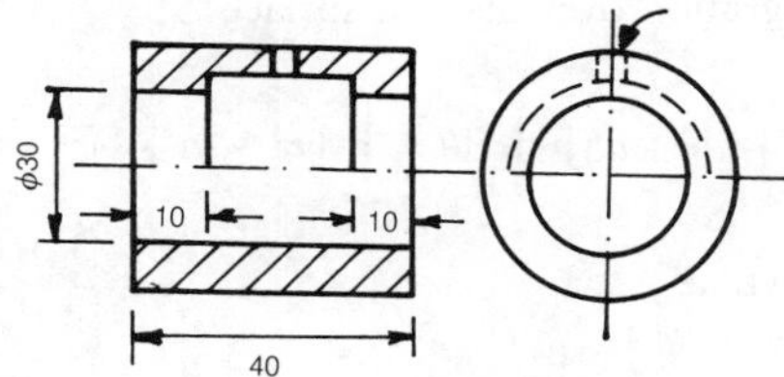

Figure E11.3

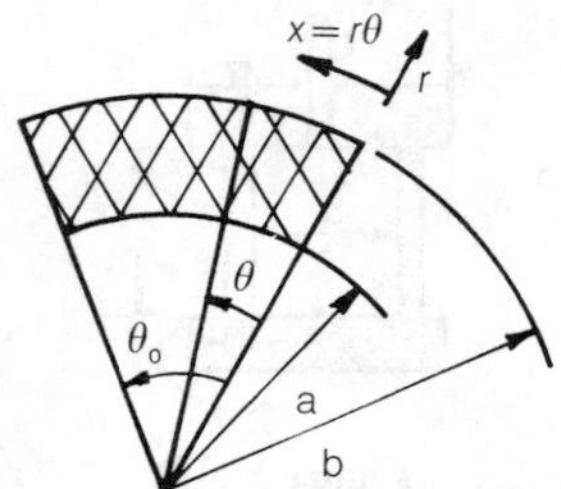

Figure E11.4

Example 11.4

A fixed Michell pad bearing for the main shaft of a ship consists of six pads as in Figure E11.4 of angle $\theta_0 = 50^\circ$ between inner radius $a = 200$ mm and outer radius $b = 300$ mm. The oil has viscosity 50 cP and the minimum film thickness must not be less than 0.1 mm where the film thickness at the trailing edge is 0.5 mm greater than the one at the leading edge owing to the pad inclination. Determine the load-carrying capacity for speed of rotation 300 rpm assuming short bearing in the radial direction. Determine the optimum h_1/h_0 value.

Solution

Neglecting $\partial/\partial x$ (or $\partial/\partial\theta$) terms owing to short bearing approximation, the Reynolds equation (11.11) becomes, for $y = r$, $x = r\theta$,

$$\frac{\partial}{\partial r}\left(h^3\frac{\mathrm{d}p}{\mathrm{d}r}\right) = 6\mu U\frac{\mathrm{d}h}{\mathrm{d}x} = k \tag{a}$$

But $h = h_0 + (h_1 - h_0)\theta/\theta_0$, $\mathrm{d}h/\mathrm{d}x = \mathrm{d}h/\mathrm{d}(r) = h_1 - h_0/r\theta_0$ and $U = 2\pi N r$, therefore

$$k = 12\mu N\mu(h_1 - h_0)\theta/\theta_0$$

Integrating twice equation (a) yields

$$p = \frac{k}{2h^3}r^2 + \frac{c_1 r}{h^3} + c_2 \tag{b}$$

where for $r = a$ and b the pressure $p = 0$. Equation (b) for $r = a$, $p = 0$ and $r = b$, $p = 0$ yields

$$c_1 = -\frac{k(b-a)}{2}, \quad c_2 = \frac{kab}{2h^3}$$

$$p = \frac{k}{2h^3}[r^2 - (b+a)r + ab]$$

The load is found by integration over the pad surface

$$W = \int_a^b\int_0^{\theta_0}\frac{k}{2h^3}[r^2 - (b+a)r + ab]r\,\mathrm{d}r\,\mathrm{d}\theta = \int_a^b[r^3 - (a+b)r^2 + abr]\,\mathrm{d}r\int_0^{\theta_0}\frac{k}{2h^3}\mathrm{d}\theta$$

The first integral has the value

$$I_1 = \left[\frac{r^4}{4} - (b+a)\frac{r^3}{3} + abr^2/2\right]_a^b = \frac{b^4 - a^4}{4} - \frac{b+a}{3}(b^3 - a^3) + \frac{ab}{2}(b^2 - a^2)$$

The second integral

$$I_2 = \int_0^{\theta_0} \frac{k}{2h^3} d\theta = \frac{\theta_0 k}{2(h_1 - h_0)} \int_{h_0}^{h_1} \frac{dh}{h^3} = \frac{\theta_0 k(h_1 + h_0)}{2h_0^2 h_1^2}$$

Therefore,
$W = I_1 I_2$. The numerical values are:

$$h_0 = 0.1E - 6m,\ h_1 = 0.6E - m,\ \theta_0 = 50° = 0.87\,\text{rad},\ \mu = 50E - 3\,\text{Pa.s},\ N = 5\,\text{rps}$$

$$k = 12 \times 5 \times 50E - 3 \times (0.6E - 3 - 0.1E - 3)/0.87 = 1.72E - 3$$

$$I_1 = \frac{0.3^2 - 0.2^2}{4} - \frac{0.3 + 0.2}{3}(0.3^3 - 0.2^3) + \frac{0.2 \times 0.3}{2}(0.3^2 - 0.2^2)$$

$$I_2 = \frac{0.87 \times 5.17E - 3 \times (0.1E - 3 + 0.6E - 3)}{2 \times (0.1E - 3 \times 0.6E - 3)^2} = 3.14E6$$

Then, the load-carrying capacity will be, per pad,

$$W = I_1 I_2 = 5143\,\text{N}$$

For six pads $W_6 = 32{,}482$ N

Repeating the calculation for $h_1/h_0 = 1.5, 2, 3, 4$ we obtain for the load-carrying capacity per pad,

h_1/h_0	1.5	2.0	2.5	3.0	4.0	5.0
W	18,797	25,377	28,434	30,076	31,721	32,482

It can be concluded that h_1/h_0 must have the value about 2.5. Above that, no substantial gain is obtained in the load-carrying capacity.

Example 11.5

A 6000-kW steam turbine moves a ship propeller at 180 rpm through a main shaft. The propeller has an efficiency of 80%, and the maximum speed of the ship is 50 km/h.

A thrust bearing is used on the main shaft to take the thrust load. It consists of 50° segments as shown in Figure 11.24, but between radii $a = 200$ and $b = 520$ mm. The oil used is SAE 40 maintained at 70 °C and surface roughness, manufacturing tolerances and oil filtration do not allow for minimum film thickness less than 80 μm. Determine the number of six-pad such thrust bearings required to sustain the load, assuming a rectangular pad and a Kingsbury and Needs solution. Determine then the power loss.

Solution

The pad geometry is

$$L = b - a = 520 - 200 = 320\,\text{mm},\ B = R\theta_0 = \frac{a + b}{2}\theta_0 = -\frac{200 + 520}{2} 50 \frac{3.14}{180} = 314\,\text{mm}$$

The mean peripheral velocity is

$$U = \omega R = 2\pi N(a + b)/2 = 2\pi(180/60)(0.200 + 0.520)/2 = 6.78\,\text{m/s}$$

From Figure 11.4, the oil viscosity is $\mu = 25$ mPa.s.

The leakage factor (Figure 11.10), for $L/B \cong 1$, is $K_L = 0.4$. Therefore, the load-carrying capacity of one pad is (equation 11.16),

$$W = \frac{6\mu U B^2 L \phi(p_{av}) K_L}{h_0^2} = \frac{6 \times 25E-3 \times 6.78 \times 0.314^2 \times 0.320 \times 0.026 \times 0.4}{(80E-6)^2} = 52{,}000\ \text{N}$$

Each complete six-pad bearing can take load $6 \times 52{,}000 = 3.12E5$ N. The thrust F of the ship is found from the power

$$P = 6E6W = FU = F50{,}000/3600$$
$$F = 6E6 \times 3600/50{,}000 = 0.42\,E6N$$

The required number of thrust bearing stages of six pads each is

$$1.15E6 \times 3.12E5 = 1.38$$

Two stages are sufficient.

The power loss, from equation (11.19), assuming $\phi(F_R) = 0.77$, is

$$P_L = \frac{25E-6 \times 6.8^2 \times 0.32 \times 0.314 \times 0.77}{80E-6} = 1118\ \text{W} = 1.12\ \text{kW}$$

for each pad. For the two stages, 12 pads, for safety reasons.

$$P_{Tot} = 6P_L = 6 \times 1.12 = 6.70\ \text{kW}$$

A more accurate estimate would be obtained taking into account that two stages (instead of 1.4) were selected, therefore the minimum oil film thickness will be somewhat greater than 80 μm. Equation (11.17) will give this value which must be then used in equation (11.19).

The bearing average pressure will be, for six pads:

$$p_{av} = \frac{1.08E6}{6 \times 0.320 \times 0.314} = 1.8\ \text{MPa}$$

Example 11.6

Optimize the design of Example 11.5 for minimum power loss.

Solution

Design parameters which can be altered are the radii a and b retaining the two-stage, six-pad design. The functions $\phi(F_R)$ and $\phi(p_{av})$ will be kept constant. The average pressure per pad is

$$p_{av} = \frac{F}{2 \times 6 \times LVB} = \frac{2F}{12(b-a)(b+a)\theta_0} = \frac{F}{6(b+a)(b-a)\theta_0}$$

The peripheral velocity $U = 2\pi NR = 2\pi N(a+b)/2 = \pi N(a+b)$. The minimum oil film thickness

$$h_0 = \left[\frac{6\mu\pi N(a+b)\theta_0(a+b)/2}{F/6(b+a)(b-a)\theta_0}(p_{av})K_L\right]^{\frac{1}{2}} \qquad \text{(a)}$$

From Figure 11.10, the relationship of K_L to L/B can be approximated as

$$L/B = 4.375K_L^2 + 0.3 \quad \text{or} \quad K_L = [L/B - 0.30]^{\frac{1}{2}}/4.375$$

Therefore,

$$h_0 = \left[\frac{18\mu\pi N(a+b)^3(b-a)_0^2}{F}\phi(p_{av})[[(b-a)2/(a+b)\theta_0 - 0.3]^{\frac{1}{2}}/4.375]\right]^{\frac{1}{2}}$$

The power loss is

$$P_L = \frac{[\pi N(a+b)]^2 (b-a)_1 \theta_0 (a+b)/2}{h_0} \phi(F_R)$$

$$P_L = \frac{2\pi^2 \mu N^2 (a+b)^3 (b-a)\theta_0(F_R)}{h_0} \tag{b}$$

where h_0 is given by equation (a).

Here, from Example 11.5, $\mu = 25E - 3$ Pa.s, $N = 6$ rps and $\phi(F_R) = 0.77$, $\phi(p_{av}) = 0.026$.

The program OPTIMUM (Appendix 5.A) will be used with starting values $a = 0.200$, $b = 0.520$ of Example 11.5 and the objective function to be minimized will be P_L as in equation (b), above.

Limits must be imposed on the design variables, a_{max}, $b_{max} = 100$ mm is specified. The initial guess $a = 0.2$ m and $b = 0.52$ m is specified.

Lines 660–680 include the definition of the objective function. To maintain the minimum film thickness 80 μm, a penalty function is defined.

$$F = P_L + 10^{10}(80 \times 10^{-6} - h_0)^2 \quad \text{for} \quad h_0 < 80 \times 10^{-6}$$

$$F = P_L \quad \text{for} \quad h_0 > 80 \times 10^{-6}$$

Running the program with Steepest Descent and Golden Section selections, the initial and final values of the design variables a and b and the resulting power loss are:

a(m)	b(m)	P_L(W)	h_{min}(μm)
0.200	0.520	16,420	87
0.100	0.479	11,364	57.8

A 30% reduction of the power loss was obtained. The penalty function definition in program OPTIMUM, Appendix 5.A, follows.

```
630 REM ***********************************************************

640 REM OBJECTIVE FUNCTION DEF
650 REM ***********************************************************

660 A = H(1): B = H(2): AP = A + B: AM = B - A
662 FO = 1.08E6
663 TO = 50*3.14159/180
665 HO = 18*25E - 3*3.14*6*TO^2*AP^3*AM*.026*
      (SQR(AM*2/TO/AP - .3)/4.375)/FO
666 HO = SQR(HO)
670 F = 2*3.14^2*25E - 3*TO*36*AP^3*AM*77/HO
672 PRINT "PL=";F;"HO=";HO
674 IF HO > 80E - 6 THEN 680
677 F = F + 1E10*(80E - 6 - HO)

680 RETURN
```

Example 11.7

An electric generator rotor weighs 600,000 N and rotates at 3600 rpm. It is supported by two 180°

sleeve bearings. The journal diameter is 400 mm and the oil used is light oil SAE 30. The oil filter allows particles 60 μm and the surface roughnesses are 5 μm.
Determine:

(a) The bearing width for a load safety factor 2.
(b) The operating parameters for normal operation.

Assume bearing of infinite length.

Solution
The load per bearing is 600,000/2 = 300,000 N. The clearance ratio is selected $c_r/r = 0.001$. Therefore, the clearance is $c_r = 0.001r = 0.001 \times 200 = 0.2$ mm. The minimum oil film thickness

$$h_{min} = 2 \times (5 + 5) + 60 = 80\ \mu m$$
$$h_{min}/c = 80/200 = 0.4$$
$$\varepsilon = 1 - h_{min}/c = 1 - 0.4 = 0.6$$

Now, a bearing temperature will be selected and then a corresponding viscosity from Figure 11.4. It will be verified later.

(a) The Sommerfeld Number yields the bearing length, with double load,

$$L = SW/\mu ND(r/c)^2$$

The Sommerfeld Number is obtained from equation (11.25) for infinitely long, 180° journal bearing.

For a temperature of 60°C, $\mu = 12$ mPa.s and a Sommerfeld Number 0.300 with equation (11.25). Then, $L = 0.578$ m and $L/D = 0.578/0.400 = 1.445$.

The Petroff torque

$$T_P = \pi^2 12E - 3 \times 60 \times 0.578 \times 0.4^3/2 \times 0.0002 = 657\ \text{Nm}$$

The correction $\phi(\varepsilon)$ is (equation 11.33)

$$\phi(0.6) = (1 + 2 \times 0.6^2)/[(2 + 0.6^2)(1 - 0.6^2)^{\frac{1}{2}}] = 0.91$$

Therefore, the frictional torque will be

$$T_J = T_P = 0.91 \times 657.7 = 598\ \text{Nm}$$

The heat generation

$$\dot{W} = T_J\omega = T_J 2\pi N = 2\pi \times 60 \times 598 = 225,440\ \text{W}$$

The oil flow for the infinite-long bearing is (equation 11.37),

$$Q_L = \pi c_r DLN(1 - \varepsilon^2)/(2 + \varepsilon^2) = 2.36 \times 10^{-3}\ \text{m}^3/\text{s}$$

The heat balance requires (equation 11.39),

$$\dot{W} = \rho c Q_L \Delta T$$

For $\rho c = 1.36\ \text{Nm/m}^3\ °\text{C}$,

$$\Delta T = \dot{W}/\rho c Q_L = 76\ °\text{C}$$

For an environmental temperature $T_0 = 20$ °C, the oil temperature is 96 °C. But it was assumed initially 60 °C. Therefore, we repeat the calculation for several temperatures. The results follow

T_0(°C)	μ(mPa.s)	ΔT	$T_e+\Delta T$
60	13	76	96
70	9	52.6	72.6
80	6.6	38.6	58.6

For $T_0 = 70\,°\mathrm{C}$, a reasonable agreement is obtained. Repeating the calculation for $T_0 = 70\,°\mathrm{C}$, $\mu = 9 \times 10^{-3}$ Pa.s,

$$L = 0.836, \quad \dot{W} = 225\,\mathrm{kW}, \quad Q_L = 3.42 \times 10^{-3}\,\mathrm{m^3/s}$$

(b) For the operating load of 300 kN per bearing, the computation is repeated to yield, with the length $L = 836$ mm found previously, operating temperature = 52 °C and eccentricity ratio = 0.05, power loss 260 kW. Unloading the bearing resulted in very small eccentricity which might present instability problems. This indicates that using a high safety factor for the load in bearings does not necessarily result in a safer bearing – the load-carrying capacity will be sufficient but problems of another nature might be encountered.

Example 11.8

In a circumferential groove bearing, the oil is supplied through a circumferential groove of enough depth to ensure uniform pressure around the journal and the flow is along the journal axis.

For such a bearing of diameter $D = 100$ mm, length of the bearing lands $L = 50$ mm, carrying a load of 30 kN at 3000 rpm, determine the operating parameters, eccentricity ratio, oil flow, exit oil temperature and power loss. The SAE 20 oil is supplied at 40 °C and pressure 20 N/mm², and the clearance ratio c_r/r is 0.001.

Consider the bearing that consists of two equal short bearings of 360° oil film.

Solution

Figure 11.4 gives for an initial estimate of viscosity at 40 °C for the SAE 20 oil, $\mu = 43$ mPa.s.

The Sommerfeld Number is for each bearing half,

$$S = \frac{\mu N L D}{W}\left(\frac{r_c}{c}\right)^2 = \frac{43E-3 \times 50 \times 0.050 \times 0.100}{30{,}000} 1000^2 = 0.358$$

The ratio $L/D = 50/100 = 0.5$.

For the bearing,

$$(L/D)^2 S = (1-\varepsilon^2)^{3/2}/2\pi^2\varepsilon$$

Solving for ε by iteration, $\varepsilon = (1-\varepsilon^2)^{3/2}/2\pi^2(L/D)S$, we obtain $\varepsilon = 0.422$.

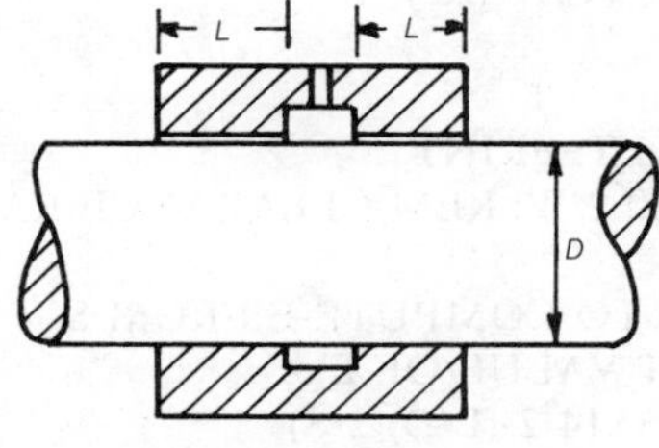

Figure E11.8

The axial oil flow is independent of the bearing rotation, provided that cavitation does not take place because the Reynolds equation is linear and the principle of superposition holds. We can consider the axial flow as a channel flow. For a differential angle $d\theta$ at a film thickness $h(\theta) = c(1 + \varepsilon \cos \theta)$ the flow is

$$dq = \frac{h^3 R\, d\theta}{12\mu L} p$$

The total flow around the bearing, for each half is

$$Q = \frac{2\pi r c^3}{12L} \int_0^{360°} (1 + \varepsilon \cos \theta)^3 \, d\theta$$

But $(1 + \varepsilon \cos \theta)^3 \, d\theta = \theta + 3\varepsilon^2\theta/2$. Therefore for $c = 0.001r = 0.050$ mm.

$$Q = \frac{2\pi p r^3 c}{12\mu L}(1 + 1.5\varepsilon^2) = \frac{2\pi \times 20E - 6 \times 0.100 \times 0.050^2 \times 10^{-6}}{12 \times 43E - 3 \times 0.050}(1 + 1.5 \times 0.422^2)$$

$$Q = 1.38\ 10^{-4}\ \mathrm{m^3/s}$$

The Petroff torque, equation (11.3),

$$T_P = \pi^2 \mu N L D^3/2c = \pi^2 \times 43E - 3 \times 50 \times 0.050 \times 0.100^3/2 \times 0.050E - 3 = 10.6\ \mathrm{Nm}$$

Heat generation (equation 11.40)

$$\dot{W} = T_P \omega \phi(\varepsilon) = T_P 2\pi N[1 + \varepsilon^2/2(1 - \varepsilon^2)](L/D)^2/(1 - \varepsilon^2)^{\frac{1}{2}} = 1019\ \mathrm{W}$$

Temperature rise, equation (4.40), for $\rho c = 1.36E6$

$$T - T_0 \dot{W}/\rho c Q = 23\ °\mathrm{C}$$

The exit temperature is $T = T_0 + 23 = 40 + 23 = 63$ °C.

Repeating the calculation for oil temperature 59 °C and so on until the exit temperature matches the assumed one, this leads to an exit temperature of 55 °C with the following operating parameters: $S = 0.460$, $\varepsilon = 0.36$, $Q = 4.51E - 4\ \mathrm{m^3/s}$, $\dot{W} = 6173$ W, $T = 50$ °C.

The computer program which does the computations follows.

```
10 REM EXAMPLE 11.8
20 PI = 3.14159
30 E = .999: REM A GUESS FOR ITERATION
40 T = 40 + DT
50 W = 15000:N = 50:L = .05:D = .1:C = .05E - 3: REM DATA
60 REM COMPUTATION OF VISCOSITY AT TEMPERATURE T...SAE 20 OIL
   ASSUME D
70 X1 = 53.2*EXP (63.3/(313 + DT))
80 X2 = X1 - 77.14:ET = EXP (X2)
90 ET = ET*6895
100 PRINT "Temperature "; T
110 PRINT "Viscosity"; ET: PRINT
120 S = ET*N*L*D*1000^2/W: REM CLEARANCE RATIO .001
130 PRINT "S = "; S
140 REM ITERATION TO COMPUTE E FROM S
150 EP = E: REM LAST VALUE OF E
160 E = (1 - E^2)^1.5/(2.*3.14^2*(L/D)^2*S)
170 IF ABS (E - EP) > 0.1 THEN 150: REM KEEP ITERATING FOR E
180 PRINT "E =  = "; E
```

```
190 Q = 2*PI*160E6*D/2*C^3/(12*ET*L)*(1 + 1.5*E^2)
200 PRINT "Q =  "; Q
210 P = PI^2*ET*N*L*D^3/2/C
220 PRINT "P = "; P
230 WP = P*20*PI*N*(1 + E^2/(2*(1 – E^2)))*(L/D)^2/SQR (1 – E^2)
240 PRINT "WP = "; WP
250 DT = WP/(1.36E6*Q)
260 PRINT "DT = "; DT
270 FLASH
300 NORMAL
310 PRINT
320 IF ABS (T – 40 – DT) > 0.1 THEN 40
330 PRINT "SOLUTION IS THE LAST SET OF RESULTS"
350 GOTO 180
```

```
Temperature      40
Viscosity        .0416964516

S = .694940861
E = .0263804748
Q = 2.77364829E – 04
P = 10.2881697
WP = 8680.34548
DT = 23.0381045

Temperature      63.0381045
Viscosity        .0170083735

S = .283472893
E = .0477063476
Q = 8.25883324E – 04
P = 4.19664088
WP = 4302.82242
DT = 3.83085591

Temperature      43.8308559
Viscosity        .0355650789

S = 592751315
E = .356307493
Q = 4.49230692E – 04
P = 6.84703651
WP = 6173.84106
DT = 10.1052513

SOLUTION IS THE LAST SET OF RESULTS

E = .356307493
Q = 4.49230692E – 04
P = 6.84703651
WP = 6173.84106
DT = 10.1052513
```

Example 11.9

The rotor of a turbo-compressor weighs 100 kN and it is supported by two tilting pad bearings with four pads each. The light turbine oil is equivalent to SAE 5 oil and it is externally cooled to

50 °C and fed to the bearings by axial grooves. Design the proper bearing and determine the operating characteristics. The rotating speed is 3600 rpm. Select clearance ratio 0.002.

Solution

To start, an L/D ratio 1 is selected. From engineering experience, (Figure 11.20a), a value of average pressure 2 MPa is selected. Each half carries a 50 kN load. Therefore,

$$P_m = W/LD, \quad 2E6 = 50E3/LD, \quad LD = 0.025$$

Because $L/D = 1$, $D^2 = 0.025$, $D > 158$ mm. The clearance ratio is selected as 0.002.

For diameters 160, 170, 180, ..., mm, the load-carrying capacity will be computed. Then, the one which will be equal to the given bearing load will be selected.

We assume an average oil film temperature 70 °C and Figure 11.4 gives $\mu = 8.1$ mPa.s. The clearance is $c_r = (c_r/r)r = 0.002\,r$.

The minimum film thickness $h_{min} = 5 + 0.04D$ and the eccentricity ratio is then

$$\varepsilon = 1 - h_{min}/c_r$$

Figure 11.26, for the given ε yields the Sommerfeld Number S. Then the load is

$$W = \frac{\mu NLD}{S}\left(\frac{r}{c_r}\right)^2$$

The procedure was programed with following results:

At $D = 185$ mm, the load-carrying capacity is $W = 51{,}979$ N, at $\varepsilon = 0.933$ and $S = 0.08$. The clearance is $c_r = 0.002 \times 0.105 = 0.370$ mm.

The find the thermal equilibrium, the power loss must be estimated. For $\varepsilon = 0.935$ Figure 11.17 gives the equivalent Sommerfeld Number for the full sleeve bearing $S = 0.14$. From Figure 11.18, the friction parameter $f(r/c) = 0.9$. Therefore,

$$f = 0.9 \times 2 \times 10^{-3} = 1.8E - 3.$$

The power loss

$$W = fW2\pi ND/2 = 1.8E - 50{,}000 \times 3.14 \times 60 \times 0.180 = 3216\ W.$$

The oil flow parameter (from Figure 11.19), $Q/RcNL = 4.8$. Therefore, $Q = 4.8 \times 0.100 \times 0.0004 \times 60 \times 0.200 = 9.2E - 4\ \text{m}^3/\text{s}$.

The temperature rise is

$$T - T_0 = \dot{W}/\rho c Q = 3261/1.36E6 \times 2.3E - 3 = 2.65°\text{C}$$

The oil film temperature is then 52.6 °C and not 70 °C as assumed. The computation is repeated until agreement is obtained. The final results are:

$D = 160$ mm; $L = 160$ mm; load capacity, $W = 57{,}600$ N; $W = 3125$ W; $Q = 5.83E - 4\ \text{m}^3/\text{s}$; T = 54 °C.

The program run and listing follow.

```
50 T=70
60 PRINT "T="; T; "ENTER VISCOSITY ";: INPUT ET
90 N=3600/60
100 INPUT "ENTER DIAMETER, m"; D
105 L=D
115 CR=.002
120 C=CR*(D/2)
```

```
130 HMTN = (5 + .04*1000*D)/1F6
140 F = 1 – HMTN/C
150 PRINT "E = "; F; "ENTER SOMMERFELD NO";: INPUT S
160 W = FT*N*L*D*(1/CR)^2/S
170 PRINT "DIAMETER D = "; D; "LOAD W = "; W
200 INPUT "MORE ITERATIONS (Y/N)"; X$
210 IF X$ = "Y" THEN 100
220 INPUT "ENTER EQ. SOM NO FOR SLEEVE BRG"; SO
230 INPUT "ENTER FRICTION PARAMETER"; FR
240 F = FR*CR
250 WP = F*W*3.14*N*D
260 INPUT "ENTER OIL FLOW PARAMETER"; QL
270 Q = QL*D/2*CR*D/2*N*L
280 DT = WP/1.36F6/Q
290 PRINT "WP = "; WP
300 PRINT "Q = "; Q
310 PRINT "T = "; 50 + DT
320 INPUT "MORE ITERATIONS (Y/N)"; X$
330 IF X$ = "Y" THEN 60
1285 T = 50 + DT

    RUN
    T = OENTER VISCOSITY 8.1F – 3
    ENTER DIAMETER, m.160
    F = .92875 ENTER SOMMERFELD NO 2.08
    DIAMETER D = .16 LOAD W = 38880.0001
    MORE ITERATIONS (Y/N) Y
    ENTER DIAMETER, m .185
    F = .932972973 ENTER SOMMERFELD NO 2.08
    DIAMETER D = .185 LOAD W = 51979.3188
    MORE ITERATIONS (Y/N) N
    ENTER FQ. SOM NO FOR SLEEVE BRG .014
    ENTER FRICTION PARAMETER .9
    ENTER OIL FLOW PARAMETER 4.75
    WP = 3261.03065
    Q = 9.02256563F – 04
    T = 52.6575774
    MORE ITERATIONS (Y/N) Y
    T = 52.6 ENTER VISCOSITY 2.12E – 3
    ENTER DIAMETER, m .185
    F = .932972973 ENTER SOMMERFELD NO 7.08
    DIAMETER D = .185 LOAD W = 77006.2501
    MORE ITERATIONS (Y/N) Y
    ENTER DIAMETER, m.160
    F = .92875 ENTER SOMMERFELD NO 2.08
    DIAMETER D = .16 LOAD W = 57600.0001
    MORE ITERATIONS (Y/N) N
    ENTER EQ. SOM NO FOR SLEEVE BRG .014
    ENTER FRICTION PARAMETER .9
    ENTER OIL FLOW PARAMETER 4.75
    WP = 3125.32992
```

Q = 5.83680001E − 04
T = 53.9371517
MORE ITERATIONS (Y/N) N

Example 11.10

The hydroelectric generator shown has a thrust load of 800 kN. The thrust bearing is of the hydrostatic type with a capillary compensator and a supply oil pressure 1000 MPa. Assuming a compensator of diameter $d_r = 1$ mm, $l = 200$ mm, oil SAE 5 supplied at 50 °C, and a land over recess diameter ratio $L/b = 0.2$, design the bearing for a minimum film thickness of 60 μm.

Solution

The width of the bearing is the periphery of the land, Figure 11.33(b)

$$2W = \pi(b + L)$$

For 50 °C, oil SAE 5 has viscosity $7E - 3$ Pa.s. Then, from equation (11.59), the load carrying capacity is

$$W = \frac{p_s\left(b + \frac{L}{2}\right)w}{1 + h^3 \cdot 128 l_c w / 3\pi d^4 L}$$

For $L = 0.2b$, the bearing recess b will be computed as follows:

$$W = \frac{p_s b^2 \times 1.1\pi \times 1.2/2}{1 + h^3 \times 128 \times l_c \times 1.2b/(2 \times 3 \times d^4 \times 0.2b)} = \frac{1.32\pi p_s b^2/2}{1 + 128 l_c h^3/d^4}$$

Therefore

$$b = \frac{2W(1 + 256 l_c h^3/d_r^4)}{1.31\pi p_s} = \frac{2 \times 800E3(1 + 128 \times 0.2 \times 60^3 E - 18/0.001^4)}{1.31\pi \times 1{,}000E6}$$

Finally, $b = 500$ mm. The other design parameters are:

$$D = 500 + 2 \times 0.2 \times 500 = 705 \text{ mm}, \quad L = 0.2b = 100 \text{ mm}, \quad w = 1.2\ b/2 = 300 \text{ mm}$$

The restriction coefficient

$$k = 3\pi d_r^4 L/128 L_c w = 3\pi \times 0.01^4 \times 0.100/128 \times 0.200 \times 1.2 \times 0.500/2 = 7.75E - 14$$

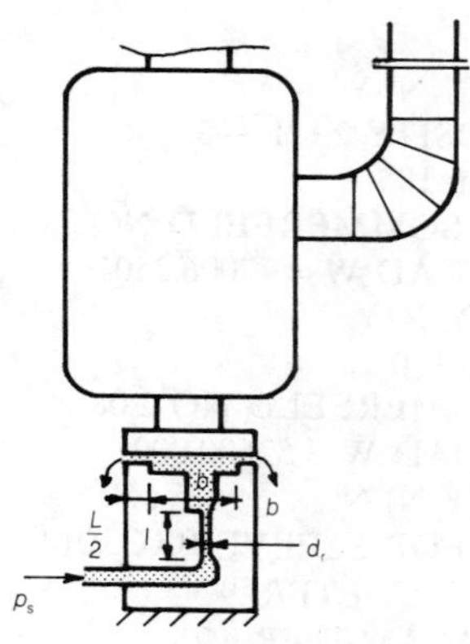

Figure E11.10

The intermediate pressure

$$p_r = p_s/(1 + h^3/k) = 1000E6/(1 + 60E - 6^3/7.75E - 14) = 2.64\,\text{MPa}$$

The oil flow

$$q = \pi d_r^4(p_s - p_r)/128\, l_c = \pi \times 0.001^4(10 - 2.64) \times 10^6/128 \times 7E - 3 = 2.64 \times 10^{-3}\,\text{m}^3/\text{s}$$

The pump power, for a pump efficiency $\eta = 75\%$:

$$W = p\dot{q}/\eta = 10E6 \times 1.28E - 4/0.75 = 1719\,\text{W}.$$

REFERENCES

British Petroleum, *Lubrication Theory and Application.* 1969. London: BP Trading Co.

Cameron, A., 1966. *The Principles of Lubrication.* New York: Wiley.

Dimarogonas, A. D., 1976. *Vibration Engineering.* St Paul, New York: West Publisher.

Dimarogonas, A. D., 1979. *History of Technology* (in Greek), Patras.

Dimarogonas, A. D., Paipetis, S. A., 1983. *Analytical Methods in Rotor Dynamics.* London: Elsevier.

Fox, R., McDonald, A. 1978. *Introduction to Fluid Mechanics.* New York: Wiley.

Huebner, K. H., 1975. *The Finite Element Method for Engineering.* New York: J. Wiley.

Pinkus, O., Sternlicht, B., 1961. *Theory of Hydrodynamic Lubrication.* New York: McGraw-Hill.

Pugh, B., 1970. *Practical Lubrication.* London: Butterworths.

Raimondi, A. A., Boyd, J., 1958. 'A solution for the finite journal bearing and its application to analysis and design', *Trans. ASLE,* **1**, No. 1, pp. 159–209.

Trumpler, P. R., 1966. *Design of Film Bearings,* New York: Macmillan.

Wills, J. G., 1980. *Lubrication Fundamentals.* New York: Dekker.

Vogelpohl, G., 1958. *Betriebssichere Gleitlager,* Berlin: Springer.

PROBLEMS

11.1 An oil sample was tested for viscosity. At 100 °C a kinematic viscosity of 100 Saybolt Universal seconds and a specific weight of 0.895 were measured. Find the kinematic and absolute viscosity on SI and English units and the nearest SAE grade.

11.2 The viscosity of an oil sample was measured at 80 °C. It had the value 8.8 mPa.s. Find the nearest SAE grade and the viscosity at 0 °C.

11.3 SAE 15W oil is defined as having a viscosity of 5000 cP at − 18 °C (0 °F). Draw on Figure 11.4 the viscosity–temperature curve for the SAE 15W oil, by linear interpolation of the parameters c_1 and c_2 in the approximation $\log\mu = c_1 \exp(B/T) + c_2$.

11.4 Determine the viscosity of the ISO VG 32 oil at a temperature of 100 °C.

11.5 Determine the viscosity of the 10W-60 oil at a temperature of 50 °C.

11.6 The wheel bearing for an overhead crane is rated at 200 kN load and it will be made porous for self-lubrication. Select the proper material and lubricant on the basis of allowable PV values keeping the ratio $L/D = 1$.

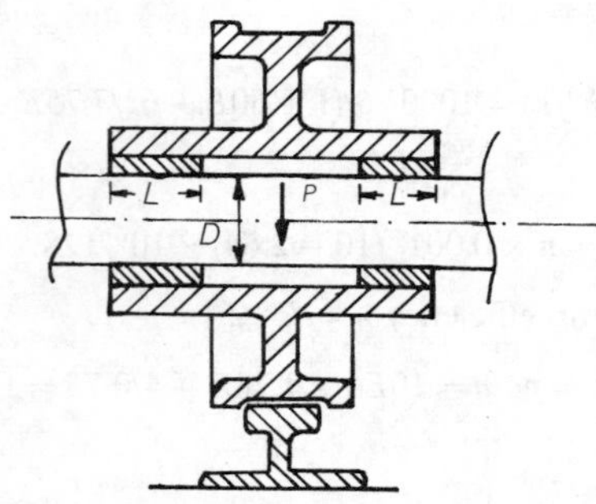

Figure P11.6

11.7 The operation of a hydraulic system is controlled by a crank and follower mechanism. The maximum force transmitted through the crank journal is 2000 N. Design a proper porous bearing, keeping $L/D = 1$, for $\omega = 50$ rad/sec.

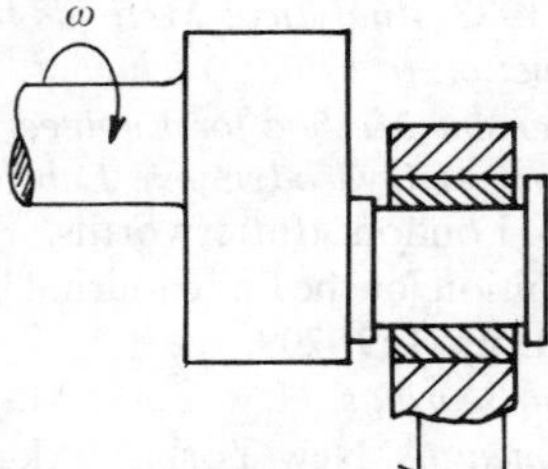

Figure P11.7

11.8 A corner crane is supporting its vertical column by a combination journal and thrust bearing. The vertical force $F = 50$ kN and the horizontal $F = 30$ kN. Design a proper porous journal and thrust bearing. The thrust bearing must have an internal recess for lubricant accumulation of the order of one-half its external diameter.

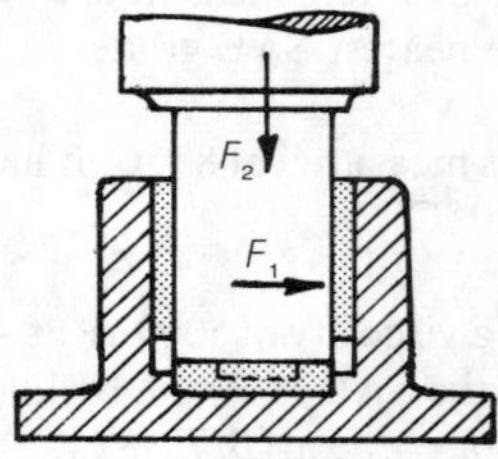

Figure P11.8

11.9 Design a porous bush for the footstep bearing shown in Figure P11.8 if the journal must have a diameter of 60 mm, for strength reasons, and the horizontal load is 2000 N. The speed of rotation is very low.

11.10 A swinging garage door has dimensions width 4 m and height 2.5 m, weighs 1000 N and the distance of the symmetrically-placed hinges is 1.60 m. Design porous bushes for the hinges assuming that the door is uniform and that it might sometimes be swinging at a rotating speed of $N = 2$ rps.

Problems 11.11 to 11.15
Design plastic bushes for the bearings of the Problems 11.6 to 11.10.

11.16 Determine the Petroff power losses for the bearing shown for oil SAE 10 entering at 40 °C, and speed of rotation $= 3200$ rpm.

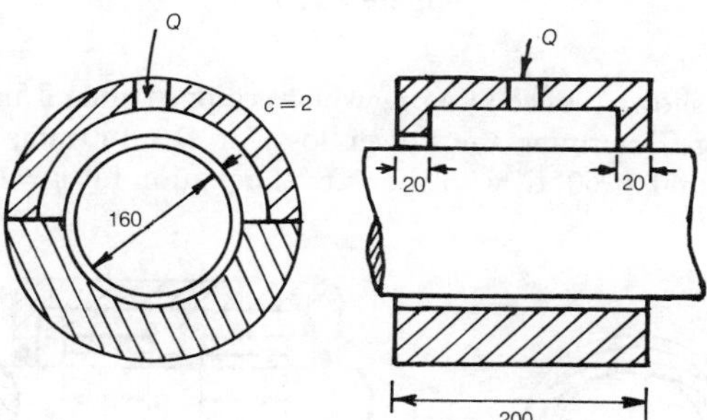

Figure P11.16

11.17 The bearing shown is supplied by oil from four equally-spaced axial grooves. The diameter is 140 mm, length $L = 180$ mm, radial clearance 0.2 mm, the oil SAE 5, oil temperature 60 °C, and speed of rotation 3000 rpm. Determine the Petroff power losses.

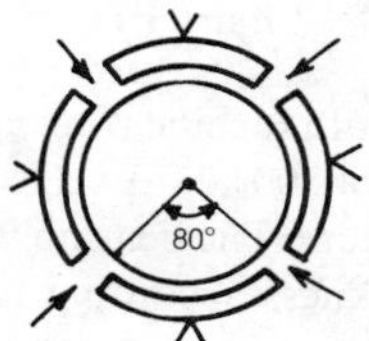

Figure P11.17

11.18 Determine the Petroff equation for the partial bearing of angle ϕ.

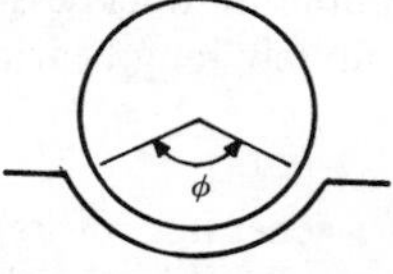

Figure P11.18

11.19 The bearing shown has two supply circumferential grooves and the central circumferential groove is for return, together with the side leakage. If the diameter is 120 mm, the length of each land is $L = 30$ mm, the oil SAE 20 at 50 °C, the clearance ratio is $c_r/r = 0.001$ and the speed of rotation $N = 45$ rps, determine the power loss for concentric bearing. Then modify the length and diameter, keeping the average bearing pressure constant, for minimum power loss.

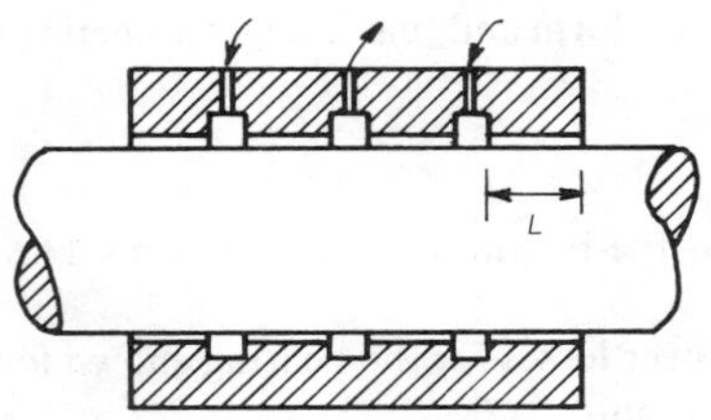

Figure P11.19

11.20 The ring fed bearing shown is oiled by a ring which is dipped into oil and by the rotation it carries oil into the bearing. Determine the power loss for $D = 140$ mm, $L = 200$ mm, $l = 600$ mm, $c_r/r = 0.001$, oil SAE 40 at 60 °C, with the Petroff equation for $n = 1000$ rpm.

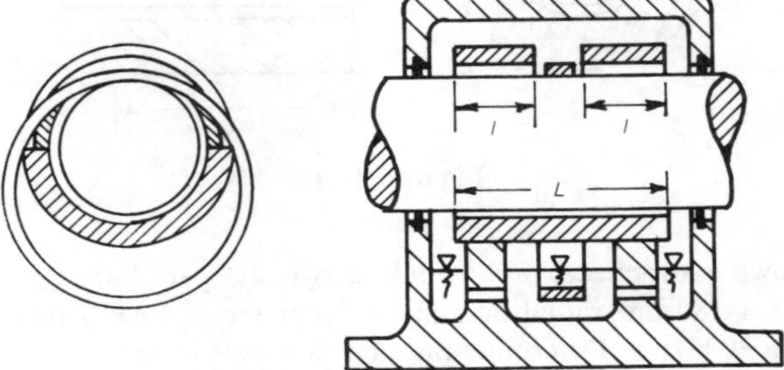

Figure P11.20

11.21 The thrust bearing for a high-pressure turbine rotor is mounted on a shaft of 360-mm diameter and it must support a thrust load of 300 kN at a speed of 3600 rpm. Turbine oil, equivalent to SAE 5 is supplied at 60 °C temperature. The bearing is of the Michell fixed pad type with a taper of 0.5% and a minimum film thickness according to Trumpler's equation [Section 11.6.6(c)] must be maintained. Design a six pad bearing for this purpose. Determine then the bearing operating parameters. Use an infinite length solution and assume that the space between pads is 6% of the annular area.

11.22 A turbocompressor operates at 6000 rpm and the 60 kN thrust is taken by a five fixed pad Michell type pad bearing of inner diameter $D_1 = 100$ mm, outer diameter $D_2 = 200$ mm, oil supply of 10 SAE at 50 °C. Determine the minimum film thickness, oil film temperature, and oil flow, for a taper of the pads 0.8% assuming infinitely long bearing and 5% of the annular space is space between pads.

11.23 The large Hoover Dam water turbine-generator rotors each weigh 1,800,000 lb. Since the shaft is vertical, the entire weight must be supported by the thrust bearing. The speed is 150 rpm. The bearing is immersed in an oil bath which contains the necessary water cooling coils. SAE 10 oil is circulated by the action of the rotating collar and is supplied to the bearing pads at 120 °F. Assuming that the inside diameter of the pads may not be less than 28 in, the number of pads is eight, and 5% of the annular area is space between pads, calculate the proportions of this bearing and the horsepower loss. Show that all the criteria for satisfactory operation are met by your design and make a sketch to scale showing all important dimensions of the bearing surfaces. (Trumpler, 1966).

11.24 The trans-Atlantic passenger liner S. S. Maasdam (Holland–America Line) built in 1952 has a single screw driven by a propeller shaft 75 cm in diameter. Using full-turbine power of 8500 hp (shaft horsepower or shp) and a shaft speed of 80 rpm, the ship speed is approximately 400 miles in 24 h. Now, the effective or two-rope horsepower (ehp) which establishes the propulsive thrust is related to shp by a 'propulsive coefficient' (PC) which is defined as the ratio ehp/shp. Assume PC to be 0.70. Propose a design of tilting-pad thrust bearing for the propeller shaft, making an engineering sketch of the main features, and presenting calculations to verify that your design meets the criteria of satisfactory operation. Use SAE 20 oil delivered to the bearings at 90 °F. Consider forward shift speed only, ignoring the problem of 'reverse'. (Trumpler, 1966).

11.25 The pad shown is known as stepped pad bearing. Determine the load-carrying capacity of such bearing for given geometry, speed and oil properties, assuming infinite pad width.

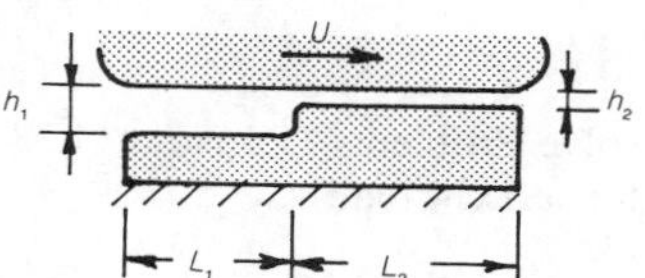

Figure P11.25

11.26 The spindle shaft bearing of a heavy lathe has extreme operating conditions load 6000 N at 450 rpm and 2800 N at 3200 rpm. It is lubricated by circulating SAE 30 oil, supplied at 45 °C and the bearing has an L/D ratio of 0.5.

Determine the bearing dimensions for the worse loading case, selecting appropriate minimum film thickness 20 μm and clearance ratio $c_r/r = 0.002$, by the short bearing approximation. Also determine the rest of the operating parameters.

11.27 The bearing of Problem 11.19 can be considered as three short bearings. With a short bearing approximation, determine the load-carrying capacity, oil film temperature, supply oil flow at atmospheric pressure and power losses for a minimum film thickness of 25 μm and clearance ratio according to the Vogelpohl equation.

11.28 Solve Problem 11.27 for oil supplied at a pressure of 0.5 MPa.

11.29 A four-axial groove bearing of diameter $D = 60$ mm, length 20 mm, clearance ratio $c_r/r = 0.001$, supplied by SAE 20 oil at 45 °C is carrying a load of 3000 N at 3000 rpm. Using the short bearing approximation, determine the operating characteristics.

11.30 The last pass of a rolling mill has a sleeve bearing loaded by 40 kN and an L/D ratio of 0.5. The clearance ratio is $c_r/r = 0.001$ and the minimum film thickness should be at least 10 μm. The speed of rotation is 4400 rpm. Determine the dimensions and operating characteristics of the bearing based on the short bearing approximation.

11.31 Solve Problem 11.30 for an L/D ratio of 2 with the infinitely-long bearing approximation.

11.32 Solve the Problem 11.20 for a 60-kN load with the infinitely-long bearing approximation. If the outer surface of the bearing enclosure is 0.5 m^2 with natural cooling, determine the equilibrium temperature of the bearing.

11.33 For the bearing of Problem 11.16, determine the operating conditions assuming a long 180° bearing approximation for a load 80 kN.

11.34 Determine the Sommerfeld Number–eccentricity ratio relation for an infinitely-long partial bearing of angle φ (Figure P11.18).

11.35 For Problem 11.34, determine the power loss and oil flow.

11.36 Solve Problem 11.26 for full finite bearing. If available, compare results with Problem 11.26.

11.37 Refer to Problem 11.36 and solve Problem 11.30 for full finite bearing. Compare with result of Problem 11.26.

11.38 Refer to Problem 11.36 and solve Problems 11.20 and 11.32 for full finite bearing. Compare results with Problem 11.20.

11.39 Refer to Problem 11.36 and solve Problems 11.16 and 11.33 for full finite bearing. Compare the results with Problem 11.16 for a 5 kN load.

11.40 Refer to Problem 11.36 and solve Problem 11.31 for full finite bearing. Compare results with Problem 11.26.

11.41 Design a hydrostatic pad bearing for the thrust bearing of Problem 11.21 with a constant pressure supply of 10 MPa and the same oil film thickness. Compare results with Problem 11.21, if available.

Problems 11.42 to 11.44
Perform the same routine for Problems 11.22, 11.23, and 11.24, as done for Problem 11.41.

11.45 Design an externally pressurized pad bearing for Problem 11.8 with air supplied at 1.5 MPa pressure.

11.46 Use the FINLUB program (see Appendix 11.A) to solve the Problems 11.26 and 11.36 and compare the results.

11.47 Use the FINLUB program to solve the Problems 11.30 and 11.37 and compare the results.

11.48 Use the FINLUB program to solve the Problems 11.32 and 11.38 and compare the results.

11.49 Perform the same routine for Problems 11.33 and 11.39, as done for Problem 11.48.

11.50 Perform the same routine for Problems 11.31 and 11.40, as done for Problem 11.48.

APPENDIX 11.A
FINLUB: A FINITE-ELEMENT PROGRAM FOR FLUID LUBRICATION

```
10 REM ************************************
20 REM *                                  *
30 REM *             FINLUB               *
40 REM *                                  *
50 REM ************************************
60 REM           COPYRIGHT 1985
70 REM by Professor Andrew D. Dimarogonas, Washington Univ., St. Louis, Mo.
```

```
80 REM  All rights reserved.  Unauthorized reproduction, disssemination,
90 REM selling or use is strictly prohibited.
100 REM  This listing is for reference purpose only.
110 REM
130 REM Solution of the lubrication equation for slider bearings
140 REM with the FEM and computation of the bearing design parameters
150 REM
160 REM          ADD / 1-12-84 / Revised 7-30-87
170 REM
180 CLS:KEY OFF:LOCATE 25,1
190 PRINT"Dimarogonas,A.D, Computer Aided Machine Design, Prentice-
Hall,1987";
200 FOR I=1 TO 14:COLOR I:LOCATE I,22+I:PRINT "FINLUB";
210 LOCATE I,50-I:PRINT "FINLUB";:NEXT I:COLOR 14
220 LOCATE 8,10:PRINT"Solution of the lubrication equation for slider
bearings";
230 LOCATE 22,25:INPUT"Hit RETURN to continue";X$
270 REM x(iar,N), y(iar,N):REM .......NODE COORDINATES OF ARC iar
280 REM ni(iar,IN), nj(iar,N), nm(iar,N):....REM ELEMENT NODES
290 REM nf(iar,N): REM ...............NODES WIT FIXED PRESSURE
300 REM tt(iar,N):REM ................FIXED PRESSURES AT ABOVE NODES
310 REM GK(N,N) ......................STRIFFNESS MATRIX
320 REM GF(N) ........................FORCE VECTOR
330 REM D(iar,N) .....................SOLUTION VECTOR
350 X0 = 41: Y0 = 101:PI=3.14159:ISCR=8:          REM PLOTTING PARAMETERS
360 REM ..............................NO OF NODES NN, ELEMENTS NE, ARCS NAR
370 REM ..............................BEARING RADIUS RA,
380 REM ..............................VELOCITY U=OM*RA, RADIAL CLEARANCE
CL,
390 REM ..............................VISCOSITY MU, ATTITUDE ANGLE FI,
400 REM ..............................ECCENTRICITY ec
402 REM ..............................ARC:ENTRANCE,EXIT ANGLES B1,B2,LENGTH
BL
403 REM ..............................Misalignment:Xmis,Ymis,FIxm,FIym
404 DIM MENU1$(25),DA(300),BI(300),DI(300),B1(10),B2(10),BL(10),PR(10)
405 DIM X(2,300),Y(2,300),NI(2,300),NJ(2,300),NM(2,300),NF(2,100),TT(2,100)
406 DIM A(300,12),GF(300),D(2,300),L$(300),NE(2),ND(2),SMAX(2),NFNODES(100)
410 READ RA, OM,  CL, MU,    EC, FA,    VL ,B1(1),B2(1),BL(1),NAR
420 DATA 100,377, .4,.005E-6,.2, -45,1500, -85,85,    200,   2
440 NMENUS=7:FOR IM=1 TO NMENUS:READ MENU1$(IM):NEXT IM
450 DATA "Stop       ","Load File ","Plot Mesh","boundary"
460 DATA "Analyze    ","Post Proc ","New Data"
470 CLS:COLOR 15:PRINT"BEARING DATA-hit ENTER for default data":PRINT:COLOR
14
480 PRINT"Bearing Radius              ";RA;:INPUT RA$:IF RA$>"" THEN
RA=VAL(RA$)
490 PRINT"Bearing Ang. Velocity     ";OM;:INPUT OM$:IF OM$>"" THEN
OM=VAL(OM$)
500 PRINT"Bearing Clearance          ";CL;:INPUT CL$:IF CL$>"" THEN
CL=VAL(CL$)
510 PRINT"Oil Dynamic Viscosity     ";MU;:INPUT MU$:IF MU$>"" THEN
MU=VAL(MU$)
520 PRINT"Bearing Ecc. (guess)      ";EC;:INPUT EC$:IF EC$>"" THEN
EC=VAL(EC$)
521 PRINT" Att. Angle,deg.(guess) ";FA;:INPUT FA$:IF FA$>"" THEN
FA=VAL(FA$)
522 PRINT"Bearing Vertical Load    ";VL;:INPUT VL$:IF VL$>"" THEN
VL=VAL(VL$)
531 PRINT"Number of arcs             ";NAR;;:INPUT N$:IF N$>"" THEN
NAR=VAL(N$)
532 PRINT:COLOR 15:FOR IAR=1 TO NAR:PRINT"Arc no ";IAR:COLOR 14
540 PRINT"Entrance Angle,deg.";B1(IAR);:INPUT B$:IF B$>"" THEN
B1(IAR)=VAL(B$)
550 PRINT"Exit Angle,deg.    ";B2(IAR);:INPUT B$:IF B$>"" THEN
B2(IAR)=VAL(B$)
560 PRINT"Length             ";BL(IAR);:INPUT B$:IF B$>"" THEN
BL(IAR)=VAL(B$)
561 PRINT"Preload            ";PR(IAR);:INPUT P$:IF P$>"" THEN
PR(IAR)=VAL(P$)
562 NEXT IAR
570 INPUT"Are the data correct (Y/N) ";DAT$:IF DAT$="n" OR DAT$="N" THEN
470
```

```
630 P0=0:REM ..........................ambient pressure
640 SCREEN ISCR:   XSC=3:YSC=1:U=OM*RA: VIND=2:XS=0
670 FOR I=1 TO NMENUS:LOCATE I+2,70:PRINT MENU1$(I);:NEXT I:  ICOUNT=0
690 LOCATE 1,70:PRINT"MAIN MENU";:LINE(530,0)-(639,8),,B:GOTO 780
710 X$=INKEY$:XS=0:IF X$="" THEN 710
730 IF LEN(X$)=1 THEN XS=ASC(X$):GOTO 840
740 X$=RIGHT$(X$,1)
750 IF X$><"H" AND X$<>"P" THEN 810
760 IF X$="H" AND VIND>1 THEN VIND=VIND-1
770 IF X$="P" AND VIND<NMENUS THEN VIND=VIND+1
780 IF VIND>1 THEN LOCATE VIND+1,69:PRINT " ";MENU1$(VIND-1);
790 LOCATE VIND+2,69:PRINT ">";MENU1$(VIND);
800 IF VIND<NMENUS THEN LOCATE VIND+3,69:PRINT " ";MENU1$(VIND+1);
810 GOTO 710
840 IF XS=13 AND VIND=1 THEN SCREEN 0:END
850 IF XS=13 AND VIND=2 THEN GOSUB 2200:REM             load file from disc
860 IF XS=13 AND VIND=4 THEN GOSUB 3030:REM             define boundary cond
870 IF XS=13 AND VIND=5 THEN GOSUB 1080:REM             solve
880 IF XS=13 AND VIND=6 THEN GOSUB 2560:REM             post processing
890 IF XS=13 AND VIND=3 THEN GOSUB  970:REM             plot mesh
900 IF XS=13 AND VIND=7 THEN GOTO 470  :REM             change data
910 IF XS=13 THEN VIND=VIND+1
920 FOR ICLEAR=14 TO 24:LOCATE ICLEAR,65:PRINT SPC(14);:NEXT ICLEAR
930 IF XS=13 AND VIND>7 THEN VIND=1:GOTO 670
940 LOCATE 25,1:PRINT"Back to the main menu. Use arrows to select,then
RETURN";
950 GOTO 670:          END
970 REM ******************************************************* PLOT
ELEMENTS
975 CLS:IAR=1:IF NAR>1 THEN INPUT"Arc number ";IAR
980 NE=NE(IAR):FOR I=1 TO NE
985 YSC=160/YMAX(IAR): XSC=500/XMAX(IAR)
990 I1 = NI(IAR,I):I2 = NJ(IAR,I): I3 = NM(IAR,I)
1000 X1=XSC*X(IAR,I1):Y1=Y(IAR,I1)*YSC:X2=XSC*X(IAR,I2):Y2=Y(IAR,I2)*YSC
1020 X3=XSC*X(IAR,I3):Y3=Y(IAR,I3)*YSC:LINE(0,0)-(0,190):LINE (0,0)-(550,0)
1040 LINE(X1,Y1)-(X2,Y2):LINE-(X3,Y3):LINE-(X1,Y1): NEXT I:RETURN
1080 ' BEARAN ************************************ BEARING ANALYSIS ROUTINE
1085 SCREEN 0:CLS:COLOR 15:PRINT"Iterating for equilibrium":PRINT:COLOR 14
1090 GOSUB 1150:FX0=FX:FY0=FY:STP$=INKEY$:IF STP$="s" OR STP$="S" THEN 1140
1100 IF ABS(FY)+ABS(FX)<VL/1000 THEN  GOSUB 1150:GOTO 1125
1110 FA=FA+.01:GOSUB 1150:DFYA=(FY-FY0)/.01:DFXA=(FX-FX0)/.01:FA=FA-.01
1111 EC=EC+.01:GOSUB 1150:DFYE=(FY-FY0)/.01:DFXE=(FX-FX0)/.01:EC=EC-.01
1115 DENOM=DFXE*DFYA-DFYE*DFXA:DFE=(-FX0*DFYA+FY0*DFXA)/DENOM
1116 DFA=(-FY0*DFXE+FX0*DFYE)/DENOM
1120 FA=FA+DFA:EC=EC+DFE:GOTO 1090
1121 REM .............. derivatives
1125 COLOR 15:PRINT"iterating for design data...":COLOR 14
1126 FX0=FX:FY0=FY:DX=CL/1000:DY=DX:SFI=SIN(FI):CFI=COS(FI)
1127 REM ..............  x=x+dx , y=y+dy ...............
1128 DE=CFI*DX+SFI*DY:DF=-SFI*DX/EC+DY/(EC*CFI):EC=EC+DE:FI=FI+DF:GOSUB
1150
1129 EC=EC-DE:FI=FI-DF:FXP1P1=FX:FYP1P1=FY
1130 REM  ................ x=x-dx , y=y+dy ................
1131 DE=-CFI*DX+SFI*DY:DF=+SFI*DX/EC+DY/(EC*CFI):EC=EC+DE:FI=FI+DF:GOSUB
1150
1132 EC=EC-DE:FI=FI-DF:FXM1P1=FX:FYM1P1=FY
1133 REM  ............... x=x+dx  , y=y-dy ...............
1134 DE=CFI*DX-SFI*DY:DF=-SFI*DX/EC-DY/(EC*CFI):EC=EC+DE:FI=FI+DF:GOSUB
1150
1135 EC=EC-DE:FI=FI-DF:FXP1M1=FX:FYP1M1=FY
1136 REM  ............... x=x-dx  , y=y-dy ................
1137 DE=-CFI*DX-SFI*DY:DF=+SFI*DX/EC-DY/(EC*CFI):EC=EC+DE:FI=FI+DF:GOSUB
1150
1138 EC=EC-DE:FI=FI-DF:FXM1M1=FX:FYM1M1=FY
1139 REM ............... stiffness parameters ............
1140 KXX=(FXP1P1-FXM1P1+FXP1M1-FXM1M1)/(4*DX)
1141 KYX=(FYP1P1-FYM1P1+FYP1M1-FYM1M1)/(4*DX)
1142 KYY=(FYP1P1-FYP1M1+FYM1P1-FYM1M1)/(4*DX)
1143 KXY=(FXP1P1-FXP1M1+FXM1P1-FXM1M1)/(4*DX)
1144 REM .................damping parameters ............
1145 XVEL=.001:GOSUB 1150:CXX=(FX-FX0)/XVEL:CYX=(FY-FY0)/XVEL:XVEL=0
```

```
1146 YVEL=.011:GOSUB 1150:CXY=(FX-FX0)/YVEL:CYY=(FY-FY0)/YVEL:YVEL=0
1149 IEND$="1":GOSUB 1150:XS=13:GOSUB 1840:SCREEN ISCR:RETURN
1150 ' ****************************************************************
1170 ' *  Subroutine bearc(R,Om,Cl,Mu,Fa,Ec,Nar,B1,B2,BL,Fx,Fy,iend$)  *
1180 ' *  Returns vertical force Fx and horizontal Fy for a fixed pad    *
1190 ' *  If iend$="any(iar)thing", plots pressures and prints results
*
1200 ' ****************************************************************
1210 SCREEN 0:COLOR 14:FI=PI*FA/180:FX=0:FY=0
1215 FOR IAR=1 TO NAR:    ND=ND(IAR):NE=NE(IAR): NN=ND:BW=BW(IAR):W1=BW+1
1220 B1=B1(IAR):B2=B2(IAR):XS=RA*PI/180*(B2-B1)/(XMAX(IAR)-XMIN(IAR))
1225 YS=BL(IAR)/(YMAX(IAR)-YMIN(IAR)):NF=NFNODES(IAR)
1260 ' ........stiffness matrix.......
1270 FOR I=1 TO NN:GF(I)=0:FOR J=1 TO W1:A(I,J)=0:NEXT J:NEXT I
1280 FOR I = 1 TO NE:          NI = NI(IAR,I):NJ = NJ(IAR,I):NM = NM(IAR,I)
1300 XI=RA*PI*B1/180+(X(IAR,NI)-XMIN(IAR))*XS:YI=(Y(IAR,NI)-YMIN(IAR))*YS
1310 XJ=RA*PI*B1/180+(X(IAR,NJ)-XMIN(IAR))*XS:YJ=(Y(IAR,NJ)-YMIN(IAR))*YS
1320 XM=RA*PI*B1/180+(X(IAR,NM)-XMIN(IAR))*XS:YM=(Y(IAR,NM)-YMIN(IAR))*YS
1330 DL = (XJ * YM + XI * YJ + YI * XM - YI * XJ - YJ * XM - XI * YM)/2
1340 BI=YJ-YM:CI=XM-XJ:BJ=YM-YI:CJ=XI-XM:BM=YI-YJ:CM=XJ-XI:BL=BL(IAR)
1360 XH=(XI+XJ+XM)/3:YH=(YI+YJ+YM)/3:ECC=EC-PR(IAR):ECY=ECC*SIN(FI)+YMIS
1361 A1=FIXM*(YH-BL/2):A2=FIYM*(XH-BL/2)
1362 ECX=ECC*COS(FI)+XMIS:ECY=ECY+A2:ECX=ECX+A1:FIMIS=ATN(ECY/ECX)
1363 ECMIS=SQR(ECY^2+ECX^2):H=CL-ECMIS*COS(XH/RA-FIMIS):KC=H^3
1370 LM = KC / (4*DL):IA = NI: IF NJ < NI THEN IA = NJ
1380 JA = ABS (NI - NJ) + 1:A(NI,1) = A(NI,1) + LM * (BI * BI + CI * CI)
1390 A(IA,JA) = A(IA,JA) + LM * (BI * BJ + CI * CJ)
1400 JA = NJ: IF NM < NJ THEN JA = NM
1410 MA = ABS (NJ - NM) + 1: A(NJ,1) = A(NJ,1) + LM * (BJ * BJ + CJ * CJ)
1420 A(JA,MA) = A(JA,MA) + LM * (BJ * BM + CJ * CM)
1430 MA = NM: IF NI < NM THEN MA = NI
1440 IA = ABS (NM - NI) + 1: A(NM,1) = A(NM,1) + LM * (BM * BM + CM * CM)
1450 A(MA,IA) = A(MA,IA) + LM * (BM * BI + CM * CI)
1460 REM ..........................................FORCE VECTOR
1470 TH=XH/RA:DH=(ECMIS-PR(IAR))/RA*SIN(XH/RA-FIMIS):Q=6*U*MU*DH
1475 Q=Q + 6*MU*2*(SIN(TH-FIMIS)*XVEL+COS(TH-FIMIS)*YVEL)
1480 GF(NI)=GF(NI)+Q*DL/3:GF(NJ)=GF(NJ)+Q*DL/3
1490 GF(NM)=GF(NM)+Q*DL/3:       NEXT I
1500 REM ................................FIXED PRESSURE BOUNDARY
CONDITIONS
1510 FOR I = 1 TO NF:IT = NF(IAR,I):    S1 = IT - BW: IF S1 < 1 THEN S1 = 1
1530 S2 = IT + BW: IF S2 > NN THEN S2 = NN
1540 FOR S = S1 TO S2:TA = 1 + ABS (IT - S): TI = IT: IF S < IT THEN TI = S
1560 GF(S) = GF(S) - A(TI,TA) * TT(IAR,IT):NEXT S
1570 J2 = IT: J1 = IT - BW: IF J1 < 1 THEN J1 = 1
1580 IF IT < BW + 1 THEN J2 = BW + 1
1590 FOR J = 1 TO BW:A(IT,J) = 0:NEXT J:FOR J = J1 TO IT:A(J,IT-J+1)=0:NEXT
J
1620 A(IT,1) = 1: GF(IT) = TT(IAR,IT):NEXT I:N=NN
1650 GOSUB 1920 :'..................................solve linear equations
1660 IF IEND$="" THEN 1720
1670 IF NX(IAR)*NY(IAR)=0 THEN PRINT"no node nx(iar),ny(iar) data": GOTO
1710
1680 PRINT "Ix ..SOLUTION .. node pressures.....Iy>>>>>>":PRINT"ARC ";IAR;"
"
1685 SMAX=0:FOR IX= 1 TO NX(IAR):FOR IY=1 TO NY(IAR):I=(IX-1)*NY(IAR)+IY
1690 IF D(IAR,I)>SMAX THEN SMAX=D(IAR,I)
1700 PRINT USING" #.#^^^^";D(IAR,I);:NEXT IY:PRINT:NEXT IX:INPUT"Hit
ENTER";XENT$
1710 SMAX(IAR)=SMAX
1720 REM ........................................INTEGRATION
1730 FOR I = 1 TO NE
1740 NI=NI(IAR,I):NJ=NJ(IAR,I):NM=NM(IAR,I):B1=B1(IAR):B2=B2(IAR)
1750 X1=RA*PI*B1/180+(X(IAR,NI)-XMIN(IAR))*XS:Y1=(Y(IAR,NI)-YMIN(IAR))*YS
1760 X2=RA*PI*B1/180+(X(IAR,NJ)-XMIN(IAR))*XS:Y2=(Y(IAR,NJ)-YMIN(IAR))*YS
1770 X3=RA*PI*B1/180+(X(IAR,NM)-XMIN(IAR))*XS:Y3=(Y(IAR,NM)-YMIN(IAR))*YS
1771 XG=(X1+X2+X3)/3:AREA=(X2*Y3+X3*Y1+X1*Y2-X2*Y1-X3*Y2-X1*Y3)/2
1772 P=(D(IAR,NI)+D(IAR,NJ)+D(IAR,NM))/3
1790 TH=XG/RA:IF P<0 THEN P=0
1800 CT = COS(TH):ST = SIN(TH)
1810 FX = FX + P*CT*AREA:FY = FY + P*ST*AREA
```

```
1820 NEXT I:   NEXT IAR :FX=FX-VL:COLOR 14
1830 PRINT "Fx=";FX+VL;"  Fy=";FY;"  e=";EC;"  Fa=";FA:IF IEND$="" THEN
1910
1835 RETURN
1840 REM ************************ COMPUTATION OF BEARING DIMENSIONLESS
DATA
1850 W=SQR(VL^2+FX*FX+FY*FY):CLS:COLOR 15:PRINT"RESULTS:":COLOR 14
1855 SO =(2*RA*BL(1))*(U/RA/6.28)* MU/W*(RA/CL)^2:REM       =mu*N*(r/c)^2/Pm
1856 PRINT:PRINT:PRINT "Sommerfeld Number              ";SO
1861 PRINT "Vertical Load                ";VL:PRINT "Horizontal Load              ";FY
1864 PRINT "Linear properties:";TAB(40);"  ";
1865 PRINT" Dimensionless Kij*C/W or Cij*C*om/W"
1866 PRINT "Spring Constant Kxx      ";KXX;TAB(50);KXX*CL/W
1867 PRINT "Spring Constant Kxy      ";KXY;TAB(50);KXY*CL/W
1868 PRINT "Spring Constant Kyx      ";KYX;TAB(50);KYX*CL/W
1869 PRINT "Spring Constant Kyy      ";KYY;TAB(50);KYY*CL/W
1870 PRINT "Damping Constant Cxx     ";CXX;TAB(50);CXX*CL*OM/W
1871 PRINT "Damping Constant Cxy     ";CXY;TAB(50);CXY*CL*OM/W
1872 PRINT "Damping Constant Cyx     ";CYX;TAB(50);CYX*CL*OM/W
1875 PRINT "Damping Constant Cyy     ";CYY;TAB(50);CYY*CL*OM/W
1889 PRINT "Relative Eccentricity    ";EC/CL
1890 PRINT "Attitude Angle,deg,      ";FA:BL=BL(1)
1900 PRINT "Length/Diameter ratio    ";BL/RA/2:INPUT"Hit
ENTER";XENT$:IEND$=""
1910 RETURN
1920 REM BANDED SYMMMETRIC MATRIX SYSTEM OF LINEAR EQUATIONS.GAUSS METHOD
1930 FD = N:MS = BW:N1 = FD - 1:FOR K = 1 TO N1
1970 C = A(K,1):K1 = K  + 1:NI = K1 + MS - 2:L = NI: IF FD < NI THEN L = FD
1990 FOR J = 2 TO MS: DI(J) = A(K,J):NEXT J
2000 FOR J=K1 TO L: K2=J-K+1:A(K,K2)=A(K,K2)/C:NEXT J:GF(K)=GF(K)/C
2020 FOR I =K1 TO L:K2 = I - K1 + 2:C = DI(K2)
2040 FOR J = I TO L:K2 = J - I + 1: K3 = J - K + 1
2060 A(I,K2) = A(I,K2) - C * A(K,K3):NEXT J:GF(I) = GF(I) - C * GF(K)
2070 NEXT I:NEXT K:GF(FD) = GF(FD) / A(FD,1)
2100 FOR I=1 TO N1:K=FD-I:K1=K+1:NI=K1+MS-2:L=NI:IF NI > FD THEN L=FD
2130 FOR J=K1 TO L:K2=J-K+1:GF(K)=GF(K)-A(K,K2)*GF(J):NEXT J:NEXT I
2170 FOR I = 1 TO FD:D(IAR,I) = GF(I):NEXT I:
RETURN
2200 REM ********************************************** file retrieval
2201 CLS: SCREEN ISCR:PRINT"Available mesh files":FILES"mesh*.*":LOCATE
21,1
2202 PRINT"Load ";NAR;"mesh files. You can make them with the AUTOMESH
program"
2203 PRINT"The geometric dimensions will be readjusted automatically for
data"
2204 PRINT"entered above.Use any convenient scale to prepare the meshes."
2220 FOR IAR=1 TO NAR:ND=0:NE=0:NDIF=0
2230 LOCATE 10+IAR,1:PRINT"Arc No ";IAR;"  Data file  ";:INPUT FILNA$
2240 IF FILNA$="" THEN  2230
2250 ICOUNT=0:XMAX(IAR)=0:YMAX(IAR)=0:XMIN(IAR)=1E+30:YMIN(IAR)=1E+30
2260 OPEN FILNA$ FOR INPUT AS #1
2270 ICOUNT=ICOUNT+1:INPUT#1,X$
2280 IF X$="filend" OR X$="FILEND" THEN 2520
2300 IF X$="triangular" OR X$="TRIANGULAR" THEN  E$="2"
2330 SYM$="":ISYM=1
2360 IF ICOUNT>1000 THEN 2530
2370 SYM$=MID$(X$,ISYM,1):ISYM=ISYM+1:IF SYM$="" AND ISYM<10 THEN 2360
2380 IF ASYM=10 THEN 2270
2390 IF SYM$<>"n" AND SYM$<>"N" THEN 2400
2391 ND=ND+1:INPUT#1,I,X(IAR,ND),Y(IAR,ND)
2392 IF X(IAR,ND)>XMAX(IAR) THEN XMAX(IAR)=X(IAR,ND)
2393 IF X(IAR,ND)<XMIN(IAR) THEN XMIN(IAR)=X(IAR,ND)
2394 IF Y(IAR,ND)>YMAX(IAR) THEN YMAX(IAR)=Y(IAR,ND)
2395 IF Y(IAR,ND)<YMIN(IAR) THEN YMIN(IAR)=Y(IAR,ND)
2400 IF SYM$<>"e" AND SYM$<>"E" THEN  2270
2410 NE=NE+1 :ES=VAL(E$)
2430 IF E$="2" THEN INPUT#1,I,NI(IAR,NE),NJ(IAR,NE),NM(IAR,NE)
2450 IF ABS(NI(IAR,NE)-NJ(IAR,NE))>NDIF THEN NDIF=ABS(NI(IAR,NE)-
NJ(IAR,NE))
2460 IF ES>1 AND ABS(NI(IAR,NE)-NM(IAR,NE))>NDIF THEN NDIF=ABS(NI(IAR,NE)-
NM(IAR,NE))
```

```
2470 IF ES>1 AND ABS(NJ(IAR,NE)-NM(IAR,NE))>NDIF THEN NDIF=ABS(NJ(IAR,NE)-
NM(IAR,NE))
2510 GOTO 2270
2520 CLOSE#1 :GOTO 2540
2530 PRINT"bad file";
2540 NE(IAR)=NE:ND(IAR)=ND:BW(IAR)=NDIF+2
2550 NEXT IAR:RETURN
2560 REM ****************************************************** PLOT
RESULTS
2570 LOCATE 1,1
2580 INPUT"Do you have QUICKBASIC 3.0 and an EGA card (Y/N)  ";S$
2590 IF S$="n" OR S$="N" THEN 2860
2600 SCREEN 8:IF ISCR=9 THEN SCREEN 9
2601 FOR IAR=1 TO NAR:NE=NE(IAR):SMAX=SMAX(IAR)
2610 CLS:FOR IC=1 TO 8:ICOL=IC:IF IC=2 OR IC=3 THEN ICOL=IC+2
2620 IF IC=5 OR IC=4 THEN ICOL=IC-2:XSC=500/XMAX(IAR):YMAX=160/YMAX(IAR)
2630 LOCATE 2*IC,73:PRINT INT(IC/8*100);:LOCATE 2*IC,78:PRINT"%";
2640 LINE(560,16*IC+6)-(570,16*IC+16),ICOL,BF:  NEXT IC :PMAX=0
2660 FOR I = 1 TO NE:                I1=NI(IAR,I):I2=NJ(IAR,I):I3=NM(IAR,I)
2670 PRESS=(D(IAR,I1)+D(IAR,I2)+D(IAR,I3))/3
2680 X1 =XSC* X(IAR,I1):Y1 = Y(IAR,I1)*YSC:LOCATE 22,1:PRINT" Y     ";
2690 X2=XSC*X(IAR,I2):Y2=Y(IAR,I2)*YSC:IF PRESS>PMAX THEN PMAX=PRESS
2700 X3 =XSC* X(IAR,I3):Y3 = Y(IAR,I3)*YSC:XG=(X1+X2+X3)/3:YG=(Y1+Y2+Y3)/3
2710 LINE(0,0)-(0,190):LINE (0,0)-(600,0)
2720 ICOL=8:IF PRESS<0 THEN LINE(XG-2,YG)-(XG+2,YG),7
2730 IF PRESS>.125*SMAX THEN ICOL=1
2740 IF PRESS>.25*SMAX THEN ICOL=4
2750 IF PRESS>.375*SMAX THEN ICOL=5
2760 IF PRESS>.5*SMAX THEN ICOL=2
2770 IF PRESS>.675*SMAX THEN ICOL=3
2780 IF PRESS>.75*SMAX THEN ICOL=6
2790 IF PRESS>.825*SMAX THEN ICOL=7
2800 LINE(X1,Y1)-(X2,Y2),ICOL:LINE-(X3,Y3),ICOL:LINE-(X1,Y1),ICOL
2810 IF ICOL<8 THEN PAINT (XG,YG),ICOL
2820 IF ICOL=8 THEN LINE(X1,Y1)-(X2,Y2),1:LINE-(X3,Y3),1:LINE-(X1,Y1),1
2830 NEXT I: LOCATE 22,10:PRINT "Max. pressure=";PMAX;:LOCATE
2,55:PRINT"X";
2840 LOCATE 25,1:INPUT"Hit return to continue";CON$:   NEXT IAR
2860 VIND=VIND+1:XS=0:CLS:RETURN
2870 REM *********************************SET NODE BOUNDARY CONDITIONS
2880 FOR IER=22 TO 24:LOCATE IER,68:PRINT SPC(11);:NEXT IER
2882 LOCATE 22,58:PRINT"Set boundary to Po (Y/N)";:INPUT BSET$
2884 IF BSET$="Y" OR BSET$="y" THEN GOSUB 3500:RETURN
2890 NFNODES(IAR)=NFNODES(IAR)+1:NF(IAR,NFNODES(IAR))=INODE
2900 NF(IAR,NFNODES(IAR))=INODE:TT(IAR,NFNODES(IAR))=P0:NF=NFNODES(IAR)
2920 XR=X(IAR,INODE)*XSC:YR=Y(IAR,INODE)*YSC
2930 LINE (XR-2,YR-2)-(XR+2,YR+2),1,BF:RETURN
2940 REM ******************************LOCKING TO THE NEAREST NODE
2950 DIST=0:PREDIST=1E+30::FOR I=1 TO ND
2960 DIST=SQR((X(IAR,I)-XHAIR)^2+(Y(IAR,I)-YHAIR)^2)
2970 IF DIST<PREDIST THEN INODE=I:PREDIST=DIST
2980 NEXT I
2990 LOCATE 25,1:PRINT SPC(78);
3000 LOCATE 25,1:PRINT "Nearest Node is  ";INODE;
3010 XLOCK=X(IAR,INODE):YLOCK=Y(IAR,INODE)
3020 RETURN
3030 REM ************************************ Assigning boundary
conditions
3035 FOR IAR=1 TO NAR:GOSUB 970
3040 LOCATE 25,1:PRINT" Move cross with keyboard arrows   ";
3050 LOCATE 14,68:PRINT"KEYBOARD:";:LOCATE 15,68:PRINT"V-elocity";
3070 LOCATE 16,68:PRINT"P-ressure ";:LOCATE 17,68:PRINT"N-ear Node";
3090 LOCATE 18,68:PRINT"E-xit MENU";:LOCATE 19,68:PRINT"F-aster   ";
3110 LOCATE 20,68:PRINT"S-lower   ";:LOCATE 21,68:PRINT"M-ore Arcs";
3120 XSTEP=10/XSC:YSTEP=10/YSC:XHAIR=1/XSC:YHAIR=1/YSC
3130 X$="aa":GOTO 3270
3140 X$=INKEY$:XS=0
3150 IF X$="" THEN 3140
3160 IF LEN(X$)>1  THEN 3270
3170 IF X$="f" OR X$="F" OR X$="s" OR X$="S" THEN LOCATE 25,1:PRINT "step
x,y =";XSTEP,YSTEP;"        ";
```

```
3180 IF X$="f"  THEN XSTEP=XSTEP*2:YSTEP=XSTEP:ZSTEP=XSTEP
3190 IF X$="s"  THEN XSTEP=XSTEP/2:YSTEP=XSTEP:ZSTEP=XSTEP
3200 IF X$="F"  THEN XSTEP=XSTEP*2:YSTEP=XSTEP:ZSTEP=XSTEP
3210 IF X$="S"  THEN XSTEP=XSTEP/2:YSTEP=XSTEP:ZSTEP=XSTEP
3220 IF X$="v" OR X$="V" THEN GOSUB 3470
3230 IF X$="p" OR X$="P" THEN GOSUB 2870
3240 IF X$="n" OR X$="N" THEN GOSUB 2940:GOTO 3270
3250 IF X$="e" OR X$="E" THEN GOTO 3465
3255 IF X$="m" OR X$="M" THEN GOTO 3460
3260 GOTO 3140
3270 X$=RIGHT$(X$,1)
3280 FOR I=5 TO 500 STEP 20:PSET(I,YHAIR*YSC),0:NEXT I
3290 FOR I=6 TO 190 STEP 10:PSET(XSC*XHAIR,I),0:NEXT I
3300 IF XLOCK>0 OR YLOCK>0 THEN XHAIR=XLOCK:YHAIR=YLOCK
3310 XLOCK=0:YLOCK=0:IF X$="H" THEN YHAIR=YHAIR-YSTEP
3330 IF X$="P" THEN YHAIR=YHAIR+YSTEP
3340 IF X$="K" THEN XHAIR=XHAIR-XSTEP
3360 IF X$="M" THEN XHAIR=XHAIR+XSTEP
3370 IF XHAIR<0 THEN XHAIR=0
3380 IF YHAIR<0 THEN YHAIR=0
3390 IF YHAIR>190 THEN YHAIR=190
3400 IF XHAIR>500 THEN XHAIR=500
3410 LOCATE 25,55:PRINT SPC(23);
3420 LOCATE 25,35:PRINT "x= ";XHAIR;:LOCATE 25,45:PRINT "y= ";YHAIR;
3430 FOR I=5 TO 500 STEP 20:PSET(I,YHAIR*YSC):NEXT I
3440 FOR I=6 TO 190 STEP 10:PSET(XSC*XHAIR,I):NEXT I
3450 GOTO 3140
3460 NEXT IAR
3465 VIND=VIND+1:RETURN
3470 ' ********************** velocity boundary conditions
3480 LOCATE 25,1:PRINT"Velocity B.C. not available...             ";
3490 RETURN
3500 ' ****************************************set boundary pressure to P0
3502 LOCATE 23,1:PRINT"Enter no of nodes in peripheral-x and axial-y ";
3503 PRINT"directions nx,ny ";:INPUT NX(IAR),NY(IAR):NX=NX(IAR):NY=NY(IAR)
3510 FOR J=1 TO NY(IAR):NF(IAR,J)=J:TT(IAR,NF(IAR,J))=P0
3520 NF(IAR,NY+J)=(NX-1)*NY+J:TT(IAR,NF(IAR,NY+J))=P0:NEXT J
3530 FOR I=2 TO NX-1:II=2*NY+I-1:NF(IAR,II)=(I-
1)*NY+1:TT(IAR,NF(IAR,II))=P0
3540 II=2*NY+NX-3+I:NF(IAR,II)=NY*I
3550 TT(IAR,NF(IAR,II))=P0:NEXT I:NFNODES(IAR)=2*(NX(IAR)+NY(IAR))-4
3560 FOR ISQ=1 TO NFNODES(IAR):X=X(IAR,NF(IAR,ISQ)):Y=Y(IAR,NF(IAR,ISQ))
3570 X=X*XSC:Y=Y*YSC:LINE (X-4,Y-2)-(X+4,Y+2),3,BF:NEXT ISQ    :RETURN
```

CHAPTER TWELVE

DESIGN OF CONTACT ELEMENTS

12.1 DRY CONTACTS

The stress distribution in highly-loaded contacts, lubricated or dry, closely follows the results of the theory developed for contacts between cylindrical and spherical surfaces. This is a rather complicated topic in the general theory of elasticity and some useful results will be presented here.

Consider the contact for two spheres of diameters d_1 and d_2 under a compressive load P. Initially, the contact is a point but due to elastic deformation, there is a circular area of contact with diameter $2a$ depending on the load. Herz studied this problem and proved that the normal stresses over the contact area have an elliptical distribution (Figure 12.1), and that the radius of the contact circle is

$$\alpha = \left[(3P/8) \frac{(1-\nu_1^2)/E_1 + (1-\nu^2)/E_2}{(1/d_1) + (1/d_2)} \right]^{\frac{1}{3}} \tag{12.1}$$

The maximum normal stress (p) is at the center of the contact circle and has a value of

$$p = 3P/2\pi a = 3p_m/2 \tag{12.2}$$

when p_m is the average pressure $P/\pi a^2$.

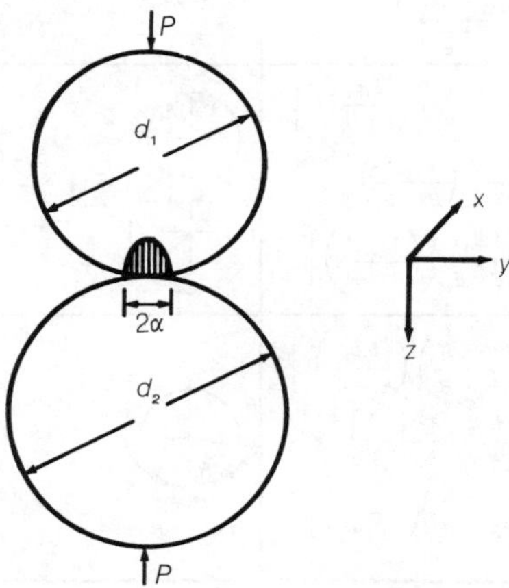

Figure 12.1 Contact of two spheres

The distribution of stresses along the center-line is shown in Figure 12.2 for a typical application. The maximum equivalent stress according to the maximum shear stress criterion appears below the contact surface along the center-line and has a value of, approximately, $0.3p_{max}$.

Solutions for different contact situations, spheres and cylinders are given in Figure 12.3, where the maximum pressure p_0, the radius of the contact circle a (for spheres) or the semiwidth of the contact area (for cylinders) and the approach δ of the centers are given. Cylinder loads are assumed per unit length.

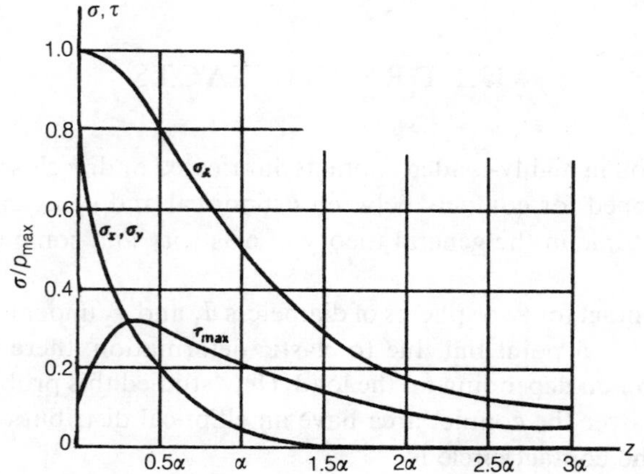

Figure 12.2 Contact stress distribution

Spheres	Cylinders
$p_o = 0.616\left[\frac{P}{R^2}\left(\frac{E_1E_2}{E_1+E_2}\right)^2\right]^{1/3}$ $a = 0.880\left[PR\left(\frac{1}{E_1}+\frac{1}{E_2}\right)\right]^{1/3}$ $\delta = 0.775\left[\frac{P^2}{R}\left(\frac{1}{E_1}+\frac{1}{E_2}\right)^2\right]^{1/3}$	$p_o = 0.591\left[\frac{P_1E_1E_2}{R(E_1+E_2)}\right]^{1/2}$ $a = 1.076\left[\frac{P_1R(E_1+E_2)}{E_1E_2}\right]^{1/2}$ $\delta = \frac{0.579P_1}{E}\left(\frac{1}{3}+\ln\frac{2R}{a}\right)$
$p_o = 0.616\left[P\left(\frac{1}{R_1}+\frac{1}{R_2}\right)^2\left(\frac{E_1E_2}{E_1+E_2}\right)^2\right]^{1/3}$ $a = 0.880\left[\frac{PR_1R_2}{(R_1+R_2)}\left(\frac{1}{E_1}+\frac{1}{E_2}\right)\right]^{1/3}$ $\delta = 0.775\left[P^2\left(\frac{1}{E}+\frac{1}{E_2}\right)^2\left(\frac{1}{R_1}+\frac{1}{R_2}\right)\right]^{1/3}$	$p_o = 0.591\left[\frac{P_1E_1E_2}{(E_1+E_2)}\left(\frac{1}{R_1}+\frac{1}{R_2}\right)\right]^{1/2}$ $a = 1.076\left[\frac{P_1R_1R_2}{(R_1+R_2)}\left(\frac{1}{E_1}+\frac{1}{E_2}\right)\right]^{1/2}$
$p_o = 0.616\left[P\left(\frac{1}{R_1}-\frac{1}{R_2}\right)^2\left(\frac{E_1E_2}{E_1-E_2}\right)^2\right]^{1/3}$ $a = 0.880\left[\frac{PR_1R_2}{(R_2-R_1)}\left(\frac{1}{E}+\frac{1}{E_2}\right)\right]^{1/3}$ $\delta = 0.775\left[P^2\left(\frac{1}{E_1}+\frac{1}{E_2}\right)\left(\frac{1}{R_1}-\frac{1}{R_2}\right)\right]^{1/3}$	$p_o = 0.591\left[\frac{P_1E_1E_2}{(E_1+E_2)}\left(\frac{1}{R_1}-\frac{1}{R_2}\right)\right]^{1/2}$ $a = 1.076\left[\frac{P_1R_1R_2}{(R_2-R_1)}\left(\frac{1}{E_1}+\frac{1}{E_2}\right)\right]^{1/2}$

Figure 12.3 Contact stress and displacements for $\nu = 0.3$.

Rolling contacts are associated with repeated loading during each rolling pass. Fatigue failures usually occur in the form of surface cracks which result in pitting, spalling and similar forms of wear. For this reason, the fatigue surface strength must now be defined.

If fatigue strength is not available from experiments, we can use the approximation that the maximum equivalent shear stress is about $0.3p_0$. For contact of cylinders, from Figure 12.3, we obtain for maximum pressure

$$p_0 = 2F/\pi a w \tag{12.3}$$

where w is the cylinder length and F the force ($P = F/w$). If the maximum stress is called 'fatigue surface strength' S_{He} and the fatigue shear strength is $S_{He}/2$, then

$$\tau_{max} = 0.3\, p_0 = 0.2 S_{fe} = S_e/2, \quad S_{He} = 1.66 S_e \tag{12.4}$$

Therefore, a fatigue surface strength of 66% greater than the fatigue strength is predicted. Because of friction, in dry contacts, experiments show that surface fatigue strength is at least $2S_e$ but for lubricated contacts $1.66S_e$ should not be exceeded.

Surface strength, however, is influenced by the material strength near the surface, which in turn is highly influenced by surface treatments. Buckingham, based on experiments, has suggested for steels the formula (Shigley 1972)

$$S_{He} = 400 H_B - 10{,}000 \tag{12.5}$$

where S_{He} is given in lb/in^2 and H_B is the Brinnell Hardness Number.

12.2 ELASTOHYDRODYNAMIC LUBRICATION

The conditions of contact in such applications as gear teeth, cams and roller bearings, frequently result in such high contact pressures that hydrodynamic lubrication would appear to be impossible.

High contact pressures can cause appreciable local elastic deformation of the contacting surfaces with consequent change in pressure distribution. A further complication is that the viscosity of the lubricant will also be affected by the high temperatures and pressures in the oil film, and also by the time of contact, that is, the time the oil film is subjected to load.

Figure 12.4 shows the film shape and pressure distribution between two disks in pure rolling, as in Figure 12.5, as predicted by elastohydrodynamic analysis. It will be noted that the high load elastically deforms the contacting area to localized restriction near the outlet. The pressure distribution closely follows that expected for dry elastic solids in contact over most of the region, but has a sharp rise at the exit side.

Considering the disks to be rigid solids and the lubricant to be incompressible and of constant viscosity, Martin (Pugh 1970) derived the following expression for the minimum film thickness, by solving the Reynolds equation presented in Chapter 11,

$$h_{min} = \frac{k\mu(U_1 + U_2)R}{W} \tag{12.6}$$

where μ = absolute viscosity, U_1 = peripheral speed of surface 1, U_2 = peripheral speed of

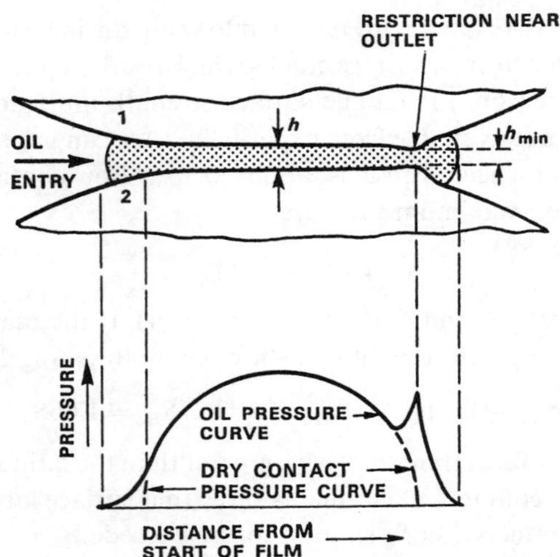

Figure 12.4 Boundary lubrication. (Courtesy BP 1969)

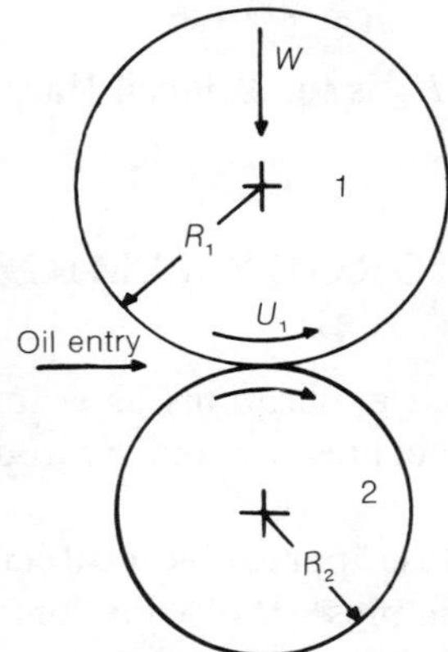

Figure 12.5 Boundary lubrication of two rotating cylinders

surface 2, R = mutual radius of curvature of roller pair = $R_1R_2/(R_1 + R_2)$, W = load per unit width, and k is a constant.

This equation suggests an oil film thickness less than the height of the asperities of the surface roughness, indicating boundary lubrication. Therefore, a different mechanism must be investigated.

Experiments show that the load can be carried for long periods without surface failure or appreciably rapid wear. This suggests that the mechanism of lubrication must be of a hydrodynamic nature, i.e. the mating surfaces must normally be separated by a continuous oil film undisturbed by metal surface irregularities.

Thus, the development of the oil film in this type of lubrication conditions cannot be explained by lubrication theory alone and the oil film thickness predicted by the Martin equation above agrees with experimental results for very light loads only. At higher loads, the elastic deformation of the disks renders the oil film thickness a function not only of the geometry but also of the pressure which deforms the material of the disk. Moreover, even moderate loads can result in high pressures, of the order of 1000 MPa. As discussed in Chapter 11, viscosity at such pressures increases drastically, changing the behavior of the oil film. Thus, the oil film thickness depends on three groups of parameters. Expressing the above influences in the semi-empirical equation, modifying equation (12.6),

$$\frac{h_{\min}}{R} = 2.6\left(\frac{\mu_0 U}{E'R}\right)^{0.7} (\alpha E')^{0.6} \left(\frac{W}{RE'}\right)^{-0.13} \tag{12.7}$$

The first term expresses the hydrodynamic influence, as also discussed in Chapter 11. The second term represents the change of viscosity with pressure with an exponential law $\mu = \mu_0 \exp \alpha(p - p_0)$ where α is a lubricant property, of 13 to 35 mm^2/N. The third term refers to the elastic deformation, where E' is the apparent modulus of elasticity which depends on the elastic properties of the materials of the two disks

$$\frac{1}{E'} = \left[\frac{1-\nu_1^2}{E_1} + \frac{1-\nu_2^2}{E_2}\right] \Big/ 2 \tag{12.8}$$

where E_1 and E_2 are the elasticity moduli, and ν_1 and ν_2 are the Poisson ratios of the two materials. Similarly, R is a geometric parameter combining the two radii $1/R = (1/R_1) + (1/R_2)$.

This type of lubrication is called 'elastohydrodynamic' or 'thin film' lubrication.

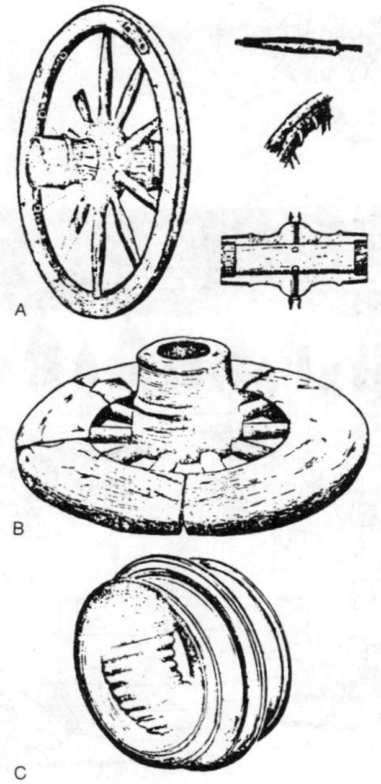

Figure 12.6 A Roman roller bearing. (Reproduced from Singer, *History of Technology*, Cambridge University Press)

12.3 ROLLING BEARINGS

A typical application of lubricated contacts under elastohydrodynamic lubrication is a type of bearing where rolling members are introduced between two rubbing surfaces, making use of the fact that the resistance to motion in this case is much less than in pure sliding. This concept is known from antiquity. Figure 12.7 shows the transportation of a colossus on rollers.

Rolling bearings have been known since the Roman times. Figure 12.6 shows such a bearing of Nero's time found in Italy, where a wheel of a chariot revolves around the axle

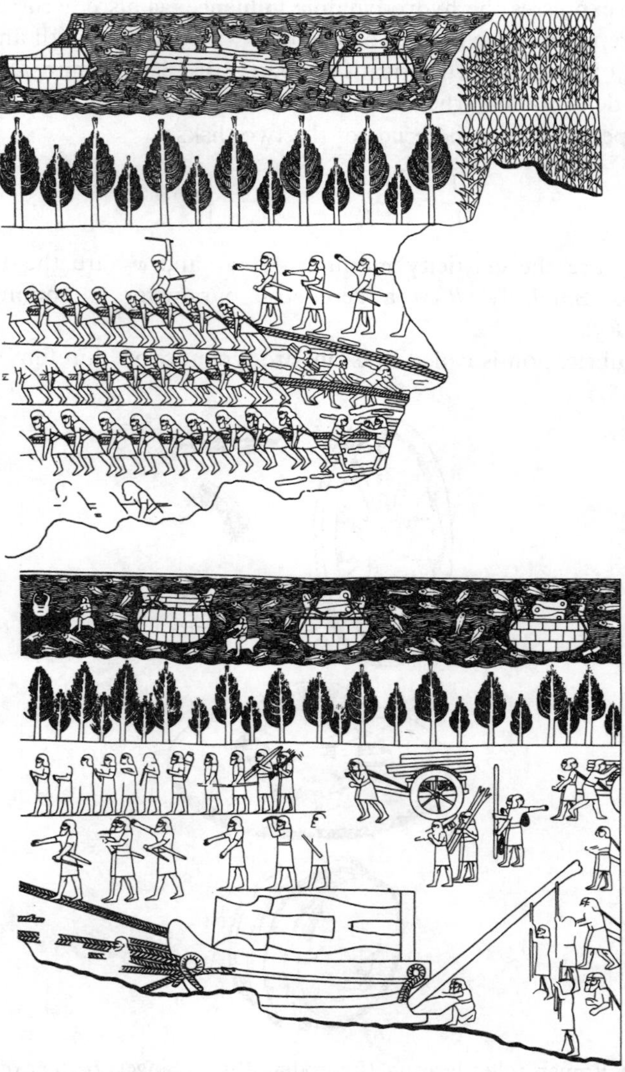

Figure 12.7 Transportation of an ancient colossus with rollers

through a roller bearing. Such bearings, described by Leonardo da Vinci, were re-invented at the turn of the century and they have been very widely used ever since.

12.3.1 Rolling bearing types

Figure 12.8(a) shows the classification of anti-friction bearings.

Each type of bearing has characteristic features which make it particularly suitable for certain applications. However, it is not possible to lay down rules for the selection of bearing types since several factors must be considered and assessed relative to each other.

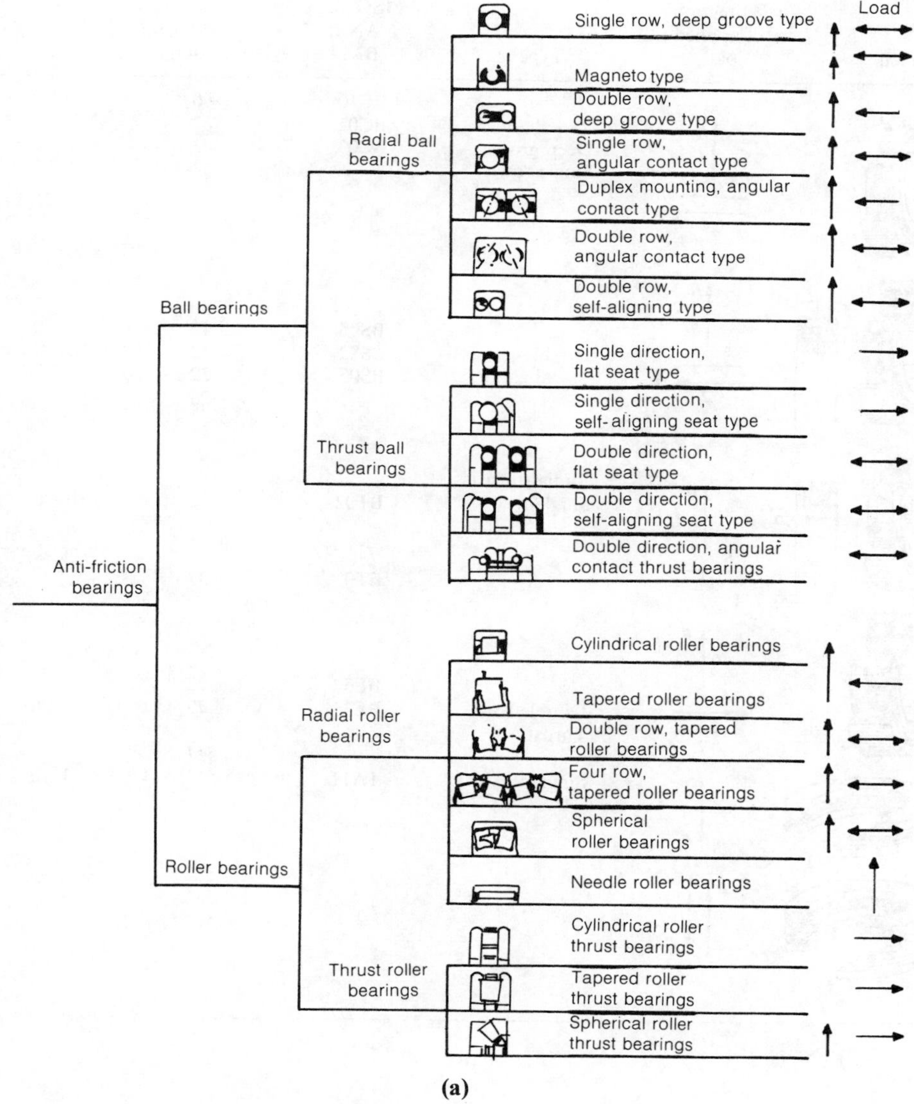

(a)

Deep groove ball bearings are normally selected for small diameter shafts. Deep groove ball bearings, cylindrical roller bearings and spherical roller bearings are considered for shafts of large diameter.

If radial space is limited then bearings with small sectional height must be selected, e.g. needle roller and cage assemblies, needle roller bearings without or with inner ring, certain series of deep groove ball bearings and spherical roller bearings.

Sketch	Direction of applied load	Type	ISO/R 15 ANSI B3.14	International	U_{max} (m/s)
	R 1; A; 0.7 A	Radial Deep groove Single-row	BC10 BC02 BC03	60 62 63 64	10–30
α	R 1; A; 0.2 A	Double-row self-aligning	BS02 BS22 BS03	12 13 22 23	10–20
α	R 1.4; A 0.7	Radial-thrust Single-row	BT02	32	10–20
			ET03	33	
	R 2.3; A; 0.7 A	Double-row Thrust	BE32 BE33	72 73	10–20
	A	Single-thrust	TA11	511, 512 513, 514	5–10
	A; 1; A			522 523 562	

(b)

Where axial space is limited and particularly narrow bearings are required, then some series of single-row cylindrical roller bearings and deep groove ball bearings can be used for radial and combined loads. For axial loads, needle roller and cage thrust assemblies, needle roller thrust bearings and some series of thrust ball bearings may be used.

Sketch	Direction of applied load	Type	ISO/R 15 ANSI B3.14	Interna- tional	U_{max} (m/s)
		Radial			
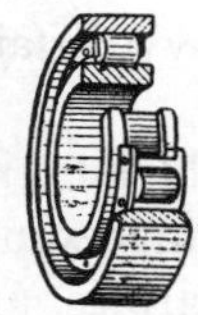	R 1.7	Without lips on outer race	RN02 RN03	N	10–20
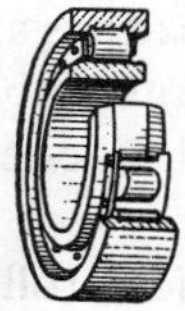	R 1.7	Without lips on inner race		NU	10–20
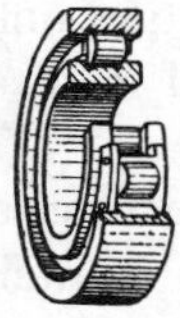	R 1.7; A, A	With one lip on inner race		NJ	10–20
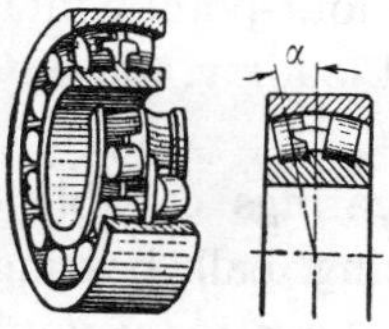 α	R 2; A 0.2 A	Double-row self-aligning	SD22 SC22 SL24 Also 30, 31	213 222 223 230	10–20
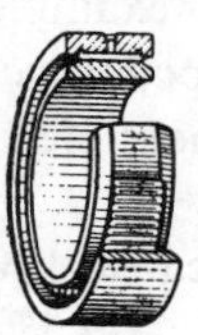	R	Needle with two lips on outer race			5–10
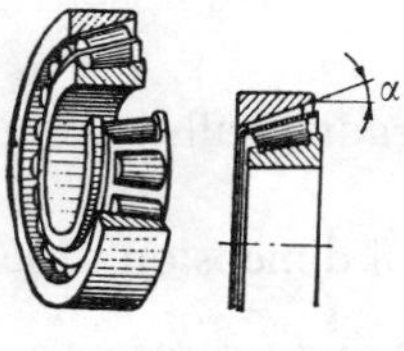 α	R 1.9; A 0.7	Radial-thrust Single-row tapered	2FB 3CC 3DC 4DB 5DD	302, 303 313 322 323 329	5–15

(c)

Figure 12.8 Classification of rolling bearings

In Figure 12.8(b, c), the capacity of antifriction bearings to support loads of radial and axial direction is indicated relative to similar deep groove ball bearings.

Generally, roller bearings can carry greater loads than ball bearings of the same external dimensions. Ball bearings are mostly used to carry light and medium loads, whilst roller bearings are often the only choice for heavy loads and large diameter shafts.

Thrust ball bearings are only suitable for moderate, purely axial loads. Single-direction thrust ball bearings can carry axial loads in one direction and double-direction thrust ball bearings can carry axial loads in both directions. Cylindrical roller and needle roller thrust bearings (with or without washers) can accommodate heavy axial loads in one direction. Spherical roller thrust bearings, in addition to very heavy axial loads, can also carry a certain amount of simultaneously acting radial load.

The most important feature affecting the ability of a bearing to carry an axial load is its angle of contact, α. The greater this angle the more suitable is the bearings for axial loading.

Single and double row angular contact ball bearings, and taper roller bearings, are mainly used for combined loads. Deep groove ball bearings and spherical roller bearings are also used. Self-aligning ball bearings and cylindrical roller bearings can also be used to a limited extent for carrying combined loads. Four-point contact ball bearings and spherical roller thrust bearings should only be considered where axial loads predominate.

Single-row angular contact ball bearings, taper roller bearings, cylindrical roller bearings and spherical roller thrust bearings can carry axial loads in one direction only. Where the direction of load varies, two such bearings arranged to carry axial loads in opposite directions must be used.

Where the axial component of the combined load is large, a separate thrust bearing can be provided for carrying the axial load independently of the radial load. Suitable radial bearings may also be used to carry axial loads only, e.g. deep groove or four-point contact ball bearings. To ensure that these bearings are only subjected to axial loading, the outer rings must have radial clearance in their housings.

Where the shaft can be misaligned relative to the housing, bearings capable of accommodating such misalignment are required, namely self-aligning ball bearings, spherical roller bearings and spherical roller thrust bearings. Misalignment can, for example, be caused by shaft deflection under load, when the bearings are fitted in housings positioned on separate bases and at a large distance from one another or when it is impossible to machine the housing seatings at one setting.

In Figure 12.8(b, c), bearing classification specified by ISO and ANSI is indicated. ISO specifies for antifriction bearings a designation defined as follows (ISO R300):

(i) The first four digits refer to the bore diameter in millimeters.
(ii) The following three letters indicate bearing type. For example, B = radial ball bearings, R = cylindrical roller bearings, T = thrust ball and roller bearing.
(iii) Then, two digits designate the geometric proportions in the form of dimension series.

In the interests of both users and manufacturers of rolling bearings and for reasons of price, quality and interchangeability, a relatively limited number of bearing sizes is desirable. Accordingly, ISO has laid down *Dimension Plans* for the boundary dimensions of metric rolling bearings (ISO/R 15, ISO/R 355 and ISO/R 104).

In the *ISO Dimension Plans* for every standard bearing bore size d, a progressive series of

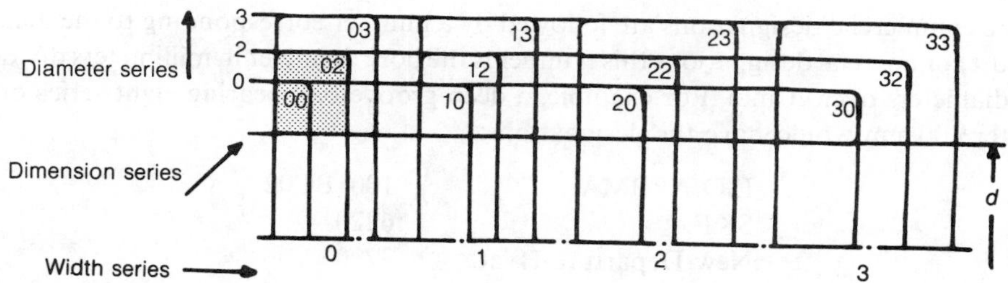

Figure 12.9 Dimension series of rolling bearings

standardized outside diameters D is specified (Diameter Series 8, 9, 0, 1, 2, 3 and 4 in ascending size order). For each Diameter Series, a series of bearing widths B is also specified (Width Series 0, 1, 2, 3, 4, 5 and 6 in ascending order of bearing width). The Width Series for radial bearings corresponds to the Height Series for thrust bearings.

The combination of a Diameter Series with a Width or Height Series is called a Dimension Series. Each Dimension Series has a number consisting of two figures: the first figure indicates the Width (or Height) Series and the second the Diameter Series.

National or international standards have not yet been drawn up for all types of needle roller bearings, but certain dimensions have been established by usage and interchangeability of needle roller bearings is thus largely assured. The same applies to spherical plain bearings.

Relative proportions for width and diameter series are shown in Figure 12.9.

The dimensional, form and running accuracy of rolling bearings has been standardized by ISO. In addition to normal tolerances, ISO recommendations specify closer tolerances for higher-precision bearings.

ISO designation has been adapted by ANSI and the Anti-Friction Bearing Manufacturers Association (AFBMA) through its Annular Bearing Engineers Committee (ABEC).

Anti-friction bearing manufacturers however are still using different designations with the most universally accepted being the ones of Swedish–German manufacturers which are now converting rapidly to ISO designations.

Therefore, the ISO/ABEC designation will be used here.

It must be stressed that not all the dimension series are available. That is, not all combinations of diameter and width series are in common use. The most popular ones, and readily available, are indicated in Figure 12.8, both with the ISO/AFBMA and the most common commercial designation. In general, diameter series 0, 1, 2, 3 and width series 0, 1, 2, 3 also are the most used. The older SAE designation specified:

Series	SAE	ISO/AFBMA
Extra light	100	01
Light	200	02
Medium	300	03
Heavy	400	04

The commercial designations are followed by a number corresponding to the diameter. For most commercial designations this number is the bore diameter in millimeters divided by 5, for diameters over 20 mm. For example, a deep groove ball bearing eight series of bore diameter 100 mm would have the designations:

ISO/AFBMA:	100 BC02
SKF:	6220
New Departure-Hyatt:	3220

Many manufacturers provide conversion tables in their catalogs.

Manufacturers are also using prefixes and suffixes to indicate other features of the bearing design, such as sealing, mounting, etc.

Anti-friction bearings have, by their nature, very close tolerances. ABEC has specified five basic grades of precision, 1, 3, 5, 7, 9 with higher grade indicating increasing precision. Grade 3 is common while higher grades have progressively smaller tolerances and they are used for precision machinery. Tables for such tolerances can be found in manufacturers ' catalogs or in AFBMA publications.

12.3.2 Antifriction bearing design database

Using the program DESIGNDB (Appendix 2.A), developed in Chapter 2, a database can be established for standard bearing dimensions from antifriction bearing manufacturers and ISO catalogs.

12.3.3 Fatigue load rating

Fatigue life of materials in the high cycle fatigue range is described by a power law (see Chapter 6)

$$\sigma^m N = \text{constant} \tag{12.9}$$

where σ is the applied stress, N is the number of cycles to failure and m is a constant.

Antifriction bearings are rated on the basis of the dynamic loading that they can sustain, purely radial or thrust for the respective bearings at 10^6 revolutions, according to ISO recommendations. If a bearing is designed for a different life, a different stress range must be applied. For all situations, since stress is proportional to the load,

$$P^p N = P_0^p N_0 \tag{12.10}$$

where $P_0 = C$ is the load carrying capacity at $N_0 = 10^6$ revolutions. Therefore, the life of the bearing, in N revolutions, for applied load P is

$$N/N_0 = (C/P)^p \tag{12.11}$$

Here the exponent p is evaluated from experiments. Most bearing manufacturers specify $p = 3$ for ball and $p = 10/3$ for roller bearings. If n rpm is the speed of rotation, the bearing life in hours will be, for a probability of failure less than 10%,

$$L_{10} = (10^6/60n)(C/P)^p \tag{12.12}$$

Table 12.1 Working life of rolling bearings (SKF General Catalogue 1977)

Class of machine	Life in working hours (L_h)
Instruments and apparatus that are used only seldom:	
Demonstration apparatus; mechanisms for operating sliding doors	500
Machines used for short periods or intermittently and whose breakdown would not have serious consequences:	
Handtools; lifting tackle in workshops; hand-operated machines generally; agricultural machines; cranes in erecting shops; domestic machines	4,000–8,000
Machines working intermittently and whose breakdown would have serious consequences:	
Auxiliary machinery in power stations; conveyor plant for flow production; lifts; cranes for piece goods; machine tools used infrequently	8,000–12,000
Machines for use 8 hours per day and not always fully utilized:	
Stationary electric motors; general-purpose gear units	12,000–20,000
Machines for use 8 hours per day and fully utilized:	
Machines for the engineering industry generally; cranes for bulk goods; ventilating fans; countershafts	20,000–30,000
Machines for continuous use 24 hours per day:	
Separators; compressors; pumps; mine hoists; stationary electric machines; machines in continuous operation on board naval vessels	40,000–60,000
Machines required to work with a high degree of reliability 24 hours per day:	
Pulp and papermaking machinery; public power plants; mine pumps; waterworks; machines in continuous operation on board merchant ships	100,000–200,000

Standard bearing ratings C are given by bearings' manufacturers for 10% probability of failure.

SKF suggests (Table 12.1), machinery lives, if such specification for the designed bearing is not available.

The fatigue load rating C in kg-force was established based on elastohydrodynamic equations, such as (12.7), in the following forms.

For ball bearings:

$$C = 10\lambda(i\cos\alpha)^{0.7} z^{2/3} d^{1/8} \quad \text{for } d_b < 25.4\,\text{mm} \tag{12.13}$$

$$36.47\lambda(i\cos\alpha)^{0.7} z^{2/3} d^{1.4} \quad \text{for } d_b > 25.4\,\text{mm} \tag{12.14}$$

For roller bearings:

$$C = 56.2\lambda(il_r\cos\alpha)^{7/9} z^{3/4} d_r^{29/27} \tag{12.15}$$

where i is the number of rows of rolling elements, α is the angle of contact defined in Figure 12.8, z is the number of rolling elements, d_b is the diameter of rolling balls (cm), d_r is the diameter of rollers (cm), l_r is the length of contact of rollers (cm), and λ is a constant which depends on the factor $\zeta = d_b\cos\alpha/d_m$ for balls, and is $d_r\cos\alpha/d_m$ for rollers, where d_m is the diameter of the circle of the ball or roller centers.

The constant λ is given in Table 12.2 for different types of bearings.

At the application stage in design, fatigue load rating C is determined from manufacturers' catalogs. However, when the bearing geometry is known, C can be computed from equations (12.13 to 12.15). Sometimes, this is particularly useful for optimization and computer aided design studies.

The bearing load rating C is determined under prescribed conditions and needs adjustment for particular design situations. To this end the following revised life equation has been established by ISO

$$L_{na} = a_1 a_2 a_3 \left(\frac{C}{P}\right)^p \tag{12.16}$$

or simply

$$L = a_1 a_2 a_3 L_{10} \tag{12.17}$$

where L_{na} = adjusted rating life in millions of revolutions (the index n represents the difference between the specified reliability and 100%); a_1 = life adjustment factor for reliability, a_2 = life adjustment factor for material, and a_3 = life adjustment factor for operating conditions.

A calculation of the adjusted rating life presupposes that the operating conditions are well defined and that the bearing loads can be accurately calculated, i.e. the calculation should consider the load spectrum, shaft deflection etc.

For the generally-accepted reliability of 90% and for conventional materials and normal operating conditions, $a_1 = a_2 = a_3 = 1$, and the two life equations become identical.

The a_1 factor for reliability is used to determine lives other than the L_{10} life, i.e. lives which are attained or exceeded with a probability greater than 90%. Values of a_1 are given in Table 12.3.

It must be noted that the values of a_1 given by AFBMA (Table 12.3), are not the same as the ones of Chapter 6 because they are not based on the standard distribution but on the Weibull distribution.

Conventional material is considered to be the steel used for rolling bearings at the time the original fatigue rating equation was established by ISO. The improvements made to the standard steels used by different manufacturers have resulted in a value of the a_2 factor which is greater than 1. Even higher values of factor a_2 can be applied when special steels, e.g. electro-slag refined (ESR) or vacuum arc remelted (VAR) steels are used.

Values of a_2 are given in manufacturers' catalogs and it is 1 for standard steels and 2 to 3 for special steels. AFBMA suggests the value 3 for bearings of good quality.

Values of a_3 are related to bearing lubrication provided that bearing operating temperatures are not excessive. The efficacy of lubrication is primarily determined by the degree of surface separation in the rolling contacts of the bearing. Under the clean conditions

Table 12.2 Constant λ

Bearing	λ
Single row	$0.476 + 2.5a - 13a^2 + 17.5a^3$
Double row	$0.451 + 2.5a - 13a^2 + 17.5a^3$
Self-aligned	$0.176 + 1.5a - 2a^2 - a^3$
	$(a = \zeta - 0.05)$

Table 12.3 Reliability factor

Reliability (%)	a_1	L_n
90	1	L_{10}
95	0.62	L_5
96	0.53	L_4
97	0.44	L_3
98	0.33	L_2
99	0.21	L_1

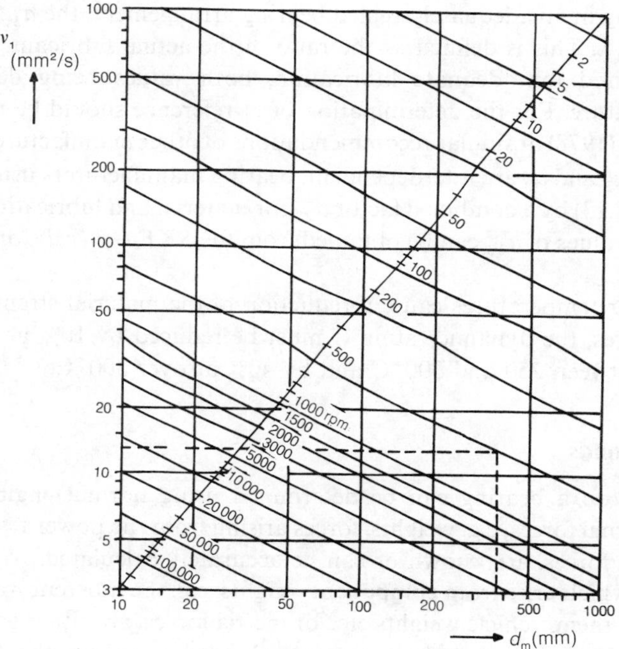

Figure 12.10 Speed coefficient, SKF bearings

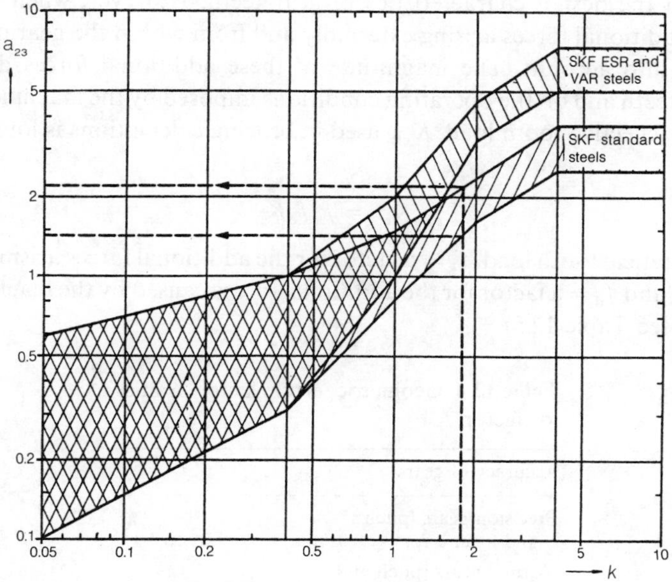

Figure 12.11 Bearing material coefficient, SKF bearings

normally prevailing in an adequately-sealed bearing arrangement, the a_3 factor is based on the viscosity ratio k. This is defined as the ratio of the actual lubricant viscosity ν to the viscosity ν_1 required for adequate lubrication, both values being determined at the operating temperature. For the determination of ν_1 reference should be made to the *SKF General Catalogue* (1978) or similar recommendations of other manufacturers (Figure 12.10).

Since factors a_2 and a_3 are interdependent, bearing manufacturers usually replace them in the equation (12.13) by a combined factor a_{23} for material and lubrication. Under normal clean conditions, values of a_{23} can be obtained from the *SKF General Catalogue* (1978) (see Figure 12.11).

Finally, higher temperatures cause a reduction of the material strength. Therefore, at higher temperatures, the dynamic rating C must be reduced by 10% at between 150 and 200 °C, 25% at between 250 and 300 °C and by 40% at over 300 °C.

12.3.4 Bearing loads

The loads acting on a bearing can be determined using normal engineering principles provided the external forces, e.g. weights, forces arising from the power transmitted or work done, and inertial forces, are known or can be accurately calculated.

Those loads which arise from component weights, e.g. the inherent weight of shafts and parts mounted on them, vehicle weights etc., or inertia forces, are either generally known or can be calculated. However, when determining the loads arising from the working loads (roll forces, cutting forces in machine tools), shock loads or additional dynamic loads, e.g. as a result of unbalance, it is often necessary to estimate the load from experience gained with earlier machines and bearing arrangements.

For example, the theoretical tooth forces on a gear can be calculated knowing the power transmitted and the design characteristics of gear teeth. However, when calculating the bearing loads additional forces arising externally and from within the gearing combination must be taken into account. The magnitude of these additional forces depends on the accuracy of the teeth and on the operating conditions imposed by the machines connected to the gearing. The effective tooth load, K_{eff} used in bearing calculations is found by means of the equation:

$$K_{eff} = f_k f_d K \tag{12.18}$$

where K = theoretical tooth load, f_k = a factor for the additional forces arising in the gearing (see Table 12.4), and f_d = a factor for the additional forces caused by the machines connected to the gearing (see Table 12.5).

Table 12.4 Geometric error factor: values for factor f_k

Accuracy of gears	f_k
Precision gears (pitch and form errors <0.02 mm)	1.05 to 1.1
Normal gears (pitch and form errors 0.02 to 0.1 mm)	1.1 to 1.3

Table 12.5 Factor for operating conditions: values for factor f_d

Operating conditions, type of machine	f_d
Machines working without shock loads, e.g. electrical machines, turbines	1.0 to 1.2
Reciprocating engines, depending on degree of unbalance	1.2 to 1.5
Pronounced shock loads, e.g. rolling mills	1.5 to 3

As a further example, belt drives may be considered. Here the loads acting on the bearings are derived from the effective belt pull which is dependent on the power being transmitted. To calculate the bearing loads the effective belt pull must be multiplied by a factor f. The factor f depends on the type of belt used, its initial tension and the additional dynamic forces. Appropriate values of factor f are:

V-belts	2 to 2.5
Plain belts with tension pulley	2.5 to 3
Plain belts	4 to 5

The larger values are suitable when the distance between the shafts is short, the peripheral speeds are slow, or the operating conditions are not known with high precision.

12.3.5 Combined loads

Rated loads of antifriction bearings are either purely radial or thrust. Experimental investigations have shown that most bearings, as indicated in Figure 12.8, can take both radial and thrust loads. The equivalent bearing load, to compare with the load rating C, is the maximum of

$$P = YF_a + VXF_r \quad \text{and} \quad VF_r \tag{12.19}$$

where P = equivalent dynamic bearing load, F_r = actual radial bearing load, F_a = actual axial bearing load, X = radial load factor for the bearing, and Y = axial load factor for the bearing. $V = 1$ if the inner ring rotates and 2 if the outer ring rotates.

Data required for the calculation of the equivalent dynamic bearing load are given in the bearing tables.

In the case of single row radial bearings, an additional axial load does not influence the equivalent bearing load P until its magnitude is such that the ratio F_a/F_r exceeds a specified value e given in bearing tables.

The bearing types are too numerous to allow a complete presentation of bearing tables here. For reference only, Table 12.6 gives the combined load factors X and Y for several types of bearings and Table 12.7 gives static and dynamic capacity of several types of bearings for particular Dimension Series.

Table 12.6 Combined loads factors

Bearing no.	$\frac{F_a}{F_r} \leqslant e$		$\frac{F_a}{F_r} > e$		e
	X	Y	X	Y	
Deep groove ball bearings					
Series EL, R, 160, 60, 62, 63, 64, EE, RLS, RMS					
$\frac{F_a}{C_0} = 0.025$	1	0	0.56	2	0.22
$= 0.04$				1.8	0.24
$= 0.07$				1.6	0.27
$= 0.13$				1.4	0.31
$= 0.25$				1.2	0.37
$= 0.5$				1	0.44
Angular contact ball bearings					
Series 72 B, 73 B	1	0	0.35	0.57	1.14
Series 72 BG, 73 BG					
A pair of bearings mounted in tandem	1	0	0.35	0.57	1.14
A pair of bearings mounted back-to-back or face-to-face	1	0.55	0.57	0.93	1.14
Series ALS, AMS	1	0	0.39	0.76	0.80
Series 32 A, 33 A	1	0.73	0.62	1.17	0.86

Bearing no.	$\frac{F_a}{F_r} \leqslant e$		$\frac{F_a}{F_r} > e$		e
	X	Y	X	Y	
Self-aligning ball bearings (contd.)					
RM 3–RM 6	1	1.8	0.65	2.8	0.34
7– 10		2.1		3.3	0.29
11– 14		2.4		3.8	0.26
15– 18		2.7		4.2	0.23
20– 48		2.9		4.5	0.21
Spherical roller bearings					
23944–239/670	1	3.7	0.67	5.5	0.18
239/710–239/950		4		6	0.17
23024C–23068CA	1	2.9	0.67	4.4	0.23
23072CA–230/500CA		3.3		4.9	0.21
24024C–24080CA	1	2.3	0.67	3.5	0.29
24084CA–240/500CA		2.4		3.6	0.28
23120C–23128C	1	2.4	0.67	3.6	0.28
23130C–231/500CA		2.3		3.5	0.29
24122C–24128C	1	1.9	0.67	2.9	0.35
24130C–24172CA		1.8		2.7	0.37
24176CA–241/500CA		1.9		2.9	0.35
22205C–22207C	1	2.1	0.67	3.1	0.32
08C– 09C		2.5		3.7	0.27
10C– 20C		2.9		4.4	0.23
22C– 44C		2.6		3.9	0.26
48 – 64		2.4		3.6	0.28
23218C–23220C	1	2.2	0.67	3.3	0.31
22C– 64CA		2		3	0.34

		Self-aligning ball bearings			
135, 126, 127, 108, 129	1	1.8	0.65	2.8	0.34
1200–1203		2		3.1	0.31
04– 05		2.3		3.6	0.27
06– 07		2.7		4.2	0.23
08– 09	1	2.9	0.65	4.5	0.21
10– 12		3.4		5.2	0.19
13– 22		3.6		5.6	0.17
24– 30		3.3		5	0.2
2200–2204		1.3		2	0.5
05– 07		1.7		2.6	0.37
08– 09		2		3.1	0.31
10– 13	1	2.3	0.65	3.5	0.28
14– 20		2.4		3.8	0.26
21– 22		2.3		3.5	0.28
1300–1303		1.8		2.8	0.34
04– 05	1	2.2	0.65	3.4	0.29
06– 09		2.5		3.9	0.25
10– 22		2.8		4.3	0.23
2301		1		1.6	0.63
2302–2304	1	1.2	0.65	1.9	0.52
05– 10		1.5		2.3	0.43
11– 18		1.6		2.5	0.39
RL 4–RL 6		2.1		3.3	0.29
7– 8		2.3		3.6	0.27
9– 11		2.7		4.2	0.23
12– 14	1	2.9	0.65	4.5	0.21
15– 18		3.4		5.2	0.19
20– 36		3.6		5.6	0.17
38– 48		4.2		6.5	0.15

21304–21305		2.8		4.2	0.24
06– 10	1	3.2	0.67	4.8	0.21
11– 19		3.4		5	0.2
20– 22		3.7		5.5	0.18
22308C–22310C		1.8		2.7	0.37
11C– 15C	1	1.9	0.67	2.9	0.35
16C– 40C		2		3	0.34
44 – 56		1.9		2.9	0.35
		Taper roller bearings			
30203–30204				1.75	0.34
05– 08	1	0	0.4	1.6	0.37
09– 22				1.45	0.41
24– 30				1.35	0.44
32206–32208				1.6	0.37
09– 22	1	0	0.4	1.45	0.41
24– 30				1.35	0.44
30302–30303				2.1	0.28
04– 07	1	0	0.4	1.95	0.31
08– 24				1.75	0.34
31305–31318	1	0	0.4	0.73	0.82
32303				2.1	0.28
32304–32307	1	0	0.4	1.95	0.31
08– 24				1.75	0.34

Table 12.7 Dynamic capacity of rolling bearings

Boundary dimensions d	D	B	Basic load ratings dynamic C	static C_0
mm			N(1 N = 0.225 lbf)	
17	26	5	1 320	915
	35	8	4 650	2 800
	35	10	4 650	2 800
	40	12	7 350	4 500
	47	14	10 400	6 550
	62	17	17 600	11 800
20	32	7	2 040	1 400
	42	8	5 400	3 400
	42	12	7 200	4 500
	47	14	9 800	6 200
	52	15	12 200	7 800
	72	19	23 600	16 600
25	37	7	2 280	1 700
	47	8	5 350	4 000
	47	12	8 650	5 600
	52	15	10 300	6 950
	62	17	17 300	11 400
	80	21	27 500	19 600
30	42	7	2 280	1 800
	55	9	8 650	5 850
	55	13	10 200	6 800
	62	16	15 000	10 000
	72	19	21 000	14 600
	90	23	33 500	24 000
35	47	7	2 360	2 000
	62	9	9 500	6 950
	62	14	12 200	8 500
	72	17	19 600	13 700
	80	21	25 500	18 000
	100	25	42 500	31 000

Boundary dimensions d	D	B	Basic load ratings dynamic C	static C_0
mm			N(1 N = 0.225 lbf)	
20	47	14	7 650	3 150
	47	18	9 650	3 800
	52	15	9 650	3 900
	52	21	13 700	5 400
25	52	15	9 300	4 000
	52	18	9 650	4 150
	62	17	13 700	5 850
	62	24	18 600	7 500
30	62	16	12 000	5 600
	62	20	11 800	5 500
	72	19	16 300	7 500
	72	27	24 000	10 000
35	72	17	12 000	6 300
	72	23	16 600	7 800
	80	21	19 300	9 500
	80	31	30 000	12 900
40	80	18	14 600	8 000
	80	23	17 300	9 000
	90	23	22 800	11 800
	90	33	34 500	15 600
45	85	19	16 600	9 000
	85	23	17 600	10 000
	100	25	29 000	15 300
	100	36	41 500	19 300
50	90	20	17 300	10 000
	90	23	17 600	10 600
	110	27	33 500	17 000
	110	40	49 000	23 600
55	100	21	20 400	12 500

Boundary dimensions d	D	B	Basic load ratings dynamic C	static C_0
mm			N(1 N = 0.225 lbf)	
15	35	11	8 150	4 250
17	40	12	9 800	5 200
	40	16	14 000	8 150
	47	14	15 600	8 650
20	47	14	13 400	7 350
	47	18	18 300	10 800
	52	15	20 400	11 600
25	52	15	15 300	8 800
	52	18	20 800	12 900
	62	17	26 000	15 000
	62	24	38 000	24 500
30	55	13	14 300	8 500
	62	16	20 400	12 000
	62	20	29 000	19 000
	72	19	34 000	20 000
	72	27	45 500	29 000
	90	23	55 000	34 000
35	62	14	19 000	11 600
	72	17	29 000	17 600
	72	23	43 000	29 000
	80	21	43 000	27 000
	80	31	57 000	38 000
	100	25	68 000	44 000
40	68	15	21 200	13 400
	80	18	38 000	24 000
	80	23	51 000	34 500
	90	23	51 000	32 500
	90	33	73 500	51 000
	110	27	88 000	57 000

Boundary dimensions d	D	T	Basic load ratings dynamic C	static C_0
mm			N(1 N = 0.225 lbf)	
20	42	15	20 800	15 600
	47	15.25	23 600	16 600
	52	16.25	29 000	20 000
	52	22.25	37 500	28 500
22	44	15	21 600	16 300
25	47	15	23 200	18 300
	52	16.25	26 500	19 300
	52	22	40 000	32 500
	62	18.25	38 000	26 500
	62	18.25	32 500	23 200
	62	25.25	51 000	39 000
28	52	16	27 000	21 600
30	55	17	30 500	24 500
	62	17.25	34 500	25 500
	62	21.25	43 000	34 000
	62	25	55 000	45 500
	72	20.75	48 000	34 000
	72	20.75	40 500	29 000
	72	28.75	65 500	52 000
32	58	17	31 500	26 000
35	62	18	36 500	30 500
	72	18.25	44 000	32 500
	72	24.25	56 000	45 000
	72	28	72 000	62 000
	80	22.75	62 000	45 500
	80	22.75	52 000	38 000
	80	32.75	81 500	65 500
40	68	19	45 000	40 000

Boundary dimensions d	D	H	Basic load ratings dynamic C	static C_0
mm			N(1 N = 0.225 lbf)	
20	35	10	9 800	16 000
	40	14	15 300	25 000
25	42	11	12 200	22 800
	47	15	19 300	34 000
	52	18	26 000	45 000
	60	24	40 000	63 000
30	47	11	12 900	26 500
	52	16	19 600	37 500
	60	21	31 000	57 000
	70	28	52 000	90 000
35	52	12	13 400	30 000
	62	18	27 000	53 000
	68	24	38 000	69 500
	80	32	63 000	112 000
40	60	13	18 000	40 000
	68	19	31 000	64 000
	78	26	50 000	98 000
	90	36	80 000	146 000
45	65	14	18 600	45 000
	73	20	31 500	68 000
	85	28	55 000	110 000
	100	39	93 000	173 000
50	70	14	19 600	50 000
	78	22	36 000	61 500
	95	31	67 000	137 000
	110	43	106 000	204 000
55	78	16	23 600	62 000
	90	25	49 000	110 000
	105	35	86 500	180 000

40	52	7	2 450	2 200
	68	9	10 200	7 800
	68	15	12 900	9 300
	80	18	23 600	16 600
	90	23	31 500	22 400
	110	27	49 000	36 500
45	58	7	4 650	3 800
	75	10	12 000	9 300
	75	16	16 300	12 200
	65	19	25 500	18 600
	100	25	40 500	30 000
	120	29	58 500	45 500
50	65	7	4 600	4 250
	80	10	12 500	10 000
	80	16	16 600	13 200
	90	20	27 000	19 600
	110	27	47 500	30 000
	130	31	67 000	52 000
55	72	9	6 400	5 600
	90	11	15 000	12 200
	90	18	21 600	17 000
	100	21	33 500	25 000
	120	29	55 000	41 500
	140	33	76 500	63 000
60	78	10	6 700	6 100
	95	11	15 300	13 200
	95	18	22 800	18 300
	110	22	36 500	28 000
	130	31	63 000	48 000
	150	35	83 000	69 500
65	85	10	9 000	8 300
	100	11	16 300	14 600
	100	18	23 600	19 600
	120	23	43 000	34 000
	140	33	71 000	56 000
	160	37	91 500	78 000
70	90	10	9 300	9 150
	110	13	21 600	19 000
	110	20	29 000	24 500
	125	24	47 500	37 500
	150	35	80 000	63 000
	180	42	110 000	104 000
	100	25	20 400	12 500
	120	29	39 000	21 600
	120	43	57 000	28 000
60	110	22	23 200	14 300
	110	28	26 000	15 600
	130	31	44 000	25 500
	130	46	67 000	32 500
65	120	23	23 600	15 600
	120	31	33 500	20 000
	140	33	47 500	28 000
	140	48	73 500	38 000
70	125	24	26 500	17 300
	125	31	34 000	21 200
	150	35	57 000	34 000
	150	51	83 000	44 000
75	130	25	29 000	19 600
	130	31	34 000	22 000
	160	37	61 000	36 500
	160	55	93 000	51 000
80	140	26	30 500	21 600
	140	33	37 500	24 500
	170	39	68 000	40 500
	170	68	104 000	57 000
85	150	28	37 500	26 000
	150	36	45 000	29 000
	180	41	75 000	46 500
	180	60	108 000	61 000
90	160	30	43 000	29 000
	160	40	54 000	35 500
	190	43	90 000	54 000
	190	64	116 000	68 000
95	170	32	49 000	34 000
	170	43	64 000	42 500
100	180	34	53 000	35 500
	180	46	75 000	50 000
	215	47	110 000	59 500
110	200	32	67 000	49 000
	200	53	96 500	63 000
	240	50	127 000	66 500
45	75	16	26 500	17 600
	85	19	40 000	25 500
	85	23	54 000	37 500
	100	25	65 500	41 500
	100	36	95 000	67 000
	120	29	104 000	69 500
50	80	16	26 500	17 600
	90	20	42 500	27 500
	90	23	56 000	40 500
	110	27	80 000	52 000
	110	40	110 000	80 000
	130	31	127 000	86 500
55	90	18	31 000	21 200
	100	21	51 000	34 000
	100	25	67 000	48 000
	120	29	100 000	67 000
	120	43	134 000	98 000
	140	33	129 000	86 500
60	95	18	32 000	22 400
	110	22	62 000	43 000
	110	28	68 000	68 000
	130	31	112 000	76 500
	130	46	153 000	114 000
	150	35	153 000	106 000
65	100	18	32 000	22 800
	120	23	72 000	51 000
	120	31	106 000	81 500
	140	33	125 000	85 000
	140	43	173 000	129 000
	160	37	180 000	127 000
70	110	20	48 000	34 000
	125	24	72 000	51 000
	125	31	106 000	81 500
	150	35	146 000	102 000
	150	51	204 000	160 000
	180	42	224 000	163 000
75	115	20	49 000	36 000
	130	25	88 000	63 000
	130	31	120 000	95 000
	160	37	176 000	125 000
	160	55	250 000	200 000
	190	45	240 000	173 000
	80	19.75	51 000	38 000
	80	24.75	64 000	50 000
	80	32	88 000	78 000
	90	25.25	73 500	56 000
	90	25.25	63 000	47 500
	90	35.25	100 000	83 000
45	75	20	50 000	44 000
	80	26	72 000	64 000
	85	20.75	57 000	44 000
	85	24.75	68 000	56 000
	85	32	91 500	81 500
	100	27.25	91 500	72 000
	100	27.25	78 000	60 000
	100	38.25	120 000	102 000
50	80	20	52 000	48 000
	80	24	58 500	56 000
	90	21.75	64 000	52 000
	90	24.75	69 500	57 000
	90	32	98 000	90 000
	110	29.25	108 000	83 000
	110	29.25	91 500	69 500
	110	42.25	146 000	127 000
55	90	23	69 500	64 000
	95	30	95 000	86 500
	100	22.75	76 500	61 000
	100	26.75	90 000	75 000
	100	35	118 000	106 000
	120	31.5	122 000	96 500
	120	31.5	104 000	80 000
	120	45.5	170 000	146 000
60	95	23	71 000	67 000
	110	23.75	83 000	65 500
	110	29.75	108 000	91 500
	110	38	143 000	134 000
	130	33.5	143 000	116 000
	130	33.5	122 000	96 500
	130	48.5	196 000	173 000
65	100	23	71 000	68 000
	100	27	83 000	83 000
	110	34	122 000	116 000
	120	48	137 000	265 000
60	85	17	27 500	71 000
	95	26	50 000	118 000
	110	35	90 000	196 000
	130	51	143 000	265 000
65	90	18	28 500	78 000
	100	27	51 000	127 000
	115	36	90 000	196 000
	140	56	166 000	360 000
70	95	18	32 500	88 000
	105	27	51 000	127 000
	125	40	102 000	232 000
	150	60	180 000	400 000
75	100	19	34 000	98 000
	110	27	52 000	134 000
	135	44	118 000	270 000
	160	65	208 000	500 000
80	105	19	34 500	102 000
	115	28	57 000	153 000
	140	44	122 000	290 000
	170	68	224 000	550 000
85	110	19	34 500	106 000
	125	31	72 000	190 000
	150	49	140 000	340 000
	180	72	236 000	610 000
90	120	22	39 000	120 000
	135	35	91 500	240 000
	155	50	153 000	400 000
	190	77	250 000	670 000
105	135	25	57 000	173 000
	150	38	102 000	280 000
	170	55	183 000	480 000
	210	85	305 000	865 000
110	145	25	58 500	190 000
	160	38	118 000	365 000
	190	63	204 000	570 000
120	155	25	68 000	250 000
	170	39	118 000	380 000
	210	70	240 000	710 000
	250	102	345 000	1 080 000

Table 12.7 *(Contd.)*

Boundary dimensions d	D	B	Basic load ratings dynamic C	static C_0
mm			N(1 N = 0.225 lbf)	
75	95	10	9 650	9 800
	115	13	22 000	20 000
	115	20	30 500	26 000
	130	25	51 000	40 500
	160	37	86 500	72 000
	190	45	118 000	114 000
80	100	10	9 500	9 800
	125	14	25 500	23 600
	125	22	36 500	31 500
	140	26	54 000	45 000
	170	39	95 000	80 000
	200	48	125 000	125 000
85	110	13	14 600	15 000
	130	14	26 000	25 000
	130	22	38 000	33 500
	150	28	64 000	53 000
	180	41	102 000	90 000
	210	52	134 000	134 000
90	115	13	15 000	15 600
	140	16	32 000	29 000
	140	24	45 000	39 000
	160	30	73 500	62 000
	190	30	110 000	98 000
	225	54	140 000	146 000
95	145	16	32 500	31 500
	145	24	46 500	41 500
	170	32	83 000	69 500
	200	45	118 000	110 000
100	125	13	15 300	17 000
	150	16	34 000	32 500
	150	24	46 500	41 500
	180	34	95 000	78 000
	215	47	134 000	132 000
105	160	18	40 000	38 000
	160	26	56 000	51 000
	190	36	102 000	143 000
	225	49	140 000	143 000

Boundary dimensions d	D	B	Basic load ratings dynamic C	static C_0
mm			N(1 N = 0.225 lbf)	
110	140	16	21 600	23 600
	170	19	44 000	42 500
	170	28	63 000	57 000
	200	38	112 000	100 000
	240	50	156 000	166 000
120	150	16	22 400	25 000
	180	19	47 500	47 500
	180	28	65 500	61 000
	215	40	112 000	100 000
	260	55	160 000	166 000
130	165	18	29 000	32 500
	200	22	61 000	61 000
	200	33	81 500	78 000
	230	40	120 000	112 000
	280	58	176 000	193 000
140	175	18	30 500	35 500
	210	22	62 000	64 000
	210	33	85 000	83 000
	250	42	127 000	122 000
	300	62	196 000	224 000
150	190	20	37 500	43 000
	225	24	71 000	73 500
	225	35	96 500	96 500
	270	45	134 000	137 000
	320	65	212 000	250 000
160	200	20	38 000	45 500
	240	25	76 500	80 000
	240	38	110 000	112 000
	290	48	143 000	146 000
	340	68	212 000	250 000
170	215	22	47 500	56 000
	260	28	91 500	96 500
	260	42	129 000	134 000
	310	52	163 000	180 000

Boundary dimensions d	D	B	Basic load ratings dynamic C	static C_0
mm			N(1 N = 0.225 lbf)	
80	125	22	60 000	44 000
	140	26	96 500	68 000
	140	33	134 000	106 000
	170	39	176 000	125 000
	170	58	250 000	200 000
	200	48	275 000	200 000
85	130	22	62 000	46 500
	150	28	110 000	78 000
	150	36	153 000	122 000
	180	41	204 000	146 000
	180	60	285 000	228 000
	210	52	310 000	223 000
90	140	24	73 500	56 000
	160	30	134 000	100 000
	160	40	186 000	150 000
	190	43	220 000	160 000
	190	64	300 000	240 000
	225	54	345 000	260 000
95	145	24	75 000	58 500
	170	32	150 000	112 000
	170	43	208 000	170 000
	200	45	250 000	190 000
	200	67	360 000	300 000
	240	55	375 000	280 000
100	150	24	76 500	60 000
	180	34	166 000	125 000
	180	46	236 000	193 000
	215	47	290 000	220 000
	215	73	425 000	355 000
	250	58	415 000	320 000
105	160	26	90 000	71 000
	190	36	183 000	137 000
	225	49	335 000	255 000
	260	60	455 000	345 000

Boundary dimensions d	D	T	Basic load ratings dynamic C	static C_0
mm			N(1 N = 0.225 lbf)	
	120	24.75	98 000	78 000
	120	32.75	129 000	112 000
	120	41	166 000	153 000
	140	36	166 000	134 000
	140	36	140 000	112 000
	140	51	224 000	200 000
70	110	25	86 500	83 000
	110	31	110 000	109 000
	125	26.25	108 000	88 000
	125	33.25	134 000	118 000
	125	41	173 000	163 000
	150	38	190 000	153 000
	150	38	160 000	127 000
	150	54	250 000	228 000
75	115	25	90 000	88 000
	115	31	112 000	118 000
	125	37	150 000	146 000
	130	27.25	120 000	100 000
	130	33.25	137 000	120 000
	130	41	176 000	170 000
	160	40	208 000	173 000
	160	40	176 000	143 000
	160	58	285 000	265 000
80	125	29	116 000	116 000
	125	36	143 000	153 000
	130	37	153 000	153 000
	140	28.25	127 000	104 000
	140	35.25	160 000	137 000
	140	46	212 000	208 000
	170	42.5	232 000	190 000
	170	42.5	193 000	153 000
	170	61.5	320 000	290 000

Boundary dimensions d	D	H	Basic load ratings dynamic C	static C_0
mm			N(1 N = 0.225 lbf)	
130	170	30	80 000	285 000
	190	45	156 000	500 000
	225	75	255 000	765 000
	270	110	430 000	1 430 000
140	180	31	81 500	305 000
	200	46	160 000	530 000
	240	80	285 000	915 000
150	190	31	83 000	325 000
	215	50	173 000	585 000
	250	80	290 000	980 000
160	200	31	86 600	345 000
	225	51	176 000	610 000
	270	87	340 000	1 180 000
170	215	34	104 000	415 000
	240	55	208 000	720 000
	280	87	345 000	1 270 000
180	225	34	104 000	430 000
	250	56	212 000	750 000
	300	95	375 000	1 370 000

12.3.6 Fluctuating loads

If loads of different magnitudes $P_1, P_2, \ldots, P_n$ act at respective numbers of revolutions N_1, $N_2, \ldots, N_n$ (equation 12.10), the approximate mean load for the remaining bearing life should not exceed, according to the Palmgren–Miner rule (see Chapter 7),

$$P = [(P_1^p N_1 + P_2^p N_2 + \cdots)/(N_1 + N_2 + \cdots)]^{1/p} \tag{12.20}$$

12.3.7 Static loads

When a bearing under load stands still, makes slow oscillating movements, or operates at very slow speeds, the load-carrying capacity is determined not by the fatigue of the material but by the permanent deformation at the points of contact between the rolling elements and the raceways. This also applies to rotating bearings which are subjected to heavy shock loads during a fraction of a revolution. Generally, loads equivalent to the basic static load rating C_0 can be accommodated without detriment to the running characteristics of the bearing.

Loads comprising radial and axial components must be converted into an equivalent static bearing load. The equivalent static bearing load is defined as that radial load (for thrust bearings the axial load) which if applied would cause the same permanent deformation in the bearing as the actual loads. It is obtained by means of the general equation:

$$P_0 = X_0 F_r + Y_0 F_a \tag{12.21}$$

where P_0 = equivalent static bearing load, F_r = actual radial load, F_a = actual axial load, X_0 = radial load factor for the bearing and Y_0 = axial load factor for the bearing.

All data necessary for the calculation of the equivalent static bearing load are given in bearing tables, such as the Tables 12.6 and 12.7.

The required static load rating C_0 of a bearing can be determined by means of the equation:

$$C_0 = s_0 P_0 \tag{12.22}$$

where C_0 = basic static load rating N, P_0 = equivalent static bearing load N, and s_0 = static safety factor

At elevated temperatures, reduced hardness of the bearing material influences the static load-carrying capacity. In this case, the bearing manufacturer must be consulted.

The following values of s_0 for a few typical applications can be used as a guide when determining the requisite basic static load rating of bearings which make occasional oscillating movements: variable-pitch propeller blades for aircraft, $s_0 > 0.5$; weir and sluice gate installations, $s_0 > 1$; moving bridges, $s_0 > 1.5$; crane hooks for large cranes without significant additional dynamic forces, small cranes for bulk goods with fairly large additional dynamic forces, $s_0 > 1.6$; generally, for spherical roller thrust bearings, $s_0 > 2$.

Where there are large fluctuations in the applied load and particularly when heavy shock loads occur during part of a revolution, it is essential to establish that the static load rating of the bearing is adequate. Severe shock loads can cause large unevenly distributed indentations round the raceways which will seriously affect the running of the bearing. In addition, shock loads cannot usually be accurately calculated. Deformation of the housing resulting in an unfavorable load distribution in the bearing may also arise.

If the heaviest load to which the bearing is subjected acts during several revolutions of the bearing, the raceways will be evenly deformed and damaging indentations will not be produced.

It follows that, dependent on the operating conditions, the heaviest load to act on a bearing should never exceed a certain value determined by the static safety factor s_0. Generally, the following minimum values can be used for s_0: applications where smooth, vibration-free running is assured, $s_0 = 0.5$; average working conditions with normal demands on quiet running, $s_0 = 1.0$; pronounced shock loads, $s_0 = 1.5$ to 2; high demands on quiet running, $s_0 = 2$.

Generally, for spherical roller thrust bearings, $s_0 > 2$.

For bearings which rotate very slowly and where the life is short, in terms of number of revolutions, the static load rating must also be taken into account. In such cases the life equation can be misleading in giving an apparently permissible load which far exceeds the basic static rating.

12.3.8 Lubrication of antifriction bearings

In rolling bearings, the lubricant has a number of functions to perform. Firstly, it must minimize the damage that results from the small degree of sliding which inevitably occurs. Secondly, the lubricant must protect the highly-polished bearing surfaces from attack by the ambient environment, particularly if moisture is present. A corrosion-inhibiting additive is therefore required. Thirdly, the lubricant should prevent dust and other contaminants from entering the bearing and causing premature failure. Finally, the lubricant may be required to act as a coolant for bearings operating at high speeds and under high loads.

Greases were the earliest and are still the most widely used lubricants for rolling bearings. As shown in Figure 12.12, the primary function of a grease pack is as a storehouse for the very small amount of oil required for lubrication. Among the advantages of grease as a lubricant are that it is an inexpensive and reliable system requiring a minimum of maintenance over a long period. The bearing is self-contained with a minimum of sealing problems. Further, grease very successfully prevents the ingress of foreign matter.

Under carefully-controlled conditions, grease lubrication has been successfully used up to $d_m N$ values of 0.6×10^6 ($d_m N$ = pitch diameter in millimeters $\times$ rotational speed in rpm).

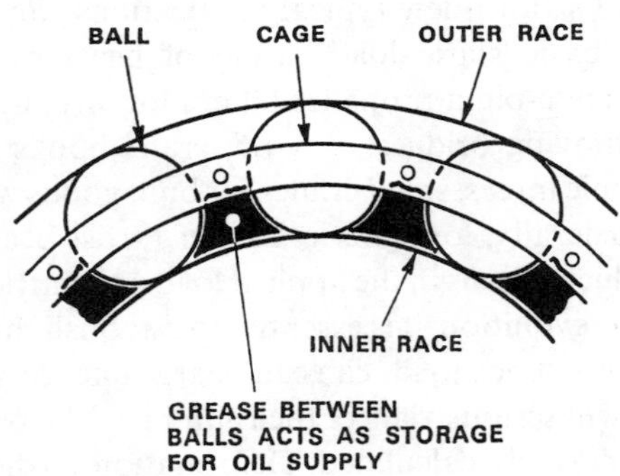

Figure 12.12 Grease lubrication of rolling bearings. (Timken Co. 1983)

At higher $d_m N$ values, the coolant property of the lubricant becomes important and some method of external oil feed must be provided. $d_m N$ is related to peripheral velocity (Figure 12.8), as $U \approx d_m n/20{,}000$.

Oil is also used since its fluidity facilitates penetration to all parts of the bearing where sliding friction is occurring.

Figure 12.13 shows the basic components of ball and roller bearings; these are the outer and inner races or rings, the rolling elements (balls or rollers) and the cage or separator which locates and separates the rolling elements. During relative motion between the outer and inner ring, two distinct modes of contact occur, rolling contact between the rolling elements and the races, and sliding contact between the cage pockets and the rolling elements. In ball bearings designed to have a high degree of osculation, such as deep-groove ball bearings, and in bearings functioning under severe conditions of load and speed, some slip must also occur between the rolling elements and the raceways of the rings. The primary function of the lubricant must be to reduce metallic interaction in these regions of contact in order to minimize wear and limit the frictional moment of the complete bearing.

In general, the frictional moment of a complete bearing will be composed of rolling resistance and hydrodynamic losses in the load-bearing components, sliding resistance and hydrodynamic losses in the cage system and bulk churning of the lubricant within and without the bearing assembly. Frictional torque follows the trend shown in Figure 12.14.

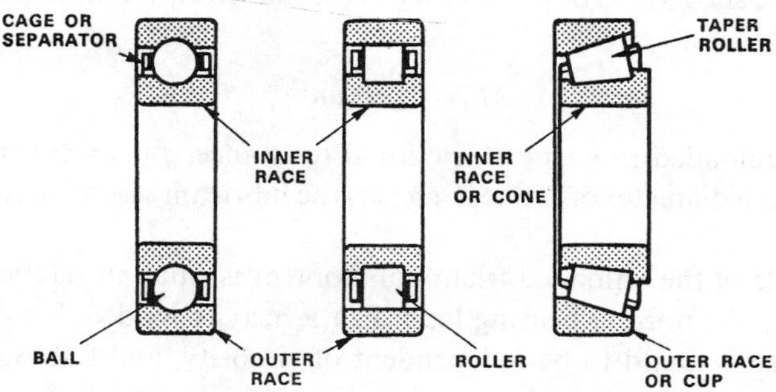

Figure 12.13 Details of rolling bearings. (Timken Co. 1983)

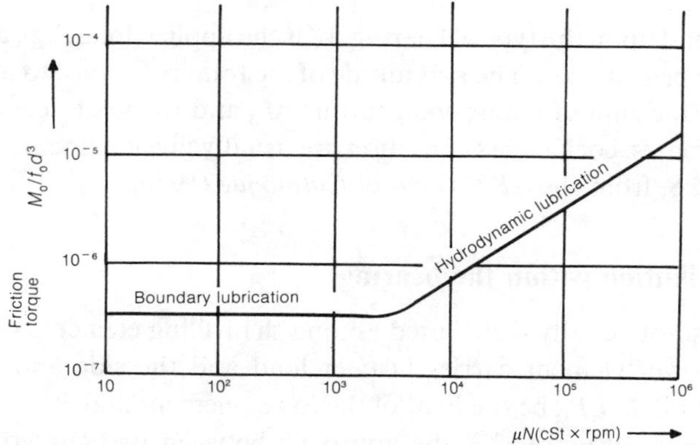

Figure 12.14 Friction torque of rolling bearings. (Timken Co. 1983)

Table 12.8 Friction coefficient for rolling bearings

Bearing type	f
Deep groove ball bearings	0.0015
Self-aligning ball bearings	0.0010
Angular contact ball bearings,	
single row	0.0020
double row	0.0024
Cylindrical roller bearings	0.0011
Needle roller bearings	0.0025
Spherical roller bearings	0.0018
Taper roller bearings	0.0018
Thrust ball bearings	0.0013
Cylindrical roller thrust bearings	0.0040
Needle roller thrust bearings	0.0040
Spherical roller thrust bearings	0.0018

At low rotational speeds the frictional torque is constant, suggesting boundary lubrication. At higher speeds and/or viscosities, involving values of μN greater than about 3×10 cSt·rev/min, a change occurs and the torque then becomes a function of both viscosity and speed, suggesting the predominance of hydrodynamic lubrication. For these conditions, the frictional torque shows a certain similarity to the Petroff equation and can be expressed in the form:

$$M_0 = f_0 d^3 (\mu n)^{2/3} \tag{12.23}$$

where M_0 is the unloaded moment of frictional resistance, f_0 is a function of the type of bearing, d is the pitch diameter of the bearing, μ is the lubricant viscosity, and n is the speed of rotation.

The magnitude of the unloaded frictional moment is small in relation to the frictional moment caused by the normal bearing load. For a heavily-loaded, low-speed bearing, the frictional resistance is found to be independent of viscosity, and follows the equation

$$M_1 = f_1 W d \left(\frac{W}{C_0} \right)^c \tag{12.24}$$

where f_1 and c depend upon the type of bearing, W is the applied load (kg), and C_0 is the static load capacity of the bearing (kg). The magnitude of the total resistance to motion of a loaded bearing is given by the sum of the no-load torque M_0 and the load-dependent torque M_1.

For rough estimates, coefficients of friction are usually given in manufacturers' catalogs, such as in Table 12.8, from the *SKF General Catalogue* (1978).

12.3.9 Load distribution within the bearing

The bearing load is not equally distributed among the rolling elements of the bearing. It is obvious that the lower element carries heavier load and the side and upper ones carry virtually no load at all. Let P_0 be the load of the lower element and $P_1, P_2, \ldots,$ the loads on the side elements. From Figure 12.3, the approach between two curved surfaces is $cP^{2/3}$, where c is a constant. The individual deflections therefore will be $\delta_0 = cP_0^{2/3}$, $\delta_1 = cP_1^{2/3}$, etc.

Therefore, because $\delta_i = \delta_0 \cos i\theta$,

$$\left(\frac{P_i}{P_0}\right)^{2/3} = \frac{\delta_i}{\delta_0} = \cos i\theta \tag{12.25}$$

The resulting vertical force, equal to the load W, is the sum of the vertical components

$$W = P_0 + 2P_1 \cos\theta + 2P_2 \cos 2\theta + ,\ldots. \tag{12.26}$$

Using equation (12.25),

$$W = P_0\left[1 + 2\sum_{i=1}^{n} (\cos i\theta)^{5/2}\right] \tag{12.27}$$

where n is the number of balls up to the horizontal position of one side only. If all the balls carry equal load, the load would be $P_0 n$. Therefore, there is a load concentration factor, with a similar meaning to stress concentration factor (Figure 12.15).

$$K_t = z\left(1 + 2\sum_{i=1}^{n} \cos^{5/2} i\theta\right)\bigg/ n \approx 4.36 \tag{12.28}$$

It is a function of the number of rolling elements z with limiting values 1 for one element only and 4.36 for infinite number of elements.

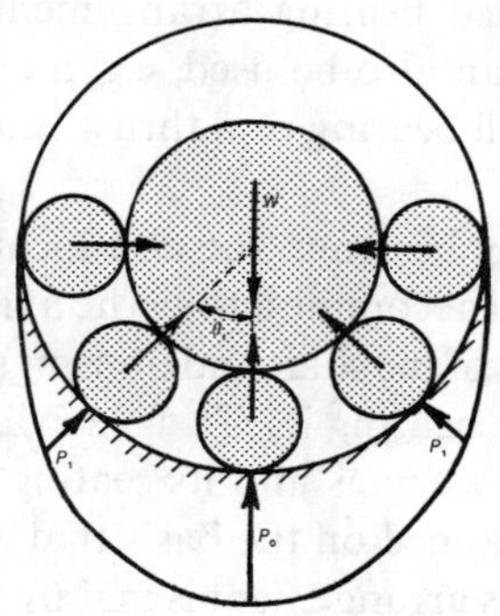

Figure 12.15 Rigid outer race. Unequal load distribution

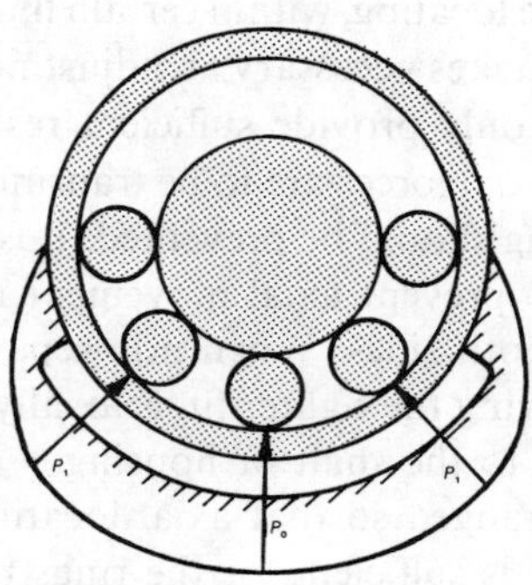

Figure 12.16 Flexible outer race for optimum load distribution

It appears that the capacity of the bearing is not fully utilized if $K_t > 1$, as usually happens. The designer can reduce the factor K_t by allowing for resilience under the heavily-loaded rolling elements, as for example with a recess at the lower part of the bearing (Figure 12.16).

12.4 APPLICATION OF ROLLING BEARINGS

12.4.1 Locating rolling bearings

A rotating shaft generally requires two bearings to support and locate it radially and axially relative to the stationary part of the machine, e.g. the housing. Normally, only one of the bearings (the locating bearing) is used to fix the position of the shaft axially, whilst the other bearings (the non-locating bearing) is free to move axially.

Axial location of the shaft is necessary in both directions and the locating bearing must be axially secured on the shaft and in the housing to limit lateral movement. In addition to locating the shaft axially the locating bearing is generally also required to provide radial support and bearings which are able to carry combined loads are then necessary, e.g. deep groove ball bearings, spherical roller bearings and double row or paired single row angular contact ball bearings. A combined bearing arrangement, with radial and axial location provided by separate bearings, can also be used, e.g. a cylindrical roller bearing mounted alongside a four-point contact ball bearing or a thrust bearing having radial freedom in the housing.

To avoid cross-location of the bearings, the non-locating bearing, which provides only radial support, must be capable of accommodating the axial displacements which arise from the differential thermal expansion of the shaft and housings. The axial displacements must be compensated for either within the bearing itself – as in cylindrical roller bearings and needle roller bearings – or between the bearings and its seating on the shaft or in the housing.

The bearings can also be arranged on the basis that axial location is provided by each bearing in one direction only. This arrangement is mainly used for short shafts supported by deep groove ball bearings, angular contact ball bearings, taper roller bearings or cylindrical roller bearings.

Some locating arrangements are shown in Figure 12.17. In some cases, such as conical roller bearings, both bearings are locating, within certain limits, because of the nature of such bearings. Wear of the bearings makes necessary an adjustment to maintain close clearances.

Interference fits in general only provide sufficient resistance to axial movement of a bearing on its seating when no axial forces are to be transmitted and the only requirement is that lateral movement of the ring should be prevented. Positive axial location or locking is necessary in all other cases. To prevent axial movement in either direction of a locating bearing it must be located at both sides. When non-separable bearings are non-locating bearings, only one ring, that having the tighter fit, is axially located; the other ring must be free to move axially in relation to the shaft or housing.

Where the bearings are arranged so that axial locating of the shaft is given by each bearing in one direction only it is sufficient for the rings to be located at one side only.

Bearings having interference fits are generally mounted against a shoulder on the shaft or

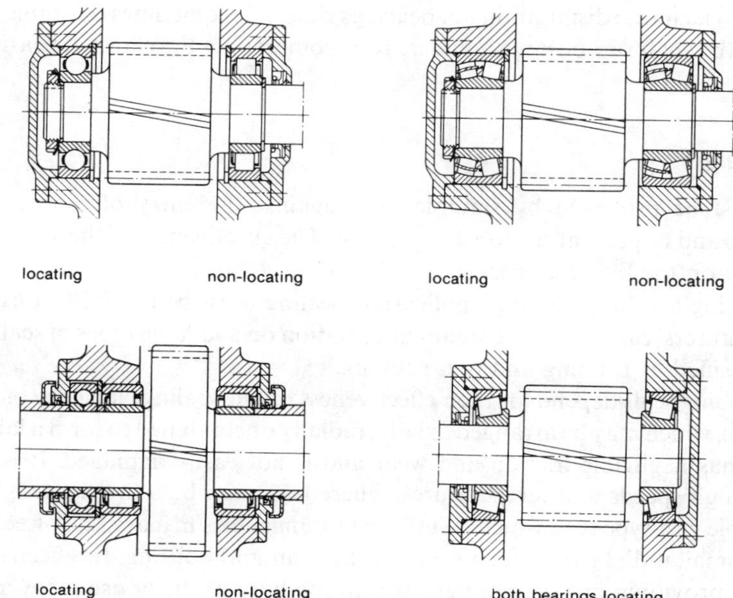

Figure 12.17 Axial support of rolling bearings. (Courtesy SKF General Catalogue, 1977)

in the housing. The inner ring is normally secured in place by means of a lock nut and locking washer or by an end plate attached by set screws to the shaft end; the outer ring is normally retained by the housing end cover, but the threaded ring screwed into the housing bore is sometimes used.

Instead of shaft or housing abutment shoulders, it is frequently convenient to use spacing sleeves or collars between the bearing rings or a bearing ring and the adjacement component, e.g. a gear. On shafts, location can also be achieved using a split collar which sits in a groove in the shaft, and is retained by either a solid outer ring which can be slid over it, or by the inner ring of the bearing itself.

Axial location of rolling bearings by means of snap rings can save space, assist rapid mounting and dismounting and simplify machining of shafts and housings. The dimensions of components adjacent to the bearings (shafts and housing shoulders, spacing collars etc.) must be such that the surface against which the face of the bearing rings about is sufficiently large. It is also necessary to ensure that rotating parts of the bearing do not come into contact with stationary components. The bearing tables give suitable dimensions for these features.

At corner points, suitable fillets must be provided, conforming with the appropriate bearing dimensions.

The better the blending of the cylindrical shaft with the shoulder, the more favorable is the stress distribution. For very heavily-loaded shafts, a larger fillet or more gradual change in diameter may be required. In this case, a spacing collar should be provided between the bearing and the shoulder to ensure a sufficiently large abutment surface for the inner ring. The spacing collar must be relieved adjacent to the shaft shoulder so that it does not contact the shaft fillet.

In order to facilitate dismounting of bearings rings it is sometimes desirable that slots be machined in the shaft or housing shoulders to accommodate the claws of bearing extractor tools.

12.4.2 Sealing

Bearings must be protected by suitable seals against the entry of moisture and other contaminants and to prevent the loss of lubricant. The effectiveness of the sealing can have a decisive effect on the life of a bearing.

In designing a rolling bearing application, sealing must be carefully considered.

Manufacturers' catalogs also contain information on available types of seals. In general, seals are classified as rubbing and non-rubbing.

Non-rubbing seals depend for their effectiveness on the sealing efficiency of narrow gaps (Figure 12.18), which may be arranged axially, radially or combined to form a labyrinth. This type of seal has negligible friction and wear and is not easily damaged. It is particularly suitable for high speeds and temperatures, where leaks can be a fire hazard.

The simple gap-type seal, which is sufficient for machines in a dry, dust-free atmosphere, comprises a small radial gap formed between the shaft and housing. Its effectiveness can be improved by providing one or more grooves in the bore of the housing cover. The grease emerging through the gap fills the grooves and helps to prevent the entry of contaminants. With oil lubrication and horizontal shafts, right- or left-hand helical grooves can be provided in the shaft or the seal bore. These serve to return any oil which may tend to leak from the housing. However, with this arrangement it is essential that the direction of rotation does not vary.

Rubbing seals (Figure 12.19), rely for their effectiveness essentially on the elasticity of the material exerting and maintaining a certain pressure at the sealing surface. The choice of seal and the required quality of the sealing surface depend on the peripheral speed.

Felt washers are mainly used with grease lubrication, e.g. in plummet-blocks. They provide a simple seal suitable for peripheral speeds up to 4 m/s and temperatures of about

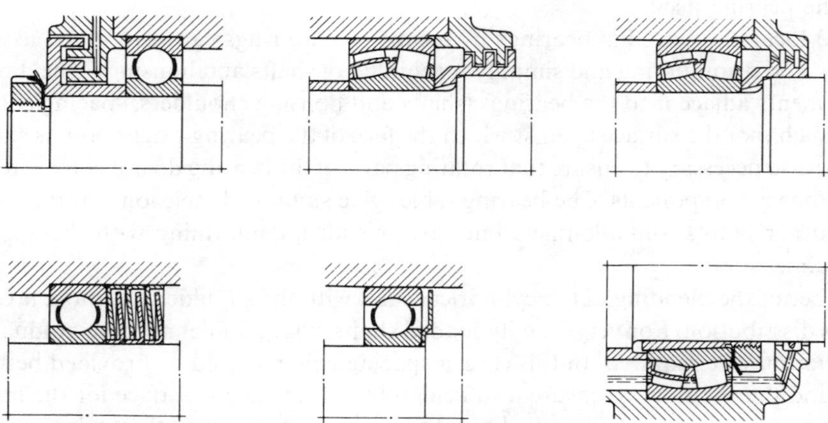

Figure 12.18 Labyrinth sealing. (Courtesy SKF General Catalogue, 1977)

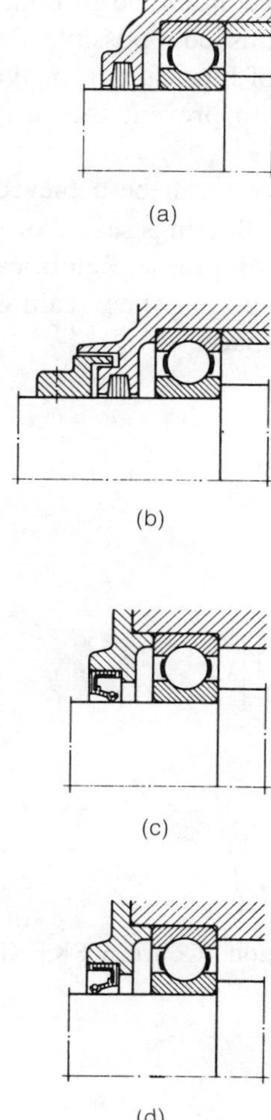

Figure 12.19 Rolling bearing seals. (Courtesy SKF General Catalogue, 1977)

100 °C. The effectiveness of the seal is considerably improved if the felt washer is supplemented by a simple labyrinth ring. The felt washers or strips should be soaked in oil at about 80 °C before assembly.

Where greater demands are made on the effectiveness of the rubbing seal, particularly for oil-lubricated bearings, lip seals are often used in preference to felt seals. A wide range of proprietary lip-type seals is available in the form of ready-to-install units comprising a seal of synthetic rubber or plastic material normally enclosed in a sheet metal casing. They are suitable for higher peripheral speeds than felt washers. As a general guide, at peripheral speeds of over 4 m/s the sealing surface should be ground, and above 8 m/s hardened or hard-chrome plated and fine ground or polished if possible. Referring to Figure 12.19, if the main requirement is to prevent leakage of lubricant from the bearing then the lip should face inwards (c); if the main purpose is to prevent the entry of dirt, then the lip should face outwards (d).

Simple space-saving arrangements can be achieved by using bearings incorporating seals or shields at one or both sides. Bearings sealed or shielded at both sides are supplied, lubricated with the correct quantity of grease. Relubrication is not normally required and they are primarily intended for applications where sealing is otherwise inadequate or where it cannot be provided for reasons of space.

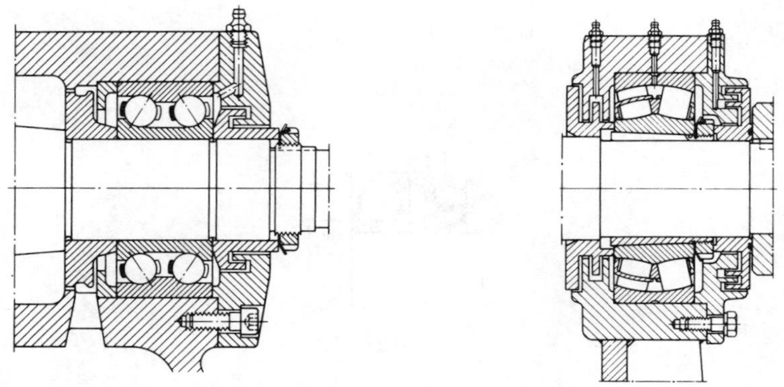

Figure 12.20 Oil lubrication. (Courtesy SKF General Catalogue, 1977)

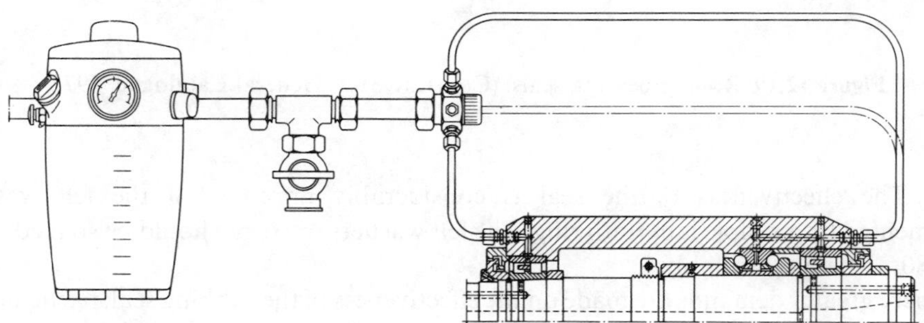

Figure 12.21 Forced oil lubrication. (Courtesy SKF General Catalogue, 1977)

12.4.3 Lubricant application

Under normal conditions, grease lubrication is usually applied. Grease is more easily retained than oil and also prevents the entrance of moisture and dust. Typical grease application is shown in Figure 12.20.

At higher speeds of rotation, oil lubrication is preferable (Figure 12.21). Additional reasons sometimes might be the necessity to remove heat generated at high rate or the existence of other elements oil-lubricated, such as gears, near the bearing. Manufacturers usually give application ranges for grease or oil lubrication.

12.5 COMPUTER SELECTION OF ANTIFRICTION BEARINGS

Using the Procedure DESIGNDB (p. 47 and Appendix 2.A), a database with the standard antifriction bearings has been established.

The procedure to retrieve the file and select the proper bearing follows (explanations are in italics)

```
    Procedure AFBEARINGS
    local  FR       'radial load
           FA       'axial load
           LH       'hours of operation
           NRPM     'revolutions per min
           A1       'reliability factor
           A2       'materials factor
           A3       'temperature, speed factor
           V        'outer ring rotation
           STATIC   'static service factor
           d        'inner diameter: real var
    begin
      input FR, FA, LH, A1, A2, A3, V, STATIC
10    input d
      input IT$  ':= 1 ball bearings
                 ':= 2 self aligning ball bearings
                 ':= 3 angular contact ball bearings
                 ':= 4 roller bearings
                 ':= 5 needle bearings
                 ':= 6 two roller self aligning bearings
                 ':= 7 conical roller bearings
                 ':= 8 axial ball bearings
      if IT$:= "1" then do                    (*Deep groove ball bearings*)
        FNAME$:= "DGBALL"
        repeat
```

```
  LRNDF file FNAME$
until real(F1$) > := d
D:= real(F2$); B:= real(F3$); C:= real(F4$); CO:= real(F5$)
N1:= real(F6$); N2:= real(F7$); TYPE$:= F8$
FOC:= FA/CO
IF FOC < .04 THEN do
  E:= .22
  IF FA/FR < E THEN do
    X:= 1; Y:= 0
  else
    X:= .56; Y:= 2
else IF FOC < .07 THEN do
  E:= .24
  IF FA/FR < E THEN do
    X:= 1; Y:= 0
  else
    X:= .56; Y:= 1.8
else IF FOC < .13 THEN do
  E:= .27
  IF FA/FR < E THEN do
    X:= 1; Y:= 0
  else
    X:= .56; Y:= 1.6
else IF FOC < .25 THEN do
  E:= .31
  IF FA/FR < E THEN do
    X:= 1; Y:= 0
  else
    X:= .56; Y:= 1.4
else IF FOC < .5 THEN do
  E:= .37
  IF FA/FR < E THEN do
    X:= 1; Y:= 0
  else
    X:= .56; Y:= 1.2
else
  E:= .44
  IF FA/FR < E THEN do
    X:= 1; Y:= 0
  else
    X:= .56; Y:= 1
P:= X*V*FR + Y*FA
PEXP:= -3
L10:= LH*60*NRPM*1E-6
CLOAD:= P*(L10/(A1*A2*A3))^PEXP
IF CLOAD > C THEN do
```

```
20      RECORD%RECORD%+1
        LRNDF FILE$
        IF real(F1$) > d THEN do
            print "Design not feasible. Increase diameter"
                  "or select heavier type of bearing"
            go to 10                                    (*Static test*)
        else
            PO:= .6*FR + .5*FA
            COLOAD:= STATIC*PO
            IF COLOAD > CO THEN do
                go to 20
            else
                print "Design is feasible"
                      "Deep groove bearing type: "TYPE$
                      "Safety margin:"
                      "    dynamic:"C/CLOAD"
                      "    static: "CO/COLOAD
    else                                        (*Other types of bearings*)
    end.
```

This procedure is included in the shaft design program SHAFTDES, Appendix 14.D, of Chapter 14.

EXAMPLES

Example 12.1

An overhead crane is rolling on four wheels of diameter $D = 400$ mm on a rail of width $w = 60$ mm. The rail is made of carbon steel AISI 1020 and the wheel of cast steel ASTM A27/GR 60–30. Determine the maximum load that each wheel can carry for continuous strength.

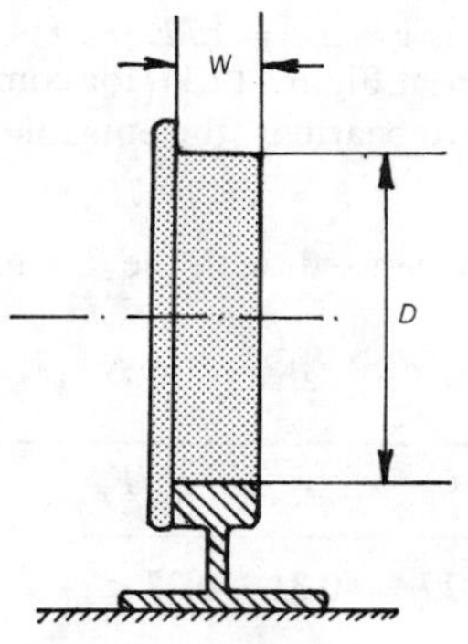

Figure E12.1

Solution

If F is the unknown force, the maximum pressure at the contact zone will be (Figure 12.3), for $E = 2.1E11$ N/m^2,

$$p = 0.591\ (FE/2Rw)^{\frac{1}{2}}$$

Solving for F_2,

$$F = (p/0.591)^2\ (2Rw/E)$$

The surface strength, for dry contact, is $2S_e$. From the material tables in Chapter 6, the fatigue strengths of AISI 1020 carbon steel and the ASTM A27/6R 60–30 cast steel are 190 and 170 N/mm^2, respectively. Therefore, the value of 170 N/mm^2 will be used. The design fatigue strength will be

$$S_e = C_p C_S C_R C_H S_n / K_f = 0.8 \times 0.7 \times 0.83 \times 1 \times 170/1 = 79\ N/mm^2$$

for 99% reliability. The surface strength is then $S_{fe} = 2S_e = 158\ N/mm^2$. The allowable load is

$$F = (158/0.591)^2 (2 \times 200 \times 60/2.IE5) = 8168\ N$$

Example 12.2

Two disks of equal diameter $D = 200$ mm, as in Figure 12.5, are used in a rotating drum, carrying load of 50 kN and rotating at 800 rpm. Their material is hardened steel and the oil used in SAE 30 at 80 °C. For this oil, $\alpha = 0.02\ mm^2/N$. On account of surface roughness and oil filtering solution, the minimum oil film thickness should not be less than 1 mm. Compute the width of the discs.

Solution

For steel, $E = 2.E5\ N/mm^2$, $\nu = 0.3$. Therefore, $E' = 2.31E5$. The equivalent radius $R = R_1 R_2 / (R_1 + R_2) = 100/2 = 50$ mm. The oil viscosity of SAE 30 at 80 °C is $\mu = 12E - 3\ N{\cdot}s/mm^2$. If the load is $P = 5000$ N, then the load W per unit width is $W = P/w$, where w is the unknown width. Solving equation (12.7) for w, because $U_1 = U_2 = 2\pi Rn/60 = 2 \times 100 \times 800/60 = 8333$ mm/s $= U$, then substitution yields, $w = 13$ mm.

Example 12.3

The bearing on the upper left of Figure 12.17 is used on a construction crane. Analysis of the system yielded for this bearing, horizontal load 3500 N, vertical load 6500 N and axial load 2000 N. The rotating speed is 2000 rpm and the desired bearing life is 6000 hours. Select a deep groove ball bearing of common steel grade with a reliability of 99%. In the load computation, shock loads have been accounted for. Strength calculations for the shaft require that its diameter is at least 60 mm. SAE 30 oil is used and it is expected to operate at 80 °C.

Solution

The total radial load is $F_r = (3500^2 + 6500^2)^{\frac{1}{2}} = 7380$ N. The reliability factor $a_1 = 0.21$ (Table 12.2). For the factor a_{23}, the required kinematic viscosity from Figure 12.10 is $\nu_1 = 12\ mm^2/s$. The oil viscosity of SAE 30 oil at 80 °C is $1.2E - 2\ N{\cdot}s/mm^2$. For a density of 890 kg/m^3 the kinematic viscosity is $\nu = \mu/\rho = 1.2E - 2/890 = 1.35E - 5\ m^2/s = 13.5\ mm^2/s$. The ratio $k = \nu/\nu_1 = 13.5/12 = 1.125$. From Figure 12.11, for common steel grade, $a_{23} = 1.25$ is a mid-range value. For radial deep groove bearings, the equivalent dynamic load will be

$$P = XF_r + YF_a$$

An iteration scheme will now be used, with the dynamic load required which is computed from equation (12.13).

$$C = P(L_n/a_1 a_2 a_3)^{\frac{1}{3}} = P(60\,NL_h/10^6\,a_1 a_2 a_3)^{\frac{1}{3}}$$

d	Series	C_0, N	F_a/C_0	e	F_a/F_r	X	Y	P	C	C_{brg}
100	61,820	17,000	0.117	0.31	0.27 < e	1	0	7,380	115,000	15,300
repeat										
100	6,320	132,000	0.015	0.22	0.27	0.56	2	8,132	127,000	134,000
repeat										
100	6,220	78,000	0.025	0.22	0.27	0.56	2	8,132	127,000	95,000
stop										

Therefore, bearing 6320 is selected with $d = 100$, $D = 180$, width $B = 34$ mm.

The static bearing capacity must now be checked. Equation (12.18) must be used. For small cranes, $S_0 = 1.6$. The equivalent static load is then,

$$P_0 = 0.6F_r + 0.5\,F_a = 5428 \text{ N}$$

and the bearing static capacity should be

$$C_0 = P_0 s_0 = 5428 \times 1.6 = 8685 \text{ N}$$

which is much lower than the bearing static capacity of 132,000 N.

Example 12.4

The wheel of a crane carries a maximum load of 50,000 N and revolves at 40 rpm when the crane is in motion. It is expected that only 30% of its operating life of 10,000 h will have the maximum load while another 30% of the time it will carry a 60% load and the rest of the time the wheel carries the dead weight of the crane (18,000 N). Bending of the shaft requires that the diameter should be at least 50 mm. Grease lubrication will be used at normal temperature and 90% reliability will suffice. Select appropriate rolling bearings.

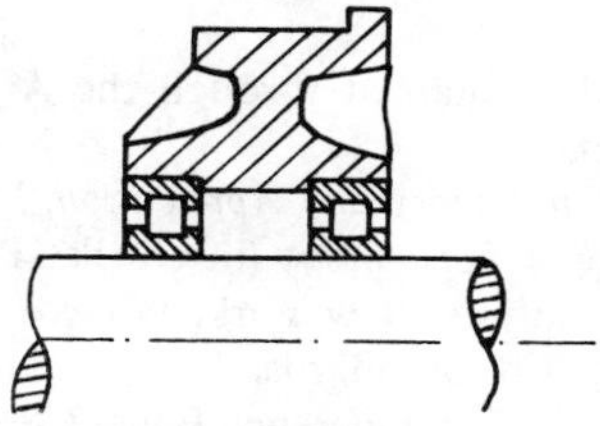

Figure E12.4

Solution

For the diameter of 50 mm, the 50 kN load looks heavy and roller bearings are initially selected. Each bearing carries half load. Because load is not constant, an equivalent load will be computed with equation (12.20). Since the number of revolutions is proportional to the percentage of operating time,

$$P = [50{,}000^p \times 0.30 + 3000^p \times 0.30 + 8000^p \times 0.40/(0.3 + 0.3 + 0.4)]^{1/p}$$

For $p = 10/3$, $P = 37{,}042$ N. Operating factors are $a_1 = 1$ (90% reliability specified), $a_{23} = 1$ (normal steel and adequate lubrication in conjunction with low speed). Therefore, for $f_k = 1$ (no gears), $f_d = 1$ (no machines connected), the effective load is 37,042 N. The required dynamic capacity of the bearing will be, by equation (12.14),

$$\text{C} = 37{,}042(60 \times 40 \times 10{,}000/1E6)^{\frac{3}{10}} = 96{,}107 \text{ N}$$

Bearing NU231 has a capacity of 110,000 N and a static capacity of 80,000 N. The dynamic capacity is adequate. The static capacity for $s_0 = 1.6$.

$$C_0 = P_0 s_0 = 37{,}042 \times 1.6 = 59{,}267 \text{ N}$$

which is less than the bearing capacity of 80,000 N. Therefore, bearing NU231 is selected with $d = 50$, $D = 110$, $B = 40$ mm.

Example 13.5

A cylindrical roller bearing SKF NU318 made of standard steel is to rotate at $n = 500$ rpm under a constant radial load of 25 kN. The kinematic viscosity of the oil to be used at the expected temperature is 38 mm²/s. What is the expected life for reliabilities 90% and 98%? From bearing tables, $d = 90$, $D = 190$, $B = 43$ mm, $C = 220{,}000$ N, $C_0 = 160{,}000$ N.

Solution
Since the load acts only in the radial direction, $P = F_r = 25{,}000$ N. The mean diameter $d_m = (d + D)/2 = (90 + 190)/2 = 140$ mm. From Figure 12.10, $\nu_1 = 21$ mm²/s. Therefore, $k = \nu/\nu_1 = 38/21 = 1.8$. The life adjustment factor is then $a_{23} = 1.4$ to 2.2. Therefore:
For reliability of 90%, $a_1 = 1$,

$$L = 1 \times (1.4 \text{ to } 2.2)(220{,}000/25{,}000)^{\frac{10}{3}} = 1970 \text{ to } 3030 \times 10^6 \text{ revolutions.}$$

For reliability of 98%, $a_1 = 0.33$,

$$L = 0.33(1.4 \text{ to } 2.2)(22{,}000/25{,}000)^{\frac{10}{3}} = 650 \text{ to } 1000 \times 10^6 \text{ revolutions.}$$

REFERENCES AND FURTHER READING

ANSI B3.14. Standard. This can be obtained through the Anti-Friction Bearings Manufacturers Association, AFBMA #20 Standard.
British Petroleum, 1969. *Lubrication Theory and Application*, London: BP Trading Co.
Dimarogonas, A. D., 1979. *History of Technology* (in Greek). Patras.
Harris, T., 1966. *Rolling Bearing Analysis.* New York: Wiley.
ISO Technical Committee 4 (ISO/TC4) Standards.
Mechanical Drives. 1978. Machine Design Reference Issue. Cleveland: Penton/IPC.
Niemann, G. 1965. *Maschinenelemente.* Munich: Springer.
Palmgren, A., 1959. *Ball and Roller Bearing Engineering.* Philadelphia: SKF Industries.
Pugh, B., 1970. *Practical Lubrication.* London: Butterworths.
Reshetov, D. N., 1978. *Machine Design*, Moscow: Mir Publishers.
Shigley, J. 1972. *Mechanical Engineering Design.* New York: McGraw-Hill.
SKF General Catalogue. 1978. Sweden.
Timken Bearing Selection Handbook. 1983. Timken Co.
Wills, J. G., 1980. *Lubrication Fundamentals.* New York: Dekker.

PROBLEMS

12.1 A large railroad car is designed for transporting large electric generators. The total weight of the car and generator is 500 ton. The wheels have a diameter of 600 mm and the width of the rail is 50 mm. The fatigue strength of the material is 200 N/mm². Each axle has two wheels. Determine the required number of axles for that car. The car will be moving very slowly and no impact loads should be considered.

12.2 A cement kiln drum is rotated by way of rolling a ring of diameter $D = 2$ m on two rollers of diameter $d = 0.40$ m. The width of the rollers is 120 mm and the material of the ring and the rollers is cast-steel ASTM A148/GR 80–50. Determine for each such station along the kiln drum how much load it can take.

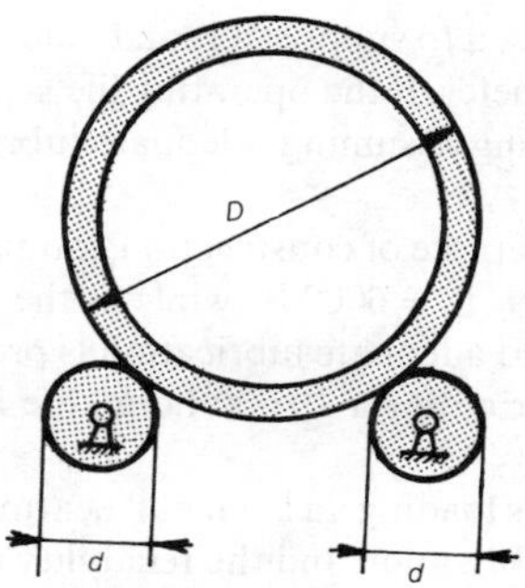

Figure P12.2

12.3 A generator rotor weighs 120 tons and is turned on a lathe supported on two pairs of rollers at the two ends. The rotor diameter at the rolling station is 1 m and the rollers have a diameter of 250 mm. The rotor material is Cr–Mo–V which has a much higher fatigue strength than the carbon steel material of the rollers which is 280 N/mm^2. Determine the required width of the rollers.

12.4 A cam and follower mechanism, as shown, operates at 3200 rpm. The diameter of the follower is 40 mm and the radius of the curvature of the cam is 10 mm at the point of maximum acceleration, which yields a maximum transmitted force of 6000 N. The lubricant used is SAE 10 and the temperature is kept at below 50 °C. The width of the follower is 20 mm. Determine the minimum oil film thickness.

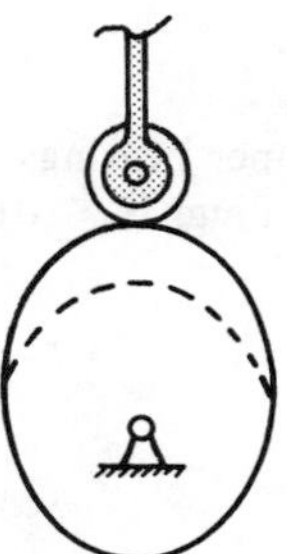

Figure P12.4

12.5 In Problem 12.4, if the material of the cam and follower is carbon steel AISI 1020, determine the safety factor for fatigue failure.

12.6 A deep groove ball bearing of size 6014 is used in a gearing application with moderate shocks. The lubricant is SAE 80 at temperature of 80 °C. The axial load is 600 N and the radial 1200 N. For a reliability of 90%, determine the bearing life in hours. The speed of rotation is 3000 rpm.

12.7 A taper roller bearing will be used for the main shaft of a lathe which rotates at a maximum speed of 2000 rpm and maximum loads, radial, 20,000 N, axial 12,000 N. The oil used is SAE 30 at temperatures which do not exceed 40 °C. The shaft diameter is at least 80 mm. Select the bearing for a reliability of 98% and 6,000 hours of life.

12.8 A deep groove ball bearing is used to support a radial load of 12,000 N, axial load of 5,000 N and no limit is set on the shaft diameter. If the operating life is 5000 hours and the desired reliability is 90%, select the proper bearing assuming adequate lubrication for $n = 1000$ rpm.

12.9 A taper roller bearing used in a piece of construction equipment has two modes of operation: In the forward mode, $E_a = 6000$ N, $F_r = 6000$ N, while in the backward mode, $F_a = 0$, $F_r = 900$ N. The speed of rotation is low and adequate lubrication is provided. Standard reliability of 90% is adequate. Select a proper bearing for a 10,000 hours life for the worst mode, $n = 33$ rpm.

12.10 The thrust load in an axial fan is loading an axial ball bearing with 68,000 N. The diameter has no limitation, there is adequate lubrication and the reliability required is 98%. The bearing life is to be 10,000 hours. Select the proper bearing for $n = 100$ rpm.

12.11 The bearing in Problem 12.6 was overloaded for 1000 hours by a factor of 1.5. Determine the remaining life under the design load.

12.12 Refer to Problem 12.11 and perform the same routine for the bearing in Problem 12.7.

12.13 Refer to Problem 12.11 and perform the same routine for the bearing in Problem 12.8.

12.14 The bearing in Problem 12.9 operates for 80% of its life in the forward mode and 20% in the backward mode. Select the proper bearing.

12.15 The bearing of Problem 12.10, owing to a malfunction, operates for 500 hours with erratic shock load. It was estimated that the shock factor was at least 2. Determine the life remaining after this incident, assuming quiet operation.

Problems 12.16 to 12.20

For the shafts shown, select the proper bearings at both ends. Make a sketch showing the proper mounting, lubrication and sealing. $H = 10{,}000$ h, $a_1 = a_2 = a_3 = 1$, s_0.

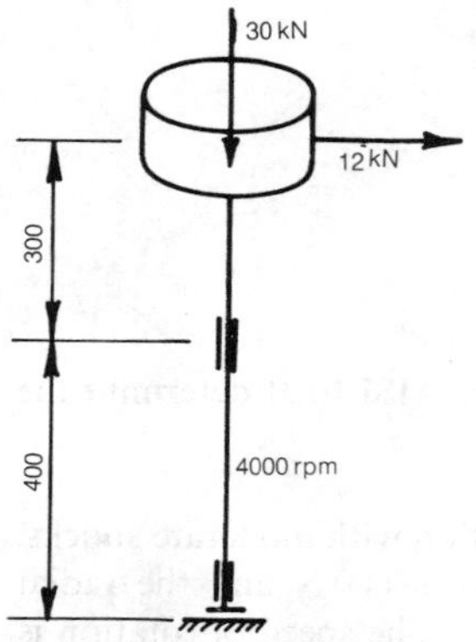

Figure P12.16

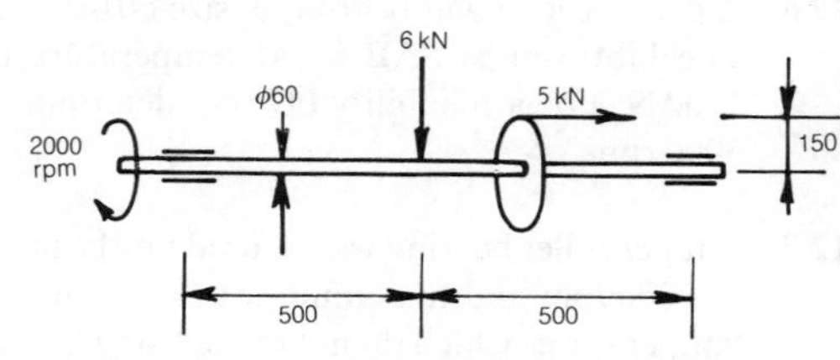

Figure P12.17

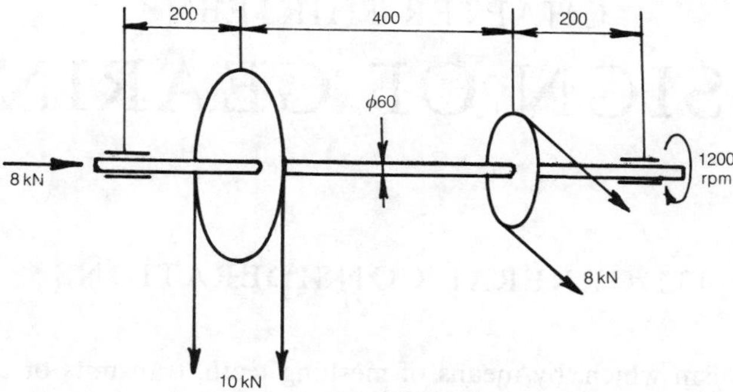

Figure P12.18

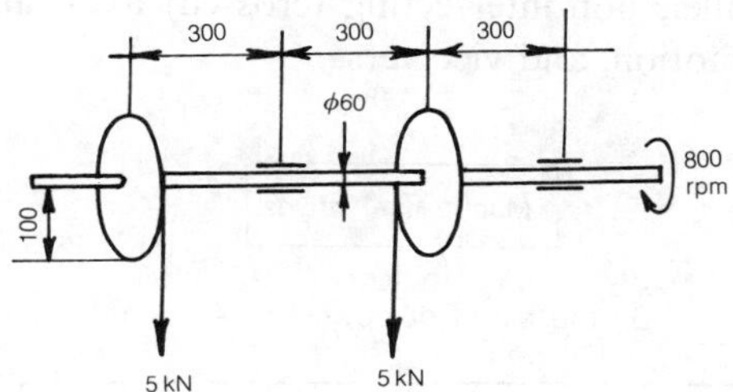

Figure P12.19

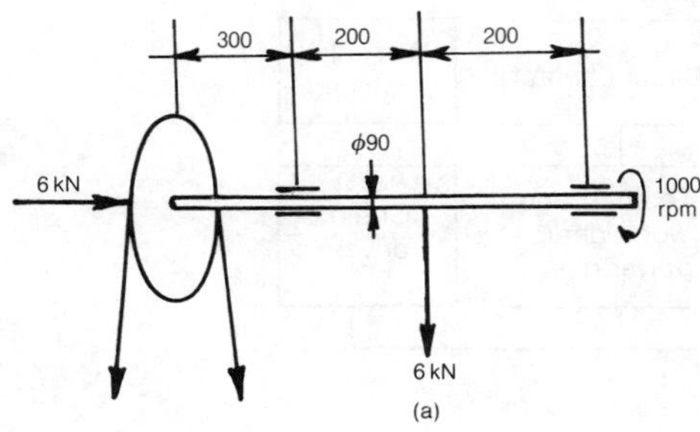

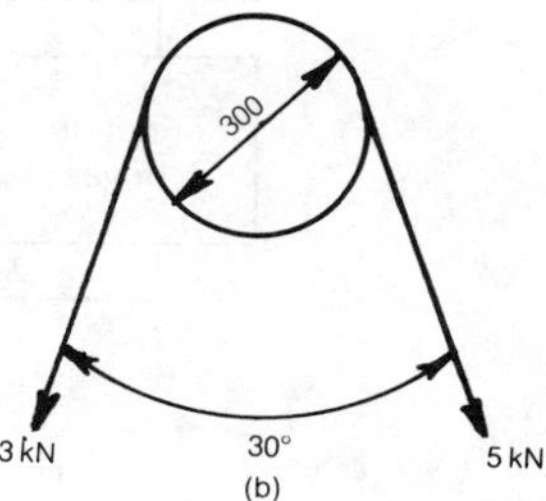

Figure P12.20

CHAPTER THIRTEEN

DESIGN OF GEARING

13.1 GENERAL CONSIDERATIONS

A gear is a mechanism which, by means of meshing teeth, transmits or converts motion, changing the angular velocity and torque between two moving systems. It is a member of the family of mechanical drives (Figure 13.1).

Toothed gearings convert and transmit rotary motion between shafts with parallel, intersecting, and non-parallel, non-intersecting (crossed) axes, and also convert rotary motion into translational motion, and vice versa.

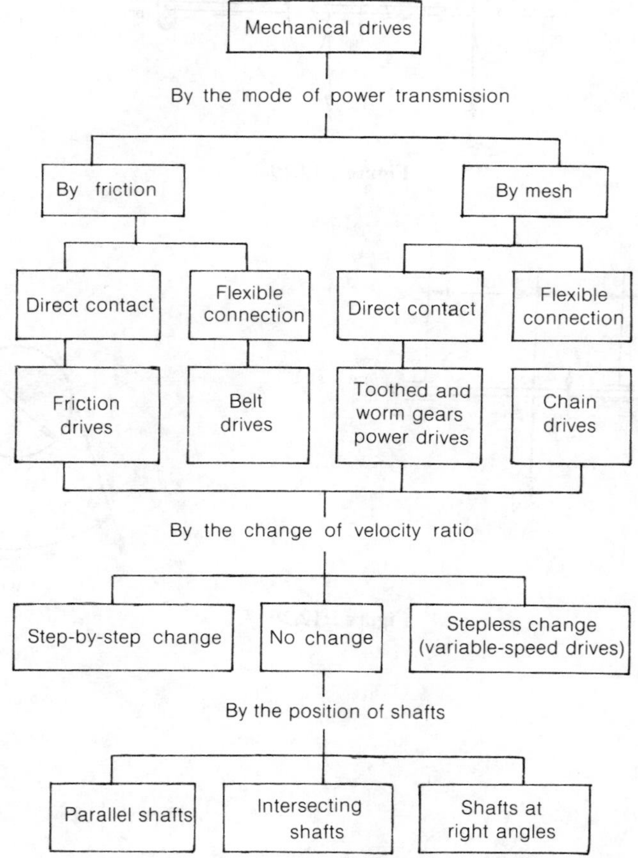

Figure 13.1 Classification of mechanical drives

Toothed gearing between parallel shafts is accomplished by spur, helical and herringbone (double-helical) gears (Figures 13.2a, b, c, d). Gearing between intersecting axes is usually accomplished by straight and spiral bevel gears (Figures 13.2f and h), and less often by skew bevel gears (Figure 13.2g). Gearing to convert rotary motion into translational motion or vice versa is accomplished by a rack and pinion (Figure 13.2e).

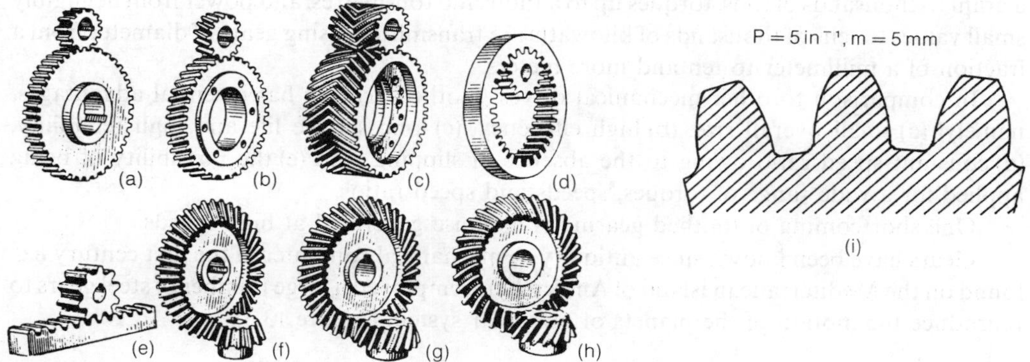

Figure 13.2 Gear drives. (Reshetov 1978)

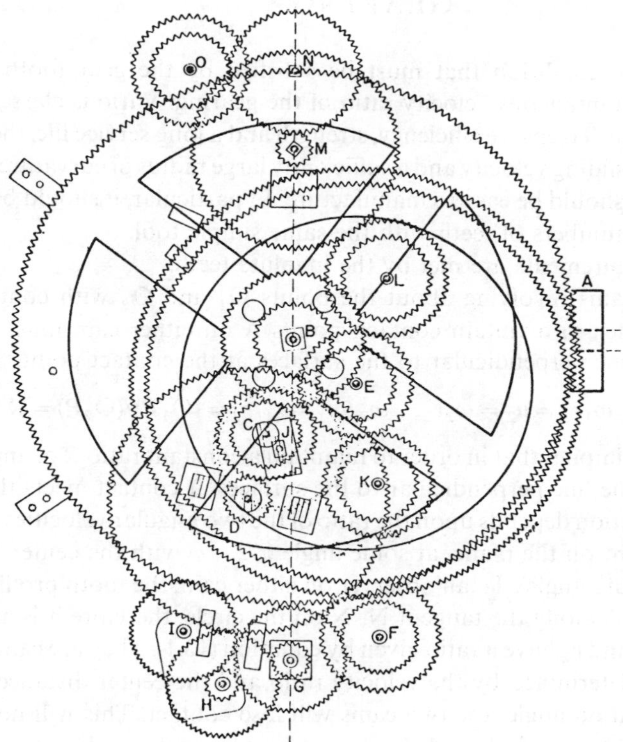

Figure 13.3 The Antikythyra Computer. (Courtesy the late Derek de Solla Price, Yale University)

Crossed helical gears, worm gears and hypoid gears are used to transmit rotation between shafts with non-parallel, crossed axes.

Toothed gearing is the most widely used and most important form of mechanical drive. Gears are used in many fields and under a wide range of conditions, from watches and instruments to the heaviest and most powerful machinery. Peripheral forces from decimals of a gram to thousands of tons, torques up to a thousand ton-metres, and power from negligibly small values to tens of thousands of kilowatts are transmitted, using gears of diameter from a fraction of a millimeter to ten and more meters.

In comparison to other mechanical drives, toothed gearing has essential advantages, namely: (a) small overall size, (b) high efficiency, (c) long service life and high reliability, (d) constant speed ratio owing to the absence of slipping, and (e) the possibility of being applied for a wide range of torques, speeds and speed ratios.

One shortcoming of toothed gearing is its noise generated at high speeds.

Gears have been known since antiquity. A mechanical computer of the first century B.C. found on the Mediterranean island of Antikythyra, employed a large number of steel gears to reproduce the motion of the planets of our solar system (Figure 13.3).

13.2 GEOMETRY AND KINEMATICS OF INVOLUTE GEARING

The basic kinematic condition that must be satisfied by the gear tooth profiles is the constancy of the instantaneous velocity ratio of the gearing. Various classes of curves can meet this requirement. To ensure efficiency, strength and a long service life, the profile should also provide for low sliding velocity and a sufficiently large radius of curvature at the points of contact. The profile should be easy to manufacture. In particular, it should be feasible to cut gears with various numbers of teeth with the same simple tool.

All of these requirements are met by the involute teeth.

Consider two cams revolving about the points O_1 and O_2 with contact at point A (Figure 13.4). In order to maintain contact, points A on either cam must have the same velocity along the line perpendicular to the surfaces at the contact point. Therefore,

$$u_1 = \omega_1 r_1 = u_2 = \omega_2 r_2, \quad \omega_2/\omega_1 = r_1/r_2 = (O_1P)/(O_2P) = R \tag{13.1}$$

Equation (13.1) implies that in order to maintain a constant ratio R of angular velocities, the point P where the line perpendicular to the surfaces at contact meets the center-line is constant and its location depends upon the ratio of the two angular velocities. Moreover, this line must be constant on the plane, at some angle $\pi/2 - \varphi$ with the center-line. Angle φ is known as the 'pressure angle'. To an observer on either cam, the tooth profile appears to be traced by the point A along the tangent N–N on the circle, therefore it is an involute. The circles with radii r_1 and r_2 have a ratio given by equation (13.1). They are called 'basic circles' and they are fully determined by the velocity ratio and the center distance.

After some rotation angle, the two cams will lose contact. This will not happen if the involute profiles are repeated around the cam at equal peripheral distances. This forms an involute gear.

Conventionally, the smaller gear is called the 'pinion' and the larger, simply the 'gear'. Indices 1 and 2 will correspond in the following to pinion and gear respectively.

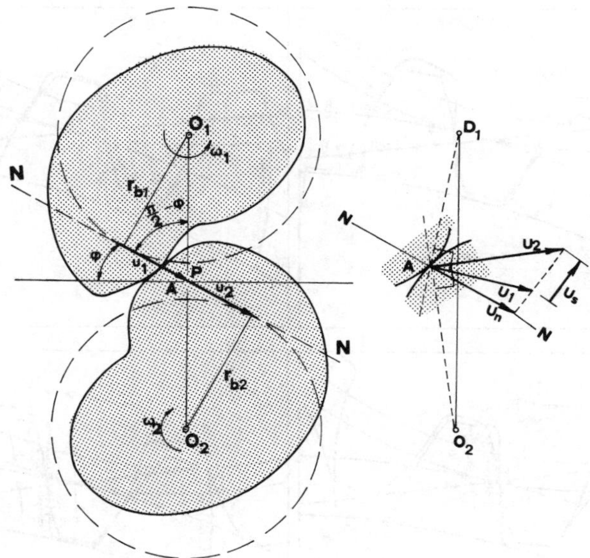

Figure 13.4 The fundamental law of gearing

The meshing of gears is equivalent to the rolling without slipping of circles of diameters $d_1 = O_1P$ and $d_2 = O_2P$ (Figure 13.4) called the 'rolling' or 'pitch diameters'. The circles are known as the rolling or pitch circles. As the straight line NN rolls without slipping around the base circles of diameters $d_{b1} = d_1 \cos\varphi$ and $d_{b2} = d_2 \cos\varphi$ (where φ is the pressure angle), points of this straight line describe involutes on each of the gears. The gears themselves have considerable slipping on the mating surfaces, as one can see from Figure 13.5, observing carefully the mating points as the gears rotate and their relative position.

The elements of toothed gearing have been standardized. The basic parameter is the module m of the teeth, proportional to the circular pitch p along the rolling circles (pitch circles), i.e. $m = p/\pi$.

On the circumference, pitch p is a multiple of number π and is therefore inconvenient to employ as a basic parameter of gearing. Therefore, in the English system the diametral pitch $P = \pi/p$ is used.

The circular pitch p is the distance between like profiles of adjacent teeth measured along an arc of the pitch circle of the gear: $d = N/P$ (where N is the number of teeth).

In a similar manner, the circular module $m = d/N$ and $m(\text{mm}) = 25.4/P$ (in.).

The following are the modules of the most widely-used range (those of the first series are preferable to those of the second):

1st series	1	1.25	1.5	2	2.5	3	4	5	
2nd series	1.125	1.375	1.75	2.25	2.75	3.5	4.5	5.5	
1st series	6	8	10	12	16	20	25	32	40
2nd series	7	9	11	14	18	22	28	36	45

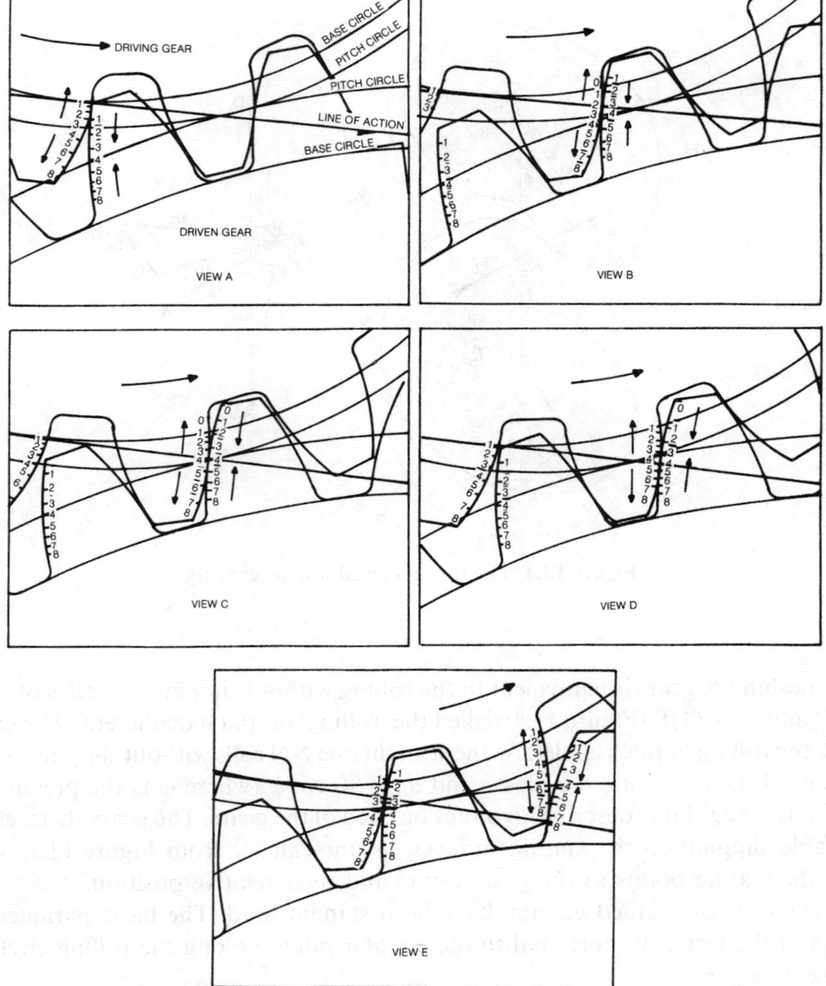

Figure 13.5 Meshing of spur gears, contact line. (Courtesy Mobil Oil Corp.)

Additional modules that can be used for reducing gears are: 1.6, 3.15, 6.3 and 12.5 mm.

As the number of teeth is increased to infinity, the gear becomes a gear rack and the involute tooth profile, a straight line, which is convenient in manufacture and inspection. The basic rack, Figure 13.6, completely determines the tooth profiles of all normal gearing, enabling all gears to mesh with one another. The parameters of the initial contour have been standardized and the profile angle is usually $\varphi = 20°$. The working depth of the teeth is $h = 2.25$ m; the radial clearance between the bottom of tooth space and the top of the tooth of the mating gear is $c = 0.25$ m (or up to 0.35 m in cutting gears with shaper cutters); the fillet radius at the bottom of the tooth of spur and helical gears is, according to most international standards, $r = 0.38$ m.

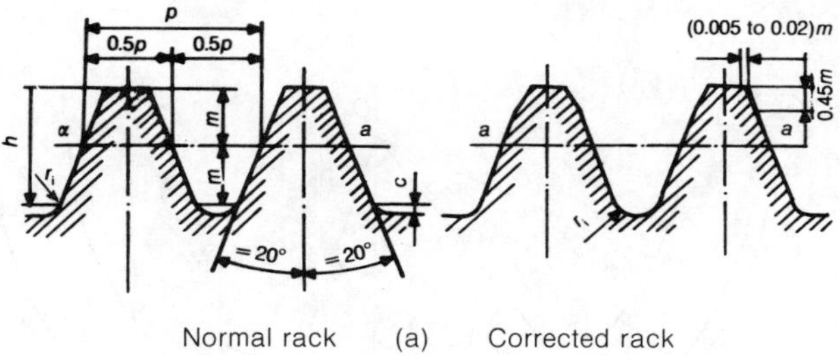

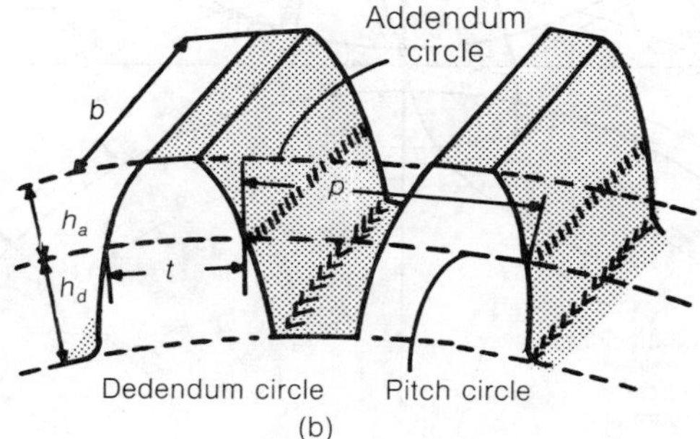

Figure 13.6 Forms of spur gear teeth

Table 13.1 Geometric relationships for spur gears

Parameter	Formula
Center-to-center distance a	$a = \dfrac{(N_1 + N_2)m}{2} = 0.5(N_1 + N_2)m$
Whole depth of teeth h	$h = 2.25m$
Addendum h_a	$h_a = m$
Radial clearance c	$c = 0.25m$
Pitch diameter d	$d_1 = mN_1$ $d_2 = mN_2$
Rolling circle diameter d	d_1 and d_2
Outside (addendum circle) diameter d_a	$d_{a1} = d_1 + 2m$ $d_{a2} = d_2 + 2m$
Foot (dedendum circle) diameter d_f	$d_{f1} = d_1 - 2m - 2c$ $d_{f2} = d_2 - 2m - 2c$

To reduce the force of impact in high-speed gearing when the teeth come into and go out of mesh and to reduce noise, the faces of the teeth are modified by flanking which is a deliberate deviation from the involute profile at the top of the tooth (over a portion of the face) into the body of the tooth. This procedure is known as 'correction'.

The principal geometric relationships for uncorrected gearing are listed in Table 13.1. The names and designations of the elements of toothed gearing are illustrated in Figures 13.6 and13.7.

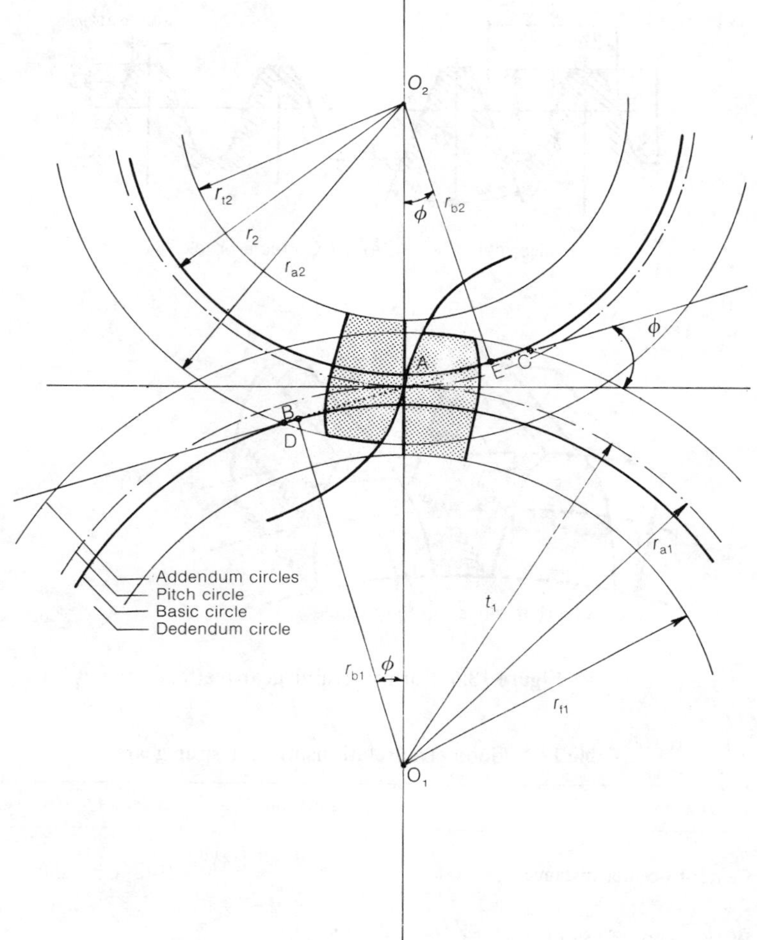

Figure 13.7 Geometry of spur gears

Gear ratios $\omega_1/\omega_2 = N_2/N_1$ for spur gears might only have values which are ratios of two integers.

The basic principle of the involute curve implies that contact between teeth will always take place along the pressure line NN. On the other hand, no contact can take place outside the addendum circle of the gear. Therefore, contact will take place on the portion of the pressure line which is inside both addendum circles, segment BC on the pressure line (Figure 13.7). The length of this segment is of particular importance for the operation of the gears. The longer this segment is, the more teeth are in contact at the same time. The sum of the arcs determined by the angles AO_2B and AO_2C is the arc of contacts, which divided by the circular pitch will yield the number of teeth in contact. A measure of this is the contact ratio, the ratio of the length (BC) divided by the circular pitch $p = \pi m$. From Figure 13.7,

$$m_c = (BC)/\pi m = [(BE - AE) + (CD - AD)]/\pi m \qquad (13.2)$$

For standard teeth, the radius of the addendum circle is $r+m$. (BE) and (CD) are computed from the triangles O_2BE and O_1CD. Using also the relations $r_1 = mN_1/2$, $r_2 = mN_2/2$, $r_b = R\cos\varphi$, the contact ratio is obtained:

$$m_c = [[(N_2+2)^2 - N_2^2\cos^2\varphi]^{\frac{1}{2}} - N_2\sin\varphi + [(N_1+2)^2 - N_1^2\cos^2\varphi]^{\frac{1}{2}} - N\sin\varphi]/\pi m \tag{13.3}$$

It is evident that contact ratio depends on the pressure angle and the number of teeth of the two gears. In fact, for the most commonly-used pressure angles, the contact ratio is related to pressure angle as follows:

Pressure angle	Contact ratio
14.5°	1.7 to 2.5
20°	1.45 to 1.85
25°	1.2 to 1.5

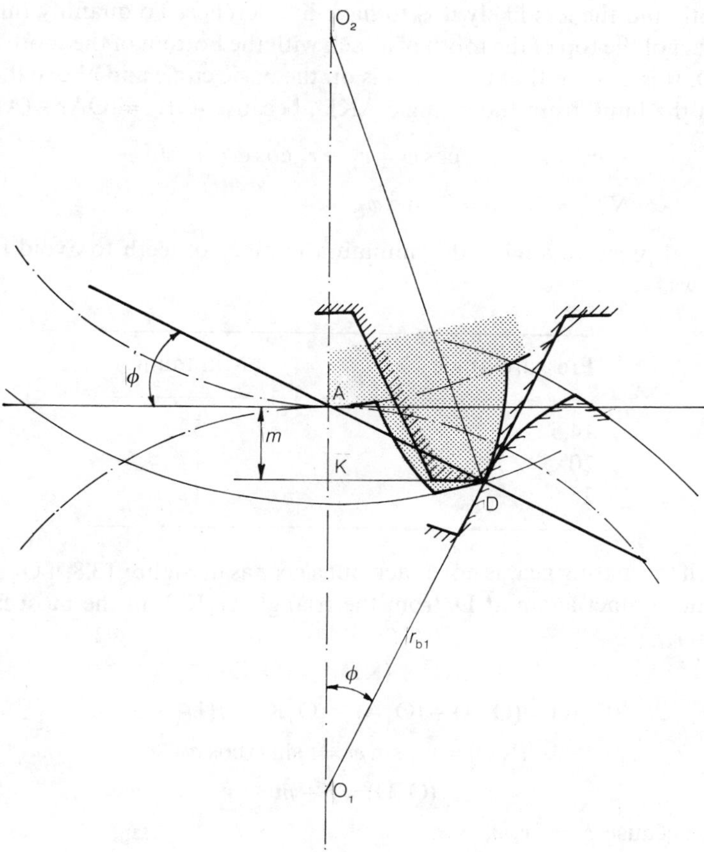

Figure 13.8 Generation of the involute

The higher values correspond to higher number of teeth of the gear, in the limit to the basic rack. The lower values correspond to equal diameters, gear ratio 1.

The basic circle has diameter $d_b = d\cos\varphi$. Therefore, the distance between the basic and the pitch circles is $r - r_b = r(1 - \cos\varphi)$. If the size of the teeth is such that the tooth and the contact extend below the basic circle, since involute does not exist below the basic circle, a non-involute shape will continue up to the dedendum circle. A radial direction is usually selected since the involute is perpendicular to the basic circle and so is the radius. At this portion of the tooth, the basic law of the gearing does not hold and therefore the speed ratio is not constant. This means change in velocity, acceleration and additional dynamic loads which wear quickly this portion of the tooth. The tooth becomes thinner at the high stress area and breakage eventually will occur. This phenomenon is known as 'interference' and the weakening of the tooth as undercutting, situations which obviously must be avoided. Gear displacement (correction) and additional machining of the top of the tooth, Figure 13.6a, have been used for this purpose. However, the main measure to avoid interference is to keep the number of teeth above a minimum value. Indeed, the distance between pitch and basic circle depends only on the pressure angle and not the number of teeth. The depth of the tooth depends on the number of teeth for a given diameter. The larger the number of teeth, the smaller the tooth and the less likely it is to have interference. To quantify this Figure 13.8 shows the contact of the top of the tooth of a rack with the bottom of the tooth of the mating pinion, point D. It is evident that if point D is on the basic circle and above there will be no interference. In the limit, from the triangle AKD, because (OK) = (OA) – (AK) = $r - m$,

$$m = r_1 - r_{b1}\cos\varphi = r_1 - r_1\cos\varphi = r_1\sin\varphi \tag{13.4}$$

$$N_{min} = 2r_1/m = 2/\sin^2\varphi \tag{13.5}$$

For the usual pressure angles, the minimum number of teeth to avoid interference is related as follows:

Pressure angle	Minimum
14.5°	32
20°	17
25°	12

In general, if the mating gear is not a rack but a gear as in Figure 13.8, if O_2 is the center of the gear and the contact again at D, from the triangle O_1KD, in the most favorable case where $r_1 = r_2 = r$,

$$(O_2K)^2 + (KD)^2 = (O_2D)^2$$

$$(O_2K) = (O_2A) + (O_1A) - (O_1K) = r(1 + \sin^2\varphi)$$

$$(KD) = r_{b1}\sin\varphi = r\sin\varphi\cos\varphi$$

$$(O_2D) = r + m$$

Therefore, because $N = 2r/m$,

$$N_{min} = 2/\{[1 + 3\sin^2\varphi]^{\frac{1}{2}} - 1\} \tag{13.6}$$

This relation yields minimum number of teeth 23, 12, 9, for pressure angles 14.5°, 20°, 25°, respectively.

For certain applications, such as for pumping of liquids, a small number of teeth is required. For such purposes, unconventional tooth profiles have been used, such as cycloidal shapes of different forms.

13.3 MECHANISMS OF FAILURE

Tooth breakage is the most dangerous kind of gear failure. It is the result of high overloads of either impact or static action, repeated overloads causing low cycle fatigue, or multiple repeated loads leading to fatigue of the material.

Cracks are usually formed at the root of the teeth on the side of the stretched fibers where the highest tensile stresses occur together with local stresses due to the shape of the teeth. Fracture occurs mainly at a cross-section through the root of the tooth.

Fatigue pitting of the surface layers of the gear teeth (Figure 13.9a) is the most serious and widespread kind of tooth damage that may happen to gears which are enclosed, well lubricated and protected against dirt.

As a result of the meshing of the teeth, the contact stresses at each point of the working surface of the teeth vary in a zero–plus cycle (with stresses of the same sign), while the stresses in the surface layers vary according to an alternating cycle, which, however, is unsymmetrical. Fatigue cracks usually originate at the surface where stress concentration occurs due to the micro-irregularities. If the hardened layer is comparatively thin, and also at high contact stresses, the cracks may originate deeper in the material.

Abrasive wear (Figure 13.9b) is the principal reason for the failure of open gearing and the closed gearing of machinery operating in environments polluted by abrasive materials.

Wear is non-uniform along the tooth profile because of the unequal sliding (rubbing) velocities and contact stresses. Wear increases dynamic loads and noise, weakens the teeth and, finally, leads to tooth breakage.

Seizing of the teeth consists in localized molecular cohesion of the contacting surfaces due to the action of the high pressure and the rupture of the film of lubrication between them. As the contact is broken, particles of one of the surfaces are torn out by the other surface. These particles then score the rubbing surfaces of the teeth.

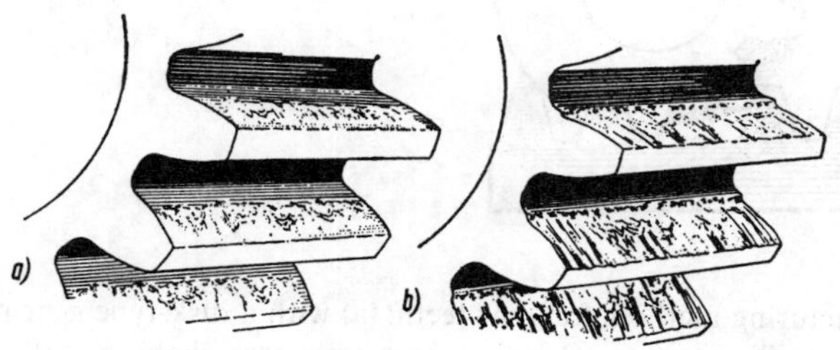

Figure 13.9 Gear teeth wear. (Decker)

13.4 MATERIALS AND MANUFACTURE

Requirements of materials for gears are similar to some extent to the ones for antifriction bearings. This is due to the high contact stresses inherent in gearing as in antifriction bearings. In addition, owing to the beam-like shape of the teeth and the repeated character of the load, considerable fatigue bending stresses are encountered. These stresses however, are usually small as compared with the contact stresses and the respective material strengths. Therefore, homogeneous materials yield gears with low bending stresses if contact stresses are near the limit. Surface hardening therefore can increase the load-carrying capacity without exceeding the fatigue strength of the bulk material. For this reason, the principal materials used for load-carrying gears are steels which can be heat-treated and surface-hardened. Cast irons are also used for gears, mostly for low speed and high sizes, because cast iron has considerable hardness and by nature has much more contact than fatigue strength in tension. Plastics and bronzes are used to a lesser extent.

Gear teeth can be cut either by direct or generating processes.

Direct cutting is performed with shape cutters (Figure 13.10), which must have the form of the space between teeth. Because this space is different for different number of teeth even for the same pitch, each cutter is assigned a range of modules and cutting is therefore approximate. In addition, the process is not productive and it is used for low accuracy gears. The advantage is that it can be performed in usual milling machines in a very simple way.

Generating processes (Figures 13.10c, d) use screw-type cutters with profiles identical to the basic rack corresponding to the pitch and module of the gear. The process is continuous and has definite advantages:

1. it is faster and more productive;
2. it is more accurate because the correct involute is generated regardless of the number of teeth;

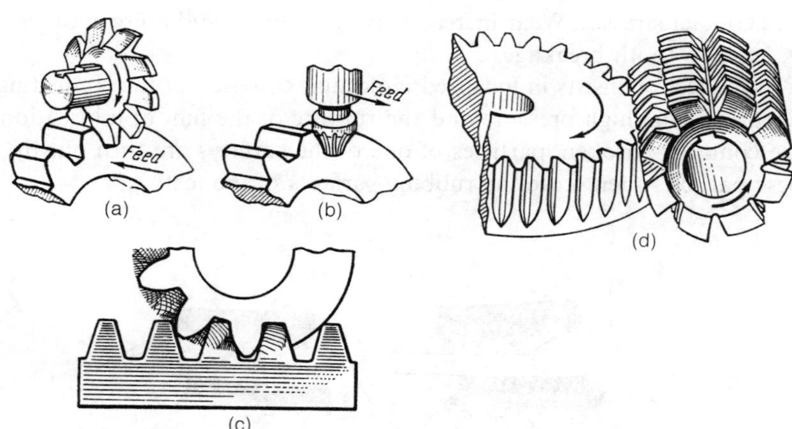

Figure 13.10 Manufacturing method, for gear teeth: (a) with a disk-type gear milling cutter; (b) with an end-mill type gear milling cutter; (c) with a rack-type gear shaping cutter; (d) with a gear hob. (Reshetov)

3. it requires only one tool for every module.

For the above reasons, generating processes are used exclusively for mass production of gears.

13.5 STRESS ANALYSIS OF GEAR TEETH

As the gears rotate, the point of contact between mating teeth moves along the profile, as one can observe in Figure 13.5. If the friction force is neglected, it can be assumed that the interaction force is perpendicular to the tooth profile all the times. It will be assumed at this point that only one tooth at a time carries the load, owing to the cutting errors and the high tooth stiffness. Multiple contact will be discussed later (p. 618).

Under the single contact assumption, the worst case is when the load is applied at the top of the tooth (Figure 13.11).

The normal force can be analyzed in tangential and radial components (Figure 13.11),

$$F_t = F_n \cos\varphi \tag{13.7}$$

$$F_r = F_n \sin\varphi = F_t \tan\varphi \tag{13.8}$$

The tangential force is known from the system parameters, power P, angular velocity ω, pinion diameter d_1:

$$F_t = 2P/d_1\omega \tag{13.9}$$

We shall deal first with the case in which the force is applied to the top of the tooth and the arm of this force is maximum, assuming that the load is transmitted by a single pair of teeth.

The force is resolved into two components: one which bends the tooth $F_n \cos\varphi$ and one which compresses the tooth $F_n \sin\varphi$, where φ is the angle at the top of the tooth. The gear tooth is treated as a cantilever beam. The dangerous cross-section is at the root of the tooth in the zone of maximum stress concentration. Lewis proposed that this is the point B where a parabola from point A is tangent to the tooth profile, Figure 13.11a.

Since fatigue cracks and failure begin on the side of the teeth subjected to tensile stress, the strength is determined for this side. The normal stress in the dangerous section is

$$\sigma = F_t h_c/(bt^2/6) \tag{13.10}$$

where t = thickness of the tooth in the dangerous section and, h_c = calculation arm of the force.

This can be written as the **Lewis equation**

$$\sigma = F_t/mbY \tag{13.11}$$

where

$$Y = t^2/6mh_c \tag{13.12}$$

is the **Lewis Form Factor**.

The maximum fillet stress is $\sigma_{max} = K_t F_t/mYb$, where K_t is the stress concentration factor

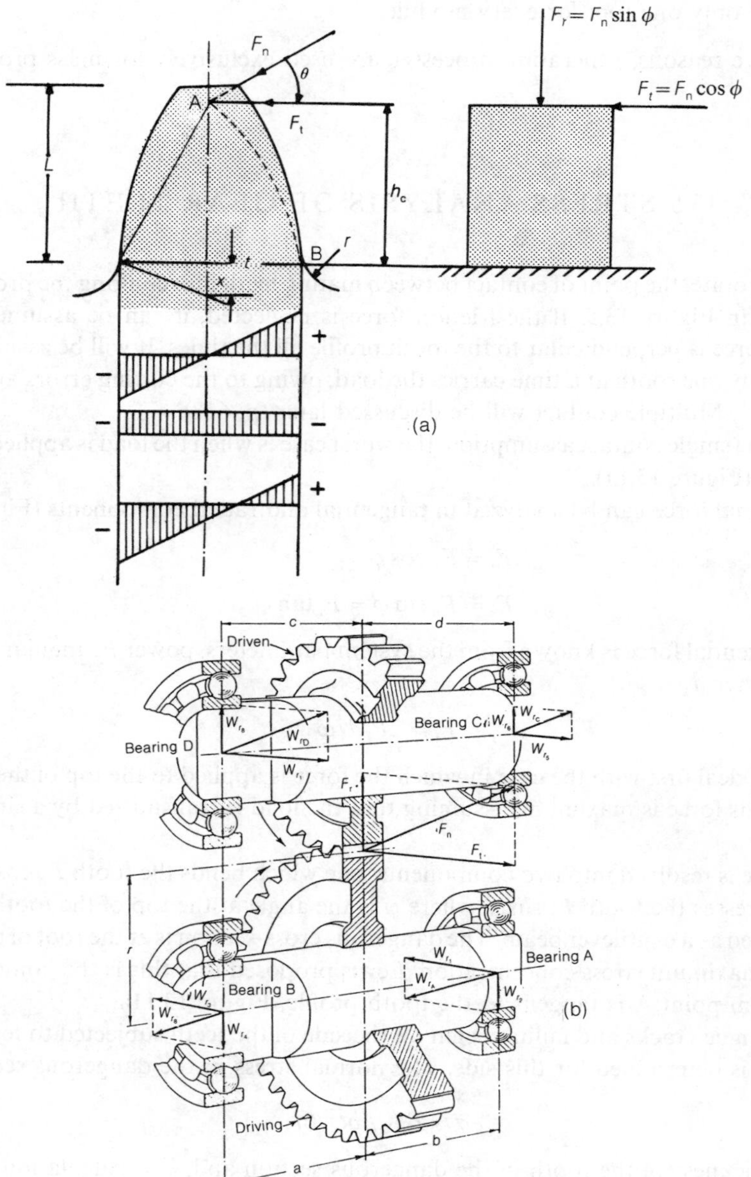

Figure 13.11 Beam strength analysis of spur gears. (b), Dudley 1962

which has been determined with photoelastic methods, Figure 13.12. This factor accounts also for the normal force which is omitted in the Lewis Equation.

The stress concentration factor depends on the radius of the fillet and, consequently, is a function of the number of teeth.

(a) Hardened gears for which the effective stress concentration factor K_f is usually taken equal to the theoretical factor K_t.

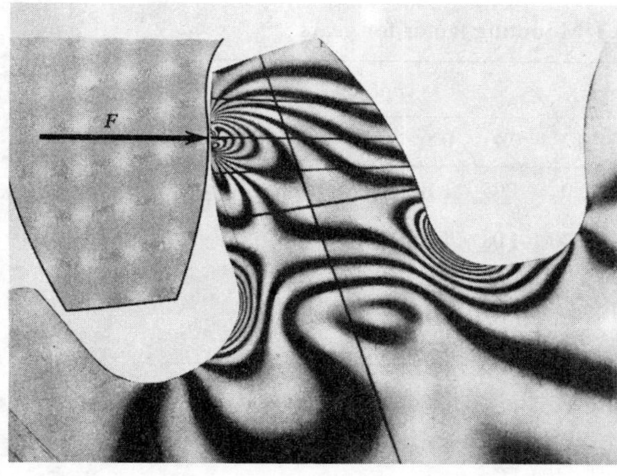

$$K_t = 0.22 + \left(\frac{t}{r}\right)^{0.2}\left(\frac{t}{L}\right)^{0.4} \quad \phi = 14.5$$

$$K_t = 0.18 + \left(\frac{t}{r}\right)^{0.15}\left(\frac{t}{L}\right)^{0.45} \quad \phi = 20$$

$$K_t = 0.14 + \left(\frac{t}{r}\right)^{0.11}\left(\frac{t}{L}\right)^{0.6} \quad \phi = 25$$

Figure 13.12 Stress concentration at the root of the tooth. (Dolan and Broghamer 1942)

(b) Structurally-improved and normalized gears for which $K_f = 0.9K_t$, this being taken conditionally into consideration by a corresponding reduction of the safety factor.

The values of the rated load and the allowable material strength must be adjusted to the realistic service conditions. AGMA (American Gears Manufacturers Association) suggest a modification of the Lewis equation in the form

$$\sigma = \frac{F_t}{mbJ}(K_v K_o K_m) \tag{13.13}$$

where the Lewis form factor Y has been modified to a factor J to include the stress concentration factor. Furthermore, K_v is the velocity factor, owing to additional dynamic loads at tooth engagement as a result of the cutting inaccuracies. It depends on peripheral velocity v and manufacturing errors. The following values are recommended:

$K_v = 1$ for high precision gears
$K_v = 78/[78 + (v)^{\frac{1}{2}}]$ for ground gears
$K_v = 50/[50 + (v)^{\frac{1}{2}}]$ for non-ground gears, where v is in ft/min.

K_o is the overloading factor depending on the uniformity of the load. If simulation results are not available, the overload factor can be taken as in Table 13.2.

K_m is the mounting factor which can be taken from Table 13.3 depending on the mounting, the accuracy and the rigidity of the shaft.

The endurance strength (S_e) is adjusted to the service conditions with the derating

Table 13.2 Shock factors

Source of power	Driven machine		
	Uniform	Moderate shock	Heavy shock
Uniform	1.00	1.25	1.75
Light shock	1.25	1.50	2.00
Medium shock	1.50	1.75	2.25

Table 13.3 Mounting factor for gears

Mounting	b/d_1 0.2	0.4	0.8	1.2	1.6	2.0
Symmetrically between bearings	1.0	1.05	1.10	1.25	1.40	(1.55)
Near one end very rigid shaft	1.00	1.08	1.20	1.40	(1.70)	—
Near one end flexible shaft	1.10	1.20	1.40	1.70	(2.0)	—
Cantilever pinion	1.25	1.55	(20)	—	—	—

equation

$$S_e = S'_n C_s C_f C_R C_t C_m \tag{13.14}$$

where: C_s = size factor = 1.0 for $m < 0.2$ ($P > 5$) and 0.85 for $m > 0.2$ ($P < 5$); C_f = surface finish derating factor (Chapter 7) and applies to the surface finish at the root of the tooth (Figure 7.7 p. 284), C_R = reliability factor (Chapter 7), C_t = temperature factor. For temperature below 70 °C = 1; for higher temperatures, for T(°C),

$$C_t = 344/(273 + T) \tag{13.15}$$

C_m = stress cycle factor. For reversed bending of the gear teeth, $C_m = 1$. For one way bending it is $\sigma = \sigma_m = \sigma_r$ and the equivalent stress range is, with the **Soderberg Criterion**, for $\sigma_m = \sigma_r = S_y/2$

$$S_y/N = S_y/(\sigma_r/S_e + \sigma_m/S_y)$$

$$C_m = N = 2/(1 + S_e/S_y) = 1.4 \text{ for gear steels}$$

Therefore, the design equation will be

$$\frac{F_t}{mbJ}(K_v K_o K_m) \leqslant S'_n C_s C_f C_r C_t C_m \tag{13.16}$$

Since the unknown is usually the module $m = 1/P$ and some of the parameters depend on it (such as J), this equation must be solved by iteration.

The geometry factor J incorporates into the Lewis form factor the stress concentration, a concept not well-known in Lewis's time. Recommended values by AGMA are given in Figure 13.13. There, for gears of regular accuracy, the load is assumed on the top of the tooth, as the most conservative location, and one tooth in contact is assumed.

For gears of high accuracy, simultaneous contact of more than one teeth can be assured and the geometry factor depends also on the number of teeth of the gear.

For computational purposes, the following expressions approximate the values of Figure 13.13 to engineering accuracy:

$\varphi = 20°$, regular gears, $J = 0.32[1 - (1.14/N_1^{0.546})]$
$\varphi = 20°$, high-accuracy gears, $J = 0.56[1 - 0.38/N_2 - (0.88/N_2)][1 - (0.26/N_1 - 5.5/N_1)]$

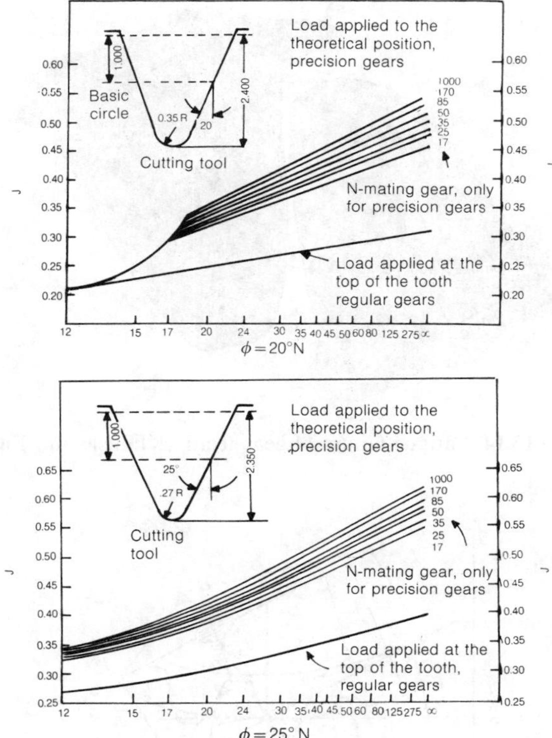

Figure 13.13 Lewis factor for spur gears. (Courtesy AGMA)

$\varphi = 25°$, regular gears, $J = 0.39[1 - (1.14/N_1^{0.546})]$
$\varphi = 25°$, high-accuracy gears, $J = 0.63[1 - (0.38/N_1 - 0.88/N_1)][1-(0.26/N_2^{1/2})-(5.5/N_2)]$.

13.6 DESIGN FOR SURFACE STRENGTH

It is observed in service that pitting usually originates at pitch point. Therefore, this is the place to expect surface fatigue failure (Figure 13.14). Maximum contact stresses in pressing together two cylinders which are in contact along elements of the cylindrical surfaces are given by the **Herz formula**

$$\sigma_H = \left[\frac{E}{2\pi(1-\nu^2)}\frac{F_n}{R_{eq}}\right]^{\frac{1}{2}} \tag{13.17}$$

where F_n = load normal to the surface, $E = 2E_1E_2/(E_1 + E_2)$ = equivalent modulus of elasticity of the material, E_1 and E_2 = moduli of elasticity of the pinion and gear materials (if they are of the same material, $E = E_1 = E_2$), ν = Poisson's ratio, and R_{eq} = equivalent radius of curvature = $R_1R_2/(R_1 + R_2)$.

In the equation for the equivalent radius of curvature, the plus sign refers to external, and

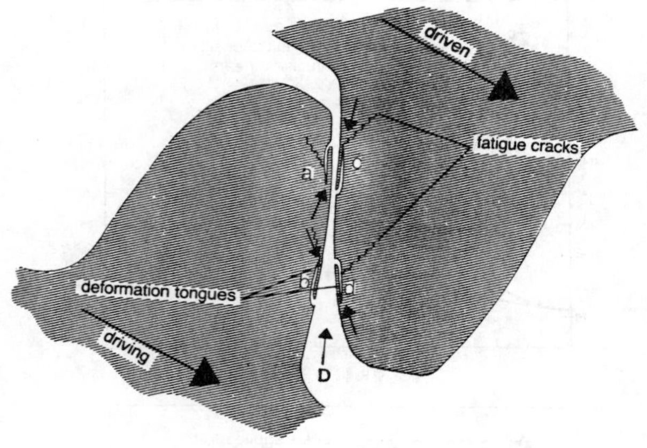

Figure 13.14 Fatigue failure of gear teeth. (Klingele and Engel)

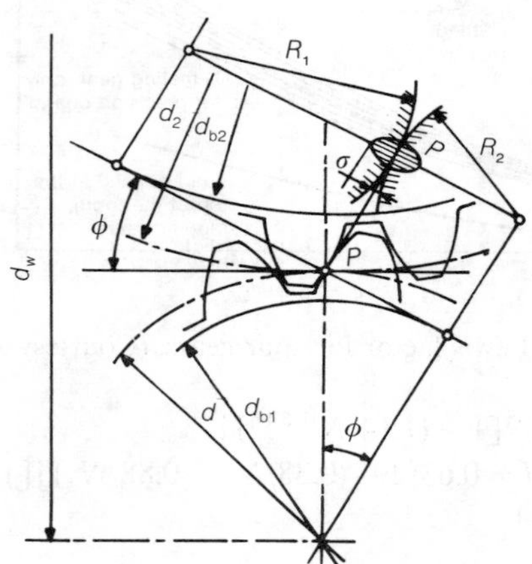

Figure 13.15 Contact stresses

the minus to internal gearing. For rack and pinion gearing, $R = R_1$ and $R_2 = \infty$.

From basic involute geometry, the radii of curvature applicable in the Herz formula are the lengths of the generating tangent NN between the involute and the point of contact of NN with the basic circle (Figures 13.15 and 13.16),

$$R_1 = (d_1 \sin \varphi)/2, \quad R_2 = (d_2 \sin \varphi)/2$$

For $\nu = 0.3$, equating to the fatigue strength S_H and solving for F_n,

$$F_n = d_1 bQK \tag{13.18}$$

where $Q = 2d_2/(d_1 + d_2) = 2N_2/(N_1 + N_2)$, a parameter which depends on the number of teeth only; $K = S_H^2 \sin \varphi(1/E_1 + 1/E_2)/1.4$ is a parameter which depends on the materials and the pressure angle φ. This equation is associated with the name of Earl Buckingham who derived it.

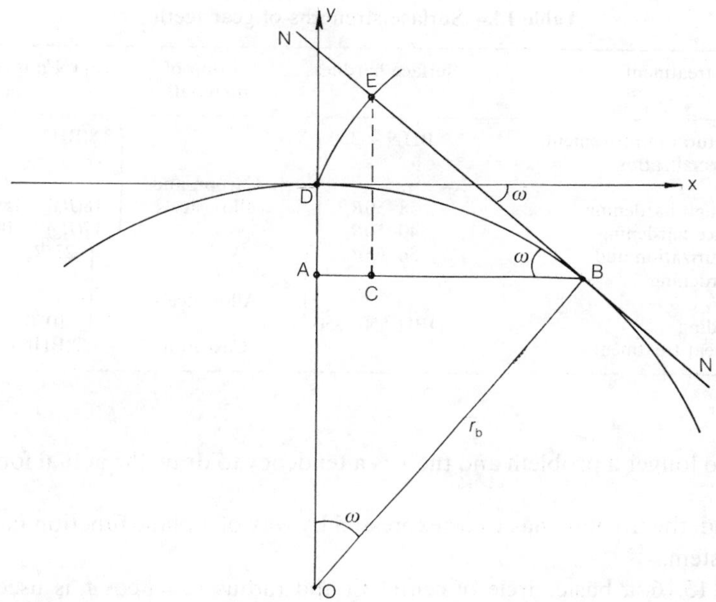

Figure 13.16 Curvature of the involute

Going back to the Herz equation and adjusting the load to the service conditions as in the case of the beam strength of teeth,

$$\sigma_H = C_p(F_t K_v K_o K_m / b d_1 I)^{\frac{1}{2}} \tag{13.19}$$

where $C_p = 0.591[E_1 E_2/(E_1 + E_2)]^{\frac{1}{2}}$ is a material factor and $I = (\sin 2\varphi)(N_2/N_1)/4(1 + N_2/N_1)$ is a geometry factor.

In this approach, recommended by AGMA, the maximum contact stress must be compared with the contact strength, adjusted for the service conditions:

$$\sigma_H < S_H C_L C_H C_T C_R = S_{He} \tag{13.20}$$

where C_L = life factor = 1 for $n > 10^7$ cycles and for shorter life, $C_L = 2.16 - 0.17\log_{10}(n)$. C_H is a hardness factor equal to 1 for $N_2/N_1 < 2$ and for similar hardness of both gears. For very different materials, $\text{BHN}_1/\text{BHN}_2 > 1.2$, where $C_H = 1 + 0.0014(N_2/N_1 - 2)(\text{BHN}_1/\text{BHN}_2)$. The remaining components in equation (13.20) are: C_R = reliability factor, C_T = temperature factor (C_t in equation 13.15) and S_H = contact fatigue strength.

If test results are not available, contact fatigue strength can be estimated on the basis of surface hardness (Table 13.4).

13.7 COMPUTER AIDED GRAPHICS OF SPUR GEARS

In designing standard spur gears, conventional techniques were used for the representation of the teeth because drawing them was a tedious effort if done by hand. With the advent of

Table 13.4 Surface strengths of gear teeth

Heat treatment	Surface hardness	Group of materials	S_H(N/mm^2)
Structural improvement, normalization	BHN < 350	Carbon and alloy steels	2.8(BHN) − 70
Through hardening	38–50R_c		18(R_c) + 150
Surface hardening	40–50R_c		17(R_c) + 200
Carburization and hardening	56–65R_c	Alloy steels	23(R_c)
Nitriding	DPH 550–750		1050
No heat treatment	—	Cast iron	2(BHN)

CAD, this is no longer a problem and there is a tendency to draw the actual form of toothed gearing.

To this end, the involute has to be expressed by way of a plane function in a convenient coordinate system.

In Figure 13.16, a basic circle of center O and radius $r_b = r\cos\phi$ is used to draw an involute from point D by way of rolling the line NN. If after some rolling angle ω the point of contact of the rolling line is at B, the length (BE) will equal the arc(BD) $= r_b\omega$. In reference to coordinate system x, y,

$$x = (AB) - (BC) = r_b \sin\omega - \omega r_b \cos\omega = r\cos\varphi\cos\omega(\tan\omega - \omega) \tag{13.21}$$

$$y = (EC) - (DA) = (EC) - (OD) + (OA) = r\cos\varphi\cos\omega[(\omega\tan\omega + 1) - (1/\cos\omega)] \tag{13.22}$$

The limits of the involute will be imposed by the addendum and dedendum circles, that is between points E_1 and E_2 where $OE_1 = r - 1.25m$, $OE_2 = r + m$, for standard tooth profiles. This gives the limits of the rolling angle:

$$\begin{aligned} &\omega_{max}\colon r_b(1+\omega^2)^{\frac{1}{2}} = r + m; \quad \omega_{max} = [(1+2/N)^2/\cos^2\varphi - 1]^{\frac{1}{2}} \\ &\omega_{min}\colon r_b[1+\omega^2]^{\frac{1}{2}} = r - 1.25m; \quad \omega_{min} = [(1-2.5/N)^2/\cos^2\phi]^{\frac{1}{2}} - 1 \end{aligned} \tag{13.23}$$

There is an obvious limitation, that $\omega > 0$. This means that the dedendum circle cannot be smaller than the basic circle. In case of smaller number of teeth than the one necessary to avoid interference, the involute is continued as a straight line along the radius DO. This happens when the quantity in the root becomes negative. This happens when $N < 2.5/(1 - \cos\phi)$. The procedure to draw a complete spur gear profile follows. The procedure GEARPLOT provides to the procedure SOLID the modelling of a solid spur gear with flat sides (explanations are in italics):

```
Procedure GEARPLOT(D, m, theta, B)
LOCAL N, IN, IN1, I, J, INface Integers: var
R, F, Xo, Yo, Xtop, Ytop, wmax, wmin, Xinv, Yinv, rota, Xtemp, Ytemp
      XR(15), YR(15) real: var
      PI + 3.14159 real: const
```

```
GLOBAL X, Y, Z(N*15), PD(N + 2, 15*N) real: var
begin
                                              (*Draw one side of the tooth*)
 N:= d/m
 r:= d/2
                                                          (*Draw root arc*)
 FOR F:= PI to 3*PI/2 step PI/10 do
   IN:= IN + 1
   Xo:= .38*m
   Yo:= (r - 1.25*m) - r*SIN(theta) + .38*m
   X(IN):= Xo + .38*m*COS(F)
   Y(IN):= Yo + .38*m*SIN(F)
                              (*Draw straight line part of profile, if it exists*)
 IF X(IN) < 0 then do
   IN = IN + 1
   X(IN):= 0
   Y(IN):= 0
 Xtop:= X(IN)
                                                          (*Draw involute*)
 wmax:= SQR((1 + 2/N)^2/COS(theta)^2 - 1)
 IF N < 2.5/(1 - COS(theta) then do
   wmin = 0
 else
   wmin:= SQR((1 - 2.5/N)^2/COS(theta)^2 - 1)
 FOR F = wmin to wmax STEP (wmax - wmin)/10 do
   Xinv:= r*COS(theta)*COS(F)*(TAN(F) - F)
   Yinv:= r*COS(theta)*COS(F)*(F*TAN(F) + 1 - 1/COS(F))
   IF Xinv > Xroot then do
     IN:= IN + 1
     X(IN):= Xinv
     Y(IN):= Yinv
                                                (*Draw other face of tooth*)
                         (*Rotate by quarter of pitch angle, axis of tooth on y-axis*)
 rota:= PI*m/(4*r)
 FOR i:= 1 TO IN do
   Xtemp:= X(I)*COS(rota) + Y(I)*SIN(rota)
   Ytemp:= - X(I)*SIN(rota) + Y(I)*COS(rota)
   X(I):= Xtemp
   Y(I):= Ytemp
                                                     (*Mirror the other half*)
 FOR I:= 1 TO IN
   Y(IN + I):= Y(IN - I + 1)
   X(IN + 1):= - X(IN - I + 1)
 IN:= 2*IN
                                                (*Draw the rest of the gear*)
```

```
rota:= PI*m/r
IN1:= IN
FOR I:= 1 TO N - 1 do
                                        (*Rotate the first tooth and add*)
  FOR J:= 1 TO IN1
    XR(J):= X(J)*COS(I*rota) + Y(J)*SIN(I*rota)
    YR(J):= - X(J)*SIN(I*rota) + Y*COS(rota)
    IN:= IN + 1
    X(IN):= XR(J)
    Y(IN):= YR(J)
                                        (*Draw other face of gear*)
INface:= IN
FOR I:= 1 TO IN do
  IN = IN + 1
  X(IN):= X(I)
  Y(IN):= Y(I)
  Z(I):= 0
  Z(IN):= B
                                        (*Define polygons for gear faces*)
  PD(1, 1 + I):= I
  PD(2, 1 + I):= IN
PD(1, 1):= INface + 2                   (*Number of nodes per facet*)
PD(2, 1):= INface + 2
PD(1, INface + 2):= 1
PD(2, IN + 2):= INface + 1
                                        (*Define polygons for top of teeth*)
FOR I:= 1 TO N
  PD(2 + I, 1):= 5
  PD(2 + I, 2):= (I - 1)*IN1 + IN1/2
  PD(2 + I, 3):= (I - 1)*IN1 + IN1/2 + 1
  PD(2 + I, 4):= INface + PD(2 + I, 2)
  PD(2 + I, 5):= INface + PD(2 + I, 3)
  PD(2 + I, 6):= PD(2 + I, 2)
                                        (*Plot gear*)
SOLID
end.
```

13.8 COMPUTER AIDED DESIGN OF SPUR GEARS

For the design of spur gears, two design equations are available.

The AGMA/Lewis equation:

$$F_t K_v K_o K_m / mJb \leqslant S'_n C_s C_f C_r C_t C_m \tag{13.24}$$

The AGMA/Buckingham equation:

$$[F_t K_v K_o K_m / b d_1 I]^{\frac{1}{2}} \leqslant S_H C_L C_H C_T C_R / C_P \tag{13.25}$$

Design data are usually the speed of the two gears, the power transmitted and sometimes the center-to-center distance $a = (d_1 + d_2)/2$ is given, if the design of the associated parts or space limitations demand it.

It is evident that the Buckingham equation does not include the module, owing to the fact that Herz stresses depend on the radius of curvature and not the size of the tooth. Substituting the moment $M_1 = F_t d_1/2$ and noting that $d_1 = 2a/(1 + R)$, where R is the given gear ratio N_2/N_1, the Buckingham equation can be solved for a to yield

$$a^2 = M_1(1 + R)^2/2bIK \tag{13.26}$$

where $K = [S_H C_L C_H C_T C_R/C_P]^2/K_v K_o K_m$.

The constant K depends only on the operating conditions and general design features, except for the velocity constant K_v which has a weak dependence on a since $v = \omega_1 d_1/2 = \omega_1 a/(1 + R)$. Therefore, equation (13.26) has to be solved by iteration.

At this point, the selection of b seems arbitrary. There are two ways to assign a value to it:

(a) From design experience, it is recommended that the width factors b/a are between 0.3 and 0.4 for structurally improved gears and 0.25 and 0.315 for hardened gears, asymmetrically located on the shaft. If the gears are symmetrically located, which means better contact, the width factor b/a can be as high as 0.5. In terms of this factor, equation (13.26) can be written as

$$a^3 = M_1(1 + R)^2/2(b/a)IK \tag{13.27}$$

(b) The width b can be used as a design variable for optimization, under the constraints of the width ratio.

In cases where the center distance a is given, the material has to be selected properly to give values of b within the recommendations for the width ratio.

For fatigue strength of the tooth, the AGMA/Lewis equation is used. In terms of known parameters, it can be rewritten as

$$m = M_1 a(1 + R)K_o K_m K_v/bJSC_R C_t C_s C_f S_n \tag{13.28}$$

This equation gives the module of the gearing. The resulting number of teeth $N_1 = md_1$ must be checked to be above the interference level, 17 teeth for pressure angle 20°.

In equation (13.28) factor J depends on the number of teeth N_1 and N_2, therefore on the module. Consequently, equation (13.28) has to be solved by iteration.

Only the smallest gear is used for fatigue strength calculations, if the two gears are made of the same material. Otherwise, both gears must be used and the larger module, if different, should be adopted.

In the definition of the form factor J, distinction is made between common and precision gears. This is because only in precision gears one may assume that more than one tooth at one time are in simultaneous contact. In common gears, even if kinematically the contact ratio is greater than 1, as it always happens, owing to inaccuracies and the high rigidity of teeth, load sharing among teeth cannot be warranted.

In precision gears, even if the contact ratio is not an integer, it is accounted for because fatigue is a cumulative process and unloading of the teeth for a fraction of their loading time contributes to the fatigue life. Only in slow gearing where a small total number of cycles is expected, must the design for strength be based on yield strength. In this case, the contact ratio should be always taken 1 and, in principle, for precision gearing the integer part of the contact ratio, though precision gears are not used for slow applications.

For gearing optimization, the whole system has to be used for objective function, including shafts, bearings, couplings etc. As a first approximation, however, the volume of the gear blanks should be minimized. To this end, the objective function is set-up in the form

$$P(m, a, b) = \pi(d_1^2 + d_2^2)b/4$$

subject to the following constraints:

(i) The Lewis equation.
(ii) The Buckingham equation.
(iii) The limits on minimum number of teeth.
(iv) The limits on the width ratio b/a.

A free parameter is the material. Several feasible designs from different materials must be found first and their total cost compared.

The procedure SPURGEARS which follows, computes the main dimensions for a spur gear drive.

```
Procedure SPURGEARS(BA, SE, SHE, R, Power, RPM1, Ko, Km, Precision,
FId, A, m)
LOCAL N1, N2 integers:var
        OM1, M1, V, Kv, K, I, Apr, B, F1, F2, F3, F4, FI, J, mprev real:var
        PI:= 3.14159 real:const
GLOBAL BA, SE, SHE, R, Power, RPM1, Ko, Km, A, M, FId real:var
        Precision integer:var
begin
  FI:= FID*PI/180
  Om1:= 2*PI*RPM1/60
  M1:= 9550*Power/RPM1
  repeat
    V:= A/(1 + R)*Om1
    IF Precision:= 1 then do                          (*High precision gears*)
      Kv:= 1
    else, IF Precision:= 2 then do                    (*Ground gears*)
      Kv:= SQR((78 + sqr(V))/78)
    else, if Precision:= 3 then do                    (*Non-ground gears*)
      Kv:= (50 + SQR(V))/50
    else, end.
    K:= (SHE/Kv)^2
    I:= SIN(FI)*R/(4*(1 + R))
    Apr:= A
```

```
        A:= [M1*(1 + R)^2/(2*BA*I*K)]^(1/3)
        until ABS(A – Apr) < 1.E – 5
      B:= A*BA
      m:= 1
      repeat
        N1:= m*2*A/(1 + R)
        N2:= R*N1
        F1:= .56*(1 – .38/SQR(N2) – .88/N2)*(1 – .26/SQR(N1) – 5.5/N1)
        F2:= F1*.63/.56
        F3:= .32*(1 – 1.14/N1^.56)
        F4:= F3*.39/.32
        IF Precision:= 1 or 2 then do
          IF FId:= 20 then do
            J:= F1
          else, IF FId:= 25 then do
            J:= F2
          else, end.
        else, IF Precision:= 3 then do
          IF FId:= 20 then do
            J:= F3
          else, IF FId:= 25 then do
            J:= F4
          else, end.
        else, end.
        mprev:= m
        m:= (M1*A*(1 + R)/(B*J*Se))*Ko*Kv*Km
      until ABS(m-mprev) < 1.E – 5
                                                        (*Print results*)
    end.
```

13.9 HELICAL GEARS

In spur gears, each tooth is parallel to the axis of rotation. The involute geometry ensures a constant speed transmission ratio. However, inaccuracies in the tooth geometry, nonuniform rotation or errors in misalignment and center distance force the gearing to deviate from the perfect involute geometry. This leads to non-constant speed ratio. Abrupt variation of the speed means acceleration and dynamic forces. This reduces the life of gearing and introduces noise and vibration. To overcome this, the generatrix of the tooth is inclined in respect to the axis of rotation by an angle ψ, forming a helix on the cylindrical surface of the gear. In this way, many teeth are in contact at the same time, equalizing the errors and giving longer life and quiet operation. In addition, breakage of one tooth is not detrimental to the gearing operation because many teeth are engaged at the same time.

Looking at a helical gearing, a circular pitch p, module $m = p/\pi$ and pressure angle φ are defined at a section perpendicular to the axis of rotation in the gear plane. At a plane A–A perpendicular to the generatrix of the tooth, another tooth profile appears, called 'normal', with geometry p_n, $m_n = p_n/\pi$ and φ_n. From the system geometry (Figure 13.17),

$$F_n = F_t \cos\varphi \tag{13.29}$$

$$\tan\varphi_n = \tan\varphi\cos\psi \tag{13.30}$$

$$m_n = m\cos\psi \tag{13.31}$$

The section of the plane A–A with the cylindrical pitch surface of the gear is obviously an ellipse. At point B however, there is a circle tangent to the ellipse, the closest circle, with a radius from the theory of conic sections

$$d_n = d\cos^2\psi \tag{13.32}$$

The operation of the gearing can be considered as equivalent with a gear in the plane A–A having diameter d_n, pitch p_n, module m_n and pressure angle φ_n. The number of teeth of this hypothetical (virtual) gear is

$$N_n = d_n/P_n = d/p\cos^3\psi = N/\cos^3\psi \tag{13.33}$$

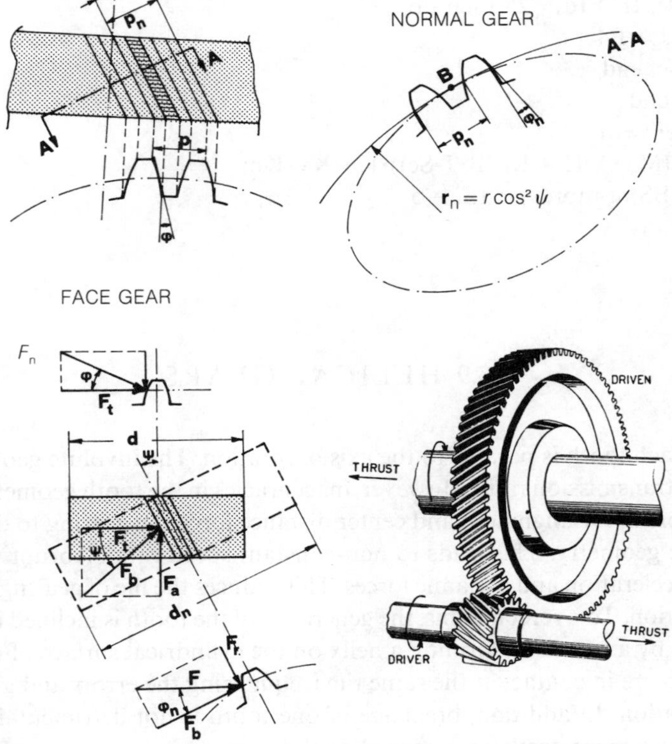

Figure 13.17 Helical gearing. (Courtesy Mobil Oil Co.)

The normal force F at the tooth contact has a space orientation and can be analyzed into a radial force F_r and a force F_D at a plane tangent to the gear cylinder surface at the point of contact (pitch point). In turn, force F_D can be analyzed into the tangential force F_t and axial force F_a. The tangential force is again,

$$F_t = P/(d/2)\omega \tag{13.34}$$

The analysis of the contact force F into the components F_r, F_t and F_a yields (Figure 13.18),

$$\left.\begin{aligned} F_r F_t \tan\varphi &= F_D \tan\varphi_n \\ F_a &= F_t \tan\psi \\ F_b &= F_t/\cos\psi \\ F &= F_b/\cos\varphi_n = F_t/\cos\psi\cos\varphi_n \\ F_r &= F_t \tan_n/\cos\psi \end{aligned}\right\} \tag{13.35}$$

Standard pitch/module are specified usually at the normal gear, because of the methods of cutting the teeth. Beam strength and surface strength is referred to the normal gear. There is one difference however. In the spur gear, the whole width of the tooth is loaded while in helical gears the situation is more complicated. The greater the helix angle ψ, the greater number of helical teeth are in contact. Therefore, the number of teeth in contact must be

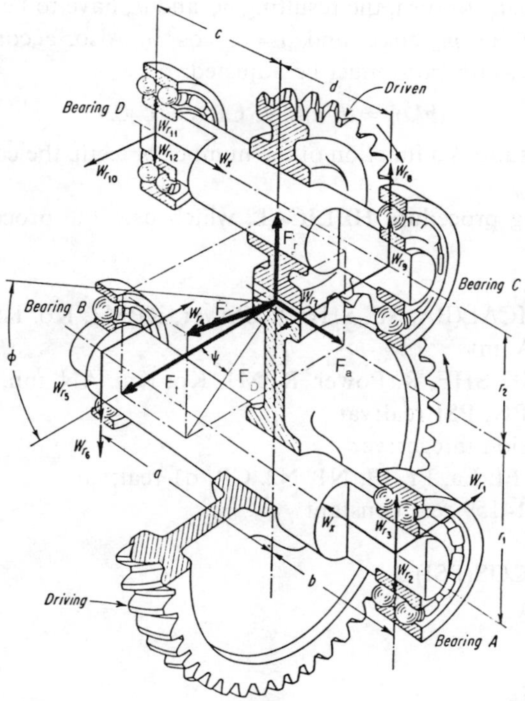

Figure 13.18 Forces in helical gear drives. (Dudley 1962)

proportional to the contact ratio (CR) and inversely proportional to $\cos\psi$. Therefore, the design equations for helical gear will be:

Beam strength:

$$\frac{F_t}{\cos\psi\, bJm} K_v K_o K_m \leqslant S'_n C_s C_f C_r C_t C_n \tag{13.36}$$

Contact fatigue strength:

$$\sigma_H = C_P [F_t K_v K_o K_m (\cos\psi/\mathrm{CR})/bdI \cos\psi]^{\frac{1}{2}} \leqslant S_H C_L C_H C_T C_R \tag{13.37}$$

AGMA suggests a 7% reduction of the mounting factor K_m and 5% reduction of the contact ratio CR owing to better performance of helical gearing to mounting and profile errors. They also issue standards with recommended geometry factors. The design procedure is similar to the spur gear design. In fact, one can use the same procedure with a new definition of the design parameters.

To this end, the tangential force F_t must be such that it will equal F_t of the normal gear. Since the tangential force is proportional to the torque and inversely proportional to diameter,

$$F_{tn} = F_t/\cos\psi = 2M_1/d_1 \cos^3\psi$$

Therefore, if the power P, proportional to M_1 is divided by $\cos^3\psi$, it will yield with the dimensions of the normal gear the right tangential force for the computation of the module and the center distance a_n. In turn, the resulting m_n and a_n have to be transformed into the helical gear quantities $m = m_n/\cos\psi$ and $a = a_n/\cos^2\psi$. Also, according to the AGMA recommendations, the width ratio must be adjusted:

$$(b/a)_n = (b/a)(0.95\,\mathrm{CR}/\cos\psi)0.93$$

Since the contact ratio is a function of the number of teeth, the computation should be iterative.

The corresponding procedure HELICAL, which uses the procedure SPURGEARS, follows:

```
Procedure HELICAL(BA, SE, SHE, R, Power, RPM1, Ko, Km,
Precision, FId, A, m)
GLOBAL BA, SE, SHE, R, Power, RPM1, Ko, Km, AN, mn,
         FId, PN, PSI real:var
         Precision integer:var
LOCAL mprev, Ft, Fa, Fr, FI, N1, N2, CR, d1 real:var
         PI:= 3.14159 real:constant
begin
  PN:= Power/COS(PSI)^3
  Km:= .97*Km
  N1:= 20
  N2:= R*20
  FI:= FId*PI/180
```

```
m:= .999
repeat
  CR:= [SQR((N2 + 2)^2 – N2^2*COS(FI)^2) – N2*SIN(FI)
        + SQR((N1 + 2)^2 – N1^2*COS(FI)^2) – N1*SIN(FI)]/(2*PI)
  BA:= BA*.95*CR/COS(PSI)
  mprev:= m
  SPURGEARS(BA, SE, SHE, R, PN, RPM1, Ko, Km,
  Precision, FId, AN, mn)
  A:= AN/COS(PSI)^2                                  (*Face gear*)
  m:= mn/COS(PSI)
  d1:= 2*A/(1 + R)
  N1:= m*d1
  N2:= N1*R
until ABS(m-mprev) < 0.01

Ft:= Power*2/(9550*RPM1*d1)               (*Computation of forces*)
Fr:= Ft*TAN(FI)                                      (*Tangential*)
Fa:= Ft*TAN(PSI)                                         (*Radial*)
end.                                                      (*Axial*)
```

13.10 BEVEL GEARS

Bevel gears are used to transmit rotation between intersecting or nearly intersecting shafts. They may have straight, skew, spiral or other curvilinear teeth (Figure 13.19).

Straight bevel gears are used for low peripheral velocities up to 2 or 3 m/s. For higher velocities curved-tooth bevel gears are used because the gears mesh more smoothly and they have a higher load-carrying capacity.

Bevel gears intersecting at an angle γ operate by rolling of two pitch cones with angles γ_1 and γ_2 for the pinion and gear respectively with

$$\gamma_1 + \gamma_2 = \gamma \tag{13.38}$$

Each of the two gears is formed by the pitch cone of angle γ_i, $i = 1$ or 2, and front and back cones intersecting with the pitch cone at right angles, having the same axis of symmetry and angle $\pi/2 - \gamma_i$.

Again, operation of the straight tooth can be modelled (Figure 13.20), by assigning a virtual spur gear having the same tooth center on the axis of symmetry.

The tooth of a bevel gear has variable thickness because of the geometry. The virtual spur gear has teeth of the same shape of the bevel gear teeth at some location. Usually, the shape of the tooth at the back cone (larger thickness) is standard and the strength calculations are performed with a virtual spur gear with tooth thickness equal to the one of the bevel gear at the middle of the width of the tooth. The basic geometric relationships are for the pinion gear 1:

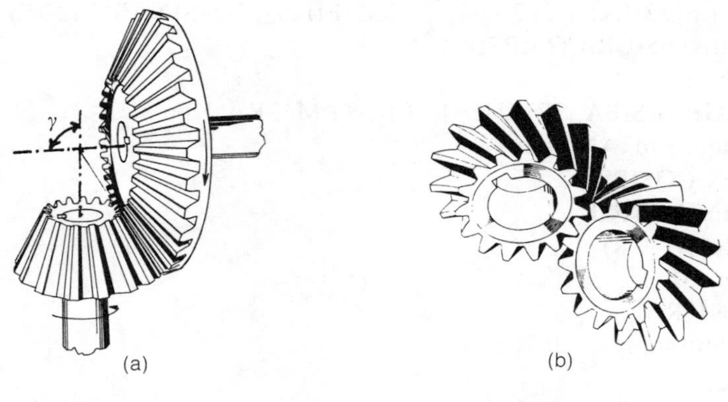

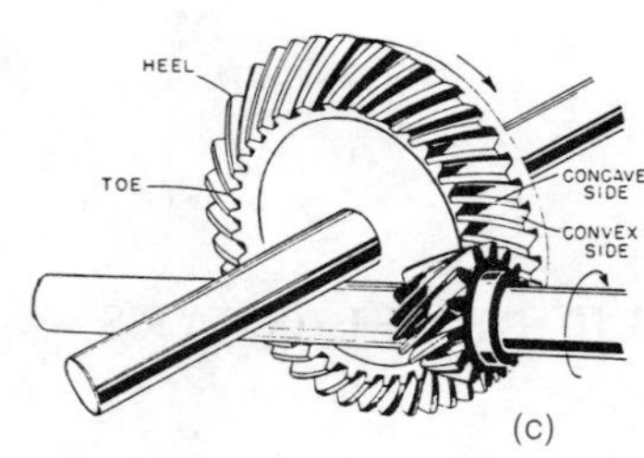

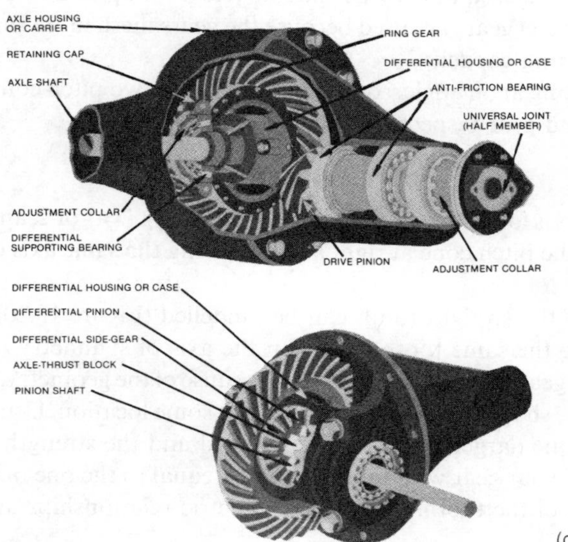

Figure 13.19 Conical gear drives. (Courtesy Mobil Oil Co.)

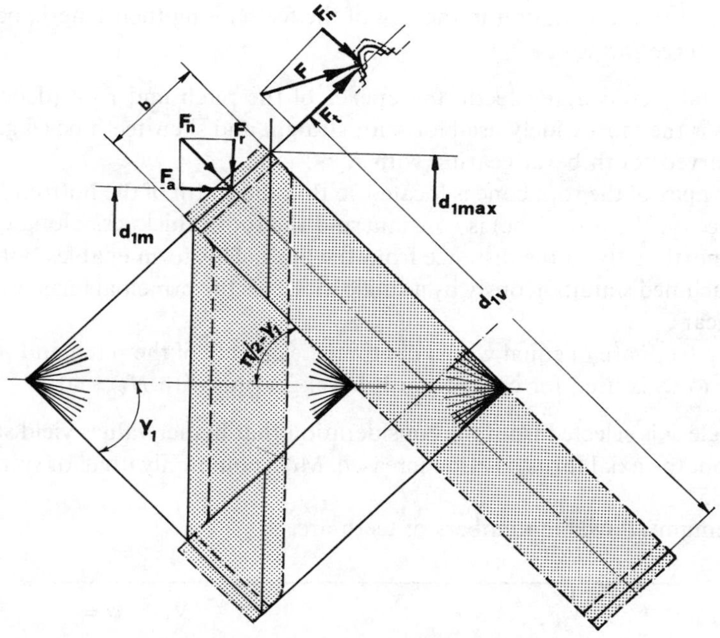

Figure 13.20 Geometry of conical gearing

	Bevel gear	Virtual gear
Maximum pitch diameter	d_1	$d_{1v} = d_1/\cos\gamma$
Pitch at maximum pitch diameter	p_1	p_1
Module at maximum pitch diameter	m_1	m_1
Mean diameter	$d_{1m} = d_1 - b\sin\gamma_1$	$d_{1v} = d_{1m}/\cos\gamma_1$
Pitch at mean diameter	p_{1m}	p_{1m}
Module at mean diameter	m_{1m}	m_{1m}
Width	b	b
Number of teeth	$N_1 = d_{1m}/m_{1m}$	$N_{1v} = N_1/\cos\gamma_1$
Pressure angle	φ	φ

The basic force relationships are, at the mean diameter,

	Bevel gear	Virtual gear
Tangential force	$F_t = P/(d_{1m}/2)\omega_1$	F_t
Transmitted force	$F = F_t/\cos\varphi$	F
Normal force	$F_n = F\sin\varphi = F_t\tan\varphi$	F_n
Axial force	$F_a = F_n\sin\gamma_1 = F_t\tan\varphi\sin\gamma_1$	0
Radial force	$F_r = F_n\cos\gamma_1 = F_t\tan\varphi\cos\gamma_1$	F_n

With respect to the variation in the size of the teeth along their length, bevel gears are classified into three forms:

Form I. Normally converging teeth; the apexes of the pitch and root (dedendum) cones coincide. This is the most widely used form for straight and skew teeth bevel gears. It is also applied for curved-tooth bevel gearing with $N_1 < 25$.
Form II. The apex of the root cone is located so that the width of the bottom land (width of the tooth space on the root cone) is constant and the tooth thickness along the pitch cone increases proportionally to the distance from the apex. This form enables both sides of the teeth to be machined simultaneously by a single tool. It is the principal form used for curved-tooth bevel gears.
Form III. The teeth are of equal whole depth; the elements of the pitch and root cones are parallel. This form is used for curved-tooth bevel gearing with $N_2 > 40$.

The spiral angle ψ is selected from the consideration that higher values yield smoother gear engagement but the axial thrust is also increased. Most commonly used for spiral bevel gears is $20° < \psi < 35°$.

The minimum allowable numbers of teeth are:

	N_1 at $\psi =$		
N_2/N_1	0° to 15°	20° to 25°	30° to 40°
1	17	17	17
1.6	15	15	14
2	13	12	11
>3.15	12	10	8

The load is non-uniformly distributed along the length of the teeth in bevel gearing. The elastic deflections at various sections of the teeth are not the same but are proportional to the distance from the axes of the gears or from the common apex of the pitch cones.

At the same time, the specific rigidity of geometrically similar teeth does not depend on their linear dimensions (module) since the effect of increasing the arm of the load is compensated by the increase in the height of the tooth cross-section. It is usually assumed, therefore, that the unit load is distributed along the length of straight or skew teeth of bevel gears proportionally to the elastic deflections, i.e. according to a trapezoid.

The point of application of the resultant load (the center of gravity of the trapezoidal load diagram) is displaced from the middle of the length of the teeth toward the large end of the teeth. However, to simplify calculations, they are carried out for the middle cross-section (located at the middle along the length of the teeth). This increases the factor of safety to some extent.

The design equation will be for the beam strength,

$$\frac{F_t}{m_{1v}bJ} K_v K_o K_m \leqslant S_n C_s C_f C_r C_t C_m \tag{13.39}$$

Geometry factor J and all the derating factors must be selected on the basis of the diameter d_{1mv} and the number of teeth N_{1v} of the virtual spur gear. AGMA has issued standards for the values of the geometry factor for several pressure, shaft and spiral angles. For the last, i.e. for spiral bevel gears, the virtual gear is a helical one and the force relationships are:

$$F_a = F_t(\tan\varphi_n \sin\gamma_1 \pm \sin\psi\cos\gamma_1)/\cos\psi \qquad (13.40)$$
$$F_r = F_t(\tan\varphi_n \cos\gamma_1 + \sin\psi\sin\gamma_1)/\cos\psi$$

where ψ is the helix angle. The + sign applies to a driving pinion with right-hand spiral rotating clockwise or left-hand spiral rotating counterclockwise, as viewed from the end cone.

Again, $\tan\varphi_n = \tan\varphi\cos\psi$.

The design equation for fatigue surface strength will be again,

$$\sigma_H = C_P[F_t K_v K_o K_m / b d_{1v} I_v]^{\frac{1}{2}} \leqslant S_H C_L C_H C_T C_R \qquad (13.41)$$

The two equations are sufficient to yield the gear width b and the module m_{1v}. The module at the maximum pitch diameter (on the back cone) is

$$m_1 = m_{1v} d_1/d_{1m} = m_{1v} d_1/(d_1 - b\sin\gamma_1) \qquad (13.42)$$

which is usually assigned a standard value and then all dimensions have to be adjusted to the exact value of the maximum module m_1.

AGMA suggests different values for the tooth geometry factors I and J which can be approximated as

$$J_{\text{bevel}} = J_{\text{spur}}(100 + N_2)240$$
$$I_{\text{bevel}} = I_{\text{spur}}(1 - 100/N_1^2)1.04$$

Since the computation for spur, helical and bevel gears is similar, these procedures have been coded in program GEARDES, Appendix 13.B.

13.11 CROSSED GEARING

In gearing with intersecting shafts and bevel gears one of the gears is mounted on an overhanging shaft. This is not a sound design because it is associated with high deflection and wear of the gear. This can be corrected with gearing with non-parallel and non-intersecting shafts because both shafts can extend beyond the gearing in both directions. Smoothness of operation is typical of such gearing, but the gears also can have higher sliding velocities, wear and friction losses.

In crossed helical gearing (Figure 13.21), the teeth have initial point contact under conditions of considerable sliding velocities. Therefore, the load-carrying capacity of this gearing is limited.

The diameters of the pitch and rolling cylinders of uncorrected gears (Figure 13.21) are

$$d_1 = m_{t1} N_1 = m_n N_1/\cos\psi_1 \quad \text{and} \quad d_2 = m_{t2} N_2 = m_n N_2/\cos\psi_2 \qquad (13.43)$$

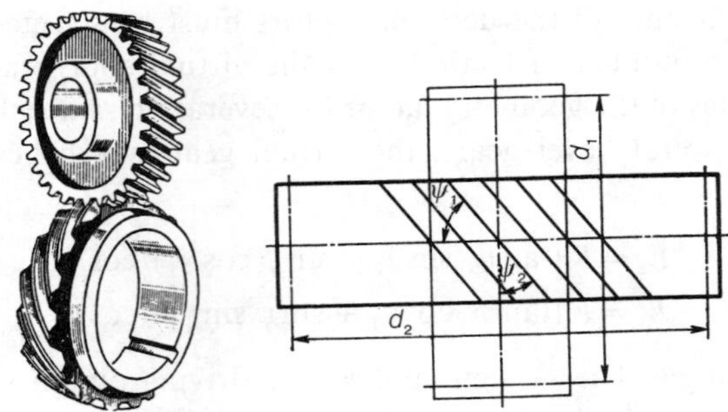

Figure 13.21 Crossed helical gearing

where m_{t1} and m_{t2} = face modules of the pinion and gear, and ψ_1 and ψ_2 = helix angles of the pinion and gear teeth.

The distance of the axes is

$$a = 0.5(d_1 + d_2)$$

The speed ratio is

$$R = n_1/n_2 = \omega_1/\omega_2 = N_2/N_1$$

Expressing N_1 and N_2 in terms of the pitch diameters, we obtain

$$R = (d_2/d_1) \tan \psi_1 \tag{13.44}$$

In power drives requiring a sufficiently high efficiency, the helix angles are selected equal or approximately the same. In certain auxiliary gearing, particularly in some instruments, it proves convenient to have a pinion and gear with the same pitch diameter. Then the required speed ratio is obtained by proper selection of the helix angles of the teeth.

Hypoid gearing consists of spiral bevel gears with crossed axes (Figure 13.19c). The speed ratio of such gears is usually in the range from 2 to 10. In addition to the mentioned advantages of gearing with non-parallel, non-intersecting axes, hypoid gearing also has a high load-carrying capacity. This is owing, primarily, to the fact that in hypoid gearings, in contrast to crossed helical gearings, tooth contact close to linear contact is achieved with optimal size and shape of the tooth bearing contact pattern. In this respect, hypoid gearing is similar to curved-tooth bevel gearing. The sliding velocity in hypoid gearing is substantially less than in crossed helical gearing. For the same diameter of gear and speed ratio, the diameter of a hypoid pinion is larger than in bevel gearing. Besides, the mating teeth in hypoid gearing run-in well and are not subject to appreciable distortion because the sliding action along the working surfaces is sufficiently uniform. Since several pairs of teeth are in mesh simultaneously, hypoid gearing can be employed in high-precision mechanisms, for example in the indexing gear trains of precision gear-cutting machines.

Hypoid gearing has found wide application in automobiles and other transportation machinery.

Contact stress and beam strength calculations for hypoid gearing can be carried out as for bevel gearing having the same diameters, face widths and large end modules.

13.12 WORM GEARS

Worm gearing (Figures 13.22 and 13.23), is one of the oldest gear drives, invented by Archimedes. It consists of a worm, i.e. a screw with trapezoidal, or approximately trapezoidal, threads, and a worm gear which is a toothed gear with teeth of special shape obtained by geometric interaction with the threads of the worm.

Worm gearing is one kind of crossed helical gearing. Worm gearing however, has initial contact along a curved line, giving better load distribution and higher strength.

Other advantages are: (a) high-speed reduction, (b) smooth and silent operation, (c) small drive size, many times 1/10th of ordinary gear drives.

Disadvantages of worm gear drives are: (a) low efficiency and (b) expensive antifriction materials necessary for the wormgear.

The speed ratio R of worm gearing is determined from the condition that for each revolution of the worm, the wormgear turns through a number of teeth equal to the number of starts, or threads, N_1 on the worm. Thus,

$$R = n_1/n_2 = \omega_1/\omega_2 = N_2/N_1 \tag{13.45}$$

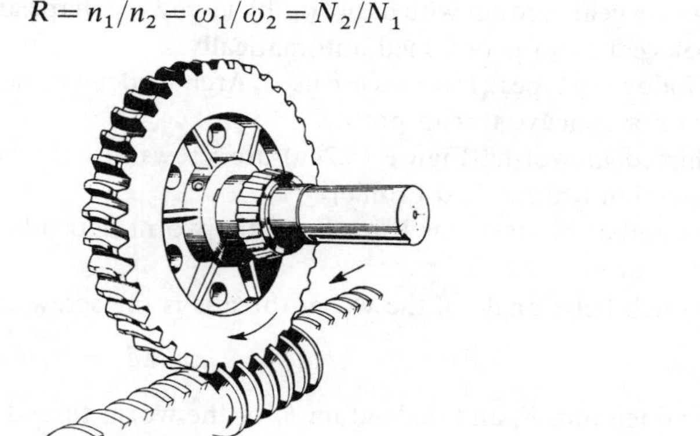

Figure 13.22 Worm-gear system. (Courtesy Mobil Oil Co.)

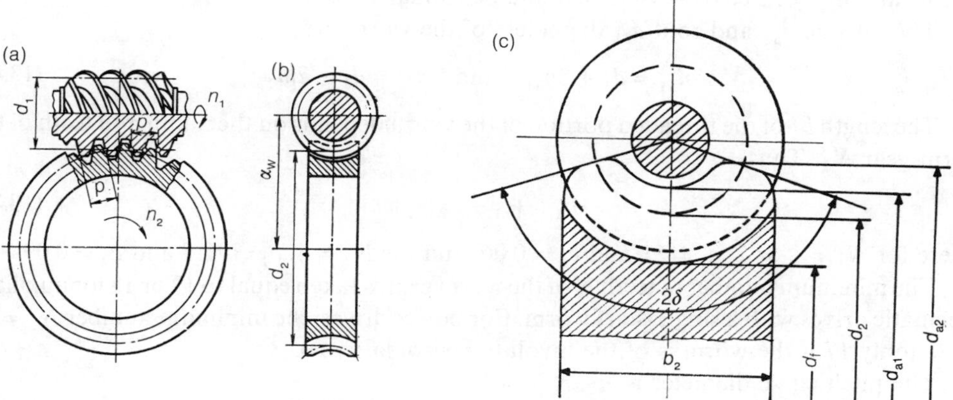

Figure 13.23 Geometry of the worm-gear

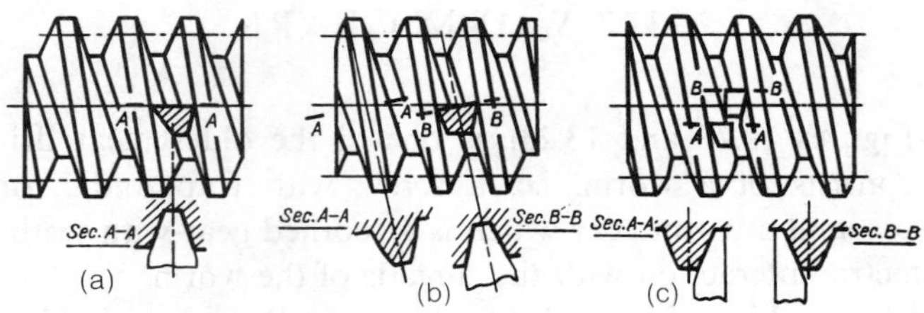

Figure 13.24 Geometry of worm thread

It must be pointed out that now the speed ratio is no longer equal to the ratio of the diameters.

Owing to low efficiency, worm gear drives are used for low and medium power transmission up to 50 kW, and for torques up to 500 kNm. The speed ratio usually ranges from 8 to 100 and, in special cases, up to 1000.

Geometrical calculations for worm gearing are similar to ones for ordinary toothed gearing.

The worm gears are cut with a tool (hob) having the shape and size of the worm, therefore the proper gear form is obtained automatically.

The following types of worms are used: Archimedean basic rack, involute helicoidal and worms with a concave thread profile.

Archimedean worms (Figure 13.24a), are screws with threads which are straight-sided in an axial section (trapezoidal profile).

The number of starts, or threads, of the worm depends upon the speed ratio and is usually 1, 2 or 4.

The pitch helix angle of the worm threads is the screw angle

$$\tan \psi = h/\pi d_1 = N_1 \pi m/\pi d_1 = N_1 m/d_1 \tag{13.46}$$

The addendum h_a and dedendum h_f of the worm threads are

$$h_{a1} = h_{a1}^* m \quad \text{and} \quad h_{f1} = h_{f1}^* m \tag{13.47}$$

where the addendum factor $h_{a1}^* = 1$ and the dedendum factor $h_{f1}^* = 1.2$ for Archimedean worms and $h_{f1}^* = 2.2 \cos \psi - 1$ for involute helicoidal worms.

The outside d_{a1} and root d_{f1} diameters of the worm are

$$d_{a1} = d_1 + 2h_{a1} \quad \text{and} \quad d_{f1} = d_1 - 2h_{f1} \tag{13.48}$$

The length b_1 of the threaded portion of the worm depends on the number of teeth of the worm gear N_2. Thus,

$$b_1 > (c_1 + c_2 N_2)m \tag{13.49}$$

where for $N_1 = 1$ or 2, $c_1 = 11$ and $c_2 = 0.06$; and for $N_1 = 4$, $c_1 = 12.5$ and $c_2 = 0.09$.

The minimum number of teeth N of the worm gear is taken equal to 17 or 18 for auxiliary kinematic drives with a single-start worm. For power drives, the minimum number $N_2 = 26$ to 28 (only 17 if the worm is of the involute helicoidal type).

The pitch circle diameter is

$$d_2 = mN_2 \tag{13.50}$$

The throat diameter d_{a2} and the root diameter d_{f2} of the worm gear are

$$d_{a2} = d_2 + 2h_{a1} \quad \text{and} \quad d_{f2} = d_2 - 2h_{f1} \tag{13.51}$$

In multiple-start drives, the effective field of engagement is less than for single-start worm gearing and therefore the outside diameter and face width of the worm gear is taken less than for the corresponding single-start drive (with the same values of d_2, d_{a2} and m). The maximum outside diameter should be

$$d_{am2} < d_{a2} + 6m/(N_1 + 2) \tag{13.52}$$

The face width of the worm gear is assigned in accordance with the outside diameter of the worm (Figure 13.23c): for $N_1 = 1$ or 2, $b_2 < 0.75d_{a1}$; for $N_1 = 4$, $b_2 < 0.67d_{a1}$.

The conditional worm gear face angle 2δ, used in strength calculations, is found from the points of intersection of a circle of diameter $d_{a1} - 0.5m$ with the end faces (contour) of the worm gear. Thus,

$$\sin\delta = \frac{b_2}{d_{a1} - 0.5m} \tag{13.53}$$

The center distance is equal to one-half of the sum of pitch diameters of the worm and worm gear. Thus,

$$a = (d_1 + d_2)/2 = (N_1 + N_2)m/2 \tag{13.54}$$

In designing worm gearing, it is necessary to plan for axial adjustment of the worm gear in assembly so that it can be made to coincide with the axial plane of the worm.

In contrast to rolling of spur gear surfaces, worm gears work primarily on sliding contact. To increase the load-carrying capacity, a geometry providing an oil wedge, as in the bearings, must be selected.

Sliding cannot be totally avoided, therefore the principal causes of worm gearing failure are surface damages, tooth wear and seizing.

Seizing is especially dangerous if the worm gear is made of hard materials such as hard bronzes or cast iron. In this case, seizing is severe, with damage to the surfaces followed by rapid wear of the teeth by particles of the worm gear material welded onto the threads of the worm. With worm gears of softer materials, seizing of a less dangerous type is observed: the material of the worm gear (bronze) is 'smeared' on the worm.

Fatigue pitting is observed mainly in gearing with a worm gear made of a seizure-resistant bronze. Such pitting, as a rule, occurs only on the worm gear.

Breakage can be observed mainly after severe wear and usually only the teeth of the worm gear are broken.

In power drives, the worms are made, as a rule, of steel heat-treated to considerable hardness. The most durable gearing has the worm made of a case-hardening steel having a hardness of 56–63 R after heat-treatment. Also extensively used are worms of medium-carbon steels surface or through hardened to 45–55 R.

Force calculations for the worm are based on the screw equations while for the worm gear the situation is similar to the one for helical gears.

Owing to geometry if the worm and gear axes are perpendicular, (Figure 13.26), the

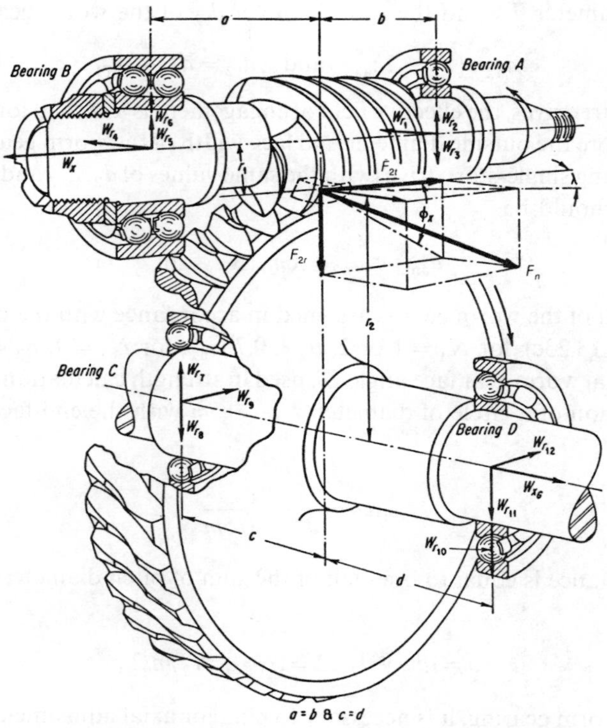

Figure 13.25 Forces in worm-gear systems. (Dudley 1962)

tangential force of the one equals the axial force of the other. Therefore (Figure 13.25),

$$\left.\begin{aligned} F_{2\mathrm{a}} &= F_{1\mathrm{t}} = P_1/(d_1/2)\omega_1 \\ F_{2\mathrm{t}} &= F_{1\mathrm{a}} = P_2/(d_2/2)\omega_2 \end{aligned}\right\} \tag{13.55}$$

To compute the other forces, the friction force must be taken into account. It is proportional to the normal force F_{n} with a direction opposing the sliding direction. For coefficient of friction f,

$$\left.\begin{aligned} F_{2\mathrm{t}} &= F_{1\mathrm{a}} = F_{\mathrm{n}} \cos\varphi_{\mathrm{n}} \cos\psi - fF_{\mathrm{n}} \sin\psi \\ F_{1\mathrm{t}} &= F_{2\mathrm{a}} = F_{\mathrm{n}} \cos\varphi_{\mathrm{n}} \sin\psi + fF_{\mathrm{n}} \cos\psi \end{aligned}\right\} \tag{13.56}$$

$$F_{1\mathrm{r}} = F_{2\mathrm{r}} = F_{\mathrm{n}} \sin\varphi_{\mathrm{n}}$$

Therefore,

$$\frac{F_{2\mathrm{t}}}{F_{1\mathrm{t}}} = \frac{\cos\varphi_{\mathrm{n}} \cos\psi - f \sin\psi}{\cos\varphi_{\mathrm{n}} \sin\psi + f \cos\psi} \tag{13.57}$$

From Figure 13.26, the sliding velocity u_{s} is analyzed into the worm and gear tangential velocities u_1 and u_2 respectively. It is $\tan\psi = u_2/u_1$.

The efficiency of the system is the ratio of the worm tangential force without friction to

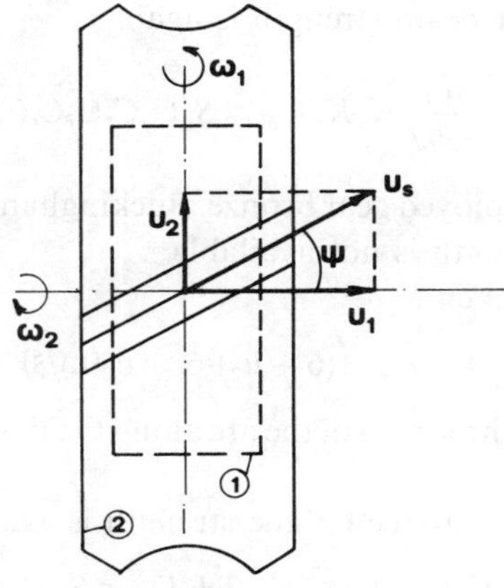

Figure 13.26 Kinematics of worm-gears

the one with friction. Therefore

$$n = \frac{\cos \varphi_n \sin \psi}{\cos \varphi_n \sin \psi + f \cos \psi} \tag{13.58}$$

The coefficient of friction f depends on sliding velocity. For usual applications AGMA recommends values

$$f = 0.123[1 - 0.23 \log_{10}(197 u_s)] \tag{13.59}$$

where $u_s = u_1 \cos \psi$ (m/s). Therefore f depends on the helix angle

$$f = 0.123[1 - 0.28 \log_{10}(197 u_1 \cos \psi)] \tag{13.60}$$

This suggests also that the helix angle plays an important role on efficiency. Optimization yields values for optimum ψ on about 40° but the improvement in efficiency is small above $\psi = 25°$. Since lower ψ yields higher gear ratio, values of between 25° and 30° are usually selected.

If the gear is driving, the efficiency will be the ratio of F without and with friction:

$$n' = \frac{\cos \varphi_n \cos \psi}{\cos \varphi_n \cos \psi - f \sin \psi} \tag{13.61}$$

For drives of reverse driving, conditions of self-locking schould not exist, i.e. ψ must be large locked. For a given coefficient of friction the helix angle must have small values if self-locking is desired. This is the case, for example, in drives where safety demands it, such as in elevators. For drives of reverse driving, conditions of self-locking should not exist, i.e. ψ must be large enough.

Strength of worm gearing follows the same lines as for helical gears, with emphasis on surface fatigue strength.

The worm is usually much stronger than the gear and calculations are performed on the worm gear.

The design equation for beam strength is again

$$\frac{F_t}{mbJ} K_v K_o K_m \leqslant S_n C_c C_f C_r C_t C_m \tag{13.62}$$

For the most usually employed gear bronze, Buckingham suggested the value of 24 ksi or 170 N/mm^2, if material strength is not available.

The velocity factor is taken as

$$K_v = (6 + u_2)/6, \quad u_2(\text{m/s}) \tag{13.63}$$

The face width b is here the length of the arc along the line of contact on the pitch cylinder of the worm.

The design equation for surface fatigue strength is again

$$\sigma_H = C_P[F_{2t} K_v K_o K_m / bd_2 I]^{\frac{1}{2}} < S_H C_L C_H C_T C_R \tag{13.64}$$

Design of worms is a little different than the other gears. Strength calculations are based primarily on the gear since the high sliding inherently present requires antifriction materials which have relatively low strength.

The diameters are not in immediate relation to the gear ratio. Therefore, the CAD procedures for the other gears are not applicable.

The design procedure starts with an estimate of the ratio (b/d_2) to be optimized later. In view of this ratio, the Buckingham/AGMA equation, applied to the gear, yields upon solution for d_2,

$$d_2^3 = 2M_1 K_o K_m K_v / R(b/d_2) I (S_H C_L C_H C_t C_R / C_P)^2 \tag{13.65}$$

The Lewis/AGMA equation for fatigue strength of the gear is solved also for d,

$$d_2^3 = 2M_1 N_1 K_v K_o K_m / R(b/d_2) J S_n C_s C_f C_r C_t C_m \tag{13.66}$$

Obviously, the larger of the two values is selected.

The module is now $m = d_2/N_2$, while $N_2 = RN_1$; N_1 is the number of starts of the worm. Again, the normal module is adjusted to available sizes.

The diameter of the worm cannot be computed from the results available up to this point. Usually, it is computed from considerations of strength of the carrying shaft, since the forces can be computed with the gear geometry already computed. For preliminary design calculations, a ratio d_2/d_1 between 3 and 6 is suggested. Now, the face width to the worm diameter ratio can be computed. In general, it will not be within the limits given above. In this case, an iteration needs to be performed, modifying the (b/d_2) ratio until an acceptable (b/d_1) ratio is achieved. This iteration is necessary also for the reason that some of the parameters are functions of the results and need to be adjusted along the computation. Rough initial values are chosen for these parameters.

The coefficient of friction is recommended by AGMA to be a function of the sliding velocity, which can be approximated as, equation (13.60),

$$f = 0.123[1 - 0.28 \log_{10}(197 u_1 \cos\psi)] \tag{13.67}$$

The procedure is similar to the ones for the other types of gear and it has been included in program WORMDES in Appendix 13.C.

13.13 HEAT CAPACITY AND DESIGN OF WORM-GEAR DRIVES

Worm gearing operates at relatively low efficiency and generates a substantial amount of heat. But the heating of the oil to a temperature exceeding the limiting value $T = 95\,°C$ leads to the danger of seizing and rapid wear.

The amount of heat by worm gearing operating continuously with an efficiency η and transmitting the power P is

$$W_H = (1 - \eta)P \tag{13.68}$$

The heat removed by convection from the free surface of the housing and by conduction through the foundation plate or frame is

$$W'_H = h_t(T - T_0)A(1 + \xi) \tag{13.69}$$

where $A =$ free surface of the housing from which heat is removed to cool the drive (included is 50% of the surface of the fins) (m^2), T and $T_0 =$ temperatures of the oil and of the surrounding air (°C), $h_t =$ heat transfer coefficient, and $\xi =$ factor taking into account heat transfer to the foundation plate or frame of the machine and amounting up to 0.3 when the housing seating surface is large.

The heat balance, $W_H = W'_H$, can be employed to find the working temperature, or the power P that can be transmitted by the worm gearing, complying with the condition that the oil temperature does not exceed T_{max}. Thus,

$$\begin{aligned} T &= T_0 + (1 - \eta)P/h_t A(1 + \xi) \\ P &= h_t(T_{max} - T_0)A(1 + \xi)/(1 - \eta) \end{aligned} \tag{13.70}$$

If the equilibrium temperature is high, additional cooling surface must be provided. This is done by adding fins to the housing, mechanical ventilation, coils with cooling fluid in the oil bath, or by cooling facilities.

The fins are designed for adequate air flow over them. Owing to the fact that heated air rises, fins should be arranged vertically for natural ventilation. For mechanical ventilation, the fins are arranged along the stream of air blown by the fan, usually horizontally.

Mechanical ventilation is accomplished by a fan mounted on the worm shaft. Air cooling is much simpler and less expensive than water cooling and is consequently more widely employed.

EXAMPLES

Example 13.1

A spur gear reducer was measured to have a pinion with 20 teeth and external diameter 110 mm and a gear with 40 teeth and 210 mm external diameter. The distance of the centers was measured to be 150 mm. It is supposed that the gears have standard involute teeth. Determine the module, pitch diameters, the contact ratio and if interference will be observed.

Solution

For standard teeth, the addendum circle has diameter $d + 2m$. For the pinion,

$$d_1 + 2m = 110$$

Because $d_1 = mN_1$, $mN_1 + 2m = 110$. Therefore, $m = 110/(N_1 + 2) = 5$ mm.
The pitch diameters $d_1 = mN_1 = 100$ mm, $d_2 = mN_2 = 200$ mm.
The contact ratio for $\varphi = 20°$, from equation (13.3),

$$m_c = [[(40+2)^2 - 40^2 \cos^2 20°]^{\frac{1}{2}} - 40 \sin 20° + [(20+2)^2 - 20^2 \cos^2 20°]^{\frac{1}{2}} - 20 \sin 20]/5\pi = 1.615$$

From equation (13.5), the minimum number of teeth to avoid interference, the most unfavorable case, is 17, for $\varphi = 20°$ pressure angle. Therefore, no interference will be encountered.

Example 13.2

The speed reducer shown has two standard spur gears of 20° involute angle, transmits power 25 kW at 1500 rpm and a gear speed 500 rpm. The center distance of the two shafts was measured, approximately, to be 320 mm. Determine the forces loading the shafts and the bearings owing to the gearing.

Solution

The gear ratio is $R = 1500/500 = 3$. Therefore, the pinion diameter is $d_1 = 2a/(1 + R) = 160$ mm. The torque is $M_1 = 9550P/n_1 = 159.16$ Nm. The tangential force $F_t = 2M_1/d_1 = 3980$ N.

The radial force is $F_r = F_t \tan 20° = 1448$ N. Owing to symmetry, these forces which are the same for pinion and gear are equally divided at the bearings. The bearing forces are then indicated in Figure E13.2.

Example 13.3

Plot the faces of two meshing gears of standard 20° involute profile, having a module 5 mm and number of teeth $N_1 = 17$, $N_2 = 34$.

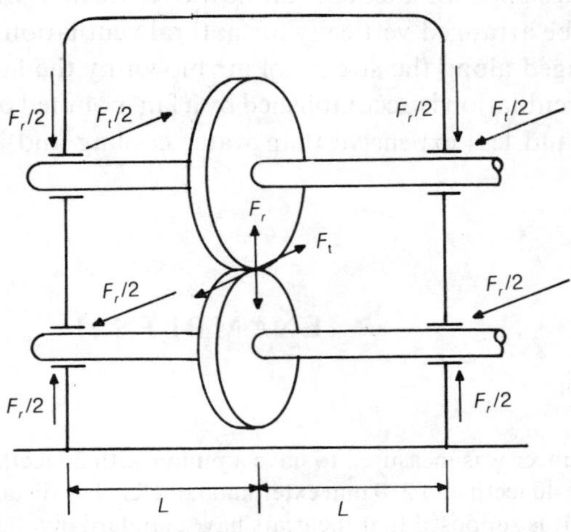

Figure E13.2

Solution
The program GEARPLOT is used. The resulting plot is shown in Figure E13.3.

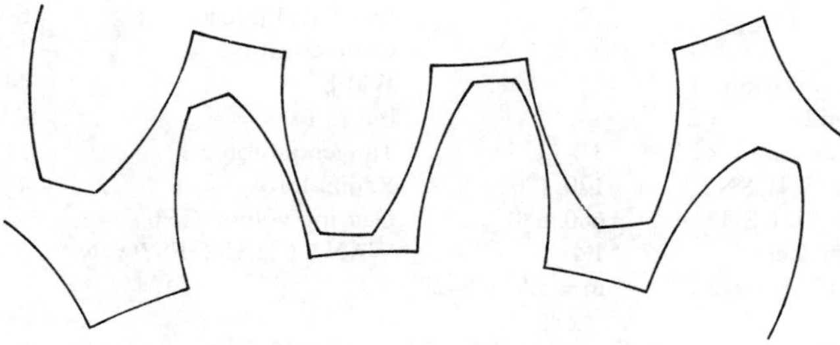

Figure E13.3

Example 13.4
An electric motor of rated power 12 kW at 1785 rpm runs an air blower at 714 rpm with one stage of standard 20 involute spur gearing. Ninety percent reliability will suffice and no shocks are expected in the motor with light shocks in the blower. The gears are to last for at least 1 million revolutions at a temperature only slightly exceeding the one of the environment. The gears will be made of AISI 1015 steel surface hardened to 270 BHN and they will be machined but not ground. Design the proper gearing.

Solution
The gear ratio is $R = 1785/714 = 2.5$. From Chapter 6, AISI 1015 carbon steel has a fatigue strength of 170 MPa. For a 270 BHN, the contact fatigue strength is (Table 13.4)

$$S_H = 2.8 \times 270 - 70 = 680 \text{ MPa}$$

The calculations are performed by the program GEARDES, Appendix 13.B. The interactive computer session follows:

```
GEAR DESIGN PROGRAM

Menu:

SPUR GEARS         1
HELICAL GEARS      2
BEVEL GEARS        3

Your selection? 1              Spur gears

DATA                                          RESULTS                  Pinion     Gear
                                              ------------------------------------------
Driving power, kW                 12          No of teeth              28         70
Pinion speed, RPM                 1785        Diameter                 112        280
Gear ratio                        2.5         BHN                      220        220
No of cycles                      1E6         Young mod                210000     210000
Pinion BHN, Young mod E           220, 2.1E5  Fatigue strength         170        170
Gear BHN, Young mod E             220, 2.1E5  Surface strength         680        680
Operating Temperature, deg C      70          SYSTEM PARAMETERS:
Involute angle (14.5, 20, 25)     20          Center distance          196
Service Factor Ko                 1.25        Modul                    4
```

Width/a ratio	.25	Pressure angle	20
Reliability %	90	Diametral pitch	6.35
Precision:		Contact ratio	1.62033522
High precision ⟨1⟩		Width	44.4592355
Ground ⟨2⟩		b/a ratio	.25
Non-ground ⟨3⟩	3	Tangential force	1146.45858
Fatigue strength SN1, SN2	170, 170	Radial force	417.047059
Surface strength SH1, SH2	680, 680	Gearing volume/1E6	3.17399106
Cycle asymmetry factor	1.4	WANT CHANGES?(Y/N)	N
Iterating...A = 177.836942	m = 3.57879427		

The 28 teeth of the gear result in operation without interference. The contact ratio is 1.6, in the acceptable range.

The overall design is judged acceptable.

Example 13.5

An electric motor of rated power 12 kW at 1785 rpm runs an air blower at 714 rpm with one stage of standard 20 involute spur gearing. Ninety percent reliability will suffice and no shocks are expected in the motor with light shocks in the blower. The gears are to last for at least 1 million revolutions at a temperature only slightly exceeding that of the environment. The gears will be made of AISI 1015 steel surface hardened to 270 BHN and they will be machined but not ground. Design a helical gearing of 30° helix angle. Compare the resulting gearing with the spur gearing of Example 13.4 which was designed for the same application.

Solution

The gear ratio is $R = 1785/714 = 2.5$. From Chapter 6, AISI 1015 carbon steel has a fatigue strength of 170 MPa. For a 270 BHN, the contact fatigue strength is (Table 13.4),

$$S_H = 2.8 \times 270 - 70 = 680 \text{ MPa}$$

The calculations are performed by the program GEARDES, Appendix 13.B. The interactive computer session follows:

GEAR DESIGN PROGRAM

Menu:

SPUR GEARS	1
HELICAL GEARS	2
BEVEL GEARS	3

Your selection ?2

DATA	Helical gears	RESULTS	Pinion	Gear
Driving power, kW	12	No of teeth	24	60
Pinion speed, RPM	1785	Diameter	83.12	207.81
Gear ratio	2.5	BHN	220	220
No of cycles	1E6	Young mod	210000	210000
Pinion BHN, Young mod E	220, 2, 1E5	Fatigue strength	170	170
Gear BHN, Young mod E	220, 2, 1E5	Surface strength	680	680
Operating Temperature, deg C	70	SYSTEM PARAMETERS:		

Involute angle (14.5, 20, 25)	20	Center distance	145.469979
Helix angle, deg	30	Modul	3.46357094
Service Factor Ko	1.25	Pressure angle	20
Width/a ratio	.25	Normal modul	3
Reliability %	90	Helix angle	30
Precision:		Diametral pitch	7.33347187
High precision ⟨1⟩		Contact ratio	1.59255386
Ground ⟨2⟩		Width	46.8002546
Non-ground ⟨3⟩	3	b/a ratio	.25
Fatigue strength SN1, SN2	170, 170	Tangential force	2058.9544
Surface strength SH1, SH2	680, 680	Radial force	748.985518
Cycle asymmetry factor	1.4	Axial force	1188.00928
Iterating... A = 187.017351	m = 2.78729453	Gearing volume/1E6	1.84046065
		WANT CHANGES?(Y/N)	N

The 24 teeth of the pinion result in operation without interference. The contact ratio is 1.6, in the acceptable range.

The overall design is judged acceptable.

In comparison with the spur gearing of Example 13.4, the helical gearing has 24 teeth, instead of 28, and center distance 145.5 instead of 196 mm, resulting in smaller gears. In fact, the volume is 1.8E6 mm which compared with the 3.17E6 of the spur gearing is almost one-half. However, the radial and tangential forces are greater, 749 and 2058 compared with the 417 and 1116 N of the spur gearing, respectively. In addition, an axial force of 1188 N is now loading the shafts.

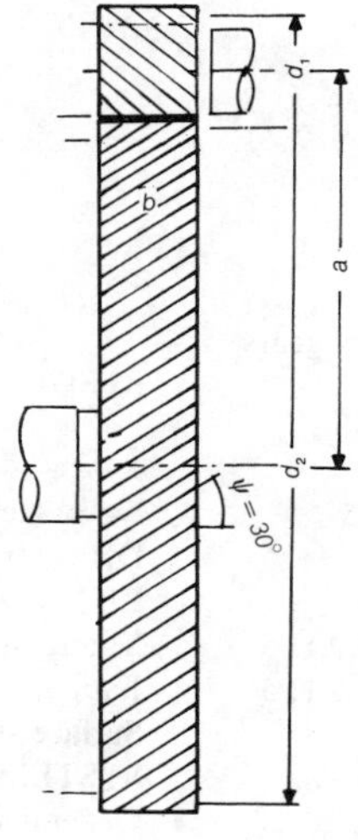

Figure E13.5

Example 13.6

An electric motor of rated power 12 kW at 1785 rpm runs an air blower at 714 rpm with one stage of standard 20 involute bevel gearing at right angles. Ninety percent reliability will suffice and no shocks are expected in the motor with light shocks in the blower. The gears are to last for at least 1 million revolutions at a temperature only slightly exceeding that of the environment. The gears will be made of AISI 1015 steel surface-hardened to 270 BHN and they will be machined but not ground. Design the proper bevel gearing with straight teeth.

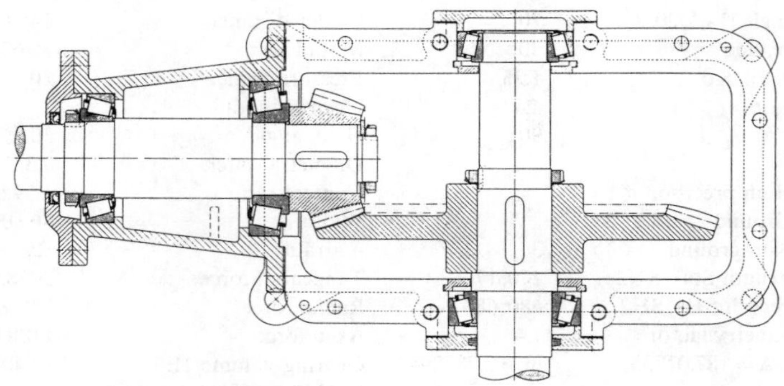

Figure E13.6

Solution

The shaft angle is 90°. The gear ratio is $R = 1785/714 = 2.5$. From Chapter 6, AISI 1015 carbon steel has a fatigue strength of 170 MPa. For a 270 BHN, the contact fatigue strength is (Table 13.4),

$$S_H = 2.8 \times 270 - 70 = 680 \text{ MPa}$$

The calculations are performed by the program GEARDES, Appendix 13.B. The interactive computer session follows:

```
GEAR DESIGN PROGRAM

Menu:

SPUR GEARS            1
HELICAL GEARS         2
BEVEL GEARS           3

Your selection ?3

                                  Bevel gears
DATA                                               RESULTS
                                                                        Pinion       Gear
Driving power, kW                 12               ------------------------------------------
Pinion speed, RPM                 1785             No of teeth          22           55
Gear ratio                        2.5              Diameter             68.62        171.55
No of cycles                      1E6              BHN                  220          220
Pinion BHN, Young mod E           220, 2.1E5       Young mod            210000       210000
Gear BHN, Young mod E             220, 2.1E5       Fatigue strength     170          170
                                                   Surface strength     680          680
Operating Temperature, deg C      70               SYSTEM PARAMETERS:
Shaft angle, deg                  90               Generatrix length    105.138741
Involute angle (14.5, 20, 25)     20               Shaft angle          90
Helix angle, deg                  0                Modul                5
Service Factor Ko                 1.25             Pressure angle       20
Width/a ratio                     .25              Diametral pitch      5.08
Reliability %                     90               Contact ratio        1.57601579
Precision:                                         Width                42.7271033
        High precision <1>                         b/a ratio            .25
        Ground         <2>                         Tangential force     2999.36699
        Non-ground     <3>        3                Radial force         680.664891
Fatigue strength SN1, SN2         170, 170         Axial force          852.73047
Surface strength SH1, SH2         600, 600         Gearing volume/1E6   1.14511891
Cycle asymmetry factor            1.4              WANT CHANGES?(Y/N)     N
```

The 22 teeth of the pinion result in operation without interference. The contact ratio is 1.57, in the acceptable range.

The overall design is judged acceptable.

Example 13.7

An electric motor of rated power 25 kW at 3550 rpm will be connected to the main shaft of a freight elevator rotating at 120 rpm with one stage of standard 20° worm-gear speed reducer. Ninety-nine percent reliability is required and no shocks are expected in the motor with light shocks on the low speed side. The worm gears are to last for at least 1 million revolutions at a temperature of 80 °C. The worm will be made of AISI 1015 steel surface hardened to 270 BHN. The gear will be cast of phosphor-bronze with a fatigue strength 60 MPa and contact stress of 165 MPa. They will be machined and ground. The worm must have one lead. Design the worm-gear drive.

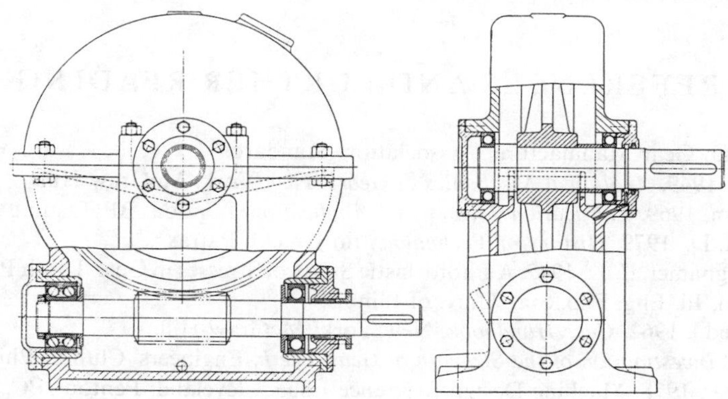

Figure E13.7

Solution

The gear ratio is $R = 3550/120 = 4.58$. From Chapter 6, AISI 1015 carbon steel has a fatigue strength of 170 MPa. For a 270 BHN, the contact fatigue strength is (Table 13.4),

$$S_H = 2.8 \times 270 - 70 = 680\ \text{MPa}$$

The calculations are performed by the program WORMDES, Appendix 13.C. The interactive computer session follows:

DATA		RESULTS	Worm	Gear
Driving power, kW	25	No of teeth	1	28
Worm speed, RPM	3550	Diameter	46.9	234.53
Gear ratio	28.9	BHN	270	120
Worm no of leads	1	Young mod	210000	115000
No of cycles	1E7	Fatigue strength	170	60
Worm BHN, Young mod E	270, 2.1E5	Surface strength	680	165
Gear BHN, Young mod E	120, 1.15E5	Modul	8.11532945	
Operating Temperature, deg C	80	Pressure angle	20	
Involute angle (14.5, 20, 25)	20	Helix angle	9.8205775	
Service Factor Ko	1.25	Diametral pitch	3.1298791	
Width/a ratio	.2	Contact ratio	1.17284908	
Reliability %	99	Width	25.3464568	

DATA		
Precision:		
High precision ⟨1⟩		
Ground ⟨2⟩		
Non-ground ⟨3⟩	2	
Fatigue strength SN1, SN2	170, 60	
Surface strength SH1, SH2	680, 165	
Cycle asymmetry factor	1.4	

RESULTS	Worm	Gear
b/a ratio	.108072017	
Tangential force	2867.5502	
Radial force	4679.71177	
Axial force/worm	2815.23957	
Gearing volume/1E6	1.13822534	
EFFICIENCY	.757762148	
WANT CHANGES?(Y/N)	N	

The 28 teeth of the gear result in operation without interference. The contact ratio is 1.17, in the acceptable range.

The overall design is judged acceptable.

REFERENCES AND FURTHER READING

AGMA, American Gear Manufacturers Association, Standards.

Buckingham, E., 1949. *Analytical Mechanics of Gears.* New York: McGraw-Hill.

British Petroleum, 1969. *Lubrication Theory and Application.* London: BP Trading Co.

Dimarogonas, A. D., 1979. *History of Technology* (in Greek), Patras.

Dolan, T. J., Broghamer, E. L., 1942. A Photoelastic Study of Stresses in Gear Tooth Profiles, Bull. No. 335. Urbana, Ill. Eng. Exp. Sta., Univ. of Illinois.

Dudley, D. W. (ed.), 1962. *Gear Handbook.* New York: McGraw-Hill.

Lewis, W., 1892. *Investigation of the Strength of Gear Teeth.* Engineers' Club of Philadelphia.

Mechanical Drives, 1978. Machine Design Reference Issue. Cleveland: Penton/IPC.

Niemann, G., 1965. *Maschinenelemente.* Munich: Springer.

Pugh, B., 1970. *Practical Lubrication.* London: Butterworths.

Reshetov, D. N., 1978. *Machine Design.* Moscow: Mir Publishers.

Shigley, J. E., 1972. *Mechanical Engineering Design.* 2nd edn. New York: McGraw-Hill.

Wills, J. G., 1980. *Lubrication Fundamentals.* New York: Dekker.

PROBLEMS

13.1 A standard 20° involute pair of gears have number of teeth 17 and 40 with module $m = 6$ mm. Determine the geometric characteristics of the gearing, diameters, center distance and contact ratio. Draw the main lines in the vicinity of two mating teeth and verify graphically the computed contact ratio.

13.2 A standard 20° involute pair of gears has center distance $a = 300$ mm, gear ratio $R = 4$ with module $m = 6$ mm. Determine the geometric characteristics of the gearing, diameters, numbers of teeth and contact ratio. Draw the main lines in the vicinity of two mating teeth and verify graphically the computed contact ratio.

13.3 A standard 20° involute pair of gears has addendum diameters 135 and 385 mm and numbers of teeth 25 and 75, respectively. Determine the geometric characteristics of the gearing, diameters, center distance and contact ratio. Draw the main lines in the vicinity of two mating teeth and verify graphically the computed contact ratio.

13.4 A standard 20° involute pair of gears has pitch diameter of the pinion 96 mm, gear ratio 3 with module $m = 8$ mm. Determine the geometric characteristics of the gearing, diameters, center distance and contact ratio. Draw the main lines in the vicinity of two mating teeth and verify graphically the computed contact ratio.

13.5 Two standard 20° involute pairs of gears (Figure P13.5), have common geometric axes of rotation and the same center distance $a = 125$, approximately, while the speed ratio $n_1/n_2 = 6.25$ and the module $m = 4$. Determine the geometric characteristics of the gearing, diameters, center distance and contact ratio. Draw the main lines in the vicinity of two mating teeth and verify graphically the computed contact ratio.

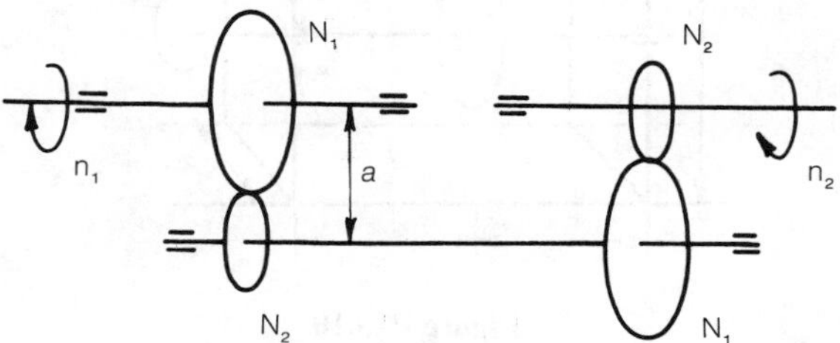

Figure P13.5

Problems 13.6 to 13.9

The standard 20° involute spur gears shown (Figures P13.6 to P13.9) respectively, transmit power $P = 16$ kW at 3500 rpm. The gear ratio is $R = 4$. Determine the gear interaction forces and the bearing loads. Draw force and bending moment diagrams for the shafts.

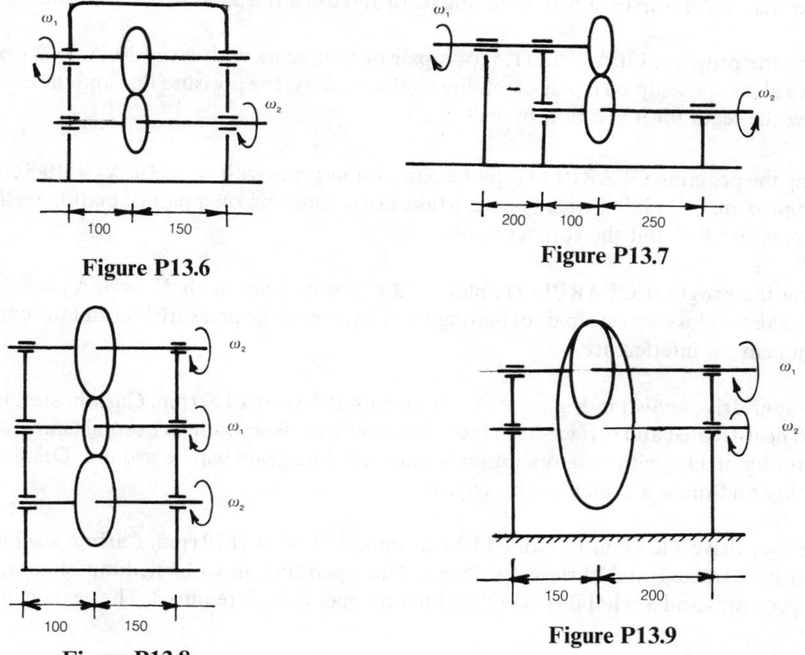

Figure P13.6

Figure P13.7

Figure P13.8

Figure P13.9

13.10 In the system shown in Figure P13.10, the lower gear rotating at 1750 rpm transmits power $P = 20$ kW. The second gear rotates at 875 rpm and also transmits power 12 kW to the third gear at 730 rpm while transmitting the balance 8 kW through its shaft. Determine the gear forces and the bearing reactions.

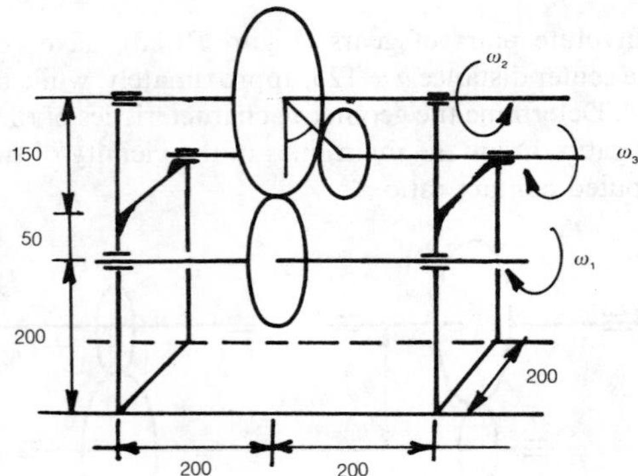

Figure P13.10

13.11 Using the program GEARPLOT (Appendix 13.A), plot a pair of spur gears with $N_1 = 12$, $N_2 = 16$, $m = 10$ mm. Make also a close-up on a pair of mating teeth and draw the pressure line and the contact limits. Comment on the possibility of interference.

13.12 Using the program GEARPLOT, plot a pair of spur gears with $N_1 = 16$, $N_2 = 16$, $m = 8$ mm. Make also a close-up on a pair of mating teeth and draw the pressure line and the contact limits.

13.13 Using the program GEARPLOT, plot a pair of spur gears with $N_1 = 20$, $N_2 = 20$, $m = 10$ mm. Make also a close-up on a pair of mating teeth and draw the pressure line and the contact limits. Draw the teeth for a rotation by half pitch.

13.14 Using the program GEARPLOT, plot a pair of spur gears with $N_1 = 16$, $N_2 = 1000$, $m = 10$ mm, to approximate a pinion-rack system. Make also a close-up on a pair of mating teeth and draw the pressure line and the contact limits.

13.15 Using the program GEARPLOT, plot a pair of spur gears with $N_1 = 8$, $N_2 = 8$, $m = 10$ mm. Make also a close-up on a pair of mating teeth and draw the pressure line and the contact limits. Comment on interference.

13.16 The spur drive shown in Figure P13.6, transmits 15 kW at 1450 rpm. Carbon steel is used SAE 1020 heat-treated and surface-hardened. The operation is smooth, at normal temperatures and a reliability of 99% with long operation is required. The gears will be ground. Design the proper gearing and draw a sketch of the system.

13.17 The spur drive shown in Figure P13.7 transmits 15 kW at 1870 rpm. Carbon steel is used SAE 1020 heat-treated and surface-hardened. The operation is with medium shocks, at normal temperatures and a reliability of 99% with long operation is required. The gears will be ground.

Design the proper gearing and draw a sketch of the system. Then, redesign for 100 hours' operation and comment on the difference.

13.18 The spur gear drive shown in Figure P13.8 transmits 15 kW at 1450 rpm. Carbon steel is used SAE 1020 heat-treated and surface-hardened. The operation is smooth, at normal temperatures and a reliability of 99% with long operation is required. The gears will be ground. Design the proper gearing and draw a sketch of the system. Redesign for 99.99% reliability and comment on the difference.

13.19 A spur gear drive is shown in Figure P13.9. Carbon steel is used SAE 1020 heat-treated and surface-hardened. The operation is smooth, at normal temperatures and a reliability of 99% with long operation is required. The gears will be ground. Design the proper gearing and draw a sketch of the system.

13.20 The dual drive shown in Figure P13.10 will have spur gears made of carbon steel SAE 1020 heat-treated and surface-hardened. The operation is smooth, at normal temperatures and a reliability of 99% with long operation is required. The spur gears will be ground. Design the proper gearing and draw a sketch of the system.

13.21 A spur gear pair is to be designed for the second gear of an automotive transmission with gear ratio $R = 1.5$ and input power 50 kW at 3500 rpm. It is required that the center distance should be $a = 90$ mm, approximately. For 1000 hours' operation and 99.9% reliability, determine the required material properties. Select a proper material and specify the required heat treatment.

13.22 A spur gear pair is to be designed for the power pack of an overhead crane with gear ratio $R = 3$ and input power 12 kW at 1200 rpm. It is required that the module is 5 mm and the gear width $b = 40$ mm. For 4000 hours' operation and 99.9% reliability, determine the required material properties. Select a proper material and specify the required heat treatment.

13.23 A spur gear pair is to be designed for the drive of a machine tool with gear ratio $R = 8$ and input power 20 kW at 3500 rpm for occasional operation equally in forward and backward motion totalling no more than 50 hours. Because it will mesh with forward and backward gears, the pinion diameter should be $d = 80$ mm, approximately. For 99% reliability, determine the required material properties. Select a proper material and specify the required heat treatment.

13.24 For the system of Problem 13.24, determine the width b if the material for pinion and gear is SAE 1020 with BHN = 300.

Problems 13.25 to 13.29

Design helical gears for the systems of problems P13.6 to P13.10, with helix angles $\psi = 30°$. Determine the bearing loads and compare with the ones for spur gears.

Problems 13.30 to 13.34

Design helical gears for the systems of problems P13.21 to P13.24. Compare solutions with spur gears. Helix angle is 25°.

13.35 A bevel gear drive must transmit at right angles 20 kW at $n = 600$ rpm, gear ratio $R = 3$ and minimum pinion teeth 20. Design the gears for medium shocks, 2000 hours' operation at 99% reliability with cast iron material.

13.36 A bevel gear drive must transmit at right angles 20 kW at $n = 2500$ rpm, gear ratio $R = 2.5$ and module $m = 8$. Design the gears for medium shocks, 2000 hours' operation at 99% reliability with carbon steel SAE 1015 material.

13.37 A bevel gear drive must transmit at right angles 20 kW at $n = 600$ rpm, gear ratio $R = 3$ and maximum pinion diameter $d = 90$ mm. Design the gears for medium shocks, 2000 hours' operation at 99% reliability. Specify material with the required heat and surface treatment.

13.38 A bevel gear drive must transmit at right angles 20 kW at $n = 600$ rpm, gear ratio $R = 3$. The tooth width must not exceed 40 mm. Design the gears for medium shocks, 2000 hours' operation at 99% reliability with cast iron material.

13.39 A vertical drill must transmit at right angles torque 400 Nm at $n = 500$ rpm to the vertical gear from a horizontal pinion rotating at 1500 rpm. Design the gears for medium shocks, 2000 hours' operation at 99% reliability with cast iron material.

13.40 Design a worm-gear drive for input power 15 kW at 2000 rpm, with center distance $a = 200$ mm, transmission ratio $R = 15$, number of worm leads $N_1 = 2$ and module $m = 5$. The worm must be made of carbon steel heat-treated and surface-hardened. A gear bronze must be selected for the gear. Determine the efficiency and the bearing forces making reasonable assumptions for the enclosure dimensions for a self-contained drive. Make a sketch for the general arrangement, if the worm is on top of the gear.

13.41 Design a worm-gear drive for input power 22 hp at 3700 rpm with approximate diameters $d_1 = 50$, $d_2 = 150$ mm, transmission ratio $R = 15$ and number of worm leads $N_1 = 2$. The worm must be made of carbon steel heat-treated and surface-hardened. A gear bronze must be selected for the gear. Determine the efficiency and the bearing forces making reasonable assumptions for the enclosure dimensions for a self-contained drive. Make a sketch for the general arrangement, if the worm is on top of the gear.

13.42 Design a worm-gear drive for input power 12 kW at 3000 rpm with worm diameter 50 mm, transmission ratio $R = 15$, number of worm leads $N_1 = 2$ and module $m = 10$. The worm must be made of carbon steel heat-treated and surface-hardened. A gear bronze must be selected for the gear. Determine the efficiency and the bearing forces making reasonable assumptions for the enclosure dimensions for a self-contained drive. Make a sketch for the general arrangement, if the worm is on top of the gear.

13.43 Design a worm-gear drive for input power 20 kW at 3550 rpm with transmission ratio $R = 30$, number of worm leads $N_1 = 2$, number of gear teeth $N_2 = 60$. The worm must be made of carbon steel heat-treated and surface-hardened. A gear bronze must be selected for the gear. Determine the efficiency and the bearing forces making reasonable assumptions for the enclosure dimensions for a self-contained drive. Make a sketch for the general arrangement, if the worm is on top of the gear.

13.44 Design a worm-gear drive for output power 10 kW at 200 rpm with center distance $a = 200$ mm, transmission ratio $R = 15$, number of worm leads $N_1 = 2$ and gear teeth $N_2 = 30$. The worm must be made of carbon steel heat-treated and surface-hardened. A gear bronze must be slected for the gear. Determine the efficiency and the bearing forces making reasonable assumptions for the enclosure dimensions for a self-contained drive. Make a sketch for the general arrangement, if the worm is on top of the gear.

Problems 13.45 to 13.49

For the worm-gear drives of Problems 13.40 to 13.44, determine the operating temperature. If it is higher than 70 °C, provide proper surface cooling.

APPENDIX 13.A
GEARPLOT: AN INVOLUTE GEAR TEETH DRAWING PROGRAM

```
10 REM ************************************************
20 REM *                                              *
30 REM *                  GEARPLOT                    *
40 REM *                                              *
50 REM ************************************************
60 REM          COPYRIGHT 1987
70 REM by Professor Andrew D. Dimarogonas, Washington University,St.Louis.
80 REM All rights reserved.  Unauthorized reproduction, dissemination or
90 REM selling is strictly prohibited.  This listing is for personal use.
100 REM
110 REM
120 REM
130 REM Plotting of involute gears
140 REM
150 REM
160 REM        ADD/11-20-85
170 REM ************************************************************
180 CLS:KEY OFF
190 FOR I=1 TO 14:COLOR I:LOCATE I,22+I:PRINT "GEARPLOT";
191 LOCATE 25,1:PRINT"   Dimarogonas,A.D., Computer Aided Machine Design,
Prentice-Hall, London 1988";
200 LOCATE I,50-I:PRINT "GEARPLOT";:NEXT I:COLOR 14
210 LOCATE 8,20:PRINT"Plotting a pair of involute spur gears";
220 REM The head and foot dimensions is, by default, m and 1.2 m
respectively.
230 LOCATE 16,20:INPUT"modul(mm)                                ";ML
240 LOCATE 17,20:INPUT"Number of teeth of the pinion gear  ";N1
250 LOCATE 18,20:INPUT"Number of teeth of the driven gear  ";N2
260 LOCATE 19,20:INPUT"Scale factor for plotting           ";SC
270 LOCATE 20,20:INPUT"Angle of involute              (deg) ";F
280 LOCATE 21,20:INPUT"Angle of rotation              (deg) ";FROT
290 LOCATE 23,20:INPUT"Hit RETURN for plotting             ";X$
300 IP=5:SCREEN 2:XSC=SC*2:YSC=SC:Y0=100/YSC:X0=300/XSC-ML*N1/2
310 F=F*3.14159/180
320 Z1=N1:F0=CL+FROT:F0=F0*3.14159/180:GOSUB 410
330 Z1=N2
340 WA=(N1+N2)*ML/2
350 X0=X0+WA*COS(CL)
360 Y0=Y0+WA*SIN(CL)
370 F0=3.14159-AG*N2/N1+CL-F0
380 GOSUB 410
390 LOCATE 25,1:INPUT"Hit RETURN to end";X$
400 SCREEN 0:COLOR 14:END
410 REM ................................Gear plotting routine
420 R1=Z1/2*ML:RB=R1*COS(F)
430 FOR T0=F0 TO 6.283+F0 STEP 6.283/Z1
440 RI=R1-ML*1.2
450 XI=X0+RI*COS(T0)
460 YI=Y0+RI*SIN(T0)
470 XL=XI:YL=YI
480 TD=SQR((R1+ML)^2-RB*RB)/RB
490 XC=X0+RB*COS(T0):YC=Y0+RB*SIN(T0)
500 REM ................plotting left straight segment below basic circle
510 X1=XI*XSC:Y1=YI*YSC:X2=XC*XSC:Y2=YC*YSC:GOSUB 1060
520 FOR T=0 TO TD STEP TD/IP
530 XA=X0+RB*COS(T0+T)
540 YA=Y0 +RB*SIN(T0+T)
550 AB=RB*T:XB=XA+AB*SIN(T0+T)
560 YB=YA-AB*COS(T0+T)
570 X9=X2:Y9=Y2
```

```
580 X2=XB*XSC:Y2=YB*YSC:X1=X9:Y1=Y9:GOSUB 1060
590 REM .....................................plotting left involute
curve
600 NEXT T
610 XP=X2:YP=Y2
620 TG=SQR(R1*R1-RB*RB)/RB
630 PI=3.14159:AM=2*PI/Z1
640 OC=RB
650 CD=SQR((XB-XC)^2+(YB-YC)^2)
660 OD=SQR((XB-X0)^2+(YB-Y0)^2)
670 CA=(OD^2+OC^2-CD^2)/(2*OC*OD)
680 SA=SQR(1-CA*CA):TA=SA/CA
690 AD=SA:IF SA<.1 THEN 710
700 AD=ATN(TA)
710 XG=XA+AB*SIN(T0+TG)
720 YG=YA-AB*COS(T0+TG)
730 OG=SQR((XG-X0)^2+(YG-Y0)^2)
740 CG=SQR((XC-XG)^2+(YC-YG)^2)
750 CA=(OC^2+OG^2-CG^2)/(2*OC*OG)
760 SA=SQR(1-CA*CA):TA=SA/CA
770 AG=SA:IF AG<.1 THEN 790
780 AG=ATN(TA)
790 AH=AG+PI/Z1:AF=AH+AG
800 XI=X0+RI*COS(T0+AF)
810 YI=Y0+RI*SIN(T0+AF)
820 X=X0+RB*COS(T0+AF)
830 Y=Y0+RB*SIN(T0+AF)
840 REM ...............plotting right straight segment below basic circle
850 X1=XI*XSC:Y1=YI*YSC:X2=X*XSC:Y2=Y*YSC:GOSUB 1060
860 FOR T=0 TO TD STEP TD/IP
870 XF=RB*COS(T0+AF-T)+X0
880 YF=RB*SIN(T0+AF-T)+Y0
890 XB=XF-RB*T*SIN(T0+AF-T)
900 YB=YF+RB*T*COS(T0+AF-T)
910 X9=X2:Y9=Y2
920 REM .................................plotting right involute curve
930 X2=XB*XSC:Y2=YB*YSC:X1=X9:Y1=Y9:GOSUB 1060
940 NEXT T
950 REM .....................................plotting outer circle
960 X1=X2:Y1=Y2:X2=XP:Y2=YP:GOSUB 1060
970 TI=SQR((XL-XI)^2+(YL-YI)^2)/RI
980 TT=AM-TI
990 XN=X0+RI*COS(T0+AF+TT)
1000 YN=Y0+RI*SIN(T0+AF+TT)
1010 REM ....................................plotting inner circle
1020 X1=XN*XSC:Y1=YN*YSC:X2=XI*XSC:Y2=YI*YSC:GOSUB 1060
1030 NEXT T0
1040 RETURN
1050 REM
1060 REM ................................plotting with clipping
1070 REM
1080 XM=630:YM=199
1090 IF X1>XM THEN 1180
1100 IF Y1>YM THEN 1180
1110 IF Y2>YM THEN 1180
1120 IF X2>XM THEN 1180
1130 IF X1<0 THEN  1180
1140 IF X2<0 THEN  1180
1150 IF Y2<0 THEN  1180
1160 IF Y1<0 THEN  1180
1170 LINE (X1,Y1)-(X2,Y2)
1180 RETURN
```

APPENDIX 13.B
GEARDES: A GEAR DESIGN PROGRAM

```
10 REM *********************************
20 REM *                               *
30 REM *         GEARDES               *
40 REM *                               *
50 REM *********************************
```

```
60 REM      copyright  1987
70 REM by professor Andrew D.Dimarogonas,Washington University,St. Louis
72 REM  All rights reserved.Unauthorised
reproduction,dissemination,selling,
74 REM or use is strictly prohibited.This listing is for reference purpose
only.
80 REM
90 REM
95 REM A gear design program using the lewis/AGMA equation
100 REM for fatigue and the Buckingham/AGMA equation for surface strength.
101 CLS:KEY OFF
102 FOR I=1 TO 14:COLOR I:LOCATE I,22+I:PRINT "GEARDES";
103 LOCATE I,50-I:PRINT "GEARDES";:NEXT I:COLOR 14
104 LOCATE 8,16:PRINT"A spur, helical, bevel gear design program ";
106 LOCATE 23,28:INPUT"Hit ENTER to continue";X$:CLS
107 REM
110 DIM MD(30)
120 FOR IM=1 TO 29:READ MD(IM):NEXT IM
130 DATA 1,1.25,1.5,2,2.5,3,4,5,6,8,10
135 DATA 12,16,20,25,32,40,45,50,55,60,65,70,75,80,85,90,95,100
140 MD(30)=200
150 CM=1.4:pi=3.14159
160 READ P,RP,R,N,H1,E1,H2,E2,TC,FD,KM
170 READ KO,BA,RA,CS,CF,PR,PH,GH,PN,GN,n1
180 DATA 30,1000,2,1E5,200,2.1E5,200,2.1E5,60,20,1.3,1.25
185 DATA .25,90,1,1,3,600,600,150,150,20
200 REM for  default data GOTO 600
210 CLS :COLOR 14
212 PRINT "GEAR DESIGN PROGRAM         "
213 PRINT "       Menu:": PRINT
214 PRINT "SPUR GEARS           1"
215 PRINT "HELICAL GEARS        2"
216 PRINT "BEVEL GEARS          3"
217 PRINT :LOCATE 6,40
218 PRINT "Your selection  ";:INPUT B1:if b1=0 then 217
219 LOCATE 4,40
220 IF B1 = 1 THEN PRINT "Spur gears"
221 IF B1 = 2 THEN PRINT "Helical gears"
222 IF B1 = 3 THEN PRINT "Bevel gears"
224 locate 8,1:PRINT"DATA:                          (default)":PRINT
350 PRINT"Driving power , kw             ";P,: INPUT P$:IF P$>"" THEN
P=VAL(P$)
360 PRINT"Pinion   speed,RPM             ";RP,:INPUT RP$:IF RP$>"" THEN
RP=VAL(RP$)
370 PRINT"Gear   ratio                   ";R,: INPUT R$:IF R$>"" THEN
R=VAL(R$)
375 PRINT"Pinion Number of Teeth         ";N1,:INPUT NW$:IF NW$>"" THEN
N1=VAL(NW$)
380 PRINT"No of cycles                   ";N,: INPUT N$:IF N$>"" THEN
N=VAL(N$)
390 PRINT"Pinion BHN                     ";H1,:INPUT H1$:IF H1$>"" THEN
H1=VAL(H1$)
395 PRINT"Pinion Young mod E             ";E1,:INPUT E1$:IF E1$>"" THEN
E1=VAL(E1$)
400 PRINT"Gear  BHN                      ";H2,:INPUT H2$:IF H2$>"" THEN
H2=VAL(H2$)
405 PRINT"Gear  Young mod E              ";E2,:INPUT E2$:IF E2$>"" THEN
E2=VAL(E2$)
410 PRINT"Operating temperature,deg C ";TC,:INPUT TC$:IF TC$>"" THEN
TC=VAL(TC$)
440 PRINT"Involute angle (14.5,20,25) ";FD,:INPUT FD$:IF FD$>"" THEN
FD=VAL(FD$)
445 on b1 goto 470,448,450
448 PRINT"Helix angle, deg               ";th,:INPUT th$:IF th$>"" THEN
th=VAL(th$)
449 goto 470
450 PRINT"Shaft angle, deg               ";gg,:INPUT gg$:IF gg$>"" THEN
gg=VAL(gg$)
470 KM=1:pa=th*pi/180:ga=gg*pi/360:if b1>1 and pa+ga=0 then 445
480 PRINT"Service Factor Ko              ";KO,:INPUT KO$:IF KO$>"" THEN
KO=VAL(KO$)
490 PRINT"Width/a ratio                  ";BA,:INPUT BA$:IF BA$>"" THEN
BA=VAL(BA$)
```

```
500 PRINT"Reliability %                         ";RA,:INPUT RA$:IF RA$>"" THEN
RA=VAL(RA$)
510 PRINT"Size factor Cs                        ";CS,:INPUT CS$:IF CS$>"" THEN
CS=VAL(CS$)
520 PRINT"Surface finish factor Cf              ";CF,:INPUT CF$:IF CF$>"" THEN
CF=VAL(CF$)
530 PRINT"Precision: High precision <1>,Ground <2>,Non-ground
<3>";PR;:INPUT PR$
540 IF PR$>"" THEN PR=VAL(PR$)
570 PRINT"fatigue strength, pinion      ";PN,:INPUT PN$:IF PN$>"" THEN
PN=VAL(PN$)
572 PRINT"fatigue strength,gear         ";GN,:INPUT GN$: IF GN$>"" THEN
GN=VAL(GN$)
580 PRINT"Surface strength,pinion       ";PH,:INPUT PH$:IF PH$>"" THEN
PH=VAL(PH$)
585 PRINT"Surface strength, gear        ";GH,:INPUT GH$:IF GH$>"" THEN
GH=VAL(GH$)
586 PRINT"Cycle asymmetry factor        ";CM,:INPUT CM$:IF CM$>"" THEN
CM=VAL(CM$)
587 PRINT:INPUT"Are the data correct (Y/N) ";XDAT$:IF XDAT$="N" OR
XDAT$="n" THEN 210
688 cr=.042*(.4343*log(100-ra)+2)^1.5+.704
    if pr=1 then cs=1
    if pr<>2 then 693
    if sn<500 then cf=.9
    if sn>=500 then cf=1.22-pn/1550
693 if pr=3 then cf=.874-pn/2275
    co=cos(pa):rp=rp*co^2:ba=ba/co
    if b1<>3 then 695
    g1=2*ga/(1+r):rp=rp*cos(g1):rem  bevel gears
695 FI = FD * 3.14/180
    REM COMPUTATION OF CV
    CV=1:km=1
    if pr=1 then km=1.25+.00075*b:if km>1.8 then km=1.8
    if pr=2 then km=1.54+.00087*b:if km>2.2 then km=2.2
    if pr=3 then km=2.2
700 for iter=1 to 10:M1 = 9550 * P /RP * (1000)
720 IX = R * SIN(2 * FI) / (4 * (1+R))
730 IF N1 = 0 THEN N1 = 20
740 if b1=2 then IX = .95 * CR * IX
741 if gl=0 or b>gl then 750
742 if b1=3 then ix=1.04*ix*(1-100/n1^2):rem bevel gears
750 IF IX < .06 THEN IX = .06
760 CL = 1:cs=1:if mo>5 then cs=.85
770 IF N < 1E+07 THEN CL = 2.16 - .17 * LOG(N) / LOG(10)
780 CH = 1
790 IF H1 / H2 > 1.2 THEN CH = 1 + .0014 * (R-2) * H1 / H2
800 CT = 344 / (273 + TC):if tc<70 then ct=1
810 CP = .591 * SQR(E1 * E2 /(E1 + E2)):aold=a
    if b1=3 then cp=1.23*cp:rem   bevel gears
    if ph=0 then 815:rem  Given  a  ,compute the material
    if aind=1 then 816
    a1=((m1*(1+r)^2)/(2*ba*ix*(ph*cl*ch*ct*cr/cp)^2)*ko*km*cv):a1=a1^.333
    a2=((m1*(1+r)^2)/(2*ba*ix*(gh*cl*ch*ct*cr/cp)^2)*ko*km*cv):a2=a2^.333
    a=a1:if a2>a1 then a=a2
    goto 816
815 locate 23,1:print spc(78);:locate 23,1
    print "Center Distance a ";a$;:Input a$:a=val(a$):if a=0 then 815
    ph=sqr(m1*(1+r)^2/(2*ba*ix*a^3)*ko*km*cv)*cp/(cl*ch*ct*cr)
    gh=sqr(m1*(1+r)^2/(2*ba*ix*a^3)*ko*km*cv)*cp/(cl*ch*ct*cr)
816 d1=2*a/(1+r)
    b=a*ba
    if pr=1 then km=1.25+.75*b
    if pr=2 then km=1.54+.875*b
    if pr=3 then km=2.5
    if b1=2 then km=.93*km
840 V = 3.14*rp * D1 / 304
850 VP = CV
860 IF PR = 2 THEN CV = SQR((78 + SQR(V)) / 78)
870 IF PR = 3 THEN CV = (50 + SQR(V)) / 50
880 IF PR = 1 THEN cv=1
```

```
900 IF MO = 0 THEN MO = 10
910 N1=d1/mo: N2 = N1 * R
920 B = BA * a
930 IF FD = 20 AND PR<3 THEN JX=.56*(1-.38/SQR(N2)-.88/N2)*(1-.26/SQR(N1)-
5.5 / N1)
940 IF FD = 20 AND PR > 2 THEN JX = .32 * (1 - 1.14 / N1 ^ .56)
950 IF FD = 25 AND PR < 3 THEN JX = .63 * (1 - .38 / SQR(N2) - .88 / N2) *
(1 - .26 / SQR(N1) - 5.5 / N1)
960 IF FD = 25 AND PR > 2 THEN JX = .39 * (1 - 1.14 / N1 ^ .56)
965 IF JX<.1 THEN JX=.1
980 if b1=2 then JX = JX * (1 + 4.5 * PA * (PA - .7) * (PA - .87)): REM
HELICAL GEARS
    if b1=3 then JX = JX * (100+n2)/240:rem       bevel gears
990 REM FATIGUE STRENGTH
1000 mp =  M1*(1+r)* KO * KM * CV / CR / CT / CM / (b*A * JX * GN * CS *
CF)
     mg =  M1*(1+r)* KO * KM * CV / CR / CT / CM / (b*A * JX * GN * CS *
CF)
     rem  select then larger modul
     mo=mg:if mp>mg then mo=mp
     ml=mo
     rem contact ratio
1030 MC = ( SQR ((N2 + 2) ^ 2 - N2 ^ 2 * COS(FI) ^ 2) - N2 * SIN(FI) + SQR
((N1 + 2) ^ 2 - N1 ^ 2 * COS (FI) ^ 2) - N1 * SIN (FI)) / 6.28
     if abs(ap-a)/a>0.001 then 1035
     if abs(ml-mo)/mo<.001 then 1060
1035 ap=a:if b1=3 then gl=d1/2/sin(g1): rem   bevel gears
1040 LOCATE 25,1:PRINT "ITERATING....a =";a;"   m=";mo;"        ";
     rem if b1>1 then 700:rem iterate for solution dependent factors
1060 FOR IM = 1 TO 29
1070 IF MO > MD(IM) AND MO < MD(IM + 1) THEN MO = MD(IM + 1)
1080 NEXT IM
1090 IF MO = 200 THEN PRINT "MODUL OUT OF RANGE": STOP
     rem transform back from virtual to face gear
     mo=mo/co:n1=n1*co^3:rem helical gears
     if b1=3 then ba=ba*co:b=b*co
     n1=n1*cos(g1):rem bevel gears
     n1=int(n1):n2=r*n1:d1=mo*n1:d2=mo*n2:if aind=0 then a=(d1+d2)/2
1180 MC=(SQR((N2+2)^2-N2^2*COS(FI)^2)-N2*SIN(FI)+SQR((N1+2)^2-
N1^2*COS(FI)^2)- N1 * SIN (FI)) / 6.28
1185 next iter
1190 CLS
1200 PRINT " RESULTS"
1210 PRINT "                                    PINION       GEAR"
1220 PRINT "----------------------------------------------------------------"
1230 PRINT "No of teeth          ",N1,N2
1240 D1 = D1 * COS (G1):D2 = D2 * COS (G1)
1250 PRINT "Diameter             ",int(d1),int(d2)
1260 PRINT "BHN                  ", H1,H2
1270 PRINT "Young modulus        ",E1,E2
1280 PRINT "Fatigue strength     ",PN,GN
1290 PRINT "Surface strength     ",PH,GH
1295 PRINT
1300 PRINT "SYSTEM PARAMETERS: "
1310 PRINT "Center distance      ";(d1+d2)/2
1320 PRINT "Modul                ";MO
1330 PRINT "Pressure angle       ";FD
1340 if b1=2 then PRINT "Helix angle          ";180 * PA / 3.14
1345 if b1=3 then PRINT "Shaft angle          ";gg
1350 PRINT "Diameter pitch       ";1 / MO * 25.4
1360 PRINT "Contact ratio        ";MC
1365 LOCATE 12,40:PRINT "Width                ";a * BA;
1380 LOCATE 13,40:PRINT "b/a ratio            ";BA;
1390 FT = 2 * M1 / D1
1400 LOCATE 14,40:PRINT "Tangential force    ";FT;
1405 FR=FT *  tan(FI) * COS(g1)
1410 LOCATE 15,40:PRINT "Radial force         ";FR;
1430 FA = FT * tan (PA)
1440 if b1<3 then LOCATE 16,40:PRINT "Axial force          ";FA;
1445 if b1=3 then LOCATE 16,40:PRINT "Axial force          ";FA*sin(g1)
1450 B =BA * a:VOLUME=3.14 * B *(D1 ^2 + D2 ^ 2) / 4 /1000000!
```

```
1460 LOCATE 17,40:PRINT "Gearing volume/1E6 ";VOLUME
1490 INPUT "Do you wand design variations  ? ( Y/N )";X$
1500 IF X$ = "N" OR X$="n" THEN GOTO 1600
1510 INPUT "<1> fatigue strength SN,<2> surface strength SH,<3> center
distance a : choice is :";IIS
1520 aind=0:ON IIS GOTO 1530,1550,1570
1530 INPUT "New fatigue strength Sn pinion,gear   ";PN,GN
1540 GOTO 700
1550 INPUT "New surface strength SH pinion,gear   ";PH,GH
1560 GOTO 700
1570 print "a= ";a;"   New center distance a       ";:input a:aind=1
1580 GOTO 700
1590 GOTO 1510
1600 END
```

APPENDIX 13.C
WORMDES: A WORM-GEAR DESIGN PROGRAM

```
10 REM *********************************
20 REM *                               *
30 REM *          WORMDES              *
40 REM *                               *
50 REM *********************************
60 REM       copyright  1987
70 REM by professor Andrew D.Dimarogonas,Washington University,St. Louis
72 REM  All rights reserved.Unauthorised
reproduction,dissemination,selling,
74 REM or use is strictly prohibited.This listing is for reference purpose
only.
80 REM
90 REM
95 REM A worm gear design program using the lewis/AGMA equation
100 REM for fatigue and the Buckingham/AGMA equation for surface strength.
101 CLS:KEY OFF
102 FOR I=1 TO 14:COLOR I:LOCATE I,22+I:PRINT "WORMDES";
103 LOCATE I,50-I:PRINT "WORMDES";:NEXT I:COLOR 14
104 LOCATE 8,26:PRINT"A worm-gear design program ";
105 locate 25,1:print" Dimarogonas, A.D., Computer Aided Machine Design,
Prentice-Hall, 1988";
106 LOCATE 23,28:INPUT"Hit ENTER to continue";X$:CLS
107 REM
110 DIM MD(30)
120 FOR IM=1 TO 29:READ MD(IM):NEXT IM
130 DATA 1,1.25,1.5,2,2.5,3,4,5,6,8,10
135 DATA 12,16,20,25,32,40,45,50,55,60,65,70,75,80,85,90,95,100
140 MD(30)=200
150 CM=1.4
160 READ P,RP,R,N,H1,E1,H2,E2,TC,FD,KM
170 READ KO,BA,RA,CS,CF,PR,PH,GH,PN,GN
180 DATA 30,1000,5,1E5,200,2.1E5,200,2.1E5,100,20,1.3,1.25
185 DATA .25,90,1,1,3,600,600,150,150
190 NW=2:FF=.05
195 BA=.2:R=15
196 GH=150:GN=30
200 REM for  default data GOTO 600
210 CLS
340 PRINT"DATA:                          (default)":PRINT
350 PRINT"Driving power , kw             ";P,: INPUT P$:IF P$>"" THEN
P=VAL(P$)
360 PRINT"Worm     speed,RPM             ";RP,:INPUT RP$:IF RP$>"" THEN
RP=VAL(RP$)
370 PRINT"Gear   ratio                   ";R,: INPUT R$:IF R$>"" THEN
R=VAL(R$)
375 PRINT"Worm no of leads               ";NW,:INPUT NW$:IF NW$>"" THEN
NW=VAL(NW$)
380 PRINT"No of cycles                   ";N,: INPUT N$:IF N$>"" THEN
N=VAL(N$)
390 PRINT"Worm  BHN                      ";H1,:INPUT H1$:IF H1$>"" THEN
H1=VAL(H1$)
395 PRINT"Worm  Young mod E              ";E1,:INPUT E1$:IF E1$>"" THEN
```

```
E1=VAL(E1$)
400 PRINT"Gear  BHN                      ";H2,:INPUT H2$:IF H2$>"" THEN
H2=VAL(H2$)
405 PRINT"Gear  Young mod E              ";E2,:INPUT E2$:IF E2$>"" THEN
E2=VAL(E2$)
410 PRINT"Operating temperature,deg C ";TC,:INPUT TC$:IF TC$>"" THEN
TC=VAL(TC$)
440 PRINT"Involute angle (14.5,20,25) ";FD,:INPUT FD$:IF FD$>"" THEN
FD=VAL(FD$)
470 KM=1
480 PRINT"Service Factor Ko              ";KO,:INPUT KO$:IF KO$>"" THEN
KO=VAL(KO$)
490 PRINT"Width/a ratio                  ";BA,:INPUT BA$:IF BA$>"" THEN
BA=VAL(BA$)
500 PRINT"Reliability %                  ";RA,:INPUT RA$:IF RA$>"" THEN
RA=VAL(RA$)
510 PRINT"Size factor Cs                 ";CS,:INPUT CS$:IF CS$>"" THEN
CS=VAL(CS$)
520 PRINT"Surface finish factor Cf       ";CF,:INPUT CF$:IF CF$>"" THEN
CF=VAL(CF$)
530 PRINT"Precision: High precision <1>,Ground <2>,Non-ground
<3>";PR;:INPUT PR$
540 IF PR$>"" THEN PR=VAL(PR$)
570 PRINT"fatigue strength, worm      ";PN,:INPUT PN$:IF PN$>"" THEN
PN=VAL(PN$)
572 PRINT"fatigue strength,gear       ";GN,:INPUT GN$: IF GN$>"" THEN
GN=VAL(GN$)
580 PRINT"Surface strength,worm       ";PH,:INPUT PH$:IF PH$>"" THEN
PH=VAL(PH$)
585 PRINT"Surface strength, gear      ";GH,:INPUT GH$:IF GH$>"" THEN
GH=VAL(GH$)
590 PRINT"Cycle asymmetry factor      ";CM,:INPUT CM$:IF CM$>"" THEN
CM=VAL(CM$)
595 PRINT:INPUT"Are the data correct (Y/N) ";XDAT$:IF XDAT$="N" OR
XDAT$="n" THEN 210
600 IF RA>=50 THEN CR=1
610 IF RA>=90! THEN CR =.897
620 IF RA>=99.9 THEN CR=.753
630 IF RA>=99.99 THEN CR=.702
640 IF RA>=99.999 THEN CR=.65
660 REM
670 FI = FD * 3.14/180
680 REM COMPUTATION OF CV
690 CV=1
700 M1 = 9550 * P / RP * (1000)
720 IX = R * SIN(2 * FI) / (4 * (1+R))
730 IF N1 = 0 THEN N1 = 100
740 IX = .95 * CR * IX
750 IF IX < .06 THEN IX = .06
760 CL = 1
770 IF N < 1E+07 THEN CL = 2.16 - .17 * LOG(N) / LOG(10)
780 CH = 1
790 IF H1 / H2 > 1.2 THEN CH = 1 + .0014 * (R-2) * H1 / H2
800 CT = 344 / (273 + T)
810 CP = .591 * SQR(E1 * E2 /(E1 + E2))
820 DD = ((2 * M1) / (R * BA * IX * (GH * CL * CH / (CT * CR * CP)) ^ 2)*
KO * KM * CV) ^ .333
830 KM = 1.25
840 V = 3.14 * D1 / 304
850 VP = CV
860 IF PR = 2 THEN CV = SQR((78 + SQR(V)) / 78)
870 IF PR = 3 THEN CV = (50 + SQR(V)) / 50
880 IF PR = 1 THEN 900
890 IF ABS(CV - VP) > .01 THEN 700
900 IF MO = 0 THEN MO = 10
910 N1 = NW : N2 = N1 * R
920 B = BA * D2
930 IF FD = 20 AND PR < 3 THEN JX = .56 * (1 - .38 / SQR(N2) - .88 / N2) *
(1 - .26 / SQR(N1) - 5.5 / N1)
940 IF FD = 20 AND PR > 2 THEN JX = .32 * (1 - 1.14 / N1 ^ .56)
950 IF FD = 25 AND PR < 3 THEN JX = .63 * (1 - .38 / SQR(N2) - .88 / N2) *
(1 - .26 / SQR(N1) - 5.5 / N1)
```

```
960 IF FD = 25 AND PR > 2 THEN JX = .39 * (1 - 1.14 / N1 ^ .56)
965 IF JX<.1 THEN JX=.1
980 JX = JX * (1 + 4.5 * PA * (PA - .7) * (PA - .87)): REM HELICAL GEARS
990 REM FATIGUE STRENGTH
1000 GG = (2 * M1 * N2 * KO * KM * CV * CR * CT * CM / (R * BA * JX * GN *
CS * CF)) ^ .333
1010 D2 = DD : IF GG > DD THEN D2 = GG
1020 MO = D2 / N2
1030 MC = ( SQR ((N2 + 2) ^ 2 - N2 ^ 2 * COS(FI) ^ 2) - N2 * SIN(FI) + SQR
((N1 + 2) ^ 2 - N1 ^ 2 * COS (FI) ^ 2) - N1 * SIN (FI)) / 6.28
1040 LOCATE 25,1:PRINT "ITERATING...";"D2=";D2;
1060 FOR IM = 1 TO 29
1070 IF MO > MD(IM) AND MO < MD(IM + 1) THEN MO = MD(IM + 1)
1080 NEXT IM
1090 IF MO = 200 THEN PRINT "MODUL OUT OF RANGE": STOP
1100 D1 = .2 * D2
1105 PA = ATN (MO * N1 / D1)
1110 A = (D1 + D2) / 2
1120 BB = BA * D2
1130 IF BB < .55 * D1 THEN 1150
1140 BA = .95 * BA: GOTO 700
1150 IF BB > .45 * D1 THEN 1180
1160 BA = 1.05 * BA
1175 GOTO 700
1180 MC = ( SQR ((N2 + 2) ^ 2 - N2 ^ 2 * COS (FI) ^ 2) - N2 * SIN (FI) +
SQR ((N1 + 2) ^ 2 - N1 ^ 2 * COS (FI) ^ 2)- N1 * SIN (FI)) / 6.28
1190 CLS
1200 PRINT " RESULTS"
1210 PRINT "                                   PINION      GEAR"
1220 PRINT "-------------------------------------------------------------"
1230 PRINT "No of teeth          ",N1,N2
1240 D1 = D1 * COS (G1):D2 = D2 * COS (G1)
1250 PRINT "Diameter             ",D1,D2
1260 PRINT "BHN                  ", H1,H2
1270 PRINT "Young mod            ",E1,E2
1280 PRINT "Fatigue strength     ",PN,GN
1290 PRINT "Surface strength     ",PH,GH
1295 PRINT
1300 PRINT "SYSTEM PARAMETERS: "
1310 PRINT " Center distance     ";A
1320 PRINT "Modul                ";MO
1330 PRINT "Pressure angle       ";FD
1340 PRINT "Helix angle          ";180 * PA / 3.14
1350 PRINT "Diameter pitch       ";1 / MO * 25.4
1360 PRINT "Contact ratio        ";MC
1365 LOCATE 12,40:PRINT "Width                    ";D2 * BA;
1380 LOCATE 13,40:PRINT "b/a ratio                ";BA;
1390 FT = 2 * M1 / D1
1400 LOCATE 14,40:PRINT "Tangential force   ";FT;
1405 FR=FT * SIN (FI) / ( COS (FI) * SIN (PA) + FF * COS(PA))
1410 LOCATE 15,40:PRINT "Radial force           ";FR;
1430 FA = FT * ( COS (FI) * COS (PA) - FF * SIN (PA)) / (COS (FI) * COS
(PA) + FF * SIN (PA))
1440 LOCATE 16,40:PRINT "Axial force/worm    ";FA;
1450 B =BA * D2:VOLUME=3.14 * B *(D1 ^2 + D2 ^ 2) / 4 /1000000!
1460 LOCATE 17,40:PRINT "Gearing volume/1E6 ";VOLUME
1470 EE = ( COS (FI) - FF * TAN (PA)) / ( COS (FI) + FF / TAN (PA))
1480 PRINT "EFFICIENCY           ";EE:PRINT
1490 INPUT "Do you wand design variations  ? ( Y/N )";X$
1500 IF X$ = "N" OR X$="n" THEN GOTO 1600
1510 INPUT "<1> fatigue strength SN,<2> surface strength SH,<3> no of
starts : choice is :";IIS
1520 ON IIS GOTO 1530,1550,1570
1530 INPUT "New fatigue strength Sn worm,gear   ";PN,GN
1540 GOTO 720
1550 INPUT "New surface strength SH worm,gear   ";PH,GH
1560 GOTO 720
1570 INPUT "New number of starts NW    ";NW
1580 GOTO 720
1590 GOTO 1510
1600 END
```

CHAPTER FOURTEEN

DESIGN OF AXISYMMETRIC ELEMENTS

14.1 DYNAMICS OF ROTARY MOTION

Rotary motion is, by its nature, very advantageous for use in machines. Almost invariably, machines have members with rotary motion.

The main advantage of the rotary motion as compared with linear motion is the absence of speed variation or reversal, associated with high acceleration (or deceleration), inertial loads and finally inertial stresses.

However, this is not always the case. Beyond the centrifugal inertial loads and stresses, there are factors that change the speed of rotation and cause acceleration and dynamic loads, such as start-up and coasting-down, intermitted loads, accidental or intended braking, etc.

The static torque for a transmitted power P at rotating speed n is

$$T_o(\text{Nm}) = 7026P\,(\text{hp})/n\,(\text{rpm}) \tag{14.1}$$

$$T_o(\text{Nm}) = 9549P\,(\text{kW})/n\,(\text{rpm})$$

If the transmitted power is P (watts) and the angular velocity is ω (rad/sec), then

$$T_o = P/\omega\,(\text{Nm}) \tag{14.2}$$

Dynamic loads can be computed in most cases.

In absence of the required information for analytical evaluation for dynamic loads, for preliminary design the static torque is increased to account for these loads with the empirical formula

$$T_d = T_o(c_1 + c_2) \tag{14.3}$$

where c_1 depends on the driving machine and c_2 depends on the driven machine (Table 14.1).

Starting torques are not included in the factors of Table 14.1 and they must be accounted for separately because sometimes they can reach very high values.

Factors determining the starting torque are the start-up time and the inertia of the system. The starting torque is also limited by the starting characteristics of the driving machine. For most driving machines, the maximum torque is known. Since in many cases this torque is high, the design objective is to increase the start-up time in order to have lower starting torque. Then, the starting torque, assumed constant, is for constant acceleration motion

$$T_{su} = \lambda J\omega/t + T_f \tag{14.4}$$

Table 14.1 Service factors for rotating machinery

Rotating machine	
Driving	c_1
Electric motors	0.5
Turbomachinery	0.75
Reciprocating engines	
4-cylinder	1.5–2.8
6-cylinder	1–2
Driven	c_2
Machine tools	3
Rolling mills	1.6–2
Blowers	1.5
Textile machinery	1.6–2.2
Rotary pumps–compressors	1.5
Reciprocating pumps–compressors	2.2–3.2
Paper mills	1.8–3
Mixing machines	1.8–3
Wood-processing machines	2–3
Electric generators	1.1–3
Hoists	1.5
Elevators	2
Cranes	3
Food processing machines	1.3–1.5
Construction machinery	1.5–2

where J is the rotating system inertia, ω the final angular velocity of rotation, t the start-up time, T_f the constant friction torque and λ a factor expressing the deviation from the assumed constant acceleration. If no measurements are available, λ must be taken between 1.2 and 2.

This approach yields acceptable results for compact machines designed with high safety factors. In recent years, computer simulation is extensively used yielding very accurate results provided that the torque-speed characteristics for the driving machine and the resisting torque are known. In most cases, knowledge of this information can be assumed and then we proceed with the system modeling and simulation which gives detailed information for the dynamic loads for the system.

A rotary dynamic system consists, in its simplest modes, on two rigid rotors, prime mover and driven machine, connected by way of an elastic shaft with negligible mass compared with the rotors and behaving as a rotary linear spring (Figure 14.1).

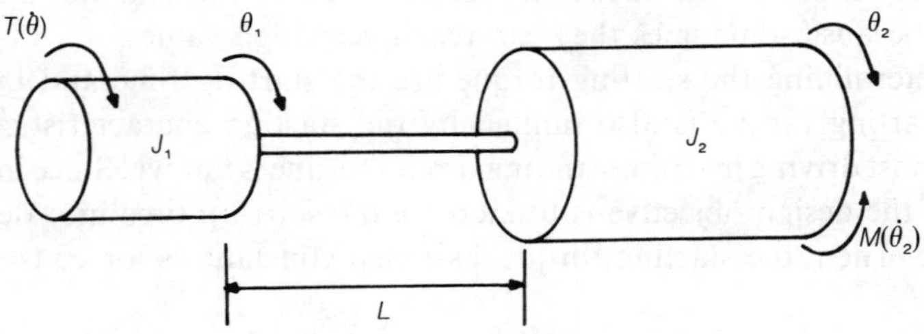

Figure 14.1 A two rotor system

For a straight shaft, the rotary spring has constant

$$k = T/\Delta\varphi = GI_p/L \tag{14.5}$$

where G is the shear modulus, I_p the polar moment of inertia, L the length of the shaft, T the static torque and $\Delta\varphi$ the resulting angular twist of the shaft.

If the shaft is a compound of sections with rotary spring constants $k_1, k_2, \ldots, k_n$, then an equivalent spring will be constant, since the twists are additive,

$$\frac{1}{k} = \frac{1}{k_1} + \frac{1}{k_2} + \cdots + \frac{1}{k_n} \tag{14.6}$$

If now, the motor and machine rotors have polar moments of inertia J_1 and J_2 and at time t they have rotated from an arbitrary initial position by angles θ_1 and θ_2 respectively, then the twist of the shaft will be $\theta_2 - \theta_1$ and the reaction torque of the twisted shaft will have magnitude $k(\theta_2 - \theta_1)$. Applying Newton's Law in respect to rotation for the rotors 1 and 2 yields

$$\left.\begin{aligned} J_1\ddot{\theta}_1 &= T(\dot{\theta}_1) + k(\theta_2 - \theta_1) \\ J_2\ddot{\theta}_2 &= -k(\theta_2 - \theta_1) - M(\dot{\theta}_2) \end{aligned}\right\} \tag{14.7}$$

or

$$\left.\begin{aligned} J_1\ddot{\theta}_1 + k\theta_1 - k\theta_2 &= T(\dot{\theta}_1) \\ J_2\ddot{\theta}_2 - k\theta_1 + k\theta_2 &= -M(\dot{\theta}_2) \end{aligned}\right\} \tag{14.8}$$

T is the torque applied by the motor which is a function of the speed of rotation $\omega_1 = \dot{\theta}_1$, and M is the resisting torque of the driven machine, usually an ascending function of the speed $\omega_2 = \dot{\theta}_2$.

Typical such functions are usually available from motor manufacturers and they have forms indicated in Figure 14.2. For induction motors Figure 14.3 gives the torque versus speed of rotation for most common speeds and capacities.

Solution of equations (14.8) is performed usually by computer in the following manner: Define the quantities

$$\left.\begin{aligned} x_1 &= \dot{\theta}_1, \quad x_2 = \theta_1 \\ x_3 &= \dot{\theta}_2, \quad x_4 = \theta_2 \end{aligned}\right\} \tag{14.9}$$

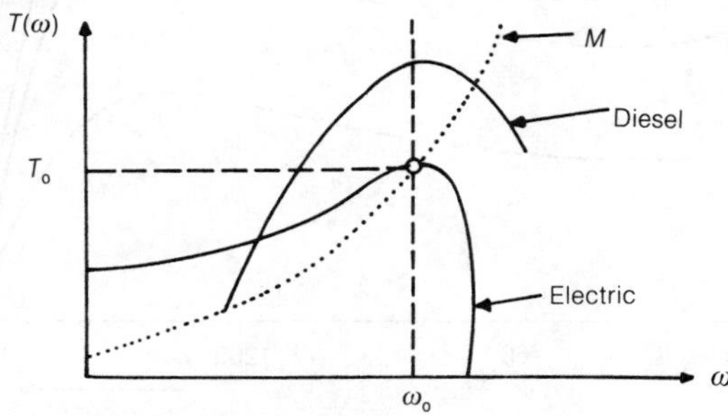

Figure 14.2 Torque characteristics of prime movers

Then, equations (14.8) are equivalent to

$$\left.\begin{aligned} \dot{x}_1 &= [T(x_1) + k(x_4 - x_2)]/J_1 \\ \dot{x}_2 &= x_1 \\ \dot{x}_3 &= -k(x_4 - x_2)/J_2 - M(x_3)/J_2 \\ \dot{x}_4 &= x_3 \end{aligned}\right\} \qquad (14.10)$$

In equations (14.10) there are four unknown functions of time x_1, x_2, x_3, x_4 and at any time if their values are known, their rate of change can be computed from the system (14.10). We make use of the fact that at some time, taken arbitrarily as time 0, we know the values of x_1, x_2, x_3, x_4, that is of $\dot{\theta}_1 = \omega_1, \theta_1, \dot{\theta}_2 = \omega_2, \theta_2$. At start-up all these angles and angular velocities are zero. Therefore, equations (14.10) yield the derivatives which can be used to calculate the new values of x_1, x_2, x_3, x_4 after a small elapsed time Δt. The simplest scheme would be to use the **Taylor expansion** for each function $x(t)$ as

$$x(t + \Delta t) = x(t) + \Delta t \frac{\mathrm{d}x(t)}{\mathrm{d}t} + \frac{\Delta t^2}{2!} \frac{\mathrm{d}^2 x(t)}{\mathrm{d}t^2} + \cdots \qquad (14.11)$$

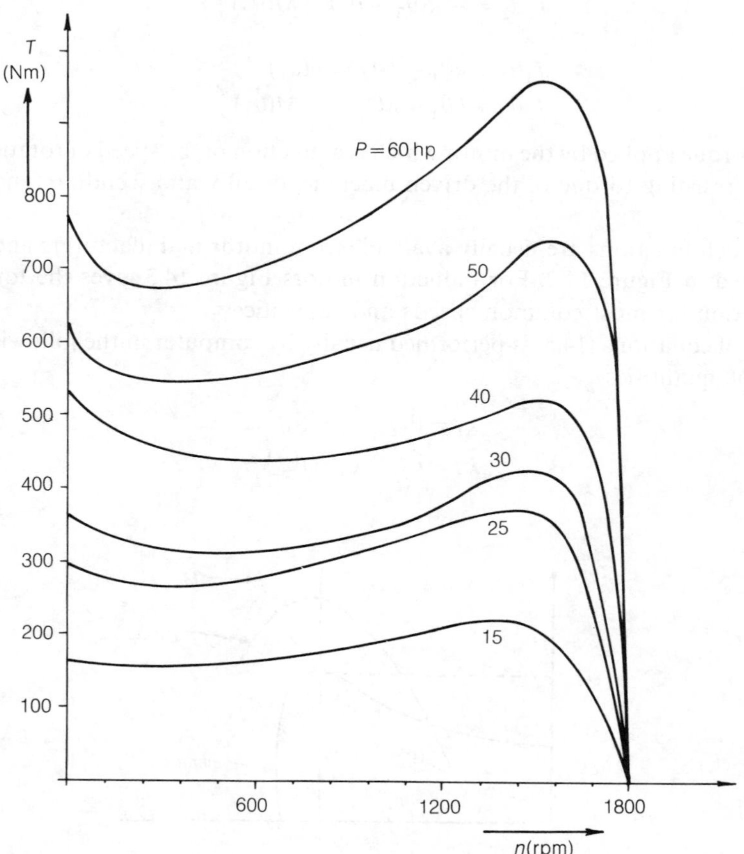

Figure 14.3a Torque characteristics of 1800 rpm, four-pole/60-Hz asynchronous motors

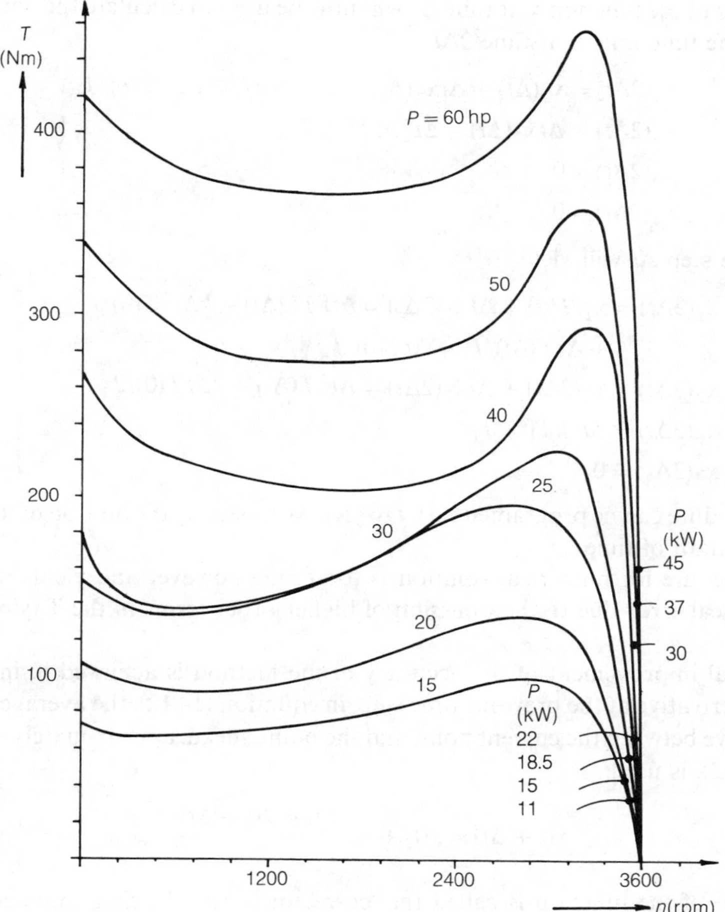

Figure 14.3b Torque characteristics of 3600 rpm, two-pole/60-Hz asynchronous motors

Retaining only the first two terms yields

$$\left.\begin{aligned} x(t+\Delta t) &= x(t) + \Delta t\, \mathrm{d}x(t)/\mathrm{d}t \\ x(t+\Delta t) &= x(t) + \Delta t \dot{x}(t) \end{aligned}\right\} \tag{14.12}$$

This formula can be applied successively for each time t to compute the values of the functions x in equations (14.10) at time $t + \Delta t$

$$\left.\begin{aligned} x_1(\Delta t) &= \Delta t\, T(0)/J_1 \\ x_2(\Delta t) &= \Delta t \dot{x}_2(0) = 0 \\ x_3(\Delta t) &= \Delta t \dot{x}_3(0) = 0 \\ x_4(\Delta t) &= \Delta t \dot{x}_4(0) = 0 \end{aligned}\right\} \tag{14.13}$$

In general $x(\Delta t) = x(0) + \dot{x}(0)\Delta t$.

The values of the function x at time Δt will now be used to calculate the derivatives and the value of the functions x at time $2\Delta t$:

$$\left.\begin{aligned} x_1(2\Delta t) &= x_1(\Delta t) + \Delta t x_1(\Delta t) = T(\Delta t)\Delta t/J_1 + \Delta t T(0)/J_1 \\ x_2(2\Delta t) &= \Delta t x_2(\Delta t) = \Delta t[\Delta t T(0)/J_1] \\ x_3(2\Delta t) &= 0 \\ x_4(2\Delta t) &= 0 \end{aligned}\right\} \tag{14.14}$$

One more step Δt will yield

$$\left.\begin{aligned} x_1(3\Delta t) &= x_1(2\Delta t) + \Delta t\dot{x}_1(2\Delta t) = \Delta t[T(2\Delta t) - k\Delta t^2 T(0)/J_1]/J_1 \\ &\quad + \Delta t T(\Delta t)/J_1 + \Delta t T(0)/J \\ x_2(3\Delta t) &= x_2(2\Delta t) + \Delta t\dot{x}_2(2\Delta t) = \Delta t^2 T(\Delta t) + 2\Delta t T(0)/J_1 \\ x_3(3\Delta t) &= \Delta t^3 k T(0)/J_1 \\ x_4(3\Delta t) &= 0 \end{aligned}\right\} \tag{14.15}$$

This procedure can be programed easily to yield successively the motion of the system for any desired length of time.

This procedure for numerical solution is too crude however and yields progressively higher numerical error due to the omission of higher-order terms in the Taylor expansion (14.11).

Substantial improvement of the accuracy of the method is achieved if, instead of the value of the derivative at the previous time step, in equation (14.12), the average of the value of the derivative between the current point and the point ahead, approximately estimated by equation (14.12), is used:

$$x(t + \Delta t) = x(t) + \Delta t\frac{\dot{x}(t) + \dot{x}(t + \Delta t)}{2} \tag{14.16}$$

This value of the function is called the 'corrector' while the first value computed by equation (14.12) is the 'predictor'. The value of the predictor is used only to compute the

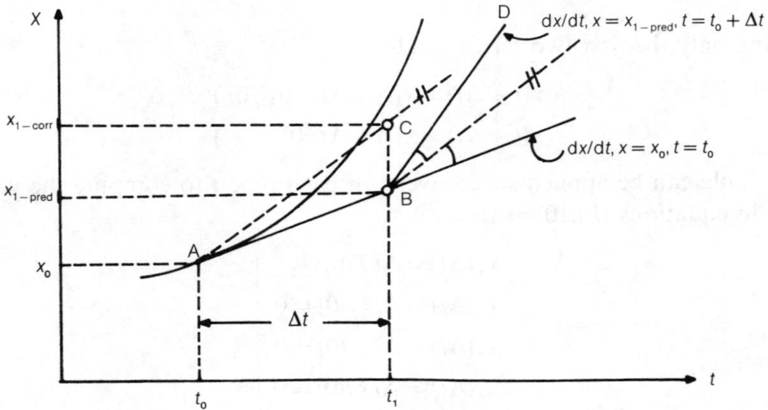

Figure 14.4 Euler integration of differential equations

derivative used in equation (14.16) for the step ahead. A geometric interpretation of this procedure is shown in Figure 14.4.

The value x is known at some value of time t. The value of x after some time step t is computed. Point B has x given by equations (14.12). This value of the function x is used to compute the derivative BD. A straight line from point A, dichotomous of the angle formed by the two derivatives yields point C with the line $t_1 = t_0 + \Delta t$, the corrector, equation (14.16). The point t_1 is used now to compute one more point ahead, and so on.

This procedure has been coded in the program SIMUL (Appendix 14.B).

14.2 PRELIMINARY SHAFT DESIGN

14.2.1 Fundamentals

Shafts are solids of revolution intended for transmitting a torque along their axes and for supporting rotating machine components. Since the transmission of torque is associated with the development of forces applied to the shafts, such as forces acting on the teeth of gears, belt tension, etc., shafts are usually subjected to transverse forces and bending moments in addition to the torque. Shafts transmitting forces and moments between their two ends without intermediate loads are known as axles (Reshetov 1980, Shigley 1972).

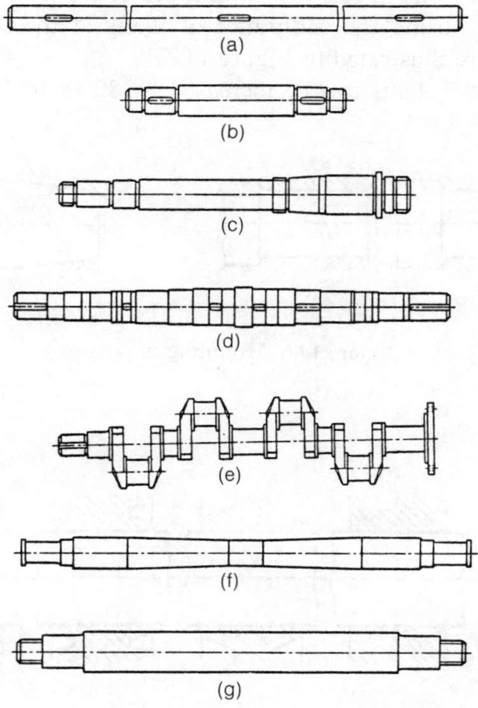

Figure 14.5 Different rotor designs. (Reshetov 1980)

Shafts can be classified with respect to their purpose, as transmission shafts carrying drive members, such as toothed gears, pulleys, chain sprockets and clutches (Figures 14.5a and b), and as main shafts of machines and other special shafts which, in addition to drive members, carry the operating members of engines or machine tools. Such members may include turbine wheels and disks, cranks, etc. (Figures 14.5c, d, e).

With respect to the shape of their geometric axes, shafts are classified as 'straight shafts' and 'crankshafts'. Crankshafts (Figure 14.5e), are used to convert reciprocating motion into rotary or vice versa. They combine the functions of ordinary shafts with those of cranks in slider-crank mechanisms. In a separate group are shafts with a geometric axis of variable shape, Figure 14.5(f, g).

The supporting sections of shafts are called 'journals' (Figure 14.6).

As to their shape, straight shafts can be plain (constant-diameter) shafts and stepped shafts. The diameter of a shaft along its length is determined by the load distribution, bending moments and torques, axial loads and conditions imposed by the manufacturing and assembly processes used.

Equal-strength shafts are more preferable than shafts of constant cross-section. Usually they are of stepped design. Such a shape is convenient in manufacture and assembly. The shoulders of the shaft can carry large axial forces.

Shafts may be hollow in design. A hollow shaft with a hole-to-outside-diameter ratio of 0.75 is lighter by about 50% than a solid shaft of equal strength and rigidity.

The endurance of shafts is influenced by stress concentration. For this reason, special design and processing methods to raise the endurance of shafts are employed.

Design methods for raising the endurance of shafts at the seating (fit) surfaces by reducing edge pressure are illustrated in Figure 14.7.

The endurance limit of shafts can be increased by 80 to 100% by strengthening the

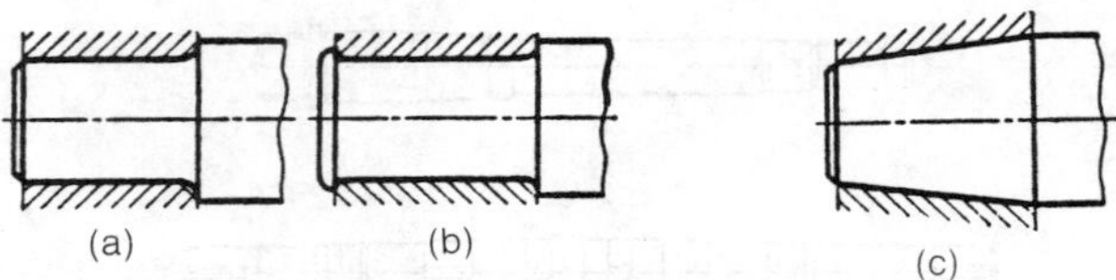

Figure 14.6 Mounting of pins

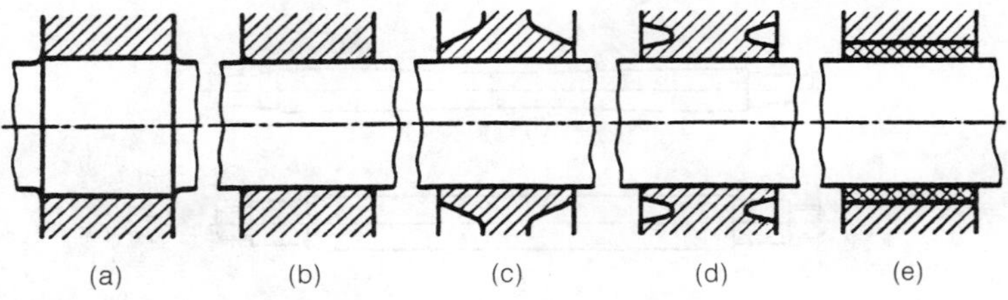

Figure 14.7 Mounting of hubs

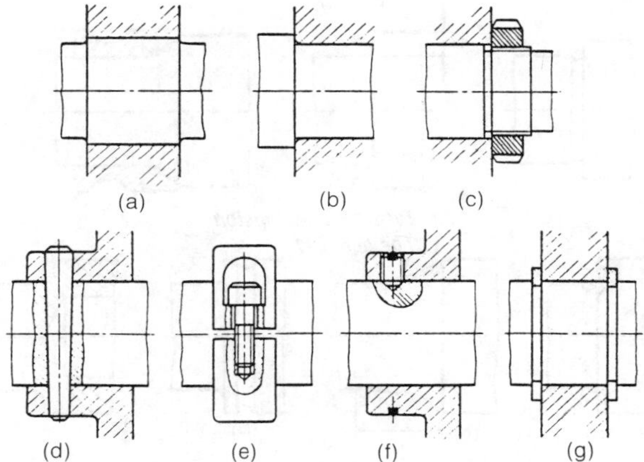

Figure 14.8 Securing hubs for axial motion

surfaces for seating hubs by work (strain) hardening (using a roll or ball burnishing process). This is effective for shafts up to 500 or 600 mm in diameter. Such hardening procedures and shot peening are widely used.

Axial loads acting on shaft or on components mounted on the shafts are transmitted as follows (Figure 14.8):

(1) *Heavy loads.* Transmission by having the loaded components bear against shoulders of the shaft, by mounting the components or locating rings with an interference fit (Figures 14.8a, b).
(2) *Medium loads.* Transmission by means of nuts, pins passing directly through the loaded components or through locating rings, or by clamp joints (Figures 14.8c, d, e).
(3) *Light loads.* Transmission by means of set screws holding either the component or a locating ring, by clamp joints or by snap locating rings (Figures 14.8e, f, g).

Two-way axial securing of a shaft is not compulsory if it is held in place by a constant force which prevents displacement (for instance, the weight of the components in the case of heavy vertical shafts).

The transition between two shaft steps of different diameters is usually designed with a fillet of a single radius or a transition surface (Figure 14.9). The radius of the fillet is to be taken as smaller than that of the edge round or radial dimension of the chamfer of the components to be mounted on the shaft step against the shoulder.

It is desirable to have the fillet radius of heavily stressed shafts larger or equal to $0.1d$.

By designing fillets of optimal shape that extend over a considerable length of the shaft, it is possible to practically eliminate stress concentration. In Figure 14.9 the optimal shapes for shafts subject to bending, torsion and tension are shown. The transition zone extends over a length of the shaft equal to its diameter. For this reason it is possible to use such fillets only in rare cases, for example for torsion shafts (i.e. shafts serving as springs and operating in torsion), for the free sections of heavily stressed shafts, etc.

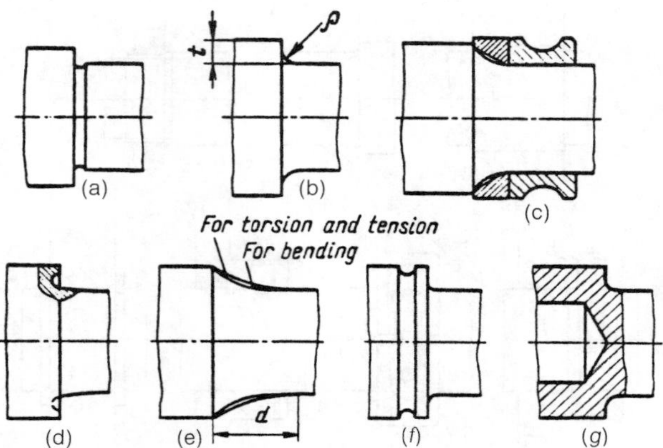

Figure 14.9 Shaft fillets for low stress concentration. (Reshetov 1980)

An effective method for increasing the strength of shafts in their transition zones is the removal of low-stressed material by providing load-relief grooves (Figure 14.9f) or by drilling an axial hole in the larger step (Figure 14.9g). This yields a more uniform stress distribution and reduces stress concentration.

Strain hardening of shaft fillets (by rolling or, for large shafts, by shot peening) can raise the load-carrying capacity of shafts by 50 to 100%.

14.2.2 Shaft materials

The selection of the material and heat treatment for shafts and axles depends upon their criteria of working capacity, including the criteria for the journals with their supports.

Because of their high strength, high modulus of elasticity, capacity to be strengthened by hardening and ease of obtaining the required bar stock or blanks by mill rolling, carbon and alloy steels are the main materials for manufacturing shafts and axles.

Shafts and axles of high rigidity which will not be heat-treated are made mainly of low carbon steels, such as AISI 1015 and 1020. The great majority of shafts are made of medium-carbon and alloy steels, grades 1040, 1045, which can be heat-treated. Alloy steels, grades 5115–5120, etc. are steels for highly-stressed shafts of critical machinery. Shafts of these steels are usually subjected to structural improvement, hardening followed by high-temperature tempering or induction surface-hardening followed by low-temperature tempering (spline shafts).

High-speed shafts running in sleeve bearings require journals of especially high hardness. They are made of case hardening steels or nitriding steels. Chromium-plated shafts have high wear resistance. In automotive engineering, chromium plating of the crankpins and main journals of crankshafts lengthens their service life up to regrinding by 3 to 5 times.

High-strength heat-treated steels can be efficiently employed for shafts whose dimensions are designed for rigidity only when dictated by requirements of the durability of splines and other surfaces subject to wear.

14.2.3 Design criteria

Shafts and rotating axles are usually designed as beams on hinged supports. This assumption corresponds with sufficient accuracy to the actual conditions for shafts running in antifriction bearings and having a single bearing in each support (Figure 14.10a). For shafts running in antifriction bearings and having two bearings in each support (Figure 14.10b), the support reaction is carried mainly by the bearing on the side of the loaded span. The outer bearings are subject to much less load and if they are not installed up against the inner bearings they may be subject to a support reaction in the opposite direction. For this reason it is more accurate to have the nominal hinged supports coincide with the middle of the inner bearings or located one-third of the distance between the bearings of each support, closer to the inner bearings. More precise calculations for such shafts should take into account the combined behavior of the shaft with the bearings as for multiple-support beams on elastic supports.

For shafts running in non-self-aligning sleeve bearings (Figure 14.10c), the pressure along the length of the bearings is asymmetrically distributed due to deformation of the shaft. The nominal hinged support should be located at the distance (0.25 to 0.3) l from the inner face of the bearing, but not over one-half of the shaft diameter from the bearing face on the side of the loaded span. More precise calculations for these shafts should take their combined behavior with the bearings into consideration.

Forces are transmitted to a shaft through the components mounted on it: toothed gears, chain sprockets, pulleys, couplings, etc. For the simplest calculations it is assumed that the components transmit forces and torques to the shaft at the middle of their width, and design is based on the corresponding cross-sections. Actually, forces of interaction between hubs and shafts are distributed along the length of the hubs, and the latter function in conjunction with the shafts (Figure 14.10d). It is more accurate to regard the moments as being in cross-

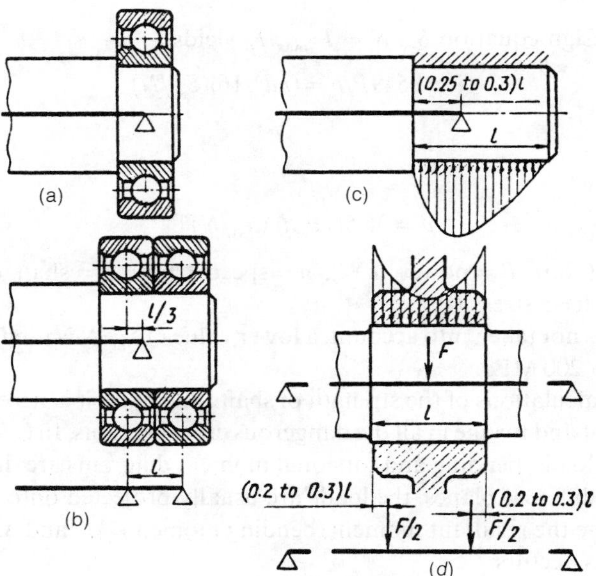

Figure 14.10 Load distribution at hubs and supports. (Reshetov 1980)

sections at distance of (0.2 to 0.3) l from the faces of the hub, where l is the hub length, and to assume that the concentrated forces of interaction between the hub and shaft are in the same cross-sections. The smaller distances from the design cross-sections to the hub faces are selected for interference fits and rigid hubs; the larger distances, for clearance fits and to less rigid hubs.

Of the various strength criteria for most shafts of high-speed machinery, endurance is of prime significance. Fatigue failures make up to 40 or 50% of the cases in which shafts become inoperative. Low-cycle fatigue may develop in operation with high overloads. For slow-speed shafts of normalized, structurally improved or hardened steels with high-temperature tempering, the limiting criterion may also turn out to be the static load capacity at peak loads (absence of excessive permanent deformation). And finally, the criterion for the shafts of brittle or low-ductility materials (cast iron or low-tempered steel), is the resistance to brittle fracture.

Rigidity criteria of shafts are determined by operating conditions of gears and bearings, and vibrations.

14.2.4 Strength calculations

The main loads acting on shafts are forces due to power transmission.

Forces constant in magnitude and direction cause constant stresses in stationary axles, and stresses that vary in an alternating symmetrical cycle in rotating axles and in shafts. Constant loads rotating together with the axles and shafts, due for instance, to unbalanced rotating components, cause constant stresses.

For preliminary design, torsion calculations should be used. This is because the length dimensions of the shaft have not yet been determined so that it is impossible to compute the bending moments.

The torsion design equation $S_{sy}/N = Tr_{max}/I_p$ yields

$$T = 9549P/n = (\pi d^3/16)(S_{sy}/N) \tag{14.17}$$

from which

$$d = 36.5[(P/n)/(S_{sy}/N)]^{\frac{1}{3}} \tag{14.18}$$

where T = torque (Nm), P = power (kW), n = speed (rpm), d = shaft diameter (m), and S_{sy}/N = allowable shear stress (Pa).

Since bending is not taken into account, a lower value of S_{sy}/N is used. It is often selected in the range 120 to 200 MPa.

For the main calculations of the strength of shafts and axles it is necessary to determine the bending moment and torque in all the dangerous cross-sections. In calculations for shafts subject to complex loads, bending and torsional moment diagrams are drawn. If the shaft is subject to loads in different planes, the loads are usually projected onto two perpendicular planes. To determine the resultant moment, bending moments M_x and M_y in perpendicular planes are added as vectors:

$$M_b = [M_x^2 + M_y^2]^{\frac{1}{2}} \tag{14.19}$$

The several stresses are computed and the failure criterion is the applied to yield the design equation.

The dangerous cross-section is determined by the moment diagrams, size of the shaft cross-sections and stress concentration. With some practice, the location of the dangerous cross-section can be readily determined without any calculations. In many cases, calculations are carried out for many sections.

The equivalent stress in the dangerous cross-section is

$$\sigma_{eq} = [\sigma^2 + a^2\tau]^{\frac{1}{2}} \leqslant S_y/N \tag{14.20}$$

where σ and τ are normal and shear stresses.

Expressing the stresses in terms of the moments and assuming $a = 4$ (according to the maximum shear theory of failure), we can write

$$\sigma_{eq} = (32/\pi d[M_b^2 + T^2]^{\frac{1}{3}} \leqslant S_y/N \tag{14.21}$$

In calculations based on the static strength at overloads, M_b and T are to be regarded as the nominal moments multiplied by the service factor.

Owing to their convenience, static strength calculations based on nominal stresses are efficiently applied in design to determine the diameters of axles and shafts with subsequent checking calculations based on endurance. Here, the calculations are usually carried out to the nominal moments and the service factors are taken from Table 14.1. In precise calculations the stresses are increased. Strengthening techniques enable the allowable stresses to be raised substantially.

The diameter of an axle subject to bending is determined from equation (14.20), taking $T = 0$. Thus,

$$d = [32M_b/(\pi S_y/N)]^{\frac{1}{3}} \tag{14.22}$$

The diameter of a shaft subject to both bending and torsion is

$$d = [32(M_b^2 + T^2)^{\frac{1}{2}}/(\pi S_y/N)]^{\frac{1}{2}} \tag{14.23}$$

14.2.5 Fatigue strength

Fatigue analysis takes into consideration the character of stress variation, static and fatigue strengths of the materials, stress concentration, the scale factor, surface conditions and surface hardening as presented in Chapter 7. Known values for calculations must include the constant (M_m and T_m) and variable (M_r and T_r) stress components (see Section 7.2).

The equivalent static stress from the **Goodman Diagram** (see Example 7.1) on p. 297, is

$$\sigma_{eq} = \sigma_m + \varepsilon\sigma_r \tag{14.24}$$

where ε depends on the magnitude of the asymmetry factor $r = \sigma_r/\sigma_m$. More precisely, for

$$r = \sigma_r/\sigma_m > r_D, \quad \varepsilon = S_u/S_e$$

and for

$$r = \sigma_r/\sigma_m < r_D, \quad \varepsilon = 1 \tag{14.25}$$

where $r_D = (S_e/S_u)(S_u - S_y)/(S_y - S_e)$.

In terms of equation (14.20), the asymmetry factor is, relating stresses to moments and torques,

$$r = \frac{\sigma_r}{\sigma_m} = [M_r^2 + T_r^2]^{\frac{1}{2}}[M_m^2 + T_m^2]^{-\frac{1}{2}} \tag{14.26}$$

Equation 14.23 now becomes

$$d = \left\{\frac{32N}{\pi S_y}[(M_m^2 + T_m^2)^{\frac{1}{2}} + \varepsilon(M_r^2 + T_r^2)^{\frac{1}{2}}]\right\}^{\frac{1}{3}} \tag{14.27}$$

This design equation must be used iteratively because in the determination of the fatigue strength some factors depend on the diameter.

Stress concentration factors are used in conjunction with the nature of the material, as explained in Chapter 7. In short, fatigue stress concentration factors are always used if variable stresses are present. On the steady stresses, stress concentration factors are used only for brittle materials.

In equation (14.27), the range for the torque T_r must be used with caution because torsion on the Goodman Diagram appears as a horizontal line. Therefore, equation (14.27) can be used for small values of the torque range only, which is almost always the case.

Under average conditions, the fatigue safety factor is $N_f = 1.5$ to 2.5. If the diameters of shafts are determined from conditions of sufficient rigidity, the value of N_f may be substantially greater. If the design loads have been determined with enough accuracy, and with accurate calculations and homogeneous materials, a lower endurance safety factor can be chosen. The lowest advisable value is $N_f = 1.3$.

If a shaft operates under conditions of non-steady-state loading and it is necessary to completely utilize all strength margins, calculations are carried out on the basis of the **Miner's Rule** (equation 7.8 on p. 291 and the fatigue curve, equation 7.3, on p. 282). Thus,

$$\sigma_{eq} = \left[\left(\sum_i n_{ri}\right)\Big/ N_0 a\right]^{1/m} \tag{14.28}$$

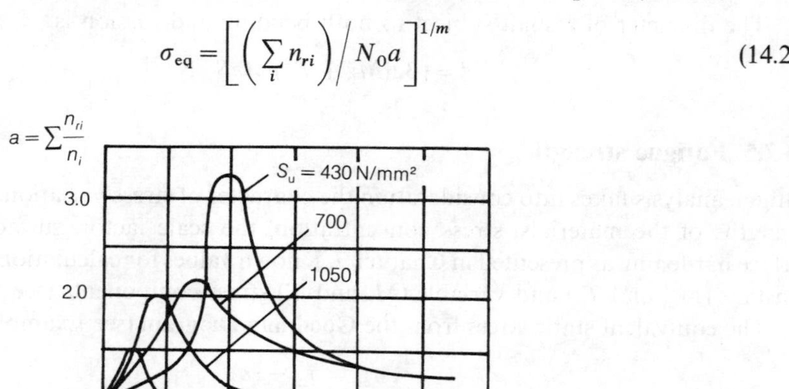

Figure 14.11 Miner's ratio for cummulative fatigue damage. (From Reshetov, by permission of Mir Publishers, Moscow)

where N_0 = number of cycles up to the inflection point of the fatigue curve, taken equal to (3 to 5) $\times 10^6$ for shafts of small cross-section and 10^7 for large shafts, n_{ri} = total number of loading cycles at stress, m = exponent of the fatigue curve, usually taken equal to 9 (for shafts with press-fitted components, $m = 6$), and σ_{ri} = alternating stress in the shaft at the maximum load applied during an appreciable length of time.

The Miner's fraction a is shown in Figure 14.11. It is greater, in particular, for soft steels. In general, a is greater than 1 in the zone $\sigma_2 > \sigma_1 > S_e$, and reaches a maximum in the zone $\sigma_2/\sigma_1 = 1.1$ to 1.2. If experimental data are not available, it is advisable to take a equal to 1, which gives a slight increase in the factor of safety and more conservative design.

14.3 DESIGN OF SHAFTS FOR RIGIDITY

The required rigidity of shafts subject to bending is determined mainly by conditions of proper operation of the drive and bearings.

The deflection of shafts has a small effect on belt and chain drives and therefore they are not usually checked for rigidity. The elastic displacement (deflection) of shafts carrying toothed gears, leads to angular displacement of meshing gears with respect to each other and consequently load concentration along the face width of the meshing teeth. It also leads to separation of the shaft axes which is unfavorable for gearing. In the case of involute gearing, it shortens the line of contact between the meshing teeth.

The rigidity of shafts running in non-self-aligning sleeve bearings should be sufficient to ensure the required uniform pressure distribution along the length of the bearing.

The rigidity of shafts running in ball bearings must be such that the balls do not jam due to misalignment of the rings if self-aligning bearings are not used.

Empirical rules are available for checking the allowable deflections and angles of inclination of the elastic lines of shafts. Thus, in speed gearboxes, the maximum deflection of shafts carrying toothed gears should not exceed 0.0002 to 0.0003 of the distance between the supports; the angle of the mutual inclination of shafts with meshing gears should be less than 0.001 radian. The angle of inclination in a radial ball bearing can be 0.5° (under the condition that the bearing life is reduced by only 20% and if the bearing unit is manufactured with ordinary accuracy). The same values for roller bearings with short cylindrical rollers and for tapered roller bearings are 0.2 and 0.15°. The maximum deflection of the shafts of induction motors should not exceed 0.1 of the air gap. These relationships are, however, of a specific nature and cannot, of course, be generalized.

The flexural rigidity (stiffness) of shafts is also important from the point of view of vibrations.

Deflections and angles of inclination of the elastic lines of shafts are determined by the conventional methods of the strength of material. For simple cases of design it is expedient to employ design formulas, treating the shaft as a beam of constant cross-section of the equivalent diameter (Figure 14.12). The computer methods of Chapter 4 must be used for shafts of complex geometry.

The required torsional rigidity of shafts is determined by various criteria.

Static elastic angular deformation of kinematic trains can affect the accuracy of machine performance, for instance that of precision engine lathes and gear-cutting machines, dividing machines, etc. The elastic deformations of the drives for low-speed machinery may cause

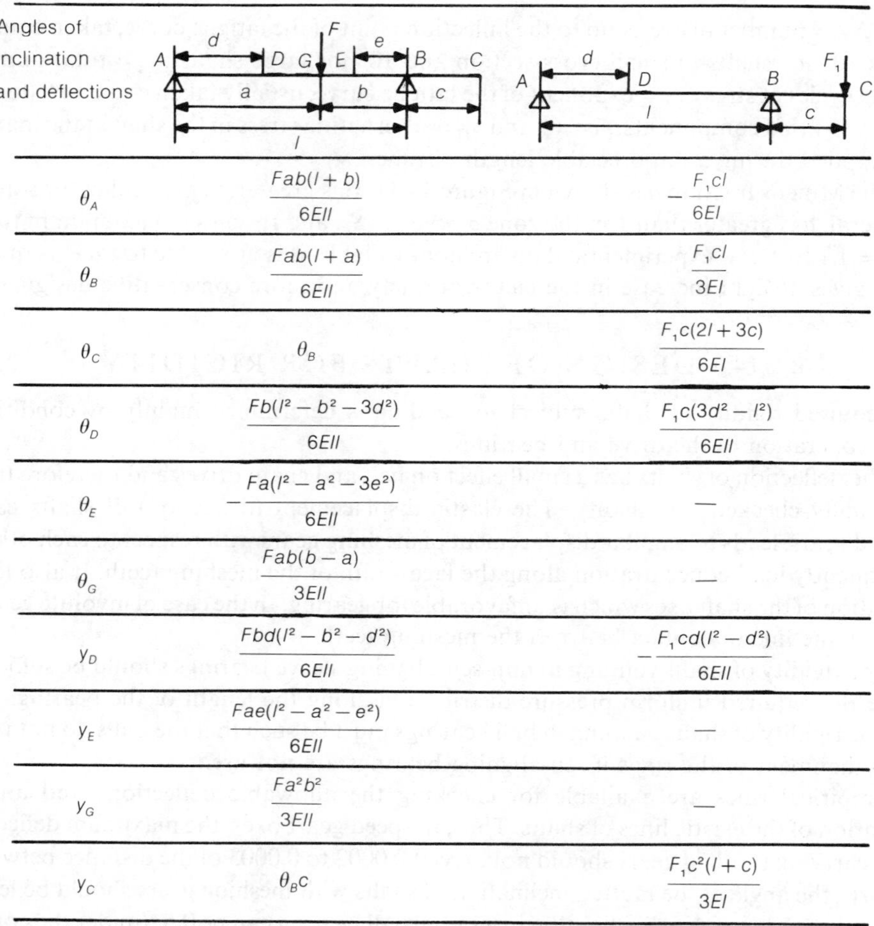

Angles of inclination and deflections		
θ_A	$\frac{Fab(l+b)}{6EIl}$	$-\frac{F_1cl}{6EI}$
θ_B	$\frac{Fab(l+a)}{6EIl}$	$\frac{F_1cl}{3EI}$
θ_C	θ_B	$\frac{F_1c(2l+3c)}{6EI}$
θ_D	$\frac{Fb(l^2-b^2-3d^2)}{6EIl}$	$\frac{F_1c(3d^2-l^2)}{6EIl}$
θ_E	$-\frac{Fa(l^2-a^2-3e^2)}{6EIl}$	—
θ_G	$\frac{Fab(b-a)}{3EIl}$	—
y_D	$\frac{Fbd(l^2-b^2-d^2)}{6EIl}$	$-\frac{F_1cd(l^2-d^2)}{6EIl}$
y_E	$\frac{Fae(l^2-a^2-e^2)}{6EIl}$	—
y_G	$\frac{Fa^2b^2}{3EIl}$	—
y_C	θ_Bc	$\frac{F_1c^2(l+c)}{3EI}$

Figure 14.12 Beam formulas for shaft deformations. (Reshetov 1980)

stick-slip motion. For this reason, for example, the angles of twist of long feed shafts of heavy machine tools are limited to values of the order of 0.5° per meter length. The elastic deformations of divided drives powered from a single motor and used for traversing overhead cranes, portals, cross-members of heavy machine tools, etc. may result in jamming of the slideways. In the transmission shafts of the mechanisms for traversing travelling cranes, angles of twist range from 0.15 to 0.20° per meter of length.

Insufficient torsional rigidity of pinion gears cut integral with their shafts may lead to load concentration along the face width of the teeth.

Torsional rigidity is of special significance for shafts for which torsional vibrations are dangerous, for example in the drives of piston engines. It can prevent resonance vibrations and increase the service life of the gearing.

For most shafts, torsional rigidity is of no essential importance and there is no necessity to check the rigidity of such shafts. An allowable angle of twist equal to 1/4° per meter length, frequently given in the literature, has become obsolete and cannot be technically substantiated. In certain cases it is exceeded by many times. This is especially true for small-diameter shafts because the stress is inversely proportional to the cube of the shaft diameter and also to the angle of twist per unit length, to the fourth power. For instance, the angles of twist of the propeller shafts of automobiles (30 to 50 mm in diameter) may reach several degrees per meter of length. The angle of twist of the cylindrical portion of length L of a shaft subject to the torque T is

$$\Delta\varphi = \frac{TL}{GJ} = cT \tag{14.29}$$

where G = shear modulus, J = polar moment of inertia of the shaft cross-section, and c = torsional compliance of the cylindrical portion of the shaft.

For a portion weakened by keyways, a factor of reduction in rigidity, k, is introduced into the right-hand member of the equations:

$$k = [1 - (4nh/d)]^{-1}, \quad \Delta\varphi = ckT \tag{14.30}$$

where h = depth of the keyway and n = factor equal to 0.5 for one key, 1 for two keys at an angle of 90°, 1.2 at an angle of 180°, and 0.4 for two tangent keys at an angle of 120°.

Peterson gives much useful information on this subject (see References).

14.4 VIBRATION OF ROTATING SHAFTS

A considerable part of the system is involved in the vibrations observed in machines, including the main kinematic train and the basic parts.

The individual vibrations of the separate transmission shafts, such as the shafts of the gearbox, play no essential role in the dynamics of a machine and are therefore not considered separately. In contrast, the vibrations of main shafts with their attached assemblies and supports (turbine rotors, crankshafts of piston engines, machine tool spindles with workpieces, etc.) may be the governing factor in design.

Of main practical importance in shaft design is the determination of the natural frequency of vibration to avoid resonance of vibrations, i.e. an increase in the amplitude of vibrations when the frequency of the exciting forces coincides with the natural frequency of vibrations. Observed in shafts are transverse or bending vibrations, angular or torsional vibrations and also combined bending and torsional vibrations.

The most widely employed are calculations of the fundamental frequencies of vibrations since these vibrations are usually dangerous because most machines operate near the lowest critical speed. Heavy or very high-speed machinery, such as turbomachinery, might operate near the second or the third.

The fundamental frequency of natural vibration of a shaft with a concentrated mass, taking into consideration the dead weight of the shaft, is most readily found if the reduced mass of the shaft is added to the concentrated mass. The reduction factor for transverse vibrations of an overhanging axle of constant cross-section with a mass at its end is 0.235. For a two-support shaft or axle with the mass at the middle it is 0.48. For torsional vibrations of a shaft with one end fixed and with a disk at the other end, it is 0.33.

14.5 COMPUTER AIDED DESIGN OF SHAFTS

Strength of rigidity analysis of shafts of complicated geometry and loading, multiple supports, etc., can be performed with one of the structural programs with prismatic beam elements presented in Chapter 4. In particular, the program TMSTAT (Appendix 4.A) is very convenient.

Recall that the program performs static analysis of a multinode beam in one plane. Therefore, all loads have to be vector-analyzed into the vertical and horizontal planes. At every position on the shaft, the resulting loads and displacements have to be vector added to yield load and displacement at this position. This analysis has to be performed separately for the static and dynamic loading. Then, equivalent loads must be computed and compared with the allowed stress of the chosen material.

Distributed loads are usually equally lumped at the nearest nodes. In most cases this is adequate and also it simplifies stress analysis since with concentrated masses, the maxima of shear forces and bending moments exist at the nodes, therefore analysis between nodes is not necessary for stress analysis.

For deflection analysis this is not true but placing some nodes in the vicinity of the place where the maximum deflection is expected, is usually adequate.

Using the program TMSTAT as an analysis subroutine, a program SHAFTDES incorporates the above procedure (Appendix 14.D). Applied torques are computed separately, something usually very small. If substantial loads exist, the shaft has to be tested with the program TMSTABIL for buckling. Shafts with continuously variable cross-section can be analyzed by way of taking enough nodes and considering constant section properties between nodes. A constant strength along the shaft length is the aim of an optimum design.

Vibration analysis of shafts can be conveniently performed with computer methods, in particular **transfer matrix methods**.

A rotating shaft, due to unbalanced masses or forces repeating every revolution, is set to harmonic motion, i.e. vibration of the form $y = y_0 \cos \omega t$, where y_0 is the vibration amplitude, variable along the shaft, and ω is the known rotation frequency. Since acceleration is the second derivative of displacement in respect to time, at every position along the shaft there will be an acceleration $a = -y_0\omega^2 \cos \omega t$. At time $t = 2\pi/\omega$, $\cos \omega t = 1$ and there is a force on a mass m equal to $ma = -my_0\omega^2$. Therefore, if at any node i there is a mass m, there will be an inertia force $-my_0\omega^2$ where y_0 is the nodal displacement. Similarly, if a disk with inertia J is at node i there will be present a moment $-J\theta\omega^2$ on the disk. The free-body diagram is shown in Figure 14.13. Equilibrium of the forces at this node yields:

$$
\begin{aligned}
M_i^+ &= J_i \theta_i^- \omega^2 + M_i^- \\
V_i^+ &= m_i y_i^- \omega^2 + V_i^-
\end{aligned}
\tag{14.31}
$$

Since deflection and slope does not change at the node, the transfer matrix is

$$
[P] = \begin{Bmatrix}
1 & 0 & 0 & 0 \\
0 & 1 & 0 & 0 \\
0 & J_i\omega^2 & 1 & 0 \\
m_i\omega^2 & 0 & 0 & 1
\end{Bmatrix}
\tag{14.32}
$$

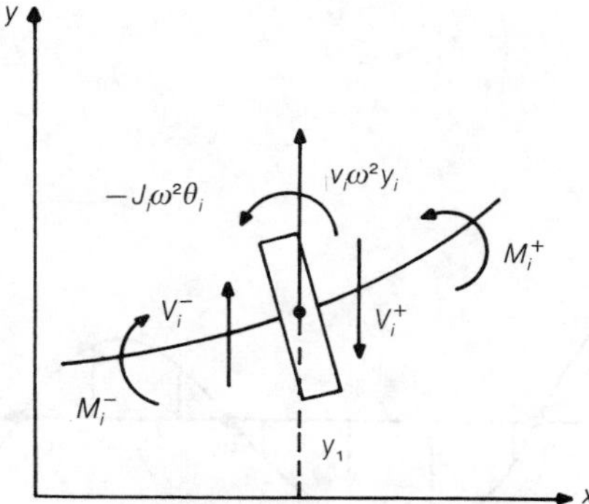

Figure 14.13 Equilibrium of a rotating disk

It is apparent that all inertial forces and moments are functions of the displacements and slopes and therefore no constant forces exist in the transfer matrix. Static loads do not affect the dynamic response in linear systems, therefore are not considered in dynamic analysis. Consequently, equation (4.11) which will result from the transfer matrix multiplication from one end to the other, will have no constant terms and they will be, for free-free boundary conditions

$$\left.\begin{aligned} d_{11}y_1 + d_{12}\theta_1 + d_{13}y_n + d_{14}\theta_n = 0 \\ d_{21}y_1 + d_{22}\theta_1 + d_{23}y_n + d_{24}\theta_n = 0 \\ d_{31}y_1 + d_{32}\theta_1 + d_{33}y_n + d_{34}\theta_n = 0 \\ d_{41}y_1 + d_{42}\theta_1 + d_{43}y_n + d_{44}\theta_n = 0 \end{aligned}\right\} \qquad (14.33)$$

This is a homogeneous system of linear equations and has solutions (that is: the system can have harmonic motion of frequency ω) when the determinant of the coefficients is zero:

$$\det[D] = 0 \qquad (14.34)$$

The values of ω which satisfy equation (14.34) are the shaft natural frequencies. The simpler way to determine them is to plot the function $\det[D(\omega)]$. The points of intersection of this function with the ω-axis are the natural frequencies.

What we call critical speeds might be sometimes a little different but for present purposes it will suffice to say that for low-speed machinery applications they are practically the same with the shaft natural frequencies.

Computation of the critical speeds with the transfer matrix method involves the computation of the determinant $\mathrm{d}(\omega) = \det[D(\omega)]$ for successive values of the frequency ω, $\omega + \Delta\omega$, $\omega + 2\Delta\omega$, $\omega + 3\Delta\omega, \ldots$, (Figure 14.14), where $\Delta\omega$ is a small increment. If the values of the determinant in two successive steps change sign, that means that the critical speed is in this interval. To implement this, at every step, the product $\mathrm{d}(\omega)\mathrm{d}(\omega + \Delta\omega)$ is formed. If this product is positive, the algorithm proceeds to the next step. If it is negative, that is there is a change of sign, the critical speed has been located and the algorithm proceeds for the next critical speed. The accuracy of the determination of the value of the critical speed is $\Delta\omega$, which can be enhanced with a variety of methods. The most simple is to search with a smaller step

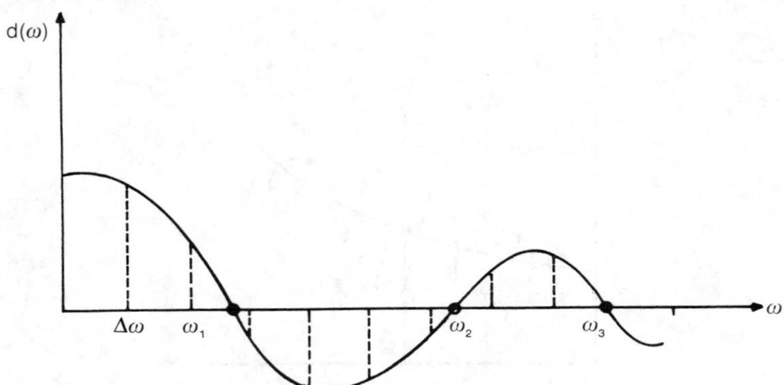

Figure 14.14 Frequency determinant

and a linear interpolation

$$\omega_c = \omega + \Delta\omega d(\omega)/[d(\omega) + d(\omega + \Delta\omega)] \tag{14.35}$$

Going back to equation (4.11) and setting the free–free boundary conditions, that is $M_1 = V_1 = M_n = V_n = 0$, the determinant of the coefficients has value

$$\det[D(\omega)] = d_{31}d_{42} - d_{41}d_{32} \tag{14.36}$$

At every critical speed, the solution of equation (14.33) yields the vibration amplitudes. Since there is no constant term, one of the displacements must be arbitrarily chosen, say $y_1 = 1$. Then the rest of the unknown displacements can be readily computed. In particular, the initial slope from the third equation (14.33), for $y_1 = 1$, is $\theta_1 = -d_{31}/d_{32}$.

The procedure ROTORDYN for the determination of the critical speeds of a rotating shaft follows (explanations are in italics).

```
  Procedure ROTORDYN
  GLOBAL NN, arrays MASS, INERTIA, SPRINGY, SPRINGT(1..NN)
           arrays EI, LENGTH(1..NN − 1)
           arrays FIELD, POINT, A(1..5,1..5)
           array V(1..5) real:var
  LOCAL NN int:var
          OM, DOM, OMmax, Det, Dprevious real:var
begin
  input NN
  input DOM, OMmax                    (*Increment, maximum value of omega*)
  FOR I:= 1 TO NN − 1 do
    input
  MASS(I), INERTIA(I), SPRINGY(I), SPRINGT(I), EI(I), LENGTH(I)
                                                       (*Right end data*)
    input MASS(NN), INERTIA(NN), SPRINGY(NN), SPRINGT(NN)
  repeat
                                                       (*Forward sweep*)
```

```
set unit diagonal matrix A
FOR I:= 1 TO NN − 1
   MATPOINT(I)
   MATFIELD(I)
   MATMULT(POINT, A, ATEMP, 5, 5, 5)
   MATMULT(FIELD, ATEMP, A, 5, 5, 5)
                                                         (*End station*)
MATPOINT(NN)
MATMULT(POINT, A, ATEMP, 5, 5, 5)
                                          (*Apply boundary conditions*)
                                                          (*Free–free*)
Dprevious:= Det
Det:= a(3, 1)*a(4, 2) − a(4, 1)*a(3, 2)
IF Dprevious*Det < 0 then do
   CRITICAL:= OM + DOM*dprevious/(Dprevious + Det)
   WRITELN "Critical speed is"; CRITICAL
                                  (*Computation of the vibration mode*)
   V(1):= 1                                             (*Arbitrarily*)
   V(2):= − a(3, 1)/a(3, 2)        (*From the third part of equation 4.11*)
   V(3):= 0; V(4):= 0
                                             (*Computation of results*)
   FOR I:= 1 TO NN − 1 do
   WRITELN I, (V(J), J:= 1 TO 4)                   (*Results at node I*)
   MATPOINT(I)
   MATFIELD(I)
   MATMULT(POINT, V, VTEMP, 5, 5, 1)
   MATMULT(FIELD, VTEMP, V, 5, 5, 1)
                                                         (*End station*)
MATPOINT(NN)
MATMULT(POINT, TEMP, V, 5, 5, 1)
WRITELN NN, (V(J), J:= 1 TO 4)                    (*Results at node NN*)
else
OM:= OM + DOM
until OM > OMmax
end.
```

This procedure is coded in program ROTORDYN, Appendix 14.A. The program computes also the response at a given speed to harmonic forces, using 5 × 5 matrices as explained in Chapter 4.

Rotating shafts can exhibit torsional vibration also. The computation is very similar, with properly chosen transfer matrices and state vector. The latter includes only two components: rotation and torque. Therefore, the transfer matrices will be 2 × 2. Since for a cylindrical shaft the torsional angle of twist is $\Delta\varphi = TL/GJ_p$, where T is the torque, L the length, G the shear modulus and J_p the polar moment of inertia of the section, the corresponding matrices will be:

State vector Field matrix Point matrix

$$\{V\}=\begin{Bmatrix}\theta\\T\end{Bmatrix},\quad [L]=\begin{bmatrix}1 & L/GI_p\\0 & 1\end{bmatrix},\quad [P]=\begin{bmatrix}1 & 0\\J_p\ \omega^2 & 1\end{bmatrix}$$

where J_p is the mass moment of inertia of the disk at the node.

Upon multiplication from left to right, a 2×2 matrix equation will result:

$$\left.\begin{aligned}\theta_n &= a_{11}\theta_1 + a_{12}T_1\\ T_n &= a_{21}\theta_1 + a_{22}T_1\end{aligned}\right\} \tag{14.37}$$

For a free–free shaft $T_1 = T_n = 0$ and the value of the characteristic determinant is simply $\det[D(\omega)] = a_{21}$. To find the vibration mode, the initial vector at station 1 is set to $\{1\ 0\}$. The procedure is exactly the same as the ROTORDYN only using the appropriate 2×2 matrices.

In the above discussion, both for bending and torsional vibration of shafts, the masses and inertias were considered concentrated at the nodes. For most practical purposes this procedure is accurate enough and simple. The mass and inertia of each element is equally divided to the end nodes, where they are added to the mass and inertia at the node, if any (a disk, for example). Transfer matrices for continuous elements can be found in Dimarogonas, *Vibration Engineering* (see References).

14.6 DISKS OF REVOLUTION

Many axisymmetric machine members have the form of disks, attached usually to shafts, or the form of compound gears, pulleys, etc. (Figure 14.15), which can be modeled as a succession of concentric rings. Disks of general shape can also be modeled by way of rings (Figure 14.16).

Such structures can be analyzed with the finite-element methods of Chapter 4. The transfer matrix method can also very conveniently be used.

To this end, a state vector $\{z\} = \{p\ u\}$ is defined at the interface between two successive

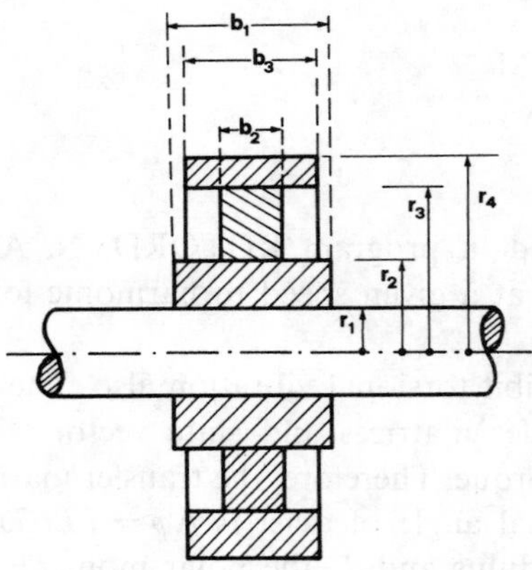

Figure 14.15 A compound hub

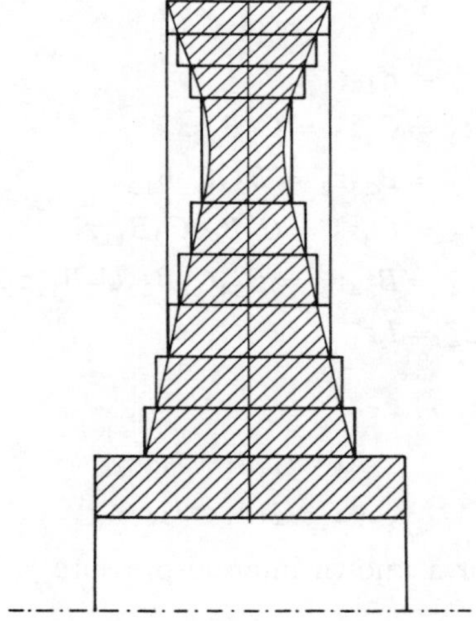

Figure 14.16 Modeling of a continuous hub

rings (Figure 14.15). The state vectors at the inner and outer surface of the ring are related by way of equations (8.49) and (8.50) in Chapter 8. They can be rewritten in transfer matrix form, solving for p, u at the outer surface,

$$p_2 = -\frac{\alpha_{11}}{\alpha_{12}} p_1 + \frac{1}{\alpha_{12}} u_1$$

$$u_2 = \left(\alpha_{21} - \frac{\alpha_{11}}{\alpha_{12}} \alpha_{22} \right) p_1 + \frac{\alpha_{22}}{\alpha_{12}} u_1 \tag{14.38}$$

$$z = \{p\ u\} \quad B = \begin{bmatrix} -\alpha_{11}/\alpha_{12} & 1/\alpha_{12} \\ \alpha_{21} - \alpha_{11}\alpha_{22}/\alpha_{12} & \alpha_{22}/\alpha_{12} \end{bmatrix}$$

Matrix $[B]$ is the transfer matrix for the ring, relating the state vectors z_1 and z_2 at the inner and outer nodes, respectively.

$$\{z_2\} = [B_1]\{z_1\} \tag{14.39}$$

If the width is different, the pressure as we cross the interface will change, while the force per unit length of circumference remains constant. This yields the point matrix at the node,

$$p_2^+ b_2 = p_2^- b_1$$

Therefore,

$$\{z^+\} = [C]\{z^-\}, \quad [C] = \begin{bmatrix} b_1/b_2 & 0 \\ 0 & 1 \end{bmatrix} \tag{14.40}$$

Successive multiplication of transfer matrices from the inner to the outer surface of the disk will yield

$$
\begin{aligned}
z_2^- &= B_{12}z_1^- \\
z_2^+ &= C_2z_2^- = C_2B_{12}z_1^- \\
z_3^- &= B_{23}z_2^+ = B_{23}C_2B_{12}z_1^- \\
z_3^+ &= C_3z_3^- = C_3B_{23}C_2B_{12}z_1^- \\
z_4^- &= B_{34}z_3^+ = B_{34}C_3B_{23}C_2B_{12}z_1^- \\
z_4^+ &= Lz_1^-
\end{aligned}
\tag{14.41}
$$

In explicit form

$$
\begin{aligned}
p_4 &= l_{11}p_1 + l_{12}u_1 \\
u_4 &= l_{21}p_1 + l_{22}u_1
\end{aligned}
\tag{14.42}
$$

The boundary conditions, for a known internal pressure p, for example, and no outside pressure, $p_1 = p$, $p_4 = 0$,

$$u_1 = (-l_{11}/l_{12})p \tag{14.43}$$

yields the unknown component of the state vector at the inner node. Successive application of equation (14.41) will give the state vector at each node. Stresses can be subsequently computed using equation (8.42).

This procedure, essentially the same with TMSTAT, Appendix 4A, was coded for shrink fits of compound wheels in program SHRINKFIT, Appendix 14.C.

Rotating disks are loaded also with inertial loads due to the rotation, called centrifugal loads. Unlike the situation in shafts, these loads can produce substantial stresses and displacements in disks of large diameter. To account for these loads, a centrifugal D'Alembert force is added to the volume differential (Figure 8.12a) of magnitude $\omega^2\gamma g^{-1}\rho\, dV$, where ρ is the radius at the differential volume. Following the integration sequence of Chapter 8, an additional term is added to the expression (8.48), namely

$$u_t = \frac{\gamma}{Eg}\omega^2\rho\left(\frac{3+\nu}{8}\right)\left[(r_1^2 + r_2^2)(1-\nu) + \frac{r_1^2r_2^2}{\rho^2}(1+\nu) - \frac{1-\nu^2}{3+\nu}\rho^2\right] \tag{14.44}$$

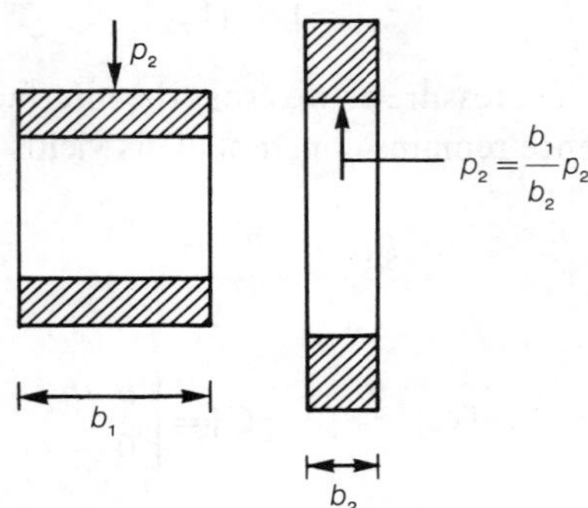

Figure 14.17 Change in hub width

and to the expressions for the stresses equation (8.42),

$$\sigma_r = \frac{\gamma}{g}\omega^2\left(\frac{3+\nu}{8}\right)\left(r_1^2 + r_2^2 - \frac{r_1^2 r_2^2}{\rho^2} - \rho^2\right)$$

$$\sigma_\theta = \frac{\gamma}{g}\omega^2\left(\frac{3+\nu}{8}\right)\left[r_1^2 + r_2^2 + \frac{r_1^2 r_2^2}{\rho^2} - \frac{1+3\nu}{3+\nu}\rho^2\right] \tag{14.45}$$

Then we can write

$$\left.\begin{aligned} u(r_1) = u_1 = a_{11}p_1 + a_{12}p_2 + u_{1t} \\ u(r_2) = u_2 = a_{21}p_1 + a_{22}p_2 + u_{2t} \end{aligned}\right\} \tag{14.46}$$

where from equation (14.44),

$$u_{2t} = (\rho_0\omega^2/E)(3+\nu)r_2[(r_1^2 + r_2^2)(1-\nu) + r_1^2(1+\nu) - (1-\nu^2)r_2^2/(3+\nu)]/8$$

$$u_{1t} = (\rho_0\omega^2/E)(3+\nu)r_1[(r_1^2 + r_2^2)(1-\nu) + r_2^2(1+\nu) - (1-\nu^2)r_1^2/(3+\nu)]/8$$

where ρ_0 is the material density γ/g.

Equation (14.46) will be solved for p_2 and u_2 to yield the transfer matrix. Since now there are constant terms, a 3×3 field matrix will be used. The first four elements are the same with the ones of the 2×2 matrix of equation (14.38). The complete matrix is

$$[B] = \begin{bmatrix} (-a_{11}/a_{12}) & (1/a_{12}) & (-u_{1t}/a_{12}) \\ (a_{21} - a_{22}a_{11}/a_{12}) & (a_{22}/a_{12}) & (-a_{22}/a_{12})u_{1t} + u_{2t}) \\ 0 & 0 & 1 \end{bmatrix}$$

The computation will be performed now with 3×3 matrices, in the same way. The program SHRINKFIT implements this procedure. If no inertia terms are to be used, the angular velocity is set to zero and the results are identical with the previous procedure.

EXAMPLES

Example 14.1

The rotor of a 11 kW, 1500 rpm rated motor has rotary inertia in respect to the axis of rotation $J_1 = 0.5\ \text{kg m}^2$ and drives a pump with a rotor of $J_2 = 0.5\ \text{kg m}^2$ through a stub-shaft of length 2 m and diameter 5 cm connected through a flexible coupling with torsional spring constant 1000 Nm/rad. The reaction of the pump is proportional to the speed of rotation. Find the time needed to reach 99% of the rotating speed using the program SIMUL. The motor torque has the form indicated in Figure 14.3(a). Then comment on the influence of changing the pump with another one having a moment inertia about the axis of rotation $J_2 = 1\ \text{kg m}^2$.

Solution

The operation torque $T_o = P/\omega_0$, where P the power and ω_0 the angular velocity of rotation. The rated $\omega_0 = 1500 \times 2\pi/60 = 157$ rad/sec and from Figure 14.3a, the actual speed of rotation is $\omega = 0.975\omega_0 = 153.2$ rad/sec, at maximum torque.

The linear reaction moment will be

$$M(\omega) = \lambda\omega$$

But for $\omega = \omega_0$ it must be $M = T_o$, therefore

$$\lambda = T_o/\omega_0 = 56.18/178 = 0.315 \text{ Nm sec}$$

The torsional spring constant for the connecting shaft is GI_p/L. For steel $G = 1.05E11$ Pa, therefore,

$$k_{Ts} = G\pi d^4/32\text{L} = 1.05E11 \times \pi \times 0.05^4/32 \times 2 = 3213 \text{ Nm/rad}$$

The coupling flexibility must be added to the shaft flexibility,

$$1/k_T = 1/k_{Tc} + 1/k_{Ts} = (1/1000) + (1/32213)$$

which gives $k_T = 970$ Nm/rad.

The program SIMUL yielded a start-up overload factor 1.8. A subsequent run with $J_2 = 1 \text{ kg m}^2$ resulted in an overload factor of 2.32. The two runs follow. The torque versus speed curve for the motor (Figure 14.3a), was approximated by way of six points connected with linear segments.

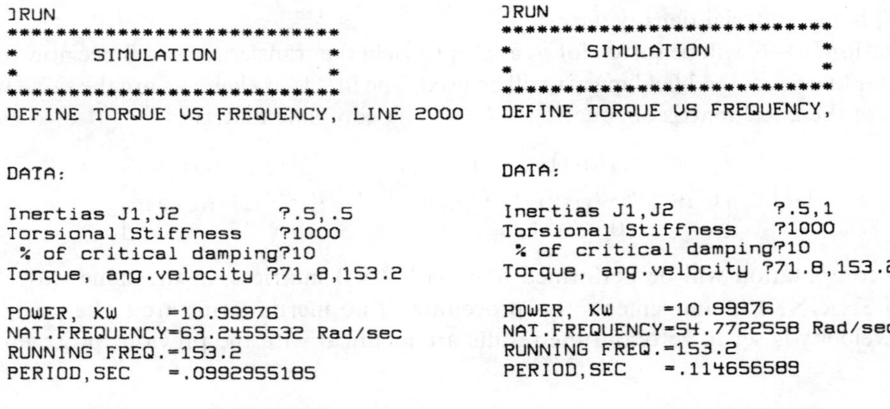

```
]RUN
****************************
*      SIMULATION          *
*                          *
****************************
DEFINE TORQUE VS FREQUENCY, LINE 2000


DATA:

Inertias J1,J2          ?.5,.5
Torsional Stiffness    ?1000
 % of critical damping?10
Torque, ang.velocity ?71.8,153.2

POWER, KW       =10.99976
NAT.FREQUENCY=63.2455532 Rad/sec
RUNNING FREQ.=153.2
PERIOD,SEC     =.0992955185

Maximum torque =129.420156
Starting time =1.469
Service factor=1.80250913
```

```
]RUN
****************************
*      SIMULATION          *
*                          *
****************************
DEFINE TORQUE VS FREQUENCY,


DATA:

Inertias J1,J2          ?.5,1
Torsional Stiffness    ?1000
 % of critical damping?10
Torque, ang.velocity ?71.8,153.2

POWER, KW       =10.99976
NAT.FREQUENCY=54.7722558 Rad/sec
RUNNING FREQ.=153.2
PERIOD,SEC     =.114656589

Maximum torque =166.834435
Starting time =2.694
Service factor=2.32359938
```

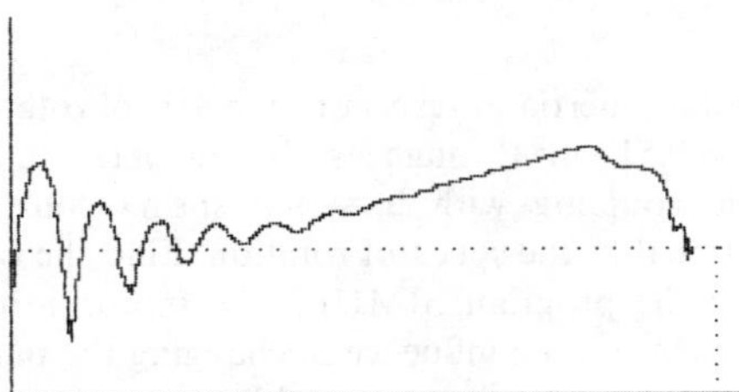

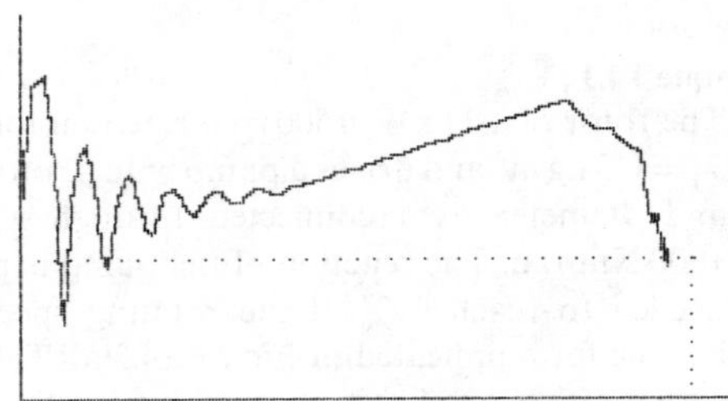

Figure E14.1

Example 14.2

The shaft of Example 14.1 is made of AISI 1015 carbon steel. Find the diameter.

Solution

The torque loading is non-reversing, and there is no bending to induce alternating stresses. From

Chapter 6, for AISI 1015 steel, the fatigue strength is 273 MPa and the yield strength 430 MPa. Since the loading is steady, the design will be based on the yield strength.

The service factor was found in Example 14.1 to be equal to 1.8. No stress concentration will be used because the material is ductile. Therefore, for $S_{sy} = S_y/2 = 215$ MPa equation (14.18) yields,

$$d = 36.5[(11/1462)/(215E6/1.80)]^{\frac{1}{3}} = 0.014 \text{ m or } 14 \text{ mm}.$$

Example 14.3

The shaft shown rotates at 930 rpm. It carries a helical gear at B with helix angle $\psi = 35°$, standard involute teeth with $\varphi = 20°$ and diameter 250 mm. On the right end D carries a flywheel of mass $m = 200$ kg. Ball bearings at A and C support the shaft. The operation is intermittent and with heavy torsional impacts equalized by the flywheel. The power of 30 kW is supplied at the left end. Determine the shaft loads and draw force and moment diagrams.

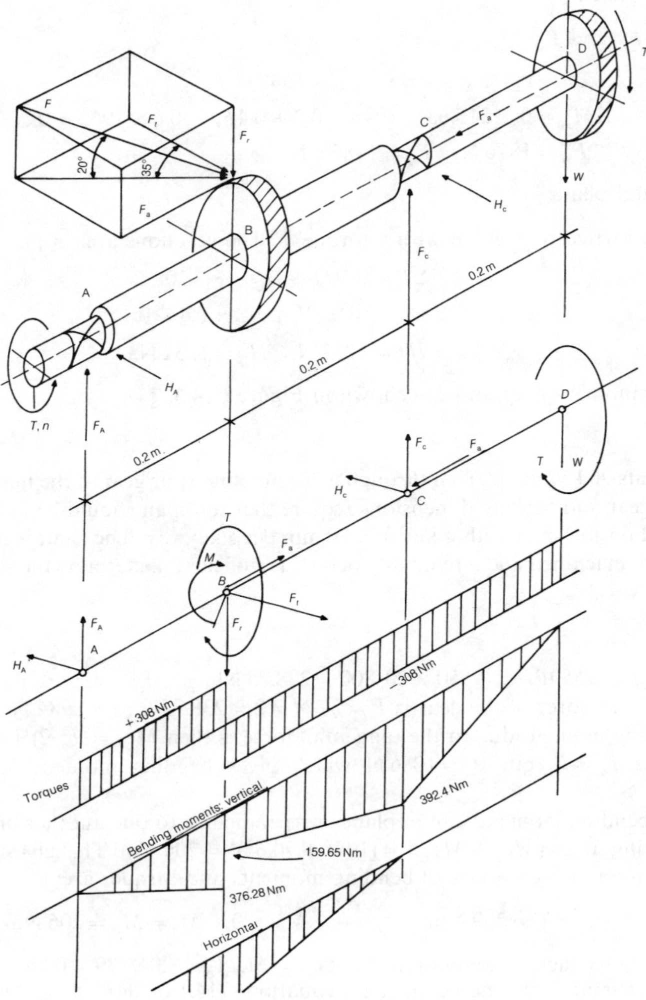

Figure E14.3

Solution

The flywheel weights $mg = 200 \times 9.81 = 1962$ N. The angular velocity is $2\pi n/60 = 2\pi \times 930/60 = 97.4$ rad/sec.

The torque between A and B is $T = P/\omega = 30{,}000/97.4 = 308$ Nm.

The flywheel supplies power to keep the shaft running and it is assumed that the maximum power supplied equals the system power, therefore the torque of 308 Nm continues to point D.

The forces on the gear are:

$$F_t = 2T/d = 2 \times 308/0.25 = 2464 \text{ N}$$
$$F_r = F_t \tan \varphi'/\cos \psi = 1095 \text{ N}$$
$$F_a = F_t \tan \psi = 1725 \text{ N}$$

The axial force is support at the bearing at C and gives bending moment $M = F_a d/2 = 215.6$ Nm.

In the vertical plane:

The reactions F_A and F_C:

$$\Sigma F = 0: \quad F_A - 1095 + F_C - 1962 = 0$$
$$\Sigma M_A = 0: \quad 215.63 - 1095 \times 0.2 + 0.4F_C - 0.6 \times 1962 = 0$$
$$F_A = 1876 \text{ N}, \quad F_C = 1180.6 \text{ N}.$$

In the horizontal plane:

The force F_t is carried to point B with a torque T. The reactions are:

$$\Sigma F = 0: H_A + H_C + F_t = 0$$
$$\Sigma M_A = 0: 0.2F_t + H_C \times 0.4 = 0$$
$$H_A = 1232 \text{ N}, \quad H_C = 1232 \text{ N}.$$

The force and moment diagrams are shown in Figure E14.3.

Example 14.4

A shaft transmits 20 kW at 800 rpm through a 20° involute spur gear at the mid-span of diameter 160 mm. The gear and bearing dimensions require that the span should be at least 260 mm. The gear is secured on the shaft with a key in a 10 mm deep keyway. The shaft is made of hot-rolled AISI 1015 steel, machined and a reliability of 90% is sufficient. Determine the shaft diameter for a service factor $N = 2$.

Solution

The torque is $T = 9550P/n = 9550 \times 20/800 = 238.75$ Nm.

The tangential force of the gear is $F = 2T/d = 2 \times 238.75/0.16 = 2984$ N.

The bending moment due to the tangential force is then $M_{b1} = (F_t/2)(L/2) = 194$ Nm. The radial force is $F_r = F_t \tan 20° = 1086$ N which gives bending moment $M_{b2} = (F_r/2)(L/2) = 70.6$ Nm.

The two bending moments are in planes perpendicular to one another and they are added vectorially, giving $M_b = (M_{b1} + M_{b2})^{\frac{1}{2}} = (194^2 + 70.6^2)^{\frac{1}{2}} = 206$ Nm. This gives reversing stresses. Then, the mean and range values of bending moments and torques are:

$$T_m = T = 239 \text{ Nm}, \quad T_r = 0, \quad M_m = 0, \quad M_r = M_b = 206 \text{ Nm}$$

The asymmetry factor is, equation (14.26), $r = M_r/T_m = 206/239 = 0.86$.

A first estimate of the diameter, equation (14.18), for $S_{sy} = 20$ MPa gives $d = 36.5[(20/800)/(20E6/2)]^{\frac{1}{3}} = 0.050$ m or 50 mm.

For AISI 1015 steel $S_u = 527$, $S_y = 430$ and $S'_n = 270$ MPa. From Appendix A (see p. 706), the stress concentration factor for the keyway is $K_t = 2.15$. From Figure 7.7 the notch sensitivity factor for notch radius 2 mm is $q = 0.75$. Therefore, the fatigue stress concentration factor is, equation (7.4), $K_f = 1 + q(K_t - 1) = 1 + 0.75(2.15 - 1) = 1.86$.

The surface finish factor $C_f = 0.80$ and the size factor $C_s = 0.70$ (Figure 7.8). For 90% reliability, Figure 6.24 gives $C_R = 1/1.1 = 0.90$. Therefore, from equation (7.3),

$$S_e = 0.8 \times 0.7 \times 0.9 \times 273/1.86 = 74\text{ MPa}$$

The asymmetry factor of the Goodman diagram is $r_D = (S_e/S_u)(S_u - S_y)/(S_y - S_e) = (74/527)(527 - 430)/(430 - 74) = 0.0382 < r$, therefore $\varepsilon = S_u/S_e = 527/74 = 7.12$. Then, equation (14.27) yields the diameter

$$d^3 = \frac{32N}{\pi S_y}(T_m + \varepsilon M_r) = \frac{32 \times 2}{43 \times 10^6}(239 + 7.12 \times 206) = 8.08E - 5 \text{ and then } d = 43.3\text{ mm}$$

Example 14.5

A high-pressure turbine rotor running at 3600 rpm has four stages sharing equally the production of 50,000 kW power. The shaft material is AISI 12Cr steel with $S_u = 830$ MPa, $S_y = 540$ MPa $S_e = 320$ MPa. Due to high temperatures and creep, a service factor of 2.5 must be used. Design the shaft for 99.9% reliability. Compute the maximum deflection owing to the shaft's own weight. The wheels weigh 3, 3.55, 4.1 and 4.5 kN from left to right, equally spaced between the two end bearings on a rotor of total length of 5 m. The power is transmitted to the right and the thrust is negligible.

Solution

The power per wheel is 50,000/4 = 12,500 kW. The torque on each wheel is then $T = 9550P/n = 9550 \times 12{,}500/3660 = 33{,}159$ Nm. The transition from one diameter to the other is assumed smooth without stress concentration.

For 99.9% reliability Figure 6.24 gives safety factor 1.4 and $C_R = 1/1.4 = 0.714$. The surface finish factor for machined shaft $C_f = 0.72$ (Figure 7.7) and the size factor $C_s = 0.6$ (Figure 7.8). These are tentative values since the exact diameters are not yet known.

Therefore, $S_e = 0.72 \times 0.6 \times 0.714 \times 320 = 98.7$ MPa.

The program SHAFTDES (Appendix 14.D) is used to yield the diameters and deflections. The run follows.

```
ENTER DATA:
Number of elements                                   ? 5
Node   1 : Enter FX,FY,MX,MY                          ? 0,0,0,0
           Support <S>, spring <P>, free <F>          ? s
Node   2 : Enter FX,FY,MX,MY                          ? 0,3000,0,0
           Support <S>, spring <P>, free <F>          ? f
Node   3 : Enter FX,FY,MX,MY                          ? 0,3550,0,0
           Support <S>, spring <P>, free <F>          ? f
Node   4 : Enter FX,FY,MX,MY                          ? 0,4100,0,0
           Support <S>, spring <P>, free <F>          ? f
Node   5 : Enter FX,FY,MX,MY                          ? 0,4500,0,0
           Support <S>, spring <P>, free <F>          ? f
Node   6 : Enter FX,FY,MX,MY                          ? 0,0,0,0
           Support <S>, spring <P>, free <F>          ? s
Element   1 :Length,QloadX,QloadY                      ? 1,0,0
           Element torque,thrust                      ? 0,0
Element   2 :Length,QloadX,QloadY                      ? 1,0,0
           Element torque,thrust                      ? 33000,0
Element   3 :Length,QloadX,QloadY                      ? 1,0,0
           Element torque,thrust                      ? 66000,0
Element   4 :Length,QloadX,QloadY                      ? 1,0,0
           Element torque,thrust                      ? 99000,0
```

(Contd.)

```
Element  5 :Length,QloadX,QloadY                        ? 1,0,0
          Element torque,thrust                        ? 132000,0
Enter Material Data:
                                    Young Modulus      ? 2.1e11
                                    Fatigue strength ? 98.7e6
                                    Yield Strength     ? 540e6
                                    Specific weight    ? 7.8
                                    Service Factor     ? 2.5


Hit ENTER to continue...
?
VERTICAL PASS...
DATA:  5  ELEMENTS,    6   NODES

  LENGTH     DIAMETER  FORCE        LOAD         SPRING
-------------------------------------------------------------------
   100.E-02  232.E-03    0.E+00     0.E+00     0.E+00 SUPPORT
   100.E-02  232.E-03  300.E+01     0.E+00     0.E+00
   100.E-02  232.E-03  355.E+01     0.E+00     0.E+00
   100.E-02  232.E-03  410.E+01     0.E+00     0.E+00
   100.E-02  232.E-03  450.E+01     0.E+00     0.E+00
     0.E+00  232.E-03    0.E+00     0.E+00     0.E+00 SUPPORT
computing...
SOLUTION ...
STATION DEFLECTION   SLOPE         MOMENT          SHEAR          REACTION
--------------------------------------------------------------------------
  1   615.E-08   623.E-06     0.E+00      0.E+00   707.E+01
  2   589.E-06   504.E-06 -707.E+01 -707.E+01
  3   953.E-06   200.E-06 -111.E+02 -407.E+01
  4   964.E-06 -181.E-06 -117.E+02 -520.E+00
  5   608.E-06 -511.E-06 -808.E+01  358.E+01
  6   703.E-08 -646.E-06    0.E+00  123.E-03   808.E+01
Hit ENTER to continue?
RESULTS:
NODE     D(-)         D(+)
------------------------------------------------------
  1      232.E-03     116.E-03    ...bearing
  2      117.E-03     147.E-03
  3      147.E-03     168.E-03
  4      168.E-03     185.E-03
  5      184.E-03     727.E-04
  6        0.E+00     727.E-04    ...bearing
Hit ENTER to continue...?
Do you want bearing calculations (Y/N) ? n
```

Figure E14.5

Example 14.6

For the shaft of Example 14.5.
Determine:

(a) The lowest three critical speeds.
(b) The vibration amplitude at stage 3 if a bucket of mass 1.2 kg at a distance 400 mm from the axis of rotation, at the same stage, brakes. The radii of gyration of the turbine wheels are 200, 245, 304, 350 mm respectively.

Solution

The mass moments of inertia of the disks about the diameter are mr_g^2, giving 12.23, 21.7, 38.6, 56.2 kg m^2. The program ROTORDYN was used.

To compute the dynamic response the inertial force due to the broken bucket and the opposite bucket, is $F = m\omega^2 R = 1.2 \times 377^2 \times 0.4 = 68{,}220$ N.

The input and results follow.

```
               Number of Elements      0                  ? 5
               Modulus of Elasticity   0                  ? 2.1e11
               Material Density        0                  ? 7800
Element data                                node data
El/No
    length     diameter       force         spring         mass     inertia
 1  ? 1          ? .116       ?             ?              ?        ?
 2  ? 1          ? .147       ?             ?              ? 300    ? 12.23
 3  ? 1          ? .168       ?             ?              ? 355    ? 21.7
 4  ? 1          ? .184       ?             ?              ? 410    ? 38.6
 5  ? 1          ? .072       ?             ?              ? 450    ? 56.2
 right end....                ?             ?              ?        ?


Enter no of solid supports  0                    ? 2
Enter numbers of support nodes:

Support  1   at node  0                          ? 1
Support  2   at node  0                          ? 6

Are data correct (Y/N)  ? y

NO      DEFLECTION  SLOPE            MOMENT          SHEAR
-------------------------------------------------------------
 1  3.583269E-03    1                0               0
 2  .9027623        .697528          -1129142        -1129176
 3  1.455792        .3804634         -1931499        -810636.8
 4  1.71525         .1345685         -2114876        -191388
 5  1.769814        -.0161648        -1452203        657639.7
 6  4.639804E-03    -2.640544        -958.125        1452120
Critical speed no  1  is:    31.04138

NO      DEFLECTION  SLOPE            MOMENT          SHEAR
-------------------------------------------------------------
 1  9.741997E-03    1                0               0
 2  .7360631        .178863          -3065627        -3066001
 3  .5934403        -.4696201        -3201044        -159180.3
 4 -2.371725E-02    -.7105829        -645644.3       2666074
 5 -.738457         -.6831437        1591555         2534681
 6 -3.605425E-03    1.445629         1972.25         -1175502
Critical speed no  2  is:    103.8481

NO      DEFLECTION  SLOPE            MOMENT          SHEAR
-------------------------------------------------------------
 1  1.861254E-02    1                0               0
 2  .5030243        -.5477457        -5781301        -5784963
 3 -.2457744        -.4763166        7181863         1.367608E+07
 4 -.3068268        .3991337         8295455         2213277
 5  .3089413        .6287316         -4506687        -1.443911E+07
Critical speed no  3  is:    325.0391

               Number of Elements      5                  ?
               Modulus of Elasticity   2.1E+11            ?
               Material Density        7800               ?
Element data                                node data
El/No
    length     diameter       force         spring         mass     inertia
 1  ? 1          ? .116       ?             ?              ?        ?
 2  ? 1          ? .147       ?             ?              ? 300    ? 12.23
 3  ? 1          ? .168       ?             ?              ? 355    ? 21.7
 4  ? 1          ? .184       ? 68220       ?              ? 410    ? 38.6
 5  ? 1          ? .072       ?             ?              ? 450    ? 56.2
 right end....                ?             ?              ?        ?
```

(Contd.)

```
Enter no of solid supports  2                               ?
Enter numbers of support nodes:

Support  1   at node  1                                     ?
Support  2   at node  6                                     ?


NO     DEFLECTION  SLOPE            MOMENT            SHEAR
-----------------------------------------------------------
 1  6.93006E-05    3.41765E-03     0                 0
 2  1.577497E-03  -2.315221E-03   -21417.29         -21434.12
 3 -1.282854E-03  -1.304796E-03    35198.41          60669.51
 4 -1.093457E-03   1.281318E-03    11326.2          -19819.67
 5  5.228928E-04   1.507973E-03   -13039.51         -31435.21
 6  2.156817E-03   3.678525E-03    2195.838          3189.082
```

Figure E14.6

Example 14.7

A compound wheel has the shape shown in Figure E14.7. It is made of steel forging with allowable yield strength $S_y/N = 120$ MPa. The wheel will be shrink fitted on a shaft of inner diameter 40 mm and outer diameter 100 mm to transmit torque 500 Nm and axial force 10 kN. The coefficient of friction is $f = 0.2$. Determine the required fit characteristics.

Solution

The program SHRINKFIT, Appendix 14.C will be used. Input and results follow.

JRUN
Enter hollow shaft inner dia ? 10
Enter number of ring elements ? 4
Ring no 1: enter d(inn), w, E, w? 100, 60, 2.1E5, .3
Ring no 2: enter d(inn), w, E, w? 150, 20, 2.1E5, .3
Ring no 3: enter d(inn), w, E, w? 200, 30, 2.1E5, .3
Ring no 4: enter d(inn), w, E, w? 250, 50, 2.1E5, .3
Enter outer diameter ? 270
Enter Torque, Axial Force ? 5E5, 1E4
Enter friction coefficient ? .2
Enter allowable yield strength ? 120
RESULTS:

Minimum required interference 1.058179E − 04
Maximum allowed interference 9.20593264E − 04
Minimum interference pressure 5.30516925
Maximum interference pressure 46.1538461
Maximum stress 120

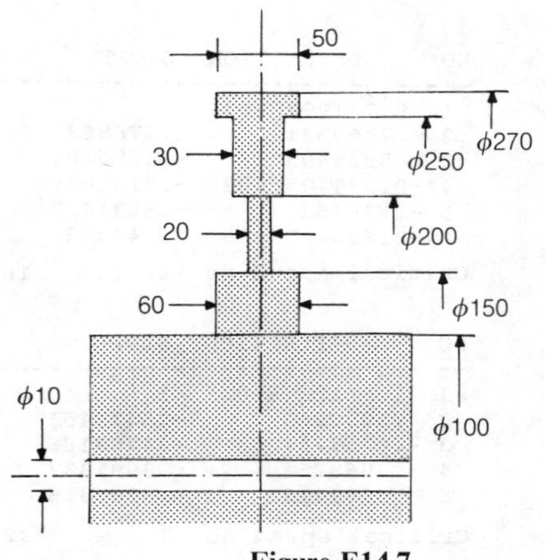

Figure E14.7

Example 14.8

For the compound wheel of Example 14.7, determine the required fit properties if the wheel rotates at 3600 rpm. Find then the maximum speed for which an interference fit for the given loads is feasible.

Solution

For $n = 3600$ rpm it is $\omega = 2\pi n/60 = 377$ rad/s. Re-running the program SHRINKFIT for the angular velocity at hand, the required fit is found, appearing in the output.

Repeated runs are then performed until a point is reached where the minimum required interference exceeds the maximum allowed by the material strength. This point is at 1525 rad/s or 14,560 rpm.

REFERENCES AND FURTHER READING

British Petroleum, 1969. *Lubrication Theory and Applications.* London: BP Trading Co.
Burr, A. H., 1982. *Mechanical Analysis and Design*, New York: Elsevier.
Dimarogonas, A. D., Paipetis, S. A., 1983. *Analytical Methods in Rotor Dynamics.* London: Elsevier Applied Science Publishers.
Dimarogonas, A. D., 1976. *Vibration Engineering.* St Paul, New York: West Publishers.
Duggan, T. V., Byrne, J., 1977. *Fatigue as a Design Criterion.* London: Macmillan.
Haenchen, R., Decker, K.-H., 1967. *Neue Festigkeitsberechnung fuer den Maschinenbau.* Munich: C. Hanser Verlag.
Huebner, K. H., 1975. *The Finite Element Method for Engineering.* New York: Wiley.
Niemann, G., 1965. *Maschinenelemente.* Berlin: Springer.
Peterson, R. E., 1974. *Stress Concentration Factors.* New York: Wiley.
Pinkus, O., Sternlicht, B., 1961. *Theory of Hydrodynamic Lubrication.* New York: McGraw-Hill.
Reshetov, D. N., 1980. *Machine Design.* Moscow: Mir Publishers.
Shigley, J. E., 1972. *Mechanical Engineering Design.* New York: McGraw-Hill.
Tondl, A., 1965. *Some Problems in Rotor Dynamics* (translation from Czech), London: Chapman and Hall.

PROBLEMS

14.1 Classify the shafts on the basis of: shape, rigidity, in-span conditions.

14.2 Discuss modes of failure of shafts.

14.3 Discuss advantages and disadvantages of solid and hollow shafts.

14.4 Identify stress raisers in shafts.

14.5 In which cases are heat treatment and surface hardening applied to shafts?

14.6 A 30 kW, 3000 rpm rated motor with a rotor of 1.45 kg m^2 moment of inertia is moving a wood shaw at constant torque corresponding to the rated motor torque through a flexible coupling of torsional spring constant 650 Nm/rad. The moment of inertia of the rotary shaw is 0.4 kg m^2. Determine the maximum starting overload if the system damping is 7% of the critical.

14.7 In Problem 14.6, the shaw is suddenly blocked. Determine the maximum torque on the shaft, (a) assuming that all kinetic energy is transformed into energy of elastic deformation of the flexible coupling, (b) using program SIMUL.

14.8 A centrifugal pump reacts with a torque which is proportional to the square of the rotating speed. The rotor of such a pump has inertia about the axis of rotation 0.8 kg m^2. It is connected to an induction motor rated 15 kW at 1500 rpm with a rotor of 2.1 kg m^2 moment of inertia. The connecting shaft has length 100 m and diameter 60 mm. The system damping is 5% of critical. Determine the starting overload.

14.9 The pump of Problem 14.8 was connected to a diesel engine which has an unbalanced torsional vibration 10% of the rated torque. Modify the torque function of program SIMUL and determine the resulting torsional vibration of the pump rotor.

14.10 The shaw of Problem 14.6 has 12 blades and the one is broken. Determine the resulting torsional vibration of the motor modifying the derivative evaluation routine of program SIMUL to have zero reacting torque for 1/12 of every rotation.

14.11 The stub-shaft connecting a 250 hp, 875 rpm motor to the gearbox of a steel rolling machine is made of AISI 1040 carbon steel. A key is used at the coupling. Determine the shaft diameter. Is reliability important?

14.12 An automobile engine delivers maximum torque at 110 hp at 3200 rpm. At low gear, the transmission ratio is 4.5 and the efficiency 87%. At high gear, respectively $\eta = 1$ and $\eta = 97\%$. Determine the main axle diameter if it is made of carbon steel AISI 1035.

14.13 A 25-hp, 1750-rpm electric motor is driving a lathe with intermittent operation and light shocks. Determine the axle diameter if no appreciable bending moment is applied and a spline is used at each end. Material AISI 1040 carbon steel.

14.14 In the automobile of Problem 14.12, the differential has a gear ratio 4.7. Determine the diameter of each wheel axle.

14.15 A 35-hp, 2500-rpm diesel engine of a boat is connected to the propeller with a shaft of 40-mm diameter. Select a proper stainless steel material.

Problems 14.16 to 14.25

The shaft of Figures P14.16 to P14.25 rotates with angular velocity ω (rad/sec). It is driven from left with power P_0 (kW) which is transmitted partly through two pulleys 1 and 2, shown schematically, powers P_1 and P_2 respectively, and the remaining power is transmitted to a driven machine at the right end with a very flexible coupling. Assuming no shear forces and bending moments transmitted to left and right ends, draw the shear force, torque and bending moment diagrams in the vertical plane.

The data are given in the table below. H_1 and H_2 are horizontal forces at pulleys 1 and 2 not shown in Figures 14.16 to 14.25. For all problems, $P_0 = 8$ kW.

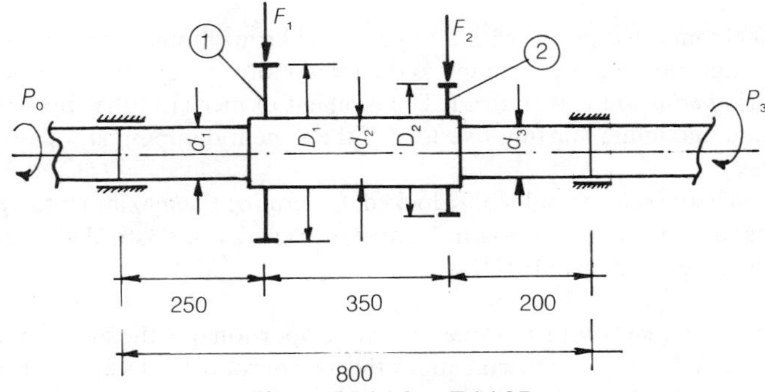

Figure P14.16 to P14.25

Problem	ω(rad/s)	P_1(kW)	P_2(kW)	F_1(N)	F_2(N)	H_1(N)	H_2(N)
14.16	200	4	4	1000	2000	—	—
14.17	200	4	2	1000	2000	2000	2000
14.18	—	—	—	2000	2000	2000	2000
14.19	200	2	4	—	—	—	—
14.20	200	2	4	2000	2000	—	—
14.21	200	2	4	—	—	2000	2000
14.22	200	6	—	3000	1000	—	—
14.23	400	4	4	1000	2000	—	—
14.24	400	4	2	1000	2000	2000	2000
14.25	400	—	—	2000	2000	2000	2000

Problems 14.26 to 14.35

For the shafts of Figures P14.16 to P14.25 determine the shaft diameters, selecting proper materials. The reliability should be 99%, the surfaces ground and the fillet radii should not exceed 3 mm. The service factor should be 2.1. Select appropriate antifriction bearings.

APPENDIX 14.A
ROTORDYN: A PROGRAM FOR DYNAMIC ANALYSIS OF ROTATING SHAFTS

```
10 KEY OFF:CLS
20 REM ***************************************************
30 REM *                                                 *
40 REM *              PROGRAM ROTORDYN                   *
50 REM *                                                 *
60 REM ***************************************************
70 REM         COPYRIGHT 1987
80 REM by Professor Andrew D. Dimarogonas, Washington University,St.Louis.
90 REM
100 REM      TRANSFER MATRIX  ANALYSIS
110 REM
120 REM         OF ROTATING SHAFTS
130 REM
140 REM            BY ADD 6/1/84      Revised 5/17/87
150 REM
160 DIM D(200,6),DAT(200,6):REM                DATA MATRIX
170 DIM P(5,5),S(5,5),A(5,5),G(5,5):REM        AUXILIARY TRANSFER MATRICES
180 DIM V(5,1),W(5,1),VMAX(200):REM            STATE VECTORS
190 DIM isu(20) :REM                            SUPPORT MATRIX
200 DIM RES(200,4):REM                         RESULTS MATRIX
210 REM                                        label
220 FOR I=1 TO 14:COLOR I:LOCATE I,22+I:PRINT "ROTORDYN";
230 LOCATE I,50-I:PRINT "ROTORDYN";:NEXT I:COLOR 14
240 LOCATE 8,12:PRINT"A Transfer Matrix Program for Rotor Dynamic";
250 PRINT " analysis";:LOCATE 21,1
252 locate 25,1:print" Dimarogonas,A.D., Computer Aided Machine Design,
Prentice-Hall,1988";
260 LOCATE 23,28:INPUT"Hit return to continue";X$
270 CLS:NMENUS=8
280 DIM L$(100)
290 IPLOT=0:SCREEN 2
300 DIM MENU1$(25)
310 FOR IM=1 TO NMENUS:READ MENU1$(IM):NEXT IM
320 DATA "Stop        ","Clear scrn","New        ","Save File ","Load File "
330 DATA "Analysis    ","Plot Rotor","Make Model"
340 XS=0
350 FOR I=1 TO NMENUS:LOCATE I+2,70:PRINT MENU1$(I);:NEXT I
360 ICOUNT=0
370 LOCATE 1,70:PRINT"MAIN MENU";
```

```
380 LINE(530,0)-(639,8),1,B
390 IF VIND=0 THEN VIND=5:GOTO 470
400 X$=INKEY$:XS=0
410 IF X$="" THEN 400
420 IF LEN(X$)=1 THEN XS=ASC(X$):GOTO 530
430 X$=RIGHT$(X$,1)
440 IF X$><"H" AND X$<>"P" THEN 500
450 IF X$="H" AND VIND>1 THEN VIND=VIND-1
460 IF X$="P" AND VIND<NMENUS THEN VIND=VIND+1
470 IF VIND>1 THEN LOCATE VIND+1,69:PRINT " ";MENU1$(VIND-1);
480 LOCATE VIND+2,69:PRINT ">";MENU1$(VIND);
490 IF VIND<NMENUS THEN LOCATE VIND+3,69:PRINT " ";MENU1$(VIND+1);
500 LOCATE 19,70:PRINT"                 ";
510 LOCATE 21,65:PRINT"                    ";
520 LOCATE 22,65:PRINT"                    ";
530 IF XS=13 AND VIND=1 THEN SCREEN 0:END
540 IF XS=13 AND VIND=2  THEN CLS:GOTO 340:'         clear screen
550 IF XS=13 AND VIND=3 THEN GOSUB  680:REM          start a new drawing
560 IF XS=13 AND VIND=4 THEN GOSUB 1060:REM          save file on disc
570 IF XS=13 AND VIND=5 THEN GOSUB 720:GOSUB 1280:REM load file from disc
580 IF XS=13 AND VIND=6 THEN GOSUB 2000:REM          Dynamic Analysis
590 IF XS=13 AND VIND=7 THEN GOSUB 3430:REM          plot rotor
600 IF XS=13 AND VIND=8 THEN GOSUB 1290:REM          make rotor model
610 REM
620 GOTO 350
630 END
640 REM
650 REM ***************************************** end of main program
660 REM
670 REM
680 REM ***************************************** start a new rotor
690 LOCATE 22,65:INPUT"SURE (Y/N)";X$:IF X$="n" OR X$="N" THEN RETURN
700 ND=0:NP=0:NC=0:RETURN
710 REM*********************************************************
720 REM                   file retrieval
730 REM*********************************************************
740 LOCATE 19,68:PRINT"                 ";
750 LOCATE 18,70:PRINT"Data file  ":LOCATE 19,68:INPUT FILNA$
760 LOCATE 19,68:PRINT SPC(10);
770 LOCATE 18,65:PRINT"New or Add (N/A)":LOCATE 19,68:INPUT F$
780 IF F$<>"a" AND F$<>"A" AND F$<>"n" AND F$<>"N" THEN 740
790 IF FILNA$="" THEN 750
800 IF F$="n" OR F$="N" THEN N=0:NS=0:N0=0
810 ICOUNT=0
820 IF F$="a" OR F$="A" THEN NS0=NS:N0=N
830 OPEN FILNA$ FOR INPUT AS #1
840 ICOUNT=ICOUNT+1:INPUT#1,X$
850 IF X$="filend" OR X$="FILEND" THEN 1020
860 SYM$=""
870 ISYM=1:IF ICOUNT>1000 THEN 1040
880 SYM$=MID$(X$,ISYM,1):ISYM=ISYM+1:IF SYM$="" AND ISYM<10 THEN 870
890 IF ISYM=10 THEN 840
900 IF SYM$<>"n" AND SYM$<>"N" THEN GOTO 940
910 REM
920 INPUT#1,I,DAT(I+N0,3),DAT(I+N0,4),DAT(I+N0,5),DAT(I+N0,6)
930 GOTO 840
940 IF SYM$<>"e" AND SYM$<>"E" THEN  970
950 INPUT#1,I,DAT(I+N0,1),DAT(I+N0,2)
960 GOTO 840
970 IF SYM$="P" OR SYM$="p" THEN INPUT#1,N,E,DENSITY:GOTO 840
980 IF SYM$<>"s" AND SYM$<>"S" THEN 1040
990 NS=NS+1
1000 INPUT#1,ISX:isu(NS)=ISX+N0
1010 GOTO 840
1020 REM
1030 CLOSE#1 :GOTO 1050
1040 PRINT"bad file";ICOUNT,ISYM
1050 N=N+N0:RETURN
1060 REM *********************************STORING TRANSFORMED DATA ON DISC
1070 LOCATE 19,68:PRINT"                 "
1080 LOCATE 18,70:PRINT"New file  ":LOCATE 19,68:INPUT FILNA$
```

```
1090 CLOSE
1100 OPEN FILNA$ FOR OUTPUT AS #1
1110 REM ** general properties
1120 PRINT#1,"PROPERTIES,";N;",";E;",";DENSITY
1130 REM **DATA           element properties
1140 FOR I=1 TO N
1150 PRINT#1,"ELEMENT,";I;",";DAT(I,1);",";DAT(I,2):NEXT I
1160 REM **DATA           node properties
1170 FOR I=1 TO N+1
1180 PRINT#1,"NODE,";I;
1190 FOR J=3 TO 6
1200 PRINT#1,",";DAT(I,J);
1210 NEXT J:PRINT#1," ":NEXT I
1220 FOR I=1 TO NS
1230 PRINT#1,"SUPPORT";",";isu(I)
1240 NEXT I
1250 PRINT#1,"FILEND"
1260 CLOSE#1
1270 RETURN
1280 REM
1290 REM ********************************************** creating rotor
model
1300 SCREEN 0:COLOR 15
1310 LOCATE 2,15:PRINT "Number of Elements       ";N;
1320 COLOR 14:LOCATE 2,60:INPUT N$:COLOR 15
1330 IF N$>"" THEN N=VAL(N$)
1340 LOCATE 3,15:PRINT "Modulus of Elasticity ";E;
1350 COLOR 14:LOCATE 3,60:INPUT E$:COLOR 15
1360 IF E$>"" THEN E=VAL(E$)
1370 LOCATE 4,15:PRINT "Material Density        ";DENSITY
1380 COLOR 14:LOCATE 4,60:INPUT D$:COLOR 15
1390 IF D$>"" THEN DENSITY=VAL(D$)
1400 REM ****************************************************** ELEMENT
DATA
1410 COLOR 15:PRINT "Element data",,,:COLOR 12:PRINT"node data";:LOCATE 6,1
1420 COLOR 15:PRINT"El/";:COLOR 12:PRINT"No";:LOCATE 7,1:COLOR 15
1430 PRINT "     length ","diameter  ",
1440 COLOR 12: PRINT"force","spring","mass";"     inertia";
1450 N1 = N - 1:IC=1  :COLOR 15
1460 FOR I = 1 TO N
1470 LOCATE 7+IC,1:PRINT I;
1480 FOR J=1 TO 6:LOCATE 7+IC,13*(J-1)+6:IF DAT(I,J)>0 THEN PRINT DAT(I,J);
1490 COLOR 14
1500 LOCATE 7+IC,13*(J-1)+5:INPUT D$:IF D$>"" THEN DAT(I,J)=VAL(D$)
1510 COLOR 15
1520 NEXT J:IF DAT(I,1)=0 OR DAT(I,2)=0 THEN PRINT"zero????";:GOTO 1470
1530 IC=IC+1:IF IC=10 THEN IC=1:FOR K=8 TO 18:LOCATE K,1:PRINT
SPC(70);:NEXT K
1540 NEXT I
1550 LOCATE 7+IC:PRINT" right end....";
1560 FOR J=3 TO 6:LOCATE 7+IC,13*(J-1)+6:IF DAT(N+1,J)>0 THEN PRINT
DAT(N+1,J);
1570 COLOR 14
1580 LOCATE 7+IC,13*(J-1)+5:INPUT D$:IF D$>"" THEN DAT(N+1,J)=VAL(D$)
1590 COLOR 15:NEXT J
1600 REM ******************************* SET "STIFFENESS" OF RIGID
SUPPORTS
1610 REM
1620 N2=INT((N+1)/2):EI=E*DAT(N2,2)
1630 LL = N *  DAT(N2,1):KS=(48 * EI) / LL ^ 3 * 1000
1640 LOCATE 19,1:COLOR 15
1650 PRINT"Enter no of solid supports ";NS;TAB(50);:COLOR 14:INPUT NS$
1660 IF NS$>"" THEN NS=VAL(NS$)
1670 COLOR 15:PRINT"Enter numbers of support nodes:":PRINT
1680 FOR I = 1 TO NS:PRINT "Support "; I;" at node ";
1690 PRINT isu(I);TAB(50);:COLOR 14:INPUT IS$:IF IS$>"" THEN
isu(I)=VAL(IS$)
1700 COLOR 15:NEXT I
1710 PRINT:INPUT"Are data correct (Y/N)  ";X$
1720 IF X$="N" OR X$="n" THEN 1410
1730 REM ************************************ FINDING MAXIMUM VALUES
```

```
1740 CLS:LTOT=0:DMAX=0:YMAX=0:FMAX=0
1750 FOR I=1 TO N+1
1760 DIA=DAT(I,2)
1770 LTOT=LTOT+DAT(I,1):IF DIA>DMAX THEN DMAX=DIA
1780 NEXT I
1790 REM                          LUMPING SHAFT MASS AND INERTIA TO END
NODES
1800 FOR I=1 TO N+1 :FOR J=1 TO 6:D(I,J)=DAT(I,J):NEXT J:IF I=N+1 THEN 1890
1810 KLUMP=KLUMP+DAT(I,4)
1820 MLUMP=3.14159*DAT(I,2)^2/4*DAT(I,1)*DENSITY
1830 ILUMP=3.14159*DAT(I,2)^4/64*DAT(I,1)*DENSITY
1840 D(I,6)=DAT(I,6)+ILUMP/2
1850 D(I+1,6)=DAT(I+1,6)+ILUMP/2
1860 D(I,5)=DAT(I,5)+MLUMP/2
1870 D(I+1,5)=DAT(I+1,5)+MLUMP/2
1880 MTOT=MTOT+DAT(I,5)+MLUMP
1890 NEXT I
1900 GOSUB 3430:REM                                         PLOTTING SHAFT
GEOMETRY
1910 KSHAFT=48*E*3.14*D(N/2,2)^4/64/LTOT^3
1920 MTOT=MTOT+D(N+1,5)
1930 KLUMP=0:FOR ISUP=1 TO NS
1940 INODE=isu(ISUP)
1950 D(INODE,4)=D(INODE,4)+100*KSHAFT
1960 KLUMP=KLUMP+D(INODE,4)
1970 NEXT ISUP
1980 LOCATE 23,1:INPUT"Hit return to continue";X$
1990 RETURN
2000 CLS
2010 SCREEN 0
2020 PRINT"*****************************************************"
2030 PRINT"*                                                   *"
2040 PRINT"*                ROTOR DYNAMICS                     *"
2050 PRINT"*                                                   *"
2060 PRINT"*****************************************************"
2070 COLOR 14,0
2080 PRINT:PRINT:PRINT"      MENU:"
2090 PRINT:PRINT
2100 PRINT"   <R>  dynamic response"
2110 PRINT
2120 PRINT"   <C>  critical speeds"
2130 PRINT
2140 PRINT"   <D>  data change"
2150 PRINT
2160 PRINT"   <M>  main menu "
2170 PRINT:PRINT"Enter your selection :";
2180 INPUT X$
2190 CLS
2200 PRINT
2210 IF X$="m" OR X$="M" THEN SCREEN 2:RETURN
2220 IF X$="d" OR X$="D" THEN SCREEN 2:ICR=0:RETURN
2230 IF X$="c" OR X$="C" OR X$="r" OR X$="R" GOTO 2250
2240  GOTO 2000
2250 REM
2260 IF X$="C" OR X$="c" THEN 2300
2270 PRINT"Enter angular velocity, rad/sec : ";
2280 INPUT  OMROT :PRINT
2290 OM=OMROT:GOTO 2390
2300 KSHAFT=48*E*3.14*D(INT((N+1)/2),2)^4/64/LTOT^3
2310 KTOT=(KSHAFT*KLUMP)/(KSHAFT+KLUMP)
2320 IF X$="c" OR X$="C" THEN OM=.1:DOM=SQR(KTOT/MTOT)/10:ICR=0
2330 LOCATE 20,1:PRINT"To increse Omega step, press  I ";
2340 LOCATE 21,1:PRINT"To decrease,           press  D";
2350 LOCATE 22,1:PRINT"To stop,               press  S";
2360 LOCATE 1,1:PRINT"Computing Characteristic Determinant...."
2370 INCR$=INKEY$:IF INCR$="i" OR INCR$="I" THEN DOM=2*DOM:INCR$=""
2380 IF INCR$="d" OR INCR$="D" THEN DOM=DOM/2:INCR$=""
2390 LOCATE 10,1:PRINT "Omega= ";OM;:LOCATE 10,55:PRINT "Omega step ";DOM;
2400 S$=INKEY$:IF S$="s" OR S$="S" THEN S$="":GOTO 2860
2410 FOR I = 1 TO 5: FOR J = 1 TO 5: A(I,J) = 0: NEXT J: NEXT I
2420 FOR I = 1 TO 5: A(I,I) = 1: NEXT I
```

```
2430 FOR M = 1 TO N + 1: F1 = D(M,3):D1 = D(M,2):M1 = D(M,5):J1=D(M,6)
2440 L1 = D(M,1):K1 = D(M,4)
2450 IF M = N + 1 THEN GOTO 2470
2460 GOSUB 3240
2470 GOSUB 3310
2480 FOR I = 1 TO 5: FOR J = 1 TO 5: G(I,J) = A(I,J): NEXT J: NEXT I
2490 GOSUB 3340: REM A = P * G
2500 IF M = N + 1 THEN GOTO 2530
2510 FOR I = 1 TO 5: FOR J = 1 TO 5: G(I,J) = A(I,J): NEXT J: NEXT I
2520 GOSUB 3390: REM A = S * G
2530 NEXT M
2540 F1 = D(N,3):K1 = D(N,4): GOSUB 3310
2550 FOR I = 1 TO 5: FOR J = 1 TO 5: G(I,J) = A(I,J): NEXT J: NEXT I
2560 GOSUB 3340
2570 FOR I = 1 TO 5: V(I,1) = 0: NEXT I
2580 DPREV=D9
2590 D9 = A(3,1) * A(4,2) - A(4,1) * A(3,2)
2600 D9=D9/E:IF X$="c" OR X$="C" THEN 2650
2610 DELTA=A(3,1)*A(4,2)-A(4,1)*A(3,2):IF DELTA=0 THEN DELTA=1E-30
2620 V(1,1)=(-A(3,5)*A(4,2)+A(4,5)*A(3,2))/DELTA
2630 V(2,1)=(-A(3,1)*A(4,5)+A(4,1)*A(3,5))/DELTA
2640 INDEX$="result":GOTO 2670
2650 V(2,1) =1
2660 V(1,1) = - A(4,2) / A(4,1)
2670 V(3,1)=0:V(4,1)=0:V(5,1)= 1
2680 IF INDEX$="result"THEN 2770
2690 IF X$="R" OR X$="r" THEN 2330
2700 COLOR 15:LOCATE 10,20:PRINT "Determinant = ";D9;:LOCATE 10,55:COLOR 14
2710 PRINT"Omega Step ";DOM;
2720 IF DPREV*D9>=0 THEN 2830
2730 OCR=OM+D9/(DPREV-D9)*DOM  :ICR=ICR+1
2740 IF X$="c" OR X$="C" THEN PRINT"Critical speed no ";ICR;" is:  ";OCR
2750 IF X$="r" OR X$="R" THEN PRINT"Speed of Rotation is:  ";OMROT
2760 OM=OCR:INDEX$="result":GOTO 2410
2770 REM
2780 D9=0:INDEX$=""
2790 GOSUB 2890:REM                                           printing
response
2800 GOSUB 3700:REM                                           plotting
response
2810 LOCATE 23,50:INPUT"Hit RETURN to continue";XX$
2820 DPREV=0:OM=OM+DOM/10:SCREEN 0
2830 IF X$="r" OR X$="R" THEN 2850
2835 IF ICR=0 THEN IF X$="c" OR X$="C" THEN OM=OM+DOM:GOTO 2330
2840 IF ICR<N+1-NS THEN IF X$="c" OR X$="C" THEN OM=OM+DOM:GOTO 2330
2850 PRINT"Hit return to continue....";:INPUT Y$
2860 X$="":OM=.0001:GOTO 2000
2870 END
2880 REM
2890 REM                                              COMPUTATION OF
RESPONSE
2900 LOCATE 1,1
2910 PRINT"NO     DEFLECTION  SLOPE          MOMENT          SHEAR"
2920 PRINT"---------------------------------------------------------------"
2930 FOR IP=1 TO 4:VMAX(IP)=0:NEXT IP
2940 FOR M = 1 TO N
2950 PRINT M;
2960 FOR IP=1 TO 4:RES(M,IP)=V(IP,1):VA=ABS(V(IP,1))
2970 IF VA>VMAX(IP) THEN VMAX(IP)=VA
2980 PRINT V(IP,1);TAB(15*IP+4);
2990 NEXT IP
3000 PRINT
3010 F1 = D(M,3):K1 = D(M,4):M1=D(M,5):J1=D(M,6)
3020 IF F1=0 THEN F1=1
3030 IF M = N + 1 THEN GOTO 3060
3040 L1 = D(M,1): D1 = D(M,2)
3050 GOSUB 3240
3060 GOSUB 3310
3070 FOR I = 1 TO 5:W(I,1) = V(I,1): NEXT I
3080 FOR I = 1 TO 5:V(I,1) = 0: FOR K = 1 TO 5: V(I,1) = P(I,K) * W(K,1) +
V(I,1): NEXT K: NEXT I
```

```
3090 IF M = N + 1 THEN GOTO 3120
3100 FOR I = 1 TO 5:W(I,1) = V(I,1): NEXT I
3110 FOR I = 1 TO 5:V(I,1) = 0: FOR K = 1 TO 5: V(I,1) = S(I,K) * W(K,1) +
V(I,1): NEXT K:NEXT I
3120 NEXT M
3130 PRINT M;
3140 FOR IP=1 TO 4:RES(M,IP)=V(IP,1):VA=ABS(V(IP,1))
3150 IF VA>VMAX(IP) THEN VMAX(IP)=VA
3160 PRINT V(IP,1);TAB(15*IP+4);
3170 NEXT IP
3180 YMAX=VMAX(1):TMAX=VMAX(2):MMAX=VMAX(3):VMAX=VMAX(4)
3190 PRINT
3200 PRINT
3210 IF X$="R" THEN 3230
3220 INPUT"Hit RETURN for plotting mode...";XX$
3230 RETURN
3240 REM FIELD MATRIX - EULER BEAM
3250 FOR I = 1 TO 5: S(I,I) = 1: NEXT I
3260 I1 = 3.14159 * D1 ^ 4 / 64: S(1,2) = L1
3270 S(3,4) = L1:S(1,3) = L1 ^ 2 / (2 * I1 * E)
3280 S(2,4) = S(1,3):S(2,3) = L1 / (I1 * E)
3290 S(1,4) = S(1,3) * L1 / 3
3300 RETURN
3310 REM POINT MATRIX: FORCE AND SPRING
3320 FOR I = 1 TO 5: P(I,I) = 1: NEXT I
3330 P(3,2)=-KT+J1*OM^2: P(4,5) = F1: P(4,1) = - K1 + M1 * OM ^ 2: RETURN
3340 REM A = P * G
3350 FOR I = 1 TO 5: FOR J = 1 TO 5
3360 C = 0
3370 FOR K = 1 TO 5: C = C + P(I,K) * G(K,J):NEXT K: A(I,J) = C: NEXT
J:NEXT I
3380 RETURN
3390 REM A=S*G
3400 FOR I = 1 TO 5: FOR J = 1 TO 5: A(I,J) = 0
3410 FOR K = 1 TO 5: A(I,J) = A(I,J) + S(I,K) * G(K,J): NEXT K: NEXT J:
NEXT I
3420 RETURN
3430 REM ********************************************* PLOTTING BEAM AND
LOADS
3440 REM
3450 SCREEN 2:CLS:XSC=500/LTOT:YSC=XSC/2.5:LSTART=0
3460 FOR I=1 TO N+1
3470 DIA=DAT(I,2):DIAM=4*SQR(DAT(I,6)/(DAT(I,5)+.000001))+.0000001
3480 LM=4*DAT(I,5)/3.14/DIAM^2/DENSITY
3490 IF DIA=0 THEN DIA=DIAPR
3500 DIAPR=DIA
3510 X1=10+LSTART:Y1=50-DIA*YSC/2:X2=10+LSTART+DAT(I,1)*XSC:Y2=50+DIA*YSC/2
3520 DY=DIAM*YSC/2
3530 LINE(X1,Y1)-(X2,Y2),7,BF:IF DAT(I,3)=0 THEN 3570
3540 LINE (X1-1,Y1-DY)-(X1+1,Y1-20-DY),1,BF
3550 LINE (X1,Y1-DY)-(X1+3,Y1-5-DY),1
3560 LINE -(X1-3,Y1-5-DY),2:LINE -(X1,Y1-DY),1
3570 Y2=Y2+DIAM*YSC/2
3580 LINE(X1-LM/2*XSC,50-DIAM*YSC/2)-(X1+LM/2*XSC,50+DIAM*YSC/2),7,BF
3590 FOR II=1 TO NS:IF isu(II)<>I THEN GOTO 3610
3600 LINE(X1,Y2)-(X1-3,Y2+5),3:LINE -(X1+3,Y2+5),3:LINE -(X1,Y2),3
3610 NEXT II
3620 LSTART=LSTART+DAT(I,1)*XSC
3630 NEXT I
3640 FOR II=1 TO NS:IF isu(II)><N+1 THEN  3660
3650 LINE(X2,Y2)-(X2-3,Y2+5),3:LINE -(X2+3,Y2+5),3:LINE -(X2,Y2),3
3660 NEXT II
3670 LOCATE 9,50:PRINT "d-max=";DMAX;
3680 LOCATE 18,50:PRINT "y-max=";VMAX(1);
3690 RETURN
3700 REM ************************************************ plot results
3710 GOSUB 3430
3720 LSTART=0:LOCATE 1,1
3730 IF X$="c" OR X$="C" THEN PRINT"Critical Speed No ";ICR;" is ";OM;"
rad/sec"
3740 IF X$="r" OR X$="R" THEN PRINT"Speed of Rotation is ";OMROT;"
rad/sec";
```

```
3750 FOR I=1 TO N
3760 X1=10+LSTART:Y1=120+RES(I,1)*20/YMAX:TSC=20/YMAX/XSC
3770 X2=10+LSTART+D(I,1)*XSC:Y2=120+RES(I+1,1)*20/YMAX
3780 SL1=RES(I,2)*TSC:SL2=RES(I+1,2)*TSC:PSET(X1,Y1)
3790 V=-RES(I,4):MOM=RES(I,3):VMAX=ABS(VMAX):MMAX=ABS(MMAX)
3800 FOR X=X1 TO X2:S=(X-X1)/(X2-X1):F1=1-3*S^2+2*S^3:F2=3*S^2-2*S^3
3810 F3=(X2-X1)*(S-2*S^2+S^3):F4=(X2-X1)*(-S^2+S^3)
3820 Y=(F1*Y1+F2*Y2+F3*SL1+F4*SL2)
3830 LINE -(X,Y),3:NEXT X:FOR X=X1 TO X2
3840 NEXT X
3850 LINE (X1,120)-(X2,120),7
3860 LSTART=LSTART+D(I,1)*XSC
3870 NEXT I
3880 RETURN
```

APPENDIX 14.B
SIMUL: A PROGRAM FOR DYNAMIC SIMULATION OF THE ROTATION OF A TWO-ROTOR SYSTEM

```
10 REM ****************************************************
20 REM *                                                  *
30 REM *                    SIMUL                         *
40 REM *                                                  *
50 REM ****************************************************
60 REM            COPYRIGHT 1987
70 REM by Andrew D. Dimarogonas, Washington Univ.,St Louis,Mo.
80 REM
90 REM            by ADD  26/5/86
100 REM
110 REM
120 REM Simulation of the torsional response of a motor-flexible shaft
130 REM -driven machine system
140 CLS:KEY OFF
150 FOR I=1 TO 14:COLOR I:LOCATE I,22+I:PRINT "SIMUL";
160 LOCATE I,50-I:PRINT "SIMUL";:NEXT I:COLOR 14
170 LOCATE 8,15:PRINT"A motor-shaft-driven machine simulation program";
180 LOCATE 23,28:INPUT"Hit ENTER to continue";X$:CLS
190 DIM DP(10),X(10),XE(10),DE(10)
200 N  = 4:CLS:COLOR 14
201 PRINT"            ENTER DATA:":PRINT:PRINT:PRINT
205 READ J1,J2,KT,T0,RO
206 DATA 10,10,1000,100,150
222 PRINT"Polar Moment of Inertia of motor                        ";J1;:INPUT
J1$:IF J1$>"" THEN J1=VAL(J1$)
223 PRINT"Polar Moment of Inertia of driven rotor                 ";J2;:INPUT
J2$:IF J2$>"" THEN J2=VAL(J2$)
224 PRINT"Torsional spring constant of shaft and coupling         ";KT;:INPUT
KT$:IF KT$>"" THEN KT=VAL(KT$)
225 PRINT"Nominal Motor Torque                                    ";T0;:INPUT
T0$:IF T0$>"" THEN T0=VAL(T0$)
226 PRINT"Running speed, rad/sec                                  ";RO;:INPUT
RO$:IF RO$>"" THEN RO=VAL(RO$)
230 C = .5 * SQR (4 * (J1 + J2) * KT)
240 C = C / 2
250 TOSC=50/T0
260 OMSC=50/RO
270 LL = T0 / RO
280 OM = SQR ((J1 + J2) * KT / J1 / J2)
290 CLS:SCREEN 2:LOCATE 15,1:PRINT "POWER, KW  =";RO * T0 / 1000;
300 LOCATE 16,1:PRINT "NAT.FREQUENCY=";OM;"Rad/sec";TAB(50);OM/6.28;" Hz";
310 LOCATE 17,1:PRINT "RUNNING FREQ.=";RO;TAB(50);RO/6.28;" Hz";
320 PE = 6.28 / OM
330 LOCATE 18,1:PRINT "PERIOD SEC  =";PE;
340 DT = PE /100
350 T = 0
360 LINE(20,0)-(630,0):LINE -(630,100):LINE -(20,100):LINE -(20,0)
370 FOR I=0 TO 20:LINE(15,5*I)-(20,5*I):NEXT I:LINE(1,1)-(20,1)
380 LINE (1,50)-(20,50):LINE (1,100)-(20,100)
390 FOR I=1 TO 20:LINE(30*I+20,100)-(30*I+20,105):NEXT I
```

```
400 LOCATE 1,1:PRINT"2";:LOCATE 13,1:PRINT "0";:LOCATE 7,1:PRINT"1";
410 TS =PE/DT:LOCATE 14,50:PRINT "Time: 1 div= ";3*TS;" sec";
420 GOSUB 880
430 X1 = XE(1):X2 = XE(2):X3 = XE(3):X4 = XE(4)
440 TT = T
450 LOCATE 24,1:PRINT"          Time              speed             shaft torque
motor torque";
460 LOCATE 25,1: PRINT USING " #########.######";TT, X3,(KT * (X2 -
X4)),TR,
470 X=TS*T:Y=X3*OMSC:GOSUB 540
480 X = TS * T:Y =KT * (X2 - X4)*TOSC:GOSUB 540
490 X = TS * T:Y =TR*TOSC:GOSUB 540
500 IF X3 > .99 * RO THEN 520
510 GOTO 420
520 END
530 REM ********************************************* PLOTTING ROUTINE
540 IF Y < 0 THEN Y = 0
550 IF Y > 99 THEN Y = 99
560 PSET (X+20,100-Y)
565 IF X>600 THEN T=0
570 RETURN
580 REM *********************************************DERIVATIVES EVALUATION
590 X1 = XE(1):X2 = XE(2):X3 = XE(3):X4 = XE(4)
600 IF X3 < RO THEN TR = 2 * T0 - X3 * T0 / RO
610 IF X3 > = RO THEN TR = T0 - (X3 - RO) * T0 / (.2 * RO)
620 D1 = (TR + KT * (X4 - X2)) / J1
630 D2 = X1
640 D3 = - KT * (X4  - X2) / J2 - LL * X3 / J2
650 D3 = D3 - C * (X3 - X1) / J2
660 D4 = X3
670 DE(1) = D1:DE(2) = D2:DE(3) = D3:DE(4) = D4
680 RETURN
690 REM *************************************************RUNGE-KUTTA METHOD
700 REM RETAIN CURRENT VALUES OF VECTOR
710 FOR I = 1 TO N:YE(I) = XE(I):NEXT I
720 GOSUB 580
730 FOR I = 1 TO N:K1(I) = DT * DE(I):NEXT I
740 REM HALF STEP
750 T = T + DT / 2
760 FOR I = 1 TO N:XE(I) = XE(I) + K1(I) / 2:NEXT I
770 GOSUB 580
780 FOR I = 1 TO N:K2(I) = DT * DE(I):NEXT I
790 FOR I = 1 TO N:XE(I) = YE(I) + K2(I) / 2:NEXT I
800 GOSUB 580
810 FOR I = 1 TO N:K3(I) = DT * DE(I):NEXT I
820 T = T + DT / 2
830 FOR I = 1 TO N:XE(I) = YE(I) + K3(I):NEXT I
840 GOSUB 580
850 FOR I = 1 TO N:K4(I) = DT * DE(I):NEXT I
860 FOR I = 1 TO N:XE(I) = YE(I) + (K1(I) + 2 * K2(I) + 2 * K3(I) + K4(I))
/ 6:NEXT I
870 RETURN
880 REM ********************************************MODIFIED EULER METHOD
890 REM PREDICTOR
900 REM RETAIN VECTOR AT POINT K
910 FOR I = 1 TO N:YE(I) = XE(I):NEXT I
920 GOSUB 580
930 REM PREDICTOR
940 FOR I = 1 TO N
950 XE(I) = XE(I) + DT * DE(I)
960 DP(I) = DE(I)
970 NEXT I
980 REM CORRECTOR
990 T = T + DT
1000 FOR IT = 1 TO NT
1010 GOSUB 580
1020 FOR I = 1 TO N
1030 XE(I) = YE(I) + (DE(I) + DP(I)) / 2 * DT
1040 NEXT I
1050 NEXT IT:LOCATE 20,1
1060 RETURN
```

APPENDIX 14.C
SHRINKFIT: A PROGRAM FOR THE DESIGN OF SHRINKFITS WITH COMPOUND HUBS

```
10 REM ************************************
20 REM *                                  *
30 REM *            SHRINKFIT             *
40 REM *                                  *
50 REM ************************************
60 REM          COPYRIGHT 1985
70 REM by Professor Andrew D. Dimarogonas, Washington Univ., St. Louis, Mo.
80 REM  All rights reserved.  Unauthorized reproduction, disssemination,
90 REM selling or use is strictly prohibited.
100 REM  This listing is for reference purpose only.
110 REM
120 REM
130 REM Shrinkfit of a compound rim
140 REM
150 REM
160 REM          ADD / 1-12--84
170 REM
180 CLS:KEY OFF
190 LOCATE 25,1:PRINT" Andrew Dimarogonas, Computer Aided Machine Design,
Prentice-Hall, London 1987";
200 FOR I=1 TO 14:COLOR I:LOCATE I,22+I:PRINT "SHRINKFIT";
210 LOCATE I,50-I:PRINT "SHRINKFIT";:NEXT I:COLOR 14
220 LOCATE 8,17:PRINT"Design of the shrink fit of a compound wheel";
230 DIM A(2,2),B(2,2),C(2,2),L(2,2),L0(2,2)
240 DIM D(10),W(10), E(10), NN(10)
250 PI = 3.14159:LOCATE 19,25:INPUT"Hit RETURN to continue";X$
260 CLS:PRINT"                    DATA:"
270 INPUT"Number of rings                                  ";N:FOR I=1 TO N
280 INPUT"inner dia, width, Young Mod., Poisson ratio
";D(I),W(I),E(I),NN(I)
290 NEXT I
300 INPUT"Outer diameter                                   ";D(N+1)
310 INPUT"Transmitted torque,axial force                   ";MO,FO
320 INPUT"Coefficient of friction                          ";FF
330 INPUT"Allowable stress at fit                          ";SA
340 REM FORWARD SWEEP
350 REM INITIALIZE TRANSFER MATRIX
360 FOR I = 1 TO 2: FOR J = 1 TO 2: L(I,J) = 0: NEXT J
370 L(I,I) = 1:NEXT I
380 FOR I = 1 TO N
390 W(0) = W(1):E(0) = E(1):NN(0) = NN(1)
400 GOSUB 750
410 FOR I1 = 1 TO 2: FOR I2 = 1 TO 2: L0(I1,I2) = 0: FOR I3 = 1 TO 2
420 L0(I1,I2) = L0(I1,I2) + B(I1,I3) * L(I3,I2): NEXT I3:NEXT I2:NEXT I1
430 FOR I1 = 1 TO 2: FOR I2 = 1 TO 2: L(I1,I2) = L0(I1,I2):NEXT I2: NEXT I1
440 IF I = N THEN 490
450 GOSUB 860
460 FOR I1 = 1 TO 2: FOR  I2 = 1 TO 2: L0(I1,I2) = 0: FOR I3 = 1 TO 2
470 L0(I1,I2) = L0(I1,I2) + L(I1,I3) * C(I3,I2): NEXT I3: NEXT I2:NEXT I1
480 FOR I1 = 1 TO 2: FOR I2 = 1 TO 2: L(I1,I2) = L0(I1,I2):NEXT I2: NEXT I1
490 NEXT I
500 REM REQUIRED FIT PRESSURE
510 PR = (2* MO / D(1) + FO) / (FF * PI * W(1) * D(1))
520 REM ...................................................INITIAL VECTOR
530 P1 = PR: U1 = - L(1,1) / L (1,2) * PR
540 I = 1: GOSUB 750
550 P2 = B(1,1) * P1 + B(1,2) * U1
560 REM .................................MAXIMUM TANGENTIAL STRESSES
570 SM = PR * (D(1) ^ 2 + D(2) ^ 2) / (D(2) ^2 - D(1) ^ 2)
580 REM ...........................................combined stresses
590 SC = SQR (SM ^ 2 + 4 * PR ^ 2)
600 REM ................................Minimum interference = U1
610 E1 = U1
620 REM ............................Maximum interference for strength
630 E2 = U1 * SA / SM :PRINT:PRINT
640 PRINT "  RESULTS:":PRINT
```

```
650 PRINT "Minimum interference           =";E1
660 PRINT
670 PRINT "Maximum interference           =";E2
680 PRINT
690 PRINT "Min. interference pressure  =";PR
700 PRINT
710 PRINT "Max  interference pressure  =";PR * SA / SM
720 PRINT
730 PRINT "With maximum stress           =";SA
740 END
750 REM .............................................Field transfer
matrix
760 R1 = D(I) / 2:NU = NN(I): R2 = D(I + 1) ./ 2
770 A(1,1)=((1+N(I))*(D(I)/2)^3 +(1-N(I))*R1*R2^2) /(E(I)*(R2 ^ 2 - R1 ^
2))
780 A(1,2) = ( - (1 + NU)*R1 * R2 ^ 2 - (1 - NU) * R1*R2^2)/(E(I)*(R2^2-
R1^2))
790 A(2,2)=(-(1+NU)*R2^3 - ( 1 - NU) * R1 ^ 2 * R2) / (E(I) * (R2 ^ 2 -R1
^2))
800 A(2,1)=((1+NU)*R1^2*R2+(1 - NU) * R1 ^ 2 * R2) / (E(I) * (R2 ^ 2 - R1 ^
1))
810 B(1,1) = - A(1,1) / A(1,2)
820 B(1,2) = 1 / A(1,2)
830 B(2,1) = A(2,1) - A(1,1) * A(2,2) / A(1,2)
840 B(2,2) = A(2,2) / A (1,2)
850 RETURN
860 REM
870 REM .............................................Point transfer
matrix
880 C(1,1) = W(I + 1) / W(I): C(1,2) = 0:C(2,1) = 0:C(2,2) = 1
890 RETURN
```

APPENDIX 14.D
SHAFTDES: A SHAFT DESIGN AND ANTIFRICTION BEARING APPLICATION PROGRAM

```
10 REM **************************************************
20 REM *                                                *
30 REM *                  SHAFTDES                      *
40 REM *                                                *
50 REM **************************************************
60 REM          COPYRIGHT 1985
70 REM by Andrew D. Dimarogonas, University of Patras, Greece. All rights
reserved. Unauthorized reproduction, dissemination, selling or use is
strictly prohibited. This listing is for reference purpose only.
80 REM
90 REM             by ADD  26/5/86
100 REM
110 REM
120 REM Design of shafts on antifriction bearings for given external loads
130 REM Material properties are provided by the user
140 REM The transfer matrix method in two orthogonal planes is used
150 REM The program TMSTAT is used
151 CLS:KEY OFF
152 FOR I=1 TO 14:COLOR I:LOCATE I,22+I:PRINT "SHAFTDES";
153 LOCATE I,50-I:PRINT "SHAFTDES";:NEXT I:COLOR 14
154 LOCATE 8,10:PRINT"A shaft design program using the transfer matrix
method";
156 LOCATE 23,28:INPUT"Hit ENTER to continue";X$:CLS
160 REM
170 DIM FX(20),FY(20),MX(20),MY(20):REM          NODAL FORCES AND MOMENTS
180 DIM QX(20),QY(20):REM                        FIELD DISTRIBUTED LOADS
190 DIM MMX(20),MPX(20),MMY(20),MPY(20):REM      MOMENTS X,Y AND +,-
200 DIM VMX(20),VPX(20),VMY(20),VPY(20):REM      SHEAR X,Y AND +,-
210 DIM TOR(20),THR(20):REM                      ELEMENT TORQUE,THRUST
220 DIM MM(20),VM(20),DM(20):REM                 NODE FORCES,DIA (-)
230 DIM MP(20),VP(20),DP(20):REM                 NODE FORCES,DIA (+)
240 DIM D(20,10):REM                             PARAMETER MATRIX
```

```
250 DIM P(5,5),S(5,5),A(5,5),G(5,5):REM              AUXILIARY TRANSFER MATRICES
260 DIM V(5,1),W(5,1):REM                            STATE VECTORS
265 DIM KX(20),KY(20),KTX(20),KTY(20):REM            SPRING CONSTANTS
270 DIM ISU(10) :REM                                 SUPPORT MATRIX
271 DIM BDIA(10):REM                                 BEARING DIAMETERS
272 DIM BFR(10):REM                                  BEARING RADIAL LOADS
273 DIM BDAT(6,30),BTYP$(30):REM                    BEARING DATA FROM FILE
280 PI=3.14159
290 CLS
300 PRINT"ENTER DATA:"
310 PI=3.14159
320 REM GOTO 790 for example data
330 PRINT"Number of elements                                    ";
340 INPUT N
350 NP1=N+1
360 N1 = N - 1:FOR I = 1 TO NP1
370 PRINT"Node ";I;
380 PRINT": Enter FX,FY,MX,MY                          ";
390 INPUT FX(I),FY(I),MX(I),MY(I)
400 PRINT"          Support <S>, spring <P>, free <F>     ";
410 INPUT X$
420 IF X$="F" OR X$="f" THEN 510
430 IF X$="S" OR X$="s" THEN 460
440 IF X$="P" OR X$="p" THEN 490
450 GOTO 400
460 ISUPP=ISUPP+1:ISU(ISUPP)=I
470 NS=ISUPP
480 GOTO 510
490 PRINT"          Enter spring constants KX,KY,KTX,KTY ";
500 INPUT KX(I),KY(I),KTX(I),KTY(I)
510 NEXT I
520 FOR I=1 TO N
530 PRINT"Element ";I;
540 PRINT          ":Length,QloadX,QloadY                   " ;
550 INPUT D(I,1),QX(I),QY(I)
560 PRINT"          Element torque,thrust                   ";
570 INPUT TOR(I),THR(I)
580 IF TOR(I)>TORMAX THEN TORMAX=TOR(I)
590 NEXT I
600 PRINT"Enter Material Data: "
610 PRINT"                         Young Modulus     ";
620 INPUT EL
630 PRINT"                         Fatigue strength ";
640 INPUT SE
650 PRINT"                         Yield Strength    ";
660 INPUT SY
670 PRINT"                         Specific weight   ";
680 INPUT SPECW
690 PRINT"                         Service Factor    ";
700 INPUT SEF
710 PRINT"Dead weight in y-direction (Y/N)                      ";
720 INPUT X$
730 IF X$="N" THEN 770
740 FOR I=1 TO N
750 QY(I)=QY(I)-PI*D(I,2)^2/4*SPECW
760 NEXT I
770 REM
780 GOTO 870
790 READ N,FY(2),FY(3),FY(4),FY(5):DATA 5,1000,1000,1000,1000
800 READ NS,ISU(1),ISU(2):DATA 2,1,6
810 READ TOR(2),TOR(3),TOR(4),TOR(5):DATA 30000,60000,90000,120000
820 READ D(1,1),D(2,1),D(3,1),D(4,1),D(5,1):DATA .8,.5,.5,.5 ,.8
830 TORMAX=120000!
840 REM HORIZONTAL PASS
850 READ EL,GS,SY,SE:DATA 2.1E11,1.E11,220E6,120E6
860 READ SEF:DATA 2
870 FOR I=1 TO N +1
875 QY(I)=QY(I)+PI*D(I,2)^2/4*SPECW
880 DP(I)=(32*TORMAX*5/3.14159/SY)^.333
890 DM(I)=DP(I)
900 D(I,2)=DM(I)
910 IF DP(I)>DM(I) THEN D(I,2)=DP(I)
```

```
915 IF D(I,2)<D(I,1)/20 THEN D(I,2)=D(I,1)/20
920 NEXT I
930 REM SET "STIFFNESS" OF RIGID SUPPORTS
940 N2 = INT (N / 2):EI = D(N2,2)^4*EL*PI/64
950 LL = N * D(N2,1):KS=(48 * EI) / LL ^ 3 * 100
960 REM INPUT PREPARATION FOR TRANSFER MATRIX SWEEP
970 FOR I=1 TO N+1
980 D(I,3)=FX(I):D(I,4)=QX(I):D(I,5)=KX(I):D(I,6)=MX(I):D(I,7)=KTX(I)
990 NEXT I:PAS$="x"
1000 COLOR 15:PRINT "PASS ";IPASS+1:COLOR 28:PRINT"HORIZONTAL
PASS....":COLOR 14
1010 GOSUB 1300  :REM PRINT DATA FOR HORIZONTAL PASS
1020 GOSUB 1430  :REM DO HORIZONTAL PASS
1030 COLOR 28:PRINT"Hit ENTER to continue...":COLOR 14:INPUT X$
1040 FOR M=1 TO N+1
1050 MMX(M)=MMY(M):VMX(M)=VMY(M)
1060 VPX(M)=VPY(M):MPX(M)=MPY(M)
1070 I=M
1080 D(I,3)=FY(I):D(I,4)=QY(I):D(I,5)=KY(I):D(I,6)=MY(I):D(I,7)=KTY(I)
1090 NEXT M:PAS$="y"
1100 COLOR 28:PRINT"VERTICAL PASS...":COLOR 14
1110 GOSUB 1300:REM PRINT DATA FOR VERTICAL PASS
1120 GOSUB 1430:color 28:REM COMPUTE VERTICAL PASS
1125 input"Hit ENTER to continue";cont$:cls:if ipass=0 then color 14:goto
1270
1130 COLOR 28:PRINT "RESULTS:":COLOR 14
1140 PRINT"NODE   D(-)        D(+) "
1150 PRINT"-----------------------------------------------"
1160 FOR I=1 TO N +1
1170 MM(I)=SQR(MMX(I)^2+MMY(I)^2)
1180 VM(I)=SQR(VMX(I)^2+VMY(I)^2)
1190 VP(I)=SQR(VPX(I)^2+VPY(I)^2)
1200 MP(I)=SQR(MPX(I)^2+MPY(I)^2)
1210 DM(I)=(32*SEF*SQR(MM(I)^2+TOR(I)^2)/(PI*SY))^.333:IF DM(I)=0 THEN
DM(I)=DM(I+1)
1220 DP(I)=(32*SEF*SQR(MP(I)^2+TOR(I+1)^2)/(PI*SY))^.333:IF DP(I)=0 THEN
DP(I)=DP(I-1)
1230 D(I,2)=DP(I):XPRINT$=""
1231 FOR II=1 TO NS:IIS = ISU(II):IF I<>IIS THEN 1235
1232 DIAM=DP(I):IF DP(I)=0 THEN DIAM=DM(I)
1233 IF DP(I)>DM(I) AND DM(I)>0 THEN DIAM=DM(I)
1234
BDIA(II)=DIAM:BFR(II)=SQR(REACTX(I)^2+REACTY(I)^2):XPRINT$="...bearing"
1235 NEXT II
1240 PRINT I;"  ";:PRINT USING "####.^^^^   ";DM(I),DP(I);:PRINT XPRINT$
1250 NEXT I
1260 COLOR 28:PRINT"Hit ENTER to continue...";:INPUT X$:COLOR 14
1270 IPASS=IPASS+1
1280 IF IPASS<2 THEN 970
1281 INPUT"Do you want bearing calculations (Y/N) ";X$
1282 IF X$="Y" OR X$="y" THEN GOSUB 3000
1290 END
1300 REM *************DATA PRINTOUT*****************************
1310 PRINT "DATA: ";N;" ELEMENTS,  ";N+1;"  NODES"
1320 PRINT
1330 PRINT "  LENGTH    DIAMETER  FORCE      LOAD      SPRING"
1340 PRINT "----------------------------------------------------------------
--"
1350 FOR I=1 TO N+1: FOR J = 1 TO 5
1360 A$ = " "
1370 FOR II = 1 TO NS: IIS = ISU(II)
1380 IF I = IIS THEN A$= " SUPPORT"
1390 NEXT II
1400 PRINT USING " ####.^^^^";D(I,J);:NEXT J:PRINT A$
1410 NEXT I
1420 RETURN
1430 REM**************TRANSFER MATRIX SWEEP**********************
1440 COLOR 28:PRINT"computing...":COLOR 14
1450 REM
1460 REM FORWARD SWEEP OF TRANSFER MATRICES
1470 REM
1480 REM SET A(5,5) TO IDENTIFY
```

```
1490 FOR I = 1 TO 5:FOR J=1 TO 5:A(I,J)=0:NEXT J: A(I,I) =1:NEXT I
1500 REM COMPUTE AND MULTIPLY TM FROM LEFT TO RIGHT
1510 FOR M = 1 TO N + 1
1520 F1 = D(M,3):K1=D(M,5)
1530 IF M > N THEN GOTO 1560
1540 L1 = D(M,1):D1 = D(M,2)^4*EL*PI/64:Q1 = D (M,4)
1550 GOSUB 2100
1560 GOSUB 2210
1570 FOR I = 1 TO 5: FOR J = 1 TO 5:G(I,J) = A(I,J): NEXT J: NEXT I
1580 GOSUB 2280:REM A=P*G
1590 IF M>N THEN GOTO 1620
1600 FOR I = 1 TO 5:FOR J = 1 TO 5:G(I,J) = A(I,J):NEXT J: NEXT I
1610 GOSUB 2330: REM A=S*G
1620 NEXT M
1630 FOR I = 1 TO 5:V(I,1) = 0:NEXT I
1640 REM SOLVE FOR UNKNOWN END CONDITIONS, HERE, FREE-FREE BEAM IS ASSUMED
WITH V0=VL=0, MO=VL=0, UNKNOWN XO,XL,FO,FL
1650 D9 = A(3,1) * A(4,2) - A(4,1) * A(3,2)
1660 V(1,1) = (-A(4,2) * A(3,5) + A(3,2) * A(4,5)) / D9
1670 V(2,1) = (-A(3,1) * A(4,5) + A(4,1) * A(3,5)) / D9
1680 V(5,1) = 1 :V(3,1)=0:V(4,1)=0
1690 REM RESULTS, PROCEED FROM LEFT TO RIGHT MULTIPLYING WITH TM'S AND
PRINTING STATE VECTOR AT EACH STATION
1700  PRINT "SOLUTION ..."
1710 PRINT"STATION DEFLECTION   SLOPE        MOMENT        SHEAR
REACTION"
1720 PRINT"-------------------------------------------------------------------
----"
1730 FOR M = 1 TO N + 1
1740 IF M > N THEN GOTO 1820
1750 PRINT M;
1760 PRINT USING " ####.^^^^";V(1,1);V(2,1);
1770 PRINT USING " ####.^^^^";V(3,1);V(4,1);
1780 FOR IIS = 1 TO NS: IF ISU(IIS) < > M THEN 1810
1790 SR =KS* V(1,1):REACTY(M)=SR:IF PAS$="x" THEN REACTX(M)=SR
1800 PRINT USING " ####.^^^^"; SR;
1810 NEXT IIS
1820 F1 = D(M,3):K1 = D(M,5)
1830 IF M > N THEN GOTO 1870
1840 L1 = D(M,1):D1 = D(M,2)^4*PI*EL/64:Q1= D(M,A)
1850 MMY(M)=V(3,1):VMY(M)=V(4,1)
1860 GOSUB 2100
1870 GOSUB 2210
1880 FOR I = 1 TO 5: W(I,1) = V(I,1):NEXT I
1890 FOR I = 1 TO 5: V(I,1) = 0: FOR K = 1 TO 5: V(I,1) = P(I,K) * W(K,1) +
V(I,1): NEXT K:NEXT I
1900 MPY(M)=V(3,1):VPY(M)=V(4,1)
1910 IF M > N THEN GOTO 1950
1920 FOR I = 1 TO 5: W(I,1) = V(I,1):NEXT I
1930 FOR I = 1 TO 5: V(I,1) = 0:FOR K = 1 TO 5: V(I,1) =S(I,K) * W(K,1) +
V(I,1):NEXT K: NEXT I
1940 PRINT
1950 NEXT M
1960 M = M-1
1970 MPY(N+1)=0:VPY(N+1)=0
1980 REM PRINT RESULTS AT END STATION
1990 PRINT M;
2000 PRINT USING " ####.^^^^";V(1,1);V(2,1);
2010 PRINT USING " ####.^^^^";V(3,1);V(4,1);
2020 FOR IIS= 1 TO NS: IF ISU(IIS) < > N + 1 THEN 2050
2030 SR =KS * V(1,1):REACTY(M)=SR:IF PAS$="x" THEN REACTX(M)=SR
2040 PRINT USING " ####.^^^^"; SR
2050 NEXT IIS
2060 RETURN
2070 REM ********************TRANSFER MATRICES********************
2080 REM TRANSFER MATRIX SUBROUTINES
2090 REM
2100 REM FIELD MATRIX - EULER BEAM
2110 FOR I = 1 TO 5: S(I,I) = 1: NEXT I
2120 EI = D1:S(1,2) = L1
2130 S(3,4) =L1:S(1,3) = L1 ^ 2 / (2 * EI)
2140 S(2,4) =S(1,3):S(2,3) = L1 / EI
```

```
2150 S(1,4) =S(1,3) * L1 /3
2160 S(1,5) =Q1 * L1 ^ 4 / (24 * EI)
2170 S(2,5) =S(1,5) * 40 / L1
2180 S(3,5) = Q1 * L1 ^ 2 / 2
2190 S(4,5) = S(3,5) * 20 /L1
2200 RETURN
2210 REM POINT MATRIX:FORCE, DISTRIBUTED LOAD AND SPRING
2220 FOR I = 1 TO 5: FOR J= 1 TO  5: P(I,J) = 0: NEXT J: NEXT I
2230 FOR I = 1 TO 5: P(I,I) = 1:NEXT I
2240 FOR I = 1 TO NS:SI = ISU(I)
2250 IF M = SI THEN K1 = KS
2260 NEXT I
2270 P(4,5) = F1: P(4,1) = - K1: RETURN
2280 REM A = P * G
2290 FOR I = 1 TO 5: FOR J = 1 TO 5
2300 C = 0
2310 FOR K = 1 TO 5:C = C + P(I,K) * G(K,J):NEXT K:A(I,J) = C: NEXT J:NEXT
I
2320 RETURN
2330 REM A = S * G
2340 FOR I = 1 TO 5: FOR J = 1 TO 5: A(I,J) = 0
2350 FOR K=1 TO 5: A(I,J)=A(I,J)+S(I,K)*G(K,J):NEXT K:NEXT J: NEXT I:RETURN
3000 REM ********************************  Antifriction bearing
calculations
3010 CLS
3020 PRINT"              ANTIFRICTION BEARING SELECTION"
3030 PRINT"                             Data:"
3040 PRINT"node      bearing No      diameter       radial load    axial load"
3050 PRINT"__________________________________________________________________"
3060 FOR I=1 TO NS:IIS=ISU(I)
3070 PRINT IIS;TAB(10);I;TAB(20);:PRINT USING"
####.^^^^";BDIA(I),BFR(I),thr(i)
3080 NEXT I:PRINT
3090 INPUT"Enter No of bearing which takes the thrust ";ITHRUST
3095 IF ITHRUST>NS OR ITHRUST<1 THEN 3090
3100 PRINT"Thrust load at bearing                 ";ITHRUST;" is
";THR(ISU(ITHRUST))
3110 INPUT"Length Units you use, <1> m, <2> mm, <3> in   ";ULEN
3120 INPUT"Force Units you use,  <1> N, <2> kg, <3> lb   ";UFOR
3130 PRINT"DESIGN DATA:"
3140 INPUT"Hours of operation                    ";LH
3150 INPUT"Revolutions per minute RPM            ";NRPM
3160 INPUT"Reliability factor a1                 ";A1
3170 INPUT"Materials factor a2                   ";A2
3180 INPUT"Temperature and speed factor a3       ";A3:V=1
3190 INPUT"Ring rotating, <I>nner, <O>uter   ";R$:IF R$="o" OR R$="O" THEN
V=1.2
3200 INPUT"Static Service Factor                 ";SST
3202 INPUT"Are above data correct (Y/N)          ";X$
3204 IF X$="n" OR X$="N" THEN CLS:GOTO 3010
3206 FOR I=1 TO NS
3208 IF ULEN=1 THEN BDIA(I)=BDIA(I)*1000
3209 IF ULEN=3 THEN BDIA(I)=BDIA(I)*25.4
3210 IF UFOR=2 THEN BFR(I)=BFR(I)*9.810001
3212 IF UFOR=3 THEN BFR(I)=BFR(I)*9.810001/2.2
3214 IF UFOR=2 THEN THR(I)=THR(I)*9.810001
3215 IF UFOR=3 THEN THR(I)=THR(I)*9.810001/2.2
3216 BDIA(I)=INT(BDIA(I)):NEXT I
3218 FOR I=1 TO NS:CLS
3220 PRINT"           DESIGN OF BEARING No ";I;" , AT NODE ";ISU(I)
3230 PRINT"           DATA:"
3240 PRINT"                 Diameter >     ";BDIA(I);" mm"
3250 PRINT"                 Radial Load = ";BFR(I);" N"
3260 PRINT"                 Axial Load =  ";THR(I);" N"
3270 PRINT"           TYPE:  BALL     <1> deep groove       <2> self
alligning"
3280 PRINT"                           <3> angular contact <4> axial"
3290 PRINT"                  ROLLER   <5> single roller     <6> double,self
algn"
3300 PRINT"                           <7> conical,roller  <8> needle"
3310 INPUT"           Enter type ";TS%:IF TS%>2 THEN PRINT"Only 1 or
2":GOTO 3310
```

```
3320 IF TS%=1 THEN NF$="deepgr.bal":PEXP=1/3
3330 IF TS%=2 THEN NF$="selfal.bal":PEXP=1/3
3332 BDIAM=BDIA(I):N1=2:fr=bfr(i):fa=thr(i)
3340 N2=N1+9:GOSUB 3550
3341 IDIA=0:FOR IR=1 TO 10:IF BDAT(1,IR)>=BDIAM THEN IDIA=1:IN1=IR:IR=10
3342 NEXT IR:IF IDIA=0 AND BDAT(1,1)<600 AND N1<200 THEN N1=N1+10:GOTO 3340
3344 N1=N1+IN1-1:N2=N1+20:GOSUB 3550:itest=0:bdiam=bdat(1,1)
3350 itest=itest+1:D=BDAT(1,itest):cap0=bdat(5,itest):ON TS% GOSUB
5100,5200
3355 P=X*V*BFR(I)+Y*THR(I):cload0=sst*p
3356 L10=LH*60*NRPM*.000001:CLOAD=P*(L10/(A1*A2*A3))^PEXP
3358 DNEW=BDAT(1,ITEST):IF DNEW>BDIAM THEN GOTO 3400
3360 CAPAC=BDAT(4,ITEST):IF CLOAD>CAPAC AND ITEST<20 THEN 3350:REM heavier
brg
3370 IF CAPAC>=CLOAD AND ITEST<20 AND cap0>cload0 THEN 3450:' ...design OK
3400 locate 15,1:PRINT"Design not feasible for this diameter and bearing
type"
3405 PRINT"Diameter= ";BDIAM;"  Dynamic Load= ";CLOAD;" Capacity= ";CAPAC
3406 print"                 Static Load= ";cload0;" Capacity= ";cap0
3410 INPUT"Do you want,<1> greater diameter, <2> different type ";X$
3420 IF X$="2" THEN CLS:GOTO 3220
3430 IF X$<>"1" THEN 3410
3440 INPUT"Enter new diameter ";BDIAM:GOTO 3340
3450 cls:PRINT:PRINT"BEARING DESIGN:"
3455 print"                    Bearing No                    ";i
3456 print"                    At Node                       ";isu(i)
3460 PRINT"                    ISO Number                    ";BTYP$(ITEST)
3462 PRINT"                    Inner Diameter (mm)           ";BDAT(1,ITEST)
3464 PRINT"                    Outer Diameter (mm)           ";BDAT(2,ITEST)
3465 PRINT"                    Width (mm)                    ";BDAT(3,ITEST)
3466 PRINT"                    Dynamic Capacity (N)          ";BDAT(4,ITEST)
3467 PRINT"                    Dynamic Load (N)              ";cload
3468 PRINT"                    Static Capacity (N)           ";BDAT(5,ITEST)
3469 PRINT"                    Static Load (N)               ";cload0
3530 PRINT:INPUT"Hit ENTER to continue";X$
3540 NEXT I
3545 RETURN
3550 REM READ FROM FILE
3560 REM ****************************************** READING RECORD OF
FILE
3600 OPEN NF$ AS #1 LEN=256
3610 FOR IGL = N1 TO N2 :IG=IGL-N1+1
3620 FIELD #1, 10 AS A1$, 10 AS A2$, 10 AS A3$, 10 AS A4$, 10 AS A5$, 10 AS
A6$,10 AS A7$, 10 AS A8$, 10 AS A9$, 10 AS A10$, 10 AS A11$, 10 AS A12$
3630 CODE%=IGL
3640 GET #1, CODE%
3650 BDAT(1,IG) =VAL( A1$):BDAT(2,IG) = VAL(A2$)
3660 BDAT(3,IG) = VAL(A3$):BDAT(4,IG) = VAL(A4$)
3670 BDAT(5,IG) = VAL(A5$):BTYP$(IG) = A6$
3680 LOCATE 25,1:PRINT "Record ";IGL;" d= ";BDAT(1,IG);"
Capacity=";BDAT(4,IG);
3720 NEXT IGL
3730 CLOSE #1
3740 RETURN
5100 REM ******************** X,Y coefficients for deep groove ball
bearings
5101 foc=af/cap0
5102 if foc=<.04 then e=.22: x=.56: y=2:if (fa/fr) < e then x=1:y=0
5103 if foc=<.07 and foc>.04 then e=.24:x=.56:y=1.8:if (fa/fr)<e then
x=1:y=0
5104 if foc=<.13 and foc>.07 then e=.27:x=.56:y=1.6:if (fa/fr)<e then
x=1:y=0
5105 if foc=<.25 and foc>.13 then e=.31:x=.56:y=1.4:if (fa/fr)<e then
x=1:y=0
5106 if foc=<.5 and foc>.25 then e=.27:x=.56:y=1.2:if (fa/fr)<e then
x=1:y=0
5107 if foc>.5 then e=.44:x=.56:y=1:if (fa/fr)<e then x=1:y=0
5110 RETURN
5200 REM ****************** X,Y coefficients for self alligned ball
bearings
5205 tp=val(btyp$(itest))
5206 if tp>=1300 and tp<= 1322 then tp=tp-100
```

```
5207 if tp>=2301 and tp<= 2318 then tp=tp-100
5210 e=.31:y=2.0:if fa/fr>=e then y=3.1:REM ...for all other types,
tentative
5215 if tp>1203 and tp<= 1205 then e=.27:y=2.3:if fa/fr>=e then y=3.6
5220 if tp>1205 and tp<= 1207 then e=.23:y=2.7:if fa/fr>=e then y=4.2
5225 if tp>1207 and tp<= 1209 then e=.21:y=2.9:if fa/fr>=e then y=4.5
5230 if tp>1209 and tp<= 1212 then e=.19:y=3.4:if fa/fr>=e then y=5.2
5235 if tp>1212 and tp<= 1222 then e=.17:y=3.6:if fa/fr>=e then y=5.6
5240 if tp>1222 and tp<= 1230 then e=.2 :y=3.3:if fa/fr>=e then y=5
5245 if tp>2200 and tp<= 2204 then e=.50:y=1.3:if fa/fr>=e then y=2
5250 if tp>2204 and tp<= 2207 then e=.37:y=1.7:if fa/fr>=e then y=2.6
5255 if tp>2207 and tp<= 2209 then e=.31:y=2.0:if fa/fr>=e then y=3.1
5260 if tp>2209 and tp<= 2213 then e=.28:y=2.3:if fa/fr>=e then y=3.5
5265 if tp>2213 and tp<= 2220 then e=.26:y=2.4:if fa/fr>=e then y=3.8
5270 if tp>2220 and tp<= 2222 then e=.28:y=2.3:if fa/fr>=e then y=3.5
5280 x=1:if fa/fr>e then x=.65
5290 RETURN
```

APPENDIX A

Table A.1 Theoretical stress concentration factors

	σ_0	$K_t = \sigma_{max}/\sigma_0$
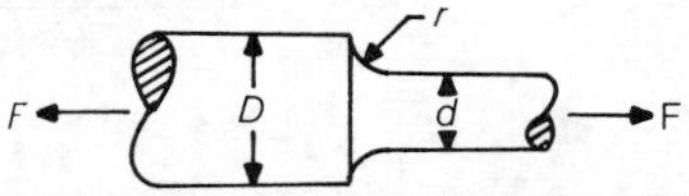	$4F/\pi d^2$	$1 + \dfrac{(r/d)^{-0.36-0.2(D/d)}}{5 + 0.12/(D/d - 1)}$
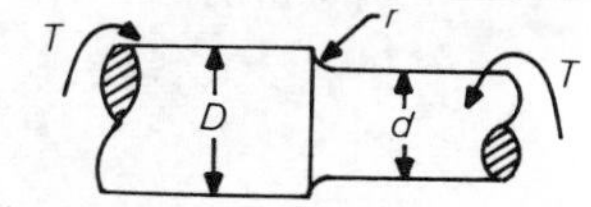	$16T/\pi d^3$	$1 + \dfrac{(r/d)^{-0.36-0.2(D/d)}}{13 + 03/(D/d - 1)}$
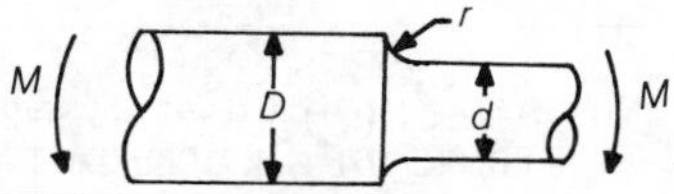	$32M/\pi d^3$	$1 + \dfrac{(r/d)^{-0.73-0.42(D/d-1)}}{5 + 4.38/(D/d - 1)^{0.16}}$
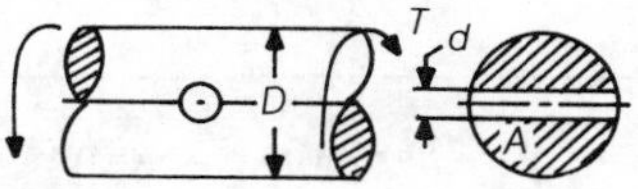	$T\Big/\left(\dfrac{\pi D^3}{16} - \dfrac{dD^2}{6}\right)$	$1 + 1.47(d/D)^{-0.197}$
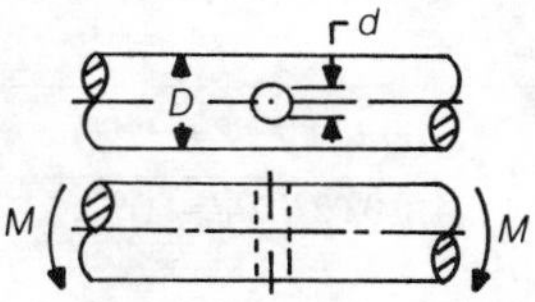	$\dfrac{M}{\pi D^3/32 - dD^2/6}$	$1 + 0.65(d/D)^{-0.275}$
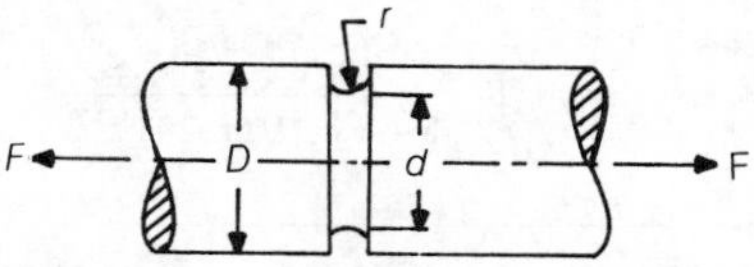	$4F/\pi d^2$	$1 + \dfrac{(r/d)^{-0.511-(D/d-1)0.34}}{3 + 0.507(D/d - 1)^{-0.42}}$
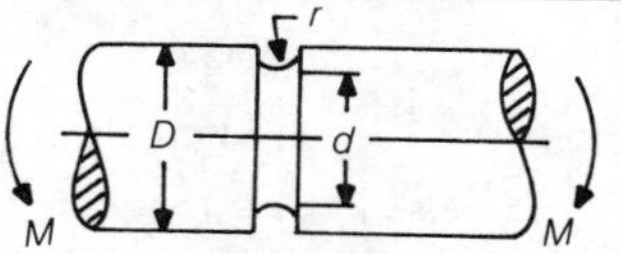	$\dfrac{32M}{\pi d^3}$	$1 + \dfrac{(r/d)^{-0.59-(D/d-1)0.184}}{5 + 0.0812/(D/d - 1)}$
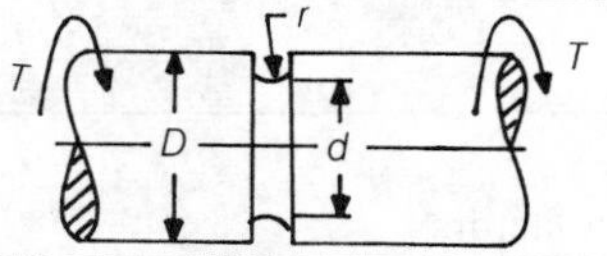	$\dfrac{16T}{\pi d^3}$	$1 + \dfrac{(r/d)^{-0.609-(D/d-1)\times 0.146}}{5 + 3.73(D/d - 1)^{-0.252}}$
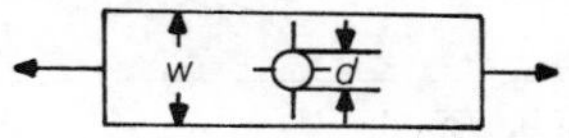	$F/(w - d)t$	$1 + \dfrac{(d/w)^{-0.179}}{0.9}$
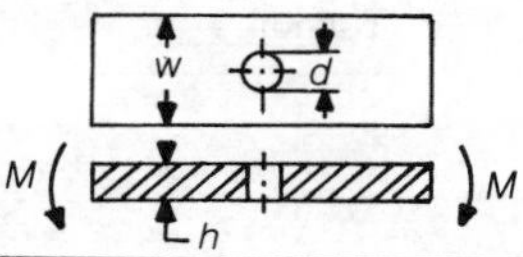	$\dfrac{6M}{(w - d)h^2}$	$1 + \dfrac{(d/w)^{-0.21-0.09(d/h)^{0.3}}}{0.954 + 0.966(d/h)^{0.65}}$

Table A.1 (*Contd.*)

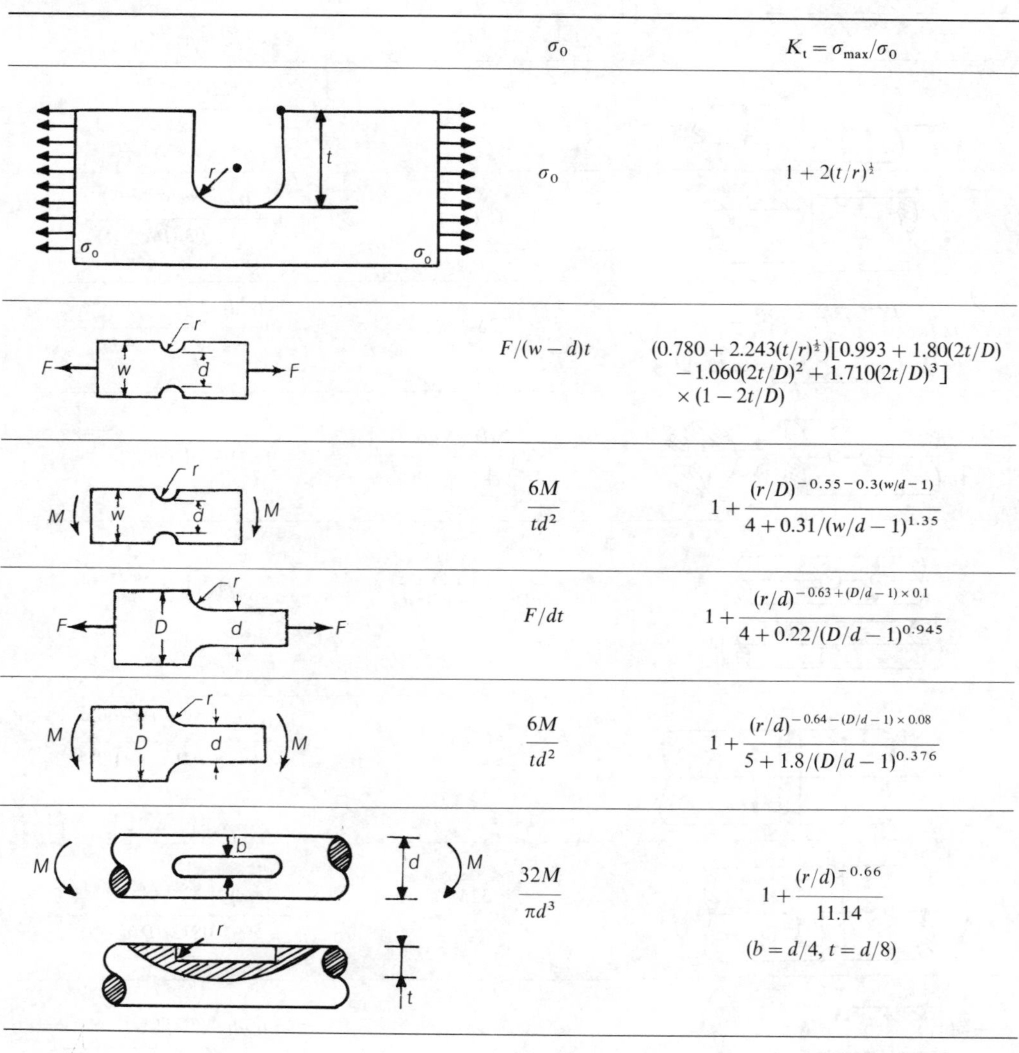

	σ_0	$K_t = \sigma_{max}/\sigma_0$
	σ_0	$1 + 2(t/r)^{\frac{1}{2}}$
	$F/(w-d)t$	$(0.780 + 2.243(t/r)^{\frac{1}{2}})[0.993 + 1.80(2t/D) - 1.060(2t/D)^2 + 1.710(2t/D)^3] \times (1 - 2t/D)$
	$\dfrac{6M}{td^2}$	$1 + \dfrac{(r/D)^{-0.55-0.3(w/d-1)}}{4 + 0.31/(w/d-1)^{1.35}}$
	F/dt	$1 + \dfrac{(r/d)^{-0.63+(D/d-1)\times 0.1}}{4 + 0.22/(D/d-1)^{0.945}}$
	$\dfrac{6M}{td^2}$	$1 + \dfrac{(r/d)^{-0.64-(D/d-1)\times 0.08}}{5 + 1.8/(D/d-1)^{0.376}}$
	$\dfrac{32M}{\pi d^3}$	$1 + \dfrac{(r/d)^{-0.66}}{11.14}$ ($b = d/4$, $t = d/8$)

Table A.2 Fatigue stress concentration factors

	K_f	
	Loading mode	
Heat treatment	Bending	Torsion
Annealed (less than 200 BHN)	1.3	1.3
Quenched and drawn (over 200 BHN)	1.6	1.6

Table A.2 (*Contd.*)

	K_f	
Annealed (less than 200 BHN)	1.6	1.3
Quenched and drawn (over 200 BHN)	2.0	1.6

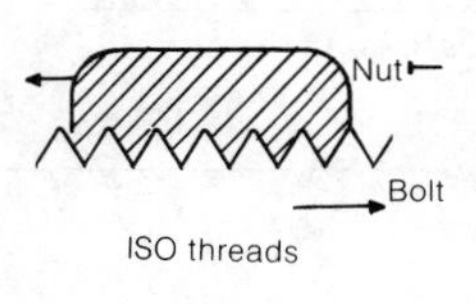

ISO threads

$K_f = k_1 k_2$

Bolt	Cut	Rolled	Annealed	Surf. Hard.	Core hard.
k_1	1	1.2	1.3	1.4	1.6
Nut	Steel, cut		Steel, rolled	Al	Bronze
k_2	1		1.05	1.1	1.15

APPENDIX B

Table B.1 Stress intensity factors for cracks

Geometry and loading	$K = f(a/b)\sigma(\pi a),\ \lambda = a/b$
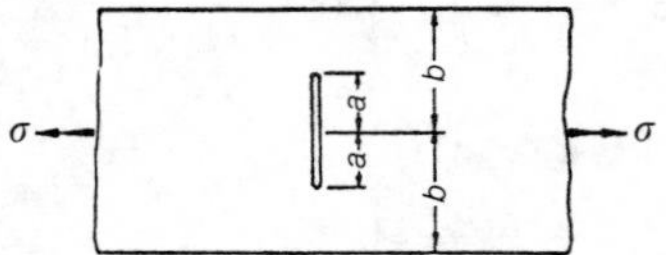	$f(a/b) = (\lambda) = (1 - 0.5\lambda + 0.37\lambda^2 - 0.044\lambda^3)/(1-\lambda)^{\frac{1}{2}}$
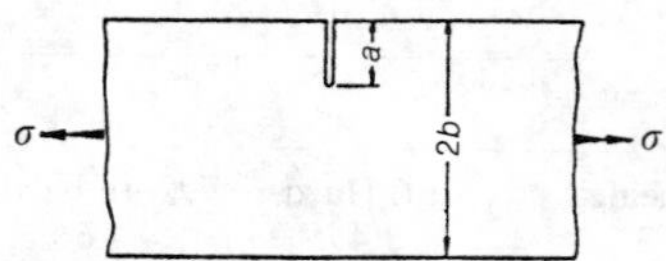	$f(\lambda) = \left(\frac{2}{\pi\lambda}\right)^{\frac{1}{2}} \frac{0.752 + 2.02\lambda + 0.37(1 - \sin(\pi\lambda/2))}{\cos(\pi\lambda/2)}$
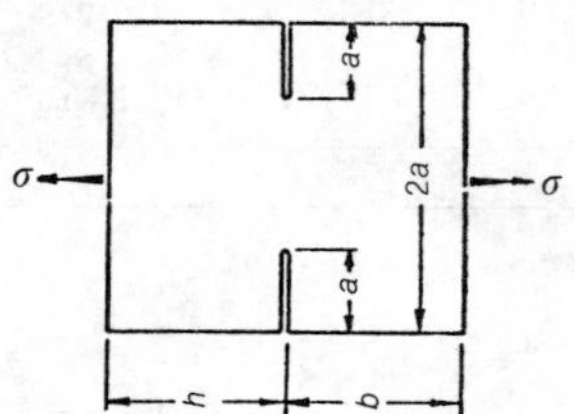	$f(\lambda) = (1 + 0.122\cos^4(\pi\lambda/2))[(2/\pi\lambda)\tan(\pi\lambda/2)]^{\frac{1}{2}}$
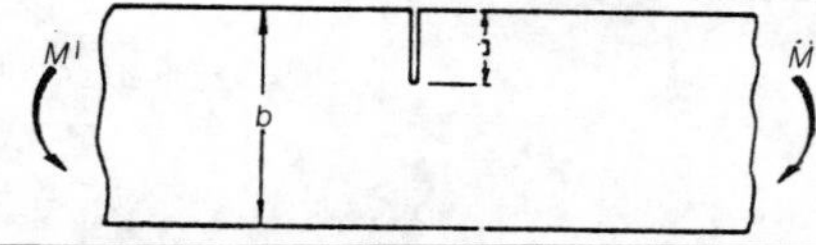	$f(\lambda) = [(2/\pi\lambda)\tan(\pi\lambda/2)]^{\frac{1}{2}} \frac{0.923 + 0.199(1 - \sin(\pi\lambda/2))^4}{\cos(\pi\lambda/2)}$
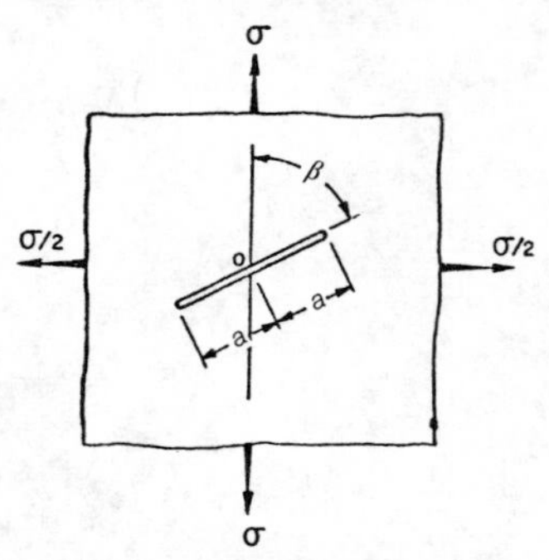	$K_1 = \frac{1}{2}\sigma(\alpha)^{\frac{1}{2}}[1 + \sin^2\beta]^{\frac{1}{2}},\ K_2 = \frac{1}{2}\sigma(\alpha)^{\frac{1}{2}}\sin\beta\cos\beta$
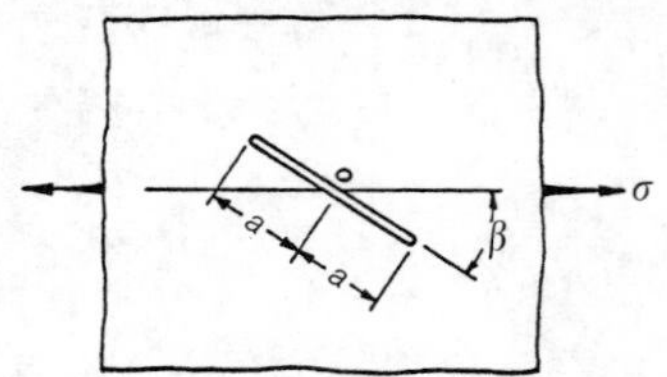	$K_1 = \frac{1}{2}\sigma(\alpha)^{\frac{1}{2}}\sin^2\beta,\ K_2 = \sigma(\alpha)^{\frac{1}{2}}\sin\beta\cos\beta$

Table B.1 (*Contd.*)

Geometry and loading	$K = \mathrm{f}(a/b)\sigma(\pi a),\ \lambda = a/b$
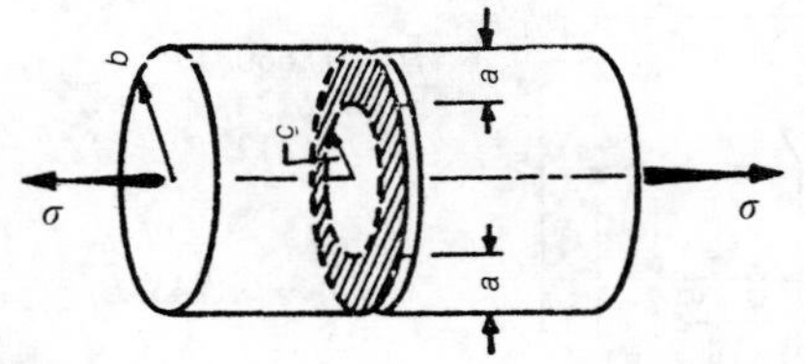	$K = \dfrac{\sigma(\pi\alpha)^{\frac{1}{2}}}{2}\left(\dfrac{c}{b}\right)^{\frac{1}{2}}\left(1 + \dfrac{1c}{2b} + \dfrac{3c^2}{8b^2} - 0.363\dfrac{c^3}{b^3} + 0.731\dfrac{c^4}{b^4}\right)$
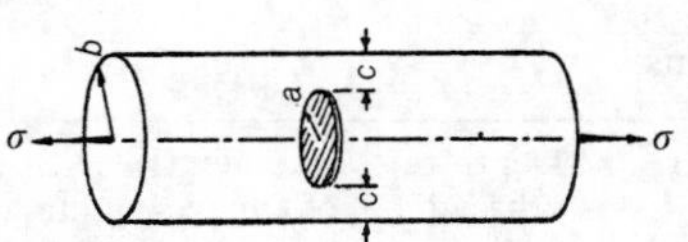	$K = \dfrac{\sigma(\pi b)^{\frac{1}{2}}}{(\alpha)^{\frac{1}{2}}E(k)}(\alpha^2 \sin^2\phi + b^2\cos^2\phi)^{\frac{1}{4}}$ $\xi = \dfrac{2}{\pi}\sigma(\pi d)^{\frac{1}{2}}, E(\xi) =$ **elliptic integral**
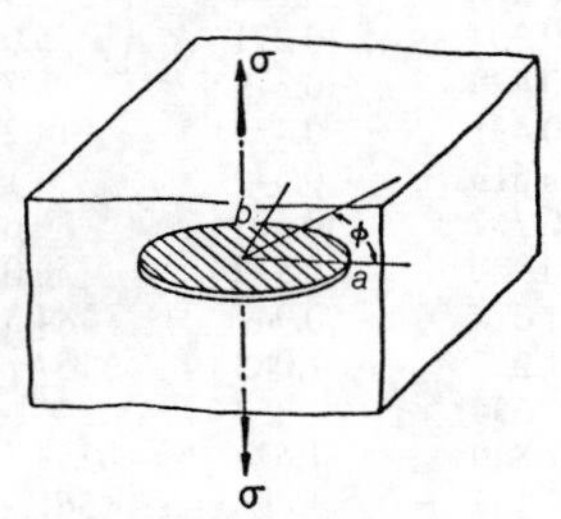	$K = \dfrac{\sigma(\pi\alpha)^{\frac{1}{2}}}{(1 - a^2/b^2)}\left(\dfrac{c}{b}\right)^{\frac{1}{2}}\left[\dfrac{2}{\pi}\left(1 + \dfrac{1a}{2b} - \dfrac{5a^2}{8b^2}\right) + 0.268\dfrac{c^3}{b^3}\right]$

APPENDIX C
SOME STANDARDS FOR MACHINE ELEMENTS

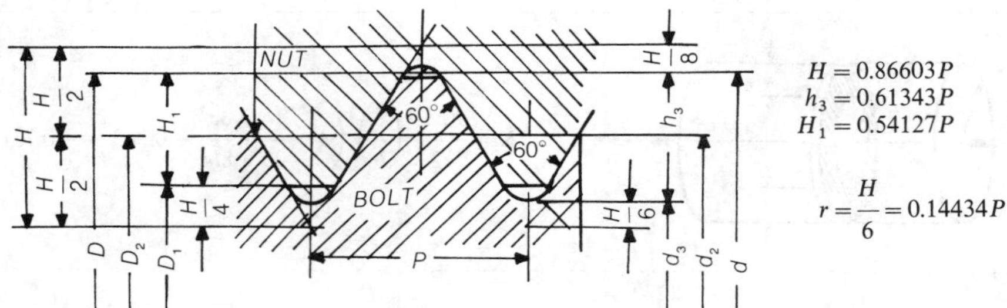

Table C.1 Bolt dimensions

	Size	Pitch (P)	Major diameter ($d = D$)	Pitch diameter ($d_2 = D_2$)	Minor diameter: Bolt (d_3)	Minor diameter: Nut (d_1)	Depth of thread (h_3)	Max. depth of engage-ment (H_1)	Stress area, A_s (mm^2)
Coarse series	M2.5	0.45	2.5	2.208	1.948	2.013	0.276	0.244	3.39
	M3	0.5	3	2.675	2.387	2.459	0.307	0.271	5.03
	M4	0.7	4	3.545	3.141	3.242	0.429	0.379	8.78
	M5	0.8	5	4.480	4.019	4.134	0.491	0.433	14.2
	M6	1	6	5.350	4.773	4.918	0.613	0.541	20.1
	M8	1.25	8	7.188	6.466	6.647	0.767	0.677	36.6
	M10	1.5	10	9.026	8.160	8.376	0.920	0.812	58.0
	M12	1.75	12	10.863	9.853	10.106	1.074	0.947	84.3
	M16	2	16	14.701	13.546	13.835	1.227	1.083	157
	M20	2.5	20	18.376	16.933	17.294	1.534	1.353	245
	M24	3	24	22.051	20.320	20.752	1.840	1.624	353
	M30	3.5	30	27.727	25.706	26.211	2.147	1.894	561
	M33	3.5	33	30.727	28.706	29.211	2.147	1.894	694
	M36	4	36	33.402	31.093	31.67	2.454	2.165	817
Fine series	M8 × 1	1	8	7.350	6.773	6.918	0.613	0.541	39.2
	M10 × 1.25	1.25	10	9.188	8.466	8.647	0.767	0.677	61.2
	M12 × 1.25	1.25	12	11.188	10.466	10.647	0.767	0.677	92.1
	M16 × 1.5	1.5	16	15.026	14.16	14.376	0.920	0.812	167
	M20 × 1.5	1.5	20	19.026	18.16	18.376	0.920	0.812	272
	M24 × 2	2	24	22.701	21.546	21.835	1.227	1.083	384
	M30 × 2	2	30	28.701	27.546	27.835	1.227	1.083	621
	M36 × 3	3	36	34.051	32.32	35.752	1.840	1.624	865

Stress area $A_s = \frac{\pi}{4}\left(\frac{d_2 + d_3}{2}\right)^2$

$h = 1.866p$
$f = 0.5p + a$
$f_1 = 0.5p + 2a - b$
$t = 0.5p + a - b$
$c = 0.25p$

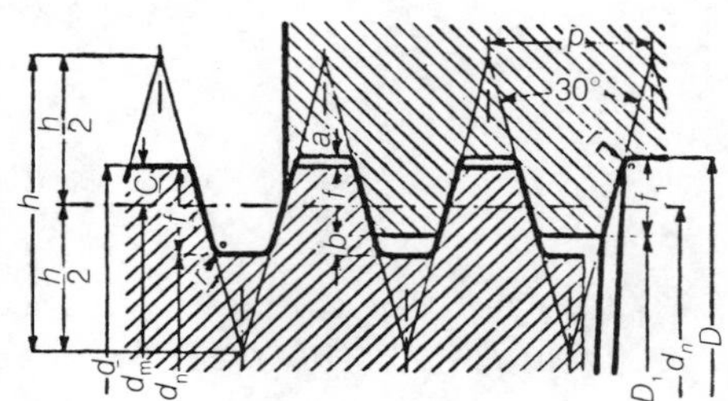

Table C.1 (*Contd.*)

d (mm)	d_n (mm)	A_s (mm^2)	f (mm)	r (mm)	D (mm)	D_1 (mm)	f_1 (mm)	r_1(1) (mm)	p (mm)	d_m (mm)	t (mm)	a (mm)	b (mm)
10	**6.5**	**33**	**1.75**	**0.25**	**10.5**	**7.5**	**1.5**	**0.20**	**3**	**8.5**	**1.25**	**0.25**	**0.5**
12	**8.5**	**57**	**1.75**	**0.25**	**12.5**	**9.5**	**1.5**	**0.20**	**3**	**10.5**	**1.25**	**0.25**	**0.5**
14	**9.5**	**71**	**2.25**	**0.25**	**14.5**	**10.5**	**2**	**0.20**	**4**	**12**	**1.75**	**0.25**	**0.5**
16	**11.5**	**104**	**2.25**	**0.25**	**16.5**	**12.5**	**2**	**0.20**	**4**	**14**	**1.75**	**0.25**	**0.5**
18	**13.5**	**143**	**2.25**	**0.25**	**18.5**	**14.5**	**2**	**0.20**	**4**	**16**	**1.75**	**0.25**	**0.5**
20	**15.5**	**189**	**2.25**	**0.25**	**20.5**	**16.5**	**2**	**0.20**	**4**	**18**	**1.75**	**0.25**	**0.5**
(22)	16.5	214	2.75	0.25	22.5	18	2.25	0.20	5	19.5	2	0.25	0.75
25	**19.5**	**299**	**2.75**	**0.25**	**25.5**	**21**	**2.25**	**0.20**	**5**	**22.5**	**2**	**0.25**	**0.75**
(28)	22.5	398	2.75	0.25	28.5	24	2.25	0.20	5	25.5	2	0.25	0.75
30	**23.5**	**434**	**3.25**	**0.25**	**30.5**	**25**	**2.75**	**0.20**	**6**	**27**	**2.5**	**0.25**	**0.75**
(32)	25.5	511	3.25	0.25	32.5	27	2.75	0.20	6	29	2.5	0.25	0.75
35	**28.5**	**638**	**3.25**	**0.25**	**35.5**	**30**	**2.75**	**0.20**	**6**	**32**	**2.6**	**0.25**	**0.75**
(38)	30.5	731	3.75	0.25	38.5	32	3.25	0.20	7	34.5	3	0.25	0.75
40	**32.5**	**830**	**3.75**	**0.25**	**40.5**	**34**	**3.25**	**0.20**	**7**	**36.5**	**3**	**0.25**	**0.75**
(42)	34.5	935	3.75	0.25	42.5	36	3.25	0.20	7	38.5	3	0.25	0.75
45	**36.5**	**1 046**	**4.25**	**0.25**	**45.5**	**38**	**3.75**	**0.20**	**8**	**41**	**3.5**	**0.25**	**0.75**
(48)	39.5	1 225	4.25	0.25	48.5	41	3.75	0.20	8	44	3.5	0.25	0.75
50	**41.5**	**1 353**	**4.25**	**0.25**	**50.5**	**43**	**3.75**	**0.20**	**8**	**46**	**3.5**	**0.25**	**0.75**
(55)	45.5	1 626	4.75	0.25	55.5	47	4.25	0.20	9	50.5	4	0.25	0.75
60	**50.5**	**2 003**	**4.75**	**0.25**	**60.5**	**52**	**4.25**	**0.20**	**9**	**55.5**	**4**	**0.25**	**0.75**
(65)	54.5	2 333	5.25	0.25	65.5	56	4.75	0.20	10	60	4.5	0.25	0.75
70	**59.5**	**2 781**	**5.25**	**0.25**	**70.5**	**61**	**4.75**	**0.20**	**10**	**65**	**4.5**	**0.25**	**0.75**
(75)	64.5	3 267	5.25	0.25	75.5	66	4.75	0.20	10	70	4.5	0.25	0.75
80	**69.5**	**3 794**	**5.25**	**0.25**	**80.6**	**71**	**4.75**	**0.20**	**10**	**75**	**4.5**	**0.25**	**0.75**

Rectangular keys

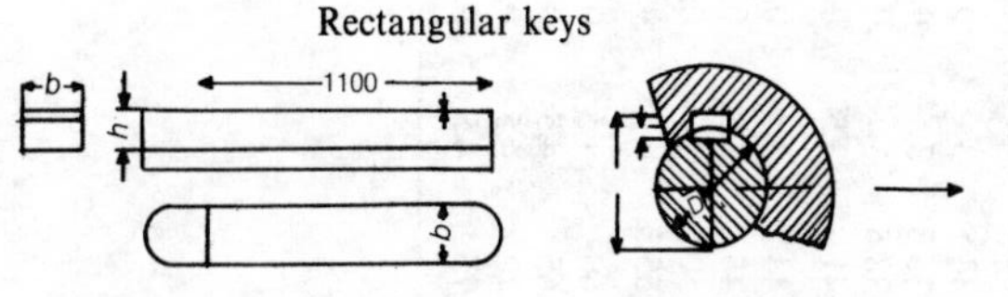

Table C.2 Key dimensions

D (mm)	b (mm)	h (mm)	z (mm)	l (mm)																t (mm)	t_1 (mm)
10 to 12	4	4	0.3	10	12	15	18	20	25	30										2.5	D + 1.5
12 to 17	5	5	0.3	10	12	15	18	20	25	30	35	40								3	D + 2
17 to 22	6	6	0.3		12	15	18	20	25	30	35	40	45	50						3.5	D + 2.5
22 to 30	8	7	0.3					20	25	30	35	40	45	50	60	70				4	D + 3
30 to 38	10	8	0.3						25	30	35	40	45	50	60	70	80	90		4.5	D + 3.5
38 to 44	12	8	0.3	30	35	40	45	50	60	70	80	90	100	120						4.5	D + 3.5
44 to 50	14	9	0.4		35	40	45	50	60	70	80	90	100	120	140					5	D + 4
50 to 58	16	10	0.4				45	50	60	70	80	90	100	120	140	160	180			5	D + 5
58 to 68	18	11	0.4					50	60	70	80	90	100	120	140	160	180	200		6	D + 5
68 to 78	20	12	0.4						60	70	80	90	100	120	140	160	180	200	220	6	D + 6
78 to 92	24	14	0.4	70	80	90	100	120	140	160	180	200	220	250	280					7	D + 7
92 to 110	28	16	0.5		80	90	100	120	140	160	180	200	220	250	280	300				8	D + 8
110 to 130	32	18	0.5			90	100	120	140	160	180	200	220	250	280	300	350			9	D + 9
130 to 150	36	20	0.5				100	120	140	160	180	200	220	250	280	300	350	400		10	D + 10
150 to 170	40	22	0.5					120	140	160	180	200	220	250	280	300	350	400		11	D + 11
170 to 200	45	25	0.5							160	180	200	220	250	280	300	350	400		13	D + 12
200 to 230	50	28	0.5								180	200	220	250	280	300	350	400		14	D + 14

Table C.2 (*Contd.*)

Woodroof keys

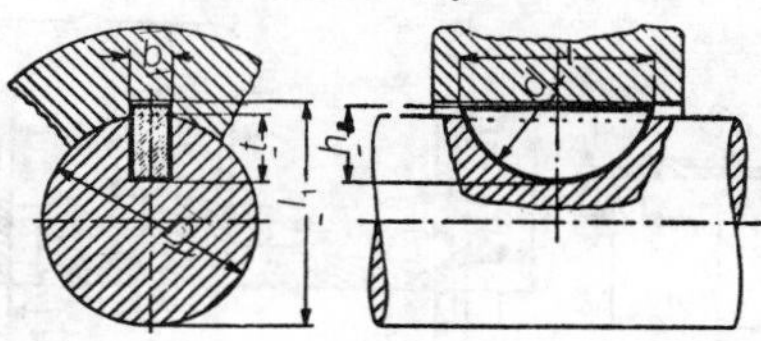

D (mm)	$b \times h$ (mm) × (mm)	t (mm)	t_1 (mm)
3 to 4	1 × 1.4	0.9	D + 0.6
4 to 5	1.5 × 1.4	0.9	D + 0.6
	1.5 × 2.6	2.1	
5 to 7	2 × 2.6	1.8	D + 0.9
	2 × 3.7	2.9	
7 to 9	2.5 × 3.7	2.9	D + 0.9
9 to 13	3 × 3.7	2.5	D + 1.3
	3 × 5	3.8	
	3 × 6.5	5.3	
13 to 17	4 × 5	3.8	D + 1.4
	4 × 6.5	5.3	
	4 × 7.5	6.3	
17 to 22	5 × 6.5	4.9	D + 1.8
	5 × 7.5	5.9	
	5 × 9	7.4	
	5 × 10	8.4	

D (mm)	$b \times h$ (mm) × (mm)	t (mm)	t_1 (mm)
22 to 28	6 × 9	7.4	D + 1.8
	6 × 10	8.4	
	6 × 11	9.4	
	6 × 13	11.4	
28 to 38	8 × 11	9.5	D + 1.7
	8 × 13	11.5	
	8 × 15	13.5	
	8 × 16	14.5	
	8 × 17	15.5	
38 to 48	10 × 16	14	D + 2.2
	10 × 17	15	
	10 × 19	17	
	10 × 24	22	
48 to 58	12 × 19	16.5	D + 2.7
	12 × 24	21.5	

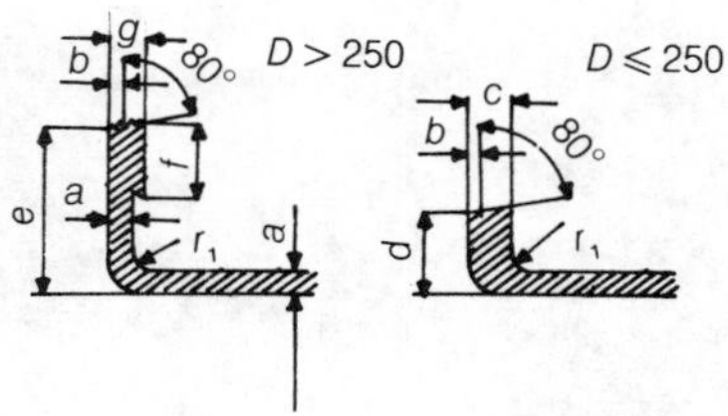

Table C.3 Dimensions of slider bearings

	a	b	c	d	e	f	g	r	r_1	r_3
25 to 50	0.5	0.5	1	7	—	—	—	2.5	1.5	1.5
50 to 75	1	1	1.5	9	—	—	—	4	2.5	4
75 to 100	1.8	1	2	10	—	—	—	4	2.5	4
00 to 150	2	1	2.5	12	—	—	—	6	4	6
50 to 200	3	1.5	3.5	15	—	—	—	10	4	6
00 to 250	3.5	1.5	4	17	—	—	—	10	6	10
50 to 300	4	1.5	—	—	22	7	6.5	15	6	10
00 to 400	5	2	—	—	28	9	7.5	18	10	15
00 to 500	5.5	2	—	—	35	10	8.5	22	15	20
00 to 600	6	2	—	—	40	12	9.5	25	15	22

Table C.3 (*Contd.*)

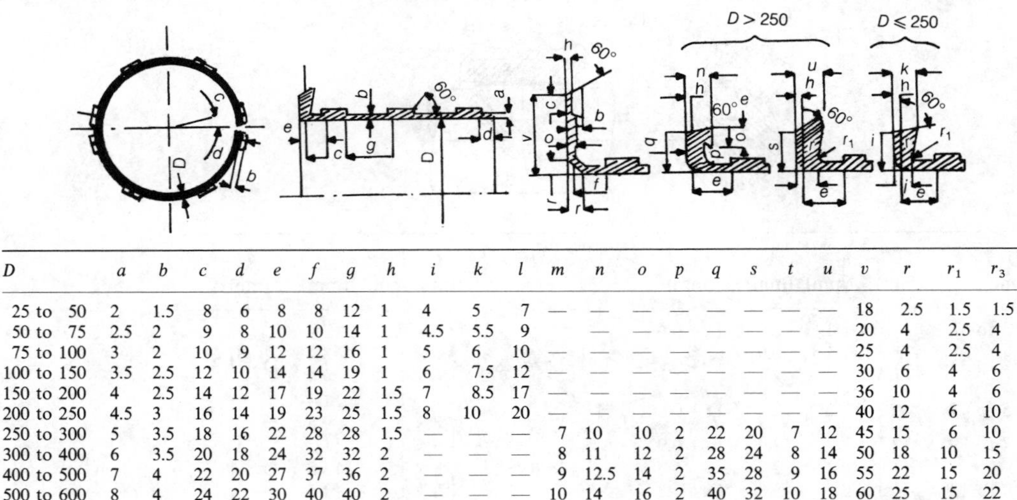

D	a	b	c	d	e	f	g	h	i	k	l	m	n	o	p	q	s	t	u	v	r	r_1	r_3
25 to 50	2	1.5	8	6	8	8	12	1	4	5	7	—	—	—	—	—	—	—	—	18	2.5	1.5	1.5
50 to 75	2.5	2	9	8	10	10	14	1	4.5	5.5	9		—	—	—	—	—	—	—	20	4	2.5	4
75 to 100	3	2	10	9	12	12	16	1	5	6	10	—	—	—	—	—	—	—	—	25	4	2.5	4
100 to 150	3.5	2.5	12	10	14	14	19	1	6	7.5	12	—	—	—	—	—	—	—	—	30	6	4	6
150 to 200	4	2.5	14	12	17	19	22	1.5	7	8.5	17	—	—	—	—	—	—	—	—	36	10	4	6
200 to 250	4.5	3	16	14	19	23	25	1.5	8	10	20	—	—	—	—	—	—	—	—	40	12	6	10
250 to 300	5	3.5	18	16	22	28	28	1.5	—	—	—	7	10	10	2	22	20	7	12	45	15	6	10
300 to 400	6	3.5	20	18	24	32	32	2	—	—	—	8	11	12	2	28	24	8	14	50	18	10	15
400 to 500	7	4	22	20	27	37	36	2	—	—	—	9	12.5	14	2	35	28	9	16	55	22	15	20
500 to 600	8	4	24	22	30	40	40	2	—	—	—	10	14	16	2	40	32	10	18	60	25	15	22

Table C.4 Hot-rolled steel sections

x	h	b	$s = r_1$	t	r_2	F (cm²)	G (kg/m)	U (m²/m)	$x-x$: J_x (cm⁴)	$x-x$: W_x (cm³)	$x-x$: i_x (cm)	$y-y$: J_y (cm⁴)	$y-y$: W_y (cm³)	$y-y$: $i_y = i_1$ (min) (cm)	S_x (cm³)	s_x (cm)
80	80	42	3.9	5.9	2.3	7.57	5.94	0.304	77.8	19.5	3.20	6.29	3.00	0.91	11.4	6.84
100	100	50	4.5	6.8	2.7	10.6	8.34	0.370	171	34.2	4.01	12.2	4.88	1.07	19.9	8.57
120	120	58	5.1	7.7	3.1	14.2	11.1	0.439	328	54.7	4.81	21.5	7.41	1.23	31.8	10.3
140	140	66	5.7	8.6	3.4	18.2	14.3	0.502	573	81.9	5.61	35.2	10.7	1.40	47.7	12.0
160	160	74	6.3	9.5	3.8	22.8	17.9	0.575	935	117	6.40	54.7	14.8	1.55	68.0	13.7
180	180	82	6.9	10.4	4.1	27.9	21.9	0.640	1 450	161	7.20	81.3	19.8	1.71	93.4	15.5
200	200	90	7.5	11.3	4.5	33.4	26.2	0.709	2 140	214	8.00	117	26.0	1.87	125	17.2
220	220	98	8.1	12.2	4.9	39.5	31.1	0.775	3 060	278	8.80	162	33.1	2.02	162	18.9
240	240	106	8.7	13.1	5.2	46.1	36.2	0.844	4 250	354	9.59	221	41.7	2.20	206	20.6
260	260	113	9.4	14.1	5.6	53.3	41.9	0.906	5 740	442	10.4	288	51.0	2.32	257	22.3
280	280	119	10.1	15.2	6.1	61.0	47.9	0.966	7 590	542	11.1	364	61.2	2.45	316	24.0
300	300	125	10.8	16.2	6.5	69.0	54.2	1.03	9 800	653	11.9	451	72.2	2.56	381	25.7
320	320	131	11.5	17.3	6.9	77.7	61.0	1.09	12 510	782	12.7	555	84.7	2.67	457	27.4
340	340	137	12.2	18.3	7.3	86.7	68.0	1.15	15 700	923	13.5	674	98.4	2.80	540	29.1
360	360	143	13.0	19.5	7.8	97.0	76.1	1.21	19 610	1 090	14.2	818	114	2.90	638	30.7
380	380	149	13.7	20.5	8.2	107	84.0	1.27	24 010	1 260	15.0	975	131	3.02	741	32.4
400	400	155	14.4	21.6	8.6	118	92.4	1.33	29 210	1 460	15.7	1 160	149	3.13	857	34.1
425	425	163	15.3	23.0	9.2	132	104	1.41	36 970	1 740	16.7	1 440	176	3.30	1 020	36.2
450	450	170	16.2	24.3	9.7	147	115	1.48	45 850	2 040	17.7	1 730	203	3.43	1 200	38.3
475	475	178	17.1	25.6	10.3	163	128	1.55	56 480	2 380	18.6	2 090	235	3.60	1 400	40.4
500	500	185	18.0	27.0	10.8	179	141	1.63	68 740	2 750	19.6	2 480	268	3.72	1 620	42.4
550	550	200	19.0	30.0	11.9	212	166	1.80	99 180	3 610	21.6	3 490	349	4.02	2 120	46.8
600	600	215	21.6	32.4	13.0	254	199	1.92	139 000	4 630	23.4	4 670	434	4.30	2 730	50.9

Table C.4 (*Contd.*)

						F	G	U	x − x			y − y						y − y	
C	h	b	s	$t = {}^*r_1$	r_2	(cm²)	(kg/m)	(m²/m)	J_x (cm⁴)	W_x (cm³)	i_x (cm)	J_y (cm⁴)	W_y (cm³)	i_y (cm)	S_x (cm³)	s_x (cm)	e_y (cm)	x_M (cm)
30 × 15	30	15	4	4.5	2	2.21	1.74	0.103	2.53	1.69	1.07	0.38	0.39	0.42	—	—	0.52	0.74
30	30	33	5	7	3.5	5.44	4.27	0.174	6.39	4.26	1.08	5.33	2.68	0.99	—	—	1.31	2.22
40 × 20	40	20	5	5.5*	2.5	3.66	2.87	0.142	7.58	3.79	1.44	1.14	0.86	0.56	—	—	0.67	1.01
40	40	35	5	7	3.5	6.21	4.87	0.199	14.1	7.05	1.50	6.68	3.08	1.04	—	—	1.33	2.32
50 × 25	50	25	5	6	3	4.92	3.86	0.181	16.8	6.73	1.85	2.49	1.48	0.71	—	—	0.81	1.34
50	50	38	5	7	3.5	7.12	5.59	0.232	26.4	10.6	1.92	9.12	3.75	1.13	—	—	1.37	2.47
60	60	30	6	6	3	6.46	5.07	0.215	31.6	10.5	2.21	4.51	2.16	0.84	—	—	0.91	1.50
65	65	42	5.5	7.5	4	9.03	7.09	0.273	57.5	17.7	2.52	14.1	5.07	1.25	—	—	1.42	2.60
80	80	45	6	8	4	11.0	8.64	0.312	106	26.5	3.10	19.4	6.36	1.33	15.9	6.65	1.45	2.67
100	100	50	6	8.5	4.5	13.5	10.6	0.372	206	41.2	3.91	29.3	8.49	1.47	24.5	8.42	1.55	2.93
120	120	55	7	9	4.5	17.0	13.4	0.434	364	60.7	4.62	43.2	11.1	1.59	36.3	10.0	1.60	3.03
140	140	60	7	10	5	20.4	16.0	0.489	605	86.4	5.45	62.7	14.8	1.75	51.4	11.8	1.75	3.37
160	160	65	7.5	10.5	5.5	24.0	18.8	0.546	925	116	6.21	85.3	18.3	1.89	68.8	13.3	1.84	3.56
180	180	70	8	11	5.5	28.0	22.0	0.611	1 350	150	6.95	114	22.4	2.02	89.6	15.1	1.92	3.75
200	200	75	8.5	11.5	6	32.2	25.3	0.661	1 910	191	7.70	148	27.0	2.14	114	16.8	2.01	3.94
220	220	80	9	12.5	6.5	37.4	29.4	0.718	2 690	245	8.48	197	33.6	2.30	146	18.5	2.14	4.20
240	240	85	9.5	13	6.5	42.3	33.2	0.775	3 600	300	9.22	248	39.6	2.42	179	20.1	2.23	4.39
260	260	90	10	14	7	48.3	37.9	0.834	4 820	371	9.99	317	47.7	2.56	221	21.8	2.36	4.66
280	280	95	10	15	7.5	53.3	41.8	0.890	6 280	448	10.9	399	57.2	2.74	266	23.6	2.53	5.02
300	300	100	10	16	8	58.8	46.2	0.950	8 030	535	11.7	495	67.8	2.90	316	25.4	2.70	5.41
320	320	100	14	17.5	8.75	75.8	59.5	0.982	10 870	679	12.1	597	80.6	2.81	413	26.3	2.60	4.82
350	350	100	14	16	8	77.3	60.6	1.047	12 840	734	12.9	570	75.0	2.72	459	28.6	2.40	4.45
380	380	102	13.5	16	8	80.4	63.1	1.110	15 760	829	14.0	615	78.7	2.77	507	31.1	2.38	4.58
400	400	110	14	18	9	91.5	71.8	1.182	20 350	1 020	14.9	846	102	3.04	618	32.9	2.65	5.11

C > 300

C ⩽ 300

T	h	b	$s=t$ $=r_1$	r_2	r_3	F (cm²)	G (kg/m)	U (m²/m)	e_x (cm)	$x-x$ J_x (cm⁴)	W_x (cm³)	i_x (cm)	$y-y$ J_y (cm⁴)	W_y (cm³)	$i_y = i_1$ (cm)	d (mm)	w_1 (mm)	w_2 (mm)
20	20	20	3	1.5	1	1.12	0.88	0.075	0.58	0.38	0.27	0.58	0.20	0.20	0.42	3.2	—	—
25	25	25	3.5	2	1	1.64	1.29	0.094	0.73	0.87	0.49	0.73	0.43	0.34	0.51	3.2	15	14
30	30	30	4	2	1	2.26	1.77	0.114	0.85	1.72	0.80	0.87	0.87	0.58	0.62	4.3	17	17
35	35	35	4.5	2.5	1	2.97	2.33	0.133	0.99	3.10	1.23	1.04	1.57	0.90	0.73	4.3	19	19
40	40	40	5	2.5	1	3.77	2.96	0.153	1.12	5.28	1.84	1.18	2.58	1.29	0.83	6.4	21	22
45	45	45	5.5	3	1.5	4.67	3.67	0.171	1.26	8.13	2.51	1.32	4.01	1.78	0.93	6.4	24	25
50	50	50	6	3	1.5	5.66	4.44	0.191	1.39	12.1	3.36	1.46	6.06	2.42	1.03	6.4	30	30
60	60	60	7	3.5	2	7.94	6.23	0.229	1.66	23.8	5.48	1.73	12.2	4.07	1.24	8.4	34	35
70	70	70	8	4	2	10.6	8.32	0.268	1.94	44.5	8.79	2.05	22.1	6.32	1.44	11	38	40
80	80	80	9	4.5	2	13.6	10.7	0.307	2.22	73.7	12.8	2.33	37.0	9.25	1.65	11	45	45
90	90	90	10	5	2.5	17.1	13.4	0.345	2.48	119	18.2	2.64	58.5	13.0	1.85	13	50	50
100	100	100	11	5.5	3	20.9	16.4	0.383	2.74	179	24.6	2.92	88.3	17.7	2.05	13	60	60
120	120	120	13	6.5	3	29.6	25.2	0.459	3.28	366	42.0	3.51	178	29.7	2.45	17	70	70
140	140	140	15	7.5	4	39.9	21.3	0.537	3.80	660	46.7	4.07	330	47.2	2.88	21	80	75

T	h	b	s	t	F (cm²)	G (kg/m)	U (m²/m)
20	20	20	3	3	1.11	0.871	0.080
25	25	25	3.5	3.5	1.63	1.28	0.100
30	30	30	4	4	2.24	1.76	0.120
35	35	35	4.5	4.5	2.95	2.31	0.140
40	40	40	5	5	3.75	2.94	0.160

L	a	s	F (cm²)	G (kg/m)	U (m²/m)
$20 \times \frac{3}{4}$	20	$\frac{3}{4}$	$\frac{1.11}{1.44}$	$\frac{0.871}{1.13}$	0.080
$25 \times \frac{3}{4}$	25	$\frac{3}{4}$	$\frac{1.41}{1.84}$	$\frac{1.11}{1.44}$	0.100
$30 \times \frac{3}{4}$	30	$\frac{3}{4}$	$\frac{1.71}{2.24}$	$\frac{1.34}{1.76}$	0.120
35×4	35	4	2.64	2.07	0.140
$40 \times \frac{4}{5}$	40	$\frac{4}{5}$	$\frac{3.04}{3.75}$	$\frac{2.39}{2.94}$	0.160
45×5	45	5	4.25	3.34	0.180
50×5	50	5	4.75	3.73	0.200

Table C.4 (*Contd.*)

					F	G	U						$x-x=y-y$			$\xi-\xi$		$\eta-\eta$		
L	a	s	r_1	r_2	(cm²)	(kg/m)	(m²/m)	$e_x = e_y$ (cm)	w (cm)	v_1 (cm)	v_2 (cm)	$J_x = J_y$ (cm⁴)	$W_x = W_y$ (cm³)	$i_x = i_y$ (cm)	J_ξ (cm⁴)	i_ξ (cm)	J_η (cm⁴)	W_η (cm)	i_η (cm)
20 × 3	20	3	3.5	2	1.12	0.88	0.077	0.60	1.41	0.85	0.70	0.39	0.28	0.59	0.62	0.74	0.15	0.18	0.37
4		4			1.45	1.14		0.64		0.90	0.71	0.48	0.35	0.58	0.77	0.73	0.19	0.21	0.36
25 × 3	25	3	3.5	2	1.42	1.12	0.097	0.73	1.77	1.03	0.87	0.79	0.45	0.75	1.27	0.95	0.31	0.30	0.47
4		4			1.85	1.45		0.76		1.08	0.89	1.01	0.58	0.74	1.61	0.93	0.40	0.37	0.47
5		5			2.26	1.77		0.80		1.13	0.91	1.18	0.69	0.72	1.87	0.91	0.50	0.44	0.47
30 × 3	30	3	5	2.5	1.74	1.36	0.116	0.84	2.12	1.18	1.04	1.41	0.65	0.90	2.24	1.14	0.57	0.48	0.57
4		4			2.27	1.78		0.89		1.24	1.05	1.81	0.86	0.89	2.85	1.12	0.76	0.61	0.58
5		5			2.78	2.18		0.92		1.30	1.07	2.16	1.04	0.88	3.41	1.11	0.91	0.70	0.57
35 × 3	35	3	5	2.5	2.04	1.60	0.136	0.96	2.47	1.36	1.23	2.29	0.90	1.06	3.63	1.34	0.95	0.70	0.68
4		4			2.67	2.10		1.00		1.41	1.24	2.96	1.18	1.05	4.68	1.33	1.24	0.88	0.68
5		5			3.28	2.57		1.04		1.47	1.25	3.56	1.45	1.04	5.63	1.31	1.49	1.10	0.68
6		6			3.87	3.04		1.08		1.53	1.27	4.14	1.71	1.04	6.50	1.30	1.77	1.16	0.68
40 × 3	40	3	6	3	2.35	1.84	0.155	1.07	2.83	1.52	1.40	3.45	1.48	1.21	5.45	1.52	1.44	0.95	0.78
4		4			3.08	2.42		1.12		1.58	1.40	4.48	1.56	1.21	7.09	1.52	1.86	1.18	0.78
5		5			3.79	2.97		1.16		1.64	1.42	5.43	1.91	1.20	8.64	1.51	2.22	1.35	0.77
6		6			4.48	3.52		1.20		1.70	1.43	6.33	2.26	1.19	9.98	1.49	2.67	1.57	0.77
45 × 4	45	4	7	3.5	3.49	2.74	0.174	1.23	3.18	1.75	1.57	6.43	1.97	1.36	10.2	1.71	2.68	1.53	0.88
5		5			4.30	3.38		1.28		1.81	1.58	7.83	2.43	1.35	12.4	1.70	3.25	1.80	0.87
6		6			5.09	4.00		1.32		1.87	1.59	9.16	2.88	1.34	14.5	1.69	3.83	2.05	0.87
7		7			5.86	4.60		1.36		1.92	1.61	10.4	3.31	1.33	16.4	1.67	4.39	2.29	0.87
50 × 4	50	4	7	3.5	3.89	3.06	0.194	1.36	3.54	1.92	1.75	8.97	2.46	1.52	14.2	1.91	3.73	1.94	0.98
5		5			4.80	3.77		1.40		1.98	1.76	11.0	3.05	1.51	17.4	1.90	4.59	2.32	0.98
6		6			5.69	4.47		1.45		2.04	1.77	12.8	3.61	1.50	20.4	1.89	5.24	2.57	0.98
7		7			6.56	5.15		1.49		2.11	1.78	14.6	4.15	1.49	23.1	1.88	6.02	2.85	0.98
8		8			7.41	5.82		1.52		2.16	1.80	16.3	4.68	1.48	25.7	1.86	6.87	3.19	0.98
9		9			8.24	6.47		1.56		2.21	1.82	17.9	5.20	1.47	28.1	1.85	7.67	3.47	0.98
55 × 5	55	5	8	4	5.32	4.18	0.213		3.89	2.15	1.93	14.7	3.70	1.66	23.3	2.09	6.11	2.84	1.07
6		6			6.31	4.95		1.56		2.21	1.94	17.3	4.40	1.66	27.4	2.08	7.24	3.28	1.07
8		8			8.23	6.46		1.64		2.32	1.97	22.1	5.72	1.64	34.8	2.06	9.35	4.03	1.07
10		10			10.1	7.90		1.72		2.43	2.00	26.3	6.97	1.62	41.4	2.02	11.3	4.65	1.07
60 × 5	60	5	8	4	5.82	4.57	0.223	1.64	4.24	2.32	2.11	19.4	4.45	1.82	30.7	2.30	8.03	3.46	1.17
6		6			6.91	5.42		1.69		2.39	2.11	22.8	5.29	1.82	36.1	2.29	9.43	3.95	1.17
8		8			9.03	7.09		1.77		2.50	2.14	29.1	6.88	1.80	46.1	2.26	12.1	4.84	1.17
10		10			11.1	8.69		1.85		2.62	2.17	34.9	8.41	1.78	55.1	2.23	14.6	5.57	1.17
65 × 6	65	6	9	4.5	7.53	5.91	0.252	1.80	4.60	2.55	2.28	29.2	6.21	1.97	46.3	2.48	12.1	4.74	1.27
7		7			8.70	6.83		1.85		2.62	2.29	33.4	7.18	1.96	53.0	2.47	13.8	5.27	1.27
8		8			9.85	7.73		1.89		2.67	2.31	37.5	8.13	1.95	59.4	2.46	15.6	5.84	1.27
9		9			11.0	8.62		1.93		2.73	2.32	41.3	9.04	1.94	65.4	2.44	17.2	6.30	1.27
11		11			13.2	10.3		2.00		2.83	2.36	48.8	10.8	1.91	76.8	2.42	20.7	7.31	1.27

ANSWERS TO SELECTED PROBLEMS

2.1 Shaft $41.949 < d < 41.974$
Hole $42 < D < 42.039$

2.2 (a) $H9/d9$, (b) $H8/f7$
(c) $H7p6$, (d) $H7/p6$

2.5 (a) $29.978 < d < 29.992$; $30 < D < 30.014$
(b) $71.981 < d < 72$; $72 < D < 72.019$

2.6 (a) $59.681 < d < 59857$; $60 < D < 60.176$
(b) $40.002 < d < 40.017$; $40 < D < 40.025$

2.7 Bearing $49.946 < d < 49.973$,
$50 < D < 50.041$
Sleave $60.031 < d < 60.048$
$60 < D < 60.028$

2.8 $50.002 < d < 50.019$; $50 < D < 50.027$

2.9 (a) $40.002 < d < 40.017$; $40 < D < 40.025$
(b) $29.995 < d < 29.798$; $30 < D < 30.035$

2.10 $19.783 < d < 19.905$; $20 < D < 20.122$

4.11 222 MPa, 0.0260 rad

4.12 $F_{\max} = 10^5$ N; $M_{\max} = 3.12 \times 10^4$ Nm

4.13 967 kN

4.16 Max stress $= R^2 g\delta/2t$
$\delta =$ density

4.17 Hemisph: $s^t = (R/2t)(H + h)\delta$
Bottom: $s = (R/2\mathrm{t})(H + R)\delta$
Cylindrical: $s_t = (hR/t)\delta$
Upper: $s_m = (RH/2t + 2R^2/3t)\delta$

4.18 $s_t = (HR/t)\delta$

4.19 $s = hD(H - h)\delta/2Ht\cos(\phi/2)$

4.20 $s = (Rp/t)(1 - R/d)$

4.30 251 MPa

4.31 $s_{x-\max} = 2.060$ MPa

4.32 $s_{x-\max} = 0.07$ MPa

4.33 162 Mpa; 0.4626 mm

4.34 192.2 MPa

4.35 138.84 Mpa

5.1 $P(b, h) = (6/bh^2 + K(b^2 + g^2 - 100)^2$

5.2 $P(d, h, t) = \pi dht + \pi d^2 t/4 + K(g_1^2 + g_2^2)$

5.3 $I(i_1, i_2) = \pi\tau br_{14}(1 + (i_{14} + 1)/i_{12})/2$
$= (i_2^4 + 1 + 100)/(i_1 i_2)^2/(i_1 i_2)^2$

5.4 1.587 m

5.5 No minimum

5.6 $d = 1.37$ m; $h = 3.16$ m

5.7 $L = 2.238$ *m*; $d_1 = 26$ mm; $L_1 = 1.96$ m

5.8 No maximum

5.9 $r_2 = 0.428$ m

5.10 $b = 5.63$; $h = 8.25$ m

5.11 $d = 2.47$; $h = 4.12$; $t = 0.00051$ m

5.12 $i_1 = 1.74$, $i_2 = 2.03$, $i_3 = 2.83$

5.13 $x = 1.59$ m

5.14 $L = 0.04$; $d = 1.94$ m

5.15 $d = 1.46$ m; $h = 3.13$ m; $t = 0.00017$ m

5.16 $L = 2.238$ m; $d_1 = 26.5$ mm; $L_1 = 1.96$ m

5.17 $r_1 = 30$ mm; $r_2 = 425$ mm

	a	b	c	d	$f_{\max}$
5.21	0.21	0.19	0.26		0.25
5.22	2.65	0.5	4.47		0.14
5.23	1.53	0.34	−1.51		0.19
5.24	0.48	1.54	0.15		0.16
5.25	1.52	−0.35	−1.51	−0.004	0.19

6.1 $u_e = s_y^2/2E$; $u_p = s_y\delta/100$

6.2 $U_p = S_y \delta V/100$

6.3 $u_e = (0.557 S_y)^2/4$;
$u_p = (0.577 S_y)^2(1 - r_0^2/R^2)$

6.10 3.02

6.11 2.57

6.12 2.97

6.13 3.28

6.14 4.68

6.15 > 4.68

6.16 1.25

6.17 5.84 kW

6.18 (Add data: Rel = 99%; $L_{th} = L_{pr}$) 1.34

6.19 1.35%

6.21 DET 1.053 MPa; MSST 1.072 MPa

6.22 DET 2.49 MPa; MSST 2.67 MPa

6.23 DET 4.31 MPa; MSST 4.95 MPa

6.24 DET 8.78 MPa; MSST 7.88 MPa

6.25 DET 361 MPa; MSST 391 MPa

6.26 DET 4810; MSST 4970 MPa

6.27 DET 35.2 MPa; MSST 35.5 MPa

6.28 9.78 kW

6.29 0.5

7.1 27.2 kN

7.2 20.8 kN

7.3 472.5 Nm

7.4 455 Nm

7.5 9050 Nm

7.6 124.5 Nm

7.7 4.457 Nm
7.8 2860 Nm
7.9 363 N
7.10 4543 N
7.11 8.8 kN
7.12 7069 N
7.13 17.781 Nm
7.14 136.8 Nm
7.15 2551 Nm
7.16 40.725 Nm
7.17 1.665 Nm
7.18 980 Nm
7.19 110 N
7.20 1438 N
7.21 49.3 kN; 1.0018×10^6 cycles
7.22 49,198 N; 240,555 cycles
7.23 750 Nm; 66,700 cycles
7.24 640.6 Nm; 1.23×10^6 cycles
7.25 6807 cycles
7.26 381 cycles
7.27 1195 cycles
7.28 4 years
7.29 16.8 cycles
7.30 140 years
8.1 7 mm
8.2 7 mm
8.3 The second
8.4 The second
8.5 12 mm
8.6 9.86
8.7 1 mm
8.8 6.6 mm
8.9 12.7 mm
8.10 4 mm
8.11 14
8.12 3 mm
8.13 $t = 9.47$; $w = 173$; $s = 35$ mm
8.14 99 mm
8.15 ASTM 284; grade D steel
8.16 12.7 mm dia
8.17 67.1 mm; $N = 11.25$
8.18 $N_{tension} = 0.116$; $N_{shear} = 0.18$
8.19 $n = 2.21$, $T = 41.41$ Nm
8.20 8 mm, 4.125 Nm
8.21 M12, 21.8 kN, 64.34 kN
8.22 M14, 79.15 Nm
8.23 2 mm
8.24 (a) 2.88 kN; 4.956 Nm
(b) 7 mm; (c) 2.42
8.25 (a) 6.66 kN
(b) 3.5 N/mm radial 0.35×10^6 N/m axial
(c) 4720 N; 22.7 mm
8.26 (a) 19 mm; (b) 0.148; (c) self lock
8.27 (a) 8.5 mm; 22.25 Nm; 0.214
(b) 11 mm
8.28 91.47 Nm; 0.417; 7.95
8.29 16 mm; 0.026; self locking
8.30 34 mm; 0.0129; 628.85 Nm; 114.4 Nm
8.31 1
8.32 84,514 N
8.33 24 mm
8.34 9.6 mm
8.35 $S_y = 36.3$ MPa
8.36 4 mm
8.37 0.29 mm
8.38 3.6 mm
8.39 3 mm
8.40 3.42 mm
8.41 1 mm
8.42 3.64 mm
8.43 15.5 mm
8.44 5.86 mm
8.45 9.4 mm
8.46 2.8
8.47 15200 N
8.48 (a) 25; (b) 3
8.49 3.5
8.50 2
8.51 460.6 Nm; 0.90; 111.8°C
8.52 $H5/n5$
8.53 5720 Nm; 0.43
8.54 $H5/n5$; 391 kN
8.55 Not feasible
9.1 $d = 31$ mm
9.2 $d = 9.37$ mm
9.3 $L = 15$ m
9.4 $d = 16$ mm
9.4 $d = 16$ mm
9.5 $d = 50$ mm
9.6 3.6
9.7 1.24
9.8 0.6
9.9 1.71
9.10 0.75
9.11 385 kN
9.12 420 kN
9.13 701 kN
9.14 3.3
9.15 2.19
9.16 $D = 93$ mm; $d = 10$ mm
9.17 $D = 0.32$ m; $d = 0.035$ m
9.18 $d = 3.5$ mm; $D = 26$ mm
9.19 0.03 MPa, 0.04 MPa
9.20 $d = 2$ mm; $D = 24.6$ mm; $N_a = 26$
9.21 $d = 15$ mm, $L = 1780$ mm
9.22 Yes
9.23 $d = 3.3$ mm, $N_a = 2$
9.24 0.3 mm
9.25 6.5 mm
9.26 20 × 840 × 840 mm plate
9.27 $d = 20$ mm
9.28 $s = 5$ mm; $h = 5$ mm

9.29 8.23 Hz
9.30 $R = 45$ mm; $h = 72$ mm
9.31 $A = 0.17\ \text{m}^2$
9.32 $b = 2.84$ m; $TR = 35$; 143; 899
9.33 $A = 0.028\ \text{m}^2$; $t = 0.1$ m
9.34 4 squares 50×50 mm; $t = 0.12$ m
10.1 108.5°C
10.2 6.11°C
10.3 0.80°C
10.4 601 RPM
10.5 14 RPM
10.6 9.8 mm
10.7 3342 rev
10.8 $k = 0.096 \times 10^{-4}$
10.9 585 hours
10.10 56.5 sec
10.11 $h = 3.34$ m
10.12 cw: 629 NM; 134 mm
ccw: 66.6 Nm; 14.7 mm
10.13 3478 N; 0.44 mm
10.14 101 Nm
10.15 1391 N; 0.44 m
10.16 184 mm; 13122 N
10.17 6
10.18 2 calipers; 4 pads; 160×160 mm
10.19 $b = 34$ mm
10.20 $b = 76$ mm
10.21 $b = 16$ mm; $h = 20$ mm
10.22 $b = 4.4$ mm
10.23 (a) Front 9303 N; rear 4653 N
(b) Front 2466 N; rear 1233 N
10.24 $b = 117$ mm; $F = 3656$ N
10.25 52 mm; $F = 1654$ N
10.26 $b = 95$ mm; $F = 2418$ N
10.27 $b = 74$ mm; $f = 662$ N
10.28 $b = 44.6$ mm; $F = 1547$ N
10.29 $b = 32$ mm; $F = 1575$ N
10.30 3075 hours
10.31 2354 hours
10.32 (a) $t = 5$ mm; $b = 88$ mm; dia 280/1400 m
(b) 31,347 hours; (c) 383 N
10.33 4472 hours; 1271 N
10.34 select $t = 5$ mm, then $b = 95$ mm
10.35 14 770 hours
10.36 $b = 17$ mm; $t = 11$ mm (A)
7×10^6 hours; 194.7 N
10.37 (a) 2 A-section V belts
(b) 1 A-section V belt
11.1 2.62 reyn; SAE 60
11.2 8.8 mPas; SAE 20
11.3 $\log(\mu_{20}) = 157.7$
$\log(\mu_{120}) = 149.2$
T°C; viscosity Pass
11.4 4 mPas
11.5 901 mPas
11.6 $D = 70$ mm; $L = 70$ mm
11.7 $D = 10$ mm; $L = 70$ mm
11.8 $D = 64$ mm; $L = 23.4$ mm
11.9 $L = 1.6$ mm
11.10 $D = 7$ mm; $L = 7$ mm
11.11 $D = 45$ mm; $L = 45$ mm
11.12 $L = D = 20$ mm
11.13 $D = 100$ mm; $L = 60$ mm
11.14 $L = 6$ mm
11.15 $L = D = 10$ mm
11.16 285 W
11.17 609 W
11.18 $T = \mu\pi RN\phi L/2c$
11.19 D small
11.20 26 W
11.21 $h_0 = 145\ \mu$m; pad 330×354 mm
79 kW
11.22 70 μm; 70.2°C; $4.12 \times 10^{-2}\ \text{m}^3/\text{sec}$
11.23 $R_1 = 14$ in; $R_2 = 84.9$ in
37.7 kW
11.24 $b = 1.12$ m; $L = 3.36$ m; $n = 8$ pads
11.25 $W = (6\mu UL/h_2^2)(m-1)s$
$/[m^3 + s/(1-s)]/Lt$
11.26 $D = 113$ mm; $c = 113\ \mu$m; $L = 66.5$ mm
11.27 $W = 8.8$ kN; $T = 50$°C;
$Q = 11 \times 10^{-6}\ \text{m}^3/\text{sec}$; $P = 1283$ W
11.28 10.6 kN
11.29 $\varepsilon = 0.66$; $c = 30\ \mu$m
11.30 $D = 60$ mm; $L = 30$ mm; $\varepsilon = 0.067$;
$c = 30\ \mu$m
11.31 $D = 89$ mm; $L = 178$ mm; $\varepsilon = 0.77$;
$c = 0.4$ mm
11.32 $T_0 + 0.73$°C
11.33 $L = 200$; $D = 160$ mm; $\varepsilon = 0.98$; $P = 63.3$ kW
11.36 $p_m = 13$ MPa; $D = 0.97$ m; $L = 0.485$ m
11.37 $p = 2.28$ MPa; $D = 60$ mm
11.38 $\varepsilon = 0.92$
11.39 $\varepsilon = 0.915$, $P = 50$ kW
11.40 $L = 226$ mm; $R = 56$ mm; $\varepsilon = 0.4$; $c = 56\ \mu$m
11.41 6 pads; $l = w = 100$ mm
11.42 With $d = l_c = 10$ mm; $Q = 0.0157\ \text{m}^3/\text{s}$;
$p_s = 26.9$ MPa
11.43 With $l_c = d = 0.25$ in:
$p_s = 3550\ \text{lb/in}^2$
11.44 $p_s = 193$ MPa; $Q = 3.63\ \text{m}^3/\text{sec}$
11.45 With $l_c = d = 10$ mm:
$L = 285$ mm; $d = 10$ mm
12.1 55 axles
12.2 58.9 kN
12.3 Roller width 508 mm
12.3 $h_{min} = 63\ \mu$m
12.5 $N < 1$, needs surface hardening
12.6 14,455 hours
12.7 $D = 240$; $d = 110$; $B = 50$ mm; tap roller
12.8 $d = 70$; $D = 180$; $B = 42$ mm;
deep grove bb
12.9 $d = 30$; $D = 55$; $B = 17$ mm; taper roller

12.10 $d = 105$; $D = 210$; $B = 85$ mm; thr bb
12.11 756 to 11,076 hours
12.12 2580 hours
12.13 2580 hours
12.14 $d = 30$; $D = 55$; $B = 13$ mm; deep grove bb
12.15 3333 hours
12.16 $d = 110$; $D = 240$; $B = 50$ mm; deep groove bb
12.17 $d = 55$; $D = 120$; $B = 29$ mm; deep gr bb
12.18 $d = 50$; $D = 110$; $B = 29.25$; tap roller
12.19 $d = 55$; $D = 72$; $B = 9$; deep groove bb
12.20 $d = 85$; $D = 150$; $B = 28$; deep groove bb
13.1 $m_c = 1.52$
13.2 $m_c = 1.59$
13.3 $m_c = 1.61$
13.4 $m_c = 1.46$
13.5 $n_c = 1.53$

	d_1	d_2	m	b
13.16	60	240	2.5	45
13.17	67	270	2.5	52
13.18	62	250	2.5	39
	76	304	2	47 (99.99%)
13.19	66	264	2	49.5
13.20	100	300	4	20
13.21	97	146	1.25	37
13.22	AISI	1010		
13.23	38	110	0.65	52
13.24	40	320	2	54
13.25	38	153	0.6	29
13.30	86	129	0.68	32
13.31	60	182	0.53	36

13.32 ASTM A159-G100 cast iron
13.33 80 640 0.38 47
13.34 SAE 1020
13.35 ASTM A159-G3000
13.36 SAE 1015
13.37 AISI 1010
13.38 ASTM A159-G3000
13.39 ASTM A159-G3500
13.46 1697°C
13.47 181.4°C
13.48 66.2°C
13.49 73°C
14.6 0.69
14.7 (a) 101; (b) 1.4
14.8 0.93
14.9 0.95
14.10 0.94
14.11 65 mm
14.12 35 mm
14.13 28 mm
14.14 58 mm
14.15 AISI 403

INDEX

ABEC (Annular Bearing Engineers' Committee), 565
abrasive wear, 492
  of gear teeth, 605
addendum circle, 601
adhesive wear, of gear teeth, 48
aesthetics, design, 8
AFBMA (Anti-Friction Bearing Manufacturers Association), 565
AGMA (American Gear Manufacturers Association), 409, 613, 616, 617, 622, 627, 642
air springs, 415
AISI (American Iron and Steel Institute), 22
algorithm, 32
alloys,
  nonferrous, 248
  steel, 237–47
Angel, I., 87
ANSI (American National Standards Institute), 22, 592, 565
ANSYS, 138
Antikythyra computer, 597
area of contact, 438
artificial intelligence, 21
ASME (American Society of Mechanical Engineers), 22, 297
ASTM (American Society for Testing and Materials), 22
AUTOMESH, 80, 99, 77, 133, 145, 147
automatic mesh generation, 75
axisymmetric solids, 133, 655
axles, 661
AWS (American Welding Society), 331

Babbitt, 229, 248, 493
base circle, involute gears, 598
BASIC, 31
basic hole system of dimensioning, 26
beams,
  computer analysis, 111–21
  curved, bending, 312–5
  straight, bending, 312–5
bearing,
  clearance, 517
  design, 515–8
  friction torque, 512
  heat balance, 513
  hydrostatic, 521–6
  infinitely long, 507
  journal, 506, 515, 517
  leakage, 505
  materials, 492–4, 516
  numerical solutions, 526
  oil flow, 514
  selection, 515
  short, 511
  sliding, 502, 713–4
  stability, 518
  thrust, 503
bearing stress,
  in pinned joints, 322, 352
  in rolling contacts, 557
bearings, rolling,
  dimension series, 565
  equivalent load, 571
  fatigue life, 566
  fitting and installation, 582
  history, 560
  lubrication, 568, 570, 578–80
  series, 564
  types, 561
Belleville washer springs, 412
belts, 455–69
  design, 463
  efficiency, 459
  flat, 465, 467
  geometry, 461
  materials, 465
  optimum design, 464
  pull factor, 458
  types, 456
  V, 466–8
benchmark tests, 30
bending of, beams, curved, 312–5
  gear teeth, 607
  shafts, 667
  welded joints, 341, 347
bevel gears, 623–7
bolts and screws,
  design for maximum fatigue strength, 336–41
  fatigue strength, 353
  grade (class) designation, 326
  initial tightening, 336

load analysis (effect of elasticity), 339
load distribution among threads, 354
materials and manufacture, 325
standard dimensioning, 710–1
stress concentration factors, 354, 707
thread locking, 325
boundary lubrication, 507, 508, 579
BP, 543, 592, 642, 687
brakes,
band, 443
cylindrical, 450
conical, 448
disk, 445
drum, 443
heat considerations, 455
materials used in, 444, 454
optimum design, 453
self-energizing, 452
brasses, 248
brazing, 329
brittle,
materials, 214, 278, 285
fracture, 293
bronzes, 248
Buckingham, Earl 612, 617, 642
buckling,
column, 398–401
of cylindrical shells, 397, 401
end conditions, 400
Euler equation, 400
of helical compression springs, 416
Johnson formula, 402
of power screws, 292
torsional, 401, 416, 398
BUCKLING procedure, 402
BUCKLING, program, 432
Burr, A., 687

cadmium, 326
CAD, 29
CAMD, 28
Cameron, A., 543
cast iron, 231–2
properties, (table), 233–6
clipping, 59
Clifford, J., 45
clutches, 443–55
air, 449
band, 443
computer aided design, 453
conical, 448
cylindrical, 450
disk, 445
materials for, 444, 454
optimum design and CAD, 453
safety, 453
thermal considerations, 455
codes, 21
cold-rolling, effect on fatigue strength, 286, 355
combined loads,
in bolts and power screws, 299–301
in rolling bearings, 571
in shafts, 230–2
COMLOAD, 264, 268, 273
computer,
aided design, 35, 54–6
aided drafting, 55
aided machine design, 28, 34
graphics, 56
programming, 32
systems, 29
utilization, 31
computed aided,
design of shafts, 672
design of spur gears, 616
graphics of spur gears, 613
inspection in gears, 601
machine design, 28, 34
selection of antifriction bearings, 587
stability analysis, 403
congruency principle, 296
constraints,
equality, 181
inequality, 182
contact ratio, in gear teeth, 603
contact stresses, 555
in gear teeth, 611
copper alloys, 248, 249
corrosion, fretting, 492
correction, in gears, 601
cost, 192
critical speed of shafts, 671
creep, 218
crossed gears, 627–8
cumulative fatigue damage, 291

Date, C., 45
da Vinci, L., 12
database, 21, 28, 36, 41
DeGarmo, E., 164, 221
Decker, K-H., 252, 264, 286, 425
Derek de Sola Price, 597
dedendum circle, 601
design,
aesthetics, 8
alternatives, 4
analysis, 7
continuation, 2
database, 21
inversion, 4
for minimum cost, 192
objectives, 7

preliminary, 5
probabilistic, 252
standards, 21
strategies, 2
synthesis, 5
DESIGNDB, 38, 43, 47
Dimarogonas, A., 147, 201, 297, 425, 543, 592, 642, 687
dimensioning 22, 23
disk of revolution, 676
distribution, normal (Gaussian), 254
drawing, mechanical, 65
Dolan, T., 609, 642
ductile materials, 214, 215, 278, 285
Dudley, D., 621, 632, 642
Duggan, T., 297, 687

eccentric loading,
of axially loaded members, 309
of pinned joints, 341
of welded joints, 347
efficiency,
belts, 459
of screws, 324
of worm gears, 633
elastohydrodynamic lubrication, 557, 579
Encarnacao, J., 45, 87
energy release rate, 295
Engel, L., 283, 491, 612
externally pressurized bearings, 521–5
expert systems, 28, 41
Euler, 399, 456

face width of,
bevel gears, 625, 627
helical gears, 622
spur gears, 617
worm gears, 631
Fadeeva, V., 147
failure,
mechanisms, in gears, 605
modes of, 107
probability of, 254–7
Faux, I., 87
fatigue,
crack propagation, 202
damage, cumulative, 291
derating factors, 282–6
failure, 108, 280–6
fracture, 107, 283
low cycle, 292
load rating, 566
Feodosyev, Y., 215, 260–4, 425
Fiacco, A., 201
files,
random, 38
sequential, 37
FINFRAME, 125, 141, 159, 165
finite element analysis,
commercial codes, 138
for lubrication, 526
method, 118
pre/post-processing, 137
prismatic, 121
system 124, 135
FINLUB, 545, 548
FINSTRES, 133, 145, 147, 170
fits, 23–28, 332, 676
FLATBELTS, 467, 485
Foley, J., 87
FORTRAN, 31
foundations, machine, 418
fracture,
brittle, 293
mechanics, 294
toughness, 295, 296
frames, machine, 405
fretting corrosion, 492
friction,
bands, 443
materials, 444, 578
sliding, 438, 490
temperatures, 440
and wear 442, 490
friction, coefficients of,
in belts, 465
in brakes and clutches, 454
in journal bearings, 499, 513
in rolling-element bearings, 580
in screws, 325
in worm gears, 632

Galerkin principle, 527
gaskets, influence of, in bolted connections, 340
Gauss elimination, 124
Gaussian distribution, 254
GEARDES, 637–41, 648
GEARPLOT, 614, 647
golden section, 189
Goodman line, 288–90, 667
Gottfried, S., 201, 202
graphics,
plane, 56
3-D, 63
grease lubrication of rolling bearings, 578
Griffith, 294
gyration, radius of, 402

hardness, 216
Haenchen, R., 286, 297, 687
Harrington, S., 87
Harris, T., 592

heat balance of bearings, 513
heat capacity, worm gears, 635
heat treatment, 225
heating and heat dissipation,
  of bearings, 513
  of brakes and clutches, 455
  of worm gear housings, 635
helical gears, 619–23
HELICSP, program, 433
Herz, 439
herringbone gears, 597
Hertz contact stresses,
  in gear teeth, 611
  in rolling-element bearings, 555
Heskett, J., 17
hidden lines, 68
high cycle fatigue, 282
Holzer, 112
Horowitz, E., 45
hot rolled sections, 715–8
hub, 676
Hubka, V., 17
Huebner, K., 543, 687
hydrostatic bearings, 521
hydrodynamic lubrication, 500, 507
hypoid gears, 628

interference fits, *see* fits
interference of involute gear teeth, 604
internal gears, 465, 495
involute gear teeth, 598–605
Irwin, 295
ISO (International Standards Organization), 45, 592

Jacobi method, 264
Johnson, R., 201, 455
Johnson's column formula, 402
joints, 317–359
journal, 662

Keenan, J., 45
keys (shaft), 712–713
keyways, fatigue stress concentration factors for,
  706–7
kilopond, kp, 377
kinematic viscosity, *see* viscosity
Klingele, K., 283, 491, 612
knowledgebase, 36, 41
Knox, C., 45, 87
Kverneland, K., 264

Lagrange interpolation, 189
Lang, J., 45
Laplace equation for membrane stresses, 128
leaf spring, 411
Leonardo da Vinci, 12

Lewis, W., 642
Lewis equation, 607, 609
Lewis form factor, 607, 609
life considerations,
  in gear teeth (surface fatigue), 611–3
  in rolling-element bearings, 567, 568
lock nuts and lock washers, 325
low cycle fatigue, 292
lubricant,
  additives, 492
  greases, 578
  solid, 440
lubrication,
  of gears, 635
  of journal bearings 506–21
  of rolling bearings, 576, 586
lubrication,
  boundary, 507, 558, 579
  elastohydrodynamic, 557, 579
  hydrodynamic, 500, 507
  hydrostatic, 521
  thin film, 507

Machine Design (journal), 264
machine design methodology, 1
machine design process, 1
machine frames, 405–9
machine foundations, 418
magnesium alloys, 249
malleable cast iron, 232
material,
  brittle, 214, 278, 285
  ductile, 214, 215, 278, 285
  machining, 222, 224
  processing, 221
  rigidity indices, 390
  section, 228
  tables, 230–49
materials,
  for bolts and screws, 325
  for clutches and brakes, 444, 454
  for gears, 606
  for journal bearings, 292–4, 516
  mechanical properties and uses (tables), 233–49
  physical properties of metals (table), 230
  for rivets, 321
  for rolling-element bearings, 568
  for shafts, 664
  for springs, 246
mean value, 253
mechanical drives, 596
membrane stresses, 128
mesh, *see* automatic mesh generation
Metals Handbook, 264
Miner–Palmgren cumulative fatigue damage, 291
Minschke, C., 201

Mirolubov, I., 297
Mobil Oil Corp., 620, 624
modelling, 109
module, 399
modulus,
  of elasticity (table), 230
  of rigidity (shear modulus, table), 230
Mohr circles, 258–263
molybdenum bisulphide, 440
moment of inertia, numerical evaluation, 344
Monte Carlo method, 187
mounting,
  of gears, 610
  of rolling bearings, 582
  of hubs, 662

natural frequency, 419
Neuber, 286
Newton, 258
Newton–Raphson method, 256
Niemann, G., 289, 425, 592, 642, 687
notch sensitivity, 285–6
numerical methods,
  finite element, 118–27
  Gauss elimination, 124
  Jacobi, 264, 274
  Monte Carlo, 187
  Newman, W., 87
  Newton–Raphson, 256, 258
  steepest descent, 185
  transfer matrix, 111–118

objective function, 179
Ocvirk, short bearing approximation, 511
oil grooves in bearings, 515
optimization, 36, 177
optimum design, 177, 194, 196, 197, 359, 417, 453, 463, 532, 618
OPTIMUM, 192, 195–7, 200, 205
Orlov, P., 5, 18, 392, 394, 395, 396, 425

pad bearings, 504, 507, 520
Pahl, G., 17
Pao, Y., 45, 87
Palmgren, A., 592
Palmgren–Miner, cumulative fatigue damage, 291
partial bearing, 388
penalty function, 183
Peterson, 286, 687
Petroff equation, 499
photoelastic method for stress concentration, 280
physical properties of materials (table), 230
pinion, definition of, 598
Pincus, O., 543, 687
Pinnifarina, S., 8, 17
pitch circle of gears, 601
pitch cone of bevel gears, 623
pitch diameter of gears, 599
pitting, 492
plotting,
  primitives, 57
  circular arcs, 58
  polygons, 58
poise, 495
Poisson ratio (table), 230
porous bearings, 494
Popov, E., 265
post-buckling behavior, 400
pre-and post-processing, 137
preferred numbers, 22
pressure,
  angle, in gears, 599
  distribution in bearings, 502
  lubricated journal bearings, 516
principal stresses, planes, directions, 259, 263
probabilistic design, 252
PROBE, 139
processes, material, 221
procedures, 33
Pugh, B., 592, 642
pull factor for belts, 458

quality control, 23
quality, machining, 24

rack cutter for gears, 606
radius of gyration, 402
Raimondi, A., 543
Rasdorf, W., 45
rational sections, 393
Reclaitis, G., 201
reflection, 62
reliability, 252–258
  factor in fatigue, 283
  of rolling bearings, 568
reyn (unit of viscosity), 495
Reynolds, 590, 526
Reshetov, D., 265, 592, 642, 665, 668, 687
Rich, E., 45
rigidity, 389, 669
rigidity indices, 390, 392
Roark, R. J., 425
Rockey, K., 148
Rockwell hardness, 216
Rodenacker, W., 17
Rogers, D., 88
rolling diameter of gears, 599
Rooke, D., 297
rotating machinery, 656
rotation, 61
ROTORDYN, 674, 684, 689
roughness, 438

running-in of bearings, 490

SAE (Society of Automotive Engineers), 326
safety factor, 250
  differential method, 252
  empirical, 251
  probabilistic, 252–7
SAFFAC program, 257, 266, 267, 272
Sandler, B., 17
Sandor, G., 9, 17
Saybolt viscosity, 496
scaling, 60
scoring, 492
scuffing, 492
sections, rational 393
seizure, 492
sealing, rolling bearings, 584
self-energizing brakes and clutches, 452
self-locking power screws, 324
self-locking worm gears, 633
service factor, 252
shaft,
  critical speeds, 673
  design, preliminary, 661
  fatigue strength, 667
  keys and keyways, 706–13
SHAFTDES program, 672, 683, 698
Shigley, J., 265, 297, 592, 642, 687
SHRINKFIT, 678, 686, 697
Siddal, J., 201
Sih, G., 295, 296, 297
Simitsis, G., 425
simulation, 661, 695, 697
size, influence on fatigue strength, 283
SKF, 592
sleeve bearings, see bearing, journal
SOLID program, procedure, 86, 91
solid modeling, 79
Soderberg, fatigue criterion, 287, 610
Sommerfeld number, 510
soldering, 329
springs,
  air, 415
  beam, 411
  Belleville washer, 412
  classification, 410
  coil, helical, 413
  coil, helical, torsional, 413, 416
  constant, 390, 411
  leaf, 412
  materials, 246
  rubber, 414
  ring, 411
  tapered beam, 411
  torsion bars, 414–6
spur gear, 601, 618
standard deviation, 253
stability, 395, 518
stability analysis, 404
standards, 21
steel, 237–47
  properties of (tables), 238–47
steepest descent, 185
Stokes, 496
strain energy,
  release rate, 295
    of materials (table), 295
strength,
  fatigue, 280
  of gear teeth, surface, 615
  properties of materials (tables), 238–47
  theories of, 258–62
  ultimate, 213
  yield, 212
stress,
  complex state of, 260–2
  concentration, 277
  in gear teeth, 607
  intensity factor, 295
  redistribution, 278
  three-dimensional state of, 264
stress concentration factors,
  fatigue, 706–7
  theoretical K_t, 277–80, 705–6
  for welded joints, 357
stress intensity factor, 295
  (tables), 708–9
surface,
  coatings, 440
  fatigue failure,
    of gear teeth, 611
    of rolling-element bearings, 566
  finish, 225, 284
  machining, 225
  members, 127
  treatments for fatigue-strengthening, 228, 286
systematic design, 12

Tada, H., 297
Teflon bearings, 494
tensile test, 212
theories of failure, 260
three-dimensional state of stress, 256
thermal capacity, *see* Heat capacity
thermal expansion coefficient, values of (table), 230
thrust bearings, *see* bearing, thrust
tilting pad bearings, 504, 520
Timoshenko, S., 425
Timken, 592
Titanium alloys, properties of (tables), 249
TMSTAT program, 116, 139, 155, 672
TMSTABIL program, 404, 435

tolerances, 23–28
Tondl, A., 687
torque, of motors, 657
torsion, 310
torsion bar springs, 415
toughness, fracture, 295
  of materials (table), 296
transfer matrix,
  field, 114
  method, 111, 672
  point, 114
transformations, 61, 63
transition temperature, 294
translation, 612
transmission ratio, for belts, 458
transmissibility, 419
Trumpler, R. R., 543

Ullman, J., 45
undercutting of gear teeth, 604

V-belts, 466
VBELTS program, 468, 487
vibration of rotating shafts, 671
virtual gear, 620, 625
viscosity, 494–8
  index, 496
  kinematic, 496
  pressure effect on, 497
  temperature effect on, 498
viscous flow, 398
Vogelpohl, G., 543

wall thickness, castings, 407
waviness of surfaces, 438
wear, 442, 492
  analytical approach to, 442
  coefficient, 442
  failure, 108
  types, 491, 492
Weaver, W., 148
Weibull distribution, 568
welding,
  electrodes, 331
  fatigue considerations, 356
  processes, 328
welds, strength, 352, 357
whip/whirl in journal bearings, 518
Wills, Jr., 543, 592, 642
Winston, P., 45
Woeler, fatigue curve, 280
Woodruff, key, 713
worm gears, 629–635
WORMDES program, 641, 652

yield strength, 212
Young's modulus, 229

zooming, 63
Zienkiewicz, O., 148